Applied Calculus

FOR BUSINESS, ECONOMICS, LIFE SCIENCES, AND SOCIAL SCIENCES

EXPONENTIAL AND LOGARITHMIC FUNCTIONS

$b > 1$

Exponential function
$f(x) = b^x$

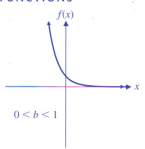

$0 < b < 1$

Exponential function
$f(x) = b^x$

$b > 1$

Logarithmic function
$f(x) = \log_b x$

REPRESENTATIVE POLYNOMIAL FUNCTIONS (DEGREE > 2)

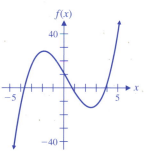

Third-degree polynomial
$f(x) = x^3 - x^2 - 14x + 11$

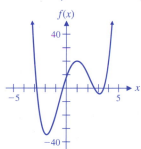

Fourth-degree polynomial
$f(x) = x^4 - 3x^3 - 9x^2 + 23x + 8$

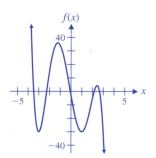

Fifth-degree polynomial
$f(x) = -x^5 - x^4 + 14x^3 + 6x^2 - 45x - 3$

REPRESENTATIVE RATIONAL FUNCTIONS

$$f(x) = \frac{x-3}{x-2}$$

$$f(x) = \frac{8}{x^2 - 4}$$

$$f(x) = x + \frac{1}{x}$$

GRAPH TRANSFORMATIONS

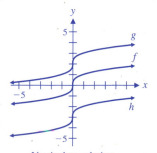

Vertical translation
$g(x) = f(x) + 2$
$h(x) = f(x) - 3$

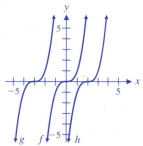

Horizontal translation
$g(x) = f(x + 3)$
$h(x) = f(x - 2)$

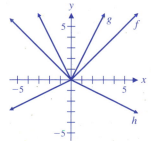

Expansion, contraction, and reflection
$g(x) = 2f(x)$
$h(x) = -0.5f(x)$

Applied Calculus

FOR BUSINESS, ECONOMICS, LIFE SCIENCES, AND SOCIAL SCIENCES

Eighth Edition

Raymond A. Barnett

Merritt College

Michael R. Ziegler

Marquette University

Karl E. Byleen

Marquette University

Pearson Education, Inc.
Upper Saddle River, New Jersey 07458

Library of Congress Cataloging-in-Publication Data

Barnett, Raymond A.
    Applied calculus for business, economics, life sciences, and social sciences. —8th ed.
    / Raymond A. Barnett, Michael R. Ziegler, Karl E. Byleen.

    p.   cm.
    Includes indexes.
    ISBN 0-13-065589-9
    1. Calculus. I. Title: Applied calculus. II. Ziegler, Michael R. III. Byleen, Karl. E. IV, Title

    QA303.2 .B2828 2003   CIP
    515—dc21   2001052343

Acquisitions Editor: *Quincy McDonald*
Editor in Chief: *Sally Yagan*
Production Editor: *Lynn Savino Wendel*
Vice President/Director of Production and Manufacturing: *David W. Riccardi*
Senior Managing Editor: *Linda Mihatov Behrens*
Assistant Managing Editor: *Bayani Mendoza De Leon*
Executive Managing Editor: *Kathleen Schiaparelli*
Manufacturing Buyer: *Alan Fischer*
Manufacturing Manager: *Trudy Pisciotti*
Executive Marketing Manager: *Patrice Lumumba Jones*
Marketing Assistant: *Rachel Beckman*
Assistant Editor: *Vince Jansen*
Editorial Assistant: *Joanne Wendelken*
Art Director: *Heather Scott*
Interior Designer: *John Christiana*
Art Editor: *Thomas Benfatti*
Creative Director: *Carole Anson*
Director of Creative Services: *Paul Belfanti*
Cover Designer: *Laura Ierardi*
Cover Photo Credits: *Corbis Digital Stock*
Art Studios: *Scientific Illustrators*
                Artworks:

                Senior Manager: *Patricia Burns*
                Production Manager: *Ronda Whitson*
                Manager, Production Technologies: *Matt Haas*
                Project Coordinator: *Jessica Einsig*
                Illustrators: *Jonathan Derk, Brandon Gilbert, Mark Landis*
                Art Quality Assurance: *Timothy Nguyen, Stacy Smith, Pamela Taylor*

© 2003, 2000, 1997, 1994, 1991, 1988, 1985, 1982 by Pearson Education, Inc.
Pearson Education, Inc.
Upper Saddle River, New Jersey 07458

Printed in the United States of America

10 9 8 7 6 5 4 3 2 1

ISBN 0-13-065589-9

Pearson Education Ltd.
Pearson Education Australia PTY, Limited
Pearson Education Singapore, Pte. Ltd.
Pearson Education North Asia Ltd.
Prentice-Hall Canada, Ltd.
Pearson Educacion de Mexico, S. A. de C. V.
Pearson Education–Japan, Inc.
Pearson Education Malaysia, Pte. Ltd.

Contents

Preface   xi

Chapter Dependencies

PART ONE   A LIBRARY OF ELEMENTARY FUNCTIONS*

1 A Beginning Library of Elementary Functions

2 Additional Elementary Functions

PART TWO   CALCULUS

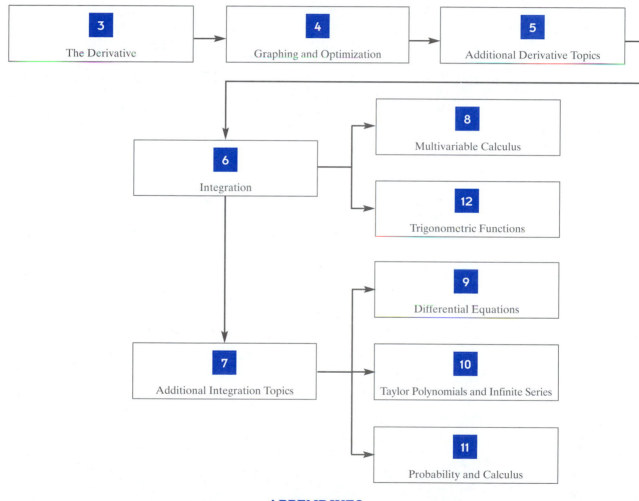

3 The Derivative

4 Graphing and Optimization

5 Additional Derivative Topics

6 Integration

8 Multivariable Calculus

12 Trigonometric Functions

9 Differential Equations

7 Additional Integration Topics

10 Taylor Polynomials and Infinite Series

11 Probability and Calculus

APPENDIXES

A Self-Test; Basic Algebra Review

B Special Topics

---

*Selected topics from Part One may be referred to as needed in Part Two or reviewed systematically before starting Part Two.

Preface

The eighth edition of *Applied Calculus for Business, Economics, Life Sciences, and Social Sciences* is designed for a one- or two-term course in calculus and for students who have had $1\frac{1}{2}$–2 years of high school algebra or the equivalent. The choice and independence of topics make the text readily adaptable to a variety of courses (see the Chapter Dependency Chart on page vii). It is one of five books in the authors' college mathematics series.

Improvements in this edition evolved out of the generous response from a large number of users of the last and previous editions as well as survey results from instructors, mathematics departments, course outlines, and college catalogs. Fundamental to a book's growth and effectiveness is classroom use and feedback. Now in its eighth edition, *Applied Calculus for Business, Economics, Life Sciences, and Social Sciences* has had the benefit of having a substantial amount of both.

■ Emphasis and Style

The text is **written for student comprehension.** Great care has been taken to write a book that is mathematically correct and accessible to students. Emphasis is on computational skills, ideas, and problem solving rather than mathematical theory. Most derivations and proofs are omitted except where their inclusion adds significant insight into a particular concept. General concepts and results are usually presented only after particular cases have been discussed.

■ Examples and Matched Problems

Over 370 completely worked examples are used to introduce concepts and to demonstrate problem-solving techniques. Many examples have multiple parts, significantly increasing the total number of worked examples. Each example is followed by a similar **matched problem for the student to work** while reading the material. This actively involves the student in the learning process. The answers to these matched problems are included at the end of each section for easy reference.

■ Exploration and Discussion

Every section contains **Explore–Discuss** problems interspersed at appropriate places to encourage the student to think about a relationship or process before a result is stated, or to investigate additional consequences of a development in the text. **Verbalization** of mathematical concepts, results, and processes is encouraged in these Explore–Discuss problems, as well as in some matched problems, and in some problems in almost every exercise set. The Explore–Discuss material also can be used as in-class or out-of-class **group activities.** In addition, at the end of every chapter, we have included two special **chapter group activities** that involve several of the concepts discussed in the chapter. Problems in the exercise sets that require verbalization are indicated by color problem numbers.

■ ## Exercise Sets

The book contains over 5,000 problems. Many problems have multiple parts, significantly increasing the total number of problems. Each exercise set is designed so that an average or below-average student will experience success and a very capable student will be challenged. Exercise sets are mostly divided into A (routine, easy mechanics), B (more difficult mechanics), and C (difficult mechanics and some theory) levels.

■ ## Applications

A major objective of this book is to give the student substantial experience in **modeling and solving real-world problems.** Enough applications are included to convince even the most skeptical student that mathematics is really useful (see the Applications Index inside the back cover). Worked examples involving applications are identified by ⛵. **Almost every exercise set contains application problems,** usually divided into business and economics, life science, and social science groupings. An instructor with students from all three disciplines can let them choose applications from their own field of interest; if most students are from one of the three areas, then special emphasis can be placed there. Most of the applications are simplified versions of actual real-world problems taken from professional journals and books. No specialized experience is required to solve any of the applications.

■ ## Internet Connections

The Internet provides a wealth of material that can be related to this book, from sources for the data in application problems to interactive exercises that provide additional insight into various mathematical processes. Every section of the book contains Internet connections identified by **www** . Links to the related web sites can be found at the **PH Companion Website** discussed later in this preface: www.prenhall.com/barnett

■ ## Technology

The generic term **graphing utility** is used to refer to any of the various graphing calculators or computer software packages that might be available to a student using this book. (See the description of the software accompanying this book later in this Preface.) Although **access to a graphing utility is not assumed,** it is likely that many students will want to make use of one of these devices. To assist these students, **optional graphing utility activities** are included in appropriate places in the book. These include brief discussions in the text, examples or portions of examples solved on a graphing utility, problems for the student to solve, and a **group activity that involves the use of technology** at the end of each chapter. Beginning with the group activity at the end of Chapter 1, and continuing throughout the text, **linear regression** on a graphing utility is used at appropriate points to illustrate **mathematical modeling with real data.** All the optional graphing utility material is clearly identified by either 🖩 or 🖩 and can be omitted without loss of continuity, if desired.

■ Graphs

All graphs are computer-generated to ensure mathematical accuracy. Graphing utility screens displayed in the text are actual output from a graphing calculator.

■ Additional Pedagogical Features

Annotation of examples and developments, in small color type, is found throughout the text to help students through critical stages (see Sections 1-1 and 3-1). **Think boxes** (dashed boxes) are used to enclose steps that are usually performed mentally (see Sections 1-1 and 3-4). **Boxes** are used to highlight important definitions, results, and step-by-step processes (see Sections 1-1 and 3-1). **Caution** statements appear throughout the text where student errors often occur (see Sections 3-1 and 3-2). **Functional use of color** improves the clarity of many illustrations, graphs, and developments, and guides students through certain critical steps (see Sections 1-1 and 3-1). **Boldface type** is used to introduce new terms and highlight important comments. **Chapter review** sections include a review of all important terms and symbols and a comprehensive review exercise. **Answers to most review exercises,** keyed to appropriate sections, are included in the back of the book. Answers to all other odd-numbered problems are also in the back of the book.

■ Content

The text begins with the development of a library of elementary functions in Chapters 1 and 2, including their properties and uses. We encourage students to investigate mathematical ideas and processes **graphically** and **numerically,** as well as **algebraically.** This development lays a firm foundation for studying mathematics both in this book and in future endeavors. Depending on the syllabus for the course and the background of the students, some or all of this material can be covered at the beginning of a course, or selected portions can be referred to as needed later in the course.

The material in Part Two (Calculus) consists of **differential calculus** (Chapters 3–5), **integral calculus** (Chapters 6 and 7), and a variety of **additional calculus topics** (Chapters 8–12). In general, Chapters 3–6 must be covered in sequence; however, certain sections can be omitted or given brief treatments, as pointed out in the discussion that follows (see the Chapter Dependency Chart on page vii).

Chapter 3 introduces the **derivative.** The first two sections cover **limits, continuity,** and the **limit properties** that are essential to understanding the definition of the derivative in Section 3-3. The remaining sections of the chapter develop **rules of differentiation** (including the chain rule for power forms) and introduce **applications** of derivatives in business and economics. The interplay between graphical, numerical, and algebraic concepts is emphasized here and throughout the text.

Chapter 4 focuses on **graphing** and **optimization.** The first two sections cover first-derivative and second-derivative graph properties, while emphasizing **polynomial graphing. Rational function** graphing is covered in Section 4-3. In a course that does not include graphing rational functions, this section can be omitted or given a brief treatment. Optimization is covered in Section 4-4, including examples and problems involving end-point solutions.

The first three sections of Chapter 5 extend the derivative concepts discussed in Chapters 3 and 4 to **exponential and logarithmic functions** (including the general form of the chain rule). This material is required for all the remaining chapters. **Implicit differentiation** is introduced in Section 5-4 and applied to **related rate problems** in Section 5-5. These topics are not referred to elsewhere in the text and can be omitted.

Chapter 6 introduces **integration.** The first two sections cover **antidifferentiation** techniques essential to the remainder of the text. Section 6-3 discusses some applications involving **differential equations** that can be omitted. Sections 6-4 and 6-5 discuss the **definite integral** in terms of **Riemann sums,** including **approximations** with various types of sums and some **simple error estimation.** As before, the interplay between graphical, numeric, and algebraic properties is emphasized. These two sections also are required for the remaining chapters in the text.

Chapter 7 covers **additional integration topics** and is organized to provide maximum flexibility for the instructor. The first section extends the **area** concepts introduced in Chapter 6 to the area between two curves and related applications. Section 7-2 covers three more **applications** of integration, and Sections 7-3 and 7-4 deal with additional **techniques of integration.** Any or all of the topics in Chapter 7 can be omitted.

The first five sections of Chapter 8 deal with **differential multivariable calculus** and can be covered any time after Section 5-3 has been completed. Section 8-6 requires the **integration** concepts discussed in Chapter 6.

After introducing the basic concepts and terminology used in the study of **differential equations** in Section 9-1, **separable differential equations** (Section 9-2) and **first-order linear differential equations** (Section 9-3) and related applications are covered thoroughly. All the growth laws introduced in Section 6-3 are covered again—this time with more emphasis on student recognition of the relevant growth law.

Chapter 10 begins by discussing the approximation of functions by **Taylor polynomials. Infinite series** follow naturally from the presentation, eliminating the need for a lengthy treatment of series of constants and all the associated convergence tests. The operations that can be performed on Taylor series and the effects of these operations on the interval of convergence are carefully discussed, since this is the way Taylor series are used in most real-world applications.

After discussing **improper integrals** in Section 11-1, the remainder of Chapter 11 discusses properties of **continuous probability density functions,** including the **uniform, exponential, and normal probability distributions.** The presentation, though comprehensive, assumes no previous experience with probability.

Chapter 12 provides brief coverage of **trigonometric functions** that can be incorporated into the course, if desired. Section 12-1 provides a review of basic trigonometric concepts. Section 12-2 can be covered any time after Section 5-3 has been completed. Section 12-3 requires the material in Chapter 6.

Appendix A contains a **self-test** and a **concise review of basic algebra** that also may be covered as part of the course or referred to as needed. Appendix B contains additional topics that can be covered in conjunction with certain sections in the text, if desired.

■ Supplements for the Student

1. A **Student Solutions Manual and Visual Calculus** by Garret J. Etgen and David Schneider is available through your book store. The manual includes detailed solutions to all odd-numbered problems and all review exercises. *Visual Calculus* by David Schneider contains over twenty routines that provide additional insight into the topics discussed in the text. Although this software has much of the computing power of standard calculus software packages, it is primarily a teaching tool that focuses on understanding mathematical concepts, rather than on computing. These routines incorporate graphics whenever possible to illustrate topics such as secant lines; tangent lines; velocity; optimization; the relationship between the graphs of f, f', f''; and the various approaches to approximating definite integrals. All the routines in this software package are menu-driven and very easy to use. The software will run on DOS or Windows platforms.

2. The **PH Companion Website,** designed to complement and expand upon the text, offers a variety of teaching and learning tools, including links to related websites, practice work for students, and the ability for instructors to monitor and evaluate students' work on the website. For more information, contact your local Prentice Hall representative.
   www.prenhall.com/barnett

3. **CourseCompass/Blackboard/WebCT** offers Course compatible content including Excel Projects, Quizzes, Chapter Destinations, Lecture Notes, and Graphing Calculator Help.

■ Supplements for the Instructor

For a summary of all available supplementary materials and detailed information regarding examination copy requests and orders, see page xix.

1. **PH Custom Test, a menu-driven random test system** for either Windows or Macintosh is available to instructors. The test system has been greatly expanded and now offers **on-line testing.** Carefully constructed algorithms use random-number generators to produce different, yet equivalent, versions of each of these problems. In addition, the system incorporates a unique **editing function** that allows the instructor to create additional problems, or alter any of the existing problems in the test, using a full set of mathematical notation. The test system offers **free-response, multiple-choice, and mixed exams.** An almost unlimited number of quizzes, review exercises, chapter tests, midterms, and final examinations, each different from the other, can be generated quickly and easily. At the same time, the system will produce answer keys, student worksheets, and a gradebook for the instructor, if desired.

2. A **Test Item File,** prepared by Laurel Technical Services, provides a hard copy of the test items available in PH Custom Test.

3. An **Instructor's Solutions Manual** provides detailed solutions to the problems not solved in the Student Solutions Manual. This manual is available to instructors without charge.

4. A **Student Solutions Manual and Visual Calculus** by Garret J. Etgen and David Schneider (see Supplements for the Student) is available to instructors.

5. The **PH Companion Website,** designed to complement and expand upon the text, offers a variety of interactive teaching and learning tools, including

links to related websites, practice work for students, and the ability for instructors to monitor and evaluate students' work on the website. For more information, contact your local Prentice Hall representative. www.prenhall.com/barnett

6. **CourseCompass/Blackboard/WebCT** offers Course compatible content including Excel Projects, Quizzes, Chapter Destinations, Lecture Notes, and Graphing Calculator Help. **CourseCompass** is the perfect course management solution that combines quality Pearson Education content with state-of-the-art Blackboard technology! It is a dynamic, interactive online course management tool powered by Blackboard. This exciting product allows you to teach with market-leading Pearson Education content in an easy-to-use customizable format. **Blackboard 5**^{SM} is a comprehensive and flexible e-Learning software platform that delivers a course management system, customizable institution-wide portals, online communities, and an advanced architecture that allows for Web-based integration of multiple administrative systems. **WebCT** is one of the most popular Web course platforms in higher education today. It is the first destination site for the higher education marketplace to offer both teaching and learning resources and a community of peers across course and institutional boundaries.

■ **Error Check**

Because of the careful checking and proofing by a number of mathematics instructors (acting independently), the authors and publisher believe this book to be substantially error-free. For any errors remaining, the authors would be grateful if they were sent to: Karl E. Byleen, 9322 W. Garden Court, Hales Corners, WI 53130; or, by e-mail, to: byleen@execpc.com

■ **Acknowledgments**

In addition to the authors, many others are involved in the successful publication of a book.

We wish to thank our colleagues who have provided input on earlier editions:

Chris Boldt, Bob Bradshaw, Celeste Carter, Bruce Chaffee, Robert Chaney, Dianne Clark, Charles E. Cleaver, Barbara Cohen, Richard L. Conlon, Catherine Cron, Madhu Deshpande, John Dickerson, Kenneth A. Dodaro, Michael W. Ecker, Jerry R. Ehman, Lucina Gallagher, Joel Haack, Martha M. Harvey, Sue Henderson, Lloyd R. Hicks, Louis F. Hoelzle, Paul Hutchins, K. Wayne James, Robert H. Johnston, Robert Krystock, James T. Loats, Frank Lopez, Roy H. Luke, Mel Mitchell, Michael Montano, Ronald Persky, Shala Peterman, Kenneth A. Peters, Jr., Tom Plavchak, Bob Prielipp, Stephen Rodi, Arthur Rosenthal, Sheldon Rothman, Elaine Russell, Daniel E. Scanlon, George R. Schriro, Arnold L. Schroeder, Hari Shanker, Larry Small, Joan Smith, Steven Terry, Delores A. Williams, Caroline Woods, Charles W. Zimmerman, and Pat Zrolka.

We also express our thanks to:

Hossein Hamedani, Carolyn Meitler, Greg Rosik, and Caroline Woods for providing a careful and thorough check of all the mathematical calculations in the book, the *Student Solutions Manual,* and the *Instructor's Solutions Manual* (a tedious but extremely important job).

Garret Etgen, Hossein Hamedani, Carolyn Meitler, and David Schneider for developing the supplemental manuals that are so important to the success of a text.

Jeanne Wallace for accurately and efficiently producing most of the manuals that supplement the text.

George Morris and his staff at Scientific Illustrators for their effective illustrations and accurate graphs.

All the people at Prentice Hall who contributed their efforts to the production of this book, especially Quincy McDonald, our acquisitions editor, and Lynn Savino Wendel, our production editor.

Producing this new edition with the help of all these extremely competent people has been a most satisfying experience.

R. A. Barnett
M. R. Ziegler
K. E. Byleen

■ Ordering Information

When requesting examination copies or placing orders for this text or any of the related supplementary material listed below, please refer to the corresponding ISBN numbers.

TITLE	ISBN NUMBER
Applied Calculus for Business, Economics, Life Sciences, and Social Sciences, Eighth Edition	0-13-065589-9
Test Item File to accompany Applied Calculus, Eighth Edition	0-13-065934-7
Computer-generated test system for Applied Calculus, Eighth Edition	
TestGen EQ WIN/Mac CD	0-13-065937-1
Instructor's Solutions Manual to accompany Applied Calculus, Eighth Edition	0-13-065580-5
Student Solutions Manual and Visual Calculus	0-13-092192-0
WebCT, Ninth Edition	0-13-035318-3
Blackboard, Ninth Edition	0-13-035317-5
Course Compass	0-13-096914-1

A Library of Elementary Functions

A Beginning Library of Elementary Functions

INTRODUCTION

The function concept is one of the most important ideas in mathematics. The study of mathematics beyond the elementary level requires a firm understanding of a basic list of elementary functions, their properties, and their graphs. See the inside front cover of this book for a list of the functions that form our library of elementary functions. Most functions in the list will be introduced to you by the end of Chapter 2 and should become a part of your mathematical toolbox for use in this and most future courses or activities that involve mathematics. A few more elementary functions may be added to these in other courses, but the functions listed inside the front cover are more than sufficient for all the applications in this text.

Section 1-1

Functions

❑ Cartesian Coordinate System
❑ Graphing: Point-by-Point
❑ Definition of a Function
❑ Functions Specified by Equations
❑ Function Notation
❑ Applications

After a brief review of the Cartesian (rectangular) coordinate system in the plane and point-by-point graphing, we discuss the concept of function, one of the most important ideas in mathematics.

❑ CARTESIAN COORDINATE SYSTEM

Recall that to form a **Cartesian** or **rectangular coordinate system,** we select two real number lines, one horizontal and one vertical, and let them cross through their origins as indicated in Figure 1. Up and to the right are the usual choices for the positive directions. These two number lines are called the **horizontal axis** and the **vertical axis,** or, together, the **coordinate axes.** The horizontal axis is usually referred to as the **x axis** and the vertical axis as the **y axis,** and each is labeled accordingly. Other labels may be used in certain situations. The coordinate axes divide the plane into four parts called **quadrants,** which are numbered counterclockwise from I to IV (see Fig. 1).

FIGURE 1 The Cartesian (rectangular) coordinate system

Now we want to assign *coordinates* to each point in the plane. Given an arbitrary point P in the plane, pass horizontal and vertical lines through the point (Fig. 1). The vertical line will intersect the horizontal axis at a point with coordinate a, and the horizontal line will intersect the vertical axis at a point with coordinate b. These two numbers written as the **ordered pair** (a, b) form the **coordinates** of the point P. The first coordinate, a, is called the **abscissa** of P; the second coordinate, b, is called the **ordinate** of P. The abscissa of Q in Figure 1 is -5, and the ordinate of Q is 5. The coordinates of a point can also be referenced in terms of the axis labels. The **x coordinate** of R in Figure 1 is 10, and the **y coordinate** of R is -10. The point with coordinates $(0, 0)$ is called the **origin.**

The procedure we have just described assigns to each point P in the plane a unique pair of real numbers (a, b). Conversely, if we are given an ordered pair of real numbers (a, b), then, reversing this procedure, we can determine a unique point P in the plane. Thus:

There is a one-to-one correspondence between the points in a plane and the elements in the set of all ordered pairs of real numbers.

This is often referred to as the **fundamental theorem of analytic geometry.**

❑   GRAPHING: POINT BY POINT

The fundamental theorem of analytic geometry allows us to look at algebraic forms geometrically and to look at geometric forms algebraically. We begin by considering an algebraic form, an equation in two variables:

$$y = 9 - x^2 \tag{1}$$

A **solution** to equation (1) is an ordered pair of real numbers (a, b) such that

$$b = 9 - a^2$$

The **solution set** for equation (1) is the set of all these ordered pairs.

To find a solution to equation (1), we replace x with a number and calculate the value of y. For example, if $x = 2$, then $y = 9 - 2^2 = 5$, and the ordered pair $(2, 5)$ is a solution of equation (1). Similarly, if $x = -3$, then $y = 9 - (-3)^2 = 0$, and $(-3, 0)$ is a solution. Since any real number substituted for x in equation (1) will produce a solution, the solution set must have an infinite number of elements. We use a rectangular coordinate system to provide a geometric representation of this set.

The **graph of an equation** is the graph of all the ordered pairs in its solution set. To **sketch the graph of an equation,** we plot enough points from its solution set in a rectangular coordinate system so that the total graph is apparent and then connect these points with a smooth curve. This process is called **point-by-point plotting.**

Example 1   ⇔   **Point-by-Point Plotting**   Sketch a graph of $y = 9 - x^2$.

SOLUTION   Make up a table of solutions—that is, ordered pairs of real numbers that satisfy the given equation. For easy mental calculation, choose integer values for x.

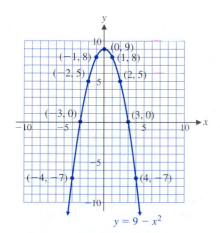

x	-4	-3	-2	-1	0	1	2	3	4
y	-7	0	5	8	9	8	5	0	-7

$$y = 9 - x^2$$

FIGURE 2

After plotting these solutions, if there are any portions of the graph that are unclear, plot additional points until the shape of the graph is apparent. Then join all the plotted points with a smooth curve as shown in Figure 2. Arrowheads are used to indicate that the graph continues beyond the portion shown here with no significant changes in shape.   ■

*Matched Problem 1**   ⇔   Sketch a graph of $y = x^2 - 4$ using point-by-point plotting.   ■

*Answer to matched problems are found near the end of each section, before the exercise set.

To graph the equation $y = -x^3 + 3x$, we use point-by-point plotting to obtain

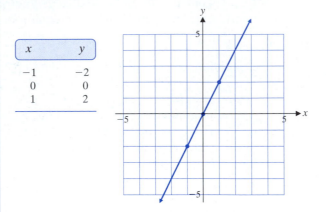

x	y
-1	-2
0	0
1	2

(A) Do you think this is the correct graph of the equation? If so, why? If not, why?

(B) Add points on the graph for $x = -2, -1.5, -0.5, 0.5, 1.5,$ and 2.

(C) Now, what do you think the graph looks like? Sketch your version of the graph, adding more points as necessary.

 (D) Graph this equation on a graphing utility and compare it with your graph from part (C).

The icon in the margin is used throughout this book to identify optional graphing utility activities that are intended to give you additional insight into the concepts under discussion. You may have to consult the manual for your **www** graphing utility or the companion website (see Preface) for the details necessary to carry out these activities. For example, to graph the equation in Explore–Discuss 1 on most graphing utilities, you first have to enter the equation (Fig. 3A) and the window variables (Fig. 3B).

As Explore–Discuss 1 illustrates, the shape of a graph may not be "apparent" from your first choice of points on the graph. One of the objectives of this chapter is to provide you with a library of basic equations that will aid you in sketching graphs. For example, the curve in Figure 2 is called a *parabola*. Notice that if we fold the paper along the y axis, the right side will match the left side. We say that the graph is *symmetric with respect to the y axis* and call the y axis the *axis of the parabola*. Later in this chapter, we will see that all parabolas have graphs with similar properties. Identifying a given equation as one whose graph will be a parabola simplifies graphing the equation by hand.

(A)

(B)

FIGURE 3

❑   DEFINITION OF A FUNCTION

Central to the concept of function is correspondence. You have already had experiences with correspondences in daily living. For example:

To each person there corresponds an annual income.

To each item in a supermarket there corresponds a price.

To each student there corresponds a grade-point average.

To each day there corresponds a maximum temperature.

For the manufacture of x items there corresponds a cost.

For the sale of x items there corresponds a revenue.

To each square there corresponds an area.

To each number there corresponds its cube.

One of the most important aspects of any science is the establishment of correspondences among various types of phenomena. Once a correspondence is known, predictions can be made. A cost analyst would like to predict costs for various levels of output in a manufacturing process; a medical researcher would like to know the correspondence between heart disease and obesity; a psychologist would like to predict the level of performance after a subject has repeated a task a given number of times; and so on.

What do all the examples above have in common? Each describes the matching of elements from one set with the elements in a second set. Consider the tables of the cube, square, and square root given in Tables 1–3.

TABLE 1	
DOMAIN	RANGE
Number	*Cube*

TABLE 2	
DOMAIN	RANGE
Number	*Square*

TABLE 3	
DOMAIN	RANGE
Number	*Square root*

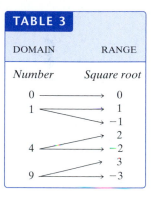

Tables 1 and 2 specify functions, but Table 3 does not. Why not? The definition of the term *function* will explain.

Definition of a Function

A **function** is a rule (process or method) that produces a correspondence between two sets of elements such that to each element in the first set there corresponds one and only one element in the second set.

The first set is called the **domain,** and the set of corresponding elements in the second set is called the **range.**

Tables 1 and 2 specify functions, since to each domain value there corresponds exactly one range value (for example, the cube of -2 is -8 and no other number). On the other hand, Table 3 does not specify a function, since to at least one domain value there corresponds more than one range value (for example, to the domain value 9 there corresponds -3 and 3, both square roots of 9).

Consider the set of students enrolled in a college and the set of faculty members of that college. Suppose we define a correspondence between the two sets by saying that a student corresponds to a faculty member if the student is currently enrolled in a course taught by that faculty member. Is this correspondence a function? Discuss.

❑ FUNCTIONS SPECIFIED BY EQUATIONS

$x - 1$

x

FIGURE 4

Most of the domains and ranges included in this book will be (infinite) sets of real numbers, and the rules associating range values with domain values will be equations in two variables. Consider, for example, the equation for the area of a rectangle with width 1 inch less than its length (Fig. 4). If x is the length, then the area y is given by

$$y = x(x - 1) \qquad x \geq 1$$

For each **input** x (length), we obtain an **output** y (area). For example:

If	$x = 5,$	then	$y = \mathbf{5}(\mathbf{5} - 1) = 5 \cdot 4 = 20.$
If	$x = 1,$	then	$y = \mathbf{1}(\mathbf{1} - 1) = 1 \cdot 0 = 0.$
If	$x = \sqrt{5},$	then	$y = \sqrt{\mathbf{5}}(\sqrt{\mathbf{5}} - 1) = 5 - \sqrt{5}$
			$\approx 2.76.$

The input values are domain values, and the output values are range values. The equation (a rule) assigns each domain value x a range value y. The variable x is called an *independent variable* (since values can be "independently" assigned to x from the domain), and y is called a *dependent variable* (since the value of y "depends" on the value assigned to x). In general, any variable used as a placeholder for domain values is called an **independent variable;** any variable that is used as a placeholder for range values is called a **dependent variable.**

When does an equation specify a function?

Functions Defined by Equations

If in an equation in two variables, we get exactly one output (value for the dependent variable) for each input (value for the independent variable), then the equation defines a function.

If we get more than one output for a given input, the equation does not define a function.

Example 2 ⇔ **Functions and Equations** Determine which of the following equations specify functions with independent variable x.

(A) $4y - 3x = 8,$ x a real number   (B) $y^2 - x^2 = 9,$ x a real number

SOLUTION (A) Solving for the dependent variable y, we have

$$4y - 3x = 8 \tag{2}$$

$$4y = 8 + 3x$$

$$y = 2 + \tfrac{3}{4}x$$

Since each input value x corresponds to exactly one output value ($y = 2 + \frac{3}{4}x$), we see that equation (2) specifies a function.

(B) Solving for the dependent variable y, we have

$$y^2 - x^2 = 9 \tag{3}$$
$$y^2 = 9 + x^2$$
$$y = \pm\sqrt{9 + x^2}$$

Since $9 + x^2$ is always a positive real number for any real number x and since each positive real number has two square roots,* to each input value x there corresponds two output values ($y = -\sqrt{9 + x^2}$ and $y = \sqrt{9 + x^2}$). For example, if $x = 4$, then equation (3) is satisfied for $y = 5$ and for $y = -5$. Thus, equation (3) does not specify a function. ∎

Matched Problem 2 ⇨   Determine which of the following equations specify functions with independent variable x.

(A) $y^2 - x^4 = 9$, x a real number

(B) $3y - 2x = 3$, x a real number

∎

Since the graph of an equation is the graph of all the ordered pairs that satisfy the equation, it is very easy to determine whether an equation specifies a function by examining its graph. The graphs of the two equations we considered in Example 2 are shown in Figure 5.

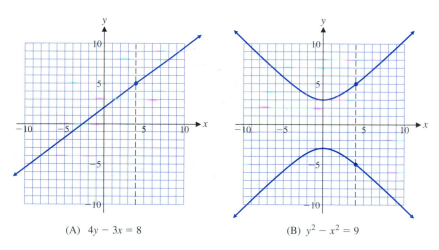

(A) $4y - 3x = 8$ (B) $y^2 - x^2 = 9$

FIGURE 5

In Figure 5A notice that any vertical line will intersect the graph of the equation $4y - 3x = 8$ in exactly one point. This shows that to each x value there corresponds exactly one y value and confirms our conclusion that this equation specifies a function. On the other hand, Figure 5B shows that there exist vertical lines that intersect the graph of $y^2 - x^2 = 9$ in two points. This indicates that there exist x values to which there correspond two different y values and verifies our conclusion that this equation does not specify a function. These observations are generalized in Theorem 1.

*Recall that each positive real number N has two square roots: $\sqrt{N}$, the principal square root, and $-\sqrt{N}$, the negative of the principal square root (see Appendix A-7).

THEOREM 1   Vertical-Line Test for a Function

An equation defines a function if each vertical line in the coordinate system passes through at most one point on the graph of the equation.

If any vertical line passes through two or more points on the graph of an equation, then the equation does not define a function.

Explore–Discuss 3

The definition of a function specifies that to each element in the domain there corresponds one and only one element in the range.

(A) Give an example of a function such that to each element of the range there correspond exactly two elements of the domain.

(B) Give an example of a function such that to each element of the range there corresponds exactly one element of the domain.

In Example 2, the domains were explicitly stated along with the given equations. In many cases, this will not be done. Unless stated to the contrary, we shall adhere to the following convention regarding domains and ranges for functions specified by equations:

> ### Agreement on Domains and Ranges
>
> If a function is specified by an equation and the domain is not indicated, then we assume that the domain is the set of all real number replacements of the independent variable (inputs) that produce real values for the dependent variable (outputs). The range is the set of all outputs corresponding to input values.
>
> In many applied problems the domain is determined by practical considerations within the problem (see Example 7).

Example 3   **Finding a Domain**   Find the domain of the function specified by the equation $y = \sqrt{4 - x}$, assuming that x is the independent variable.

Solution   For y to be real, $4 - x$ must be greater than or equal to 0; that is,

$$4 - x \geq 0$$
$$-x \geq -4$$
$$x \leq 4 \qquad \text{Sense of inequality reverses when both sides are divided by } -1.$$

Thus,

Domain: $x \leq 4$ (inequality notation) or $(-\infty, 4]$ (interval notation)

Matched Problem 3   Find the domain of the function specified by the equation $y = \sqrt{x - 2}$, assuming x is the independent variable.

❑   FUNCTION NOTATION

We have just seen that a function involves two sets, a domain and a range, and a rule of correspondence that enables us to assign to each element in the domain exactly one element in the range. We use different letters to denote names for numbers; in essentially the same way, we will now use different letters to denote names for functions. For example, f and g may be used to name the functions specified by the equations $y = 2x + 1$ and $y = x^2 + 2x - 3$:

$$f: \quad y = 2x + 1$$
$$g: \quad y = x^2 + 2x - 3 \qquad\qquad (4)$$

If x represents an element in the domain of a function f, then we frequently use the symbol

$$\boldsymbol{f(x)}$$

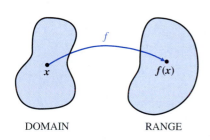

DOMAIN   RANGE

FIGURE 6

in place of y to designate the number in the range of the function f to which x is paired (Fig. 6). This symbol does *not* represent the product of f and x. The symbol $f(x)$ is read as "f of x," "f at x," or "the value of f at x." Whenever we write $y = f(x)$, we assume that the variable x is an independent variable and that both y and $f(x)$ are dependent variables.

Using function notation, we can now write functions f and g in (4) in the form

$$f(x) = 2x + 1 \qquad \text{and} \qquad g(x) = x^2 + 2x - 3$$

Let us find $f(3)$ and $g(-5)$. To find $f(3)$, we replace x with 3 wherever x occurs in $f(x) = 2x + 1$ and evaluate the right side:

$$f(x) = 2x + 1$$
$$f(\mathbf{3}) = 2 \cdot \mathbf{3} + 1$$
$$ = 6 + 1 = 7 \qquad \text{For input 3, the output is 7.}$$

Thus,

$$f(3) = 7 \qquad \text{The function } f \text{ assigns the range value 7 to the domain value 3.}$$

To find $g(-5)$, we replace x by -5 wherever x occurs in $g(x) = x^2 + 2x - 3$ and evaluate the right side:

$$g(x) = x^2 + 2x - 3$$
$$g(-\mathbf{5}) = (-\mathbf{5})^2 + 2(-\mathbf{5}) - 3$$
$$ = 25 - 10 - 3 = 12 \qquad \text{For input } -5, \text{ the output is 12.}$$

Thus,

$$g(-5) = 12 \qquad \text{The function } g \text{ assigns the range value 12 to the domain value } -5.$$

It is very important to understand and remember the definition of $f(x)$:

The Symbol $f(x)$

For any element x in the domain of the function f, the symbol $\boldsymbol{f(x)}$ represents the element in the range of f corresponding to x in the domain of f. If x is an input value, then $f(x)$ is the corresponding output value. If x is an element that is not in the domain of f, then f is *not defined at* x and $f(x)$ *does not exist.*

Example 4 ⇔ **Function Evaluation** If

$$f(x) = \frac{12}{x - 2} \qquad g(x) = 1 - x^2 \qquad h(x) = \sqrt{x - 1}$$

then:

(A) $f(6) \boxed{= \frac{12}{6 - 2}}^{*} = \frac{12}{4} = 3$

(B) $g(-2) \boxed{= 1 - (-2)^2} = 1 - 4 = -3$

(C) $h(-2) \boxed{= \sqrt{-2 - 1}} = \sqrt{-3}$

But $\sqrt{-3}$ is not a real number. Since we have agreed to restrict the domain of a function to values of x that produce real values for the function, -2 is not in the domain of h and $h(-2)$ does not exist.

(D) $f(0) + g(1) - h(10) \boxed{= \frac{12}{0 - 2} + (1 - 1^2) - \sqrt{10 - 1}}$

$$= \frac{12}{-2} + 0 - \sqrt{9}$$

$$= -6 - 3 = -9$$

■

Matched Problem 4 ⇔ Use the functions in Example 4 to find:

(A) $f(-2)$ (B) $g(-1)$ (C) $h(-8)$ (D) $\dfrac{f(3)}{h(5)}$

■

Example 5 ⇔ **Finding Domains** Find the domains of functions f, g, and h:

$$f(x) = \frac{12}{x - 2} \qquad g(x) = 1 - x^2 \qquad h(x) = \sqrt{x - 1}$$

Solution   *Domain of f:* $12/(x - 2)$ represents a real number for all replacements of x by real numbers except for $x = 2$ (division by 0 is not defined). Thus, $f(2)$ does not exist, and the domain of f is the set of all real numbers except 2. We often indicate this by writing

$$f(x) = \frac{12}{x - 2} \qquad x \neq 2$$

Domain of g: The domain is R, the set of all real numbers, since $1 - x^2$ represents a real number for all replacements of x by real numbers.

Domain of h: The domain is the set of all real numbers x such that $\sqrt{x - 1}$ is a real number—that is, such that

$$x - 1 \geq 0$$
$$x \geq 1 \qquad \text{or} \qquad [1, \infty)$$

■

---

*Dashed boxes are used throughout the book to represent steps that are usually performed mentally.

Matched Problem 5 ⟿ Find the domains of functions F, G, and H:

$$F(x) = x^2 - 3x + 1 \qquad G(x) = \frac{5}{x + 3} \qquad H(x) = \sqrt{2 - x}$$ ■

In addition to evaluating functions at specific numbers, it is important to be able to evaluate functions at expressions that involve one or more variables. For example, the **difference quotient**

$$\frac{f(x + h) - f(x)}{h} \qquad x \text{ and } x + h \text{ in the domain of } f, h \neq 0$$

is studied extensively in calculus.

Explore–Discuss 4

Let x and h be real numbers.

(A) If $f(x) = 4x + 3$, which of the following is true?

(1) $f(x + h) = 4x + 3 + h$
(2) $f(x + h) = 4x + 4h + 3$
(3) $f(x + h) = 4x + 4h + 6$

(B) If $g(x) = x^2$, which of the following is true?

(1) $g(x + h) = x^2 + h$
(2) $g(x + h) = x^2 + h^2$
(3) $g(x + h) = x^2 + 2hx + h^2$

(C) If $M(x) = x^2 + 4x + 3$, describe the operations that must be performed to evaluate $M(x + h)$.

Example 6 ⟿ **Using Function Notation**   For $f(x) = x^2 - 2x + 7$, find:

(A) $f(a)$ (B) $f(a + h)$ (C) $\dfrac{f(a + h) - f(a)}{h}$

SOLUTION   (A) $f(a) = a^2 - 2a + 7$

(B) $f(a + h) = (a + h)^2 - 2(a + h) + 7$

$$= a^2 + 2ah + h^2 - 2a - 2h + 7$$

(C) $\dfrac{f(a + h) - f(a)}{h} = \dfrac{(a^2 + 2ah + h^2 - 2a - 2h + 7) - (a^2 - 2a + 7)}{h}$

$$= \frac{2ah + h^2 - 2h}{h} = \frac{h(2a + h - 2)}{h} = 2a + h - 2$$ ■

Matched Problem 6 ⟿ Repeat Example 6 for $f(x) = x^2 - 4x + 9$. ■

 ❑ APPLICATIONS

We now turn to the important concepts of **break-even** and **profit-loss** analysis, which we will return to a number of times in this book. Any manufacturing company has **costs, C**, and **revenues, R**. The company will have a **loss** if $R < C$, will **break even** if $R = C$, and will have a **profit** if $R > C$. Costs include **fixed**

costs such as plant overhead, product design, setup, and promotion; and **variable costs,** which are dependent on the number of items produced at a certain cost per item. In addition, **price–demand** functions, usually established by financial departments using historical data or sampling techniques, play an important part in profit–loss analysis. We will let x, the number of units manufactured and sold, represent the independent variable. Cost functions, revenue functions, profit functions, and price–demand functions are often stated in the following forms, where $a, b, m,$ and n are constants determined from the context of a particular problem:

Cost Function

$$C = (\text{fixed costs}) + (\text{variable costs})$$
$$= a + bx$$

Price–Demand Function

$$p = m - nx \qquad \text{x is the number of items that can be sold at \$p per item.}$$

Revenue Function

$$R = (\text{number of items sold}) \times (\text{price per item})$$
$$= xp = x(m - nx)$$

Profit Function

$$P = R - C$$
$$= x(m - nx) - (a + bx)$$

Example 7 and Matched Problem 7 explore the relationships among the algebraic definition of a function, the numerical values of the function, and the graphical representation of the function. The interplay among algebraic, numeric, and graphic viewpoints is an important aspect of our treatment of functions and their use. In Example 7, we also see how a function can be used to describe data from the real world, a process that is often referred to as *mathematical modeling.* The material in this example will be returned to in subsequent sections so that we can analyze it in greater detail and from different points of view.

Example 7 **Price–Demand and Revenue Modeling** A manufacturer of a popular automatic camera wholesales the camera to retail outlets throughout the United States. Using statistical methods, the financial department in the company produced the price–demand data in Table 4, where p is the wholesale

TABLE 4	
PRICE–DEMAND	
x(MILLIONS)	p($)
2	87
5	68
8	53
12	37

TABLE 5	
REVENUE	
x (MILLIONS)	$R(x)$ (MILLION $)
1	90
3	
6	
9	
12	
15	

price per camera at which x million cameras are sold. Notice that as the price goes down, the number sold goes up.

Using special analytical techniques (regression analysis), an analyst arrived at the following price–demand function that models the Table 4 data:

$$p(x) = 94.8 - 5x \qquad 1 \leq x \leq 15 \qquad (5)$$

(A) Plot the data in Table 4. Then sketch a graph of the price–demand function in the same coordinate system.

(B) What is the company's revenue function for this camera, and what is the domain of this function?

(C) Complete Table 5, computing revenues to the nearest million dollars.

(D) Plot the data in Table 5. Then sketch a graph of the revenue function using these points.

(E) Plot the revenue function on a graphing utility.

SOLUTION   (A)

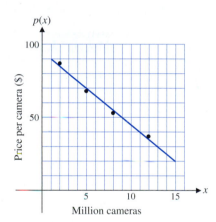

FIGURE 7 Price–demand

In Figure 7, notice that the model approximates the actual data in Table 4, and it is assumed that it gives realistic and useful results for all other values of x between 1 million and 15 million.

(B) $R(x) = xp(x) = x(94.8 - 5x)$ million dollars
     Domain: $1 \leq x \leq 15$
[Same domain as the price–demand function, equation (5).]

(C)

TABLE 5	
REVENUE	
x (MILLIONS)	$R(x)$ (MILLION $)
1	90
3	239
6	389
9	448
12	418
15	297

(D)

(E)

Matched Problem 7 The financial department in Example 7, using statistical techniques, produced the data in Table 6, where $C(x)$ is the cost in millions of dollars for manufacturing and selling x million cameras.

TABLE 6	
Cost Data	
x (MILLIONS)	$C(x)$ (MILLION $)
1	175
5	260
8	305
12	395

Using special analytical techniques (regression analysis), an analyst produced the following cost function to model the data:

$$C(x) = 156 + 19.7x \qquad 1 \le x \le 15 \tag{6}$$

(A) Plot the data in Table 6. Then sketch a graph of equation (6) in the same coordinate system.

(B) What is the company's profit function for this camera, and what is its domain?

(C) Complete Table 7, computing profits to the nearest million dollars.

TABLE 7	
Profit	
x (MILLIONS)	$P(x)$ (MILLION $)
1	−86
3	
6	
9	
12	
15	

(D) Plot the points from part (C). Then sketch a graph of the profit function through these points.

(E) Plot the profit function on a graphing utility.

Answers to Matched Problems   **1.**

2. (A) Does not specify a function
   (B) Specifies a function

3. $x \geq 2$ (inequality notation) or $[2, \infty)$ (interval notation)

4. (A) -3 (B) 0 (C) Does not exist   (D) 6

5. Domain of F: R; domain of G: all real numbers except -3;
   domain of H: $x \leq 2$ (inequality notation) or $(-\infty, 2]$ (interval notation)

6. (A) $a^2 - 4a + 9$ (B) $a^2 + 2ah + h^2 - 4a - 4h + 9$

   (C) $2a + h - 4$

7. (A)

(B) $P(x) = R(x) - C(x) = x(94.8 - 5x) - (156 + 19.7x)$;
   domain: $1 \leq x \leq 15$

(C)

TABLE 7

PROFIT

x(MILLIONS)	$P(x)$ (MILLION \$)
1	-86
3	24
6	115
9	115
12	25
15	-155

(D)

 (E)

Exercise 1-1

A *Indicate whether each table in Problems 1–6 specifies a function.*

1.

DOMAIN	RANGE
3 ⟶ 0	
5 ⟶ 1	
7 ⟶ 2	

2.

DOMAIN	RANGE
−1 ⟶ 5	
−2 ⟶ 7	
−3 ⟶ 9	

3.

DOMAIN	RANGE
3 ⟶ 5	
6	
4 ⟶ 7	
5 ⟶ 8	

4.

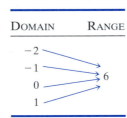

DOMAIN	RANGE
8 ⟶ 0	
9 ⟶ 1	
2	
10 ⟶ 3	

5.

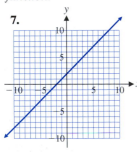

DOMAIN	RANGE
3	
6 ⟶ 5	
9	
12 ⟶ 6	

6.

DOMAIN	RANGE
−2	
−1	
0 ⟶ 6	
1	

Indicate whether each graph in Problems 7–12 specifies a function.

7.

8.

9.

10.

11.

12.

If $f(x) = 2x - 3$ and $g(x) = x^2 + 2x$, find each of the expressions in Problems 13–30.

13. $f(2)$ **14.** $f(1)$ **15.** $f(-1)$

16. $f(-2)$ **17.** $g(3)$ **18.** $g(1)$

19. $f(0)$ **20.** $f(\frac{1}{3})$ **21.** $g(-3)$

22. $g(-2)$ **23.** $f(1) + g(2)$ **24.** $g(1) + f(2)$

25. $g(2) - f(2)$ **26.** $f(3) - g(3)$ **27.** $g(3) \cdot f(0)$

28. $g(0) \cdot f(-2)$ **29.** $\dfrac{g(-2)}{f(-2)}$ **30.** $\dfrac{g(-3)}{f(2)}$

In Problems 31–38, use the following graph of a function f to determine x or y to the nearest integer, as indicated. Some problems may have more than one answer.

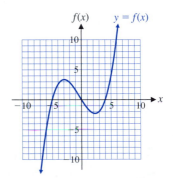

31. $y = f(-5)$ **32.** $y = f(4)$ **33.** $y = f(5)$

34. $y = f(-2)$ **35.** $0 = f(x)$ **36.** $3 = f(x), x < 0$

37. $-4 = f(x)$ **38.** $4 = f(x)$

B *In Problems 39–48, find the domain of each function.*

39. $F(x) = 2x^3 - x^2 + 3$

40. $H(x) = 7 - 2x^2 - x^4$

41. $f(x) = \dfrac{x - 2}{x + 4}$

42. $g(x) = \dfrac{x + 1}{x - 2}$

43. $F(x) = \dfrac{x + 2}{x^2 + 3x - 4}$

44. $G(x) = \dfrac{x - 7}{x^2 + x - 6}$

45. $g(x) = \sqrt{7 - x}$

46. $f(x) = \sqrt{5 + x}$

47. $G(x) = \dfrac{1}{\sqrt{7 - x}}$

48. $F(x) = \dfrac{1}{\sqrt{5 + x}}$

49. Two people are discussing the function

$$f(x) = \frac{x^2 - 4}{x^2 - 9}$$

and one says to the other, "$f(2)$ exists but $f(3)$ does not." Explain what they are talking about.

50. Referring to the function in Problem 49, do $f(-2)$ and $f(-3)$ exist? Explain.

The verbal statement "function f multiplies the square of the domain element by 3 and then subtracts 7 from the result" and the algebraic statement "$f(x) = 3x^2 - 7$" define the same function. In Problems 51–54, translate each verbal definition of a function into an algebraic definition.

51. Function g subtracts 5 from twice the cube of the domain element.

52. Function f multiplies the domain element by -3 and adds 4 to the result.

53. Function G multiplies the square root of the domain element by 2 and subtracts the square of the domain element from the result.

54. Function F multiplies the cube of the domain element by -8 and adds 3 times the square root of 3 to the result.

In Problems 55–58, translate each algebraic definition of the function into a verbal definition.

55. $f(x) = 2x - 3$

56. $g(x) = -2x + 7$

57. $F(x) = 3x^3 - 2\sqrt{x}$

58. $G(x) = 4\sqrt{x} - x^2$

Determine which of the equations in Problems 59–68 specify functions with independent variable x. For those that do, find the domain. For those that do not, find a value of x to which there corresponds more than one value of y.

59. $4x - 5y = 20$

60. $3y - 7x = 15$

61. $x^2 - y = 1$

62. $x - y^2 = 1$

63. $x + y^2 = 10$

64. $x^2 + y = 10$

65. $xy - 4y = 1$

66. $xy + y - x = 5$

67. $x^2 + y^2 = 25$

68. $x^2 - y^2 = 16$

69. If $F(t) = 4t + 7$, find:

$$\frac{F(3 + h) - F(3)}{h}$$

70. If $G(r) = 3 - 5r$, find:

$$\frac{G(2 + h) - G(2)}{h}$$

71. If $Q(x) = x^2 - 5x + 1$, find:

$$\frac{Q(2 + h) - Q(2)}{h}$$

72. If $P(x) = 2x^2 - 3x - 7$, find:

$$\frac{P(3 + h) - P(3)}{h}$$

C *In Problems 73–80, find and simplify:*

$$\frac{f(a + h) - f(a)}{h}$$

73. $f(x) = 4x - 3$

74. $f(x) = -3x + 9$

75. $f(x) = 4x^2 - 7x + 6$

76. $f(x) = 3x^2 + 5x - 8$

77. $f(x) = x^3$

78. $f(x) = x^3 - x$

79. $f(x) = \sqrt{x}$

80. $f(x) = \dfrac{1}{x}$

Problems 81–84 refer to the area A and perimeter P of a rectangle with length l and width w (see the figure).

$$A = lw$$
$$P = 2l + 2w$$

81. The area of a rectangle is 25 square inches. Express the perimeter $P(w)$ as a function of the width w, and state the domain of this function.

82. The area of a rectangle is 81 square inches. Express the perimeter $P(l)$ as a function of the length l, and state the domain of this function.

83. The perimeter of a rectangle is 100 meters. Express the area $A(l)$ as a function of the length l, and state the domain of this function.

84. The perimeter of a rectangle is 160 meters. Express the area $A(w)$ as a function of the width w, and state the domain of this function.

Applications

Business & Economics

85. *Price–demand.* A company manufactures memory chips for microcomputers. Its marketing research department, using statistical techniques, collected the data shown in Table 8, where p is the wholesale price per chip at which x million chips can be sold. Using special analytical techniques (regression analysis), an analyst produced the following price–demand function to model the data:

$$p(x) = 75 - 3x \qquad 1 \leqslant x \leqslant 20$$

TABLE 8

PRICE–DEMAND

x(MILLIONS)	p($)
1	72
4	63
9	48
14	33
20	15

Plot the data points in Table 8, and sketch a graph of the price–demand function in the same coordinate system. What would be the estimated price per chip for a demand of 7 million chips? For a demand of 11 million chips?

86. *Price–demand.* A company manufactures "notebook" computers. Its marketing research department, using statistical techniques, collected the data shown in Table 9, where p is the wholesale price per computer at which x thousand computers can be sold. Using special analytical techniques (regression analysis), an analyst produced the following price–demand function to model the data:

$$p(x) = 2,000 - 60x \qquad 1 \leqslant x \leqslant 25$$

TABLE 9

PRICE–DEMAND

x(THOUSANDS)	p($)
1	1,940
8	1,520
16	1,040
21	740
25	500

Plot the data points in Table 9, and sketch a graph of the price–demand function in the same coordinate system. What would be the estimated price per computer for a demand of 11 thousand computers? For a demand of 18 thousand computers?

87. *Revenue.*

   (A) Using the price–demand function

$$p(x) = 75 - 3x \qquad 1 \leqslant x \leqslant 20$$

   from Problem 85, write the company's revenue function and indicate its domain.

   (B) Complete Table 10, computing revenues to the nearest million dollars.

TABLE 10

REVENUE

x(MILLIONS)	$R(x)$ (MILLION $)
1	72
4	
8	
12	
16	
20	

   (C) Plot the points from part (B) and sketch a graph of the revenue function through these points. Choose millions for the units on the horizontal and vertical axes.

88. *Revenue.*

   (A) Using the price–demand function

$$p(x) = 2,000 - 60x \qquad 1 \leqslant x \leqslant 25$$

   from Problem 86, write the company's revenue function and indicate its domain.

   (B) Complete Table 11, computing revenues to the nearest thousand dollars.

TABLE 11

REVENUE

x(THOUSANDS)	$R(x)$ (THOUSAND $)
1	1,940
5	
10	
15	
20	
25	

(C) Plot the points from part (B) and sketch a graph of the revenue function through these points. Choose thousands for the units on the horizontal and vertical axes.

89. *Profit.* The financial department for the company in Problems 85 and 87 established the following cost function for producing and selling x million memory chips:

$$C(x) = 125 + 16x \text{ million dollars}$$

(A) Write a profit function for producing and selling x million memory chips, and indicate its domain.

(B) Complete Table 12, computing profits to the nearest million dollars.

TABLE 12	
PROFIT	
x (MILLIONS)	$P(x)$ (MILLION $)
1	−69
4	
8	
12	
16	
20	

(C) Plot the points in part (B) and sketch a graph of the profit function through these points.

90. *Profit.* The financial department for the company in Problems 86 and 88 established the following cost function for producing and selling x thousand "notebook" computers:

$$C(x) = 4,000 + 500x \text{ thousand dollars}$$

(A) Write a profit function for producing and selling x thousand "notebook" computers, and indicate the domain of this function.

(B) Complete Table 13, computing profits to the nearest thousand dollars.

TABLE 13	
PROFIT	
x (THOUSANDS)	$P(x)$ (THOUSAND $)
1	−2,560
5	
10	
15	
20	
25	

(C) Plot the points in part (B) and sketch a graph of the profit function through these points.

91. *Packaging.* A candy box is to be made out of a piece of cardboard that measures 8 by 12 inches. Equal-sized squares x inches on a side will be cut out of each corner, and then the ends and sides will be folded up to form a rectangular box.

(A) Express the volume of the box $V(x)$ in terms of x.

(B) What is the domain of the function V (determined by the physical restrictions)?

(C) Complete Table 14.

TABLE 14	
VOLUME	
x	$V(x)$
1	
2	
3	

(D) Plot the points in part (C) and sketch a graph of the volume function through these points.

92. *Packaging.* Refer to Problem 91.

(A) Table 15 shows the volume of the box for some values of x between 1 and 2. Use these values to estimate to one decimal place the value of x between 1 and 2 that would produce a box with a volume of 65 cubic inches.

TABLE 15	
VOLUME	
x	$V(x)$
1.1	62.524
1.2	64.512
1.3	65.988
1.4	66.976
1.5	67.5
1.6	67.584
1.7	67.252

(B) Describe how you could refine this table to estimate x to two decimal places.

(C) Carry out the refinement you described in part (B) and approximate x to two decimal places.

93. *Packaging.* Refer to Problems 91 and 92.

(A) Examine the graph of $V(x)$ from Problem 91D and discuss the possible locations of other values of x that would produce a box with a volume of 65 cubic inches. Construct a table like Table 15 to estimate any such value to one decimal place.

(B) Refine the table you constructed in part (A) to provide an approximation to two decimal places.

94. *Packaging.* A parcel delivery service will only deliver packages with length plus girth (distance around) not exceeding 108 inches. A rectangular shipping box with square ends x inches on a side is to be used.

(A) If the full 108 inches is to be used, express the volume of the box $V(x)$ in terms of x.

(B) What is the domain of the function V (determined by the physical restrictions)?

(C) Complete Table 16.

TABLE 16

VOLUME

x	$V(x)$
5	
10	
15	
20	
25	

(D) Plot the points in part (C) and sketch a graph of the volume function through these points.

Life Sciences

95. *Muscle contraction.* In a study of the speed of muscle contraction in frogs under various loads, noted British biophysicist and Nobel prize winner A.W. Hill determined that the weight w (in grams) placed on the muscle and the speed of contraction v (in centimeters per second) are approximately related by an equation of the form

$$(w + a)(v + b) = c$$

where a, b, and c are constants. Suppose that for a certain muscle, $a = 15$, $b = 1$, and $c = 90$. Express v as a function of w. Find the speed of contraction if a weight of 16 grams is placed on the muscle.

Social Sciences

96. *Politics.* The percentage s of seats in the House of Representatives won by Democrats and the percentage v of votes cast for Democrats (when expressed as decimal fractions) are related by the equation

$$5v - 2s = 1.4 \quad 0 < s < 1, \quad 0.28 < v < 0.68$$

(A) Express v as a function of s, and find the percentage of votes required for the Democrats to win 51% of the seats.

(B) Express s as a function of v, and find the percentage of seats won if Democrats receive 51% of the votes.

Elementary Functions: Graphs and Transformations

- ❑ A BEGINNING LIBRARY OF ELEMENTARY FUNCTIONS
- ❑ VERTICAL AND HORIZONTAL SHIFTS
- ❑ REFLECTIONS, EXPANSIONS, AND CONTRACTIONS
- ❑ PIECEWISE-DEFINED FUNCTIONS

The functions

$$g(x) = x^2 - 4 \qquad h(x) = (x-4)^2 \qquad k(x) = -4x^2$$

all can be expressed in terms of the function $f(x) = x^2$ as follows:

$$g(x) = f(x) - 4 \qquad h(x) = f(x-4) \qquad k(x) = -4f(x)$$

In this section we will see that the graphs of functions g, h, and k are closely related to the graph of function f. Insight gained by understanding these relationships will help us analyze and interpret the graphs of many different functions.

❑   A BEGINNING LIBRARY OF ELEMENTARY FUNCTIONS

As you progress through this book, and most any other mathematics course beyond this one, you will repeatedly encounter a relatively small list of elementary functions. We will identify these functions, study their basic properties, and include them in a library of elementary functions (see the inside front cover). This library will become an important addition to your mathematical toolbox and can be used in any course or activity where mathematics is applied.

Figure 1 shows six basic functions that you will encounter frequently. You should know the definition, domain, and range of each, and be able to recognize their graphs. For Figure 1B, recall the definition of *absolute value:*

$$|x| = \begin{cases} -x & \text{if } x < 0 \\ x & \text{if } x \geq 0 \end{cases}$$

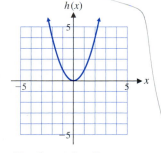

(A) **Identity function**	(B) **Absolute value function**	(C) **Square function**
$f(x) = x$	$g(x) = |x|$	$h(x) = x^2$
Domain: R	Domain: R	Domain: R
Range: R	Range: $[0, \infty)$	Range: $[0, \infty)$

FIGURE 1 Some basic functions and their graphs
Note: Letters used to designate these functions may vary from context to context; R is the set of all real numbers.

(continued)

(D) **Cube function**
$m(x) = x^3$
Domain: R
Range: R

(E) **Square root function**
$n(x) = \sqrt{x}$
Domain: $[0, \infty)$
Range: $[0, \infty)$

(F) **Cube root function**
$p(x) = \sqrt[3]{x}$
Domain: R
Range: R

FIGURE 1 *(continued)*

❑   VERTICAL AND HORIZONTAL SHIFTS

If a new function is formed by performing an operation on a given function, then the graph of the new function is called a **transformation** of the graph of the original function. For example, graphs of both $y = f(x) + k$ and $y = f(x + h)$ are transformations of the graph of $y = f(x)$.

Explore–Discuss 1

Let $f(x) = x^2$.

(A) Graph $y = f(x) + k$ for $k = -4, 0$, and 2 simultaneously in the same coordinate system. Describe the relationship between the graph of $y = f(x)$ and the graph of $y = f(x) + k$ for k any real number.

(B) Graph $y = f(x + h)$ for $h = -4, 0$, and 2 simultaneously in the same coordinate system. Describe the relationship between the graph of $y = f(x)$ and the graph of $y = f(x + h)$ for h any real number.

Example 1 ➭ **Vertical and Horizontal Shifts**

(A) How are the graphs of $y = |x| + 4$ and $y = |x| - 5$ related to the graph of $y = |x|$? Confirm your answer by graphing all three functions simultaneously in the same coordinate system.

(B) How are the graphs of $y = |x + 4|$ and $y = |x - 5|$ related to the graph of $y = |x|$? Confirm your answer by graphing all three functions simultaneously in the same coordinate system.

SOLUTION   (A) The graph of $y = |x| + 4$ is the same as the graph of $y = |x|$ shifted upward 4 units, and the graph of $y = |x| - 5$ is the same as the graph of $y = |x|$ shifted downward 5 units. Figure 2 confirms these conclusions. [It appears that the graph of $y = f(x) + k$ is the graph of $y = f(x)$ shifted up if k is positive and down if k is negative.]

(B) The graph of $y = |x + 4|$ is the same as the graph of $y = |x|$ shifted to the left 4 units, and the graph of $y = |x - 5|$ is the same as the graph of $y = |x|$ shifted to the right 5 units. Figure 3 confirms these conclusions. [It appears that the graph of $y = f(x + h)$ is the graph of $y = f(x)$ shifted right if h is negative and left if h is positive—the opposite of what you might expect.]

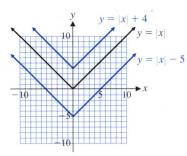

FIGURE 2 Vertical shifts   **FIGURE 3** Horizontal shifts

Matched Problem 1 ↩ (A) How are the graphs of $y = \sqrt{x} + 5$ and $y = \sqrt{x} - 4$ related to the graph of $y = \sqrt{x}$? Confirm your answer by graphing all three functions simultaneously in the same coordinate system.

(B) How are the graphs of $y = \sqrt{x + 5}$ and $y = \sqrt{x - 4}$ related to the graph of $y = \sqrt{x}$? Confirm your answer by graphing all three functions simultaneously in the same coordinate system.

Comparing the graphs of $y = f(x) + k$ with the graph of $y = f(x)$, we see that the graph of $y = f(x) + k$ can be obtained from the graph of $y = f(x)$ by **vertically translating** (shifting) the graph of the latter upward k units if k is positive and downward $|k|$ units if k is negative. Comparing the graphs of $y = f(x + h)$ with the graph of $y = f(x)$, we see that the graph of $y = f(x + h)$ can be obtained from the graph of $y = f(x)$ by **horizontally translating** (shifting) the graph of the latter h units to the left if h is positive and $|h|$ units to the right if h is negative.

Example 2 ↩ **Vertical and Horizontal Translations (Shifts)**   The graphs in Figure 4 are either horizontal or vertical shifts of the graph of $f(x) = x^2$. Write appropriate equations for functions $H, G, M,$ and N in terms of f.

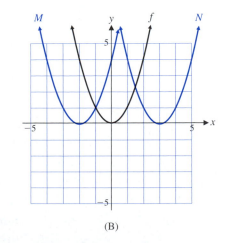

(A)   (B)

FIGURE 4 Vertical and horizontal shifts

SOLUTION Functions H and G are vertical shifts given by

$$H(x) = x^2 + 2 \qquad G(x) = x^2 - 4$$

Functions M and N are horizontal shifts given by

$$M(x) = (x + 2)^2 \qquad N(x) = (x - 3)^2$$

∎

Matched Problem 2 ✍ The graphs in Figure 5 are either horizontal or vertical shifts of the graph of $f(x) = \sqrt[3]{x}$. Write appropriate equations for functions H, G, M, and N in terms of f.

∎

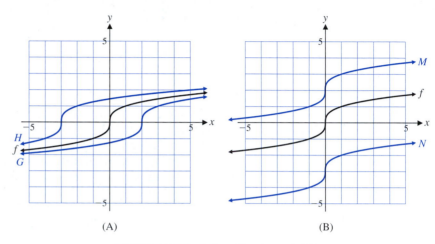

FIGURE 5 Vertical and horizontal shifts

❑ REFLECTIONS, EXPANSIONS, AND CONTRACTIONS

We now investigate how the graph of $y = Af(x)$ is related to the graph of $y = f(x)$ for different real numbers A.

Explore–Discuss 2

(A) Graph $y = Ax^2$ for $A = 1, 4$, and $\frac{1}{4}$ simultaneously in the same coordinate system.

(B) Graph $y = Ax^2$ for $A = -1, -4$, and $-\frac{1}{4}$ simultaneously in the same coordinate system.

(C) Describe the relationship between the graph of $h(x) = x^2$ and the graph of $G(x) = Ax^2$ for A any real number.

Comparing $y = Af(x)$ to $y = f(x)$, we see that the graph of $y = Af(x)$ can be obtained from the graph of $y = f(x)$ by multiplying each ordinate value of the latter by A. The result is a **vertical expansion** of the graph of $y = f(x)$ if $A > 1$, a **vertical contraction** of the graph of $y = f(x)$ if $0 < A < 1$, and a **reflection in the x axis** if $A = -1$. If A is a negative number other than -1, then the result is a combination of a reflection in the x axis and either a vertical expansion or a vertical contraction.

Example 3   ⇨   **Reflections, Expansions, and Contractions**

(A) How are the graphs of $y = 2|x|$ and $y = 0.5|x|$ related to the graph of $y = |x|$? Confirm your answer by graphing all three functions simultaneously in the same coordinate system.

(B) How is the graph of $y = -2|x|$ related to the graph of $y = |x|$? Confirm your answer by graphing both functions simultaneously in the same coordinate system.

SOLUTION   (A) The graph of $y = 2|x|$ is a vertical expansion of the graph of $y = |x|$ by a factor of 2, and the graph of $y = 0.5|x|$ is a vertical contraction of the graph of $y = |x|$ by a factor of 0.5. Figure 6 confirms this conclusion.

(B) The graph of $y = -2|x|$ is a reflection in the x axis and a vertical expansion of the graph of $y = |x|$. Figure 7 confirms this conclusion.

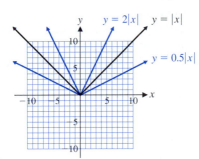

FIGURE 6 Vertical expansion and contraction

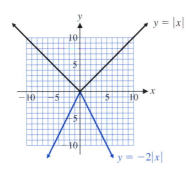

FIGURE 7 Reflection and vertical expansion

Matched Problem 3   ⇨   (A) How are the graphs of $y = 2x$ and $y = 0.5x$ related to the graph of $y = x$? Confirm your answer by graphing all three functions simultaneously in the same coordinate system.

(B) How is the graph of $y = -0.5x$ related to the graph of $y = x$? Confirm your answer by graphing both functions in the same coordinate system.

The various transformations considered above are summarized in the following box for easy reference:

Graph Transformations (Summary)

Vertical Translation:

$y = f(x) + k$ $\begin{cases} k > 0 & \text{Shift graph of } y = f(x) \text{ up } k \text{ units.} \\ k < 0 & \text{Shift graph of } y = f(x) \text{ down } |k| \text{ units.} \end{cases}$

Horizontal Translation:

$y = f(x + h)$ $\begin{cases} h > 0 & \text{Shift graph of } y = f(x) \text{ left } h \text{ units.} \\ h < 0 & \text{Shift graph of } y = f(x) \text{ right } |h| \text{ units.} \end{cases}$

(continued)

> **Graph Transformations (Summary) (continued)**
>
> **Reflection:**
>
> $y = -f(x)$ Reflect the graph of $y = f(x)$ in the x axis.
>
> **Vertical Expansion and Contraction:**
>
> $$y = Af(x) \begin{cases} A > 1 & \text{Expand graph of } y = f(x) \text{ vertically} \\ & \text{by multiplying each ordinate value by } A. \\ 0 < A < 1 & \text{Contract graph of } y = f(x) \text{ vertically} \\ & \text{by multiplying each ordinate value by } A. \end{cases}$$

Explore–Discuss 3

Use a graphing utility to explore the graph of $y = A(x + h)^2 + k$ for various values of the constants A, h, and k. Discuss how the graph of $y = A(x + h)^2 + k$ is related to the graph of $y = x^2$.

Example 4 ➾ **Combining Graph Transformations** Discuss the relationship between the graphs of $y = -|x - 3| + 1$ and $y = |x|$. Confirm your answer by graphing both functions simultaneously in the same coordinate system.

Solution   The graph of $y = -|x - 3| + 1$ is a reflection in the x axis, a horizontal translation of 3 units to the right, and a vertical translation of 1 unit upward of the graph of $y = |x|$. Figure 8 confirms this description.

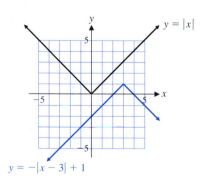

FIGURE 8 Combined transformations

Matched Problem 4 ➾ The graph of $y = G(x)$ in Figure 9 involves a reflection and a translation of the graph of $y = x^3$. Describe how the graph of function G is related to the graph of $y = x^3$ and find an equation of the function G.

FIGURE 9 Combined transformations

PIECEWISE-DEFINED FUNCTIONS

Earlier we noted that the absolute value of a real number x is defined as

$$|x| = \begin{cases} -x & \text{if } x < 0 \\ x & \text{if } x \geq 0 \end{cases}$$

Notice that this function is defined by different rules for different parts of its domain. Functions whose definitions involve more than one rule are called **piecewise-defined functions.** Graphing one of these functions involves graphing each rule over the appropriate portion of the domain (Fig. 10). In Figure 10C, notice that an open dot is used to show that the point $(0, -2)$ is not part of the graph and a solid dot is used to show that $(0, 2)$ is part of the graph.

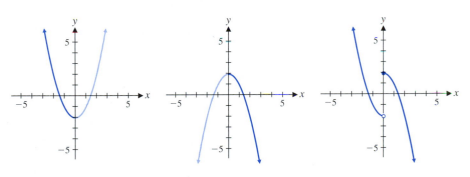

(A) $y = x^2 - 2$ (B) $y = 2 - x^2$ (C) $y = \begin{cases} x^2 - 2 & \text{if } x < 0 \\ 2 - x^2 & \text{if } x \geq 0 \end{cases}$

FIGURE 10 Graphing a piecewise-defined function

As the next example illustrates, piecewise-defined functions occur naturally in many applications.

Example 5 **Natural Gas Rates**   Easton Utilities uses the rates shown in Table 1 to compute the monthly cost of natural gas for each customer. Write a piecewise definition for the cost of consuming x CCF (cubic hundred feet) of natural gas and graph the function.

WWW

TABLE 1
CHARGES PER MONTH
$0.7866 per CCF for the first 5 CCF
$0.4601 per CCF for the next 35 CCF
$0.2508 per CCF for all over 40 CCF

SOLUTION   If $C(x)$ is the cost, in dollars, of using x CCF of natural gas in one month, then the first line of Table 1 implies that

$$C(x) = 0.7866x \quad \text{if } 0 \leqslant x \leqslant 5$$

Note that $C(5) = 3.933$ is the cost of 5 CCF. If $5 < x \leqslant 40$, then $x - 5$ represents the amount of gas that cost $0.4601 per CCF, $0.4601(x - 5)$ represents the cost of this gas, and the total cost is

$$C(x) = 3.933 + 0.4601(x - 5)$$

If $x > 40$, then

$$C(x) = 20.0365 + 0.2508(x - 40)$$

where $20.0365 = C(40)$, the cost of the first 40 CCF. Combining all these equations, we have the following piecewise definition for $C(x)$:

$$C(x) = \begin{cases} 0.7866x & \text{if } 0 \leqslant x \leqslant 5 \\ 3.933 + 0.4601(x - 5) & \text{if } 5 < x \leqslant 40 \\ 20.0365 + 0.2508(x - 40) & \text{if } 40 < x \end{cases}$$

To graph C, first note that each rule in the definition of C represents a transformation of the identity function $f(x) = x$. Graphing each transformation over the indicated interval produces the graph of C shown in Figure 11.

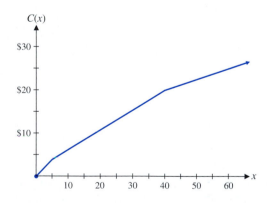

FIGURE 11 Cost of purchasing x CCF of natural gas

Matched Problem 5   **Natural Gas Rates**   Trussville Utilities uses the rates shown in Table 2 to compute the monthly cost of natural gas for residential customers. Write a piecewise definition for the cost of consuming x CCF of natural gas and graph the function.

www

TABLE 2		
CHARGES PER MONTH		
$0.7675 per CCF for the first 50 CCF		
$0.6400 per CCF for the next 150 CCF		
$0.6130 per CCF for all over 200 CCF		

Answers to Matched Problems

1. (A) The graph of $y = \sqrt{x} + 5$ is the same as the graph of $y = \sqrt{x}$ shifted upward 5 units, and the graph of $y = \sqrt{x} - 4$ is the same as the graph of $y = \sqrt{x}$ shifted downward 4 units. The figure confirms these conclusions.

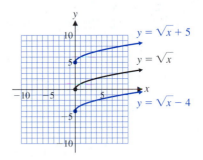

(B) The graph of $y = \sqrt{x + 5}$ is the same as the graph of $y = \sqrt{x}$ shifted to the left 5 units, and the graph of $y = \sqrt{x - 4}$ is the same as the graph of $y = \sqrt{x}$ shifted to the right 4 units. The figure confirms these conclusions.

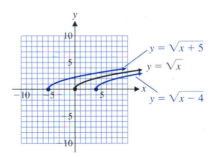

2. $H(x) = \sqrt[3]{x + 3}, G(x) = \sqrt[3]{x - 2}, M(x) = \sqrt[3]{x + 2}, N(x) = \sqrt[3]{x - 3}$

3. (A) The graph of $y = 2x$ is a vertical expansion of the graph of $y = x$, and the graph of $y = 0.5x$ is a vertical contraction of the graph of $y = x$. The figure confirms these conclusions.

(B) The graph of $y = -0.5x$ is a vertical contraction and a reflection in the x axis of the graph of $y = x$. The figure confirms this conclusion.

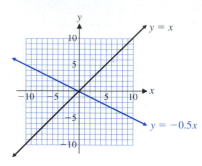

4. The graph of function G is a reflection in the x axis and a horizontal translation of 2 units to the left of the graph of $y = x^3$. An equation for G is $G(x) = -(x + 2)^3$.

5. $C(x) = \begin{cases} 0.7675x & \text{if } 0 \le x \le 50 \\ 38.375 + 0.64(x - 50) & \text{if } 50 < x \le 200 \\ 134.375 + 0.613(x - 200) & \text{if } 200 < x \end{cases}$

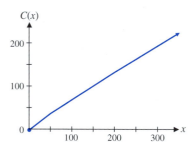

Exercise 1-2

A *Without looking back in the text, indicate the domain and range of each of the functions in Problems 1–8.*

1. $f(x) = 2x$
2. $g(x) = -0.3x$
3. $h(x) = -0.6\sqrt{x}$
4. $k(x) = 4\sqrt{x}$
5. $m(x) = 3|x|$
6. $n(x) = -0.1x^2$
7. $r(x) = -x^3$
8. $s(x) = 5\sqrt[3]{x}$

Graph each of the functions in Problems 9–20 using the graphs of functions f and g below.

9. $y = f(x) + 2$
10. $y = g(x) - 1$
11. $y = f(x + 2)$
12. $y = g(x - 1)$
13. $y = g(x - 3)$
14. $y = f(x + 3)$
15. $y = g(x) - 3$
16. $y = f(x) + 3$
17. $y = -f(x)$
18. $y = -g(x)$
19. $y = 0.5g(x)$
20. $y = 2f(x)$

B *In Problems 21–28, indicate how the graph of each function is related to the graph of one of the six basic functions in Figure 1 (at the beginning of this section). Sketch a graph of each function.*

21. $g(x) = -|x + 3|$
22. $h(x) = -|x - 5|$
23. $f(x) = (x - 4)^2 - 3$
24. $m(x) = (x + 3)^2 + 4$
25. $f(x) = 7 - \sqrt{x}$
26. $g(x) = -6 + \sqrt[3]{x}$
27. $h(x) = -3|x|$
28. $m(x) = -0.4x^2$

  Check your descriptions and graphs in Problems 21–28 by graphing each function on a graphing utility.

Each graph in Problems 29–36 is the result of applying a sequence of transformations to the graph of one of the six basic functions in Figure 1 (at the beginning of this section). Identify the basic function and describe the transformation verbally. Write an equation for the given graph.

29.

30.

31.

32.

33.

34.

35.

36.

Check your equations in Problems 29–36 by graphing each on a graphing utility.

In Problems 37–42, the graph of the function g is formed by applying the indicated sequence of transformations to the given function f. Find an equation for the function g and graph g using $-5 \le x \le 5$ and $-5 \le y \le 5$.

37. The graph of $f(x) = \sqrt{x}$ is shifted 2 units to the right and 3 units down.

38. The graph of $f(x) = \sqrt[3]{x}$ is shifted 3 units to the left and 2 units up.

39. The graph of $f(x) = |x|$ is reflected in the x axis and shifted to the left 3 units.

40. The graph of $f(x) = |x|$ is reflected in the x axis and shifted to the right 1 unit.

41. The graph of $f(x) = x^3$ is reflected in the x axis and shifted 2 units to the right and down 1 unit.

42. The graph of $f(x) = x^2$ is reflected in the x axis and shifted to the left 2 units and up 4 units.

Graph each function in Problems 43–48.

43. $f(x) = \begin{cases} 2 - 2x & \text{if } x < 2 \\ x - 2 & \text{if } x \ge 2 \end{cases}$

44. $g(x) = \begin{cases} x + 1 & \text{if } x < -1 \\ 2 + 2x & \text{if } x \ge -1 \end{cases}$

45. $h(x) = \begin{cases} x - 2 & \text{if } x < 0 \\ x + 1 & \text{if } x \ge 0 \end{cases}$

46. $k(x) = \begin{cases} x + 1 & \text{if } x < 0 \\ 2 + x & \text{if } x \ge 0 \end{cases}$

47. $m(x) = \begin{cases} -0.5x - 1 & \text{if } x < -2 \\ 4 - x^2 & \text{if } -2 \le x \le 2 \\ 0.5x - 1 & \text{if } x > 2 \end{cases}$

48. $n(x) = \begin{cases} 2x + 7 & \text{if } x < -2 \\ x^2 - 1 & \text{if } -2 \le x \le 2 \\ -2x + 7 & \text{if } x > 2 \end{cases}$

Check Problems 43–48 by graphing on a graphing utility.

C *Each of the graphs in Problems 49–54 involves a reflection in the x axis and/or a vertical expansion or contraction of one of the basic functions in Figure 1 (at the beginning of this section). Identify the basic function, and describe the transformation verbally. Write an equation for the given graph.*

49.

50.

51.

52.

53.

54.

 Check your equations in Problems 49–54 by graphing each on a graphing utility.

Changing the order in a sequence of transformations may change the final result. Investigate each pair of transformations in Problems 55–60 to determine if reversing their order can produce a different result. Support your conclusions with specific examples and/or mathematical arguments.

55. Vertical shift; horizontal shift

56. Vertical shift; reflection in y axis

57. Vertical shift; reflection in x axis

58. Vertical shift; vertical expansion

59. Horizontal shift; reflection in y axis

60. Horizontal shift; vertical contraction

Applications

Business & Economics

61. *Price–demand.* A retail chain sells CD players. The retail price $p(x)$ (in dollars) and the weekly demand x for a particular model are related by

$$p(x) = 115 - 4\sqrt{x} \qquad 9 \le x \le 289$$

(A) Describe how the graph of function p can be obtained from the graph of one of the basic functions in Figure 1 (at the beginning of this section).

(B) Sketch a graph of function p using part (A) as an aid.

62. *Price–supply.* The manufacturers of the CD players in Problem 61 are willing to supply x players at a price of $p(x)$ as given by the equation

$$p(x) = 4\sqrt{x} \qquad 9 \le x \le 289$$

(A) Describe how the graph of function p can be obtained from the graph of one of the basic functions in Figure 1 (at the beginning of this section).

(B) Sketch a graph of function p using part (A) as an aid.

63. *Hospital costs.* Using statistical methods, the financial department of a hospital arrived at the cost equation

$$C(x) = 0.00048(x - 500)^3 + 60,000 \qquad 100 \le x \le 1,000$$

where $C(x)$ is the cost in dollars for handling x cases per month.

(A) Describe how the graph of function C can be obtained from the graph of one of the basic functions in Figure 1 (at the beginning of this section).

(B) Sketch a graph of function C using part (A) and a graphing utility as aids.

64. *Price–demand.* A company manufactures and sells in-line skates. Their financial department has established the price–demand function

$$p(x) = 190 - 0.013(x - 10)^2 \qquad 10 \le x \le 100$$

where $p(x)$ is the price at which x thousand pairs of skates can be sold.

(A) Describe how the graph of function p can be obtained from the graph of one of the basic functions in Figure 1 (at the beginning of this section).

(B) Sketch a graph of function p using part (A) and a graphing utility as aids.

65. *Electricity rates.* Table 3 shows the electricity rates charged by Monroe Utilities in the summer months. The base is a fixed monthly charge, independent of the kWh (kilowatthours) used during the month.

(A) Write a piecewise definition of the monthly charge $S(X)$ for a customer who uses x kWh in a summer month.

(B) Graph $S(X)$.

WWW

TABLE 3

SUMMER (JULY–OCTOBER)

Base charge, $8.50
First 700 kWh or less at 0.0650/kWh
Over 700 kWh at 0.0900/kWh

66. *Electricity rates.* Table 4 shows the electricity rates charged by Monroe Utilities in the winter months.

(A) Write a piecewise definition of the monthly charge $W(x)$ for a customer who uses x kWh in a winter month.

(B) Graph $W(x)$.

WWW

TABLE 4

WINTER (NOVEMBER–JUNE)

Base charge, $8.50
First 700 kWh or less at 0.0650/kWh
Over 700 kWh at 0.0530/kWh

67. *State income tax.* Table 5 shows a recent state income tax schedule for married couples filing a joint return in the state of Kansas.

(A) Write a piecewise definition for the tax due $T(x)$ on an income of x dollars.

(B) Graph $T(x)$.

(C) Find the tax due on a taxable income of $40,000. Of $70,000.

WWW

TABLE 5

KANSAS STATE INCOME TAX

SCHEDULE I—MARRIED FILING JOINT
If taxable income is:

Over	But Not Over	Tax Due Is:
$0	$30,000	3.50% of taxable income
$30,000	$60,000	$1,050 plus 6.25% of excess over $30,000
$60,000		$2,925 plus 6.45% of excess over $60,000

68. *State income tax.* Table 6 shows a recent state income tax schedule for individuals filing a return in the state of Kansas.

(A) Write a piecewise definition for the tax due $T(x)$ on an income of x dollars.

(B) Graph $T(x)$.

(C) Find the tax due on a taxable income of $20,000. Of $35,000.

(D) Would it be better for a married couple in Kansas with two equal incomes to file jointly or separately? Discuss.

WWW

TABLE 6

KANSAS STATE INCOME TAX

SCHEDULE II—SINGLE, HEAD OF HOUSE-HOLD, OR MARRIED FILING SEPARATE
If taxable income is:

Over	But Not Over	Tax Due Is:
$0	$15,000	3.50% of taxable income
$15,000	$30,000	$525 plus 6.25% of excess over $15,000
$30,000		$1,462.50 plus 6.45% of excess over $30,000

Life Sciences

69. *Physiology.* A good approximation of the normal weight of a person 60 inches or taller but not taller than 80 inches is given by $w(x) = 5.5x - 220$, where x is height in inches and $w(x)$ is weight in pounds.

(A) Describe how the graph of function w can be obtained from the graph of one of the basic functions in Figure 1 (at the beginning of this section).

(B) Sketch a graph of function w using part (A) as an aid.

70. *Physiology.* The average weight of a particular species of snake is given by $w(x) = 463x^3$, $0.2 \le x \le 0.8$, where x is length in meters and $w(x)$ is weight in grams.

(A) Describe how the graph of function w can be obtained from the graph of one of the basic functions in Figure 1 (at the beginning of this section).

(B) Sketch a graph of function w using part (A) as an aid.

Social Sciences

71. *Safety research.* Under ideal conditions, if a person driving a vehicle slams on the brakes and skids to a stop, the speed of the vehicle $v(x)$ (in miles per hour) is given approximately by $v(x) = C\sqrt{x}$, where x is the length of skid marks (in feet) and C is a constant that depends on the road conditions and the weight of the vehicle. For a particular vehicle, $v(x) = 7.08\sqrt{x}$ and $4 \leqslant x \leqslant 144$.

(A) Describe how the graph of function v can be obtained from the graph of one of the basic functions in Figure 1 (at the beginning of this section).

(B) Sketch a graph of function v using part (A) as an aid.

72. *Learning.* A production analyst has found that on the average it takes a new person $T(x)$ minutes to perform a particular assembly operation after x performances of the operation, where $T(x) = 10 - \sqrt[3]{x}, 0 \leqslant x \leqslant 125$.

(A) Describe how the graph of function T can be obtained from the graph of one of the basic functions in Figure 1 (at the beginning of this section).

(B) Sketch a graph of function T using part (A) as an aid.

Section 1-3

Linear Functions and Straight Lines

- ❑ INTERCEPTS
- ❑ LINEAR FUNCTIONS, EQUATIONS, AND INEQUALITIES
- ❑ GRAPHS OF $Ax + By = C$
- ❑ SLOPE OF A LINE
- ❑ EQUATIONS OF LINES: SPECIAL FORMS
- ❑ APPLICATIONS

In this section we will add another important class of functions to our basic list of elementary functions. These functions are called *linear functions* and include the identity function $f(x) = x$ as a special case. We will investigate the relationship between linear functions and the solutions to linear equations and inequalities. (A detailed treatment of algebraic solutions to linear equations and inequalities can be found in Appendix A-8.) Finally, we will review the concept of slope and some of the standard equations of straight lines. These new tools will be applied to a variety of significant applications, including cost and price–demand functions.

❑   **INTERCEPTS**

Figure 1 illustrates the graphs of three functions f, g, and h.

(A) $f(x) = -2x + 1$ (B) $g(x) = 3 + 2x - x^2$ (C) $h(x) = x^3 - 2x$

FIGURE 1 Graphs of several functions

If the graph of a function f crosses the x axis at a point with x coordinate a, then a is called an **x intercept** of f. If the graph of f crosses the y axis at a point with y coordinate b, then b is called the **y intercept.** It is common practice to refer to both the numbers a and b and the points $(a, 0)$ and $(0, b)$ as the x and y intercepts. If the y intercept exists, then 0 must be in the domain of f and the y intercept is simply $f(0)$. Thus, the graph of a function can have at most one y intercept. The x intercepts are all real solutions or roots of $f(x) = 0$, which may vary from none to an unlimited number. In Figure 1, function f has one y intercept and one x intercept; function g has one y intercept and two x intercepts; and function h has one y intercept and three x intercepts.

❑ LINEAR FUNCTIONS, EQUATIONS, AND INEQUALITIES

In Figure 1, the graph of $f(x) = -2x + 1$ is a straight line, and because of this, we choose to call this type of function a *linear function*. In general:

Linear and Constant Functions

A function f is a **linear function** if

$$f(x) = mx + b \qquad m \neq 0$$

where m and b are real numbers. The **domain** is set of all real numbers, and the **range** is the set of all real numbers. If $m = 0$, then f is called a **constant function,**

$$f(x) = b$$

which has the set of all real numbers as its **domain** and the constant b as its **range.**

Since $mx + b, m \neq 0$, is a first-degree polynomial, linear functions are also called **first-degree functions.** Figure 2 shows the graphs of two linear functions f and g, and a constant function h.

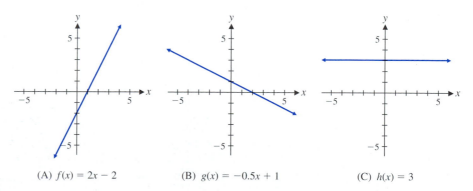

(A) $f(x) = 2x - 2$ (B) $g(x) = -0.5x + 1$ (C) $h(x) = 3$

FIGURE 2 Two linear functions and a constant function

In Section 1-2 we saw that the graph of the identity function $f(x) = x$ is a straight line. The graph of a linear function $g(x) = mx + b$, $m \neq 0$, is the same as the graph of $f(x) = x$ reflected in the x axis if m is negative, vertically expanded or contracted by a factor of $|m|$, and shifted up or down $|b|$ units. In general, it can be shown that:

The graph of a linear function is a straight line that is neither horizontal nor vertical.

The graph of a constant function is a horizontal straight line.

What about vertical lines? Recall from Section 1-1 that the graph of a function cannot contain two points with the same x coordinate and different y coordinates. Since *all* points on a vertical line have the same x coordinate, the graph of a function can never be a vertical line. Later in this section we will discuss equations of vertical lines, but these equations never define functions.

Explore–Discuss 1

(A) Is it possible for a linear function to have two x intercepts? No x intercepts? If either of your answers is yes, give an example.

(B) Is it possible for a linear function to have two y intercepts? No y intercept? If either of your answers is yes, give an example.

(C) Discuss the possible number of x and y intercepts for a constant function.

Example 1 ⇔ **Intercepts, Equations, and Inequalities**

(A) Graph $f(x) = \frac{3}{2}x - 4$ in a rectangular coordinate system.

(B) Find the x and y intercepts algebraically to two decimal places.

(C) Graph $f(x) = \frac{3}{2}x - 4$ in a standard viewing window.

(D) Find the x and y intercepts to two decimal places using trace and zoom or an appropriate built-in routine in your graphing utility.

(E) Solve $\frac{3}{2}x - 4 \le 0$ graphically to two decimal places using parts (A) and (B) or (C) and (D).

SOLUTION   (A) Graph in a rectangular coordinate system:

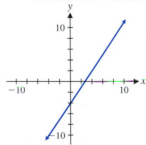

(B) Finding intercepts algebraically:

y intercept: $f(0) = \frac{3}{2}(0) - 4 = -4$

x intercept: $f(x) = 0$

$$\frac{3}{2}x - 4 = 0$$

$$\frac{3}{2}x = 4$$

$$x = \frac{8}{3} \approx 2.67$$

 (C) Graph in a graphing utility:

(D) Finding intercepts graphically in a graphing utility:

x intercept: 2.67   y intercept: −4

(E) Solving $\frac{3}{2}x - 4 \leq 0$ graphically using parts (A) and (B) or (C) and (D):
The linear inequality $\frac{3}{2}x - 4 \leq 0$ holds for those values of x for which the
graph of $f(x) = \frac{3}{2}x - 4$ in the figure in part (A) or (C) is at or below the
x axis. This happens for x less than or equal to the x intercept found in
parts (B) or (D). Thus, the solution set for the linear inequality is $x \leq 2.67$
or $(-\infty, 2.67]$.

Matched Problem 1 ⇨ (A) Graph $f(x) = -\frac{4}{3}x + 5$ in a rectangular coordinate system.

(B) Find the x and y intercepts algebraically to two decimal places.

(C) Graph $f(x) = -\frac{4}{3}x + 5$ in a standard viewing window.

(D) Find the x and y intercepts to two decimal places using trace and zoom
or an appropriate built-in routine in your graphing utility.

(E) Solve $-\frac{4}{3}x + 5 \geq 0$ graphically to two decimal places using parts (A)
and (B) or (C) and (D).

❑ **GRAPHS OF $Ax + By = C$**

We now investigate graphs of linear, or first-degree, equations in two variables:

$$Ax + By = C \tag{1}$$

where A and B are not both 0. Depending on the values of A and B, this equation defines a linear function, a constant function, or no function at all. If $A \neq 0$
and $B \neq 0$, then equation (1) can be written as

$$y = -\frac{A}{B}x + \frac{C}{B} \qquad \text{Linear function (slanted line)} \tag{2}$$

which is in the form $f(x) = mx + b, m \neq 0$, and hence is a linear function. If
$A = 0$ and $B \neq 0$, then equation (1) can be written as

$$0x + By = C$$

$$y = \frac{C}{B} \qquad \text{Constant function (horizontal line)} \tag{3}$$

which is in the form $g(x) = b$, and hence is a constant function. If $A \neq 0$ and
$B = 0$, then equation (1) can be written as

$$Ax + 0y = C$$

$$x = \frac{C}{A} \qquad \text{Not a function (vertical line)} \tag{4}$$

We can see that the graph of (4) is a vertical line, since the equation is satisfied
for any value of y as long as x is the constant C/A. Hence, this form does not
define a function.

The following theorem is a generalization of the preceding discussion:

THEOREM 1 **Graph of a Linear Equation in Two Variables**

In a Cartesian plane, the graph of any equation of the form

Standard Form $\quad Ax + By = C$ $\qquad\qquad\qquad\qquad$ (5)

where A, B, and C are real constants (A and B not both 0), is a straight line. Every straight line in a Cartesian plane coordinate system is the graph of an equation of this type.

Vertical and horizontal lines have particularly simple equations, which are special cases of equation (5):

$\qquad$ Horizontal line with y intercept $C/B = b$: $\quad y = b$

$\qquad$ Vertical line with x intercept $C/A = a$: $\quad x = a$

Explore–Discuss 2

Graph the following three special cases of $Ax + By = C$ in the same coordinate system:

(A) $3x + 2y = 6$ $\qquad$ (B) $0x - 3y = 12$ $\qquad$ (C) $2x + 0y = 10$

Which cases define functions? Explain why, or why not.
$\qquad$ Graph each case in the same viewing window using a graphing utility. (Check your manual on how to graph vertical lines.)

Sketching the graphs of equations of either form

$\qquad Ax + By = C \qquad$ or $\qquad y = mx + b$

is very easy, since the graph of each equation is a straight line. All that is necessary is to plot any two points from the solution set and use a straightedge to draw a line through these two points. The x and y intercepts are usually the easiest to find.

Example 2 ⇔ **Sketching Graphs of Lines**

(A) Graph $x = -4$ and $y = 6$ simultaneously in the same rectangular coordinate system. Also, graph in a graphing utility.

(B) Write the equations of the vertical and horizontal lines that pass through the point $(7, -5)$.

(C) Graph the equation $2x - 3y = 12$ by hand. Also, graph in a graphing utility.

SOLUTION $\quad$ (A) Graphing $x = -4$ and $y = 6$:

By hand

In a graphing utility

(B) Horizontal line through $(7, -5)$: $y = -5$
Vertical line through $(7, -5)$: $x = 7$

(C) Graphing $2x - 3y = 12$:   For the hand-drawn graph, find the intercepts by first letting $x = 0$ and solving for y and then letting $y = 0$ and solving for x. Then draw a line through the intercepts. To graph in a graphing utility, solve the equation for y in terms of x and enter the result.

By hand

$y_1 = (2/3)x - 4$

In a graphing utility

Matched Problem 2

(A) Graph $x = 5$ and $y = -3$ simultaneously in the same rectangular coordinate system. Also, graph in a graphing utility.

(B) Write the equations of the vertical and horizontal lines that pass through the point $(-8, 2)$.

(C) Graph the equation $3x + 4y = 12$ by hand. Also, graph in a graphing utility.

□ SLOPE OF A LINE

If we take two points $P_1(x_1, y_1)$ and $P_2(x_2, y_2)$, on a line, then the ratio of the change in y to the change in x as the point moves from point P_1 to point P_2 is called the **slope** of the line. In a sense, slope provides a measure of the "steepness" of a line relative to the x axis. The change in x is often called the **run** and the change in y the **rise.**

Slope of a Line

If a line passes through two distinct points $P_1(x_1, y_1)$ and $P_2(x_2, y_2)$, then its slope is given by the formula

$$m = \frac{y_2 - y_1}{x_2 - x_1} \qquad x_1 \neq x_2$$

$$= \frac{\text{vertical change (rise)}}{\text{horizontal change (run)}}$$

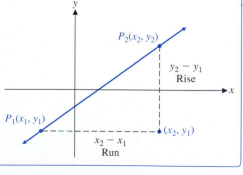

For a horizontal line, y does not change; hence, its slope is 0. For a vertical line, x does not change; hence, $x_1 = x_2$ and its slope is not defined. In general, the slope of a line may be positive, negative, 0, or not defined. Each case is illustrated geometrically in Table 1.

TABLE 1 GEOMETRIC INTERPRETATION OF SLOPE		
LINE	SLOPE	EXAMPLE
Rising as x moves from left to right	Positive	
Falling as x moves from left to right	Negative	
Horizontal	0	
Vertical	Not defined	

In using the formula to find the slope of the line through two points, it does not matter which point is labeled P_1 or P_2, since changing the labeling will change the sign in both the numerator and denominator of the slope formula, resulting in equivalent expressions. In addition, it is important to note that the definition of slope does not depend on the two points chosen on the line as long as they are distinct. This follows from the fact that the ratios of corresponding sides of similar triangles are equal.

Example 3 ➾ **Finding Slopes**   Sketch a line through each pair of points, and find the slope of each line.

(A) $(-3, -2), (3, 4)$ (B) $(-1, 3), (2, -3)$
(C) $(-2, -3), (3, -3)$ (D) $(-2, 4), (-2, -2)$

SOLUTION   (A)

$$m = \frac{4 - (-2)}{3 - (-3)} = \frac{6}{6} = 1$$

(B)

$$m = \frac{-3 - 3}{2 - (-1)} = \frac{-6}{3} = -2$$

(C)

$$m = \frac{-3 - (-3)}{3 - (-2)} = \frac{0}{5} = 0$$

(D)

$$m = \frac{-2 - 4}{-2 - (-2)} = \frac{-6}{0}$$

Slope is not defined

Matched Problem 3 ➾   Find the slope of the line through each pair of points.

(A) $(-2, 4), (3, 4)$ (B) $(-2, 4), (0, -4)$
(C) $(-1, 5), (-1, -2)$ (D) $(-1, -2), (2, 1)$

❑   EQUATIONS OF LINES: SPECIAL FORMS

Let us start by investigating why $y = mx + b$ is called the *slope–intercept form* for a line.

Explore–Discuss 3

(A) Graph $y = x + b$ for $b = -5, -3, 0, 3,$ and 5 simultaneously in the same coordinate system. Verbally describe the geometric significance of b.

(B) Graph $y = mx - 1$ for $m = -2, -1, 0, 1,$ and 2 simultaneously in the same coordinate system. Verbally describe the geometric significance of m.

(C) Using a graphing utility, explore the graph of $y = mx + b$ for different values of m and b.

As you can see from the exploration above, constants m and b in $y = mx + b$ have special geometric significance, which we now explicitly state.

If we let $x = 0$, then $y = b$, and we observe that the graph of $y = mx + b$ crosses the y axis at $(0, b)$. The constant b is the *y intercept*. For example, the y intercept of the graph of $y = -4x - 1$ is -1.

To determine the geometric significance of m, we proceed as follows: If $y = mx + b$, then by setting $x = 0$ and $x = 1$, we conclude that $(0, b)$ and $(1, m + b)$ lie on its graph (a line). Hence, the slope of this graph (line) is given by:

$$\text{Slope} = \frac{y_2 - y_1}{x_2 - x_1} = \frac{(m + b) - b}{1 - 0} = m$$

Thus, m is the slope of the line given by $y = mx + b$.

Slope–Intercept Form

The equation

$$y = mx + b \qquad m = \text{Slope}, b = y \text{ intercept} \tag{6}$$

is called the **slope–intercept form** of an equation of a line.

Example 4   ➭   **Using the Slope–Intercept Form**

(A) Find the slope and y intercept, and graph $y = -\frac{2}{3}x - 3$.

(B) Write the equation of the line with slope $\frac{2}{3}$ and y intercept -2.

SOLUTION   (A) Slope $= m = -\frac{2}{3}$ (B) $m = \frac{2}{3}$ and $b = -2$;
     y intercept $= b = -3$ thus, $y = \frac{2}{3}x - 2$

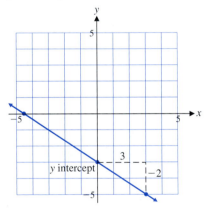

Matched Problem 4   ➭   Write the equation of the line with slope $\frac{1}{2}$ and y intercept -1. Graph.

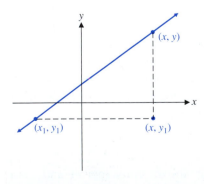

FIGURE 3

Suppose that a line has slope m and passes through a fixed point (x_1, y_1). If the point (x, y) is any other point on the line (Fig. 3), then

$$\frac{y - y_1}{x - x_1} = m$$

That is,

$$y - y_1 = m(x - x_1) \tag{7}$$

We now observe that (x_1, y_1) also satisfies equation (7) and conclude that equation (7) is an equation of a line with slope m that passes through (x_1, y_1).

> ### Point–Slope Form
>
> An equation of a line with slope m that passes through (x_1, y_1) is
>
> $$y - y_1 = m(x - x_1) \qquad\qquad (7)$$
>
> which is called the **point–slope form** of an equation of a line.

The point–slope form is extremely useful, since it enables us to find an equation for a line if we know its slope and the coordinates of a point on the line or if we know the coordinates of two points on the line.

Example 5 ➯ **Using the Point–Slope Form**

(A) Find an equation for the line that has slope $\frac{1}{2}$ and passes through $(-4, 3)$. Write the final answer in the form $Ax + By = C$.

(B) Find an equation for the line that passes through the two points $(-3, 2)$ and $(-4, 5)$. Write the resulting equation in the form $y = mx + b$.

SOLUTION (A) Use $y - y_1 = m(x - x_1)$. Let $m = \frac{1}{2}$ and $(x_1, y_1) = (-4, 3)$. Then

$$\begin{aligned}
y - 3 &= \tfrac{1}{2}[x - (-4)] \\
y - 3 &= \tfrac{1}{2}(x + 4) \qquad\qquad \color{blue}{\text{Multiply by 2.}} \\
2y - 6 &= x + 4 \\
-x + 2y &= 10 \qquad \text{or} \qquad x - 2y = -10
\end{aligned}$$

(B) First, find the slope of the line by using the slope formula:

$$m = \frac{y_2 - y_1}{x_2 - x_1} = \frac{5 - 2}{-4 - (-3)} = \frac{3}{-1} = -3$$

Now use $y - y_1 = m(x - x_1)$ with $m = -3$ and $(x_1, y_1) = (-3, 2)$:

$$\begin{aligned}
y - 2 &= -3[x - (-3)] \\
y - 2 &= -3(x + 3) \\
y - 2 &= -3x - 9 \\
y &= -3x - 7
\end{aligned}$$

■

Matched Problem 5 ➯ (A) Find an equation for the line that has slope $\frac{2}{3}$ and passes through $(6, -2)$. Write the resulting equation in the form $Ax + By = C, A > 0$.

(B) Find an equation for the line that passes through $(2, -3)$ and $(4, 3)$. Write the resulting equation in the form $y = mx + b$.

■

The various forms of the equation of a line that we have discussed are summarized in Table 2 for convenient reference.

TABLE 2
EQUATIONS OF A LINE

Standard form	$Ax + By = C$	A and B not both 0
Slope–intercept form	$y = mx + b$	Slope: m; y intercept: b
Point–slope form	$y - y_1 = m(x - x_1)$	Slope: m; point: (x_1, y_1)
Horizontal line	$y = b$	Slope: 0
Vertical line	$x = a$	Slope: undefined

❑   APPLICATIONS

We will now see how equations of lines occur in certain applications.

Example 6 ⇔   **Cost Equation**   The management of a company that manufactures roller skates has fixed costs (costs at 0 output) of $300 per day and total costs of $4,300 per day at an output of 100 pairs of skates per day. Assume that cost C is linearly related to output x.

(A) Find the slope of the line joining the points associated with outputs of 0 and 100; that is, the line passing through $(0, 300)$ and $(100, 4,300)$.

(B) Find an equation of the line relating output to cost. Write the final answer in the form $C = mx + b$.

(C) Graph the cost equation from part (B) for $0 \leqslant x \leqslant 200$.

Solution   (A) $m = \dfrac{y_2 - y_1}{x_2 - x_1} = \dfrac{4,300 - 300}{100 - 0} = \dfrac{4,000}{100} = 40$

(B) We must find an equation of the line that passes through $(0, 300)$ with slope 40. We use the slope–intercept form:

$$C = mx + b$$
$$C = 40x + 300$$

(C)

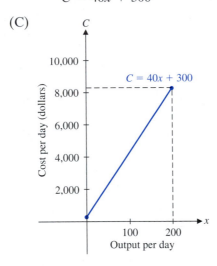

In Example 6, the **fixed cost** of $300 per day covers plant cost, insurance, and so on. This cost is incurred whether or not there is any production. The **variable cost** is $40x$, which depends on the day's output. Note that the slope 40 is the cost of producing one pair of skates; that is, the cost per unit output.

Matched Problem 6 ⇔   Answer parts (A) and (B) in Example 6 for fixed costs of $250 per day and total costs of $3,450 per day at an output of 80 pairs of skates per day.   ∎

Example 7 ⇔   **Price–Demand**   At the beginning of the twenty-first century, the world demand for crude oil was about 75 million barrels per day and the price of a barrel fluctuated between $20 and $40. Suppose that the daily demand for crude oil is 76.1 million barrels when the price is $25.52 per barrel and this demand drops to 74.9 million barrels when the price rises to $33.68. Assuming a linear relationship between the demand x and the price p, find a linear

www

function in the form $p = ax + b$ that models the price–demand relationship for crude oil. Use this model to predict the demand if the price rises to $39.12 per barrel.

SOLUTION   Find the equation of the line that passes through $(76.1, 25.52)$ and $(74.9, 33.68)$. We first find the slope of the line:

$$m = \frac{33.68 - 25.52}{74.9 - 76.1} = \frac{8.16}{-1.2} = -6.8$$

Use the point–slope form the find the equation of the line:

$$p - p_1 = m(x - x_1)$$
$$p - 25.52 = -6.8(x - 76.1)$$
$$p - 25.52 = -6.8x + 517.48$$
$$p = -6.8x + 543$$

To find the demand when the price is $39.12 per barrel, we solve the equation $p = 39.12$ for x:

$$p = 39.12$$
$$-6.8x + 543 = 39.12$$
$$-6.8x = -503.88$$
$$x = \frac{-503.88}{-6.8} = 74.1 \text{ million barrels per day}$$

■

Matched Problem 7 ⇨   The daily supply for crude oil also varies with the price. Suppose that the daily supply is 73.4 million barrels when the price is $23.84 and this supply rises to 77.4 million barrels when the price rises to $34.24. Assuming a linear relationship between the supply x and the price p, find a linear function in the form $p = ax + b$ that models the price–supply relationship for crude oil. Use this model to predict the supply if the price drops to $20.98 per barrel. ■

Figure 4 shows the price–demand function from Example 7 and the price–supply function from Matched Problem 7 graphed on the same axes. In a free competitive market, the price of a product is determined by the relationship betwen supply and demand. The price tends to stabilize at the point of intersection of the demand and supply functions. This point is called the **equilibrium point,** the corresponding price is called the **equilibrium price,** and the common value of supply and demand is called the **equilibrium quantity.**

FIGURE 4

The equilibrium point for the supply and demand functions in Figure 4 is found as follows:

$$2.6x - 167 = -6.8x + 543$$
$$9.4x = 710$$
$$x = \frac{710}{9.4} = 75.532 \text{ million barrels} \qquad \textcolor{blue}{\text{Equilibrium quantity}}$$
$$p = 2.6(75.532) - 167 = \$29.38 \qquad \textcolor{blue}{\text{Equilibrium price}}$$
$$p = -6.8(75.532) + 543 = \$29.38 \qquad \textcolor{blue}{\text{Check}}$$

Answers to Matched Problems   **1.** (A)

(B) x intercept: 3.75; y intercept: 5

(C)

(D) x intercept: 3.75; y intercept: 5   (E) $x \leq 3.75$ or $(-\infty, 3.75]$

2. (A)

By hand

In graphing utility

(B) Horizontal line: $y = 2$; vertical line: $x = -8$

(C)

By hand

In graphing utility

3. (A) 0   (B) −4   (C) Not defined   (D) 1

4. $y = \frac{1}{2}x - 1$

5. (A) $2x - 3y = 18$ (B) $y = 3x - 9$
6. (A) $m = 40$ (B) $C = 40x + 250$

7. $p(x) = 2.6x - 167$; 72.3 million barrels

Exercise 1-3

A *Problems 1–4 refer to graphs (A)–(D).*

(A)

(B)

(C)

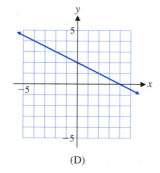

(D)

1. Identify the graph(s) of linear functions with a negative slope.

2. Identify the graph(s) of linear functions with a positive slope.

3. Identify the graph(s) of any constant functions. What is the slope of the graph?

4. Identify any graphs that are not the graphs of functions. What can you say about their slopes?

In Problems 5–8, sketch a graph of each equation in a rectangular coordinate system.

5. $y = 2x - 3$ **6.** $y = \frac{x}{2} + 1$

7. $2x + 3y = 12$ **8.** $8x - 3y = 24$

In Problems 9–12, find the slope and y intercept of the graph of each equation.

9. $y = 3x + 1$ **10.** $y = \frac{x}{5} - 2$

11. $y = -\frac{3}{7}x - 6$ **12.** $y = 0.7x + 5$

In Problems 13–16, write an equation of the line with the indicated slope and y intercept.

13. Slope $= -2$ **14.** Slope $= \frac{3}{4}$
y intercept $= 3$ y intercept $= -5$

15. Slope $= \frac{4}{3}$ **16.** Slope $= -5$
y intercept $= -4$ y intercept $= 9$

B *Sketch a graph of each equation or pair of equations in Problems 17–22 in a rectangular coordinate system.*

17. $y = -\frac{2}{3}x - 2$ **18.** $y = -\frac{3}{2}x + 1$

19. $3x - 2y = 10$ **20.** $5x - 6y = 15$

21. $x = 3$; $y = -2$ **22.** $x = -3$; $y = 2$

 Check your graphs for Problems 17–22 by graphing each in a graphing utility.

In Problems 23–26, find the slope of the graph of each equation.

23. $4x + y = 3$ **24.** $5x - y = -2$

25. $3x + 5y = 15$ **26.** $2x - 3y = 18$

27. (A) Graph $f(x) = 1.2x - 4.2$ in a rectangular coordinate system.
(B) Find the x and y intercepts algebraically to one decimal place.
(C) Graph $f(x) = 1.2x - 4.2$ in a graphing utility.

(D) Find the x and y intercepts to one decimal place using trace and zoom or an appropriate built-in routine in your graphing utility.

(E) Using the results of parts (A) and (B) or (C) and (D), find the solution set for the linear inequality

$$1.2x - 4.2 > 0$$

28. (A) Graph $f(x) = -0.8x + 5.2$ in a rectangular coordinate system.

(B) Find the x and y intercepts algebraically to one decimal place.

(C) Graph $f(x) = -0.8x + 5.2$ in a graphing utility.

(D) Find the x and y intercepts to one decimal place using trace and zoom or an appropriate built-in routine in your graphing utility.

(E) Using the results of parts (A) and (B) or (C) and (D), find the solution set for the linear inequality

$$-0.8x + 5.2 < 0$$

In Problems 29–32, write the equations of the vertical and horizontal lines through each point.

29. $(4, -3)$ **30.** $(-5, 6)$

31. $(-1.5, -3.5)$ **32.** $(2.6, 3.8)$

In Problems 33–38, write the equation of the line through each indicated point with the indicated slope. Write the final answer in the form $y = mx + b$.

33. $m = -4; (2, -3)$ **34.** $m = -6; (-4, 1)$

35. $m = \frac{3}{2}; (-4, -5)$ **36.** $m = \frac{4}{3}; (-6, 2)$

37. $m = 0; (-1.5, 4.6)$ **38.** $m = 0; (3.1, -2.7)$

In Problems 39–46, find the slope of the line that passes through the given points.

39. $(2, 5)$ and $(5, 7)$ **40.** $(1, 2)$ and $(3, 5)$

41. $(-2, -1)$ and $(2, -6)$ **42.** $(2, 3)$ and $(-3, 7)$

43. $(5, 3)$ and $(5, -3)$ **44.** $(1, 4)$ and $(0, 4)$

45. $(-2, 5)$ and $(3, 5)$ **46.** $(2, 0)$ and $(2, -3)$

In Problems 47–54, write an equation of the line through each indicated pair of points. Write the final answer in the form $Ax + By = C$. Indicate whether the equation defines a linear function, a constant function, or neither.

47. $(2, 5)$ and $(5, 7)$ **48.** $(1, 2)$ and $(3, 5)$

49. $(-2, -1)$ and $(2, -6)$ **50.** $(2, 3)$ and $(-3, 7)$

51. $(5, 3)$ and $(5, -3)$ **52.** $(1, 4)$ and $(0, 4)$

53. $(-2, 5)$ and $(3, 5)$ **54.** $(2, 0)$ and $(2, -3)$

55. Discuss the relationship among the graphs of the lines with equation $y = mx + 2$, where m is any real number.

56. Discuss the relationship among the graphs of the lines with equation $y = -0.5x + b$, where b is any real number.

C 57. (A) Graph the following equations in the same coordinate system:

$$3x + 2y = 6 \qquad 3x + 2y = 3$$
$$3x + 2y = -6 \qquad 3x + 2y = -3$$

(B) From your observations in part (A), describe the family of lines obtained by varying C in $Ax + By = C$ while holding A and B fixed.

58. (A) Graph the following two equations in the same coordinate system:

$$3x + 4y = 12 \qquad 4x - 3y = 12$$

(B) Graph the following two equations in the same coordinate system:

$$2x + 3y = 12 \qquad 3x - 2y = 12$$

(C) From your observations in parts (A) and (B), describe the apparent relationship of the graphs of

$$Ax + By = C \qquad \text{and} \qquad Bx - Ay = C$$

59. Describe the relationship between the graphs of $f(x) = mx + b$ and $g(x) = |mx + b|, m \neq 0$, and illustrate with examples. Is $g(x)$ always, sometimes, or never a linear function?

60. Describe the relationship between the graphs of $f(x) = mx + b$ and $g(x) = m|x| + b, m \neq 0$, and illustrate with examples. Is $g(x)$ always, sometimes, or never a linear function?

Applications

Business & Economics

61. *Simple interest.* If $P (the principal) is invested at an interest rate of r, then the amount A that is due after t years is given by

$$A = Prt + P$$

If $100 is invested at 6% ($r = 0.06$), then

$$A = 6t + 100, \quad t \geq 0$$

(A) What will $100 amount to after 5 years? After 20 years?

(B) Sketch a graph of $A = 6t + 100$ for $0 \leq t \leq 20$.

(C) Find the slope of the graph and interpret verbally.

62. *Simple interest.* Use the simple interest formula from Problem 61. If $1,000 is invested at 7.5% ($r = 0.075$), then $A = 75t + 1,000, t \geq 0$.

(A) What will $1,000 amount to after 5 years? After 20 years?

(B) Sketch a graph of $A = 75t + 1{,}000$ for $0 \le t \le 1{,}000$.

(C) Find the slope of the graph and interpret verbally.

63. *Cost function.* The management of a company that manufactures surfboards has fixed costs (at 0 output) of $200 per day and total costs of $3,800 per day at a daily output of 20 boards.

(A) Assuming the total cost per day, $C(x)$, is linearly related to the total output per day, x, write an equation for the cost function.

(B) What are the total costs for an output of 12 boards per day?

(C) Graph the cost function for $0 \le x \le 20$.

64. *Cost function.* Repeat Problem 63 if the company has fixed costs of $300 per day and total costs per day at an output of 20 boards of $5,100.

65. *Price–demand function.* A manufacturing company is interested in introducing a new power mower. Its market research department gave the management the price–demand forecast listed in Table 3.

TABLE 3

PRICE–DEMAND

DEMAND x	WHOLESALE PRICE ($) $p(x)$
0	200
2,400	160
4,800	120
7,800	70

(A) Plot these points, letting $p(x)$ represent the price at which x number of mowers can be sold (demand). Label the horizontal axis x.

(B) Note that the points in part (A) lie along a straight line. Find an equation for the price–demand function.

(C) What would be the price for a demand of 3,000 units?

(D) Write a brief verbal interpretation of the slope of the line found in part (B).

66. *Depreciation.* Office equipment was purchased for $20,000 and is assumed to have a scrap value of $2,000

after 10 years. If its value is depreciated linearly (for tax purposes) from $20,000 to $2,000:

(A) Find the linear equation that relates value (V) in dollars to time (t) in years.

(B) What would be the value of the equipment after 6 years?

(C) Graph the equation for $0 \le t \le 10$.

(D) Write a brief verbal interpretation of the slope of the line found in part (A).

67. *Equilibrium point.* At a price of $2.50 per bushel, the
www annual U.S. supply and demand for corn are 8.5 and 9.8 million bushels, respectively. When the price rises to $3.30, the supply increases to 10.5 million bushels while the demand decreases to 7.8 million bushels.

(A) Assuming that the price–supply and the price–demand equations are linear, find equations for each.

(B) Find the equilibrium point for the U.S. corn market.

68. *Equilibrium point.* At a price of $5.50 per bushel, the
www annual U.S. supply and demand for soybeans are 2.4 and 2.9 million bushels, respectively. When the price rises to $7.30, the supply increases to 2.8 million bushels while the demand decreases to 2.4 million bushels.

(A) Assuming that the price–supply and the price–demand equations are linear, find equations for each.

(B) Find the equilibrium point for the U.S. soybean market.

www *Merck & Co., Inc. is the world's largest pharmaceutical company. Problems 69 and 70 refer to the data in Table 4 taken from the company's annual reports.*

TABLE 4

SELECTED FINANCIAL DATA FOR MERCK & CO., INC. (BILLION $)

	1995	1996	1997	1998	1999
Sales	16.7	19.8	23.6	26.9	32.7
Income	3.3	3.8	4.6	5.2	5.9

69. *Sales analysis.* A mathematical model for Merck's sales is given by

$$f(x) = 16.12 + 3.91x$$

where $x = 0$ corresponds to 1995.

(A) Complete the following table. Round values of $f(x)$ to one decimal place.

x	Sales	$f(x)$
0	16.7	
1	19.8	
2	23.6	
3	26.9	
4	32.7	

(B) Sketch the graph of f and the sales data on the same axes.

(C) Use the modeling equation to estimate the sales in 2005. In 2010.

(D) Write a brief verbal description of the company's sales from 1995 to 1999.

70. *Sales analysis.* A mathematical model for Merck's income is given by

$$f(x) = 3.24 + 0.66x$$

where $x = 0$ corresponds to 1995.

(A) Complete the following table. Round values of $f(x)$ to one decimal place.

x	Income	$f(x)$
0	3.3	
1	3.8	
2	4.6	
3	5.2	
4	5.9	

(B) Sketch the graph of f and the income data on the same axes.

(C) Use the modeling equation to estimate the income in 2005. In 2010.

(D) Write a brief verbal description of the company's income from 1995 to 1999.

71. *Energy consumption.* Analyzing data from the U.S. Energy Department for the period between 1920 and 1960 reveals that coal consumption as a percentage of all energy consumed (wood, coal, petroleum, natural gas, hydro, and nuclear) decreased almost linearly. Percentages for this period are given in the table.

Year	Consumption (%)
1920	72
1930	60
1940	50
1950	37
1960	22

(A) Let x represent the number of years since 1900. Find the equation of the linear function f satisfying $f(20) = 72$ and $f(60) = 22$. Graph the data and this function on the same axes. Does this function provide a good model for this data?

(B) Use $f(x)$ to estimate (to the nearest 1%) the percent of coal consumption in 1927. In 1953.

(C) If we assume that $f(x)$ continues to provide a good description of the percentage of coal consumption after 1960, when would $f(x)$ indicate that the percentage of coal consumption has reached 0? Did this really happen? (Consult

some references if you are not certain.) If not, what are some reasons for the percentage of coal consumption to level off or even to increase at some point in time after 1960?

72. *Petroleum consumption.* Analyzing data from the U.S. Energy Department for the period between 1920 and 1960 reveals that petroleum consumption as a percentage of all energy consumed (wood, coal, petroleum, natural gas, hydro, and nuclear) increased almost linearly. Percentages for this period are given in the table.

Year	Consumption (%)
1920	11
1930	22
1940	29
1950	37
1960	44

(A) Let x represent the number of years since 1900. Find the equation of the linear function f satisfying $f(20) = 11$ and $f(60) = 44$. Graph the data and this function on the same axes. Does this function provide a good model for this data?

(B) Use $f(x)$ to estimate (to the nearest 1%) the percent of petroleum consumption in 1932. In 1956.

(C) If we assume that $f(x)$ continues to provide a good description of the percentage of petroleum consumption after 1960, when would this percentage reach 100%? Is this likely to happen? Explain.

Life Sciences

73. *Nutrition.* In a nutrition experiment, a biologist wants to prepare a special diet for the experimental animals. Two food mixes, A and B, are available. If mix A contains 20% protein and mix B contains 10% protein, what combination of each mix will provide exactly 20 grams of protein? Let x be the amount of A used and let y be the amount of B used. Then write a linear equation relating x, y, and 20. Graph this equation for $x \geq 0$ and $y \geq 0$.

74. *Ecology.* As one descends into the ocean, pressure increases linearly. The pressure is 15 pounds per square inch on the surface and 30 pounds per square inch 33 feet below the surface.

(A) If p is the pressure in pounds and d is the depth below the surface in feet, write an equation that expresses p in terms of d. [*Hint*: Find an equation of the line that passes through $(0, 15)$ and $(33, 30)$.]

(B) What is the pressure at 12,540 feet (the average depth of the ocean)?

(C) Graph the equation for $0 \le d \le 12{,}540$.

(D) Write a brief verbal interpretation of the slope of the line found in part (A).

Social Sciences

75. *Psychology.* In an experiment on motivation, J. S. Brown trained a group of rats to run down a narrow passage in a cage to obtain food in a goal box. Using a harness, he then connected the rats to an overhead wire that was attached to a spring scale. A rat was placed at different distances d (in centimeters) from the goal box, and the pull p (in grams) of the rat toward the food was measured. Brown found that the relationship between these two variables was very close to being linear and could be approximated by the equation

$$p = -\tfrac{1}{5}d + 70 \qquad 30 \le d \le 175$$

(See J. S. Brown, *Journal of Comparative and Physiological Psychology,* 1948, 41:450–465.)

(A) What was the pull when $d = 30$? When $d = 175$?

(B) Graph the equation.

(C) What is the slope of the line?

Section 1-4

Quadratic Functions

- ❑ QUADRATIC FUNCTIONS, EQUATIONS, AND INEQUALITIES
- ❑ PROPERTIES OF QUADRATIC FUNCTIONS AND THEIR GRAPHS
- ❑ APPLICATIONS

If the degree of a linear function is increased by one, we obtain a *second-degree function,* usually called a *quadratic function,* another basic function that we will need in our library of elementary functions. We will investigate relationships between quadratic functions and the solutions to quadratic equations and inequalities. (A detailed treatment of algebraic solutions to quadratic equations can be found in Appendix A-9.) Other important properties of quadratic functions will also be investigated, including maximum and minimum properties. We will then be in a position to solve important practical problems such as finding production levels that will produce maximum revenue or maximum profit.

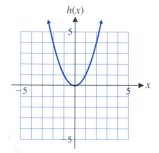

FIGURE 1 Square function $h(x) = x^2$

❑ QUADRATIC FUNCTIONS, EQUATIONS, AND INEQUALITIES

The graph of the square function $h(x) = x^2$ is shown in Figure 1. Notice that the graph is symmetric with respect to the y axis and that $(0, 0)$ is the lowest point on the graph. Let's explore the effect of applying a sequence of basic transformations to the graph of h.

Explore–Discuss 1

Indicate how the graph of each function is related to the graph of $h(x) = x^2$. Find the highest or lowest point, whichever exists, on each graph.

(A) $f(x) = (x - 3)^2 - 7 = x^2 - 6x + 2$

(B) $g(x) = 0.5(x + 2)^2 + 3 = 0.5x^2 + 2x + 5$

(C) $m(x) = -(x - 4)^2 + 8 = -x^2 + 8x - 8$

(D) $n(x) = -3(x + 1)^2 - 1 = -3x^2 - 6x - 4$

Graphing the functions in Explore–Discuss 1 produces figures similar in shape to the graph of the square function in Figure 1. These figures are called *parabolas.* The functions that produced these parabolas are examples of the important class of *quadratic functions,* which we now define.

Quadratic Functions

If a, b, and c are real numbers with $a \neq 0$, then the function

$$f(x) = ax^2 + bx + c$$

is a **quadratic function** and its graph is a **parabola.**

Since the expression $ax^2 + bx + c$ represents a real number for all real number replacements of x,

The domain of a quadratic function is the set of all real numbers.

We will discuss methods for determining the range of a quadratic function later in this section. Typical graphs of quadratic functions are illustrated in Figure 2.

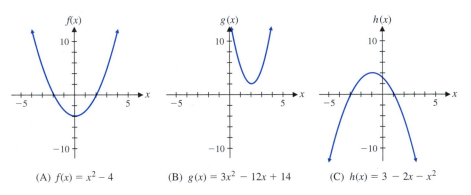

(A) $f(x) = x^2 - 4$ (B) $g(x) = 3x^2 - 12x + 14$ (C) $h(x) = 3 - 2x - x^2$

FIGURE 2 Graphs of quadratic functions

Example 1 ⇔   **Intercepts, Equations, and Inequalities**

(A) Sketch a graph of $f(x) = -x^2 + 5x + 3$ in a rectangular coordinate system.

(B) Find x and y intercepts algebraically to two decimal places.

(C) Graph $f(x) = -x^2 + 5x + 3$ in a standard viewing window.

(D) Find the x and y intercepts to two decimal places using trace and zoom or an appropriate built-in routine in your graphing utility.

(E) Solve the quadratic inequality $-x^2 + 5x + 3 \geqslant 0$ graphically to two decimal places using the results of parts (A) and (B) or (C) and (D).

(F) Solve the equation $-x^2 + 5x + 3 = 4$ graphically to two decimal places using the intersection routine in your graphing utility.

SOLUTION   (A) Hand-sketching a graph of f:

(B) Finding intercepts algebraically:

y intercept: $f(0) = -(0)^2 + 5(0) + 3 = 3$

x intercepts: $f(x) = 0$

$-x^2 + 5x + 3 = 0$ *Quadratic equation*

$x = \dfrac{-b \pm \sqrt{b^2 - 4ac}}{2a}$ *Quadratic formula (see Appendix A-9)*

$x = \dfrac{-(5) \pm \sqrt{5^2 - 4(-1)(3)}}{2(-1)}$

$= \dfrac{-5 \pm \sqrt{37}}{-2} = -0.54$ or 5.54

(C) Graphing in a graphing utility:

(D) Finding intercepts graphically using a graphing utility:

x intercept: -0.54 x intercept: 5.54 y intercept: 3

(E) Solving $-x^2 + 5x + 3 \geq 0$ graphically: The quadratic inequality

$$-x^2 + 5x + 3 \geq 0$$

holds for those values of x for which the graph of $f(x) = -x^2 + 5x + 3$ in the figures in parts (A) and (C) is at or above the x axis. This happens for

x between the two x intercepts [found in part (B) or (D)], including the two x intercepts. Thus, the solution set for the quadratic inequality is $-0.54 \leq x \leq 5.54$ or $[-0.54, 5.54]$.

(F) Solving the equation $-x^2 + 5x + 3 = 4$ using a graphing utility:

$-x^2 + 5x + 3 = 4$ at $x = 0.21$ $-x^2 + 5x + 3 = 4$ at $x = 4.79$

Matched Problem 1 ⇔ (A) Sketch a graph of $g(x) = 2x^2 - 5x - 5$ in a rectangular coordinate system.

(B) Find x and y intercepts algebraically to two decimal places.

(C) Graph $g(x) = 2x^2 - 5x - 5$ in a standard viewing window.

(D) Find the x and y intercepts to two decimal places using trace and zoom or an appropriate built-in routine in your graphing utility.

(E) Solve $2x^2 - 5x - 5 \geq 0$ graphically to two decimal places using the results of parts (A) and (B) or (C) and (D).

(F) Solve the equation $2x^2 - 5x - 5 = -3$ graphically to two decimal places using trace and zoom or an appropriate built-in routine in your graphing utility.

Explore–Discuss 2   How many x intercepts can the graph of a quadratic function have? How many y intercepts? Explain your reasoning.

❑ PROPERTIES OF QUADRATIC FUNCTIONS
   AND THEIR GRAPHS

Many useful properties of the quadratic function can be uncovered by transforming

$$f(x) = ax^2 + bx + c \qquad a \neq 0$$

into the **standard form**

$$f(x) = a(x - h)^2 + k$$

The process of *completing the square* (see Appendix A-9) is central to the transformation. We illustrate the process through a specific example and then generalize the results.

Consider the quadratic function given by

$$f(x) = -2x^2 + 16x - 24 \tag{1}$$

We use completing the square to transform this function into standard form:

$$f(x) = -2x^2 + 16x - 24$$

Factor the coefficient of x^2 out of the first two terms.

$$= -2(x^2 - 8x) - 24$$

$$= -2(x^2 - 8x + \textbf{?}) - 24$$

Add 16 to complete the square inside the parentheses. Because of the -2 outside the parentheses, we have actually added -32, so we must add 32 to the outside.

$$= -2(x^2 - 8x + \textbf{16}) - 24 + \textbf{32}$$

$$= -2(x - 4)^2 + 8$$

The transformation is complete and can be checked by multiplying out.

Thus,

$$f(x) = -2(x - 4)^2 + 8 \tag{2}$$

If $x = 4$, then $-2(x - 4)^2 = 0$ and $f(4) = 8$. For any other value of x, the negative number $-2(x - 4)^2$ is added to 8, making it smaller. (Think about this.) Therefore,

$$f(4) = 8$$

is the *maximum value* of $f(x)$ for all x—a very important result! Furthermore, if we choose any two x values that are the same distance from 4, we will obtain the same function value. For example, $x = 3$ and $x = 5$ are each one unit from $x = 4$ and their function values are

$$f(3) = -2(3 - 4)^2 + 8 = 6$$
$$f(5) = -2(5 - 4)^2 + 8 = 6$$

Thus, the vertical line $x = 4$ is a line of symmetry. That is, if the graph of equation (1) is drawn on a piece of paper and the paper is folded along the line $x = 4$, then the two sides of the parabola will match exactly. All these results are illustrated by graphing equations (1) and (2) and the line $x = 4$ simultaneously in the same coordinate system (Fig. 3).

From the preceding discussion, we see that as x moves from left to right, $f(x)$ is increasing on $(-\infty, 4]$, and decreasing on $[4, \infty)$, and that $f(x)$ can assume no value greater than 8. Thus,

Range of f: $y \leq 8$ or $(-\infty, 8]$

In general, the graph of a quadratic function is a parabola with line of symmetry parallel to the vertical axis. The lowest or highest point on the parabola, whichever exists, is called the **vertex.** The maximum or minimum value of a quadratic function always occurs at the vertex of the parabola. The line of symmetry through the vertex is called the **axis** of the parabola. In the example above, $x = 4$ is the axis of the parabola and $(4, 8)$ is its vertex.

Applying the graph transformation properties discussed in Section 1-2 to the transformed equation,

$$f(x) = -2x^2 + 16x - 24$$
$$= -2(x - 4)^2 + 8$$

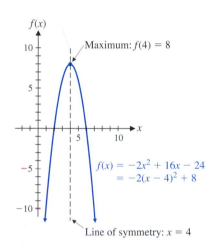

Maximum: $f(4) = 8$

$f(x) = -2x^2 + 16x - 24$
$= -2(x - 4)^2 + 8$

Line of symmetry: $x = 4$

FIGURE 3 Graph of a quadratic function

we see that the graph of $f(x) = -2x^2 + 16x - 24$ is the graph of $g(x) = x^2$ vertically expanded by a factor of 2, reflected in the x axis, and shifted to the right 4 units and up 8 units, as shown in Figure 4.

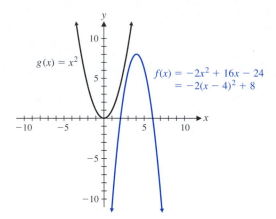

FIGURE 4 Graph of f is the graph of g transformed

Note the important results we have obtained from the standard form of the quadratic function f:

- The vertex of the parabola
- The axis of the parabola
- The maximum value of $f(x)$
- The range of the function f
- The relationship between the graph of $g(x) = x^2$ and the graph of $f(x) = -2x^2 + 16x - 24$

Now, let us explore the effects of changing the constants $a, h,$ and k on the graph of $f(x) = a(x - h)^2 + k$.

Explore–Discuss 3

(A) Let $a = 1$ and $h = 5$. Graph $f(x) = a(x - h)^2 + k$ for $k = -4, 0,$ and 3 simultaneously in the same coordinate system. Explain the effect of changing k on the graph of f.

(B) Let $a = 1$ and $k = 2$. Graph $f(x) = a(x - h)^2 + k$ for $h = -4, 0,$ and 5 simultaneously in the same coordinate system. Explain the effect of changing h on the graph of f.

(C) Let $h = 5$ and $k = -2$. Graph $f(x) = a(x - h)^2 + k$ for $a = 0.25, 1,$ and 3 simultaneously in the same coordinate system. Graph function f for $a = 1, -1,$ and -0.25 simultaneously in the same coordinate system. Explain the effect of changing a on the graph of f.

(D) Discuss parts (A)–(C) using a graphing utility and a standard viewing window.

The preceding discussion is generalized for all quadratic functions in the following box:

Properties of a Quadratic Function and Its Graph

Given a quadratic function and the standard form obtained by completing the square:

$$f(x) = ax^2 + bx + c \qquad a \neq 0$$
$$= a(x - h)^2 + k$$

We summarize general properties as follows:

1. The graph of f is a parabola:

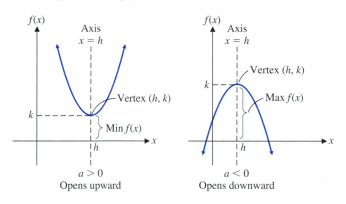

2. Vertex: (h, k) (parabola increases on one side of the vertex and decreases on the other)

3. Axis (of symmetry): $x = h$ (parallel to y axis)

4. $f(h) = k$ is the minimum if $a > 0$ and the maximum if $a < 0$

5. Domain: All real numbers
   Range: $(-\infty, k]$ if $a < 0$ or $[k, \infty)$ if $a > 0$

6. The graph of f is the graph of $g(x) = ax^2$ translated horizontally h units and vertically k units.

Example 2 ⇨ **Analyzing a Quadratic Function**   Given the quadratic function

$$f(x) = 0.5x^2 - 6x + 21$$

(A) Find the standard form for f.

(B) Find the vertex and the maximum or minimum. State the range of f.

(C) Describe how the graph of function f can be obtained from the graph of $g(x) = x^2$ using transformations discussed in Section 1-2.

(D) Sketch a graph of function f in a rectangular coordinate system.

(E) Graph function f using a suitable viewing window.

(F) Find the vertex and the maximum or minimum graphically using the minimum routine. State the range of f.

SOLUTION   (A) Complete the square to find the standard form:

$$f(x) = 0.5x^2 - 6x + 21$$
$$= 0.5(x^2 - 12x + \ ?\ ) + 21$$
$$= 0.5(x^2 - 12x + 36) + 21 - 18$$
$$= 0.5(x - 6)^2 + 3$$

(B) From the standard form, we see that $h = 6$ and $k = 3$. Thus, vertex: $(6, 3)$; minimum: $f(6) = 3$; range: $y \geq 3$ or $[3, \infty)$.

(C) The graph of $f(x) = 0.5(x - 6)^2 + 3$ is the same as the graph of $g(x) = x^2$ vertically contracted by a factor of 0.5, and shifted to the right 6 units and up 3 units.

(D) Graph in a rectangular coordinate system:

 (E) Graph in a graphing utility:

 (F) Finding the vertex, minimum, and range graphically using a graphing utility:

Vertex: $(6, 3)$; minimum: $f(6) = 3$; range: $y \geq 3$ or $[3, \infty)$.

Matched Problem 2   Given the quadratic function: $f(x) = -0.25x^2 - 2x + 2$

(A) Find the standard form for f.

(B) Find the vertex and the maximum or minimum. State the range of f.

(C) Describe how the graph of function f can be obtained from the graph of $g(x) = x^2$ using transformations discussed in Section 1-2.

(D) Sketch a graph of function f in a rectangular coordinate system.

(E) Graph function f using a suitable viewing window.

(F) Find the vertex and the maximum or minimum graphically using the minimum routine. State the range of f.

❑ APPLICATIONS

Example 3   **Maximum Revenue**   This is a continuation of Example 7 in Section 1-1. Recall that the financial department in the company that produces an automatic camera arrived at the following price–demand function and the corresponding revenue function:

$$p(x) = 94.8 - 5x \qquad \text{Price–demand function}$$
$$R(x) = xp(x) = x(94.8 - 5x) \qquad \text{Revenue function}$$

where $p(x)$ is the wholesale price per camera at which x million cameras can be sold and $R(x)$ is the corresponding revenue (in million dollars). Both functions have domain $1 \leq x \leq 15$.

(A) Find the output to the nearest thousand cameras that will produce the maximum revenue. What is the maximum revenue to the nearest thousand dollars? Solve the problem algebraically by completing the square.

(B) What is the wholesale price per camera (to the nearest dollar) that produces the maximum revenue?

(C) Graph the revenue function using an appropriate viewing window.

(D) Find the output to the nearest thousand cameras that will produce the maximum revenue. What is the maximum revenue to the nearest thousand dollars? Solve the problem graphically using the maximum routine.

SOLUTION   (A) Algebraic solution:

$$R(x) = x(94.8 - 5x)$$
$$= -5x^2 + 94.8x$$
$$= -5(x^2 - 18.96x + \textbf{?})$$
$$= -5(x^2 - 18.96x + 89.8704) + 449.352$$
$$= -5(x - 9.48)^2 + 449.352$$

The maximum revenue of 449.352 million dollars ($449,352,000) occurs when $x = 9.480$ million cameras (9,480,000 cameras).

(B) Finding the wholesale price per camera: Use the price–demand function for an output of 9.480 million cameras:

$$p(x) = 94.8 - 5x$$
$$p(9.480) = 94.8 - 5(9.480)$$
$$= \$47$$

(C) Graph in a graphing utility:

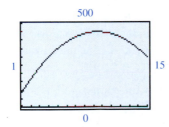

(D) Graphical solution using a graphing utility:

An output of 9.480 million cameras (9,480,000 cameras) will produce a maximum revenue of 449.352 million dollars ($449,352,000). ■

Matched Problem 3   ⤶   The financial department in Example 3, using statistical and analytical techniques (see Matched Problem 7 in Section 1-1), arrived at the cost function

$$C(x) = 156 + 19.7x \quad \textit{Cost function}$$

where $C(x)$ is the cost (in million dollars) for manufacturing and selling x million cameras.

(A) Using the revenue function from Example 3 and the cost function above, write an equation for the profit function.

(B) Find the output to the nearest thousand cameras that will produce the maximum profit. What is the maximum profit to the nearest thousand dollars? Solve the problem algebraically by completing the square.

(C) What is the wholesale price per camera (to the nearest dollar) that produces the maximum profit?

(D) Graph the profit function using an appropriate viewing window.

(E) Find the output to the nearest thousand cameras that will produce the maximum profit. What is the maximum profit to the nearest thousand dollars? Solve the problem graphically using the maximum routine.

Example 4 ⇨ **Break-Even Analysis** Use the revenue function from Example 3 and the cost function from Matched Problem 3:

$$R(x) = x(94.8 - 5x) \qquad \text{Revenue function}$$
$$C(x) = 156 + 19.7x \qquad \text{Cost function}$$

Both have domain $1 \leqslant x \leqslant 15$.

(A) Sketch the graphs of both functions in the same coordinate system.

(B) **Break-even points** are the production levels at which $R(x) = C(x)$. Find the break-even points algebraically to the nearest thousand cameras.

(C) Plot both functions simultaneously in the same viewing window.

(D) Find the break-even points graphically to the nearest thousand cameras using the intersection routine.

(E) Recall that a loss occurs if $R(x) < C(x)$ and a profit occurs if $R(x) > C(x)$. For what outputs (to the nearest thousand cameras) will a loss occur? A profit?

SOLUTION (A) Sketch of functions:

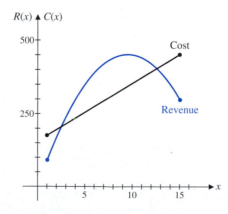

(B) Find x such that $R(x) = C(x)$:

$$x(94.8 - 5x) = 156 + 19.7x$$
$$-5x^2 + 75.1x - 156 = 0$$

$$x = \frac{-75.1 \pm \sqrt{75.1^2 - 4(-5)(-156)}}{2(-5)}$$ Quadratic formula

$$= \frac{-75.1 \pm \sqrt{2,520.01}}{-10}$$

$$x = 2.490 \quad \text{or} \quad 12.530$$

The company breaks even at $x = 2.490$ and 12.530 million cameras.

(C) Graph in a graphing utility:

(D) Graphical solution:

The company breaks even at $x = 2.490$ and 12.530 million cameras.

(E) Use the results from parts (A) and (B) or (C) and (D):

Loss: $1 \leqslant x < 2.490$ or $12.530 < x \leqslant 15$
Profit: $2.490 < x < 12.530$

Matched Problem 4   Use the profit equation from Matched Problem 3:

$$P(x) = R(x) - C(x)$$
$$= -5x^2 + 75.1x - 156 \quad \text{Profit function}$$

Domain: $1 \leqslant x \leqslant 15$

(A) Sketch a graph of the profit function in a rectangular coordinate system.

(B) Break-even points occur when $P(x) = 0$. Find the break-even points algebraically to the nearest thousand cameras.

(C) Plot the profit function in an appropriate viewing window.

(D) Find the break-even points graphically to the nearest thousand cameras using an appropriate built-in routine.

(E) A loss occurs if $P(x) < 0$, and a profit occurs if $P(x) > 0$. For what outputs (to the nearest thousand cameras) will a loss occur? A profit?

Answers to Matched Problems

1. (A) $g(x)$

(B) x intercepts: $-0.77, 3.27$; y intercept: -5

(C)

(D) x intercepts: $-0.77, 3.27$; y intercept: -5

(E) $x \le -0.77$ or $x \ge 3.27$; or $(-\infty, -0.77]$ or $[3.27, \infty)$

(F) $x = -0.35, 2.85$

2. (A) $f(x) = -0.25(x + 4)^2 + 6$.

(B) Vertex: $(-4, 6)$; maximum: $f(-4) = 6$; range: $y \le 6$ or $(-\infty, 6]$

(C) The graph of $f(x) = -0.25(x + 4)^2 + 6$ is the same as the graph of $g(x) = x^2$ vertically contracted by a factor of 0.25, reflected in the x axis, and shifted 4 units to the left and 6 units up.

(D)

(E)

(F) Vertex: $(-4, 6)$; maximum: $f(-4) = 6$; range: $y \le 6$ or $(-\infty, 6]$

3. (A) $P(x) = R(x) - C(x) = -5x^2 + 75.1x - 156$

(B) $P(x) = R(x) - C(x) = -5(x - 7.51)^2 + 126.0005$; output of 7.510 million cameras will produce a maximum profit of 126.001 million dollars.

(C) $p(7.510) = \$57$

(D)

(E)

An output of 7.51 million cameras will produce a maximum profit of 126.001 million dollars. (Notice that maximum profit does not occur at the same output where maximum revenue occurs.)

4. (A) $P(x)$

(B) $x = 2.490$ or 12.530 million cameras

(C)

Exercise 1-4

A *In Problems 1–4, complete the square and find the standard form of each quadratic function.*

1. $f(x) = x^2 - 4x + 3$ **2.** $g(x) = x^2 - 2x - 5$

3. $m(x) = -x^2 + 6x - 4$ **4.** $n(x) = -x^2 + 8x - 9$

In Problems 5–8, write a brief verbal description of the relationship between the graph of the indicated function (from Problems 1–4) and the graph of $y = x^2$.

5. $f(x) = x^2 - 4x + 3$ **6.** $g(x) = x^2 - 2x - 5$

7. $m(x) = -x^2 + 6x - 4$ **8.** $n(x) = -x^2 + 8x - 9$

9. Match each equation with a graph of one of the functions $f, g, m,$ or n in the figure.

(A) $y = -(x + 2)^2 + 1$ (B) $y = (x - 2)^2 - 1$
(C) $y = (x + 2)^2 - 1$ (D) $y = -(x - 2)^2 + 1$

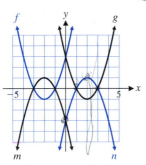

Figure for 9

10. Match each equation with a graph of one of the functions $f, g, m,$ or n in the figure.

(A) $y = (x - 3)^2 - 4$
(B) $y = -(x + 3)^2 + 4$
(C) $y = -(x - 3)^2 + 4$
(D) $y = (x + 3)^2 - 4$

Figure for 10

For the functions indicated in Problems 11–14, find each of the following to the nearest integer by referring to the graphs for Problems 9 and 10.

(A) Intercepts   (B) Vertex
(C) Maximum or minimum (D) Range
(E) Increasing interval   (F) Decreasing interval

11. Function n in the figure for Problem 9

12. Function m in the figure for Problem 10

13. Function f in the figure for Problem 9

14. Function g in the figure for Problem 10

In Problems 15–18, find each of the following:

(A) Intercepts   (B) Vertex
(C) Maximum or minimum (D) Range

15. $f(x) = -(x - 3)^2 + 2$

16. $g(x) = -(x + 2)^2 + 3$

17. $m(x) = (x + 1)^2 - 2$

18. $n(x) = (x - 4)^2 - 3$

B *In Problems 19–22, write an equation for each graph in the form $y = a(x - h)^2 + k$, where a is either 1 or −1 and h and k are integers.*

19.

20.

21.

22.

In Problems 23–28, find the standard form for each quadratic function. Then find each of the following:

    (A) Intercepts   (B) Vertex

    (C) Maximum or minimum   (D) Range

23. $f(x) = x^2 - 8x + 12$

24. $g(x) = x^2 - 6x + 5$

25. $r(x) = -4x^2 + 16x - 15$

26. $s(x) = -4x^2 - 8x - 3$

27. $u(x) = 0.5x^2 - 2x + 5$

28. $v(x) = 0.5x^2 + 4x + 10$

 29. Let $f(x) = 0.3x^2 - x - 8$. Solve each equation graphically to two decimal places.

    (A) $f(x) = 4$ (B) $f(x) = -1$ (C) $f(x) = -9$

30. Let $g(x) = -0.6x^2 + 3x + 4$. Solve each equation graphically to two decimal places.

    (A) $g(x) = -2$ (B) $g(x) = 5$ (C) $g(x) = 8$

31. Explain under what graphical conditions a quadratic function has exactly one real zero.

32. Explain under what graphical conditions a quadratic function has no real zeros.

C *In Problems 33–36, first write each function in standard form; then find each of the following (to two decimal places):*

    (A) Intercepts   (B) Vertex

    (C) Maximum or minimum   (D) Range

33. $g(x) = 0.25x^2 - 1.5x - 7$

34. $m(x) = 0.20x^2 - 1.6x - 1$

35. $f(x) = -0.12x^2 + 0.96x + 1.2$

36. $n(x) = -0.15x^2 - 0.90x + 3.3$

 Solve Problems 37–42 graphically to two decimal places using a graphing utility.

37. $2 - 5x - x^2 = 0$

38. $7 + 3x - 2x^2 = 0$

39. $1.9x^2 - 1.5x - 5.6 < 0$

40. $3.4 + 2.9x - 1.1x^2 \geq 0$

41. $2.8 + 3.1x - 0.9x^2 \leq 0$

42. $1.8x^2 - 3.1x - 4.9 > 0$

43. Given that f is a quadratic function with minimum $f(x) = f(2) = 4$, find the axis, vertex, range, and x intercepts.

44. Given that f is a quadratic function with maximum $f(x) = f(-3) = -5$, find the axis, vertex, range, and x intercepts.

In Problems 45–48:

    (A) Graph f and g in the same coordinate system.

    (B) Solve $f(x) = g(x)$ algebraically to two decimal places.

    (C) Solve $f(x) > g(x)$ using parts (A) and (B).

    (D) Solve $f(x) < g(x)$ using parts (A) and (B).

45. $f(x) = -0.4x(x - 10)$
$g(x) = 0.3x + 5$
$0 \leq x \leq 10$

46. $f(x) = -0.7x(x - 7)$
$g(x) = 0.5x + 3.5$
$0 \leq x \leq 7$

47. $f(x) = -0.9x^2 + 7.2x$
$g(x) = 1.2x + 5.5$
$0 \leq x \leq 8$

48. $f(x) = -0.7x^2 + 6.3x$
$g(x) = 1.1x + 4.8$
$0 \leq x \leq 9$

49. Give a simple example of a quadratic function that has no real zeros. Explain how its graph is related to the x axis.

50. Give a simple example of a quadratic function that has exactly one real zero. Explain how its graph is related to the x axis.

 Solve each problem algebraically, then check on a graphing utility, using built-in zero and maximum/minimum routines.

Business & Economics

51. *Tire mileage.* An automobile tire manufacturer collected the data in the table relating tire pressure x (in pounds per square inch) and mileage (in thousands of miles):

x	Mileage
28	45
30	52
32	55
34	51
36	47

A mathematical model for the data is given by

$$f(x) = -0.518x^2 + 33.3x - 481$$

(A) Complete the following table. Round values of $f(x)$ to one decimal place.

x	Mileage	$f(x)$
28	45	
30	52	
32	55	
34	51	
36	47	

(B) Sketch the graph of f and the mileage data in the same coordinate system.

(C) Use values of the modeling function rounded to two decimal places to estimate the mileage for a tire pressure of 31 pounds per square inch. For 35 pounds per square inch.

(D) Write a brief description of the relationship between tire pressure and mileage.

52. *Automobile production.* The table shows the retail market share of passenger cars from Ford Motor Company as a percentage of the U.S. market.

WWW

Year	Market Share
1975	23.6%
1980	17.2%
1985	18.8%
1990	20.0%
1995	20.7%

A mathematical model for this data is given by

$$f(x) = 0.04x^2 - 0.8x + 22$$

where $x = 0$ corresponds to 1975.

(A) Complete the following table.

x	Market Share	$f(x)$
0	23.6	
5	17.2	
10	18.8	
15	20.0	
20	20.7	

(B) Sketch the graph of f and the market share data in the same coordinate system.

(C) Use values of the modeling function f to estimate Ford's market share in 2000. In 2005.

(D) Write a brief verbal description of Ford's market share from 1975 to 1995.

53. *Revenue.* Refer to Problems 85 and 87, Exercise 1-1. We found that the marketing research department for the company that manufactures and sells memory chips for microcomputers established the following price–demand and revenue functions:

$$p(x) = 75 - 3x \qquad \text{Price–demand function}$$
$$R(x) = xp(x) = x(75 - 3x) \quad \text{Revenue function}$$

where $p(x)$ is the wholesale price in dollars at which x million chips can be sold, and $R(x)$ is in millions of dollars. Both functions have domain $1 \leqslant x \leqslant 20$.

(A) Sketch a graph of the revenue function in a rectangular coordinate system.

(B) Find the output that will produce the maximum revenue. What is the maximum revenue?

(C) What is the wholesale price per chip (to the nearest dollar) that produces the maximum revenue?

54. *Revenue.* Refer to Problems 86 and 88, Exercise 1-1. We found that the marketing research department for the company that manufactures and sells "notebook" computers established the following price–demand and revenue functions:

$$p(x) = 2{,}000 - 60x \qquad \text{Price–demand function}$$
$$R(x) = xp(x) \qquad\qquad \text{Revenue function}$$
$$ = x(2{,}000 - 60x)$$

where $p(x)$ is the wholesale price in dollars at which x thousand computers can be sold, and $R(x)$ is in thousands of dollars. Both functions have domain $1 \leqslant x \leqslant 25$.

(A) Sketch a graph of the revenue function in a rectangular coordinate system.

(B) Find the output (to the nearest hundred computers) that will produce the maximum revenue. What is the maximum revenue to the nearest thousand dollars?

(C) What is the wholesale price per computer (to the nearest dollar) that produces the maximum revenue?

55. *Break-even analysis.* Use the revenue function from Problem 53 in this exercise and the cost function from Problem 89, Exercise 1-1:

$$R(x) = x(75 - 3x) \qquad \text{Revenue function}$$
$$C(x) = 125 + 16x \qquad \text{Cost function}$$

where x is in millions of chips, and $R(x)$ and $C(x)$ are in millions of dollars. Both functions have domain $1 \leqslant x \leqslant 20$.

(A) Sketch a graph of both functions in the same rectangular coordinate system.

(B) Find the break-even points to the nearest thousand chips.

(C) For what outputs will a loss occur? A profit?

56. *Break-even analysis.* Use the revenue function from Problem 54 in this exercise and the cost function from Problem 90, Exercise 1-1:

$$R(x) = x(2,000 - 60x) \quad \text{Revenue function}$$
$$C(x) = 4,000 + 500x \quad \text{Cost function}$$

where x is thousands of computers, and $C(x)$ and $R(x)$ are in thousands of dollars. Both functions have domain $1 \leq x \leq 25$.

(A) Sketch a graph of both functions in the same rectangular coordinate system.

(B) Find the break-even points.

(C) For what outputs will a loss occur? Will a profit occur?

57. *Profit–loss analysis.* Use the revenue function from Problem 53 in this exercise and the cost function from Problem 89, Exercise 1-1:

$$R(x) = x(75 - 3x) \quad \text{Revenue function}$$
$$C(x) = 125 + 16x \quad \text{Cost function}$$

where x is in millions of chips, and $R(x)$ and $C(x)$ are in millions of dollars. Both functions have domain $1 \leq x \leq 20$.

(A) Form a profit function P, and graph R, C, and P in the same rectangular coordinate system.

(B) Discuss the relationship between the intersection points of the graphs of R and C and the x intercepts of P.

(C) Find the x intercepts of P to the nearest thousand chips. Find the break-even points to the nearest thousand chips.

(D) Refer to the graph drawn in part (A). Does the maximum profit appear to occur at the same output level as the maximum revenue? Are the maximum profit and the maximum revenue equal? Explain.

(E) Verify your conclusion in part (D) by finding the output (to the nearest thousand chips) that produces the maximum profit. Find the maximum profit (to the nearest thousand dollars), and compare with Problem 53B.

58. *Profit–loss analysis.* Use the revenue function from Problem 54 in this exercise and the cost function from Problem 90, Exercise 1-1:

$$R(x) = x(2,000 - 60x) \quad \text{Revenue function}$$
$$C(x) = 4,000 + 500x \quad \text{Cost function}$$

where x is thousands of computers, and $R(x)$ and $C(x)$ are in thousands of dollars. Both functions have domain $1 \leq x \leq 25$.

(A) Form a profit function P, and graph R, C, and P in the same rectangular coordinate system.

(B) Discuss the relationship between the intersection points of the graphs of R and C and the x intercepts of P.

(C) Find the x intercepts of P to the nearest hundred computers. Find the break-even points.

(D) Refer to the graph drawn in part (A). Does the maximum profit appear to occur at the same output level as the maximum revenue? Are the maximum profit and the maximum revenue equal? Explain.

(E) Verify your conclusion in part (D) by finding the output that produces the maximum profit. Find the maximum profit and compare with Problem 54B.

Life Sciences

59. *Medicine.* The French physician Poiseuille was the first to discover that blood flows faster near the center of an artery than near the edge. Experimental evidence has shown that the rate of flow v (in centimeters per second) at a point x centimeters from the center of an artery (see the figure) is given by

$$v = f(x) = 1,000(0.04 - x^2) \quad 0 \leq x \leq 0.2$$

Find the distance from the center that the rate of flow is 20 centimeters per second. Round answer to two decimal places.

Figure for 59 and 60

60. *Medicine.* Refer to Problem 59. Find the distance from the center that the rate of flow is 30 centimeters per second. Round answer to two decimal places.

Important Terms and Symbols

1-1 *Functions.* Cartesian or rectangular coordinate system; coordinate axes; quadrants; ordered pairs; coordinates; abscissa; ordinate; fundamental theorem of analytic geometry; solution and solution set for an equation in two variables; graph of an equation in two variables; point-by-point plotting; function; domain; range; functions specified by equations; input; output; independent variable; dependent variable; vertical-line test; function notation; difference quotient; cost function; price–demand function; revenue function; profit function

$$f(x); \quad C = a + bx; \quad p = m - nx;$$
$$R = xp; \quad P = R - C$$

1-2 *Elementary Functions: Graphs and Transformations.* Six basic functions: identity, absolute value, square, cube, square root, cube root; transformations; horizontal shift or translation; vertical shift or translation; reflection in the *x* axis; vertical expansion; vertical contraction; piecewise-defined function; discontinuity

$$f(x) = x; \quad g(x) = |x|; \quad h(x) = x^2;$$
$$m(x) = x^3; \quad n(x) = \sqrt{x}; \quad p(x) = \sqrt[3]{x}$$

See the inside front cover for the graphs of these functions.

1-3 *Linear Functions and Straight Lines.* *x* intercepts; *y* intercept; linear function; constant function; first-degree

function; graph of a linear function; graph of a constant function; graphs of linear equations in two variables; standard form for the equation of a line; equations of vertical lines and horizontal lines; slope; slope–intercept form; point–slope form; equilibrium point; equilibrium price; equilibrium quantity

$$f(x) = mx + b, m \neq 0; \quad f(x) = b; \quad x = a;$$
$$Ax + By = C; y = mx + b; y - y_1 = m(x - x_1);$$
$$m = \frac{y_2 - y_1}{x_2 - x_1}, x_1 \neq x_2$$

1-4 *Quadratic Functions.* Quadratic function; parabola; finding intercepts; standard form; vertex; axis; maximum or minimum; range; break-even points

$$f(x) = ax^2 + bx + c = a(x - h)^2 + k, \quad a \neq 0$$

Review Exercise

Work through all the problems in this chapter review and check your answers in the back of the book. Answers to all review problems are there along with section numbers in italics to indicate where each type of problem is discussed. Where weaknesses show up, review appropriate sections in the text.

A

1. Use point-by-point plotting to sketch a graph of $y = 5 - x^2$. Use integer values for *x* from -3 to 3.

2. Indicate whether each graph specifies a function:

(A)

(B)

(C)

(D)

3. For $f(x) = 2x - 1$ and $g(x) = x^2 - 2x$, find:
   (A) $f(-2) + g(-1)$ (B) $f(0) \cdot g(4)$
   (C) $\dfrac{g(2)}{f(3)}$ (D) $\dfrac{f(3)}{g(2)}$

4. Use the graph of function *f* in the figure to determine (to the nearest integer) *x* or *y* as indicated.
   (A) $y = f(0)$ (B) $4 = f(x)$
   (C) $y = f(3)$ (D) $3 = f(x)$
   (E) $y = f(-6)$ (F) $-1 = f(x)$

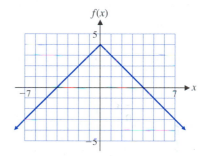

Figure for 4

5. Sketch a graph of each of the functions in parts (A)–(D) using the graph of function *f* in the figure below.
   (A) $y = -f(x)$ (B) $y = f(x) + 4$
   (C) $y = f(x - 2)$ (D) $y = -f(x + 3) - 3$

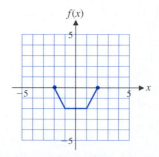

Figure for 5

6. Refer to the figure below.

Figure for 6

(A) Identify the graphs of any linear functions with positive slopes.

(B) Identify the graphs of any linear functions with negative slopes.

(C) Identify the graphs of any constant functions. What are their slopes?

(D) Identify any graphs that are not graphs of functions. What can you say about their slopes?

7. Write an equation in the form $y = mx + b$ for a line with slope $-\frac{2}{3}$ and y intercept 6.

8. Write the equations of the vertical line and the horizontal line that pass through $(-6, 5)$.

9. Sketch a graph of $2x - 3y = 18$. What are the intercepts and slope of the line?

10. Complete the square and find the standard form for the quadratic function

$$f(x) = -x^2 + 4x$$

Then write a brief verbal description of the relationship between the graph of f and the graph of $y = x^2$.

11. Match each equation with a graph of one of the functions $f, g, m,$ or n in the figure.

(A) $y = (x - 2)^2 - 4$ (B) $y = -(x + 2)^2 + 4$
(C) $y = -(x - 2)^2 + 4$ (D) $y = (x + 2)^2 - 4$

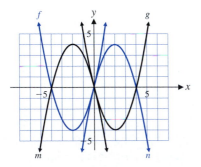

Figure for 11

12. Referring to the graph of function f in the figure for Problem 11 and using known properties of quadratic functions, find each of the following to the nearest integer:

(A) Intercepts   (B) Vertex
(C) Maximum or minimum (D) Range
(E) Increasing interval (F) Decreasing interval

B

13. Indicate which of the following equations define a linear function or a constant function:

(A) $2x - 3y = 5$ (B) $x = -2$
(C) $y = 4 - 3x$ (D) $y = -5$

(E) $x = 3y + 5$ (F) $\dfrac{x}{2} - \dfrac{y}{3} = 1$

14. Find the domain of each function:

(A) $f(x) = \dfrac{2x - 5}{x^2 - x - 6}$ (B) $g(x) = \dfrac{3x}{\sqrt{5 - x}}$

15. The function g is defined by $g(x) = 2x - 3\sqrt{x}$. Translate into a verbal definition.

16. Find the standard form for $f(x) = 4x^2 + 4x - 3$ and then find the intercepts, the vertex, the maximum or minimum, and the range.

17. Describe the graphs of $x = -3$ and $y = 2$. Graph both simultaneously in the same rectangular coordinate system.

18. Let $f(x) = 0.4x\,(x + 4)(2 - x)$.

(A) Sketch a graph of f on graph paper by first plotting points using odd integer values of x from -3 to 3. Then complete the graph using a graphing utility.

(B) Discuss the number of solutions of each of the following equations:

$$f(x) = 3 \qquad f(x) = 2 \qquad f(x) = 1$$

(C) Approximate the solutions of each equation in part (B) to two decimal places.

19. Find $\dfrac{f(2 + h) - f(2)}{h}$ for $f(x) = 3 - 2x$.

20. Find $\dfrac{f(a + h) - f(a)}{h}$ for $f(x) = x^2 - 3x + 1$.

21. Explain how the graph of $m(x) = -|x - 4|$ is related to the graph of $y = |x|$.

22. Explain how the graph of $g(x) = 0.3x^3 + 3$ is related to the graph of $y = x^3$.

23. The following graph is the result of applying a sequence of transformations to the graph of $y = x^2$. Describe the transformations verbally and write an equation for the graph.

Figure for 23

24. The graph of a function f is formed by vertically expanding the graph of $y = \sqrt{x}$ by a factor of 2, and shifting it to the left 3 units and down 1 unit. Find an equation for function f and graph it for $-5 \leqslant x \leqslant 5$ and $-5 \leqslant y \leqslant 5$.

25. Sketch the graph of

$$f(x) = \begin{cases} -x - 2 & \text{for } x < 0 \\ 0.2x^2 & \text{for } x \geqslant 0 \end{cases}$$

26. Write the equation of a line through each indicated point with the indicated slope. Write the final answer in the form $y = mx + b$.

(A) $m = -\frac{2}{3}; (-3, 2)$ (B) $m = 0; (3, 3)$

27. Write the equation of the line through the two indicated points. Write the final answer in the form $Ax + By = C$.

(A) $(-3, 5), (1, -1)$ (B) $(-1, 5), (4, 5)$
(C) $(-2, 7), (-2, -2)$

28. Write an equation for the graph shown in the form $y = a(x - h)^2 + k$, where a is either -1 or $+1$ and h and k are integers.

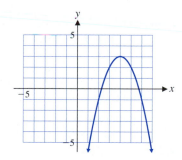

Figure for 28

29. Given $f(x) = -0.4x^2 + 3.2x + 1.2$, find the following algebraically (to one decimal place) without referring to a graph:

(A) Intercepts   (B) Vertex
(C) Maximum or minimum   (D) Range

30. Graph $f(x) = -0.4x^2 + 3.2x + 1.2$ in a graphing utility and find the following (to one decimal place)

using trace and zoom or an appropriate built-in routine:

(A) Intercepts   (B) Vertex
(C) Maximum or minimum   (D) Range

C

31. The following graph is the result of applying a sequence of transformations to the graph of $y = \sqrt[3]{x}$. Describe the transformations verbally, and write an equation for the graph.

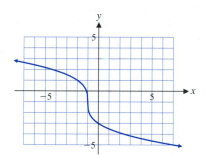

Figure for 31

32. Graph

$$y = mx + b \quad \text{and} \quad y = -\frac{1}{m}x + b$$

simultaneously in the same coordinate system for b fixed and several different values of $m, m \neq 0$. Describe the apparent relationship between the graphs of the two equations.

33. Find and simplify $\dfrac{f(x + h) - f(x)}{h}$ for each of the following functions:

(A) $f(x) = \sqrt{x}$ (B) $f(x) = \dfrac{1}{x}$

34. Given $G(x) = 0.3x^2 + 1.2x - 6.9$, find the following algebraically (to one decimal place) without the use of a graph:

(A) Intercepts   (B) Vertex
(C) Maximum or minimum   (D) Range
(E) Increasing and decreasing intervals

35. Graph $G(x) = 0.3x^2 + 1.2x - 6.9$ in a standard viewing window. Then find each of the following (to one decimal place) using trace or zoom or an appropriate built-in routine:

(A) Intercepts   (B) Vertex
(C) Maximum or minimum   (D) Range
(E) Increasing and decreasing intervals

Applications

Business & Economics

36. *Linear depreciation.* A computer system was purchased by a small business for $12,000 and, for tax purposes, is assumed to have a salvage value of $2,000 after 8 years. If its value is depreciated linearly from $12,000 to $2,000:

   (A) Find the linear equation that relates the value V in dollars to the time t in years. Then graph the equation in a rectangular coordinate system.

   (B) What would be the value of the system after 5 years?

37. *Compound interest.* If $1,000 is invested at $100r\%$ compounded annually, at the end of 3 years it will grow to $A = 1,000(1 + r)^3$.

   (A) Graph the equation in a graphing utility for $0 \leqslant r \leqslant 0.25$; that is, for money invested at between 0% and 25% compounded annually.

   (B) At what rate of interest would $1,000 have to be invested to amount to $1,500 in 3 years? Solve for r graphically (to four decimal places) using trace and zoom or an appropriate built-in routine.

38. *Markup.* A sporting goods store sells a tennis racket that cost $130 for $208 and court shoes that cost $50 for $80.

   (A) If the markup policy of the store for items that cost over $10 is assumed to be linear and is reflected in the pricing of these two items, write an equation that relates retail price R to cost C.

   (B) What would be the retail price of a pair of in-line skates that cost $120?

   (C) What would be the cost of a pair of cross-country skis that had a retail price of $176?

   (D) What is the slope of the graph of the equation found in part (A)? Interpret the slope relative to the problem.

39. *Demand.* The table shows the annual per capita consumption of eggs in the United States.

WWW

Year	Number of Eggs
1980	271
1985	255
1990	233
1995	236
2000	256

A mathematical model for this data is given by

$$f(x) = 0.28x^2 - 6.5x + 274$$

where $x = 0$ corresponds to 1980.

   (A) Complete the following table.

x	Consumption	$f(x)$
0	271	
5	255	
10	233	
15	236	
20	256	

   (B) Graph $y = f(x)$ and the data in the same coordinate system.

   (C) Use the modeling function f to estimate the per capita egg consumption in 2005. In 2010.

   (D) Based on the information in the table, write a brief description of egg consumption from 1980 to 2000.

40. *Electricity rates.* The table shows the electricity rates charged by Easton Utilities in the summer months.
WWW

   (A) Write a piecewise definition of the monthly charge $S(x)$ (in dollars) for a customer who uses x kWh in a summer month.

   (B) Graph $S(x)$.

ENERGY CHARGE (JUNE–SEPTEMBER)
$3.00 for the first 20 kWh or less
5.70¢ per kWh for the next 180 kWh
3.46¢ per kWh for the next 800 kWh
2.17¢ per kWh for all over 1,000 kWh

41. *Equilibrium point.* At a price of $1.90 per bushel, WWW the annual U.S. supply and demand for barley are 410 million bushels and 455 million bushels, respectively. When the price rises to $2.70 per bushel, the supply increases to 430 million bushels and the demand decreases to 415 million bushels.

   (A) Assuming that the price–supply and the price–demand equations are linear, find equations for each.

   (B) Find the equilibrium point for the U.S. barley market.

42. *Break-even analysis.* A video production company is planning to produce an instructional videotape. The producer estimates that it will cost $84,000 to shoot the video and $15 per unit to copy and distribute the tape. The wholesale price of the tape is $50 per unit.

   (A) Write cost and revenue equations, and graph both simultaneously in a rectangular coordinate system.

(B) Determine algebraically when $R = C$. Then, with the aid of part (A), determine when $R < C$ and $R > C$.

(C) Using a graphing utility, determine when $R = C$, $R < C$, and $R > C$.

43. *Break-even analysis.* The research department in a company that manufactures AM/FM clock radios established the following price–demand, cost, and revenue functions:

$$p(x) = 50 - 1.25x \qquad \textcolor{blue}{\textit{Price–demand function}}$$
$$C(x) = 160 + 10x \qquad \textcolor{blue}{\textit{Cost function}}$$
$$R(x) = xp(x)$$
$$\qquad = x(50 - 1.25x) \qquad \textcolor{blue}{\textit{Revenue function}}$$

where x is in thousands of units, and $C(x)$ and $R(x)$ are in thousands of dollars. All three functions have domain $1 \le x \le 40$.

(A) Graph the cost function and the revenue function simultaneously in the same coordinate system.

(B) Determine algebraically when $R = C$. Then, with the aid of part (A), determine when $R < C$ and $R > C$ to the nearest unit.

(C) Determine algebraically the maximum revenue (to the nearest thousand dollars) and the output (to the nearest unit) that produces the maximum revenue. What is the wholesale price of the radio (to the nearest dollar) at this output?

44. *Profit–loss analysis.* Use the cost and revenue functions from Problem 43.

(A) Write a profit function and graph it in a graphing utility.

(B) Determine graphically when $P = 0$, $P < 0$, and $P > 0$ to the nearest unit.

(C) Determine graphically the maximum profit (to the nearest thousand dollars) and the output (to the nearest unit) that produces the maximum profit. What is the wholesale price of the radio (to the nearest dollar) at this output? [Compare with Problem 43C.]

45. *Construction.* A construction company has 840 feet of chain-link fence that is used to enclose storage areas for equipment and materials at construction sites. The supervisor wants to set up two identical rectangular storage areas sharing a common fence (see the figure):

Figure for 45

Assuming that all fencing is used:

(A) Express the total area $A(x)$ enclosed by both pens as a function of x.

(B) From physical considerations, what is the domain of the function A?

(C) Graph function A in a rectangular coordinate system.

(D) Use the graph to discuss the number and approximate locations of values of x that would produce storage areas with a combined area of 25,000 square feet.

(E) Approximate graphically (to the nearest foot) the values of x that would produce storage areas with a combined area of 25,000 square feet.

(F) Determine algebraically the dimensions of the storage areas that have the maximum total combined area. What is the maximum area?

Life Sciences

46. *Air pollution.* On an average summer day in a large city, the pollution index at 8:00 AM is 20 parts per million, and it increases linearly by 15 parts per million each hour until 3:00 PM. Let $P(x)$ be the amount of pollutants in the air x hours after 8:00 AM.

(A) Express $P(x)$ as a linear function of x.

(B) What is the air pollution index at 1:00 PM?

(C) Graph the function P for $0 \le x \le 7$.

(D) What is the slope of the graph? (The slope is the amount of increase in pollution for each additional hour of time.)

Social Sciences

47. *Psychology—sensory perception.* One of the oldest studies in psychology concerns the following question: Given a certain level of stimulation (light, sound, weight lifting, electric shock, and so on), how much should the stimulation be increased for a person to notice the difference? In the middle of the nineteenth century, E. H. Weber (a German physiologist) formulated a law that still carries his name: If Δs is the change in stimulus that will just be noticeable at a stimulus level s, then the ratio of Δs to s is a constant:

$$\frac{\Delta s}{s} = k$$

Hence, the amount of change that will be noticed is a linear function of the stimulus level, and we note that the greater the stimulus, the more it takes to notice a difference. In an experiment on weight lifting, the constant k for a given person was found to be $\frac{1}{30}$.

(A) Find Δs (the difference that is just noticeable) at the 30 pound level. At the 90 pound level.

(B) Graph $\Delta s = s/30$ for $0 \le s \le 120$.

(C) What is the slope of the graph?

Group Activity 1 *Introduction to Regression Analysis*

In real-world applications collected data may be assembled in table form and then examined to find a function to model the data. A very powerful mathematical tool called *regression analysis* is frequently used for this purpose. Several of the modeling functions stated in examples and exercises in Chapter 1 were constructed using *regression techniques* and a graphing utility.

Regression analysis is the process of fitting a function to a set of data points. This process is also referred to as **curve fitting.** In this group activity, we will restrict our attention to **linear regression**—that is, to fitting data with linear functions. Other types of regression analysis will be discussed in the next chapter.

At this time, we will not be interested in discussing the underlying mathematical methods used to construct a particular regression equation. Instead, we will concentrate on the mechanics of using a graphing utility to apply regression techniques to data sets.

In Example 7, Section 1-1, we were given the data in Table 1 and the corresponding modeling function

$$p(x) = 94.8 - 5x \tag{1}$$

where p is the wholesale price of a camera and x is the demand in millions.

TABLE 1

PRICE–DEMAND

x (MILLIONS)	p ($)
2	87
5	68
8	53
12	37

Now we want to see how the function p was determined using linear regression on a graphing utility. On most graphing utilities, this process can be broken down into three steps:

Step 1. Enter the data set.

Step 2. Compute the desired regression equation.

Step 3. Graph the data set and the regression equation in the same viewing window.

The details for carrying out each step vary greatly from one graphing utility to another. Consult your manual. If you are using a Texas Instruments graphing calculator or a spreadsheet on a computer, you may also want to consult the supplements for this book (see Preface).

Figure 1 shows the results of applying this process to the data in Table 1 on a graphing calculator, and Figure 2 shows the same process on a spreadsheet.

FIGURE 1 Linear regression on a graphing calculator

Price-Demand	
x (millions)	p ($)
2	87
5	68
8	53
12	37
Slope	y intercept
-4.9634703	94.7534247

p = -4.9634703x + 94.7534247

FIGURE 2 Linear regression on a spreadsheet

Notice how well the line graphed in Figures 1 and 2 appears to fit this set of data. At this time, we will not discuss any mathematical techniques for determining how well a curve fits a given set of data. Instead, we will simply assume that, in some sense, the linear regression line for a given data set is the "best fit" for that data.

Most graphing utilities compute coefficients of regression equations to many more decimal places than we will need for our purposes. Normally, when a regression equation is written, the coefficients are rounded. For example, if we round each coefficient in Figure 1 or 2 to one decimal place, we will obtain the modeling function f we used in Example 7, Section 1-1 [see equation (1)].

(A) If you have not already done so, carry out steps 1 to 3 for the data in Table 1. Write a brief summary of the details for performing these steps on your graphing utility and keep it for future reference. (Some of the optional exercises in subsequent chapters will require you to compute regression equations.)

(B) Each of the following examples and exercises from Chapter 1 contains a data set and an equation that models the data. In each case, use a graphing utility to compute a linear regression equation for the data. Plot the data and graph the equation in the same viewing rectangle. Discuss the difference between the graphing utility's regression equation and the modeling function stated in the exercise.

   ■ Section 1-1, Matched Problem 7
   ■ Exercise 1-1, Problems 85 and 86
   ■ Exercise 1-3, Problems 69 and 70

Group Activity 2   Mathematical Modeling in Business

A company manufactures and sells mountain bikes. The management would like to have price–demand and cost functions for break-even and profit–loss analysis. Price–demand and cost functions are often established by collecting appropriate data at different levels of output, and then finding a model in the form of a basic elementary function (from our library of elementary functions) that "closely fits" the collected data. The financial department, using statistical techniques, arrived at the price–demand and cost data in Tables 1 and 2, where p is the wholesale price of a bike for a demand of x hundred bikes, $0 \leq x \leq 220$, and C is the cost (in hundreds of dollars) of producing and selling x bikes, $0 \leq x \leq 220$.

TABLE 1	
PRICE-DEMAND	
x (HUNDREDS)	p ($)
0	525
64	370
125	270
185	130

TABLE 2	
COST	
x (HUNDREDS)	C (HUNDRED $)
0	8,470
30	13,510
120	19,140
180	22,580
220	28,490

(A) *Building a Mathematical Model for Price–Demand.* Plot the data in Table 1 and observe that the relationship between p and x appears to be linear.

   1. If you have completed Group Activity 1, use your graphing utility to find the linear regression line for the data in Table 1. Graph the line and the data in the same viewing window.

   2. The linear regression line found in part 1 is a mathematical model for price–demand and is given by

   $$p(x) = -2.09x + 519 \qquad \text{Price–demand equation}$$

   Graph the data points from Table 1 and the price–demand equation in the same rectangular coordinate system.

   3. The linear regression line defines a linear function. Interpret the slope of the line. Discuss its domain and range. Using the mathematical model, determine the price for a demand of 10,000 bikes. For a demand of 20,000 bikes.

(B) *Building a Mathematical Model for Cost.* Plot the data in Table 2 in a rectangular coordinate system. Which type of function appears to best fit the data?

   1. If you have completed Group Activity 1, use your graphing utility to find the linear regression line for the data in Table 2. Graph the line and the data in the same viewing window.

   2. The linear regression line found in part 1 is a mathematical model for cost and is given by

   $$C(x) = 81x + 9,498 \qquad \text{Cost equation}$$

   Graph the data points from Table 2 and the cost equation in the same rectangular coordinate system.

   3. Interpret the slope of the cost equation function. Discuss its domain and range. Using the mathematical model, determine the cost for an output and sales of 10,000 bikes. For an output and sales of 20,000 bikes.

(C) *Break-Even and Profit–Loss Analysis.* Refer to the price–demand equation in part (A) and the cost equation in part (B). Write an equation for the revenue function and state its domain. Write an equation for the profit function and state its domain.

   1. Graph the revenue function and the cost function simultaneously in the same rectangular coordinate system. Determine algebraically at what outputs (to the nearest unit) the break-even points occur.

Determine the outputs where costs exceed revenues and where revenues exceed costs.

2. Graph the revenue function and the cost function simultaneously in the same viewing window. Determine graphically at what outputs (to the nearest unit) break-even occurs, costs exceed revenues, and revenues exceed costs.

3. Graph the profit function in a rectangular coordinate system. Determine algebraically at what outputs (to the nearest unit) the break-even points occur. Determine where profits occur and where losses occur. At what output and price will a maximum profit occur? Does the maximum revenue and maximum profit occur for the same output? Discuss.

4. Graph the profit function in a graphing utility. Determine graphically at what outputs (to the nearest unit) break-even occurs, losses occur, and profits occur. At what output and price will a maximum profit occur? Does the maximum revenue and maximum profit occur for the same output? Discuss.

Additional Elementary Functions

INTRODUCTION

In this chapter we add the following four general classes of functions to the beginning library of elementary functions started in Chapter 1: polynomial, rational, exponential, and logarithmic. This expanded library of elementary functions should take care of most of your function needs in this and many other courses. The linear and quadratic functions studied in Chapter 1 are special cases of the more general class of functions called *polynomials*. ⬤▭

Section 2-1

Polynomial and Rational Functions

- ❑ POLYNOMIAL FUNCTIONS
- ❑ POLYNOMIAL ROOT APPROXIMATION
- ❑ REGRESSION POLYNOMIALS
- ❑ RATIONAL FUNCTIONS
- ❑ APPLICATION

❑ POLYNOMIAL FUNCTIONS

In Chapter 1 you were introduced to the basic functions

$$f(x) = b \qquad\qquad \text{Constant function}$$
$$f(x) = ax + b \quad a \neq 0 \qquad \text{Linear function}$$
$$f(x) = ax^2 + bx + c \quad a \neq 0 \quad \text{Quadratic function}$$

as well as some special cases of

$$f(x) = ax^3 + bx^2 + cx + d \qquad a \neq 0 \qquad \text{Cubic function}$$

Most of the earlier applications we considered, including cost, revenue, profit, loss, and packaging applications, made use of these functions. Notice the evolving pattern going from the constant function to the cubic function—the terms in each equation are of the form ax^n, where n is a nonnegative integer and a is a real number. All these functions are special cases of the general class of functions called *polynomial functions*:

Polynomial Function

A **polynomial function** is a function of the form

$$f(x) = a_n x^n + a_{n-1} x^{n-1} + \cdots + a_1 x + a_0$$

for n a nonnegative integer, called the **degree** of the polynomial. The coefficients $a_0, a_1, \ldots, a_n$ are real numbers with $a_n \neq 0$. The **domain** of a polynomial function is the set of all real numbers.

The shape of the graph of a polynomial function is connected to the degree of the polynomial. The shapes of odd-degree polynomial functions have something in common, and the shapes of even-degree polynomial functions have something in common. Figure 1 shows graphs of representative polynomial functions from degrees 1 to 6, and suggests some general properties of graphs of polynomial functions.

Notice that the odd-degree polynomial graphs start negative, end positive, and cross the x axis at least once. The even-degree polynomial graphs start positive, end positive, and may not cross the x axis at all. In all cases in Figure 1, the coefficient of the highest-degree term was chosen positive. If any leading

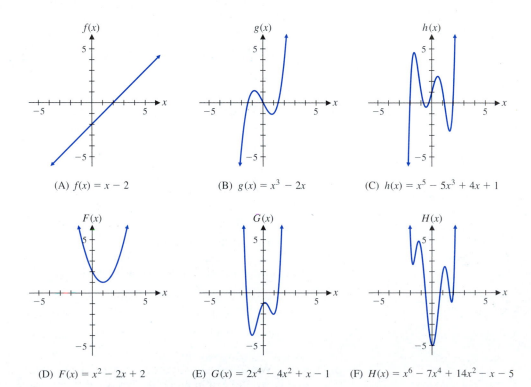

(A) $f(x) = x - 2$ (B) $g(x) = x^3 - 2x$ (C) $h(x) = x^5 - 5x^3 + 4x + 1$

(D) $F(x) = x^2 - 2x + 2$ (E) $G(x) = 2x^4 - 4x^2 + x - 1$ (F) $H(x) = x^6 - 7x^4 + 14x^2 - x - 5$

FIGURE 1 Graphs of polynomial functions

coefficient had been chosen negative, then we would have a similar graph but flipped over.

The graph of a polynomial function is **continuous,** with no holes or breaks. That is, the graph can be drawn without removing a pen from the paper. Also, the graph of a polynomial has no sharp corners. Figure 2 shows the graphs of two functions, one that is not continuous, and the other that is continuous, but with a sharp corner. Neither function is a polynomial.

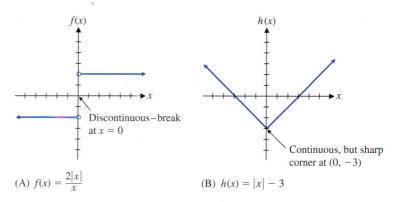

(A) $f(x) = \dfrac{2|x|}{x}$ (B) $h(x) = |x| - 3$

FIGURE 2 Discontinuous and sharp-corner functions

Figure 1 gives examples of polynomial functions with graphs containing the maximum number of *turning points* possible for a polynomial of that degree. A **turning point** on a continuous graph is a point that separates an increasing portion from a decreasing portion, or vice versa. In general, it can be shown that:

> **The graph of a polynomial function of positive degree n can have at most $(n - 1)$ turning points and can cross the x axis at most n times.**

Explore–Discuss 1

(A) What is the least number of turning points an odd-degree polynomial function can have? An even-degree polynomial function?

(B) What is the maximum number of x intercepts the graph of a polynomial function of degree n can have?

(C) What is the maximum number of real solutions an nth-degree polynomial equation can have?

(D) What is the least number of x intercepts the graph of a polynomial function of odd degree can have? Of even degree?

(E) What is the least number of real solutions a polynomial function of odd degree can have? Of even degree?

We now compare the graphs of two polynomial functions relative to points close to the origin and then "zoom out" to compare points distant from the origin. Compare the graphs in Figure 3.

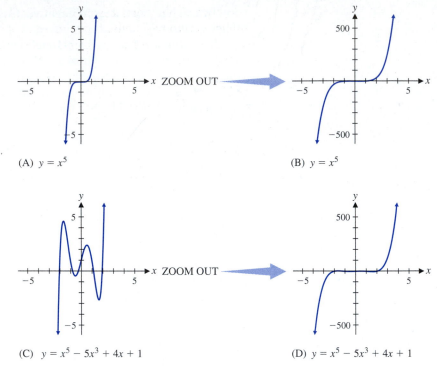

(A) $y = x^5$ ZOOM OUT ⟶   (B) $y = x^5$

(C) $y = x^5 - 5x^3 + 4x + 1$ ZOOM OUT ⟶   (D) $y = x^5 - 5x^3 + 4x + 1$

FIGURE 3 Close and distant comparisons

Figure 3 clearly shows that the highest-degree term in the polynomial dominates all other terms combined in the polynomial. As we "zoom out," the graph of $y = x^5 - 5x^3 + 4x + 1$ looks more and more like the graph of $y = x^5$. This is a general property of polynomial functions.

Explore–Discuss 2

Compare the graphs of $y = x^6$ and $y = x^6 - 7x^4 + 14x^2 - x - 5$ in the following two viewing windows:

(A) $-5 \le x \le 5, -5 \le y \le 5$
(B) $-5 \le x \le 5, -500 \le y \le 500$

❑ POLYNOMIAL ROOT APPROXIMATION

An x intercept of a function f is also called a **zero*** of the function and a **root** of the equation $f(x) = 0$. Approximating the zeros of a function with a graphing utility is a simple matter, provided that we can find a window that shows where the graph of the function touches or crosses the x axis. But how can we be sure that a particular window shows all the zeros? For polynomial functions, Theorem 1 can be used to locate a window that contains all the zeros of the

---

*Only real numbers can be x intercepts. Functions may have roots or zeros that are not real numbers, but these will not appear as x intercepts. Since we have agreed to discuss only functions whose domains are sets of real numbers, we will not consider any complex zeros or roots that are not real numbers.

polynomial. Theorem 1 is due to A. L. Cauchy (1789–1857), a French mathematician who has many theorems and concepts associated with his name, and illustrates that classical mathematical results are often useful tools in modern applications.

THEOREM 1   Locating Zeros of a Polynomial

If r is a zero of the polynomial

$$P(x) = x^n + a_{n-1}x^{n-1} + a_{n-2}x^{n-2} + \cdots + a_1 x + a_0$$

then*

$$|r| < 1 + \max\{|a_{n-1}|, |a_{n-2}|, \ldots, |a_1|, |a_0|\}$$

—————■

In a polynomial, the coefficient of the term containing the highest power of x is often referred to as the **leading coefficient.** Notice that Theorem 1 requires that this leading coefficient must be a 1.

Example 1   **Approximating the Zeros of a Polynomial**   Approximate (to two decimal places) the real zeros of

$$P(x) = 2x^4 - 5x^3 - 4x^2 + 3x + 6$$

SOLUTION   Since the leading coefficient of $P(x)$ is 2, Theorem 1 cannot be applied to $P(x)$. But it can be applied to

$$Q(x) = \tfrac{1}{2}P(x) = \tfrac{1}{2}(2x^4 - 5x^3 - 4x^2 + 3x + 6)$$
$$= x^4 - \tfrac{5}{2}x^3 - 2x^2 + \tfrac{3}{2}x + 3$$

Since multiplying a function by a positive constant expands or contracts the graph (see Section 1-2), but does not change the x intercepts, $P(x)$ and $Q(x)$ have the same zeros. Thus, Theorem 1 implies that any zero, r, of $P(x)$ must satisfy

$$|r| < 1 + \max\{|-\tfrac{5}{2}|, |-2|, |\tfrac{3}{2}|, |3|\} = 1 + 3 = 4$$

and all zeros of $P(x)$ must be between -4 and 4. Graphing $P(x)$ (Fig. 4A), we see two real zeros of $P(x)$ and we can be certain that there are no zeros outside this window. Using a built-in approximation routine, the real zeros of $P(x)$ (to two decimal places) are 1.15 (Fig. 4B) and 2.89 (Fig. 4C). Note also that $P(x)$ has three turning points in this window, the maximum allowable for a fourth-degree polynomial. Thus, there cannot be any turning points outside this window.

(A)

(B)

(C)

FIGURE 4

■

—————
*Note: $\max\{|a_{n-1}|, |a_{n-2}|, \ldots, |a_1|, |a_0|\}$ is the largest number in the list $|a_{n-1}|, |a_{n-2}|, \ldots, |a_0|$.

Matched Problem 1   Approximate (to two decimal places) the real zeros of

$$P(x) = 3x^3 + 12x^2 + 9x + 4$$

 ❑   REGRESSION POLYNOMIALS

In Group Activity 1 at the end of Chapter 1, we saw that regression techniques can be used to fit a straight line to a set of data. Linear functions are not the only ones that can be applied in this manner. Most graphing utilities have the ability to fit a variety of curves to a given set of data. We will discuss polynomial regression models in this section and other types of regression models in later sections.

Example 2   **Estimating the Weight of a Fish**   Using the length of a fish to estimate its weight is of interest to both scientists and sport anglers. The data in Table 1 give the average weights of lake trout for certain lengths. Use the data and regression techniques to find a polynomial model that can be used to estimate the weight of a lake trout for any length. Estimate (to the nearest ounce) the weights of lake trout of lengths 39, 40, 41, 42, and 43 inches, respectively.

WWW

TABLE 1			
LAKE TROUT			
LENGTH (IN.)	WEIGHT (OZ)	LENGTH (IN.)	WEIGHT (OZ)
x	y	x	y
10	5	30	152
14	12	34	226
18	26	38	326
22	56	44	536
26	96		

SOLUTION   The graph of the data in Table 1 (Fig. 5A) indicates that a linear regression model would not be appropriate in this case. And, in fact, we would not expect a linear relationship between length and weight. Instead, it is more likely that the weight would be related to the cube of the length. We use a cubic regression polynomial to model the data (Fig. 5B). (Consult your manual for the details of calculating regression polynomials on your graphing utility.) Figure 5C adds the graph of the polynomial model to the graph of the data. The graph in Figure 5C shows that this cubic polynomial does provide a good fit for the data. (We will have more to say about the choice of functions and the accuracy of the fit provided by regression analysis later in the book.) Figure 5D shows the estimated weights for the lengths requested.

(A)

(B)

(C)

(D)

FIGURE 5

Matched Problem 2   The data in Table 2 gives the average weights of pike for certain lengths. Use a cubic regression polynomial to model the data. Estimate (to the nearest ounce) the weights of pike of lengths 39, 40, 41, 42, and 43 inches, respectively.

WWW

TABLE 2
PIKE

LENGTH (IN.)	WEIGHT (OZ)	LENGTH (IN.)	WEIGHT (OZ)
x	y	x	y
10	5	30	108
14	12	34	154
18	26	38	210
22	44	44	326
26	72	52	522

❑ **RATIONAL FUNCTIONS**

Just as rational numbers are defined in terms of quotients of integers, *rational functions* are defined in terms of quotients of polynomials. The following equations define rational functions:

$$f(x) = \frac{1}{x} \qquad g(x) = \frac{x-2}{x^2 - x - 6} \qquad h(x) = \frac{x^3 - 8}{x}$$
$$p(x) = 3x^2 - 5x \qquad q(x) = 7 \qquad r(x) = 0$$

Rational Function

A **rational function** is any function of the form

$$f(x) = \frac{n(x)}{d(x)} \qquad d(x) \neq 0$$

where $n(x)$ and $d(x)$ are polynomials. The **domain** is the set of all real numbers such that $d(x) \neq 0$. We assume that $n(x)/d(x)$ is reduced to lowest terms.

Example 3 ➦ **Domain and Intercepts**   Find the domain and intercepts for the rational function

$$f(x) = \frac{x-2}{x+1}$$

SOLUTION   *Domain:*   The denominator is 0 at $x = -1$. Therefore, the domain is the set of all real numbers except -1. The graph of f cannot cross the vertical line $x = -1$.

x intercepts: Find x such that $f(x) = 0$. This happens only if $x - 2 = 0$, that is, at $x = 2$. Thus, 2 is the only x intercept.

y intercept: The y intercept is

$$f(0) = \frac{0-2}{0+1} = -2$$

Matched Problem 3 ⇔ Find the domain and intercepts for the rational function: $g(x) = \dfrac{2x}{x-2}$

In the next example we investigate the graph of $f(x) = (x-2)/(x+1)$ near the *point of discontinuity, $x = -1$*, and the behavior of the graph as x increases or decreases without bound. Using this information, we can complete a sketch of the graph of function f with little additional trouble. The investigation uncovers some characteristic features of graphs of rational functions.

Example 4 ⇔ **Graph of a Rational Function**   Given function f from Example 3:

$$f(x) = \frac{x-2}{x+1}$$

(A) Investigate the graph of f near the point of discontinuity, $x = -1$.
(B) Investigate the graph of f as x increases or decreases without bound.
(C) Sketch a graph of function f.

SOLUTION   (A) *Let x approach -1 from the left:*

x	$f(x)$
-2	4
-1.1	31
-1.01	301
-1.001	3,001
-1.0001	30,001
-1.00001	300,001

We see that $f(x)$ increases without bound as x approaches -1 from the left. Symbolically,

$$f(x) \to \infty \quad \text{as} \quad x \to -1^-$$

Let x approach -1 from the right:

x	$f(x)$
0	-2
-0.9	-29
-0.99	-299
-0.999	$-2,999$
-0.9999	$-29,999$
-0.99999	$-299,999$

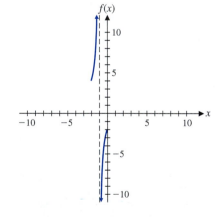

We see that $f(x)$ decreases without bound as x approaches -1 from the right. Symbolically,

$$f(x) \to -\infty \quad \text{as} \quad x \to -1^+$$

The vertical line $x = -1$ is called a *vertical asymptote*. The graph of f gets closer to this line as x gets closer to -1. Sketching the vertical asymptote provides an aid to drawing the graph of f near the asymptote (Fig. 6).

FIGURE 6 Graph near vertical asymptote $x = -1$

(B) Divide each term in the numerator and denominator of $f(x)$ by x, the highest power of x to occur in the numerator and denominator:

$$f(x) = \frac{x-2}{x+1} = \frac{1 - \dfrac{2}{x}}{1 + \dfrac{1}{x}}$$

As x increases or decreases without bound, $2/x$ and $1/x$ approach 0 and $f(x)$ gets closer and closer to 1. The horizontal line $y = 1$ is called a *horizontal asymptote.* The graph of $y = f(x)$ gets closer to this line as x decreases or increases without bound. But how does the graph of $y = f(x)$ approach the horizontal line $y = 1$? Does it approach from above? From below? Or from both? To answer these questions we investigate table values as follows:

Let x approach ∞:

x	$f(x)$
10	0.72727
100	0.97030
1,000	0.99700
10,000	0.99970
100,000	0.99997

The graph of $y = f(x)$ approaches the line $y = 1$ from below as x increases without bound.

Let x approach −∞:

x	$f(x)$
−10	1.33333
−100	1.03030
−1,000	1.00300
−10,000	1.00030
−100,000	1.00003

The graph of $y = f(x)$ approaches the line $y = 1$ from above as x decreases without bound.

Sketching the horizontal asymptote first provides an aid to drawing the graph of f as x moves away from the origin (Fig. 7).

(C) It is now easy to complete the sketch of the graph of f using the intercepts from Example 3 and filling in any points of uncertainty (Fig. 8).

FIGURE 7 Graph of $y = f(x)$ near horizontal asymptote $y = 1$

x	$f(x)$
−4	2
−2	4
0	−2
2	0

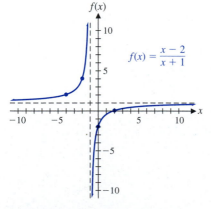

$$f(x) = \frac{x - 2}{x + 1}$$

FIGURE 8 Sketch of the rational function f

Matched Problem 4 ↩   Given function g from Matched Problem 3: $g(x) = \dfrac{2x}{x - 2}$

    (A) Investigate the graph of g near the point of discontinuity, $x = 2$.

    (B) Investigate the graph of g as x increases or decreases without bound.

    (C) Sketch a graph of function g.

    ■

Graphing rational functions is considerably aided by locating vertical and horizontal asymptotes first, if they exist. The following general procedures are suggested by Example 4:

Vertical and Horizontal Asymptotes and Rational Functions

Given the rational function

$$f(x) = \frac{n(x)}{d(x)}$$

where $n(x)$ and $d(x)$ are polynomials without common factors:

1. If a is a real number such that $d(a) = 0$, then the line $x = a$ is a **vertical asymptote** of the graph of $y = f(x)$.

2. **Horizontal asymptotes,** if any exist, can be found by dividing each term of the numerator $n(x)$ and denominator $d(x)$ by the highest power of x that appears in the numerator and denominator, and then proceeding as in Example 4.

Example 5 ↩   **Graphing Rational Functions**   Given the rational function:

$$f(x) = \frac{3x}{x^2 - 4}$$

    (A) Find intercepts and equations for any vertical and horizontal asymptotes.

    (B) Using the information from part (A) and additional points as necessary, sketch a graph of f for $-7 \leqslant x \leqslant 7$ and $-7 \leqslant y \leqslant 7$.

SOLUTION   (A) *x intercepts:* $f(x) = 0$ only if $3x = 0$, or $x = 0$. Thus, the only x intercept is 0.

y intercept:

$$f(0) = \frac{3 \cdot 0}{0^2 - 4} = \frac{0}{-4} = 0$$

Thus, the y intercept is 0.

Vertical asymptotes:

$$f(x) = \frac{3x}{x^2 - 4} = \frac{3x}{(x - 2)(x + 2)}$$

The denominator is 0 at $x = -2$ and $x = 2$; hence, $x = -2$ and $x = 2$ are vertical asymptotes.

Horizontal asymptotes: Divide each term in the numerator and denominator by x^2, the highest power of x in the numerator and denominator:

$$f(x) = \frac{3x}{x^2 - 4} = \frac{\dfrac{3x}{x^2}}{\dfrac{x^2}{x^2} - \dfrac{4}{x^2}} = \frac{\dfrac{3}{x}}{1 - \dfrac{4}{x^2}}$$

As x increases or decreases without bound, the numerator tends to 0 and the denominator tends to 1; thus, $f(x)$ tends to 0. The line $y = 0$ is a horizontal asymptote.

(B) Use the information from part (A) and plot additional points as necessary to complete the graph, as shown in Figure 9.

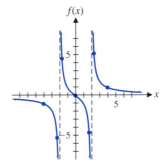

x	$f(x)$
-4	-1
-2.3	-5.3
-1.7	4.6
0	0
1.7	-4.6
2.3	5.3
4	1

FIGURE 9

Matched Problem 5   Given the rational function: $g(x) = \dfrac{3x + 3}{x^2 - 9}$

(A) Find all intercepts and equations for any vertical and horizontal asymptotes.

(B) Using the information from part (A) and additional points as necessary, sketch a graph of g for $-10 \leq x \leq 10$ and $-10 \leq y \leq 10$.

❑ APPLICATION

Rational functions occur naturally in many types of applications.

Example 6   **Employee Training**   A company that manufactures computers has established that, on the average, a new employee can assemble $N(t)$ components per day after t days of on-the-job training, as given by

$$N(t) = \frac{50t}{t + 4} \qquad t \geq 0$$

Sketch a graph of N, $0 \leq t \leq 100$, including any vertical or horizontal asymptotes. What does $N(t)$ approach as t increases without bound?

SOLUTION   *Vertical asymptotes:*   None for $t \geq 0$

Horizontal asymptote:

$$N(t) = \frac{50t}{t + 4} = \frac{50}{1 + \dfrac{4}{t}}$$

$N(t)$ approaches 50 as t increases without bound. Thus, $y = 50$ is a horizontal asymptote.

Sketch of graph: The graph is shown in the margin.

$N(t)$ approaches 50 as t increases without bound. It appears that 50 components per day would be the upper limit that an employee would be expected to assemble.

Matched Problem 6 ⇨ Repeat Example 6 for: $N(t) = \dfrac{25t + 5}{t + 5}$ $t \geq 0$

Answers to Matched Problems

1. -3.19

2.
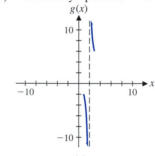

3. Domain: all real numbers except 2; x intercept: 0; y intercept: 0

4. (A) Vertical asymptote: $x = 2$ (B) Horizontal asymptote: $y = 2$

(C)

5. (A) x intercept: -1; y intercept; $-\frac{1}{3}$;
Vertical asymptotes: $x = -3$ and $x = 3$;
Horizontal asymptote: $y = 0$ (B)

6. No vertical asymptotes for $t \geq 0$; $y = 25$ is a horizontal asymptote. $N(t)$ approaches 25 as t increases without bound. It appears that 25 components per day would be the upper limit that an employee would be expected to assemble.

Exercise 2-1

A *For each polynomial function in Problems 1–6, find the following:*

(A) *Degree of the polynomial*
(B) *Maximum number of turning points of the graph*
(C) *Maximum number of x intercepts of the graph*
(D) *Minimum number of x intercepts of the graph*
(E) *Maximum number of y intercepts of the graph*
(F) *Minimum number of y intercepts of the graph*

1. $f(x) = ax^2 + bx + c, a \neq 0$

2. $f(x) = ax + b, a \neq 0$

3. $f(x) = ax^5 + bx^4 + cx^3 + dx^2 + ex + f, a \neq 0$

4. $f(x) = ax^4 + bx^3 + cx^2 + dx + e, a \neq 0$

5. $f(x) = ax^6 + bx^5 + cx^4 + dx^3 + ex^2 + fx + g, a \neq 0$

6. $f(x) = ax^3 + bx^2 + cx + d, a \neq 0$

Each graph in Problems 7–14 is the graph of a polynomial function. Answer the following questions for each graph:

(A) *How many turning points are on the graph?*
(B) *What is the minimum degree of a polynomial function that could have the graph?*
(C) *Is the leading coefficient of the polynomial negative or positive?*

7. **8.**

9. **10.**

11. **12.**

13. **14.**

B *For each rational function in Problems 15–20:*

(A) *Find the intercepts for the graph.*
(B) *Determine the domain.*
(C) *Find any vertical or horizontal asymptotes for the graph.*

(D) *Sketch any asymptotes as dashed lines. Then sketch a graph of* $y = f(x)$ *for* $-10 \leqslant x \leqslant 10$ *and* $-10 \leqslant y \leqslant 10$.

(E) *Graph* $y = f(x)$ *in a standard viewing window using a graphing utility.*

15. $f(x) = \dfrac{x+2}{x-2}$ **16.** $f(x) = \dfrac{x-3}{x+3}$

17. $f(x) = \dfrac{3x}{x+2}$ **18.** $f(x) = \dfrac{2x}{x-3}$

19. $f(x) = \dfrac{4-2x}{x-4}$ **20.** $f(x) = \dfrac{3-3x}{x-2}$

21. How does the graph of $f(x) = 2x^4 - 5x^2 + x + 2$ compare to the graph of $y = 2x^4$ as we "zoom out" (see Fig. 3)?

22. How does the graph of $f(x) = x^3 - 2x + 2$ compare to the graph of $y = x^3$ as we "zoom out"?

23. How does the graph of $f(x) = -x^5 + 4x^3 - 4x + 1$ compare to the graph of $y = -x^5$ as we "zoom out"?

24. How does the graph of $f(x) = -x^5 + 5x^3 + 4x - 1$ compare to the graph of $y = -x^5$ as we "zoom out"?

25. Compare the graph of $y = 2x^4$ to the graph of $y = 2x^4 - 5x^2 + x + 2$ in the following two viewing windows:

(A) $-5 \leqslant x \leqslant 5, -5 \leqslant y \leqslant 5$
(B) $-5 \leqslant x \leqslant 5, -500 \leqslant y \leqslant 500$

26. Compare the graph of $y = x^3$ to the graph of $y = x^3 - 2x + 2$ in the following two viewing windows:

(A) $-5 \leqslant x \leqslant 5, -5 \leqslant y \leqslant 5$
(B) $-5 \leqslant x \leqslant 5, -500 \leqslant y \leqslant 500$

27. Compare the graph of $y = -x^5$ to the graph of $y = -x^5 + 4x^3 - 4x + 1$ in the following two viewing windows:

(A) $-5 \leqslant x \leqslant 5, -5 \leqslant y \leqslant 5$
(B) $-5 \leqslant x \leqslant 5, -500 \leqslant y \leqslant 500$

28. Compare the graph of $y = -x^5$ to the graph of $y = -x^5 + 5x^3 - 5x + 2$ in the following two viewing windows:

(A) $-5 \leqslant x \leqslant 5, -5 \leqslant y \leqslant 5$
(B) $-5 \leqslant x \leqslant 5, -500 \leqslant y \leqslant 500$

In Problems 29–34, use Theorem 1 to find an interval on the x axis that must contain all the zeros of each polynomial. Then approximate each real zero to two decimal places.

29. $2x^3 - x^2 - 7x + 3$

30. $3x^3 + 10x^2 + 6x - 2$

31. $x^4 + 2x^3 - 3x^2 - 4x + 1$

32. $x^4 - 3x^3 - 4x^2 + 3x + 1$

33. $x^5 - 12x^4 + 7x^3 + 15$

34. $x^5 + 14x^4 - 10x^2 - 15$

35. Graph the line $y = 0.5x + 3$. Choose any two distinct points on this line and find the linear regression model for the data set consisting of the two points you chose. Experiment with other lines of your choice. Discuss the relationship between a linear regression model for two points and the line that goes through the two points.

36. Graph the parabola $y = x^2 - 5x$. Choose any three distinct points on this parabola and find the quadratic regression model for the data set consisting of the three points you chose. Experiment with other parabolas of your choice. Discuss the relationship between a quadratic regression model for three non-collinear points and the parabola that goes through the three points.

C *For each rational function in Problems 37–42:*

(A) *Find any intercepts for the graph.*

(B) *Find any vertical and horizontal asymptotes for the graph.*

(C) *Sketch any asymptotes as dashed lines. Then sketch a graph of f for* $-10 \leqslant x \leqslant 10$ *and* $-10 \leqslant y \leqslant 10$.

(D) *Graph the function in a standard viewing window using a graphing utility.*

37. $f(x) = \dfrac{2x^2}{x^2 - x - 6}$ **38.** $f(x) = \dfrac{3x^2}{x^2 + x - 6}$

39. $f(x) = \dfrac{6 - 2x^2}{x^2 - 9}$ **40.** $f(x) = \dfrac{3 - 3x^2}{x^2 - 4}$

41. $f(x) = \dfrac{-4x}{x^2 + x - 6}$ **42.** $f(x) = \dfrac{5x}{x^2 + x - 12}$

43. Write an equation for the lowest-degree polynomial function with the graph and intercepts shown in the figure.

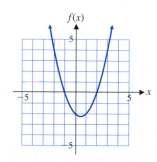

Figure for 43

44. Write an equation for the lowest-degree polynomial function with the graph and intercepts shown in the figure.

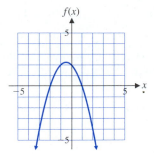

Figure for 44

45. Write an equation for the lowest-degree polynomial function with the graph and intercepts shown in the figure.

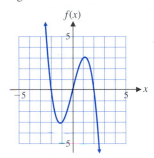

Figure for 45

46. Write an equation for the lowest-degree polynomial function with the graph and intercepts shown in the figure.

Figure for 46

Applications

Business & Economics

47. *Average cost.* A company manufacturing snowboards has fixed costs of $200 per day and total costs of $3,800 per day at a daily output of 20 boards.

(A) Assuming that the total cost per day, $C(x)$, is linearly related to the total output per day, x, write an equation for the cost function.

(B) The average cost per board for an output of x boards is given by $\overline{C}(x) = C(x)/x$. Find the average cost function.

(C) Sketch a graph of the average cost function, including any asymptotes, for $1 \leq x \leq 30$.

(D) What does the average cost per board tend to as production increases?

48. *Average cost.* A company manufacturing surfboards has fixed costs of $300 per day and total costs of $5,100 per day at a daily output of 20 boards.

(A) Assuming that the total cost per day, $C(x)$, is linearly related to the total output per day, x, write an equation for the cost function.

(B) The average cost per board for an output of x boards is given by $\overline{C}(x) = C(x)/x$. Find the average cost function.

(C) Sketch a graph of the average cost function, including any asymptotes, for $1 \leq x \leq 30$.

(D) What does the average cost per board tend to as production increases?

49. *Replacement time.* An office copier has an initial price of $2,500. A service contract costs $200 for the first year and increases $50 per year thereafter. It can be shown that the total cost of the copier after n years is given by

$$C(n) = 2,500 + 175n + 25n^2$$

The average cost per year for n years is given by $\overline{C}(n) = C(n)/n$.

(A) Find the rational function $\overline{C}$.

(B) Sketch a graph of $\overline{C}$ for $2 \leq n \leq 20$.

(C) When is the average cost per year at a minimum, and what is the minimum average annual cost? [*Hint*: Refer to the sketch in part (B) and evaluate $\overline{C}(n)$ at appropriate integer values until a minimum value is found.] The time when the average cost is minimum is frequently referred to as the **replacement time** for the piece of equipment.

(D) Graph the average cost function $\overline{C}$ in a graphing utility and use trace and zoom or an appropriate built-in routine to find when the average annual cost is at a minimum.

50. *Minimum average cost.* Financial analysts in a company that manufactures audio CD players arrived at the following daily cost equation for manufacturing x CD players per day:

$$C(x) = x^2 + 2x + 2{,}000$$

The average cost per unit at a production level of x players per day is $\overline{C}(x) = C(x)/x$.

(A) Find the rational function $\overline{C}$.

(B) Sketch a graph of $\overline{C}$ for $5 \leqslant x \leqslant 150$.

(C) For what daily production level (to the nearest integer) is the average cost per unit at a minimum, and what is the minimum average cost per player (to the nearest cent)? [*Hint*: Refer to the sketch in part (B) and evaluate $\overline{C}(x)$ at appropriate integer values until a minimum value is found.]

(D) Graph the average cost function $\overline{C}$ in a graphing utility and use trace and zoom or an appropriate built-in routine to find the daily production level (to the nearest integer) at which the average cost per player is at a minimum. What is the minimum average cost to the nearest cent?

51. *Minimum average cost.* A consulting firm, using statistical methods, provided a veterinary clinic with the cost equation

$$C(x) = 0.00048(x - 500)^3 + 60{,}000$$

$$100 \leqslant x \leqslant 1{,}000$$

where $C(x)$ is the cost in dollars for handling x cases per month. The average cost per case is given by $\overline{C}(x) = C(x)/x$.

(A) Write the equation for the average cost function $\overline{C}$.

(B) Graph $\overline{C}$ on a graphing utility.

(C) Use trace and zoom or an appropriate built-in routine to find the monthly caseload for the minimum average cost per case. What is the minimum average cost per case?

52. *Minimum average cost.* The financial department of a hospital, using statistical methods, arrived at the cost equation

$$C(x) = 20x^3 - 360x^2 + 2{,}300x - 1{,}000$$

$$1 \leqslant x \leqslant 12$$

where $C(x)$ is the cost in thousands of dollars for handling x thousand cases per month. The average cost per case is given by $\overline{C}(x) = C(x)/x$.

(A) Write the equation for the average cost function $\overline{C}$.

(B) Graph $\overline{C}$ on a graphing utility.

(C) Use trace and zoom or an appropriate built-in routine to find the monthly caseload for the minimum average cost per case. What is the minimum average cost per case to the nearest dollar?

53. *Equilibrium point.* A particular CD is sold through a chain of stores in a city. A marketing company has established price–demand and price–supply tables for selected prices for this CD (Tables 3 and 4), where x is the daily number of CDs that people are willing to buy and the store is willing to sell at a price of p dollars per CD.

TABLE 3	
PRICE–DEMAND	
x	$p = D(x)$ ($)
25	19.50
100	14.25
175	10.00
250	8.25

TABLE 4	
PRICE–SUPPLY	
x	$p = S(x)$ ($)
25	2.10
100	3.80
175	8.50
250	15.70

(A) Use a linear regression equation to model the data in Table 3 and a quadratic regression equation to model the data in Table 4.

(B) Find the point of intersection of the two equations from part (A). Write the equilibrium price to the nearest cent and the equilibrium quantity to the nearest unit.

54. *Equilibrium point.* Repeat Problem 53 with the tables at the top of the next page, except use a quadratic regression model for the data in Table 5 and a linear regression model for the data in Table 6.

TABLE 5

PRICE–DEMAND

x	$p = D(x)$ ($)
0	24
40	23
65	20
115	11

TABLE 6

PRICE–SUPPLY

x	$p = S(x)$ ($)
0	5
40	10
65	12
115	16

57. *Physiology.* In a study on the speed of muscle contraction in frogs under various loads, researchers W. O. Fems and J. Marsh found that the speed of contraction decreases with increasing loads. In particular, they found that the relationship between speed of contraction v (in centimeters per second) and load x (in grams) is given approximately by

$$v(x) = \frac{26 + 0.06x}{x} \qquad x \geq 5$$

(A) What does $v(x)$ approach as x increases?

(B) Sketch a graph of function v.

Life Sciences

55. *Health care.* Table 7 shows the total national expenditures (in billion dollars) and the per capita expenditures (in dollars) for selected years since 1970.

(A) Let x represent the number of years since 1970 and find a cubic regression polynomial for the total national expenditures.

(B) Use the polynomial model from part (A) to estimate the total national expenditures (to the nearest tenth of a billion) for 2010.

56. *Health care.* Refer to Table 7.

(A) Let x represent the number of years since 1970 and find a cubic regression polynomial for the per capita expenditures.

(B) Use the polynomial model from part (A) to estimate the per capita expenditures (to the nearest dollar) for 2010.

Social Sciences

58. *Learning theory.* In 1917, L. L. Thurstone, a pioneer in quantitative learning theory, proposed the rational function

$$f(x) = \frac{a(x + c)}{(x + c) + b}$$

to model the number of successful acts per unit time that a person could accomplish after x practice sessions. Suppose that for a particular person enrolled in a typing class,

$$f(x) = \frac{55(x + 1)}{(x + 8)} \qquad x \geq 0$$

where $f(x)$ is the number of words per minute the person is able to type after x weeks of lessons.

(A) What does $f(x)$ approach as x increases?

(B) Sketch a graph of function f, including any vertical or horizontal asymptotes.

www   **TABLE 7**

NATIONAL HEALTH EXPENDITURES

DATE	TOTAL EXPENDITURES (BILLION $)	PER CAPITA EXPENDITURES ($)
1970	73.2	341
1975	132.9	592
1980	247.3	1,052
1985	428.2	1,666
1990	699.4	2,689
1995	993.7	3,638

59. *Marriage.* Table 8 shows the marriage and divorce rates per 1,000 population for selected years since 1960.

www

TABLE 8		
MARRIAGES AND DIVORCES (PER 1,000 POPULATION)		
DATE	MARRIAGES	DIVORCES
1960	8.5	2.2
1965	9.3	2.5
1970	10.6	3.5
1975	10.0	4.8
1980	10.6	5.2
1985	10.1	5.0
1990	9.8	4.7
1995	8.9	4.4

www

(A) Let x represent the number of years since 1960 and find a cubic regression polynomial for the marriage rate.

(B) Use the polynomial model from part (A) to estimate the marriage rate (to one decimal place) for 2005. For 2010.

60. *Divorce.* Refer to Table 8.

(A) Let x represent the number of years since 1950 and find a cubic regression polynomial for the divorce rate.

(B) Use the polynomial model from part (A) to estimate the divorce rate (to one decimal place) for 2005.

Section 2-2

Exponential Functions

- ❑ EXPONENTIAL FUNCTIONS
- ❑ BASE e EXPONENTIAL FUNCTIONS
- ❑ GROWTH AND DECAY APPLICATIONS
- ❑ COMPOUND INTEREST
- ❑ CONTINUOUS COMPOUND INTEREST

This section introduces the important class of functions called *exponential functions.* These functions are used extensively in modeling and solving a wide variety of real-world problems, including growth of money at compound interest; growth of populations of people, animals, and bacteria; radioactive decay; and learning associated with the mastery of such devices as a new computer or an assembly process in a manufacturing plant.

❑ EXPONENTIAL FUNCTIONS

We start by noting that

$$f(x) = 2^x \qquad \text{and} \qquad g(x) = x^2$$

are not the same function. Whether a variable appears as an exponent with a constant base or as a base with a constant exponent makes a big difference. The function g is a quadratic function, which we have already discussed. The function f is a new type of function called an *exponential function.* In general:

> ### *Exponential Function*
>
> The equation
>
> $$f(x) = b^x \qquad b > 0, b \neq 1$$
>
> defines an **exponential function** for each different constant b, called the **base.** The **domain** of f is the set of all real numbers, and the **range** of f is the set of all positive real numbers.

We require the base b to be positive to avoid imaginary numbers such as $(-2)^{1/2} = \sqrt{-2} = i\sqrt{2}$. We exclude $b = 1$ as a base, since $f(x) = 1^x = 1$ is a constant function, which we have already considered.

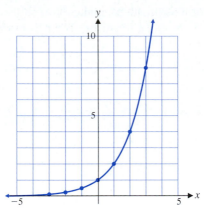

FIGURE 1 $y = 2^x$

Asked to hand-sketch graphs of equations such as $y = 2^x$ or $y = 2^{-x}$, many students would not hesitate at all. [*Note*: $2^{-x} = 1/2^x = (1/2)^x$.] They would probably make up tables by assigning integers to x, plot the resulting points, and then join these points with a smooth curve as in Figure 1. The only catch is that we have not defined 2^x for all real numbers. From Appendix A-7, we know what $2^5, 2^{-3}, 2^{2/3}, 2^{-3/5}, 2^{1.4}$, and $2^{-3.14}$ mean (that is, 2^p, where p is a rational number), but what does

$$2^{\sqrt{2}}$$

mean? The question is not easy to answer at this time. In fact, a precise definition of $2^{\sqrt{2}}$ must wait for more advanced courses, where it is shown that

$$2^x$$

names a positive real number for x any real number, and that the graph of $y = 2^x$ is indeed as indicated in Figure 1.

It is useful to compare the graphs of $y = 2^x$ and $y = 2^{-x}$ by plotting both on the same set of coordinate axes, as shown in Figure 2A. The graph of

$$f(x) = b^x \qquad b > 1 \text{ (Fig. 2B)}$$

looks very much like the graph of $y = 2^x$, and the graph of

$$f(x) = b^x \qquad 0 < b < 1 \text{ (Fig. 2B)}$$

looks very much like the graph of $y = 2^{-x}$. Note that in both cases the x axis is a horizontal asymptote for the graphs.

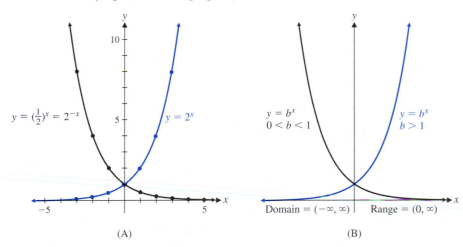

(A)   (B)

FIGURE 2 Exponential functions

The graphs in Figure 2 suggest the following important general properties of exponential functions, which we state without proof:

Basic Properties of the Graph of $f(x) = b^x, b > 0, b \neq 1$

  1. All graphs will pass through the point $(0, 1)$. $b^0 = 1$ for any permissible base b.

  2. All graphs are continuous curves, with no holes or jumps.
  3. The x axis is a horizontal asymptote.
  4. If $b > 1$, then b^x increases as x increases.
  5. If $0 < b < 1$, then b^x decreases as x increases.

The use of a calculator with the key $\boxed{y^x}$, or its equivalent, makes the graphing of exponential functions almost routine. Example 1 illustrates the process.

Example 1 ⇔ **Graphing Exponential Functions** Sketch a graph of $y = (\frac{1}{2})4^x$, $-2 \leqslant x \leqslant 2$.

SOLUTION   Use a calculator to create the table of values shown. Plot these points, and then join them with a smooth curve as in Figure 3.

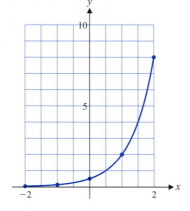

x	y
−2	0.031
−1	0.125
0	0.50
1	2.00
2	8.00

FIGURE 3 Graph of $y = (\frac{1}{2})4^x$

Matched Problem 1 ⇔ Sketch a graph of $y = (\frac{1}{2})4^{-x}, -2 \leqslant x \leqslant 2$.

Explore–Discuss 1

Graph the functions $f(x) = 2^x$ and $g(x) = 3^x$ on the same set of coordinate axes. At which values of x do the graphs intersect? For which values of x is the graph of f above the graph of g? Below the graph of g? Are the graphs close together as x increases without bound? Are the graphs close together as x decreases without bound? Discuss.

Exponential functions, whose domains include irrational numbers, obey the familiar laws of exponents discussed in Appendix A-7 for rational exponents. We summarize these exponent laws here and add two other important and useful properties.

Exponential Function Properties

For a and b positive, $a \neq 1$, $b \neq 1$, and x and y real:

1. Exponent laws:

$$a^x a^y = a^{x+y} \qquad \frac{a^x}{a^y} = a^{x-y} \qquad \frac{4^{2y}}{4^{5y}} = 4^{2y-5y} = 4^{-3y}$$

$$(a^x)^y = a^{xy} \qquad (ab)^x = a^x b^x \qquad \left(\frac{a}{b}\right)^x = \frac{a^x}{b^x}$$

2. $a^x = a^y$ if and only if $x = y$ If $7^{5t+1} = 7^{3t-3}$, then
$5t + 1 = 3t - 3$, and $t = -2$.

3. For $x \neq 0$,

$a^x = b^x$ if and only if $a = b$ If $a^5 = 2^5$, then $a = 2$.

❑   BASE e EXPONENTIAL FUNCTIONS

Of all the possible bases b we can use for the exponential function $y = b^x$, which ones are the most useful? If you look at the keys on a calculator, you will probably see $\boxed{10^x}$ and $\boxed{e^x}$. It is clear why base 10 would be important, because our number system is a base 10 system. But what is e, and why is it included as a base? It turns out that base e is used more frequently than all other bases combined. The reason for this is that certain formulas and the results of certain processes found in calculus and more advanced mathematics take on their simplest form if this base is used. This is why you will see e used extensively in expressions and formulas that model real-world phenomena. In fact, its use is so prevalent that you will often hear people refer to $y = e^x$ as the exponential function.

The base e is an irrational number, and like π, it cannot be represented exactly by any finite decimal fraction. However, e can be approximated as closely as we like by evaluating the expression

$$\left(1 + \frac{1}{x}\right)^x \tag{1}$$

for sufficiently large x. What happens to the value of expression (1) as x increases without bound? Think about this for a moment before proceeding. Maybe you guessed that the value approaches 1, because

$$1 + \frac{1}{x}$$

approaches 1, and 1 raised to any power is 1. Let us see if this reasoning is correct by actually calculating the value of the expression for larger and larger values of x. Table 1 summarizes the results.

TABLE 1

x	$\left(1 + \dfrac{1}{x}\right)^x$
1	2
10	2.593 74 ...
100	2.704 81 ...
1,000	2.716 92 ...
10,000	2.718 14 ...
100,000	2.718 27 ...
1,000,000	2.718 28 ...
⋮	⋮

Interestingly, the value of expression (1) is never close to 1, but seems to be approaching a number close to 2.7183. In fact, as x increases without bound, the value of expression (1) approaches an irrational number that we call *e*. The irrational number *e* to 12 decimal places is

$$e = 2.718\ 281\ 828\ 459$$

Compare this value of *e* with the value of e^1 from a calculator. Exactly who discovered the constant *e* is still being debated. It is named after the great Swiss mathematician Leonhard Euler (1707–1783).

Exponential Function with Base e

Exponential functions with base *e* and base $1/e$, respectively, are defined by

$$y = e^x \quad \text{and} \quad y = e^{-x}$$

Domain: $(-\infty, \infty)$

Range: $(0, \infty)$

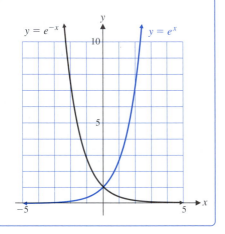

Explore–Discuss 2 Graph the functions $f(x) = e^x$, $g(x) = 2^x$, and $h(x) = 3^x$ on the same set of coordinate axes. At which values of *x* do the graphs intersect? For positive values of *x*, which of the three graphs lies above the other two? Below the other two? How does your answer change for negative values of *x*?

❑ GROWTH AND DECAY APPLICATIONS

Most exponential growth and decay problems are modeled using base *e* exponential functions. We present two applications here and many more in Exercise 2-2.

Example 2 ➭ **Exponential Growth** Cholera, an intestinal disease, is caused by a cholera bacterium that multiplies exponentially by cell division as given approximately by

$$N = N_0 e^{1.386t}$$

where *N* is the number of bacteria present after *t* hours and N_0 is the number of bacteria present at the start ($t = 0$). If we start with 25 bacteria, how many bacteria (to the nearest unit) will be present:

(A) In 0.6 hour?   (B) In 3.5 hours?

SOLUTION   Substituting $N_0 = 25$ into the preceding equation, we obtain

$$N = 25e^{1.386t}$$ *The graph is shown in Figure 4.*

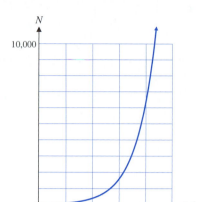

FIGURE 4

(A) Solve for N when $t = 0.6$:

$$N = 25e^{1.386(0.6)}$$ *Use a calculator.*

$$= 57 \text{ bacteria}$$

(B) Solve for N when $t = 3.5$:

$$N = 25e^{1.386(3.5)}$$ *Use a calculator.*

$$= 3,197 \text{ bacteria}$$

Matched Problem 2   Refer to the exponential growth model for cholera in Example 2. If we start with 55 bacteria, how many bacteria (to the nearest unit) will be present:

(A) In 0.85 hour?   (B) In 7.25 hours?

Example 3   **Exponential Decay**   Cosmic-ray bombardment of the atmosphere produces neutrons, which in turn react with nitrogen to produce radioactive carbon-14 (^{14}C). Radioactive ^{14}C enters all living tissues through carbon dioxide, which is first absorbed by plants. As long as a plant or animal is alive, ^{14}C is maintained in the living organism at a constant level. Once the organism dies, however, ^{14}C decays according to the equation

$$A = A_0 e^{-0.000124t}$$

where A is the amount present after t years and A_0 is the amount present at time $t = 0$. If 500 milligrams of ^{14}C is present in a sample from a skull at the time of death, how many milligrams will be present in the sample in:

(A) 15,000 years?   (B) 45,000 years?

Compute answers to two decimal places.

SOLUTION   Substituting $A_0 = 500$ in the decay equation, we have

$$A = 500e^{-0.000124t}$$ *See the graph in Figure 5.*

FIGURE 5

(A) Solve for A when $t = 15,000$:

$$A = 500e^{-0.000124(15,000)} \qquad \textit{Use a calculator.}$$
$$= 77.84 \text{ milligrams}$$

(B) Solve for A when $t = 45,000$:

$$A = 500e^{-0.000124(45,000)} \qquad \textit{Use a calculator.}$$
$$= 1.89 \text{ milligrams}$$

Matched Problem 3 Refer to the exponential decay model in Example 3. How many milligrams of ^{14}C would have to be present at the beginning in order to have 25 milligrams present after 18,000 years? Compute the answer to the nearest milligram.

Explore–Discuss 3 (A) On the same set of coordinate axes, graph the three decay equations $A = A_0e^{-0.35t}$, $t \geq 0$, for $A_0 = 10, 20,$ and 30.

(B) Identify any asymptotes for the three graphs in part (A).

(C) Discuss the long-term behavior for the equations in part (A).

Example 4 **Depreciation** Table 2 gives the market value of a minivan (in dollars) x years after its purchase. Find an exponential regression model of the form $y = ab^x$ for this data set. Estimate the purchase price of the van. Estimate the value of the van 10 years after its purchase. Round answers to the nearest dollar.

TABLE 2	
x	VALUE ($)
1	12,575
2	9,455
3	8,115
4	6,845
5	5,225
6	4,485

SOLUTION   Enter the data into a graphing utility (Fig. 6A) and find the exponential regression equation (Fig. 6B). The estimated purchase price is $y_1(0) = \$14,910$. The data set and the regression equation are graphed in Figure 6C. Using the trace feature, we see that the estimated value after 10 years is $\$1,959$.

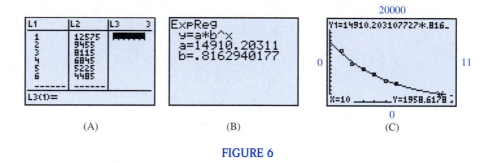

| (A) | (B) | (C) |

FIGURE 6

Matched Problem 4   Table 3 gives the market value of a luxury sedan (in dollars) x years after its purchase. Find an exponential regression model of the form $y = ab^x$ for this data set. Estimate the purchase price of the sedan. Estimate the value of the sedan 10 years after its purchase. Round answers to the nearest dollar.

TABLE 3	
x	VALUE ($)
1	23,125
2	19,050
3	15,625
4	11,875
5	9,450
6	7,125

❑ COMPOUND INTEREST

We now turn to the growth of money at compound interest. The fee paid to use another's money is called **interest.** It is usually computed as a percent (called **interest rate**) of the principal over a given period of time. If, at the end of a payment period, the interest due is reinvested at the same rate, then the interest earned as well as the principal will earn interest during the next payment period. Interest paid on interest reinvested is called **compound interest,** and may be calculated using the following compound interest formula:

Compound Interest

If a **principal P (present value)** is invested at an annual **rate r** (expressed as a decimal) compounded m times a year, then the **amount A (future value)** in the account at the end of t years is given by

$$A = P\left(1 + \frac{r}{m}\right)^{mt}$$

[*Note:* P could be replaced by A_0, but convention dictates otherwise.]

Example 5 **Compound Growth**   If $1,000 is invested in an account paying 10% compounded monthly, how much will be in the account at the end of 10 years? Compute the answer to the nearest cent.

Solution   We use the compound interest formula as follows:

$$A = P\left(1 + \frac{r}{m}\right)^{mt}$$

$$= 1,000\left(1 + \frac{0.10}{12}\right)^{(12)(10)} \qquad \textit{Use a calculator.}$$

$$= \$2,707.04$$

The graph of

$$A = 1,000\left(1 + \frac{0.10}{12}\right)^{12t}$$

for $0 \le t \le 20$ is shown in Figure 7.

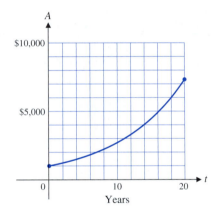

FIGURE 7

Matched Problem 5   If you deposit $5,000 in an account paying 9% compounded daily, how much will you have in the account in 5 years? Compute the answer to the nearest cent.

Explore–Discuss 4   Suppose that $1,000 is deposited in a savings account at an annual rate of 5%. Guess the amount in the account at the end of 1 year if interest is compounded (1) quarterly, (2) monthly, (3) daily, (4) hourly. Use the compound interest formula to compute the amounts at the end of 1 year to the nearest cent. Discuss the accuracy of your initial guesses.

 ❑ CONTINUOUS COMPOUND INTEREST

Returning to the compound interest formula,

$$A = P\left(1 + \frac{r}{m}\right)^{mt}$$

suppose that the principal P, the annual rate r, and the time t are held fixed, and the number of compounding periods per year m is increased without bound. Will the amount A increase without bound, or will it tend to some limiting value?

Starting with $P = \$100$, $r = 0.08$, and $t = 2$ years, we construct Table 4 for several values of m with the aid of a calculator. Notice that the largest gain appears in going from annual to semiannual compounding. Then, the gains slow down as m increases. It appears that A gets closer and closer to \$117.35 as m gets larger and larger.

TABLE 4

COMPOUNDING FREQUENCY	m	$A = 100\left(1 + \dfrac{0.08}{m}\right)^{2m}$
Annually	1	\$116.6400
Semiannually	2	116.9859
Quarterly	4	117.1659
Weekly	52	117.3367
Daily	365	117.3490
Hourly	8,760	117.3510

It can be shown that

$$P\left(1 + \frac{r}{m}\right)^{mt}$$

gets closer and closer to Pe^{rt} as the number of compounding periods m gets larger and larger. The latter is referred to as the **continuous compound interest formula,** a formula that is widely used in business, banking, and economics.

Continuous Compound Interest Formula

If a principal P is invested at an annual rate r (expressed as a decimal) compounded continuously, then the amount A in the account at the end of t years is given by

$$A = Pe^{rt}$$

Example 6

Compounding Daily and Continuously   What amount will an account have after 2 years if \$5,000 is invested at an annual rate of 8%:

(A) Compounded daily?   (B) Compounded continuously?

Compute answers to the nearest cent.

SOLUTION   (A) Use the compound interest formula

$$A = P\left(1 + \frac{r}{m}\right)^{mt}$$

with $P = 5{,}000$, $r = 0.08$, $m = 365$, and $t = 2$:

$$A = 5{,}000\left(1 + \frac{0.08}{365}\right)^{(365)(2)} \qquad \text{Use a calculator.}$$

$$= \$5{,}867.45$$

(B) Use the continuous compound interest formula

$$A = Pe^{rt}$$

with $P = 5{,}000$, $r = 0.08$, and $t = 2$:

$$A = 5{,}000e^{(0.08)(2)} \qquad \textcolor{blue}{\textit{Use a calculator.}}$$
$$= \$5{,}867.55$$

Matched Problem 6 What amount will an account have after 1.5 years if $8,000 is invested at an annual rate of 9%:

(A) Compounded weekly?   (B) Compounded continuously?

Compute answers to the nearest cent.

The formulas for simple interest, compound interest, and continuous compound interest are summarized in the box for convenient reference.

Interest Formulas

Simple interest $\qquad\qquad\qquad A = P(1 + rt)$

Compound interest $\qquad\qquad A = P\left(1 + \dfrac{r}{m}\right)^{mt}$

Continuous compound interest $\quad A = Pe^{rt}$

Answers to Matched Problems

1.

2. (A) 179 bacteria   (B) 1,271,659 bacteria

3. 233 mg   **4.** Purchase price: $30,363; value after 10 yr: $2,864

5. $7,841.13   **6.** (A) $9,155.23   (B) $9,156.29

Exercise 2-2

A

1. Match each equation with the graph of f, g, h, or k in the figure.

   (A) $y = 2^x$ (B) $y = (0.2)^x$
   (C) $y = 4^x$ (D) $y = (\frac{1}{3})^x$

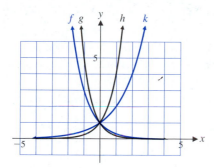

2. Match each equation with the graph of f, g, h, or k in the figure.

   (A) $y = (\frac{1}{4})^x$ (B) $y = (0.5)^x$

   (C) $y = 5^x$ (D) $y = 3^x$

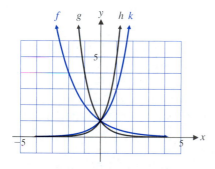

Graph each function in Problems 3–14 over the indicated interval.

3. $y = 5^x; [-2, 2]$ 4. $y = 3^x; [-3, 3]$

5. $y = (\frac{1}{5})^x = 5^{-x}; [-2, 2]$ 6. $y = (\frac{1}{3})^x = 3^{-x}; [-3, 3]$

7. $f(x) = -5^x; [-2, 2]$ 8. $g(x) = -3^{-x}; [-3, 3]$

9. $y = -e^{-x}; [-3, 3]$ 10. $y = -e^x; [-3, 3]$

11. $y = 100e^{0.1x}; [-5, 5]$ 12. $y = 10e^{0.2x}; [-10, 10]$

13. $g(t) = 10e^{-0.2t}; [-5, 5]$ 14. $f(t) = 100e^{-0.1t}; [-5, 5]$

Simplify each expression in Problems 15–20.

15. $(4^{3x})^{2y}$ 16. $10^{3x-1}10^{4-x}$ 17. $\dfrac{e^{x-3}}{e^{x-4}}$

18. $\dfrac{e^x}{e^{1-x}}$ 19. $(2e^{1.2t})^3$ 20. $(3e^{-1.4x})^2$

B *In Problems 21–28, describe the transformations that can be used to obtain the graph of g from the graph of f (see Section 1-2).*

21. $g(x) = -2^x; f(x) = 2^x$ 22. $g(x) = 2^{x-2}; f(x) = 2^x$

23. $g(x) = 3^{x+1}; f(x) = 3^x$ 24. $g(x) = -3^x; f(x) = 3^x$

25. $g(x) = e^x + 1; f(x) = e^x$

26. $g(x) = e^x - 2; f(x) = e^x$

27. $g(x) = 2e^{-(x+2)}; f(x) = e^{-x}$

28. $g(x) = 0.5e^{-(x-1)}; f(x) = e^{-x}$

 Check the answers to Problems 21–28 by graphing each pair of functions in the same viewing window of a graphing utility.

29. Use the graph of f shown in the figure to sketch the graph of each of the following.

   (A) $y = f(x) - 1$ (B) $y = f(x + 2)$
   (C) $y = 3f(x) - 2$ (D) $y = 2 - f(x - 3)$

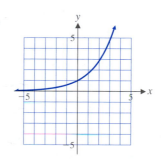

Figure for 29 and 30

30. Use the graph of f shown in the figure to sketch the graph of each of the following.

   (A) $y = f(x) + 2$ (B) $y = f(x - 3)$
   (C) $y = 2f(x) - 4$ (D) $y = 4 - f(x + 2)$

In Problems 31–40, graph each function over the indicated interval.

31. $f(t) = 2^{t/10}; [-30, 30]$

32. $G(t) = 3^{t/100}; [-200, 200]$

33. $y = -3 + e^{1+x}; [-4, 2]$

34. $y = 2 + e^{x-2}; [-1, 5]$

35. $y = e^{|x|}; [-3, 3]$

36. $y = e^{-|x|}; [-3, 3]$

37. $C(x) = \dfrac{e^x + e^{-x}}{2}; [-5, 5]$

38. $M(x) = e^{x/2} + e^{-x/2}; [-5, 5]$

39. $y = e^{-x^2}$; $[-3, 3]$

40. $y = 2^{-x^2}$; $[-3, 3]$

41. Find all real numbers a such that $a^2 = a^{-2}$. Explain why this does not violate the second exponential function property in the box on page 99.

42. Find real numbers a and b such that $a \neq b$ but $a^4 = b^4$. Explain why this does not violate the third exponential function property in the box on page 99.

Solve each equation in Problems 43–48 for x.

43. $10^{2-3x} = 10^{5x-6}$ **44.** $5^{3x} = 5^{4x-2}$

45. $4^{5x-x^2} = 4^{-6}$ **46.** $7^{x^2} = 7^{2x+3}$

47. $5^3 = (x+2)^3$ **48.** $(1-x)^5 = (2x-1)^5$

C *Solve each equation in Problems 49–52 for x. (Remember: $e^x \neq 0$ and $e^{-x} \neq 0$.)*

49. $(x-3)e^x = 0$ **50.** $2xe^{-x} = 0$

51. $3xe^{-x} + x^2e^{-x} = 0$ **52.** $x^2e^x - 5xe^x = 0$

Graph each function in Problems 53–56 over the indicated interval.

53. $h(x) = x(2^x)$; $[-5, 0]$ **54.** $m(x) = x(3^{-x})$; $[0, 3]$

55. $N = \dfrac{100}{1 + e^{-t}}$; $[0, 5]$ **56.** $N = \dfrac{200}{1 + 3e^{-t}}$; $[0, 5]$

 In Problems 57–60, approximate the real zeros of each function to two decimal places.

57. $f(x) = 4^x - 7$ **58.** $f(x) = 5 - 3^{-x}$

59. $f(x) = 2 + 3x + 10^x$ **60.** $f(x) = 7 - 2x^2 + 2^{-x}$

Applications

Business & Economics

61. *Finance.* Suppose that $2,500 is invested at 7% compounded quarterly. How much money will be in the account in:

  (A) $\frac{3}{4}$ year?   (B) 15 years?

  Compute answers to the nearest cent.

62. *Finance.* Suppose that $4,000 is invested at 6% compounded weekly. How much money will be in the account in:

  (A) $\frac{1}{2}$ year?   (B) 10 years?

  Compute answers to the nearest cent.

63. *Money growth.* If you invest $7,500 in an account paying 8.35% compounded continuously, how much money will be in the account at the end of:

  (A) 5.5 years?   (B) 12 years?

64. *Money growth.* If you invest $5,250 in an account paying 7.45% compounded continuously, how much money will be in the account at the end of:

  (A) 6.25 years?   (B) 17 years?

65. *Finance.* A person wishes to have $15,000 cash for a new car 5 years from now. How much should be placed in an account now, if the account pays 6.75% compounded weekly? Compute the answer to the nearest dollar.

66. *Finance.* A couple just had a baby. How much should they invest now at 5.5% compounded daily in order to have $40,000 for the child's education 17 years from now? Compute the answer to the nearest dollar.

67. *Money growth.* BanxQuote operates a network of Web **www** sites providing real-time market data from leading financial providers. The following rates for 12-month certficates of deposit were taken from the Web sites:

  (A) Stonebridge Bank, 6.93% compounded monthly

  (B) DeepGreen Bank, 6.96% compounded daily

  (C) Provident Bank, 6.80% compounded continuously

  Compute the value of $10,000 invested in each account at the end of 1 year.

68. *Money growth.* Refer to Problem 67. The following **www** rates for 30-month certificates of deposit were also taken from BanxQuote Web sites:

  (A) Oriental Bank & Trust, 6.50% compounded quarterly

  (B) BMW Bank of North America, 6.36% compounded monthly

  (C) BankFirst Corporation, 6.30% compounded daily

  Compute the value of $10,000 invested in each account at the end of 2.5 years.

69. *Present value.* A promissory note will pay $50,000 at maturity $5\frac{1}{2}$ years from now. How much should you be willing to pay for the note now if money is worth 8% compounded continuously?

70. *Present value.* A promissory note will pay $30,000 at maturity 10 years from now. How much should you be willing to pay for the note now if money is worth 7% compounded continuously?

71. *Advertising.* A company is trying to introduce a new product to as many people as possible through television advertising in a large metropolitan area with 2 million possible viewers. A model for the number of people N (in millions) who are aware of the product after t days of advertising was found to be

$$N = 2(1 - e^{-0.037t})$$

Graph this function for $0 \leq t \leq 50$. What value does N tend to as t increases without bound?

72. *Learning curve.* People assigned to assemble circuit boards for a computer manufacturing company undergo on-the-job training. From past experience it was found that the learning curve for the average employee is given by

$$N = 40(1 - e^{-0.12t})$$

where N is the number of boards assembled per day after t days of training. Graph this function for $0 \leq t \leq 30$. What is the maximum number of boards an average employee can be expected to produce in 1 day?

73. *Sports salaries.* Table 5 gives the average salary (in thousands of dollars) for players in the National Hockey League (NHL) and the National Basketball Association (NBA) in selected years since 1990.

(A) Let x represent the number of years since 1990 and find an exponential regression model ($y = ab^x$) for the NHL's average salary data. Estimate the average salary (to the nearest thousand dollars) for 1998 and 2010.

(B) The average salary in the NHL for 1998 was $1,167,000. How does this compare with the value estimated from the model in part (A)? How does this 1998 salary information affect the estimated salary in 2010? Explain.

WWW

TABLE 5

AVERAGE SALARIES (THOUSAND $)

YEAR	NHL	NBA
1990	211	750
1991	271	900
1992	368	1,100
1993	467	1,300
1994	562	1,700
1995	733	1,900
1996	892	2,000

74. *Sports salaries.* Refer to Table 5.

(A) Let x represent the number of years since 1990 and find an exponential regression model ($y = ab^x$) for the NBA's average salary data. Estimate the average salary (to the nearest thousand dollars) for 1997 and 2010.

(B) The average salary in the NBA for 1997 was $2,200,000. How does this compare with the estimated value from the model in part (A)? How does this 1997 salary information affect the estimated salary in 2010? Explain.

Life Sciences

75. *Marine biology.* Marine life is dependent upon the microscopic plant life that exists in the photic zone, a zone that goes to a depth where about 1% of the surface light remains. In some waters with a great deal of sediment, the photic zone may go down only 15 to 20 feet. In some murky harbors, the intensity of light d feet below the surface is given approximately by

$$I = I_0 e^{-0.23d}$$

What percentage of the surface light will reach a depth of:

(A) 10 feet?   (B) 20 feet?

76. *Marine biology.* Refer to Problem 75. Light intensity I relative to depth d (in feet) for one of the clearest bodies of water in the world, the Sargasso Sea in the West Indies, can be approximated by

$$I = I_0 e^{-0.00942d}$$

where I_0 is the intensity of light at the surface. What percent of the surface light will reach a depth of:

(A) 50 feet?   (B) 100 feet?

77. *HIV/AIDS epidemic.* The Joint United Nation Program on HIV/AIDS reported that prior to 1999 HIV had infected 47.3 million people worldwide and estimated that the disease will continue to spread at an annual rate of 11.6% compounded continuously. Let 1998 be year 0 and assume this rate does not change.

(A) Write an equation that models the worldwide growth of HIV, starting in 1998.

(B) Based on the model, how many cases (to the nearest thousand) should we expect by the end of 2005? Of 2010?

(C) Sketch a graph of this growth equation from 1998 to 2010.

78. *HIV/AIDS epidemic.* The Joint United Nations Program on HIV/AIDS reported that prior to 1999 13.9 million people worldwide had died of AIDS and estimated that the total number of deaths will continue to grow at an annual rate of 16.5% compounded continuously. Let 1998 be year 0 and assume that this rate does not change.

(A) Write an equation that models total worldwide deaths due to AIDS, starting in 1998.

(B) Based on the model, how many total deaths (to the nearest hundred thousand) should we expect by the end of 2005? Of 2010?

(C) Sketch a graph of this growth equation from 1998 to 2010.

Social Sciences

79. *World population growth.* It took from the dawn of humanity to 1830 for the population to grow to the first billion people, just 100 more years (by 1930) for the second billion, and 3 billion more were added in only 60 more years (by 1990). In 1995, the estimated world population was 5.7 billion. In 1994, the World Bank estimated the world population would be growing at an annual rate of 1.14% compounded continuously until 2030.

(A) Write an equation that models the world population growth, letting 1995 be year 0.

(B) Based on the model, what is the expected world population (to the nearest hundred million) in 2010? In 2030?

(C) Sketch a graph of the equation found in part (A). Cover the years from 1995 through 2030.

80. *Population growth in Ethiopia.* In 1995, the estimated population in Ethiopia was 88 million people. In 1994, the World Bank estimated the population would grow at an annual rate of 1.67% compounded continuously until 2030.

(A) Write an equation that models the population growth in Ethiopia, letting 1995 be year 0.

(B) Based on the model, what is the expected population in Ethiopia (to the nearest million) in 2010? In 2030?

(C) Sketch a graph of the equation found in part (A). Cover the years from 1995 through 2030.

81. *Internet growth.* The number of Internet hosts grew very rapidly from 1994 to 2000 (Table 6).

(A) Let x represent the number of years since 1994. Find an exponential regression model $(y = ab^x)$ for this data set and estimate the number of hosts in 2010 (to the nearest million).

(B) Discuss the implications of this model if the number of Internet hosts continues to grow at this rate.

WWW

TABLE 6

INTERNET HOSTS (MILLIONS)

YEAR	HOSTS
1994	2.4
1995	4.9
1996	9.5
1997	16.1
1998	29.7
1999	43.2
2000	72.4

Source: Internet Software Consortium

82. *Life expectancy.* Table 7 shows the life expectancy (in years) at birth for residents of the United States from 1970 to 1995. Let x represent years since 1970. Find an exponential regression model for this data and use it to estimate the life expectancy for a person born in 2010.

WWW

TABLE 7

YEAR OF BIRTH	LIFE EXPECTANCY
1970	70.8
1975	72.6
1980	73.7
1985	74.7
1990	75.4
1995	75.9
1997	76.5

Section 2-3   Logarithmic Functions

- ❑ INVERSE FUNCTIONS
- ❑ LOGARITHMIC FUNCTIONS
- ❑ PROPERTIES OF LOGARITHMIC FUNCTIONS
- ❑ CALCULATOR EVALUATION OF LOGARITHMS
- ❑ APPLICATION

Find the exponential function keys $\boxed{10^x}$ and $\boxed{e^x}$ on your calculator. Close to these keys you will find $\boxed{\text{LOG}}$ and $\boxed{\text{LN}}$ keys. The latter represent *logarithmic functions,* and each is closely related to the exponential function it is near. In fact, the exponential function and the corresponding logarithmic function are said to be *inverses* of each other. In this section we will develop the concept of inverse functions and use it to define a logarithmic function as the inverse of an exponential function. We will then investigate basic properties of logarithmic functions, use a calculator to evaluate them for particular values of x, and apply them to real-world problems.

Logarithmic functions are used in modeling and solving many types of problems. For example, the decibel scale is a logarithmic scale used to measure sound intensity, and the Richter scale is a logarithmic scale used to measure the strength of the force of an earthquake. An important business application has to do with finding the time it takes money to double if it is invested at a certain rate compounded a given number of times a year or compounded continuously. This requires the solution of an exponential equation, and logarithms play a central role in the process.

❑ INVERSE FUNCTIONS

Look at the graphs of $f(x) = \dfrac{x}{2}$ and $g(x) = \dfrac{|x|}{2}$ in Figure 1:

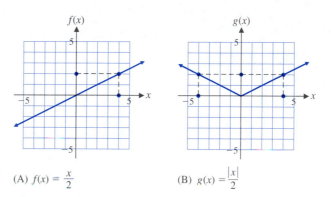

(A) $f(x) = \dfrac{x}{2}$ (B) $g(x) = \dfrac{|x|}{2}$

FIGURE 1

Because both f and g are functions, each domain value corresponds to exactly one range value. For which function does each range value correspond to exactly one domain value? This is the case only for function f. Note that for the range value 2, the corresponding domain value is 4. For function g the range value 2 corresponds to both -4 and 4. Function f is said to be *one-to-one*. In general:

┌───┐

One-to-One Functions

A function f is said to be **one-to-one** if each range value corresponds to exactly one domain value.

└───┘

It can be shown that any continuous function that is either increasing or decreasing for all domain values is one-to-one. If a continuous function increases for some domain values and decreases for others, it cannot be one-to-one. Figure 1 shows an example of each case.

┌──────────────────────┐
│ Explore–Discuss 1 │
└──────────────────────┘

Graph $f(x) = 2^x$ and $g(x) = x^2$. For a range value of 4, what are the corresponding domain values for each function? Which of the two functions is one-to-one? Explain why.

Starting with a one-to-one function f we can obtain a new function called the *inverse* of f as follows:

> ### Inverse of a Function
>
> If f is a one-to-one function, then the **inverse** of f is the function formed by interchanging the independent and dependent variables for f. Thus, if (a, b) is a point on the graph of f, then (b, a) is a point on the graph of the inverse of f.
>
> [*Note:*   If f is not one-to-one, then f **does not have an inverse.**]

A number of important functions in any library of elementary functions are the inverses of other basic functions in the library. In this course, we are interested in the inverses of exponential functions, called *logarithmic functions.*

❑   LOGARITHMIC FUNCTIONS

If we start with the exponential function f defined by

$$y = 2^x \tag{1}$$

and interchange the variables, we obtain the inverse of f:

$$x = 2^y \tag{2}$$

We call the inverse the **logarithmic function with base 2,** and write

$$y = \log_2 x \qquad \text{if and only if} \qquad x = 2^y$$

We can graph $y = \log_2 x$ by graphing $x = 2^y$, since they are equivalent. Any ordered pair of numbers on the graph of the exponential function will be on the graph of the logarithmic function if we interchange the order of the components. For example, $(3, 8)$ satisfies equation (1) and $(8, 3)$ satisfies equation (2). The graphs of $y = 2^x$ and $y = \log_2 x$ are shown in Figure 2. Note that if we fold the paper along the dashed line $y = x$ in Figure 2, the two graphs match exactly. The line $y = x$ is a line of symmetry for the two graphs.

FIGURE 2

Exponential function		Logarithmic function	
x	$y = 2^x$	$x = 2^y$	y
-3	$\frac{1}{8}$	$\frac{1}{8}$	-3
-2	$\frac{1}{4}$	$\frac{1}{4}$	-2
-1	$\frac{1}{2}$	$\frac{1}{2}$	-1
0	1	1	0
1	2	2	1
2	4	4	2
3	8	8	3

$$\begin{bmatrix} \text{Ordered} \\ \text{Pairs} \\ \text{reversed} \end{bmatrix}$$

In general, since the graphs of all exponential functions of the form $f(x) = b^x, b \neq 1, b > 0$, are either increasing or decreasing (see Section 2-2), exponential functions have inverses.

> ### Logarithmic Functions
>
> The inverse of an exponential function is called a **logarithmic function.** For $b > 0$ and $b \neq 1$,
>
> Logarithmic form   Exponential form
> $$y = \log_b x \qquad \text{is equivalent to} \qquad x = b^y$$
>
> The **log to the base b of x** is the exponent to which b must be raised to obtain x. [*Remember*:   A logarithm is an exponent.] The **domain** of the logarithmic function is the set of all positive real numbers, which is also the range of the corresponding exponential function; and the **range** of the logarithmic function is the set of all real numbers, which is also the domain of the corresponding exponential function. Typical graphs of an exponential function and its inverse, a logarithmic function, are shown in the figure in the margin.

The following examples involve converting logarithmic forms to equivalent exponential forms, and vice versa.

Example 1 ➭   **Logarithmic–Exponential Conversions**   Change each logarithmic form to an equivalent exponential form:

(A) $\log_5 25 = 2$ (B) $\log_9 3 = \frac{1}{2}$ (C) $\log_2(\frac{1}{4}) = -2$

Solution   (A) $\log_5 25 = 2$ is equivalent to $25 = 5^2$
(B) $\log_9 3 = \frac{1}{2}$ is equivalent to $3 = 9^{1/2}$
(C) $\log_2(\frac{1}{4}) = -2$ is equivalent to $\frac{1}{4} = 2^{-2}$ ■

Matched Problem 1 ➭   Change each logarithmic form to an equivalent exponential form:

(A) $\log_3 9 = 2$ (B) $\log_4 2 = \frac{1}{2}$ (C) $\log_3(\frac{1}{9}) = -2$ ■

Example 2 ➭   **Exponential–Logarithmic Conversions**   Change each exponential form to an equivalent logarithmic form:

(A) $64 = 4^3$ (B) $6 = \sqrt{36}$ (C) $\frac{1}{8} = 2^{-3}$

Solution   (A) $64 = 4^3$ is equivalent to $\log_4 64 = 3$
(B) $6 = \sqrt{36}$ is equivalent to $\log_{36} 6 = \frac{1}{2}$
(C) $\frac{1}{8} = 2^{-3}$ is equivalent to $\log_2(\frac{1}{8}) = -3$ ■

Matched Problem 2 ➭   Change each exponential form to an equivalent logarithmic form:

(A) $49 = 7^2$ (B) $3 = \sqrt{9}$ (C) $\frac{1}{3} = 3^{-1}$ ■

To gain a little deeper understanding of logarithmic functions and their relationship to the exponential functions, we consider a few problems where we want to find x, b, or y in $y = \log_b x$, given the other two values. All values are chosen so that the problems can be solved exactly without a calculator.

Example 3 ➭   **Solutions of the Equation $y = \log_b x$**   Find y, b, or x, as indicated.

(A) Find y: $y = \log_4 16$ (B) Find x: $\log_2 x = -3$
(C) Find y: $y = \log_8 4$ (D) Find b: $\log_b 100 = 2$

Solution   (A) $y = \log_4 16$ is equivalent to $16 = 4^y$. Thus,

$$y = 2$$

(B) $\log_2 x = -3$ is equivalent to $x = 2^{-3}$. Thus,

$$x = \frac{1}{2^3} = \frac{1}{8}$$

(C) $y = \log_8 4$ is equivalent to

$$4 = 8^y \qquad \text{or} \qquad 2^2 = 2^{3y}$$

Thus,

$$3y = 2$$
$$y = \tfrac{2}{3}$$

(D) $\log_b 100 = 2$ is equivalent to $100 = b^2$. Thus,

$$b = 10 \qquad \text{Recall that } b \text{ cannot be negative.}$$

Matched Problem 3   Find y, b, or x, as indicated.

(A) Find y: $y = \log_9 27$ (B) Find x: $\log_3 x = -1$
(C) Find b: $\log_b 1{,}000 = 3$

❑   PROPERTIES OF LOGARITHMIC FUNCTIONS

Logarithmic functions have many powerful and useful properties. We list eight basic properties in Theorem 1.

THEOREM 1   **Properties of Logarithmic Functions**

If b, M, and N are positive real numbers, $b \neq 1$, and p and x are real numbers, then:

1. $\log_b 1 = 0$ **5.** $\log_b MN = \log_b M + \log_b N$

2. $\log_b b = 1$ **6.** $\log_b \dfrac{M}{N} = \log_b M - \log_b N$

3. $\log_b b^x = x$ **7.** $\log_b M^p = p \log_b M$

4. $b^{\log_b x} = x, \quad x > 0$ **8.** $\log_b M = \log_b N$ if and only if $M = N$

The first four properties in Theorem 1 follow directly from the definition of a logarithmic function. Here we will sketch a proof of property 5. The other properties are established in a similar way. Let

$$u = \log_b M \qquad \text{and} \qquad v = \log_b N$$

Or, in equivalent exponential form,

$$M = b^u \qquad \text{and} \qquad N = b^v$$

Now, see if you can provide reasons for each of the following steps:

$$\log_b MN = \log_b b^u b^v = \log_b b^{u+v} = u + v = \log_b M + \log_b N$$

Example 4   **Using Logarithmic Properties**

(A) $\log_b \dfrac{wx}{yz}$ $= \log_b wx - \log_b yz$

$= \log_b w + \log_b x - (\log_b y + \log_b z)$

$= \log_b w + \log_b x - \log_b y - \log_b z$

(B) $\log_b(wx)^{3/5} = \frac{3}{5}\log_b wx = \frac{3}{5}(\log_b w + \log_b x)$

Matched Problem 4 ➾ Write in simpler logarithmic forms, as in Example 4.

(A) $\log_b \dfrac{R}{ST}$　　(B) $\log_b\left(\dfrac{R}{S}\right)^{2/3}$

The following examples and problems, although somewhat artificial, will give you additional practice in using basic logarithmic properties.

Example 5 ➾ **Solving Logarithmic Equations**　Find x so that:

$$\tfrac{3}{2}\log_b 4 - \tfrac{2}{3}\log_b 8 + \log_b 2 = \log_b x$$

SOLUTION
$$\tfrac{3}{2}\log_b 4 - \tfrac{2}{3}\log_b 8 + \log_b 2 = \log_b x$$
$$\log_b 4^{3/2} - \log_b 8^{2/3} + \log_b 2 = \log_b x \qquad \text{Property 7}$$
$$\log_b 8 - \log_b 4 + \log_b 2 = \log_b x$$
$$\log_b \frac{8 \cdot 2}{4} = \log_b x \qquad \text{Properties 5 and 6}$$
$$\log_b 4 = \log_b x$$
$$x = 4 \qquad \text{Property 8}$$

Matched Problem 5 ➾ Find x so that:　$3\log_b 2 + \tfrac{1}{2}\log_b 25 - \log_b 20 = \log_b x$

Example 6 ➾ **Solving Logarithmic Equations**　Solve: $\log_{10} x + \log_{10}(x + 1) = \log_{10} 6$

SOLUTION
$$\log_{10} x + \log_{10}(x + 1) = \log_{10} 6$$
$$\log_{10}[x(x + 1)] = \log_{10} 6 \qquad \text{Property 5}$$
$$x(x + 1) = 6 \qquad \text{Property 8}$$
$$x^2 + x - 6 = 0 \qquad \text{Solve by factoring.}$$
$$(x + 3)(x - 2) = 0$$
$$x = -3, 2$$

We must exclude $x = -3$, since the domain of the function $\log_{10}(x + 1)$ is $x > -1$ or $(-1, \infty)$; hence, $x = 2$ is the only solution.

Matched Problem 6 ➾ Solve: $\log_3 x + \log_3(x - 3) = \log_3 10$

Explore–Discuss 2

Discuss the relationship between each of the following pairs of expressions. If the two expressions are equivalent, explain why. If they are not, give an example.

(A) $\log_b M - \log_b N$;　$\dfrac{\log_b M}{\log_b N}$

(B) $\log_b M - \log_b N$;　$\log_b \dfrac{M}{N}$

(C) $\log_b M + \log_b N$;　$\log_b MN$

(D) $\log_b M + \log_b N$;　$\log_b(M + N)$

❑ CALCULATOR EVALUATION OF LOGARITHMS

Of all possible logarithmic bases, the base e and the base 10 are used almost exclusively. Before we can use logarithms in certain practical problems, we need to be able to approximate the logarithm of any positive number either to base 10 or to base e. And conversely, if we are given the logarithm of a number to base 10 or base e, we need to be able to approximate the number. Historically, tables were used for this purpose, but now calculators make computations faster and far more accurate.

Common logarithms (also called **Briggsian logarithms**) are logarithms with base 10. **Natural logarithms** (also called **Napierian logarithms**) are logarithms with base e. Most calculators have a key labeled "log" (or "LOG") and a key labeled "ln" (or "LN"). The former represents a common (base 10) logarithm and the latter a natural (base e) logarithm. In fact, "log" and "ln" are both used extensively in mathematical literature, and whenever you see either used in this book without a base indicated, they will be interpreted as follows:

Logarithmic Notation

Common logarithm: $\log x = \log_{10} x$
Natural logarithm: $\ln x = \log_e x$

Finding the common or natural logarithm using a calculator is very easy. On some calculators, you simply enter a number from the domain of the function and press $\boxed{\text{LOG}}$ or $\boxed{\text{LN}}$. On other calculators, you press either $\boxed{\text{LOG}}$ or $\boxed{\text{LN}}$, enter a number from the domain, and then press $\boxed{\text{ENTER}}$. Check the user's manual for your calculator.

Example 7 ⮑   **Calculator Evaluation of Logarithms**   Use a calculator to evaluate each to six decimal places:

(A) $\log 3{,}184$ (B) $\ln 0.000\,349$ (C) $\log(-3.24)$

SOLUTION   (A) $\log 3{,}184 = 3.502\,973$ (B) $\ln 0.000\,349 = -7.960\,439$

(C) $\log(-3.24) = $ Error*   *−3.24 is not in the domain of the log function.* ■

Matched Problem 7   Use a calculator to evaluate each to six decimal places:

(A) $\log 0.013\,529$ (B) $\ln 28.693\,28$ (C) $\ln(-0.438)$ ■

We now turn to the second problem mentioned above: Given the logarithm of a number, find the number. We make direct use of the logarithmic–exponential relationships, which follow from the definition of logarithmic function given at the beginning of this section.

Logarithmic–Exponential Relationships

$\log x = y$ is equivalent to $x = 10^y$
$\ln x = y$ is equivalent to $x = e^y$

---

*Some calculators use a more advanced definition of logarithms involving complex numbers and will display an ordered pair of real numbers as the value of $\log(-3.24)$. You should interpret such a result as an indication that the number entered is not in the domain of the logarithm function as we have defined it.

Example 8 ⇔ **Solving $\log_b x = y$ for x**   Find x to four decimal places, given the indicated logarithm:

(A) $\log x = -2.315$ (B) $\ln x = 2.386$

SOLUTION   (A) $\log x = -2.315$ *Change to equivalent exponential form.*

$x = 10^{-2.315}$ *Evaluate with a calculator.*

$= 0.0048$

(B) $\ln x = 2.386$ *Change to equivalent exponential form.*

$x = e^{2.386}$ *Evaluate with a calculator.*

$= 10.8699$

Matched Problem 8 ⇔ Find x to four decimal places, given the indicated logarithm:

(A) $\ln x = -5.062$ (B) $\log x = 2.0821$

Example 9 ⇔ **Solving Exponential Equations**   Solve for x to four decimal places:

(A) $10^x = 2$ (B) $e^x = 3$ (C) $3^x = 4$

SOLUTION   (A) $10^x = 2$ *Take common logarithms of both sides.*

$\log 10^x = \log 2$ *Property 3*

$x = \log 2$ *Use a calculator.*

$= 0.3010$

(B) $e^x = 3$ *Take natural logarithms of both sides.*

$\ln e^x = \ln 3$ *Property 3*

$x = \ln 3$ *Use a calculator.*

$= 1.0986$

(C) $3^x = 4$ *Take either natural or common logarithms of both sides. (We choose commmon logarithms.)*

$\log 3^x = \log 4$ *Property 7*

$x \log 3 = \log 4$ *Solve for x.*

$x = \dfrac{\log 4}{\log 3}$ *Use a calculator.*

$= 1.2619$

Exponential equations can also be solved graphically by graphing both sides of an equation and finding the points of intersection. Figure 3 illustrates this approach for the equations in Example 9.

(A) $y_1 = 10^x$ (B) $y_1 = e^x$ (C) $y_1 = 3^x$
    $y_2 = 2$ $y_2 = 3$ $y_2 = 4$

FIGURE 3 Graphical solution of exponential equations

Matched Problem 9 ⇔ Solve for x to four decimal places:

(A) $10^x = 7$ (B) $e^x = 6$ (C) $4^x = 5$

Discuss how you could find $y = \log_5 38.25$ using either natural or common logarithms on a calculator. [*Hint*: Start by rewriting the equation in exponential form.]

❑ APPLICATION

A convenient and easily understood way of comparing different investments is to use their **doubling times**—the length of time it takes the value of an investment to double. Logarithm properties, as you will see in Example 10, provide us with just the right tool for solving some doubling-time problems.

Example 10 **Doubling Time for an Investment** How long (to the next whole year) will it take money to double if it is invested at 20% compounded annually?

SOLUTION We use the compound interest formula discussed in Section 2-2:

$$A = P\left(1 + \frac{r}{m}\right)^{mt} \qquad \text{Compound interest}$$

The problem is to find t, given $r = 0.20$, $m = 1$, and $A = 2p$; that is,

$$2P = P(1 + 0.2)^t$$
$$2 = 1.2^t \qquad \text{Solve for } t \text{ by taking the natural or common}$$
$$1.2^t = 2 \qquad \text{logarithm of both sides (we choose the}$$
$$\ln 1.2^t = \ln 2 \qquad \text{natural logarithm).}$$
$$t \ln 1.2 = \ln 2 \qquad \text{Property 7}$$
$$t = \frac{\ln 2}{\ln 1.2} \qquad \text{Use a calculator.}$$
$$= 3.8 \text{ years} \qquad [\text{Note: } (\ln 2)/(\ln 1.2) \ne \ln 2 - \ln 1.2]$$
$$\approx 4 \text{ years} \qquad \text{To the next whole year}$$

FIGURE 4 $y_1 = 1.2^x$, $y_2 = 2$

When interest is paid at the end of 3 years, the money will not be doubled; when paid at the end of 4 years, the money will be slightly more than doubled.

Example 10 can also be solved graphically by graphing both sides of the equation $2 = 1.2^t$, and finding the intersection point (Fig. 4).

Matched Problem 10 How long (to the next whole year) will it take money to triple if it is invested at 13% compounded annually?

It is interesting and instructive to graph the doubling times for various rates compounded annually. We proceed as follows:

$$A = P(1 + r)^t$$
$$2P = P(1 + r)^t$$
$$2 = (1 + r)^t$$
$$(1 + r)^t = 2$$
$$\ln(1 + r)^t = \ln 2$$
$$t \ln(1 + r) = \ln 2$$
$$t = \frac{\ln 2}{\ln(1 + r)}$$

FIGURE 5

Figure 5 shows the graph of this equation (doubling time in years) for interest rates compounded annually from 1 to 70% (expressed as decimals). Note the dramatic change in doubling time as rates change from 1 to 20% (from 0.01 to 0.20).

Answers to Matched Problems

1. (A) $9 = 3^2$ (B) $2 = 4^{1/2}$ (C) $\frac{1}{9} = 3^{-2}$

2. (A) $\log_7 49 = 2$ (B) $\log_9 3 = \frac{1}{2}$ (C) $\log_3(\frac{1}{3}) = -1$

3. (A) $y = \frac{3}{2}$ (B) $x = \frac{1}{3}$ (C) $b = 10$

4. (A) $\log_b R - \log_b S - \log_b T$ (B) $\frac{2}{3}(\log_b R - \log_b S)$ **5.** $x = 2$

6. $x = 5$

7. (A) $-1.868\,734$ (B) $3.356\,663$ (C) Not defined

8. (A) 0.0063 (B) 120.8092

9. (A) 0.8451 (B) 1.7918 (C) 1.1610

10. 9 yr

Exercise 2-3

A *For Problems 1–6, rewrite in equivalent exponential form.*

1. $\log_3 27 = 3$ **2.** $\log_2 32 = 5$ **3.** $\log_{10} 1 = 0$

4. $\log_e 1 = 0$ **5.** $\log_4 8 = \frac{3}{2}$ **6.** $\log_9 27 = \frac{3}{2}$

For Problems 7–12, rewrite in equivalent logarithmic form.

7. $49 = 7^2$ **8.** $36 = 6^2$ **9.** $8 = 4^{3/2}$

10. $9 = 27^{2/3}$ **11.** $A = b^u$ **12.** $M = b^x$

For Problems 13–24, evaluate without a calculator.

13. $\log_{10} 1$ **14.** $\log_e 1$ **15.** $\log_e e$

16. $\log_{10} 10$ **17.** $\log_{0.2} 0.2$ **18.** $\log_{13} 13$

19. $\log_{10} 10^3$ **20.** $\log_{10} 10^{-5}$ **21.** $\log_2 2^{-3}$

22. $\log_3 3^5$ **23.** $\log_{10} 1{,}000$ **24.** $\log_6 36$

For Problems 25–30, write in terms of simpler logarithmic forms, as in Example 4.

25. $\log_b \dfrac{P}{Q}$ **26.** $\log_b FG$ **27.** $\log_b L^5$

28. $\log_b w^{15}$ **29.** $\log_b \dfrac{p}{qrs}$ **30.** $\log_b PQR$

B *For Problems 31–42, find x, y, or b without a calculator.*

31. $\log_3 x = 2$ **32.** $\log_2 x = 2$

33. $\log_7 49 = y$ **34.** $\log_3 27 = y$

35. $\log_b 10^{-4} = -4$ **36.** $\log_b e^{-2} = -2$

37. $\log_4 x = \frac{1}{2}$ **38.** $\log_{25} x = \frac{1}{2}$

39. $\log_{1/3} 9 = y$ **40.** $\log_{49}(\frac{1}{7}) = y$

41. $\log_b 1{,}000 = \frac{3}{2}$ **42.** $\log_b 4 = \frac{2}{3}$

For Problems 43–54, write in terms of simpler logarithmic forms, going as far as you can with logarithmic properties (see Example 4).

43. $\log_b \dfrac{x^5}{y^3}$ **44.** $\log_b(x^2 y^3)$

45. $\log_b \sqrt[3]{N}$ **46.** $\log_b \sqrt[5]{Q}$

47. $\log_b(x^2 \sqrt[3]{y})$ **48.** $\log_b \sqrt[3]{\dfrac{x^2}{y}}$

49. $\log_b(50 \cdot 2^{-0.2t})$ **50.** $\log_b(100 \cdot 1.06^t)$

51. $\log_b[P(1 + r)^t]$ **52.** $\log_e Ae^{-0.3t}$

53. $\log_e 100e^{-0.01t}$ **54.** $\log_{10}(67 \cdot 10^{-0.12x})$

Find x in Problems 55–62.

55. $\log_b x = \frac{2}{3}\log_b 8 + \frac{1}{2}\log_b 9 - \log_b 6$

56. $\log_b x = \frac{2}{3}\log_b 27 + 2\log_b 2 - \log_b 3$

57. $\log_b x = \frac{3}{2}\log_b 4 - \frac{2}{3}\log_b 8 + 2\log_b 2$

58. $\log_b x = 3\log_b 2 + \frac{1}{2}\log_b 25 - \log_b 20$

59. $\log_b x + \log_b(x - 4) = \log_b 21$

60. $\log_b(x + 2) + \log_b x = \log_b 24$

61. $\log_{10}(x - 1) - \log_{10}(x + 1) = 1$

62. $\log_{10}(x + 6) - \log_{10}(x - 3) = 1$

Graph Problems 63 and 64 by converting to exponential form first.

63. $y = \log_2(x - 2)$ **64.** $y = \log_3(x + 2)$

65. Explain how the graph of the equation in Problem 63 can be obtained from the graph of $y = \log_2 x$ using a simple transformation (see Section 1-2).

66. Explain how the graph of the equation in Problem 64 can be obtained from the graph of $y = \log_3 x$ using a simple transformation (see Section 1-2).

67. What are the domain and range of the function defined by $y = 1 + \ln(x + 1)$?

68. What are the domain and range of the function defined by $y = \log(x - 1) - 1$?

For Problems 69 and 70, evaluate to five decimal places using a calculator.

69. (A) log 3,527.2   (B) log 0.006 913 2
     (C) ln 277.63   (D) ln 0.040 883

70. (A) log 72.604   (B) log 0.033 041
     (C) ln 40,257   (D) ln 0.005 926 3

For Problems 71 and 72, find x to four decimal places.

71. (A) $\log x = 1.1285$ (B) $\log x = -2.0497$
     (C) $\ln x = 2.7763$ (D) $\ln x = -1.8879$

72. (A) $\log x = 2.0832$ (B) $\log x = -1.1577$
     (C) $\ln x = 3.1336$ (D) $\ln x = -4.3281$

For Problems 73–80, solve each equation to four decimal places.

73. $10^x = 12$ **74.** $10^x = 153$

75. $e^x = 4.304$ **76.** $e^x = 0.3059$

77. $1.03^x = 2.475$ **78.** $1.075^x = 1.837$

79. $1.005^{12t} = 3$ **80.** $1.02^{4t} = 2$

 Check Problems 73–80 by solving graphically.

Graph Problems 81–88 using a calculator and point-by-point plotting. Indicate increasing and decreasing intervals.

81. $y = \ln x$ **82.** $y = -\ln x$

83. $y = |\ln x|$ **84.** $y = \ln |x|$

85. $y = 2\ln(x + 2)$ **86.** $y = 2\ln x + 2$

87. $y = 4\ln x - 3$ **88.** $y = 4\ln(x - 3)$

 Check Problems 81–88 by graphing on a graphing utility.

C

89. Explain why the logarithm of 1 for any permissible base is 0.

90. Explain why 1 is not a suitable logarithmic base.

91. Write $\log_{10} y - \log_{10} c = 0.8x$ in an exponential form that is free of logarithms.

92. Write $\log_e x - \log_e 25 = 0.2t$ in an exponential form that is free of logarithms.

93. Let $p(x) = \ln x$, $q(x) = \sqrt{x}$, and $r(x) = x$. Use a graphing utility to draw graphs of all three functions in the same viewing window for $1 \le x \le 16$. Discuss what it means for one function to be larger than another on an interval, and then order the three functions from largest to smallest for $1 < x \le 16$.

94. Let $p(x) = \log x$, $q(x) = \sqrt[3]{x}$. and $r(x) = x$. Use a graphing utility to draw graphs of all three functions in the same viewing window for $1 \le x \le 16$. Discuss what it means for one function to be smaller than another on an interval, and then order the three functions from smallest to largest for $1 < x \le 16$.

Applications

Business & Economics

95. *Doubling time.* In its first 10 years the Gabelli
www Growth Fund produced an average annual return of 21.36%. Assume that money invested in this fund continues to earn 21.36% compounded annually. How long will it take money invested in this fund to double?

96. *Doubling time.* In its first 10 years the Janus Flexi-
www ble Income Fund produced an average annual return of 9.58%. Assume that money invested in this fund continues to earn 9.58% compounded annually. How long will it take money invested in this fund to double?

97. *Investing.* How many years (to two decimal places) will it take $1,000 to grow to $1,800 if it is invested at 6% compounded quarterly? Compounded continuously?

98. *Investing.* How many years (to two decimal places) will it take $5,000 to grow to $7,500 if it is invested at 8% compounded semiannually? Compounded continuously?

99. *Investment.* A newly married couple wishes to have $30,000 in 6 years for the down payment on a house. At what rate of interest compounded continuously (to three decimal places) must $20,000 be invested now to accomplish this goal?

100. *Investment.* The parents of a newborn child want to have $60,000 for the child's college education 17 years from now. At what rate of interest compounded continuously (to three decimal places) must a grandparent's gift of $20,000 be invested now to achieve this goal?

101. *Supply and demand.* A cordless screwdriver is sold through a national chain of discount stores. A marketing company established price–demand and price–supply tables (Tables 1 and 2), where x is the number of screwdrivers people are willing to buy and the store is willing to sell each month at a price of p dollars per screwdriver.

TABLE 1

PRICE–DEMAND

x	$p = D(x)$ ($)
1,000	91
2,000	73
3,000	64
4,000	56
5,000	53

TABLE 2

PRICE–SUPPLY

x	$p = S(x)$ ($)
1,000	9
2,000	26
3,000	34
4,000	38
5,000	41

(A) Find a logarithmic regression model ($y = a + b \ln x$) for the data in Table 1. Estimate the demand (to the nearest unit) at a price level of $50.

(B) Find a logarithmic regression model ($y = a + b \ln x$) for the data in Table 2. Estimate the supply (to the nearest unit) at a price level of $50.

(C) Does a price level of $50 represent a stable condition, or is the price likely to increase or decrease? Explain.

102. *Equilibrium point.* Use the models constructed in Problem 101 to find the equilibrium point. Write the equilibrium price to the nearest cent and the equilibrium quantity to the nearest unit.

Life Sciences

103. *Sound intensity: decibels.* Because of the extraordinary range of sensitivity of the human ear (a range of over 1,000 million millions to 1), it is helpful to use a logarithmic scale, rather than an absolute scale, to measure sound intensity over this range. The unit of measure is called the *decibel,* after the inventor of the telephone, Alexander Graham Bell. If we let N be the number of decibels, I the power of the sound in question (in watts per square centimeter), and I_0 the power of sound just below the threshold of hearing (approximately 10^{-16} watt per square centimeter), then

$$I = I_0 10^{N/10}$$

Show that this formula can be written in the form

$$N = 10 \log \frac{I}{I_0}$$

104. *Sound intensity: decibels.* Use the formula in Problem 105 (with $I_0 = 10^{-16}$ W/cm²) to find the decibel ratings of the following sounds:

(A) Whisper: 10^{-13} W/cm²
(B) Normal conversation: 3.16×10^{-10} W/cm²
(C) Heavy traffic: 10^{-8} W/cm²
(D) Jet plane with afterburner: 10^{-1} W/cm²

105. *Agriculture.* Table 3 shows the yield (in bushels per acre) and the total production (in millions of bushels) for corn in the United States for selected years since 1950. Let x represent years since 1900. Find a logarithmic regression model ($y = a + b \ln x$) for the yield. Estimate (to one decimal place) the yield in 2010.

TABLE 3   www

UNITED STATES CORN PRODUCTION

YEAR	x	YIELD(BUSHELS PER ACRE)	TOTAL PRODUCTION (MILLION BUSHELS)
1950	50	37.6	2,782
1960	60	55.6	3,479
1970	70	81.4	4,802
1980	80	97.7	6,867
1990	90	115.6	7,802
2000	100	139.6	10,192

106. *Agriculture.* Refer to Table 3. Find a logarithmic regression model ($y = a + b \ln x$) for the total production. Estimate (to the nearest million) the production in 2010.

Social Sciences

107. *World population.* If the world population is now 5.8 billion people and if it continues to grow at an annual rate of 1.14% compounded continuously, how long (to the nearest year) will it take before there is only 1 square yard of land per person? (The Earth contains approximately 1.68×10^{14} square yards of land.)

108. *Archaeology: carbon-14 dating.* The radioactive carbon-14 (^{14}C) in an organism at the time of its death decays according to the equation

$$A = A_0 e^{-0.000124t}$$

where t is time in years and A_0 is the amount of ^{14}C present at time $t = 0$. (See Example 3 in Section 2-2.) Estimate the age of a skull uncovered in an archaeological site if 10% of the original amount of ^{14}C is still present. [*Hint:* Find t such that $A = 0.1A_0$.]

Important Terms and Symbols

2-1 *Polynomial and Rational Functions.* Polynomial function; degree; continuity; turning point; root; zero; leading coefficient; rational function; points of discontinuity; vertical and horizontal asymptotes

$$f(x) = a_n x^n + a_{n-1} x^{n-1} + \cdots + a_1 x + a_0, a_n \neq 0;$$

$$f(x) = \frac{n(x)}{d(x)}, d(x) \neq 0$$

2-2 *Exponential Functions.* Exponential function; base; basic graphs; horizontal asymptote; basic properties; irrational number e; exponential function with base e; exponential growth; exponential decay; compound interest; principal (present value); amount (future value); continuous compound interest

$$f(x) = b^x, b > 0, b \neq 1; \quad y = e^x; \quad N = N_0 e^{kt};$$

$$A = A_0 e^{-kt}; \quad A = P\left(1 + \frac{r}{m}\right)^{mt}; \quad A = Pe^{rt}$$

2-3 *Logarithmic Functions.* Inverse functions; one-to-one functions; logarithmic function; base; equivalent exponential form; properties; common logarithm; natural logarithm; calculator evaluation; solving logarithmic and exponential equations; doubling time

$$y = \log_b x \text{ is equivalent to } x = b^y;$$
$$\log_b x, b > 0, b \neq 1; \quad \log x; \quad \ln x$$

Review Exercise

Work through all the problems in this chapter review and check your answers in the back of the book. Answers to all review problems are there along with section numbers in italics to indicate where each type of problem is discussed. Where weaknesses show up, review appropriate sections in the text.

A

1. Write in logarithmic form using base e: $u = e^v$

2. Write in logarithmic form using base 10: $x = 10^y$

3. Write in exponential form using base e: $\ln M = N$

4. Write in exponential form using base 10: $\log u = v$

Simplify Problems 5 and 6.

5. $\dfrac{5^{x+4}}{5^{4-x}}$ 6. $\left(\dfrac{e^u}{e^{-u}}\right)^u$

Solve Problems 7–9 for x exactly without using a calculator.

7. $\log_3 x = 2$ 8. $\log_x 36 = 2$

9. $\log_2 16 = x$

Solve Problems 10–12 for x to three decimal places.

10. $10^x = 143.7$ 11. $e^x = 503,000$

12. $\log x = 3.105$ 13. $\ln x = -1.147$

For each polynomial function in Problems 14 and 15, find the following:

(A) The degree of the polynomial

(B) The maximum number of turning points of the graph

(C) The maximum number of x intercepts of the graph

(D) The minimum number of x intercepts of the graph

(E) The maximum number of y intercepts of the graph

(F) The minimum number of y intercepts of the graph

14. $p(x) = ax^3 + bx^2 + cx + d, a \neq 0$

15. $p(x) = ax^4 + bx^3 + cx^2 + dx + e, a \neq 0$

Each graph in Problems 16 and 17 is the graph of a polynomial function. Answer the following questions for each graph:

(A) How many turning points are on the graph?

(B) What is the minimum degree of a polynomial function that could have the graph?

(C) Is the leading coefficient of the polynomial negative or positive?

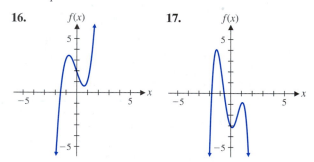

16.   17.

B *For each rational function in Problems 18 and 19:*

(A) Find the intercepts for the graph.

(B) Determine the domain.

(C) Find any vertical or horizontal asymptotes for the graph.

(D) Sketch any asymptotes as dashed lines. Then sketch a graph of f for $-10 \leq x \leq 10$ and $-10 \leq y \leq 10$.

 (E) Graph $y = f(x)$ in a standard viewing window using a graphing utility.

18. $f(x) = \dfrac{x + 4}{x - 2}$ 19. $f(x) = \dfrac{3x - 4}{2 + x}$

Solve Problems 20–27 for x exactly without using a calculator.

20. $\log(x + 5) = \log(2x - 3)$

21. $2 \ln(x - 1) = \ln(x^2 - 5)$

22. $9^{x-1} = 3^{1+x}$ **23.** $e^{2x} = e^{x^2-3}$

24. $2x^2 e^x = 3xe^x$ **25.** $\log_{1/3} 9 = x$

26. $\log_x 8 = -3$ **27.** $\log_9 x = \frac{3}{2}$

Solve Problems 28–37 for x to four decimal places.

28. $x = 3(e^{1.49})$ **29.** $x = 230(10^{-0.161})$

30. $\log x = -2.0144$ **31.** $\ln x = 0.3618$

32. $35 = 7(3^x)$ **33.** $0.01 = e^{-0.05x}$

34. $8{,}000 = 4{,}000(1.08^x)$ **35.** $5^{2x-3} = 7.08$

36. $x = \log_2 7$ **37.** $x = \log_{0.2} 5.321$

38. How does the graph of $f(x) = x^4 - 4x^2 + 1$ compare to the graph of $y = x^4$ as we "zoom out"?

39. Compare the graphs of $y = x^4$ and $y = x^4 - 4x^2 + 1$ in the following two viewing windows:

   (A) $-5 \le x \le 5, -5 \le y \le 5$

   (B) $-5 \le x \le 5, -500 \le y \le 500$

40. Let $p(x) = 2x^4 - 11x^3 - 15x^2 - 14x - 16$. Approximate the real zeros of $p(x)$ to two decimal places.

41. Let $f(x) = e^x - 1$ and $g(x) = \ln(x + 2)$. Find all points of intersection for the graphs of f and g. Round answers to two decimal places.

Simplify Problems 42 and 43.

42. $e^x(e^{-x} + 1) - (e^x + 1)(e^{-x} - 1)$

43. $(e^x - e^{-x})^2 - (e^x + e^{-x})(e^x - e^{-x})$

Applications

Business & Economics

The two formulas below will be of use in some of the problems that follow:

$$A = P\left(1 + \frac{r}{m}\right)^{mt} \quad \text{Compound interest}$$

$$A = Pe^{rt} \quad \text{Continuous compound interest}$$

54. *Money growth.* Provident Bank of Cincinnati, Ohio **www** recently offered a certificate of deposit that paid 6.59% compounded continuously. If a $5,000 CD earns this rate for 5 years, how much will it be worth?

55. *Money growth.* Capital One Bank of Glen Allen, **www** Virginia recently offered a certificate of deposit that paid 6.58% compounded daily. If a $5,000 CD earns this rate for 5 years, how much will it be worth?

56. *Money growth.* How long will it take for money invested at 6.59% compounded continuously to triple?

Graph Problems 44–46 over the indicated interval. Indicate increasing and decreasing intervals.

44. $y = 2^{x-1}; [-2, 4]$ **45.** $f(t) = 10e^{-0.08t}; t \ge 0$

46. $y = \ln(x + 1); (-1, 10]$

C

47. Noting that $\pi = 3.141\,592\,654\ldots$ and $\sqrt{2} = 1.414\,213\,562\ldots$, explain why the calculator results shown here are obvious. Discuss similar connections between the natural logarithmic function and the exponential function with base e.

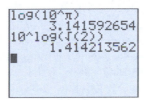

Solve Problems 48–51 exactly without using a calculator.

48. $\log x - \log 3 = \log 4 - \log(x + 4)$

49. $\ln(2x - 2) - \ln(x - 1) = \ln x$

50. $\ln(x + 3) - \ln x = 2 \ln 2$

51. $\log 3x^2 = 2 + \log 9x$

52. Write $\ln y = -5t + \ln c$ in an exponential form free of logarithms. Then solve for y in terms of the remaining variables.

53. Explain why 1 cannot be used as a logarithmic base.

57. *Money growth.* How long will it take for money invested at 6.58% compounded daily to double?

58. *Minimum average cost.* The financial department of a company that manufactures in-line skates has fixed costs of $300 per day and total costs of $4,300 per day at an output of 100 pairs of skates per day. Assume that the cost $C(x)$ is linearly related to output x.

   (A) Find an expression for the cost function $C(x)$ and the average cost function $\overline{C}(x) = C(x)/x$.

(B) Sketch a graph of the average cost function for $5 \leq x \leq 200$.

(C) Identify any asymptotes.

(D) What does the average cost approach as production increases?

59. *Minimum average cost.* The cost $C(x)$ in thousands of dollars for operating a hospital for a year is given by

$$C(x) = 20x^3 - 360x^2 + 2,300x - 1,000$$

where x is the number of cases per year (in thousands). The average cost function $\overline{C}$ is given by $\overline{C}(x) = C(x)/x$.

(A) Write an equation for the average cost function.

(B) Graph the average cost function for $1 \leq x \leq 12$.

(C) Use trace and zoom or a built-in routine to find the number of cases per year the hospital should handle to have the minimum average cost. What is the minimum average cost?

60. *Equilibrium point.* A company is planning to introduce a 10-piece set of nonstick cookware. A marketing company established price–demand and price–supply tables for selected prices (Tables 1 and 2), where x is the number of cookware sets people are willing to buy and the company is willing to sell each month at a price of p dollars per set.

TABLE 1	
PRICE–DEMAND	
x	$p = D(x)$ ($)
985	330
2,145	225
2,950	170
4,225	105
5,100	50

TABLE 2	
PRICE–SUPPLY	
x	$p = S(x)$ ($)
985	30
2,145	75
2,950	110
4,225	155
5,100	190

(A) Find a quadratic regression model for the data in Table 1. Estimate the demand at a price level of $180.

(B) Find a linear regression model for the data in Table 2. Estimate the supply at a price level of $180.

(C) Does a price level of $180 represent a stable condition, or is the price likely to increase or decrease? Explain.

(D) Use the models in parts (A) and (B) to find the equilibrium point. Write the equilibrium price to the nearest cent and the equilibrium quantity to the nearest unit.

61. *Telecommunications.* According to the Telecommunications Industry Association, wireless telephone subscriptions grew from about 4 million in 1990 to over 76 million in 1999 (Table 3). Let x represent years since 1990.

www

TABLE 3	
WIRELESS TELEPHONE SUBSCRIBERS	
YEAR	MILLION SUBSCRIBERS
1990	4
1991	6
1992	9
1993	13
1994	19
1995	28
1996	38
1997	49
1998	60
1999	76

(A) Find an exponential regression model ($y = ab^x$) for this data. Estimate (to the nearest million) the number of subscribers in 2000. In 2010.

(B) The actual number of subscribers in 2000 was approximately 97 million. How does this compare with the estimate in part (A)? What effect will this additional 2000 information have on the estimate for 2010?

Life Sciences

62. *Medicine.* One leukemic cell injected into a healthy mouse will divide into 2 cells in about $\frac{1}{2}$ day. At the end of the day these 2 cells will divide into 4. This doubling continues until 1 billion cells are formed; then the animal dies with leukemic cells in every part of the body.

(A) Write an equation that will give the number N of leukemic cells at the end of t days.

(B) When, to the nearest day, will the mouse die?

63. *Marine biology.* The intensity of light entering water is reduced according to the exponential equation

$$I = I_0 e^{-kd}$$

where I is the intensity d feet below the surface, I_0 is the intensity at the surface, and k is the coefficient of extinction. Measurements in the Sargasso Sea in the West Indies have indicated that half of the surface light reaches a depth of 73.6 feet. Find k (to five decimal places), and find the depth (to the nearest foot) at which 1% of the surface light remains.

64. *Agriculture.* The total U.S. corn consumption (in millions of bushels) is shown in Table 4 for selected years since 1975. Let x represent years since 1900.

www

TABLE 4

CORN CONSUMPTION

YEAR	x	TOTAL CONSUMPTION (MILLION BUSHELS)
1975	75	522
1980	80	659
1985	85	1,152
1990	90	1,373
1995	95	1,690

(A) Find a logarithmic regression model $(y = a + b \ln x)$ for the data. Estimate (to the nearest million bushels) the total consumption in 1996 and in 2010.

(B) The actual consumption in 1996 was 1,583 million bushels. How does this compare with the estimated consumption in part (A)? What effect will this additional 1996 information have on the estimate for 2010? Explain.

Social Sciences

65. *Population growth.* Many countries have a population growth rate of 3% (or more) per year. At this rate, how many years (to the nearest tenth of a year) will it take a population to double? Use the annual compounding growth model $P = P_0(1 + r)^t$.

66. *Population growth.* Repeat Problem 64 using the continuous compounding growth model $P = P_0 e^{rt}$.

67. *Medicare.* The annual expenditures for Medicare (in billions of dollars) by the U.S. government for selected years since 1980 are shown in Table 5. Let x represent years since 1980.

www

TABLE 5

MEDICARE EXPENDITURES

YEAR	BILLION $
1980	37
1985	72
1990	111
1995	181

(A) Find an exponential regression model $(y = ab^x)$ for the data. Estimate (to the nearest billion) the total expenditures in 2010.

(B) When will the total expenditures reach 500 billion dollars?

Group Activity 1   *Comparing the Growth of Exponential and Polynomial Functions, and Logarithmic and Root Functions*

(A) An exponential function such as $f(x) = 2^x$ increases extremely rapidly for large values of x, more rapidly than any polynomial function. Show that the graphs of $f(x) = 2^x$ and $g(x) = x^2$ intersect three times. The intersection points divide the x axis into four regions. Describe which function is greater than the other relative to each region.

(B) A logarithmic function such as $r(x) = \ln x$ increases extremely slowly for large values of x, more slowly than a function like $s(x) = \sqrt[3]{x}$. Sketch graphs of both functions in the same coordinate system for $x > 0$, and determine how many times the two graphs intersect. Describe which function is greater than the other relative to the regions determined by the intersection points.

Group Activity 2   *Comparing Regression Models*

We have used polynomial, exponential, and logarithmic regression models to fit curves to data sets. And there are other equations that can be used for curve fitting. (The TI-83 Plus graphing calculator has 10 equations on its STAT-CALC menu.) How can we determine which equation provides the best fit for a given set of data? There are two principal ways to select models. The first is to use information about the type of data to help make a choice. For example, we expect the weight of a fish to be related to the cube of its length. And we expect most populations to grow exponentially, at least over the short term. The second method for choosing between equations involves developing a measure of how close an equation fits a given data set. This is best introduced through an example. Consider the data set in Figure 1, where L1 represents the x coordinates and L2 represents the y coordinates. The graph of this data set is shown in Figure 2. Suppose that we arbitrarily choose the equation $y_1 = 0.6x + 2$ to model the data (Fig. 3).

FIGURE 1   FIGURE 2   FIGURE 3 $y_1 = 0.6x + 2$

To measure how well the graph of y_1 fits the data, we examine the difference between the y coordinates in the data set and the corresponding y coordinates on the graph of y_1 (L3 in Figs. 4 and 5). Each of these differences is called a **residual.** The most commonly accepted measure of the fit provided by a given model is the **sum of the squares of the residuals (SSR).** Computing this quantity is a simple matter on a graphing calculator (Fig. 6) or a spreadsheet (Fig. 7).

FIGURE 4   FIGURE 5 +L2; ■ L3   FIGURE 6 Two ways to calculate SSR

	A	B	C	D	E
1		Data Set			
2	x	y	y1=0.6x + 2	Residual	Residual^2
3	2	2	3.2	-1.2	1.44
4	4	3	4.4	-1.4	1.96
5	6	8	5.6	2.4	5.76
6	8	5	6.8	-1.8	3.24
7				SSR	12.4

FIGURE 7

(A) Find the linear regression model for the data in Figure 1, compute the SSR for this equation, and compare it with the one we computed for y_1.

TABLE 1	
ANNUAL ADVERTISING EXPENDITURES, 1950–1995	
x(YEARS)	y(BILLION $)
0	5.7
5	9.2
10	12.0
15	15.3
20	19.6
25	27.9
30	53.6
35	94.8
40	128.6
45	160.9

It turns out that among all possible linear polynomials, **the linear regression model minimizes the sum of the squares of the residuals.** For this reason, the linear regression model is often called the **least squares line.** A similar statement can be made for polynomials of any fixed degree. That is, the quadratic regression model minimizes the SSR over all quadratic polynomials, the cubic regression model minimizes the SSR over all cubic polynomials, and so on. The same statement cannot be made for exponential or logarithmic regression models. Nevertheless, the SSR can still be used to compare exponential, logarithmic, and polynomial models.

(B) Find the exponential and logarithmic regression models for the data in Figure 1, compute their SSRs, and compare with the linear model.

(C) National annual advertising expenditures for selected years since 1950 are shown in Table 1, where x is years since 1950 and y is total expenditures in billions of dollars. Which regression model would fit the data best: a quadratic model, a cubic model, or an exponential model? Use the SSRs to support your choice.

Calculus

The Derivative

INTRODUCTION

How do algebra and calculus differ? The two words *static* and *dynamic* probably come as close as any in expressing the difference between the two disciplines. In algebra, we solve equations for a particular value of a variable—a static notion. In calculus, we are interested in how a change in one variable affects another variable—a dynamic notion.

Parts (A)–(C) of the figure at the top of the next page illustrate three basic problems in calculus. It may surprise you to learn that all three problems—as different as they appear—are related mathematically. The solutions to these problems and the discovery of their relationship required the creation of a new kind of mathematics. Isaac Newton (1642–1727) of England and Gottfried Wilhelm von Leibniz (1646–1716) of Germany, simultaneously and independently, developed this new mathematics, called **the calculus**—it was an idea whose time had come.

In addition to solving the problems described in the figure, calculus will enable us to solve many other important problems. Until fairly recently, calculus was used primarily in the physical sciences, but now people in many other disciplines are finding it a useful tool.

(A) Find the equation of the tangent line at (x_1, y_1) given $y = f(x)$

(B) Find the instantaneous velocity of a falling object

(C) Find the indicated area bounded by $y = g(x)$, $x = a$, $x = b$, and the x axis

---

Section 3-1 | # Introduction to Limits

❑ FUNCTIONS AND GRAPHS: BRIEF REVIEW
❑ LIMITS
❑ LIMIT EVALUATION
❑ LIMITS OF DIFFERENCE QUOTIENTS

Basic to the study of calculus is the concept of *limit*. This concept helps us to describe, in a precise way, the behavior of $f(x)$ when x is close to but not equal to a particular value c. In this section we develop a combined numerical, graphical, and algebraic approach for evaluating limits. Our discussion concentrates on concept development and understanding rather than on formal mathematical detail.

❑ FUNCTIONS AND GRAPHS: BRIEF REVIEW

The graph of the function $y = f(x) = x + 2$ is the graph of the set of all ordered pairs $(x, f(x))$. For example, if $x = 2$, then $f(2) = 4$ and $(2, f(2)) = (2, 4)$ is a point on the graph of f. Figure 1 shows $(-1, f(-1))$, $(1, f(1))$, and $(2, f(2))$ plotted on the graph of f. Notice that the domain values -1, 1, and 2 are associated with the x axis, and the range values $f(-1) = 1$, $f(1) = 3$, and $f(2) = 4$ are associated with the y axis.

Given x, it is sometimes useful to be able to read $f(x)$ directly from the graph of f. Example 1 reviews this process.

FIGURE 1

Example 1 ⇔ **Finding Values of a Function from Its Graph** Complete the table below using the given graph of the function g.

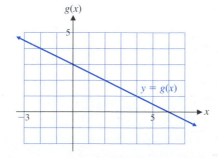

x	$g(x)$
-2	
1	
3	
4	

SOLUTION   To determine $g(x)$, proceed vertically from the x value on the x axis to the graph of g, then horizontally to the corresponding y value, $g(x)$, on the y axis (as indicated by the dashed lines):

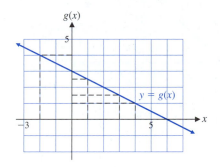

x	$g(x)$
-2	4.0
1	2.5
3	1.5
4	1.0

Matched Problem 1   ☞   Complete the table below using the given graph of the function h.

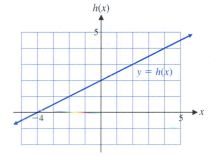

x	$h(x)$
-2	
-1	
0	
1	
2	
3	
4	

❑ LIMITS

We introduce the important notion of *limit* through two examples, after which the limit concept is defined.

Example 2   ☞   **Analyzing a Limit**   Let $f(x) = x + 2$. Discuss the behavior of $f(x)$ numerically, graphically, and algebraically when x is chosen closer and closer to 2, but not equal to 2.

SOLUTION   To investigate the behavior of $f(x)$ numerically for x near 2, we construct a table of values (Table 1).

TABLE 1

x	1.5	1.8	1.9	1.99	1.999 → 2 ← 2.001	2.01	2.1	2.2	2.5
$f(x)$	3.5	3.8	3.9	3.99	3.999 → 4 ← 4.001	4.01	4.1	4.2	4.5

The numerical values in the table indicate that if x is close to 2, the associated function value $f(x)$ is close to 4.

(A)   (B)

www FIGURE 2

To investigate the behavior of $f(x)$ graphically, we draw a graph of f for x near 2 (Fig. 2A). The thin vertical lines in Figure 2A represent values of x that are approaching 2 from either side. The corresponding horizontal lines identify the value of $f(x)$ associated with each value of x. The graph indicates that the values of $f(x)$ approach 4 as x approaches 2 from either side of 2. Figure 2B shows the graph of f and a table of values of x and $f(x)$ on a Texas Instruments TI-83 set in the graph-table mode. If you trace along the graph, the coordinates of the cursor are highlighted in the table. This clearly illustrates the important relationship between the graph of a function and the numeric values of the function.

Finally, to investigate the behavior algebraically, we simply note that if x approaches 2 then $x + 2$ must approach $2 + 2 = 4$. Collecting all this evidence, we conclude that

$$\lim_{x \to 2}(x + 2) = 4 \qquad \text{or} \qquad x + 2 \to 4 \quad \text{as} \quad x \to 2$$

Also note that $f(2) = 4$. Thus, the value of the function at 2 and the limit of the function at 2 are the same. That is,

$$\lim_{x \to 2}(x + 2) = f(2)$$

Graphically, this means that there is no break or hole in the graph of f at $x = 2$.

Matched Problem 2 ⇨ Let $f(x) = x + 1$.

(A) Complete the following table:

x	0.9	0.99	0.999	$\to$ 1 $\leftarrow$	1.001	1.01	1.1
$f(x)$	?	?	?	$\to$? $\leftarrow$?	?	?

(B) Graph $f(x) = x + 1$.

(C) Use the table, the graph, and the algebraic expression $x + 1$ to find

$$\lim_{x \to 1}(x + 1)$$

(D) Use a similar approach to find

$$\lim_{x \to 0}(x + 1) \quad \text{and} \quad \lim_{x \to 3}(x + 1)$$

The results found in Example 2 and Matched Problem 2 were fairly obvious. The next example is a little less obvious.

Example 3 ⇨ **Analyzing a Limit**   Let

$$g(x) = \frac{x^2 - 4}{x - 2} \qquad x \neq 2$$

Even though the function is not defined when $x = 2$ (both the numerator and denominator are 0), we can still ask how $g(x)$ behaves when x is near 2 but not equal to 2. Can you guess what happens to $g(x)$ as x approaches 2? The numerator tending to 0 is a force pushing the fraction toward 0. The denominator tending to 0 is another force pushing the fraction toward larger values. How do these two forces balance out?

SOLUTION   Proceeding as before, we construct a table of values of $g(x)$ (Table 2):

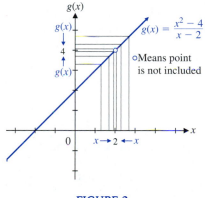

FIGURE 3

TABLE 2

x	1.5	1.8	1.9	1.99	1.999 $\to$ 2 $\leftarrow$ 2.001	2.01	2.1	2.2	2.5
$g(x)$	3.5	3.8	3.9	3.99	3.999 $\to$ 4 $\leftarrow$ 4.001	4.01	4.1	4.2	4.5

Notice that these values agree with the values of $f(x) = x + 2$ in Table 1. This suggests that there is an algebraic relationship between f and g:

$$g(x) = \frac{x^2 - 4}{x - 2} = \frac{(x - 2)(x + 2)}{x - 2} = x + 2 \qquad x \neq 2$$

Thus, we see that $g(x) = f(x)$ for all x, except $x = 2$. This implies that the graph of g is the same as the graph of f (Fig. 2), except that the graph of g has a hole at the point with coordinates $(2, 4)$, as shown in Figure 3.

Since the behavior of $(x^2 - 4)/(x - 2)$ for x near 2 but not equal to 2 is the same as the behavior of $x + 2$ for x near 2 but not equal to 2, we have

$$\lim_{x \to 2} \frac{x^2 - 4}{x - 2} = \lim_{x \to 2}(x + 2) = 4$$

We see that the limit of the function g at 2 exists even though the function is not defined there (the graph has a hole at $x = 2$).

Matched Problem 3 ⇨   Proceed as in Example 3 to find $\lim_{x \to 1} \dfrac{x^2 - 1}{x - 1}$.

REMARK

If you investigate the graph of $g(x) = (x^2 - 4)/(x - 2)$ (from Example 3) with a graphing utility, you may not see a hole in the graph at $x = 2$ (see Fig. 4). This

FIGURE 4 $g(x) = \dfrac{x^2 - 4}{x - 2}$ for $-3 \leqslant x \leqslant 5$ **FIGURE 5** $g(x) = \dfrac{x^2 - 4}{x - 2}$ for $-1 \leqslant x \leqslant 5$

is due to the difference between the coordinates of a point in the plane (an infinite set) and the coordinates of a pixel on the screen (a finite set). On many graphing utilities, a simple way to make this hole visible is to choose the window parameters so that $x = 2$ is the midpoint of the graphing interval (see Fig. 5). Notice that the graphing utility does not display a y coordinate corresponding to $x = 2$, since $g(2)$ is not defined. See Problems 75 and 76 in Exercise 3-1 for another way to choose window parameters so that holes in graphs will be visible.

We now present an informal definition of the important concept of limit. A precise definition is not needed for our discussion, but one is given in the footnote.*

Limit

We write

$$\lim_{x \to c} f(x) = L \qquad \text{or} \qquad f(x) \to L \quad \text{as} \quad x \to c$$

if the functional value $f(x)$ is close to the single real number L whenever x is close to, but not equal to, c (on either side of c).

[*Note*: The existence of a limit at c has nothing to do with the value of the function at c. In fact, c may not even be in the domain of f (see Example 3). However, the function must be defined on both sides of c.]

The next example involves the **absolute value function:**

$$f(x) = |x| = \begin{cases} -x & \text{if } x < 0 \\ x & \text{if } x \geqslant 0 \end{cases} \qquad \begin{aligned} f(-2) &= |-2| = -(-2) = 2 \\ f(3) &= |3| = 3 \end{aligned}$$

The graph of f is shown in Figure 6.

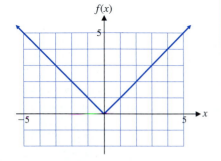

FIGURE 6 $f(x) = |x|$

---

*To make the informal definition of limit precise, the use of the word *close* must be made more precise. This is done as follows: We write $\lim_{x \to c} f(x) = L$ if for each $e > 0$, there exists a $d > 0$ such that $|f(x) - L| < e$ whenever $0 < |x - c| < d$. This definition is used to establish particular limits and to prove many useful properties of limits that will be helpful to us in finding particular limits. [Even though intuitive notions of limit existed for a long time, it was not until the nineteenth century that a precise definition was given by the German mathematician Karl Weierstrass (1815–1897).]

Example 4 ➭ **Analyzing a Limit**   Let $h(x) = |x|/x$. Explore the behavior of $h(x)$ for x near 0 but not equal to 0 using a table and a graph. Find $\lim_{x \to 0} h(x)$, if it exists.

SOLUTION   The function h is defined for all real numbers except 0. For example,

$$h(-2) = \frac{|-2|}{-2} = \frac{2}{-2} = -1$$

$$h(0) = \frac{|0|}{0} = \frac{0}{0} \qquad \text{Not defined}$$

$$h(2) = \frac{|2|}{2} = \frac{2}{2} = 1$$

In general, $h(x)$ is -1 for all negative x and 1 for all positive x. Table 3 and Figure 7 illustrate the behavior of $h(x)$ for x near 0.

TABLE 3

x	-2	-1	-0.1	-0.01	-0.001	$\to$	0	$\leftarrow$	0.001	0.01	0.1	1	2
$h(x)$	-1	-1	-1	-1	-1	$\to -1$	$\neq 1 \leftarrow$		1	1	1	1	1

FIGURE 7

When x is near 0 (on either side of 0), is $h(x)$ near one specific number? The answer is "No," because $h(x)$ is -1 for $x < 0$ and 1 for $x > 0$. Consequently, we say that

$$\lim_{x \to 0} \frac{|x|}{x} \quad \text{does not exist}$$

Thus, neither $h(x)$ nor the limit of $h(x)$ exists at $x = 0$. However, the limit from the left and the limit from the right both exist at 0, but they are not equal. (We discuss this further below.) ∎

Matched Problem 4 ➭ Graph

$$h(x) = \frac{x - 2}{|x - 2|}$$

and find $\lim_{x \to 2} h(x)$, if it exists. ∎

In Example 2 we found it helpful to examine the values of the function $f(x)$ as x approached 2 from the left and then from the right. In Example 4 we saw that the values of the function $h(x)$ approached two different numbers, depending on the direction of approach, and it was natural to refer to these values as "the limit from the left" and "the limit from the right." These experiences suggest that the notion of **one-sided limits** will be very useful when discussing basic limit concepts.

We write

$$\lim_{x \to c^-} f(x) = K$$

$x \to c^-$ is read "x approaches c from the left" and means $x \to c$ and $x < c$.

and call K the **limit from the left** (or **left-hand limit**) if $f(x)$ is close to K whenever x is close to c, but to the left of c, on the real number line. We write

$$\lim_{x \to c^+} f(x) = L$$

$x \to c^+$ is read "x approaches c from the right" and means $x \to c$ and $x > c$.

and call L the **limit from the right** (or **right-hand limit**) if $f(x)$ is close to L whenever x is close to c, but to the right of c, on the real number line.

We now make the following important observation:

On the Existence of a Limit

For a limit to exist, the limit from the left and the limit from the right must exist and be equal.

In Example 4,

$$\lim_{x \to 0^-} \frac{|x|}{x} = -1 \qquad \text{and} \qquad \lim_{x \to 0^+} \frac{|x|}{x} = 1$$

Since the left- and right-hand limits are not the same,

$$\lim_{x \to 0} \frac{|x|}{x} \text{ does not exist}$$

Example 5 ⇔ **Analyzing Limits Graphically** Given the graph of the function f shown in Figure 8, we discuss the behavior of $f(x)$ for x near (A) -1, (B) 1, and (C) 2.

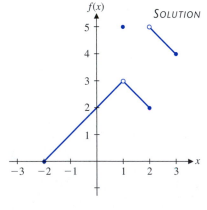

FIGURE 8

SOLUTION (A) Since we have only a graph to work with, we use vertical and horizontal lines to relate the values of x and the corresponding values of $f(x)$. For any x near -1 on either side of -1, we see that the corresponding value of $f(x)$, determined by a horizontal line, is close to 1.

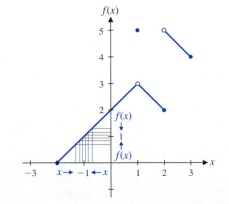

$$\lim_{x \to -1^-} f(x) = 1$$

$$\lim_{x \to -1^+} f(x) = 1$$

$$\lim_{x \to -1} f(x) = 1$$

$$f(-1) = 1$$

(B) Again, for any x near but not equal to 1, the vertical and horizontal lines indicate that the corresponding value of $f(x)$ is close to 3. However, the open dot at $(1, 3)$, indicates that $f(1) \neq 3$.

$$\lim_{x \to 1^-} f(x) = 3$$

$$\lim_{x \to 1^+} f(x) = 3$$

$$\lim_{x \to 1} f(x) = 3$$

$$f(1) = 5$$

(C) The abrupt break in the graph at $x = 2$ indicates that the behavior near $x = 2$ is more complicated than in the two preceding cases. If x is close to 2 on the left side of 2, the corresponding horizontal line intersects the y axis at a point close to 2. If x is close to 2 on the right side of 2, the corresponding horizontal line intersects the y axis at a point close to 5. Thus, this is a case where the one-sided limits are different.

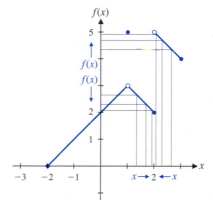

$$\lim_{x \to 2^-} f(x) = 2$$

$$\lim_{x \to 2^+} f(x) = 5$$

$$\lim_{x \to 2} f(x) \text{ does not exist}$$

$$f(2) = 2$$

Matched Problem 5   Given the graph of the function f shown in Figure 9, discuss the following, as we did in Example 5:

(A) Behavior of $f(x)$ for x near 0

(B) Behavior of $f(x)$ for x near 1

(C) Behavior of $f(x)$ for x near 3

FIGURE 9

❑ LIMIT EVALUATION

Tables and graphs are very useful tools for investigating limits, especially if something unusual happens at the point in question. However, many of the limits encountered in calculus are routine and can be evaluated quickly, using a little algebraic simplification, some intuition, and basic properties of limits. The following list of properties of limits forms the basis for this approach.

THEOREM 1 Properties of Limits

Let f and g be two functions, and assume that

$$\lim_{x\to c} f(x) = L \qquad \lim_{x\to c} g(x) = M$$

where L and M are real numbers (both limits exist). Then

1. $\lim_{x\to c} [f(x) + g(x)] = \lim_{x\to c} f(x) + \lim_{x\to c} g(x) = L + M$

2. $\lim_{x\to c} [f(x) - g(x)] = \lim_{x\to c} f(x) - \lim_{x\to c} g(x) = L - M$

3. $\lim_{x\to c} kf(x) = k \lim_{x\to c} f(x) = kL \qquad$ for any constant k

4. $\lim_{x\to c} [f(x) \cdot g(x)] = \left[\lim_{x\to c} f(x)\right]\left[\lim_{x\to c} g(x)\right] = LM$

5. $\lim_{x\to c} \dfrac{f(x)}{g(x)} = \dfrac{\lim_{x\to c} f(x)}{\lim_{x\to c} g(x)} = \dfrac{L}{M} \qquad$ if $M \neq 0$

6. $\lim_{x\to c} \sqrt[n]{f(x)} = \sqrt[n]{\lim_{x\to c} f(x)} = \sqrt[n]{L} \qquad L > 0$ for n even

Explore–Discuss 1

The properties listed in Theorem 1 can be paraphrased in brief verbal statements. For example, property 1 simply states that *the limit of a sum is equal to the sum of the limits*. Write brief verbal statements for the remaining properties in Theorem 1.

Example 6 **Using Limit Properties** Find $\lim_{x\to 3} (x^2 - 4x)$.

SOLUTION First, note the following obvious limit:

$$\lim_{x\to 3} x = 3$$

(All this says is that if x approaches 3, then x approaches 3.) Now we make use of this limit and the limit properties:

$$\lim_{x\to 3}(x^2 - 4x) = \lim_{x\to 3} x^2 - \lim_{x\to 3} 4x \qquad \text{Property 2}$$

$$= \left(\lim_{x\to 3} x\right) \cdot \left(\lim_{x\to 3} x\right) - 4\lim_{x\to 3} x \qquad \text{Properties 3 and 4}$$

$$= 3 \cdot 3 - 4 \cdot 3 = -3$$

With a little practice you will soon be able to omit the steps in the dashed boxes and simply write

$$\lim_{x\to 3} (x^2 - 4x) = 3 \cdot 3 - 4 \cdot 3 = -3$$

Matched Problem 6 Find $\lim_{x\to -2} (x^2 + 5x)$.

What happens if we try to evaluate a limit like the one in Example 6, but with x approaching an unspecified number, such as c? Proceeding as we did in Example 6, we have

$$\lim_{x \to c} (x^2 - 4x) = c \cdot c - 4 \cdot c = c^2 - 4c$$

If we let $f(x) = x^2 - 4x$, we have

$$\lim_{x \to c} f(x) = \lim_{x \to c} (x^2 - 4x) = c^2 - 4c = f(c)$$

That is, this limit can be evaluated simply by evaluating the function f at c. It would certainly simplify the process of evaluating limits if we could identify the functions for which

$$\lim_{x \to c} f(x) = f(c) \tag{1}$$

since we could use this fact to evaluate the limit. It turns out that there are many functions that satisfy equation (1). We postpone a detailed discussion of these functions until the next section. For now, we note that if

$$f(x) = a_n x^n + a_{n-1} x^{n-1} + \cdots + a_0$$

is a polynomial function, the properties in Theorem 1 imply that

$$\lim_{x \to c} f(x) = \lim_{x \to c} (a_n x^n + a_{n-1} x^{n-1} + \cdots + a_0)$$

$$= a_n c^n + a_{n-1} c^{n-1} + \cdots + a_0 = f(c)$$

And if

$$r(x) = \frac{n(x)}{d(x)}$$

is a rational function, where $n(x)$ and $d(x)$ are polynomials, $d(c) \neq 0$, then using property 5 and the fact that polynomials $n(x)$ and $d(x)$ satisfy equation (1), we have

$$\lim_{x \to c} r(x) = \lim_{x \to c} \frac{n(x)}{d(x)} = \frac{\lim_{x \to c} n(x)}{\lim_{x \to c} d(x)} = \frac{n(c)}{d(c)} = r(c)$$

These results are summarized in Theorem 2.

THEOREM 2   **Limits of Polynomial and Rational Functions**

1. $\lim_{x \to c} f(x) = f(c)$ f any polynomial function

2. $\lim_{x \to c} r(x) = r(c)$ r any rational function with a nonzero denominator at $x = c$

Example 7 ⇔   **Evaluating Limits**   Find each limit:

(A) $\lim_{x \to 2} (x^3 - 5x - 1)$ (B) $\lim_{x \to -1} \sqrt{2x^2 + 3}$ (C) $\lim_{x \to 4} \dfrac{2x}{3x + 1}$

SOLUTION   (A) $\lim_{x \to 2} (x^3 - 5x - 1) = 2^3 - 5 \cdot 2 - 1 = -3$ Theorem 2

(B) $\lim_{x \to -1} \sqrt{2x^2 + 3} = \sqrt{\lim_{x \to -1} (2x^2 + 3)}$ Property 6

$= \sqrt{2(-1)^2 + 3}$ Theorem 2

$= \sqrt{5}$

$$\text{(C)} \lim_{x \to 4} \frac{2x}{3x + 1} = \frac{2 \cdot 4}{3 \cdot 4 + 1} \qquad \textit{Theorem 2}$$

$$= \frac{8}{13}$$

■

Matched Problem 7 ⇔   Find each limit:

(A) $\lim_{x \to -1} (x^4 - 2x + 3)$ (B) $\lim_{x \to 2} \sqrt{3x^2 - 6}$ (C) $\lim_{x \to -2} \frac{x^2}{x^2 + 1}$

■

It is important to note that there are restrictions on some of the limit properties. In particular, if

$$\lim_{x \to c} f(x) = 0 \qquad \text{and} \qquad \lim_{x \to c} g(x) = 0$$

then finding

$$\lim_{x \to c} \frac{f(x)}{g(x)} \tag{2}$$

may present some difficulties, since limit property 5 (the limit of a quotient) does not apply when $\lim_{x \to c} g(x) = 0$. We often have to use algebraic manipulation or other devices to determine the outcome. Recall from Examples 3 and 4 that

$$\lim_{x \to 2} \frac{x^2 - 4}{x - 2} = \lim_{x \to 2} \frac{(x - 2)(x + 2)}{x - 2} = \lim_{x \to 2} (x + 2) = 4$$

and

$$\lim_{x \to 0} \frac{|x|}{x} \text{ does not exist}$$

From these two examples, it is clear that knowing only that $\lim_{x \to c} f(x) = 0$ and $\lim_{x \to c} g(x) = 0$ is not enough to determine limit (2). Depending on the choice of functions f and g, the limit (2) may or may not exist. Consequently, if we are given limit (2) and $\lim_{x \to c} f(x) = 0$ and $\lim_{x \to c} g(x) = 0$, then limit (2) is said to be **indeterminate,** or, more specifically, a **0/0 indeterminate form.**

CAUTION

The expression 0/0 does not represent a real number and should never be used as the value of a limit. If a limit is a 0/0 indeterminate form, further investigation is always required to determine whether the limit exists and to find its value, if it does exist.

Explore–Discuss 2

Use algebraic, numerical, and/or graphical techniques to analyze each of the following indeterminate forms:

(A) $\lim_{x \to 1} \frac{x - 1}{x^2 - 1}$ (B) $\lim_{x \to 1} \frac{(x - 1)^2}{x^2 - 1}$ (C) $\lim_{x \to 1} \frac{x^2 - 1}{(x - 1)^2}$

❑   LIMITS OF DIFFERENCE QUOTIENTS

Let the function f be defined in a open interval containing the number a. One of the most important limits in calculus is the limit of the **difference quotient:**

$$\lim_{h \to 0} \frac{f(a + h) - f(a)}{h} \tag{3}$$

If

$$\lim_{h \to 0} [f(a + h) - f(a)] = 0$$

as it often does, then limit (3) is an indeterminate form. The following examples illustrate some useful techniques for evaluating limits of difference quotients.

Example 8 ➭ **Limit of a Difference Quotient**   Find the following limit for $f(x) = 4x - 5$.

$$\lim_{h \to 0} \frac{f(3 + h) - f(3)}{h}$$

SOLUTION

$$\lim_{h \to 0} \frac{f(3 + h) - f(3)}{h} = \lim_{h \to 0} \frac{[4(3 + h) - 5] - [4(3) - 5]}{h}$$

$$= \lim_{h \to 0} \frac{12 + 4h - 5 - 12 + 5}{h}$$

$$= \lim_{h \to 0} \frac{4h}{h} = \lim_{h \to 0} 4 = 4$$

Since this is a 0/0 indeterminate form and property 5 in Theorem 1 does not apply, we proceed with algebraic simplification.

■

Matched Problem 8 ➭ Find the following limit for $f(x) = 7 - 2x$: $\displaystyle\lim_{h \to 0} \frac{f(4 + h) - f(4)}{h}$.

■

Explore–Discuss 3

The following is an incorrect solution to Example 8 with the invalid statements indicated by $\neq$. Explain why each $\neq$ is used.

$$\lim_{h \to 0} \frac{f(3 + h) - f(3)}{h} \neq \lim_{h \to 0} \frac{4(3 + h) - 5 - 4(3) - 5}{h}$$

$$= \lim_{h \to 0} \frac{-10 + 4h}{h}$$

$$\neq \lim_{h \to 0} \frac{-10 + 4}{1} = -6$$

Example 9 ➭ **Limit of a Difference Quotient**   Find the following limit for $f(x) = |x + 5|$.

$$\lim_{h \to 0} \frac{f(-5 + h) - f(-5)}{h}$$

SOLUTION

$$\lim_{h \to 0} \frac{f(-5 + h) - f(-5)}{h} = \lim_{h \to 0} \frac{|(-5 + h) + 5| - |-5 + 5|}{h}$$

$$= \lim_{h \to 0} \frac{|h|}{h} \text{ does not exist}$$

Since this is a 0/0 indeterminate form and property 5 in Theorem 1 does not apply, we proceed with algebraic simplification. See Example 4.

■

Matched Problem 9 ➷ Find the following limit for $f(x) = |x - 1|$: $\lim\limits_{h \to 0} \dfrac{f(1 + h) - f(1)}{h}$.

Example 10 ➷ **Limit of a Difference Quotient** Find the following limit for $f(x) = \sqrt{x}$.

$$\lim_{h \to 0} \frac{f(2 + h) - f(2)}{h}$$

SOLUTION

$$\lim_{h \to 0} \frac{f(2 + h) - f(2)}{h} = \lim_{h \to 0} \frac{\sqrt{2 + h} - \sqrt{2}}{h}$$

This is $0/0$ indeterminate form, so property 5 in Theorem 1 does not apply. Rationalizing the numerator will be of help.

$$= \lim_{h \to 0} \frac{\sqrt{2 + h} - \sqrt{2}}{h} \cdot \frac{\sqrt{2 + h} + \sqrt{2}}{\sqrt{2 + h} + \sqrt{2}} \qquad (A - B)(A + B) = A^2 - B^2$$

$$= \lim_{h \to 0} \frac{2 + h - 2}{h(\sqrt{2 + h} + \sqrt{2})}$$

$$= \lim_{h \to 0} \frac{1}{\sqrt{2 + h} + \sqrt{2}}$$

$$= \frac{1}{\sqrt{2} + \sqrt{2}} = \frac{1}{2\sqrt{2}}$$

Matched Problem 10 ➷ Find the following limit for $f(x) = \sqrt{x}$: $\lim\limits_{h \to 0} \dfrac{f(3 + h) - f(3)}{h}$.

Answers to Matched Problems

1.

x	-2	-1	0	1	2	3	4
$h(x)$	1.0	1.5	2.0	2.5	3.0	3.5	4.0

2. (A)

x	0.9	0.99	0.999 → 1 ← 1.001	1.01	1.1
$f(x)$	1.9	1.99	1.999 → 2 ← 2.001	2.01	2.1

(B)

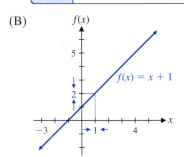

$f(x) = x + 1$

(C) $\lim\limits_{x \to 1} (x + 1) = 2$

(D) $\lim\limits_{x \to 0} (x + 1) = 1$;
$\lim\limits_{x \to 3} (x + 1) = 4$

3. 2

4.

$h(x) = \dfrac{x - 2}{|x - 2|}$

$\lim\limits_{x \to 2} \dfrac{x - 2}{|x - 2|}$ does not exist

5. (A) $\lim_{x \to 0^-} f(x) = 0$

  $\lim_{x \to 0^+} f(x) = 0$

  $\lim_{x \to 0} f(x) = 0$

    $f(0) = 0$

(B) $\lim_{x \to 1^-} f(x) = 1$

  $\lim_{x \to 1^+} f(x) = 2$

  $\lim_{x \to 1} f(x)$ does not exist

    $f(1)$ not defined

(C) $\lim_{x \to 3^-} f(x) = 3$

  $\lim_{x \to 3^+} f(x) = 3$

  $\lim_{x \to 3} f(x) = 3$

    $f(3) = 2$

6. -6 **7.** (A) 6 (B) $\sqrt{6}$ (C) $\frac{4}{5}$ **8.** -2 **9.** Does not exist
10. $1/(2\sqrt{3})$

Exercise 3-1

A *In Problems 1–10, find each limit, if it exists.*

1. $\lim_{x \to 3} 4x$

2. $\lim_{x \to -2} 3x$

3. $\lim_{x \to -4} (x + 5)$

4. $\lim_{x \to 5} (x - 3)$

5. $\lim_{x \to 2} x(x - 4)$

6. $\lim_{x \to -1} x(x + 3)$

7. $\lim_{x \to -3} \dfrac{x}{x + 5}$

8. $\lim_{x \to 4} \dfrac{x - 2}{x}$

9. $\lim_{x \to 1} \sqrt{5x + 4}$

10. $\lim_{x \to 0} \sqrt{16 - 7x}$

Given $\lim_{x \to 3} f(x) = 5$ *and* $\lim_{x \to 3} g(x) = 9,$ *find the indicated limits in Problems 11–20.*

11. $\lim_{x \to 3} [f(x) - g(x)]$

12. $\lim_{x \to 3} [f(x) + g(x)]$

13. $\lim_{x \to 3} 4g(x)$

14. $\lim_{x \to 3} (-2) f(x)$

15. $\lim_{x \to 3} \dfrac{f(x)}{g(x)}$

16. $\lim_{x \to 3} [f(x) \cdot g(x)]$

17. $\lim_{x \to 3} \sqrt{f(x)}$

18. $\lim_{x \to 3} \sqrt{g(x)}$

19. $\lim_{x \to 3} \dfrac{f(x) + g(x)}{2f(x)}$

20. $\lim_{x \to 3} \dfrac{g(x) - f(x)}{3g(x)}$

In Problems 21–24, sketch a possible graph of a function that satisfies the given conditions.

21. $f(0) = 1$; $\lim_{x \to 0^-} f(x) = 3$; $\lim_{x \to 0^+} f(x) = 1$

22. $f(1) = -2$; $\lim_{x \to 1^-} f(x) = 2$; $\lim_{x \to 1^+} f(x) = -2$

23. $f(-2) = 2$; $\lim_{x \to -2^-} f(x) = 1$; $\lim_{x \to -2^+} f(x) = 1$

24. $f(0) = -1$; $\lim_{x \to 0^-} f(x) = 2$; $\lim_{x \to 0^+} f(x) = 2$

In Problems 25–28, use the graph of the function f shown below to estimate the indicated limits and function values.

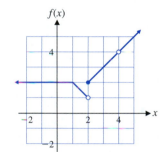

$f(x)$

25. (A) $\lim_{x \to 0^-} f(x)$ (B) $\lim_{x \to 0^+} f(x)$

  (C) $\lim_{x \to 0} f(x)$ (D) $f(0)$

26. (A) $\lim_{x \to 1^-} f(x)$ (B) $\lim_{x \to 1^+} f(x)$

  (C) $\lim_{x \to 1} f(x)$ (D) $f(1)$

27. (A) $\lim_{x \to 2^-} f(x)$ (B) $\lim_{x \to 2^+} f(x)$

  (C) $\lim_{x \to 2} f(x)$ (D) $f(2)$

  (E) Is it possible to redefine $f(2)$ so that $\lim_{x \to 2} f(x) = f(2)$? Explain.

28. (A) $\lim_{x \to 4^-} f(x)$ (B) $\lim_{x \to 4^+} f(x)$

  (C) $\lim_{x \to 4} f(x)$ (D) $f(4)$

  (E) Is it possible to define $f(4)$ so that $\lim_{x \to 4} f(x) = f(4)$? Explain.

In Problems 29–32, use the graph of the function g shown below to estimate the indicated limits and function values.

29. (A) $\lim\limits_{x \to 1^-} g(x)$ (B) $\lim\limits_{x \to 1^+} g(x)$

   (C) $\lim\limits_{x \to 1} g(x)$ (D) $g(1)$

   (E) Is it possible to define $g(1)$ so that $\lim\limits_{x \to 1} g(x) = g(1)$? Explain.

30. (A) $\lim\limits_{x \to 2^-} g(x)$ (B) $\lim\limits_{x \to 2^+} g(x)$

   (C) $\lim\limits_{x \to 2} g(x)$ (D) $g(2)$

31. (A) $\lim\limits_{x \to 3^-} g(x)$ (B) $\lim\limits_{x \to 3^+} g(x)$

   (C) $\lim\limits_{x \to 3} g(x)$ (D) $g(3)$

   (E) Is it possible to redefine $g(3)$ so that $\lim\limits_{x \to 3} g(x) = g(3)$? Explain.

32. (A) $\lim\limits_{x \to 4^-} g(x)$ (B) $\lim\limits_{x \to 4^+} g(x)$

   (C) $\lim\limits_{x \to 4} g(x)$ (D) $g(4)$

B *In Problems 33–52, use limit properties, algebraic simplification, tables of values, and/or graphs to find each limit, if it exists.*

33. $\lim\limits_{x \to -3} \dfrac{x^2 - 9}{x + 3}$ **34.** $\lim\limits_{x \to -5} \dfrac{x^2 - 25}{x + 5}$

35. $\lim\limits_{x \to 1^+} \dfrac{|x - 1|}{x - 1}$ **36.** $\lim\limits_{x \to 3^-} \dfrac{x - 3}{|x - 3|}$

37. $\lim\limits_{x \to 1^-} \dfrac{|x - 1|}{x - 1}$ **38.** $\lim\limits_{x \to 3^+} \dfrac{x - 3}{|x - 3|}$

39. $\lim\limits_{x \to 1} \dfrac{|x - 1|}{x - 1}$ **40.** $\lim\limits_{x \to 3} \dfrac{x - 3}{|x - 3|}$

41. $\lim\limits_{x \to 1} \dfrac{x - 2}{x^2 - 2x}$ **42.** $\lim\limits_{x \to 1} \dfrac{x + 3}{x^2 + 3x}$

43. $\lim\limits_{x \to 2} \dfrac{x - 2}{x^2 - 2x}$ **44.** $\lim\limits_{x \to -3} \dfrac{x + 3}{x^2 + 3x}$

45. $\lim\limits_{x \to 2} \dfrac{x^2 - x - 6}{x + 2}$ **46.** $\lim\limits_{x \to 3} \dfrac{x^2 + x - 6}{x + 3}$

47. $\lim\limits_{x \to -2} \dfrac{x^2 - x - 6}{x + 2}$ **48.** $\lim\limits_{x \to -3} \dfrac{x^2 + x - 6}{x + 3}$

49. $\lim\limits_{x \to 3} \left(\dfrac{x}{x + 3} + \dfrac{x - 3}{x^2 - 9} \right)$

50. $\lim\limits_{x \to 2} \left(\dfrac{1}{x + 2} + \dfrac{x - 2}{x^2 - 4} \right)$

51. $\lim\limits_{x \to 0} \left(\sqrt{x^2 + 9} - \dfrac{x^2 + 3x}{x} \right)$

52. $\lim\limits_{x \to 1} \left(\dfrac{x^2 - 1}{x - 1} + \sqrt{x^2 + 3} \right)$

Compute the following limit for each function in Problems 53–60:

$$\lim\limits_{h \to 0} \dfrac{f(2 + h) - f(2)}{h}$$

53. $f(x) = 3x + 1$ **54.** $f(x) = 5x - 1$

55. $f(x) = x^2 + 1$ **56.** $f(x) = x^2 - 2$

57. $f(x) = \sqrt{x} - 2$ **58.** $f(x) = 1 + \sqrt{x}$

59. $f(x) = |x - 2| - 3$ **60.** $f(x) = 2 + |x - 2|$

61. Use a table of values to investigate the behavior of the following limits:

   (A) $\lim\limits_{x \to 0^-} \dfrac{1}{x}$ (B) $\lim\limits_{x \to 0^+} \dfrac{1}{x}$

   Write a brief verbal description of each limit.

62. Use a table of values to investigate the behavior of the following limits:

   (A) $\lim\limits_{x \to 0^-} \dfrac{1}{x^2}$ (B) $\lim\limits_{x \to 0^+} \dfrac{1}{x^2}$

   Write a brief verbal description of each limit.

C

63. Let f be defined by

$$f(x) = \begin{cases} 1 + mx & \text{if } x \le 1 \\ 4 - mx & \text{if } x > 1 \end{cases}$$

where m is a constant.

   (A) Graph f for $m = 1$, and find

$$\lim\limits_{x \to 1^-} f(x) \quad \text{and} \quad \lim\limits_{x \to 1^+} f(x)$$

   (B) Graph f for $m = 2$, and find

$$\lim\limits_{x \to 1^-} f(x) \quad \text{and} \quad \lim\limits_{x \to 1^+} f(x)$$

   (C) Find m so that

$$\lim\limits_{x \to 1^-} f(x) = \lim\limits_{x \to 1^+} f(x)$$

   and graph f for this value of m.

   (D) Write a brief verbal description of each graph. How does the graph in part (C) differ from the graphs in parts (A) and (B)?

64. Let f be defined by

$$f(x) = \begin{cases} -3m + 0.5x & \text{if } x \leqslant 2 \\ 3m - x & \text{if } x > 2 \end{cases}$$

where m is a constant.

(A) Graph f for $m = 0$, and find

$$\lim_{x \to 2^-} f(x) \qquad \text{and} \qquad \lim_{x \to 2^+} f(x)$$

(B) Graph f for $m = 1$, and find

$$\lim_{x \to 2^-} f(x) \qquad \text{and} \qquad \lim_{x \to 2^+} f(x)$$

(C) Find m so that

$$\lim_{x \to 2^-} f(x) = \lim_{x \to 2^+} f(x)$$

and graph f for this value of m.

(D) Write a brief verbal description of each graph. How does the graph in part (C) differ from the graphs in parts (A) and (B)?

Find each limit in Problems 65–68, where a is a real constant.

65. $\displaystyle\lim_{h \to 0} \frac{(a + h)^2 - a^2}{h}$

66. $\displaystyle\lim_{h \to 0} \frac{[3(a + h) - 2] - (3a - 2)}{h}$

67. $\displaystyle\lim_{h \to 0} \frac{\sqrt{a + h} - \sqrt{a}}{h}, \quad a > 0$

68. $\displaystyle\lim_{h \to 0} \frac{\dfrac{1}{a + h} - \dfrac{1}{a}}{h}, \quad a \neq 0$

In Problems 69–74, use a table of values to estimate each limit to three decimal places.

69. $\displaystyle\lim_{x \to 1} \frac{x^{10} - 1}{x - 1}$

70. $\displaystyle\lim_{x \to 1} \frac{x^{15} - 1}{x - 1}$

71. $\displaystyle\lim_{x \to 0} \frac{2^x - 1}{x}$

72. $\displaystyle\lim_{x \to 0} \frac{3^x - 1}{x}$

73. $\displaystyle\lim_{x \to 0} (1 + x)^{1/x}$

74. $\displaystyle\lim_{x \to 0} (1 + 2x)^{1/x}$

 Problems 75–80 require the use of a graphing utility. When investigating limits with a graphing utility, it is helpful to choose Xmin and Xmax so that coordinates displayed on the screen have finite decimal expansions. Problems 75 and 76 describe a simple way to accomplish this.

75. Let Xmin $= 0$, and use trial and error to find an integer n so that if Xmax $= n$, then the x coordinates of points displayed on the screen are integers.

76. Given any value for Xmin, let

$$\text{Xmax} = \text{Xmin} + h \cdot n$$

where n is the integer discovered in Problem 75. Discuss the nature of the x coordinates of points displayed on the screen if $h = 0.5$. If $h = 0.1$.

In Problems 77–80, graph each function and use zoom and trace to investigate the left- and right-hand limits at the indicated value(s) of c. Use the ideas discussed in Problems 75 and 76 to choose Xmin and Xmax.

77. $f(x) = \dfrac{x^4 - 10x^2 + 24}{4 - x^2}; c = -2, c = 2$

78. $f(x) = \dfrac{x^4 - 12x^2 + 27}{x^2 - 9}; c = -3, c = 3$

79. $f(x) = \dfrac{x^3 - 9x}{|x^2 - 9|}; c = -3, c = 3$

80. $f(x) = \dfrac{4x - x^3}{|x^2 - 4|}; c = -2, c = 2$

Applications

Business & Economics

81. *Telephone rates.* A 10-10 long-distance telephone service charges \$0.99 for the first 20 minutes or less of a state-to-state call and \$0.07 for each additional minute or fraction thereof.

(A) Write a piecewise definition of the charge $F(x)$ for a state-to-state long-distance call lasting x minutes.

(B) Graph $F(x)$ for $0 < x \leqslant 40$.

(C) Find $\displaystyle\lim_{x \to 20^-} F(x)$, $\displaystyle\lim_{x \to 20^+} F(x)$, and $\displaystyle\lim_{x \to 20} F(x)$, whichever exist.

82. *Telephone rates.* A second 10-10 long-distance
www telephone service charges \$0.09 per minute or fraction thereof for calls lasting 10 minutes or more

and \$0.18 per minute or fraction thereof for calls lasting less than 10 minutes.

(A) Write a piecewise definition of the charge $G(x)$ for a state-to-state long-distance call lasting x minutes.

(B) Graph $G(x)$ for $0 < x \leqslant 40$.

(C) Find $\displaystyle\lim_{x \to 10^-} G(x)$, $\displaystyle\lim_{x \to 10^+} G(x)$, and $\displaystyle\lim_{x \to 10} G(x)$, whichever exist.

83. *Telephone rates.* Refer to Problems 81 and 82. Write a brief verbal comparison of these two services for calls lasting 20 minutes or less.

84. *Telephone rates.* Refer to Problems 81 and 82. Write a brief verbal comparison of these two services for calls lasting more than 20 minutes.

www *A company sells custom embroidered apparel and promotional products. The table shows the volume discounts offered by the company, where x is the volume of a purchase in dollars. Problems 85 and 86 deal with two different interpretations of this discount method.*

VOLUME DISCOUNT (EXCLUDING TAX)	
Volume (x)	Discount Amount
$300 \leq x < $1,000	3%
$1,000 \leq x < $3,000	5%
$3,000 \leq x < $5,000	7%
$5,000 \leq x	10%

85. *Volume discount.* Assume that the volume discounts in the table apply to the entire purchase. That is, if the volume x satisfies $300 \leq x < $1,000, the entire purchase is discounted 3%. If the volume x satisfies $1,000 \leq x < $3,000 the entire purchase is discounted 5% and so on.

(A) If x is the volume of a purchase before the discount is applied, write a piecewise definition for the discounted price $D(x)$ of this purchase.

(B) Use one-sided limits to investigate the limit of $D(x)$ as x approaches $1,000. As x approaches $3,000.

86. *Volume discount.* Assume that the volume discounts in the table apply only to that portion of the volume in each interval in the table. That is, the discounted price for a $4,000 purchase would be computed as follows:

$$300 + 0.97(700) + 0.95(2,000) + 0.93(1,000) = 3,809$$

(A) If x is the volume of a purchase before the discount is applied, write a piecewise definition for the discounted price $P(x)$ of this purchase.

(B) Use one-sided limits to investigate the limit of $P(x)$ as x approaches $1,000. As x approaches $3,000.

(C) Compare this discount method with the one in Problem 85. Does one always produce a lower price than the other? Discuss.

Life Sciences

87. *Pollution.* A state charges polluters an annual fee of **www** $20 per ton for each ton of pollutant emitted into the atmosphere, up to a maximum of 4,000 tons. No fees are charged for emissions beyond the 4,000-ton limit. Write a piecewise definition of the fees $F(x)$ charged for the emission of x tons of pollutant in a year. What is the limit of $F(x)$ as x approaches 4,000 tons? As x approaches 8,000 tons?

88. *Pollution.* Refer to Problem 87. The fee per ton of **www** pollution is given by $A(x) = F(x)/x$. Write a piecewise definition of $A(x)$. What is the limit of $A(x)$ as x approaches 4,000 tons? As x approaches 8,000 tons?

Social Sciences

89. *Voter turnout.* Statisticians often use piecewise-defined functions to predict outcomes of elections. For the functions f and g given below, find the limit of each function as x approaches 5 and as x approaches 10.

$$f(x) = \begin{cases} 0 & \text{if } x \leq 5 \\ 0.8 - 0.08x & \text{if } 5 < x < 10 \\ 0 & \text{if } 10 \leq x \end{cases}$$

$$g(x) = \begin{cases} 0 & \text{if } x \leq 5 \\ 0.8x - 0.04x^2 - 3 & \text{if } 5 < x < 10 \\ 1 & \text{if } 10 \leq x \end{cases}$$

Section 3-2

Limits and Continuity

- ❑ CONTINUITY
- ❑ CONTINUITY PROPERTIES
- ❑ INFINITE LIMITS
- ❑ SOLVING INEQUALITIES USING CONTINUITY PROPERTIES

Theorem 2 in Section 3-1 states that if f is a polynomial function or a rational function with a nonzero denominator at $x = c$, then

$$\lim_{x \to c} f(x) = f(c) \tag{1}$$

Functions that satisfy equation (1) are said to be *continuous* at $x = c$. A firm understanding of continuous functions is essential for sketching and analyzing graphs. We will also see that continuity properties provide a simple and efficient method for solving inequalities, a tool that we will use extensively in later sections.

❑ CONTINUITY

Compare the graphs shown in Figure 1, which were discussed in Examples 2 to 4 in Section 3-1. Notice that two of the graphs are broken; that is, they cannot be drawn without lifting a pen off the paper. Informally, a function is *continuous over an interval* if its graph over the interval can be drawn without removing a pen from the paper. A function whose graph is broken (disconnected) at $x = c$ is said to be *discontinuous* at $x = c$. Function f (Fig. 1A) is continuous for all x. Function g (Fig. 1B) is discontinuous at $x = 2$, but is continuous over any interval that does not include 2. Function h (Fig. 1C) is discontinuous at $x = 0$, but is continuous over any interval that does not include 0.

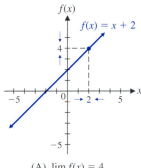

(A) $\lim\limits_{x \to 2} f(x) = 4$
$f(2) = 4$

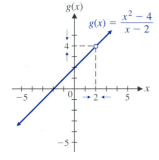

(B) $\lim\limits_{x \to 2} g(x) = 4$
$g(2)$ is not defined

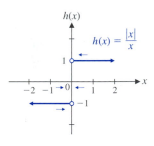

(C) $\lim\limits_{x \to 0} h(x)$ does not exist
$h(0)$ is not defined

FIGURE 1

Most graphs of natural phenomena are continuous, whereas many graphs in business and economics applications have discontinuities. Figure 2A illustrates temperature variation over a 24-hour period—a continuous phenomenon. Figure 2B illustrates warehouse inventory over a 1-week period—a discontinuous phenomenon.

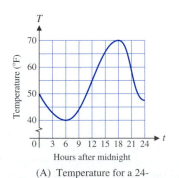

(A) Temperature for a 24-hour period

(B) Inventory in a warehouse during 1 week

FIGURE 2

(A) Write a brief verbal description of the temperature variation illustrated in Figure 2A, including estimates of the high and low temperatures during this period and the times at which they occurred.

(B) Write a brief verbal description of the changes in inventory illustrated in Figure 2B, including estimates of the changes in inventory and the times at which these changes occurred.

The preceding discussion leads to the following formal definition of continuity:

Continuity

A function f is **continuous at the point $x = c$** if

1. $\lim_{x \to c} f(x)$ exists   **2.** $f(c)$ exists   **3.** $\lim_{x \to c} f(x) = f(c)$

A function is **continuous on the open interval*** **(a, b)** if it is continuous at each point on the interval.

_____

*See Appendix A-8 for a review of interval notation.

If one or more of the three conditions in the definition fails, the function is **discontinuous** at $x = c$.

Sketch a graph of a function that is discontinuous at a point because it fails to satisfy condition 1 in the definition of continuity. Repeat for conditions 2 and 3.

Example 1 ⌇ **Continuity of a Function Defined by a Graph** Use the definition of continuity to discuss the continuity of the function whose graph is shown in Figure 3.

FIGURE 3

SOLUTION We begin by identifying the points of discontinuity. Examining the graph, we see breaks and/or holes in the graph at $x = -4, -2, 1,$ and 3. Now we must determine which conditions in the definition of continuity are not satisfied at each of these points. In each case, we find the value of the function and the limit of the function at the point in question.

Discontinuity at $x = -4$:

$$\lim_{x \to -4^-} f(x) = 2$$

$$\lim_{x \to -4^+} f(x) = 3$$

Since the one-sided limits are different, the limit does not exist (Section 3-1).

$$\lim_{x \to -4} f(x) \text{ does not exist}$$

$$f(-4) = 3$$

Thus, f is not continuous at $x = -4$ because condition 1 is not satisfied.

Discontinuity at $x = -2$:

$$\lim_{x \to -2^-} f(x) = 5$$

$$\lim_{x \to -2^+} f(x) = 5$$

The hole at $(-2, 5)$ indicates that 5 is not the value of f at -2. Since there is no solid dot elsewhere on the vertical line $x = -2$, $f(-2)$ is not defined.

$$\lim_{x \to -2} f(x) = 5$$

$$f(-2) \text{ does not exist}$$

Thus, f is not continuous at $x = -2$ because condition 2 is not satisfied.

Discontinuity at $x = 1$:

$$\lim_{x \to 1^-} f(x) = 4$$

$$\lim_{x \to 1^+} f(x) = 1$$

$$\lim_{x \to 1} f(x) \text{ does not exist}$$

$$f(1) \text{ does not exist}$$

This time, f is not continuous at $x = 1$ because both conditions 1 and 2 are not satisfied.

Discontinuity at $x = 3$:

$$\lim_{x \to 3^-} f(x) = 3$$

$$\lim_{x \to 3^+} f(x) = 3$$

The solid dot at $(3, 1)$ indicates that $f(3) = 1$.

$$\lim_{x \to 3} f(x) = 3$$

$$f(3) = 1$$

Conditions 1 and 2 are satisfied, but f is not continuous at $x = 3$ because condition 3 is not satisfied.

Having identified and discussed all points of discontinuity, we can now conclude that f is continuous except at $x = -4, -2, 1$, and 3. ■

Matched Problem 1 ⇦ Use the definition of continuity to discuss the continuity of the function whose graph is shown in Figure 4.

FIGURE 4

■

For functions defined by equations, it is also important to be able to locate points of discontinuity by examining the equation.

Example 2 ⇨ **Continuity of Functions Defined by Equations** Using the definition of continuity, discuss the continuity of each function at the indicated point (s).

(A) $f(x) = x + 2$ at $x = 2$ (B) $g(x) = \dfrac{x^2 - 4}{x - 2}$ at $x = 2$

(C) $h(x) = \dfrac{|x|}{x}$ at $x = 0$ and at $x = 1$

SOLUTION (A) f is continuous at $x = 2$, since

$$\lim_{x \to 2} f(x) = 4 = f(2) \quad \textit{See Figure 1A on page 149.}$$

(B) g is not continuous at $x = 2$, since $g(2) = 0/0$ is not defined (see Fig. 1B on page 149).

(C) h is not continuous at $x = 0$, since $h(0) = |0|/0$ is not defined; also, $\lim_{x \to 0} h(x)$ does not exist.

h is continuous at $x = 1$, since

$$\lim_{x \to 1} \frac{|x|}{x} = 1 = h(1) \quad \textit{See Figure 1C on page 149.}$$

Matched Problem 2 ⇨ Using the definition of continuity, discuss the continuity of each function at the indicated point (s).

(A) $f(x) = x + 1$ at $x = 1$ (B) $g(x) = \dfrac{x^2 - 1}{x - 1}$ at $x = 1$

(C) $h(x) = \dfrac{x - 2}{|x - 2|}$ at $x = 2$ and at $x = 0$

We can also talk about one-sided continuity, just as we talked about one-sided limits. For example, a function is said to be **continuous on the right** at $x = c$ if $\lim_{x \to c^+} f(x) = f(c)$ and **continuous on the left** at $x = c$ if $\lim_{x \to c^-} f(x) = f(c)$. A function is **continuous on the closed interval [a, b]** if it is continuous on the open interval (a, b) and is continuous on the right at a and continuous on the left at b.

Figure 5A illustrates a function that is continuous on the closed interval $[-1, 1]$. Figure 5B illustrates a function that is continuous on a half-closed interval $[0, \infty)$.

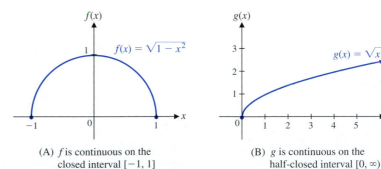

(A) f is continuous on the closed interval $[-1, 1]$

(B) g is continuous on the half-closed interval $[0, \infty)$

FIGURE 5 Continuity on closed and half-closed intervals

❑   CONTINUITY PROPERTIES

www   Functions have some useful **general continuity properties:**

> **If two functions are continuous on the same interval, then their sum, difference, product, and quotient are continuous on the same interval, except for values of x that make a denominator 0.**

These properties, along with Theorem 1 below, enable us to determine intervals of continuity for some important classes of functions without having to look at their graphs or use the three conditions in the definition.

THEOREM 1   **Continuity Properties of Some Specific Functions**

(A) A constant function $f(x) = k$, where k is a constant, is continuous for all x.

$f(x) = 7$ is continuous for all x.

(B) For n a positive integer, $f(x) = x^n$ is continuous for all x.

$f(x) = x^5$ is continuous for all x.

(C) A polynomial function is continuous for all x.

$2x^3 - 3x^2 + x - 5$ is continuous for all x.

(D) A rational function is continuous for all x except those values that make a denominator 0.

$\dfrac{x^2 + 1}{x - 1}$ is continuous for all x except x = 1, a value that makes the denominator 0.

(E) For n an odd positive integer greater than 1, $\sqrt[n]{f(x)}$ is continuous wherever $f(x)$ is continuous.

$\sqrt[3]{x^2}$ is continuous for all x.

(F) For n an even positive integer, $\sqrt[n]{f(x)}$ is continuous wherever $f(x)$ is continuous and nonnegative.

$\sqrt[4]{x}$ is continuous on the interval $[0, \infty)$.

Parts C and D of Theorem 1 are the same as Theorem 2 in Section 3-1. They are repeated here to emphasize their importance.

Example 3 ⇨   **Using Continuity Properties**   Using Theorem 1 and the general properties of continuity, determine where each function is continuous.

(A) $f(x) = x^2 - 2x + 1$ (B) $f(x) = \dfrac{x}{(x + 2)(x - 3)}$

(C) $f(x) = \sqrt[3]{x^2 - 4}$ (D) $f(x) = \sqrt{x - 2}$

SOLUTION   (A) Since f is a polynomial function, f is continuous for all x.

(B) Since f is a rational function, f is continuous for all x except -2 and 3 (values that make the denominator 0).

(C) The polynomial function $x^2 - 4$ is continuous for all x. Since $n = 3$ is odd, f is continuous for all x.

(D) The polynomial function $x - 2$ is continuous for all x and nonnegative for $x \geq 2$. Since $n = 2$ is even, f is continuous for $x \geq 2$, or on the interval $[2, \infty)$. ■

Matched Problem 3 ☞ Using Theorem 1 and the general properties of continuity, determine where each function is continuous.

(A) $f(x) = x^4 + 2x^2 + 1$ (B) $f(x) = \dfrac{x^2}{(x+1)(x-4)}$

(C) $f(x) = \sqrt{x - 4}$ (D) $f(x) = \sqrt[3]{x^3 + 1}$ ■

❑   INFINITE LIMITS

A function is discontinuous at any point c where $\lim_{x \to c} f(x)$ does not exist. For example, if the one-sided limits are different at $x = c$, the limit does not exist and the function is discontinuous at $x = c$ (see Fig. 3 on page 150). Another situation where a limit may fail to exist involves functions whose values become very large as x approaches c. The special symbol ∞ is often used to describe this type of behavior. To illustrate this case, consider the functions

$$f(x) = \frac{1}{x} \quad \text{and} \quad g(x) = \frac{1}{x^2}$$

The graph of f in Figure 6A indicates that the values of $f(x)$ are very large negative numbers if x is near 0 on the left and very large positive numbers if x is near 0 on the right. In neither case does the limit exist. However, it is

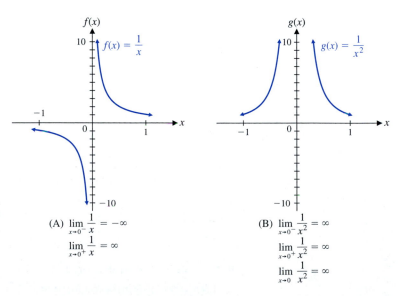

(A) $\lim\limits_{x \to 0^-} \dfrac{1}{x} = -\infty$

$\lim\limits_{x \to 0^+} \dfrac{1}{x} = \infty$

(B) $\lim\limits_{x \to 0^-} \dfrac{1}{x^2} = \infty$

$\lim\limits_{x \to 0^+} \dfrac{1}{x^2} = \infty$

$\lim\limits_{x \to 0} \dfrac{1}{x^2} = \infty$

FIGURE 6

convenient to use the special symbol ∞ to describe the nature of the graph at $x = 0$. Thus, we write

$$\lim_{x \to 0^-} f(x) = -\infty \qquad \text{and} \qquad \lim_{x \to 0^+} f(x) = \infty$$

Since these two statements represent different types of behavior, we cannot write a single limit statement to describe the nature of the graph at $x = 0$.

The graph of g in Figure 6B shows that the values of $g(x)$ are very large positive numbers when x is near 0 on either side of 0. Once again, the limit of $g(x)$ as x approaches 0 does not exist, but we can describe the behavior of the graph of g near 0 by writing

$$\lim_{x \to 0} g(x) = \infty$$

For both functions f and g, the line $x = 0$ (the vertical axis) is a *vertical asymptote*. These ideas are summarized in the following box.

www

> ## Vertical Asymptotes
>
> If the limit of a function f fails to exist as x approaches c from the left because the values of $f(x)$ are becoming very large positive numbers (or very large negative numbers), we say that*
>
> $$\lim_{x \to c^-} f(x) = \infty \quad (\text{or} -\infty)$$
>
> If this happens as x approaches c from the right, we say that
>
> $$\lim_{x \to c^+} f(x) = \infty \quad (\text{or} -\infty)$$
>
> If both one-sided limits exhibit the same behavior, we say that
>
> $$\lim_{x \to c} f(x) = \infty \quad (\text{or} -\infty)$$
>
> If any of the above hold, the line $x = c$ is a **vertical asymptote** for the graph of $y = f(x)$.
>
> ---
>
> *The precise definition of this limit statement is as follows: $\lim_{x \to c^-} f(x) = \infty$ if for each $N > 0$, there exists a $d > 0$ such that $f(x) > N$ whenever $c - d < x < c$. Similar statements can be made for limits from the right, unrestricted limits, and limits involving very large negative numbers.

Example 4   **Limits at Points of Discontinuity**   For the function

$$f(x) = \frac{1 - x}{x^4 - x^2}$$

use ∞ and $-\infty$, as appropriate, to describe the behavior at each point of discontinuity, and identify all vertical asymptotes.

SOLUTION   First, we factor the denominator and identify the points of discontinuity:

$$x^4 - x^2 = x^2(x^2 - 1) = x^2(x - 1)(x + 1)$$

Thus, f is discontinuous at $x = -1, 0$, and 1. We use a numerical approach to investigate the behavior of $f(x)$ near each of these discontinuities.

Behavior at $x = -1$ [values of $f(x)$ rounded to the nearest integer]:

TABLE 1

x	-1.01	-1.001	$-1.0001 \rightarrow$	-1	$\leftarrow -0.9999$	-0.999	-0.99
$f(x)$	98	998	$9,998 \rightarrow \infty$		$-\infty \leftarrow -10,002$	$-1,002$	-102

Examining the values of $f(x)$ near $x = -1$ (Table 1), we see that the values of $f(x)$ are large positive numbers for x near -1 on the left and large negative numbers for x near -1 on the right. Thus, we must use one-sided limits to describe the behavior at $x = -1$:

$$\lim_{x \to -1^-} \frac{1-x}{x^4 - x^2} = \infty \quad \text{and} \quad \lim_{x \to -1^+} \frac{1-x}{x^4 - x^2} = -\infty$$

This shows that the line $x = -1$ is a vertical asymptote for the graph of $y = f(x)$.
Behavior at $x = 0$ [values of $f(x)$ rounded to the nearest integer]:

TABLE 2

x	-0.01	-0.001	$\rightarrow 0 \leftarrow$	0.001	0.01
$f(x)$	$-10,101$	$-1,001,001 \rightarrow$	$-\infty \leftarrow$	$-999,001$	$-9,901$

Since the values of $f(x)$ for x near 0, on either side of 0, are very large negative numbers (Table 2), we can use a single limit statement to describe the behavior at $x = 0$:

$$\lim_{x \to 0} \frac{1-x}{x^4 - x^2} = -\infty$$

Thus, the line $x = 0$ (the y axis) is a vertical asymptote for the graph of $y = f(x)$.
Behavior at $x = 1$ [values of $f(x)$ rounded to three decimal places]:

TABLE 3

x	0.9	0.99	$0.999 \rightarrow$	1	$\leftarrow 1.001$	1.01	1.1
$f(x)$	-0.650	-0.513	$-0.501 \rightarrow$	-0.5	$\leftarrow -0.499$	-0.488	-0.394

The values in Table 3 suggest that $\lim_{x \to 1} f(x)$ exists. We confirm this by using algebraic simplification (notice that this limit is an indeterminate form):

$$\lim_{x \to 1} \frac{1-x}{x^4 - x^2} = \lim_{x \to 1} \frac{1-x}{x^2(x+1)(x-1)} \qquad \frac{0}{0} \text{ indeterminate form}$$

$$= \lim_{x \to 1} \frac{-1}{x^2(x+1)} \qquad \frac{1-x}{x-1} = -1, x \neq 1$$

$$= -0.5$$

Since this limit exists, there is no vertical asymptote at $x = 1$.
The graph of $y = f(x)$ (Fig. 7) illustrates the behavior indicated by the following limit statements.

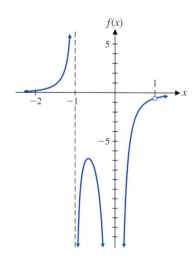

FIGURE 7 $f(x) = \dfrac{1-x}{x^4 - x^2}$

$$\lim_{x \to -1^-} f(x) = \infty \qquad \lim_{x \to 0} f(x) = -\infty$$

$$\lim_{x \to -1^+} f(x) = -\infty \qquad \lim_{x \to 1} f(x) = -0.5$$

We will have much more to say about sketching and analyzing graphs involving asymptotes in Section 4-3. ∎

Matched Problem 4 ✏ For the function

$$f(x) = \frac{x - 3}{x^2 - 4x + 3}$$

use ∞ and $-\infty$, as appropriate, to describe the behavior at each point of discontinuity, and identify all vertical asymptotes. ∎

CAUTION

Figure 8A shows the graph of the function f from Example 4 on a graphing utility. It appears that the graphing utility has also drawn the vertical asymptote at $x = -1$, but this is not the case. As we saw in Example 4, points close to -1 on the left have large positive y coordinates, while points close to -1 on the right have large negative y coordinates. For the given x range on this particular graphing utility, there are no points on the screen with x coordinate -1. The graphing utility simply connected the last point to the left of -1 with the first point to the right of -1. Since these points are not visible on the screen, this gives the appearance of a vertical asymptote. Figure 8B shows the graph of the same function with a much larger y range. Now these two points are visible and, in fact, the graph appears to be continuous at $x = -1$, which we know is not true. When you graph functions with vertical asymptotes on a graphing utility, you should proceed as we did in Example 4 to identify the asymptotes first. You cannot depend on the graphing utility to identify asymptotes.

(A)

(B)

FIGURE 8

❑ SOLVING INEQUALITIES USING CONTINUITY PROPERTIES

One of the basic tools for analyzing graphs in calculus is a special line graph called a *sign chart*. We will make extensive use of these charts in later sections. In the following discussion, we use continuity properties to develop a simple and efficient procedure for constructing sign charts.

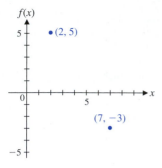

FIGURE 9

Suppose that a function f is continuous over the interval $(1, 8)$ and $f(x) \neq 0$ for any x in $(1, 8)$. Also suppose that $f(2) = 5$, a positive number. Is it possible for $f(x)$ to be negative for any x in the interval $(1, 8)$? The answer is "no." If $f(7)$ were -3, for example, as shown in Figure 9, how would it be possible to join the points $(2, 5)$ and $(7, -3)$ with the graph of a continuous function without crossing the x axis between 1 and 8 at least once? [Crossing the x axis would violate our assumption that $f(x) \neq 0$ for any x in $(1, 8)$.] Thus, we conclude that $f(x)$ must be positive for all x in $(1, 8)$. If $f(2)$ were negative, then, using the same type of reasoning, $f(x)$ would have to be negative over the entire interval $(1, 8)$.

In general, **if f is continuous and $f(x) \neq 0$ on the interval (a, b), then $f(x)$ cannot change sign on (a, b).** This is the essence of Theorem 2.

THEOREM 2   Sign Properties on an Interval (a, b)

If f is continuous on (a, b) and $f(x) \neq 0$ for all x in (a, b), then either $f(x) > 0$ for all x in (a, b) or $f(x) < 0$ for all x in (a, b).

Theorem 2 provides the basis for an effective method of solving many types of inequalities. Example 5 illustrates the process.

Example 5 ⇔ **Solving an Inequality**   Solve: $\dfrac{x + 1}{x - 2} > 0$.

SOLUTION   We start by using the left side of the inequality to form the function f:

$$f(x) = \frac{x + 1}{x - 2}$$

FIGURE 10

The rational function f is discontinuous at $x = 2$, and $f(x) = 0$ for $x = -1$ (a fraction is 0 when the numerator is 0 and the denominator is not 0). We plot $x = 2$ and $x = -1$, which we call *partition numbers,* on a real number line (Fig. 10). (Note that the dot at 2 is open, because the function is not defined at $x = 2$.) The partition numbers 2 and -1 determine three open intervals: $(-\infty, -1)$, $(-1, 2)$, and $(2, \infty)$. The function f is continuous and nonzero on each of these intervals. From Theorem 2 we know that $f(x)$ does not change sign on any of these intervals. Thus, we can find the sign of $f(x)$ on each of these intervals by selecting a **test number** in each interval and evaluating $f(x)$ at that number. Since any number in each subinterval will do, we choose test numbers that are easy to evaluate: $-2, 0$, and 3. Table 4 shows the results.

The sign of $f(x)$ at each test number is the same as the sign of $f(x)$ over the interval containing that test number. Using this information, we construct a **sign chart** for $f(x)$:

Test Numbers	
x	$f(x)$
-2	$\frac{1}{4}$ (+)
0	$-\frac{1}{2}$ (−)
3	4 (+)

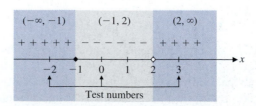

Now using the sign chart, we can easily write the solution for the given nonlinear inequality:

$$f(x) > 0 \quad \text{for} \quad \begin{array}{l} x < -1 \quad \text{or} \quad x > 2 \qquad \text{Inequality notation} \\ (-\infty, -1) \cup (2, \infty) \qquad \text{Interval notation} \end{array}$$

Most of the inequalities we encounter will involve strict inequalities ($>$ or $<$). If it is necessary to solve inequalities of the form $\geq$ or $\leq$, we simply include the endpoint x of any interval if f is defined at x and $f(x)$ satisfies the given inequality. For example, referring to the sign chart in Example 5, the solution of the inequality

$$\frac{x+1}{x-2} \geq 0 \quad \text{is} \quad \begin{array}{l} x \leq -1 \quad \text{or} \quad x > 2 \qquad \text{Inequality notation} \\ (-\infty, -1] \cup (2, \infty) \qquad \text{Interval notation} \end{array}$$

In general, given a function f, we will call all values x such that f is discontinuous at x or $f(x) = 0$ **partition numbers. Partition numbers determine open intervals where $f(x)$ does not change sign.** By using a test number from each interval, we can construct a sign chart for $f(x)$ on the real number line. It is then an easy matter to determine where $f(x) < 0$ or $f(x) > 0$; that is, to solve the inequality $f(x) < 0$ or $f(x) > 0$.

We summarize the procedure for constructing sign charts in the following box:

Constructing Sign Charts

Given a function f,

Step 1. Find all partition numbers. That is,

(A) Find all numbers where f is discontinuous. (Rational functions are discontinuous for values of x that make a denominator 0.)

(B) Find all numbers where $f(x) = 0$. (For a rational function, this occurs where the numerator is 0 and the denominator is not 0.)

Step 2. Plot the numbers found in step 1 on a real number line, dividing the number line into intervals.

Step 3. Select a test number in each open interval determined in step 2, and evaluate $f(x)$ at each test number to determine whether $f(x)$ is positive ($+$) or negative ($-$) in each interval.

Step 4. Construct a sign chart using the real number line in step 2. This will show the sign of $f(x)$ on each open interval.

[*Note*: From the sign chart, it is easy to find the solution for the inequality $f(x) < 0$ or $f(x) > 0$.]

Matched Problem 5 ⇔ Solve: $\dfrac{x^2 - 1}{x - 3} < 0$.

Let $y_1 = (x + 1)/(x - 2)$ and $y_2 = y_1/|y_1|$. Figure 11 shows the graph of y_2 on a graphing utility. Discuss the relationship between this graph and the sign chart constructed in the solution of Example 5.

FIGURE 11

Answers to Matched Problems

1. f is not continuous at $x = -3, -1, 2$, and 4.

$x = -3$: $\lim_{x \to -3} f(x) = 3$, but $f(-3)$ does not exist

$x = -1$: $f(-1) = 1$, but $\lim_{x \to -1} f(x)$ does not exist

$x = 2$: $\lim_{x \to 2} f(x) = 5$, but $f(2) = 3$

$x = 4$: $\lim_{x \to 4} f(x)$ does not exist, and $f(4)$ does not exist

2. (A) f is continuous at $x = 1$, since $\lim_{x \to 1} f(x) = 2 = f(1)$.

   (B) g is not continuous at $x = 1$, since $g(1)$ is not defined.

   (C) h is not continuous at $x = 2$ for two reasons: $h(2)$ does not exist and $\lim_{x \to 2} h(x)$ does not exist.

   h is continuous at $x = 0$, since $\lim_{x \to 0} h(x) = -1 = h(0)$.

3. (A) Since f is a polynomial function, f is continuous for all x.

   (B) Since f is a rational function, f is continuous for all x except -1 and 4 (values that make the denominator 0).

   (C) The polynomial function $x - 4$ is continuous for all x and nonnegative for $x \geqslant 4$. Since $n = 2$ is even, f is continuous for $x \geqslant 4$, or on the interval $[4, \infty)$.

   (D) The polynomial function $x^3 + 1$ is continuous for all x. Since $n = 3$ is odd, f is continuous for all x.

4. f has a vertical asymptote at $x = 1$, since $\lim_{x \to 1^-} f(x) = -\infty$ and $\lim_{x \to 1^+} f(x) = \infty$; f is discontinuous at $x = 3$, since $f(3)$ does not exist, but there is no vertical asymptote at $x = 3$, since $\lim_{x \to 3} f(x) = 0.5$.

5. $-\infty < x < -1$ or $1 < x < 3$; $(-\infty, -1) \cup (1, 3)$

A *In Problems 1–6, sketch a possible graph of a function that satisfies the given conditions at $x = 1$, and discuss the continuity of f at $x = 1$.*

1. $f(1) = 2$ and $\lim_{x \to 1} f(x) = 2$

2. $f(1) = -2$ and $\lim_{x \to 1} f(x) = 2$

3. $f(1) = 2$ and $\lim_{x \to 1} f(x) = -2$

4. $f(1) = -2$ and $\lim_{x \to 1} f(x) = -2$

5. $f(1) = -2$, $\lim_{x \to 1^-} f(x) = 2$, and $\lim_{x \to 1^+} f(x) = -2$

6. $f(1) = 2$, $\lim_{x \to 1^-} f(x) = 2$, and $\lim_{x \to 1^+} f(x) = -2$

In Problems 7–10, sketch a possible graph of a function that is continuous for all x except x = 1 and satisfies the given conditions at x = 1.

7. $\lim\limits_{x \to 1^-} f(x) = -\infty$ **8.** $\lim\limits_{x \to 1} f(x) = \infty$

9. $\lim\limits_{x \to 1^-} f(x) = -\infty$ and $\lim\limits_{x \to 1^+} f(x) = \infty$

10. $\lim\limits_{x \to 1^-} f(x) = \infty$ and $\lim\limits_{x \to 1^+} f(x) = -\infty$

Use Theorem 1 to determine where each function in Problems 11–16 is continuous.

11. $f(x) = 3x - 4$ **12.** $h(x) = 4 - 2x$

13. $g(x) = \dfrac{3x}{x + 2}$ **14.** $k(x) = \dfrac{2x}{x - 4}$

15. $m(x) = \dfrac{x + 1}{(x - 1)(x + 4)}$

16. $n(x) = \dfrac{x - 2}{(x - 3)(x + 1)}$

B *Problems 17–20 refer to the function f shown in the graph. Use the graph to estimate limits as outlined below.*

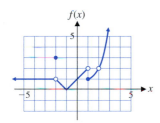

For each value of c,

(A) *Find* $\lim\limits_{x \to c^-} f(x), \lim\limits_{x \to c^+} f(x), \lim\limits_{x \to c} f(x),$ *and* $f(c)$.

(B) *Is f continuous at x = c? Explain.*

17. $c = 1$ **18.** $c = 2$

19. $c = -2$ **20.** $c = -1$

21. Given the following function f,

$$f(x) = \begin{cases} 2 & \text{if } x \text{ is an integer} \\ 1 & \text{if } x \text{ is not an integer} \end{cases}$$

  (A) Graph f.   (B) $\lim\limits_{x \to 2} f(x) = ?$ (C) $f(2) = ?$

  (D) Is f continuous at $x = 2$?

  (E) Where is f discontinuous?

22. Given the following function g,

$$g(x) = \begin{cases} -1 & \text{if } x \text{ is an even integer} \\ 1 & \text{if } x \text{ is not an even integer} \end{cases}$$

  (A) Graph g.   (B) $\lim\limits_{x \to 1} g(x) = ?$ (C) $g(1) = ?$

  (D) Is g continuous at $x = 1$?

  (E) Where is g discontinuous?

In Problems 23–32, use $-\infty$ *or* ∞ *where appropriate to describe the behavior at each point of discontinuity, and identify all vertical asymptotes.*

23. $f(x) = \dfrac{1}{x + 3}$ **24.** $g(x) = \dfrac{x}{4 - x}$

25. $h(x) = \dfrac{x^2 + 4}{x^2 - 4}$ **26.** $k(x) = \dfrac{x^2 - 9}{x^2 + 9}$

27. $F(x) = \dfrac{x^2 - 4}{x^2 + 4}$ **28.** $G(x) = \dfrac{x^2 + 9}{9 - x^2}$

29. $H(x) = \dfrac{x^2 - 2x - 3}{x^2 - 4x + 3}$ **30.** $K(x) = \dfrac{x^2 + 2x - 3}{x^2 - 4x + 3}$

31. $T(x) = \dfrac{8x - 16}{x^4 - 8x^3 + 16x^2}$

32. $S(x) = \dfrac{6x + 9}{x^4 + 6x^3 + 9x^2}$

In Problems 33–38, solve each inequality using a sign chart. Express answers in inequality and interval notation.

33. $x^2 - x - 12 < 0$ **34.** $x^2 - 2x - 8 < 0$

35. $x^2 + 21 > 10x$ **36.** $x^2 + 7x > -10$

37. $\dfrac{x^2 + 5x}{x - 3} > 0$ **38.** $\dfrac{x - 4}{x^2 + 2x} < 0$

 In Problems 39–44, use a graphing utility to approximate the partition numbers of each function $f(x)$ to two decimal places. Then solve the following inequalities:

  (A) $f(x) > 0$ (B) $f(x) < 0$

Express answers in interval notation.

39. $f(x) = x^3 - 3x^2 - 2x + 5$

40. $f(x) = x^3 + 3x^2 - 4x - 8$

41. $f(x) = x^4 - 6x^2 + 3x + 5$

42. $f(x) = x^4 - 4x^2 - 2x + 2$

43. $f(x) = \dfrac{3 + 6x - x^3}{x - 1}$

44. $f(x) = \dfrac{x^3 - 5x + 1}{x + 1}$

Use Theorem 1 to determine where each function in Problems 45–52 is continuous. Express the answer in interval notation.

45. $\sqrt{x - 6}$ **46.** $\sqrt{7 - x}$

47. $\sqrt[3]{5 - x}$ **48.** $\sqrt[3]{x - 8}$

49. $\sqrt{x^2 - 9}$ **50.** $\sqrt{4 - x^2}$

51. $\sqrt{x^2 + 1}$ **52.** $\sqrt[3]{x^2 + 2}$

In Problems 53–58, graph f, locate all points of discontinuity, and discuss the behavior of f at these points.

53. $f(x) = \begin{cases} 1 + x & \text{if } x < 1 \\ 5 - x & \text{if } x \geqslant 1 \end{cases}$

54. $f(x) = \begin{cases} x^2 & \text{if } x \le 1 \\ 2x & \text{if } x > 1 \end{cases}$

55. $f(x) = \begin{cases} 1 + x & \text{if } x \le 2 \\ 5 - x & \text{if } x > 2 \end{cases}$

56. $f(x) = \begin{cases} x^2 & \text{if } x \le 2 \\ 2x & \text{if } x > 2 \end{cases}$

57. $f(x) = \begin{cases} -x & \text{if } x < 0 \\ 1 & \text{if } x = 0 \\ x & \text{if } x > 0 \end{cases}$

58. $f(x) = \begin{cases} 1 & \text{if } x < 0 \\ 0 & \text{if } x = 0 \\ 1 + x & \text{if } x > 0 \end{cases}$

 In Problems 59–62, locate all points of discontinuity of f, using a graphing utility as an aid, and discuss the behavior of f at these points. [Hint: Select Xmin and Xmax so that the suspected point of discontinuity is the midpoint of the graphing interval (Xmin, Xmax).]

59. $f(x) = x + \dfrac{|2x - 4|}{x - 2}$

60. $f(x) = x + \dfrac{|3x + 9|}{x + 3}$

61. $f(x) = \dfrac{x^2 - 1}{|x| - 1}$

62. $f(x) = \dfrac{x^3 - 8}{|x| - 2}$

C

63. Use the graph of the function g to answer the following questions:

(A) Is g continuous on the open interval $(-1, 2)$?

(B) Is g continuous from the right at $x = -1$? That is, does $\lim_{x \to -1^+} g(x) = g(-1)$?

(C) Is g continuous from the left at $x = 2$? That is, does $\lim_{x \to 2^-} g(x) = g(2)$?

(D) Is g continuous on the closed interval $[-1, 2]$?

$g(x) = -x^2 + 2x + 2$

Figure for 63

64. Use the graph of the function f to answer the following questions:

(A) Is f continuous on the open interval $(0, 3)$?

(B) Is f continuous from the right at $x = 0$? That is, does $\lim_{x \to 0^+} f(x) = f(0)$?

(C) Is f continuous from the left at $x = 3$? That is, does $\lim_{x \to 3^-} f(x) = f(3)$?

(D) Is f continuous on the closed interval $[0, 3]$?

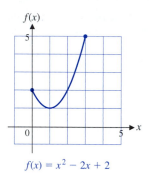

$f(x) = x^2 - 2x + 2$

Figure for 64

*Problems 65 and 66 refer to the **greatest integer function**, which is denoted by $[x]$ and is defined as follows:*

$$[x] = \text{greatest integer} \le x$$

For example,

$$[-3.6] = \text{greatest integer} \le -3.6 = -4$$
$$[2] = \text{greatest integer} \le 2 = 2$$
$$[2.5] = \text{greatest integer} \le 2.5 = 2$$

The graph of $f(x) = [x]$ is shown. There, we can see that

$$\begin{aligned}
[x] &= -2 &&\text{for} && -2 \le x < -1 \\
[x] &= -1 &&\text{for} && -1 \le x < 0 \\
[x] &= 0 &&\text{for} && 0 \le x < 1 \\
[x] &= 1 &&\text{for} && 1 \le x < 2 \\
[x] &= 2 &&\text{for} && 2 \le x < 3
\end{aligned}$$

and so on.

$f(x) = [x]$

Figure for 65 and 66

65. (A) Is *f* continuous from the right at $x = 0$?

(B) Is *f* continuous from the left at $x = 0$?

(C) Is *f* continuous on the open interval $(0, 1)$?

(D) Is *f* continuous on the closed interval $[0, 1]$?

(E) Is *f* continuous on the half-closed interval $[0, 1)$?

66. (A) Is *f* continuous from the right at $x = 2$?

(B) Is *f* continuous from the left at $x = 2$?

(C) Is *f* continuous on the open interval $(1, 2)$?

(D) Is *f* continuous on the closed interval $[1, 2]$?

(E) Is *f* continuous on the half-closed interval $[1, 2)$?

In Problems 67–70, sketch a possible graph for a function f that is continuous for all real numbers and satisfies the given conditions. Find the x intercepts for f.

67. $f(x) < 0$ on $(-\infty, -5)$ and $(2, \infty)$; $f(x) > 0$ on $(-5, 2)$

68. $f(x) > 0$ on $(-\infty, -4)$ and $(3, \infty)$; $f(x) < 0$ on $(-4, 3)$

69. $f(x) < 0$ on $(-\infty, -6)$ and $(-1, 4)$; $f(x) > 0$ on $(-6, -1)$ and $(4, \infty)$

70. $f(x) > 0$ on $(-\infty, -3)$ and $(2, 7)$; $f(x) < 0$ on $(-3, 2)$ and $(7, \infty)$

71. The function $f(x) = 2/(1 - x)$ satisfies $f(0) = 2$ and $f(2) = -2$. Is *f* equal to 0 anywhere on the interval $(-1, 3)$? Does this contradict Theorem 2? Explain.

72. The function $f(x) = 6/(x - 4)$ satisfies $f(2) = -3$ and $f(7) = 2$. Is *f* equal to 0 anywhere on the interval $(0, 9)$? Does this contradict Theorem 2? Explain.

73. The function *f* is continuous and never 0 on the interval $(0, 4)$, and continuous and never 0 on the interval $(4, 8)$. Also, $f(2) = 3$ and $f(6) = -3$. Discuss the validity of the following statement and illustrate your conclusions with graphs: Either $f(4) = 0$ or *f* is discontinuous at $x = 4$.

74. The function *f* is continuous and never 0 on the interval $(-3, 1)$, and continuous and never 0 on the interval $(1, 4)$. Also, $f(-2) = -3$ and $f(3) = 4$. Discuss the validity of the following statement and illustrate your conclusions with graphs: Either $f(1) = 0$ or *f* is discontinuous at $x = 1$.

Applications

Business & Economics

75. *Postal rates.* First-class postage in 2001 was $0.34 for the first ounce (or any fraction thereof) and $0.21 for each additional ounce (or fraction thereof).

(A) Write a piecewise definition of the first-class postage $P(x)$ for a letter weighing *x* ounces.

(B) Graph $P(x)$ for $0 < x \leqslant 5$.

(C) Is $P(x)$ continuous at $x = 4.5$? At $x = 4$? Explain.

76. *Telephone rates.* A long-distance telephone service charges $0.15 for the first minute (or any fraction thereof) and $0.10 for each additional minute (or fraction thereof).

(A) Write a piecewise definition of the charge $R(x)$ for a state-to-state long-distance call lasting *x* minutes.

(B) Graph $R(x)$ for $0 < x \leqslant 6$

(C) Is $R(x)$ continuous at $x = 3.5$? At $x = 3$? Explain.

77. *Postal rates.* Discuss the differences between the function $Q(x) = 0.34 + 0.21[\![x]\!]$ and the function $P(x)$ defined in Problem 75.

78. *Telephone rates.* Discuss the differences between the function $S(x) = 0.15 + 0.1[\![x]\!]$ and the function $R(x)$ defined in Problem 76.

79. *Natural gas rates.* Table 5 shows the rates for natural gas charged by the Middle Tennessee Natural Gas Utility District during the summer months. The customer charge is a fixed monthly charge, independent of the amount of gas used during the month.

(A) Write a piecewise definition of the monthly charge $S(x)$ for a customer who uses *x* therms* in a summer month.

www

TABLE 5	
SUMMER (MAY–SEPTEMBER)	
Base charge	$5.00
First 50 therms	0.63 per therm
Over 50 therms	0.45 per therm

*A British thermal unit (Btu) is the amount of heat required to raise the temperature of 1 pound of water 1 degree Fahrenheit and a therm is 100,000 Btu.

(B) Graph $S(x)$.

(C) Is $S(x)$ continuous at $x = 50$? Explain.

80. *Natural gas rates.* Table 6 shows the rates for natural gas charged by the Middle Tennessee Natural Gas Utility District during the winter months. The customer charge is a fixed monthly charge, independent of the amount of gas used during the month.

(A) Write a piecewise definition of the monthly charge $S(x)$ for a customer who uses x therms in a winter month.

(B) Graph $S(x)$.

(C) Is $S(x)$ continuous at $x = 5$? At $x = 50$? Explain.

www

TABLE 6	
WINTER (OCTOBER–APRIL)	
Base charge	$5.00
First 5 therms	0.69 per therm
Next 45 therms	0.65 per therm
Over 50 therms	0.63 per therm

81. *Income.* A personal computer salesperson receives a base salary of $1,000 per month and a commission of 5% of all sales over $10,000 during the month. If the monthly sales are $20,000 or more, the salesperson is given an additional $500 bonus. Let $E(s)$ represent the person's earnings during the month as a function of the monthly sales s.

(A) Graph $E(s)$ for $0 \le s \le 30,000$.

(B) Find $\lim_{s \to 10,000} E(s)$ and $E(10,000)$.

(C) Find $\lim_{s \to 20,000} E(s)$ and $E(20,000)$.

(D) Is E continuous at $s = 10,000$? At $s = 20,000$?

82. *Equipment rental.* An office equipment rental and leasing company rents typewriters for $10 per day (and any fraction thereof) or for $50 per 7 day week. Let $C(x)$ be the cost of renting a typewriter for x days.

(A) Graph $C(x)$ for $0 \le x \le 10$.

(B) Find $\lim_{x \to 4.5} C(x)$ and $C(4.5)$.

(C) Find $\lim_{x \to 8} C(x)$ and $C(8)$.

(D) Is C continuous at $x = 4.5$? At $x = 8$?

Life Sciences

83. *Animal supply.* A medical laboratory raises its own rabbits. The number of rabbits $N(t)$ available at any time t depends on the number of births and deaths.

When a birth or death occurs, the function N generally has a discontinuity, as shown in the figure.

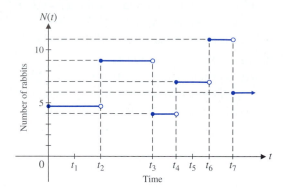

Figure for 83

(A) Where is the function N discontinuous?

(B) $\lim_{t \to t_5} N(t) = ?$; $N(t_5) = ?$

(C) $\lim_{t \to t_3} N(t) = ?$; $N(t_3) = ?$

Social Sciences

84. *Learning.* The graph might represent the history of a particular person learning the material on limits and continuity in this book. At time t_2, the student's mind goes blank during a quiz. At time t_4, the instructor explains a concept particularly well, and suddenly, a big jump in understanding takes place.

(A) Where is the function p discontinuous?

(B) $\lim_{t \to t_1} p(t) = ?$; $p(t_1) = ?$

(C) $\lim_{t \to t_2} p(t) = ?$; $p(t_2) = ?$

(D) $\lim_{t \to t_4} p(t) = ?$; $p(t_4) = ?$

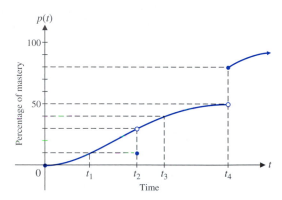

Figure for 84

Section 3-3

The Derivative

- ❑ RATE OF CHANGE
- ❑ SLOPE
- ❑ THE DERIVATIVE

We now use the concept of limit to solve the two important problems illustrated in Figure 1. The solution of these two apparently unrelated problems involves a common concept called *the derivative*.

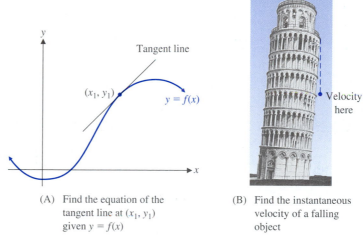

(A) Find the equation of the tangent line at (x_1, y_1) given $y = f(x)$

(B) Find the instantaneous velocity of a falling object

FIGURE 1 Two basic problems of calculus

❑ RATE OF CHANGE

Let us start by considering a simple example.

Example 1

Revenue Analysis   The revenue (in dollars) from the sale of x plastic planter boxes is given by

$$R(x) = 20x - 0.02x^2 \qquad 0 \leq x \leq 1{,}000$$

which is graphed in Figure 2.

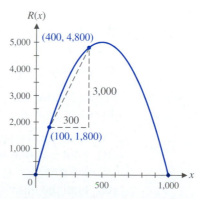

FIGURE 2 $R(x) = 20x - 0.02x^2$

(A) What is the change in revenue if production is changed from 100 planters to 400 planters?

(B) What is the average change in revenue for this change in production?

SOLUTION (A) The change in revenue is given by

$$R(400) - R(100) = 20(400) - 0.02(400)^2 - [20(100) - 0.02(100)^2]$$
$$= 4,800 - 1,800 = \$3,000$$

Thus, increasing production from 100 planters to 400 planters will increase revenue by $3,000.

(B) To find the average change in revenue, we divide the change in revenue by the change in production:

$$\frac{R(400) - R(100)}{400 - 100} = \frac{3,000}{300} = \$10$$

Thus, the average change in revenue is $10 per planter when production is increased from 100 to 400 planters. ∎

Matched Problem 1 Refer to the revenue function in Example 1.

(A) What is the change in revenue if production is changed from 600 planters to 800 planters?

(B) What is the average change in revenue for this change in production? ∎

In general, if we are given a function $y = f(x)$ and if x is changed from a to $a + h$, then y will change from $f(a)$ to $f(a + h)$. The *average rate of change is the ratio of the change in y to the change in x.*

Average Rate of Change

For $y = f(x)$, the **average rate of change from $x = a$ to $x = a + h$** is

$$\frac{f(a + h) - f(a)}{(a + h) - a} = \frac{f(a + h) - f(a)}{h} \qquad h \neq 0 \qquad (1)$$

As we noted in Section 3-1, the mathematical expression (1) is called the **difference quotient.** The preceding discussion shows that the difference quotient can be interpreted as an average rate of change. The next example illustrates another interpretation of this quotient: velocity of a moving object.

Example 2 **Velocity** A small steel ball dropped from a tower will fall a distance of y feet in x seconds, as given approximately by the formula (from physics)

$$y = f(x) = 16x^2$$

Figure 3 shows the position of the ball on a coordinate line (positive direction down) at the end of 0, 1, 2, and 3 seconds.

(A) Find the average velocity from $x = 2$ seconds to $x = 3$ seconds.

(B) Find and simplify the average velocity from $x = 2$ seconds to $x = 2 + h$ seconds, $h \neq 0$.

(C) Find the limit of the expression from part B as $h \to 0$, if it exists.

(D) Discuss possible interpretations of the limit from part C.

0 ← Position at start ($x = 0$ second)
16 ← Position at $x = 1$ second [$y = 16(1^2) = 16$ feet]

64 ← Position at $x = 2$ seconds [$y = 16(2^2) = 64$ feet]

144 ← Position at $x = 3$ seconds [$y = 16(3^2) = 144$ feet]

Ground

FIGURE 3 *Note:*   Positive y direction is down.

SOLUTION   (A) Recall the formula $d = rt$, which can be written in the form

$$r = \frac{d}{t} = \frac{\text{Distance covered}}{\text{Elapsed time}} = \text{Average velocity}$$

For example, if a person drives from San Francisco to Los Angeles (a distance of about 420 miles) in 7 hours, then the average velocity is

$$r = \frac{d}{t} = \frac{420}{7} = 60 \text{ miles per hour}$$

Sometimes the person will be traveling faster and sometimes slower, but the average velocity is 60 miles per hour. In our present problem, the average velocity of the steel ball from $x = 2$ seconds to $x = 3$ seconds is

$$\begin{aligned}
\text{Average velocity} &= \frac{\text{Distance covered}}{\text{Elapsed time}} \\
&= \frac{f(3) - f(2)}{3 - 2} \\
&= \frac{16(3)^2 - 16(2)^2}{1} = 80 \text{ feet per second}
\end{aligned}$$

Thus, we see that if $y = f(x)$ is the position of the falling ball, then the average velocity is simply the average rate of change of $f(x)$ with respect to time x. And we have another interpretation of the difference quotient (1).

(B) Proceeding as in part A,

$$\begin{aligned}
\text{Average velocity} &= \frac{\text{Distance covered}}{\text{Elapsed time}} \\
&= \frac{f(2 + h) - f(2)}{h} \qquad \text{\color{blue}Difference quotient} \\
&= \frac{16(2 + h)^2 - 16(2)^2}{h} \qquad \text{\color{blue}Simplify this 0/0} \\
& \qquad\qquad\qquad\qquad\qquad \text{\color{blue}indeterminate form.} \\
&= \frac{64 + 64h + 16h^2 - 64}{h} \\
&= \frac{h(64 + 16h)}{h} = 64 + 16h \qquad h \neq 0
\end{aligned}$$

Notice that if $h = 1$, the average velocity is 80 feet per second, our result in part A.

(C) The limit of the average velocity expression from part B as $h \to 0$ is

$$\lim_{h \to 0} \frac{f(2 + h) - f(2)}{h} = \lim_{h \to 0}(64 + 16h)$$

$$= 64 \text{ feet per second}$$

(D) The average velocity over smaller and smaller time intervals approaches 64 feet per second. This limit can be interpreted as the velocity of the ball at the *instant* when the ball has been falling for exactly 2 seconds. Thus, 64 feet per second is referred to as the **instantaneous velocity** at $x = 2$ seconds. And we have solved the one of the basic problems of calculus (see Fig. 1B). ■

Matched Problem 2 ✏ For the falling steel ball in Example 2, find

(A) The average velocity from $x = 1$ second to $x = 2$ seconds

(B) The average velocity (in simplified form) from $x = 1$ second to $x = 1 + h$ seconds, $h \neq 0$

(C) The instantaneous velocity at $x = 1$ second ■

Explore–Discuss 1

Recall the revenue function in Example 1: $R(x) = 20x - 0.02x^2$. Find

$$\lim_{h \to 0} \frac{R(100 + h) - R(100)}{h}$$

Discuss possible interpretations of this limit.

The ideas introduced in Example 2 are not confined to average velocity but can be applied to the average rate of change of any function.

Instantaneous Rate of Change

For $y = f(x)$, the **instantaneous rate of change at $x = a$** is

$$\lim_{h \to 0} \frac{f(a + h) - f(a)}{h} \tag{2}$$

if the limit exists.

The adjective *instantaneous* is often omitted with the understanding that the phrase **rate of change** always refers to the instantaneous rate of change and not the average rate of change. Similarly, **velocity** always refers to the instantaneous rate of change of distance with respect to time.

❑ SLOPE

So far our interpretations of the difference quotient have been numerical in nature. Now we want to consider a geometric interpretation. A line through two points on the graph of a function is called a **secant line**. If $(a, f(a))$ and $(a + h, f(a + h))$ are two points on the graph of $y = f(x)$, we can use the slope formula from Section 1-3 to find the slope of the secant line through these points (see Fig. 4).

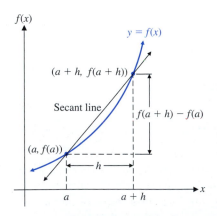

FIGURE 4 Secant line

$$\textbf{Slope of secant line} = \frac{f(a + h) - f(a)}{(a + h) - a}$$

$$= \frac{f(a + h) - f(a)}{h} \qquad \textit{Difference quotient}$$

Thus, the difference quotient can be interpreted as both the average rate of change and the slope of the secant line.

Example 3 ⇨ **Slope of a Secant Line**   Given $f(x) = x^2$,

(A) Find the slope of the secant line for $a = 1$, and $h = 2$ and 1, respectively. Graph $y = f(x)$ and the two secant lines.

(B) Find and simplify the slope of the secant line for $a = 1$ and h any nonzero number.

(C) Find the limit of the expression in part B.

(D) Discuss possible interpretations of the limit in part C.

SOLUTION   (A) For $a = 1$ and $h = 2$, the secant line goes through $(1, f(1)) = (1, 1)$ and $(3, f(3)) = (3, 9)$, and its slope is

$$\frac{f(1 + 2) - f(1)}{2} = \frac{3^2 - 1^2}{2} = 4$$

For $a = 1$ and $h = 1$, the secant line goes through $(1, f(1)) = (1, 1)$ and $(2, f(2)) = (2, 4)$, and its slope is

$$\frac{f(1 + 1) - f(1)}{1} = \frac{2^2 - 1^2}{1} = 3$$

The graphs of $y = f(x)$ and the two secant lines are shown in Figure 5.

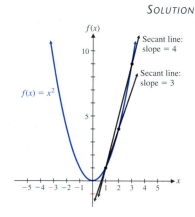

FIGURE 5 Secant lines

(B) For $a = 1$ and h any nonzero number, the secant line goes through $(1, f(1)) = (1, 1)$ and $(1 + h, f(1 + h)) = (1 + h, (1 + h)^2)$, and its slope is

$$\frac{f(1 + h) - f(h)}{h} = \frac{(1 + h)^2 - 1^2}{h}$$

$$= \frac{1 + 2h + h^2 - 1}{h}$$

$$= \frac{h(2 + h)}{h}$$

$$= 2 + h \qquad h \neq 0$$

(C) The limit of the secant line slope from part B is

$$\lim_{h \to 0} \frac{f(1 + h) - f(h)}{h} = \lim_{h \to 0} (2 + h)$$

$$= 2$$

(D) In part C, we saw that the limit of the slopes of the secant lines through the point $(1, f(1))$ is 2. If we graph the line through $(1, f(1))$ with slope 2 (Fig. 6), this line appears to be the limit of the secant lines. The slope obtained from the limit of slopes of secants lines is called the *slope of the graph* at $x = 1$ and the line through the point $(1, f(1))$ with this slope is called the *tangent line*. And we have solved another basic problem of calculus (see Fig. 1A). ■

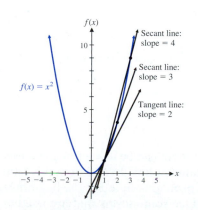

FIGURE 6 Tangent line

Matched Problem 3 ➯ Given $f(x) = x^2$,

(A) Find the slope of the secant line for $a = 2$, and $h = 2$ and 1, respectively.

(B) Find and simplify the slope of the secant line for $a = 2$ and h any non-zero number.

(C) Find the limit of the expression in part B.

(D) Find the slope of the graph and the slope of the tangent line at $a = 2$. ∎

The ideas introduced in the preceding example are summarized in the following box.

> ### Slope of a Graph
>
> Given $y = f(x)$, the **slope of the graph** at the point $(a, f(a))$ is given by
>
> $$\lim_{h \to 0} \frac{f(a + h) - f(a)}{h} \qquad (3)$$
>
> provided that the limit exists. The slope of the graph is also the **slope of the tangent line** at the point $(a, f(a))$.

From plane geometry we know that a line tangent to a circle is a line that passes through one and only one point of the circle (see Fig. 7A). Although this definition cannot be extended to graphs of functions in general, the visual relationship between graphs of functions and their tangent lines is similar to the circle case (see Fig. 7B). Limit (3) provides both a mathematically sound definition for the concept of tangent line and a method for approximating the slope of the tangent line.

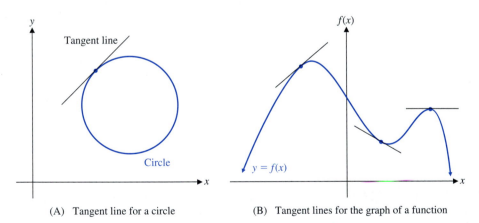

(A) Tangent line for a circle

(B) Tangent lines for the graph of a function

FIGURE 7

❑ THE DERIVATIVE

We have seen that the limit of a difference quotient can be interpreted as a rate of change, as a velocity, or as the slope of a tangent line. In addition, this limit provides solutions to two of the three basic problems stated at the beginning of this chapter. We are now ready to introduce some terms that are used to refer to this limit. To follow customary practice, we use x in place of a and think of the difference quotient

$$\frac{f(x + h) - f(x)}{h}$$

as a function of h, with x held fixed as h tends to 0.

The Derivative

For $y = f(x)$, we define the **derivative of f at x,** denoted by $f'(x)$, to be

$$f'(x) = \lim_{h \to 0} \frac{f(x + h) - f(x)}{h} \text{ if the limit exists}$$

If $f'(x)$ exists for each x in the open interval (a, b), then f is said to be **differentiable** over (a, b).

(Differentiability from the left or from the right is defined using $h \to 0^-$ or $h \to 0^+$, respectively, in place of $h \to 0$ in the definition above.)

The process of finding the derivative of a function is called **differentiation.** That is, the derivative of a function is obtained by **differentiating** the function.

Interpretations of the Derivative

The derivative of a function f is a new function f'. The domain of f' is a subset of the domain of f. The derivative has various applications and interpretations, including the following:

1. *Slope of the tangent line.* For each x in the domain of f', $f'(x)$ is the slope of the line tangent to the graph of f at the point $(x, f(x))$.
2. *Instantaneous rate of change.* For each x in the domain of f', $f'(x)$ is the instantaneous rate of change of $y = f(x)$ with respect to x.
3. *Velocity.* If $f(x)$ is the position of a moving object at time x, then $v = f'(x)$ is the velocity of the object at that time.

Example 4 illustrates the **two-step process** that we use to find derivatives in this section. In subsequent sections, we develop rules for finding derivatives that do not involve limits. However, it is important that you master the limit process in order to fully comprehend and appreciate the various applications we will consider.

Example 4   **Finding a Derivative**   Find $f'(x)$, the derivative of f at x, for $f(x) = 4x - x^2$.

SOLUTION   To find $f'(x)$, we use a two-step process:

Step 1. Form the difference quotient and simplify:

$$\frac{f(x + h) - f(x)}{h} = \frac{[4(x + h) - (x + h)^2] - (4x - x^2)}{h}$$

$$= \frac{4x + 4h - x^2 - 2xh - h^2 - 4x + x^2}{h}$$

$$= \frac{4h - 2xh - h^2}{h}$$

$$= \frac{h(4 - 2x - h)}{h}$$

$$= 4 - 2x - h \qquad h \neq 0$$

Step 2.   Find the limit of the difference quotient:

$$f'(x) = \lim_{h \to 0} \frac{f(x + h) - f(x)}{h}$$

$$= \lim_{h \to 0} (4 - 2x - h) = 4 - 2x$$

Thus, if $f(x) = 4x - x^2$, then $f'(x) = 4 - 2x$. The derivative f' is a new function derived from the function f.

■

Matched Problem 4   ➯   Find $f'(x)$, the derivative of f at x, for $f(x) = 8x - 2x^2$.

■

Example 5   ➯   **Finding Tangent Line Slopes**   In Example 4 we started with the function
WWW   specified by $f(x) = 4x - x^2$ and found the derivative of f at x to be $f'(x) = 4 - 2x$. Thus, the slope of a tangent line to the graph of f at any point $(x, f(x))$ on the graph is

$$m = f'(x) = 4 - 2x$$

(A) Find the slope of the graph of f at $x = 0$, $x = 2$, and $x = 3$.
(B) Graph $y = f(x) = 4x - x^2$, and use the slopes found in part (A) to make a rough sketch of the tangent lines to the graph at $x = 0$, $x = 2$, and $x = 3$.

SOLUTION   (A) Using $f'(x) = 4 - 2x$, we have

$$f'(0) = 4 - 2(0) = 4 \qquad \textit{Slope at x = 0}$$
$$f'(2) = 4 - 2(2) = 0 \qquad \textit{Slope at x = 2}$$
$$f'(3) = 4 - 2(3) = -2 \qquad \textit{Slope at x = 3}$$

(B)

Matched Problem 5   ➯   In Matched Problem 4, we started with the function specified by $f(x) = 8x - 2x^2$. Using the derivative found there:

(A) Find the slope of the graph of f at $x = 1$, $x = 2$, and $x = 4$.
(B) Graph $y = f(x) = 8x - 2x^2$, and use the slopes from part (A) to make a rough sketch of the tangent lines to the graph at $x = 1$, $x = 2$, and $x = 4$.

■

Explore–Discuss 2

In Example 4 we found that the derivative of $f(x) = 4x - x^2$ is $f'(x) = 4 - 2x$, and in Example 5 we graphed $f(x)$ and several tangent lines.

(A) Graph f and f' on the same set of axes.

(B) The graph of f' is a straight line. Is it a tangent line for the graph of f? Explain.

(C) Find the x intercept for the graph of f'. What is the slope of the line tangent to the graph of f for this value of x? Write a verbal description of the relationship between the slopes of the tangent lines of a function and the x intercepts of the derivative of the function.

REMARK: DERIVATIVES AND TECHNOLOGY

Refer to the values of $f'(x)$ computed in Example 5A. Some graphing utilities, including most graphing calculators, have a built-in **numerical differentiation** routine that will approximate numerically the value of $f'(x)$ for any given value of x. Figure 8A illustrates this approach on the TI-83 graphing calculator. Other graphing utilities use **symbolic differentiation** to find an algebraic formula for the derivative, and then evaluate this formula at the indicated values of x. Figure 8B illustrates this approach in Maple V, a very powerful mathematical software package.

```
nDeriv(Y1,X,0)
                    4
nDeriv(Y1,X,2)
                    0
nDeriv(Y1,X,3)
                   -2
```

```
• f:=4*x-x^2;d:=diff(f,x);
                    2
       f := 4 x - x
       d := 4 - 2 x
• x:=0;d;x:=2;d;x:=3;d;
       x := 0
             4
       x := 2
             0
       x := 3
            -2
```

(A) Numerical differentiation:
    y1=4x-x^2

(B) Symbolic differentiation:
    f:=4*x-x^2

FIGURE 8

Differentiation is as basic to calculus as addition is to algebra, and it is important that you learn to perform this operation for yourself, without using technology. However, you certainly can use numerical differentiation on a graphing utility to check your work.

Example 6   **Finding a Derivative**   Find $f'(x)$, the derivative of f at x, for $f(x) = \sqrt{x} + 2$.

SOLUTION   To find $f'(x)$, we find the following limit using the two-step process.

$$\lim_{h \to 0} \frac{f(x + h) - f(x)}{h}$$

Step 1.   Form the difference quotient and simplify:

$$\frac{f(x + h) - f(x)}{h} = \frac{(\sqrt{x + h} + 2 - (\sqrt{x} + 2)}{h}$$

$$= \frac{\sqrt{x + h} - \sqrt{x}}{h}$$

Since this is a 0/0 indeterminate form, we change the form by rationalizing the numerator:

$$\frac{\sqrt{x+h}-\sqrt{x}}{h} \cdot \frac{\sqrt{x+h}+\sqrt{x}}{\sqrt{x+h}+\sqrt{x}} = \frac{x+h-x}{h(\sqrt{x+h}+\sqrt{x})}$$

$$= \frac{h}{h(\sqrt{x+h}+\sqrt{x})}$$

$$= \frac{1}{\sqrt{x+h}+\sqrt{x}} \qquad h \neq 0$$

Step 2. Find the limit of the difference quotient:

$$f'(x) = \lim_{h \to 0} \frac{f(x+h)-f(x)}{h}$$

$$= \lim_{h \to 0} \frac{1}{\sqrt{x+h}+\sqrt{x}}$$

$$= \frac{1}{\sqrt{x}+\sqrt{x}} = \frac{1}{2\sqrt{x}} \qquad x > 0$$

Thus, the derivative of $f(x) = \sqrt{x} + 2$ is $f'(x) = 1/(2\sqrt{x})$, a new function. The domain of f is $[0, \infty)$. Since $f'(0)$ is not defined, the domain of f' is $(0, \infty)$, a subset of the domain of f. ■

Matched Problem 6 ⇨ Find $f'(x)$ for $f(x) = \sqrt{x} + 4$. ■

Example 7 ⇨ **Sales Analysis** The total sales of a company (in millions of dollars) t months from now are given by $S(t) = \sqrt{t} + 2$. Find $S(25)$ and $S'(25)$, and interpret. Use these results to estimate the total sales after 26 months and after 27 months. Check the values of $S(25)$ and $S'(25)$ on a graphing utility.

SOLUTION The total sales function S has the same form as the function f in Example 6— only the letters used to represent the function and the independent variable have been changed. It follows that S' and f' also have the same form:

$$S(t) = \sqrt{t} + 2 \qquad f(x) = \sqrt{x} + 2$$

$$S'(t) = \frac{1}{2\sqrt{t}} \qquad f'(x) = \frac{1}{2\sqrt{x}}$$

Evaluating S and S' at $t = 25$, we have

$$S(25) = \sqrt{25} + 2 = 7 \qquad S'(25) = \frac{1}{2\sqrt{25}} = 0.1$$

Figure 9 shows a check of these results on a graphing utility. Thus, 25 months from now the total sales are $7 million and are increasing at the rate of $0.1 million ($100,000) per month. If this instantaneous rate of change of sales remained constant, the sales would grow to $7.1 million after 26 months, $7.2 million after 27 months, and so on. Even though $S'(t)$ is not a constant function in this case, these values provide useful estimates of the total sales. ■

```
25→X
              25
Y1
               7
nDeriv(Y1,X,25)
      .1000000001
```

FIGURE 9 $y_1 = \sqrt{x} + 2$

Matched Problem 7   The total sales of a company (in millions of dollars) t months from now are given by $S(t) = \sqrt{t} + 4$. Find $S(12)$ and $S'(12)$, and interpret. Use these results to estimate the total sales after 13 months and after 14 months. (Use the derivative found in Matched Problem 6.) Check with a graphing utility.

Refer to Example 7. It is instructive to compare the estimates of total sales obtained by using the derivative with the corresponding exact values of $S(t)$:

Exact values   Estimated values

$$S(26) = \sqrt{26} + 2 = 7.099\ldots \approx 7.1$$
$$S(27) = \sqrt{27} + 2 = 7.196\ldots \approx 7.2$$

For this function, the estimated values provide very good approximations to the exact values of $S(t)$. For other functions, the approximation might not be as accurate.

Using the instantaneous rate of change of a function at a point to estimate values of the function at nearby points is a simple but important application of the derivative.

❑ NONEXISTENCE OF THE DERIVATIVE

The existence of a derivative at $x = a$ depends on the existence of a limit at $x = a$, that is, on the existence of

$$f'(a) = \lim_{h \to 0} \frac{f(a + h) - f(a)}{h} \tag{4}$$

If the limit does not exist at $x = a$, we say that the function f is **nondifferentiable at $x = a$, or $f'(a)$ does not exist.**

Explore–Discuss 3

Let $f(x) = |x - 1|$.

(A) Graph f.

(B) Complete the following table:

h	−0.1	−0.01	−0.001 → 0 ←		0.001	0.01	0.1
$\dfrac{f(1 + h) - f(1)}{h}$	?	?	?	→ ? ←	?	?	?

(C) Find the following limit, if it exists.

$$\lim_{h \to 0} \frac{f(1 + h) - f(1)}{h}$$

(D) Use the results of parts (A)–(C) to discuss the existence of $f'(1)$.

Repeat parts (A)–(D) for $g(x) = \sqrt[3]{x - 1}$.

How can we recognize the points on the graph of f where $f'(a)$ does not exist? It is impossible to describe all the ways that the limit in (1) can fail to

(A) Not continuous at $x = a$

(B) Graph has sharp corner at $x = a$

(C) Vertical tangent at $x = a$

(D) Vertical tangent at $x = a$

FIGURE 10 The function f is nondifferentiable at $x = a$.

exist. However, we can illustrate some common situations where $f'(a)$ does fail to exist (see Fig. 10):

1. If the graph of f has a hole or a break at $x = a$, then $f'(a)$ does not exist (Fig. 10A).

2. If the graph of f has a sharp corner at $x = a$, then $f'(a)$ does not exist and the graph has no tangent line at $x = a$ (Fig. 10B). (In Fig. 10B, the left- and right-hand derivatives exist but are not equal.)

3. If the graph of f has a vertical tangent line at $x = a$, then $f'(a)$ does not exist (Fig. 10C and D).

Answers to Matched Problems

1. (A) $-\$1,600$ (B) $-\$8$ per planter
2. (A) 48 ft/s   (B) $32 + 16h$ (C) 32 ft/s
3. (A) 6, 5   (B) $4 + h$ (C) 4   (D) Both are 4
4. $f'(x) = 8 - 4x$
5. (A) $f'(1) = 4, f'(2) = 0, f'(4) = -8$

   (B)

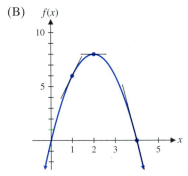

6. $f'(x) = 1/(2\sqrt{x + 4})$
7. $S(12) = 4, S'(12) = 0.125$; 12 months from now the total sales are \$4 million and are increasing at the rate of \$0.125 million (\$125,000) per month. The estimated total sales are \$4.125 million after 13 months and \$4.25 million after 14 months.

Exercise 3-3

A *In Problems 1 and 2, find the indicated quantity for*
y = f(x) = 5 − x², and interpret in terms of the graph
shown below.

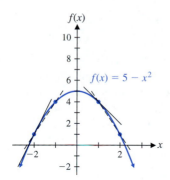

1. (A) $\dfrac{f(2) - f(1)}{2 - 1}$ (B) $\dfrac{f(1 + h) - f(1)}{h}$

   (C) $\lim\limits_{h \to 0} \dfrac{f(1 + h) - f(1)}{h}$

2. (A) $\dfrac{f(-1) - f(-2)}{-1 - (-2)}$ (B) $\dfrac{f(-2 + h) - f(-2)}{h}$

   (C) $\lim\limits_{h \to 0} \dfrac{f(-2 + h) - f(-2)}{h}$

3. Find the indicated quantities for $f(x) = 3x^2$.

   (A) The average rate of change of $f(x)$ if x changes from 1 to 4.

   (B) The slope of the secant line through the points $(1, f(1))$ and $(4, f(4))$ on the graph of $y = f(x)$.

   (C) The slope of the secant line through the points $(1, f(1))$ and $(1 + h, f(1 + h)), h \neq 0$. Simplify your answer.

   (D) The slope of the graph at $(1, f(1))$.

   (E) The instantaneous rate of change of $y = f(x)$ with respect to x at $x = 1$.

   (F) The slope of the tangent line at $(1, f(1))$.

   (G) The equation of the tangent line at $(1, f(1))$.

4. Find the indicated quantities for $f(x) = 3x^2$.

   (A) The average rate of change of $f(x)$ if x changes from 2 to 5.

   (B) The slope of the secant line through the points $(2, f(2))$ and $(5, f(5))$ on the graph of $y = f(x)$.

   (C) The slope of the secant line through the points $(2, f(2))$ and $(2 + h, f(2 + h)), h \neq 0$. Simplify your answer.

   (D) The slope of the graph at $(2, f(2))$.

   (E) The instantaneous rate of change of $y = f(x)$ with respect to x at $x = 2$.

   (F) The slope of the tangent line at $(2, f(2))$.

   (G) The equation of the tangent line at $(2, f(2))$.

In Problems 5–12, use the two-step process to find $f'(x)$
and then find $f'(1), f'(2)$, and $f'(3)$.

5. $f(x) = 4$ **6.** $f(x) = -3$

7. $f(x) = 2x - 3$ **8.** $f(x) = 4x + 3$

9. $f(x) = 2 - x^2$ **10.** $f(x) = 2x^2 + 5$

11. $f(x) = 3 + \dfrac{2}{x}$ **12.** $f(x) = \dfrac{5}{x}$

B *Problems 13 and 14 refer to the graph of*
y = f(x) = x² + x shown.

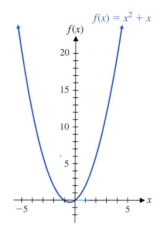

13. (A) Find the slope of the secant line joining $(1, f(1))$ and $(3, f(3))$.

   (B) Find the slope of the secant line joining $(1, f(1))$ and $(1 + h, f(1 + h))$.

   (C) Find the slope of the tangent line at $(1, f(1))$.

   (D) Find the equation of the tangent line at $(1, f(1))$.

14. (A) Find the slope of the secant line joining $(2, f(2))$ and $(4, f(4))$.

   (B) Find the slope of the secant line joining $(2, f(2))$ and $(2 + h, f(2 + h))$.

   (C) Find the slope of the tangent line at $(2, f(2))$.

   (D) Find the equation of the tangent line at $(2, f(2))$.

In Problems 15 and 16, suppose an object moves along
the y axis so that its location is $y = f(x) = x^2 + x$ at
time x (y is in meters and x is in seconds). Find

15. (A) The average velocity (the average rate of change of y with respect to x) for x changing from 1 to 3 seconds

   (B) The average velocity for x changing from 1 to $1 + h$ seconds

   (C) The instantaneous velocity at $x = 1$ second

16. (A) The average velocity (the average rate of change of y with respect to x) for x changing from 2 to 4 seconds

(B) The average velocity for x changing from 2 to $2 + h$ seconds

(C) The instantaneous velocity at $x = 2$ seconds

In Problems 17–22, find $f'(x)$ using the two-step limiting process. Then find $f'(1), f'(2),$ and $f'(3)$.

17. $f(x) = 5x - 2x^2$ **18.** $f(x) = 3x - 4x^2$

19. $f(x) = 4 - 2\sqrt{x}$ **20.** $f(x) = 6\sqrt{x} - 7$

21. $f(x) = \dfrac{1}{x + 3}$ **22.** $f(x) = \dfrac{1}{2 - x}$

Problems 23–30 refer to the function F in the graph shown. Use the graph to determine whether $F'(x)$ exists at each indicated value of x.

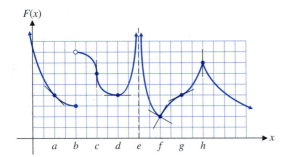

$F(x)$

23. $x = a$ **24.** $x = b$ **25.** $x = c$

26. $x = d$ **27.** $x = e$ **28.** $x = f$

29. $x = g$ **30.** $x = h$

31. Given $f(x) = x^2 - 4x$,

(A) Find $f'(x)$.

(B) Find the slopes of the tangent lines to the graph of f at $x = 0, 2,$ and 4.

(C) Graph f, and sketch in the tangent lines at $x = 0, 2,$ and 4.

32. Given $f(x) = x^2 + 2x$,

(A) Find $f'(x)$.

(B) Find the slopes of the tangent lines to the graph of f at $x = -2, -1,$ and 1.

(C) Graph f, and sketch in the tangent lines at $x = -2, -1,$ and 1.

33. If an object moves along a line so that it is at $y = f(x) = 4x^2 - 2x$ at time x (in seconds), find the instantaneous velocity function $v = f'(x)$, and find the velocity at times $x = 1, 3,$ and 5 seconds (y is measured in feet).

34. Repeat Problem 33 with $f(x) = 8x^2 - 4x$.

In Problems 35–38, use zoom and trace on a graphing utility to find points on the graph of $y = f(x)$ near $(0, f(0))$. Then use secant line slopes to approximate

$f'(0)$ *to two decimal places. (If your graphing utility has a numerical differentiation routine that will approximate the value of a derivative, use it to check your answers.)*

35. $f(x) = 2^x$

36. $f(x) = 3^x$

37. $f(x) = \sqrt{2 + 2x - x^2}$

38. $f(x) = \sqrt{3 - 2x - x^2}$

39. Let $f(x) = x^2, g(x) = x^2 - 3,$ and $h(x) = x^2 + 1$.

(A) How are the graphs of these functions related? How would you expect the derivatives of these functions to be related?

(B) Use the two-step process to find the derivative of $m(x) = x^2 + C$, where C is any real number constant.

40. Let $f(x) = 2x, g(x) = 2x - 1,$ and $h(x) = 2x + 2$.

(A) How are the graphs of these functions related? How would you expect the derivatives of these functions to be related?

(B) Use the two-step process to find the derivative of $m(x) = 2x + C$, where C is any real number constant.

41. (A) Give a geometric explanation of the following statement: If $f(x) = C$ is a constant function, then $f'(x) = 0$.

(B) Use the two-step process to verify the statement in part (A).

42. (A) Give a geometric explanation of the following statement: If $f(x) = mx + b$ is a linear function, then $f'(x) = m$.

(B) Use the two-step process to verify the statement in part (A).

C *In Problems 43–46, sketch the graph of f and determine where f is nondifferentiable.*

43. $f(x) = \begin{cases} 2x & \text{if } x < 1 \\ 2 & \text{if } x \geq 1 \end{cases}$

44. $f(x) = \begin{cases} 2x & \text{if } x < 2 \\ 6 - x & \text{if } x \geq 2 \end{cases}$

45. $f(x) = \begin{cases} x^2 + 1 & \text{if } x < 0 \\ 1 & \text{if } x \geq 0 \end{cases}$

46. $f(x) = \begin{cases} 2 - x^2 & \text{if } x \leq 0 \\ 2 & \text{if } x > 0 \end{cases}$

In Problems 47–52, determine whether f is differentiable at $x = 0$ by considering

$$\lim_{h \to 0} \frac{f(0 + h) - f(0)}{h}$$

47. $f(x) = |x|$ **48.** $f(x) = 1 - |x|$

49. $f(x) = x^{1/3}$ **50.** $f(x) = x^{2/3}$

51. $f(x) = \sqrt{1 - x^2}$ **52.** $f(x) = \sqrt{1 + x^2}$

53. A ball dropped from a balloon falls $y = 16x^2$ feet in x seconds. If the balloon is 576 feet above the ground when the ball is dropped, when does the ball hit the ground? What is the impact velocity of the ball at the instant it hits the ground?

54. Repeat Problem 53 if the balloon is 1,024 feet above the ground when the ball is dropped.

Applications

Business & Economics

55. *Revenue.* The revenue (in dollars) from the sale of x car seats for infants is given by

$$R(x) = 60x - 0.025x^2 \qquad 0 \leqslant x \leqslant 2,400$$

(A) Find the average change in revenue if production is changed from 1,000 car seats to 1,050 car seats.

(B) Use the two-step process to find $R'(x)$.

(C) Find the revenue and the instantaneous rate of change of revenue at a production level of 1,000 car seats and write a brief verbal interpretation of these results.

56. *Profit.* The profit (in dollars) from the sale of x car seats for infants is given by

$$P(x) = 45x - 0.025x^2 - 5,000 \qquad 0 \leqslant x \leqslant 2,400$$

(A) Find the average change in profit if production is changed from 800 car seats to 850 car seats.

(B) Use the two-step process to find $P'(x)$.

(C) Find the profit and the instantaneous rate of change of profit at a production level of 800 car seats and write a brief verbal interpretation of these results.

57. *Sales analysis.* The total sales of a company (in millions of dollars) t months from now are given by

$$S(t) = 2\sqrt{t} + 10$$

(A) Use the two-step process to find $S'(t)$.

(B) Find $S(15)$ and $S'(15)$. Write a brief verbal interpretation of these results.

(C) Use the results in part (B) to estimate the total sales after 16 months and after 17 months.

58. *Sales analysis.* The total sales of a company (in millions of dollars) t months from now are given by

$$S(t) = 2\sqrt{t} + 6$$

(A) Use the two-step process to find $S'(t)$.

(B) Find $S(10)$ and $S'(10)$. Write a brief verbal interpretation of these results.

(C) Use the results in part (B) to estimate the total sales after 11 months and after 12 months.

59. *Mineral Production.* The U.S. production of zinc (in **WWW** thousands of metric tons) is given approximately by

$$p(t) = 14t^2 - 6.6t + 602.4$$

where t is time in years and $t = 0$ corresponds to 1995.

(A) Use the two-step process to find $p'(t)$.

(B) Find the annual production in 2010, the instantaneous rate of change of production in 2010, and write a brief verbal interpretation of these results.

60. *Mineral Consumption.* The U.S. consumption of **WWW** copper (in thousands of metric tons) is given approximately by

$$p(t) = 27t^2 - 75t + 6,015$$

where t is time in years and $t = 0$ corresponds to 1990.

(A) Use the two-step process to find $p'(t)$.

(B) Find the annual production in 2010, the instantaneous rate of change of production in 2010, and write a brief verbal interpretation of these results.

61. *Electricity consumption.* Table 1 gives the retail sales of electricity (in billion kilowatthours) for the residential and commercial sectors in the United States during the 1990s.

WWW

TABLE 1

RETAIL ELECTRICITY SALES

YEAR	RESIDENTIAL	COMMERCIAL
1990	924	751
1991	955	766
1992	936	761
1993	995	795
1994	1,008	820
1995	1,043	863
1996	1,082	887
1997	1,075	928
1998	1,128	969
1999	1,146	983

(A) Let x represent time (in years) with $x = 0$ corresponding to 1990 and let y represent the corresponding residential sales. Enter the appropriate data set in a graphing utility and find a quadratic regression equation for the data.

(B) If $y = R(x)$ denotes the regression equation found in part (A), find $R(20)$ and $R'(20)$, and write a brief verbal interpretation of these results. Round answers to the nearest tenth of a billion.

62. *Electricity consumption.* Refer to the data in Table 1.

www

(A) Let x represent time (in years) with $x = 0$ corresponding to 1990 and let y represent the corresponding commercial sales. Enter the appropriate data set in a graphing utility and find a quadratic regression equation for the data.

(B) If $y = C(x)$ denotes the regression equation found in part (A), find $C(20)$ and $C'(20)$, and write a brief verbal interpretation of these results. Round answers to the nearest tenth of a billion.

Life Sciences

63. *Air pollution.* The ozone level (in parts per billion) on a summer day in a metropolitan area is given by

$$P(t) = 80 + 12t - t^2$$

where t is time in hours and $t = 0$ corresponds to 9 AM.

(A) Use the two-step process to find $P'(t)$.

(B) Find $P(3)$ and $P'(3)$. Write a brief verbal interpretation of these results.

64. *Medicine.* The body temperature (in degrees Fahrenheit) of a patient t hours after being given a fever-reducing drug is given by

$$F(t) = 98 + \frac{4}{t + 1}$$

(A) Use the two-step process to find $F'(t)$.

(B) Find $F(3)$ and $F'(3)$. Write a brief verbal interpretation of these results.

Social Sciences

65. *Infant mortality.* The number of infant deaths per 100,000 births for males in the United States is given approximately by

$$f(t) = 0.011t^2 - t + 29.8$$

where t is time in years and $t = 0$ corresponds to 1960.

(A) Use the two-step process to find $f'(t)$.

(B) Find the number of male infant deaths in 2000, the instantaneous rate of change of the number of male infant deaths in 2000, and write a brief verbal interpretation of these results.

66. *Infant mortality.* The number of infant deaths per 100,000 births for females in the United States is given approximately by

$$f(t) = 0.008t^2 - 0.74t + 23$$

where t is time in years and $t = 0$ corresponds to 1960.

(A) Use the two-step prrocess to find $f'(t)$.

(B) Find the number of female infant deaths in 2000, the instantaneous rate of change of the number of female infant deaths in 2000, and write a brief verbal interpretation of these results.

Section 3-4   Derivatives of Constants, Power Forms, and Sums

- ❑ Derivative of a Constant Function
- ❑ Power Rule
- ❑ Derivative of a Constant Times a Function
- ❑ Derivatives of Sums and Differences
- ❑ Applications

In the preceding section, we defined the derivative of f at x as

$$f'(x) = \lim_{h \to 0} \frac{f(x + h) - f(x)}{h}$$

if the limit exists, and we used this definition and a two-step process to find the derivatives of several functions. In this and the next two sections, we develop some rules based on this definition that will enable us to determine the derivatives of a rather large class of functions without having to go through the two-step process each time.

Before starting on these rules, we list some symbols that are widely used to represent derivatives.

> ### Derivative Notation
>
> Given $y = f(x)$, then
>
> $$f'(x) \qquad y' \qquad \frac{dy}{dx}$$
>
> all represent the derivative of f at x.

Each of these symbols for the derivative has its particular advantage in certain situations. All of them will become familiar to you after a little experience.

❑   ## DERIVATIVE OF A CONSTANT FUNCTION

Suppose

$$f(x) = C \qquad C \text{ a constant} \qquad \textit{A constant function}$$

Geometrically, the graph of $f(x) = C$ is a horizontal straight line with slope 0 (see Fig. 1); hence, we would expect $f'(x) = 0$. We will show that this is actually the case using the definition of the derivative and the two-step process introduced earlier. We want to find

$$f'(x) = \lim_{h \to 0} \frac{f(x + h) - f(x)}{h} \qquad \textit{Definition of } f'(x)$$

Step 1. $\quad \dfrac{f(x + h) - f(x)}{h} = \dfrac{C - C}{h} = \dfrac{0}{h} = 0 \qquad h \neq 0$

Step 2. $\quad \lim_{h \to 0} 0 = 0$

Thus,

$$f'(x) = 0$$

We conclude that

The derivative of any constant function is 0.

FIGURE 1

> ### Derivative of a Constant Function Rule
>
> If $y = f(x) = C$, then
>
> $$f'(x) = 0$$
>
> Also, $y' = 0$ and $dy/dx = 0$.
>
> [*Note*: When we write $C' = 0$ or $\dfrac{d}{dx} C = 0$, we mean that $y' = \dfrac{dy}{dx} = 0$ where $y = C$.]

Example 1 ⇨   **Differentiating Constant Functions**

(A) If $f(x) = 3$, then $f'(x) = 0$.   (B) If $y = -1.4$, then $y' = 0$.

(C) If $y = \pi$, then $\dfrac{dy}{dx} = 0$.   (D) $\dfrac{d}{dx} 23 = 0$

■

Matched Problem 1 ⇨   Find

(A) $f'(x)$ for $f(x) = -24$ (B) y' for $y = 12$

(C) $\dfrac{dy}{dx}$ for $y = -\sqrt{7}$ (D) $\dfrac{d}{dx}(-\pi)$

■

❑   POWER RULE

A function of the form $f(x) = x^k$, where k is a real number, is called a **power function.** The elementary functions (see the inside front cover) listed below are examples of power functions:

$$f(x) = x \qquad h(x) = x^2 \qquad m(x) = x^3 \tag{1}$$
$$n(x) = \sqrt{x} \qquad p(x) = \sqrt[3]{x}$$

Explore–Discuss 1

(A) It is clear that the functions f, h, and m in (1) are power functions. Explain why the functions n and p are also power functions.

(B) The domain of a power function depends on the power. Discuss the domain of each of the following power functions:

$$r(x) = x^4 \qquad s(x) = x^{-4} \qquad t(x) = x^{1/4}$$
$$u(x) = x^{-1/4} \qquad v(x) = x^{1/5} \qquad w(x) = x^{-1/5}$$

The definition of the derivative and the two-step process introduced in the preceding section can be used to find the derivatives of many power functions. For example, it can be shown that

If $f(x) = x^2$,   then $f'(x) = 2x$.
If $f(x) = x^3$,   then $f'(x) = 3x^2$.
If $f(x) = x^4$,   then $f'(x) = 4x^3$.
If $f(x) = x^5$,   then $f'(x) = 5x^4$.

Notice the pattern in these derivatives. In each case, the power in f becomes the coefficient in f' and the power in f' is 1 less than the power in f. In general, for any positive integer n:

If $f(x) = x^n$,   then $f'(x) = nx^{n-1}$. \tag{2}

In fact, more advanced techniques can be used to show that (2) holds for *any* real number n. We will assume this general result for the remainder of this book.

Power Rule

If $y = f(x) = x^n$, where n is a real number, then

$$f'(x) = nx^{n-1}$$

Also, $y' = nx^{n-1}$ and $dy/dx = nx^{n-1}$.

Explore–Discuss 2

(A) Write a verbal description of the power rule.

(B) If $f(x) = x$, what is $f'(x)$? Discuss how this derivative can be obtained from the power rule.

Example 2   ⇨   **Differentiating Power Functions**

WWW

(A) If $f(x) = x^5$,   then $f'(x) = 5x^{5-1} = 5x^4$.
(B) If $y = x^{25}$,   then $y' = 25x^{25-1} = 25x^{24}$.

(C) If $y = x^{-3}$, then $\dfrac{dy}{dx} = -3x^{-3-1} = -3x^{-4} = -\dfrac{3}{x^4}$.

(D) $\dfrac{d}{dx}x^{5/3} = \dfrac{5}{3}x^{(5/3)-1} = \dfrac{5}{3}x^{2/3}$.

■

Matched Problem 2 ➭ Find

(A) $f'(x)$ for $f(x) = x^6$ (B) y' for $y = x^{30}$

(C) $\dfrac{dy}{dx}$ for $y = x^{-2}$ (D) $\dfrac{d}{dx}x^{3/2}$

■

In some cases, properties of exponents must be used to rewrite an expression before the power rule is applied.

Example 3 ➭ **Differentiating Power Functions**

(A) If $f(x) = 1/x^4$, we can write $f(x) = x^{-4}$ and

$$f'(x) = -4x^{-4-1} = -4x^{-5} \quad \text{or} \quad \dfrac{-4}{x^5}$$

(B) If $y = \sqrt{x}$, we can write $y = x^{1/2}$ and

$$y' = \dfrac{1}{2}x^{(1/2)-1} = \dfrac{1}{2}x^{-1/2} \quad \text{or} \quad \dfrac{1}{2\sqrt{x}}$$

(C) $\dfrac{d}{dx}\dfrac{1}{\sqrt[3]{x}} = \dfrac{d}{dx}x^{-1/3} = -\dfrac{1}{3}x^{(-1/3)-1} = -\dfrac{1}{3}x^{-4/3} \quad \text{or} \quad \dfrac{-1}{3\sqrt[3]{x^4}}$

■

Matched Problem 3 ➭ Find

(A) $f'(x)$ for $f(x) = \dfrac{1}{x}$ (B) y' for $y = \sqrt[3]{x^2}$ (C) $\dfrac{d}{dx}\dfrac{1}{\sqrt{x}}$

■

❑ **DERIVATIVE OF A CONSTANT TIMES A FUNCTION**

Let $f(x) = ku(x)$, where k is a constant and u is differentiable at x. Then, using the two-step process, we have the following:

Step 1. $\dfrac{f(x+h)-f(x)}{h} = \dfrac{ku(x+h)-ku(x)}{h} = k\left[\dfrac{u(x+h)-u(x)}{h}\right]$

Step 2. $\displaystyle\lim_{h\to 0}\dfrac{f(x+h)-f(x)}{h} = \lim_{h\to 0}k\left[\dfrac{u(x+h)-u(x)}{h}\right]$ $\displaystyle\lim_{x\to c}kg(x)=k\lim_{x\to c}g(x)$

$\displaystyle = k\lim_{h\to 0}\left[\dfrac{u(x+h)-u(x)}{h}\right]$ Definition of $u'(x)$

$= ku'(x)$

Thus,

The derivative of a constant times a differentiable function is the constant times the derivative of the function.

> *Constant Times a Function Rule*
>
> If $y = f(x) = ku(x)$, then
>
> $\quad f'(x) = ku'(x)$
>
> Also,
>
> $\quad y' = ku' \qquad \dfrac{dy}{dx} = k\dfrac{du}{dx}$

Example 4 ⟹ **Differentiating a Constant Times a Function**

(A) If $f(x) = 3x^2$, then $f'(x) = 3 \cdot 2x^{2-1} = 6x$.

(B) If $y = \dfrac{x^3}{6} = \dfrac{1}{6}x^3$, then $\dfrac{dy}{dx} = \dfrac{1}{6} \cdot 3x^{3-1} = \dfrac{1}{2}x^2$.

(C) If $y = \dfrac{1}{2x^4} = \dfrac{1}{2}x^{-4}$, then $y' = \dfrac{1}{2}(-4x^{-4-1}) = -2x^{-5}$ or $\dfrac{-2}{x^5}$.

(D) $\dfrac{d}{dx}\dfrac{0.4}{\sqrt{x^3}} = \dfrac{d}{dx}\dfrac{0.4}{x^{3/2}} = \dfrac{d}{dx}0.4x^{-3/2} = 0.4\left[-\dfrac{3}{2}x^{(-3/2)-1}\right]$

$\qquad\qquad\qquad = -0.6x^{-5/2}$ or $-\dfrac{0.6}{\sqrt{x^5}}$

Matched Problem 4 ⟹ Find

(A) $f'(x)$ for $f(x) = 4x^5$ (B) $\dfrac{dy}{dx}$ for $y = \dfrac{x^4}{12}$

(C) y' for $y = \dfrac{1}{3x^3}$ (D) $\dfrac{d}{dx}\dfrac{0.9}{\sqrt[3]{x}}$

❑ DERIVATIVES OF SUMS AND DIFFERENCES

Let $f(x) = u(x) + v(x)$, where $u'(x)$ and $v'(x)$ exist. Then, using the two-step process, we have the following:

Step 1. $\dfrac{f(x+h) - f(x)}{h} = \dfrac{[u(x+h) + v(x+h)] - [u(x) + v(x)]}{h}$

$\qquad = \dfrac{u(x+h) + v(x+h) - u(x) - v(x)}{h}$

$\qquad = \dfrac{u(x+h) - u(x)}{h} + \dfrac{v(x+h) - v(x)}{h}$

Step 2. $\displaystyle\lim_{h\to0}\frac{f(x+h)-f(x)}{h}=\lim_{h\to0}\left[\frac{u(x+h)-u(x)}{h}+\frac{v(x+h)-v(x)}{h}\right]$

$$\lim_{x\to c}\,[g(x)+h(x)]=\lim_{x\to c}g(x)+\lim_{x\to c}h(x)$$

$$=\lim_{h\to0}\frac{u(x+h)-u(x)}{h}+\lim_{h\to0}\frac{v(x+h)-v(x)}{h}$$

$$=u'(x)+v'(x)$$

Thus,

> **The derivative of the sum of two differentiable functions is the sum of the derivatives.**

Similarly, we can show that:

> **The derivative of the difference of two differentiable functions is the difference of the derivatives.**

Together, we then have the **sum and difference rule** for differentiation:

Sum and Difference Rule

If $y=f(x)=u(x)\pm v(x)$, then

$$f'(x)=u'(x)\pm v'(x)$$

Also,

$$y'=u'\pm v'\qquad\frac{dy}{dx}=\frac{du}{dx}\pm\frac{dv}{dx}$$

[*Note:* This rule generalizes to the sum and difference of any given number of functions.]

With this and the other rules stated previously, we will be able to compute the derivatives of all polynomials and a variety of other functions.

Example 5 ➯ **Differentiating Sums and Differences**

(A) If $f(x)=3x^2+2x$, then

$$f'(x)=(3x^2)'+(2x)'=3(2x)+2(1)=6x+2$$

(B) If $y=4+2x^3-3x^{-1}$, then

$$y'=(4)'+(2x^3)'-(3x^{-1})'=0+2(3x^2)-3(-1)x^{-2}=6x^2+3x^{-2}$$

(C) If $y=\sqrt[3]{x}-3x$, then

$$\frac{dy}{dx}=\frac{d}{dx}x^{1/3}-\frac{d}{dx}3x=\frac{1}{3}x^{-2/3}-3=\frac{1}{3x^{2/3}}-3$$

(D) $\dfrac{d}{dx}\left(\dfrac{5}{3x^2} - \dfrac{2}{x^4} + \dfrac{x^3}{9}\right) = \dfrac{d}{dx}\dfrac{5}{3}x^{-2} - \dfrac{d}{dx}\,2x^{-4} + \dfrac{d}{dx}\dfrac{1}{9}x^3$

$\qquad\qquad\qquad\qquad = \dfrac{5}{3}(-2)x^{-3} - 2(-4)x^{-5} + \dfrac{1}{9}\cdot 3x^2$

$\qquad\qquad\qquad\qquad = -\dfrac{10}{3x^3} + \dfrac{8}{x^5} + \dfrac{1}{3}x^2$

∎

Matched Problem 5   Find

(A) $f'(x)$ for $f(x) = 3x^4 - 2x^3 + x^2 - 5x + 7$

(B) y' for $y = 3 - 7x^{-2}$

(C) $\dfrac{dy}{dx}$ for $y = 5x^3 - \sqrt[4]{x}$

(D) $\dfrac{d}{dx}\left(-\dfrac{3}{4x} + \dfrac{4}{x^3} - \dfrac{x^4}{8}\right)$

∎

   ❑   **APPLICATIONS**

Example 6   **Instantaneous Velocity**   An object moves along the y axis (marked in feet) so that its position at time x (in seconds) is

$$f(x) = x^3 - 6x^2 + 9x$$

(A) Find the instantaneous velocity function v.

(B) Find the velocity at $x = 2$ and $x = 5$ seconds.

(C) Find the time(s) when the velocity is 0.

SOLUTION   (A) $v = f'(x) = (x^3)' - (6x^2)' + (9x)' = 3x^2 - 12x + 9$

(B) $f'(2) = 3(2)^2 - 12(2) + 9 = -3$ feet per second

$\qquad f'(5) = 3(5)^2 - 12(5) + 9 = 24$ feet per second

(C) $v = f'(x) = 3x^2 - 12x + 9 = 0$

$\qquad\qquad\qquad 3(x^2 - 4x + 3) = 0$

$\qquad\qquad\qquad 3(x - 1)(x - 3) = 0$

$\qquad\qquad\qquad\qquad\qquad x = 1, 3$

Thus, $v = 0$ at $x = 1$ and $x = 3$ seconds.

∎

Matched Problem 6   Repeat Example 6 for $f(x) = x^3 - 15x^2 + 72x$.

Example 7   **Tangents**   Let $f(x) = x^4 - 6x^2 + 10$.

(A) Find $f'(x)$.

(B) Find the equation of the tangent line at $x = 1$.

(C) Find the values of x where the tangent line is horizontal.

SOLUTION   (A) $f'(x) = (x^4)' - (6x^2)' + (10)'$

$$= 4x^3 - 12x$$

(B) $y - y_1 = m(x - x_1)$ $y_1 = f(x_1) = f(1) = (1)^4 - 6(1)^2 + 10 = 5$

$y - 5 = -8(x - 1)$ $m = f'(x_1) = f'(1) = 4(1)^3 - 12(1) = -8$

$y = -8x + 13$ Tangent line at $x = 1$

(C) Since a horizontal line has 0 slope, we must solve $f'(x) = 0$ for x:

$$f'(x) = 4x^3 - 12x = 0$$
$$4x(x^2 - 3) = 0$$
$$4x(x + \sqrt{3})(x - \sqrt{3}) = 0$$
$$x = 0, -\sqrt{3}, \sqrt{3}$$

Matched Problem 7   Repeat Example 7 for $f(x) = x^4 - 8x^3 + 7$.

REMARK

In Example 7 we used algebraic techniques to solve the equation $f'(x) = 0$. A graphing utility also can be used to approximate the solutions to equations of this form. Figure 2 illustrates this process for the equation in Example 7C. Note that the graphing utility gives decimal approximations to $-\sqrt{3}$ and $\sqrt{3}$.

FIGURE 2 $y_1 = 4x^3 - 12x$

In business and economics, the rates at which certain quantities are changing often provide useful insight into various economic systems. A manufacturer, for example, is interested not only in the total cost $C(x)$ at certain production levels, but is also interested in the rate of change of costs at various production levels.

www   In economics, the word **marginal** refers to a rate of change, that is, to a derivative. Thus, if

$C(x)$ = total cost of producing x items

then

$C'(x)$ = **marginal cost**

= instantaneous rate of change of total cost $C(x)$ with respect to the number of items produced at a production level of x items

Example 8   **Marginal Cost**   Suppose that the total cost $C(x)$ (in thousands of dollars) for manufacturing x sailboats per year is given by the function

$$C(x) = 575 + 25x - 0.25x^2 \qquad 0 \le x \le 50$$

(See Fig. 3).

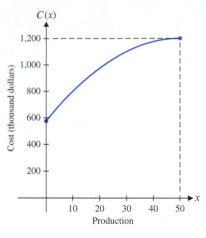

FIGURE 3

(A) Find the marginal cost at a production level of x boats per year.

(B) Find the marginal cost at a production level of 40 boats per year, and interpret the results.

SOLUTION (A) The marginal cost at a production level of x boats is

$$C'(x) = (575)' + (25x)' - (0.25x^2)' = 25 - 0.5x$$

(B) The marginal cost at a production level of 40 boats is

$$C'(40) = 25 - 0.5(40) = 5 \quad \text{or} \quad \$5,000 \text{ per boat}$$

At a production level of 40 boats per year, the total cost is increasing at the rate of \$5,000 per boat. ∎

Matched Problem 8 ⇨ Suppose that the total cost $C(x)$ (in thousands of dollars) for manufacturing x sailboats per year is given by the function

$$C(x) = 500 + 24x - 0.2x^2 \qquad 0 \le x \le 50$$

(A) Find the marginal cost at a production level of x boats per year.

(B) Find the marginal cost at a production level of 35 boats per year, and interpret the results. ∎

The interpretation of marginal cost given in Example 8B is the usual way we use the derivative to describe the manner in which a quantity is changing. However, there is a special way to interpret the rate of change of a total cost function that is used extensively in economics. Referring to the results in Example 8B, if the total cost is increasing at the rate of \$5,000 per boat when 40 boats are produced per year, and if production is increased to 41 boats per year, the total cost will increase by approximately \$5,000. But the increase in total cost is just the cost of producing the next boat. Thus, the cost of producing the 41st boat must be approximately \$5,000, the marginal cost at the production level of 40 boats per year. These ideas are summarized in the following box:

> ## *Marginal Cost Function*
>
> If $C(x)$ is the total cost of producing x items, the marginal cost function $C'(x)$ approximates the cost of producing one more item at a production level of x items.

Example 9   **Marginal Cost**   Refer to the total cost function given in Example 8:

$$C(x) = 575 + 25x - 0.25x^2 \qquad 0 \leqslant x \leqslant 50$$

(A) Use the marginal cost function to approximate the cost of producing the 31st boat.

(B) Use the total cost function to find the exact cost of producing the 31st boat.

SOLUTION   (A) From Example 8, $C'(x) = 25 - 0.5x$. Thus,

$$C'(30) = 25 - 0.5(30) = 10 \quad \text{or} \quad \$10,000$$

The cost of producing the 31st boat is approximately $10,000.

(B) The exact cost of producing the 31st boat is

$$\begin{pmatrix} \text{total cost of} \\ \text{producing} \\ \text{31 boats} \end{pmatrix} - \begin{pmatrix} \text{total cost of} \\ \text{producing} \\ \text{30 boats} \end{pmatrix}$$

$$= C(31) \quad - \quad C(30)$$
$$= 1,109.75 - 1,100 = 9.750 \quad \text{or} \quad \$9,750$$

Notice that the marginal cost of $10,000 per boat is a close approximation to this exact cost. ■

Matched Problem 9   Refer to the cost function given in Matched Problem 8:

$$C(x) = 500 + 24x - 0.2x^2 \qquad 0 \leqslant x \leqslant 50$$

(A) Use the marginal cost function to approximate the cost of producing the 41st boat.

(B) Use the total cost function to find the exact cost of producing the 41st boat. ■

REMARK

A derivative can always be interpreted as an instantaneous rate of change, as we did in Example 8B. The interpretation of the marginal cost function as the approximate cost of producing the next item, as in Example 9A, is a special case that applies to total cost functions. This interpretation also applies to total revenue functions and total profit functions, but not to most of the other economic functions we will consider.

Answers to Matched Problems   **1.** All are 0.

2. (A) $6x^5$ (B) $30x^{29}$ (C) $-2x^{-3} = -2/x^3$ (D) $\frac{3}{2}x^{1/2}$

3. (A) $-x^{-2}$, or $-1/x^2$ (B) $\frac{2}{3}x^{-1/3}$, or $2/(3\sqrt[3]{x})$
   (C) $-\frac{1}{2}x^{-3/2}$, or $-1/(2\sqrt{x^3})$

4. (A) $20x^4$ (B) $x^3/3$ (C) $-x^{-4}$, or $-1/x^4$
   (D) $-0.3x^{-4/3}$, or $-0.3/\sqrt[3]{x^4}$

5. (A) $12x^3 - 6x^2 + 2x - 5$ (B) $14x^{-3}$, or $14/x^3$
   (C) $15x^2 - \frac{1}{4}x^{-3/4}$, or $15x^2 - 1/(4x^{3/4})$
   (D) $3/(4x^2) - (12/x^4) - (x^3/2)$

6. (A) $v = 3x^2 - 30x + 72$ (B) $f'(2) = 24$ ft/sec; $f'(5) = -3$ ft/sec
   (C) $x = 4$ and $x = 6$ sec

7. (A) $f'(x) = 4x^3 - 24x^2$ (B) $y = -20x + 20$ (C) $x = 0$ and $x = 6$

8. (A) $C'(x) = 24 - 0.4x$
   (B) $C'(35) = 10$ or $10,000$ per boat; at a production level of 35 boats, total cost is increasing at the rate of $10,000$ per boat.

9. (A) $C'(40) = 8$ or $8,000$; the cost of producing the 41st boat is approximately $8,000$.
   (B) $C(41) - C(40) = 7.8$ or $7,800$

Exercise 3-4

A *Find the indicated derivatives in Problems 1–18.*

1. $f'(x)$ for $f(x) = 7$ **2.** $\dfrac{d}{dx}3$

3. $\dfrac{dy}{dx}$ for $y = x^9$ **4.** y' for $y = x^6$

5. $\dfrac{d}{dx}x^3$ **6.** $g'(x)$ for $g(x) = x^5$

7. y' for $y = x^{-4}$ **8.** $\dfrac{dy}{dx}$ for $y = x^{-8}$

9. $g'(x)$ for $g(x) = x^{8/3}$
10. $f'(x)$ for $f(x) = x^{9/2}$

11. $\dfrac{dy}{dx}$ for $y = \dfrac{1}{x^{10}}$ **12.** y' for $y = \dfrac{1}{x^{12}}$

13. $f'(x)$ for $f(x) = 5x^2$ **14.** $\dfrac{d}{dx}(-2x^3)$

15. y' for $y = 0.4x^7$

16. $f'(x)$ for $f(x) = 0.8x^4$

17. $\dfrac{d}{dx}\left(\dfrac{x^3}{18}\right)$ **18.** $\dfrac{dy}{dx}$ for $y = \dfrac{x^5}{25}$

Problems 19–24 refer to functions f and g that satisfy $f'(2) = 3$ and $g'(2) = -1$. In each problem, find $h'(2)$ for the indicated function h.

19. $h(x) = 4f(x)$ **20.** $h(x) = 5g(x)$

21. $h(x) = f(x) + g(x)$ **22.** $h(x) = g(x) - f(x)$
23. $h(x) = 2f(x) - 3g(x) + 7$
24. $h(x) = -4f(x) + 5g(x) - 9$

B *Find the indicated derivatives in Problems 25–48.*

25. $\dfrac{d}{dx}(2x^{-5})$ **26.** y' for $y = -4x^{-1}$

27. $f'(x)$ for $f(x) = \dfrac{4}{x^4}$ **28.** $\dfrac{dy}{dx}$ for $y = \dfrac{-3}{x^6}$

29. $\dfrac{d}{dx}\dfrac{-1}{2x^2}$ **30.** y' for $y = \dfrac{1}{6x^3}$

31. $f'(x)$ for $f(x) = -3x^{1/3}$

32. $\dfrac{dy}{dx}$ for $y = -8x^{1/4}$

33. $\dfrac{d}{dx}(2.4x^2 - 3.5x + 4)$

34. y' for $y = 3x^2 + 4x - 7$

35. $\dfrac{dy}{dx}$ for $y = 3x^5 - 2x^3 + 5$

36. $f'(x)$ for $f(x) = 2.6x^3 - 6.7x + 5.2$

37. $\dfrac{d}{dx}(3x^{-4} + 2x^{-2})$

38. y' for $y = 2x^{-3} - 4x^{-1}$

39. $\dfrac{dy}{dx}$ for $y = \dfrac{1}{2x} - \dfrac{2}{3x^3}$

40. $f'(x)$ for $f(x) = \dfrac{3}{4x^3} + \dfrac{1}{2x^5}$

41. $\dfrac{d}{dx}(3x^{2/3} - 5x^{1/3})$

42. $\dfrac{d}{dx}(8x^{3/4} + 4x^{-1/4})$

43. $\dfrac{d}{dx}\left(\dfrac{3}{x^{3/5}} - \dfrac{6}{x^{1/2}}\right)$

44. $\dfrac{d}{dx}\left(\dfrac{5}{x^{1/5}} - \dfrac{8}{x^{3/2}}\right)$

45. $\dfrac{d}{dx}\dfrac{1}{\sqrt[3]{x}}$

46. y' for $y = \dfrac{10}{\sqrt[5]{x}}$

47. $\dfrac{dy}{dx}$ for $y = \dfrac{1.2}{\sqrt{x}} - 3.2x^{-2} + x$

48. $f'(x)$ for $f(x) = 2.8x^{-3} - \dfrac{0.6}{\sqrt[3]{x^2}} + 7$

For Problems 49–52, find

(A) $f'(x)$

(B) The slope of the graph of f at $x = 2$ and $x = 4$

(C) The equations of the tangent lines at $x = 2$ and $x = 4$

(D) The value(s) of x where the tangent line is horizontal

49. $f(x) = 6x - x^2$

50. $f(x) = 2x^2 + 8x$

51. $f(x) = 3x^4 - 6x^2 - 7$

52. $f(x) = x^4 - 32x^2 + 10$

If an object moves along the y axis (marked in feet) so that its position at time x (in seconds) is given by the indicated function in Problems 53–56, find

(A) The instantaneous velocity function $v = f'(x)$

(B) The velocity when $x = 0$ and $x = 3$ seconds

(C) The time(s) when $v = 0$

53. $f(x) = 176x - 16x^2$

54. $f(x) = 80x - 10x^2$

55. $f(x) = x^3 - 9x^2 + 15x$

56. $f(x) = x^3 - 9x^2 + 24x$

 Problems 57–64 require the use of a graphing utility. For each problem, find $f'(x)$ and approximate (to two decimal places) the value(s) of x where the graph of f has a horizontal tangent line.

57. $f(x) = x^2 - 3x - 4\sqrt{x}$

58. $f(x) = x^2 + x - 10\sqrt{x}$

59. $f(x) = 3\sqrt[3]{x^4} - 1.5x^2 - 3x$

60. $f(x) = 3\sqrt[3]{x^4} - 2x^2 + 4x$

61. $f(x) = 0.05x^4 + 0.1x^3 - 1.5x^2 - 1.6x + 3$

62. $f(x) = 0.02x^4 - 0.06x^3 - 0.78x^2 + 0.94x + 2.2$

63. $f(x) = 0.2x^4 - 3.12x^3 + 16.25x^2 - 28.25x + 7.5$

64. $f(x) = 0.25x^4 - 2.6x^3 + 8.1x^2 - 10x + 9$

65. Let $f(x) = ax^2 + bx + c, a \neq 0$. Recall that the graph of $y = f(x)$ is a parabola. Use the derivative $f'(x)$ to derive a formula for the x coordinate of the vertex of this parabola.

66. Now that you know how to find derivatives, explain why it is no longer necessary for you to memorize the formula for the x coordinate of the vertex of a parabola.

67. Give an example of a cubic polynomial function that has

(A) No horizontal tangents

(B) One horizontal tangent

(C) Two horizontal tangents

68. Can a cubic polynomial function have more than two horizontal tangents? Explain.

C *In Problems 69–72, find each derivative.*

69. $f'(x)$ for $f(x) = \dfrac{10x + 20}{x}$

70. $\dfrac{dy}{dx}$ for $y = \dfrac{x^2 + 25}{x^2}$

71. $\dfrac{d}{dx}\dfrac{x^4 - 3x^3 + 5}{x^2}$

72. y' for $y = \dfrac{2x^5 - 4x^3 + 2x}{x^3}$

In Problems 73 and 74, use the definition of the derivative and the two-step process to verify each statement.

73. $\dfrac{d}{dx}x^3 = 3x^2$

74. $\dfrac{d}{dx}x^4 = 4x^3$

75. The domain of the power function $f(x) = x^{1/3}$ is the set of all real numbers. Find the domain of the derivative $f'(x)$. Discuss the nature of the graph of $y = f(x)$ for any x values excluded from the domain of $f'(x)$.

76. The domain of the power function $f(x) = x^{2/3}$ is the set of all real numbers. Find the domain of the derivative $f'(x)$. Discuss the nature of the graph of $y = f(x)$ for any x values excluded from the domain of $f'(x)$.

Applications

Business & Economics

77. *Marginal cost.* The total cost (in dollars) of producing x tennis rackets per day is

$$C(x) = 800 + 60x - 0.25x^2 \qquad 0 \leqslant x \leqslant 120$$

(A) Find the marginal cost at a production level of x rackets.

(B) Find the marginal cost at a production level of 60 rackets, and interpret the result.

(C) Find the actual cost of producing the 61st racket, and compare this cost with the result found in part (B).

(D) Find $C'(80)$, and interpret the result.

78. *Marginal cost.* The total cost (in dollars) of producing x portable radios per day is

$$C(x) = 1{,}000 + 100x - 0.5x^2 \qquad 0 \leqslant x \leqslant 100$$

(A) Find the marginal cost at a production level of x radios.

(B) Find the marginal cost at a production level of 80 radios, and interpret the result.

(C) Find the actual cost of producing the 81st radio, and compare this cost with the result found in part (B).

(D) Find $C'(50)$, and interpret the result.

 79. *Motor vehicle production.* Annual limousine production in the United States for selected years is given in Table 1.

TABLE 1	
YEAR	LIMOUSINE PRODUCTION
1980	2,500
1985	6,500
1990	4,400
1995	3,400
2000	5,700

(A) Let x represent time (in years) since 1980, and let y represent the corresponding U.S. production of limousines. Enter the data in a graphing utility and find a cubic regression equation for the data.

(B) If $y = L(x)$ denotes the regression equation found in part (A), find $L(12)$ and $L'(12)$ (to the nearest hundred), and write a brief verbal interpretation of these results.

(C) Repeat part (B) for $L(18)$ and $L'(18)$.

 80. *Motor vehicle operation.* The total number of registered limousine operators in the United States for selected years is given in Table 2.

www

TABLE 2	
YEAR	LIMOUSINE OPERATORS
1985	4,000
1990	7,000
1995	8,900
2000	11,000

(A) Let x represent time (in years) since 1980 and let y represent the corresponding number of registered limousine operators in the United States. Enter the data in a graphing utility and find a quadratic regression equation for the data.

(B) If $y = L(x)$ denotes the regression equation found in part (A), find $L(18)$ and $L'(18)$ (to the nearest hundred), and write a brief verbal interpretation of these results.

81. *Marginal cost.* The total cost (in dollars) of producing x microwave ovens per week is shown in the figure. Which is greater, the approximate cost of producing the 101st oven or the approximate cost of producing the 401st oven? Does this graph represent a manufacturing process that is becoming more efficient or less efficient as production levels increase? Explain.

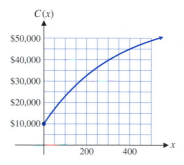

Figure for 81

82. *Marginal cost.* The total cost (in dollars) of producing x electric stoves per week is shown in the figure. Which is greater, the approximate cost of producing the 101st stove or the approximate cost of producing the 401st stove? Does this graph represent a manufacturing process that is becoming more efficient or less efficient as production levels increase? Explain.

Figure for 82

83. *Advertising.* Using past records, it is estimated that a marine manufacturer will sell $N(x)$ power boats after spending $x thousand on advertising, as given by

$$N(x) = 1,000 - \frac{3,780}{x} \qquad 5 \le x \le 30$$

See the figure.

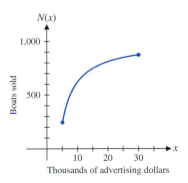

Figure for 83

(A) Find $N'(x)$.

(B) Find $N'(10)$ and $N'(20)$. Write a brief verbal interpretation of these results.

84. *Price–demand equation.* Suppose that in a given gourmet food store, people are willing to buy x pounds of chocolate candy per day at $p per quarter pound, as given by the price–demand equation

$$x = 10 + \frac{180}{p} \qquad 2 \le p \le 10$$

This function is graphed in the figure. Find the demand and the instantaneous rate of change of demand with respect to price when the price is $5. Write a brief verbal interpretation of these results.

Figure for 84

Life Sciences

85. *Medicine.* A person x inches tall has a pulse rate of y beats per minute, as given approximately by

$$y = 590x^{-1/2} \qquad 30 \le x \le 75$$

What is the instantaneous rate of change of pulse rate at the

(A) 36-inch level?   (B) 64-inch level?

86. *Ecology.* A coal-burning electrical generating plant emits sulfur dioxide into the surrounding air. The concentration $C(x)$, in parts per million, is given approximately by

$$C(x) = \frac{0.1}{x^2}$$

where x is the distance from the plant in miles. Find the instantaneous rate of change of concentration at

(A) $x = 1$ mile   (B) $x = 2$ miles

Social Sciences

87. *Learning.* Suppose that a person learns y items in x hours, as given by

$$y = 50\sqrt{x} \qquad 0 \le x \le 9$$

(see the figure). Find the rate of learning at the end of

(A) 1 hour   (B) 9 hours

Figure for 87

88. *Learning.* If a person learns y items in x hours, as given by

$$y = 21\sqrt[3]{x^2} \qquad 0 \le x \le 8$$

find the rate of learning at the end of

(A) 1 hour (B) 8 hours

Section 3-5	**Derivatives of Products and Quotients**

- ❑ DERIVATIVES OF PRODUCTS
- ❑ DERIVATIVES OF QUOTIENTS

The derivative rules discussed in the preceding section added substantially to our ability to compute and apply derivatives to many practical problems. In this and the next sections we add a few more rules that will increase this ability even further.

❑ DERIVATIVES OF PRODUCTS

In Section 3-4 we found that the derivative of a sum is the sum of the derivatives. Is the derivative of a product the product of the derivatives?

Explore–Discuss 1

Let $F(x) = x^2$, $S(x) = x^3$, and $f(x) = F(x)S(x) = x^5$. Which of the following is $f'(x)$?

(A) $F'(x)S'(x)$ (B) $F(x)S'(x)$

(C) $F'(x)S(x)$ (D) $F(x)S'(x) + F'(x)S(x)$

Comparing the various expressions computed in Explore–Discuss 1, we see that the derivative of a product is not the product of the derivatives, but appears to involve a slightly more complicated form.

Using the definition of the derivative and the two-step process, it can be shown that

> **The derivative of the product of two functions is the first function times the derivative of the second function plus the second function times the derivative of the first function.**

That is,

www

Product Rule

If

$$y = f(x) = F(x)S(x)$$

and if $F'(x)$ and $S'(x)$ exist, then

$$f'(x) = F(x)S'(x) + S(x)F'(x)$$

Also,

$$y' = FS' + SF' \qquad \frac{dy}{dx} = F\frac{dS}{dx} + S\frac{dF}{dx}$$

Example 1 ➭ **Differentiating a Product**   Use two different methods to find $f'(x)$ for
$$f(x) = 2x^2(3x^4 - 2).$$

SOLUTION   **Method 1.**   Use the product rule:
$$f'(x) = 2x^2(3x^4 - 2)' + (3x^4 - 2)(2x^2)' \qquad \textit{First times derivative}$$
$$= 2x^2(12x^3) + (3x^4 - 2)(4x) \qquad\qquad \textit{of second plus second}$$
$$= 24x^5 + 12x^5 - 8x \qquad\qquad\qquad\quad \textit{times derivative of first}$$
$$= 36x^5 - 8x$$

Method 2.   Multiply first; then take derivatives:
$$f(x) = 2x^2(3x^4 - 2) = 6x^6 - 4x^2$$
$$f'(x) = 36x^5 - 8x$$

Matched Problem 1 ➭ Use two different methods to find $f'(x)$ for $f(x) = 3x^3(2x^2 - 3x + 1)$.

At this point, all the products we encounter can be differentiated by either of the methods illustrated in Example 1. In the next and later sections we will see that there are situations where the product rule must be used. Unless instructed otherwise, you should use the product rule to differentiate all products in this section to gain experience with the use of this important differentiation rule.

Example 2 ➭ **Tangent Lines**   Let $f(x) = (2x - 9)(x^2 + 6)$.

(A) Find the equation of the line tangent to the graph of $f(x)$ at $x = 3$.
(B) Find the value(s) of x where the tangent line is horizontal.

SOLUTION   (A) First, find $f'(x)$:
$$f'(x) = (2x - 9)(x^2 + 6)' + (x^2 + 6)(2x - 9)'$$
$$= (2x - 9)(2x) + (x^2 + 6)(2)$$

Now, find the equation of the tangent line at $x = 3$:
$$y - y_1 = m(x - x_1) \qquad y_1 = f(x_1) = f(3) = -45$$
$$y - (-45) = 12(x - 3) \qquad m = f'(x_1) = f'(3) = 12$$
$$y = 12x - 81 \qquad\qquad \textit{Tangent line at } x = 3$$

(B) The tangent line is horizontal at any value of x such that $f'(x) = 0$, so
$$f'(x) = (2x - 9)2x + (x^2 + 6)2 = 0$$
$$6x^2 - 18x + 12 = 0$$
$$x^2 - 3x + 2 = 0$$
$$(x - 1)(x - 2) = 0$$
$$x = 1, 2$$

The tangent line is horizontal at $x = 1$ and at $x = 2$.

Matched Problem 2 ➭ Repeat Example 2 for $f(x) = (2x + 9)(x^2 - 12)$.

As Example 2 illustrates, the way we write $f'(x)$ depends on what we want to do with it. If we are interested only in evaluating $f'(x)$ at specified values of x, the form in part (A) is sufficient. However, if we want to solve $f'(x) = 0$, we must multiply and collect like terms, as we did in part (B).

❑ DERIVATIVES OF QUOTIENTS

As is the case with a product, the derivative of a quotient of two functions is not the quotient of the derivatives of the two functions.

Let $T(x) = x^5$, $B(x) = x^2$, and

$$f(x) = \frac{T(x)}{B(x)} = \frac{x^5}{x^2} = x^3$$

Which of the following is $f'(x)$?

(A) $\dfrac{T'(x)}{B'(x)}$ (B) $\dfrac{T'(x)B(x)}{[B(x)]^2}$ (C) $\dfrac{T(x)B'(x)}{[B(x)]^2}$

(D) $\dfrac{T'(x)B(x)}{[B(x)]^2} - \dfrac{T(x)B'(x)}{[B(x)]^2} = \dfrac{T'(x)B(x) - T(x)B'(x)}{[B(x)]^2}$

The expressions in Explore–Discuss 2 suggest that the derivative of a quotient leads to a more complicated quotient than you might expect.

In general, if $T(x)$ and $B(x)$ are any two differentiable functions and

$$f(x) = \frac{T(x)}{B(x)}$$

it can be shown that

$$f'(x) = \frac{B(x)T'(x) - T(x)B'(x)}{[B(x)]^2}$$

Thus,

The derivative of the quotient of two functions is the bottom function times the derivative of the top function minus the top function times the derivative of the bottom function, all over the bottom function squared.

www

Quotient Rule

If

$$y = f(x) = \frac{T(x)}{B(x)}$$

and if $T'(x)$ and $B'(x)$ exist, then

$$f'(x) = \frac{B(x)T'(x) - T(x)B'(x)}{[B(x)]^2}$$

Also,

$$y' = \frac{BT' - TB'}{B^2} \qquad \frac{dy}{dx} = \frac{B\dfrac{dT}{dx} - T\dfrac{dB}{dx}}{B^2}$$

Example 3 ⇔ **Differentiating Quotients**

(A) If $f(x) = \dfrac{x^2}{2x - 1}$, find $f'(x)$.   (B) If $y = \dfrac{x^2 - x}{x^3 + 1}$, find y'.

(C) Find $\dfrac{d}{dx}\dfrac{x^2 - 3}{x^2}$ by using the quotient rule and also by splitting the fraction into two fractions.

SOLUTION   (A) $f'(x) = \dfrac{(2x - 1)(x^2)' - x^2(2x - 1)'}{(2x - 1)^2}$

$= \dfrac{(2x - 1)(2x) - x^2(2)}{(2x - 1)^2}$

$= \dfrac{4x^2 - 2x - 2x^2}{(2x - 1)^2}$

$= \dfrac{2x^2 - 2x}{(2x - 1)^2}$

The bottom times the derivative of the top minus the top times the derivative of the bottom, all over the square of the bottom

(B) $y' = \dfrac{(x^3 + 1)(x^2 - x)' - (x^2 - x)(x^3 + 1)'}{(x^3 + 1)^2}$

$= \dfrac{(x^3 + 1)(2x - 1) - (x^2 - x)(3x^2)}{(x^3 + 1)^2}$

$= \dfrac{2x^4 - x^3 + 2x - 1 - 3x^4 + 3x^3}{(x^3 + 1)^2}$

$= \dfrac{-x^4 + 2x^3 + 2x - 1}{(x^3 + 1)^2}$

(C) **Method 1.** Use the quotient rule:

$$\frac{d}{dx}\frac{x^2 - 3}{x^2} = \frac{x^2\dfrac{d}{dx}(x^2 - 3) - (x^2 - 3)\dfrac{d}{dx}x^2}{(x^2)^2}$$

$$= \frac{x^2(2x) - (x^2 - 3)2x}{x^4}$$

$$= \frac{2x^3 - 2x^3 + 6x}{x^4} = \frac{6x}{x^4} = \frac{6}{x^3}$$

Method 2. Split into two fractions:

$$\frac{x^2 - 3}{x^2} = \frac{x^2}{x^2} - \frac{3}{x^2} = 1 - 3x^{-2}$$

$$\frac{d}{dx}(1 - 3x^{-2}) = 0 - 3(-2)x^{-3} = \frac{6}{x^3}$$

Comparing methods 1 and 2, we see that it often pays to change an expression algebraically before blindly using a differentiation formula. ■

Matched Problem 3 ⇨ Find

(A) $f'(x)$ for $f(x) = \dfrac{2x}{x^2 + 3}$ (B) y' for $y = \dfrac{x^3 - 3x}{x^2 - 4}$

(C) $\dfrac{d}{dx} \dfrac{2 + x^3}{x^3}$ two ways

Explore–Discuss 3

Explain why $\neq$ is used below, and then find the correct derivative.

$$\frac{d}{dx} \frac{x^3}{x^2 + 3x + 4} \neq \frac{3x^2}{2x + 3}$$

Example 4 ⇨ **Sales Analysis** The total sales S (in thousands of games) for a home video game t months after the game is introduced are given by

$$S(t) = \frac{125t^2}{t^2 + 100}$$

(A) Find $S'(t)$.

(B) Find $S(10)$ and $S'(10)$. Write a brief verbal interpretation of these results. Check on a graphing utility.

(C) Use the results from part (B) to estimate the total sales after 11 months.

SOLUTION (A) $S'(t) = \dfrac{(t^2 + 100)(125t^2)' - 125t^2(t^2 + 100)'}{(t^2 + 100)^2}$

$= \dfrac{(t^2 + 100)(250t) - 125t^2(2t)}{(t^2 + 100)^2}$

$= \dfrac{250t^3 + 25,000t - 250t^3}{(t^2 + 100)^2}$

$= \dfrac{25,000t}{(t^2 + 100)^2}$

(B) $S(10) = \dfrac{125(10)^2}{10^2 + 100} = 62.5$ and $S'(10) = \dfrac{25,000(10)}{(10^2 + 100)^2} = 6.25$

The total sales after 10 months are 62,500 games, and sales are increasing at the rate of 6,250 games per month. Figure 1 shows a check of these results on a graphing utility.

(C) The total sales will increase by approximately 6,250 games during the next month. Thus, the estimated total sales after 11 months are $62,500 + 6,250 = 68,750$ games.

FIGURE 1 $y_1 = \dfrac{125\,x^2}{x^2 + 100}$

Matched Problem 4 ⇨ Refer to Example 4. Suppose that the total sales S (in thousands of games) t months after the game is introduced are given by

$$S(t) = \frac{150t}{t + 3}$$

(A) Find $S'(t)$.

(B) Find $S(12)$ and $S'(12)$. Write a brief verbal interpretation of these results. Check on a graphing utility.

(C) Use the results from part (B) to estimate the total sales after 13 months.

Answers to Matched Problems

1. $30x^4 - 36x^3 + 9x^2$

2. (A) $y = 84x - 297$
   (B) $x = -4, x = 1$

3. (A) $\dfrac{(x^2 + 3)2 - (2x)(2x)}{(x^2 + 3)^2} = \dfrac{6 - 2x^2}{(x^2 + 3)^2}$

   (B) $\dfrac{(x^2 - 4)(3x^2 - 3) - (x^3 - 3x)(2x)}{(x^2 - 4)^2} = \dfrac{x^4 - 9x^2 + 12}{(x^2 - 4)^2}$

   (C) $-\dfrac{6}{x^4}$

4. (A) $S'(t) = \dfrac{450}{(t + 3)^2}$

   (B) $S(12) = 120; S'(12) = 2$. After 12 months, the total sales are 120,000 games, and sales are increasing at the rate of 2,000 games per month.

   (C) 122,000 games

Exercise 3-5

The answers to most of the problems in this exercise set contain both an unsimplified form and a simplified form of the derivative. When checking your work, first check that you applied the rules correctly, and then check that you performed the algebraic simplification correctly. Unless instructed otherwise, when differentiating a product, use the product rule rather than performing the multiplication first.

A *In Problems 1–18, find $f'(x)$ and simplify.*

1. $f(x) = 2x^3(x^2 - 2)$

2. $f(x) = 5x^2(x^3 + 2)$

3. $f(x) = (x - 3)(2x - 1)$

4. $f(x) = (3x + 2)(4x - 5)$

5. $f(x) = \dfrac{x}{x - 3}$ 6. $f(x) = \dfrac{3x}{2x + 1}$

7. $f(x) = \dfrac{2x + 3}{x - 2}$ 8. $f(x) = \dfrac{3x - 4}{2x + 3}$

9. $f(x) = (x^2 + 1)(2x - 3)$

10. $f(x) = (3x + 5)(x^2 - 3)$

11. $f(x) = (0.4x + 2)(0.5x - 5)$

12. $f(x) = (0.5x - 4)(0.2x + 1)$

13. $f(x) = \dfrac{x^2 + 1}{2x - 3}$ 14. $f(x) = \dfrac{3x + 5}{x^2 - 3}$

15. $f(x) = (x^2 + 2)(x^2 - 3)$

16. $f(x) = (x^2 - 4)(x^2 + 5)$

17. $f(x) = \dfrac{x^2 + 2}{x^2 - 3}$ 18. $f(x) = \dfrac{x^2 - 4}{x^2 + 5}$

Problems 19–24 refer to functions f and g that satisfy $f(1) = 3, f'(1) = 2, g(1) = 5,$ and $g'(1) = -3$. In each problem, find $h'(1)$ for the indicated function h.

19. $h(x) = f(x)g(x)$ 20. $h(x) = \dfrac{f(x)}{g(x)}$

21. $h(x) = \dfrac{g(x)}{f(x)}$ 22. $h(x) = f(x)f(x)$

23. $h(x) = \dfrac{1}{f(x)}$ 24. $h(x) = \dfrac{1}{g(x)}$

B *In Problems 25–32, find the indicated derivatives and simplify.*

25. $f'(x)$ for $f(x) = (2x + 1)(x^2 - 3x)$

26. y' for $y = (x^3 + 2x^2)(3x - 1)$

27. $\dfrac{dy}{dx}$ for $y = (2.5x - x^2)(4x + 1.4)$

28. $\dfrac{d}{dx}[(3 - 0.4x^3)(0.5x^2 - 2x)]$

29. y' for $y = \dfrac{5x - 3}{x^2 + 2x}$

30. $f'(x)$ for $f(x) = \dfrac{3x^2}{2x - 1}$

31. $\dfrac{d}{dx} \dfrac{x^2 - 3x + 1}{x^2 - 1}$

32. $\dfrac{dy}{dx}$ for $y = \dfrac{x^4 - x^3}{3x - 1}$

In Problems 33–36, find $f'(x)$ and find the equation of the line tangent to the graph of f at $x = 2$.

33. $f(x) = (1 + 3x)(5 - 2x)$

34. $f(x) = (7 - 3x)(1 + 2x)$

35. $f(x) = \dfrac{x - 8}{3x - 4}$ **36.** $f(x) = \dfrac{2x - 5}{2x - 3}$

In Problems 37–40, find $f'(x)$ and find the value(s) of x where $f'(x) = 0$.

37. $f(x) = (2x - 15)(x^2 + 18)$

38. $f(x) = (2x - 3)(x^2 - 6)$

39. $f(x) = \dfrac{x}{x^2 + 1}$ **40.** $f(x) = \dfrac{x}{x^2 + 9}$

In Problems 41–44, find $f'(x)$ two ways: by using the product or quotient rule and by simplifying first.

41. $f(x) = x^3(x^4 - 1)$ **42.** $f(x) = x^4(x^3 - 1)$

43. $f(x) = \dfrac{x^3 + 9}{x^3}$ **44.** $f(x) = \dfrac{x^4 + 4}{x^4}$

C *In Problems 45–56, find each derivative and simplify.*

45. $f'(x)$ for $f(x) = (2x^4 - 3x^3 + x)(x^2 - x + 5)$

46. $\dfrac{dy}{dx}$ for $y = (x^2 - 3x + 1)(x^3 + 2x^2 - x)$

47. $\dfrac{d}{dx} \dfrac{3x^2 - 2x + 3}{4x^2 + 5x - 1}$

48. y' for $y = \dfrac{x^3 - 3x + 4}{2x^2 + 3x - 2}$

49. $\dfrac{dy}{dx}$ for $y = 9x^{1/3}(x^3 + 5)$

50. $\dfrac{d}{dx}[(4x^{1/2} - 1)(3x^{1/3} + 2)]$

51. $f'(x)$ for $f(x) = \dfrac{6\sqrt[3]{x}}{x^2 - 3}$

52. y' for $y = \dfrac{2\sqrt{x}}{x^2 - 3x + 1}$

53. $\dfrac{d}{dx} \dfrac{x^3 - 2x^2}{\sqrt[3]{x^2}}$

54. $\dfrac{dy}{dx}$ for $y = \dfrac{x^2 - 3x + 1}{\sqrt[4]{x}}$

55. $f'(x)$ for $f(x) = \dfrac{(2x^2 - 1)(x^2 + 3)}{x^2 + 1}$

56. y' for $y = \dfrac{2x - 1}{(x^3 + 2)(x^2 - 3)}$

Problems 57–60 refer to a function of the form $f(x) = [u(x)]^n$, where $u(x)$ is a differentiable function.

57. Use the product rule to show that if $n = 2$, then $f'(x) = 2u(x)u'(x)$.

58. Use the product rule and the result from Problem 57 to show that $f'(x) = 3[u(x)]^2 u'(x)$ if $n = 3$.

59. Based on the results in Problems 57 and 58, write a formula for $f'(x)$ for any n.

60. Use the quotient rule to find $f'(x)$ if $n = -1$. Does this agree with your formula in Problem 59?

In Problems 61–64, approximate (to two decimal places) the value (s) of x where the graph of f has a horizontal tangent line.

61. $f(x) = (x^2 + 4)(x^2 - 2x)$

62. $f(x) = (x^2 + 3)(x^2 + 4x)$

63. $f(x) = \dfrac{x^3 + 17x - 2}{x^2 + 1}$ **64.** $f(x) = \dfrac{x^3 + 15x - 1}{x^2 + 1}$

Applications

Business & Economics

65. *Sales analysis.* The total sales S (in thousands of CDs) for a compact disk are given by

$$S(t) = \dfrac{90t^2}{t^2 + 50}$$

where t is the number of months since the release of the CD.

(A) Find $S'(t)$.

(B) Find $S(10)$ and $S'(10)$. Write a brief verbal interpretation of these results.

(C) Use the results from part (B) to estimate the total sales after 11 months.

66. *Sales analysis.* A communications company has installed a cable television system in a city. The total number N (in thousands) of subscribers t months after the installation of the system is given by

$$N(t) = \frac{200t}{t + 5}$$

(A) Find $N'(t)$.

(B) Find $N(15)$ and $N'(15)$. Write a brief verbal interpretation of these results.

(C) Use the results from part (B) to estimate the total number of subscribers after 16 months.

67. *Price–demand equation.* According to classical economic theory, the demand x for a quantity in a free market decreases as the price p increases (see the figure). Suppose that the number x of CD players people are willing to buy per week from a retail chain at a price of $\$p$ is given by

$$x = \frac{4,000}{0.1p + 1} \qquad 10 \le p \le 70$$

Figure for 67 and 68

(A) Find dx/dp.

(B) Find the demand and the instantaneous rate of change of demand with respect to price when the price is $40. Write a brief verbal interpretation of these results.

(C) Use the results from part (B) to estimate the demand if the price is increased to $41.

68. *Price–supply equation.* Also according to classical economic theory, the supply x for a quantity in a free market increases as the price p increases (see the figure). Suppose that the number x of CD

players a retail chain is willing to sell per week at a price of $\$p$ is given by

$$x = \frac{100p}{0.1p + 1} \qquad 10 \le p \le 70$$

(A) Find dx/dp.

(B) Find the supply and the instantaneous rate of change of supply with respect to price when the price is $40. Write a brief verbal interpretation of these results.

(C) Use the results from part (B) to estimate the supply if the price is increased to $41.

Life Sciences

69. *Medicine.* A drug is injected into the bloodstream of a patient through her right arm. The concentration of the drug (in milligrams per cubic centimeter) in the bloodstream of the left arm t hours after the injection is given by

$$C(t) = \frac{0.14t}{t^2 + 1}$$

(A) Find $C'(t)$.

(B) Find $C'(0.5)$ and $C'(3)$, and interpret the results.

70. *Drug sensitivity.* One hour after x milligrams of a particular drug are given to a person, the change in body temperature $T(x)$, in degrees Fahrenheit, is given approximately by

$$T(x) = x^2\left(1 - \frac{x}{9}\right) \qquad 0 \le x \le 7$$

The rate $T'(x)$ at which T changes with respect to the size of the dosage, x, is called the *sensitivity* of the body to the dosage.

(A) Find $T'(x)$ using the product rule.

(B) Find $T'(1)$, $T'(3)$, and $T'(6)$.

Social Sciences

71. *Learning.* In the early days of quantitative learning theory (around 1917), L. L. Thurstone found that a given person successfully accomplished $N(x)$ acts after x practice acts, as given by

$$N(x) = \frac{100x + 200}{x + 32}$$

(A) Find the instantaneous rate of change of learning, $N'(x)$, with respect to the number of practice acts x.

(B) Find $N'(4)$ and $N'(68)$.

Chain Rule: Power Form

- CHAIN RULE: POWER RULE
- COMBINING RULES OF DIFFERENTIATION

In this section we develop a rule for differentiating powers of functions—a special case of the very important *chain rule,* which we will return to in Chapter 5. Also, for the first time, we will encounter some product forms that cannot be simplified by multiplication and must be differentiated by the power rule.

❑ CHAIN RULE: POWER RULE

We have already made extensive use of the power rule,

$$\frac{d}{dx} x^n = nx^{n-1} \tag{1}$$

Now we want to generalize this rule so that we can differentiate functions of the form $[u(x)]^n$, where $u(x)$ is a differentiable function. Is rule (1) still valid if we replace x with a function $u(x)$?

Explore–Discuss 1

Let $u(x) = 2x^2$ and $f(x) = [u(x)]^3 = 8x^6$. Which of the following is $f'(x)$?

(A) $3[u(x)]^2$ (B) $3[u'(x)]^2$ (C) $3[u(x)]^2 u'(x)$

The calculations in Explore–Discuss 1 show that we cannot generalize the power rule simply by replacing x with $u(x)$ in equation (1).

How can we find a formula for the derivative of $[u(x)]^n$, where $u(x)$ is an arbitrary differentiable function? Let's begin by considering the derivatives of $[u(x)]^2$ and $[u(x)]^3$ to see if a general pattern emerges. Since $[u(x)]^2 = u(x)u(x)$, we use the product rule to write

$$\frac{d}{dx}[u(x)]^2 = \frac{d}{dx}[u(x)u(x)]$$
$$= u(x)u'(x) + u(x)u'(x)$$
$$= 2u(x)u'(x) \tag{2}$$

Since $[u(x)]^3 = [u(x)]^2 u(x)$, we now use the product rule and the result in equation (2) to write

$$\frac{d}{dx}[u(x)]^3 = \frac{d}{dx}\{[u(x)]^2 u(x)\}$$
$$= [u(x)]^2 \frac{d}{dx}u(x) + u(x)\frac{d}{dx}[u(x)]^2 \qquad \text{Use equation (2) to substitute for}$$
$$= [u(x)]^2 u'(x) + u(x)[2u(x)u'(x)] \qquad \frac{d}{dx}[u(x)]^2.$$
$$= 3[u(x)]^2 u'(x)$$

Continuing in this fashion, it can be shown that

$$\frac{d}{dx}[u(x)]^n = n[u(x)]^{n-1}u'(x) \qquad n \text{ a positive integer} \tag{3}$$

Using more advanced techniques, formula (3) can be established for all real numbers n. Thus, we have the **general power rule:**

> ### General Power Rule
>
> **www**
>
> If $u(x)$ is a differentiable function, n is any real number, and
>
> $$y = f(x) = [u(x)]^n$$
>
> then
>
> $$f'(x) = n[u(x)]^{n-1}u'(x)$$
>
> This rule is often written more compactly as
>
> $$y' = nu^{n-1}u' \quad \text{or} \quad \frac{d}{dx}u^n = nu^{n-1}\frac{du}{dx} \quad \text{where } u = u(x)$$

The general power rule is a special case of a very important and useful differentiation rule called the **chain rule.** In essence, the chain rule will enable us to differentiate a composition form $f[g(x)]$ if we know how to differentiate $f(x)$ and $g(x)$. We defer a complete discussion of the chain rule until Chapter 5.

Example 1   ⇨   **Differentiating Power Forms**   Find $f'(x)$:

(A) $f(x) = (3x + 1)^4$ (B) $f(x) = (x^3 + 4)^7$

(C) $f(x) = \dfrac{1}{(x^2 + x + 4)^3}$ (D) $f(x) = \sqrt{3 - x}$

SOLUTION

(A) $f(x) = (3x + 1)^4$ Let $u = 3x + 1, n = 4.$

$f'(x) = 4(3x + 1)^3(3x + 1)'$ $nu^{n-1}\dfrac{du}{dx}$

$\qquad = 4(3x + 1)^3 3$ $\dfrac{du}{dx} = 3$

$\qquad = 12(3x + 1)^3$

(B) $f(x) = (x^3 + 4)^7$ Let $u = (x^3 + 4), n = 7.$

$f'(x) = 7(x^3 + 4)^6(x^3 + 4)'$ $nu^{n-1}\dfrac{du}{dx}$

$\qquad = 7(x^3 + 4)^6 3x^2$ $\dfrac{du}{dx} = 3x^2$

$\qquad = 21x^2(x^3 + 4)^6$

(C) $f(x) = \dfrac{1}{(x^2 + x + 4)^3} = (x^2 + x + 4)^{-3}$ Let $u = x^2 + x + 4, n = -3.$

$f'(x) = -3(x^2 + x + 4)^{-4}(x^2 + x + 4)'$ $nu^{n-1}\dfrac{du}{dx}$

$\qquad = -3(x^2 + x + 4)^{-4}(2x + 1)$ $\dfrac{du}{dx} = 2x + 1$

$\qquad = \dfrac{-3(2x + 1)}{(x^2 + x + 4)^4}$

(D) $f(x) = \sqrt{3 - x} = (3 - x)^{1/2}$ Let $u = 3 - x, n = \frac{1}{2}.$

$f'(x) = \dfrac{1}{2}(3 - x)^{-1/2}(3 - x)'$ $nu^{n-1}\dfrac{du}{dx}$

$\qquad = \dfrac{1}{2}(3 - x)^{-1/2}(-1)$ $\dfrac{du}{dx} = -1$

$\qquad = -\dfrac{1}{2(3 - x)^{1/2}} \quad \text{or} \quad -\dfrac{1}{2\sqrt{3 - x}}$

Matched Problem 1 ⇔ Find $f'(x)$:

(A) $f(x) = (5x + 2)^3$ (B) $f(x) = (x^4 - 5)^5$

(C) $f(x) = \dfrac{1}{(x^2 + 4)^2}$ (D) $f(x) = \sqrt{4 - x}$

Notice that we used two steps to differentiate each function in Example 1. First, we applied the general power rule; then we found du/dx. As you gain experience with the general power rule, you may want to combine these two steps. If you do this, be certain to multiply by du/dx. For example,

$$\frac{d}{dx}(x^5 + 1)^4 = 4(x^5 + 1)^3 5x^4 \qquad \textit{Correct}$$

$$\frac{d}{dx}(x^5 + 1)^4 \neq 4(x^5 + 1)^3 \qquad \textit{du/dx = 5x⁴ is missing}$$

If we let $u(x) = x$, then $du/dx = 1$, and the general power rule reduces to the (ordinary) power rule discussed in Section 3-4. Compare the following:

$$\frac{d}{dx} x^n = nx^{n-1} \qquad \textit{Yes—power rule}$$

$$\frac{d}{dx} u^n = nu^{n-1}\frac{du}{dx} \qquad \textit{Yes—general power rule}$$

$$\frac{d}{dx} u^n \neq nu^{n-1} \qquad \textit{Unless u(x) = x + k so that du/dx = 1}$$

❑ COMBINING RULES OF DIFFERENTIATION

The following examples illustrate the use of the general power rule in combination with other rules of differentiation.

Example 2 ⇔ **Tangent Lines**   Find the equation of the line tangent to the graph of f at $x = 2$ for $f(x) = x^2\sqrt{2x + 12}$. Check with a graphing utility.

SOLUTION

$$f(x) = x^2\sqrt{2x + 12}$$
$$= x^2(2x + 12)^{1/2} \qquad \textit{Apply the product rule.}$$

$$f'(x) = x^2\frac{d}{dx}(2x + 12)^{1/2} + (2x + 12)^{1/2}\frac{d}{dx}x^2 \qquad \begin{array}{l}\textit{Use the general}\\\textit{power rule to}\\\textit{differentiate}\end{array}$$

$$= x^2\left[\frac{1}{2}(2x + 12)^{-1/2}\right](2) + (2x + 12)^{1/2}(2x) \qquad \begin{array}{l}\textit{(2x + 12)}^{1/2}\textit{ and}\\\textit{the ordinary}\end{array}$$

$$= \frac{x^2}{\sqrt{2x + 12}} + 2x\sqrt{2x + 12} \qquad \begin{array}{l}\textit{power rule to}\\\textit{differentiate x}^2\textit{.}\end{array}$$

$$f'(2) = \frac{4}{\sqrt{16}} + 4\sqrt{16} = 1 + 16 = 17$$

$$f(2) = 4\sqrt{16} = 16$$

$$(x_1, y_1) = (2, f(2)) = (2, 16) \qquad \textit{Point}$$

$$m = f'(2) = 17 \qquad \textit{Slope}$$

$$y - 16 = 17(x - 2) \qquad \textit{y – y₁ = m(x – x₁)}$$

$$y = 17x - 18 \qquad \textit{Tangent line}$$

Figure 1A checks the values of $f(2)$ and $f'(2)$, and Figure 1B shows the graphs of $y = f(x)$ and the tangent line at $x = 2$. Note that we graphed the equation $y = 17x - 18$ in order to check our work. We did not use a built-in routine for graphing tangent lines. Such a routine would produce the same graph, but it would not tell us if we have the correct equation for the tangent line.

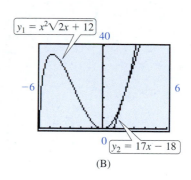

(A)

(B)

FIGURE 1

Matched Problem 2 ⇔ Find $f'(x)$ and the equation of the line tangent to the graph of f at $x = 3$ for $f(x) = x\sqrt{15 - 2x}$. Check with a graphing utility.

Example 3 ⇔ **Tangent Lines**   Find the value(s) of x where the tangent line is horizontal for

$$f(x) = \frac{x^3}{(2 - 3x)^5}$$

SOLUTION   Use the quotient rule:

$$f'(x) = \frac{(2 - 3x)^5 \dfrac{d}{dx} x^3 - x^3 \dfrac{d}{dx}(2 - 3x)^5}{[(2 - 3x)^5]^2}$$

Use the ordinary power rule to differentiate x^3 and the general power rule to differentiate $(2 - 3x)^5$.

$$= \frac{(2 - 3x)^5 3x^2 - x^3 5(2 - 3x)^4 (-3)}{(2 - 3x)^{10}}$$

$$= \frac{(2 - 3x)^4 3x^2 [(2 - 3x) + 5x]}{(2 - 3x)^{10}}$$

$$= \frac{3x^2 (2 + 2x)}{(2 - 3x)^6} = \frac{6x^2 (x + 1)}{(2 - 3x)^6}$$

Since a fraction is 0 when the numerator is 0 and the denominator is not, we see that $f'(x) = 0$ at $x = -1$ and $x = 0$. Thus, the graph of f will have horizontal tangent lines at $x = -1$ and $x = 0$.

Matched Problem 3 ⇔ Find the value(s) of x where the tangent line is horizontal for

$$f(x) = \frac{x^3}{(3x - 2)^2}$$

Example 4 ⇔ **Combining Differentiation Rules**   Starting with the function f in Example 3, write f as a product and then differentiate.

SOLUTION $$f(x) = \frac{x^3}{(2 - 3x)^5} = x^3 (2 - 3x)^{-5}$$

$$f'(x) = x^3 \frac{d}{dx}(2 - 3x)^{-5} + (2 - 3x)^{-5} \frac{d}{dx} x^3$$

$$= x^3(-5)(2 - 3x)^{-6}(-3) + (2 - 3x)^{-5} 3x^2$$

$$= 15x^3(2 - 3x)^{-6} + 3x^2(2 - 3x)^{-5}$$

At this point, we have an unsimplified form for $f'(x)$. This may be satisfactory for some purposes, but not for others. For example, if we need to solve the equation $f'(x) = 0$, we must simplify algebraically:

$$f'(x) = \frac{15x^3}{(2 - 3x)^6} + \frac{3x^2}{(2 - 3x)^5} = \frac{15x^3}{(2 - 3x)^6} + \frac{3x^2(2 - 3x)}{(2 - 3x)^6}$$

$$= \frac{15x^3 + 3x^2(2 - 3x)}{(2 - 3x)^6} = \frac{3x^2(5x + 2 - 3x)}{(2 - 3x)^6}$$

$$= \frac{3x^2(2 + 2x)}{(2 - 3x)^6} = \frac{6x^2(1 + x)}{(2 - 3x)^6}$$

■

Matched Problem 4 ⇔ Refer to the function f in Matched Problem 3. Write f as a product and then differentiate.

■

As Example 4 illustrates, any quotient can be converted to a product and differentiated by the product rule. However, if the derivative must be simplified, it is usually easier to use the quotient rule. (Compare the algebraic simplifications in Example 4 with those in Example 3.) There is one special case where using negative exponents is the preferred method—a fraction whose numerator is a constant.

Example 5 ⇔ **Alternative Methods of Differentiation**   Find $f'(x)$ two ways for

$$f(x) = \frac{4}{(x^2 + 9)^3}$$

Solution   **Method 1.**   Use the quotient rule:

$$f'(x) = \frac{(x^2 + 9)^3 \frac{d}{dx} 4 - 4 \frac{d}{dx}(x^2 + 9)^3}{[(x^2 + 9)^3]^2}$$

$$= \frac{(x^2 + 9)^3(0) - 4[3(x^2 + 9)^2(2x)]}{(x^2 + 9)^6}$$

$$= \frac{-24x(x^2 + 9)^2}{(x^2 + 9)^6} = \frac{-24x}{(x^2 + 9)^4}$$

Method 2.   Rewrite as a product, and use the general power rule:

$$f(x) = \frac{4}{(x^2 + 9)^3} = 4(x^2 + 9)^{-3}$$

$$f'(x) = 4(-3)(x^2 + 9)^{-4}(2x)$$

$$= \frac{-24x}{(x^2 + 9)^4}$$

Which method do you prefer?

■

Matched Problem 5 ⇔ Find $f'(x)$ two ways for $f(x) = \dfrac{5}{(x^3 + 1)^2}$.

Answers to Matched Problems   **1.** (A) $15(5x + 2)^2$ (B) $20x^3(x^4 - 5)^4$ (C) $-4x/(x^2 + 4)^3$
(D) $-1/(2\sqrt{4 - x})$

2. $f'(x) = \sqrt{15 - 2x} - \dfrac{x}{\sqrt{15 - 2x}}; y = 2x + 3$ **3.** $x = 0, x = 2$

4. $-6x^3(3x - 2)^{-3} + 3x^2(3x - 2)^{-2} = \dfrac{3x^2(x - 2)}{(3x - 2)^3}$ **5.** $\dfrac{-30x^2}{(x^3 + 1)^3}$

Exercise 3-6

The answers to many of the problems in this exercise set contain both an unsimplified form and a simplified form of a derivative. When checking your work, first check that you applied the rules correctly, and then check that you performed the algebraic simplification correctly.

A *In Problems 1–6, replace the ? with an expression that will make the indicated equation valid.*

1. $\dfrac{d}{dx}(3x + 4)^4 = 4(3x + 4)^3$ ___?___

2. $\dfrac{d}{dx}(5 - 2x)^6 = 6(5 - 2x)^5$ ___?___

3. $\dfrac{d}{dx}(4 - 2x^2)^3 = 3(4 - 2x^2)^2$ ___?___

4. $\dfrac{d}{dx}(3x^2 + 7)^5 = 5(3x^2 + 7)^4$ ___?___

5. $\dfrac{d}{dx}(1 + 2x + 3x^2)^7 = 7(1 + 2x + 3x^2)^6$ ___?___

6. $\dfrac{d}{dx}(4 - 3x - 2x^2)^8 = 8(4 - 3x - 2x^2)^7$ ___?___

In Problems 7–20, find $f'(x)$ using the general power rule and simplify.

7. $f(x) = (2x + 5)^3$ **8.** $f(x) = (3x - 7)^5$

9. $f(x) = (5 - 2x)^4$ **10.** $f(x) = (9 - 5x)^2$

11. $f(x) = (4 + 0.2x)^5$ **12.** $f(x) = (6 - 0.5x)^4$

13. $f(x) = (3x^2 + 5)^5$ **14.** $f(x) = (5x^2 - 3)^6$

15. $f(x) = (x^3 - 2x^2 + 2)^8$ **16.** $f(x) = (2x^2 + x + 1)^7$

17. $f(x) = (2x - 5)^{1/2}$ **18.** $f(x) = (4x + 3)^{1/2}$

19. $f(x) = (x^4 + 1)^{-2}$ **20.** $f(x) = (x^5 + 2)^{-3}$

In Problems 21–24, find $f'(x)$ and the equation of the line tangent to the graph of f at the indicated value of x. Find the value(s) of x where the tangent line is horizontal.

21. $f(x) = (2x - 1)^3$; $x = 1$

22. $f(x) = (3x - 1)^4$; $x = 1$

23. $f(x) = (4x - 3)^{1/2}$; $x = 3$

24. $f(x) = (2x + 8)^{1/2}$; $x = 4$

B *In Problems 25–44, find dy/dx using the general power rule and simplify.*

25. $y = 3(x^2 - 2)^4$ **26.** $y = 2(x^3 + 6)^5$

27. $y = 2(x^2 + 3x)^{-3}$ **28.** $y = 3(x^3 + x^2)^{-2}$

29. $y = \sqrt{x^2 + 8}$ **30.** $y = \sqrt[3]{3x - 7}$

31. $y = \sqrt[3]{3x + 4}$ **32.** $y = \sqrt{2x - 5}$

33. $y = \sqrt[4]{0.8x + 3.6}$ **34.** $y = \sqrt[5]{1.6x - 4.5}$

35. $y = (x^2 - 4x + 2)^{1/2}$ **36.** $y = (2x^2 + 2x - 3)^{1/2}$

37. $y = \dfrac{1}{2x + 4}$ **38.** $y = \dfrac{1}{3x - 7}$

39. $y = \dfrac{1}{(x^3 + 4)^5}$ **40.** $y = \dfrac{1}{(x^2 - 3)^6}$

41. $y = \dfrac{1}{4x^2 - 4x + 1}$ **42.** $y = \dfrac{1}{2x^2 - 3x + 1}$

43. $y = \dfrac{4}{\sqrt{x^2 - 3x}}$ **44.** $y = \dfrac{3}{\sqrt[3]{x - x^2}}$

In Problems 45–50, find $f'(x)$, and find the equation of the line tangent to the graph of f at the indicated value of x.

45. $f(x) = x(4 - x)^3$; $x = 2$

46. $f(x) = x^2(1 - x)^4$; $x = 2$

47. $f(x) = \dfrac{x}{(2x - 5)^3}$; $x = 3$

48. $f(x) = \dfrac{x^4}{(3x - 8)^2}$; $x = 4$

49. $f(x) = x\sqrt{2x + 2}$; $x = 1$

50. $f(x) = x\sqrt{x - 6}$; $x = 7$

In Problems 51–56, find $f'(x)$, and find the value(s) of x where the tangent line is horizontal.

51. $f(x) = x^2(x - 5)^3$

52. $f(x) = x^3(x - 7)^4$

53. $f(x) = \dfrac{x}{(2x + 5)^2}$

54. $f(x) = \dfrac{x - 1}{(x - 3)^3}$

55. $f(x) = \sqrt{x^2 - 8x + 20}$ **56.** $f(x) = \sqrt{x^2 + 4x + 5}$

In Problems 57–62, approximate (to two decimal places) the value(s) of x where the graph of f has a horizontal tangent line.

57. $f(x) = x(x - 1)(x^2 - 5)$

58. $f(x) = x(x - 2)(x + 3)(x - 4)$

59. $f(x) = (x^3 - 2x^2)(x^2 + 1)$

60. $f(x) = (x^3 + 3x^2)(x^2 + 4)$

61. $f(x) = \sqrt{x^4 - 6x^2 + x + 12}$

62. $f(x) = \sqrt{x^4 - 4x^3 + 4x + 20}$

C *In Problems 63–74, find each derivative and simplify.*

63. $\dfrac{d}{dx}[3x(x^2 + 1)^3]$

64. $\dfrac{d}{dx}[2x^2(x^3 - 3)^4]$

65. $\dfrac{d}{dx}\dfrac{(x^3 - 7)^4}{2x^3}$

66. $\dfrac{d}{dx}\dfrac{3x^2}{(x^2 + 5)^3}$

67. $\dfrac{d}{dx}[(2x - 3)^2(2x^2 + 1)^3]$

68. $\dfrac{d}{dx}[(x^2 - 1)^3(x^2 - 2)^2]$

69. $\dfrac{d}{dx}(4x^2\sqrt{x^2 - 1})$

70. $\dfrac{d}{dx}(3x\sqrt{2x^2 + 3})$

71. $\dfrac{d}{dx}\dfrac{2x}{\sqrt{x - 3}}$

72. $\dfrac{d}{dx}\dfrac{x^2}{\sqrt{x^2 + 1}}$

73. $\dfrac{d}{dx}\sqrt{(2x - 1)^3(x^2 + 3)^4}$

74. $\dfrac{d}{dx}\sqrt{\dfrac{4x + 1}{2x^2 + 1}}$

Applications

Business & Economics

75. *Marginal cost.* The total cost (in hundreds of dollars) of producing x calculators per day is

$$C(x) = 10 + \sqrt{2x + 16} \qquad 0 \leq x \leq 50$$

(see the figure).

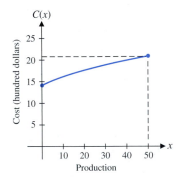

Figure for 75

(A) Find the marginal cost at a production level of x calculators.

(B) Find $C'(24)$ and $C'(42)$, and interpret the results.

76. *Marginal cost.* The total cost (in hundreds of dollars) of producing x cameras per week is

$$C(x) = 6 + \sqrt{4x + 4} \qquad 0 \leq x \leq 30$$

(A) Find the marginal cost at a production level of x cameras.

(B) Find $C'(15)$ and $C'(24)$, and interpret the results.

77. *Price–supply equation.* The number x of stereo speakers a retail chain is willing to sell per week at a price of $\$p$ is given by

$$x = 80\sqrt{p + 25} - 400 \qquad 20 \leq p \leq 100$$

(see the figure).

Figure for 77 and 78

(A) Find dx/dp.

(B) Find the supply and the instantaneous rate of change of supply with respect to price when the price is $\$75$. Write a brief verbal interpretation of these results.

78. *Price–demand equation.* The number x of stereo speakers people are willing to buy per week from a retail chain at a price of $\$p$ is given by

$$x = 1{,}000 - 60\sqrt{p + 25} \qquad 20 \leq p \leq 100$$

(see the figure on the preceding page).

(A) Find dx/dp.

(B) Find the demand and the instantaneous rate of change of demand with respect to price when the price is $75. Write a brief verbal interpretation of these results.

79. *Compound interest.* If $1,000 is invested at an annual interest rate r compounded monthly, the amount in the account at the end of 4 years is given by

$$A = 1{,}000(1 + \tfrac{1}{12}r)^{48}$$

Find the rate of change of the amount A with respect to the interest rate r.

80. *Compound interest.* If $100 is invested at an annual interest rate r compounded semiannually, the amount in the account at the end of 5 years is given by

$$A = 100(1 + \tfrac{1}{2}r)^{10}$$

Find the rate of change of the amount A with respect to the interest rate r.

Life Sciences

81. *Bacteria growth.* The number y of bacteria in a certain colony after x days is given approximately by

$$y = (3 \times 10^{6})\left[1 - \frac{1}{\sqrt[3]{(x^2 - 1)^2}}\right]$$

Find dy/dx.

82. *Pollution.* A small lake in a resort area became contaminated with harmful bacteria because of excessive septic tank seepage. After treating the lake with a bactericide, the Department of Public Health estimated the bacteria concentration (number per cubic centimeter) after t days to be given by

$$C(t) = 500(8 - t)^2 \qquad 0 \leq t \leq 7$$

(A) Find $C'(t)$ using the general power rule.

(B) Find $C'(1)$ and $C'(6)$, and interpret the results.

Social Sciences

83. *Learning.* In 1930, L. L. Thurstone developed the
WWW following formula to indicate how learning time T depends on the length of a list n:

$$T = f(n) = \frac{c}{k} n \sqrt{n - a}$$

where a, c, and k are empirical constants. Suppose that for a particular person, time T (in minutes) for learning a list of length n is

$$T = f(n) = 2n\sqrt{n - 2}$$

(A) Find dT/dn.

(B) Find $f'(11)$ and $f'(27)$, and interpret the results.

Section 3-7

Marginal Analysis in Business and Economics

❑   MARGINAL COST, REVENUE, AND PROFIT
❑   APPLICATION
❑   MARGINAL AVERAGE COST, REVENUE, AND PROFIT

❑   MARGINAL COST, REVENUE, AND PROFIT

WWW

One important use of calculus in business and economics is in *marginal analysis.* We introduced the concept of *marginal cost* earlier. There is no reason to stop there. Economists also talk about *marginal revenue* and *marginal profit.* Recall that the word *marginal* refers to an instantaneous rate of change—that is, a derivative. Thus, we define the following:

Marginal Cost, Revenue, and Profit

If x is the number of units of a product produced in some time interval, then

$$\text{total cost} = C(x) \qquad \text{total revenue} = R(x)$$
$$\textbf{marginal cost} = C'(x) \qquad \textbf{marginal revenue} = R'(x)$$
$$\text{total profit} = P(x) = R(x) - C(x)$$
$$\textbf{marginal profit} = P'(x) = R'(x) - C'(x)$$
$$= (\text{marginal revenue}) - (\text{marginal cost})$$

Marginal cost (or revenue or profit) is the instantaneous rate of change of cost (or revenue or profit) relative to production at a given production level.

It is important to remember that whenever we refer to a cost function $C(x)$ it is understood that $C(x)$ represents the *total cost* of producing x items. To find the exact cost of producing a particular item, we use the difference of two successive values of $C(x)$:

$$\text{Total cost of producing } x + 1 \text{ items} = C(x + 1)$$
$$\text{Total cost of producing } x \text{ items} = C(x)$$
$$\text{Exact cost of producing the } (x + 1)\text{st item} = C(x + 1) - C(x)$$

As we noted in Section 3-4, the marginal cost function can be used to approximate this exact cost. To see why, we return to the definition of a derivative:

$$C'(x) = \lim_{h \to 0} \frac{C(x + h) - C(x)}{h} \qquad \text{Marginal cost}$$

$$C'(x) \approx \frac{C(x + h) - C(x)}{h} \qquad h \neq 0$$

$$C'(x) \approx C(x + 1) - C(x) \qquad h = 1$$

Thus, we see that the marginal cost $C'(x)$ approximates $C(x + 1) - C(x)$, the exact cost of producing the $(x + 1)$st item. These observations are summarized below and illustrated in Figure 1.

FIGURE 1 $C'(x) \approx C(x + 1) - C(x)$

Marginal Cost and Exact Cost

If $C(x)$ is the total cost of producing x items, the marginal cost function approximates the exact cost of producing the $(x + 1)$st item:

$$\underset{\text{Marginal cost}}{C'(x)} \approx \underset{\text{Exact cost}}{C(x + 1) - C(x)}$$

Similar interpretations can be made for total revenue functions and total profit functions.

Example 1 **Cost Analysis** A company manufactures fuel tanks for automobiles. The total weekly cost (in dollars) of producing x tanks is given by

$$C(x) = 10,000 + 90x - 0.05x^2$$

(A) Find the marginal cost function.

(B) Find $C'(500)$, and discuss the various interpretations of this result.

(C) Find the exact cost of producing the 501st tank, and discuss the relationship between this result and the marginal cost found in part (B).

Check all calculations on a graphing utility.

SOLUTION   (A) $C'(x) = 90 - 0.1x$

(B) $C'(500) = 90 - 0.1(500) = \$40$

The standard interpretation of a derivative certainly applies to marginal cost functions. Thus, at a production level of 500 tanks per week, total production costs are increasing at the rate of $40 per tank. The special interpretation of a marginal cost function also applies. That is, $40 is the approximate cost of producing the 501st tank.

(C) $C(501) = 10,000 + 90(501) - 0.05(501)^2$

$\qquad = \$42,539.95$ *Total cost of producing 501 tanks per week*

$C(500) = 10,000 + 90(500) - 0.05(500)^2$

$\qquad = \$42,500.00$ *Total cost of producing 500 tanks per week*

$C(501) - C(500) = 42,539.95 - 42,500.00$

$\qquad\qquad\qquad = \$39.95$ *Exact cost of producing the 501st tank*

Comparing this result with the marginal cost found in part (B), we see that the marginal cost does provide a good approximation to the exact cost of producing the 501st tank.

See Figure 2 for a check of these calculations.

```
hDeriv(Y1,X,500)
                  40
500→X:Y1
              42500
501→X:Y1
          42539.95
```

FIGURE 2 $y_1 = 10,000 + 90x - 0.05x^2$

Matched Problem 1   A company manufactures automatic transmissions for automobiles. The total weekly cost (in dollars) of producing x transmissions is given by

$$C(x) = 50,000 + 600x - 0.75x^2$$

(A) Find the marginal cost function.

(B) Find $C'(200)$, and discuss the various interpretations of this result.

(C) Find the exact cost of producing the 201st transmission, and discuss the relationship between this result and the marginal cost found in part (B).

Check all calculations on a graphing utility.

$C(x)$

Slope = $C'(500)$ = 40

Slope = $C'(200)$ = 70

FIGURE 3 $C(x) = 10,000 + 90x - 0.05x^2$

It is instructive to compute both the approximate cost $C'(x)$ and the exact cost $C(x + 1) - C(x)$ for comparison purposes. However, in actual practice the marginal cost is used much more frequently than the exact cost. One reason for this is that the marginal cost is easily visualized when examining the graph of the total cost function. Figure 3 shows the graph of the cost function discussed in Example 1 with tangent lines added at $x = 200$ and $x = 500$. The graph clearly shows that as production increases, the slope of the tangent line decreases. Thus, the cost of producing the next tank also decreases, a desirable characteristic of a total cost function. We will have much more to say about graphical analysis in Chapter 4.

❑ APPLICATION

We now want to discuss how price, demand, revenue, cost, and profit are tied together in typical applications. Although either price or demand can be used as the independent variable in a price–demand equation, it is common practice to use demand as the independent variable when marginal revenue, cost, and profit are also involved.

Explore–Discuss 1

The market research department of a company used test marketing to determine the demand for a new radio (Table 1).

TABLE 1

DEMAND x	PRICE p
3,000	$7
6,000	$4

(A) Assuming that the relationship between price p and demand x is linear, find the price–demand equation and write the result in the form $x = f(p)$. Graph the equation and find the domain of f. Discuss the effect of price increases and decreases on demand.

(B) Solve the equation found in part (A) for p, obtaining an equation of the form $p = g(x)$. Graph this equation and find the domain of g. Discuss the effect of price increases and decreases on demand.

Example 2 ⇌

Production Strategy The market research department of a company recommends that the company manufacture and market a new transistor radio. After suitable test marketing, the research department presents the following **price–demand equation:**

$$x = 10,000 - 1,000p \qquad \text{x is demand at price p.} \qquad (1)$$

Or, solving (1) for p,

$$p = 10 - 0.001x \qquad (2)$$

where x is the number of radios retailers are likely to buy at $\$p$ per radio.
　　The financial department provides the following **cost function:**

$$C(x) = 7,000 + 2x \qquad (3)$$

where $\$7,000$ is the estimate of fixed costs (tooling and overhead) and $\$2$ is the estimate of variable costs per radio (materials, labor, marketing, transportation, storage, etc.).

(A) Find the domain of the function defined by the price–demand equation (2).

(B) Find the marginal cost function $C'(x)$ and interpret.

(C) Find the revenue function as a function of x, and find its domain.

(D) Find the marginal revenue at $x = 2,000, 5,000$, and $7,000$. Interpret these results.

(E) Graph the cost function and the revenue function in the same coordinate system, find the intersection points of these two graphs, and interpret the results.

(F) Find the profit function and its domain, and sketch its graph.

(G) Find the marginal profit at $x = 1,000, 4,000$, and $6,000$. Interpret these results.

SOLUTION   (A) Since price p and demand x must be nonnegative, we have $x \geqslant 0$ and

$$p = 10 - 0.001x \geqslant 0$$
$$10 \geqslant 0.001x$$
$$10,000 \geqslant x$$

Thus, the permissible values of x are $0 \leqslant x \leqslant 10,000$.

(B) The marginal cost is $C'(x) = 2$. Since this is a constant, it costs an additional \$2 to produce one more radio at any production level.

(C) The **revenue** is the amount of money R received by the company for manufacturing and selling x radios at \$$p$ per radio and is given by

$$R = (\text{number of radios sold})(\text{price per radio}) = xp$$

In general, the revenue R can be expressed as a function of p by using equation (1) or as a function of x by using equation (2). As we mentioned earlier, when using marginal functions, we will always use the number of items x as the independent variable. Thus, the **revenue function** is

$$R(x) = xp = x(10 - 0.001x) \quad \text{Using equation (2)} \tag{4}$$
$$= 10x - 0.001x^2$$

Since equation (2) is defined only for $0 \leqslant x \leqslant 10,000$, it follows that the domain of the revenue function is $0 \leqslant x \leqslant 10,000$.

(D) The **marginal revenue** is

$$R'(x) = 10 - 0.002x$$

For production levels of $x = 2,000, 5,000,$ and $7,000$, we have

$$R'(2,000) = 6 \qquad R'(5,000) = 0 \qquad R'(7,000) = -4$$

This means that at production levels of 2,000, 5,000, and 7,000, the respective approximate changes in revenue per unit change in production are \$6, \$0, and $-\$4$. That is, at the 2,000 output level, revenue increases as production increases; at the 5,000 output level, revenue does not change with a "small" change in production; and at the 7,000 output level, revenue decreases with an increase in production.

(E) When we graph $R(x)$ and $C(x)$ in the same coordinate system, we obtain Figure 4. The intersection points are called the **break-even points** because revenue equals cost at these production levels—the company neither makes nor loses money, but just breaks even. The break-even points are obtained as follows:

$$C(x) = R(x)$$
$$7,000 + 2x = 10x - 0.001x^2$$
$$0.001x^2 - 8x + 7,000 = 0$$
$$x^2 - 8,000x + 7,000,000 = 0 \quad \text{Solve using the quadratic formula (see Appendix A-9).}$$

$$x = \frac{8,000 \pm \sqrt{8,000^2 - 4(7,000,000)}}{2}$$
$$= \frac{8,000 \pm \sqrt{36,000,000}}{2}$$
$$= \frac{8,000 \pm 6,000}{2}$$
$$= 1,000, \quad 7,000$$

$$R(1{,}000) = 10(1{,}000) - 0.001(1{,}000)^2 = 9{,}000$$
$$C(1{,}000) = 7{,}000 + 2(1{,}000) = 9{,}000$$
$$R(7{,}000) = 10(7{,}000) - 0.001(7{,}000)^2 = 21{,}000$$
$$C(7{,}000) = 7{,}000 + 2(7{,}000) = 21{,}000$$

Thus, the break-even points are $(1{,}000, 9{,}000)$ and $(7{,}000, 21{,}000)$, as shown in Figure 4. Further examination of the figure shows that cost is greater than revenue for production levels between 0 and 1,000 and also between 7,000 and 10,000. Consequently, the company incurs a loss at these levels. On the other hand, for production levels between 1,000 and 7,000, revenue is greater than cost and the company makes a profit.

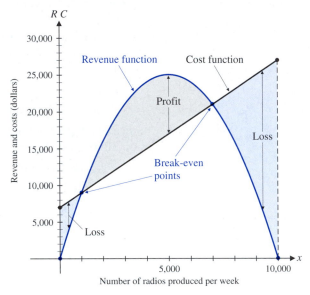

FIGURE 4

(F)   The **profit function** is

$$
\begin{aligned}
P(x) &= R(x) - C(x) \\
&= (10x - 0.001x^2) - (7{,}000 + 2x) \\
&= -0.001x^2 + 8x - 7{,}000
\end{aligned}
$$

The domain of the cost function is $x \geq 0$ and the domain of the revenue function is $0 \leq x \leq 10{,}000$. Thus, the domain of the profit function is the set of x values for which both functions are defined; that is, $0 \leq x \leq 10{,}000$. The graph of the profit function is shown in Figure 5. Notice that the x coordinates of the break-even points in Figure 4 are the x intercepts of the profit function. Furthermore, the intervals where cost is greater than revenue and where revenue is greater than cost correspond, respectively, to the intervals where profit is negative and the intervals where profit is positive.

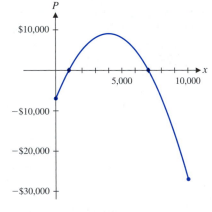

FIGURE 5

(G)   The **marginal profit** is

$$P'(x) = -0.002x + 8$$

For production levels of 1,000, 4,000, and 6,000, we have

$$P'(1{,}000) = 6 \qquad P'(4{,}000) = 0 \qquad P'(6{,}000) = -4$$

FIGURE 6
$y_1 = 7,000 + 2x$
$y_2 = 10x - 0.001x^2$

www

This means that at production levels of 1,000, 4,000, and 6,000, the respective approximate changes in profit per unit change in production are $6, $0, and −$4. That is, at the 1,000 output level, profit will be increased if production is increased; at the 4,000 output level, profit does not change for "small" changes in production; and at the 6,000 output level, profits will decrease if production is increased. It seems the best production level to produce a maximum profit is 4,000. ■

Break-even points can also be located by using graphical approximation on a graphing utility. The break-even points for Example 2 are shown in Figure 6.

Example 2 warrants careful study, since a number of important ideas in economics and calculus are involved. In the next chapter, we will develop a systematic procedure for finding the production level (and, using the demand equation, the selling price) that will maximize profit.

Matched Problem 2   Refer to the revenue and profit functions in Example 2.

(A) Find $R'(3,000)$ and $R'(6,000)$, and interpret the results.

(B) Find $P'(2,000)$ and $P'(7,000)$, and interpret the results. ■

Explore–Discuss 2

Let

$$C(x) = 12,000 + 5x \qquad \text{and} \qquad R(x) = 9x - 0.002x^2$$

Explain why $\neq$ is used below. Then find the correct expression for the profit function.

$$P(x) = R(x) - C(x) \neq 9x - 0.002x^2 - 12,000 + 5x$$

❑   MARGINAL AVERAGE COST, REVENUE, AND PROFIT

Sometimes, it is desirable to carry out marginal analysis relative to **average cost (cost per unit), average revenue (revenue per unit), and average profit (profit per unit).** The relevant definitions are summarized in the following box:

Marginal Average Cost, Revenue, and Profit

If x is the number of units of a product produced in some time interval, then

Cost per unit: **average cost** $= \overline{C}(x) = \dfrac{C(x)}{x}$

marginal average cost $= \overline{C}'(x) = \dfrac{d}{dx}\overline{C}(x)$

Revenue per unit: average revenue $= \overline{R}(x) = \dfrac{R(x)}{x}$

marginal average revenue $= \overline{R}'(x) = \dfrac{d}{dx}\overline{R}(x)$

Profit per unit: average profit $= \overline{P}(x) = \dfrac{P(x)}{x}$

marginal average profit $= \overline{P}'(x) = \dfrac{d}{dx}\overline{P}(x)$

Example 3 **Cost Analysis** A small machine shop manufactures drill bits used in the petroleum industry. The shop manager estimates that the total daily cost (in dollars) of producing x bits is

$$C(x) = 1{,}000 + 25x - 0.1x^2$$

(A) Find $\overline{C}(x)$ and $\overline{C}'(x)$.

(B) Find $\overline{C}(10)$ and $\overline{C}'(10)$, and interpret.

(C) Use the results in part (B) to estimate the average cost per bit at a production level of 11 bits per day.

SOLUTION (A) $\overline{C}(x) = \dfrac{C(x)}{x} = \dfrac{1{,}000 + 25x - 0.1x^2}{x}$

$= \dfrac{1{,}000}{x} + 25 - 0.1x$ Average cost function

$\overline{C}'(x) = \dfrac{d}{dx}\overline{C}(x) = -\dfrac{1{,}000}{x^2} - 0.1$ Marginal average cost function

(B) $\overline{C}(10) = \dfrac{1{,}000}{10} + 25 - 0.1(10) = \$124$

$\overline{C}'(10) = -\dfrac{1{,}000}{10^2} - 0.1 = -\$10.10$

At a production level of 10 bits per day, the average cost of producing a bit is \$124, and this cost is decreasing at the rate of \$10.10 per bit.

(C) If production is increased by 1 bit, then the average cost per bit will decrease by approximately \$10.10. Thus, the average cost per bit at a production level of 11 bits per day is approximately \$124 − \$10.10 = \$113.90. ∎

Matched Problem 3 Consider the cost function for the production of radios from Example 2:

$$C(x) = 7{,}000 + 2x$$

(A) Find $\overline{C}(x)$ and $\overline{C}'(x)$.

(B) Find $\overline{C}(100)$ and $\overline{C}'(100)$, and interpret.

(C) Use the results in part (B) to estimate the average cost per radio at a production level of 101 radios. ∎

A student produced the following solution to Matched Problem 3:

$$C(x) = 7,000 + 2x \quad \text{Cost}$$
$$C'(x) = 2 \quad \text{Marginal cost}$$
$$\frac{C'(x)}{x} = \frac{2}{x} \quad \text{"Average" of the marginal cost}$$

Explain why the last function is not the same as the marginal average cost function.

_____

CAUTION

1. The marginal average cost function must be computed by first finding the average cost function and then finding its derivative. As Explore–Discuss 3 illustrates, reversing the order of these two steps produces a different function that does not have any useful economic interpretations.

2. Recall that the marginal cost function has two interpretations: the usual interpretation of any derivative as an instantaneous rate of change, and the special interpretation as an approximation to the exact cost of the $(x + 1)$st item. This special interpretation does not apply to the marginal average cost function. Referring to Example 3, it would be incorrect to interpret $\overline{C}'(10) = -\$10.10$ to mean that the average cost of the next bit is approximately $-\$10.10$. In fact, the phrase "average cost of the next bit" does not even make sense. Averaging is a concept applied to a collection of items, not to a single item.

   These remarks also apply to revenue and profit functions.

Answers to Matched Problems

1. (A) $C'(x) = 600 - 1.5x$
   (B) $C'(200) = 300$. At a production level of 200 transmissions, total costs are increasing at the rate of $300 per transmission. Also, the approximate cost of producing the 201st transmission is $300.
   (C) $C(201) - C(200) = \$299.25$. The marginal cost from part (B) provides a good approximation to this exact cost.

2. (A) $R'(3,000) = 4$. At a production level of 3,000, a unit increase in production will increase revenue by approx. $4.
   $R'(6,000) = -2$. At a production level of 6,000, a unit increase in production will decrease revenue by approx. $2.
   (B) $P'(2,000) = 4$. At a production level of 2,000, a unit increase in production will increase profit by approx. $4.
   $P'(7,000) = -6$. At a production level of 7,000, a unit increase in production will decrease profit by approx. $6.

3. (A) $\overline{C}(x) = \dfrac{7,000}{x} + 2$; $\overline{C}'(x) = -\dfrac{7,000}{x^2}$
   (B) $\overline{C}(100) = \$72$; $\overline{C}'(100) = -\$0.70$. At a production level of 100 radios, the average cost per radio is $72, and this average cost is decreasing at a rate of $0.70 per radio.
   (C) Approx. $71.30.

Business & Economics

1. *Cost analysis.* The total cost (in dollars) of producing x food processors is

$$C(x) = 2,000 + 50x - 0.5x^2$$

(A) Find the exact cost of producing the 21st food processor.

(B) Use the marginal cost to approximate the cost of producing the 21st food processor.

2. *Cost analysis.* The total cost (in dollars) of producing x electric guitars is

$$C(x) = 1,000 + 100x - 0.25x^2$$

(A) Find the exact cost of producing the 51st guitar.

(B) Use the marginal cost to approximate the cost of producing the 51st guitar.

3. *Cost analysis.* The total cost (in dollars) of manufacturing x auto body frames is

$$C(x) = 60,000 + 300x$$

(A) Find the average cost per unit if 500 frames are produced.

(B) Find the marginal average cost at a production level of 500 units, and interpret the results.

(C) Use the results from parts (A) and (B) to estimate the average cost per frame if 501 frames are produced.

4. *Cost analysis.* The total cost (in dollars) of printing x dictionaries is

$$C(x) = 20,000 + 10x$$

(A) Find the average cost per unit if 1,000 dictionaries are produced.

(B) Find the marginal average cost at a production level of 1,000 units, and interpret the results.

(C) Use the results from parts (A) and (B) to estimate the average cost per dictionary if 1,001 dictionaries are produced.

5. *Profit analysis.* The total profit (in dollars) from the sale of x skateboards is

$$P(x) = 30x - 0.3x^2 - 250 \qquad 0 \le x \le 100$$

(A) Find the exact profit from the sale of the 26th skateboard.

(B) Use the marginal profit to approximate the profit from the sale of the 26th skateboard.

6. *Profit analysis.* The total profit (in dollars) from the sale of x portable stereos is

$$P(x) = 22x - 0.2x^2 - 400 \qquad 0 \le x \le 100$$

(A) Find the exact profit from the sale of the 41st stereo.

(B) Use the marginal profit to approximate the profit from the sale of the 41st stereo.

7. *Profit analysis.* The total profit (in dollars) from the sale of x video cassettes is

$$P(x) = 5x - 0.005x^2 - 450 \qquad 0 \le x \le 1,000$$

Evaluate the marginal profit at the given values of x, and interpret the results.

(A) $x = 450$ (B) $x = 750$

8. *Profit analysis.* The total profit (in dollars) from the sale of x cameras is

$$P(x) = 12x - 0.02x^2 - 1,000 \qquad 0 \le x \le 600$$

Evaluate the marginal profit at the given values of x, and interpret the results.

(A) $x = 200$ (B) $x = 350$

9. *Profit analysis.* The total profit (in dollars) from the sale of x lawn mowers is

$$P(x) = 30x - 0.03x^2 - 750 \qquad 0 \le x \le 1,000$$

(A) Find the average profit per mower if 50 mowers are produced.

(B) Find the marginal average profit at a production level of 50 mowers and interpret.

(C) Use the results from parts (A) and (B) to esti- mate the average profit per mower if 51 mow- ers are produced.

10. *Profit analysis.* The total profit (in dollars) from the sale of x charcoal grills is

$$P(x) = 20x - 0.02x^2 - 320 \qquad 0 \leq x \leq 1,000$$

(A) Find the average profit per grill if 40 grills are produced.

(B) Find the marginal average profit at a produc- tion level of 40 grills and interpret.

(C) Use the results from parts (A) and (B) to esti- mate the average profit per grill if 41 grills are produced.

11. *Revenue analysis.* The price p (in dollars) and the de- mand x for a particular clock radio are related by the equation

$$x = 4,000 - 40p$$

(A) Express the price p in terms of the demand x and find the domain of this function.

(B) Find the revenue $R(x)$ from the sale of x clock radios. What is the domain of R?

(C) Find the marginal revenue at a production level of 1,600 clock radios and interpret.

(D) Find the marginal revenue at a production level of 2,500 clock radios and interpret.

12. *Revenue analysis.* The price p (in dollars) and the de- mand x for a particular steam iron are related by the equation

$$x = 1,000 - 20p$$

(A) Express the price p in terms of the demand x and find the domain of this function.

(B) Find the revenue $R(x)$ from the sale of x clock radios. What is the domain of R?

(C) Find the marginal revenue at a production level of 400 steam irons and interpret.

(D) Find the marginal revenue at a production level of 650 steam irons and interpret.

13. *Revenue, cost and profit.* The price–demand equa- tion and the cost function for the production of table saws are given, respectively, by

$$x = 6,000 - 30p \qquad \text{and} \qquad C(x) = 72,000 + 60x$$

where x is the number of saws that can be sold at a price of $\$p$ per saw and $C(x)$ is the total cost (in dol- lars) of producing x saws.

(A) Express the price p as a function of the demand x and find the domain of this funtion.

(B) Find the marginal cost.

(C) Find the revenue function and state its domain.

(D) Find the marginal revenue.

(E) Find $R'(1,500)$ and $R'(4,500)$, and interpret the results.

(F) Graph the cost function and the revenue func- tion on the same coordinate system for $0 \leq x \leq 6,000$. Find the break-even points, and indicate regions of loss and profit.

(G) Find the profit function in terms of x.

(H) Find the marginal profit.

(I) Find $P'(1,500)$ and $P'(3,000)$, and interpret the results.

14. *Revenue, cost, and profit.* The price–demand equa- tion and the cost function for the production of tele- vision sets are given, respectively, by

$$x = 9,000 - 30p \qquad \text{and} \qquad C(x) = 150,000 + 30x$$

where x is the number of sets that can be sold at a price of $\$p$ per set and $C(x)$ is the total cost (in dol- lars) of producing x sets.

(A) Express the price p as a function of the demand x and find the domain of this function.

(B) Find the marginal cost.

(C) Find the revenue function and state its domain.

(D) Find the marginal revenue.

(E) Find $R'(3,000)$ and $R'(6,000)$, and interpret the results.

(F) Graph the cost function and the revenue func- tion on the same coordinate system for $0 \leq x \leq 9,000$. Find the break-even points, and indicate regions of loss and profit.

(G) Find the profit function in terms of x.

(H) Find the marginal profit.

(I) Find $P'(1,500)$ and $P'(4,500)$, and interpret the results.

15. *Revenue, cost, and profit.* A company is planning to manufacture and market a new two-slice electric toaster. After conducting extensive market surveys, the research department provides the following esti- mates: a weekly demand of 200 toasters at a price of $\$16$ per toaster and a weekly demand of 300 toasters at a price of $\$14$ per toaster. The financial department estimates that weekly fixed costs will be $\$1,400$ and variable costs (cost per unit) will be $\$4$.

(A) Assume that the relationship between the price p and the demand x is linear. Use the research department's estimates to express p as a function of x and find the domain of this function.

(B) Find the revenue function in terms of x and state its domain.

(C) Assume that the cost function is linear. Use the financial department's estimates to express the cost function in terms of x.

(D) Graph the cost function and the revenue function on the same coordinate system for $0 \le x \le 1{,}000$. Find the break-even points, and indicate regions of loss and profit.

(E) Find the profit function in terms of x.

(F) Evaluate the marginal profit at $x = 250$ and $x = 475$, and interpret the results.

16. *Revenue, cost, and profit.* The company in Problem 15 is also planning to manufacture and market a four-slice toaster. For this toaster, the research department's estimates are a weekly demand of 300 toasters at a price of \$25 per toaster and a weekly demand of 400 toasters at a price of \$20. The financial department's estimates are fixed weekly costs of \$5,000 and variable costs of \$5 per toaster.

(A) Assume that the relationship between the price p and the demand x is linear. Use the research department's estimates to express p as a function of x and find the domain of this function.

(B) Find the revenue function in terms of x and state its domain.

(C) Assume that the cost function is linear. Use the financial department's estimates to express the cost function in terms of x.

(D) Graph the cost function and the revenue function on the same coordinate system for $0 \le x \le 800$. Find the break-even points, and indicate regions of loss and profit.

(E) Find the profit function in terms of x.

(F) Evaluate the marginal profit at $x = 325$ and $x = 425$, and interpret the results.

17. *Revenue, cost, and profit.* The total cost and the total revenue (in dollars) for the production and sale of x ski jackets are given, respectively, by

$$C(x) = 24x + 21{,}900 \quad \text{and} \quad R(x) = 200x - 0.2x^2$$
$$0 \le x \le 1{,}000$$

(A) Find the value of x where the graph of $R(x)$ has a horizontal tangent line.

(B) Find the profit function $P(x)$.

(C) Find the value of x where the graph of $P(x)$ has a horizontal tangent line.

(D) Graph $C(x)$, $R(x)$, and $P(x)$ on the same coordinate system for $0 \le x \le 1{,}000$. Find the break-even points. Find the x intercepts for the graph of $P(x)$.

18. *Revenue, cost, and profit.* The total cost and the total revenue (in dollars) for the production and sale of x hair dryers are given, respectively, by

$$C(x) = 5x + 2{,}340 \quad \text{and} \quad R(x) = 40x - 0.1x^2$$
$$0 \le x \le 400$$

(A) Find the value of x where the graph of $R(x)$ has a horizontal tangent line.

(B) Find the profit function $P(x)$.

(C) Find the value of x where the graph of $P(x)$ has a horizontal tangent line.

(D) Graph $C(x)$, $R(x)$, and $P(x)$ on the same coordinate system for $0 \le x \le 400$. Find the break-even points. Find the x intercepts for the graph of $P(x)$.

19. *Break-even analysis.* The price–demand equation and the cost function for the production of garden hoses are given, respectively, by

$$p = 20 - \sqrt{x} \quad \text{and} \quad C(x) = 500 + 2x$$

where x is the number of garden hoses that can be sold at a price of \$$p$ per unit and $C(x)$ is the total cost (in dollars) of producing x garden hoses.

(A) Express the revenue function in terms of x.

(B) Graph the cost function and the revenue function in the same viewing window for $0 \le x \le 400$. Use approximation techniques to find the break-even points correct to the nearest unit.

20. *Break-even analysis.* The price–demand equation and the cost function for the production of hand-woven silk scarves are given, respectively, by

$$p = 60 - 2\sqrt{x} \quad \text{and} \quad C(x) = 3{,}000 + 5x$$

where x is the number of scarves that can be sold at a price of \$$p$ per unit and $C(x)$ is the total cost (in dollars) of producing x scarves.

(A) Express the revenue function in terms of x.

(B) Graph the cost function and the revenue function in the same viewing window for $0 \le x \le 900$. Use approximation techniques to find the break-even points correct to the nearest unit.

21. *Break-even analysis.* Table 2 contains price–demand and total cost data for the production of overhead projectors, where p is the wholesale price (in dollars)

TABLE 2		
x	$p\,(\$)$	$C\,(\$)$
3,190	581	1,130,000
4,570	405	1,241,000
5,740	181	1,410,000
7,330	85	1,620,000

of a projector for an annual demand of x projectors and C is the total cost (in dollars) of producing x projectors.

(A) Find a quadratic regression equation for the price–demand data using x as the independent variable.

(B) Find a linear regression equation for the cost data using x as the independent variable. Use this equation to estimate the fixed costs and the variable costs per projector. Round answers to the nearest dollar.

(C) Find the break-even points. Round answers to the nearest integer.

(D) Find the price range for which the company will make a profit. Round answers to the nearest dollar.

22. *Break-even analysis.* Table 3 contains price–demand and total cost data for the production of treadmills, where p is the wholesale price (in dollars) of a treadmill for an annual demand of x treadmills and C is the total cost (in dollars) of producing x treadmills.

TABLE 3		
x	$p\,(\$)$	$C\,(\$)$
2,910	1,435	3,650,000
3,415	1,280	3,870,000
4,645	1,125	4,190,000
5,330	910	4,380,000

(A) Find a linear regression equation for the price–demand data using x as the independent variable.

(B) Find a linear regression equation for the cost data using x as the independent variable. Use this equation to estimate the fixed costs and the variable costs per treadmill. Round answers to the nearest dollar.

(C) Find the break-even points. Round answers to the nearest integer.

(D) Find the price range for which the company will make a profit. Round answers to the nearest dollar.

Important Terms and Symbols

3-1 *Limits.* Limit; absolute value function; limit from the left; left-hand limit; limit from the right; right-hand limit; properties of limits; limits of polynomial and rational functions; 0/0 indeterminate form; difference quotient;

$$\lim_{x \to c} f(x); \qquad \lim_{x \to c^-} f(x); \qquad \lim_{x \to c^+} f(x)$$

$$\frac{f(a + h) - f(a)}{h}; \quad \lim_{h \to 0} \frac{f(a + h) - f(a)}{h}$$

3-2 *Limits and Continuity.* Continuous at a point; continuous on an open interval; discontinuous at a point; continuous on the right; continuous on the left; continuous on a closed interval; general continuity properties; continuity properties of specific functions; vertical asymptote; solving inequalities using continuity properties; sign chart; test number; partition number

$$\lim_{x \to c} f(x) = f(c) \text{ if } f \text{ is continuous at } x = c;$$

$$\lim_{x \to c} f(x) = \infty; \quad \lim_{x \to c} f(x) = -\infty$$

3-3 *The Derivative.* Average rate of change; difference quotient; average velocity; instantaneous velocity; instantaneous rate of change; secant line; slope of a graph; slope of a tangent line; derivative; differentiation; differentiable; interpretations of the derivative; nondifferentiable;

$$f'(x) = \lim_{h \to 0} \frac{f(x + h) - f(x)}{h}$$

3-4 *Derivatives of Constants, Power Forms, and Sums.* Derivative notation; derivative of a constant function rule; power rule; constant times a function rule; sum and difference rule; marginal cost function

$$f'(x); \quad y'; \quad \frac{dy}{dx}$$

3-5 *Derivatives of Products and Quotients.* Product rule; quotient rule

3-6 *Chain Rule: Power Form.* General power rule; chain rule; combining rules of differentiation

3-7 *Marginal Analysis in Business and Economics.* Marginal cost; marginal revenue; marginal profit; exact cost; price–demand equation; cost function; revenue function; marginal revenue; break-even points; profit function; marginal profit; average cost; marginal average cost; average revenue; marginal average revenue; average profit; marginal average profit

$$C'(x); \quad \overline{C}(x); \quad \overline{C}'(x); \quad R'(x); \quad \overline{R}(x); \quad \overline{R}'(x);$$
$$P'(x); \quad \overline{P}(x); \quad \overline{P}'(x)$$

Summary of Rules of Differentiation

$$\frac{d}{dx} k = 0$$

$$\frac{d}{dx} x^n = nx^{n-1}$$

$$\frac{d}{dx} kf(x) = kf'(x)$$

$$\frac{d}{dx}[u(x) \pm v(x)] = u'(x) \pm v'(x)$$

$$\frac{d}{dx}[F(x)S(x)] = F(x)S'(x) + S(x)F'(x)$$

$$\frac{d}{dx}\frac{T(x)}{B(x)} = \frac{B(x)T'(x) - T(x)B'(x)}{[B(x)]^2}$$

$$\frac{d}{dx}[u(x)]^n = n[u(x)]^{n-1}u'(x)$$

Review Exercise

Work through all the problems in this chapter review and check your answers in the back of the book. Answers to all review problems are there along with section numbers in italics to indicate where each type of problem is discussed. Where weaknesses show up, review appropriate sections in the text.

Many of the problems in this exercise set ask you to find a derivative. Most of the answers to these problems contain both an unsimplified form and a simplified form of the derivative. When checking your work, first check that you applied the rules correctly, and then check that you performed the algebraic simplification correctly.

A **1.** Find the indicated quantities for $y = f(x) = 2x^2 + 5$:

(A) The change in y if x changes from 1 to 3

(B) The average rate of change of y with respect to x if x changes from 1 to 3

(C) The slope of the secant line through the points $(1, f(1))$ and $(3, f(3))$ on the graph of $y = f(x)$

(D) The instantaneous rate of change of y with respect to x at $x = 1$

(E) The slope of the line tangent to the graph of $y = f(x)$ at $x = 1$

(F) $f'(1)$

2. Use the two-step limiting process to find $f'(x)$ for $f(x) = -3x + 2$.

3. If $\lim_{x \to 1} f(x) = 2$ and $\lim_{x \to 1} g(x) = 4$, find

(A) $\lim_{x \to 1}(5f(x) + 3g(x))$ (B) $\lim_{x \to 1}[f(x)g(x)]$

(C) $\lim_{x \to 1}\dfrac{g(x)}{f(x)}$

In Problems 4–6, use the graph of f shown below to estimate the indicated limits and function values.

$f(x)$

4. (A) $\lim_{x \to 1^-} f(x)$ (B) $\lim_{x \to 1^+} f(x)$

(C) $\lim_{x \to 1} f(x)$ (D) $f(1)$

5. (A) $\lim_{x \to 2^-} f(x)$ (B) $\lim_{x \to 2^+} f(x)$

(C) $\lim_{x \to 2} f(x)$ (D) $f(2)$

6. (A) $\lim_{x \to 3^-} f(x)$ (B) $\lim_{x \to 3^+} f(x)$

(C) $\lim_{x \to 3} f(x)$ (D) $f(3)$

In Problems 7–9, use the graph of the function f shown in the figure to answer each question.

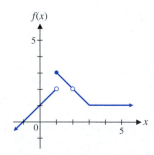

$f(x)$

7. (A) $\lim_{x \to 1} f(x) = ?$ (B) $f(1) = ?$

(C) Is f continuous at $x = 1$?

8. (A) $\lim\limits_{x \to 2} f(x) = ?$ (B) $f(2) = ?$

  (C) Is f continuous at $x = 2$?

9. (A) $\lim\limits_{x \to 3} f(x) = ?$ (B) $f(3) = ?$

  (C) Is f continuous at $x = 3$?

10. Use the two-step limiting process to find $f'(x)$ for $f(x) = 5x^2$.

11. If $f(5) = 4, f'(5) = -1, g(5) = 2,$ and $g'(5) = -3,$ find $h'(5)$ for each of the following functions:

  (A) $h(x) = 2f(x) + 3g(x)$ (B) $h(x) = f(x)g(x)$

  (C) $h(x) = \dfrac{f(x)}{g(x)}$ (D) $h(x) = [f(x)]^2$

12. Replace the **?** in the equation below with an expression that makes the equation valid.

$$\frac{d}{dx}(3x^2 + 4x + 1)^5 = 5(3x^2 + 4x + 1)^4 \underline{\;\; ? \;\;}$$

In Problems 13–24, find $f'(x)$ and simplify.

13. $f(x) = 3x^4 - 2x^2 + 1$ **14.** $f(x) = 2x^{1/2} - 3x$

15. $f(x) = 5$ **16.** $f(x) = \dfrac{1}{2x^2} + \dfrac{x^2}{2}$

17. $f(x) = \dfrac{0.5}{x^4} + 0.25x^4$

18. $f(x) = (2x - 1)(3x + 2)$

19. $f(x) = (x^2 - 1)(x^3 - 3)$

20. $f(x) = (0.2x - 1.5)(0.5x + 0.4)$

21. $f(x) = \dfrac{2x}{x^2 + 2}$ **22.** $f(x) = \dfrac{1}{3x + 2}$

23. $f(x) = (2x - 3)^3$ **24.** $f(x) = (x^2 + 2)^{-2}$

B *Problems 25 and 26 refer to the function f in the figure.*

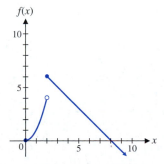

Figure for 25 and 26: $f(x) = \begin{cases} x^2 & \text{if } 0 \leqslant x < 2 \\ 8 - x & \text{if } x \geqslant 2 \end{cases}$

25. (A) $\lim\limits_{x \to 2^-} f(x) = ?$ (B) $\lim\limits_{x \to 2^+} f(x) = ?$

  (C) $\lim\limits_{x \to 2} f(x) = ?$ (D) $f(2) = ?$

  (E) Is f continuous at $x = 2$?

26. (A) $\lim\limits_{x \to 5^-} f(x) = ?$ (B) $\lim\limits_{x \to 5^+} f(x) = ?$

  (C) $\lim\limits_{x \to 5} f(x) = ?$ (D) $f(5) = ?$

  (E) Is f continuous at $x = 5$?

In problems 27–29, solve each inequality. Express the answer in interval notation. Use a graphing utility in Problem 29 to approximate partition numbers to two decimal places.

27. $x^2 - x < 12$

28. $\dfrac{x - 5}{x^2 + 3x} > 0$

29. $x^3 + x^2 - 4x - 2 > 0$

30. Let $f(x) = 0.5x^2 - 5$.

  (A) Find the slope of the secant line through $(2, f(2))$ and $(4, f(4))$.

  (B) Find the slope of the secant line through $(2, f(2))$ and $(2 + h, f(2 + h)), h \neq 0$.

  (C) Find the slope of the tangent line at $x = 2$.

In Problems 31–39, find the indicated derivative and simplify.

31. $\dfrac{dy}{dx}$ for $y = 3x^4 - 2x^{-3} + 5$

32. y' for $y = (2x^2 - 3x + 2)(x^2 + 2x - 1)$

33. $f'(x)$ for $f(x) = \dfrac{2x - 3}{(x - 1)^2}$

34. y' for $y = 2\sqrt{x} + \dfrac{4}{\sqrt{x}}$

35. $g'(x)$ for $g(x) = 1.8\sqrt[3]{x} + \dfrac{0.9}{\sqrt[3]{x}}$

36. $\dfrac{d}{dx}[(x^2 - 1)(2x + 1)^2]$ **37.** $\dfrac{d}{dx}\sqrt[3]{x^3 - 5}$

38. $\dfrac{dy}{dx}$ for $y = \dfrac{3x^2 + 4}{x^2}$ **39.** $\dfrac{d}{dx}\dfrac{(x^2 + 2)^4}{2x - 3}$

40. For $y = f(x) = x^2 + 4$, find

  (A) The slope of the graph at $x = 1$

  (B) The equation of the tangent line at $x = 1$ in the form $y = mx + b$

41. Repeat Problem 40 for $f(x) = x^3(x + 1)^2$.

In Problems 42–45, find the value(s) of x where the tangent line is horizontal.

42. $f(x) = 10x - x^2$ **43.** $f(x) = (x + 3)(x^2 - 45)$

44. $f(x) = \dfrac{x}{x^2 + 4}$ **45.** $f(x) = x^2(2x - 15)^3$

In Problems 46–48, approximate (to two decimal places) the value(s) of x where the graph of f has a horizontal tangent line.

46. $f(x) = x^4 - 2x^3 - 5x^2 + 7x$

47. $f(x) = \dfrac{x^3 - 5x + 10}{x^2 + 2}$

48. $f(x) = \dfrac{5x^4 - 40x^2}{(x^2 + 1)^2}$

49. If an object moves along the y axis (scale in feet) so that it is at $y = f(x) = 16x^2 - 4x$ at time x (in seconds), find

(A) The instantaneous velocity function

(B) The velocity at time $x = 3$ seconds

50. An object moves along the y axis (scale in feet) so that at time x (in seconds) it is at $y = f(x) = 96x - 16x^2$. Find

(A) The instantaneous velocity function

(B) The time(s) when the velocity is 0

51. Let $f(x) = x^3$, $g(x) = (x - 4)^3$, and $h(x) = (x + 3)^3$.

(A) How are the graphs of f, g, and h related? Illustrate your conclusion by graphing f, g, and h on the same coordinate axes.

(B) How would you expect the graphs of the derivatives of these functions to be related? Illustrate your conclusion by graphing f', g', and h' on the same coordinate axes.

52. Let $f(x)$ be a differentiable function and let k be a nonzero constant. For each function g, write a brief verbal description of the relationship between the graphs of f and g. Do the same for the graphs of f' and g'.

(A) $g(x) = f(x + k)$　　(B) $g(x) = f(x) + k$

In Problems 53–57, determine where f is continuous. Express the answer in interval notation.

53. $f(x) = x^2 - 4$　　　**54.** $f(x) = \dfrac{x + 1}{x - 2}$

55. $f(x) = \dfrac{x + 4}{x^2 + 3x - 4}$　　**56.** $f(x) = \sqrt[3]{4 - x^2}$

57. $f(x) = \sqrt{4 - x^2}$

In Problems 58–61, evaluate the indicated limit. Use $-\infty$ or ∞ where appropriate.

58. $\lim\limits_{x \to 3^-} \dfrac{x + 1}{3 - x}$　　**59.** $\lim\limits_{x \to 3^-} \dfrac{x + 1}{3 - x}$

60. $\lim\limits_{x \to 3} \dfrac{x + 1}{3 - x}$　　**61.** $\lim\limits_{x \to 3} \dfrac{x + 1}{(3 - x)^2}$

In Problems 62–71, find each limit, if it exists.

62. $\lim\limits_{x \to 3} \dfrac{2x - 3}{x + 5}$　　**63.** $\lim\limits_{x \to 3} (2x^2 - x + 1)$

64. $\lim\limits_{x \to 0} \dfrac{2x}{3x^2 - 2x}$　　**65.** $\lim\limits_{x \to 3} \dfrac{x - 3}{x^2 - 9}$

66. $\lim\limits_{x \to 4^-} \dfrac{|x - 4|}{x - 4}$　　**67.** $\lim\limits_{x \to 4^+} \dfrac{|x - 4|}{x - 4}$

68. $\lim\limits_{x \to 4} \dfrac{|x - 4|}{x - 4}$

69. $\lim\limits_{h \to 0} \dfrac{[(2 + h)^2 - 1] - [2^2 - 1]}{h}$

70. $\lim\limits_{h \to 0} \dfrac{f(2 + h) - f(2)}{h}$ for $f(x) = x^2 + 4$

71. $\lim\limits_{h \to 0} \dfrac{f(x + h) - f(x)}{h}$ for $f(x) = \dfrac{1}{x + 2}$

72. Let

$$f(x) = \dfrac{x^3 - 4x^2 - 4x + 16}{|x^2 - 4|}$$

Graph f and use zoom and trace to investigate the left- and right-hand limits at the indicated values of c.

(A) $c = -2$　　(B) $c = 0$　　(C) $c = 2$

In Problems 73 and 74, use the definition of the derivative and the two-step process to find $f'(x)$.

73. $f(x) = x^2 - x$　　**74.** $f(x) = \sqrt{x} - 3$

C *Problems 75–78 refer to the function f in the figure. Determine whether f is differentiable at the indicated value of x.*

75. $x = 0$　　**76.** $x = 1$　　**77.** $x = 2$　　**78.** $x = 3$

In Problems 79–82, find $f'(x)$ and simplify.

79. $f(x) = (x - 4)^4(x + 3)^3$

80. $f(x) = \dfrac{x^5}{(2x + 1)^4}$

81. $f(x) = \dfrac{\sqrt{x^2 - 1}}{x}$ **82.** $f(x) = \dfrac{x}{\sqrt{x^2 + 4}}$

83. The domain of the power function $f(x) = x^{1/5}$ is the set of all real numbers. Find the domain of the derivative $f'(x)$. Discuss the nature of the graph of $y = f(x)$ for any x values excluded from the domain of $f'(x)$.

84. Let f be defined by

$$f(x) = \begin{cases} x^2 - m & \text{if } x \leq 1 \\ -x^2 + m & \text{if } x > 1 \end{cases}$$

where m is a constant.

(A) Graph f for $m = 0$, and find

$$\lim_{x \to 1^-} f(x) \quad \text{and} \quad \lim_{x \to 1^+} f(x)$$

(B) Graph f for $m = 2$, and find

$$\lim_{x \to 1^-} f(x) \quad \text{and} \quad \lim_{x \to 1^+} f(x)$$

(C) Find m so that

$$\lim_{x \to 1^-} f(x) = \lim_{x \to 1^+} f(x)$$

and graph f for this value of m.

(D) Write a brief verbal description of each graph. How does the graph in part (C) differ from the graphs in parts (A) and (B)?

85. Let $f(x) = 1 - |x - 1|, 0 \leq x \leq 2$ (see the figure).

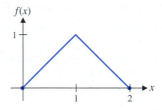

Figure for 71

(A) $\displaystyle\lim_{h \to 0^-} \dfrac{f(1 + h) - f(1)}{h} = ?$

(B) $\displaystyle\lim_{h \to 0^+} \dfrac{f(1 + h) - f(1)}{h} = ?$

(C) $\displaystyle\lim_{h \to 0} \dfrac{f(1 + h) - f(1)}{h} = ?$

(D) Does $f'(1)$ exist?

Applications

Business & Economics

86. *Natural gas rates.* The table shows the winter rates for natural gas charged by the Bay State Gas Company. The customer charge is a fixed monthly charge, independent of the amount of gas used during the month.

www

(A) Write a piecewise definition of the monthly charge $S(x)$ for a customer who uses x therms in a winter month.

(B) Graph $S(x)$.

(C) Is $S(x)$ continuous at $x = 90$? Explain.

NATURAL GAS RATES	
Monthly customer charge	$7.47
First 90 therms	$0.4000 per therm
All usage over 90 therms	$0.2076 per therm

87. *Cost analysis.* The total cost (in dollars) of producing x television sets is

$$C(x) = 10,000 + 200x - 0.1x^2$$

(A) Find the exact cost of producing the 101st television set.

(B) Use the marginal cost to approximate the cost of producing the 101st television set.

88. *Cost analysis.* The total cost (in dollars) of producing x bicycles is

$$C(x) = 5,000 + 40x + 0.05x^2$$

(A) Find the total cost and the marginal cost at a production level of 100 bicycles, and interpret the results.

(B) Find the average cost and the marginal average cost at a production level of 100 bicycles, and interpret the results.

89. *Cost analysis.* The total cost (in dollars) of producing x laser printers per week is shown in the figure. Which is greater, the approximate cost of producing the 201st printer or the approximate cost of producing the 601st printer? Does this graph represent a manufacturing process that is becoming more efficient or less efficient as production levels increase? Explain.

Figure for 89

90. *Cost analysis.* Let

$$p = 25 - 0.01x \quad \text{and} \quad C(x) = 2x + 9,000$$
$$0 \leqslant x \leqslant 2,500$$

be the price–demand equation and the cost function, respectively, for the manufacture of umbrellas.

(A) Find the marginal cost, average cost, and marginal average cost functions.

(B) Express the revenue in terms of x, and find the marginal revenue, average revenue, and marginal average revenue functions.

(C) Find the profit, marginal profit, average profit, and marginal average profit functions.

(D) Find the break-even point(s).

(E) Evaluate the marginal profit at $x = 1,000$, 1,150, and 1,400, and interpret the results.

(F) Graph $R = R(x)$ and $C = C(x)$ on the same coordinate system, and locate regions of profit and loss.

91. *Employee training.* A company producing computer components has established that on the average, a new employee can assemble $N(t)$ components per day after t days of on-the-job training, as given by

$$N(t) = \frac{40t}{t + 2}$$

(A) Find the average rate of change of $N(t)$ from 3 days to 6 days.

(B) Find the instantaneous rate of change of $N(t)$ at 3 days.

92. *Sales analysis.* Past sales records for a swimming pool manufacturer indicate that the total number of swimming pools, N (in thousands), sold during a year is given by

$$N(t) = t\sqrt{4 + t}$$

where t is the number of months since the beginning of the year. Find $N(5)$ and $N'(5)$, and interpret.

93. *Natural gas consumption.* The data in Table 1 give the U.S. consumption of natural gas in trillions of cubic feet.

WWW

TABLE 1

YEAR	NATURAL GAS CONSUMPTION
1960	12.0
1970	21.1
1980	19.9
1990	18.7
2000	21.9

(A) Let x represent time (in years), with $x = 0$ corresponding to 1960, and let y represent the corresponding U.S. consumption of natural gas. Enter the data set in a graphing utility and find a cubic regression equation for the data.

(B) If $y = N(x)$ denotes the regression equation found in part (A), find $N(50)$ and $N'(50)$, and write a brief verbal interpretation of these results.

94. *Break-even analysis.* Table 2 contains price–demand and total cost data from a bakery for the production of kringles (a Danish pastry), where p is the price (in dollars) of a kringle for a daily demand of x kringles and C is the total cost (in dollars) of producing x kringles.

TABLE 2

x	$p\,(\$)$	$C\,(\$)$
125	9	740
140	8	785
170	7	850
200	6	900

(A) Find a linear regression equation for the price–demand data using x as the independent variable.

(B) Find a linear regression equation for the cost data using x as the independent variable. Use this equation to estimate the fixed costs and the variable costs per kringle.

(C) Find the break-even points.

(D) Find the price range for which the bakery will make a profit.

In all answers, round dollar amounts to the nearest cent.

Life Sciences

95. *Pollution.* A sewage treatment plant disposes of its effluent through a pipeline that extends 1 mile toward the center of a large lake. The concentration of effluent $C(x)$, in parts per million, x meters from the end of the pipe is given approximately by

$$C(x) = 500(x + 1)^{-2}$$

What is the instantaneous rate of change of concentration at 9 meters? At 99 meters?

96. Medicine. The body temperature (in degrees Fahrenheit) of a patient t hours after being given a fever-reducing drug is given by

$$F(t) = 98 + \frac{4}{\sqrt{t+1}}$$

Find $F(3)$ and $F'(3)$. Write a brief verbal interpretation of these results.

Social Sciences

97. Learning. If a person learns N items in t hours, as given by

$$N(t) = 20\sqrt{t}$$

find the rate of learning after

(A) 1 hour   (B) 4 hours

Group Activity 1   *Minimal Average Cost*

www   If $C(x)$ is the total cost of producing x items, the marginal cost function $C'(x)$ gives the approximate cost of the next item produced, while the average cost function $\overline{C}(x)$ gives the average cost per item for the items already produced. Thus, $C'(x)$ looks forward to the next item, while $\overline{C}(x)$ looks backward at all the items produced thus far. Given this difference in viewpoint, it is somewhat surprising that there is an important relationship between these two functions. As we will see, information gained from comparing the values of $C'(x)$ and $\overline{C}(x)$ can help determine the production level that minimizes average cost.

TABLE 1		
x	$\overline{C}(x)$	$C'(x)$
100		
200		
300		
400		
500		
600		
700		
800		
900		
1,000		

(A) The total cost (in dollars) of producing x items is given by

$$C(x) = 0.01x^2 + 40x + 3{,}600$$

Find $C'(x)$ and $\overline{C}(x)$, and complete Table 1.

(B) Repeat part (A) for

$$C(x) = 0.00016x^3 - 0.12x^2 + 30x + 10{,}000$$

(C) Examine the values in the tables from parts (A) and (B), and write a brief verbal description of the behavior of each function. Does each average cost function appear to have a minimum value? What is the minimal value, and where does it occur? What relationship do you observe between the minimum average cost and the marginal cost at the production level that minimizes average cost?

(D) If you have access to a graphing utility, confirm your observations in part (C) by examining the graphs of $C'(x)$ and $\overline{C}(x)$.

(E) The following statements can help justify the relationship you observed in part (C). In each case, fill in the blank with "increase" or "decrease" and justify your choice.

1. If $C'(x) < \overline{C}(x)$ (that is, the cost of the next item is less than the average cost of the items already produced), then increasing production by 1 item will _____ the average cost.

2. If $C'(x) > \overline{C}(x)$ (that is, the cost of the next item is more than the average cost of the items already produced), then increasing production by 1 item will _____ the average cost.

(F) Discuss the validity of the following statement for an arbitrary cost function $C(x)$: If the minimum value of $\overline{C}(x)$ occurs at a production level x, then $C'(x) = \overline{C}(x)$ at that production level.

(G) We used a quadratic function and a cubic function in parts (A) and (B) to illustrate the relationship between $C'(x)$ and $\overline{C}(x)$. But linear cost functions are one of the most important types. To see why we did not choose a linear cost function, try to parallel the development above for the cost function

$$C(x) = 30x + 12{,}000$$

Do any of your findings contradict the statements in parts (E) and (F)?

Group Activity 2   *Numerical Differentiation on a Graphing Utility*

 Most graphing utilities have a built-in routine for approximating the derivative of a function, often denoted by nDeriv (check the manual for your graphing utility). For example, nDeriv(x^3, x, a) approximates the derivative of $y = x^3$ at a number a.

(A) Find nDeriv(x^3, x, a) on a graphing utility for $a = 1, 2, 3, 4,$ and 5, and compare with the corresponding values of

$$\frac{d}{dx}x^3 = 3x^2$$

(B) Enter $y_1 = x^3$, and $y_2 = 3x^2$, and $y_3 = $ nDeriv(y_1, x, x) in the equation editor of a graphing utility. Graph y_2 and y_3 for $-2 \leqslant x \leqslant 2, -2 \leqslant y \leqslant 2$, and discuss the relationship between these graphs. Use the trace feature or tables of values to support your conclusions.

Most graphing utilities use the following average of difference quotients with a fixed value of h to approximate a derivative:

$$\frac{1}{2}\left(\frac{f(x + h) - f(x)}{h} + \frac{f(x - h) - f(x)}{-h}\right) = \frac{f(x + h) - f(x - h)}{2h}$$

Thus, for a given number a and a fixed value of h,

$$\text{nDeriv}(f(x), x, a) = \frac{f(a + h) - f(a - h)}{2h} \tag{1}$$

(C) Use equation (1) to find and simplify nDeriv$(f(x), x, a)$ for $f(x) = x^3$. Compare the values of the simplified form with the values of nDeriv you computed in part (A) to see if you can determine the fixed value of h for your graphing utility. Check your manual to see if you can change this value.

(D) Let $f(x) = |x|$. What is nDeriv$(f(x), x, 0)$? Is $f(x)$ differentiable at $x = 0$?

(E) Repeat part (D) for $f(x) = 1/x^2$.

(F) When using nDeriv to approximate the derivative of a function $f(x)$, why is it important to know in advance the location of any points where $f'(x)$ does not exist?

Graphing and Optimization

INTRODUCTION

Historically, one of the primary reasons for studying calculus was the development of powerful tools for sketching graphs of functions. In fact, prior to the very recent advent of graphing utilities, hand-sketching was the only way most students could produce graphs. Now things have changed. An accurate graph of a function specified by an equation can be produced quickly and easily on a graphing utility. Does this mean that the old tools are no longer relevant? On the contrary, efficient use of graphing technology still requires a person at the helm, directing the electronic device, and this person still needs a sound foundation in graphing tools and analysis.

In the first three sections of this chapter, we develop the graphing concepts and tools necessary to analyze, discuss, and sketch by hand the graphs of functions. The functions we consider may be specified by equations or by information obtained from tables, verbal descriptions, or other graphs. In the last section, we use these graphing tools to solve optimization problems, one of the most important applications of calculus to real-world problems.

A brief review of Section 2-1 would be helpful before beginning this chapter. In particular, we will make frequent use of the following important property of polynomials:

The graph of a polynomial function of positive degree n can have at most $n - 1$ turning points and at most n real zeros or x intercepts.

Since the derivative is associated with the slope of the graph of a function at a point, we might expect that it is also associated with other properties of a graph. As we will see in this and the next section, the derivative can tell us a great deal about the shape of the graph of a function. In addition, this investigation will lead to methods for finding absolute maximum and minimum values for functions that do not require graphing. Manufacturing companies can use these methods to find production levels that will minimize cost or maximize profit. Pharmacologists can use them to find levels of drug dosages that will produce maximum sensitivity to a drug. And so on.

❑ INCREASING AND DECREASING FUNCTIONS

Explore–Discuss 1

Figure 1 shows the graph of $y = f(x)$ and a sign chart for $f'(x)$, where

$$f(x) = x^3 - 3x$$

and

$$f'(x) = 3x^2 - 3 = 3(x + 1)(x - 1)$$

Discuss the relationship between the graph of f and the sign of $f'(x)$ over each interval where $f'(x)$ has a constant sign. Also, describe the behavior of the graph of f at each partition number for f'.

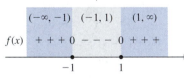

FIGURE 1

Graphs of functions generally have *rising* and *falling* sections as we scan graphs from left to right. Referring to the graph of $f(x) = x^3 - 3x$ in Figure 1, we see that on the interval $(-\infty, -1)$, the graph of f is *rising*, $f(x)$ is *increasing*,* and the slope of the graph is positive $[f'(x) > 0]$. On the other hand, on the interval $(-1, 1)$, the graph of f is *falling*, $f(x)$ is *decreasing,* and the slope of the graph is negative $[f'(x) < 0]$. Finally, on the interval $(1, \infty)$, once again the graph of f is rising, $f(x)$ is increasing, and $f'(x) > 0$. At $x = -1$ and $x = 1$, the slope of the graph is 0 $[f'(x) = 0]$ and the tangent lines are horizontal.

In general, if $f'(x) > 0$ (is positive) on the interval (a, b) (Fig. 2), then $f(x)$ increases (↗) and the graph of f rises as we move from left to right over the interval; if $f'(x) < 0$ (is negative) on an interval (a, b), then $f(x)$ decreases (↘) and the graph of f falls as we move from left to right over the interval. We summarize these important results in the box.

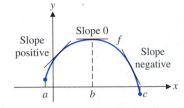

FIGURE 2

*Formally, we say that the function f is **increasing** on an interval (a, b) if $f(x_2) > f(x_1)$ whenever $a < x_1 < x_2 < b$; and f is **decreasing** on (a, b) if $f(x_2) < f(x_1)$ whenever $a < x_1 < x_2 < b$.

www

Increasing and Decreasing Functions

For the interval (a, b),

$f'(x)$	$f(x)$	GRAPH OF f	EXAMPLES
+	Increases ↗	Rises ↗	
−	Decreases ↘	Falls ↘	

Explore–Discuss 2

The graphs of $f(x) = x^2$ and $g(x) = |x|$ are shown in Figure 3. Both functions change from decreasing to increasing at $x = 0$. Discuss the relationship between the graph of each function at $x = 0$ and the derivative of the function at $x = 0$.

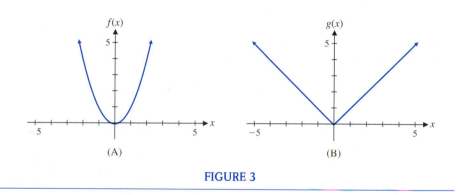

(A)   (B)

FIGURE 3

Example 1 ✎   **Finding Intervals Where a Function Is Increasing or Decreasing**
Given the function $f(x) = 8x - x^2$,

(A) Which values of x correspond to horizontal tangent lines?

(B) For which values of x is $f(x)$ increasing? Decreasing?

(C) Sketch a graph of f. Add any horizontal tangent lines.

SOLUTION   (A) $f'(x) = 8 - 2x = 0$
$$x = 4$$

Thus, a horizontal tangent line exists at $x = 4$ only.

(B) We will construct a sign chart for $f'(x)$ to determine which values of x make $f'(x) > 0$ and which values make $f'(x) < 0$. Recall from Section 3-2 that the partition numbers for a function are the points where the function is 0 or discontinuous. Thus, when constructing a sign chart for $f'(x)$, we must locate all points where $f'(x) = 0$ or $f'(x)$ is discontinuous. From part (A) we know that $f'(x) = 8 - 2x = 0$ at $x = 4$. Since $f'(x) = 8 - 2x$ is a polynomial, it is continuous for all x. Thus, 4 is the only partition number. We construct a sign chart for the intervals $(-\infty, 4)$ and $(4, \infty)$, using test numbers 3 and 5:

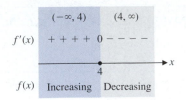

Thus, $f(x)$ is increasing on $(-\infty, 4)$ and decreasing on $(4, \infty)$.

(C)

x	$f(x)$
0	0
2	12
4	16
6	12
8	0

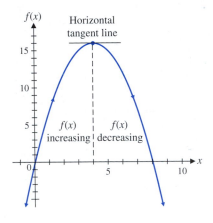

Matched Problem 1   Repeat Example 1 for $f(x) = x^2 - 6x + 10$.

As Example 1 illustrates, construction of a sign chart will play an important role in using the derivative to analyze and sketch the graph of a function f. The partition numbers for f' are central to the construction of these sign charts and also to the analysis of the graph of $y = f(x)$. We already know that if $f'(c) = 0$, the graph of $y = f(x)$ will have a horizontal tangent line at $x = c$. But the partition numbers for f' also include the numbers c where $f'(c)$ does not exist.* There are two possibilities at this type of number: $f(c)$ does not exist, or $f(c)$ exists, but the slope of the tangent line at $x = c$ is undefined.

Critical Values of f

The values of x in the domain of f where $f'(x) = 0$ or $f'(x)$ does not exist are called the **critical values** of f. The critical values of f are always partition numbers for f', but f' may have partition numbers that are not critical values.

It is important to understand that although f' may not be defined at a critical value c, f must be defined at c.

Critical values of a function f are always in the domain of f.

---

*We are assuming that $f'(c)$ does not exist at any point of discontinuity of f. There do exist functions where f' is discontinuous at $x = c$, yet $f'(c)$ exists. However, we do not consider such functions in this book.

Example 2 will illustrate the relationship between critical values and partition numbers.

Example 2 ⇔ **Partition Numbers and Critical Values**   For each function, find the partition numbers for f', the critical values for f, and determine the intervals where f is increasing and those where f is decreasing.

$$(A) \ f(x) = 1 + x^3 \quad (B) \ f(x) = (1 - x)^{1/3} \quad (C) \ f(x) = \frac{1}{x - 2}$$

Solution   (A) $f(x) = 1 + x^3 \qquad f'(x) = 3x^2 = 0$

$$x = 0$$

The only partition number for f' is $x = 0$. Since 0 is in the domain of f, $x = 0$ is also the only critical value for f.

Sign chart for $f'(x) = 3x^2$ (partition number is 0):

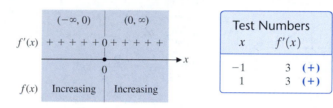

		Test Numbers	
		x	$f'(x)$
		-1	3 (+)
		1	3 (+)

The sign chart indicates that $f(x)$ is increasing on $(-\infty, 0)$ and $(0, \infty)$. Since f is continuous at $x = 0$, it follows that $f(x)$ is increasing for all x. The graph of f is shown in Figure 4.

(B) $f(x) = (1 - x)^{1/3} \qquad f'(x) = -\dfrac{1}{3}(1 - x)^{-2/3} = \dfrac{-1}{3(1 - x)^{2/3}}$

To find partition numbers for f', we note that f' is continuous for all x except for values of x for which the denominator is 0; that is, $f'(1)$ does not exist and f' is discontinuous at $x = 1$. Since the numerator is the constant -1, $f'(x) \neq 0$ for any value of x. Thus, $x = 1$ is the only partition number for f'. Since 1 is in the domain of f, $x = 1$ is also the only critical value of f. When constructing the sign chart for f' we use the abbreviation ND to note the fact that $f'(x)$ is *not defined* at $x = 1$.

Sign chart for $f'(x) = -1/[3(1 - x)^{2/3}]$ (partition number is 1):

		Test Numbers	
		x	$f'(x)$
		0	$-\frac{1}{3}$ (−)
		2	$-\frac{1}{3}$ (−)

The sign chart indicates that f is decreasing on $(-\infty, 1)$ and $(1, \infty)$. Since f is continuous at $x = 1$, it follows that $f(x)$ is decreasing for all x. Thus, **a**

FIGURE 4

$f(x)$

FIGURE 5

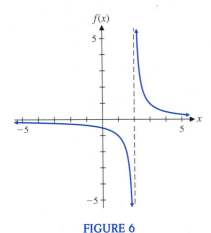

$f(x)$

FIGURE 6

continuous function can be decreasing (or increasing) on an interval containing values of x where $f'(x)$ does not exist. The graph of f is shown in Figure 5. Notice that the undefined derivative at $x = 1$ results in a vertical tangent line at $x = 1$. In general, **a vertical tangent will occur at $x = c$ if f is continuous at $x = c$ and $|f'(x)|$ becomes larger and larger as x approaches c.**

(C) $f(x) = \dfrac{1}{x - 2}$ $f'(x) = \dfrac{-1}{(x - 2)^2}$

To find the partition numbers for f', note that $f'(x) \neq 0$ for any x and f' is not defined at $x = 2$. Thus, $x = 2$ is the only partition number for f'. However, $x = 2$ is *not* in the domain of f. Consequently, $x = 2$ is not a critical value of f. This function has no critical values.

Sign chart for $f'(x) = -1/(x - 2)^2$ (partition number is 2):

	$(-\infty, 2)$	$(2, \infty)$
$f'(x)$	$- - - -$ ND $- - - -$	
	2	
$f(x)$	Decreasing	Decreasing

Test Numbers	
x	$f'(x)$
1	-1 $(-)$
3	-1 $(-)$

Thus, f is decreasing on $(-\infty, 2)$ and $(2, \infty)$. See the graph of f in Figure 6. ∎

Matched Problem 2 ⇒ For each function, find the partition numbers for f', the critical values for f, and determine the intervals where f is increasing and those where f is decreasing.

(A) $f(x) = 1 - x^3$ (B) $f(x) = (1 + x)^{1/3}$ (C) $f(x) = \dfrac{1}{x}$ ∎

Explore–Discuss 3

A student examined the sign chart in Example 2C and concluded that $f(x) = 1/(x - 2)$ is decreasing for all x except $x = 2$. However, $f(1) = -1 < f(3) = 1$, which seems to indicate that f is increasing. Discuss the difference between the correct answer in Example 2C and the student's answer. Explain why the student's description of where f is decreasing is unacceptable.

CAUTION

Example 2C illustrates two important ideas.

1. **Do not assume that all partition numbers for the derivative f' are critical values of the function f.** To be a critical value, a partition number must also be in the domain of f.

2. **The values where a function is increasing or decreasing must always be expressed in terms of open intervals that are subsets of the domain of the function.**

❑ LOCAL EXTREMA

When the graph of a continuous function changes from rising to falling, a high point, or *local maximum,* occurs; and when the graph changes from falling to rising, a low point, or *local minimum,* occurs. In Figure 7, high points occur at c_3 and c_6, and low points occur at c_2 and c_4. In general, we call $f(c)$ a **local maximum** if there exists an interval (m, n) containing c such that

$$f(x) \leq f(c) \qquad \text{for all } x \text{ in } (m, n)$$

Note that this inequality need only hold for values of x near c, hence the use of the term *local.*

FIGURE 7

The quantity $f(c)$ is called a **local minimum** if there exists an interval (m, n) containing x such that

$$f(x) \geq f(c) \qquad \text{for all } x \text{ in } (m, n)$$

The quantity $f(c)$ is called a **local extremum** if it is either a local maximum or a local minimum. A point on a graph where a local extremum occurs is also called a **turning point.** Thus, in Figure 7 we see that local maxima occur at c_3 and c_6, local minima occur at c_2 and c_4, and all four values produce local extrema. Also note that the local maximum $f(c_3)$ is not the highest point on the graph in Figure 7. Later in this chapter we consider the problem of finding the highest and lowest points on a graph. For now, we are concerned only with locating local extrema.

How can we locate local maxima and minima if we are given the equation for a function and not its graph? The key is to examine the critical values of the function. The local extrema of the function f in Figure 7 occur either at points where the derivative is 0 (c_2 and c_3) or at points where the derivative does not exist (c_4 and c_6). In other words, local extrema occur only at critical values of f. Theorem 1 shows that this is true in general.

THEOREM 1   ### Existence of Local Extrema

If f is continuous on the interval (a, b), c is a number in (a, b) and $f(c)$ is a local extremum, then either $f'(c) = 0$ or $f'(c)$ does not exist (is not defined).

Theorem 1 states that a local extremum can occur only at a critical value, but it does not imply that every critical value produces a local extremum. In Figure 7, c_1 and c_5 are critical values (the slope is 0), but the function does not have a local maximum or local minimum at either of these values.

Our strategy for finding local extrema is now clear. We find all critical values for f and test each one to see if it produces a local maximum, a local minimum, or neither.

❑ FIRST-DERIVATIVE TEST

If $f'(x)$ exists on both sides of a critical value c, the sign of $f'(x)$ can be used to determine whether the point $(c, f(c))$ is a local maximum, a local minimum, or neither. The various possibilities are summarized in the following box and illustrated in Figure 8.

www

First-Derivative Test for Local Extrema

Let c be a critical value of f [$f(c)$ defined and either $f'(c) = 0$ or $f'(c)$ not defined]. Construct a sign chart for $f'(x)$ close to and on either side of c.

SIGN CHART	$f(c)$
$f'(x)$ $---$ $+++$ / m c n / $f(x)$ Decreasing Increasing	$f(c)$ is local minimum. If $f'(x)$ changes from negative to positive at c, then $f(c)$ is a local minimum.
$f'(x)$ $+++$ $---$ / m c n / $f(x)$ Increasing Decreasing	$f(c)$ is local maximum. If $f'(x)$ changes from positive to negative at c, then $f(c)$ is a local maximum.
$f'(x)$ $---$ $---$ / m c n / $f(x)$ Decreasing Decreasing	$f(c)$ is not a local extremum. If $f'(x)$ does not change sign at c, then $f(c)$ is neither a local maximum nor a local minimum.
$f'(x)$ $+++$ $+++$ / m c n / $f(x)$ Increasing Increasing	$f(c)$ is not a local extremum. If $f'(x)$ does not change sign at c, then $f(c)$ is neither a local maximum nor a local minimum.

$f'(c) = 0$
Horizontal tangent

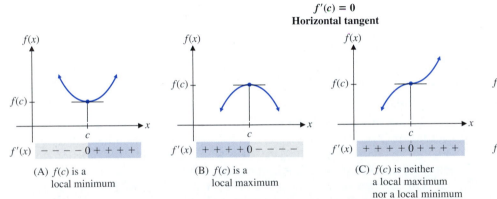

(A) $f(c)$ is a local minimum

(B) $f(c)$ is a local maximum

(C) $f(c)$ is neither a local maximum nor a local minimum

(D) $f(c)$ is neither a local maximum nor a local minimum

FIGURE 8 Local extrema

Example 3 ⇨ **Locating Local Extrema**   Given $f(x) = x^3 - 6x^2 + 9x + 1$,

(A) Find the critical values of f.   (B) Find the local maxima and minima.
(C) Sketch the graph of f.

SOLUTION   (A) Find all numbers x in the domain of f where $f'(x) = 0$ or $f'(x)$ does not exist.

$$f'(x) = 3x^2 - 12x + 9 = 0$$
$$3(x^2 - 4x + 3) = 0$$
$$3(x - 1)(x - 3) = 0$$
$$x = 1 \quad \text{or} \quad x = 3$$

$f'(x)$ exists for all x; the critical values are $x = 1$ and $x = 3$.

(B) The easiest way to apply the first-derivative test for local maxima and minima is to construct a sign chart for $f'(x)$ for all x. Partition numbers for $f'(x)$ are $x = 1$ and $x = 3$ (which also happen to be critical values for f).

Sign chart for $f'(x) = 3(x - 1)(x - 3)$:

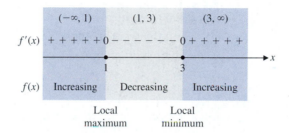

Test Numbers	
x	$f'(x)$
0	9 **(+)**
2	−3 **(−)**
4	9 **(+)**

The sign chart indicates that f increases on $(-\infty, 1)$, has a local maximum at $x = 1$, decreases on $(1, 3)$, has a local minimum at $x = 3$, and increases on $(3, \infty)$. These facts are summarized in the following table:

x	$f'(x)$	$f(x)$	GRAPH OF f
$(-\infty, 1)$	+	Increasing	Rising
$x = 1$	0	Local maximum	Horizontal tangent
$(1, 3)$	−	Decreasing	Falling
$x = 3$	0	Local minimum	Horizontal tangent
$(3, \infty)$	+	Increasing	Rising

$f'(c)$ is not defined but $f(c)$ is defined

(E) $f(c)$ is a local minimum

(F) $f(c)$ is a local maximum

(G) $f(c)$ is neither a local maximum nor a local minimum

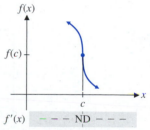

(H) $f(c)$ is neither a local maximum nor a local minimum

(C) We sketch a graph of *f* using the information from part (B) and point-by-point plotting.

x	f(x)
0	1
1	5
2	3
3	1
4	5

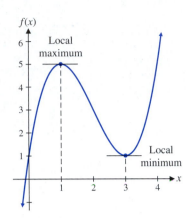

Matched Problem 3 ✐   Given $f(x) = x^3 - 9x^2 + 24x - 10$,

(A) Find the critical values of *f*.

(B) Find the local maxima and minima.

(C) Sketch a graph of *f*.

REMARK

Local extrema are easy to recognize on a graphing utility. Figure 9A shows the graph of the function in Example 3. The local maximum and minimum are clearly visible, but how do we determine their location? Two methods for approximating local extrema on a graphing utility are stated below. Consult the manual for your graphing utility for details. You might also want to see if your graphing utility will graph a numerical approximation to the derivative of a function. If it does, you do not have to find and enter the derivative in the graphing utility.

1. Graph the derivative and use built-in root approximation routines to find the critical values. Examining the graph in Figure 9B, we see that $x = 1$ is the critical value that corresponds to the local maximum in Figure 9A. The critical value for the local minimum is found in the same way.

2. Graph the function and use built-in routines that approximate local maxima and minima. Figure 9C shows that the local minimum occurs at $x = 3$ (after rounding). The local maximum is found in a similar manner.

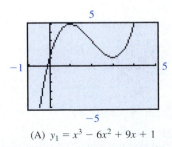

(A) $y_1 = x^3 - 6x^2 + 9x + 1$

(B) $y_2 = 3x^2 - 12x + 9$

(C) $y_1 = x^3 - 6x^2 + 9x + 1$

FIGURE 9 Approximating local extrema with a graphing utility

How can you tell if you have found all the local extrema for a function? In general, this can be a difficult question to answer. However, in the case of a polynomial function, there is an easily determined upper limit on the number of local extrema. Since the local extrema are the x intercepts of the derivative, this limit is a consequence of the number of x intercepts of a polynomial. The relevant information is summarized in the following theorem, which is stated without proof:

THEOREM 2   Intercepts and Local Extrema for Polynomial Functions

If $f(x) = a_n x^n + a_{n-1} x^{n-1} + \cdots + a_1 x + a_0$, $a_n \neq 0$, is an nth-degree polynomial, then f has at most n x intercepts and at most $n - 1$ local extrema.

Theorem 2 does not guarantee that every nth-degree polynomial has exactly $n - 1$ local extrema; it says only that there can never be more than $n - 1$ local extrema. For example, the third-degree polynomial in Example 3 has two local extrema, while the third-degree polynomial in Example 2A does not have any.

□   **ANALYZING GRAPHS**

www   In addition to providing information for hand-sketching graphs, the derivative is also an important tool for analyzing graphs and discussing the interplay between a function and its rate of change. The next two examples illustrate this process in the context of some applications to economics.

Example 4 ⇔   **Agricultural Exports and Imports**   Over the past several decades, the United States has exported more agricultural products than it has imported, maintaining a positive balance of trade in this area. However, the trade balance fluctuated considerably during this period. The graph in Figure 10 approximates the rate of change of the trade balance over a 15-year period, where $B(t)$ is the trade balance (in billions of dollars) and t is time (in years).

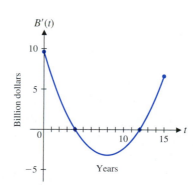

FIGURE 10 Rate of change of the balance of trade

(A) Write a brief verbal description of the graph of $y = B(t)$, including a discussion of any local extrema.

(B) Sketch a possible graph of $y = B(t)$.

SOLUTION   (A) The graph of the derivative $y = B'(t)$ contains the same essential information as a sign chart. That is, we see that $B'(t)$ is positive on $(0, 4)$, 0 at

$t = 4$, negative on $(4, 12)$, 0 at $t = 12$, and positive on $(12, 15)$. Hence, the trade balance increases for the first 4 years to a local maximum, decreases for the next 8 years to a local minimum, and then increases for the final 3 years.

(B) Without additional information concerning the actual values of $y = B(t)$, we cannot produce an accurate graph. However, we can sketch a possible graph that illustrates the important features, as shown in Figure 11. The absence of a scale on the vertical axis is a consequence of the lack of information about the values of $B(t)$.

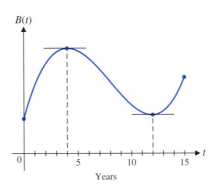

FIGURE 11 Balance of trade

Matched Problem 4   The graph in Figure 12 approximates the rate of change of the U.S. share of the total world production of motor vehicles over a 20-year period, where $S(t)$ is the U.S. share (as a percentage) and t is time (in years).

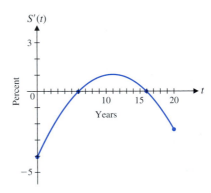

FIGURE 12

(A) Write a brief verbal description of the graph of $y = S(t)$, including a discussion of any local extrema.

(B) Sketch a possible graph of $y = S(t)$.

Example 5   **Revenue Analysis**   The graph of the total revenue $R(x)$ (in dollars) from the sale of x bookcases is shown in Figure 13.

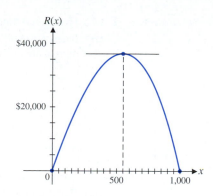

FIGURE 13 Revenue

(A) Write a brief verbal description of the graph of the marginal revenue function $y = R'(x)$, including a discussion of any x intercepts.

(B) Sketch a possible graph of $y = R'(x)$.

SOLUTION (A) The graph of $y = R(x)$ indicates that $R(x)$ increases on $(0, 550)$, has a local maximum at $x = 550$, and decreases on $(550, 1{,}000)$. Consequently, the marginal revenue function $R'(x)$ must be positive on $(0, 550)$, 0 at $x = 550$, and negative on $(550, 1{,}000)$.

(B) A possible graph of $y = R'(x)$ illustrating the information summarized in part (A) is shown in Figure 14.

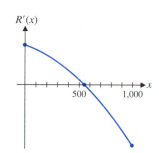

FIGURE 14 Marginal revenue

Matched Problem 5 The graph of the total revenue $R(x)$ (in dollars) from the sale of x desks is shown in Figure 15.

(A) Write a brief verbal description of the graph of the marginal revenue function $y = R'(x)$, including a discussion of any x intercepts.

(B) Sketch a possible graph of $y = R'(x)$.

FIGURE 15

Comparing Examples 4 and 5, we see that we were able to obtain more information about the function from the graph of its derivative (Example 4) than we were when the process was reversed (Example 5). In the next section we introduce some ideas that will enable us to extract additional information about the derivative from the graph of the function.

Answers to Matched Problems

1. (A) Horizontal tangent line at $x = 3$.
   (B) Decreasing on $(-\infty, 3)$;
       increasing on $(3, \infty)$
   (C)

2. (A) Partition number: $x = 0$; critical value: $x = 0$; decreasing for all x
   (B) Partition number: $x = -1$; critical value: $x = -1$; increasing for all x
   (C) Partition number: $x = 0$; no critical values; decreasing on $(-\infty, 0)$ and $(0, \infty)$

3. (A) Critical values: $x = 2, x = 4$
   (B) Local maximum at $x = 2$;
       local minimum at $x = 4$
   (C)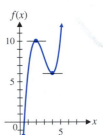

4. (A) The U.S. share of the world market decreases for 6 years to a local minimum, increases for the next 10 years to a local maximum, and then decreases for the final 4 years.
   (B)

5. (A) The marginal revenue is positive on $(0, 450)$, 0 at $x = 450$, and negative on $(450, 1{,}000)$.
   (B)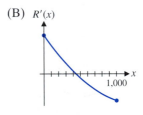

Exercise 4-1

A *Problems 1–8 refer to the graph of $y = f(x)$ below.*

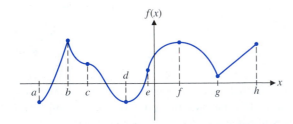

1. Identify the intervals over which $f(x)$ is increasing.

2. Identify the intervals over which $f(x)$ is decreasing.

3. Identify the intervals over which $f'(x) < 0$.

4. Identify the intervals over which $f'(x) > 0$.

5. Identify the x coordinates of the points where $f'(x) = 0$.

6. Identify the x coordinates of the points where $f'(x)$ does not exist.

7. Identify the x coordinates of the points where $f(x)$ has a local maximum.

8. Identify the x coordinates of the points where $f(x)$ has a local minimum.

In Problems 9 and 10, $f(x)$ is continuous on $(-\infty, \infty)$ and has critical values at $x = a, b, c,$ and d. Use the sign chart for $f'(x)$ to determine whether f has a local maximum, a local minimum, or neither at each critical value.

9.

10.

B

In Problems 11–24, find the intervals where $f(x)$ is increasing, the intervals where $f(x)$ is decreasing, and the local extrema.

11. $f(x) = x^2 - 8x + 8$

12. $f(x) = x^2 + 4x - 2$

13. $f(x) = -2x^2 - 16x - 25$

14. $f(x) = -3x^2 + 12x - 5$

15. $f(x) = x^3 + 4x - 5$

16. $f(x) = -x^3 - 4x + 8$

17. $f(x) = -x^3 + 9x^2 - 15$

18. $f(x) = x^3 - 12x^2 + 36x$

19. $f(x) = 2x^3 - 3x^2 - 36x$

20. $f(x) = -2x^3 + 3x^2 + 120x$

21. $f(x) = x^4 - 18x^2$

22. $f(x) = -x^4 + 50x^2$

23. $f(x) = -x^4 + 32x$

24. $f(x) = x^4 + 4x$

 In Problems 25–30, use a graphing utility to approximate the critical values of $f(x)$ to two decimal places. Find the intervals where $f(x)$ is increasing, the intervals where $f(x)$ is decreasing, and the local extrema.

25. $f(x) = x^4 + x^2 + x$

26. $f(x) = x^4 + x^2 - 9x$

27. $f(x) = x^4 - 4x^3 + 9x$

28. $f(x) = x^4 + 5x^3 - 15x$

29. $f(x) = x^4 - 2x^3 - 5x^2 + 4x$

30. $f(x) = x^4 + 2x^3 - 4x^2 - 6x$

In Problems 31–36, find the intervals where $f(x)$ is increasing, the intervals where $f(x)$ is decreasing, and sketch the graph. Add horizontal tangent lines.

31. $f(x) = 4 + 8x - x^2$

32. $f(x) = 2x^2 - 8x + 9$

33. $f(x) = x^3 - 3x + 1$

34. $f(x) = x^3 - 12x + 2$

35. $f(x) = 10 - 12x + 6x^2 - x^3$

36. $f(x) = x^3 + 3x^2 + 3x$

In Problems 37–44, $f(x)$ is continuous on $(-\infty, \infty)$. Use the given information to sketch the graph of f.

37.

x	-2	-1	0	1	2
$f(x)$	-1	1	2	3	1

38.

x	-2	-1	0	1	2
$f(x)$	1	3	2	1	-1

39.

x	-2	-1	0	2	4
$f(x)$	2	1	2	1	0

40.

x	-2	-1	0	2	3
$f(x)$	-3	0	2	-1	0

41. $f(-2) = 4, f(0) = 0, f(2) = -4;$
   $f'(-2) = 0, f'(0) = 0, f'(2) = 0;$
   $f'(x) > 0$ on $(-\infty, -2)$ and $(2, \infty);$
   $f'(x) < 0$ on $(-2, 0)$ and $(0, 2)$

42. $f(-2) = -1, f(0) = 0, f(2) = 1;$
   $f'(-2) = 0, f'(2) = 0;$
   $f'(x) > 0$ on $(-\infty, -2), (-2, 2)$, and $(2, \infty)$

43. $f(-1) = 2, f(0) = 0, f(1) = -2;$
   $f'(-1) = 0, f'(1) = 0, f'(0)$ is not defined;
   $f'(x) > 0$ on $(-\infty, -1)$ and $(1, \infty);$
   $f'(x) < 0$ on $(-1, 0)$ and $(0, 1)$

44. $f(-1) = 2, f(0) = 0, f(1) = 2;$
   $f'(-1) = 0, f'(1) = 0, f'(0)$ is not defined;
   $f'(x) > 0$ on $(-\infty, -1)$ and $(0, 1);$
   $f'(x) < 0$ on $(-1, 0)$ and $(1, \infty)$

Problems 45–50 involve functions f_1–f_6 and their derivatives g_1–g_6. Use the graphs shown in figures (A) and (B) to match each function f_i with its derivative g_j.

Figure (A) for 45–50

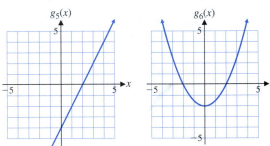

Figure (B) for 45–50

45. f_1 **46.** f_2 **47.** f_3

48. f_4 **49.** f_5 **50.** f_6

In Problems 51–56, use the given graph of $y = f'(x)$ to find the intervals where f is increasing, the intervals where f is decreasing, and the local extrema. Sketch a possible graph for $y = f(x)$.

51.

52.

53.

54.

55.

56.

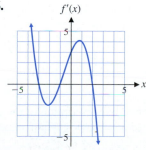

In Problems 57–60, use the given graph of $y = f(x)$ to find the intervals where $f'(x) > 0$, the intervals where $f'(x) < 0$, and the values of x for which $f'(x) = 0$. Sketch a possible graph for $y = f'(x)$.

57.

58.

59.

60.

C *In Problems 61–72, find the critical values, the intervals where $f(x)$ is increasing, the intervals where $f(x)$ is decreasing, and the local extrema. Do not graph.*

61. $f(x) = x + \dfrac{4}{x}$

62. $f(x) = \dfrac{9}{x} + x$

63. $f(x) = 1 + \dfrac{1}{x} + \dfrac{1}{x^2}$

64. $f(x) = 3 - \dfrac{4}{x} - \dfrac{2}{x^2}$

65. $f(x) = \dfrac{x^2}{x - 2}$

66. $f(x) = \dfrac{x^2}{x + 1}$

67. $f(x) = x^4(x - 6)^2$

68. $f(x) = x^3(x - 5)^2$

69. $f(x) = 3(x - 2)^{2/3} + 4$

70. $f(x) = 6(4 - x)^{2/3} + 4$

71. $f(x) = 2\sqrt{x} - x, x > 0$

72. $f(x) = x - 4\sqrt{x}, x > 0$

73. Let $f(x) = x^3 + kx$, where k is a constant. Discuss the number of local extrema and the shape of the graph of f if:

(A) $k > 0$ (B) $k < 0$ (C) $k = 0$

74. Let $f(x) = x^4 + kx^2$, where k is a constant. Discuss the number of local extrema and the shape of the graph of f if:

(A) $k > 0$ (B) $k < 0$ (C) $k = 0$

Applications

Business & Economics

75. *Profit analysis.* The graph of the total profit $P(x)$ (in dollars) from the sale of x cordless electric screwdrivers is shown in the figure.

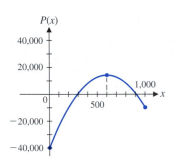

Figure for 75

(A) Write a brief verbal description of the graph of the marginal profit function $y = P'(x)$, including a discussion of any x intercepts.

(B) Sketch a possible graph of $y = P'(x)$.

76. *Revenue analysis.* The graph of the total revenue $R(x)$ (in dollars) from the sale of x cordless electric screwdrivers is shown in the figure.

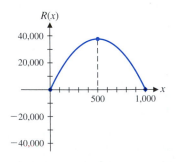

Figure for 76

(A) Write a brief verbal description of the graph of the marginal revenue function $y = R'(x)$, including a discussion of any x intercepts.

(B) Sketch a possible graph of $y = R'(x)$.

77. *Price analysis.* The graph in the figure approximates the rate of change of the price of bacon over a 70-month period, where $B(t)$ is the price of a pound of sliced bacon (in dollars) and t is time (in months).

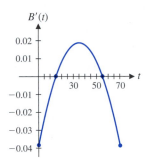

Figure for 77

(A) Write a brief verbal description of the graph of $y = B(t)$, including a discussion of any local extrema.

(B) Sketch a possible graph of $y = B(t)$.

78. *Price analysis.* The graph in the figure approximates the rate of change of the price of eggs over a 70-month period, where $E(t)$ is the price of a dozen eggs (in dollars) and t is time (in months).

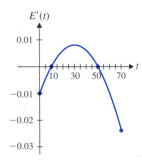

Figure for 78

(A) Write a brief verbal description of the graph of $y = E(t)$, including a discussion of any local extrema.

(B) Sketch a possible graph of $y = E(t)$.

79. *Average cost.* A manufacturer incurs the following costs in producing x toasters in one day for $0 < x < 150$: fixed costs, \$320; unit production cost, \$20 per toaster; equipment maintenance and repairs, $0.05x^2$ dollars. Thus, the cost of manufacturing x toasters in one day is given by

$$C(x) = 0.05x^2 + 20x + 320 \qquad 0 < x < 150$$

(A) What is the average cost, $\overline{C}(x)$, per toaster if x toasters are produced in one day?

(B) Find the critical values for $\overline{C}(x)$, the intervals where the average cost per toaster is decreasing, the intervals where the average cost per toaster is increasing, and the local extrema. Do not graph.

80. *Average cost.* A manufacturer incurs the following costs in producing x blenders in one day for $0 < x < 200$: fixed costs, \$450; unit production cost, \$30 per blender; equipment maintenance and repairs, $0.08x^2$ dollars.

(A) What is the average cost, $\overline{C}(x)$, per blender if x blenders are produced in one day?

(B) Find the critical values for $\overline{C}(x)$, the intervals where the average cost per blender is decreasing, the intervals where the average cost per blender is increasing, and the local extrema. Do not graph.

81. *Marginal analysis.* Show that profit will be increasing over production intervals (a, b) for which marginal revenue is greater than marginal cost. [*Hint:* $P(x) = R(x) - C(x)$]

82. *Marginal analysis.* Show that profit will be decreasing over production intervals (a, b) for which marginal revenue is less than marginal cost.

Life Sciences

83. *Medicine.* A drug is injected into the bloodstream of a patient through the right arm. The concentration of the drug in the bloodstream of the left arm t hours after the injection is approximated by

$$C(t) = \frac{0.14t}{t^2 + 1} \qquad 0 < t < 24$$

Find the critical values for $C(t)$, the intervals where the concentration of the drug is increasing, the intervals where the concentration of the drug is decreasing, and the local extrema. Do not graph.

84. *Medicine.* The concentration $C(t)$, in milligrams per cubic centimeter, of a particular drug in a patient's bloodstream is given by

$$C(t) = \frac{0.16t}{t^2 + 4t + 4} \qquad 0 < t < 12$$

where t is the number of hours after the drug is taken orally. Find the critical values for $C(t)$, the intervals where the concentration of the drug is increasing, the intervals where the concentration of the drug is decreasing, and the local extrema. Do not graph.

Social Sciences

85. *Politics.* Public awareness of a congressional candidate before and after a successful campaign was approximated by

$$P(t) = \frac{8.4t}{t^2 + 49} + 0.1 \qquad 0 < t < 24$$

where t is time (in months) after the campaign started and $P(t)$ is the fraction of people in the congressional district who could recall the candidate's (and later, congressman's) name. Find the critical values for $P(t)$, the time intervals where the fraction is increasing, the time intervals where the fraction is decreasing, and the local extrema. Do not graph.

Section 4-2	# Second Derivative and Graphs

- ❑ CONCAVITY
- ❑ INFLECTION POINTS
- ❑ SECOND-DERIVATIVE TEST
- ❑ ANALYZING GRAPHS

In Section 4-1 we saw that the derivative can be used to determine when a graph is rising and falling. Now we want to see what the *second derivative* (the derivative of the derivative) can tell us about the shape of a graph.

❑ CONCAVITY

Consider the functions

$$f(x) = x^2 \quad \text{and} \quad g(x) = \sqrt{x}$$

for x in the interval $(0, \infty)$. Since

$$f'(x) = 2x > 0 \quad \text{for } 0 < x < \infty$$

and

$$g'(x) = \frac{1}{2\sqrt{x}} > 0 \quad \text{for } 0 < x < \infty$$

both functions are increasing on $(0, \infty)$.

Explore–Discuss 1

(A) Discuss the difference in the shapes of the graphs of f and g shown in Figure 1.

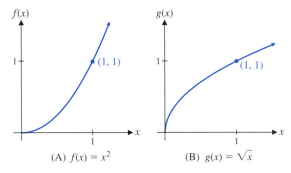

(A) $f(x) = x^2$ (B) $g(x) = \sqrt{x}$

FIGURE 1

(B) Complete the following table and discuss the relationship between the values of the derivatives of f and g and the shapes of their graphs.

x	0.25	0.5	0.75	1
$f'(x)$				
$g'(x)$				

We use the term *concave upward* to describe a graph that opens upward and *concave downward* to describe a graph that opens downward. Thus, the graph of f in Figure 1A is concave upward, and the graph of g in Figure 1B is concave downward. Finding a mathematical formulation of concavity will help us sketch and analyze graphs.

It will be instructive to examine the slopes of f and g at various points on their graphs (see Fig. 2). We can make two observations about each graph. Looking at the graph of f in Figure 2A, we see that $f'(x)$ (the slope of the tangent line) is *increasing* and that the graph lies *above* each tangent line. Looking at Figure 2B, we see that $g'(x)$ is *decreasing* and that the graph lies *below* each tangent line.

(A) $f(x) = x^2$

(B) $g(x) = \sqrt{x}$

FIGURE 2

With these ideas in mind, we state the general definition of concavity:

The graph of a function *f* is concave upward on the interval (a, b) if $f'(x)$ is *increasing* on (a, b) and is concave downward on the interval (a, b) if $f'(x)$ is *decreasing* on (a, b).

Geometrically, the graph is concave upward on (a, b) if it lies above its tangent lines in (a, b) and is concave downward on (a, b) if it lies below its tangent lines in (a, b).

How can we determine when $f'(x)$ is increasing or decreasing? In Section 4-1 we used the derivative of a function to determine when that function is increasing or decreasing. Thus, to determine when the function $f'(x)$ is increasing or decreasing, we use the derivative of $f'(x)$. The derivative of the derivative of a function is called the *second derivative* of the function. Various notations for the second derivative are given in the following box:

Second Derivative

For $y = f(x)$, the **second derivative** of f, provided that it exists, is

$$f''(x) = \frac{d}{dx}f'(x)$$

Other notations for $f''(x)$ are

$$\frac{d^2y}{dx^2} \qquad y''$$

---

www   Returning to the functions f and g discussed at the beginning of this section, we have

$$f(x) = x^2 \qquad\qquad g(x) = \sqrt{x} = x^{1/2}$$

$$f'(x) = 2x \qquad\qquad g'(x) = \frac{1}{2}x^{-1/2} = \frac{1}{2\sqrt{x}}$$

$$f''(x) = \frac{d}{dx}2x = 2 \qquad g''(x) = \frac{d}{dx}\frac{1}{2}x^{-1/2} = -\frac{1}{4}x^{-3/2} = -\frac{1}{4\sqrt{x^3}}$$

For $x > 0$, we see that $f''(x) > 0$; thus, $f'(x)$ is increasing and the graph of f is concave upward (see Fig. 2A). For $x > 0$, we also see that $g''(x) < 0$; thus, $g'(x)$ is decreasing and the graph of g is concave downward (see Fig. 2B). These ideas are summarized in the following box:

Concavity

For the interval (a, b),

$f''(x)$	$f'(x)$	GRAPH OF $y = f(x)$	EXAMPLES
$+$	Increasing	Concave upward	
$-$	Decreasing	Concave downward	

Be careful not to confuse concavity with falling and rising. As Figure 3 illustrates, a graph that is concave upward on an interval may be falling, rising, or both falling and rising on that interval. A similar statement holds for a graph that is concave downward.

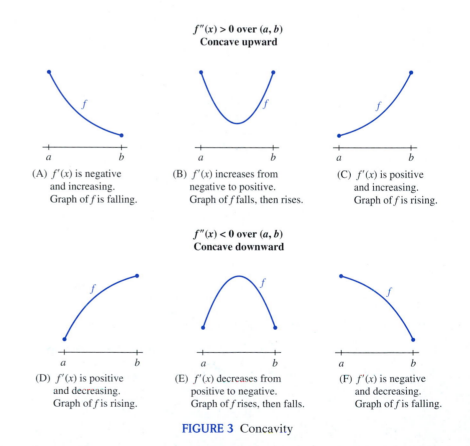

$f''(x) > 0$ over (a, b)
Concave upward

(A) $f'(x)$ is negative and increasing. Graph of f is falling.

(B) $f'(x)$ increases from negative to positive. Graph of f falls, then rises.

(C) $f'(x)$ is positive and increasing. Graph of f is rising.

$f''(x) < 0$ over (a, b)
Concave downward

(D) $f'(x)$ is positive and decreasing. Graph of f is rising.

(E) $f'(x)$ decreases from positive to negative. Graph of f rises, then falls.

(F) $f'(x)$ is negative and decreasing. Graph of f is falling.

FIGURE 3 Concavity

Example 1 ⇨ **Determining Concavity of a Graph** Let $f(x) = x^3$. Find the intervals where the graph of f is concave upward and the intervals where the graph of f is concave downward. Sketch a graph of f.

SOLUTION   To determine concavity, we must determine the sign of $f''(x)$.

$$f(x) = x^3 \qquad f'(x) = 3x^2 \qquad f''(x) = 6x$$

Sign chart for $f''(x) = 6x$ (partition number is 0):

Test Numbers	
x	$f''(x)$
-1	-6 **(−)**
1	6 **(+)**

Graph of f

FIGURE 4

Thus, the graph of f is concave downward on $(-\infty, 0)$ and concave upward on $(0, \infty)$. The graph of f (without going through other graphing details) is shown in Figure 4. ■

Matched Problem 1 ⮠ Repeat Example 1 for $f(x) = 1 - x^3$. ■

The graph in Example 1 changes from concave downward to concave upward at the point $(0, 0)$. This point is called an *inflection point*.

❑ INFLECTION POINTS

Explore–Discuss 2

Discuss the relationship between the change in concavity of each of the following functions at $x = 0$ and the second derivative at and near 0.

(A) $f(x) = x^3$ (B) $g(x) = x^{4/3}$ (C) $h(x) = x^4$

In general, an **inflection point** is a point on the graph of the function where the concavity changes (from upward to downward or from downward to upward). For the concavity to change at a point, $f''(x)$ must change sign at that point. But in Section 3-2 we saw that the partition numbers* identify the points where a function can change sign. Thus, we have the following theorem:

THEOREM 1   **Inflection Points**

If $y = f(x)$ is continuous on (a, b) and has an inflection point at $x = c$, then either $f''(c) = 0$ or $f''(c)$ does not exist.

▬▬▬▬▬▬▬■

Note that inflection points can occur only at partition numbers of f'', but not every partition number of f'' produces an inflection point. Two additional requirements must be satisfied for an inflection point:

A partition number c for f'' produces an inflection point for the graph of f only if

1. $f''(x)$ changes sign at c, and
2. c is in the domain of f.

Figure 5 on the next page illustrates several typical cases.

---

*As we did with the first derivative, we assume that if f'' is discontinuous at c, then $f''(c)$ does not exist.

$f''(x)$ $+ + + + 0 - - - -$ $f''(x)$ $- - - - 0 + + + +$ $f''(x)$ $+ + + + 0 - - - -$ $f''(x)$ $- - -$ ND $+ + +$

(A) $f'(c) > 0$ (B) $f'(c) < 0$ (C) $f'(c) = 0$ (D) $f'(c)$ is not defined

FIGURE 5 Inflection points

If $f'(c)$ exists and $f''(x)$ changes sign at $x = c$, the tangent line at an inflection point $(c, f(c))$ will always lie below the graph on the side that is concave upward and above the graph on the side that is concave downward (see Fig. 5A, B, and C).

Example 2   **Locating Inflection Points**   Find the inflection point(s) of

$$f(x) = x^3 - 6x^2 + 9x + 1$$

SOLUTION   Since inflection point(s) occur at values of x where $f''(x)$ changes sign, we construct a sign chart for $f''(x)$.

$$f(x) = x^3 - 6x^2 + 9x + 1$$
$$f'(x) = 3x^2 - 12x + 9$$
$$f''(x) = 6x - 12 = 6(x - 2)$$

Sign chart for $f''(x) = 6(x - 2)$ (partition number is 2):

FIGURE 6

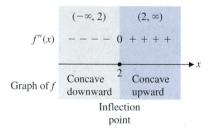

	$(-\infty, 2)$	$(2, \infty)$
$f''(x)$	$- - - -$ 0	$+ + + +$

Graph of f Concave downward   Concave upward

Inflection point

Test Numbers	
x	$f''(x)$
1	-6 **(−)**
3	6 **(+)**

From the sign chart we see that the graph of f has an inflection point at $x = 2$. The graph of f is shown in Figure 6. (See also Example 3 in Section 4-1.) ∎

Matched Problem 2   Find the inflection point(s) of $f(x) = x^3 - 9x^2 + 24x - 10$. (See the answer to Matched Problem 3 in Section 4-1 for the graph of f.) ∎

REMARK

Inflection points can be difficult to recognize on a graphing utility, but they are easily located using root approximation routines. Examining the graph of the function $f(x)$ from Example 2 on a graphing utility (Fig. 7A), it is clear that there must be an inflection point somewhere between the local maximum at $x = 1$ and the local minimum at $x = 3$. Graphing the second derivative and using a built-in root approximation routine (Fig. 7B) shows that this inflection point occurs at $x = 2$. Many graphing utilities also graph a numerical approximation to the second derivative that can be used to find inflection points. For details, consult the manual for your graphing utility.

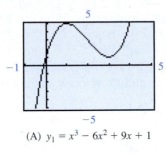

(A) $y_1 = x^3 - 6x^2 + 9x + 1$

(B) $y_2 = 6x - 12$

FIGURE 7

It is important to remember that the partition numbers of f'' are only candidates for inflection points. The function f must be defined at $x = c$, and the second derivative must change sign at $x = c$ in order for the graph to have an inflection point at $x = c$. For example, consider

$$f(x) = x^4 \qquad g(x) = \frac{1}{x}$$

$$f'(x) = 4x^3 \qquad g'(x) = -\frac{1}{x^2}$$

$$f''(x) = 12x^2 \qquad g''(x) = \frac{2}{x^3}$$

In each case, $x = 0$ is a partition number for the second derivative, but neither graph has an inflection point at $x = 0$. Function f does not have an inflection point at $x = 0$ because $f''(x)$ does not change sign at $x = 0$ (see Fig. 8A). Function g does not have an inflection point at $x = 0$ because $g(0)$ is not defined (see Fig. 8B).

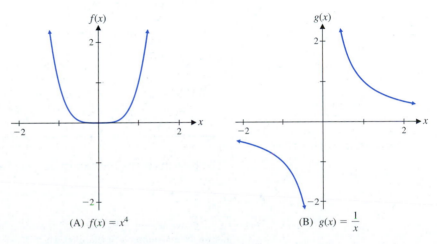

(A) $f(x) = x^4$

(B) $g(x) = \frac{1}{x}$

FIGURE 8

❑ SECOND-DERIVATIVE TEST

Now we want to see how the second derivative can be used to find local extrema. Suppose that f is a function satisfying $f'(c) = 0$ and $f''(c) > 0$. First, note that if $f''(c) > 0$, it follows from the properties of limits* that $f''(x) > 0$ in some interval (m, n) containing c. Thus, the graph of f must be concave upward in this interval. But this implies that $f'(x)$ is increasing in this interval. Since $f'(c) = 0$, $f'(x)$ must change from negative to positive at $x = c$, and $f(c)$ is a local minimum (see Fig. 9). Reasoning in the same fashion, we conclude that if $f'(c) = 0$ and $f''(c) < 0$, then $f(c)$ is a local maximum. Of course, it is possible that both $f'(c) = 0$ and $f''(c) = 0$. In this case, the second derivative cannot be used to determine the shape of the graph around $x = c$; $f(c)$ may be a local minimum, a local maximum, or neither.

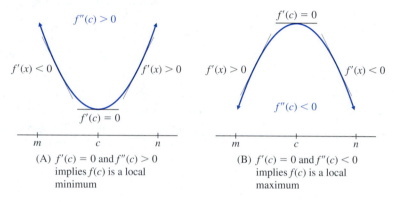

(A) $f'(c) = 0$ and $f''(c) > 0$
implies $f(c)$ is a local
minimum

(B) $f'(c) = 0$ and $f''(c) < 0$
implies $f(c)$ is a local
maximum

FIGURE 9 Second derivative and local extrema

The sign of the second derivative thus provides a simple test for identifying local maxima and minima. This test is most useful when we do not want to draw the graph of the function. If we are interested in drawing the graph and have already constructed the sign chart for $f'(x)$, the first-derivative test can be used to identify the local extrema.

Second-Derivative Test for Local Maxima and Minima

Let c be a critical value for $f(x)$.

$f'(c)$	$f''(c)$	GRAPH OF f IS:	$f(c)$	EXAMPLE
0	+	Concave upward	Local minimum	
0	−	Concave downward	Local maximum	
0	0	?	Test fails	

The first-derivative test must be used whenever $f''(c) = 0$ or $f''(c)$ does not exist.

---

*Actually, we are assuming that $f''(x)$ is continuous in an interval containing c. It is very unlikely that we will encounter a function for which $f''(c)$ exists but $f''(x)$ is not continuous in an interval containing c.

Example 3 ⇔ **Testing for Local Extrema**   Find the local maxima and minima for each function. Use the second-derivative test when it applies.

(A) $f(x) = x^3 - 6x^2 + 9x + 1$ (B) $f(x) = \frac{1}{6}x^6 - 4x^5 + 25x^4$

SOLUTION (A) Take first and second derivatives and find critical values:

$$f(x) = x^3 - 6x^2 + 9x + 1$$
$$f'(x) = 3x^2 - 12x + 9 = 3(x - 1)(x - 3)$$
$$f''(x) = 6x - 12 = 6(x - 2)$$

Critical values are $x = 1$ and $x = 3$.

$$f''(1) = -6 < 0 \qquad \textit{f has a local maximum at } x = 1.$$
$$f''(3) = 6 > 0 \qquad \textit{f has a local minimum at } x = 3.$$

(B) $$f(x) = \frac{1}{6}x^6 - 4x^5 + 25x^4$$
$$f'(x) = x^5 - 20x^4 + 100x^3 = x^3(x - 10)^2$$
$$f''(x) = 5x^4 - 80x^3 + 300x^2$$

Critical values are $x = 0$ and $x = 10$.

$$f''(0) = 0 \qquad \textit{The second-derivative test fails at both critical values, so}$$
$$f''(10) = 0 \qquad \textit{the first-derivative test must be used.}$$

Sign chart for $f'(x) = x^3(x - 10)^2$ (partition numbers are 0 and 10):

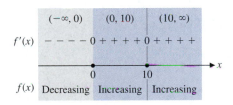

Test Numbers	
x	$f'(x)$
-1	-121 (−)
1	81 (+)
11	$1{,}331$ (+)

From the sign chart we see that $f(x)$ has a local minimum at $x = 0$ and does not have a local extremum at $x = 10$. ■

Matched Problem 3 ⇔ Find the local maxima and minima for each function. Use the second-derivative test when it applies.

(A) $f(x) = x^3 - 9x^2 + 24x - 10$ (B) $f(x) = 10x^6 - 24x^5 + 15x^4$ ■

A common error is to assume that $f''(c) = 0$ implies that $f(c)$ is not a local extremum. As Example 3B illustrates, if $f''(c) = 0$, then $f(c)$ may or may not be a local extremum. **The first-derivative test *must* be used whenever $f''(c) = 0$ or $f''(c)$ does not exist.**

❑ ANALYZING GRAPHS

In the next two examples we combine increasing/decreasing properties with concavity properties to analyze the graph of a function.

Example 4 ➯ **Analyzing a Graph** Figure 10 shows the graph of the derivative of a function *f*. Use this graph to discuss the graph of *f*. Include a sketch of a possible graph of *f*.

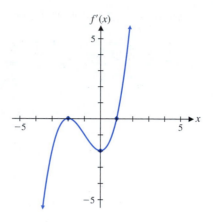

FIGURE 10

SOLUTION The sign of the derivative determines where the original function is increasing and decreasing, and the increasing/decreasing properties of the derivative determine the concavity of the original function. The relevant information obtained from the graph of f' is summarized in Table 1, and a possible graph of f is shown in Figure 11.

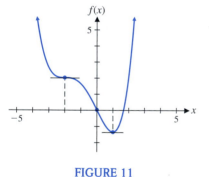

FIGURE 11

TABLE 1		
x	$f'(x)$ *(Fig. 10)*	$f(x)$ *(Fig. 11)*
$-\infty < x < -2$	Negative and increasing	Decreasing and concave upward
$x = -2$	Local maximum	Inflection point
$-2 < x < 0$	Negative and decreasing	Decreasing and concave downward
$x = 0$	Local minimum	Inflection point
$0 < x < 1$	Negative and increasing	Decreasing and concave upward
$x = 1$	x intercept	Local minimum
$1 < x < \infty$	Positive and increasing	Increasing and concave upward

Matched Problem 4 ➯ Figure 12 shows the graph of the derivative of a function *f*. Use this graph to discuss the graph of *f*. Include a sketch of a possible graph of *f*.

FIGURE 12

Example 5 ⇔   **Maximum Rate of Change**   Using past records, a company estimates that it will sell $N(x)$ items after spending $\$x$ thousand on advertising, as given by

$$N(x) = 2{,}000 - 2x^3 + 60x^2 - 450x \qquad 5 \leq x \leq 15$$

When is the rate of change of sales with respect to advertising expenditures increasing? Decreasing? What is the maximum rate of change? Graph N and N' on the same coordinate system and interpret.

SOLUTION   The rate of change of sales with respect to advertising expenditures is

$$N'(x) = -6x^2 + 120x - 450 = -6(x - 5)(x - 15)$$

To determine when this rate is increasing and decreasing, we find $N''(x)$, the derivative of $N'(x)$:

$$N''(x) = -12x + 120 = 12(10 - x)$$

The information obtained by analyzing the signs of $N'(x)$ and $N''(x)$ is summarized in Table 2 (sign charts are omitted).   ■

TABLE 2				
x	$N''(x)$	$N'(x)$	$N'(x)$	$N(x)$
$5 < x < 10$	+	+	Increasing	Increasing, concave upward
$x = 10$	0	+	Local maximum	Inflection point
$10 < x < 15$	−	+	Decreasing	Increasing, concave downward

Thus, we see that $N'(x)$, the rate of change of sales, is increasing on $(5, 10)$ and decreasing on $(10, 15)$. Both N and N' are graphed in Figure 13. An examination of the graph of $N'(x)$ shows that the maximum rate of change is $N'(10) = 150$. Notice that $N'(x)$ has a local maximum and $N(x)$ has an inflection point at $x = 10$. This value of x is referred to as the **point of diminishing returns,** since the rate of change of sales begins to decrease at this point.   ■

www

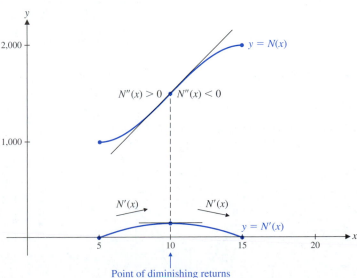

Point of diminishing returns

FIGURE 13

Matched Problem 5   Repeat Example 5 for $N(x) = 5{,}000 - x^3 + 60x^2 - 900x$, $10 \le x \le 30$.

Answers to Matched Problems

1. Concave upward on $(-\infty, 0)$
   Concave downward on $(0, \infty)$

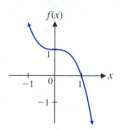

2. Inflection point at $x = 3$

3. (A) $f(2)$ is a local maximum; $f(4)$ is a local minimum

   (B) $f(0)$ is a local minimum; no local extremum at $x = 1$

4.

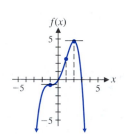

x	$f'(x)$	$f(x)$
$-\infty < x < -1$	Positive and decreasing	Increasing and concave downward
$x = -1$	Local minimum	Inflection point
$-1 < x < 1$	Positive and increasing	Increasing and concave upward
$x = 1$	Local maximum	Inflection point
$1 < x < 2$	Positive and decreasing	Increasing and concave downward
$x = 2$	x intercept	Local maximum
$2 < x < \infty$	Negative and decreasing	Decreasing and concave downward

5. $N'(x)$ is increasing on $(10, 20)$, decreasing on $(20, 30)$; maximum rate of change is $N'(20) = 300$; $x = 20$ is point of diminishing returns

Point of diminishing returns

Exercise 4-2

A   *Problems 1–8 refer to the graph of $y = f(x)$ below.*

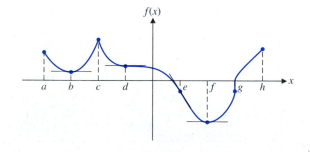

1. Identify intervals over which the graph of f is concave upward.

2. Identify intervals over which the graph of f is concave downward.

3. Identify intervals over which $f''(x) < 0$.

4. Identify intervals over which $f''(x) > 0$.

5. Identify intervals over which $f'(x)$ is increasing.

6. Identify intervals over which $f'(x)$ is decreasing.

7. Identify x coordinates of inflection points.

8. Identify x coordinates of local extrema for $f'(x)$.

In Problems 9–12, match the indicated conditions with one of the graphs (A)–(D) shown in the figure.

(A)   (B)   (C)   (D)

9. $f'(x) > 0$ and $f''(x) > 0$ on (a, b)

10. $f'(x) > 0$ and $f''(x) < 0$ on (a, b)

11. $f'(x) < 0$ and $f''(x) > 0$ on (a, b)

12. $f'(x) < 0$ and $f''(x) < 0$ on (a, b)

In Problems 13–20, find the indicated derivative for each function.

13. $f''(x)$ for $f(x) = 2x^3 - 4x^2 + 5x - 6$

14. $g''(x)$ for $g(x) = -x^3 + 2x^2 - 3x + 9$

15. $h''(x)$ for $h(x) = 2x^{-1} - 3x^{-2}$

16. $k''(x)$ for $k(x) = -6x^{-2} + 12x^{-3}$

17. d^2y/dx^2 for $y = x^2 - 18x^{1/2}$

18. d^2y/dx^2 for $y = x^3 - 24x^{1/3}$

19. y'' for $y = (x^2 + 9)^4$

20. y'' for $y = (x^2 - 16)^5$

B In Problems 21–32, find all local maxima and minima using the second-derivative test whenever it applies (do not graph). If the second-derivative test fails, use the first-derivative test.

21. $f(x) = 2x^2 - 8x + 6$

22. $f(x) = 6x - x^2 + 4$

23. $f(x) = 2x^3 - 3x^2 - 12x - 5$

24. $f(x) = 2x^3 + 3x^2 - 12x - 1$

25. $f(x) = 3 - x^3 + 3x^2 - 3x$

26. $f(x) = x^3 + 6x^2 + 12x + 2$

27. $f(x) = x^4 - 8x^2 + 10$

28. $f(x) = x^4 - 18x^2 + 50$

29. $f(x) = x^6 + 3x^4 + 2$

30. $f(x) = 4 - x^6 - 6x^4$

31. $f(x) = x + \dfrac{16}{x}$ **32.** $f(x) = x + \dfrac{25}{x}$

In Problems 33–38, find the intervals where the graph of f is concave upward, the intervals where the graph of f is concave downward, and the inflection points.

33. $f(x) = x^4 + 6x^2$

34. $f(x) = x^4 + 6x$

35. $f(x) = x^3 - 4x^2 + 5x - 2$

36. $f(x) = -x^3 - 5x^2 + 4x - 3$

37. $f(x) = -x^4 + 12x^3 - 12x + 24$

38. $f(x) = x^4 - 2x^3 - 36x + 12$

In Problems 39–44, describe the graph of f at the given point relative to the existence of a local maximum or minimum with one of the following phrases: "Local maximum," "Local minimum," "Neither," or "Unable to determine from the given information." Assume that $f(x)$ is continuous on $(-\infty, \infty)$.

39. $(2, f(2))$ if $f'(2) = 0$ and $f''(2) > 0$

40. $(4, f(4))$ if $f'(4) = 1$ and $f''(4) < 0$

41. $(-3, f(-3))$ if $f'(-3) = 0$ and $f''(-3) = 0$

42. $(-1, f(-1))$ if $f'(-1) = 0$ and $f''(-1) < 0$

43. $(6, f(6))$ if $f'(6) = 1$ and $f''(6)$ does not exist

44. $(5, f(5))$ if $f'(5) = 0$ and $f''(5)$ does not exist

In Problems 45–52, $f(x)$ is continuous on $(-\infty, \infty)$. Use the given information to sketch the graph of f.

45.

x	-4	-2	-1	0	2	4
$f(x)$	0	3	1.5	0	-1	-3

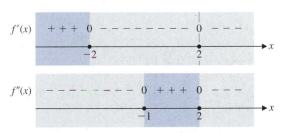

46.

x	-4	-2	-1	0	2	4
$f(x)$	0	-2	-1	0	1	3

47.

x	−3	0	1	2	4	5
$f(x)$	−4	0	2	1	−1	0

$f'(x)$ + + + ND + + + 0 − − − − − − − − 0 + + +

$f''(x)$ + + + ND − − − − − − − 0 + + + + + + + +

48.

x	−4	−2	0	2	4	6
$f(x)$	0	3	0	−2	0	3

$f'(x)$ + + + 0 − − − ND − − − 0 + + + + + + + +

$f''(x)$ − − − − − − − − ND + + + + + + + + 0 − − −

49. $f(0) = 2, f(1) = 0, f(2) = -2;$
  $f'(0) = 0, f'(2) = 0;$
  $f'(x) > 0$ on $(-\infty, 0)$ and $(2, \infty);$
  $f'(x) < 0$ on $(0, 2);$
  $f''(1) = 0;$
  $f''(x) > 0$ on $(1, \infty);$
  $f''(x) < 0$ on $(-\infty, 1)$

50. $f(-2) = -2, f(0) = 1, f(2) = 4;$
  $f'(-2) = 0, f'(2) = 0;$
  $f'(x) > 0$ on $(-2, 2);$
  $f'(x) < 0$ on $(-\infty, -2)$ and $(2, \infty);$
  $f''(0) = 0;$
  $f''(x) > 0$ on $(-\infty, 0);$
  $f''(x) < 0$ on $(0, \infty)$

51. $f(-1) = 0, f(0) = -2, f(1) = 0;$
  $f'(0) = 0, f'(-1)$ and $f'(1)$ are not defined;
  $f'(x) > 0$ on $(0, 1)$ and $(1, \infty);$
  $f'(x) < 0$ on $(-\infty, -1)$ and $(-1, 0);$
  $f''(-1)$ and $f''(1)$ are not defined;
  $f''(x) > 0$ on $(-1, 1);$
  $f''(x) < 0$ on $(-\infty, -1)$ and $(1, \infty)$

52. $f(0) = -2, f(1) = 0, f(2) = 4;$
  $f'(0) = 0, f'(2) = 0, f'(1)$ is not defined;
  $f'(x) > 0$ on $(0, 1)$ and $(1, 2);$
  $f'(x) < 0$ on $(-\infty, 0)$ and $(2, \infty);$

  $f''(1)$ is not defined;
  $f''(x) > 0$ on $(-\infty, 1);$
  $f''(x) < 0$ on $(1, \infty)$

In Problems 53–60, find local maxima, local minima, and inflection points. Sketch the graph of each function. Include tangent lines at each local extremum and inflection point.

53. $f(x) = x^3 - 6x^2 + 16$

54. $f(x) = x^3 - 9x^2 + 15x + 10$

55. $f(x) = x^3 + x + 2$

56. $f(x) = 1 - 3x - x^3$

57. $f(x) = (2 - x)^3 + 1$

58. $f(x) = (1 + x)^3 - 1$

59. $f(x) = x^3 - 12x$

60. $f(x) = 27x - x^3$

In Problems 61–64, use the graph of $y = f'(x)$ to discuss the graph of $y = f(x)$. Organize your conclusions in a table (see Example 4), and sketch a possible graph of $y = f(x)$.

61.

62.

63.

64.

Problems 65–70 require the use of a graphing utility. Approximate the x coordinates of the inflection points of f to two decimal places. Find the intervals where the graph of f is concave upward and the intervals where the graph is concave downward.

65. $f(x) = 0.25x^4 - 2x^3 + 1.5x^2 + 21x + 15$

66. $f(x) = 0.25x^4 - 3.5x^3 + 12x^2 + 21x$

67. $f(x) = x^5 + 2x^4 + 4x^2 - 5$

68. $f(x) = x^5 - 3x^4 + x^3 - x^2 + 10$

69. $f(x) = x^5 - 3x^4 - x^3 + 7x^2 - 2$

70. $f(x) = x^5 - 2x^4 - 3x^3 + 4x^2 + 4x + 5$

In Problems 71–74, assume that f, f′, and f″ are continuous for all real numbers.

71. Explain how you can locate inflection points for the graph of $y = f(x)$ by examining the graph of $y = f'(x)$.

72. Explain how you can determine where $f'(x)$ is increasing or decreasing by examining the graph of $y = f(x)$.

73. Explain how you can locate local maxima and minima for the graph of $y = f'(x)$ by examining the graph of $y = f(x)$.

74. Explain how you can locate local maxima and minima for the graph of $y = f(x)$ by examining the graph of $y = f'(x)$.

C *Find the inflection points in Problems 75–78. Do not graph.*

75. $f(x) = \dfrac{1}{x^2 + 12}$

76. $f(x) = \dfrac{x^2}{x^2 + 12}$

77. $f(x) = \dfrac{x}{x^2 + 12}$

78. $f(x) = \dfrac{x^3}{x^2 + 12}$

Business & Economics

79. *Inflation.* One commonly used measure of inflation is
www the annual rate of change of the Consumer Price Index (CPI). A newspaper headline proclaims that the rate of change of inflation for consumer prices is increasing. What does this say about the shape of the graph of the CPI?

80. *Inflation.* Another commonly used measure of infla-
www tion is the annual rate of change of the Producers Price Index (PPI). A government report states that the rate of change of inflation for producer prices is decreasing. What does this say about the shape of the graph of the PPI?

81. *Cost analysis.* A company manufactures a variety of lighting fixtures at different locations. The total cost $C(x)$ (in dollars) of producing x desk lamps per week at plant A is shown in the figure. Discuss the shape of the graph of the marginal cost function $C'(x)$ and interpret in terms of the efficiency of the production process at this plant.

Figure for 81   Production costs at plant A

82. *Cost analysis.* The company in Problem 81 produces the same lamp at another plant. The total cost $C(x)$

(in dollars) of producing x desk lamps per week at plant B is shown in the figure. Discuss the shape of the graph of the marginal cost function $C'(x)$ and interpret in terms of the efficiency of the production process at plant B. Compare the production processes at these two plants.

Figure for 82   Production costs at plant B

83. *Revenue.* The marketing research department for a computer company used a large city to test market their new product. They found that the relationship between price p (dollars per unit) and the demand x (units per week) was given approximately by

$$p = 1,296 - 0.12x^2 \qquad 0 < x < 80$$

Thus, the weekly revenue can be approximated by

$$R(x) = xp = 1,296x - 0.12x^3 \qquad 0 < x < 80$$

(A) Find the local extrema for the revenue function.

(B) Over which intervals is the graph of the revenue function concave upward? Concave downward?

84. *Profit.* Suppose that the cost equation for the company in Problem 83 is

$$C(x) = 830 + 396x$$

(A) Find the local extrema for the profit function.

(B) Over which intervals is the graph of the profit function concave upward? Concave downward?

85. Advertising. A company estimates that it will sell $N(x)$ units of a product after spending $\$x$ thousand on advertising, as given by

$$N(x) = -3x^3 + 225x^2 - 3,600x + 17,000 \qquad 10 \leqslant x \leqslant 40$$

(A) When is the rate of change of sales $N'(x)$ increasing? Decreasing?

(B) Find the inflection points for the graph of N.

(C) Graph N and N' on the same coordinate system.

(D) What is the maximum rate of change of sales?

86. Advertising. A company estimates that it will sell $N(x)$ units of a product after spending $\$x$ thousand on advertising, as given by

$$N(x) = -2x^3 + 90x^2 - 750x + 2,000 \qquad 5 \leqslant x \leqslant 25$$

(A) When is the rate of change of sales $N'(x)$ increasing? Decreasing?

(B) Find the inflection points for the graph of N.

(C) Graph N and N' on the same coordinate system.

(D) What is the maximum rate of change of sales?

87. Advertising. An automobile dealer uses television advertising to promote car sales. Using past records, the dealer arrived at the data in the table, where x is the number of ads placed monthly and y is the number of cars sold that month.

NUMBER OF ADS x	NUMBER OF CARS y
10	325
12	339
20	417
30	546
35	615
40	682
50	795

(A) Enter the data in a graphing utility and find a cubic regression equation for the number of cars sold monthly as a function of the number of ads.

(B) How many ads should the dealer place each month to maximize the rate of change of sales with respect to the number of ads, and how many cars can the dealer expect to sell with this number of ads? Round answers to the nearest integer.

88. Advertising. A music store uses radio advertising to promote sales of CDs. The store manager used past records to determine the data in the table, where x is the number of ads placed monthly and y is the number of CDs sold that month.

NUMBER OF ADS x	NUMBER OF CDS y
10	345
14	488
20	746
30	1,228
40	1,671
50	1,955

(A) Enter the data in a graphing utility and find a cubic regression equation for the number of CDs sold monthly as a function of the number of ads.

(B) How many ads should the store manager place each month to maximize the rate of change of sales with respect to the number of ads, and how many CDs can the manager expect to sell with this number of ads? Round answers to the nearest integer.

Life Sciences

89. Population growth—bacteria. A drug that stimulates reproduction is introduced into a colony of bacteria. After t minutes, the number of bacteria is given approximately by

$$N(t) = 1,000 + 30t^2 - t^3 \qquad 0 \leqslant t \leqslant 20$$

(A) When is the rate of growth $N'(t)$ increasing? Decreasing?

(B) Find the inflection points for the graph of N.

(C) Sketch the graphs of N and N' on the same coordinate system.

(D) What is the maximum rate of growth?

90. Drug sensitivity. One hour after x milligrams of a particular drug are given to a person, the change in body temperature $T(x)$, in degrees Fahrenheit, is given by

$$T(x) = x^2\left(1 - \frac{x}{9}\right) \qquad 0 \leqslant x \leqslant 6$$

The rate $T'(x)$ at which $T(x)$ changes with respect to the size of the dosage x is called the *sensitivity* of the body to the dosage.

(A) When is $T'(x)$ increasing? Decreasing?

(B) Where does the graph of T have inflection points?

(C) Sketch the graphs of T and T' on the same co-ordinate system.

(D) What is the maximum value of $T'(x)$?

Social Sciences

91. *Learning.* The time T (in minutes) it takes a person to learn a list of length n is

$$T(n) = 0.08n^3 - 1.2n^2 + 6n \qquad n \geq 0$$

(A) When is the rate of change of T with respect to the length of the list increasing? Decreasing?

(B) Where does the graph of T have inflection points? Graph T and T' on the same coordinate system.

(C) What is the minimum value of $T'(n)$?

Section 4-3

Curve Sketching Techniques: Unified and Extended

- ❑ LIMITS AT INFINITY
- ❑ VERTICAL ASYMPTOTES
- ❑ GRAPHING STRATEGY
- ❑ USING THE STRATEGY
- ❑ APPLICATION

In this section we apply, in a systematic way, all the graphing concepts discussed in the preceding three sections. Before we do this, we need to discuss the behavior of a graph as x increases or decreases without bound and also at certain points of discontinuity.

❑ LIMITS AT INFINITY

An important element in the analysis of the graph of a function is the behavior of the function as x increases or decreases without bound. Recall from Section 3-2 that we used the special symbol ∞ to describe limits that increase or decrease without bound. We write $x \to \infty$ to indicate that x is increasing without bound and $x \to -\infty$ to indicate that x is decreasing without bound.

We begin by considering power functions of the form x^p and $1/x^p$, where p is a positive real number.

Explore–Discuss 1

(A) Complete the following table:

x	100	1,000	10,000	100,000	1,000,000
x^2					
$1/x^2$					

(B) Describe verbally the behavior of x^2 as x increases without bound. Then use limit notation to describe this behavior.

(C) Repeat part (B) for $1/x^2$.

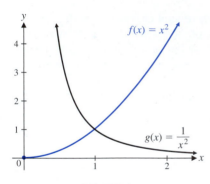

FIGURE 1

$$\lim_{x \to \infty} f(x) = \infty$$

$$\lim_{x \to \infty} g(x) = 0$$

If p is a positive real number, then x^p increases as x increases, and it can be shown that there is no upper bound on the values of x^p. We indicate this by writing

$$x^p \to \infty \quad \text{as} \quad x \to \infty \qquad \text{or} \qquad \lim_{x \to \infty} x^p = \infty$$

Since the reciprocals of very large numbers are very small numbers, it follows that $1/x^p$ approaches 0 as x increases without bound. We indicate this behavior by writing

$$\frac{1}{x^p} \to 0 \quad \text{as} \quad x \to \infty \qquad \text{or} \qquad \lim_{x \to \infty} \frac{1}{x^p} = 0$$

Figure 1 illustrates this behavior for $f(x) = x^2$ and $g(x) = 1/x^2$.

Limits of power forms as x decreases without bound behave in a similar manner, with two important differences. First, if x is negative, then x^p is not a real number for all values of p. For example, $x^{1/2} = \sqrt{x}$ is not a real number for negative values of x. Second, if x^p represents a real number for all real x, then it may approach ∞ or $-\infty$, depending on the value of p. For example,

$$\lim_{x \to -\infty} x^2 = \infty \qquad \text{but} \qquad \lim_{x \to -\infty} x^3 = -\infty$$

For the function g in Figure 1, the line $y = 0$ (the x axis) is called a *horizontal asymptote*. In general, a line $y = b$ is a **horizontal asymptote** for the graph of $y = f(x)$ if $f(x)$ approaches b as x either decreases without bound or increases without bound. Symbolically, $y = b$ is a horizontal asymptote if either

$$\lim_{x \to -\infty} f(x) = b \qquad \text{or} \qquad \lim_{x \to \infty} f(x) = b$$

In the first case, the graph of f will be close to the horizontal line $y = b$ for large (in absolute value) negative x (see Fig. 2). In the second case, the graph will be close to the horizontal line $y = b$ for large positive x.

Theorem 1 summarizes the various possibilities for limits of power functions as x increases or decreases without bound.

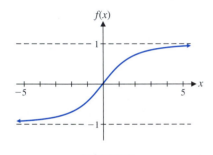

FIGURE 2

$$\lim_{x \to -\infty} f(x) = -1$$

$$\lim_{x \to \infty} f(x) = 1$$

THEOREM 1   Limits at Infinity for Power Functions

If p is a positive real number and k is any real constant, then

1. $\displaystyle\lim_{x \to -\infty} \frac{k}{x^p} = 0$ **2.** $\displaystyle\lim_{x \to \infty} \frac{k}{x^p} = 0$

3. $\displaystyle\lim_{x \to -\infty} kx^p = \pm\infty$ **4.** $\displaystyle\lim_{x \to \infty} kx^p = \pm\infty$

provided that x^p names a real number for negative values of x. The limits in 3 and 4 will be either $-\infty$ or ∞, depending on k and p.

It is important to understand that the symbol ∞ does not represent an actual number that x is approaching, but is used only to indicate that x is increasing with no upper limit on its size. And as we discussed in Section 3-2, when we say that a limit is equal to ∞, we are using ∞ to describe the behavior of a limit that does not exist. Finally, we note that **limit properties 1–6 in Theorem 1 in Section 3-1 are valid if we replace the statement $x \to c$ with $x \to \infty$ or $x \to -\infty$.**

Now we want to use the ideas discussed above to investigate limits at infinity for polynomial functions and rational functions. We begin with an example involving a polynomial function.

Example 1   ➱   **Limit of a Polynomial Function at Infinity**   Let

$$p(x) = 2x^3 - x^2 - 7x + 3$$

Discuss the limit of $p(x)$ as x approaches ∞ and as x approaches $-\infty$.

SOLUTION   Since limits of power functions of the form $1/x^p$ approach 0 as x approaches ∞ or $-\infty$, it is convenient to work with these reciprocal forms whenever possible. If we factor out the term involving the highest power of x, we can write $p(x)$ as

$$p(x) = 2x^3\left(1 - \frac{1}{2x} - \frac{7}{2x^2} + \frac{3}{2x^3}\right)$$

Using Theorem 1 and limit properties from Section 3-1, we have

$$\lim_{x\to\infty}\left(1 - \frac{1}{2x} - \frac{7}{2x^2} + \frac{3}{2x^3}\right) = 1 - 0 - 0 + 0 = 1$$

Thus, for large values of x,

$$\left(1 - \frac{1}{2x} - \frac{7}{2x^2} + \frac{3}{2x^3}\right) \approx 1$$

and

$$p(x) = 2x^3\left(1 - \frac{1}{2x} - \frac{7}{2x^2} + \frac{3}{2x^3}\right) \approx 2x^3$$

Since $2x^3 \to \infty$ as $x \to \infty$, we can conclude that

$$\lim_{x\to\infty}(2x^3 - x^2 - 7x + 3) = \lim_{x\to\infty} 2x^3\left(1 - \frac{1}{2x} - \frac{7}{2x^2} + \frac{3}{2x^3}\right) = \infty$$

Similarly, $2x^3 \to -\infty$ as $x \to -\infty$ implies that

$$\lim_{x\to-\infty}(2x^3 - x^2 - 7x + 3) = \lim_{x\to-\infty} 2x^3\left(1 - \frac{1}{2x} - \frac{7}{2x^2} + \frac{3}{2x^3}\right) = -\infty$$

Thus, we can conclude that the behavior of $p(x)$ for large values is the same as the behavior of the highest-degree term, $2x^3$. ■

 Figure 3 illustrates the behavior described in Example 1 on a graphing utility.

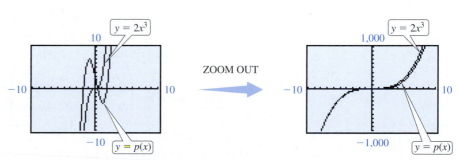

FIGURE 3 $p(x) = 2x^3 - x^2 - 7x + 3$

Matched Problem 1   ➱   Let $p(x) = -4x^4 + 2x^3 + 3x$. Discuss the limit of $p(x)$ as x approaches ∞ and as x approaches $-\infty$. ■

The ideas introduced in Example 1 are generalized to any polynomial in Theorem 2.

THEOREM 2 **Limits at Infinity for Polynomial Functions**

If

$$p(x) = a_n x^n + a_{n-1} x^{n-1} + \cdots + a_1 x + a_0 \qquad a_n \neq 0, \quad n \geqslant 1$$

then

$$\lim_{x \to \infty} p(x) = \lim_{x \to \infty} a_n x^n = \pm \infty$$

and

$$\lim_{x \to -\infty} p(x) = \lim_{x \to -\infty} a_n x^n = \pm \infty$$

Each limit will be either $-\infty$ or ∞, depending on a_n and n.

A polynomial of degree 0 is a constant function, $p(x) = a_0$, and its limit as x approaches ∞ or $-\infty$ is the number a_0. For any polynomial of degree 1 or greater, Theorem 2 states that the limit as x approaches ∞ or $-\infty$ cannot be equal to a number. Thus, it follows that **polynomials of degree 1 or greater never have horizontal asymptotes.**

Explore–Discuss 2

Write a verbal description of the relationship between a polynomial and its highest-degree term as stated in Theorem 2.

www

Since a rational function is the ratio of two polynomials, it is not surprising that reciprocals of powers of x can also be used to analyze limits at infinity for rational functions. For example, consider the rational function

$$f(x) = \frac{2x^2 - 5x + 9}{3x^2 + 4x - 8}$$

Factoring the highest-degree term out of the numerator and the denominator, we have

$$f(x) = \frac{2x^2\left(1 - \dfrac{5}{2x} + \dfrac{9}{2x^2}\right)}{3x^2\left(1 + \dfrac{4}{3x} - \dfrac{8}{3x^2}\right)} = \frac{2}{3} \cdot \frac{1 - \dfrac{5}{2x} + \dfrac{9}{2x^2}}{1 + \dfrac{4}{3x} - \dfrac{8}{3x^2}}$$

Using Theorem 1, we have

$$\lim_{x \to \infty} f(x) = \lim_{x \to \infty}\left(\frac{2}{3} \cdot \frac{1 - \dfrac{5}{2x} + \dfrac{9}{2x^2}}{1 + \dfrac{4}{3x} - \dfrac{8}{3x^2}}\right) = \frac{2}{3} \cdot \frac{1 - 0 + 0}{1 + 0 - 0} = \frac{2}{3}$$

and

$$\lim_{x \to -\infty} f(x) = \lim_{x \to -\infty}\left(\frac{2}{3} \cdot \frac{1 - \dfrac{5}{2x} + \dfrac{9}{2x^2}}{1 + \dfrac{4}{3x} - \dfrac{8}{3x^2}}\right) = \frac{2}{3} \cdot \frac{1 - 0 + 0}{1 + 0 - 0} = \frac{2}{3}$$

Thus, for this rational function, the behavior as x approaches infinity is determined by the ratio of the highest-degree term in the numerator to the highest-degree term in the denominator. Theorem 3 generalizes this to any rational function and lists the three possible outcomes. This provides us with a useful tool for analyzing the behavior of rational functions as x approaches infinity.

THEOREM 3   **Limits at Infinity and Horizontal Asymptotes for Rational Functions**

If

$$f(x) = \frac{a_m x^m + a_{m-1} x^{m-1} + \cdots + a_1 x + a_0}{b_n x^n + b_{n-1} x^{n-1} + \cdots + b_1 x + b_0} \qquad a_m \neq 0, \quad b_n \neq 0$$

then

$$\lim_{x \to \infty} f(x) = \lim_{x \to \infty} \frac{a_m x^m}{b_n x^n} \qquad \text{and} \qquad \lim_{x \to -\infty} f(x) = \lim_{x \to -\infty} \frac{a_m x^m}{b_n x^n}$$

There are three possible cases for these limits:

1. If $m < n$, then $\lim\limits_{x \to \infty} f(x) = \lim\limits_{x \to -\infty} f(x) = 0$, and the line $y = 0$ (the x axis) is a horizontal asymptote for $f(x)$.

2. If $m = n$, then $\lim\limits_{x \to \infty} f(x) = \lim\limits_{x \to -\infty} f(x) = a_m/b_n$, and the line $y = a_m/b_n$ is a horizontal asymptote for $f(x)$.

3. If $m > n$, then each limit will be ∞ or $-\infty$, depending on m, n, a_m, and b_n, and $f(x)$ does not have a horizontal asymptote.

Notice in cases 1 and 2 of Theorem 3 that the limit is the same if x approaches ∞ or $-\infty$. Thus, **a rational function can have at most one horizontal asymptote.**

Explore–Discuss 3

Case 1 in Theorem 3 can be stated verbally as follows:

If the degree of the numerator of a rational function is less than the degree of the denominator, then the x axis is a horizontal asymptote.

Write similar verbal statements for cases 2 and 3.

Example 2   **Horizontal Asymptotes for Rational Functions**   Find all horizontal asymptotes for each function:

(A) $f(x) = \dfrac{5x^3 - 2x^2 + 1}{4x^3 + 2x - 7}$ (B) $f(x) = \dfrac{3x^4 - x^2 + 1}{8x^6 - 10}$

(C) $f(x) = \dfrac{2x^5 - x^3 - 1}{6x^3 + 2x^2 - 7}$

Solution   (A) $\dfrac{a_m x^m}{b_n x^n} = \dfrac{5x^3}{4x^3} = \dfrac{5}{4}$

The line $y = \frac{5}{4}$ is a horizontal asymptote for $f(x)$ (Theorem 3, case 2).

(B) $\dfrac{a_m x^m}{b_n x^n} = \dfrac{3x^4}{8x^6} = \dfrac{3}{8x^2}$

The line $y = 0$ (the x axis) is a horizontal asymptote for $f(x)$ (Theorem 3, case 1).

(C) $\dfrac{a_m x^m}{b_n x^n} = \dfrac{2x^5}{6x^3} = \dfrac{x^2}{3}$

Thus, $f(x)$ does not have a horizontal asymptote (Theorem 3, case 3). ■

Matched Problem 2 ➯   Find all horizontal asymptotes for each function:

(A) $f(x) = \dfrac{4x^3 - 5x + 8}{2x^4 - 7}$ (B) $f(x) = \dfrac{5x^6 + 3x}{2x^5 - x - 5}$

(C) $f(x) = \dfrac{2x^3 - x + 7}{4x^3 + 3x^2 - 100}$ ■

❑ VERTICAL ASYMPTOTES

www   Theorem 3 added an important tool for locating horizontal asymptotes to our mathematical toolbox. But in Section 3-2 we discussed another type of asymptote—a vertical asymptote. Theorem 4 provides a simple and effective tool for locating vertical asymptotes.

THEOREM 4   **Locating Vertical Asymptotes**

Let $f(x) = n(x)/d(x)$, where both n and d are continuous at $x = c$. If, at $x = c$, the denominator $d(x)$ is 0 and the numerator $n(x)$ is not 0, then the line $x = c$ is a vertical asymptote for the graph of f.

[*Note:* Since a rational function is a ratio of two polynomial functions and polynomial functions are continuous for all real numbers, this theorem includes rational functions as a special case.]

If $f(x) = n(x)/d(x)$ and both $n(c) = 0$ and $d(c) = 0$, then the limit of $f(x)$ as x approaches c involves an indeterminate form and Theorem 4 does not apply:

$$\lim_{x \to c} f(x) = \lim_{x \to c} \frac{n(x)}{d(x)} \qquad \frac{0}{0} \text{ indeterminate form}$$

Algebraic simplification is often useful in this situation.

Example 3 ➯   **Locating Vertical Asymptotes**   Find all vertical asymptotes for

$$f(x) = \frac{x^2 + x - 2}{x^2 - 1}$$

SOLUTION   Let $n(x) = x^2 + x - 2$ and $d(x) = x^2 - 1$. Factoring the denominator, we see that

$$d(x) = x^2 - 1 = (x - 1)(x + 1)$$

Since $d(-1) = 0$ and $n(-1) = -2 \neq 0$, Theorem 4 tells us that the line $x = -1$ is a vertical asymptote. On the other hand, $d(1) = 0$ but $n(1) = 0$ also, so

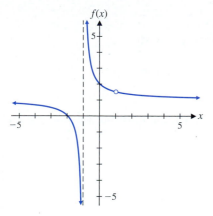

FIGURE 4

$$f(x) = \frac{x^2 + x - 2}{x^2 - 1}$$

Theorem 4 does not apply at $x = 1$. We use algebraic simplification to investigate the behavior of the function at $x = 1$:

$$\lim_{x \to 1} f(x) = \lim_{x \to 1} \frac{x^2 + x - 2}{x^2 - 1}$$

$$= \lim_{x \to 1} \frac{(x - 1)(x + 2)}{(x - 1)(x + 1)}$$

$$= \lim_{x \to 1} \frac{x + 2}{x + 1}$$

$$= \frac{3}{2}$$

Since the limit exists as x approaches 1, f does not have a vertical asymptote at $x = 1$ (see Fig. 4).

■

Matched Problem 3 ✍ Find all vertical asymptotes for $f(x) = \dfrac{x - 3}{x^2 - 4x + 3}$.

■

❑ GRAPHING STRATEGY

WWW We now have powerful tools to determine the shape of a graph of a function—even before we plot any points. We can accurately sketch the graphs of many functions using these tools and point-by-point plotting as necessary (often, very little point-by-point plotting is required). These same tools can be used to analyze a graph produced on a graphing utility or other electronic device. We organize these tools in the graphing strategy summarized in the next box.

Graphing Strategy for $y = f(x)$

Step 1. *Analyze $f(x)$.*

(A) Find the domain of f. [The domain of f is the set of all real numbers x that produce real values for $f(x)$.]

(B) Find intercepts. [The y intercept is $f(0)$, if it exists; the x intercepts are the solutions to $f(x) = 0$, if they exist.]

(C) Find asymptotes. (Use Theorems 3 and 4, if they apply; otherwise, calculate limits at points of discontinuity and as x increases and decreases without bound.)

Step 2. *Analyze $f'(x)$.* Find any critical values for $f(x)$ and any partition numbers for $f'(x)$. [Remember, every critical value for $f(x)$ is also a partition number for $f'(x)$, but some partition numbers for $f'(x)$ may not be critical values for $f(x)$.] Construct a sign chart for $f'(x)$, determine the intervals where $f(x)$ is increasing and decreasing, and find local maxima and minima.

Step 3. *Analyze $f''(x)$.* Construct a sign chart for $f''(x)$, determine where the graph of f is concave upward and concave downward, and find any inflection points.

Step 4. *Sketch the graph of f.* Draw asymptotes and locate intercepts, local maxima and minima, and inflection points. Sketch in what you know from steps 1 to 3. In regions of uncertainty, use point-by-point plotting to complete the graph.

❏ USING THE STRATEGY

Some examples will illustrate the use of the graphing strategy.

Example 4 ⇔ **Using the Graphing Strategy** Analyze the function

$$f(x) = x^4 - 2x^3$$

following the graphing strategy. State all the pertinent information, and sketch the graph of f.

SOLUTION **Step 1.** *Analyze $f(x)$.* $f(x) = x^4 - 2x^3$

(A) Domain: All real x

(B) y intercept: $f(0) = 0$

x intercept:
$$f(x) = 0$$
$$x^4 - 2x^3 = 0$$
$$x^3(x - 2) = 0$$
$$x = 0, 2$$

(C) Asymptotes: Since f is a polynomial, there are no horizontal or vertical asymptotes.

Step 2. *Analyze $f'(x)$.* $f'(x) = 4x^3 - 6x^2 = 4x^2(x - \frac{3}{2})$

Critical values for $f(x)$: 0 and $\frac{3}{2}$

Partition numbers for $f'(x)$: 0 and $\frac{3}{2}$

Sign chart for $f'(x)$:

Test Numbers

x	$f'(x)$	
-1	-10	(−)
1	-2	(−)
2	8	(+)

Thus, $f(x)$ is decreasing on $(-\infty, \frac{3}{2})$, increasing on $(\frac{3}{2}, \infty)$, and has a local minimum at $x = \frac{3}{2}$.

Step 3. *Analyze $f''(x)$.* $f''(x) = 12x^2 - 12x = 12x(x - 1)$

Partition numbers for $f''(x)$: 0 and 1

Sign chart for $f''(x)$:

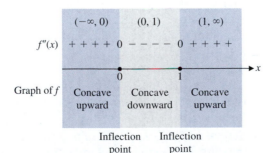

Test Numbers

x	$f''(x)$	
-1	24	(+)
$\frac{1}{2}$	-3	(−)
2	24	(+)

Thus, the graph of f is concave upward on $(-\infty, 0)$ and $(1, \infty)$, concave downward on $(0, 1)$, and has inflection points at $x = 0$ and $x = 1$.

Step 4.   *Sketch the graph of f.*

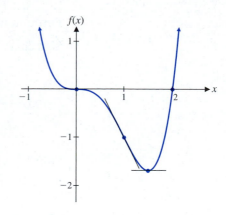

x	$f(x)$
0	0
1	−1
$\frac{3}{2}$	−$\frac{27}{16}$
2	0

Matched Problem 4 ➯   Analyze the function $f(x) = x^4 + 4x^3$ following the graphing strategy. State all the pertinent information, and sketch the graph of f.

Example 5 ➯   **Using the Graphing Strategy**   Use the graphing strategy to analyze the function $f(x) = (x - 1)/(x - 2)$. State all the pertinent information, and sketch the graph of f.

SOLUTION   **Step 1.**   *Analyze $f(x)$.* $f(x) = \dfrac{x - 1}{x - 2}$

(A) Domain: All real x, except $x = 2$

(B) y intercept: $f(0) = \dfrac{0 - 1}{0 - 2} = \dfrac{1}{2}$

x intercepts: Since a fraction is 0 when its numerator is 0 and the denominator is not 0, the x intercept is $x = 1$.

(C) Horizontal asymptote: $\dfrac{a_m x^m}{b_n x^n} = \dfrac{x}{x} = 1$

Thus, the line $y = 1$ is a horizontal asymptote.

Vertical asymptote: The denominator is 0 for $x = 2$, and the numerator is not 0 for this value. Therefore, the line $x = 2$ is a vertical asymptote.

Step 2.   *Analyze $f'(x)$.* $f'(x) = \dfrac{(x - 2)(1) - (x - 1)(1)}{(x - 2)^2} = \dfrac{-1}{(x - 2)^2}$

Critical values for $f(x)$:   None

Partition number for $f'(x)$: $x = 2$

Sign chart for $f'(x)$:

Test Numbers	
x	$f'(x)$
1	−1 (−)
3	−1 (−)

Thus, $f(x)$ is decreasing on $(-\infty, 2)$ and $(2, \infty)$. There are no local extrema.

Step 3. *Analyze* $f''(x)$. $f''(x) = \dfrac{2}{(x-2)^3}$

Partition number for $f''(x)$: $x = 2$

Sign chart for $f''(x)$:

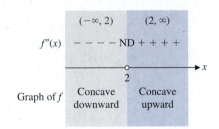

Test Numbers

x	$f''(x)$	
1	-2	$(-)$
3	2	$(+)$

Thus, the graph of f is concave downward on $(-\infty, 2)$ and concave upward on $(2, \infty)$. Since $f(2)$ is not defined, there is no inflection point at $x = 2$, even though $f''(x)$ changes sign at $x = 2$.

Step 4. *Sketch the graph of f.* Insert intercepts and asymptotes, and plot a few additional points (for functions with asymptotes, plotting additional points is often helpful). Then sketch the graph:

x	$f(x)$
-2	$\frac{3}{4}$
0	$\frac{1}{2}$
1	0
$\frac{3}{2}$	-1
$\frac{5}{2}$	3
3	2
4	$\frac{3}{2}$

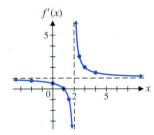

Matched Problem 5 Analyze the function $f(x) = 2x/(1 - x)$ following the graphing strategy. State all the pertinent information, and sketch the graph of f.

❑ APPLICATION

Example 6 **Average Cost** Given the cost function $C(x) = 5{,}000 + 0.5x^2$, where x is the number of items produced, use the graphing strategy to analyze the graph of the average cost function. State all the pertinent information and sketch the graph of the average cost function. Find the marginal cost function and graph it on the same set of coordinate axes.

Solution The average cost function is

$$\overline{C}(x) = \frac{5{,}000 + 0.5x^2}{x} = \frac{5{,}000}{x} + 0.5x$$

Step 1. *Analyze* $\overline{C}(x)$

(A) Domain: Since negative values of x do not make sense and $\overline{C}(0)$ is not defined, the domain is the set of positive real numbers.

(B) Intercepts: None

(C) Horizontal asymptote: $\dfrac{a_m x^m}{b_n x^n} = \dfrac{0.5x^2}{x} = 0.5x$

Thus, there is no horizontal asymptote.

Vertical asymptote: The line $x = 0$ is a vertical asymptote since the denominator is 0 and the numerator is not 0 for $x = 0$.

Oblique asymptotes: Some graphs have asymptotes that are neither vertical nor horizontal. These are called **oblique asymptotes.** If x is a large positive number, then $5{,}000/x$ is very small and

$$\overline{C}(x) = \frac{5{,}000}{x} + 0.5x \approx 0.5x$$

That is,

$$\lim_{x \to \infty} \left[\overline{C}(x) - 0.5x\right] = \lim_{x \to \infty} \frac{5{,}000}{x} = 0$$

This implies that the graph of $y = \overline{C}(x)$ approaches the line $y = 0.5x$ as x approaches ∞. This line is an oblique asymptote for the graph of $y = \overline{C}(x)$.

Step 2.   *Analyze* $\overline{C}'(x)$

$$\overline{C}'(x) = -\frac{5{,}000}{x^2} + 0.5$$

$$= \frac{0.5x^2 - 5{,}000}{x^2}$$

$$= \frac{0.5(x - 100)(x + 100)}{x^2}$$

Critical value for $\overline{C}(x)$:   100

Partition numbers for $\overline{C}'(x)$:   0 and 100

Sign chart for $\overline{C}'(x)$:

Test Numbers	
x	$\overline{C}'(x)$
50	-1.5 $(-)$
125	0.18 $(+)$

Thus, $\overline{C}(x)$ is decreasing on $(0, 100)$, increasing on $(100, \infty)$, and has a local minimum at $x = 100$.

Step 3.   *Analyze* $\overline{C}''(x)$: $\overline{C}''(x) = \dfrac{10{,}000}{x^3}$

$\overline{C}''(x)$ is positive for all positive x; therefore, the graph of $y = \overline{C}(x)$ is concave upward on $(0, \infty)$.

Step 4. *Sketch the graph of $\overline{C}$.* The graph of $\overline{C}$ is shown in Figure 5.

FIGURE 5

The marginal cost function is $C'(x) = x$. The graph of this linear function is also shown in Figure 5.

The graph in Figure 5 illustrates an important principle in economics:

The minimum average cost occurs when the average cost is equal to the marginal cost.

Matched Problem 6 Given the cost function $C(x) = 1{,}600 + 0.25x^2$, where x is the number of items produced,

(A) Use the graphing strategy to analyze the graph of the average cost function. State all the pertinent information and sketch the graph of the average cost function. Find the marginal cost function and graph it on the same set of coordinate axes. Include any oblique asymptotes.

(B) Find the minimum average cost.

Answers to Matched Problems

1. $p(x) = -4x^4\left(1 - \dfrac{1}{2x} - \dfrac{3}{4x^3}\right) \approx -4x^4$ for x large (in absolute value);

$\lim\limits_{x \to -\infty} p(x) = -\infty$, $\lim\limits_{x \to \infty} p(x) = -\infty$

2. (A) $y = 0$ (B) No horizontal asymptote   (C) $y = \frac{1}{2}$

3. $x = 1$

4. Domain: $(-\infty, \infty)$
y intercept: $f(0) = 0$; x intercepts: $-4, 0$
Asymptotes: No horizontal or vertical asymptotes
Decreasing on $(-\infty, -3)$; increasing on $(-3, \infty)$; local minimum at $x = -3$
Concave upward on $(-\infty, -2)$ and $(0, \infty)$; concave downward on $(-2, 0)$
Inflection points at $x = -2$ and $x = 0$

x	$f(x)$
-4	0
-3	-27
-2	-16
0	0

5. Domain: All real x, except $x = 1$
   y intercept: $f(0) = 0$; x intercept: 0
   Horizontal asymptote: $y = -2$;   vertical asymptote: $x = 1$
   Increasing on $(-\infty, 1)$ and $(1, \infty)$
   Concave upward on $(-\infty, 1)$; concave downward on $(1, \infty)$

x	$f(x)$
-1	-1
0	0
$\frac{1}{2}$	2
$\frac{3}{2}$	-6
2	-4
5	$-\frac{5}{2}$

6. (A) Domain: $(0, \infty)$
   Intercepts: None
   Vertical asymptote: $x = 0$;   oblique asymptote: $y = 0.25x$
   Decreasing on $(0, 80)$; increasing on $(80, \infty)$; local minimum at $x = 80$
   Concave upward on $(0, \infty)$

(B) Minimum average cost is 40 at $x = 80$.

Exercise 4-3

A *Problems 1–14 refer to the graph of $y = f(x)$ shown in the figure below.*

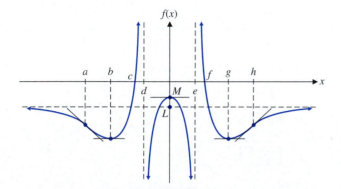

1. Identify the intervals over which $f'(x) < 0$.

2. Identify the intervals over which $f'(x) > 0$.

3. Identify the intervals over which $f(x)$ is increasing.

4. Identify the intervals over which $f(x)$ is decreasing.

5. Identify the x coordinate(s) of the point(s) where $f(x)$ has a local maximum.

6. Identify the x coordinate(s) of the point(s) where $f(x)$ has a local minimum.

7. Identify the intervals over which $f''(x) < 0$.

8. Identify the intervals over which $f''(x) > 0$.

9. Identify the intervals over which the graph of f is concave upward.

10. Identify the intervals over which the graph of f is concave downward.

11. Identify the x coordinate(s) of the inflection point(s).

12. Identify the horizontal asymptote(s).

13. Identify the vertical asymptote(s).

14. Identify the x and y intercepts.

In Problems 15 – 24, evaluate the indicated limit. Use $-\infty$ or ∞ where appropriate.

15. $\lim\limits_{x \to \infty} (-5x^3 + 7x - 9)$ **16.** $\lim\limits_{x \to \infty} (4x^3 + 5x^2 + 3)$

17. $\lim\limits_{x \to -\infty} (5x^4 + 2x^3 + 7)$ **18.** $\lim\limits_{x \to -\infty} (3x^6 - 2x^5 + 7x)$

19. $\lim\limits_{x\to\infty}\dfrac{2-3x}{1+2x}$

20. $\lim\limits_{x\to\infty}\dfrac{2x^2+5}{5x^2-2}$

21. $\lim\limits_{x\to\infty}\dfrac{3x^3+5x}{9x^2-4}$

22. $\lim\limits_{x\to\infty}\dfrac{2x-7}{5x^2+2}$

23. $\lim\limits_{x\to-\infty}\dfrac{2x^2-5x}{3x^3+7}$

24. $\lim\limits_{x\to-\infty}\dfrac{3x^3-5}{2x^2+9}$

In Problems 25–36, find all horizontal and vertical asymptotes.

25. $f(x)=\dfrac{3x-4}{x}$

26. $f(x)=\dfrac{3x}{x-4}$

27. $f(x)=\dfrac{x^2-4}{x^2+4}$

28. $f(x)=\dfrac{x^2+9}{x^2-9}$

29. $f(x)=\dfrac{x}{x^2-1}$

30. $f(x)=\dfrac{x^3}{x^2+1}$

31. $f(x)=\dfrac{x^3}{x^2-16}$

32. $f(x)=\dfrac{3x}{x^2+3}$

33. $f(x)=\dfrac{2x^2+x-3}{x^2-2x+1}$

34. $f(x)=\dfrac{5x^2-2x+1}{2x^2-5x-3}$

35. $f(x)=\dfrac{3x^2+2x+5}{2x^2+3x-20}$

36. $f(x)=\dfrac{2x^2+x-6}{x^2+4x+4}$

B *In Problems 37–46, summarize the pertinent information obtained by applying the graphing strategy and sketch the graph of $y=f(x)$.*

37. $f(x)=x^3-6x^2$

38. $f(x)=3x^2-x^3$

39. $f(x)=(x+4)(x-2)^2$

40. $f(x)=(2-x)(x+1)^2$

41. $f(x)=8x^3-2x^4$

42. $f(x)=x^4-4x^3$

43. $f(x)=\dfrac{x+3}{x-3}$

44. $f(x)=\dfrac{2x-4}{x+2}$

45. $f(x)=\dfrac{x}{x-2}$

46. $f(x)=\dfrac{2+x}{3-x}$

In Problems 47–54, use the given information to sketch the graph of f. Assume that f is continuous on its domain and that all intercepts are included in the table of values.

47. Domain: All real x; $\lim\limits_{x\to\pm\infty}f(x)=2$

x	-4	-2	0	2	4
$f(x)$	0	-2	0	-2	0

48. Domain: All real x; $\lim\limits_{x\to-\infty}f(x)=-3$; $\lim\limits_{x\to\infty}f(x)=3$

x	-2	-1	0	1	2
$f(x)$	0	2	0	-2	0

49. Domain: All real x, except $x=-2$; $\lim\limits_{x\to-2^-}f(x)=\infty$; $\lim\limits_{x\to-2^+}f(x)=-\infty$; $\lim\limits_{x\to\infty}f(x)=1$

x	-4	0	4	6
$f(x)$	0	0	3	2

50. Domain: All real x, except $x=1$; $\lim\limits_{x\to1^-}f(x)=\infty$; $\lim\limits_{x\to1^+}f(x)=\infty$; $\lim\limits_{x\to\infty}f(x)=-2$

x	-4	-2	0	2
$f(x)$	0	-2	0	0

51. Domain: All real x, except $x = -1$;
$f(-3) = 2, f(-2) = 3, f(0) = -1, f(1) = 0$;
$f'(x) > 0$ on $(-\infty, -1)$ and $(-1, \infty)$;
$f''(x) > 0$ on $(-\infty, -1)$; $f''(x) < 0$ on $(-1, \infty)$;
vertical asymptote: $x = -1$;
horizontal asymptote: $y = 1$

52. Domain: All real x, except $x = 1$;
$f(0) = -2, f(2) = 0$;
$f'(x) < 0$ on $(-\infty, 1)$ and $(1, \infty)$;
$f''(x) < 0$ on $(-\infty, 1)$;
$f''(x) > 0$ on $(1, \infty)$;
vertical asymptote: $x = 1$;
horizontal asymptote: $y = -1$

53. Domain: All real x, except $x = -2$ and $x = 2$;
$f(-3) = -1, f(0) = 0, f(3) = 1$;
$f'(x) < 0$ on $(-\infty, -2)$ and $(2, \infty)$;
$f'(x) > 0$ on $(-2, 2)$;
$f''(x) < 0$ on $(-\infty, -2)$ and $(-2, 0)$;
$f''(x) > 0$ on $(0, 2)$ and $(2, \infty)$;
vertical asymptotes: $x = -2$ and $x = 2$;
horizontal asymptote: $y = 0$

54. Domain: All real x, except $x = -1$ and $x = 1$;
$f(-2) = 1, f(0) = 0, f(2) = 1$;
$f'(x) > 0$ on $(-\infty, -1)$ and $(0, 1)$;
$f'(x) < 0$ on $(-1, 0)$ and $(1, \infty)$;
$f''(x) > 0$ on $(-\infty, -1)$, $(-1, 1)$, and $(1, \infty)$;
vertical asymptotes: $x = -1$ and $x = 1$;
horizontal asymptote: $y = 0$

55. Theorem 2 states that

$$\lim_{x \to \infty} (a_n x^n + a_{n-1} x^{n-1} + \cdots + a_0) = \pm\infty$$

What conditions must n and a_n satisfy for the limit to be ∞? For the limit to be $-\infty$?

56. Theorem 2 also states that

$$\lim_{x \to -\infty} (a_n x^n + a_{n-1} x^{n-1} + \cdots + a_0) = \pm\infty$$

What conditions must n and a_n satisfy for the limit to be ∞? For the limit to be $-\infty$?

57. Let $p(x) = x^3 - 2x^2$.

(A) Find $\lim_{x \to \infty} p'(x)$ and $\lim_{x \to \infty} p''(x)$. Describe the shape of the graph of $y = p(x)$ for large positive values of x.

(B) Find $\lim_{x \to -\infty} p'(x)$ and $\lim_{x \to -\infty} p''(x)$. Describe the shape of the graph of $y = p(x)$ for large (in absolute value) negative values of x.

58. Repeat Problem 57 for $p(x) = x^4 - 2x^3$.

C *In Problems 59 and 60, show that the line $y = x$ is an oblique asymptote for the graph of $y = f(x)$, summarize the pertinent information obtained by applying the graphing strategy, and sketch the graph of $y = f(x)$.*

59. $f(x) = x + \dfrac{1}{x}$ **60.** $f(x) = x - \dfrac{1}{x}$

In Problems 61–72, summarize the pertinent information obtained by applying the graphing strategy and sketch the graph of $y = f(x)$.

61. $f(x) = x^3 - x$

62. $f(x) = x^3 + x$

63. $f(x) = (x^2 + 3)(9 - x^2)$

64. $f(x) = (x^2 + 3)(x^2 - 1)$

65. $f(x) = (x^2 - 4)^2$

66. $f(x) = (x^2 - 1)(x^2 - 5)$

67. $f(x) = 2x^6 - 3x^5$

68. $f(x) = 3x^5 - 5x^4$

69. $f(x) = \dfrac{x}{x^2 - 4}$

70. $f(x) = \dfrac{1}{x^2 - 4}$

71. $f(x) = \dfrac{1}{1 + x^2}$

72. $f(x) = \dfrac{x^2}{1 + x^2}$

In Problems 73–80, apply steps 1–3 of the graphing strategy to $f(x)$. Use a graphing utility to approximate (to two decimal places) x intercepts, critical values, and x coordinates of inflection points. Summarize all the pertinent information.

73. $f(x) = x^4 - 5x^3 + 3x^2 + 8x - 5$

74. $f(x) = x^4 + 2x^3 - 5x^2 - 4x + 4$

75. $f(x) = x^4 - 21x^3 + 100x^2 + 20x + 100$

76. $f(x) = x^4 - 12x^3 + 28x^2 + 76x - 50$

77. $f(x) = -x^4 - x^3 + 2x^2 - 2x + 3$

78. $f(x) = -x^4 + x^3 + x^2 + 6$

79. $f(x) = 0.1x^5 + 0.3x^4 - 4x^3 - 5x^2 + 40x + 30$

80. $f(x) = x^5 + 4x^4 - 7x^3 - 20x^2 + 20x - 20$

Applications

Business & Economics

81. *Revenue.* The marketing research department for a computer company used a large city to test market their new product. They found that the relationship between price p (dollars per unit) and the demand x (units sold per week) was given approximately by

$$p = 1,296 - 0.12x^2 \qquad 0 \leq x \leq 80$$

Thus, the weekly revenue can be approximated by

$$R(x) = xp = 1,296x - 0.12x^3 \qquad 0 \leq x \leq 80$$

Graph the revenue function R.

82. *Profit.* Suppose that the cost function $C(x)$ (in dollars) for the company in Problem 81 is

$$C(x) = 830 + 396x$$

(A) Write an equation for the profit $P(x)$.

(B) Graph the profit function P.

83. *Pollution.* In Silicon Valley (California), a number of computer-related manufacturing firms were found to be contaminating underground water supplies with toxic chemicals stored in leaking underground containers. A water quality control agency ordered the companies to take immediate corrective action and to contribute to a monetary pool for testing and cleanup of the underground contamination. Suppose that the required monetary pool (in millions of dollars) for the testing and cleanup is estimated to be given by

$$P(x) = \frac{2x}{1 - x} \qquad 0 \leq x < 1$$

where x is the percentage (expressed as a decimal fraction) of the total contaminant removed.

(A) Where is $P(x)$ increasing? Decreasing?

(B) Where is the graph of P concave upward? Downward?

(C) Find any horizontal and vertical asymptotes.

(D) Find the x and y intercepts.

(E) Sketch a graph of P.

84. *Employee training.* A company producing computer components has established that on the average a new employee can assemble $N(t)$ components per day after t days of on-the-job training, as given by

$$N(t) = \frac{100t}{t + 9} \qquad t \geq 0$$

(A) Where is $N(t)$ increasing? Decreasing?

(B) Where is the graph of N concave upward? Downward?

(C) Find any horizontal and vertical asymptotes.

(D) Find the intercepts.

(E) Sketch a graph of N.

85. *Replacement time.* An office copier has an initial price of $3,200. A maintenance/service contract costs $300 for the first year and increases $100 per year thereafter. It can be shown that the total cost of the copier (in dollars) after n years is given by

$$C(n) = 3,200 + 250n + 50n^2 \qquad n \geq 1$$

(A) Write an expression for the average cost per year, $\overline{C}(n)$, for n years.

(B) Graph the average cost function found in part (A).

(C) When is the average cost per year minimum? (This is frequently referred to as the **replacement time** for this piece of equipment.)

86. *Construction costs.* The management of a manufacturing plant wishes to add a fenced-in rectangular storage yard of 20,000 square feet, using the plant building as one side of the yard (see the figure). If x is the distance (in feet) from the building to the fence parallel to the building, show that the length of the fence required for the yard is given by

$$L(x) = 2x + \frac{20,000}{x} \qquad x > 0$$

Figure for 86

(A) Graph L.

(B) What are the dimensions of the rectangle requiring the least amount of fencing?

87. *Average and marginal costs.* The total daily cost (in dollars) of producing x park benches is given by

$$C(x) = 1,000 + 5x + 0.1x^2$$

(A) Sketch the graphs of the average cost function and the marginal cost function on the same set of coordinate axes. Include any oblique asymptotes.

(B) Find the minimum average cost.

88. *Average and marginal costs.* The total daily cost (in dollars) of producing x picnic tables is given by

$$C(x) = 500 + 2x + 0.2x^2$$

(A) Sketch the graphs of the average cost function and the marginal cost function on the same set of coordinate axes. Include any oblique asymptotes.

(B) Find the minimum average cost.

89. *Minimizing average costs.* The data in the table give the total daily costs y (in dollars) of producing x pepperoni pizzas at various production levels.

NUMBER OF PIZZAS x	TOTAL COSTS y
50	395
100	475
150	640
200	910
250	1,140
300	1,450

(A) Enter the data in a graphing utility and find a quadratic regression equation for the total costs.

(B) Use the regression equation from part (A) to find the minimum average cost (to the nearest cent) and the corresponding production level (to the nearest integer).

90. *Minimizing average costs.* The data in the table give the total daily costs y (in dollars) of producing x deluxe pizzas at various production levels.

NUMBER OF PIZZAS x	TOTAL COSTS y
50	595
100	755
150	1,110
200	1,380
250	1,875
300	2,410

(A) Enter the data in a graphing utility and find a quadratic regression equation for the total costs.

(B) Use the regression equation from part (A) to find the minimum average cost (to the nearest cent) and the corresponding production level (to the nearest integer).

Life Sciences

91. *Medicine.* A drug is injected into the bloodstream of a patient through her right arm. The concentration of the drug in the bloodstream of the left arm t hours after the injection is given by

$$C(t) = \frac{0.14t}{t^2 + 1}$$

Graph C.

92. *Physiology.* In a study on the speed of muscle contraction in frogs under various loads, researchers W. O. Fems and J. Marsh found that the speed of contraction decreases with increasing loads. More precisely, they found that the relationship between speed of contraction S (in centimeters per second) and load w (in grams) is given approximately by

$$S(w) = \frac{26 + 0.06w}{w} \qquad w \geq 5$$

Graph S.

Social Sciences

93. *Psychology: retention.* An experiment on retention is conducted in a psychology class. Each student in the class is given one day to memorize the same list of 30 special characters. The lists are turned in at the end of the day, and for each succeeding day for 30 days each student is asked to turn in a list of as many of the symbols as can be recalled. Averages are taken, and it is found that

$$N(t) = \frac{5t + 20}{t} \qquad t \geq 1$$

provides a good approximation of the average number of symbols $N(t)$ retained after t days. Graph N.

Section 4-4 Optimization; Absolute Maxima and Minima

- ❑ ABSOLUTE MAXIMA AND MINIMA
- ❑ APPLICATIONS

We are now ready to consider one of the most important applications of the derivative, namely, the use of derivatives to find the *absolute maximum* or *minimum* value of a function. As we mentioned earlier, an economist may be interested in the price or production level of a commodity that will bring a maximum profit; a doctor may be interested in the time it takes for a drug to reach its maximum concentration in the bloodstream after an injection; and a city planner might be interested in the location of heavy industry in a city to produce minimum pollution in residential and business areas. Before we launch an attack on problems of this type, which are called *optimization problems,* we have to say a few words about the procedures needed to find absolute maximum and absolute minimum values of functions. We have most of the tools we need from the previous sections.

❑ ABSOLUTE MAXIMA AND MINIMA

First, what do we mean by *absolute maximum* and *absolute minimum*? We say that $f(c)$ is an **absolute maximum** of f if

$$f(c) \geq f(x)$$

for all x in the domain of f. Similarly, $f(c)$ is called an **absolute minimum** of f if

$$f(c) \leq f(x)$$

for all x in the domain of f. Figure 1 illustrates some typical examples.

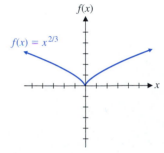

(A) No absolute maximum or minimum
One local maximum at $x = -2$
One local minimum at $x = 2$

(B) Absolute maximum at $x = 0$
No absolute minimum

(C) Absolute minimum at $x = 0$
No absolute maximum

FIGURE 1

Functions f, g, and h, along with their graphs, are shown in Figure 2.

(A) $f(x) = 1 - |x|$ (B) $g(x) = x - \dfrac{1}{x}$ (C) $h(x) = x - \dfrac{|x|}{x}$

FIGURE 2

(A) Which of these functions are continuous on $[-1, 1]$?

(B) Find the absolute maximum and the absolute minimum of each function on $[-1, 1]$, if they exist, and the corresponding values of x that produce these absolute extrema.

(C) Suppose that a function p is continuous on $[-1, 1]$ and satisfies $p(-1) = 0$ and $p(1) = 0$. Sketch a possible graph for p. Does the function you graphed have an absolute maximum? An absolute minimum? Can you modify your sketch so that p does not have an absolute maximum or an absolute minimum on $[-1, 1]$?

In many practical problems, the domain of a function is restricted because of practical or physical considerations. If the domain is restricted to some closed interval, as is often the case, then Theorem 1 can be proved.

THEOREM 1   **Extreme Value Theorem**

A function f that is continuous on a closed interval $[a, b]$ (see Section 4-1) has both an absolute maximum value and an absolute minimum value on that interval.

It is important to understand that the absolute maximum and minimum values depend on both the function f and the interval $[a, b]$. Figure 3 illustrates four cases.

(A) $[a, b] = [2, 12]$ (B) $[a, b] = [4, 10]$ (C) $[a, b] = [4, 8]$ (D) $[a, b] = [3, 11]$

FIGURE 3 Absolute extrema for $f(x) = x^3 - 21x^2 + 135x - 170$ for various closed intervals

In all four cases illustrated in Figure 3, the absolute maximum value and absolute minimum value both occur at a critical value or an endpoint. Both the absolute maximum value and the absolute minimum value are unique, but each can occur at more than one point in the interval (Fig. 3D). In general,

Absolute extrema (if they exist) must always occur at critical values or at endpoints.

Thus, to find the absolute maximum or minimum value of a continuous function on a closed interval, we simply identify the endpoints and the critical values in the interval, evaluate the function at each, and then choose the largest and smallest values out of this group.

www

> ### Steps in Finding Absolute Maximum and Minimum Values of a Continuous Function f on a Closed Interval $[a, b]$
>
> **Step 1.** Check to make certain that f is continuous over $[a, b]$.
> **Step 2.** Find the critical values in the interval (a, b).
> **Step 3.** Evaluate f at the endpoints a and b and at the critical values found in step 2.
> **Step 4.** The absolute maximum $f(x)$ on $[a, b]$ is the largest of the values found in step 3.
> **Step 5.** The absolute minimum $f(x)$ on $[a, b]$ is the smallest of the values found in step 3.

Example 1 ⇨ **Finding Absolute Extrema**   Find the absolute maximum and absolute minimum values of

$$f(x) = x^3 + 3x^2 - 9x - 7$$

on each of the following intervals:

(A) $[-6, 4]$ (B) $[-4, 2]$ (C) $[-2, 2]$

SOLUTION   (A) The function is continuous for all values of x.

$$f'(x) = 3x^2 + 6x - 9 = 3(x - 1)(x + 3)$$

Thus, $x = -3$ and $x = 1$ are critical values in the interval $(-6, 4)$. Evaluate f at the endpoints and critical values $(-6, -3, 1, \text{ and } 4)$, and choose the maximum and minimum from these:

$$f(-6) = -61 \qquad \text{Absolute minimum}$$
$$f(-3) = 20$$
$$f(1) = -12$$
$$f(4) = 69 \qquad \text{Absolute maximum}$$

(B) Interval: $[-4, 2]$

x	$f(x)$	
-4	13	
-3	20	Absolute maximum
1	-12	Absolute minimum
2	-5	

(C) Interval: $[-2, 2]$

x	$f(x)$	
-2	15	Absolute maximum
1	-12	Absolute minimum
2	-5	

The critical value $x = -3$ is not included in this table, because it is not in the interval $[-2, 2]$.

Matched Problem 1 ✏ Find the absolute maximum and absolute minimum values of

$$f(x) = x^3 - 12x$$

on each of the following intervals:

(A) $[-5, 5]$ (B) $[-3, 3]$ (C) $[-3, 1]$

Now, suppose that we want to find the absolute maximum or minimum value of a function that is continuous on an interval that is not closed. Since Theorem 1 no longer applies, we cannot be certain that the absolute maximum or minimum value exists. Figure 4 on the next page illustrates several ways that functions can fail to have absolute extrema.

In general, the best procedure to follow when the interval is not a closed interval (that is, not of the form $[a, b]$) is to sketch the graph of the function. However, one special case that occurs frequently in applications can be analyzed without drawing a graph. It often happens that f is continuous on an interval I and has only one critical value c in the interval I (here, I can be any

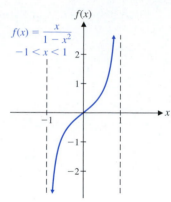

(A) No absolute extrema on $(-\infty, \infty)$:
$-1 < f(x) < 1$ for all x
$[f(x) \neq 1$ or -1 for any $x]$

(B) No absolute extrema on $(1, 2)$:
$3 < f(x) < 5$ for $x \in (1, 2)$
$[f(x) \neq 3$ or 5 for any $x \in (1, 2)]$

(C) No absolute extrema on $(-1, 1)$:
Graph has vertical asymptotes
at $x = -1$ and $x = 1$

FIGURE 4 Functions with no absolute extrema

type of interval—open, closed, or half-closed). If this is the case and if $f''(c)$ exists, we have the second-derivative test for absolute extrema given in the box below.

Second-Derivative Test for Absolute Maximum and Minimum When f Is Continuous on an Interval I and c Is the Only Critical Value in I

$f'(c)$	$f''(c)$	$f(c)$	EXAMPLE
0	+	Absolute minimum	
0	−	Absolute maximum	
0	0	Test fails	

Example 2 ⇔ **Finding an Absolute Extremum on an Open Interval** Find the absolute minimum value of

$$f(x) = x + \frac{4}{x}$$

on the interval $(0, \infty)$.

SOLUTION $\qquad f'(x) = 1 - \frac{4}{x^2} = \frac{x^2 - 4}{x^2} = \frac{(x-2)(x+2)}{x^2} \qquad f''(x) = \frac{8}{x^3}$

The only critical value in the interval $(0, \infty)$ is $x = 2$. Since $f''(2) = 1 > 0$, $f(2) = 4$ is the absolute minimum value of f on $(0, \infty)$. ∎

Matched Problem 2 ⇔ Find the absolute maximum value of

$$f(x) = 12 - x - \frac{9}{x}$$

on the interval $(0, \infty)$.

■

www

□   APPLICATIONS

Now we want to solve some applied problems that involve absolute extrema. Before beginning, we outline in the next box the steps to follow in solving this type of problem. The first step is the most difficult one. The techniques used to solve optimization problems are best illustrated through a series of examples.

Strategy for Solving Applied Optimization Problems

Step 1.   Introduce variables and a function f, including the domain I of f, and then construct a mathematical model of the form

Maximize (or minimize) $f(x)$ on the interval I

Step 2.   Find the absolute maximum (or minimum) value of $f(x)$ on the interval I and the value(s) of x where this occurs.

Step 3.   Use the solution to the mathematical model to answer the questions asked in the problem.

Example 3 ⇔ **Maximizing Revenue and Profit**   A company manufactures and sells x transistor radios per week. If the weekly cost and price–demand equations are

$$C(x) = 5,000 + 2x$$
$$p = 10 - 0.001x \qquad 0 \leq x \leq 10,000$$

find the following for each week:

(A) The maximum revenue

(B) The maximum profit, the production level that will realize the maximum profit, and the price that the company should charge for each radio to realize the maximum profit

SOLUTION   (A) The revenue received for selling x radios at $\$p$ per radio is

$$
\begin{aligned}
R(x) &= xp \\
&= x(10 - 0.001x) \\
&= 10x - 0.001x^2
\end{aligned}
$$

Thus, the mathematical model is

Maximize $R(x) = 10x - 0.001x^2 \qquad 0 \leq x \leq 10,000$
$$R'(x) = 10 - 0.002x$$
$$10 - 0.002x = 0$$
$$x = 5,000 \qquad \text{Only critical value}$$

Use the second-derivative test for absolute extrema:

$$R''(x) = -0.002 < 0 \qquad \text{for all } x$$

Thus, the maximum revenue is

$$\text{Max } R(x) = R(5,000) = \$25,000$$

(B) Profit = revenue − cost

$$
\begin{aligned}
P(x) &= R(x) - C(x)\\
&= 10x - 0.001x^2 - 5,000 - 2x\\
&= 8x - 0.001x^2 - 5,000
\end{aligned}
$$

The mathematical model is

$$
\begin{aligned}
\text{Maximize } P(x) &= 8x - 0.001x^2 - 5,000 \qquad 0 \leqslant x \leqslant 10,000\\
P'(x) &= 8 - 0.002x\\
8 - 0.002x &= 0\\
x &= 4,000\\
P''(x) &= -0.002 < 0 \qquad \text{for all } x
\end{aligned}
$$

Since $x = 4,000$ is the only critical value and $P''(x) < 0$,

$$\text{Max } P(x) = P(4,000) = \$11,000$$

Using the price–demand equation with $x = 4,000$, we find that

$$p = 10 - 0.001(4,000) = \$6$$

Thus, a maximum profit of \$11,000 per week is realized when 4,000 radios are produced weekly and sold for \$6 each. Notice that this is not the same level of production that produces the maximum revenue. ■

All the results in Example 3 are illustrated in Figure 5. We also note that profit is maximum when

$$P'(x) = R'(x) - C'(x) = 0$$

that is, when the marginal revenue is equal to the marginal cost (the rate of increase in revenue is the same as the rate of increase in cost at the 4,000 output level—notice that the slopes of the two curves are the same at this point).

Matched Problem 3 ⇔ Repeat Example 3 for

$$C(x) = 90,000 + 30x$$

$$p = 300 - \frac{x}{30} \qquad 0 \leqslant x \leqslant 9,000$$

■

Example 4 ⇔ **Maximizing Profit** The government has decided to tax the company in Example 3 \$2 for each radio produced. Taking into account this additional cost, how many radios should the company manufacture each week to maximize its weekly profit? What is the maximum weekly profit? How much should the company charge for the radios to realize the maximum weekly profit?

R C

30,000

Cost and revenue (dollars)

Revenue

20,000

Profit

Maximum
profit

Cost

Loss

10,000

Profit

Maximum
revenue

Minimum
cost

Loss

0   5,000   10,000   → x

Production (number of radios per week)

FIGURE 5

SOLUTION   The tax of $2 per unit changes the company's cost equation:

$$C(x) = \text{original cost} + \text{tax}$$
$$= 5{,}000 + 2x + 2x$$
$$= 5{,}000 + 4x$$

The new profit function is

$$P(x) = R(x) - C(x)$$
$$= 10x - 0.001x^2 - 5{,}000 - 4x$$
$$= 6x - 0.001x^2 - 5{,}000$$

Thus, we must solve the following:

$$\text{Maximize } P(x) = 6x - 0.001x^2 - 5{,}000 \qquad 0 \le x \le 10{,}000$$
$$P'(x) = 6 - 0.002x$$
$$6 - 0.002x = 0$$
$$x = 3{,}000$$
$$P''(x) = -0.002 < 0 \qquad \text{for all } x$$
$$\text{Max } P(x) = P(3{,}000) = \$4{,}000$$

Using the price–demand equation with $x = 3{,}000$, we find that

$$p = 10 - 0.001(3{,}000) = \$7$$

Thus, the company's maximum profit is $4,000 when 3,000 radios are produced and sold weekly at a price of $7.

Even though the tax caused the company's cost to increase by $2 per radio, the price that the company should charge to maximize its profit increases by only $1. The company must absorb the other $1 with a resulting decrease of $7,000 in maximum profit.

∎

Matched Problem 4 Repeat Example 4 if

$$C(x) = 90,000 + 30x$$

$$p = 300 - \frac{x}{30} \qquad 0 \leq x \leq 9,000$$

and the government decides to tax the company $20 for each unit produced. Compare the results with the results in Matched Problem 3B. ∎

Example 5 **Maximizing Yield**   A walnut grower estimates from past records that if 20 trees are planted per acre, each tree will average 60 pounds of nuts per year. If for each additional tree planted per acre (up to 15) the average yield per tree drops 2 pounds, how many trees should be planted to maximize the yield per acre? What is the maximum yield?

SOLUTION   Let x be the number of additional trees planted per acre. Then

$$20 + x = \text{total number of trees per acre}$$
$$60 - 2x = \text{yield per tree}$$
$$\text{yield per acre} = (\text{total number of trees per acre})(\text{yield per tree})$$
$$Y(x) = (20 + x)(60 - 2x)$$
$$= 1,200 + 20x - 2x^2 \qquad 0 \leq x \leq 15$$

Thus, we must solve the following:

$$\text{Maximize } Y(x) = 1,200 + 20x - 2x^2 \qquad 0 \leq x \leq 15$$
$$Y'(x) = 20 - 4x$$
$$20 - 4x = 0$$
$$x = 5$$
$$Y''(x) = -4 < 0 \qquad \text{for all } x$$

Hence,

$$\text{Max } Y(x) = Y(5) = 1{,}250 \text{ pounds per acre}$$

Thus, a maximum yield of 1,250 pounds of nuts per acre is realized if 25 trees are planted per acre. ∎

Matched Problem 5 Repeat Example 5 starting with 30 trees per acre and a reduction of 1 pound per tree for each additional tree planted. ∎

Explore–Discuss 2

In Example 5, letting x be the number of *additional* trees planted per acre produced a simple and direct solution to the problem. However, this is not the most obvious choice for a variable. Suppose that we proceed as follows:

Let x be the total number of trees planted per acre and let y be the yield per tree. Then the yield per acre is given by xy.

(A) Find y when $x = 20$ and when $x = 21$. Find the equation of the line through these two points.

(B) Use the equation from part (A) to express the yield per acre in terms of either x or y, and use this expression to solve Example 5.

(C) Compare this method of solution to the one used in Example 5 with respect to ease of comprehension and ease of computation.

Example 6   **Maximizing Area**   A farmer wants to construct a rectangular pen next to a barn 60 feet long, using all of the barn as part of one side of the pen. Find the dimensions of the pen with the largest area that the farmer can build if

(A) 160 feet of fencing material is available.

(B) 250 feet of fencing material is available.

SOLUTION   (A) We begin by constructing and labeling the figure shown in the margin. The area of the pen is

$$A = (x + 60)y$$

Before we can maximize the area, we must determine a relationship between x and y in order to express A as a function of one variable. In this case, x and y are related to the total amount of available fencing material:

$$x + y + 60 + x + y = 160$$
$$2x + 2y = 100$$
$$y = 50 - x$$

Thus,

$$A(x) = (x + 60)(50 - x)$$

Now we need to determine the permissible values of x; that is, the domain of the function A. Since the farmer wants to use all of the barn as part of one side of the pen, x cannot be negative. Since y is the other dimension of the pen, y cannot be negative. Thus,

$$y = 50 - x \geqslant 0$$
$$50 \geqslant x$$

The domain of A is $[0, 50]$. Thus, we must solve the following:

$$\text{Maximize } A(x) = (x + 60)(50 - x) \qquad 0 \leqslant x \leqslant 50$$
$$A(x) = 3{,}000 - 10x - x^2$$
$$A'(x) = -10 - 2x$$
$$-10 - 2x = 0$$
$$x = -5$$

Since $x = -5$ is not in the interval $[0, 50]$, there are no critical values in the interval. $A(x)$ is continuous on $[0, 50]$, so the absolute maximum must occur at one of the endpoints.

$$A(0) = 3{,}000 \qquad \textit{Maximum area}$$
$$A(50) = 0$$

If $x = 0$, then $y = 50$. Thus, the dimensions of the pen with largest area are 60 feet by 50 feet.

(B) If 250 feet of fencing material is available, then

$$x + y + 60 + x + y = 250$$
$$2x + 2y = 190$$
$$y = 95 - x$$

The model becomes

$$\text{Maximize } A(x) = (x + 60)(95 - x) \qquad 0 \leq x \leq 95$$
$$A(x) = 5{,}700 + 35x - x^2$$
$$A'(x) = 35 - 2x$$
$$35 - 2x = 0$$
$$x = \tfrac{35}{2} = 17.5 \qquad \textit{The only critical value}$$
$$A''(x) = -2 < 0 \qquad \text{for all } x$$
$$\text{Max } A(x) = A(17.5) = 6{,}006.25$$
$$y = 95 - 17.5 = 77.5$$

This time, the dimensions of the pen with the largest area are 77.5 feet by 77.5 feet.

77.5 ft   60 ft   77.5 ft

Matched Problem 6 Repeat Example 6 if the barn is 80 feet long.

REMARK

A graphing utility is a convenient tool for locating critical values and solutions at endpoints. The graphs of the area functions from Example 6A and B (see Fig. 6) confirm the results we obtained in our solutions.

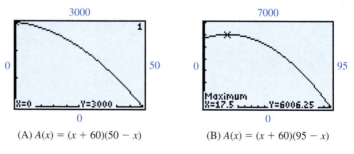

(A) $A(x) = (x + 60)(50 - x)$ (B) $A(x) = (x + 60)(95 - x)$

FIGURE 6

Example 7 **Inventory Control** A recording company anticipates that there will be a demand for 20,000 copies of a certain compact disk (CD) during the following year. It costs the company $0.50 to store a CD for one year. Each time it must make additional CDs, it costs $200 to set up the equipment. How many CDs should the company make during each production run to minimize its total storage and setup costs?

SOLUTION This type of problem is called an **inventory control problem.** One of the basic assumptions made in such problems is that the demand is uniform. For example, if there are 250 working days in a year, the daily demand would be $20{,}000 \div 250 = 80$ CDs. The company could decide to produce all 20,000

Number in storage

Average number in storage

First   Second   Third   Fourth

Production run

FIGURE 7

CDs at the beginning of the year. This would certainly minimize the setup costs but would result in very large storage costs. At the other extreme, it could produce 80 CDs each day. This would minimize the storage costs but would result in very large setup costs. Somewhere between these two extremes is the optimal solution that will minimize the total storage and setup costs. Let

x = number of CDs manufactured during each production run

y = number of production runs

It is easy to see that the total setup cost for the year is $200y$, but what is the total storage cost? If the demand is uniform, the number of CDs in storage between production runs will decrease from x to 0, and the average number in storage each day is $x/2$. This result is illustrated in Figure 7.

Since it costs $0.50 to store a CD for one year, the total storage cost is $0.5(x/2) = 0.25x$ and the total cost is

$$\text{total cost} = \text{setup cost} + \text{storage cost}$$
$$C = 200y + 0.25x$$

In order to write the total cost C as a function of one variable, we must find a relationship between x and y. If the company produces x CDs in each of y production runs, the total number of CDs produced is xy. Thus,

$$xy = 20{,}000$$
$$y = \frac{20{,}000}{x}$$

Certainly, x must be at least 1 and cannot exceed 20,000. Thus, we must solve the following:

Minimize $C(x) = 200\left(\dfrac{20{,}000}{x}\right) + 0.25x \qquad 1 \le x \le 20{,}000$

$$C(x) = \frac{4{,}000{,}000}{x} + 0.25x$$

$$C'(x) = -\frac{4{,}000{,}000}{x^2} + 0.25$$

$$-\frac{4{,}000{,}000}{x^2} + 0.25 = 0$$

$$x^2 = \frac{4{,}000{,}000}{0.25}$$

$$x^2 = 16{,}000{,}000 \qquad \textcolor{blue}{-4{,}000 \text{ is not a critical value, since}}$$
$$x = 4{,}000 \qquad\qquad \textcolor{blue}{1 \le x \le 20{,}000.}$$

$$C''(x) = \frac{8{,}000{,}000}{x^3} > 0 \qquad \text{for } x \in (1, 20{,}000)$$

Thus,

$$\text{Min } C(x) = C(4{,}000) = 2{,}000$$
$$y = \frac{20{,}000}{4{,}000} = 5$$

The company will minimize its total cost by making 4,000 CDs five times during the year.

Matched Problem 7 Repeat Example 7 if it costs $250 to set up a production run and $0.40 to store a CD for one year.

Answers to Matched Problems

1. (A) Absolute maximum: $f(5) = 65$; absolute minimum: $f(-5) = -65$
   (B) Absolute maximum: $f(-2) = 16$; absolute minimum: $f(2) = -16$
   (C) Absolute maximum: $f(-2) = 16$; absolute minimum: $f(1) = -11$

2. $f(3) = 6$

3. (A) Max $R(x) = R(4,500) = \$675,000$
   (B) Max $P(x) = P(4,050) = \$456,750$; $p = \$165$

4. Max $P(x) = P(3,750) = \$378,750$; $p = \$175$; price increases $10, profit decreases $78,000

5. Max $Y(x) = Y(15) = 2,025$ lb/acre

6. (A) 80 ft by 40 ft   (B) 82.5 ft by 82.5 ft

7. Make 5,000 CDs four times during the year

Exercise 4-4

A *Problems 1–10 refer to the graph of $y = f(x)$ shown below. Find the absolute minimum and the absolute maximum over the indicated interval.*

1. $[0, 10]$ 2. $[2, 8]$ 3. $[0, 8]$ 4. $[2, 10]$
5. $[1, 10]$ 6. $[0, 9]$ 7. $[1, 9]$ 8. $[0, 2]$
9. $[2, 5]$ 10. $[5, 8]$

In Problems 11–16, find the absolute maximum and minimum, if either exists, for each function.

11. $f(x) = x^2 - 2x + 3$ 12. $f(x) = x^2 + 4x - 3$
13. $f(x) = -x^2 - 6x + 9$ 14. $f(x) = -x^2 + 2x + 4$
15. $f(x) = x^3 + x$ 16. $f(x) = -x^3 - 2x$

B *In Problems 17–22, find the indicated extremum of each function, if it exists.*

17. Absolute maximum value of
$$f(x) = 20 - 3x - \frac{12}{x} \qquad x > 0$$

18. Absolute minimum value of
$$f(x) = 4 + x + \frac{9}{x} \qquad x > 0$$

19. Absolute minimum value of
$$f(x) = 10 + 2x + \frac{64}{x^2} \qquad x > 0$$

20. Absolute maximum value of
$$f(x) = 20 - 4x - \frac{250}{x^2} \qquad x > 0$$

21. Absolute maximum value (rounded to two decimal places) of
$$f(x) = 20 + 15x - 2x^2 - \frac{7}{x} \qquad x > 0$$

22. Absolute minimum value (rounded to two decimal places) of
$$f(x) = 20 - 7x + 4x^2 + \frac{9}{x} \qquad x > 0$$

In Problems 23–26, find the absolute maximum and minimum, if either exists, for each function on the indicated intervals.

23. $f(x) = x^3 - 6x^2 + 9x - 6$
   (A) $[-1, 5]$ (B) $[-1, 3]$ (C) $[2, 5]$

24. $f(x) = 2x^3 - 3x^2 - 12x + 24$
   (A) $[-3, 4]$ (B) $[-2, 3]$ (C) $[-2, 1]$

25. $f(x) = (x - 1)(x - 5)^3 + 1$
   (A) $[0, 3]$ (B) $[1, 7]$ (C) $[3, 6]$

26. $f(x) = x^4 - 8x^2 + 16$
   (A) $[-1, 3]$ (B) $[0, 2]$ (C) $[-3, 4]$

C *Preliminary word problems:*

27. How would you divide a 10-inch line so that the product of the two lengths is a maximum?

28. What quantity should be added to 5 and subtracted from 5 to produce the maximum product of the results?

29. Find two numbers whose difference is 30 and whose product is a minimum.

30. Find two positive numbers whose sum is 60 and whose product is a maximum.

31. Find the dimensions of a rectangle with perimeter 100 centimeters that has maximum area. Find the maximum area.

32. Find the dimensions of a rectangle of area 225 square centimeters that has the least perimeter. What is the perimeter?

Problems 33–36 refer to a rectangular area enclosed by a fence that costs $B per foot. Discuss the existence of a solution and the economical implications of each optimization problem.

33. Given a fixed area, minimize the cost of the fencing.

34. Given a fixed area, maximize the cost of the fencing.

35. Given a fixed amount to spend on fencing, maximize the enclosed area.

36. Given a fixed amount to spend on fencing, minimize the enclosed area.

Applications

Business & Economics

37. *Maximum revenue and profit.* A company manufactures and sells x television sets per month. The monthly cost and price–demand equations are

$$C(x) = 72,000 + 60x$$

$$p = 200 - \frac{x}{30} \qquad 0 \leq x \leq 6,000$$

(A) Find the maximum revenue.

(B) Find the maximum profit, the production level that will realize the maximum profit, and the price the company should charge for each television set.

(C) If the government decides to tax the company $5 for each set it produces, how many sets should the company manufacture each month to maximize its profit? What is the maximum profit? What should the company charge for each set?

38. *Maximum revenue and profit.* Repeat Problem 37 for

$$C(x) = 60,000 + 60x$$

$$p = 200 - \frac{x}{50} \qquad 0 \leq x \leq 10,000$$

39. *Maximum profit.* The table contains price–demand and total cost data for the production of radial arm saws, where p is the wholesale price (in dollars) of a saw for an annual demand of x saws and C is the total cost (in dollars) of producing x saws.

x	p	C
950	240	130,000
1,200	210	150,000
1,800	160	180,000
2,050	120	190,000

(A) Find a quadratic regression equation for the price–demand data using x as the independent variable.

(B) Find a linear regression equation for the cost data using x as the independent variable.

(C) What is the maximum profit? What is the wholesale price per saw that should be charged to realize the maximum profit? Round answers to the nearest dollar.

40. *Maximum profit.* The table contains price–demand and total cost data for the production of airbrushes, where p is the wholesale price (in dollars) of an airbrush for an annual demand of x airbrushes and C is the total cost (in dollars) of producing x airbrushes.

x	p	C
2,300	98	145,000
3,300	84	170,000
4,500	67	190,000
5,200	51	210,000

(A) Find a quadratic regression equation for the price–demand data using x as the independent variable.

(B) Find a linear regression equation for the cost data using x as the independent variable.

(C) What is the maximum profit? What is the wholesale price per airbrush that should be charged to realize the maximum profit? Round answers to the nearest dollar.

41. *Car rental.* A car rental agency rents 200 cars per day at a rate of $30 per day. For each $1 increase in rate, 5 fewer cars are rented. At what rate should the cars be rented to produce the maximum income? What is the maximum income?

42. *Rental income.* A 300-room hotel in Las Vegas is filled to capacity every night at $80 a room. For each $1 increase in rent, 3 fewer rooms are rented. If each rented room costs $10 to service per day, how much should the management charge for each room to maximize gross profit? What is the maximum gross profit?

43. *Agriculture.* A commercial cherry grower estimates from past records that if 30 trees are planted per acre, each tree will yield an average of 50 pounds of cherries per season. If for each additional tree planted per acre (up to 20), the average yield per tree is reduced by 1 pound, how many trees should be planted per acre to obtain the maximum yield per acre? What is the maximum yield?

44. *Agriculture.* A commercial pear grower must decide on the optimum time to have fruit picked and sold. If the pears are picked now, they will bring 30¢ per pound, with each tree yielding an average of 60 pounds of salable pears. If the average yield per tree increases 6 pounds per tree per week for the next 4 weeks, but the price drops 2¢ per pound per week, when should the pears be picked to realize the maximum return per tree? What is the maximum return?

45. *Manufacturing.* A candy box is to be made out of a piece of cardboard that measures 8 by 12 inches. Squares of equal size will be cut out of each corner, and then the ends and sides will be folded up to form a rectangular box. What size square should be cut from each corner to obtain a maximum volume?

46. *Packaging.* A parcel delivery service will deliver a package only if the length plus girth (distance around) does not exceed 108 inches.

Figure for 46

(A) Find the dimensions of a rectangular box with square ends that satisfies the delivery service's restriction and has maximum volume. What is the maximum volume?

(B) Find the dimensions (radius and height) of a cylindrical container that meets the delivery service's requirement and has maximum volume. What is the maximum volume?

47. *Construction costs.* A fence is to be built to enclose a rectangular area of 800 square feet. The fence along three sides is to be made of material that costs $6 per foot. The material for the fourth side costs $18 per foot. Find the dimensions of the rectangle that will allow the most economical fence to be built.

48. *Construction costs.* The owner of a retail lumber store wants to construct a fence to enclose an outdoor storage area adjacent to the store, using all of the store as part of one side of the area (see the figure). Find the dimensions that will enclose the largest area if

(A) 240 feet of fencing material is used.

(B) 400 feet of fencing material is used.

Figure for 48

49. *Inventory control.* A publishing company sells 50,000 copies of a certain book each year. It costs the company $1 to store a book for one year. Each time it must print additional copies, it costs the company $1,000 to set up the presses. How many books should the company produce during each printing in order to minimize its total storage and setup costs?

50. *Operational costs.* The cost per hour for fuel to run a train is $v^2/4$ dollars, where v is the speed of the train in miles per hour. (Note that the cost goes up as the square of the speed.) Other costs, including labor, are $300 per hour. How fast should the train travel on a 360-mile trip to minimize the total cost for the trip?

51. *Construction costs.* A freshwater pipeline is to be run from a source on the edge of a lake to a small resort community on an island 5 miles offshore, as indicated in the figure.

Figure for 51

(A) If it costs 1.4 times as much to lay the pipe in the lake as it does on land, what should x be (in miles) to minimize the total cost of the project?

(B) If it costs only 1.1 times as much to lay the pipe in the lake as it does on land, what should x be to minimize the total cost of the project? [*Note:* Compare with Problem 56.]

52. *Manufacturing costs.* A manufacturer wants to produce cans that will hold 12 ounces (approximately 22 cubic inches) in the form of a right circular cylinder. Find the dimensions (radius of an end and height) of the can that will use the smallest amount of material. Assume that the circular ends are cut out of squares, with the corner portions wasted, and the sides are made from rectangles, with no waste.

Life Sciences

53. *Bacteria control.* A recreational swimming lake is treated periodically to control harmful bacteria growth. Suppose that t days after a treatment, the concentration of bacteria per cubic centimeter is given by

$$C(t) = 30t^2 - 240t + 500 \qquad 0 \leq t \leq 8$$

How many days after a treatment will the concentration be minimal? What is the minimum concentration?

54. *Drug concentration.* The concentration $C(t)$, in milligrams per cubic centimeter, of a particular drug in a patient's bloodstream is given by

$$C(t) = \frac{0.16t}{t^2 + 4t + 4}$$

where t is the number of hours after the drug is taken. How many hours after the drug is given will the concentration be maximum? What is the maximum concentration?

55. *Laboratory management.* A laboratory uses 500 white mice each year for experimental purposes. It costs $4 to feed a mouse for one year. Each time mice are ordered from a supplier, there is a service charge of $10 for processing the order. How many mice should be ordered each time to minimize the total cost of feeding the mice and of placing the orders for the mice?

56. *Bird flights.* Some birds tend to avoid flights over large bodies of water during daylight hours. (It is speculated that more energy is required to fly over water than land, because air generally rises over land and falls over water during the day.) Suppose that an adult bird with this tendency is taken from its nesting area on the edge of a large lake to an island 5 miles offshore and is then released (see the figure).

Figure for 56

(A) If it takes 1.4 times as much energy to fly over water as land, how far up-shore (x, in miles) should the bird head to minimize the total energy expended in returning to the nesting area?

(B) If it takes only 1.1 times as much energy to fly over water as land, how far up-shore should the bird head to minimize the total energy expended in returning to the nesting area? [*Note:* Compare with Problem 51.]

57. *Botany.* If it is known from past experiments that the height (in feet) of a given plant after t months is given approximately by

$$H(t) = 4t^{1/2} - 2t \qquad 0 \leq t \leq 2$$

how long, on the average, will it take a plant to reach its maximum height? What is the maximum height?

58. *Pollution.* Two heavy industrial areas are located 10 miles apart, as indicated in the figure. If the concentration of particulate matter (in parts per million) decreases as the reciprocal of the square of the distance from the source, and area A_1 emits eight times the particulate matter as A_2, the concentration of particulate matter at any point between the two areas is given by

$$C(x) = \frac{8k}{x^2} + \frac{k}{(10 - x)^2} \qquad 0.5 \leq x \leq 9.5, \quad k > 0$$

How far from A_1 will the concentration of particulate matter between the two areas be at a minimum?

Figure for 58

Social Sciences

59. *Politics.* In a newly incorporated city, it is estimated that the voting population (in thousands) will increase according to

$$N(t) = 30 + 12t^2 - t^3 \qquad 0 \le t \le 8$$

where t is time in years. When will the rate of increase be most rapid?

60. *Learning.* A large grocery chain found that on the average, a checker can recall $P\%$ of a given price list x hours after starting work, as given approximately by

$$P(x) = 96x - 24x^2 \qquad 0 \le x \le 3$$

At what time x does the checker recall a maximum percentage? What is the maximum?

Important Terms and Symbols

4-1 *First Derivative and Graphs.* Increasing and decreasing functions; rising and falling graphs; critical values; local extremum; local maximum; local minimum; turning point; first-derivative test for local extrema

4-2 *Second Derivative and Graphs.* Concave upward; concave downward; second derivative; concavity and the second derivative; inflection point; second-derivative test for local maxima and minima; point of diminishing returns

$$f''(x); \quad \frac{d^2y}{dx^2}; \quad y''$$

4-3 *Curve Sketching Techniques: Unified and Extended.* Horizontal asymptote; limits at infinity for power functions, polynomials, and rational functions; locating vertical asymptotes; graphing strategy [analyze $f(x)$ to find domain, intercepts, horizontal and vertical asymptotes; analyze $f'(x)$ to find increasing and decreasing regions, local extrema; analyze $f''(x)$ to find concave upward and concave downward regions, inflection points]; oblique asymptote

4-4 *Optimization: Absolute Maxima and Minima.* Absolute maximum; absolute minimum; absolute extrema of a function on a closed interval; second-derivative test for absolute maximum and minimum; optimization problems; inventory control

Review Exercise

Work through all the problems in this chapter review and check your answers in the back of the book. Answers to all review problems are there along with section numbers in italics to indicate where each type of problem is discussed. Where weaknesses show up, review appropriate sections in the text.

A *Problems 1–8 refer to the graph of $y = f(x)$ below. Identify the points or intervals on the x axis that produce the indicated behavior.*

1. $f(x)$ is increasing (a, c_1) (c_3, c_6)

2. $f'(x) < 0$ (c_1, c_3) (c_6, b)

3. Graph of f is concave downward $(a, c_1)($

4. Local minima (c_3)

5. Absolute maxima (c_1) (c_6)

6. $f'(x)$ appears to be 0

7. $f'(x)$ does not exist

8. Inflection points

In Problems 9 and 10, use the given information to sketch the graph of f. Assume that f is continuous on its domain and that all intercepts are included in the given information.

9. Domain: All real x

x	-3	-2	-1	0	2	3
$f(x)$	0	3	2	0	-3	0

10. Domain: All real x;

$$f(-2) = 1, f(0) = 0, f(2) = 1;$$
$$f'(0) = 0; f'(x) < 0 \text{ on } (-\infty, 0);$$
$$f'(x) > 0 \text{ on } (0, \infty);$$
$$f''(-2) = 0, f''(2) = 0;$$
$$f''(x) < 0 \text{ on } (-\infty, -2) \text{ and } (2, \infty);$$
$$f''(x) > 0 \text{ on } (-2, 2);$$
$$\lim_{x \to -\infty} f(x) = 2; \lim_{x \to \infty} f(x) = 2$$

11. Find $f''(x)$ for $f(x) = x^4 + 5x^3$.

12. Find y'' for $y = 3x + \dfrac{4}{x}$.

B *Problems 13–16 refer to the function*

$$f(x) = x^3 - 18x^2 + 81x$$

13. Using $f(x)$,

(A) Determine the domain of f.

(B) Find any intercepts for the graph of f.

(C) Find any horizontal or vertical asymptotes for the graph of f.

14. Using $f'(x)$,

(A) Find critical values for $f(x)$.

(B) Find partition numbers for $f'(x)$.

(C) Find intervals over which $f(x)$ is increasing; decreasing.

(D) Find any local maxima and minima.

15. Using $f''(x)$,

(A) Find intervals over which the graph of f is concave upward; concave downward.

(B) Find any inflection points.

16. Use the results of Problems 13 to 15 to graph f.

Problems 17–20 refer to the function

$$y = f(x) = \frac{3x}{x + 2}$$

17. Using $f(x)$,

(A) Determine the domain of f.

(B) Find any intercepts for the graph of f.

(C) Find any horizontal or vertical asymptotes for the graph of f.

18. Using $f'(x)$,

(A) Find critical values for $f(x)$.

(B) Find partition numbers for $f'(x)$.

(C) Find intervals over which $f(x)$ is increasing; decreasing.

(D) Find any local maxima and minima

19. Using $f''(x)$,

(A) Find intervals over which the graph of f is concave upward; concave downward.

(B) Find any inflection points.

20. Use the results of Problems 17 to 19 to graph f.

21. Use the graph of $y = f'(x)$ shown below to discuss the graph of $y = f(x)$. Organize your conclusions in a table (see Example 4, Section 4-2). Sketch a possible graph for $y = f(x)$.

Figure for 21 and 22

22. Refer to the graph of $y = f'(x)$ above. Which of the following could be the graph of $y = f''(x)$?

23. Use the second-derivative test to find any local extrema for

$$f(x) = x^3 - 6x^2 - 15x + 12$$

24. Find the absolute maximum and absolute minimum, if either exists, for

$$y = f(x) = x^3 - 12x + 12 \qquad -3 \leq x \leq 5$$

25. Find the absolute minimum, if it exists, for

$$y = f(x) = x^2 + \frac{16}{x^2} \qquad x > 0$$

In Problems 26–31, evaluate the indicated limit. Use $-\infty$ or ∞ where appropriate.

26. $\lim_{x \to \infty} (-2x^3)$

27. $\lim_{x \to -\infty} (x^3 - 2x^2)$

28. $\lim_{x \to \infty} (2x^4 - 5x^3)$

29. $\lim_{x \to \infty} \frac{4x^3 + 5}{2x^3 + 2x}$

30. $\lim_{x \to \infty} \frac{4x^4 + 5}{2x^3 + 2x}$

31. $\lim_{x \to \infty} \frac{4x^3 + 5}{2x^4 + 2x}$

In Problems 32 and 33, find horizontal and vertical asymptotes, if they exist.

32. $f(x) = \frac{2x}{x^2 + 4}$

33. $f(x) = \frac{4x^3}{x^2 - 4}$

34. Let $y = f(x)$ be a polynomial function with local minima at $x = a$ and $x = b, a < b$. Must f have at least one local maximum between a and b? Justify your answer.

35. The derivative of $f(x) = x^{-1}$ is $f'(x) = -x^{-2}$. Since $f'(x) < 0$ for $x \ne 0$, is it correct to say that $f(x)$ is decreasing for all x except $x = 0$? Explain.

36. Discuss the difference between a partition number for $f'(x)$ and a critical value for $f(x)$, and illustrate with examples.

C

37. Find the absolute maximum for $f'(x)$ if

$$f(x) = 6x^2 - x^3 + 8$$

Graph f and f' on the same coordinate system for $0 \le x \le 4$.

38. Find two positive numbers whose product is 400 and whose sum is a minimum. What is the minimum sum?

39. Let $f(x) = (x - 1)^3(x + 3)$. Apply the graphing strategy discussed in Section 4-3, summarize the pertinent information, and sketch the graph.

 In Problems 40 and 41, apply steps 1–3 of the graphing strategy discussed in Section 4-3 and summarize the pertinent information. Round any approximate values to two decimal places.

40. $f(x) = x^4 + x^3 - 4x^2 - 3x + 4$

41. $f(x) = 0.25x^4 - 5x^3 + 31x^2 - 70x$

Applications

Business & Economics

42. *Price analysis.* The graph in the figure approximates the rate of change of the price of tomatoes over a 60-month period, where $p(t)$ is the price of a pound of tomatoes and t is time (in months).

(A) Write a brief verbal description of the graph of $y = p(t)$, including a discussion of local extrema and inflection points.

(B) Sketch a possible graph of $y = p(t)$.

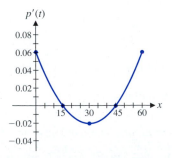

Figure for 42

43. *Maximum revenue and profit.* A company manufactures and sells x electric stoves per month. The monthly cost and price–demand equations are

$$C(x) = 350x + 50,000$$
$$p = 500 - 0.025x \qquad 0 \le x \le 20,000$$

(A) Find the maximum revenue.

(B) How many stoves should the company manufacture each month to maximize its profit? What is the maximum monthly profit? How much should the company charge for each stove?

(C) If the government decides to tax the company $20 for each stove it produces, how many stoves should the company manufacture each month to maximize its profit? What is the maximum monthly profit? How much should the company charge for each stove?

44. *Construction.* A fence is to be built to enclose a rectangular area. The fence along three sides is to be made of material that costs $5 per foot. The material for the fourth side costs $15 per foot.

(A) If the area is 5,000 square feet, find the dimensions of the rectangle that will allow the most economical fence to be built.

(B) If $3,000 is available for the fencing, find the dimensions of the rectangle that will enclose the most area.

45. *Rental income.* A 200-room hotel in Fresno is filled to capacity every night at a rate of $40 per room. For each $1 increase in the nightly rate, 4 fewer rooms are rented. If each rented room costs $8 a day to service, how much should the management charge per room in order to maximize gross profit? What is the maximum gross profit?

46. *Inventory control.* A computer store sells 7,200 boxes of floppy disks annually. It costs the store $0.20 to store a box of disks for one year. Each time it reorders disks, the store must pay a $5.00 service charge for processing the order. How many times during the year should the store order disks to minimize the total storage and reorder costs?

47. *Average cost.* The total cost of producing x garbage disposals per day is given by

$$C(x) = 4,000 + 10x + 0.1x^2$$

Find the minimum average cost. Graph the average cost and the marginal cost functions on the same coordinate system. Include any oblique asymptotes.

48. *Average cost.* The data in the table give the total daily cost y (in dollars) of producing x dozen chocolate chip cookies at various production levels.

DOZENS OF COOKIES x	TOTAL COST y
50	119
100	187
150	248
200	382
250	505
300	695

(A) Enter the data in a graphing utility and find a quadratic regression equation for the total cost.

(B) Use the regression equation from part (A) to find the minimum average cost (to the nearest cent) and the corresponding production level (to the nearest integer).

49. *Advertising.* A chain of appliance stores uses television ads to promote the sales of refrigerators. Analyzing past records produced the data in the table, where x is the number of ads placed monthly and y is the number of refrigerators sold that month.

NUMBER OF ADS x	NUMBER OF REFRIGERATORS y
10	271
20	427
25	526
30	629
45	887
48	917

(A) Enter the data in a graphing utility, set the utility to display two decimal places, and find a cubic regression equation for the number of refrigerators sold monthly as a function of the number of ads.

(B) How many ads should be placed each month to maximize the rate of change of sales with respect to the number of ads, and how many refrigerators can be expected to be sold with this number of ads? Round answers to the nearest integer.

Life Sciences

50. *Bacteria control.* If t days after a treatment, the bacteria count per cubic centimeter in a body of water is given by

$$C(t) = 20t^2 - 120t + 800 \qquad 0 \le t \le 9$$

in how many days will the count be a minimum?

Social Sciences

51. *Politics.* In a new suburb, it is estimated that the number of registered voters will grow according to

$$N = 10 + 6t^2 - t^3 \qquad 0 \le t \le 5$$

where t is time in years and N is in thousands. When will the rate of increase be maximum?

Group Activity 1   *Maximizing Profit*

A company manufactures and sells x air-conditioners per month. The monthly cost and price–demand equations are

$$C(x) = 180x + 20{,}000$$
$$p = 220 - 0.001x \qquad 0 \leqslant x \leqslant 100{,}000$$

(A) How many air conditioners should the company manufacture each month to maximize its monthly profit? What is the maximum monthly profit, and what should the company charge for each air conditioner to realize the maximum monthly profit?

(B) Repeat part (A) if the government decides to tax the company at the rate of $18 per air conditioner produced. How much revenue will the government receive from the tax on these air conditioners?

(C) Repeat part (A) if the government raises the tax to $23 per air conditioner. Discuss the effect of this tax increase on the government's tax revenue.

(D) Repeat part (A) if the government sets the tax rate at $$t$ per air conditioner. What value of t will maximize the government's tax revenue? What is the government's maximum tax revenue?

Group Activity 2   *Minimizing Construction Costs*

Two resort communities are located on separate islands in a lake, as indicated in Figure 1. A communications company wants to run underground cables from a single source on the shore to each of the islands. Since the high cost of laying cable under water is directly proportional to the amount of cable used, the company wants to position the source of the cables so that the amount of cable used is minimized.

Island 1

Island 2

5 mi

8 mi

20 mi

FIGURE 1

(A) Assume that the shoreline is straight and position a coordinate system as indicated in Figure 2. Let f be the function that represents the total length of cable used. Express f in terms of x (Fig. 2), graph f and f' (use the numerical derivative of f to graph f'), and find the value of x (in miles) that will minimize the total cable length.

FIGURE 2

(B) Repeat part (A) under the assumption that the shoreline is in the shape of the parabola $y = 0.02x^2 - 2$ (Fig. 3).

FIGURE 3

Additional Derivative Topics

INTRODUCTION

In this chapter we complete our discussion of derivatives by first looking at the differentiation of forms that involve the exponential and logarithmic functions and then considering some additional topics and applications involving all the different types of functions we have encountered thus far. You will probably find it helpful to review some of the important properties of the exponential and logarithmic functions given in Chapter 2 before proceeding further.

Section 5-1

The Constant *e* and Continuous Compound Interest

- ❑ THE CONSTANT *e*
- ❑ CONTINUOUS COMPOUND INTEREST

In Chapter 2, both the exponential function with base *e* and continuous compound interest were introduced informally. Now, with limit concepts at our disposal, we can give precise definitions of *e* and continuous compound interest.

❑ THE CONSTANT *e*

The special irrational number *e* is a particularly suitable base for both exponential and logarithmic functions. The reasons for choosing this number as a

303

base will become clear as we develop differentiation formulas for the exponential function e^x and the natural logarithmic function $\ln x$.

In precalculus treatments (Chapter 2), the number e is informally defined as an irrational number that can be approximated by the expression $[1 + (1/n)]^n$ by taking n sufficiently large. Now we will use the limit concept to formally define e as either of the following two limits:

www

The Number e

$$e = \lim_{n \to \infty} \left(1 + \frac{1}{n}\right)^n \qquad \text{or, alternatively,} \qquad e = \lim_{s \to 0}(1 + s)^{1/s}$$

$$e = 2.718\ 281\ 828\ 459\ldots$$

We will use both these limit forms. [*Note:* If $s = 1/n$, then as $n \to \infty, s \to 0$.]

The proof that the indicated limits exist and represent an irrational number between 2 and 3 is not easy and is omitted here. Many people reason (incorrectly) that the limits are 1, since "$1 + s$ approaches 1 as $s \to 0$, and 1 to any power is 1." A little experimentation with a calculator can convince you otherwise. Consider the following table of values for s and $f(s) = (1 + s)^{1/s}$ and the graph shown in Figure 1 for s close to 0.

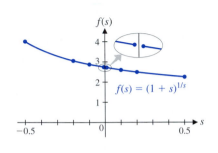

FIGURE 1

s approaches 0 from the left $\rightarrow$ 0 $\leftarrow$ s approaches 0 from the right

s	-0.5	-0.2	-0.1	$-0.01 \rightarrow 0 \leftarrow 0.01$	0.1	0.2	0.5
$(1 + s)^{1/s}$	4.0000	3.0518	2.8680	$2.7320 \rightarrow e \leftarrow 2.7048$	2.5937	2.4883	2.2500

Compute some of the table values with a calculator yourself, and also try several values of s even closer to 0. Note that the function is discontinuous at $s = 0$.

Exactly who discovered e is still being debated. It is named after the great mathematician Leonhard Euler (1707–1783), who computed e to 23 decimal places using $[1 + (1/n)]^n$.

❑ CONTINUOUS COMPOUND INTEREST

Now we will see how e appears quite naturally in the important application of compound interest. Let us start with simple interest, move on to compound interest, and then on to continuous compound interest.

If a principal P is borrowed at an annual rate of r,* then after t years at simple interest the borrower will owe the lender an amount A given by

$$A = P + Prt = P(1 + rt) \qquad \text{Simple interest} \tag{1}$$

On the other hand, if interest is compounded n times a year, the borrower will owe the lender an amount A given by

$$A = P\left(1 + \frac{r}{n}\right)^{nt} \qquad \text{Compound interest} \tag{2}$$

---

*If r is the interest rate written as a decimal, then $100r\%$ is the rate using %. For example, if $r = 0.12$, we have $100r\% = 100(0.12)\% = 12\%$. The expressions 0.12 and 12% are therefore equivalent. Unless stated otherwise, all formulas in this book use r in decimal form.

where r/n is the interest rate per compounding period and nt is the number of compounding periods. Suppose that P, r, and t in equation (2) are held fixed and n is increased. Will the amount A increase without bound, or will it tend to approach some limiting value?

Let us perform a calculator experiment before we attack the general limit problem. If $P = \$100$, $r = 0.06$, and $t = 2$ years, then

$$A = 100\left(1 + \frac{0.06}{n}\right)^{2n}$$

We compute A for several values of n in Table 1. The biggest gain appears in the first step; then the gains slow down as n increases. In fact, it appears that A might be tending to approach \$112.75 as n gets larger and larger.

www

TABLE 1

COMPOUNDING FREQUENCY	n	$A = 100\left(1 + \dfrac{0.06}{n}\right)^{2n}$
Annually	1	\$112.3600
Semiannually	2	112.5509
Quarterly	4	112.6493
Monthly	12	112.7160
Weekly	52	112.7419
Daily	365	112.7486
Hourly	8,760	112.7496

Explore–Discuss 1

(A) Suppose that \$1,000 is deposited in a savings account that earns 6% simple interest. How much will be in the account after 2 years?

(B) Suppose that \$1,000 is deposited in a savings account that earns compound interest at a rate of 6% per year. How much will be in the account after 2 years if interest is compounded annually? Semiannually? Quarterly? Weekly?

(C) How frequently must interest be compounded at the 6% rate in order to have \$1,150 in the account after 2 years?

Now we turn back to the general problem for a moment. Keeping P, r, and t fixed in equation (2), we compute the following limit and observe an interesting and useful result:

$$\lim_{n\to\infty} P\left(1 + \frac{r}{n}\right)^{nt} = P \lim_{n\to\infty}\left(1 + \frac{r}{n}\right)^{(n/r)rt}$$

Insert r/r in the exponent and let $s = r/n$. Note that $n \to \infty$ implies $s \to 0$.

$$= P \lim_{s\to 0}\left[(1 + s)^{1/s}\right]^{rt}$$

Use the limit property given in the footnote below.*

$$= P\left[\lim_{s\to 0}(1 + s)^{1/s}\right]^{rt}$$

$$\lim_{s\to 0}(1 + s)^{1/s} = e$$

$$= Pe^{rt}$$

*The following new limit property is used: If $\lim_{x\to c} f(x)$ exists, then $\lim_{x\to c}[f(x)]^p = [\lim_{x\to c} f(x)]^p$, provided that the last expression names a real number.

The resulting formula is called the **continuous compound interest formula,** a very important and widely used formula in business and economics.

> ### Continuous Compound Interest
>
> $$A = Pe^{rt}$$
>
> where
>
> P = principal
> r = annual nominal interest rate compounded continuously
> t = time in years
> A = amount at time t

Example 1 **Computing Continuously Compounded Interest**   If $100 is invested at 6% compounded continuously,* what amount will be in the account after 2 years? How much interest will be earned?

SOLUTION

$$A = Pe^{rt}$$
$$= 100e^{(0.06)(2)} \qquad \textit{6\% is equivalent to } r = 0.06.$$
$$\approx \$112.7497$$

(Compare this result with the values calculated in Table 1.) The interest earned is $112.7497 − $100 = $12.7497. ■

Matched Problem 1 What amount (to the nearest cent) will an account have after 5 years if $100 is invested at an annual nominal rate of 8% compounded annually? Semiannually? Continuously? ■

Example 2 **Graphing the Growth of an Investment**   Recently, Union Savings Bank offered a 5-year certificate of deposit (CD) that earned 5.75% compounded continuously. If $1,000 is invested in one of these CDs, graph the amount in the account relative to time for a period of 5 years.

www

SOLUTION   We want to graph

$$A = 1{,}000e^{0.0575t} \qquad 0 \le t \le 5$$

We construct a table of values using a calculator, graph the points from the table, and join the points with a smooth curve.

t	$A(\$)$
0	1,000
1	1,059
2	1,122
3	1,188
4	1,259
5	1,333

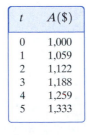

*Following common usage, we will often write "at 6% compounded continuously," understanding that this means "at an annual nominal rate of 6% compounded continuously."

Matched Problem 2 ⟿   If \$5,000 is invested in a Union Savings Bank 4-year CD that earns 5.61% compounded continuously, graph the amount in the account relative to time for a period of 4 years. ■

Example 3 ⟿   **Computing Growth Time**   How long will it take an investment of \$5,000 to grow to \$8,000 if it is invested at 5% compounded continuously?

SOLUTION   Starting with the continous compound interest formula $A = Pe^{rt}$, we must solve for *t*:

FIGURE 2
$y_1 = 5,000e^{0.05x}$
$y_2 = 8,000$

$$A = Pe^{rt}$$
$$8,000 = 5,000e^{0.05t} \quad \textcolor{blue}{\text{Divide both sides by 5,000 and reverse the equation.}}$$
$$e^{0.05t} = 1.6 \quad \textcolor{blue}{\text{Take the natural logarithm of both sides—recall}}$$
$$\ln e^{0.05t} = \ln 1.6 \quad \textcolor{blue}{\text{that } \log_b b^x = x.}$$
$$0.05t = \ln 1.6$$
$$t = \frac{\ln 1.6}{0.05}$$
$$t \approx 9.4 \text{ years}$$

    Figure 2 shows an alternative method for solving Example 3 on a graphing utility.

Matched Problem 3 ⟿   How long will it take an investment of \$10,000 to grow to \$15,000 if it is invested at 9% compounded continuously? ■

Example 4 ⟿   **Computing Doubling Time**   How long will it take money to double if it is invested at 6.5% compounded continuously?

SOLUTION   Starting with the continuous compound interest formula $A = Pe^{rt}$, we must solve for *t* given $A = 2P$ and $r = 0.065$:

FIGURE 3
$y_1 = e^{0.065x}$
$y_2 = 2$

$$2P = Pe^{0.065t} \quad \textcolor{blue}{\text{Divide both sides by } P \text{ and reverse the equation.}}$$
$$e^{0.065t} = 2 \quad \textcolor{blue}{\text{Take the natural logarithm of both sides.}}$$
$$\ln e^{0.065t} = \ln 2$$
$$0.065t = \ln 2$$
$$t = \frac{\ln 2}{0.065}$$
$$t \approx 10.66 \text{ years}$$

    Figure 3 illustrates the solution of $e^{0.065t} = 2$ from Example 4 on a graphing utility.

Matched Problem 4 ⟿   How long will it take money to triple if it is invested at 5.5% compounded continuously? ■

Explore–Discuss 2   You are considering three options for investing \$10,000: at 7% compounded annually, at 6% compounded monthly, and at 5% compounded continuously.

(A) Which option would be the best for investing \$10,000 for 8 years?

(B) How long would you need to invest your money for the third option to be the best?

Answers to Matched Problems   **1.** $146.93; $148.02; $149.18

2. $A = 5{,}000e^{0.0561t}$

t	$A(\$)$
0	5,000
1	5,289
2	5,594
3	5,916
4	6,258

3. 4.51 yr

4. 19.97 yr

Exercise 5-1

A *Use a calculator to evaluate A to the nearest cent in Problems 1 and 2.*

1. $A = \$1{,}000e^{0.1t}$ for $t = 2, 5,$ and 8

2. $A = \$5{,}000e^{0.08t}$ for $t = 1, 4,$ and 10

3. If $6,000 is invested at 10% compounded continuously, graph the amount in the account relative to time for a period of 8 years.

4. If $4,000 is invested at 8% compounded continuously, graph the amount in the account relative to time for a period of 6 years.

B *In Problems 5–10, solve for t or r to two decimal places.*

5. $2 = e^{0.06t}$

6. $2 = e^{0.03t}$

7. $3 = e^{0.1t}$

8. $3 = e^{0.25t}$

9. $2 = e^{5r}$

10. $3 = e^{10r}$

C *In Problems 11 and 12, complete each table to five decimal places using a calculator.*

11.

n	$[1 + (1/n)]^n$
10	2.593 74
100	
1,000	
10,000	
100,000	
1,000,000	
10,000,000	
$\downarrow$	$\downarrow$
∞	$e = 2.718\ 281\ 828\ 459\ldots$

12.

s	$(1 + s)^{1/s}$
0.01	2.704 81
−0.01	
0.001	
−0.001	
0.000 1	
−0.000 1	
0.000 01	
−0.000 01	
$\downarrow$	$\downarrow$
0	$e = 2.718\ 281\ 828\ 459\ldots$

13. Use a calculator and a table of values to investigate

$$\lim_{n \to \infty} (1 + n)^{1/n}$$

Do you think this limit exists? If so, what do you think it is?

14. Use a calculator and a table of values to investigate

$$\lim_{s \to 0^+} \left(1 + \frac{1}{s}\right)^s$$

Do you think this limit exists? If so, what do you think it is?

15. It can be shown that the number e satisfies the inequality

$$\left(1 + \frac{1}{n}\right)^n < e < \left(1 + \frac{1}{n}\right)^{n+1} \qquad n \geq 1$$

Illustrate this graphically by graphing

$$y_1 = (1 + 1/n)^n$$
$$y_2 = 2.718\ 281\ 828 \approx e$$
$$y_3 = (1 + 1/n)^{n+1}$$

in the same viewing window for $1 \leq n \leq 20$.

16. It can be shown that

$$e^s = \lim_{n \to \infty} \left(1 + \frac{s}{n} \right)^n$$

for any real number *s*. Illustrate this graphically for *s* = 2 by graphing

$$y_1 = (1 + 2/n)^n$$
$$y_2 = 7.389\ 056\ 099 \approx e^2$$

in the same viewing window for $1 \leqslant n \leqslant 50$.

Applications

Business & Economics

17. *Continuous compound interest.* Recently, Provident
WWW Bank offered a 10-year CD that earns 5.51% compounded continuously.

  (A) If $10,000 is invested in this CD, how much will it be worth in 10 years?

  (B) How long will it take for the account to be worth $15,000?

18. *Continuous compound interest.* Provident Bank also
WWW offers a 3-year CD that earns 5.28% compounded continuously.

  (A) If $10,000 is invested in this CD, how much will it be worth in 3 years?

  (B) How long will it take for the account to be worth $11,000?

19. *Present value.* A note will pay $20,000 at maturity 10 years from now. How much should you be willing to pay for the note now if money is worth 5.2% compounded continuously?

20. *Present value.* A note will pay $50,000 at maturity 5 years from now. How much should you be willing to pay for the note now if money is worth 6.4% compounded continuously?

21. *Continuous compound interest.* An investor bought stock for $20,000. Five years later, the stock was sold for $30,000. If interest is compounded continuously, what annual nominal rate of interest did the original $20,000 investment earn?

22. *Continuous compound interest.* A family paid $40,000 cash for a house. Fifteen years later, they sold the house for $100,000. If interest is compounded continuously, what annual nominal rate of interest did the original $40,000 investment earn?

23. *Present value.* Solving $A = Pe^{rt}$ for *P*, we obtain

$$P = Ae^{-rt}$$

which is the present value of the amount *A* due in *t* years if money earns interest at an annual nominal rate *r* compounded continuously.

  (A) Graph $P = 10,000e^{-0.08t}$, $0 \leqslant t \leqslant 50$.

  (B) $\lim_{t \to \infty} 10,000e^{-0.08t} = ?$ [Guess, using part (A).]

  [*Conclusion:* The longer the duration of time until the amount *A* is due, the smaller its present value, as we would expect.]

24. *Present value.* Referring to Problem 23, in how many years will the $10,000 have to be due in order for its present value to be $5,000?

25. *Doubling time.* How long will it take money to double if it is invested at 7% compounded continuously?

26. *Doubling time.* How long will it take money to double if it is invested at 5% compounded continuously?

27. *Doubling rate.* At what nominal rate compounded continuously must money be invested to double in 8 years?

28. *Doubling rate.* At what nominal rate compounded continuously must money be invested to double in 10 years?

29. *Growth time.* A man with $20,000 to invest decides to diversify his investments by placing $10,000 in an account that earns 7.2% compounded continuously and $10,000 in an account that earns 8.4% compounded annually. Use graphical approximation methods to determine how long it will take for his total investment in the two accounts to grow to $35,000.

30. *Growth time.* A woman invests $5,000 in an account that earns 8.8% compounded continuously and $7,000 in an account that earns 9.6% compounded annually. Use graphical approximation methods to determine how long it will take for her total investment in the two accounts to grow to $20,000.

31. *Doubling times*

(A) Show that the doubling time t (in years) at an annual rate r compounded continuously is given by

$$t = \frac{\ln 2}{r}$$

(B) Graph the doubling-time equation from part (A) for $0.02 \leqslant r \leqslant 0.30$. Are these restrictions on r reasonable? Explain.

(C) Determine the doubling times (in years, to two decimal places) for $r = 5\%$, 10%, 15%, 20%, 25% and 30%.

32. *Doubling rates*

(A) Show that the rate r that doubles an investment at continuously compounded interest in t years is given by

$$r = \frac{\ln 2}{t}$$

(B) Graph the doubling-rate equation from part (A) for $1 \leqslant t \leqslant 20$. Are these restrictions on t reasonable? Explain.

(C) Determine the doubling rates for $t = 2, 4, 6, 8, 10$, and 12 years.

Life Sciences

33. *Radioactive decay.* A mathematical model for the decay of radioactive substances is given by

$$Q = Q_0 e^{rt}$$

where

Q_0 = amount of the substance at time $t = 0$
r = continuous compound rate of decay
t = time in years
Q = amount of the substance at time t

If the continuous compound rate of decay of radium per year is $r = -0.000\,433\,2$, how long will it take a certain amount of radium to decay to half the original amount? (This period of time is the *half-life* of the substance.)

34. *Radioactive decay.* The continuous compound rate of decay of carbon-14 per year is $r = -0.000\,123\,8$. How long will it take a certain amount of carbon-14 to decay to half the original amount? (Use the radioactive decay model in Problem 33.)

35. *Radioactive decay.* A cesium isotope has a half-life of 30 years. What is the continuous compound rate of decay? (Use the radioactive decay model in Problem 33.)

36. *Radioactive decay.* A strontium isotope has a half-life of 90 years. What is the continuous compound rate of decay? (Use the radioactive decay model in Problem 33.)

Social Sciences

37. *World population.* A mathematical model for world population growth over short periods of time is given by

$$P = P_0 e^{rt}$$

where

P_0 = population at time $t = 0$
r = continuous compound rate of growth
t = time in years
P = population at time t

How long will it take the world population to double if it continues to grow at its current continuous compound rate of 1.3% per year?

38. *U. S. population.* How long will it take for the U. S. population to double if it continues to grow at a rate of 0.85% per year?

39. *Population growth.* Some underdeveloped nations have population doubling times of 50 years. At what continuous compound rate is the population growing? (Use the population growth model in Problem 37.)

40. *Population growth.* Some developed nations have population doubling times of 200 years. At what continuous compound rate is the population growing? (Use the population growth model in Problem 37.)

Section 5-2	Derivatives of Logarithmic and Exponential Functions

- ❑ Derivative Formulas for $\ln x$ and e^x
- ❑ Graphing Techniques
- ❑ Application

In this section we discuss derivative formulas for $\ln x$ and e^x. Out of all the possible choices for bases for the logarithmic and exponential functions, $\log_b x$ and b^x, it turns out (as we will see in this and the next section) that the simplest derivative formulas occur when the base b is chosen to be e.

(A) Using the graph of $f(x) = 2^x$ in Figure 1, sketch tangent lines for several values of x, estimate their slopes, and sketch a graph of $f'(x)$. Is the graph of the derivative above or below the graph of $f(x) = 2^x$?

(B) Using the graph of $g(x) = 3^x$ in Figure 2, sketch tangent lines for several values of x, estimate their slopes, and sketch a graph of $g'(x)$. Is the graph of the derivative above or below the graph of $g(x) = 3^x$?

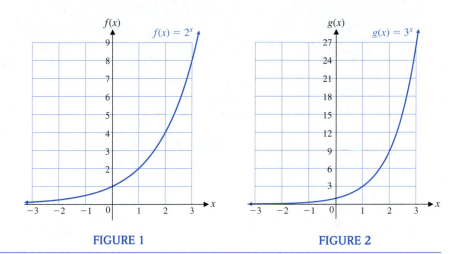

FIGURE 1   FIGURE 2

Explore–Discuss 1 led to the discovery that for $f(x) = 2^x$, the graphs of $f(x)$ and $f'(x)$ look much the same, with $f'(x)$ always below $f(x)$. Similarly, for $g(x) = 3^x$, the graphs of $g(x)$ and $g'(x)$ look much the same, with $g'(x)$ always above $g(x)$. These observations suggest that there may be an exponential function with base between 2 and 3 that is its own derivative. In this section we show that the exponential function with base $e = 2.718...$ has exactly this property—that is,

$$\frac{d}{dx}e^x = e^x$$

We will also obtain the derivative formula

$$\frac{d}{dx}\ln x = \frac{1}{x}$$

for the natural logarithm function.

❑ **DERIVATIVE FORMULAS FOR $\ln x$ AND e^x**

We are now ready to derive a formula for the derivative of

$$f(x) = \ln x = \log_e x \qquad x > 0$$

using the definition of the derivative

$$f'(x) = \lim_{h \to 0} \frac{f(x + h) - f(x)}{h}$$

and the two-step process discussed in Section 3-3.

Step 1. Simplify the difference quotient first:

$$\frac{f(x + h) - f(x)}{h} = \frac{\ln(x + h) - \ln x}{h}$$

$$= \frac{1}{h}[\ln(x + h) - \ln x] \qquad \text{Use } \ln A - \ln B = \ln \frac{A}{B}.$$

$$= \frac{1}{h}\ln \frac{x + h}{x} \qquad \qquad \text{Multiply by } 1 = x/x \text{ to change form.}$$

$$= \frac{x}{x} \cdot \frac{1}{h}\ln \frac{x + h}{x}$$

$$= \frac{1}{x}\left[\frac{x}{h}\ln\left(1 + \frac{h}{x}\right)\right] \qquad \text{Use } p \ln A = \ln A^p.$$

$$= \frac{1}{x}\ln\left(1 + \frac{h}{x}\right)^{x/h}$$

Step 2. Find the limit: Let $s = h/x$. For x fixed, if $h \to 0$, then $s \to 0$. Thus,

$$\frac{d}{dx}\ln x = \lim_{h \to 0} \frac{f(x + h) - f(x)}{h}$$

$$= \lim_{h \to 0}\left[\frac{1}{x}\ln\left(1 + \frac{h}{x}\right)^{x/h}\right] \qquad \begin{array}{l}\text{Let } s = h/x. \text{ Note that } h \to 0 \text{ im-}\\ \text{plies } s \to 0.\end{array}$$

$$= \frac{1}{x}\lim_{s \to 0}[\ln(1 + s)^{1/s}] \qquad \begin{array}{l}\text{Use the new limit property given in the}\\ \text{footnote below.*}\end{array}$$

$$= \frac{1}{x}\ln\left[\lim_{s \to 0}(1 + s)^{1/s}\right] \qquad \text{Use the definition of } e.$$

$$= \frac{1}{x}\ln e \qquad \qquad \ln e = \log_e e = 1$$

$$= \frac{1}{x}$$

Thus,

$$\frac{d}{dx}\ln x = \frac{1}{x}$$

In the next section we will show that, in general,

$$\frac{d}{dx}\log_b x = \frac{1}{\ln b}\left(\frac{1}{x}\right)$$

which is a somewhat more complicated result than the above—unless $b = e$.

In the process of finding the derivative of e^x, we will use (without proof) the fact that

$$\lim_{h \to 0} \frac{e^h - 1}{h} = 1$$

---

*The following new limit property is used: If $\lim_{x \to c} f(x)$ exists and is positive, then $\lim_{x \to c}[\ln f(x)] = \ln[\lim_{x \to c} f(x)]$.

Explore–Discuss 2	Compute

$$\frac{e^h - 1}{h}$$

for the following values of h: $-0.1, -0.01, -0.001, -0.0001, 0.0001, 0.001, 0.01,$ 0.1. Do your calculations make it reasonable to conclude that

$$\lim_{h \to 0} \frac{e^h - 1}{h} = 1?$$

Discuss.

We now apply the two-step process to the exponential function $f(x) = e^x$.

Step 1. Simplify the difference quotient first:

$$\frac{f(x + h) - f(x)}{h} = \frac{e^{x+h} - e^x}{h} \qquad \text{Use } e^{a+b} = e^a e^b.$$

$$= \frac{e^x e^h - e^x}{h} \qquad \text{Factor out } e^x.$$

$$= e^x \left(\frac{e^h - 1}{h} \right)$$

Step 2. Compute the limit of the result in step 1:

$$\frac{d}{dx} e^x = \lim_{h \to 0} \frac{f(x + h) - f(x)}{h}$$

$$= \lim_{h \to 0} e^x \left(\frac{e^h - 1}{h} \right)$$

$$= e^x \lim_{h \to 0} \left(\frac{e^h - 1}{h} \right) \qquad \text{Use the assumed limit given above.}$$

$$= e^x \cdot 1 = e^x$$

Thus,

$$\frac{d}{dx} e^x = e^x$$

In the next section we will show that

$$\frac{d}{dx} b^x = b^x \ln b$$

which is, again, a somewhat more complicated result than the above—unless $b = e$.

The two results just obtained explain why e^x is so widely used that it is sometimes referred to as *the* exponential function. These two new and important derivative formulas are restated in the box for reference.

> ### Derivatives of the Natural Logarithmic and Exponential Functions
>
> $$\frac{d}{dx}\ln x = \frac{1}{x} \qquad \frac{d}{dx}e^x = e^x$$

www

These new derivative formulas can be combined with the rules of differentiation discussed in Chapter 3 to differentiate a wide variety of functions.

Example 1 **Finding Derivatives** Find $f'(x)$ for:

(A) $f(x) = 2e^x + 3\ln x$ (B) $f(x) = \dfrac{e^x}{x^3}$

(C) $f(x) = (\ln x)^4$ (D) $f(x) = \ln x^4$

SOLUTION (A) $f'(x) = 2\dfrac{d}{dx}e^x + 3\dfrac{d}{dx}\ln x$

$$= 2e^x + 3\left(\frac{1}{x}\right) = 2e^x + \frac{3}{x}$$

(B) $f'(x) = \dfrac{x^3\dfrac{d}{dx}e^x - e^x\dfrac{d}{dx}x^3}{(x^3)^2}$ Quotient rule

$$= \frac{x^3e^x - e^x 3x^2}{x^6} = \frac{x^2e^x(x-3)}{x^6} = \frac{e^x(x-3)}{x^4}$$

(C) $\dfrac{d}{dx}(\ln x)^4 = 4(\ln x)^3\dfrac{d}{dx}\ln x$ Power rule for functions

$$= 4(\ln x)^3\left(\frac{1}{x}\right) = \frac{4(\ln x)^3}{x}$$

(D) $\dfrac{d}{dx}\ln x^4 = \dfrac{d}{dx}(4\ln x)$ Property of logarithms

$$= 4\left(\frac{1}{x}\right) = \frac{4}{x}$$

Matched Problem 1 Find $f'(x)$ for:

(A) $f(x) = 4\ln x - 5e^x$ (B) $f(x) = x^2e^x$
(C) $f(x) = \ln x^3$ (D) $f(x) = (\ln x)^3$

CAUTION

$$\frac{d}{dx}e^x \neq xe^{x-1} \qquad \frac{d}{dx}e^x = e^x$$

The power rule cannot be used to differentiate the exponential function. The power rule applies to exponential forms x^n where the exponent is a constant and the base is a variable. In the exponential form e^x, the base is a constant and the exponent is a variable.

❑ GRAPHING TECHNIQUES

Using the techniques discussed in Chapter 4, we can use first and second derivatives to gain useful information about the graphs of $y = \ln x$ and $y = e^x$. With the derivative formulas given above, we can construct Table 1.

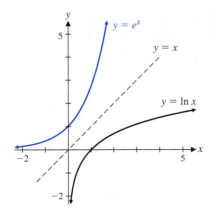

FIGURE 3 e^x is continuous on $(-\infty, \infty)$, $\ln x$ is continuous on $(0, \infty)$.

TABLE 1			
ln x		e^x	
$y = \ln x$	$x > 0$	$y = e^x$	$-\infty < x < \infty$
$y' = 1/x > 0$	$x > 0$	$y' = e^x > 0$	$-\infty < x < \infty$
$y'' = -1/x^2 < 0$	$x > 0$	$y'' = e^x > 0$	$-\infty < x < \infty$

From the table, we can see that both functions are increasing throughout their respective domains, the graph of $y = \ln x$ is always concave downward, and the graph of $y = e^x$ is always concave upward. It can be shown that $\lim_{x \to 0^+} \ln x = -\infty$ and $\lim_{x \to -\infty} e^x = 0$. Thus, the y axis is a vertical asymptote for the graph of $y = \ln x$ and the x axis is a horizontal asymptote for the graph of e^x. Both equations are graphed in Figure 3.

Notice that if we fold the page along the dashed line $y = x$, the two graphs match exactly (see Section 2-3). Also notice that both graphs are unbounded as $x \to \infty$. Comparing each graph with the graph of $y = x$ (the dashed line), we conclude that e^x grows more rapidly than x and $\ln x$ grows more slowly than x. In fact, the following limits can be established:

> ### *Growth of Exponential and Logarithmic Functions*
>
> $$\lim_{x \to \infty} \frac{x^p}{e^x} = 0, \quad p > 0 \qquad \text{and} \qquad \lim_{x \to \infty} \frac{\ln x}{x^p} = 0, \quad p > 0$$

The limits in the box indicate that e^x grows more rapidly than any positive power of x, and $\ln x$ grows more slowly than any positive power of x.

Now we will apply graphing techniques to a slightly more complicated function.

Example 2 ↩ **Graphing Strategy**   Analyze the function $f(x) = xe^x$ following the steps in the graphing strategy discussed in Section 4-3. State all the pertinent information and sketch the graph of f.

SOLUTION   **Step 1.** *Analyze $f(x)$:* $f(x) = xe^x$

(A) Domain: All real numbers
(B) y intercept: $f(0) = 0$
x intercept: $xe^x = 0$ for $x = 0$ only, since $e^x > 0$ for all x (see Fig. 3).
(C) Vertical asymptotes: None
Horizontal asymptotes: We have not developed limit techniques for functions of this type to determine the behavior of $f(x)$ as $x \to -\infty$ and

$x \to \infty$. However, the following tables of values suggest the nature of the graph of f as $x \to -\infty$ and $x \to \infty$:

x	1	5	10	$\to \infty$
$f(x)$	2.72	742.07	220,264.66	$\to \infty$

x	-1	-5	-10	$\to -\infty$
$f(x)$	-0.37	-0.03	$-0.000\ 45$	$\to 0$

Step 2. *Analyze $f'(x)$:*

$$f'(x) = x\frac{d}{dx}e^x + e^x\frac{d}{dx}x$$

$$= xe^x + e^x = e^x(x + 1)$$

Critical value for $f(x)$: -1
Partition number for $f'(x)$: -1
Sign chart for $f'(x)$:

Thus, $f(x)$ decreases on $(-\infty, -1)$, has a local minimum at $x = -1$, and increases on $(-1, \infty)$. (Since $e^x > 0$ for all x, we do not have to evaluate e^{-2} to conclude that $-e^{-2} < 0$ when using the test number -2.)

Step 3. *Analyze $f''(x)$:*

$$f''(x) = e^x\frac{d}{dx}(x + 1) + (x + 1)\frac{d}{dx}e^x$$

$$= e^x + (x + 1)e^x = e^x(x + 2)$$

Sign chart for $f''(x)$ (partition number is -2):

Thus, the graph of f is concave downward on $(-\infty, -2)$, has an inflection point at $x = -2$, and is concave upward on $(-2, \infty)$.

www   **Step 4.**   *Sketch the graph of f using the information from steps 1 to 3:*

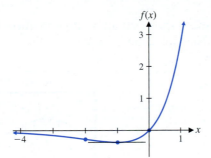

x	$f(x)$
-2	-0.27
-1	-0.37
0	0

Matched Problem 2   Analyze the function $f(x) = x \ln x$. State all the pertinent information and sketch the graph of f.

❑   APPLICATION

Example 3   **Maximizing Profit**   The market research department of a chain of pet stores test marketed their aquarium pumps (as well as other items) in several of their stores in a test city. They found that the weekly demand for aquarium pumps is given approximately by

$$p = 12 - 2 \ln x \qquad 0 < x < 90$$

where x is the number of pumps sold each week and $\$p$ is the price of one pump. If each pump costs the chain $3, how should it be priced in order to maximize the weekly profit?

SOLUTION   Although we want to find the price that maximizes the weekly profit, it will be easier first to find the number of pumps that will maximize the weekly profit. The revenue equation is

$$R(x) = xp = 12x - 2x \ln x$$

The cost equation is

$$C(x) = 3x$$

and the profit equation is

$$\begin{aligned} P(x) &= R(x) - C(x) \\ &= 12x - 2x \ln x - 3x \\ &= 9x - 2x \ln x \end{aligned}$$

Thus, we must solve the following:

$$\text{Maximize} \quad P(x) = 9x - 2x \ln x \qquad 0 < x < 90$$

$$\begin{aligned} P'(x) &= 9 - 2x\left(\frac{1}{x}\right) - 2 \ln x \\ &= 7 - 2 \ln x = 0 \\ 2 \ln x &= 7 \\ \ln x &= 3.5 \\ x &= e^{3.5} \end{aligned}$$

$$P''(x) = -2\left(\frac{1}{x}\right) = -\frac{2}{x}$$

Since $x = e^{3.5}$ is the only critical value and $P''(e^{3.5}) < 0$, the maximum weekly profit occurs when $x = e^{3.5} \approx 33$ and $p = 12 - 2 \ln e^{3.5} = \$5$.

 Figure 4 shows an alternative method for finding the maximum value of $P(x)$ in Example 3 on a graphing utility.

FIGURE 4 $y_1 = 9x - 2x \ln x$

Matched Problem 3 ⇔ Repeat Example 3 if each pump costs the chain \$3.50.

Answers to Matched Problems   **1.** (A) $(4/x) - 5e^x$ (B) $xe^x(x + 2)$ (C) $3/x$ (D) $3(\ln x)^2/x$

2. Domain: $(0, \infty)$
y intercept: None [$f(0)$ is not defined]
x intercept: $x = 1$
Increasing on (e^{-1}, ∞)
Decreasing on $(0, e^{-1})$
Local minimum at $x = e^{-1} \approx 0.368$
Concave upward on $(0, \infty)$

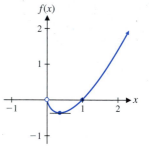

$f(x)$

x	5	10	100	$\to \infty$
$f(x)$	8.05	23.03	460.52	$\to \infty$

x	0.1	0.01	0.001	0.000 1	$\to 0$
$f(x)$	-0.23	-0.046	$-0.006\ 9$	$-0.000\ 92$	$\to 0$

3. Maximum profit occurs for $x = e^{3.25} \approx 26$ and $p = \$5.50$.

Exercise 5-2

For many of the problems in this exercise set, the answers in the back of the book include both an unsimplified form and a simplified form. When checking your work, first check that you applied the rules correctly, and then check that you performed the algebraic simplification correctly.

A *In Problems 1–8, find the indicated derivative and simplify.*

1. $f'(x)$ for $f(x) = 2e^x + 3 \ln x$

2. $g'(x)$ for $g(x) = 5e^x - 6 \ln x$

3. $\dfrac{dy}{dx}$ for $y = 4e^x - 3x^e$

4. $\dfrac{dy}{dx}$ for $y = -7e^x + 9x^e$

5. y' for $y = \ln x^3$

6. y' for $y = (\ln x)^3$

7. $\dfrac{d}{dx}(\ln x)^6$

8. $\dfrac{d}{dx}\ln x^6$

B   *In Problems 9–32, find $f'(x)$ and simplify.*

9. $f(x) = x^4 \ln x$ **10.** $f(x) = x^3 \ln x$

11. $f(x) = x^3 e^x$ **12.** $f(x) = x^4 e^x$

13. $f(x) = \dfrac{e^x}{x^2 + 9}$ **14.** $f(x) = \dfrac{e^x}{x^2 + 4}$

15. $f(x) = \dfrac{\ln x}{x^4}$ **16.** $f(x) = \dfrac{\ln x}{x^3}$

17. $f(x) = (x + 2)^3 \ln x$ **18.** $f(x) = (x - 1)^2 \ln x$

19. $f(x) = (x + 1)^3 e^x$ **20.** $f(x) = (x - 2)^3 e^x$

21. $f(x) = \dfrac{x^2 + 1}{e^x}$ **22.** $f(x) = \dfrac{x + 1}{e^x}$

23. $f(x) = x(\ln x)^3$ **24.** $f(x) = x(\ln x)^2$

25. $f(x) = (4 - 5e^x)^3$ **26.** $f(x) = (5 - \ln x)^4$

27. $f(x) = \sqrt{1 + \ln x}$ **28.** $f(x) = \sqrt{1 + e^x}$

29. $f(x) = xe^x - e^x$ **30.** $f(x) = x \ln x - x$

31. $f(x) = 2x^2 \ln x - x^2$

32. $f(x) = x^2 e^x - 2xe^x + 2e^x$

In Problems 33–36, find the equation of the line tangent to the graph of $y = f(x)$ at the indicated value of x.

33. $f(x) = e^x; \quad x = 1$

34. $f(x) = e^x; \quad x = 2$

35. $f(x) = \ln x; \quad x = e$

36. $f(x) = \ln x; \quad x = 1$

37. A student claims that the tangent line to the graph of $f(x) = e^x$ at $x = 3$ passes through the point $(2, 0)$ (see the figure). Is she correct? Will the tangent line at $x = 4$ pass through $(3, 0)$? Explain.

Figure for 37

38. A student claims that the tangent line to the graph of $g(x) = \ln x$ at $x = 3$ passes through the origin (see the figure). Is he correct? Will the tangent line at $x = 4$ pass through the origin? Explain.

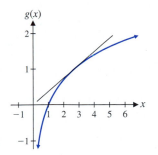

Figure for 38

Problems 39 and 40 make use of the following familiar properties of exponential and logarithmic functions (see Sections 2-2 and 2-3):

$$(a^b)^c = a^{bc} \qquad \ln ab = \ln a + \ln b$$

39. Find the derivatives of $y = e^{2x}$, $y = e^{3x}$, and $y = e^{4x}$. What do you think would be the derivative of $y = e^{kx}$ for any constant k? Justify your answer.

40. Find the derivatives of $y = \ln(2x)$, $y = \ln(3x)$, and $y = \ln(4x)$. What do you think would be the derivative of $y = \ln(kx)$ for any positive constant k? Justify your answer.

C   *In Problems 41–46, find the indicated extremum of each function for $x > 0$.*

41. Absolute maximum value of
$$f(x) = 5x - 2x \ln x$$

42. Absolute minimum value of
$$f(x) = 4x \ln x - 7x$$

43. Absolute minimum value of
$$f(x) = \dfrac{e^x}{x^2}$$

44. Absolute maximum value of
$$f(x) = \dfrac{x^3}{e^x}$$

45. Absolute maximum value of
$$f(x) = x^2(3 - \ln x)$$

46. Absolute minimum value of
$$f(x) = x^3(\ln x - 2)$$

In Problems 47–54, apply the graphing strategy discussed in Section 4-3 to f, summarize the pertinent information, and sketch the graph.

47. $f(x) = 1 - e^x$

48. $f(x) = 1 - \ln x$

49. $f(x) = x - \ln x$

50. $f(x) = e^x - x$

51. $f(x) = (3 - x)e^x$

52. $f(x) = (x - 2)e^x$

53. $f(x) = x^2 \ln x$

54. $f(x) = \dfrac{\ln x}{x}$

 Problems 55–58 require the use of a graphing utility. Approximate the critical values of $f(x)$ to two decimal places and find the intervals where $f(x)$ is increasing, the intervals where $f(x)$ is decreasing, and the local extrema.

55. $f(x) = e^x - 2x^2$

56. $f(x) = e^x + x^2$

57. $f(x) = 20 \ln x - e^x$

58. $f(x) = x^2 - 3x \ln x$

 In Problems 59–62, use graphical approximation methods to find the points of intersection of $f(x)$ and $g(x)$ (to two decimal places).

59. $f(x) = e^x; g(x) = x^4$
[Note that there are three points of intersection and that e^x is greater than x^4 for large values of x.]

60. $f(x) = e^x; g(x) = x^5$
[Note that there are two points of intersection and that e^x is greater than x^5 for large values of x.]

61. $f(x) = \ln x; g(x) = x^{1/5}$
[Note that $\ln x$ is less than $x^{1/5}$ for large values of x.]

62. $f(x) = \ln x; g(x) = x^{1/4}$
[Note that $\ln x$ is less than $x^{1/4}$ for large values of x.]

Applications

Business & Economics

63. *Maximum profit.* A national food service runs food concessions for sporting events throughout the country. Their marketing research department chose a particular football stadium to test market a new jumbo hot dog. It was found that the demand for the new hot dog is given approximately by

$$p = 5 - \ln x \qquad 5 \leq x \leq 50$$

where x is the number of hot dogs (in thousands) that can be sold during one game at a price of $\$p$. If the concessionaire pays $\$1$ for each hot dog, how should the hot dogs be priced to maximize the profit per game?

64. *Maximum profit.* On a national tour of a rock band, the demand for T-shirts is given by

$$p = 15 - 4 \ln x \qquad 1 \leq x \leq 40$$

where x is the number of T-shirts (in thousands) that can be sold during a single concert at a price of $\$p$. If the shirts cost the band $\$5$ each, how should they be priced in order to maximize the profit per concert?

65. *Maximum profit.* A regional chain of department stores has collected the data in the table, showing weekly sales of a certain brand of jeans. The same jeans have been offered at various prices, ranging from the regular price of $\$35.99$ to the lowest sale price of $\$23.99$. Each pair of this brand costs the chain $\$20.00$. Use logarithmic regression ($p = a + b \ln x$) to find the price (to the nearest cent) that will maximize profit.

PAIRS OF JEANS x	PRICE PER PAIR ($\$$) p
21,543	23.99
14,029	25.99
12,130	27.99
9,169	29.99
6,964	31.99
5,506	33.99
4,187	35.99

66. *Maximum profit.* A mail-order company specializes in the sale of athletic shoes. Its market research team has determined the weekly demand shown in the table for a particular model of basketball shoe when priced at various levels. The company is able to purchase this model from the manufacturer for $\$62$ per pair. Use logarithmic regression ($p = a + b \ln x$) to find the price (to the nearest cent) that will maximize profit.

DEMAND	PRICE PER PAIR ($)
x	p
4,312	74
3,064	79
2,499	84
2,047	89
1,823	94
1,781	99

67. *Minimum average cost.* The cost of producing x units of a product is given by

$$C(x) = 600 + 100x - 100 \ln x \qquad x \geqslant 1$$

Find the minimum average cost.

68. *Minimum average cost.* The cost of producing x units of a product is given by

$$C(x) = 1{,}000 + 200x - 200 \ln x \qquad x \geqslant 1$$

Find the minimum average cost.

69. *Maximizing revenue.* A cosmetic company is planning the introduction and promotion of a new lipstick line. The marketing research department, after test marketing the new line in a carefully selected large city, found that the demand in that city is given approximately by

$$p = 10e^{-x} \qquad 0 \leqslant x \leqslant 2$$

where x thousand lipsticks were sold per week at a price of $\$p$ each.

(A) At what price will the weekly revenue $R(x) = xp$ be maximum? What is the maximum weekly revenue in the test city?

(B) Graph R for $0 \leqslant x \leqslant 2$.

70. *Maximizing revenue.* Repeat Problem 69 using the demand equation $p = 12e^{-x}, 0 \leqslant x \leqslant 2$.

Life Sciences

71. *Blood pressure.* An experiment was set up to find a relationship between weight and systolic blood pressure in normal children. Using hospital records for 5,000 normal children, it was found that the systolic blood pressure was given approximately by

$$P(x) = 17.5(1 + \ln x) \qquad 10 \leqslant x \leqslant 100$$

where $P(x)$ is measured in millimeters of mercury and x is measured in pounds. What is the rate of change of blood pressure with respect to weight at the 40-pound weight level? At the 90-pound weight level?

72. *Blood pressure.* Graph the systolic blood pressure equation in Problem 71.

73. *Drug concentration.* The concentration of a drug in the bloodstream t hours after injection is given approximately by

$$C(t) = 4.35e^{-t} \qquad 0 \leqslant t \leqslant 5$$

where $C(t)$ is concentration in milligrams per milliliter.

(A) What is the rate of change of concentration after 1 hour? After 4 hours?

(B) Graph C.

74. *Water pollution.* The use of iodine crystals is a popular way of making small quantities of nonpotable water safe to drink. Crystals placed in a 1-ounce bottle of water will dissolve until the solution is saturated. After saturation, half of this solution is poured into a quart container of nonpotable water, and after about an hour, the water is usually safe to drink. The half-empty 1-ounce bottle is then refilled to be used again in the same way. Suppose that the concentration of iodine in the 1-ounce bottle t minutes after the crystals are introduced can be approximated by

$$C(t) = 250(1 - e^{-t}) \qquad t \geqslant 0$$

where $C(t)$ is the concentration of iodine in micrograms per milliliter.

(A) What is the rate of change of the concentration after 1 minute? After 4 minutes?

(B) Graph C for $0 \leqslant t \leqslant 5$.

Social Sciences

75. *Psychology: stimulus/response.* In psychology, the
www   Weber–Fechner law for stimulus response is

$$R = k \ln \frac{S}{S_0}$$

where R is the response, S is the stimulus, and S_0 is the lowest level of stimulus that can be detected. Find dR/dS.

76. *Psychology: learning.* A mathematical model for the average of a group of people learning to type is given by

$$N(t) = 10 + 6 \ln t \qquad t \geqslant 1$$

where $N(t)$ is the number of words per minute typed after t hours of instruction and practice (2 hours per day, 5 days per week). What is the rate of learning after 10 hours of instruction and practice? After 100 hours?

| Section 5-3 | # Chain Rule: General Form |

- ❑ COMPOSITE FUNCTIONS
- ❑ CHAIN RULE
- ❑ GENERALIZED DERIVATIVE RULES
- ❑ OTHER LOGARITHMIC AND EXPONENTIAL FUNCTIONS

| Explore–Discuss 1 |

(A) We have shown that the function $f(x) = e^x$ is its own derivative. Use the graph of $g(x) = e^{x^2}$ in Figure 1 to explain why $g(x)$ is not its own derivative.

(B) Use the graph of $h(x) = 2xe^{x^2}$ in Figure 2 to explain why $h(x)$ could be the derivative of $g(x)$. [We will show in this section that $h(x)$ is indeed the derivative of $g(x)$.]

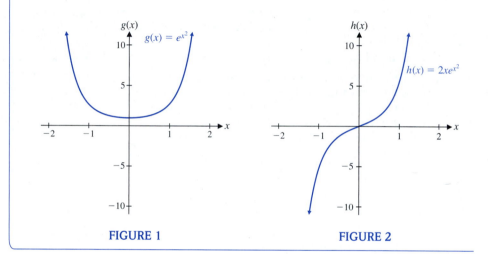

FIGURE 1   FIGURE 2

In Section 3-6 we introduced the power form of the chain rule:

$$\frac{d}{dx}[u(x)]^n = n[u(x)]^{n-1}u'(x) \qquad \text{For example,}$$

$$\frac{d}{dx}(x^2 - 3)^5 = 5(x^2 - 3)^4 \frac{d}{dx}(x^2 - 3)$$
$$= 10x(x^2 - 3)^4$$

This general power rule is a special case of one of the most important derivative rules of all—the *chain rule*—which will enable us to determine the derivatives of some fairly complicated functions in terms of derivatives of more elementary functions.

Suppose that you were asked to find the derivative of

$$m(x) = \ln(2x + 1) \qquad \text{or} \qquad n(x) = e^{3x^2 - 1}$$

We have formulas for computing derivatives of $\ln x$ and e^x, and polynomial functions in general, but not in the combinations indicated. The chain rule is used to compute derivatives of functions that are *compositions* of more elementary functions whose derivatives are known. We therefore start this section with a brief review of *composite functions*.

❑ COMPOSITE FUNCTIONS

Let us look at function m more closely:

$$m(x) = \ln(2x + 1)$$

The function m is a combination of the natural logarithm function and a linear function. To see this more clearly, let

$$y = f(u) = \ln u \quad \text{and} \quad u = g(x) = 2x + 1$$

Then we can express y as a function of x as follows:

$$y = f(u) = f[g(x)] = \ln(2x + 1) = m(x)$$

The function m is said to be the *composite* of the two simpler functions f and g. (Loosely speaking, we can think of m as a function of a function.) In general, we have the following:

WWW

> **Composite Functions**
>
> A function m is a **composite** of functions f and g if
>
> $$m(x) = f[g(x)]$$
>
> The domain of m is the set of all numbers x such that x is in the domain of g and $g(x)$ is in the domain of f.

Example 1 ⟳ **Composite Functions**   Let $f(u) = e^u$, $g(x) = 3x^2 + 1$, and $m(v) = v^{3/2}$. Find:

(A) $f[g(x)]$ (B) $g[f(u)]$ (C) $m[g(x)]$

SOLUTION   (A) $f[g(x)] = e^{g(x)} = e^{3x^2 + 1}$
(B) $g[f(u)] = 3[f(u)]^2 + 1 = 3(e^u)^2 + 1 = 3e^{2u} + 1$
(C) $m[g(x)] = [g(x)]^{3/2} = (3x^2 + 1)^{3/2}$

Matched Problem 1 ⟳   Let $f(u) = \ln u$, $g(x) = 2x^3 + 4$, and $m(v) = v^{-5}$. Find:

(A) $f[g(x)]$ (B) $g[f(u)]$ (C) $m[g(x)]$

Example 2 ⟳ **Composite Functions**   Write each function as a composition of the natural logarithm or exponential function and a polynomial.

(A) $y = \ln(x^3 - 2x^2 + 1)$ (B) $y = e^{x^2 + 4}$

SOLUTION   (A) Let

$$y = f(u) = \ln u$$
$$u = g(x) = x^3 - 2x^2 + 1$$

Check: $y = f[g(x)] = \ln[g(x)] = \ln(x^3 - 2x^2 + 1)$

(B) Let

$$y = f(u) = e^u$$
$$u = g(x) = x^2 + 4$$

Check: $y = f[g(x)] = e^{g(x)} = e^{x^2 + 4}$

Matched Problem 2   ↩   Repeat Example 2 for:

(A) $y = e^{2x^3+7}$ (B) $y = \ln(x^4 + 10)$

■

❑   CHAIN RULE

The word *chain* in the name *chain rule* comes from the fact that a function formed by composition (such as those in Example 1) involves a chain of functions—that is, a function of a function. The *chain rule* will enable us to compute the derivative of a composite function in terms of the derivatives of the functions making up the composition.

Suppose that

$$y = m(x) = f[g(x)]$$

is a composite of f and g, where

$$y = f(u) \qquad \text{and} \qquad u = g(x)$$

We would like to express the derivative dy/dx in terms of the derivatives of f and g. From the definition of a derivative (see Section 3-3), we have

$$\frac{dy}{dx} = \lim_{h \to 0} \frac{m(x + h) - m(x)}{h} \qquad \text{Substitute } m(x + h) = f[g(x + h)] \text{ and } m(x) = f[g(x)].$$

$$= \lim_{h \to 0} \frac{f[g(x + h)] - f[g(x)]}{h} \qquad \text{Multiply by } 1 = \frac{g(x + h) - g(x)}{g(x + h) - g(x)}.$$

$$= \lim_{h \to 0} \left[\frac{f[g(x + h)] - f[g(x)]}{h} \cdot \frac{g(x + h) - g(x)}{g(x + h) - g(x)} \right]$$

$$= \lim_{h \to 0} \left[\frac{f[g(x + h)] - f[g(x)]}{g(x + h) - g(x)} \cdot \frac{g(x + h) - g(x)}{h} \right] \qquad (1)$$

We recognize the second factor in equation (1) as the difference quotient for $g(x)$. To interpret the first factor as the difference quotient for $f(u)$, we let $k = g(x + h) - g(x)$. Since $u = g(x)$, we can write

$$u + k \; \underline{\underline{\;= g(x) + g(x + h) - g(x)\;}} = g(x + h)$$

Substituting in equation (1), we now have

$$\frac{dy}{dx} = \lim_{h \to 0} \left[\frac{f(u + k) - f(u)}{k} \cdot \frac{g(x + h) - g(x)}{h} \right] \qquad (2)$$

If we assume that $k = [g(x + h) - g(x)] \to 0$ as $h \to 0$, we can find the limit of each difference quotient in equation (2):

$$\frac{dy}{dx} = \left[\lim_{k \to 0} \frac{f(u + k) - f(u)}{k} \right] \left[\lim_{h \to 0} \frac{g(x + h) - g(x)}{h} \right]$$

$$= f'(u)g'(x)$$

$$= \frac{dy}{du} \frac{du}{dx}$$

This result is correct under rather general conditions, and is called the *chain rule*, but our "derivation" is superficial, because it ignores a number of hidden problems. Since a formal proof of the chain rule is beyond the scope of this book, we simply state it as follows:

www

> **Chain Rule**
>
> If $y = f(u)$ and $u = g(x)$, define the composite function
>
> $$y = m(x) = f[g(x)]$$
>
> Then
>
> $$\frac{dy}{dx} = \frac{dy}{du}\frac{du}{dx} \qquad \text{provided that } \frac{dy}{du} \text{ and } \frac{du}{dx} \text{ exist}$$
>
> Or, equivalently,
>
> $$m'(x) = f'[g(x)]g'(x) \qquad \text{provided that } f'[g(x)] \text{ and } g'(x) \text{ exist}$$

Example 3 ↪   **Using the Chain Rule**   Find dy/dx, given:

(A) $y = \ln(x^2 - 4x + 2)$ (B) $y = e^{2x^3+5}$ (C) $y = (3x^2 + 1)^{3/2}$

SOLUTION   (A) Let $y = \ln u$ and $u = x^2 - 4x + 2$. Then

$$\frac{dy}{dx} = \frac{dy}{du}\frac{du}{dx} \quad {}^*$$

$$= \frac{1}{u}(2x - 4)$$

$$= \frac{1}{x^2 - 4x + 2}(2x - 4) \quad \text{Since } u = x^2 - 4x + 2$$

$$= \frac{2x - 4}{x^2 - 4x + 2}$$

(B) Let $y = e^u$ and $u = 2x^3 + 5$. Then

$$\frac{dy}{dx} = \frac{dy}{du}\frac{du}{dx}$$

$$= e^u(6x^2)$$

$$= 6x^2 e^{2x^3+5} \quad \text{Since } u = 2x^3 + 5$$

(C) We have two methods:

Method 1.   *Chain rule: general form.* Let $y = u^{3/2}$ and $u = 3x^2 + 1$. Then

$$\frac{dy}{dx} = \frac{dy}{du}\frac{du}{dx}$$

$$= \tfrac{3}{2}u^{1/2}(6x)$$

$$= \tfrac{3}{2}(3x^2 + 1)^{1/2}(6x) \quad \text{Since } u = 3x^2 + 1$$

$$= 9x(3x^2 + 1)^{1/2} \quad \text{or} \quad 9x\sqrt{3x^2 + 1}$$

---

*After some experience with the chain rule, the steps in the dashed boxes are usually done mentally.

Method 2. *Chain rule: power form (general power rule)*

$$\frac{d}{dx}(3x^2 + 1)^{3/2} = \tfrac{3}{2}(3x^2 + 1)^{1/2}\frac{d}{dx}(3x^2 + 1) \qquad \frac{d}{dx}[u(x)]^n = n[u(x)]^{n-1}\frac{d}{dx}u(x)$$

$$= \tfrac{3}{2}(3x^2 + 1)^{1/2}(6x)$$

$$= 9x(3x^2 + 1)^{1/2} \qquad \text{or} \qquad 9x\sqrt{3x^2 + 1}$$

The general power rule stated in Section 3-6 can be derived using the chain rule as follows: Given $y = [u(x)]^n$, let $y = v^n$ and $v = u(x)$. Then

$$\frac{dy}{dx} = \frac{dy}{dv}\frac{dv}{dx}$$

$$= nv^{n-1}\frac{d}{dx}u(x)$$

$$= n[u(x)]^{n-1}\frac{d}{dx}u(x) \qquad \text{Since } v = u(x)$$

Matched Problem 3 ⇔ Find dy/dx, given:

(A) $y = e^{3x^4+6}$ (B) $y = \ln(x^2 + 9x + 4)$ (C) $y = (2x^3 + 4)^{-5}$
(Use two methods.)

The chain rule can be extended to compositions of three or more functions. For example, if $y = f(w)$, $w = g(u)$, and $u = h(x)$, then

$$\frac{dy}{dx} = \frac{dy}{dw}\frac{dw}{du}\frac{du}{dx}$$

Example 4 ⇔ **Using the Chain Rule** For $y = h(x) = e^{1+(\ln x)^2}$, find dy/dx.

SOLUTION Note that h is of the form $y = e^w$, where $w = 1 + u^2$ and $u = \ln x$. Thus,

$$\frac{dy}{dx} = \frac{dy}{dw}\frac{dw}{du}\frac{du}{dx}$$

$$= e^w(2u)\left(\frac{1}{x}\right)$$

$$= e^{1+u^2}(2u)\left(\frac{1}{x}\right) \qquad \text{Since } w = 1 + u^2$$

$$= e^{1+(\ln x)^2}(2\ln x)\left(\frac{1}{x}\right) \qquad \text{Since } u = \ln x$$

$$= \frac{2}{x}(\ln x)e^{1+(\ln x)^2}$$

Matched Problem 4 ⇔ For $y = h(x) = [\ln(1 + e^x)]^3$, find dy/dx.

❑   GENERALIZED DERIVATIVE RULES

In practice, it is not necessary to introduce additional variables when using the chain rule as we did in Examples 3 and 4. Instead, the chain rule can be used to extend the derivative rules for specific functions to general derivative rules for compositions. This is what we did above when we showed that the general power rule is a consequence of the chain rule. The same technique can be applied to functions of the form $y = e^{f(x)}$ and $y = \ln[f(x)]$ (see Problems 71 and 72 in Exercise 5-3). The results are summarized in the following box:

General Derivative Rules

$$\frac{d}{dx}[f(x)]^n = n[f(x)]^{n-1}f'(x) \tag{3}$$

$$\frac{d}{dx}\ln[f(x)] = \frac{1}{f(x)}f'(x) \tag{4}$$

$$\frac{d}{dx}e^{f(x)} = e^{f(x)}f'(x) \tag{5}$$

For power, natural logarithm, or exponential forms, we can either use the chain rule discussed earlier or these special differentiation formulas, which are based on the chain rule. Use whichever is easier for you. In Example 5 we use the general derivative rules.

Example 5 ⇨ **Using General Derivative Rules**

(A) $\dfrac{d}{dx}e^{2x} = e^{2x}\dfrac{d}{dx}2x$ Using equation (5)

$\qquad = e^{2x}(2) = 2e^{2x}$

(B) $\dfrac{d}{dx}\ln(x^2 + 9) = \dfrac{1}{x^2 + 9}\dfrac{d}{dx}(x^2 + 9)$ Using equation (4)

$\qquad\qquad = \dfrac{1}{x^2 + 9}2x = \dfrac{2x}{x^2 + 9}$

(C) $\dfrac{d}{dx}(1 + e^{x^2})^3 = 3(1 + e^{x^2})^2\dfrac{d}{dx}(1 + e^{x^2})$ Using equation (3)

$\qquad\qquad = 3(1 + e^{x^2})^2 e^{x^2}\dfrac{d}{dx}x^2$ Using equation (5)

$\qquad\qquad = 3(1 + e^{x^2})^2 e^{x^2}(2x)$

$\qquad\qquad = 6xe^{x^2}(1 + e^{x^2})^2$

Matched Problem 5 ⇨ Find:

(A) $\dfrac{d}{dx}\ln(x^3 + 2x)$ (B) $\dfrac{d}{dx}e^{3x^2+2}$ (C) $\dfrac{d}{dx}(2 + e^{-x^2})^4$

❑ OTHER LOGARITHMIC AND EXPONENTIAL FUNCTIONS

In most applications involving logarithmic or exponential functions, the number e is the preferred base. However, there are situations where it is convenient to use a base other than e. Derivatives of $y = \log_b x$ and $y = b^x$ can be obtained by expressing these functions in terms of the natural logarithmic and exponential functions.

We begin by finding a relationship between $\log_b x$ and $\ln x$ for any base $b, b > 0$ and $b \neq 1$. Some of you may prefer to remember the process, and others the formula.

$$y = \log_b x \qquad \text{Change to exponential form.}$$
$$b^y = x \qquad \text{Take the natural logarithm of both sides.}$$
$$\ln b^y = \ln x \qquad \text{Recall that } \ln b^y = y \ln b.$$
$$y \ln b = \ln x \qquad \text{Solve for } y.$$
$$y = \frac{1}{\ln b} \ln x$$

Thus,

$$\log_b x = \frac{1}{\ln b} \ln x \qquad \text{Change-of-base formula*} \tag{6}$$

Differentiating both sides of equation (6), we have

$$\frac{d}{dx} \log_b x = \frac{1}{\ln b} \frac{d}{dx} \ln x = \frac{1}{\ln b}\left(\frac{1}{x}\right)$$

Explore–Discuss 2	(A) The graphs of $f(x) = \log_2 x$ and $g(x) = \log_4 x$ are shown in Figure 3. Which graph belongs to which function?

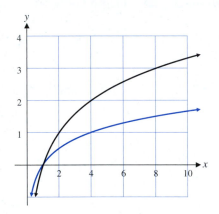

FIGURE 3

(B) Sketch graphs of $f'(x)$ and $g'(x)$.

(C) The function $f(x)$ is related to $g(x)$ in the same way that $f'(x)$ is related to $g'(x)$. What is that relationship?

Example 6 ⟹ **Differentiating Logarithmic Functions** Find $f'(x)$ for:

(A) $f(x) = \log_2 x$ (B) $f(x) = \log(1 + x^3)$

SOLUTION (A) $f(x) = \log_2 x = \dfrac{1}{\ln 2} \ln x$ *Using equation (6)*

$$f'(x) = \frac{1}{\ln 2}\left(\frac{1}{x}\right)$$

---

*Equation (6) is a special case of the **general change-of-base formula** for logarithms (which can be derived in the same way): $\log_b x = (\log_a x)/(\log_a b)$.

(B) $f(x) = \log(1 + x^3)$ 　　　Recall that $\log r = \log_{10} r$.

$$= \frac{1}{\ln 10}\ln(1 + x^3) \qquad \text{Using equation (6)}$$

$$f'(x) = \frac{1}{\ln 10}\left(\frac{1}{1 + x^3}3x^2\right) = \frac{1}{\ln 10}\left(\frac{3x^2}{1 + x^3}\right)$$

■

Matched Problem 6 ➭ Find $f'(x)$ for:

(A) $f(x) = \log x$ 　　(B) $f(x) = \log_3(x + x^2)$

■

Now we want to find a relationship between b^x and e^x for any base $b, b > 0$ and $b \neq 1$.

$$y = b^x \qquad \text{Take the natural logarithm of both sides.}$$
$$\ln y = \ln b^x$$
$$= x \ln b \qquad \text{If } \ln A = B, \text{ then } A = e^B.$$
$$y = e^{x \ln b}$$

Thus,

$$b^x = e^{x \ln b} \tag{7}$$

Differentiating both sides of equation (7), we have

$$\frac{d}{dx}b^x = e^{x \ln b}\ln b = b^x \ln b$$

Example 7 ➭ **Differentiating Exponential Functions**　Find $f'(x)$ for:

(A) $f(x) = 2^x$ 　　(B) $f(x) = 10^{x^5 + x}$

Solution 　(A) $f(x) = 2^x = e^{x \ln 2}$ 　　　　　　　Using equation (7)
　　　　　$f'(x) = e^{x \ln 2}\ln 2 = 2^x \ln 2$
　　　　(B) $f(x) = 10^{x^5 + x} = e^{(x^5 + x)\ln 10}$ 　　Using equation (7)
　　　　　$f'(x) = e^{(x^5 + x)\ln 10}(5x^4 + 1)\ln 10$
　　　　　　　$= 10^{x^5 + x}(5x^4 + 1)\ln 10$

■

Matched Problem 7 ➭ Find $f'(x)$ for:

(A) $f(x) = 5^x$ 　　(B) $f(x) = 4^{x^2 + 3x}$

■

Answers to Matched Problems　**1.** (A) $\ln(2x^3 + 4)$ 　　(B) $2(\ln u)^3 + 4$ 　　(C) $(2x^3 + 4)^{-5}$

2. (A) $y = f(u) = e^u; u = g(x) = 2x^3 + 7$
　　(B) $y = f(u) = \ln u; u = g(x) = x^4 + 10$

3. (A) $12x^3 e^{3x^4 + 6}$ 　　(B) $\dfrac{2x + 9}{x^2 + 9x + 4}$ 　　(C) $-30x^2(2x^3 + 4)^{-6}$

4. $\dfrac{3e^x[\ln(1 + e^x)]^2}{1 + e^x}$

5. (A) $\dfrac{3x^2 + 2}{x^3 + 2x}$ 　　(B) $6xe^{3x^2 + 2}$ 　　(C) $-8xe^{-x^2}(2 + e^{-x^2})^3$

6. (A) $\dfrac{1}{\ln 10}\left(\dfrac{1}{x}\right)$ 　　(B) $\dfrac{1}{\ln 3}\left(\dfrac{1 + 2x}{x + x^2}\right)$

7. (A) $5^x \ln 5$ 　　(B) $4^{x^2 + 3x}(2x + 3)\ln 4$

Exercise 5-3

A *In Problems 1–6, find $f[g(x)]$.*

1. $f(u) = u^3; g(x) = 3x^2 + 2$
2. $f(u) = u^4; g(x) = 1 - 4x^3$
3. $f(u) = \ln u; g(x) = 2x^2 + 5$
4. $f(u) = \ln u; g(x) = 3x^4 + 10$
5. $f(u) = e^u; g(x) = -x^2$
6. $f(u) = e^u; g(x) = 3x^3$

Write each composite function in Problems 7–12 in the form $y = f(u)$ and $u = g(x)$.

7. $y = (3x^2 - x + 5)^4$
8. $y = (2x^3 + x + 3)^5$
9. $y = \ln(3x^2 - 2x + 9)$
10. $y = \ln(x^4 + x^2 + 12)$
11. $y = e^{1+x+x^2}$
12. $y = e^{x^4+2x^2+5}$

In Problems 13–18, express y in terms of x. Use the chain rule to find dy/dx, and then express dy/dx in terms of x.

13. $y = u^2; u = 2 + e^x$
14. $y = u^3; u = 3 - \ln x$
15. $y = e^u; u = 2 - x^4$
16. $y = e^u; u = x^6 + 5x^2$
17. $y = \ln u; u = 4x^5 - 7$
18. $y = \ln u; u = 2 + 3x^4$

In Problems 19–52, find the derivative and simplify. (For some of these problems, the answers in the back of the book include both an unsimplified form and a simplified form. When checking your work, first check that you applied the rules correctly, and then check that you performed the algebraic simplification correctly.)

19. $\frac{d}{dx}\ln(x - 3)$
20. $\frac{d}{dw}\ln(w + 100)$
21. $\frac{d}{dt}\ln(3 - 2t)$
22. $\frac{d}{dy}\ln(4 - 5y)$
23. $\frac{d}{dx}3e^{2x}$
24. $\frac{d}{dy}2e^{3y}$
25. $\frac{d}{dt}2e^{-4t}$
26. $\frac{d}{dr}6e^{-3r}$

B

27. $\frac{d}{dx}100e^{-0.03x}$
28. $\frac{d}{dt}1{,}000e^{0.06t}$
29. $\frac{d}{dx}\ln(x + 1)^4$
30. $\frac{d}{dx}\ln(x + 1)^{-3}$

31. $\frac{d}{dx}(2e^{2x} - 3e^x + 5)$
32. $\frac{d}{dt}(1 + e^{-t} - e^{-2t})$
33. $\frac{d}{dx}e^{3x^2-2x}$
34. $\frac{d}{dx}e^{x^3-3x^2+1}$
35. $\frac{d}{dt}\ln(t^2 + 3t)$
36. $\frac{d}{dx}\ln(x^3 - 3x^2)$
37. $\frac{d}{dx}\ln(x^2 + 1)^{1/2}$
38. $\frac{d}{dx}\ln(x^4 + 5)^{3/2}$
39. $\frac{d}{dt}[\ln(t^2 + 1)]^4$
40. $\frac{d}{dw}[\ln(w^3 - 1)]^2$
41. $\frac{d}{dx}(e^{2x} - 1)^4$
42. $\frac{d}{dx}(e^{x^2} + 3)^5$
43. $\frac{d}{dx}\frac{e^{2x}}{x^2 + 1}$
44. $\frac{d}{dx}\frac{e^{x+1}}{x + 1}$
45. $\frac{d}{dx}(x^2 + 1)e^{-x}$
46. $\frac{d}{dx}(1 - x)e^{2x}$
47. $\frac{d}{dx}(e^{-x}\ln x)$
48. $\frac{d}{dx}\frac{\ln x}{e^x + 1}$
49. $\frac{d}{dx}\frac{1}{\ln(1 + x^2)}$
50. $\frac{d}{dx}\frac{1}{\ln(1 - x^3)}$
51. $\frac{d}{dx}\sqrt[3]{\ln(1 - x^2)}$
52. $\frac{d}{dt}\sqrt[5]{\ln(1 - t^5)}$

53. Describe the relationship between the graphs of $y = \ln x$ and $y = \log_b x$ for $b > 1$.
54. Describe the relationship between the graphs of $y = \ln x$ and $y = \log_b x$ for $0 < b < 1$.

C *In Problems 55–60, apply the graphing strategy discussed in Section 4-3 to f, state the pertinent information, and sketch the graph.*

55. $f(x) = 1 - e^{-x}$
56. $f(x) = 2 - 3e^{-2x}$
57. $f(x) = \ln(1 - x)$
58. $f(x) = \ln(2x + 4)$
59. $f(x) = e^{-0.5x^2}$
60. $f(x) = \ln(x^2 + 4)$

In Problems 61 and 62, express y in terms of x. Use the chain rule to find dy/dx, and express dy/dx in terms of x.

61. $y = 1 + w^2; w = \ln u; u = 2 + e^x$
62. $y = \ln w; w = 1 + e^u; u = x^2$

In Problems 63–70, find each derivative.

63. $\frac{d}{dx}\log_2(3x^2 - 1)$
64. $\frac{d}{dx}\log(x^3 - 1)$
65. $\frac{d}{dx}10^{x^2+x}$
66. $\frac{d}{dx}8^{1-2x^2}$
67. $\frac{d}{dx}\log_3(4x^3 + 5x + 7)$
68. $\frac{d}{dx}\log_5(5^{x^2-1})$

69. $\dfrac{d}{dx} 2^{x^3-x^2+4x+1}$ **70.** $\dfrac{d}{dx} 10^{\ln x}$

71. Use the chain rule to derive the formula:

$$\frac{d}{dx}\ln[f(x)] = \frac{1}{f(x)}f'(x)$$

72. Use the chain rule to derive the formula:

$$\frac{d}{dx}e^{f(x)} = e^{f(x)}f'(x)$$

73. (A) Use the graphs of $f(x) = \ln(x^2 + 1)$ and $g(x) = 1/(x^2 + 1)$ (see the figure) to explain why $g(x)$ is not the derivative of $f(x)$.

(B) Compute $f'(x)$ and sketch its graph.

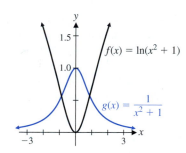

Figure for 73

74. (A) How many times is the chain rule used in computing the derivative of $f(x) = e^{\sqrt{3x}}$?

(B) Construct a function $g(x)$ such that the chain rule is used three times in computing its derivative.

75. Suppose a student reasons that the functions $f(x) = \ln[5(x^2 + 3)^4]$ and $g(x) = 4\ln(x^2 + 3)$ must have the same derivative, since he has entered $f(x), g(x), f'(x)$, and $g'(x)$ into a graphing utility, but only three graphs appear (see the figure). Is his reasoning correct? Are $f'(x)$ and $g'(x)$ the same function? Explain.

(A)

(B)

Figure for 75

76. Suppose a student reasons that the functions $f(x) = (x + 1)\ln(x + 1) - x$ and $g(x) = (x + 1)^{1/3}$ must have the same derivative, since she has entered $f(x)$, $g(x), f'(x)$, and $g'(x)$ into a graphing utility, but only three graphs appear (see the figure). Is her reasoning correct? Are $f'(x)$ and $g'(x)$ the same function? Explain.

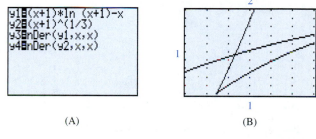

(A)

(B)

Figure for 76

Applications

Business & Economics

77. *Maximum revenue.* Suppose the price–demand equation for x units of a commodity is determined from empirical data to be

$$p = 100e^{-0.05x}$$

where x units are sold per day at a price of $\$p$ each. Find the production level and price that maximize revenue. What is the maximum revenue?

78. *Maximum revenue.* Repeat Problem 77 using the price–demand equation

$$p = 10e^{-0.04x}$$

79. *Maximum profit.* Refer to Problem 77. If the daily fixed cost is $\$400$ and the cost per unit is $\$6$, use approximation techniques to find the production level and the price that maximize profit. What is the maximum profit? [*Hint:* Graph $y = P(x)$ and $y = P'(x)$ in the same viewing window.]

80. *Maximum profit.* Refer to Problem 78. If the daily fixed cost is $\$30$ and the cost per unit is $\$0.70$, use approximation techniques to find the production level and the price that maximize profit. What is the maximum profit? [*Hint:* Graph $y = P(x)$ and $y = P'(x)$ in the same viewing window.]

81. *Maximum profit.* A mail-order company specializing in computer equipment has collected the data in the table, showing the weekly demand x for Data-Link modems at various prices p. The company purchases the modems from the manufacturer for $100 each. Use exponential regression $(p = ab^x)$ to find the price (to the nearest cent) that will maximize the weekly profit.

DEMAND	PRICE PER MODEM ($)
x	p
412	169.95
488	149.95
575	139.95
722	129.95
786	119.95

82. *Maximum profit.* The mail-order company in Problem 81 also sells 100 megabyte disks. The data in the table show the weekly demand x for these disks at various price levels p. The company purchases the disks from the manufacturer for $6 each. Use exponential regression $(p = ab^x)$ to find the price (to the nearest cent) that will maximize the weekly profit.

DEMAND	PRICE PER DISK ($)
x	p
578	16.95
942	14.95
1,218	13.95
1,758	11.95
2,198	10.95

83. *Salvage value.* The salvage value S (in dollars) of a company airplane after t years is estimated to be given by

$$S(t) = 300,000e^{-0.1t}$$

What is the rate of depreciation (in dollars per year) after 1 year? 5 years? 10 years?

84. *Resale value.* The resale value R (in dollars) of a company car after t years is estimated to be given by

$$R(t) = 20,000e^{-0.15t}$$

What is the rate of depreciation (in dollars per year) after 1 year? 2 years? 3 years?

85. *Promotion and maximum profit.* A recording company has produced a new compact disk featuring a very popular recording group. Before launching a national sales campaign, the marketing research department chose to test market the CD in a bellwether city. Their interest is in determining the length of a sales campaign that will maximize total profits. From empirical data, the research department estimates that the proportion of a target group of 50,000 persons buying the CD after t days of television promotion is given by $1 - e^{-0.03t}$. If $4 is received for each CD sold, the total revenue after t days of promotion will be approximated by

$$R(t) = (4)(50,000)(1 - e^{-0.03t}) \qquad t \geq 0$$

Television promotion costs are

$$C(t) = 4,000 + 3,000t \qquad t \geq 0$$

(A) How many days of television promotion should be used to maximize total profit? What is the maximum total profit? What percentage of the target market will have purchased the CD when the maximum profit is reached?

(B) Graph the profit function.

86. *Promotion and maximum profit.* Repeat Problem 85 using the revenue equation

$$R(t) = (3)(60,000)(1 - e^{-0.04t})$$

Life Sciences

87. *Blood pressure and age.* A research group using hospital records developed the following approximate mathematical model relating systolic blood pressure and age:

$$P(x) = 40 + 25 \ln(x + 1) \qquad 0 \leq x \leq 65$$

where $P(x)$ is pressure measured in millimeters of mercury and x is age in years. What is the rate of change of pressure at the end of 10 years? At the end of 30 years? At the end of 60 years?

88. *Biology.* A yeast culture at room temperature (68°F) is placed in a refrigerator maintaining a constant temperature of 38°F. After t hours, the temperature T of the culture is given approximately by

$$T = 30e^{-0.58t} + 38 \qquad t \geq 0$$

What is the rate of change of temperature of the culture at the end of 1 hour? At the end of 4 hours?

89. *Bacterial growth.* A single cholera bacterium divides every 0.5 hour to produce two complete cholera bacteria. If we start with a colony of 5,000 bacteria, after t hours there will be

$$A(t) = 5,000 \cdot 2^{2t}$$

bacteria. Find $A'(t)$, $A'(1)$, and $A'(5)$, and interpret the results.

90. *Bacterial growth.* Repeat Problem 89 for a starting colony of 1,000 bacteria where a single bacterium divides every 0.25 hour.

Social Sciences

91. *Sociology.* Daniel Lowenthal, a sociologist at Columbia University, made a 5-year study on the sale of popular records relative to their position in the top 20. He found that the average number of sales $N(n)$ of the nth-ranking record was given approximately by

$$N(n) = N_1 e^{-0.09(n-1)} \qquad 1 \leq n \leq 20$$

where N_1 was the number of sales of the top record on the list at a given time. Graph N for $N_1 = 1,000,000$ records.

92. *Political science.* Thomas W. Casstevens, a political scientist at Oakland University, has studied legislative turnover. He (with others) found that the number $N(t)$ of continuously serving members of an elected legislative body remaining t years after an election is given approximately by a function of the form

$$N(t) = N_0 e^{-ct}$$

In particular, for the 1965 election for the U.S. House of Representatives, it was found that

$$N(t) = 434 e^{-0.0866t}$$

What is the rate of change after 2 years? After 10 years?

Section 5-4

Implicit Differentiation

- ❑ SPECIAL FUNCTION NOTATION
- ❑ IMPLICIT DIFFERENTIATION

❑   SPECIAL FUNCTION NOTATION

The equation

$$y = 2 - 3x^2 \tag{1}$$

defines a function f with y as a dependent variable and x as an independent variable. Using function notation, we would write

$$y = f(x) \qquad \text{or} \qquad f(x) = 2 - 3x^2$$

In order to reduce to a minimum the number of symbols involved in a discussion, we will often write equation (1) in the form

$$y = 2 - 3x^2 = y(x)$$

where y is *both* a dependent variable and a function symbol. This is a convenient notation, and no harm is done as long as one is aware of the double role of y. Other examples are

$$x = 2t^2 - 3t + 1 = x(t)$$
$$z = \sqrt{u^2 - 3u} = z(u)$$
$$r = \frac{1}{(s^2 - 3s)^{2/3}} = r(s)$$

This type of notation will simplify much of the discussion and work that follows.

Until now we have considered functions involving only one independent variable. There is no reason to stop there. The concept can be generalized to functions involving two or more independent variables, and this will be done in detail in Chapter 8. For now, we will "borrow" the notation for a function involving two independent variables. For example,

$$F(x, y) = x^2 - 2xy + 3y^2 - 5$$

specifies a function F involving two independent variables.

❑ IMPLICIT DIFFERENTIATION

Consider the equation

$$3x^2 + y - 2 = 0 \tag{2}$$

and the equation obtained by solving equation (2) for y in terms of x,

$$y = 2 - 3x^2 \tag{3}$$

Both equations define the same function using x as the independent variable and y as the dependent variable. For equation (3), we can write

$$y = f(x)$$

where

$$f(x) = 2 - 3x^2 \tag{4}$$

and we have an **explicit** (clearly stated) rule that enables us to determine y for each value of x. On the other hand, the y in equation (2) is the same y as in equation (3), and equation (2) **implicitly** gives (implies though does not plainly express) y as a function of x. Thus, we say that equations (3) and (4) define the function f explicitly, and equation (2) defines f implicitly.

The direct use of an equation that defines a function implicitly to find the derivative of the dependent variable with respect to the independent variable is called **implicit differentiation.** Let us differentiate equation (2) implicitly and equation (3) directly, and compare results.

Starting with

$$3x^2 + y - 2 = 0$$

we think of y as a function of x— that is, $y = y(x)$— and write

$$3x^2 + y(x) - 2 = 0$$

Then we differentiate both sides with respect to x:

$$\frac{d}{dx}[(3x^2 + y(x) - 2)] = \frac{d}{dx}0$$

$$\frac{d}{dx}3x^2 + \frac{d}{dx}y(x) - \frac{d}{dx}2 = 0$$

$$6x + y' - 0 = 0$$

Since y is a function of x, but is not explicitly given, we simply write $\frac{d}{dx}y(x) = y'$ to indicate its derivative.

Now, we solve for y':

$$y' = -6x$$

Note that we get the same result if we start with equation (3) and differentiate directly:

$$y = 2 - 3x^2$$
$$y' = -6x$$

Why are we interested in implicit differentiation? In general, why do we not solve for y in terms of x and differentiate directly? The answer is that there are many equations of the form

$$F(x, y) = 0 \tag{5}$$

that are either difficult or impossible to solve for y explicitly in terms of x (try it for $x^2y^5 - 3xy + 5 = 0$ or for $e^y - y = 3x$, for example). But it can be shown that, under fairly general conditions on F, equation (5) will define one or more functions where y is a dependent variable and x is an independent variable. To find y' under these conditions, we differentiate equation (5) implicitly.

Explore–Discuss 1

www

(A) How many tangent lines are there to the graph in Figure 1 when $x = 0$? When $x = 1$? When $x = 2$? When $x = 4$? When $x = 6$?

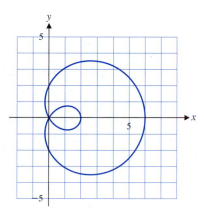

FIGURE 1

(B) Sketch the tangent lines referred to in part (A) and estimate each of their slopes.
(C) Explain why the graph in Figure 1 is not the graph of a function.

Example 1 ⇨ **Differentiating Implicitly**   Given

$$F(x, y) = x^2 + y^2 - 25 = 0 \qquad (6)$$

find y' and the slope of the graph at $x = 3$.

SOLUTION   We start with the graph of $x^2 + y^2 - 25 = 0$ (a circle, as shown in Fig. 2) so that we can interpret our results geometrically. From the graph it is clear that equation (6) does not define a function. But with a suitable restriction on the variables, equation (6) can define two or more functions. For example, the upper half and the lower half of the circle each define a function. A point on each half-circle that corresponds to $x = 3$ is found by substituting $x = 3$ into equation (6) and solving for y:

$$x^2 + y^2 - 25 = 0$$
$$(3)^2 + y^2 = 25$$
$$y^2 = 16$$
$$y = \pm 4$$

$x^2 + y^2 = 25$

FIGURE 2

Thus, the point $(3, 4)$ is on the upper half-circle, and the point $(3, -4)$ is on the lower half-circle. We will use these results in a moment. We now differentiate equation (6) implicitly, treating y as a function of x; that is, $y = y(x)$:

$$x^2 + y^2 - 25 = 0$$

$$x^2 + [y(x)]^2 - 25 = 0$$

$$\frac{d}{dx}\{x^2 + [y(x)]^2 - 25\} = \frac{d}{dx}\,0$$

$$\frac{d}{dx}x^2 + \frac{d}{dx}[y(x)]^2 - \frac{d}{dx}25 = 0 \qquad \textit{Use the chain rule.}$$

$$2x + 2[y(x)]^{2-1}y'(x) - 0 = 0$$

$$2x + 2yy' = 0 \qquad \textit{Solve for } y' \textit{ in terms of x and y.}$$

$$y' = -\frac{2x}{2y}$$

$$y' = -\frac{x}{y} \qquad \textit{Leave the answer in terms of x and y.}$$

www   We have found y' without first solving $x^2 + y^2 - 25 = 0$ for y in terms of x. And by leaving y' in terms of x and y, we can use $y' = -x/y$ to find y' for *any* point on the graph of $x^2 + y^2 - 25 = 0$ (except where $y = 0$). In particular, for $x = 3$, we found that $(3, 4)$ and $(3, -4)$ are on the graph; thus, the slope of the graph at $(3, 4)$ is

$$y'|_{(3,4)} = -\tfrac{3}{4} \qquad \textit{The slope of the graph at } (3, 4)$$

and the slope at $(3, -4)$ is

$$y'|_{(3,-4)} = -\tfrac{3}{-4} = \tfrac{3}{4} \qquad \textit{The slope of the graph at } (3, -4)$$

The symbol

$$y'|_{(a,\, b)}$$

is used to indicate that we are evaluating y' at $x = a$ and $y = b$.

The results are interpreted geometrically on the original graph as shown in Figure 3. ■

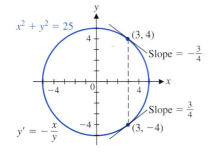

FIGURE 3

In Example 1, the fact that y' is given in terms of both x and y is not a great disadvantage. We have only to make certain that when we want **to evaluate y' for a particular value of x and y, say, (x_0, y_0), the ordered pair must satisfy the original equation.**

Matched Problem 1 ↩ Graph $x^2 + y^2 - 169 = 0$, find y' by implicit differentiation, and find the slope of the graph when $x = 5$. ■

Example 2 ↩ **Differentiating Implicitly**   Find the equation(s) of the tangent line(s) to the graph of

$$y - xy^2 + x^2 + 1 = 0$$

at the point(s) where $x = 1$.   (7)

SOLUTION   We first find y when $x = 1$:

$$y - xy^2 + x^2 + 1 = 0$$
$$y - (1)y^2 + (1)^2 + 1 = 0$$
$$y - y^2 + 2 = 0$$
$$y^2 - y - 2 = 0$$
$$(y - 2)(y + 1) = 0$$
$$y = -1, 2$$

Thus, there are two points on the graph of (7) where $x = 1$; namely, $(1, -1)$ and $(1, 2)$. We next find the slope of the graph at these two points by differentiating equation (7) implicitly:

$$y - xy^2 + x^2 + 1 = 0$$

$$\frac{d}{dx}y - \frac{d}{dx}xy^2 + \frac{d}{dx}x^2 + \frac{d}{dx}1 = \frac{d}{dx}0$$

Use the product rule and the chain rule for $\frac{d}{dx}xy^2$.

$$y' - (x \cdot 2yy' + y^2) + 2x = 0$$
$$y' - 2xyy' - y^2 + 2x = 0$$
$$y' - 2xyy' = y^2 - 2x$$
$$(1 - 2xy)y' = y^2 - 2x$$
$$y' = \frac{y^2 - 2x}{1 - 2xy}$$

Solve for y' by getting all terms involving y' on one side.

Now, find the slope at each point:

$$y'|_{(1,-1)} = \frac{(-1)^2 - 2(1)}{1 - 2(1)(-1)} = \frac{1 - 2}{1 + 2} = \frac{-1}{3} = -\frac{1}{3}$$

$$y'|_{(1,2)} = \frac{(2)^2 - 2(1)}{1 - 2(1)(2)} = \frac{4 - 2}{1 - 4} = \frac{2}{-3} = -\frac{2}{3}$$

Equation of tangent line at $(1, -1)$:   Equation of tangent line at $(1, 2)$:

$$y - y_1 = m(x - x_1) \qquad\qquad y - y_1 = m(x - x_1)$$
$$y + 1 = -\tfrac{1}{3}(x - 1) \qquad\qquad y - 2 = -\tfrac{2}{3}(x - 1)$$
$$y + 1 = -\tfrac{1}{3}x + \tfrac{1}{3} \qquad\qquad y - 2 = -\tfrac{2}{3}x + \tfrac{2}{3}$$
$$y = -\tfrac{1}{3}x - \tfrac{2}{3} \qquad\qquad y = -\tfrac{2}{3}x + \tfrac{8}{3}$$

Matched Problem 2   Repeat Example 2 for $x^2 + y^2 - xy - 7 = 0$ at $x = 1$.

Explore–Discuss 2

The slopes of the tangent lines to $y^2 + 3xy + 4x = 9$ when $x = 0$ can be found in either of the following ways: (1) by differentiating the equation implicitly; or (2) by solving for y explicitly in terms of x (using the quadratic formula), and then computing the derivative. Which of the two methods is more efficient? Explain.

Example 3   **Differentiating Implicitly**   Find x' for $x = x(t)$ defined implicitly by

$$t \ln x = xe^t - 1$$

and evaluate x' at $(t, x) = (0, 1)$.

SOLUTION   It is important to remember that x is the dependent variable and t is the independent variable. Therefore, we differentiate both sides of the equation with respect to t (using product and chain rules where appropriate), and then solve for x':

$$t \ln x = xe^t - 1$$ Differentiate implicitly with respect to t.

$$\frac{d}{dt}(t \ln x) = \frac{d}{dt}(xe^t) - \frac{d}{dt} 1$$

$$t \frac{x'}{x} + \ln x = xe^t + x'e^t$$ Clear fractions.

$$x \cdot t \frac{x'}{x} + x \cdot \ln x = x \cdot xe^t + x \cdot e^t x'$$ $x \neq 0$

$$tx' + x \ln x = x^2 e^t + xe^t x'$$ Solve for x'.

$$tx' - xe^t x' = x^2 e^t - x \ln x$$ Factor out x'.

$$(t - xe^t)x' = x^2 e^t - x \ln x$$

$$x' = \frac{x^2 e^t - x \ln x}{t - xe^t}$$

Now, we evaluate x' at $(t, x) = (0, 1)$, as requested:

$$x'|_{(0,1)} = \frac{(1)^2 e^0 - 1 \ln 1}{0 - 1e^0}$$

$$= \frac{1}{-1} = -1$$

Matched Problem 3   Find x' for $x = x(t)$ defined implicitly by

$$1 + x \ln t = te^x$$

and evaluate x' at $(t, x) = (1, 0)$.

Answers to Matched Problems   **1.** $y' = -x/y$; when $x = 5$, $y = \pm 12$, thus, $y'|_{(5,12)} = -\frac{5}{12}$ and $y'|_{(5,-12)} = \frac{5}{12}$

2. $y' = \frac{y - 2x}{2y - x}$; $y = \frac{4}{5}x - \frac{14}{5}$, $y = \frac{1}{5}x + \frac{14}{5}$ **3.** $x' = \frac{te^x - x}{t \ln t - t^2 e^x}$; $x'|_{(1,0)} = -1$

A *In Problems 1–4, find y' two ways:*

(A) *Differentiate the given equation implicitly and then solve for y'.*

(B) *Solve the given equation for y and then differentiate directly.*

1. $3x + 5y + 9 = 0$ **2.** $-2x + 6y - 4 = 0$

3. $3x^2 - 4y - 18 = 0$ **4.** $2x^3 + 5y - 2 = 0$

In Problems 5–22, use implicit differentiation to find y' and evaluate y' at the indicated point.

5. $y - 5x^2 + 3 = 0$; $(1, 2)$

6. $5x^3 - y - 1 = 0$; $(1, 4)$

7. $x^2 - y^3 - 3 = 0$; $(2, 1)$

8. $y^2 + x^3 + 4 = 0$; $(-2, 2)$

9. $y^2 + 2y + 3x = 0$; $(-1, 1)$

10. $y^2 - y - 4x = 0$; $(0, 1)$

B

11. $xy - 6 = 0$; $(2, 3)$

12. $3xy - 2x - 2 = 0$; $(2, 1)$

13. $2xy + y + 2 = 0$; $(-1, 2)$

14. $2y + xy - 1 = 0$; $(-1, 1)$

15. $x^2 y - 3x^2 - 4 = 0$; $(2, 4)$

16. $2x^3y - x^3 + 5 = 0; (-1, 3)$

17. $e^y = x^2 + y^2; (1, 0)$ 18. $x^2 - y = 4e^y; (2, 0)$

19. $x^3 - y = \ln y; (1, 1)$ 20. $\ln y = 2y^2 - x; (2, 1)$

21. $x \ln y + 2y = 2x^3; (1, 1)$

22. $xe^y - y = x^2 - 2; (2, 0)$

In Problems 23 and 24, find x' for x = x(t) defined implicitly by the given equation. Evaluate x' at the indicated point.

23. $x^2 - t^2x + t^3 + 11 = 0; (-2, 1)$

24. $x^3 - tx^2 - 4 = 0; (-3, -2)$

Problems 25 and 26 refer to the equation and graph shown in the figure .

$(x - 1)^2 + (y - 1)^2 = 1$

25. Use implicit differentiation to find the slopes of the tangent lines at the points on the graph where $x = 1.6$. Check your answers by visually estimating the slopes on the graph in the figure.

26. Find the slopes of the tangent lines at the points on the graph where $x = 0.2$. Check your answers by visually estimating the slopes on the graph in the figure.

In Problems 27–30, find the equation(s) of the tangent line(s) to the graphs of the indicated equations at the point(s) with the given value of x.

27. $xy - x - 4 = 0; x = 2$

28. $3x + xy + 1 = 0; x = -1$

29. $y^2 - xy - 6 = 0; x = 1$

30. $xy^2 - y - 2 = 0; x = 1$

31. If $xe^y = 1$, find y' in two ways: first by implicit differentiation, then by solving for y explicitly in terms of x. Which method do you prefer? Explain.

32. Explain the difficulty in solving $x^3 + y + xe^y = 1$ for y as an explicit function of x. Find the slope of the tangent line to the graph of the equation at the point $(0, 1)$.

C *In Problems 33–40, find y' and the slope of the tangent line to the graph of each equation at the indicated point.*

33. $(1 + y)^3 + y = x + 7; (2, 1)$

34. $(y - 3)^4 - x = y; (-3, 4)$

35. $(x - 2y)^3 = 2y^2 - 3; (1, 1)$

36. $(2x - y)^4 - y^3 = 8; (-1, -2)$

37. $\sqrt{7 + y^2} - x^3 + 4 = 0; (2, 3)$

38. $6\sqrt{y^3 + 1} - 2x^{3/2} - 2 = 0; (4, 2)$

39. $\ln(xy) = y^2 - 1; (1, 1)$

40. $e^{xy} - 2x = y + 1; (0, 0)$

41. Find the equation(s) of the tangent line(s) at the point(s) on the graph of the equation

$$y^3 - xy - x^3 = 2$$

where $x = 1$. Round all approximate values to two decimal places.

42. Refer to the equation in Problem 41. Find the equation(s) of the tangent line(s) at the point(s) on the graph where $y = -1$. Round all approximate values to two decimal places.

Applications

Business & Economics

For the demand equations in Problems 43–46, find the rate of change of p with respect to x by differentiating implicitly (x is the number of items that can be sold at a price of $p.)

43. $x = p^2 - 2p + 1,000$ 44. $x = p^3 - 3p^2 + 200$

45. $x = \sqrt{10,000 - p^2}$ 46. $x = \sqrt[3]{1,500 - p^3}$

Life Sciences

47. *Biophysics.* In biophysics, the equation

$$(L + m)(V + n) = k$$

is called the *fundamental equation of muscle contraction,* where m, n, and k are constants, and V is the velocity of the shortening of muscle fibers for a muscle subjected to a load of L. Find dL/dV using implicit differentiation.

48. *Biophysics.* In Problem 47, find dV/dL using implicit differentiation.

Related Rates

The workers in a union are concerned that the rate at which wages are increasing is lagging behind the rate of increase in the company's profits. An automobile dealer wants to predict how badly an anticipated increase in interest rates will decrease his rate of sales. An investor is studying the connection between the rate of increase in the Dow Jones Average and the rate of increase in the Gross Domestic Product over the past 50 years.

In each of these situations there are two quantities—wages and profits in the first instance, for example—that are changing with respect to time. We would like to discover the precise relationship between the rates of increase (or decrease) of the two quantities. We will begin our discussion of such *related rates* by considering some familiar situations in which the two quantities are distances, and the two rates are velocities.

Example 1 ↩
www

Related Rates and Motion A 26-foot ladder is placed against a wall (Fig. 1). If the top of the ladder is sliding down the wall at 2 feet per second, at what rate is the bottom of the ladder moving away from the wall when the bottom of the ladder is 10 feet away from the wall?

SOLUTION

FIGURE 1

Many people reason that since the ladder is of constant length, the bottom of the ladder will move away from the wall at the same rate that the top of the ladder is moving down the wall. This is not the case, as we will see.

At any moment in time, let x be the distance of the bottom of the ladder from the wall, and let y be the distance of the top of the ladder on the wall (see Fig. 1). Both x and y are changing with respect to time and can be thought of as functions of time; that is, $x = x(t)$ and $y = y(t)$. Furthermore, x and y are related by the Pythagorean relationship:

$$x^2 + y^2 = 26^2 \tag{1}$$

Differentiating equation (1) implicitly with respect to time t, and using the chain rule where appropriate, we obtain

$$2x\frac{dx}{dt} + 2y\frac{dy}{dt} = 0 \tag{2}$$

The rates dx/dt and dy/dt are related by equation (2); hence, this type of problem is referred to as a **related rates problem.**

Now our problem is to find dx/dt when $x = 10$ feet, given that $dy/dt = -2$ (y is decreasing at a constant rate of 2 feet per second). We have all the quantities we need in equation (2) to solve for dx/dt, except y. When $x = 10$, y can be found using equation (1):

$$10^2 + y^2 = 26^2$$
$$y = \sqrt{26^2 - 10^2} = 24 \text{ feet}$$

Substitute $dy/dt = -2$, $x = 10$, and $y = 24$ into (2); then solve for dx/dt:

$$2(10)\frac{dx}{dt} + 2(24)(-2) = 0$$

$$\frac{dx}{dt} = \frac{-2(24)(-2)}{2(10)} = 4.8 \text{ feet per second}$$

Thus, the bottom of the ladder is moving away from the wall at a rate of 4.8 feet per second.

Matched Problem 1 ➭ Again, a 26-foot ladder is placed against a wall (Fig. 1). If the bottom of the ladder is moving away from the wall at 3 feet per second, at what rate is the top moving down when the top of the ladder is 24 feet up the wall? ■

Explore–Discuss 1

(A) For which values of x and y in Example 1 is dx/dt equal to 2 (that is, the same rate at which the ladder is sliding down the wall)?

(B) When is dx/dt greater than 2? Less than 2?

Suggestions for Solving Related Rates Problems

Step 1. Sketch a figure if helpful.

Step 2. Identify all relevant variables, including those whose rates are given and those whose rates are to be found.

Step 3. Express all given rates and rates to be found as derivatives.

Step 4. Find an equation connecting the variables in step 2.

Step 5. Implicitly differentiate the equation found in step 4, using the chain rule where appropriate, and substitute in all given values.

Step 6. Solve for the derivative that will give the unknown rate.

Example 2 ➭
www

Related Rates and Motion   Suppose that two motorboats leave from the same point at the same time. If one travels north at 15 miles per hour and the other travels east at 20 miles per hour, how fast will the distance between them be changing after 2 hours?

SOLUTION   First, draw a picture, as shown in Figure 2.

FIGURE 2

All variables, x, y, and z, are changing with time. Hence, they can be thought of as functions of time; $x = x(t)$, $y = y(t)$, and $z = z(t)$, given implicitly. It now makes sense to take derivatives of each variable with respect to time. From the Pythagorean theorem,

$$z^2 = x^2 + y^2 \tag{3}$$

We also know that

$$\frac{dx}{dt} = 20 \text{ miles per hour} \qquad \text{and} \qquad \frac{dy}{dt} = 15 \text{ miles per hour}$$

We would like to find dz/dt at the end of 2 hours; that is, when $x = 40$ miles and $y = 30$ miles. To do this, we differentiate both sides of equation (3) with respect to t and solve for dz/dt:

$$2z\frac{dz}{dt} = 2x\frac{dx}{dt} + 2y\frac{dy}{dt} \tag{4}$$

We have everything we need except z. When $x = 40$ and $y = 30$, we find z from equation (3) to be 50. Substituting the known quantities into equation (4), we obtain

$$2(50)\frac{dz}{dt} = 2(40)(20) + 2(30)(15)$$

$$\frac{dz}{dt} = 25 \text{ miles per hour}$$

Thus, the boats will be separating at a rate of 25 miles per hour. ■

Matched Problem 2 ⇔ Repeat Example 2 for the situation at the end of 3 hours. ∎

Example 3 ⇔ **Related Rates and Motion** Suppose a point is moving along the graph of $x^2 + y^2 = 25$ (Fig. 3). When the point is at $(-3, 4)$, its x coordinate is increasing at the rate of 0.4 unit per second. How fast is the y coordinate changing at that moment?

Solution Since both x and y are changing with respect to time, we can think of each as a function of time:

$$x = x(t) \quad \text{and} \quad y = y(t)$$

but restricted so that

$$x^2 + y^2 = 25 \tag{5}$$

Our problem is now to find dy/dt, given $x = -3$, $y = 4$, and $dx/dt = 0.4$. Implicitly differentiating both sides of equation (5) with respect to t, we have

$$x^2 + y^2 = 25$$

$$2x\frac{dx}{dt} + 2y\frac{dy}{dt} = 0 \qquad \text{Divide both sides by 2.}$$

$$x\frac{dx}{dt} + y\frac{dy}{dt} = 0 \qquad \text{Substitute } x = -3, y = 4, \text{ and } dx/dt = 0.4, \text{ and solve for } dy/dt.$$

$$(-3)(0.4) + 4\frac{dy}{dt} = 0$$

$$\frac{dy}{dt} = 0.3 \text{ unit per second} \qquad \blacksquare$$

y

(−3, 4)

$x^2 + y^2 = 25$

FIGURE 3

Matched Problem 3 ⇔ A point is moving on the graph of $y^3 = x^2$. When the point is at $(-8, 4)$, its y coordinate is decreasing at 2 units per second. How fast is the x coordinate changing at that moment? ∎

Example 4 ⇔ **Related Rates and Business** Suppose that for a company manufacturing transistor radios, the cost, revenue, and profit equations are given by

$$C = 5{,}000 + 2x \qquad \text{Cost equation}$$
$$R = 10x - 0.001x^2 \qquad \text{Revenue equation}$$
$$P = R - C \qquad \text{Profit equation}$$

where the production output in 1 week is x radios. If production is increasing at the rate of 500 radios per week when production is 2,000 radios, find the rate of increase in:

(A) Cost   (B) Revenue   (C) Profit

Solution If production x is a function of time (it must be, since it is changing with respect to time), then C, R, and P must also be functions of time. These functions are implicitly (rather than explicitly) given. Letting t represent time in weeks, we differentiate both sides of each of the three equations above with respect to t, and then substitute $x = 2{,}000$ and $dx/dt = 500$ to find the desired rates.

(A) $C = 5,000 + 2x$ *Think:* $C = C(t)$ *and* $x = x(t)$.

$$\frac{dC}{dt} = \frac{d}{dt}(5,000) + \frac{d}{dt}(2x)$$ *Differentiate both sides with respect to t.*

$$\frac{dC}{dt} = 0 + 2\frac{dx}{dt} = 2\frac{dx}{dt}$$

Since $dx/dt = 500$ when $x = 2,000$,

$$\frac{dC}{dt} = 2(500) = \$1,000 \text{ per week}$$

Cost is increasing at a rate of \$1,000 per week.

(B) $R = 10x - 0.001x^2$

$$\frac{dR}{dt} = \frac{d}{dt}(10x) - \frac{d}{dt}0.001x^2$$

$$\frac{dR}{dt} = 10\frac{dx}{dt} - 0.002x\frac{dx}{dt}$$

$$\frac{dR}{dt} = (10 - 0.002x)\frac{dx}{dt}$$

Since $dx/dt = 500$ when $x = 2,000$,

$$\frac{dR}{dt} = [10 - 0.002(2,000)](500) = \$3,000 \text{ per week}$$

Revenue is increasing at a rate of \$3,000 per week.

(C) $P = R - C$

$$\frac{dP}{dt} = \frac{dR}{dt} - \frac{dC}{dt}$$

$$= \$3,000 - \$1,000$$ *Results from parts (A) and (B)*

$$= \$2,000 \text{ per week}$$

Profit is increasing at a rate of \$2,000 per week.

■

Matched Problem 4 ⇔   Repeat Example 4 for a production level of 6,000 radios per week.

■

Explore–Discuss 2

(A) In Example 4 suppose that $x(t) = 500t + 500$. Find the time and production level at which the profit is maximized.

(B) Suppose that $x(t) = t^2 + 492t + 16$. Find the time and production level at which the profit is maximized.

(C) Explain why it is unnecessary to know a formula for $x(t)$ in order to determine the production level at which the profit is maximized.

Answers to Matched Problems   **1.** $dy/dt = -1.25$ ft/sec   **2.** $dz/dt = 25$ mi/hr   **3.** $dx/dt = 6$ units/sec

4. (A) $dC/dt = \$1,000/\text{wk}$ (B) $dR/dt = -\$1,000/\text{wk}$

     (C) $dP/dt = -\$2,000/\text{wk}$

Exercise 5-5

A *In Problems 1–6, assume that $x = x(t)$ and $y = y(t)$. Find the indicated rate, given the other information.*

1. $y = x^2 + 2$; $dx/dt = 3$ when $x = 5$; find dy/dt

2. $y = x^3 - 3$; $dx/dt = -2$ when $x = 2$; find dy/dt

3. $x^2 + y^2 = 1$; $dy/dt = -4$ when $x = -0.6$ and $y = 0.8$; find dx/dt

4. $x^2 + y^2 = 4$; $dy/dt = 5$ when $x = 1.2$ and $y = -1.6$; find dx/dt

5. $x^2 + 3xy + y^2 = 11$; $dx/dt = 2$ when $x = 1$ and $y = 2$; find dy/dt

6. $x^2 - 2xy - y^2 = 7$; $dy/dt = -1$ when $x = 2$ and $y = -1$; find dx/dt

B

7. A point is moving on the graph of $xy = 36$. When the point is at $(4, 9)$, its x coordinate is increasing at 4 units per second. How fast is the y coordinate changing at that moment?

8. A point is moving on the graph of $4x^2 + 9y^2 = 36$. When the point is at $(3, 0)$, its y coordinate is decreasing at 2 units per second. How fast is its x coordinate changing at that moment?

9. A boat is being pulled toward a dock as indicated in the figure . If the rope is being pulled in at 3 feet per second, how fast is the distance between the dock and the boat decreasing when it is 30 feet from the dock?

Rope

4 ft

Figure for 9 and 10

10. Refer to Problem 9. Suppose that the distance between the boat and the dock is decreasing at 3.05 feet per second. How fast is the rope being pulled in when the boat is 10 feet from the dock?

11. A rock is thrown into a still pond and causes a circular ripple. If the radius of the ripple is increasing at 2 feet per second, how fast is the area changing when the radius is 10 feet? [Use $A = \pi R^2$, $\pi \approx 3.14$.]

12. Refer to Problem 11. How fast is the circumference of a circular ripple changing when the radius is 10 feet? [Use $C = 2\pi R$, $\pi \approx 3.14$.]

13. The radius of a spherical balloon is increasing at the rate of 3 centimeters per minute. How fast is the

volume changing when the radius is 10 centimeters? [Use $V = \frac{4}{3}\pi R^3$, $\pi \approx 3.14$.]

14. Refer to Problem 13. How fast is the surface area of the sphere increasing when the radius is 10 centimeters? [Use $S = 4\pi R^2$, $\pi \approx 3.14$.]

15. Boyle's law for enclosed gases states that if the volume is kept constant, the pressure P and temperature T are related by the equation

$$\frac{P}{T} = k$$

where k is a constant. If the temperature is increasing at 3 kelvin per hour, what is the rate of change of pressure when the temperature is 250 kelvin and the pressure is 500 pounds per square inch?

16. Boyle's law for enclosed gases states that if the temperature is kept constant, the pressure P and volume V of the gas are related by the equation

$$VP = k$$

where k is a constant. If the volume is decreasing by 5 cubic inches per second, what is the rate of change of the pressure when the volume is 1,000 cubic inches and the pressure is 40 pounds per square inch?

17. A 10-foot ladder is placed against a vertical wall. Suppose the bottom slides away from the wall at a constant rate of 3 feet per second. How fast is the top sliding down the wall (negative rate) when the bottom is 6 feet from the wall? [*Hint*: Use the Pythagorean theorem: $a^2 + b^2 = c^2$, where c is the length of the hypotenuse of a right triangle and a and b are the lengths of the two shorter sides.]

18. A weather balloon is rising vertically at the rate of 5 meters per second. An observer is standing on the ground 300 meters from the point where the balloon was released. At what rate is the distance between the observer and the balloon changing when the balloon is 400 meters high?

C

19. A street light is on top of a 20-foot pole. A person who is 5 feet tall walks away from the pole at the rate of 5 feet per second. At what rate is the tip of the person's shadow moving away from the pole when he is 20 feet from the pole?

20. Refer to Problem 19. At what rate is the person's shadow growing when he is 20 feet from the pole?

21. Helium is pumped into a spherical balloon at a constant rate of 4 cubic feet per second. How fast is the radius increasing after 1 minute? After 2 minutes? Is there any time at which the radius is increasing at a rate of 100 feet per second? Explain.

22. A point is moving along the x axis at a constant rate of 5 units per second. At which point is its distance from $(0, 1)$ increasing at a rate of 2 units per second? At 4 units per second? At 5 units per second? At 10 units per second? Explain.

23. A point is moving on the graph of $y = e^x + x + 1$ in such a way that its x coordinate is always increasing at a rate of 3 units per second. How fast is the y coordinate changing when the point crosses the x axis?

24. A point is moving on the graph of $x^3 + y^2 = 1$ in such a way that its y coordinate is always increasing at a rate of 2 units per second. At which point(s) is the x coordinate increasing at a rate of 1 unit per second?

Applications

Business & Economics

25. *Cost, revenue, and profit rates.* Suppose that for a company manufacturing calculators, the cost, revenue, and profit equations are given by

$$C = 90,000 + 30x \qquad R = 300x - \frac{x^3}{30}$$

$$P = R - C$$

where the production output in 1 week is x calculators. If production is increasing at a rate of 500 calculators per week when production output is 6,000 calculators, find the rate of increase (decrease) in:

(A) Cost   (B) Revenue   (C) Profit

26. *Cost, revenue, and profit rates.* Repeat Problem 25 for

$$C = 72,000 + 60x \qquad R = 200x - \frac{x^2}{30}$$

$$P = R - C$$

where production is increasing at a rate of 500 calculators per week at a production level of 1,500 calculators.

27. *Advertising.* A retail store estimates that weekly sales s and weekly advertising costs x (both in dollars) are related by

$$s = 60,000 - 40,000e^{-0.0005x}$$

The current weekly advertising costs are $2,000, and these costs are increasing at the rate of $300 per week. Find the current rate of change of sales.

28. *Advertising.* Repeat Problem 27 for

$$s = 50,000 - 20,000e^{-0.0004x}$$

29. *Price–demand.* The price p (in dollars) and demand x for a product are related by

$$2x^2 + 5xp + 50p^2 = 80,000$$

(A) If the price is increasing at a rate of $2 per month when the price is $30, find the rate of change of the demand.

(B) If the demand is decreasing at a rate of 6 units per month when the demand is 150 units, find the rate of change of the price.

30. *Price–demand.* Repeat Problem 29 for

$$x^2 + 2xp + 25p^2 = 74,500$$

Life Sciences

31. *Pollution.* An oil tanker aground on a reef is leaking oil that forms a circular oil slick about 0.1 foot thick (see the figure). To estimate the rate dV/dt (in cubic feet per minute) at which the oil is leaking from the tanker, it was found that the radius of the slick was increasing at 0.32 foot per minute ($dR/dt = 0.32$) when the radius R was 500 feet. Find dV/dt, using $\pi \approx 3.14$.

Figure for 31

Social Sciences

32. *Learning.* A person who is new on an assembly line performs an operation in T minutes after x performances of the operation, as given by

$$T = 6\left(1 + \frac{1}{\sqrt{x}}\right)$$

If $dx/dt = 6$ operations per hour, where t is time in hours, find dT/dt after 36 performances of the operation.

Important Terms and Symbols

5-1 *The Constant e and Continuous Compound Interest.* Definition of e; continuous compound interest

$$A = Pe^{rt}$$

5-2 *Derivatives of Logarithmic and Exponential Functions.* Derivative formulas for the natural logarithmic and exponential functions; graph properties of $y = \ln x$ and $y = e^x$

5-3 *Chain Rule: General Form.* Composite functions; chain rule; general derivative formulas; derivative formulas for $y = \log_b x$ and $y = b^x$

5-4 *Implicit Differentiation.* Special function notation; function explicitly defined; function implicitly defined; implicit differentiation

$$y = f(x); \quad y = y(x); \quad F(x, y) = 0; \quad y'|_{(a, b)}$$

5-5 *Related Rates.* Related rates

$$x = x(t); \quad y = y(t)$$

Additional Rules of Differentiation

$$\frac{d}{dx} \ln x = \frac{1}{x} \qquad \frac{d}{dx} e^x = e^x$$

$$\frac{d}{dx} \ln[f(x)] = \frac{1}{f(x)} f'(x) \qquad \frac{d}{dx} e^{f(x)} = e^{f(x)} f'(x)$$

$$\frac{d}{dx} \log_b x = \frac{1}{\ln b} \frac{d}{dx} \ln x = \frac{1}{\ln b}\left(\frac{1}{x}\right)$$

$$\frac{d}{dx} b^x = \frac{d}{dx} e^{x \ln b} = e^{x \ln b} \ln b = b^x \ln b$$

$$\frac{d}{dx} [f(x)]^n = n[f(x)]^{n-1} f'(x)$$

$$\frac{dy}{dx} = \frac{dy}{du} \frac{du}{dx}, \quad \frac{dy}{dx} = \frac{dy}{dw} \frac{dw}{du} \frac{du}{dx}, \quad \text{and so on}$$

Review Exercise

Work through all the problems in this chapter review and check your answers in the back of the book. Answers to all review problems are there along with section numbers in italics to indicate where each type of problem is discussed. Where weaknesses show up, review appropriate sections in the text.

A

1. Use a calculator to evaluate $A = 2,000e^{0.09t}$ to the nearest cent for $t = 5, 10,$ and 20.

Find the indicated derivatives in Problems 2–4.

2. $\dfrac{d}{dx}(2 \ln x + 3e^x)$ 3. $\dfrac{d}{dx} e^{2x-3}$

4. y' for $y = \ln(2x + 7)$

5. Let $y = \ln u$ and $u = 3 + e^x$.

   (A) Express y in terms of x.
   (B) Use the chain rule to find dy/dx, and then express dy/dx in terms of x.

6. Find y' for $y = y(x)$ defined implicitly by the equation $2y^2 - 3x^3 - 5 = 0$, and evaluate at $(x, y) = (1, 2)$.

7. For $y = 3x^2 - 5$, where $x = x(t)$ and $y = y(t)$, find dy/dt if $dx/dt = 3$ when $x = 12$.

B

8. Graph $y = 100e^{-0.1x}$.

9. Use a calculator and a table of values to investigate

$$\lim_{n \to \infty} \left(1 + \frac{2}{n}\right)^n$$

Do you think the limit exists? If so, what do you think it is?

Find the indicated derivatives in Problems 10–15.

10. $\dfrac{d}{dz} [(\ln z)^7 + \ln z^7]$ 11. $\dfrac{d}{dx} (x^6 \ln x)$

12. $\dfrac{d}{dx} \dfrac{e^x}{x^6}$ 13. y' for $y = \ln(2x^3 - 3x)$

14. $f'(x)$ for $f(x) = e^{x^3 - x^2}$

15. dy/dx for $y = e^{-2x} \ln 5x$

16. Find the equation of the line tangent to the graph of $y = f(x) = 1 + e^{-x}$ at $x = 0$. At $x = -1$.

17. Find y' for $y = y(x)$ defined implicitly by the equation $x^2 - 3xy + 4y^2 = 23$, and find the slope of the graph at $(-1, 2)$.

18. Find x' for $x = x(t)$ defined implicitly by $x^3 - 2t^2x + 8 = 0$, and evaluate at $(t, x) = (-2, 2)$.

19. Find y' for $y = y(x)$ defined implicitly by $x - y^2 = e^y$, and evaluate at $(1, 0)$.

20. Find y' for $y = y(x)$ defined implicitly by $\ln y = x^2 - y^2$, and evaluate at $(1, 1)$.

21. A point is moving on the graph of $y^2 - 4x^2 = 12$ so that its x coordinate is decreasing at 2 units per second when $(x, y) = (1, 4)$. Find the rate of change of the y coordinate.

22. A 17-foot ladder is placed against a wall. If the foot of the ladder is pushed toward the wall at 0.5 foot per second, how fast is the top rising when the foot of the ladder is 8 feet from the wall?

23. Water from a water heater is leaking onto a floor. A circular pool is created whose area is increasing at the rate of 24 square inches per minute. How fast is the radius R of the pool increasing when the radius is 12 inches? $[A = \pi R^2]$

C *In Problems 24–27, find the absolute maximum value of $f(x)$ for $x > 0$.*

24. $f(x) = 11x - 2x \ln x$

25. $f(x) = 10xe^{-2x}$

26. $f(x) = 3x - x^2 + e^{-x}$

27. $f(x) = \dfrac{\ln x}{e^x}$

In Problems 28 and 29, apply the graphing strategy discussed in Section 4-4 to f, summarize the pertinent information, and sketch the graph.

28. $f(x) = 5 - 5e^{-x}$ **29.** $f(x) = x^3 \ln x$

30. Let $y = w^3$, $w = \ln u$, and $u = 4 - e^x$.

(A) Express y in terms of x.

(B) Use the chain rule to find dy/dx, and then express dy/dx in terms of x.

Find the indicated derivatives in Problems 31–33.

31. y' for $y = 5^{x^2-1}$

32. $\dfrac{d}{dx} \log_5(x^2 - x)$

33. $\dfrac{d}{dx} \sqrt{\ln(x^2 + x)}$

34. Find y' for $y = y(x)$ defined implicitly by the equation $e^{xy} = x^2 + y + 1$, and evaluate at $(0, 0)$.

35. A rock is thrown into a still pond and causes a circular ripple. Suppose the radius is increasing at a constant rate of 3 feet per second. Show that the area does not increase at a constant rate. When is the rate of increase of the area the smallest? The largest? Explain.

36. A point moves along the graph of $y = x^3$ in such a way that its y coordinate is increasing at a constant rate of 5 units per second. Does the x coordinate ever increase at a faster rate than the y coordinate? Explain.

Applications

Business & Economics

37. *Doubling time.* How long will it take money to double if it is invested at 5% interest compounded

(A) Annually?   (B) Continuously?

38. *Continuous compound interest.* If $100 is invested at 10% interest compounded continuously, the amount (in dollars) at the end of t years is given by

$$A = 100e^{0.1t}$$

Find $A'(t)$, $A'(1)$, and $A'(10)$.

39. *Marginal analysis.* The price–demand equation for 14-cubic foot refrigerators at an appliance store is

$$p(x) = 1,000e^{-0.02x}$$

where x is the monthly demand and p is the price in dollars. Find the marginal revenue equation.

40. *Maximum revenue.* For the price–demand equation in Problem 39, find the production level and price per unit that produces the maximum revenue. What is the maximum revenue?

41. *Maximum revenue.* Graph the revenue function from Problems 39 and 40 for $0 \leq x \leq 100$.

42. *Maximum profit.* Refer to Problem 39. If the refrigerators cost the store $220 each, find the price (to the nearest cent) that maximizes the profit. What is the maximum profit (to the nearest dollar)?

43. *Maximum profit.* The data in the table show the daily demand x for cream puffs at a state fair at various price levels p. If it costs $1 to make a cream puff, use logarithmic regression ($p = a + b \ln x$) to find the price (to the nearest cent) that maximizes profit.

DEMAND x	PRICE PER CREAM PUFF ($) p
3,125	1.99
3,879	1.89
5,263	1.79
5,792	1.69
6,748	1.59
8,120	1.49

44. *Minimum average cost.* The cost of producing x units of a product is given by

$$C(x) = 200 + 50x - 50 \ln x \qquad x \geq 1$$

Find the minimum average cost.

45. *Demand equation.* Given the demand equation

$$x = \sqrt{5,000 - 2p^3}$$

find the rate of change of p with respect to x by implicit differentiation (x is the number of items that can be sold at a price of p per item).

46. *Rate of change of revenue.* A company is manufacturing a new video game and can sell all that it manufactures. The revenue (in dollars) is given by

$$R = 36x - \frac{x^2}{20}$$

where the production output in 1 day is x games. If production is increasing at 10 games per day when production is 250 games per day, find the rate of increase in revenue.

Life Sciences

47. *Drug concentration.* The concentration of a drug in the bloodstream t hours after injection is given approximately by

$$C(t) = 5e^{-0.3t}$$

where $C(t)$ is concentration in milligrams per milliliter. What is the rate of change of concentration after 1 hour? After 5 hours?

48. *Would healing.* A circular wound on an arm is healing at the rate of 45 square millimeters per day (the area of the wound is decreasing at this rate). How fast is the radius R of the wound decreasing when $R = 15$ millimeters? $[A = \pi R^2]$

Social Sciences

49. *Psychology: learning.* In a computer assembly plant, a new employee, on the average, is able to assemble

$$N(t) = 10(1 - e^{-0.4t})$$

units after t days of on-the-job training.

(A) What is the rate of learning after 1 day? After 5 days?

(B) Graph N for $0 \leq t \leq 10$.

50. *Learning.* A new worker on the production line performs an operation in T minutes after x performances of the operation, as given by

$$T = 2\left(1 + \frac{1}{x^{3/2}}\right)$$

If, after performing the operation 9 times, the rate of improvement is $dx/dt = 3$ operations per hour, find the rate of improvement in time dT/dt in performing each operation.

Group Activity 1 *Elasticity of Demand*

www

We normally expect that by lowering the price of a commodity we will increase demand, and that by raising the price we will decrease demand. To measure the sensitivity of demand to changes in price, we define the **elasticity of demand,** denoted by $E(p)$, to be the limit of the relative change in demand x divided by the relative change in price p. Therefore, if $x = f(p)$ is the price–demand equation, then

$$E(p) = \lim_{h \to 0} \frac{\dfrac{f(p + h) - f(p)}{f(p)}}{\dfrac{h}{p}}$$

$$= \lim_{h \to 0} \left[\frac{p}{f(p)} \cdot \frac{f(p + h) - f(p)}{h}\right]$$

$$= \frac{p}{f(p)} f'(p)$$

Since an increase in price normally produces a decrease in demand, we expect $E(p)$ to be negative. If $E(p) = -1$, a 10% increase in price produces a 10% decrease in demand, and we say that demand has **unit elasticity.** If

$E(p) = -4$, a 10% increase in price produces a 40% decrease in demand; we express such sensitivity of demand to price by saying that demand is **elastic** whenever $E(p) < -1$. If $E(p) = -0.25$, a 10% increase in price produces only a 2.5% decrease in demand; we express such insensitivity to price by saying that demand is **inelastic** whenever $E(p) > -1$.

(A) Consider the price–demand equation

$$x = f(p) = 10{,}000 - 500p \qquad 0 \leqslant p \leqslant 20$$

Calculate the elasticity of demand at $p = 4, 7, 10, 13,$ and 16, and describe each case as elastic, inelastic, or of unit elasticity.

(B) Graph the elasticity of demand $E(p)$ and the revenue $R(p)$ as functions of price, and graph the line $y = -1$. Use the graphs to explain the connection between elasticity and the increasing/decreasing behavior of the revenue function.

(C) Repeat part (B) for the price–demand equation

$$x = f(p) = \sqrt{144 - 2p} \qquad 0 \leqslant p \leqslant 72$$

(D) Suppose that $x = g(p)$ is any price–demand equation for which $g(p)$ is differentiable. Let the associated revenue function be $R(p) = xp = g(p)p$. Show that $R'(p) = g(p)[E(p) + 1]$. Explain why revenue increases with price if and only if demand is inelastic.

Group Activity 2   *Point of Diminishing Returns**

Table 1 shows the total number of copies of a new spreadsheet for personal computers that have been sold x months after the product has been introduced. The **point of diminishing returns** occurs at the value of x where the rate of change of sales assumes its maximum value. In Section 4-3 we saw that a cubic polynomial model can be used to describe this situation.

TABLE 1	
MONTH x	TOTAL SALES (THOUSANDS OF COPIES)
4	71
8	182
12	305
16	405
20	450

(A) Find a cubic regression model $P(x)$ for the data in Table 1, and find the point of diminishing returns (correct to two decimal places).

The cubic polynomial model provides a good description of the data for $4 \leqslant x \leqslant 20$, but not for $x > 20$. A more sophisticated model that can be used to predict the behavior for $x > 20$ is provided by a **logistic growth function** of the form

www

$$L(x) = \frac{c}{1 + ae^{-bx}} \tag{1}$$

*This group activity requires a graphing utility that supports logistic regression.

(B) Find a logistic regression model $L(x)$ for the data in Table 1, and find the point of diminishing returns (correct to two decimal places) for this model.

(C) Graph $P(x)$ and $L(x)$ for Xmin = 4, Xmax = 20, Ymin = 0, and Ymax = 500. Are there any significant differences in the descriptions of the data provided by these two functions?

(D) Repeat part (C) with Xmax increased to 30.

(E) Find $\lim_{x \to \infty} L(x)$ (correct to the nearest integer). The value of this limit is called the **carrying capacity** of the model. Compare the carrying capacity with the value of $L(x)$ at the point of diminishing returns found in part (B).

(F) Sales data for a new word processor are given in Table 2. Find a logistic regression model for the data, the point of diminishing returns (correct to two decimal places), and the carrying capacity (correct to the nearest integer).

(G) For any function of the form in equation (1), show that the carrying capacity is c, the point of diminishing returns is $(\ln a)/b$, and the value of the function at the point of diminishing returns is $c/2$.

TABLE 2

MONTH x	TOTAL SALES (THOUSANDS OF COPIES)
4	151
8	204
12	327
16	480
20	530

Integration

6 CHAPTER

INTRODUCTION

The preceding three chapters dealt with differential calculus. We now begin the development of the second main part of calculus, called *integral calculus.* Two types of integrals are introduced, the *indefinite integral* and the *definite integral,* each quite different from the other. But through the remarkable *fundamental theorem of calculus,* we show that not only are the two integral forms intimately related, but both are intimately related to differentiation. ⊂⊃

Section 6-1

Antiderivatives and Indefinite Integrals

- ❑ ANTIDERIVATIVES
- ❑ GEOMETRIC–NUMERIC LOOK AT ANTIDERIVATIVES
- ❑ ANTIDERIVATIVES AND INDEFINITE INTEGRALS: ALGEBRAIC FORMS
- ❑ ANTIDERIVATIVES AND INDEFINITE INTEGRALS: EXPONENTIAL AND LOGARITHMIC FORMS
- ❑ APPLICATIONS

Many operations in mathematics have reverses—compare addition and subtraction, multiplication and division, and powers and roots. We now know how to find the derivatives of many functions. The reverse operation, *antidifferentiation* (the reconstruction of a function from its derivative) will receive our

351

attention in this and the next two sections. A function F is an **antiderivative** of a function f if $F'(x) = f(x)$. We look at the problem geometrically, numerically, and algebraically, and use an algebraic approach to develop special antiderivative formulas in much the same way as we developed derivative formulas.

❑ ANTIDERIVATIVES

Explore–Discuss 1

(A) Find three antiderivatives of $2x$.

(B) How many antiderivatives of $2x$ exist, and how are they related to each other?

(C) What notation would you use to represent all antiderivatives of $2x$?

What is an antiderivative of x^2?

$$\frac{x^3}{3} \qquad \text{is an antiderivative of } x^2$$

since

$$\frac{d}{dx}\left(\frac{x^3}{3}\right) = x^2$$

Note also that

$$\frac{d}{dx}\left(\frac{x^3}{3} + 2\right) = x^2 \qquad \frac{d}{dx}\left(\frac{x^3}{3} - \pi\right) = x^2 \qquad \frac{d}{dx}\left(\frac{x^3}{3} + \sqrt{5}\right) = x^2$$

Hence,

$$\frac{x^3}{3} + 2 \qquad \frac{x^3}{3} - \pi \qquad \frac{x^3}{3} + \sqrt{5}$$

are also antiderivatives of x^2, since each has x^2 as a derivative. In fact, it appears that

$$\frac{x^3}{3} + C \qquad \text{for any real number } C$$

is an antiderivative of x^2, since

$$\frac{d}{dx}\left(\frac{x^3}{3} + C\right) = x^2$$

Thus, antidifferentiation of a given function does not lead to a unique function, but to an entire family of functions.

Does the expression

$$\frac{x^3}{3} + C \qquad \text{with } C \text{ any real number}$$

include all antiderivatives of x^2? Theorem 1 (which we state without proof) indicates that the answer is yes.

THEOREM 1   **On Antiderivatives**

If the derivatives of two functions are equal on an open interval (a, b), then the functions differ by at most a constant. Symbolically: If F and G are differentiable functions on the interval (a, b) and $F'(x) = G'(x)$ for all x in (a, b), then $F(x) = G(x) + k$ for some constant k.

❑ GEOMETRIC–NUMERIC LOOK AT ANTIDERIVATIVES

Explore–Discuss 2

Given $f'(x) = 1$:

(A) What does $f'(x) = 1$ tell us about graphs of any antiderivative function f?

(B) Sketch, in the same coordinate system, graphs of three different antiderivatives $y = f(x)$ satisfying $f(0) = -2, f(0) = 0$, and $f(0) = 2$. What are the geometric relationships among the graphs?

TABLE 1

PROPERTIES OF THE GRAPH OF $y = f(x)$ DETERMINED FROM THE GRAPH OF $y = f'(x)$

$f'(x)$	GRAPH OF $y = f(x)$
Positive	Increasing
Negative	Decreasing
Increasing	Concave upward
Decreasing	Concave downward

We can visualize certain characteristics of an antiderivative by using slopes. If we are given the graph of a derivative function $y = f'(x)$, we can sketch a possible graph of an antiderivative function f showing its characteristic shape. Recall the properties listed in Table 1.

Example 1 ➩ **Analyzing the Graph of $y = f(x)$ Given the Graph of $y = f'(x)$**
Given the graph of $y = f'(x)$ in Figure 1, sketch possible graphs of three antiderivative functions f such that $f(0) = -1, f(0) = 0$, and $f(0) = 2$, all using the same set of coordinate axes.

SOLUTION   Table 2 summarizes the information obtained from the graph of $y = f'(x)$.

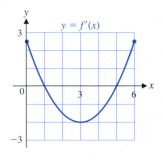

FIGURE 1

TABLE 2

x	$f'(x)$	GRAPH OF $y = f(x)$
$0 < x < 1$	Positive and decreasing	Increasing and concave downward
$x = 1$	x intercept	Local maximum
$1 < x < 3$	Negative and decreasing	Decreasing and concave downward
$x = 3$	Local minimum	Inflection point
$3 < x < 5$	Negative and increasing	Decreasing and concave upward
$x = 5$	x intercept	Local minimum
$5 < x < 6$	Positive and increasing	Increasing and concave upward

If $f(x)$ is one antiderivative of $f'(x)$, since all antiderivatives are given by $y = f(x) + C$, **the graph of any antiderivative function can be obtained from another by a vertical translation.** We know where each of the three graphs of $y = f(x)$ start and the general shape over the interval $(0, 6)$. Figure 2 shows possible graphs of three antiderivative functions having the required characteristics.

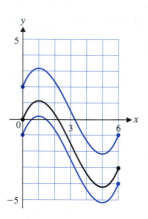

FIGURE 2

Matched Problem 1   Given the graph of $y = f'(x)$ in Figure 3, sketch possible graphs of three antiderivative functions f such that $f(0) = -1, f(0) = 0$, and $f(0) = 1$, all using the same set of coordinate axes.

FIGURE 3

□   ANTIDERIVATIVES AND INDEFINITE INTEGRALS: ALGEBRAIC FORMS

Theorem 1 states that if the derivatives of two functions are equal, then the functions differ by at most a constant. We use the symbol

$$\int f(x)\, dx$$

called the **indefinite integral,** to represent the family of all antiderivatives of $f(x)$, and write

$$\int f(x)\, dx = F(x) + C \qquad \text{if} \qquad F'(x) = f(x)$$

The symbol $\int$ is called an **integral sign,** and the function $f(x)$ is called the **integrand.** The symbol dx indicates that the antidifferentiation is performed with respect to the variable x. (We will have more to say about the symbols $\int$ and dx later in this chapter.) The arbitrary constant C is called the **constant of integration.** Referring to the preceding discussion, we can write

$$\int x^2\, dx = \frac{x^3}{3} + C \qquad \text{since} \qquad \frac{d}{dx}\left(\frac{x^3}{3} + C\right) = x^2$$

Of course, variables other than x can be used in indefinite integrals. For example,

$$\int t^2\, dt = \frac{t^3}{3} + C \qquad \text{since} \qquad \frac{d}{dt}\left(\frac{t^3}{3} + C\right) = t^2$$

or

$$\int u^2\, du = \frac{u^3}{3} + C \qquad \text{since} \qquad \frac{d}{du}\left(\frac{u^3}{3} + C\right) = u^2$$

The fact that indefinite integration and differentiation are reverse operations, except for the addition of the constant of integration, can be expressed symbolically as

$$\frac{d}{dx}\left[\int f(x)\,dx \right] = f(x) \qquad \text{\textit{The derivative of the indefinite integral of } f(x) \text{ is } f(x).}$$

and

$$\int F'(x)\,dx = F(x) + C \qquad \text{\textit{The indefinite integral of the derivative of } F(x) \text{ is } F(x) + C.}$$

Just as with differentiation, we can develop formulas and special properties that will enable us to find indefinite integrals of many frequently encountered functions. To start, we list some formulas that can be established using the definitions of antiderivative and indefinite integral, and the properties of derivatives considered in Chapter 3.

> ### Indefinite Integral Formulas and Properties
>
> For k and C constants:
>
> **1.** $\int k\,dx = kx + C$
>
> **2.** $\displaystyle\int x^n\,dx = \frac{x^{n+1}}{n+1} + C, \qquad n \neq -1$
>
> **3.** $\int kf(x)\,dx = k\int f(x)\,dx$
>
> **4.** $\int [f(x) \pm g(x)]\,dx = \int f(x)\,dx \pm \int g(x)\,dx$

We will establish formula 2 and property 3 here (the others may be shown to be true in a similar manner). To establish formula 2, we simply differentiate the right side to obtain the integrand on the left side. Thus,

$$\frac{d}{dx}\left(\frac{x^{n+1}}{n+1} + C \right) = \frac{(n+1)x^n}{n+1} + 0 = x^n \qquad n \neq -1$$

(Notice that formula 2 cannot be used when $n = -1$; that is, when the integrand is x^{-1} or $1/x$. The indefinite integral of $x^{-1} = 1/x$ will be considered later in this section.)

To establish property 3, let F be a function such that $F'(x) = f(x)$. Then

$$k\int f(x)\,dx = k\int F'(x)\,dx = k[F(x) + C_1] = kF(x) + kC_1$$

and since $[kF(x)]' = kF'(x) = kf(x)$, we have

$$\int kf(x)\,dx = \int kF'(x)\,dx = kF(x) + C_2$$

But $kF(x) + kC_1$ and $kF(x) + C_2$ describe the same set of functions, since C_1 and C_2 are arbitrary real numbers. Thus, property 3 is established.

CAUTION

It is important to remember that property 3 states that **a constant factor can be moved across an integral sign. A variable factor cannot be moved across an integral sign:**

Constant Factor   **Variable Factor**

$$\int 5x^{1/2}\,dx = 5\int x^{1/2}\,dx \qquad \int xx^{1/2}\,dx \ne x\int x^{1/2}\,dx$$

Now let us put the formulas and properties to use.

Example 2 ⇨ **Using Indefinite Integral Properties and Formulas**

(A) $\displaystyle\int 5\,dx = 5x + C$

(B) $\displaystyle\int x^4\,dx = \frac{x^{4+1}}{4+1} + C = \frac{x^5}{5} + C$

(C) $\displaystyle\int 5t^7\,dt = 5\int t^7\,dt = 5\frac{t^8}{8} + C = \frac{5}{8}t^8 + C$

(D)

$$\begin{aligned}
\int (4x^3 + 2x - 1)\,dx &= \int 4x^3\,dx + \int 2x\,dx - \int dx \\
&= 4\int x^3\,dx + 2\int x\,dx - \int dx \\
&= \frac{4x^4}{4} + \frac{2x^2}{2} - x + C
\end{aligned}$$

Property 4 can be extended to the sum and difference of an arbitrary number of functions.

$$= x^4 + x^2 - x + C$$

(E) $\displaystyle\int \frac{3\,dx}{x^2} = \int 3x^{-2}\,dx = \frac{3x^{-2+1}}{-2+1} + C = -3x^{-1} + C$

(F) $\displaystyle\int 5\sqrt[3]{u^2}\,du = 5\int u^{2/3}\,du = 5\frac{u^{(2/3)+1}}{\frac{2}{3}+1} + C$

$$= 5\frac{u^{5/3}}{\frac{5}{3}} + C = 3u^{5/3} + C$$ ∎

To check any of the results in Example 2, we differentiate the final result to obtain the integrand in the original indefinite integral. When you evaluate an indefinite integral, do not forget to include the arbitrary constant C.

Matched Problem 2 ⇨ Find each of the following:

(A) $\int dx$ (B) $\displaystyle\int 3t^4\,dt$ (C) $\displaystyle\int (2x^5 - 3x^2 + 1)\,dx$

(D) $\displaystyle\int 4\sqrt[5]{w^3}\,dw$ (E) $\displaystyle\int \left(2x^{2/3} - \frac{3}{x^4}\right)\,dx$ ∎

Example 3 ⇨ **Using Indefinite Integral Properties and Formulas**

(A) $\displaystyle\int \frac{x^3 - 3}{x^2}\,dx = \int \left(\frac{x^3}{x^2} - \frac{3}{x^2}\right)\,dx$

$$= \int (x - 3x^{-2})\,dx$$

$$= \int x\,dx - 3\int x^{-2}\,dx$$

$$= \frac{x^{1+1}}{1+1} - 3\frac{x^{-2+1}}{-2+1} + C$$

$$= \tfrac{1}{2}x^2 + 3x^{-1} + C$$

(B) $\int \left(\frac{2}{\sqrt[3]{x}} - 6\sqrt{x}\right) dx = \int (2x^{-1/3} - 6x^{1/2})\, dx$

$$= 2 \int x^{-1/3}\, dx - 6 \int x^{1/2}\, dx$$

$$= 2\,\frac{x^{(-1/3)+1}}{-\frac{1}{3}+1} - 6\,\frac{x^{(1/2)+1}}{\frac{1}{2}+1} + C$$

$$= 2\,\frac{x^{2/3}}{\frac{2}{3}} - 6\,\frac{x^{3/2}}{\frac{3}{2}} + C$$

$$= 3x^{2/3} - 4x^{3/2} + C$$

Matched Problem 3   Find each indefinite integral.

(A) $\int \frac{x^4 - 8x^3}{x^2}\, dx$ (B) $\int \left(8\sqrt[3]{x} - \frac{6}{\sqrt{x}}\right) dx$

□ ANTIDERIVATIVES AND INDEFINITE INTEGRALS:
EXPONENTIAL AND LOGARITHMIC FORMS

We now give indefinite integral formulas for e^x and $1/x$. (Recall that the form $x^{-1} = 1/x$ is not covered by formula 2, given earlier.)

Indefinite Integral Formulas

5. $\int e^x\, dx = e^x + C$ **6.** $\int \frac{1}{x}\, dx = \ln|x| + C$, $x \neq 0$

Formula 5 follows immediately from the derivative formula for the exponential function discussed in Chapter 5. Because of the absolute value, formula 6 does not follow directly from the derivative formula for the natural logarithm function. Let us show that

$$\frac{d}{dx}\ln|x| = \frac{1}{x} \qquad x \neq 0$$

We consider two cases, $x > 0$ and $x < 0$:

Case 1. $x > 0$:

$$\frac{d}{dx}\ln|x| = \frac{d}{dx}\ln x \qquad \text{Since } |x| = x \text{ for } x > 0$$

$$= \frac{1}{x}$$

Case 2. $x < 0$:

$$\frac{d}{dx}\ln|x| = \frac{d}{dx}\ln(-x) \qquad \text{Since } |x| = -x \text{ for } x < 0$$

$$= \frac{1}{-x}\frac{d}{dx}(-x)$$

$$= \frac{-1}{-x} = \frac{1}{x}$$

Thus,

$$\frac{d}{dx}\ln|x| = \frac{1}{x} \qquad x \neq 0$$

and hence,

$$\int \frac{1}{x}dx = \ln|x| + C \qquad x \neq 0$$

What about the indefinite integral of $\ln x$? We postpone a discussion of $\int \ln x \, dx$ until Section 7-3, where we will be able to find it using a technique called *integration by parts*.

Example 4 ⇨ **Exponential and Logarithmic Forms**

$$\int \left(2e^x + \frac{3}{x}\right)dx = 2\int e^x \, dx + 3\int \frac{1}{x}dx$$

$$= 2e^x + 3\ln|x| + C$$

■

Matched Problem 4 ⇨ Find: $\int \left(\frac{5}{x} - 4e^x\right)dx$

■

[*Note:* More general forms of exponential and logarithmic functions are considered in Section 6-2.]

❑ APPLICATIONS

Let us now consider some applications of the indefinite integral to see why we are interested in finding antiderivatives of functions.

Example 5 ⇨ **Curves**   Find the equation of the curve that passes through $(2, 5)$ if its slope is given by $dy/dx = 2x$ at any point x.

SOLUTION   We are interested in finding a function $y = f(x)$ such that

$$\frac{dy}{dx} = 2x \tag{1}$$

and

$$y = 5 \qquad \text{when} \qquad x = 2 \tag{2}$$

If $dy/dx = 2x$, then

$$y = \int 2x \, dx$$
$$= x^2 + C \tag{3}$$

Since $y = 5$ when $x = 2$, we determine the *particular value of C* so that

$$5 = 2^2 + C$$

Thus, $C = 1$, and

$$y = x^2 + 1$$

is the *particular antiderivative* out of all those possible from equation (3) that satisfies both equations (1) and (2) (see Fig. 4).

■

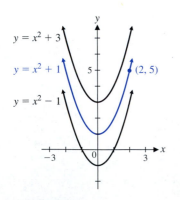

$y = x^2 + 3$
$y = x^2 + 1$
$y = x^2 - 1$
$(2, 5)$

FIGURE 4 $y = x^2 + C$

Note how Example 5 differs from Example 1, where we reconstructed a possible graph of a function using the graph of its derivative. In Example 5 we actually found an equation for the antiderivative whose graph passes through $(2, 5)$. Consequently, a precise graph of the antiderivative could be produced.

Explore–Discuss 3

Graph the derivative function $y = f'(x) = 2x$ in Example 5 and use the information from the graph to confirm the shape of the graphs of the antiderivative functions shown in Figure 4.

Matched Problem 5   ↩   Find the equation of the curve that passes through $(2, 6)$ if its slope is given by $dy/dx = 3x^2$ at any point x.   ■

In certain situations, it is easier to determine the rate at which something happens than how much of it has happened in a given length of time (for example, population growth rates, business growth rates, rate of healing of a wound, rates of learning or forgetting). If a rate function (derivative) is given and we know the value of the dependent variable for a given value of the independent variable, then—if the rate function is not too complicated—we can often find the original function by integration.

Example 6   ↩   **Cost Function**   If the marginal cost of producing x units is given by

$$C'(x) = 0.3x^2 + 2x$$

and the fixed cost is $2,000, find the cost function $C(x)$ and the cost of producing 20 units.

SOLUTION   Recall that marginal cost is the derivative of the cost function and that fixed cost is cost at a 0 production level. Thus, the mathematical problem is to find $C(x)$ given

$$C'(x) = 0.3x^2 + 2x \qquad C(0) = 2,000$$

We now find the indefinite integral of $0.3x^2 + 2x$ and determine the arbitrary integration constant using $C(0) = 2,000$:

$$C'(x) = 0.3x^2 + 2x$$
$$C(x) = \int (0.3x^2 + 2x)\, dx$$
$$= 0.1x^3 + x^2 + K \qquad \text{\textit{Since C represents the cost, we use K for the constant of integration.}}$$

But

$$C(0) = (0.1)0^3 + 0^2 + K = 2,000$$

Thus, $K = 2,000$, and the particular cost function is

$$C(x) = 0.1x^3 + x^2 + 2,000$$

We now find $C(20)$, the cost of producing 20 units:

$$C(20) = (0.1)20^3 + 20^2 + 2,000$$
$$= \$3,200$$

See Figure 5 for a geometric representation.   ■

FIGURE 5

Matched Problem 6
Find the revenue function $R(x)$ when the marginal revenue is

$$R'(x) = 400 - 0.4x$$

and no revenue results at a 0 production level. What is the revenue at a production level of 1,000 units? ∎

Example 7
Advertising An FM radio station is launching an aggressive advertising campaign in order to increase the number of daily listeners. The station currently has 27,000 daily listeners, and management expects the number of daily listeners, $S(t)$, to grow at the rate of

$$S'(t) = 60t^{1/2}$$

listeners per day, where t is the number of days since the campaign began. How long should the campaign last if the station wants the number of daily listeners to grow to 41,000?

SOLUTION
We must solve the equation $S(t) = 41,000$ for t, given that

$$S'(t) = 60t^{1/2} \quad \text{and} \quad S(0) = 27,000$$

First, we use integration to find $S(t)$:

$$S(t) = \; 60t^{1/2} \, dt$$

$$= 60 \frac{t^{3/2}}{\frac{3}{2}} + C$$

$$= 40t^{3/2} + C$$

Since

$$S(0) = 40(0)^{3/2} + C = 27,000$$

we have $C = 27,000$, and

$$S(t) = 40t^{3/2} + 27,000$$

Now we solve the equation $S(t) = 41,000$ for t:

$$40t^{3/2} + 27,000 = 41,000$$
$$40t^{3/2} = 14,000$$
$$t^{3/2} = 350$$
$$t = 350^{2/3} \qquad \textcolor{blue}{\textit{Use a calculator.}}$$
$$= 49.664\ 419\ldots$$

Thus, the advertising campaign should last approximately 50 days. ∎

A graphing utility offers an alternative approach to solving $S(t) = 41,000$ in Example 7 (see Fig. 6).

FIGURE 6 $y_1 = 40x^{1.5} + 27,000$
$y_2 = 41,000$

Matched Problem 7 ✎   The current monthly circulation of the magazine *Computing News* is 640,000 copies. Due to competition from a new magazine in the same field, the monthly circulation of *Computing News*, $C(t)$, is expected to decrease at the rate of

$$C'(t) = -6,000t^{1/3}$$

copies per month, where t is the time in months since the new magazine began publication. How long will it take for the circulation of *Computing News* to decrease to 460,000 copies per month?

■

CAUTION

1. $\displaystyle\int e^x \, dx \neq \frac{e^{x+1}}{x+1} + C$

The power rule applies only to power functions of the form x^n, where the exponent n is a real constant not equal to -1 and the base x is the variable. The function e^x is an exponential function with variable exponent x and constant base e. The correct form is

$$\int e^x \, dx = e^x + C$$

2. $\displaystyle\int x(x^2 + 2) \, dx \neq \frac{x^2}{2}\left(\frac{x^3}{3} + 2x\right) + C$

The integral of a product is not equal to the product of the integrals. The correct form is

$$\int x(x^2 + 2) \, dx = \int (x^3 + 2x) \, dx = \frac{x^4}{4} + x^2 + C$$

3. Not all elementary functions have elementary antiderivatives. But finding one when it exists can markedly simplify the solution of some problems.

Answers to Matched Problems   **1.**

2. (A) $x + C$ (B) $\frac{3}{5}t^5 + C$
(C) $(x^6/3) - x^3 + x + C$
(D) $\frac{5}{2}w^{8/5} + C$
(E) $\frac{6}{5}x^{5/3} + x^{-3} + C$

3. (A) $\frac{1}{3}x^3 - 4x^2 + C$ (B) $6x^{4/3} - 12x^{1/2} + C$ 4. $5\ln|x| - 4e^x + C$
5. $y = x^3 - 2$ 6. $R(x) = 400x - 0.2x^2; R(1,000) = \$200,000$
7. $t = (40)^{3/4} \approx 16$ mo

Exercise 6-1

A *In Problems 1–18, find each indefinite integral. (Check by differentiating.)*

1. $\int 2.7 \, dx$ 2. $\int -6.28 \, dx$

3. $\int x^8 \, dx$ 4. $\int x^{-3} \, dx$

5. $\int t^{-4} \, dt$ 6. $\int t^7 \, dt$

7. $\int 2\pi \, dt$ 8. $\int e \, dt$

9. $\int 3u^6 \, du$ 10. $\int 2u^5 \, du$

11. $\int (1 + 3u^2) \, du$ 12. $\int (8 - u^3) \, du$

13. $\int 5e^t \, dt$ 14. $\int 7.67e^t \, dt$

15. $\int \dfrac{3}{z}\, dz$

16. $\int 9s^{-1}\, ds$

17. $\int x(1 + x)\, dx$

18. $\int x^2(1 - x^2)\, dx$

Each graph in Problems 19–22 is the graph of a derivative function $y = f'(x)$. For each, sketch possible graphs of three antiderivative functions f satisfying $f(0) = -2$, $f(0) = 0$, and $f(0) = 2$, respectively, all on the same coordinate system.

19. $f'(x)$

20. $f'(x)$

21. $f'(x)$

22. $f'(x)$

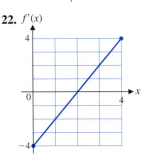

In Problems 23–32, find all the antiderivatives for each derivative.

23. $\dfrac{dy}{dx} = 200x^4$

24. $\dfrac{dx}{dt} = 42t^5$

25. $\dfrac{dP}{dx} = 24 - 6x$

26. $\dfrac{dy}{dx} = 3x^2 - 4x^3$

27. $\dfrac{dy}{du} = 2u^5 - 3u^2 - 1$

28. $\dfrac{dA}{dt} = 3 - 12t^3 - 9t^5$

29. $\dfrac{dy}{dx} = e^x + 3$

30. $\dfrac{dy}{dx} = x - e^x$

31. $\dfrac{dx}{dt} = 5t^{-1} + 1$

32. $\dfrac{du}{dv} = \dfrac{4}{v} + \dfrac{v}{4}$

B   In Problems 33 and 34, discuss the validity of each statement. If the statement is always true, explain why. If not, give a counterexample.

33. (A) If n is an integer, then $x^{n+1}/(n + 1)$ is an anti-derivative of x^n.

(B) The function $f(x) = \pi$ is an antiderivative of the function $g(x) = 0$.

34. (A) $\int \dfrac{d}{dx}(x^4 + x^2)\, dx = x^4 + x^2 + C$

(B) $\dfrac{d}{dx}\left(\int x^2\, dx \right) = x^2 + C$

In Problems 35–38, could the three graphs be antiderivatives of the same function? Explain.

35.

36.

37.

38.

In Problems 39–60, find each indefinite integral. (Check by differentiation.)

39. $\int 20x^{1/3}\, dx$

40. $\int 10u^{3/2}\, du$

41. $\int 9t^{-4}\, dt$

42. $\int 8t^{-5}\, dt$

43. $\int \dfrac{du}{\sqrt{u}}$

44. $\int \dfrac{dt}{\sqrt[3]{t}}$

45. $\int \dfrac{dx}{4x^3}$

46. $\int \dfrac{6\, dm}{m^2}$

47. $\int \dfrac{4 + u}{u}\, du$

48. $\int \dfrac{1 - y^2}{3y}\, dy$

49. $\int \left(3x^2 - \dfrac{2}{x^2} \right) dx$

50. $\int \left(4x^3 + \dfrac{2}{x^3} \right) dx$

51. $\int \left(10x^4 - \dfrac{8}{x^5} - 2 \right) dx$

52. $\int \left(\dfrac{6}{x^4} - \dfrac{2}{x^3} + 1 \right) dx$

53. $\int \left(3\sqrt{x} + \dfrac{2}{\sqrt[7]{x}} \right) dx$

54. $\int \left(\dfrac{2}{\sqrt[3]{x}} - \sqrt[3]{x^2} \right) dx$

55. $\int \left(\sqrt[3]{x^2} - \dfrac{4}{x^3} \right) dx$

56. $\int \left(\dfrac{12}{x^5} - \dfrac{1}{\sqrt[3]{x^2}} \right) dx$

57. $\int \dfrac{e^x - 3x}{4}\, dx$

58. $\int \dfrac{e^x - 3x^2}{2}\, dx$

59. $\int \dfrac{12 + 5z - 3z^3}{z^4}\, dz$

60. $\int \dfrac{(1 + z^2)^2}{z}\, dz$

Each graph in Problems 61–64 is the graph of a deriva-tive function y = f'(x). For each, sketch possible graphs of three antiderivative functions f satisfying f(0) = −2, f(0) = 0, and f(0) = 2, respectively, all on the same coordinate system.

61.

62.

63.

64.

In Problems 65–74, find the particular antiderivative of each derivative that satisfies the given condition.

65. $\dfrac{dy}{dx} = 2x - 3; y(0) = 5$

66. $\dfrac{dy}{dx} = 5 - 4x; y(0) = 20$

67. $C'(x) = 6x^2 - 4x; C(0) = 3{,}000$

68. $R'(x) = 600 - 0.6x; R(0) = 0$

69. $\dfrac{dx}{dt} = \dfrac{20}{\sqrt{t}}; x(1) = 40$

70. $\dfrac{dR}{dt} = \dfrac{100}{t^2}; R(1) = 400$

71. $\dfrac{dy}{dx} = 2x^{-2} + 3x^{-1} - 1; y(1) = 0$

72. $\dfrac{dy}{dx} = 3x^{-1} + x^{-2}; y(1) = 1$

73. $\dfrac{dx}{dt} = 4e^t - 2; x(0) = 1$

74. $\dfrac{dy}{dt} = 5e^t - 4; y(0) = -1$

75. Find the equation of the curve that passes through $(2, 3)$ if its slope is given by

$$\frac{dy}{dx} = 4x - 3$$

for each x.

76. Find the equation of the curve that passes through $(1, 3)$ if its slope is given by

$$\frac{dy}{dx} = 12x^2 - 12x$$

for each x.

C *In Problems 77–82, find each indefinite integral.*

77. $\displaystyle\int \frac{2x^4 - x}{x^3}\, dx$

78. $\displaystyle\int \frac{x^{-1} - x^4}{x^2}\, dx$

79. $\displaystyle\int \frac{x^5 - 2x}{x^4}\, dx$

80. $\displaystyle\int \frac{1 - 3x^4}{x^2}\, dx$

81. $\displaystyle\int \frac{x^2 e^x - 2x}{x^2}\, dx$

82. $\displaystyle\int \frac{1 - xe^x}{x}\, dx$

For each derivative in Problems 83–88, find an anti-derivative that satisfies the given condition.

83. $\dfrac{dM}{dt} = \dfrac{t^2 - 1}{t^2}; M(4) = 5$

84. $\dfrac{dR}{dx} = \dfrac{1 - x^4}{x^3}; R(1) = 4$

85. $\dfrac{dy}{dx} = \dfrac{5x + 2}{\sqrt[3]{x}}; y(1) = 0$

86. $\dfrac{dx}{dt} = \dfrac{\sqrt{t^3} - t}{\sqrt{t^3}}; x(9) = 4$

87. $p'(x) = -\dfrac{10}{x^2}; p(1) = 20$

88. $p'(x) = \dfrac{10}{x^3}; p(1) = 15$

In Problems 89–92, find the derivative or indefinite inte-gral as indicated.

89. $\dfrac{d}{dx}\left(\displaystyle\int x^3\, dx \right)$

90. $\dfrac{d}{dt}\left(\displaystyle\int \frac{\ln t}{t}\, dt \right)$

91. $\displaystyle\int \frac{d}{dx}(x^4 + 3x^2 + 1)\, dx$

92. $\displaystyle\int \frac{d}{du}(e^{u^2})\, du$

Applications

Business & Economics

93. *Cost function.* The marginal average cost for producing x digital sports watches is given by

$$\overline{C}'(x) = -\frac{1,000}{x^2} \qquad \overline{C}(100) = 25$$

where $\overline{C}(x)$ is the average cost in dollars. Find the average cost function and the cost function. What are the fixed costs?

94. *Renewable energy.* In 1998 the U.S. consumption of
WWW renewable energy was 6.98 quadrillion Btu (that is, 6.98×10^{15} Btu). Since 1960, consumption has been growing at a rate given by

$$f'(t) = 0.004t + 0.062$$

where t is years after 1960. Find $f(t)$ and estimate the U.S. consumption of renewable energy in 2020.

95. *Manufacturing costs.* The graph of the marginal cost function from the manufacturing of x thousand watches per month [where cost $C(x)$ is in thousands of dollars per month] is given in the figure.

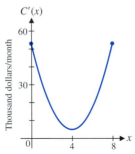

Figure for 95

(A) Using the graph shown, verbally describe the shape of the graph of the cost function $C(x)$ as x increases from 0 to 8,000 watches per month.

(B) Given the equation of the marginal cost function,

$$C'(x) = 3x^2 - 24x + 53$$

find the cost function if monthly fixed costs at 0 output are $30,000. What is the cost for manufacturing 4,000 watches per month? 8,000 watches per month?

(C) Graph the cost function for $0 \le x \le 8$. [Check the shape of the graph relative to the analysis in part (A).]

(D) Why do you think that the graph of the cost function is steeper at both ends than in the middle?

96. *Revenue.* The graph of the marginal revenue function from the sale of x digital sports watches is given in the figure.

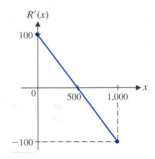

Figure for 96

(A) Using the graph shown, verbally describe the shape of the graph of the revenue function $R(x)$ as x increases from 0 to 1,000.

(B) Find the equation of the marginal revenue function (the linear function shown in the figure).

(C) Find the equation of the revenue function that satisfies $R(0) = 0$. Graph the revenue function over the interval [0, 1,000]. [Check the shape of the graph relative to the analysis in part (A).]

(D) Find the price–demand equation and determine the price when the demand is 700 units.

97. *Sales analysis.* The monthly sales of a particular personal computer are expected to decline at the rate of

$$S'(t) = -25t^{2/3}$$

computers per month, where t is time in months and $S(t)$ is the number of computers sold each month. The company plans to stop manufacturing this computer when the monthly sales reach 800 computers. If the monthly sales now ($t = 0$) are 2,000 computers, find $S(t)$. How long will the company continue to manufacture this computer?

98. *Sales analysis.* The rate of change of the monthly sales of a new home video game cartridge is given by

$$S'(t) = 500t^{1/4} \qquad S(0) = 0$$

where t is the number of months since the game was released and $S(t)$ is the number of cartridges sold each month. Find $S(t)$. When will the monthly sales reach 20,000 cartridges?

99. *Sales analysis.* Repeat Problem 97 if $S'(t) = -25t^{2/3} - 70$ and all other information remains the same. Use a graphing utility to approximate the solution to the equation $S(t) = 800$ to two decimal places.

100. *Sales analysis.* Repeat Problem 98 if $S'(t) = 500t^{1/4} + 300$ and all other information remains the

same. Use a graphing utility to approximate the solution to the equation $S(t) = 20,000$ to two decimal places.

101. *Labor costs and learning.* A defense contractor is
www starting production on a new missile control system. On the basis of data collected while assembling the first 16 control systems, the production manager obtained the following function describing the rate of labor use:

$$g(x) = 2,400x^{-1/2}$$

where $g(x)$ is the number of labor-hours required to assemble the xth unit of the control system. For example, after assembling 16 units, the rate of assembly is 600 labor-hours per unit, and after assembling 25 units, the rate of assembly is 480 labor-hours per unit. The more units assembled, the more efficient the process because of learning. If 19,200 labor-hours are required to assemble the first 16 units, how many labor-hours, $L(x)$, will be required to assemble the first x units? The first 25 units?

102. *Labor costs and learning.* If the rate of labor use in Problem 101 is

$$g(x) = 2,000x^{-1/3}$$

and if the first 8 control units require 12,000 labor-hours, how many labor-hours, $L(x)$, will be required for the first x control units? The first 27 control units?

Life Sciences

103. *Weight–height.* For an average person, the rate of change of weight W (in pounds) with respect to height h (in inches) is given approximately by

$$\frac{dW}{dh} = 0.0015h^2$$

Find $W(h)$ if $W(60) = 108$ pounds. Also, find the weight for a person who is 5 feet 10 inches tall.

104. *Wound healing.* If the area A of a healing wound changes at a rate given approximately by

$$\frac{dA}{dt} = -4t^{-3}   \quad 1 \leq t \leq 10$$

where t is time in days and $A(1) = 2$ square centimeters, what will the area of the wound be in 10 days?

Social Sciences

105. *Urban growth.* The rate of growth of the population, $N(t)$, of a newly incorporated city t years after incorporation is estimated to be

$$\frac{dN}{dt} = 400 + 600\sqrt{t}   \quad 0 \leq t \leq 9$$

If the population was 5,000 at the time of incorporation, find the population 9 years later.

106. *Learning.* A beginning high school language class was chosen for an experiment in learning. Using a list of 50 words, the experiment involved measuring the rate of vocabulary memorization at different times during a continuous 5-hour study session. It was found that the average rate of learning for the entire class was inversely proportional to the time spent studying and was given approximately by

$$V'(t) = \frac{15}{t}   \quad 1 \leq t \leq 5$$

If the average number of words memorized after 1 hour of study was 15 words, what was the average number of words learned after t hours of study for $1 \leq t \leq 5$? After 4 hours of study? (Round answer to the nearest whole number.)

Integration by Substitution

- ❑ REVERSING THE CHAIN RULE
- ❑ INTEGRATION BY SUBSTITUTION
- ❑ ADDITIONAL SUBSTITUTION TECHNIQUES
- ❑ APPLICATION

Many of the indefinite integral formulas introduced in the preceding section are based on corresponding derivative formulas studied earlier. We now consider indefinite integral formulas and procedures based on the chain rule for differentiation.

❑ REVERSING THE CHAIN RULE

Recall the chain rule:

$$\frac{d}{dx} f[g(x)] = f'[g(x)]g'(x)$$

The expression on the right is formed by taking the derivative of the outside function f and multiplying it by the derivative of the inside function g. If we recognize an integrand as a chain-rule form $f'[g(x)]g'(x)$, we can easily find an antiderivative and its indefinite integral:

Reversing the Chain Rule

$$\int f'[g(x)]g'(x) \, dx = f[g(x)] + C \tag{1}$$

Explore–Discuss 1

(A) Which of the following has e^{x^3-1} as an antiderivative?

$$x^2 e^{x^3-1} \qquad 3x^2 e^{x^3-1} \qquad 3x e^{x^3-1}$$

(B) Which of the following would have e^{x^3-1} as an antiderivative if it were multiplied by a constant factor? A variable factor?

$$3x e^{x^3-1} \qquad x^2 e^{x^3-1}$$

We are interested in finding the indefinite integral:

$$\int 3x^2 e^{x^3-1} \, dx \tag{2}$$

The integrand appears to be the chain-rule form $e^{g(x)}g'(x)$, which is the derivative of $e^{g(x)}$. Since

$$\frac{d}{dx} e^{x^3-1} = 3x^2 e^{x^3-1}$$

it follows that

$$\int 3x^2 e^{x^3-1} \, dx = e^{x^3-1} + C \tag{3}$$

How does the following indefinite integral differ from integral (2)?

$$\int x^2 e^{x^3-1} \, dx \tag{4}$$

It is missing the constant factor 3. That is, $x^2 e^{x^3-1}$ is within a constant factor of being the derivative of e^{x^3-1}. But because a constant factor can be moved across the integral sign, this causes us little trouble in finding the indefinite integral of $x^2 e^{x^3-1}$. We introduce the constant factor 3, and at the same time multiply by $\frac{1}{3}$ and move the $\frac{1}{3}$ factor outside the integral sign. This is equivalent to multiplying the integrand in integral (4) by 1.

$$\int x^2 e^{x^3-1} \, dx = \int \frac{3}{3} x^2 e^{x^3-1} \, dx$$

$$= \frac{1}{3} \int 3x^2 e^{x^3-1} \, dx = \frac{1}{3} e^{x^3-1} + C \tag{5}$$

The derivative of the right side of equation (5) is the integrand of the indefinite integral (4). You should check this.

How does the following indefinite integral differ from integral (2)?

$$\int 3x e^{x^3-1} \, dx \tag{6}$$

It is missing a variable factor x. This is more serious. As tempting as it might be, we *cannot* adjust integral (6) by introducing the variable factor x and moving $1/x$ outside the integral sign, as we did with the constant 3 in equation (5). If we could move $1/x$ across the integral sign, what would stop us from moving the entire integrand across the integral sign? Then, indefinite integration would become a trivial exercise and would not give us the results we want—antiderivatives of the integrand.

Summary of the Integral Forms Above

$$\int 3x^2 e^{x^3-1} \, dx = e^{x^3-1} + C \qquad \text{Integrand is a chain-rule form.}$$

$$\int x^2 e^{x^3-1} \, dx = \frac{1}{3} e^{x^3-1} + C \qquad \text{Integrand can be adjusted to a chain-rule form.}$$

$$\int 3x e^{x^3-1} \, dx = \; ? \qquad \text{Integrand cannot be adjusted to be a chain-rule form.}$$

CAUTION

A constant factor can be moved across an integral sign, but a variable factor cannot.

There is nothing wrong with educated guessing when looking for an antiderivative of a given function, and you are encouraged to do so. You have only to check the result by differentiation. If you are right, you go on your way; if you are wrong, you simply try another approach.

In Section 5-3, we saw that the chain rule extends the derivative formulas for x^n, e^x, and $\ln x$ to derivative formulas for $[f(x)]^n$, $e^{f(x)}$, and $\ln[f(x)]$. The chain rule can also be used to extend the indefinite integral formulas discussed in Section 6-1. Some general formulas are summarized in the following box:

> ### General Indefinite Integral Formulas
>
> **1.** $\displaystyle\int [f(x)]^n f'(x)\, dx = \frac{[f(x)]^{n+1}}{n+1} + C, n \neq -1$
>
> **2.** $\displaystyle\int e^{f(x)} f'(x)\, dx = e^{f(x)} + C$
>
> **3.** $\displaystyle\int \frac{1}{f(x)} f'(x)\, dx = \ln|f(x)| + C$

Each formula can be verified by using the chain rule to show that the derivative of the function on the right is the integrand on the left. For example,

$$\frac{d}{dx}[e^{f(x)} + C] = e^{f(x)} f'(x)$$

verifies formula 2.

Example 1 ➭ **Reversing the Chain Rule**

(A) $\displaystyle\int (3x+4)^{10}(3)\, dx = \frac{(3x+4)^{11}}{11} + C$ Formula 1 with $f(x) = 3x + 4$ and $f'(x) = 3$

Check: $\displaystyle\frac{d}{dx}\frac{(3x+4)^{11}}{11} = 11\frac{(3x+4)^{10}}{11}\frac{d}{dx}(3x+4) = (3x+4)^{10}(3)$

(B) $\displaystyle\int e^{x^2}(2x)\, dx = e^{x^2} + C$ Formula 2 with $f(x) = x^2$ and $f'(x) = 2x$

Check: $\displaystyle\frac{d}{dx}e^{x^2} = e^{x^2}\frac{d}{dx}x^2 = e^{x^2}(2x)$

(C) $\displaystyle\int \frac{1}{1+x^3}3x^2\, dx = \ln|1+x^3| + C$ Formula 3 with $f(x) = 1 + x^3$ and $f'(x) = 3x^2$

Check: $\displaystyle\frac{d}{dx}\ln|1+x^3| = \frac{1}{1+x^3}\frac{d}{dx}(1+x^3) = \frac{1}{1+x^3}3x^2$ ∎

Matched Problem 1 ➭ Find each indefinite integral.

(A) $\displaystyle\int (2x^3 - 3)^{20}(6x^2)\, dx$ (B) $\displaystyle\int e^{5x}(5)\, dx$

(C) $\displaystyle\int \frac{1}{4+x^2}2x\, dx$ ∎

❑ ᴵɴᴛᴇɢʀᴀᴛɪᴏɴ ʙʏ Sᴜʙsᴛɪᴛᴜᴛɪᴏɴ

The key step in using formulas 1, 2, and 3 is recognizing the form of the integrand. Some people find it difficult to identify $f(x)$ and $f'(x)$ in these formulas and prefer to use a *substitution* to simplify the integrand. The *method of substitution,* which we now discuss, becomes increasingly useful as one progresses in studies of integration.

We start by introducing the idea of the *differential.* We represented the derivative by the symbol dy/dx taken as a whole. We now define dy and dx as two separate quantities with the property that their ratio is still equal to $f'(x)$:

Differentials

If $y = f(x)$ defines a differentiable function, then:

1. The **differential** dx of the independent variable x is an arbitrary real number.

2. The **differential** dy of the dependent variable y is defined as the product of $f'(x)$ and dx—that is, as

   $$dy = f'(x)\, dx$$

Differentials involve mathematical subtleties that are treated carefully in advanced mathematics courses. Here, we are interested in them mainly as a bookkeeping device to aid in the process of finding indefinite integrals. We can always check the results by differentiating.

Example 2   **Differentials**

(A) If $y = f(x) = x^2$, then

$$dy \;\bigl[= f'(x)\, dx\bigr]= 2x\, dx$$

(B) If $u = g(x) = e^{3x}$, then

$$du \;\bigl[= g'(x)\, dx\bigr]= 3e^{3x}\, dx$$

(C) If $w = h(t) = \ln(4 + 5t)$, then

$$dw \;\bigl[= h'(t)\, dt\bigr]= \frac{5}{4 + 5t}\, dt$$

Matched Problem 2   (A) Find dy for $y = f(x) = x^3$.   (B) Find du for $u = h(x) = \ln(2 + x^2)$.
(C) Find dv for $v = g(t) = e^{-5t}$.

The **method of substitution** is developed through the following examples.

Example 3   **Using Substitution**   Find $\int (x^2 + 2x + 5)^5 (2x + 2)\, dx$.

SOLUTION   If

$$u = x^2 + 2x + 5$$

then the differential of u is

$$du = (2x + 2)\, dx$$

Notice that du is one of the factors in the integrand. Substitute u for $x^2 + 2x + 5$ and du for $(2x + 2)\, dx$ to obtain

$$\int (x^2 + 2x + 5)^5(2x + 2)\, dx = \int u^5\, du$$

$$= \frac{u^6}{6} + C$$

$$= \frac{1}{6}(x^2 + 2x + 5)^6 + C \qquad \text{Since } u = x^2 + 2x + 5$$

Check: $\dfrac{d}{dx}\dfrac{1}{6}(x^2 + 2x + 5)^6 = \dfrac{1}{6}(6)(x^2 + 2x + 5)^5\dfrac{d}{dx}(x^2 + 2x + 5)$

$$= (x^2 + 2x + 5)^5(2x + 2) \qquad\blacksquare$$

Matched Problem 3 ➩ Find $\int(x^2 - 3x + 7)^4(2x - 3)\, dx$ by substitution. ■

The substitution method is also called the **change-of-variable method,** since u replaces the variable x in the process. Substituting $u = f(x)$ and $du = f'(x)\, dx$ in formulas 1, 2, and 3 produces the general indefinite integral formulas in the following box:

General Indefinite Integral Formulas

4. $\displaystyle\int u^n\, du = \frac{u^{n+1}}{n + 1} + C, \qquad n \neq -1$

5. $\int e^u\, du = e^u + C$

6. $\displaystyle\int \frac{1}{u}\, du = \ln|u| + C$

These formulas are valid if u is an independent variable or if u is a function of another variable and du is its differential with respect to that variable.

The substitution method for evaluating certain indefinite integrals is outlined in the following box:

Integration by Substitution

Step 1. Select a substitution that appears to simplify the integrand. In particular, try to select u so that du is a factor in the integrand.

Step 2. Express the integrand entirely in terms of u and du, completely eliminating the original variable and its differential.

Step 3. Evaluate the new integral, if possible.

Step 4. Express the antiderivative found in step 3 in terms of the original variable.

Example 4 ➩ **Using Substitution** Use a substitution to find the following:

(A) $\int (3x + 4)^6(3)\, dx$ (B) $\int e^{t^2}(2t)\, dt$

SOLUTION (A) If we let $u = 3x + 4$, then $du = 3\, dx$, and

$$\int (3x + 4)^6 (3)\, dx = \int u^6\, du \qquad \text{Use formula 4.}$$

$$= \frac{u^7}{7} + C$$

$$= \frac{(3x + 4)^7}{7} + C \qquad \text{Since } u = 3x + 4$$

Check: $\dfrac{d}{dx} \dfrac{(3x + 4)^7}{7} = \dfrac{7(3x + 4)^6}{7} \dfrac{d}{dx}(3x + 4) = (3x + 4)^6(3)$

(B) If we let $u = t^2$, then $du = 2t\, dt$, and

$$e^{t^2}(2t)dt = e^u\, du \qquad \text{Use formula 5.}$$

$$= e^u + C$$

$$= e^{t^2} + C \qquad \text{Since } u = t^2$$

Check: $\dfrac{d}{dt} e^{t^2} = e^{t^2} \dfrac{d}{dt} t^2 = e^{t^2}(2t)$

Matched Problem 4 ⇦   Use a substitution to find each indefinite integral.

(A) $\int (2x^3 - 3)^4 (6x^2)\, dx$ (B) $\int e^{5w}(5)\, dw$

CAUTION

Integration by substitution is an effective procedure for some indefinite integrals, but not all. Substitution is not helpful for $\int e^{x^2}\, dx$ or $\int (\ln x)dx$, for example.

❑   **ADDITIONAL SUBSTITUTION TECHNIQUES**

In order to use the substitution method, **the integrand must be expressed entirely in terms of u and du.** In some cases, the integrand will have to be modified before making a substitution and using one of the integration formulas. Example 5 illustrates this process.

Example 5 ⇦   **Substitution Techniques**   Integrate:

(A) $\displaystyle\int \frac{1}{4x + 7}\, dx$ (B) $\displaystyle\int te^{-t^2}\, dt$ (C) $\displaystyle\int 4x^2 \sqrt{x^3 + 5}\, dx$

SOLUTION   (A) If $u = 4x + 7$, then $du = 4\, dx$. We are missing a factor of 4 in the integrand to match formula 6 exactly. Recalling that a constant factor can be moved across an integral sign, we proceed as follows:

$$\int \frac{1}{4x + 7}\, dx = \int \frac{1}{4x + 7} \frac{4}{4}\, dx$$

$$= \frac{1}{4} \int \frac{1}{4x + 7} 4\, dx \qquad \text{Substitute } u = 4x + 7 \text{ and } du = 4\, dx.$$

$$= \frac{1}{4} \int \frac{1}{u}\, du \qquad \text{Use formula 6.}$$

$$= \tfrac{1}{4} \ln|u| + C$$

$$= \tfrac{1}{4} \ln|4x + 7| + C \qquad \text{Since } u = 4x + 7$$

Check:

$$\frac{d}{dx}\frac{1}{4}\ln|4x+7| = \frac{1}{4}\frac{1}{4x+7}\frac{d}{dx}(4x+7) = \frac{1}{4}\frac{1}{4x+7}4 = \frac{1}{4x+7}$$

(B) If $u = -t^2$, then $du = -2t\,dt$. Proceed as in part (A):

$$\int te^{-t^2}\,dt = \int e^{-t^2}\frac{-2}{-2}t\,dt$$

$$= -\frac{1}{2}\int e^{-t^2}(-2t)\,dt \quad \text{Substitute } u = -t^2 \text{ and } du = -2t\,dt.$$

$$= -\frac{1}{2}\int e^u\,du \qquad \text{Use formula 5.}$$

$$= -\tfrac{1}{2}e^u + C$$

$$= -\tfrac{1}{2}e^{-t^2} + C \qquad \text{Since } u = -t^2$$

Check: $\dfrac{d}{dt}(-\tfrac{1}{2}e^{-t^2}) = -\tfrac{1}{2}e^{-t^2}\dfrac{d}{dt}(-t^2) = -\tfrac{1}{2}e^{-t^2}(-2t) = te^{-t^2}$

(C) $\displaystyle\int 4x^2\sqrt{x^3+5}\,dx = 4\int\sqrt{x^3+5}(x^2)\,dx$ Move the 4 across the integral sign and proceed as before.

$$= 4\int\sqrt{x^3+5}\,\frac{3}{3}(x^2)\,dx$$

$$= \frac{4}{3}\int\sqrt{x^3+5}(3x^2)\,dx \quad \text{Substitute } u = x^3+5 \text{ and } du = 3x^2\,dx.$$

$$= \frac{4}{3}\int\sqrt{u}\,du$$

$$= \frac{4}{3}\int u^{1/2}\,du \qquad \text{Use formula 4.}$$

$$= \frac{4}{3}\frac{u^{3/2}}{\frac{3}{2}} + C$$

$$= \tfrac{8}{9}u^{3/2} + C$$

$$= \tfrac{8}{9}(x^3+5)^{3/2} + C \qquad \text{Since } u = x^3+5$$

Check: $\dfrac{d}{dx}[\tfrac{8}{9}(x^3+5)^{3/2}] = \tfrac{4}{3}(x^3+5)^{1/2}\dfrac{d}{dx}(x^3+5)$

$$= \tfrac{4}{3}(x^3+5)^{1/2}(3x^2) = 4x^2\sqrt{x^3+5}$$ ■

Matched Problem 5 ⇨ Integrate:

(A) $\displaystyle\int e^{-3x}dx$ (B) $\displaystyle\int\frac{x}{x^2-9}dx$ (C) $\displaystyle\int 5t^2(t^3+4)^{-2}\,dt$ ■

Even if it is not possible to find a substitution that makes an integrand match one of the integration formulas exactly, a substitution may sufficiently simplify the integrand so that other techniques can be used.

Example 6 ⇨ **Substitution Techniques**   Find: $\displaystyle\int\frac{x}{\sqrt{x+2}}dx$

SOLUTION   Proceeding as before, if we let $u = x+2$, then $du = dx$ and

$$\int\frac{x}{\sqrt{x+2}}dx = \int\frac{x}{\sqrt{u}}du$$

Notice that this substitution is not yet complete, because we have not expressed the integrand entirely in terms of u and du. As we noted earlier, only a constant factor can be moved across an integral sign, so we cannot move x outside the integral sign (as much as we would like to). Instead, we must return to the original substitution, solve for x in terms of u, and use the resulting equation to complete the substitution:

$$u = x + 2 \quad \text{Solve for x in terms of u.}$$
$$u - 2 = x \quad \text{Substitute this expression for x.}$$

Thus,

$$\int \frac{x}{\sqrt{x+2}}\, dx = \int \frac{u-2}{\sqrt{u}}\, du \qquad \text{Simplify the integrand.}$$

$$= \int \frac{u-2}{u^{1/2}}\, du$$

$$= \int (u^{1/2} - 2u^{-1/2})\, du$$

$$= \int u^{1/2}\, du - 2\int u^{-1/2}\, du$$

$$= \frac{u^{3/2}}{\frac{3}{2}} - 2\frac{u^{1/2}}{\frac{1}{2}} + C$$

$$= \tfrac{2}{3}(x+2)^{3/2} - 4(x+2)^{1/2} + C \quad \text{Since u = x + 2}$$

$$Check: \quad \frac{d}{dx}[\tfrac{2}{3}(x+2)^{3/2} - 4(x+2)^{1/2}] = (x+2)^{1/2} - 2(x+2)^{-1/2}$$

$$= \frac{x+2}{(x+2)^{1/2}} - \frac{2}{(x+2)^{1/2}}$$

$$= \frac{x}{(x+2)^{1/2}} \qquad \blacksquare$$

Matched Problem 6 ✏ Find: $\int x\sqrt{x+1}\, dx$ ■

  ❑ APPLICATION

Example 7 ✏   **Price–Demand**   The market research department for a supermarket chain has determined that for one store the marginal price $p'(x)$ at x tubes per week for a certain brand of toothpaste is given by

$$p'(x) = -0.015e^{-0.01x}$$

Find the price–demand equation if the weekly demand is 50 tubes when the price of a tube is $2.35. Find the weekly demand when the price of a tube is $1.89.

SOLUTION $p(x) = \int -0.015e^{-0.01x}\, dx$
$$= -0.015 \int e^{-0.01x}\, dx$$

$$= -0.015 \int e^{-0.01x} \frac{-0.01}{-0.01} \, dx$$

$$= \frac{-0.015}{-0.01} \int e^{-0.01x}(-0.01) \, dx \qquad \text{Substitute } u = -0.01x \text{ and } du = -0.01 \, dx.$$

$$= 1.5 \int e^u \, du$$

$$= 1.5e^u + C$$

$$= 1.5e^{-0.01x} + C \qquad \text{Since } u = -0.01x$$

We find C by noting that

$$p(50) = 1.5e^{-0.01(50)} + C = \$2.35$$

$$C = \$2.35 - 1.5e^{-0.5} \qquad \text{Use a calculator.}$$

$$= \$2.35 - 0.91$$

$$= \$1.44$$

Thus,

$$p(x) = 1.5e^{-0.01x} + 1.44$$

To find the demand when the price is \$1.89, we solve $p(x) = \$1.89$ for x:

$$1.5e^{-0.01x} + 1.44 = 1.89$$

$$1.5e^{-0.01x} = 0.45$$

$$e^{-0.01x} = 0.3$$

$$-0.01x = \ln 0.3$$

$$x = -100 \ln 0.3 \approx 120 \text{ tubes}$$

FIGURE 1
$y_1 = 1.5e^{-0.01x} + 1.44$
$y_2 = 1.89$

A graphing utility offers an alternative approach to solving $p(x) = 1.89$ in Example 7 (see Fig. 1).

Matched Problem 7 The marginal price $p'(x)$ at a supply level of x tubes per week for a certain brand of toothpaste is given by

$$p'(x) = 0.001e^{0.01x}$$

Find the price–supply equation if the supplier is willing to supply 100 tubes per week at a price of \$1.65 each. How many tubes would the supplier be willing to supply at a price of \$1.98 each?

Explore–Discuss 2

In each of the following examples explain why $\neq$ is used; then work the problem correctly using either an appropriate substitution or another method.

1. $\displaystyle \int (x^2 + 3)^2 \, dx = \int (x^2 + 3)^2 \frac{2x}{2x} \, dx$

$$\neq \frac{1}{2x} \int (x^2 + 3)^2 (2x) \, dx$$

2. $\displaystyle \int \frac{1}{10x + 3} \, dx = \int \frac{1}{u} \, dx \qquad u = 10x + 3$

$$\neq \ln|u| + C$$

We conclude with two final cautions (the first was stated earlier, but is worth repeating):

CAUTION

1. A variable cannot be moved across an integral sign!
2. An integral must be expressed entirely in terms of u and du before applying integration formulas 4, 5, and 6.

Answers to Matched Problems

1. (A) $\frac{1}{21}(2x^3 - 3)^{21} + C$ (B) $e^{5x} + C$
   (C) $\ln|4 + x^2| + C$ or $\ln(4 + x^2) + C$, since $4 + x^2 > 0$

2. (A) $dy = 3x^2\, dx$ (B) $du = \dfrac{2x}{2 + x^2}\, dx$ (C) $dv = -5e^{-5t} dt$

3. $\frac{1}{5}(x^2 - 3x + 7)^5 + C$ 4. (A) $\frac{1}{5}(2x^3 - 3)^5 + C$ (B) $e^{5w} + C$

5. (A) $-\frac{1}{3}e^{-3x} + C$ (B) $\frac{1}{2}\ln|x^2 - 9| + C$ (C) $-\frac{5}{3}(t^3 + 4)^{-1} + C$

6. $\frac{2}{5}(x + 1)^{5/2} - \frac{2}{3}(x + 1)^{3/2} + C$ 7. $p(x) = 0.1e^{0.01x} + 1.38$; 179 tubes

---

Exercise 6-2

A *In Problems 1–40, find each indefinite integral, and check the result by differentiating.*

1. $\displaystyle\int (4x - 3)^2\,(4)\,dx$ 2. $\displaystyle\int (7x + 1)^3\,(7)\,dx$

3. $\displaystyle\int (x^2 + 1)^4\,(2x)\,dx$ 4. $\displaystyle\int (x^5 - 2)^2\,(5x^4)\,dx$

5. $\displaystyle\int e^{2x}\,(2)\,dx$ 6. $\displaystyle\int e^{x^3}\,(3x^2)\,dx$

7. $\displaystyle\int \frac{1}{1 + x^4}\,(4x^3)\,dx$ 8. $\displaystyle\int \frac{1}{1 + e^x}\,(e^x)\,dx$

B

9. $\displaystyle\int (x + 3)^{10}\,dx$ 10. $\displaystyle\int (x - 3)^{-4}\,dx$

11. $\displaystyle\int (6t - 7)^{-2}\,dt$ 12. $\displaystyle\int (5t + 1)^3\,dt$

13. $\displaystyle\int (t^2 + 1)^5\, t\,dt$ 14. $\displaystyle\int (t^3 + 4)^{-2}\, t^2\,dt$

15. $\displaystyle\int x e^{x^2}\,dx$ 16. $\displaystyle\int e^{-0.01x}\,dx$

17. $\displaystyle\int \frac{1}{5x + 4}\,dx$ 18. $\displaystyle\int \frac{x}{1 + x^2}\,dx$

19. $\displaystyle\int e^{1-t}\,dt$ 20. $\displaystyle\int \frac{3}{2 - t}\,dt$

21. $\displaystyle\int \frac{t}{(3t^2 + 1)^4}\,dt$ 22. $\displaystyle\int \frac{t^2}{(t^3 - 2)^5}\,dt$

23. $\displaystyle\int \frac{x^2}{(4 - x^3)^2}\,dx$ 24. $\displaystyle\int \frac{x}{(5 - 2x^2)^5}\,dx$

25. $\displaystyle\int x\sqrt{x + 4}\,dx$ 26. $\displaystyle\int x\sqrt{x - 9}\,dx$

27. $\displaystyle\int \frac{x}{\sqrt{x - 3}}\,dx$ 28. $\displaystyle\int \frac{x}{\sqrt{x + 5}}\,dx$

29. $\displaystyle\int x(x - 4)^9\,dx$ 30. $\displaystyle\int x(x + 6)^8\,dx$

31. $\displaystyle\int e^{2x}(1 + e^{2x})^3\,dx$ 32. $\displaystyle\int e^{-x}(1 - e^{-x})^4\,dx$

33. $\displaystyle\int \frac{1 + x}{4 + 2x + x^2}\,dx$ 34. $\displaystyle\int \frac{x^2 - 1}{x^3 - 3x + 7}\,dx$

35. $\displaystyle\int (2x + 1)e^{x^2 + x + 1}\,dx$ 36. $\displaystyle\int (x^2 + 2x)e^{x^3 + 3x^2}\,dx$

37. $\displaystyle\int (e^x - 2x)^3(e^x - 2)\,dx$

38. $\int (x^2 - e^x)^4 (2x - e^x)\,dx$

39. $\int \dfrac{x^3 + x}{(x^4 + 2x^2 + 1)^4}\,dx$ **40.** $\int \dfrac{x^2 - 1}{(x^3 - 3x + 7)^2}\,dx$

In Problems 41–46, imagine that the indicated "solutions" were given to you by a student whom you are tutoring in this class.

(A) How would you have the student check each solution?

(B) Is the solution right or wrong? If the solution is wrong, explain what is wrong and how it can be corrected.

(C) Show a correct solution for each incorrect solution, and check the result by differentiation.

41. $\int \dfrac{1}{2x - 3}\,dx = \ln|2x - 3| + C$

42. $\int \dfrac{x}{x^2 + 5}\,dx = \ln|x^2 + 5| + C$

43. $\int x^3 e^{x^4}\,dx = e^{x^4} + C$

44. $\int e^{4x-5}\,dx = e^{4x-5} + C$

45. $\int 2(x^2 - 2)^2\,dx = \dfrac{(x^2 - 2)^2}{3x} + C$

46. $\int (-10x)(x^2 - 3)^{-4}\,dx = (x^2 - 3)^{-5} + C$

C *In Problems 47–58, find each indefinite integral, and check the result by differentiating.*

47. $\int x\sqrt{3x^2 + 7}\,dx$ **48.** $\int x^2\sqrt{2x^3 + 1}\,dx$

49. $\int x(x^3 + 2)^2\,dx$ **50.** $\int x(x^2 + 2)^2\,dx$

51. $\int x^2(x^3 + 2)^2\,dx$ **52.** $\int (x^2 + 2)^2\,dx$

53. $\int \dfrac{x^3}{\sqrt{2x^4 + 3}}\,dx$ **54.** $\int \dfrac{x^2}{\sqrt{4x^3 - 1}}\,dx$

55. $\int \dfrac{(\ln x)^3}{x}\,dx$ **56.** $\int \dfrac{e^x}{1 + e^x}\,dx$

57. $\int \dfrac{1}{x^2}e^{-1/x}\,dx$ **58.** $\int \dfrac{1}{x \ln x}\,dx$

In Problems 59–64, find the antiderivative of each derivative.

59. $\dfrac{dx}{dt} = 7t^2(t^3 + 5)^6$ **60.** $\dfrac{dm}{dn} = 10n(n^2 - 8)^7$

61. $\dfrac{dy}{dt} = \dfrac{3t}{\sqrt{t^2 - 4}}$ **62.** $\dfrac{dy}{dx} = \dfrac{5x^2}{(x^3 - 7)^4}$

63. $\dfrac{dp}{dx} = \dfrac{e^x + e^{-x}}{(e^x - e^{-x})^2}$ **64.** $\dfrac{dm}{dt} = \dfrac{\ln(t - 5)}{t - 5}$

Use substitution techniques to derive the integration formulas in Problems 65 and 66. Then check your work by differentiation.

65. $\int e^{au}\,du = \dfrac{1}{a}e^{au} + C, \qquad a \neq 0$

66. $\int \dfrac{1}{au + b}\,du = \dfrac{1}{a}\ln|au + b| + C, \qquad a \neq 0$

Applications

Business & Economics

67. *Price–demand equation.* The marginal price for a weekly demand of x bottles of baby shampoo in a drug store is given by

$$p'(x) = \dfrac{-6,000}{(3x + 50)^2}$$

Find the price–demand equation if the weekly demand is 150 when the price of a bottle of shampoo is $4. What is the weekly demand when the price is $2.50?

68. *Price–supply equation.* The marginal price at a supply level of x bottles of baby shampoo per week is given by

$$p'(x) = \dfrac{300}{(3x + 25)^2}$$

Find the price–supply equation if the distributor of the shampoo is willing to supply 75 bottles a week at a price of $1.60 per bottle. How many bottles would the supplier be willing to supply at a price of $1.75 per bottle?

69. *Cost function.* The weekly marginal cost of producing x pairs of tennis shoes is given by

$$C'(x) = 12 + \dfrac{500}{x + 1}$$

where $C(x)$ is cost in dollars. If the fixed costs are $2,000 per week, find the cost function. What is the average cost per pair of shoes if 1,000 pairs of shoes are produced each week?

70. *Revenue function.* The weekly marginal revenue from the sale of x pairs of tennis shoes is given by

$$R'(x) = 40 - 0.02x + \frac{200}{x + 1} \qquad R(0) = 0$$

where $R(x)$ is revenue in dollars. Find the revenue function. Find the revenue from the sale of 1,000 pairs of shoes.

71. *Marketing.* An automobile company is ready to introduce a new line of cars with a national sales campaign. After test marketing the line in a carefully selected city, the marketing research department estimates that sales (in millions of dollars) will increase at the monthly rate of

$$S'(t) = 10 - 10e^{-0.1t} \qquad 0 \leqslant t \leqslant 24$$

t months after the national campaign has started.

(A) What will be the total sales, $S(t)$, t months after the beginning of the national campaign if we assume 0 sales at the beginning of the campaign?

(B) What are the estimated total sales for the first 12 months of the campaign?

(C) When will the estimated total sales reach $100 million? Use a graphing utility to approximate the answer to two decimal places.

72. *Marketing.* Repeat Problem 71 if the monthly rate of increase in sales is found to be approximated by

$$S'(t) = 20 - 20e^{-0.05t} \qquad 0 \leqslant t \leqslant 24$$

73. *Oil production.* Using data from the first 3 years of production as well as geological studies, the management of an oil company estimates that oil will be pumped from a producing field at a rate given by

$$R(t) = \frac{100}{t + 1} + 5 \qquad 0 \leqslant t \leqslant 20$$

where $R(t)$ is the rate of production (in thousands of barrels per year) t years after pumping begins. How many barrels of oil, $Q(t)$, will the field produce the first t years if $Q(0) = 0$? How many barrels will be produced the first 9 years?

74. *Oil production.* Assume that the rate in Problem 73 is found to be

$$R(t) = \frac{120t}{t^2 + 1} + 3 \qquad 0 \leqslant t \leqslant 20$$

(A) When is the rate of production greatest?

(B) How many barrels of oil, $Q(t)$, will the field produce the first t years if $Q(0) = 0$? How many barrels will be produced the first 5 years?

(C) How long (to the nearest tenth of a year) will it take to produce a total of a quarter of a million barrels of oil?

Life Sciences

75. *Biology.* A yeast culture is growing at the rate of $W'(t) = 0.2e^{0.1t}$ grams per hour. If the starting culture weighs 2 grams, what will be the weight of the culture, $W(t)$, after t hours? After 8 hours?

76. *Medicine.* The rate of healing for a skin wound (in square centimeters per day) is approximated by $A'(t) = -0.9e^{-0.1t}$. If the initial wound has an area of 9 square centimeters, what will its area, $A(t)$, be after t days? After 5 days?

77. *Pollution.* A contaminated lake is treated with a bactericide. The rate of increase in harmful bacteria t days after the treatment is given by

$$\frac{dN}{dt} = -\frac{2,000t}{1 + t^2} \qquad 0 \leqslant t \leqslant 10$$

where $N(t)$ is the number of bacteria per milliliter of water (since dN/dt is negative, the count of harmful bacteria is decreasing).

(A) Find the minimum value of dN/dt.

(B) If the initial count was 5,000 bacteria per milliliter, find $N(t)$ and then find the bacteria count after 10 days.

(C) When (to two decimal places) is the bacteria count 1,000 bacteria per milliliter?

78. *Pollution.* An oil tanker aground on a reef is losing oil and producing an oil slick that is radiating outward at a rate given approximately by

$$\frac{dR}{dt} = \frac{60}{\sqrt{t + 9}} \qquad t \geqslant 0$$

where R is the radius (in feet) of the circular slick after t minutes. Find the radius of the slick after 16 minutes if the radius is 0 when $t = 0$.

Social Sciences

79. *Learning.* In a particular business college, it was found that an average student enrolled in an advanced typing class progressed at a rate of $N'(t) = 6e^{-0.1t}$ words per minute per week, t weeks after enrolling in a 15-week course. If at the beginning of the course a student could type 40 words per minute, how many words per minute, $N(t)$, would the student be expected to type t weeks into the course? After completing the course?

80. *Learning.* In the same business college, it was also found that an average student enrolled in a

beginning shorthand class progressed at a rate of $N'(t) = 12e^{-0.06t}$ words per minute per week, t weeks after enrolling in a 15-week course. If at the beginning of the course a student could take dictation in shorthand at 0 words per minute, how many words per minute, $N(t)$, would the student be expected to handle t weeks into the course? After completing the course?

81. *College enrollment.* The projected rate of increase in enrollment in a new college is estimated by

$$\frac{dE}{dt} = 5{,}000(t + 1)^{-3/2} \qquad t \geqslant 0$$

where $E(t)$ is the projected enrollment in t years. If enrollment is 2,000 now ($t = 0$), find the projected enrollment 15 years from now.

Differential Equations; Growth and Decay

- ❑ DIFFERENTIAL EQUATIONS AND SLOPE FIELDS
- ❑ CONTINUOUS COMPOUND INTEREST REVISITED
- ❑ EXPONENTIAL GROWTH LAW
- ❑ POPULATION GROWTH, RADIOACTIVE DECAY, AND LEARNING
- ❑ COMPARISON OF EXPONENTIAL GROWTH PHENOMENA

In the preceding section we considered equations of the form

$$\frac{dy}{dx} = 6x^2 - 4x \qquad p'(x) = -400e^{-0.04x}$$

These are examples of *differential equations*. In general, an equation is a **differential equation** if it involves an unknown function and one or more of its derivatives. Other examples of differential equations are

$$\frac{dy}{dx} = ky \qquad y'' - xy' + x^2 = 5 \qquad \frac{dy}{dx} = 2xy$$

The first and third equations are called **first-order** equations because each involves a first derivative, but no higher derivative. The second equation is called a **second-order** equation because it involves a second derivative, and no higher derivatives. Finding solutions to different types of differential equations (functions that satisfy the equation) is the subject matter for entire books and courses on this topic. Here, we consider only a few very special but very important first-order equations that have immediate and significant applications.

We start by looking at some first-order equations geometrically, in terms of *slope fields*. We then reconsider continuous compound interest as modeled by a first-order differential equation, and from this we will be able to generalize the approach to a wide variety of other types of growth phenomena.

❑ DIFFERENTIAL EQUATIONS AND SLOPE FIELDS

We introduce the concept of *slope field* through an example. Suppose that we are given the first-order differential equation

$$\frac{dy}{dx} = 0.2y \tag{1}$$

A function f is a solution of equation (1) if $y = f(x)$ satisfies equation (1) for all values of x in the domain of f. Geometrically interpreted, equation (1) gives us the slope of a solution curve that passes through the point (x, y). For example, if $y = f(x)$ is a solution of equation (1) that passes through the point $(0, 2)$, then the slope of f at $(0, 2)$ is given by

$$\frac{dy}{dx} = 0.2(2) = 0.4$$

We indicate this by drawing a short segment of the tangent line at the point $(0, 2)$, as shown in Figure 1A. This procedure is repeated for points $(-3, 1)$ and $(2, 3)$, as also shown in Figure 1A. Assuming that the graph of f passes through all three points, we sketch an approximate graph of f in Figure 1B.

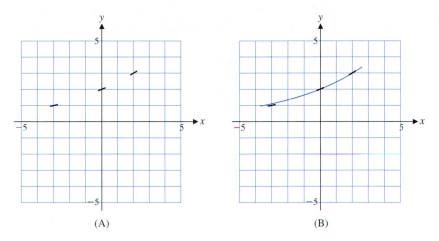

(A)   (B)

FIGURE 1

If we continue the process of drawing tangent line segments at each point in the grid in Figure 1—a task easily handled by computers but not by hand—we obtain a *slope field*. A slope field for differential equation (1), drawn by a computer, is shown in Figure 2. In general, a **slope field** for a first-order differential equation is obtained by drawing tangent line segments determined by the equation at each point in a grid. In a more advanced treatment of the subject, one can find out a lot about the shape and behavior of solution curves of first-order differential equations by looking at slope fields. Our interests are more modest here.

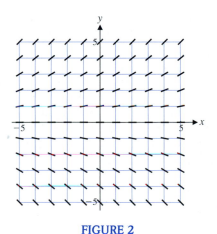

FIGURE 2

Explore–Discuss 1

(A) In Figure 1A (or a copy), draw tangent line segments for a solution curve of differential equation (1) that passes through $(-3, -1)$, $(0, -2)$, and $(2, -3)$.

(B) In Figure 1B (or a copy), sketch an approximate graph of the solution curve that passes through these three points. (Repeat the tangent line segments first.)

(C) Of all the elementary functions discussed in the first two chapters, make a conjecture as to what type of function appears to be a solution to differential equation (1).

In Explore–Discuss 1, if you guessed that solutions to equation (1) are exponential functions, you are to be congratulated. We now show that

$$y = Ce^{0.2x} \tag{2}$$

is a solution to equation (1) for any real number C. [Later in this section we show how to find equation (2) directly from equation (1).] To do this, we substitute

FIGURE 3

$y = Ce^{0.2x}$ into equation (1) to see if the left side is equal to the right side for all real x:

$$\frac{dy}{dx} = 0.2y$$

$$\text{Left side:} \quad \frac{dy}{dx} = \frac{d}{dx}(Ce^{0.2x}) = 0.2Ce^{0.2x}$$

$$\text{Right side:} \quad 0.2y = 0.2Ce^{0.2x}$$

Thus, equation (2) is a solution of equation (1) for C any real number. Which values of C will produce solution curves that pass through $(0, 2)$ and $(0, -2)$, respectively? Substituting the coordinates of each point into equation (2) and solving for C (a task left the reader), we obtain

$$y = 2e^{0.2x} \qquad \text{and} \qquad y = -2e^{0.2x} \tag{3}$$

as can easily be checked. The graphs of equations (3) are shown in Figure 3 and confirm the results in Figure 1B.

Explore–Discuss 2

(A) In Figure 3 (or a copy), sketch in an approximate solution curve that passes through $(0, 3)$ and one that passes through $(0, -3)$.

(B) Use a graphing utility to graph $y = Ce^{0.2x}$ for $C = -4, -3, -2, 2, 3, 4$, all in the same viewing window. Notice how these solution curves follow the flow of the tangent line segments in the slope field in Figure 2.

As indicated above, drawing slope fields by hand is not a task for human beings—a 20 by 20 grid would require drawing 400 tangent line segments! Repetitive tasks of this type are what computers are for. A few problems in Exercise 6-3 involve interpreting slope fields, not drawing them.

❑ **CONTINUOUS COMPOUND INTEREST REVISITED**

Let P be the initial amount of money deposited in an account, and let A be the amount in the account at any time t. Instead of assuming that the money in the account earns a particular rate of interest, suppose we say that the rate of growth of the amount of money in the account at any time t is proportional to the amount present at that time. Since dA/dt is the rate of growth of A with respect to t, we have

$$\frac{dA}{dt} = rA \qquad A(0) = P \qquad A, P > 0 \tag{4}$$

where r is an appropriate constant. We would like to find a function $A = A(t)$ that satisfies these conditions. Multiplying both sides of equation (4) by $1/A$, we obtain

$$\frac{1}{A}\frac{dA}{dt} = r$$

Now we integrate each side with respect to t:

$$\int \frac{1}{A}\frac{dA}{dt}\,dt = \int r\,dt \qquad \frac{dA}{dt}dt = A'(t)dt = dA$$

$$\int \frac{1}{A}\,dA = \int r\,dt$$

$$\ln|A| = rt + C \qquad |A| = A, \text{ since } A > 0$$
$$\ln A = rt + C$$

We convert this last equation into the equivalent exponential form

$$A = e^{rt+C} \qquad \text{Definition of logarithmic function:}$$
$$\qquad\qquad\qquad y = \ln x \text{ if and only if } x = e^y$$
$$\quad\; = e^C e^{rt} \qquad \text{Property of exponents: } b^m b^n = b^{m+n}$$

Since $A(0) = P$, we evaluate $A(t) = e^C e^{rt}$ at $t = 0$ and set it equal to P:

$$A(0) = e^C e^0 = e^C = P$$

Hence, $e^C = P$, and we can rewrite $A = e^C e^{rt}$ in the form

$$A = Pe^{rt}$$

This is the same continuous compound interest formula obtained in Section 5-1, where the principal P is invested at an annual nominal rate r compounded continuously for t years.

❑ EXPONENTIAL GROWTH LAW

In general, if the rate of change with respect to time of a quantity Q is proportional to the amount present and $Q(0) = Q_0$, then proceeding in exactly the same way as above, we obtain the following:

Exponential Growth Law

If $\dfrac{dQ}{dt} = rQ$ $\qquad$ and $\qquad$ $Q(0) = Q_0$, $\qquad$ then $\qquad$ $Q = Q_0 e^{rt}$,

where

Q_0 = amount at $t = 0$

r = continuous compound growth rate (expressed as a decimal)

t = time

Q = quantity at time t

The constant r in the exponential growth law is sometimes called the **growth constant,** or the **growth rate.** The last term can be misleading, since the rate of growth of Q with respect to time is dQ/dt, not r. Notice that if $r < 0$, then $dQ/dt < 0$ and Q is decreasing. This type of growth is called **exponential decay.**

Once we know that the rate of growth of something is proportional to the amount present, we know that it has exponential growth and we can use the results summarized in the box without having to solve the differential equation each time. The exponential growth law applies not only to money invested at interest compounded continuously, but also to many other types of problems—population growth, radioactive decay, natural resource depletion, and so on.

❑ POPULATION GROWTH, RADIOACTIVE DECAY, AND LEARNING

The world population passed 1 billion in 1804, 2 billion in 1927, 3 billion in 1960, 4 billion in 1974, 5 billion in 1987, and 6 billion in 1999, as illustrated in Figure 4. **Population growth** over certain periods of time often can be approximated by the exponential growth law described above.

FIGURE 4 World population growth

Example 1 ⇔ **Population Growth** India had a population of about 1.0 billion in 2000 ($t = 0$). Let P represent the population (in billions) t years after 2000, and assume a growth rate of 1.3% compounded continuously.

    (A) Find an equation that represents India's population growth after 2000, assuming that the 1.3% growth rate continues.

    (B) What is the estimated population (to the nearest tenth of a billion) for India in the year 2030?

    (C) Graph the equation found in part (A) from 2000 to 2030.

SOLUTION   (A) The exponential growth law applies, and we have

$$\frac{dP}{dt} = 0.013P \qquad P(0) = 1.0$$

Thus,

$$P = 1.0e^{0.013t} \tag{5}$$

    (B) Using equation (5), we can estimate the population in India in 2030 ($t = 30$):

$$P = 1.0e^{0.013(30)} = 1.5 \text{ billion people}$$

    (C) The graph is shown in Figure 5.

FIGURE 5 Population of India

Matched Problem 1 ⇔ Assuming the same continuous compound growth rate as in Example 1, what will India's population be (to the nearest tenth of a billion) in the year 2015?

Example 2 ⇔ **Population Growth** If the exponential growth law applies to Canada's population growth, at what continuous compound growth rate will the population double over the next 100 years?

SOLUTION   The problem is to find r, given $P = 2P_0$ and $t = 100$:

$$P = P_0e^{rt}$$
$$2P_0 = P_0e^{100r}$$

$$2 = e^{100r}$$

$$100r = \ln 2$$

Take the natural logarithm of both sides and reverse the equation.

$$r = \frac{\ln 2}{100}$$

$$\approx 0.0069 \quad \text{or} \quad 0.69\%$$

Matched Problem 2 If the exponential growth law applies to population growth in Nigeria, find the doubling time (to the nearest year) of the population if it grows at 2.1% per year compounded continuously.

We now turn to another type of exponential growth—**radioactive decay.** In 1946, Willard Libby (who later received a Nobel Prize in Chemistry) found that as long as a plant or animal is alive, radioactive carbon-14 is maintained at a constant level in its tissues. Once the plant or animal is dead, however, the radioactive carbon-14 diminishes by radioactive decay at a rate proportional to the amount present. Thus,

$$\frac{dQ}{dt} = rQ \qquad Q(0) = Q_0$$

and we have another example of the exponential growth law. The continuous compound rate of decay for radioactive carbon-14 has been found to be 0.000 123 8; thus, $r = -0.000\ 123\ 8$, since decay implies a negative continuous compound growth rate.

Example 3 **Archaeology** A piece of human bone was found at an archaeological site in Africa. If 10% of the original amount of radioactive carbon-14 was present, estimate the age of the bone (to the nearest 100 years).

SOLUTION Using the exponential growth law for

$$\frac{dQ}{dt} = -0.000\ 123\ 8Q \qquad Q(0) = Q_0$$

we find that

$$Q = Q_0 e^{-0.0001238t}$$

and our problem is to find t so that $Q = 0.1Q_0$ (since the amount of carbon-14 present now is 10% of the amount present, Q_0, at the death of the person). Thus,

$$0.1Q_0 = Q_0 e^{-0.0001238t}$$

$$0.1 = e^{-0.0001238t}$$

$$\ln 0.1 = \ln e^{-0.0001238t}$$

$$t = \frac{\ln 0.1}{-0.000\ 123\ 8} \approx 18{,}600 \text{ years}$$

FIGURE 6 $y_1 = e^{-0.0001238x}$; $y_2 = 0.1$

See Figure 6 for a graphical solution to Example 3.

Matched Problem 3 Estimate the age of the bone in Example 3 (to the nearest 100 years) if 50% of the original amount of carbon-14 is present.

In learning certain skills such as typing and swimming, a mathematical model often used is one that assumes there is a maximum skill attainable, say, M, and the rate of improving is proportional to the difference between that achieved, y, and that attainable, M. Mathematically,

$$\frac{dy}{dt} = k(M - y) \qquad y(0) = 0$$

We solve this type of problem using the same technique that was used to obtain the exponential growth law. First, multiply both sides of the first equation by $1/(M - y)$ to obtain

$$\frac{1}{M - y}\frac{dy}{dt} = k$$

and then integrate each side with respect to t:

$$\int \frac{1}{M - y}\frac{dy}{dt}\,dt = \int k\,dt$$

$$-\int \frac{1}{M - y}\left(-\frac{dy}{dt}\right)dt = \int k\,dt \qquad \text{Substitute } u = M - y \text{ and}$$

$$-\int \frac{1}{u}\,du = \int k\,dt \qquad\qquad du = -dy = -\frac{dy}{dt}\,dt.$$

$$-\ln|u| = kt + C \qquad \text{Substitute } M - y = u.$$

$$-\ln(M - y) = kt + C \qquad \text{Absolute value signs are not required.}$$

$$\ln(M - y) = -kt - C \qquad \text{(Why?)}$$

Change this last equation to equivalent exponential form:

$$M - y = e^{-kt-C}$$
$$M - y = e^{-C}e^{-kt}$$
$$y = M - e^{-C}e^{-kt}$$

Now, $y(0) = 0$; hence,

$$y(0) = M - e^{-C}e^0 = 0$$

Solving for e^{-C}, we obtain

$$e^{-C} = M$$

and our final solution is

$$y = M - Me^{-kt} = M(1 - e^{-kt})$$

Example 4 ↩ **Learning** For a particular person who is learning to swim, it is found that the distance y (in feet) the person is able to swim in 1 minute after t hours of practice is given approximately by

$$y = 50(1 - e^{-0.04t})$$

What is the rate of improvement (to two decimal places) after 10 hours of practice?

SOLUTION
$$y = 50 - 50e^{-0.04t}$$
$$y'(t) = 2e^{-0.04t}$$
$$y'(10) = 2e^{-0.04(10)} \approx 1.34 \text{ feet per hour of practice}$$

Matched Problem 4   In Example 4, what is the rate of improvement (to two decimal places) after 50 hours of practice?

❑   COMPARISON OF EXPONENTIAL GROWTH PHENOMENA

The graphs and equations given in Table 1 compare several widely used growth models. These are divided basically into two groups: unlimited growth and limited growth. Following each equation and graph is a short (and necessarily incomplete) list of areas in which the models are used. This only touches on a subject that has been extensively developed and which you are likely to encounter in greater depth in the future.

TABLE 1

EXPONENTIAL GROWTH

DESCRIPTION	MODEL	SOLUTION	GRAPH	USES
Unlimited growth: Rate of growth is proportional to the amount present	$\dfrac{dy}{dt} = ky$ $k, t > 0$ $y(0) = c$	$y = ce^{kt}$		• Short-term population growth (people, bacteria, etc.) • Growth of money at continuous compound interest • Price–supply curves
Exponential decay: Rate of growth is proportional to the amount present	$\dfrac{dy}{dt} = -ky$ $k, t > 0$ $y(0) = c$	$y = ce^{-kt}$		• Depletion of natural resources • Radioactive decay • Light absorption in water • Price–demand curves • Atmospheric pressure (t is altitude)
Limited growth: Rate of growth is proportional to the difference between the amount present and a fixed limit	$\dfrac{dy}{dt} = k(M - y)$ $k, t > 0$ $y(0) = 0$	$y = M(1 - e^{-kt})$		• Sales fads (for example, skateboards) • Depreciation of equipment • Company growth • Learning
Logistic growth: Rate of growth is proportional to the amount present and to the difference between the amount present and a fixed limit	$\dfrac{dy}{dt} = ky(M - y)$ $k, t > 0$ $y(0) = \dfrac{M}{1 + c}$	$y = \dfrac{M}{1 + ce^{-kMt}}$		• Long-term population growth • Epidemics • Sales of new products • Rumor spread • Company growth

Answers to Matched Problems   **1.** 1.2 billion people   **2.** 33 yr   **3.** 5,600 yr   **4.** 0.27 ft/hr

A *In Problems 1–8, find the general or particular solution, as indicated, for each differential equation.*

1. $\dfrac{dy}{dx} = 4.16x$ **2.** $\dfrac{dy}{dx} = x^{-2}$

3. $\dfrac{dy}{dx} = e^{0.5x}$ **4.** $\dfrac{dy}{dx} = \dfrac{2}{x}$

5. $\dfrac{dy}{dx} = x^2 - x; y(0) = 0$ **6.** $\dfrac{dy}{dx} = \sqrt{x}; y(0) = 0$

7. $\dfrac{dy}{dx} = -2xe^{-x^2}; y(0) = 3$

8. $\dfrac{dy}{dx} = e^{x-3}; y(3) = -5$

B *Problems 9–14 refer to the following slope fields:*

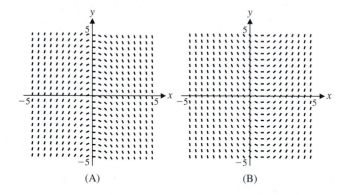

(A)   (B)

9. Which slope field is associated with the differential equation $dy/dx = x - 1$? Briefly justify your answer.

10. Which slope field is associated with the differential equation $dy/dx = -x$? Briefly justify your answer.

11. Solve the differential equation $dy/dx = x - 1$, and find the particular solution that passes through $(0, -2)$.

12. Solve the differential equation $dy/dx = -x$, and find the particular solution that passes through $(0, 3)$,

13. Graph the particular solution found in Problem 11 in the appropriate figure above (or a copy).

14. Graph the particular solution found in Problem 12 in the appropriate figure above (or a copy).

In Problems 15–20, find the general or particular solution, as indicated, for each differential equation.

15. $\dfrac{dy}{dt} = 2y$

16. $\dfrac{dy}{dt} = -3y$

17. $\dfrac{dy}{dx} = -0.5y; y(0) = 100$

18. $\dfrac{dy}{dx} = 0.1y; y(0) = -2.5$

19. $\dfrac{dx}{dt} = -8x$

20. $\dfrac{dx}{dt} = -8t$

C *Problems 21–28 refer to the following slope fields:*

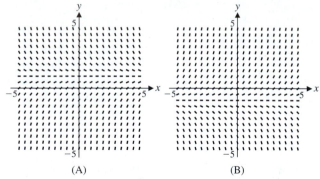

(A)   (B)

21. Which slope field is associated with the differential equation $dy/dx = 1 - y$? Briefly justify your answer.

22. Which slope field is associated with the differential equation $dy/dx = y + 1$? Briefly justify your answer.

23. Show that $y = 1 - Ce^{-x}$ is a solution of the differential equation $dy/dx = 1 - y$ for any real number C. Find the particular solution that passes through $(0, 0)$.

24. Show that $y = Ce^x - 1$ is a solution of the differential equation $dy/dx = y + 1$ for any real number C. Find the particular solution that passes through $(0, 0)$.

25. Graph the particular solution found in Problem 23 in the appropriate figure above (or a copy).

26. Graph the particular solution found in Problem 24 in the appropriate figure above (or a copy).

27. Use a graphing utility to graph $y = 1 - Ce^{-x}$ for $C = -2, -1, 1$, and 2 for $-5 \leqslant x \leqslant 5, -5 \leqslant y \leqslant 5$, all in the same viewing window. Observe how the solution curves go with the flow of the tangent line segments in the corresponding slope field shown above.

28. Use a graphing utility to graph $y = Ce^x - 1$ for $C = -2, -1, 1$, and 2 for $-5 \leqslant x \leqslant 5, -5 \leqslant y \leqslant 5$, all in the same viewing window. Observe how the solution curves go with the flow of the tangent line segments in the corresponding slope field shown above.

 In Problems 29–36, use a graphing utility to graph the given examples of the various cases in Table 1.

29. Unlimited growth:
$y = 1,000e^{0.08t}$
$0 \leqslant t \leqslant 15$
$0 \leqslant y \leqslant 3,500$

30. Unlimited growth:
$y = 5,250e^{0.12t}$
$0 \leqslant t \leqslant 10$
$0 \leqslant y \leqslant 20,000$

31. Exponential decay:
$p = 100e^{-0.05x}$
$0 \leqslant x \leqslant 30$
$0 \leqslant p \leqslant 100$

32. Exponential decay:
$p = 1,000e^{-0.08x}$
$0 \leqslant x \leqslant 40$
$0 \leqslant p \leqslant 1,000$

33. Limited growth:

$$N = 100(1 - e^{-0.05t})$$

$$0 \leq t \leq 100$$

$$0 \leq N \leq 100$$

34. Limited growth:

$$N = 1{,}000(1 - e^{-0.07t})$$

$$0 \leq t \leq 70$$

$$0 \leq N \leq 1{,}000$$

35. Logistic growth:

$$N = \frac{1{,}000}{1 + 999e^{-0.4t}}$$

$$0 \leq t \leq 40$$

$$0 \leq N \leq 1{,}000$$

36. Logistic growth:

$$N = \frac{400}{1 + 99e^{-0.4t}}$$

$$0 \leq t \leq 30$$

$$0 \leq N \leq 400$$

37. Show that the rate of logistic growth, $dy/dt = ky(M - y)$, has its maximum value when $y = M/2$.

38. Find the value of t for which the logistic function

$$y = \frac{M}{1 + ce^{-kMt}}$$

is equal to $M/2$.

39. Let $Q(t)$ denote the population of the world at time t. In 1967 the world population was 3.5 billion and increasing at 2.0% per year; in 1999 it was 6.0 billion and increasing at 1.3% per year. In which year, 1967 or 1999, was dQ/dt (the rate of growth of Q with respect to t) greater? Explain.

40. Refer to Problem 39. Explain why the world population function $Q(t)$ does not satisfy an exponential growth law.

Applications

Business & Economics

41. *Continuous compound interest.* Find the amount A in an account after t years if

$$\frac{dA}{dt} = 0.08A \quad \text{and} \quad A(0) = 1{,}000$$

42. *Continuous compound interest.* Find the amount A in an account after t years if

$$\frac{dA}{dt} = 0.12A \quad \text{and} \quad A(0) = 5{,}250$$

43. *Continuous compound interest.* Find the amount A in an account after t years if

$$\frac{dA}{dt} = rA \quad A(0) = 8{,}000 \quad A(2) = 9{,}020$$

44. *Continuous compound interest.* Find the amount A in an account after t years if

$$\frac{dA}{dt} = rA \quad A(0) = 5{,}000 \quad A(5) = 7{,}460$$

45. *Price–demand.* The marginal price dp/dx at x units of demand per week is proportional to the price p. There is no weekly demand at a price of $100 per unit $[p(0) = 100]$, and there is a weekly demand of 5 units at a price of $77.88 per unit $[p(5) = 77.88]$.

(A) Find the price–demand equation.

(B) At a demand of 10 units per week, what is the price?

(C) Graph the price–demand equation for $0 \leq x \leq 25$.

46. *Price–supply.* The marginal price dp/dx at x units of supply per day is proportional to the price p. There is

no supply at a price of $10 per unit $[p(0) = 10]$, and there is a daily supply of 50 units at a price of $12.84 per unit $[p(50) = 12.84]$.

(A) Find the price–supply equation.

(B) At a supply of 100 units per day, what is the price?

(C) Graph the price–supply equation for $0 \leq x \leq 250$.

47. *Advertising.* A company is trying to expose a new product to as many people as possible through television advertising. Suppose that the rate of exposure to new people is proportional to the number of those who have not seen the product out of L possible viewers. No one is aware of the product at the start of the campaign, and after 10 days 40% of L are aware of the product. Mathematically,

$$\frac{dN}{dt} = k(L - N) \quad N(0) = 0 \quad N(10) = 0.4L$$

(A) Solve the differential equation.

(B) What percent of L will have been exposed after 5 days of the campaign?

(C) How many days will it take to expose 80% of L?

(D) Graph the solution found in part (A) for $0 \leq t \leq 90$.

48. *Advertising.* Suppose that the differential equation for Problem 47 is

$$\frac{dN}{dt} = k(L - N) \quad N(0) = 0 \quad N(10) = 0.1L$$

(A) Interpret $N(10) = 0.1L$ verbally.

(B) Solve the differential equation.

(C) How many days will it take to expose 50% of L?

(D) Graph the solution found in part (B) for $0 \leqslant t \leqslant 300$.

Life Sciences

49. *Biology.* For relatively clear bodies of water, light intensity is reduced according to

$$\frac{dI}{dx} = -kI \qquad I(0) = I_0$$

where I is the intensity of light at x feet below the surface. For the Sargasso Sea off the West Indies, $k = 0.00942$. Find I in terms of x, and find the depth at which the light is reduced to half of that at the surface.

50. *Blood pressure.* It can be shown under certain assumptions that blood pressure P in the largest artery in the human body (the aorta) changes between beats with respect to time t according to

$$\frac{dP}{dt} = -aP \qquad P(0) = P_0$$

where a is a constant. Find $P = P(t)$ that satisfies both conditions.

51. *Drug concentration.* A single injection of a drug is administered to a patient. The amount Q in the body then decreases at a rate proportional to the amount present, and for a particular drug the rate is 4% per hour. Thus,

$$\frac{dQ}{dt} = -0.04Q \qquad Q(0) = Q_0$$

where t is time in hours.

(A) If the initial injection is 3 milliliters $[Q(0) = 3]$, find $Q = Q(t)$ satisfying both conditions.

(B) How many milliliters (to two decimal places) are in the body after 10 hours?

(C) How many hours (to two decimal places) will it take for only 1 milliliter of the drug to be left in the body?

(D) Graph the solution found in part (A).

52. *Simple epidemic.* A community of 1,000 people is assumed to be homogeneously mixed. One person who has just returned from another community has influenza. Assume that the home community has not had influenza shots and all are susceptible. One mathematical model for an influenza epidemic assumes that influenza tends to spread at a rate in direct proportion to the number who have it, N, and to the number who have not yet contracted it—in this case, $1,000 - N$. Mathematically,

$$\frac{dN}{dt} = kN(1,000 - N) \qquad N(0) = 1$$

where N is the number of people who have contracted influenza after t days. For $k = 0.0004$, it can be shown that $N(t)$ is given by

$$N(t) = \frac{1,000}{1 + 999e^{-0.4t}}$$

(A) How many people have contracted influenza after 10 days? After 20 days?

(B) How many days will it take until half the community has contracted influenza?

(C) Find $\lim_{t \to \infty} N(t)$.

(D) Graph $N = N(t)$ for $0 \leqslant t \leqslant 30$.

53. *Nuclear accident.* One of the dangerous radioactive isotopes detected after the Chernobyl nuclear accident in 1986 was cesium-137. If 93.3% of the cesium-137 emitted during the accident is still present 3 years later, find the continuous compound rate of decay of this isotope.

54. *Insecticides.* Many countries have banned the use of the insecticide DDT because of its long-term adverse effects. Five years after a particular country stopped using DDT, the amount of DDT in the ecosystem had declined to 75% of the amount present at the time of the ban. Find the continuous compound rate of decay of DDT.

Social Sciences

55. *Archaeology.* A skull from an ancient tomb was discovered and was found to have 5% of the original amount of radioactive carbon-14 present. Estimate the age of the skull. (See Example 3.)

56. *Learning.* For a particular person learning to type, it was found that the number of words per minute, N, the person was able to type after t hours of practice was given approximately by

$$N = 100(1 - e^{-0.02t})$$

See Table 1 (limited growth) for a characteristic graph. What is the rate of improvement after 10 hours of practice? After 40 hours of practice?

57. *Small group analysis.* In a study on small group dynamics, sociologists Stephan and Mischler found that when the members of a discussion group of 10 were ranked according to the number of times each participated, the number of times $N(k)$ the kth-ranked person participated was given approximately by

$$N(k) = N_1 e^{-0.11(k-1)} \qquad 1 \leqslant k \leqslant 10$$

where N_1 is the number of times the first-ranked person participated in the discussion. If, in a particular discussion group of 10 people, $N_1 = 180$, estimate how many times the sixth-ranked person participated. The 10th-ranked person.

58. *Perception.* One of the oldest laws in mathematical psychology is the Weber–Fechner law (discovered in the middle of the nineteenth century). It concerns a person's sensed perception of various strengths of stimulation involving weights, sound, light, shock, taste, and so on. One form of the law states that the rate of change of sensed sensation S with respect to stimulus R is inversely proportional to the strength of the stimulus R. Thus,

$$\frac{dS}{dR} = \frac{k}{R}$$

where k is a constant. If we let R_0 be the threshold level at which the stimulus R can be detected (the least amount of sound, light, weight, and so on, that can be detected), it is appropriate to write

$$S(R_0) = 0$$

Find a function S in terms of R that satisfies the conditions above.

59. *Rumor spread.* A group of 400 parents, relatives, and friends are waiting anxiously at Kennedy Airport for a student charter flight to return after a year in Europe. It is stormy and the plane is late. A particular parent thought he had heard that the plane's radio had gone out and related this news to some friends, who in turn passed it on to others, and so on. Sociol-

ogists have studied rumor propagation and have found that a rumor tends to spread at a rate in direct proportion to the number who have heard it, x, and to the number who have not, $P - x$, where P is the total population. Mathematically, for our case, $P = 400$ and

$$\frac{dx}{dt} = 0.001x(400 - x) \qquad x(0) = 1$$

where t is time (in minutes). From this, it can be shown that

$$x(t) = \frac{400}{1 + 399e^{-0.4t}}$$

See Table 1 (logistic growth) for a characteristic graph.

(A) How many people have heard the rumor after 5 minutes? 20 minutes?

(B) Find $\lim_{t \to \infty} x(t)$.

(C) Graph $x = x(t)$ for $0 \le t \le 30$.

60. *Rumor spread.* In Problem 59, how long (to the nearest minute) will it take for half of the group of 400 to have heard the rumor?

Section 6-4

Geometric–Numeric Introduction to the Definite Integral

❑ AREA
❑ RATE, AREA, AND DISTANCE
❑ RATE, AREA, AND TOTAL CHANGE

The first three sections of this chapter focused on the *indefinite integral*—the set of all antiderivatives of a given function. In this and the next section we introduce a related form called the *definite integral.* Our approach in this section is intuitive and informal. In the next section these concepts will be made more precise.

We introduce the *definite integral* by considering the area bounded by the graph of a function f, the x axis, and the vertical lines $x = a$ and $x = b$, as shown in Figure 1. If $f(x)$ is positive, we refer to this area as the area under the graph of $y = f(x)$ from $x = a$ to $x = b$.

We denote the shaded area in Figure 1 by the **definite integral symbol:**

$$\int_a^b f(x)\, dx = \left(\begin{array}{c} \text{area under the graph} \\ \text{from } x = a \text{ to } x = b \end{array} \right)$$

This symbol is precisely defined in Section 6-5, where we will also see how the definite integral of a function over an interval $[a, b]$ is related to the indefinite integral (antiderivative) of the function through the fundamental theorem of calculus. To keep things simple at the start, we restrict $f(x)$ to positive values—this restriction will be relaxed shortly. We also assume that all functions referred to are continuous over any interval under consideration.

Surprisingly, out of this analysis of area we will be able to find the total change in a function over an interval from its derivative. For example, given a

FIGURE 1 Area under the graph of $y = f(x)$

rate function (derivative) or a table of rate values, we will be able to determine the distance that a steel ball dropped from a bridge travels in a given time interval, or the additional costs resulting from an increase in production.

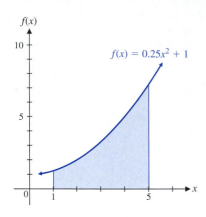

FIGURE 2 What is the shaded area?

❑ AREA

How do we find the shaded area in Figure 2? That is, how do we find the area bounded by the graph of $f(x) = 0.25x^2 + 1$, the x axis, and the vertical lines $x = 1$ and $x = 5$? [This cumbersome description is usually shortened to "the area under the graph of $f(x) = 0.25x^2 + 1$ from $x = 1$ to $x = 5$."] Our standard geometric area formulas do not apply directly, but the formula for the area of a rectangle can be used indirectly. To see how, we look at a method of approximating the area under the graph by using rectangles. This method will give us any accuracy desired, which is quite different from finding the area exactly. Our first area approximation is made by dividing the interval $[1, 5]$ on the x axis into four equal parts of length

$$\Delta x = \frac{5 - 1}{4} = 1$$

(It is customary to denote the length of the subintervals by Δx, which is read "delta x," since Δ is the Greek capital letter delta.) We then place a rectangle on each subinterval with a height determined by the function evaluated at the left endpoint of the subinterval (see Fig. 3).

Summing the areas of the rectangles in Figure 3, we obtain a **left sum** of four rectangles, denoted by L_4, as follows:

$$L_4 = f(1) \cdot 1 + f(2) \cdot 1 + f(3) \cdot 1 + f(4) \cdot 1$$
$$= 1.25 + 2.00 + 3.25 + 5 = 11.5$$

From Figure 3, since $f(x)$ is increasing, it is clear that the left sum L_4 underestimates the area, and we can write

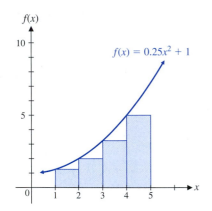

FIGURE 3 Left rectangles

$$11.5 = L_4 < \int_1^5 (0.25x^2 + 1) \, dx = \text{area}$$

Explore–Discuss 1

If $f(x)$ were decreasing over the interval $[1, 5]$, would the left sum L_4 over- or underestimate the actual area under the curve? Explain.

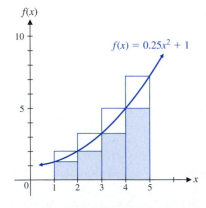

FIGURE 4 Left and right rectangles

Now suppose that we use the right endpoint of each subinterval to obtain the height of the rectangle placed on top of it. Superimposing this result on top of Figure 3, we obtain Figure 4.

Summing the areas of the higher rectangles in Figure 4, we obtain the **right sum** of the four rectangles, denoted by R_4, as follows (compare R_4 below with L_4 above and note that R_4 can be obtained from L_4 by deleting one rectangle area and adding one more):

$$R_4 = f(2) \cdot 1 + f(3) \cdot 1 + f(4) \cdot 1 + f(5) \cdot 1$$
$$= 2.00 + 3.25 + 5.00 + 7.25 = 17.5$$

From Figure 4, since $f(x)$ is increasing, it is clear that the right sum R_4 overestimates the area, and we conclude that the actual area is between 11.5 and 17.5. That is,

$$11.5 = L_4 < \int_1^5 (0.25x^2 + 1)\, dx < R_4 = 17.5 \tag{1}$$

Since the actual area lies between the left estimate and the right estimate, the average should be even closer:

$$\text{average} = \frac{L_4 + R_4}{2} = \frac{11.5 + 17.5}{2} = 14.5$$

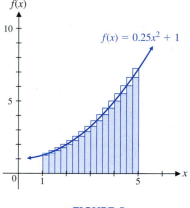

FIGURE 5

Explore–Discuss 2

If $f(x)$ in Figure 4 were decreasing over the interval $[1, 5]$, would the right sum R_4 over- or underestimate the actual area under the curve? Explain.

The first approximation of the area under the curve in (1) is fairly coarse, but the method outlined can be continued with increasingly accurate results by dividing the interval $[1, 5]$ into more and more equal subintervals. Of course, this is not a job for hand calculation but a job that computers are designed to do.* Figure 5 shows left and right rectangle approximations for 16 equal subdivisions.

For this case,

$$\Delta x = \frac{5 - 1}{16} = 0.25$$

$$L_{16} = f(1) \cdot \Delta x + f(1.25) \cdot \Delta x + \cdots + f(4.75) \cdot \Delta x$$
$$= 13.59$$

$$R_{16} = f(1.25) \cdot \Delta x + f(1.50) \cdot \Delta x + \cdots + f(5) \cdot \Delta x$$
$$= 15.09$$

Thus, we now know that the area under the curve is between 13.59 and 15.09. That is,

$$13.59 = L_{16} < \int_1^5 (0.25x^2 + 1)\, dx < R_{16} = 15.09 \tag{2}$$

$$\text{average} = \frac{L_{16} + R_{16}}{2} = \frac{13.59 + 15.09}{2} = 14.34$$

For 100 equal subdivisions, computer calculations give us

$$14.214 = L_{100} < \int_1^5 (0.25x^2 + 1)\, dx < R_{100} = 14.454 \tag{3}$$

$$\text{average} = \frac{L_{100} + R_{100}}{2} = \frac{14.214 + 14.454}{2} = 14.334$$

For an approximation process to be useful, we need an estimate of the maximum possible error produced by the process. We define the **error in an approximation** to be the absolute value of the difference between the approximated value and the actual value.

---

*The computer software that accompanies this book will perform these calculations (see the Preface).

If a and b are the coordinates of two points on a real number line, the distance between these two points is given by $|b - a|$. The following inequalities involving distance on a real number line are very useful for estimating errors in approximations.

1. If x is between a and b on a number line, then

$$|x - a| \leqslant |b - a| \qquad \text{and} \qquad |x - b| \leqslant |b - a|$$

2. If $A = \dfrac{a + b}{2}$, then

$$|x - A| \leqslant \tfrac{1}{2}|b - a|$$

Interpret each of these inequalities both verbally and geometrically on a real number line.

A function is **monotone** over an interval $[a, b]$ if it is either increasing over $[a, b]$ or decreasing over $[a, b]$. The following important observation about monotone functions forms the basis for estimating the error when a left sum or a right sum is used to approximate an area.

> **If f is a monotone function, then the area under the graph of $f(x)$ always lies between the left sum L_n and the right sum R_n for any integer n.**

Consider a positive monotone function $y = f(x)$, as shown in Figure 6. If $\int_a^b f(x)\,dx$ is the actual area under the graph of $y = f(x)$ from $x = a$ to $x = b$, this area lies between L_n and R_n. Thus, approximating this area with either L_n or R_n produces an error that is less than the absolute value of the difference between the left sum and the right sum, $|R_n - L_n|$. This difference is just the sum of the areas of the white rectangles shown in Figure 6. (Think about this.)

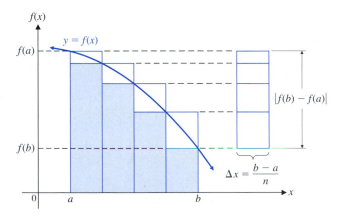

FIGURE 6 $|R_n - L_n| = |f(b) - f(a)|\Delta x$

Thus,

$$\text{error} \leqslant |R_n - L_n| = |f(b) - f(a)|\Delta x = |f(b) - f(a)|\frac{b - a}{n}$$

If the approximation is the average of left and right sums, the bound for the error is cut in half. These results are summarized in the following box for convenient reference:

> ### Error Bounds for Left and Right Sums and Their Average (Monotone Functions)
>
> If $f(x)$ is monotonic on the interval $[a, b]$ and
>
> $$I = \int_a^b f(x)\, dx, \quad L_n = \text{left sum}, \quad R_n = \text{right sum}, \quad A_n = \frac{L_n + R_n}{2}$$
>
> then the following error bounds hold:
>
> $$|I - L_n| \le |f(b) - f(a)| \frac{b - a}{n}$$
>
> $$|I - R_n| \le |f(b) - f(a)| \frac{b - a}{n}$$
>
> $$|I - A_n| \le |f(b) - f(a)| \frac{b - a}{2n}$$

These formulas not only enable us to compute error bounds for a particular approximation, L_n, R_n, or A_n, but more important, they can be used to tell us in advance how large n should be to produce an error less than a specified amount.

Example 1   ➭   **Approximating Areas**   Given the function $f(x) = 9 - 0.25x^2$, we are interested in approximating the area under $y = f(x)$ from $x = 2$ to $x = 5$.

(A) Graph the function over the interval $[0, 6]$; then draw left and right rectangles for the interval $[2, 5]$ with $n = 6$.

(B) Calculate L_6, R_6, and A_6; then calculate error bounds for each.

(C) How large should n be chosen for L_n, R_n, and A_n for the approximation of $\int_2^5 (9 - 0.25x^2)\,dx$ to be within 0.05 of the true value?

SOLUTION   (A) $\Delta x = 0.5$:

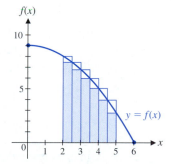

(B) $L_6 = f(2) \cdot \Delta x + f(2.5) \cdot \Delta x + f(3) \cdot \Delta x + f(3.5) \cdot \Delta x + f(4) \cdot \Delta x$
$\qquad + f(4.5) \cdot \Delta x = 18.53$

$\quad\; R_6 = f(2.5) \cdot \Delta x + f(3) \cdot \Delta x + f(3.5) \cdot \Delta x + f(4) \cdot \Delta x$
$\qquad + f(4.5) \cdot \Delta x + f(5) \cdot \Delta x = 15.91$

$\quad\; A_6 = \dfrac{L_6 + R_6}{2} = 17.22$

Error bound for L_6 and R_6:

$$\text{error} \le |f(5) - f(2)| \frac{5 - 2}{6} = |2.75 - 8|(0.5) = 2.625$$

Error bound for A_6:

$$\text{error} \leq \frac{2.625}{2} = 1.3125$$

(C) For L_n and R_n, find n such that error ≤ 0.05:

$$|f(b) - f(a)|\frac{b - a}{n} \leq 0.05$$

$$|2.75 - 8|\frac{3}{n} \leq 0.05$$

$$15.75 \leq 0.05n$$

$$n \geq \frac{15.75}{0.05} = 315$$

For A_n, find n such that error ≤ 0.05:

$$|f(b) - f(a)|\frac{b - a}{2n} \leq 0.05$$

$$|2.75 - 8|\frac{3}{2n} \leq 0.05$$

$$7.875 \leq 0.05n$$

$$n \geq \frac{7.875}{0.05} = 157.5$$

$$n \geq 158 \qquad \textcolor{blue}{\text{Round up to the next integer.}} \qquad ■$$

Matched Problem 1 ⤶ Given the function $f(x) = 8 - 0.5x^2$, we are interested in approximating the area under $y = f(x)$ from $x = 1$ to $x = 3$.

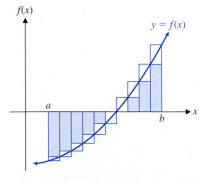

FIGURE 7 Nonmonotonic function

(A) Graph the function over the interval $[0, 4]$; then draw left and right rectangles for the interval $[1, 3]$ with $n = 4$.

(B) Calculate L_4, R_4, and A_4; then calculate error bounds for each.

(C) How large should n be chosen for L_n, R_n, and A_n for the approximation of $\int_1^3 (8 - 0.5x^2)dx$ to be within 0.5 of the true value? ■

If f is not monotonic, the area may not lie between the left sum and the right sum, and our error estimates cannot be applied to the entire interval. However, we can usually divide the interval $[a, b]$ into subintervals such that f is monotonic on each, and then proceed as above (see Fig. 7).

The discussion until now assumed that the graph of $y = f(x)$ is above the x axis; that is, $f(x)$ is positive on the interval $[a, b]$. If part (or all) of the graph lies below the x axis, then for some (or all) values of x, $f(x)$ will be negative and $f(x) \Delta x$ will be the negative of the area of a rectangle lying below the x axis (Fig. 8). In this case, the left sum and the right sum, L_n and R_n, are the sums of the areas of the rectangles above the x axis and the negatives of the areas of the rectangles below the x axis. It is customary to refer to these as **signed areas,** since the positive quantities represent areas and the negative quantities represent negatives of areas.

This observation leads to the following expanded interpretation of the *definite integral symbol:*

FIGURE 8 Signed areas

Definite Integral Symbol for Functions with Negative Values

If $f(x)$ is positive for some values of x on $[a, b]$ and negative for others, then the **definite integral symbol**

$$\int_a^b f(x)\, dx$$

represents the cumulative sum of the signed areas between the graph of $y = f(x)$ and the x axis where the areas above the x axis are counted positively and the areas below the x axis are counted negatively (see Fig. 9, where A and B are actual areas of the indicated regions).

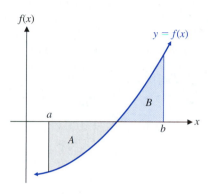

FIGURE 9 $\int_a^b f(x)\, dx = -A + B$

If a function $f(x)$ is monotonic and has both negative and positive values on the interval $[a, b]$, as in Figures 8 and 9, the process for approximating $\int_a^b f(x)\, dx$ by L_n, R_n, or A_n, including the use of the associated error bound formulas, proceeds unchanged from that used above when $f(x)$ was restricted to positive quantities.

❑ RATE, AREA, AND DISTANCE

We now turn our attention to a seemingly unrelated problem involving distance, rate, and time. In our introduction to the derivative in Chapter 3, we started with the concept

$$\text{average rate} = \frac{\text{total distance}}{\text{elapsed time}} \tag{4}$$

and then developed the concept of instantaneous rate as the limit of a difference quotient:

$$\frac{ds}{dt} = \lim_{h \to 0} \frac{s(t + h) - s(t)}{h}$$

$$= \lim_{\Delta t \to 0} \frac{s(t + \Delta t) - s(t)}{\Delta t}$$

The increment notation Δt is sometimes used in place of h in the difference quotient.

where s is the position of a moving object at time t. Now we are going to reverse the process. That is, we will start with a rate function, either in the form of an

FIGURE 10 Steel ball dropped from a bridge

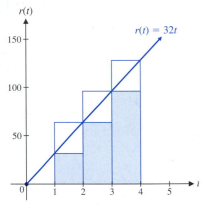

FIGURE 11 $\Delta x = 1$

equation or a table, and work back to distance. Central to the process is equation (4) in the form

$$\text{distance} = (\text{rate})(\text{time}) \tag{5}$$

A concrete example should make the process clear.

From physics it can be shown (neglecting air resistance) that a small steel ball dropped from a bridge (Fig. 10) will have an instantaneous rate of descent given approximately by

$$r(t) = 32t$$

where $r(t)$ is the rate in feet per second at the end of t seconds. (Note that the rate at which the ball falls increases with time due to gravity, as you would expect.) How far will the ball fall between the first and fourth second? To answer this question, we look at the area under the curve $r(t) = 32t$ from $t = 1$ to $t = 4$. But how do we get distance from area? To make the connection, we divide the interval $[1, 4]$ into three equal parts and estimate the area under the curve using L_3 and R_3, as shown in Figure 11.

Area of first left rectangle $= r(1)\,\Delta t$

But

$$r(1)\,\Delta t = (\text{rate at end of 1st second}) \times (1\ \text{second})$$
$$= \text{distance traveled from } t = 1 \text{ to } t = 2 \text{ at a constant rate of } r(1)$$

Since $r(t)$ is increasing over the interval $[1, 2]$, $r(1)$ is the minimum rate on this interval. Thus, $r(1)\,\Delta t$ is an underestimate of the actual distance the steel ball travels from $t = 1$ to $t = 2$.

Area of first right rectangle $= r(2)\,\Delta t$

But

$$r(2)\,\Delta t = (\text{rate at end of 2nd second}) \times (1\ \text{second})$$
$$= \text{distance traveled from } t = 1 \text{ to } t = 2 \text{ at a constant rate of } r(2)$$

Since $r(t)$ is increasing over the interval $[1, 2]$, $r(2)$ is the maximum rate on this interval. Thus, $r(2)\,\Delta t$ is an overestimate of the actual distance the steel ball travels from $t = 1$ to $t = 2$.

The same analysis applies to the second and third rectangles, and we conclude that

$$L_3 \leq \left(\begin{array}{c} \text{distance traveled} \\ \text{from } t = 1 \text{ to } t = 4 \end{array} \right) \leq R_3$$

Proceeding in the same way for arbitrary n, we can write

$$L_n \leq \left(\begin{array}{c} \text{distance traveled} \\ \text{from } t = 1 \text{ to } t = 4 \end{array} \right) \leq R_n$$

This suggests that the actual area under the rate function curve $r(t) = 32t$ from $t = 1$ to $t = 4$ represents the exact distance the steel ball travels from $t = 1$ to $t = 4$; that is,

$$\int_1^4 r(t)\, dt = \left(\begin{array}{c} \text{distance traveled} \\ \text{from } t = 1 \text{ to } t = 4 \end{array} \right)$$

In general, it can be shown that:

> ### Rate, Area, and Distance
>
> If $r = r(t)$ is a positive rate function for an object moving on a line, then
>
> $$\int_a^b r(t)\, dt = \begin{pmatrix} \text{net distance traveled} \\ \text{from } t = a \text{ to } t = b \end{pmatrix}$$
>
> $$= \begin{pmatrix} \text{total change in position} \\ \text{from } t = a \text{ to } t = b \end{pmatrix}$$

Example 2 ⇔ **Distance and Area**

(A) Using the rate function $r(t) = 32t$ from the discussion above, estimate the distance the ball falls from $t = 1$ to $t = 4$ using A_6. Calculate an error bound for this estimate.

(B) How large should n be so that the error in using A_n is not greater than 0.5 foot?

SOLUTION (A) $\Delta t = \dfrac{b - a}{n} = \dfrac{4 - 1}{6} = 0.5$

$L_6 = r(1)\,\Delta t + r(1.5)\,\Delta t + r(2)\,\Delta t + r(2.5)\,\Delta t + r(3)\,\Delta t + r(3.5)\,\Delta t$
$= 216$

$R_6 = r(1.5)\,\Delta t + r(2)\,\Delta t + r(2.5)\,\Delta t + r(3)\,\Delta t + r(3.5)\,\Delta t + r(4)\,\Delta t$
$= 264$

$A_6 = \dfrac{216 + 264}{2} = 240 \text{ feet}$ *Approximate distance traveled from $t = 1$ to $t = 4$*

$\text{Error} \leqslant |r(4) - r(1)|\dfrac{4 - 1}{2 \cdot 6} = 24 \text{ feet}$

Thus,

$$\begin{pmatrix} \text{Distance traveled} \\ \text{from } t = 1 \text{ to } t = 4 \end{pmatrix} = \int_1^4 r(t)\, dt = 240 \pm 24 \text{ feet}$$

[*Note:* $\int_1^4 r(t)\, dt = 240 \pm 24$ feet is a common and convenient way of representing $240 - 24 \leqslant \int_1^4 r(t)\, dt \leqslant 240 + 24$, and we will use this form where appropriate.]

(B) Solve the error bound inequality:

$$|r(4) - r(1)|\dfrac{4 - 1}{2n} \leqslant 0.5$$
$$\dfrac{144}{n} \leqslant 0.5$$
$$n \geqslant 288$$

Matched Problem 2 ⇔ (A) Using the rate function $r(t) = 32t$, approximate the distance traveled from $t = 2$ to $t = 5$ using A_6. Calculate an error bound for this estimate.

(B) How large should n be so that the error in using A_n is not greater than 1 foot?

□ RATE, AREA, AND TOTAL CHANGE

Recall from Chapter 3 that marginal cost (or revenue or profit) is the instantaneous rate of change of cost (or revenue or profit) relative to production at a given production level. We now investigate how total change in cost is related to marginal cost and area through a concrete example.

A company in Florida manufactures cruising power boats. Marginal costs at various production levels are given in Table 1, where x is the number of boats produced per month and the marginal cost $C'(x)$ is in thousands of dollars.

We are interested in estimating the additional cost (total change in cost) in going from a production level of 5 boats per month to 25 boats per month. We make this estimate by graphing the values in Table 1 along with left and right rectangles, as shown in Figure 12.

We assume that a continuous decreasing curve representing the marginal cost function passes through each of the Table 1 points as indicated, but we do not know the equation of this curve. (This situation occurs very frequently in real-world applications.) Nevertheless, we can still estimate the total change in cost by proceeding as we did above in finding distance, given a rate.

$$\text{Area of first left rectangle} = C'(5)\,\Delta x$$

But

$$C'(5)\,\Delta x = \begin{pmatrix} \text{approximate cost of producing} \\ \text{the next boat at an output} \\ \text{level of 5 boats} \end{pmatrix} \times 5$$
$$= \text{approximate additional cost of increasing} \\ \text{production from 5 to 10 boats per month}$$

Since $C'(x)$ is decreasing over the interval $[5, 10]$, $C'(5)$ is the maximum rate on this interval. Thus, $C'(5)\,\Delta x$ is an overestimate of the additional cost of increasing production from 5 to 10 boats per month.

$$\text{Area of first right rectangle} = C'(10)\,\Delta x$$

But

$$C'(10)\,\Delta x = \begin{pmatrix} \text{approximate cost of producing} \\ \text{the next boat at an output} \\ \text{level of 10 boats} \end{pmatrix} \times 5$$
$$= \text{approximate additional cost of increasing} \\ \text{production from 5 to 10 boats per month}$$

Since $C'(x)$ is decreasing over the interval $[5, 10]$, $C'(10)$ is the minimum rate on this interval. Thus, $C'(10)\,\Delta x$ is an underestimate of the additional cost of increasing production from 5 to 10 boats per month.

The same analysis applies to the second, third, and fourth rectangles, and we conclude that

$$R_4 \leqslant \begin{pmatrix} \text{additional cost of increasing} \\ \text{production from 5 to 25 boats} \\ \text{per month} \end{pmatrix} \leqslant L_4$$

This suggests that the actual area under the graph of the marginal cost function $y = C'(x)$—if we had the whole graph—would represent the additional cost

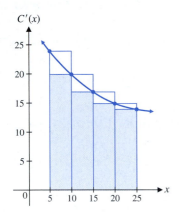

TABLE 1					
MARGINAL COSTS					
x	5	10	15	20	25
$C'(x)$	24	20	17	15	14

FIGURE 12 $\Delta x = 5$

of increasing production from 5 to 25 boats per month, and we would be able to write

$$\int_5^{25} C'(x)\,dx = \begin{pmatrix} \text{additional cost of increasing} \\ \text{production from 5 to 25 boats} \\ \text{per month} \end{pmatrix}$$

In general, it can be shown that:

Rate, Area, and Total Change

If $y = F'(x)$ is a rate function (derivative), then the cumulated sum of the signed areas between the graph of $y = F'(x)$ and the x axis from $x = a$ to $x = b$ represents the total net change in $F(x)$ from $x = a$ to $x = b$. Symbolically,

$$\int_a^b F'(x)\,dx = \begin{pmatrix} \text{total net change in } F(x) \\ \text{from } x = a \text{ to } x = b \end{pmatrix}$$

Informally, a **definite integral** produces the total change in a function from its rate of change. We are very close to a landmark result in mathematics, the *fundamental theorem of calculus,* which will be discussed in Section 6-5.

Example 3   **Change in Cost and Area**   Refer to Table 1 and Figure 12 on the preceding page, and use A_4 to estimate the additional cost (total change in cost) in going from a production level of 5 boats per month to 25 boats per month. Calculate an error bound for this estimate.

SOLUTION

$\Delta x = 5$

$L_4 = C'(5)\,\Delta x + C'(10)\,\Delta x + C'(15)\,\Delta x + C'(20)\,\Delta x$

$\quad = \$380 \text{ thousand}$ *Overestimate*

$R_4 = C'(10)\,\Delta x + C'(15)\,\Delta x + C'(20)\,\Delta x + C'(25)\,\Delta x$

$\quad = \$330 \text{ thousand}$ *Underestimate*

$A_4 = \dfrac{380 + 330}{2} = \$355 \text{ thousand}$

Error bound estimate for A_4:

$$\text{error} \leq |C'(5) - C'(25)|\frac{25 - 5}{2 \cdot 4} = \$25 \text{ thousand}$$

Thus, the additional cost (total change in cost) in going from a production level of 5 boats per month to 25 boats per month is

$$\int_5^{25} C'(x)\,dx = \$355{,}000 \pm \$25{,}000$$

Matched Problem 3   Refer to Table 1 and Figure 12 on the preceding page, and use A_2 to estimate the additional cost (total change in cost) in going from a production level of 10 boats per month to 20 boats per month. Calculate an error bound for this estimate.

Answers to Matched Problems

1. (A) $\Delta x = 0.5$:

(B) $L_4 = 12.625$, $R_4 = 10.625$, $A_4 = 11.625$;
error for L_4 and $R_4 = 2$; A_4 error bound $= 1$
(C) $n > 16$ for L_n and R_n; $n > 8$ for A_n

2. (A) $\left(\begin{array}{c}\text{Distance traveled}\\ \text{from } t = 2 \text{ to } t = 5\end{array}\right) = 336 \pm 24$ ft

(B) $n \geqslant 144$

3. $\left(\begin{array}{c}\text{Additional cost of increasing}\\ \text{production from 10 to 20 boats}\\ \text{per month}\end{array}\right) = \$172{,}500 \pm \$12{,}500$

Exercise 6-4

A *Problems 1–10 involve estimating the area under the curves in Figures A–D from $x = 1$ to $x = 4$. For each figure, divide the interval $[1, 4]$ into three equal sub-intervals.*

(A)

(B)

(C)

(D)

1. Draw in left and right rectangles for Figures A and B.

2. Draw in left and right rectangles for Figures C and D.

3. Using the results of Problem 1, compute L_3, R_3, and A_3 for Figure A and for Figure B.

4. Using the results of Problem 2, compute L_3, R_3, and A_3 for Figure C and for Figure D.

5. Replace the question marks with L_3 and R_3 as appropriate. Explain your choice.

$$? \leqslant \int_1^4 f(x)\, dx \leqslant ? \qquad ? \leqslant \int_1^4 g(x)\, dx \leqslant ?$$

6. Replace the question marks with L_3 and R_3 as appropriate. Explain your choice.

$$? \leqslant \int_1^4 u(x)\, dx \leqslant ? \qquad ? \leqslant \int_1^4 v(x)\, dx \leqslant ?$$

7. Compute error bounds for L_3, R_3, and A_3 found in Problem 3 for both figures.

8. Compute error bounds for L_3, R_3, and A_3 found in Problem 4 for both figures.

9. Using results from the preceding problems, interpret the following in geometric language:

$$\int_1^4 f(x)\, dx = 16.5 \pm 3.5$$

10. Using results from the preceding problems, interpret the following in geometric language:

$$\int_1^4 u(x)\, dx = 10.5 \pm 3.5$$

11. An arrow is shot vertically into the air with an initial rate of 128 feet per second. Neglecting air resistance, its rate of ascent $r(t)$ (in feet per second) at the end of t seconds is shown in the figure.

Figure for 11

(A) Estimate the height the arrow reaches using left and right sums and their average over four equal subdivisions. (Read values directly from the graph.) Calculate error bounds for these estimates.

(B) Explain how the areas of the rectangles used in the left or right sum approximations in part (A) are related to distance.

(C) How many equal subdivisions on the interval $[0, 4]$ should be used to have A_n approximate the height the arrow reaches within 1 foot?

12. Referring to Problem 11, use left and right sums and their average using two equal subdivisions to estimate the distance the arrow travels during the first 2 seconds of flight. Estimate error bounds for each estimate.

B *In Problems 13 and 14, discuss the validity of each statement. If the statement is always true, explain why. If not, give a counterexample.*

13. (A) The function $f(x) = x^2 - 2x$ is monotone over $[0, 2]$.

(B) The function $f(x) = x^2 - 2x$ is monotone over $[1, 3]$.

14. (A) If f is a monotone function on $[a, b]$, then the area under the graph of $f(x)$ is greater than the left sum L_n and less than the right sum R_n for any positive integer n.

(B) If the area under the graph of $f(x)$ on $[a, b]$ is equal to both the left sum L_n and the right sum R_n for some positive integer n, then f is a constant function.

Problems 15 and 16 refer to the figure below showing two parcels of land along a river.

15. You are interested in purchasing both parcels of land shown in the figure and wish to make a quick check on their combined area. There is no equation for the river frontage, so you use the average of the left and right sums of rectangles covering the area. The 1,000-foot baseline is divided into 10 equal parts. At the end of each subinterval, a measurement is made from the baseline to the river, and the results are tabulated. Let x be the distance from the left end of the baseline and let $h(x)$ be the distance from the baseline to the river at x. Estimate the combined area of both parcels using A_{10}, and calculate an error bound for this estimate. How many subdivisions of the baseline would be required so that the error in using A_n would not exceed 2,500 square feet?

x	0	100	200	300	400	500
$h(x)$	0	183	235	245	260	286

x	600	700	800	900	1,000
$h(x)$	322	388	453	489	500

16. Estimate the separate area of each parcel in Problem 15 using A_5 for each, and calculate an error bound for each estimate. The baseline length of each parcel is 500 feet.

17. A car is traveling at 75 miles per hour (110 feet per second), and the driver slams on the brakes to stop. During the 7 seconds it takes to stop, the speed of the car at the end of each second is recorded by a special instrument as follows:

t	0	1	2	3	4	5	6	7
$r(t)$	110	85	63	45	29	16	5	0

where $r(t)$ is the speed (in feet per second) at the end of t seconds, $0 \leq t \leq 7$.

(A) Estimate the distance required to stop using the average of left and right sums over seven equal subdivisions. Calculate an error bound for this estimate.

(B) How many equal subdivisions should be used on the interval $[0, 7]$ to have A_n approximate the stopping distance within 5 feet?

18. In Problem 17, estimate the distance traveled during the first 3 seconds after applying the brakes using the average of left and right sums over three equal subdivisions. Calculate an error bound for this estimate. How many equal subdivisions should be used on the interval $[0, 3]$ to have A_n approximate the distance within 10 feet?

19. The figure shows the graph of a function that is not monotonic over the interval $[0, 5]$. We wish to estimate the area under the curve over this interval.

Figure for 19

(A) Find a point P on the x axis that divides $[0, 5]$ into two parts such that $f(x)$ is monotonic over each.

(B) Divide the interval $[0, 5]$ into five equal subintervals, and sketch left and right rectangles over the interval.

(C) Describe the behavior of left and right rectangles to the left and to the right of P relative to under- and overestimating the true area.

(D) Find numbers N_1 and N_2 by combining appropriate left and right rectangle sums so that

$$N_1 \le \int_0^5 f(x)\, dx \le N_2$$

20. Find numbers N_1 and N_2 by combining appropriate left and right rectangle sums from Problem 19 so that

$$N_1 \le \int_1^4 f(x)\, dx \le N_2$$

C *Problems 21 and 22 refer to the figure below.*

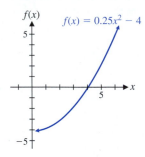

21. Approximate $\int_2^5 (0.25x^2 - 4)\, dx$ using L_6, R_6, and A_6. Compute error bounds for each. (Round all values to two decimal places.) Describe in geometric terms what the definite integral over the interval $[2, 5]$ represents.

22. Approximate $\int_1^6 (0.25x^2 - 4)\, dx$ using L_5, R_5, and A_5. Compute error bounds for each. (Round all values to two decimal places.) Describe in geometric terms what the definite integral over the interval $[1, 6]$ represents.

 For Problems 23–26, use a graphing utility to determine the intervals on which each function is monotone.

23. $f(x) = e^{-x^2}$

24. $f(x) = \dfrac{3}{1 + 2e^{-x}}$

25. $f(x) = x^4 - 2x^2 + 3$

26. $f(x) = e^{x^2}$

In Problems 27–32, the left sum L_n, the right sum R_n, or their average A_n is used to approximate the definite integral to the indicated accuracy. How large must n be chosen in each case? (Each function is monotonic over the indicated interval.)

27. $\displaystyle\int_1^3 \ln x\, dx = R_n \pm 0.1$

28. $\displaystyle\int_0^{10} \ln(x^2 + 1)\, dx = L_n \pm 0.5$

29. $\displaystyle\int_1^3 x^x\, dx = L_n \pm 0.5$

30. $\displaystyle\int_1^4 x^x\, dx = R_n \pm 0.5$

31. $\displaystyle\int_0^3 e^{-x^2}\, dx = A_n \pm 0.05$

32. $\displaystyle\int_0^3 e^{x^2}\, dx = A_n \pm 0.05$

33. If f is monotonic on the interval $[a, b]$, give an argument as to why we are justified in writing

$$\int_a^b f(x)\, dx = \lim_{n \to \infty} L_n$$

34. If f is monotonic on the interval $[a, b]$, give an argument as to why we are justified in writing

$$\int_a^b f(x)\, dx = \lim_{n \to \infty} R_n$$

Applications

Business & Economics

35. *Cost.* A company manufactures mountain bikes. The research department produced the marginal cost graph shown, where $C'(x)$ is in dollars and x is the number of bikes produced per month. Estimate the increase in cost going from a production level of 300 bikes per month to 900 bikes per month. Use the average of the left and right sums over two equal subintervals, and compute an error bound for this estimate.

Figure for 35 and 36

36. *Cost.* Referring to Problem 35, estimate the increase in cost going from a production level of 0 bikes per month to 600 bikes per month using left and right sums over two equal subintervals. Replace the question marks with the values of L_2 or R_2 as appropriate:

$$? \leq \int_0^{600} C'(x)\, dx \leq ?$$

37. *Continuous compound interest.* Ten thousand dollars is deposited in an account where the rate of change of the amount in the account is given by

$$A'(t) = 800e^{0.08t}$$

t years after the initial deposit. By how much will the account change from the end of the second year to the end of the sixth year? Estimate the amount by using left and right sums over four equal subintervals. (Calculate all quantities to the nearest dollar.) Replace the question marks with the values of L_4 or R_4 as appropriate:

$$? \leq \int_2^6 800e^{0.08t}\, dt \leq ?$$

38. *Continuous compound interest.* One thousand dollars is deposited in an account where the rate of change of the amount in the account is given by

$$A'(t) = 50e^{0.05t}$$

t years after the initial deposit. By how much will the account change from the end of the first year to the end of the fifth year? Estimate the amount by using the average of the left sum and the right sum over four equal subintervals. Calculate an error bound for this estimate. (Calculate all quantities to the nearest dollar.)

39. *Employee training.* A company producing computer components has established that, on the average, a new employee can assemble $N(t)$ components per day after t days of on-the-job training, as indicated in the following table (a new employee's productivity increases continuously with time on the job):

t	0	20	40	60	80	100	120
$N(t)$	10	51	68	76	81	84	86

Use the average of left and right sums to estimate the total number of units produced by a new employee over the first 60 days. Over the second 60 days. Use three equal subintervals for each. Calculate an error bound for each estimate.

40. *Employee training.* For a new employee in Problem 39, use left and right sums to estimate the number of units produced over the time interval $[20, 100]$. Replace the question marks with the values of L_4 or R_4 as appropriate:

$$? \leq \int_{20}^{100} N(t)\, dt \leq ?$$

41. *Revenue.* The research department for a market chain in a city established the following price–demand equation for a premium beer sold by the six-pack:

$$p = 8 - \frac{x}{50} \qquad 0 \leq x \leq 300 \qquad \text{Price–demand}$$

where x is the number of six-packs sold per day at a price of p dollars each. From the price–demand equation we obtain the revenue and marginal revenue equations and their graphs shown in the figure.

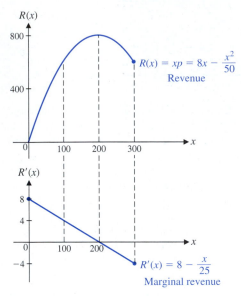

Figure for 41 and 42

(A) Interpret $\int_{100}^{200} R'(x)\, dx$ geometrically and relative to change in revenue.

(B) Approximate $\int_{100}^{200} R'(x)\, dx$ using the average of the left and right sums over four equal subintervals. Calculate an error bound for this estimate.

(C) Evaluate $R(200) - R(100)$. How does this quantity relate to part (A)?

42. *Revenue.* Refer to Problem 41.

(A) Interpret $\int_{100}^{300} R'(x)\, dx$ geometrically and relative to change in revenue. Guess what the value should be from the graph.

(B) Approximate $\int_{100}^{300} R'(x)\, dx$ using the average of left and right sums over four equal subintervals. Calculate an error bound for this estimate.

(C) Evaluate $R(300) - R(100)$. How does this quantity relate to part (A)?

Life Sciences

43. *Medicine.* The rate of healing $A'(t)$ (in square cen-
www timeters per day) for a certain type of abrasive skin wound is given approximately by the following table:

t	0	1	2	3	4	5
$A'(t)$	0.90	0.81	0.74	0.67	0.60	0.55

t	6	7	8	9	10
$A'(t)$	0.49	0.45	0.40	0.36	0.33

$A(t)$ is the area of the wound in square centimeters after t days of healing.

(A) Approximate the area of the wound that will be healed in the first 5 days of healing using the left and right sums over five equal subintervals.

(B) Replace the question marks with values of L_5 and R_5 as appropriate:

$$? \le \int_0^5 A'(t)\, dt \le ?$$

44. *Medicine.* Refer to Problem 43. Approximate the area of the wound that will be healed in the second 5 days of healing. Use the average of the left and right sums over five equal subintervals. Calculate an error bound for this estimate.

Social Sciences

45. *Learning.* During a special study on learning, a psy-
www chologist found that, on the average, the rate of learning a list of special symbols in a code, $N'(x)$, after x days of practice was given approximately by the following table values:

x	0	2	4	6	8	10	12
$N'(x)$	29	26	23	21	19	17	15

$N(x)$ is the number of special symbols learned after x days of practice. Approximate the number of code symbols learned from the sixth to the twelfth day using the average of the left and right sums over three equal subintervals. Calculate an error bound for this estimate.

46. *Learning.* For the data in Problem 45, approximate the number of code symbols learned during the first 6 days of practice using left and right sums over three equal subintervals. Replace the question marks with values of L_3 and R_3 as appropriate:

$$? \le \int_0^6 N'(x)\, dx \le ?$$

Section 6-5

Definite Integral as a Limit of a Sum; Fundamental Theorem of Calculus

- ❑ DEFINITE INTEGRAL AS A LIMIT OF A SUM
- ❑ FUNDAMENTAL THEOREM OF CALCULUS
- ❑ RECOGNIZING A DEFINITE INTEGRAL: AVERAGE VALUE

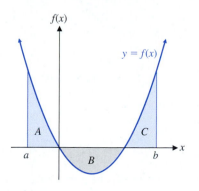

FIGURE 1

$$\int_a^b f(x)\,dx = A - B + C$$

In the preceding section we presented an intuitive and informal introduction to the definite integral and the fundamental theorem of calculus. There, we used the definite integral symbol $\int_a^b f(x)\,dx$ to represent the cumulative sums of signed areas between the graph of $y = f(x)$ and the x axis from $x = a$ to $x = b$. See Figure 1, where A, B, and C represent actual areas of the respective regions.

We also found that if $f(x)$ represents a rate, then $\int_a^b f(x)\,dx$ can be interpreted as the total net change in an antiderivative of $f(x)$ from $x = a$ to $x = b$. In particular, if $f(x) = C'(x)$, the marginal cost for producing x items, then $\int_a^b C'(x)\,dx$ represents the total change in cost going from a production level of a units to b units.

❑ DEFINITE INTEGRAL AS A LIMIT OF A SUM

We now make some of the concepts in the preceding section more precise. We start by giving a precise description of left and right sums: If we partition the interval $[a, b]$ into n equal subintervals of length $\Delta x = (b - a)/n$ with end points $a = x_0, x_1, \ldots, x_n = b$, then, using **summation notation** (see Appendix B-1):

$$L_n = \textbf{left sum} = \sum_{k=1}^{n} f(x_{k-1})\,\Delta x$$

$$= f(x_0)\,\Delta x + f(x_1)\,\Delta x + \cdots + f(x_{n-1})\,\Delta x$$

$$R_n = \textbf{right sum} = \sum_{k=1}^{n} f(x_k)\,\Delta x$$

$$= f(x_1)\,\Delta x + f(x_2)\,\Delta x + \cdots + f(x_n)\,\Delta x$$

Other rectangle sums, such as the *midpoint sum*, can also be used to approximate $\int_a^b f(x)\,dx$. With the midpoint sum, the height of each rectangle is found by evaluating $f(x)$ at the midpoint of each subinterval instead of at an endpoint. The **midpoint sum** is given as follows:

$$M_n = \textbf{midpoint sum} = \sum_{k=1}^{n} f\left(\frac{x_{k-1} + x_k}{2}\right)\,\Delta x$$

$$= f\left(\frac{x_0 + x_1}{2}\right)\,\Delta x + f\left(\frac{x_1 + x_2}{2}\right)\,\Delta x + \cdots + f\left(\frac{x_{n-1} + x_n}{2}\right)\,\Delta x$$

Figure 2 shows the use of the midpoint sum to approximate $\int_1^6 (0.25x^2 - 4)\,dx$ for $n = 5, 10,$ and 20 (all done on a computer, of course).

(A) $M_5 = -2.19$ (B) $M_{10} = -2.11$ (C) $M_{20} = -2.09$

FIGURE 2 Midpoint sums $[f(x) = 0.25x^2 - 4]$

Explore–Discuss 1

In Figure 2, let A represent the exact area below the x axis from $x = 1$ to $x = 4$, and let B represent the exact area above the x axis from $x = 4$ to $x = 6$. Make a reasonable guess for the following limit in terms of areas A and B, and explain the reasoning behind your guess.

$$\lim_{n \to \infty} M_n = \ ?$$

We are now ready to give a general definition of the *definite integral* in a form that not only covers the discussions above and in the preceding section as special cases, but also opens the concept up to many other interpretations and applications.

Definition of a Definite Integral

Let f be a continuous function defined on the closed interval $[a, b]$, and let

1. $a = x_0 < x_1 < \cdots < x_{n-1} < x_n = b$
2. $\Delta x_k = x_k - x_{k-1}$ for $k = 1, 2, \ldots, n$
3. $\Delta x_k \to 0$ as $n \to \infty$
4. $x_{k-1} \leqslant c_k \leqslant x_k$ for $k = 1, 2, \ldots, n$

Then

$$\int_a^b f(x)\, dx = \lim_{n \to \infty} \sum_{k=1}^n f(c_k)\, \Delta x_k$$
$$= \lim_{n \to \infty} [f(c_1)\, \Delta x_1 + f(c_2)\, \Delta x_2 + \cdots + f(c_n)\, \Delta x_n]$$

is called a **definite integral** of f from a to b. The **integrand** is $f(x)$, the **lower limit** is a, and the **upper limit** is b.

In the definition of a definite integral, we divide the closed interval $[a, b]$ into n subintervals of arbitrary length in such a way that the length of each subinterval $\Delta x_k = x_k - x_{k-1}$ tends to 0 as n increases without bound. From each of the n subintervals we then select a point c_k and form the sum

$$\sum_{k=1}^{n} f(c_k) \, \Delta x_k = f(c_1) \, \Delta x_1 + f(c_2) \, \Delta x_2 + \cdots + f(c_n) \, \Delta x_n$$

which is called a **Riemann sum** [named after the celebrated German mathematician Georg Riemann (1826–1866)].

Under the conditions stated in the definition, it can be shown that the limit of the Riemann sum always exists, and it is a real number. The limit is independent of the nature of the subdivisions of $[a, b]$ as long as condition 3 holds, and it is independent of the choice of c_k as long as condition 4 holds.* It is important to remember that irrespective of its original interpretation in a practical problem, **a definite integral can always be interpreted geometrically in terms of signed areas just as a derivative can always be interpreted geometrically in terms of slope.**

The left, right, and midpoint sums discussed above are particularly simple Riemann sums with subintervals all the same length and c_k chosen in a regular manner as an endpoint of a subinterval or the midpoint of the subinterval. Any one of these special sums (as well as others) can be used to approximate a definite integral. To make these approximations useful, we need formulas for error bounds that include nonmonotonic functions. We state these formulas in the following box without proof.

Error Bounds for L_n, R_n, and M_n

Let

$$I = \int_a^b f(x) \, dx$$

$$L_n = \text{left sum}, \qquad R_n = \text{right sum}, \qquad M_n = \text{midpoint sum}$$

Left and Right Sum Error Bound

If $|f'(x)| \leq B_1$ for all x on $[a, b]$, then

$$|I - L_n| \leq \frac{B_1(b - a)^2}{2n} \qquad |I - R_n| \leq \frac{B_1(b - a)^2}{2n}$$

Midpoint Error Bound

If $|f''(x)| \leq B_2$ for all x on $[a, b]$, then

$$|I - M_n| \leq \frac{B_2(b - a)^3}{24n^2}$$

---

*In this book we consider only limits of Riemann sums for which the n subintervals of $[a, b]$ have the same length. Condition 3 is then satisfied automatically.

Example 1 ⇔ **Using Estimates for Error Bounds** Let

$$I = \int_{-1}^{3} (4x - x^3)\, dx$$

For the integrand $f(x) = 4x - x^3$, use the graphs of $y = f'(x)$ and $y = f''(x)$ in Figure 3 as needed.

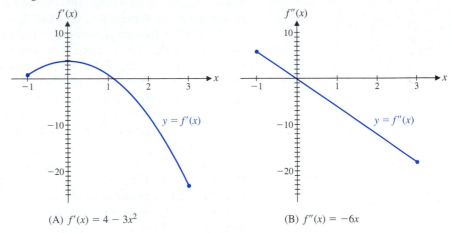

(A) $f'(x) = 4 - 3x^2$ (B) $f''(x) = -6x$

FIGURE 3

(A) How large should n be chosen for either L_n or R_n to approximate I with an error no greater than 0.5?

(B) How large should n be chosen for M_n to approximate I with an error no greater than 0.5?

SOLUTION (A) Choose n such that

$$\frac{B_1(b - a)^2}{2n} \le 0.5$$

where B_1 is any upper bound for $|f'(x)|$ on the interval $[-1, 3]$. [*Caution:* We are looking for an upper bound for $|f'(x)|$, not for $f'(x)$, on the interval $[-1, 3]$.] From the graph of $y = f'(x)$ in Figure 3A, we see that B_1 can be chosen as any number 23 or larger. We choose $B_1 = 23$. [Graphs of $y = f'(x)$ are often used to get an upper bound for $|f'(x)|$ on an interval $[a, b]$ if an upper bound is not easily obtained algebraically or is not obvious.] Thus,

$$\frac{23[3 - (-1)]^2}{2n} \le 0.5$$

$$\frac{23(8)}{n} \le 0.5$$

$$n \ge \frac{23(8)}{0.5} = 368$$

(B) Choose n such that

$$\frac{B_2(b - a)^3}{24n^2} \le 0.5$$

where B_2 is any upper bound for $|f''(x)|$ on the interval $[-1, 3]$. From the graph of $y = f''(x)$ in Figure 3B, we see that B_2 can be chosen as any number 18 or larger. We choose $B_2 = 18$. Thus,

$$\frac{18[3 - (-1)]^3}{24n^2} \le 0.5$$

$$\frac{18(64)}{24(0.5)} \le n^2$$

$$n \ge \sqrt{\frac{18(64)}{24(0.5)}} \qquad \textit{Since n is positive}$$

$$= 9.8 \quad \text{or} \quad 10$$

[*Note*:   The midpoint sum required only 10 subintervals to get the same accuracy that either the left or right sum achieved using 368 subintervals. In both cases, the accuracy may be considerably better than indicated, but these error bound formulas guarantee the results for the indicated values of n. Another approximation sum that is usually more accurate for a given value of n, *Simpson's rule*, is discussed in Group Activity 1.]

Matched Problem 1 ⇔ Given $I = \int_{-1}^{2}(x^2 - 4x)\, dx$:

(A) How large should n be chosen for either L_n or R_n to approximate I with an error no greater than 0.1?

(B) How large should n be chosen for M_n to approximate I with an error no greater than 0.1?

Explore–Discuss 2

Describe the difference between the *definite integral* $\int_{a}^{b} f(x)\, dx$ and the *indefinite integral* $\int f(x)\, dx$ in terms of what each represents.

In the next box we state several useful properties of the definite integral. Note that properties 3 and 4 parallel properties 3 and 4 given earlier in Section 6-1 for the indefinite integral.

Definite Integral Properties

1. $\int_{a}^{a} f(x)\, dx = 0$

2. $\int_{a}^{b} f(x)\, dx = -\int_{b}^{a} f(x)\, dx$

3. $\int_{a}^{b} kf(x)\, dx = k \int_{a}^{b} f(x)\, dx, \quad k$ a constant

4. $\int_{a}^{b} [f(x) \pm g(x)]\, dx = \int_{a}^{b} f(x)\, dx \pm \int_{a}^{b} g(x)\, dx$

5. $\int_{a}^{b} f(x)\, dx = \int_{a}^{c} f(x)\, dx + \int_{c}^{b} f(x)\, dx$

Most of these properties follow directly from the definition of the definite integral. Example 2 illustrates how they can be used.

Example 2 ⤳ **Using Properties of the Definite Integral** If

$$\int_0^2 x \, dx = 2 \qquad \int_0^2 x^2 \, dx = \frac{8}{3} \qquad \int_2^3 x^2 \, dx = \frac{19}{3}$$

then:

(A) $\displaystyle\int_0^2 12x^2 \, dx = 12 \int_0^2 x^2 \, dx = 12\left(\frac{8}{3}\right) = 32$

(B) $\displaystyle\int_0^2 (2x - 6x^2) \, dx = 2 \int_0^2 x \, dx - 6 \int_0^2 x^2 \, dx = 2(2) - 6\left(\frac{8}{3}\right) = -12$

(C) $\displaystyle\int_3^2 x^2 \, dx = -\int_2^3 x^2 \, dx = -\frac{19}{3}$

(D) $\displaystyle\int_5^5 3x^2 \, dx = 0$

(E) $\displaystyle\int_0^3 3x^2 \, dx = 3 \int_0^2 x^2 \, dx + 3 \int_2^3 x^2 \, dx = 3\left(\frac{8}{3}\right) + 3\left(\frac{19}{3}\right) = 27$ ∎

Matched Problem 2 ⤳ Using the same integral values given in Example 2, find:

(A) $\int_2^3 6x^2 \, dx$ (B) $\int_0^2 (9x^2 - 4x) \, dx$ (C) $\int_2^0 3x \, dx$

(D) $\int_{-2}^{-2} 3x \, dx$ (E) $\int_0^3 12x^2 \, dx$ ∎

❑ FUNDAMENTAL THEOREM OF CALCULUS

We are now ready to present one of the most important results in mathematics, the *fundamental theorem of calculus.* We start by reviewing the marginal cost interpretation of the definite integral discussed earlier.

If $C'(x)$ represents marginal cost at an output of x units, then $\int_a^b C'(x) \, dx$ represents the total change in cost as production changes from $x = a$ units to $x = b$ units (see Fig. 4). We can also get the total change in cost by using $C(b) - C(a)$, where $C(x)$ is the cost function, an antiderivative of the marginal cost function $C'(x)$. Thus, it appears that we can write

$$\int_a^b C'(x) \, dx = C(b) - C(a)$$

FIGURE 4 $\int_a^b C'(x) \, dx$ = total change in cost

In general, since $F'(x)$ is an instantaneous rate of change at x, then

$$\int_a^b F'(x) \, dx \text{ represents the total change in } F(x) \text{ from } x = a \text{ to } x = b$$

But this is also given by $F(b) - F(a)$. Thus, it appears that

$$\int_a^b F'(x) \, dx = F(b) - F(a)$$

It turns out that this remarkable result is true under rather general conditions, which we now state:

Fundamental Theorem of Calculus

If f is a continuous function on the closed interval $[a, b]$ and F is any anti-derivative of f, then

$$\int_a^b f(x)\, dx = F(x)\Big|_a^b = F(b) - F(a) \qquad F'(x) = f(x)$$

If we can find an antiderivative of $f(x)$, then we can evaluate $\int_a^b f(x)\, dx$, the limit of a Riemann sum over the interval $[a, b]$, exactly and quickly by evaluating an antiderivative of $f(x)$ at the endpoints a and b and taking the difference. Of course, if $f(x)$ is given in table or graph form without an equation, or if an antiderivative of $f(x)$ cannot be found or does not exist (a common occurrence in the real world), we must resort to the use of an approximation procedure such as those described earlier. The symbol $F(x)|_a^b$ is used to represent the **net change** in $F(x)$ from $x = a$ to $x = b$ and is included as a convenient intermediate step in the evaluation of a definite integral by the fundamental theorem.

Now you know why we studied techniques of indefinite integration before this section—so we would have methods of finding antiderivatives of large classes of elementary functions for use with the fundamental theorem. It is important to remember that:

Any antiderivative of $f(x)$ can be used in the fundamental theorem. One generally chooses the simplest antiderivative by letting $C = 0$, since any other value of C will drop out when computing the difference $F(b) - F(a)$.

A variety of definite integrals are evaluated using the fundamental theorem in the following examples.

Example 3 ⇔ **Evaluating Definite Integrals**   Evaluate: $\displaystyle\int_1^2 \left(2x + 3e^x - \frac{4}{x}\right) dx$

SOLUTION
$$\int_1^2 \left(2x + 3e^x - \frac{4}{x}\right) dx = 2\int_1^2 x\, dx + 3\int_1^2 e^x\, dx - 4\int_1^2 \frac{1}{x}\, dx$$

$$= 2\frac{x^2}{2}\Big|_1^2 + 3e^x\Big|_1^2 - 4\ln|x|\Big|_1^2$$

$$= (2^2 - 1^2) + (3e^2 - 3e^1) - (4\ln 2 - 4\ln 1)$$

$$= 3 + 3e^2 - 3e - 4\ln 2 \approx 14.24$$

■

Matched Problem 3 ⇔ Evaluate: $\displaystyle\int_1^3 \left(4x - 2e^x + \frac{5}{x}\right) dx$

■

The evaluation of a definite integral is a two-step process: First, find an antiderivative, then find the net change in that antiderivative. If *substitution techniques* are required to find the antiderivative, there are two different ways to proceed. The next example illustrates both methods.

Example 4 ⇴ **Definite Integrals and Substitution Techniques** Evaluate:

$$\int_0^5 \frac{x}{x^2 + 10}\, dx$$

Solution We will solve this problem using substitution in two different ways:

Method 1. Use substitution in an indefinite integral to find an antiderivative as a function of x; then evaluate the definite integral:

$$\int \frac{x}{x^2 + 10}\, dx = \frac{1}{2} \int \frac{1}{x^2 + 10}\, 2x\, dx \qquad \text{Substitute } u = x^2 + 10 \text{ and}$$
$$\text{du = 2x dx.}$$

$$= \frac{1}{2} \int \frac{1}{u}\, du$$

$$= \tfrac{1}{2} \ln|u| + C$$

$$= \tfrac{1}{2} \ln(x^2 + 10) + C \qquad \text{Since } u = x^2 + 10 > 0$$

We choose $C = 0$ and use the antiderivative $\tfrac{1}{2} \ln(x^2 + 10)$ to evaluate the definite integral:

$$\int_0^5 \frac{x}{x^2 + 10}\, dx = \frac{1}{2} \ln(x^2 + 10) \Big|_0^5$$

$$= \tfrac{1}{2} \ln 35 - \tfrac{1}{2} \ln 10 \approx 0.626$$

Method 2. Substitute directly in the definite integral, changing both the variable of integration and the limits of integration: In the definite integral

$$\int_0^5 \frac{x}{x^2 + 10}\, dx$$

the upper limit is $x = 5$ and the lower limit is $x = 0$. When we make the substitution $u = x^2 + 10$ in this definite integral, we must change the limits of integration to the corresponding values of u:

$$x = 5 \qquad \text{implies} \qquad u = 5^2 + 10 = 35 \qquad \text{New upper limit}$$
$$x = 0 \qquad \text{implies} \qquad u = 0^2 + 10 = 10 \qquad \text{New lower limit}$$

Thus, we have

$$\int_0^5 \frac{x}{x^2 + 10}\, dx = \frac{1}{2} \int_0^5 \frac{1}{x^2 + 10}\, 2x\, dx$$

$$= \frac{1}{2} \int_{10}^{35} \frac{1}{u}\, du$$

$$= \frac{1}{2} \left(\ln|u| \, \Big|_{10}^{35} \right)$$

$$= \tfrac{1}{2}(\ln 35 - \ln 10) \approx 0.626$$

Matched Problem 4 ⇴ Use both methods described in Example 4 to evaluate: $\displaystyle\int_0^1 \frac{1}{2x + 4}\, dx$

Example 5 ➾ **Definite Integrals and Substitution**   Use method 2 described in Example 4 to evaluate

$$\int_{-4}^{1} \sqrt{5 - t}\ dt$$

SOLUTION   If $u = 5 - t$, then $du = -dt$, and

$t = 1$	implies	$u = 5 - 1 = 4$	New upper limit
$t = -4$	implies	$u = 5 - (-4) = 9$	New lower limit

Notice that the lower limit for u is larger than the upper limit. Be careful not to reverse these two values when substituting in the definite integral.

$$\int_{-4}^{1} \sqrt{5 - t}\ dt = -\int_{-4}^{1} \sqrt{5 - t}\ (-dt)$$

$$= -\int_{9}^{4} \sqrt{u}\ du$$

$$= -\int_{9}^{4} u^{1/2}\ du$$

$$= -\left(\frac{u^{3/2}}{\frac{3}{2}} \Big|_{9}^{4} \right)$$

$$= -\left[\tfrac{2}{3}(4)^{3/2} - \tfrac{2}{3}(9)^{3/2} \right]$$

$$= -\left[\tfrac{16}{3} - \tfrac{54}{3} \right] = \tfrac{38}{3} \approx 12.667$$ ■

Matched Problem 5 ➾   Use method 2 described in Example 4 to evaluate: $\displaystyle\int_{2}^{5} \frac{1}{\sqrt{6 - t}}\ dt$ ■

Explore–Discuss 3   Explain why $\neq$ is used in each and finish each correctly:

1. $\displaystyle\int_{0}^{2} e^x\ dx = e^x \Big|_{0}^{2} \neq e^2$

2. $\displaystyle\int_{2}^{5} \frac{1}{2x + 3}\ dx \neq \frac{1}{2} \int_{2}^{5} \frac{1}{u}\ du$ $u = 2x + 3,$ $du = 2\ dx$

Example 6 ➾ **Change in Profit**   A company manufactures x television sets per month. The monthly marginal profit (in dollars) is given by

$$P'(x) = 165 - 0.1x   \qquad 0 \leq x \leq 4{,}000$$

The company is currently manufacturing 1,500 sets per month, but is planning to increase production. Find the total change in the monthly profit if monthly production is increased to 1,600 sets.

SOLUTION

$$P(1{,}600) - P(1{,}500) = \int_{1{,}500}^{1{,}600} (165 - 0.1x)\ dx$$

$$= (165x - 0.05x^2) \Big|_{1{,}500}^{1{,}600}$$

$$= [165(1{,}600) - 0.05(1{,}600)^2] - [165(1{,}500) - 0.05(1{,}500)^2]$$

$$= 136{,}000 - 135{,}000$$

$$= 1{,}000$$

Thus, increasing monthly production from 1,500 units to 1,600 units will increase the monthly profit by $1,000. ∎

Matched Problem 6 Repeat Example 6 if

$$P'(x) = 300 - 0.2x \qquad 0 \leq x \leq 3,000$$

and monthly production is increased from 1,400 sets to 1,500 sets. ∎

Example 7 **Useful Life** An amusement company maintains records for each video game it installs in an arcade. Suppose that $C(t)$ and $R(t)$ represent the total accumulated costs and revenues (in thousands of dollars), respectively, t years after a particular game has been installed and that

$$C'(t) = 2 \qquad R'(t) = 9e^{-0.5t}$$

The value of t for which $C'(t) = R'(t)$ is called the **useful life** of the game.

(A) Find the useful life of the game to the nearest year.
(B) Find the total profit accumulated during the useful life of the game.

FIGURE 5 Useful life

SOLUTION (A) $R'(t) = C'(t)$
$9e^{-0.5t} = 2$
$e^{-0.5t} = \frac{2}{9}$ *Convert to equivalent logarithmic form.*
$-0.5t = \ln\frac{2}{9}$
$t = -2\ln\frac{2}{9} \approx 3$ years

Thus, the game has a useful life of 3 years. This is illustrated graphically in Figure 5.

(B) The total profit accumulated during the useful life of the game is

$$P(3) - P(0) = \int_0^3 P'(t)\,dt$$

$$= \int_0^3 [R'(t) - C'(t)]\,dt$$

$$= \int_0^3 (9e^{-0.5t} - 2)\,dt$$

$$= \left(\frac{9}{-0.5}e^{-0.5t} - 2t\right)\Big|_0^3 \qquad Recall: \int e^{ax}\,dx = \frac{1}{a}e^{ax} + C$$

$$= (-18e^{-0.5t} - 2t)|_0^3$$

$$= (-18e^{-1.5} - 6) - (-18e^0 - 0)$$

$$= 12 - 18e^{-1.5} \approx 7.984 \quad \text{or} \quad \$7,984$$
∎

Matched Problem 7 Repeat Example 7 if $C'(t) = 1$ and $R'(t) = 7.5e^{-0.5t}$. ∎

Example 8 **Numerical Integration on a Graphing Utility** Evaluate (to three decimal places): $\int_{-1}^{2} e^{-x^2}\,dx$

SOLUTION   The integrand e^{-x^2} does not have an elementary antiderivative, so we are unable to use the fundamental theorem to evaluate the definite integral. Instead, we use a numerical integration routine that has been preprogrammed in a graphing utility (consult your user's manual for specific details). Such a routine is an approximation algorithm, more powerful than the left sum and right sum methods discussed in Section 6-4. From Figure 6,

$$\int_{-1}^{2} e^{-x^2}\, dx = 1.629$$

FIGURE 6

Matched Problem 8   Evaluate (to three decimal places): $\int_{1.5}^{4.3} \dfrac{x}{\ln x}\, dx$

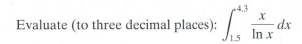

❑   RECOGNIZING A DEFINITE INTEGRAL: AVERAGE VALUE

Recall that the derivative of a function f was defined in Section 3-3 by

$$f'(x) = \lim_{h \to 0} \frac{f(x + h) - f(x)}{h}$$

This form is generally not easy to compute directly, but it is easy to recognize it in certain practical problems (slope, instantaneous velocity, rates of change, and so on). Once we know that we are dealing with a derivative, we then proceed to try to compute the derivative using derivative formulas and rules.

Similarly, evaluating a definite integral using the definition

$$\int_{a}^{b} f(x)\, dx = \lim_{n \to \infty} [f(c_1)\, \Delta x_1 + f(c_2)\, \Delta x_2 + \cdots + f(c_n)\, \Delta x_n] \qquad (1)$$

is generally not easy; but the form on the right occurs naturally in many practical problems. We can use the fundamental theorem to evaluate the definite integral (once it is recognized) if an antiderivative can be found; otherwise, we will approximate it using a rectangle sum. We will now illustrate these points by finding the *average value* of a continuous function.

Suppose that the temperature F (in degrees Fahrenheit) in the middle of a small shallow lake from 8 AM ($t = 0$) to 6 PM ($t = 10$) during the month of May is given approximately as shown in Figure 7.

How can we compute the average temperature from 8 AM to 6 PM? We know that the average of a finite number of values $a_1, a_2, \ldots, a_n$ is given by

$$\text{average} = \frac{a_1 + a_2 + \cdots + a_n}{n}$$

But how can we handle a continuous function with infinitely many values? It would seem reasonable to divide the time interval $[0, 10]$ into n equal subintervals, compute the temperature at a point in each subinterval, and then use the average of these values as an approximation of the average value of the continuous function $F = F(t)$ over $[0, 10]$. We would expect the approximations to improve as n increases. In fact, we would be inclined to define the limit of the average of n values as $n \to \infty$ as the *average value of F over* $[0, 10]$, if the limit exists. This is exactly what we will do:

$$\left(\begin{array}{c} \text{average temperature} \\ \text{for } n \text{ values} \end{array} \right) = \frac{1}{n} [F(t_1) + F(t_2) + \cdots + F(t_n)] \qquad (2)$$

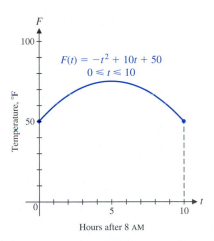

$F(t) = -t^2 + 10t + 50$
$0 \leq t \leq 10$

Temperature, °F

Hours after 8 AM

FIGURE 7

where t_k is a point in the kth subinterval. We will call the limit of equation (2) as $n \to \infty$ the *average temperature over the time interval* $[0, 10]$.

Form (2) looks sort of like form (1), but we are missing the Δt_k. We take care of this by multiplying equation (2) by $(b - a)/(b - a)$, which will change the form of equation (2) without changing its value:

$$\frac{b - a}{b - a} \cdot \frac{1}{n}[F(t_1) + F(t_2) + \cdots + F(t_n)] = \frac{1}{b - a} \cdot \frac{b - a}{n}[F(t_1) + F(t_2) + \cdots + F(t_n)]$$

$$= \frac{1}{b - a}\left[F(t_1)\frac{b - a}{n} + F(t_2)\frac{b - a}{n} + \cdots + F(t_n)\frac{b - a}{n}\right]$$

$$= \frac{1}{b - a}[F(t_1)\,\Delta t + F(t_2)\,\Delta t + \cdots + F(t_n)\,\Delta t]$$

Thus,

$$\begin{pmatrix} \text{average temperature} \\ \text{over } [a, b] = [0, 10] \end{pmatrix} = \lim_{n \to \infty}\left\{\frac{1}{b - a}[F(t_1)\,\Delta t + F(t_2)\,\Delta t + \cdots + F(t_n)\,\Delta t]\right\}$$

$$= \frac{1}{b - a}\left\{\lim_{n \to \infty}[F(t_1)\,\Delta t + F(t_2)\,\Delta t + \cdots + F(t_n)\,\Delta t]\right\}$$

Now the limit inside the braces is of form (1)—that is, a definite integral. Thus,

$$\begin{pmatrix} \text{average temperature} \\ \text{over } [a, b] = [0, 10] \end{pmatrix} = \frac{1}{b - a}\int_a^b F(t)\,dt$$

$$= \frac{1}{10 - 0}\int_0^{10}(-t^2 + 10t + 50)\,dt \qquad \text{We now evaluate the definite}$$

$$= \frac{1}{10}\left(-\frac{t^3}{3} + 5t^2 + 50t\right)\Big|_0^{10} \qquad \text{integral using the fundamen-}$$

$$= \frac{200}{3} \approx 67°F \qquad \text{tal theorem.}$$

In general, proceeding as above for an arbitrary continuous function f over an interval $[a, b]$, we obtain the following general formula:

Average Value of a Continuous Function f over $[a, b]$

$$\frac{1}{b - a}\int_a^b f(x)\,dx$$

Explore–Discuss 4

In Figure 8 the rectangle has the same area as the area under the graph of $y = f(x)$ from $x = a$ to $x = b$. Explain how the average value of $f(x)$ over the interval $[a, b]$ is related to the height of the rectangle.

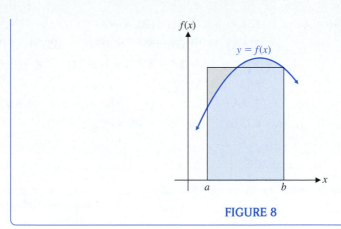

FIGURE 8

Example 9 ➷ **Average Value of a Function**   Find the average value of $f(x) = x - 3x^2$ over the interval $[-1, 2]$.

SOLUTION
$$\frac{1}{b-a} \int_a^b f(x)\, dx = \frac{1}{2-(-1)} \int_{-1}^2 (x - 3x^2)\, dx$$

$$= \frac{1}{3}\left(\frac{x^2}{2} - x^3 \right)\Big|_{-1}^2 = -\frac{5}{2}$$

■

Matched Problem 9 ➷ Find the average value of $g(t) = 6t^2 - 2t$ over the interval $[-2, 3]$.

■

Example 10 ➷ **Average Price**   Given the demand function

$$p = D(x) = 100e^{-0.05x}$$

find the average price (in dollars) over the demand interval $[40, 60]$.

SOLUTION   Average price $= \dfrac{1}{b-a} \displaystyle\int_a^b D(x)\, dx$

$$= \frac{1}{60-40} \int_{40}^{60} 100e^{-0.05x}\, dx$$

$$= \frac{100}{20} \int_{40}^{60} e^{-0.05x}\, dx \qquad \text{Use } \int e^{ax}\, dx = \frac{1}{a}e^{ax},\, a \neq 0.$$

$$= -\frac{5}{0.05} e^{-0.05x}\Big|_{40}^{60}$$

$$= 100(e^{-2} - e^{-3}) \approx \$8.55$$

■

Matched Problem 10 ➷ Given the supply equation

$$p = S(x) = 10e^{0.05x}$$

find the average price (in dollars) over the supply interval $[20, 30]$.

■

Exercise 6-5

A *Problems 1–6 refer to the following figure with the indicated areas:*

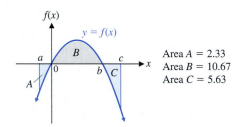

$f(x)$

$y = f(x)$

Area A = 2.33
Area B = 10.67
Area C = 5.63

1. $\int_a^0 f(x)\, dx = ?$ 2. $\int_b^c f(x)\, dx = ?$

3. $\int_a^b f(x)\, dx = ?$ 4. $\int_0^c f(x)\, dx = ?$

5. $\int_0^b \dfrac{f(x)}{10}\, dx = ?$ 6. $\int_0^b 10f(x)\, dx = ?$

Evaluate the integrals in Problems 7–20.

7. $\int_2^3 2x\, dx$ 8. $\int_1^2 3x^2\, dx$

9. $\int_3^4 5\, dx$ 10. $\int_{12}^{20} dx$

11. $\int_1^3 (2x - 3)\, dx$ 12. $\int_1^3 (6x + 5)\, dx$

13. $\int_{-3}^4 (4 - x^2)\, dx$ 14. $\int_{-1}^2 (x^2 - 4x)\, dx$

15. $\int_0^1 24x^{11}\, dx$ 16. $\int_0^2 30x^5\, dx$

17. $\int_0^1 e^{2x}\, dx$ 18. $\int_{-1}^1 e^{5x}\, dx$

19. $\int_1^{3.5} 2x^{-1}\, dx$ 20. $\int_1^2 \dfrac{dx}{x}$

B *Problems 21–24 refer to the figure for Problems 1–6 and the indicated areas.*

21. $\int_b^0 f(x)\, dx = ?$ 22. $\int_0^a f(x)\, dx = ?$

23. $\int_c^0 f(x)\, dx = ?$ 24. $\int_b^a f(x)\, dx = ?$

25. In Example 2 of Section 6-4 we used the rate function $r(t) = 32t$ and left and right sums to estimate the distance that a steel ball dropped from a bridge falls over the time interval $[1, 4]$. Find this distance using a definite integral and the fundamental theorem of calculus.

26. Repeat Problem 25 for the time interval $[2, 5]$.

Evaluate the integrals in Problems 27–44.

27. $\int_1^2 (2x^{-2} - 3)\, dx$ 28. $\int_1^2 (5 - 16x^{-3})\, dx$

29. $\int_1^4 3\sqrt{x}\, dx$ 30. $\int_4^{25} \dfrac{2}{\sqrt{x}}\, dx$

31. $\int_2^3 12(x^2 - 4)^5 x\, dx$ 32. $\int_0^1 32(x^2 + 1)^7 x\, dx$

33. $\int_3^9 \dfrac{1}{x - 1}\, dx$ 34. $\int_2^8 \dfrac{1}{x + 1}\, dx$

35. $\int_{-5}^{10} e^{-0.05x}\, dx$ 36. $\int_{-10}^{25} e^{-0.01x}\, dx$

37. $\int_1^e \dfrac{\ln t}{t}\, dt$ 38. $\int_e^{e^2} \dfrac{(\ln t)^2}{t}\, dt$

39. $\int_0^2 x\sqrt{4 - x^2}\, dx$ 40. $\int_0^4 x\sqrt{4 - x}\, dx$

41. $\int_0^1 xe^{-x^2}\, dx$ 42. $\int_0^1 xe^{x^2}\, dx$

43. $\int_{-2}^{-1} \dfrac{x^2 + 1}{x}\, dx$ 44. $\int_{-2}^{-1} \dfrac{x}{x^2 + 1}\, dx$

In Problems 45–52:

(A) Find the average value of each function over the indicated interval.

(B) Use a graphing utility to graph the function and its average value over the indicated interval in the same viewing window.

45. $f(x) = 500 - 50x$; $[0, 10]$
46. $g(x) = 2x + 7$; $[0, 5]$
47. $f(t) = 3t^2 - 2t$; $[-1, 2]$
48. $g(t) = 4t - 3t^2$; $[-2, 2]$
49. $f(x) = \sqrt[3]{x}$; $[1, 8]$
50. $g(x) = \sqrt{x + 1}$; $[3, 8]$
51. $f(x) = 4e^{-0.2x}$; $[0, 10]$
52. $f(x) = 64e^{0.08x}$; $[0, 10]$

Problems 53–58 refer to the figure below.

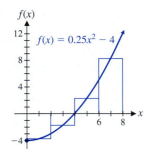

$f(x)$

$f(x) = 0.25x^2 - 4$

53. Use the midpoint sum M_4 to approximate the integral $I = \int_0^8 (0.25x^2 - 4)\, dx$. Calculate an error bound for this estimate.

54. Use the midpoint sum M_3 to approximate the integral $\int_0^6 (0.25x^2 - 4)\, dx$. Calculate an error bound for this estimate.

55. Evaluate $I = \int_0^8 (0.25x^2 - 4)\, dx$ using the fundamental theorem. Use the midpoint sum M_4 from Problem 53 to calculate the actual error $|I - M_4|$. Does this error lie within the error bound computed in Problem 53? Explain.

56. Evaluate $I = \int_0^6 (0.25x^2 - 4)\, dx$ using the fundamental theorem. Use the midpoint sum M_3 from Problem 54 to calculate the actual error $|I - M_3|$. Does this error lie within the error bound computed in Problem 54? Explain.

57. How large must n be chosen to have a midpoint sum M_n approximate $I = \int_0^8 (0.25x^2 - 4)\, dx$ with an error that does not exceed 0.005?

58. How large must n be chosen to have a midpoint sum M_n approximate $I = \int_0^6 (0.25x^2 - 4)\, dx$ with an error that does not exceed 0.05?

C *Write Problems 59–62 in the form $\int_a^b f(x)\, dx$ and evaluate using the fundamental theorem of calculus.*

59. $\lim\limits_{n\to\infty} \left[(1 - c_1^2)\, \Delta x + (1 - c_2^2)\, \Delta x + \cdots \right.$
$$\left. + (1 - c_n^2)\, \Delta x\right],$$
where $\Delta x = \dfrac{5 - 2}{n}$ and $c_k = 2 + k\dfrac{3}{n}, k = 1, 2, \ldots, n$

60. $\lim\limits_{n\to\infty} \left[(c_1^2 - 3)\, \Delta x + (c_2^2 - 3)\, \Delta x + \cdots \right.$
$$\left. + (c_n^2 - 3)\, \Delta x\right],$$
where $\Delta x = \dfrac{10 - 0}{n}$ and $c_k = 0 + k\dfrac{10}{n},$
$k = 1, 2, \ldots, n$

61. $\lim\limits_{n\to\infty} \left[(3c_1^2 - 2c_1 + 3)\, \Delta x + (3c_2^2 - 2c_2 + 3)\, \Delta x \right.$
$$\left. + \cdots + (3c_n^2 - 2c_n + 3)\, \Delta x\right],$$

where $\Delta x = \dfrac{12 - 2}{n}$ and $c_k = 2 + k\dfrac{10}{n}$,
$k = 1, 2, \ldots, n$

62. $\lim\limits_{n\to\infty} \left[(4c_1^3 + 3c_1^2 - 5)\, \Delta x + (4c_2^3 + 3c_2^2 - 5)\, \Delta x \right.$
$$\left. + \cdots + (4c_n^3 + 3c_n^2 - 5)\, \Delta x\right],$$

where $\Delta x = \dfrac{2 - 1}{n}$ and $c_k = 1 + k\dfrac{1}{n}$,
$k = 1, 2, \ldots, n$

Evaluate the integrals in Problems 63–68.

63. $\displaystyle\int_2^3 x\sqrt{2x^2 - 3}\, dx$

64. $\displaystyle\int_0^1 x\sqrt{3x^2 + 2}\, dx$

65. $\displaystyle\int_0^1 \frac{x - 1}{x^2 - 2x + 3}\, dx$

66. $\displaystyle\int_1^2 \frac{x + 1}{2x^2 + 4x + 4}\, dx$

67. $\displaystyle\int_{-1}^1 \frac{e^{-x} - e^x}{(e^{-x} + e^x)^2}\, dx$

68. $\displaystyle\int_6^7 \frac{\ln(t - 5)}{t - 5}\, dt$

Problems 69–72 refer to the following: In a more advanced treatment of exponential and logarithmic functions, the natural logarithmic function is defined in terms of a definite integral,

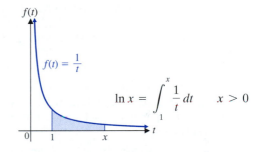

$f(t)$

$f(t) = \dfrac{1}{t}$

$$\ln x = \int_1^x \frac{1}{t}\, dt \qquad x > 0$$

69. Round all calculations to four decimal places.

(A) Approximate
$$\ln 2 = \int_1^2 \frac{1}{t}\, dt$$
using the midpoint sum with $n = 5$. Calculate an error bound.

(B) Evaluate $\ln 2$ directly on your calculator.

(C) Calculate the actual error from using the midpoint sum. [Use the results from parts (A) and (B) and error $= |\ln 2 - M_5|$.] Is this within the error bound determined in part (A)?

70. Round all calculations to four decimal places.

(A) Approximate
$$\ln 3 = \int_1^3 \frac{1}{t}\, dt$$
using the midpoint sum with $n = 10$. Calculate an error bound.

(B) Evaluate ln 3 directly on your calculator.

(C) Calculate the actual error from using the midpoint sum. [Use the results from parts (A) and (B) and error = $|\ln 3 - M_{10}|$.] Is this within the error bound determined in part (A)?

71. In using the midpoint sum in Problem 69, how large should n be chosen so that the error in approximating ln 2 does not exceed 0.0005?

72. In using the midpoint sum in Problem 70, how large should n be chosen so that the error in approximating ln 3 does not exceed 0.0005?

 Evaluate each definite integral in Problems 73–76 (to three decimal places) using a numerical integration routine.

73. $\int_{1.7}^{3.5} x \ln x \, dx$ **74.** $\int_{-1}^{1} e^{x^2} \, dx$

75. $\int_{-2}^{2} \frac{1}{1 + x^2} \, dx$ **76.** $\int_{0}^{3} \sqrt{9 - x^2} \, dx$

Applications

Business & Economics

77. *Cost.* A company manufactures mountain bikes. The research department produced the following marginal cost function:

$$C'(x) = 500 - \frac{x}{3} \qquad 0 \leqslant x \leqslant 900$$

where $C'(x)$ is in dollars and x is the number of bikes produced per month. Compute the increase in cost going from a production level of 300 bikes per month to 900 bikes per month. Set up a definite integral and evaluate.

78. *Cost.* Referring to Problem 77, compute the increase in cost going from a production level of 0 bikes per month to 600 bikes per month. Set up a definite integral and evaluate.

79. *Salvage value.* A new piece of industrial equipment will depreciate in value rapidly at first, then less rapidly as time goes on. Suppose that the rate (in dollars per year) at which the book value of a new milling machine changes is given approximately by

$$V'(t) = f(t) = 500(t - 12) \qquad 0 \leqslant t \leqslant 10$$

where $V(t)$ is the value of the machine after t years. What is the total loss in value of the machine in the first 5 years? In the second 5 years? Set up appropriate integrals and solve.

80. *Maintenance costs.* Maintenance costs for an apartment house generally increase as the building gets older. From past records, a managerial service determines that the rate of increase in maintenance costs (in dollars per year) for a particular apartment complex is given approximately by

$$M'(x) = f(x) = 90x^2 + 5{,}000$$

where x is the age of the apartment complex in years and $M(x)$ is the total (accumulated) cost of maintenance for x years. Write a definite integral that will give the total maintenance costs from the end of the

second year to the end of the seventh year after the apartment complex was built, and evaluate it.

81. *Employee training.* A company producing computer components has established that, on the average, a new employee can assemble $N(t)$ components per day after t days of on-the-job training, as indicated in the table (a new employee's productivity usually increases with time on the job up to a leveling-off point):

t	0	20	40	60	80	100	120
$N(t)$	10	51	68	76	81	84	85

(A) Find a quadratic regression equation for the data, and graph it and the data set in the same viewing window.

(B) Use the regression equation and a numerical integration routine on a graphing utility to approximate the number of units assembled by a new employee during the first 100 days on the job.

82. *Employee training.* Refer to Problem 81.

(A) Find a cubic regression equation for the data, and graph it and the data set in the same viewing window.

(B) Use the regression equation and a numerical integration routine on a graphing utility to approximate the number of units assembled by a new employee during the second 60 days on the job.

83. *Useful life.* The total accumulated costs $C(t)$ and revenues $R(t)$ (in thousands of dollars), respectively, for a coin-operated photocopying machine satisfy

$$C'(t) = \tfrac{1}{11}t \qquad \text{and} \qquad R'(t) = 5te^{-t^2}$$

where t is time in years. Find the useful life of the machine to the nearest year. What is the total profit accumulated during the useful life of the machine?

84. *Useful life.* The total accumulated costs $C(t)$ and revenues $R(t)$ (in thousands of dollars), respectively, for a coal mine satisfy

$$C'(t) = 3 \quad \text{and} \quad R'(t) = 15e^{-0.1t}$$

where t is the number of years the mine has been in operation. Find the useful life of the mine to the nearest year. What is the total profit accumulated during the useful life of the mine?

85. *Average cost.* The total cost (in dollars) of manufacturing x auto body frames is $C(x) = 60,000 + 300x$.

(A) Find the average cost per unit if 500 frames are produced. [*Hint:* Recall that $\overline{C}(x)$ is the average cost per unit.]

(B) Find the average value of the cost function over the interval $[0, 500]$.

(C) Discuss the difference between parts (A) and (B).

86. *Average cost.* The total cost (in dollars) of printing x dictionaries is $C(x) = 20,000 + 10x$.

(A) Find the average cost per unit if 1,000 dictionaries are produced.

(B) Find the average value of the cost function over the interval $[0, 1,000]$.

(C) Discuss the difference between parts (A) and (B).

87. *Cost.* The marginal cost at various levels of output per month for a company that manufactures watches is shown in the table. The output x is given in thousands of units per month, and $C(x)$ is given in thousands of dollars per month.

x	0	1	2	3	4	5	6	7	8
$C'(x)$	58	30	18	9	5	7	17	33	51

(A) Find a quadratic regression equation for the data, and graph it and the data set in the same viewing window.

(B) Use the regression equation and a numerical integration routine on a graphing utility to approximate (to the nearest dollar) the increased cost in going from a production level of 2 thousand watches per month to 8 thousand watches per month.

88. *Cost.* Refer to Problem 87.

(A) Find a cubic regression equation for the data, and graph it and the data set in the same viewing window.

(B) Use the regression equation and a numerical integration routine on a graphing utility to approximate (to the nearest dollar) the increased cost in going from a production level

of 1 thousand watches per month to 7 thousand watches per month.

89. *Supply function.* Given the supply function

$$P = S(x) = 10(e^{0.02x} - 1)$$

find the average price (in dollars) over the supply interval $[20, 30]$.

90. *Demand function.* Given the demand function

$$p = D(x) = \frac{1,000}{x}$$

find the average price (in dollars) over the demand interval $[400, 600]$.

91. *Labor costs and learning.* A defense contractor is starting production on a new missile control system. On the basis of data collected while assembling the first 16 control systems, the production manager obtained the following function for rate of labor use:

$$g(x) = 2,400x^{-1/2}$$

where $g(x)$ is the number of labor-hours required to assemble the xth unit of a control system. Approximately how many labor-hours will be required to assemble the 17th through the 25th control units? [*Hint:* Let $a = 16$ and $b = 25$.]

92. *Labor costs and learning.* If the rate of labor use in Problem 91 is

$$g(x) = 2,000x^{-1/3}$$

approximately how many labor-hours will be required to assemble the 9th through the 27th control units? [*Hint:* Let $a = 8$ and $b = 27$.]

93. *Inventory.* A store orders 600 units of a product every 3 months. If the product is steadily depleted to 0 by the end of each 3 months, the inventory on hand, I, at any time t during the year is illustrated as shown in the figure.

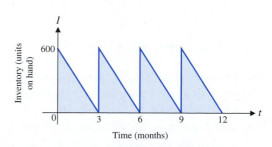

Figure for 93

(A) Write an inventory function (assume that it is continuous) for the first 3 months. [The graph is a straight line joining $(0, 600)$ and $(3, 0)$.]

(B) What is the average number of units on hand for a 3-month period?

94. Repeat Problem 93 with an order of 1,200 units every 4 months.

95. *Oil production.* Using data from the first 3 years of production as well as geological studies, the management of an oil company estimates that oil will be pumped from a producing field at a rate given by

$$R(t) = \frac{100}{t + 1} + 5 \qquad 0 \leqslant t \leqslant 20$$

where $R(t)$ is the rate of production (in thousands of barrels per year) t years after pumping begins. Approximately how many barrels of oil will the field produce during the first 10 years of production? From the end of the 10th year to the end of the 20th year of production?

96. *Oil production.* In Problem 95, if the rate is found to be

www

$$R(t) = \frac{120t}{t^2 + 1} + 3 \qquad 0 \leqslant t \leqslant 20$$

approximately how many barrels of oil will the field produce during the first 5 years of production? The second 5 years of production?

97. *Profit.* Let $R(t)$ and $C(t)$ represent the total accumulated revenues and costs (in dollars), respectively, for a producing oil well, where t is time in years. The graphs of the derivatives of R and C over a 5-year period are shown in the figure. Use the midpoint sum with $n = 5$ to approximate the total accumulated profits from the well over this 5-year period. Estimate necessary function values from the graphs.

Figure for 97 and 98

98. *Revenue.* Use the figure and the midpoint sum with $n = 5$ to approximate the average annual revenue from the oil well.

99. *Real estate.* A surveyor produced the table below by measuring the vertical distance (in feet) across a piece of real estate at 600-foot intervals, starting at 300 (see the figure). Use these values and the midpoint sum to estimate the area of the property.

x	300	900	1,500	2,100
$f(x)$	900	1,700	1,700	900

Figure for 99

100. *Real estate.* Repeat Problem 99 for the following table of measurements:

x	200	600	1,000	1,400	1,800	2,200
$f(x)$	600	1,400	1,800	1,800	1,400	600

Life Sciences

101. *Biology.* A yeast culture weighing 2 grams is removed from a refrigerator unit and is expected to grow at the rate of $W'(t) = 0.2e^{0.1t}$ grams per hour at a higher controlled temperature. How much will the weight of the culture increase during the first 8 hours of growth? How much will the weight of the culture increase from the end of the 8th hour to the end of the 16th hour of growth?

102. *Medicine.* The rate of healing for a skin wound (in square centimeters per day) is given approximately by $A'(t) = -0.9e^{-0.1t}$. The initial wound has an area of 9 square centimeters. How much will the area change during the first 5 days? The second 5 days?

103. *Temperature.* If the temperature $C(t)$ in an aquarium is made to change according to

$$C(t) = t^3 - 2t + 10 \qquad 0 \leqslant t \leqslant 2$$

(in degrees Celsius) over a 2-hour period, what is the average temperature over this period?

104. *Medicine.* A drug is injected into the bloodstream of a patient through her right arm. The concentration of the drug in the bloodstream of the left arm t hours after the injection is given by

$$C(t) = \frac{0.14t}{t^2 + 1}$$

What is the average concentration of the drug in the bloodstream of the left arm during the first hour after the injection? During the first 2 hours after the injection?

105. *Medicine: respiration.* Physiologists use a machine **WWW** called a pneumotachograph to produce a graph of the rate of flow $R(t)$ of air into the lungs (inspiration) and out of the lungs (expiration). The figure gives the graph of the inspiration phase of the breathing cycle for a person at rest. The area under this graph represents the total volume of air inhaled during the inspiration phase. Use the midpoint sum with $n = 3$ to approximate the area under the graph. Estimate the necessary function values from the graph.

Figure for 105 and 106

106. *Medicine: respiration.* Use the result obtained in Problem 105 to approximate the average volume of air in the lungs during the inspiration phase.

Social Sciences

107. *Politics.* Public awareness of a congressional candidate before and after a successful campaign was approximated by

$$P(t) = \frac{8.4t}{t^2 + 49} + 0.1 \qquad 0 \leq t \leq 24$$

where t is time in months after the campaign started and $P(t)$ is the fraction of people in the congressional district who could recall the candidate's name. What is the average fraction of people who could recall the candidate's name during the first 7 months after the campaign began? During the first 2 years after the campaign began?

108. *Population composition.* Because of various factors (such as birth rate expansion, then contraction; family flights from urban areas; and so on), the number of children in a large city was found to increase and then decrease rather drastically. If the number of children over a 6-year period was found to be given approximately by

$$N(t) = -\tfrac{1}{4}t^2 + t + 4 \qquad 0 \leq t \leq 6$$

what was the average number of children in the city over the 6-year time period? [Assume $N = N(t)$ is continuous.]

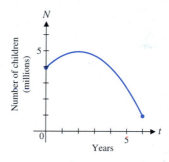

Figure for 108

Important Terms and Symbols

6-1 *Antiderivatives and Indefinite Integrals.* Antiderivative; indefinite integral; integral sign; integrand; constant of integration

6-2 *Integration by Substitution.* Reversing the chain rule; general indefinite integral formulas; differentials; method of substitution; change of variable

6-3 *Differential Equations: Growth and Decay.* Differential equation; first-order; second-order; slope field; continuous compound interest; exponential growth law; growth constant; growth rate; exponential decay; population growth; radioactive decay; unlimited growth; limited growth; logistic growth

$$\frac{dQ}{dt} = rQ; \qquad Q = Q_0 e^{rt}$$

6-4 *Geometric–Numeric Introduction to the Definite Integral.* Area under a curve; definite integral symbol; left sum; right sum; monotone functions; error bounds for left and right sums and their average; signed areas; rate, area, and distance; rate, area, and total change

$$\int_a^b f(x)\,dx; \quad L_n; \quad R_n; \quad A_n$$

6-5 *Definite Integral as a Limit of a Sum; Fundamental Theorem of Calculus.* Left sum; right sum; midpoint sum; general error bounds for left, right, and midpoint sums; definite integral as a limit of a Riemann sum; integrand; lower limit; upper limit; properties of definite integrals; fundamental theorem of calculus; useful life; recognizing a definite integral; average value of a continuous function

$$L_n; \quad R_n; \quad M_n;$$
$$\int_a^b f(x)\,dx; \qquad F(x)\Big|_a^b = F(b) - F(a);$$
$$\sum_{k=1}^n f(c_k)\,\Delta x_k; \qquad \frac{1}{b-a}\int_a^b f(x)\,dx$$

Integration Formulas and Properties

$$\int k\, dx = kx + C$$

$$\int kf(x)\, dx = k \int f(x)\, dx$$

$$\int [f(x) \pm g(x)]\, dx = \int f(x)\, dx \pm \int g(x)\, dx$$

$$\int u^n\, du = \frac{u^{n+1}}{n+1} + C, \qquad n \neq -1$$

$$\int e^u\, du = e^u + C$$

$$\int e^{au}\, du = \frac{1}{a}e^{au} + C, \qquad a \neq 0$$

$$\int \frac{1}{u}\, du = \ln|u| + C, \qquad u \neq 0$$

$$\int_a^a f(x)\, dx = 0$$

$$\int_a^b f(x)\, dx = -\int_b^a f(x)\, dx$$

$$\int_a^b kf(x)\, dx = k \int_a^b f(x)\, dx, \qquad k \text{ a constant}$$

$$\int_a^b [f(x) \pm g(x)]\, dx = \int_a^b f(x)\, dx \pm \int_a^b g(x)\, dx$$

$$\int_a^b f(x)\, dx = \int_a^c f(x)\, dx + \int_c^b f(x)\, dx$$

Review Exercise

Work through all the problems in this chapter review and check your answers in the back of the book. Answers to all review problems are there along with section numbers in italics to indicate where each type of problem is discussed. Where weaknesses show up, review appropriate sections in the text.

A *Find each integral in Problems 1–6.*

1. $\int (6x + 3)\, dx$ **2.** $\int_{10}^{20} 5\, dx$

3. $\int_0^9 (4 - t^2)\, dt$ **4.** $\int (1 - t^2)^3\, t\, dt$

5. $\int \frac{1 + u^4}{u}\, du$ **6.** $\int_0^1 xe^{-2x^2}\, dx$

In Problems 7 and 8, find the derivative or indefinite integral as indicated.

7. $\dfrac{d}{dx}\left(\int e^{-x^2}\, dx\right)$ **8.** $\int \dfrac{d}{dx}(\sqrt{4 + 5x})\, dx$

9. Find a function $y = f(x)$ that satisfies both conditions:

$$\frac{dy}{dx} = 3x^2 - 2 \qquad f(0) = 4$$

10. From the graph of $y = f'(x)$ shown in the figure, verbally describe the shape of the graph of an antiderivative function f. How would the graph of one antiderivative of $f'(x)$ differ from another?

$f'(x)$

Figure for 10 and 11

11. From the graph of $y = f'(x)$ shown in the figure, sketch possible graphs of three antiderivative functions f such that $f(0) = -2$, $f(0) = 0$, and $f(0) = 2$, all on the same set of coordinate axes.

12. Find all antiderivatives of:

(A) $\dfrac{dy}{dx} = 8x^3 - 4x - 1$ (B) $\dfrac{dx}{dt} = e^t - 4t^{-1}$

13. Approximate $\int_1^5 (x^2 + 1)\, dx$ using a midpoint sum with $n = 2$. Calculate an error bound for this approximation.

14. Evaluate the integral in Problem 13 using the fundamental theorem of calculus, and calculate the actual error $|I - M_2|$ produced in using M_2.

15. Use the table of values below and a midpoint sum with $n = 4$ to approximate $\int_1^{17} f(x)\, dx$.

x	3	7	11	15
$f(x)$	1.2	3.4	2.6	0.5

16. Find the average value of $f(x) = 6x^2 + 2x$ over the interval $[-1, 2]$.

17. Describe a rectangle that would have the same area as the area under the graph of $f(x) = 6x^2 + 2x$ from $x = -1$ to $x = 2$ (see Problem 16).

B *Use the graph and actual areas of the indicated regions in the figure to evaluate the integrals in Problems 18–25:*

$f(x)$

$y = f(x)$

Area $A = 1$
Area $B = 2$
Area $C = 2$
Area $D = 0.6$

18. $\int_a^b 5f(x)\,dx$ **19.** $\int_b^c \dfrac{f(x)}{5}\,dx$ **20.** $\int_b^d f(x)\,dx$

21. $\int_a^c f(x)\,dx$ **22.** $\int_0^d f(x)\,dx$ **23.** $\int_b^a f(x)\,dx$

24. $\int_c^b f(x)\,dx$ **25.** $\int_d^0 f(x)\,dx$

26. For the graph of $y = f'(x)$ shown, verbally describe the shape of the graph of an antiderivative function f. How would the graph of one antiderivative of $f'(x)$ differ from another?

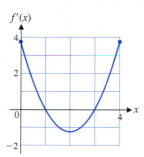

Figure for 26 and 27

27. For the graph of $y = f'(x)$ shown, sketch possible graphs of three antiderivative functions f such that $f(0) = -1, f(0) = 0$, and $f(0) = 1$, all on the same set of coordinate axes.

Problems 28–33 refer to the slope field shown in the figure:

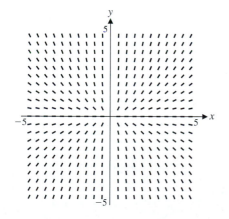

28. (A) For $dy/dx = (2y)/x$, what is the slope of a solution curve at $(2, 1)$? At $(-2, -1)$?

  (B) For $dy/dx = (2x)/y$, what is the slope of a solution curve at $(2, 1)$? At $(-2, -1)$?

29. Is the slope field shown in the figure for $dy/dx = (2x)/y$ or for $dy/dx = (2y)/x$? Explain.

30. Show that $y = Cx^2$ is a solution of $dy/dx = (2y)/x$ for any real number C.

31. Referring to Problem 30, find the particular solution of $dy/dx = (2y)/x$ that passes through $(2, 1)$. Through $(-2, -1)$.

32. Graph the two particular solutions found in Problem 31 in the slope field shown above (or a copy).

33. Use a graphing utility to graph in the same viewing window graphs of $y = Cx^2$ for $C = -2, -1, 1$, and 2 for $-5 \le x \le 5$ and $-5 \le y \le 5$.

Find each integral in Problems 34–44.

34. $\int_{-1}^1 \sqrt{1 + x}\,dx$ **35.** $\int_{-1}^0 x^2(x^3 + 2)^{-2}\,dx$

36. $\int 5e^{-t}\,dt$ **37.** $\int_1^e \dfrac{1 + t^2}{t}\,dt$

38. $\int xe^{3x^2}\,dx$ **39.** $\int_{-3}^1 \dfrac{1}{\sqrt{2 - x}}\,dx$

40. $\int_0^3 \dfrac{x}{1 + x^2}\,dx$ **41.** $\int_0^3 \dfrac{x}{(1 + x^2)^2}\,dx$

42. $\int x^3(2x^4 + 5)^5\,dx$ **43.** $\int \dfrac{e^{-x}}{e^{-x} + 3}\,dx$

44. $\int \dfrac{e^x}{(e^x + 2)^2}\,dx$

45. Find a function $y = f(x)$ that satisfies both conditions:

$$\frac{dy}{dx} = 3x^{-1} - x^{-2} \qquad f(1) = 5$$

46. Find the equation of the curve that passes through $(2, 10)$ if its slope is given by

$$\frac{dy}{dx} = 6x + 1$$

for each x.

Problems 47–50 refer to the following: A toy rocket is shot vertically into the air with an initial rate of 160 feet per second. Neglecting air resistance, its rate of ascent, $r(t)$ (in feet per second), at the end of t seconds is shown in the figure.

47. Estimate the height the rocket reaches using left and right sums and their average over five equal subdivisions. (Read values directly from the graph.) Calculate error bounds for these estimates.

48. Explain how the areas of the rectangles used in the left or right sum approximations in Problem 47 are related to distance.

49. How many equal subdivisions should be used on the interval $[0, 5]$ to have A_n approximate the height the rocket reaches within 1 foot?

50. Set up a definite integral that represents the height the rocket reaches and evaluate it. [*Hint:* First find the equation of the rate function.]

51. (A) Find the average value of $f(x) = 3\sqrt{x}$ over the interval $[1, 9]$.

　　(B) Graph $f(x) = 3\sqrt{x}$ and its average over the interval $[1, 9]$ in the same coordinate system.

C *Find each integral in Problems 52–56.*

52. $\displaystyle\int \frac{(\ln x)^2}{x}\, dx$　　**53.** $\displaystyle\int x(x^3 - 1)^2\, dx$

54. $\displaystyle\int \frac{x}{\sqrt{6 - x}}\, dx$　　**55.** $\displaystyle\int_0^7 x\sqrt{16 - x}\, dx$

56. $\displaystyle\int_{-1}^1 x(x + 1)^4\, dx$

57. Find a function $y = f(x)$ that satisfies both conditions:

$$\frac{dy}{dx} = 9x^2 e^{x^3} \qquad f(0) = 2$$

58. Solve the differential equation:

$$\frac{dN}{dt} = 0.06N \qquad N(0) = 800 \qquad N > 0$$

Problems 59–63 involve estimating the value of the definite integral

$$I = \int_0^1 e^{-x^2}\, dx$$

(The integrand does not have an elementary antiderivative.)

59. Graph $f(x) = e^{-x^2}$ over the interval $[0, 1]$.

60. Approximate I using a midpoint sum with $n = 5$.

61. Calculate $f''(x)$ and show that 2 is an upper bound for $|f''(x)|$ by graphing $|f''(x)|$ on the interval $[0, 1]$ using a graphing utility.

62. Use the results from Problem 61 to calculate an error bound for the approximation M_5 in Problem 60.

63. Using the result from Problem 62, determine how large n should be so that M_n approximates I with an error that does not exceed 0.0005.

 Graph Problems 64–67 on a graphing utility and identify each as unlimited growth, exponential decay, limited growth, or logistic growth:

64. $N = 50(1 - e^{-0.07t});\ 0 \leqslant t \leqslant 80,\ 0 \leqslant N \leqslant 60$

65. $p = 500e^{-0.03x};\ 0 \leqslant x \leqslant 100,\ 0 \leqslant p \leqslant 500$

66. $A = 200e^{0.08t};\ 0 \leqslant t \leqslant 20,\ 0 \leqslant A \leqslant 1{,}000$

67. $N = \dfrac{100}{1 + 9e^{-0.3t}};\ 0 \leqslant t \leqslant 25,\ 0 \leqslant N \leqslant 100$

 Evaluate each definite integral in Problems 68–70 (to three decimal places) using a numerical integration routine.

68. $\displaystyle\int_{-0.5}^{0.6} \frac{1}{\sqrt{1 - x^2}}\, dx$

69. $\displaystyle\int_{-2}^3 x^2 e^x\, dx$

70. $\displaystyle\int_{0.5}^{2.5} \frac{\ln x}{x^2}\, dx$

Applications

Business & Economics

71. *Cost.* A company manufactures downhill skis. The research department produced the marginal cost graph shown in the figure, where $C'(x)$ is in dollars and x is the number of pairs of skis produced per week. Estimate the increase in cost going from a production level of 200 to 600 pairs of skis per week. Use left and right sums over two equal subintervals. Replace the question marks with the values of L_2 and R_2 as appropriate:

$$? \leqslant \int_{200}^{600} C'(x)\, dx \leqslant ?$$

Figure for 71

72. *Cost.* Explain how the increase in production cost is related to the left or right rectangles used in Problem 71.

73. *Cost.* Assuming that the marginal cost function in Problem 71 is linear, find its equation and write a definite integral that represents the increase in costs going from a production level of 200 to 600 pairs of skis per week. Evaluate the definite integral.

74. *Profit and production.* The weekly marginal profit for an output of x units is given approximately by

$$P'(x) = 150 - \frac{x}{10} \qquad 0 \leq x \leq 40$$

What is the total change in profit for a production change from 10 units per week to 40 units? Set up a definite integral and evaluate it.

75. *Profit function.* If the marginal profit for producing x units per day is given by

$$P'(x) = 100 - 0.02x \qquad P(0) = 0$$

where $P(x)$ is the profit in dollars, find the profit function P and the profit on 10 units of production per day.

76. *Resource depletion.* An oil well starts out producing oil at the rate of 60,000 barrels of oil per year, but the production rate is expected to decrease by 4,000 barrels per year. Thus, if $P(t)$ is the total production (in thousands of barrels) in t years, then

$$P'(t) = f(t) = 60 - 4t \qquad 0 \leq t \leq 15$$

Write a definite integral that will give the total production after 15 years of operation and evaluate it.

77. *Inventory.* Suppose that the inventory of a certain item t months after the first of the year is given approximately by

$$I(t) = 10 + 36t - 3t^2 \qquad 0 \leq t \leq 12$$

What is the average inventory for the second quarter of the year?

78. *Price–supply.* Given the price–supply function

$$p = S(x) = 8(e^{0.05x} - 1)$$

find the average price (in dollars) over the supply interval $[40, 50]$.

79. *Employee training.* A company producing sound system equipment has found that, on the average, a new employee can assemble $N(t)$ components per day after t days of on-the-job training, as indicated in the following table:

t	0	10	20	30	40	50
$N(t)$	5	10	14	17	19	20

(A) Use the average of left and right sums, A_5, for five equal subdivisions to estimate the total number of units assembled by a new employee the first 50 days of employment. Calculate an error bound for the estimate.

📷 (B) Find a quadratic regression equation for the data, and graph it and the data set in the same viewing window.

📷 (C) Use the regression equation and a numerical integration routine on a graphing utility to approximate the number of units assembled by a new employee the first 50 days of employment.

80. *Useful life.* The total accumulated costs $C(t)$ and revenues $R(t)$ (in thousands of dollars), respectively, for a coal mine satisfy

$$C'(t) = 3 \qquad \text{and} \qquad R'(t) = 20e^{-0.1t}$$

where t is the number of years the mine has been in operation. Find the useful life of the mine to the nearest year. What is the total profit accumulated during the useful life of the mine?

81. *Marketing.* The market research department for an automobile company estimates that the sales (in millions of dollars) of a new automobile will increase at the monthly rate of

$$S'(t) = 4e^{-0.08t} \qquad 0 \leq t \leq 24$$

t months after the introduction of the automobile. What will be the total sales $S(t)$, t months after the automobile is introduced if we assume that there were 0 sales at the time the automobile entered the marketplace? What are the estimated total sales during the first 12 months after the introduction of the automobile? How long will it take for the total sales to reach $40 million?

Life Sciences

82. *Pollution.* In an industrial area, the concentration **www** $C(t)$ of particulate matter (in parts per million) during a 12-hour period is given in the figure. Use a midpoint sum with $n = 6$ to approximate the average concentration during this 12-hour period. Estimate the necessary function values from the graph.

83. *Wound healing.* The area of a small, healing surface wound changes at a rate given approximately by

$$\frac{dA}{dt} = -5t^{-2} \qquad 1 \le t \le 5$$

where t is time in days and $A(1) = 5$ square centimeters. What will the area of the wound be in 5 days?

84. *Pollution.* An environmental protection agency estimates that the rate of seepage of toxic chemicals from a waste dump (in gallons per year) is given by

$$R(t) = \frac{1,000}{(1 + t)^2}$$

where t is time in years since the discovery of the seepage. Find the total amount of toxic chemicals that seep from the dump during the first 4 years after the seepage is discovered.

85. *Population.* The population of Mexico was 100 million in 2000 and was growing at a rate of 1.5% per year compounded continuously.

WWW

(A) Assuming that the population continues to grow at this rate, estimate the population of Mexico in the year 2025.

(B) At the growth rate indicated, how long will it take the population of Mexico to double?

Social Sciences

86. *Archaeology.* The continuous compound rate of decay for carbon-14 is $r = -0.000\ 123\ 8$. A piece of animal bone found at an archaeological site contains 4% of the original amount of carbon-14. Estimate the age of the bone.

87. *Learning.* In a particular business college, it was found that an average student enrolled in a typing class progressed at a rate of $N'(t) = 7e^{-0.1t}$ words per minute t weeks after enrolling in a 15-week course. If at the beginning of the course a student could type 25 words per minute, how many words per minute, $N(t)$, would the student be expected to type t weeks into the course? After completing the course?

Group Activity 1   Simpson's Rule

Introduction to Simpson's Rule

The left sum L_n, right sum R_n, average A_n of left and right sums, and midpoint sum M_n can all be used as numerical integration devices to approximate definite integrals. If a function f is increasing, the right sum overestimates and the left sum underestimates the definite integral (Fig. 1).

$$L_n \le \int_a^b f(x)\, dx \le R_n$$

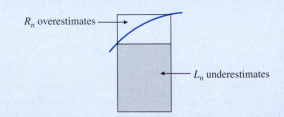

FIGURE 1

The average A_n of left and right sums is a better estimate for the definite integral of a monotone function than either L_n or R_n. It is often called a *trapezoidal sum,* and denoted by T_n, since the average area of a left rectangle and right rectangle is the area of a trapezoid (the smaller, shaded trapezoid in Fig. 2). Adding the following expressions for L_n and R_n and dividing by 2 gives the expression for $T_n = A_n$ called the **trapezoidal rule.** The midpoint sum M_n, like T_n, is also a better estimate for the definite integral of a monotone function than either the left or right sum.

$$L_n = [f(x_0) + f(x_1) + \cdots + f(x_{n-1})]\,\Delta x \qquad \text{Left sum}$$
$$R_n = [f(x_1) + f(x_2) + \cdots + f(x_n)]\,\Delta x \qquad \text{Right sum}$$
$$T_n = [f(x_0) + 2f(x_1) + \cdots + 2f(x_{n-1}) + f(x_n)]\frac{\Delta x}{2} \qquad \text{Trapezoidal rule}$$
$$M_n = \left[f\!\left(\frac{x_0 + x_1}{2}\right) + f\!\left(\frac{x_1 + x_2}{2}\right) + \cdots + f\!\left(\frac{x_{n-1} + x_n}{2}\right)\right]\Delta x \quad \text{Midpoint sum}$$

A midpoint sum rectangle has the same area as the corresponding tangent line trapezoid (the larger trapezoid in Fig. 2). It appears from Figure 2, and can be proved in general, that the trapezoidal sum error is about double the midpoint sum error when the graph of the function is concave up or concave down.

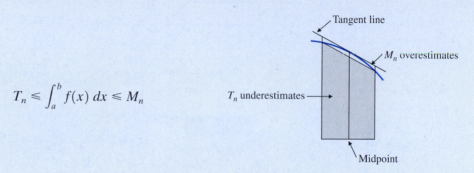

$$T_n \le \int_a^b f(x)\,dx \le M_n$$

FIGURE 2

This suggests that a weighted average of the two estimates, with the midpoint sum being counted double the trapezoidal sum, might be an even better estimate than either separately. This weighted average is called **Simpson's rule,** and is given symbolically as follows:

$$S_{2n} = \frac{2M_n + T_n}{3} \qquad \text{Simpson's rule} \tag{1}$$

The trapezoidal rule involves integrand values at $n + 1$ points (including a and b), and the midpoint sum involves integrand values at the n midpoints. So Simpson's rule involves integrand values at $2n + 1$ points (including a and b). To apply Simpson's rule, we divide the interval $[a, b]$ into $2n$ equal subdivisions, evaluate the integrand at each subdivision point, and then use n equal subdivisions for each of the sums M_n and T_n. Thus, **Simpson's rule always requires an even number of equal subdivisions of [a, b].**

(A) For $\Delta x = (b - a)/4$, show, starting with equation (1) in the form

$$S_4 = \frac{2M_2 + T_2}{3}$$

that

$$S_4 = [f(x_0) + 4f(x_1) + 2f(x_2) + 4f(x_3) + f(x_4)]\frac{\Delta x}{3}$$

Error Comparisons

One way to compare the effectiveness of a numerical integration technique is to find the error produced for a given value of n and to see what happens to the error as n is increased. We do this for the five sums discussed above, L_n, R_n, M_n, T_n, and S_{2n}, by using these sums to approximate a definite integral of known value and looking at the corresponding errors.

(B) Given the definite integral

$$I = \int_4^{12} \frac{dx}{x} = 1.099 \qquad \text{To three decimal places}$$

use the values of the integrand in Table 1 to complete Table 2. Round all answers to three decimal places.

TABLE 1

INTEGRAND VALUES

x	4	5	6	7	8	9	10	11	12
$1/x$	0.250	0.200	0.167	0.143	0.125	0.111	0.100	0.091	0.083

TABLE 2

ERROR

	$L_n - I$	$R_n - I$	$M_n - I$	$T_n - I$	$S_{2n} - I$
$n = 2$	0.401				0.002
$n = 4$			−0.009		

Table 2 suggests that the relationships between error and the change in n shown in Table 3 exist. This can be confirmed in general.

TABLE 3

ERROR AND CHANGE IN n

	FACTOR BY WHICH ERROR IS CHANGED		
CHANGE IN n	L_n, R_n	M_n, T_n	S_{2n}
$2n$	$\frac{1}{2}$	$\frac{1}{4}$	$\frac{1}{16}$
$10n$	$\frac{1}{10}$ Adds one more digit of accuracy	$\frac{1}{10^2}$ Adds two more digits of accuracy	$\frac{1}{10^4}$ Adds four more digits of accuracy

We have already noted that L_n and R_n give exact results if the integrand is a constant function. Also, M_n and T_n give exact results for constant and linear functions. What about Simpson's rule? Simpson's rule gives exact results if the integrand is any polynomial of degree 3 or less. It should now be clear why Simpson's rule is so popular. However, Simpson's rule is generally not used in numerical integration routines. More powerful algorithms exist. For example, the Gauss–Kronrod algorithm is used in a number of popular graphing utilities and gives exact results for polynomials of degree 5 or less.

(C) Use Simpson's rule to approximate the definite integral

$$I = \int_2^{10} \frac{x}{\ln x}\, dx$$

for $2n = 4$ and for $2n = 8$. Compute S_{2n} to three decimal places.

(D) Using more powerful algorithms, it is known that $I = 27.159$ to three decimal places. Compute the error $|I - S_{2n}|$ for the two estimates in part (C).

Application

We now turn to an example where Simpson's rule is used to estimate cost. (Note that it is not necessary to find a regression equation, as we would in order to use a numerical integration routine on a graphing utility.)

A company manufactures and wholesales a popular brand of sunglasses. Their financial research department collected the data in Table 4, showing marginal cost $C'(x)$ (in dollars) at selected production levels of x pairs of sunglasses per hour.

TABLE 4

MARGINAL COSTS

t	50	100	150	200	250	300	350	400	450
$C'(x)$	21.80	16.95	15.31	14.50	14.00	13.66	13.42	13.25	13.11

(E) If the company is now producing 50 pairs of sunglasses per hour and increases production to 450 pairs per hour, estimate the total increase in costs per hour using L_n and R_n with an appropriate value of n. How are L_n and R_n related to the actual increase in cost?

(F) Use Simpson's rule to estimate the increase in cost in part (E).

Group Activity 2   **Bell-Shaped Curves**

f(x)

$$f(x) = \frac{1}{\sigma\sqrt{2\pi}}\, e^{-(x-\mu)^2/2\sigma^2}$$

FIGURE 1 Normal curve

One of the most important functions in probability and statistics is the **normal probability density function** and its bell-shaped graph or **normal curve,** as shown in Figure 1.

It can be shown that the total area under the curve from $-\infty$ to ∞, for μ any real number and σ any positive real number, is always 1. Thus, the area under the curve over an interval $[a, b]$ is the percentage of the total area that is under the curve between a and b. We will interpret the normal probability density function through an example.

A manufacturer of 100-watt light bulbs tests a large sample and finds that the average life of these bulbs is 5 hundred hours ($\mu = 5$) with a *standard deviation* of 1 hundred hours ($\sigma = 1$). (Standard deviation measures the dispersion of the normal probability density function about the mean or average. A small standard deviation is associated with a tall, narrow normal curve, and a large standard deviation is related to a low, flat normal curve. You will see this below.) For $\mu = 5$ and $\sigma = 1$, the probability density function and corre-

$$f(x) = \frac{1}{\sqrt{2\pi}} e^{-(x-5)^2/2}$$

FIGURE 2

sponding normal curve are shown in Figure 2. The area under the curve between 5 hundred hours and 6 hundred hours represents the percentage of light bulbs in the manufacturing process that will have a life between 5 and 6 hundred hours; that is, the probability of a light bulb drawn at random having a life between 5 and 6 hundred hours.

(A) Write a definite integral that represents the probability of a light bulb drawn at random having a life between 5 and 6 hundred hours.

(B) Approximate the definite integral in part (A) with a midpoint sum using five equal subintervals. Explain what the result means relative to the original problem.

(C) Compute $f''(x)$. Use a graphing utility to graph $y = |f''(x)|$ over the interval $[5, 6]$, and use the graph to show that 0.4 is an upper bound for $|f''(x)|$ on this interval.

(D) Calculate an error bound for the estimate M_5 in part (B) using the result from part (C).

(E) Using the result from part (C), how large should n be chosen so that the error in using M_n is no greater than 0.00005?

(F) If the area under the normal curve from 5 to ∞ is 0.5, what is the probability of selecting a light bulb at random that has a life greater than 600 hours? Explain how you arrived at your answer.

(G) What is the probability of selecting a light bulb at random that has a life less than 500 hours? Explain how you arrived at your answer.

(H) Use a numerical integration routine on a graphing utility to find the probability of selecting a light bulb at random that has a life between 450 and 550 hours. That has a life between 350 and 650 hours.

(I) Graph normal probability density functions with $\mu = 8$ and $\sigma = 1, 2$, and 3, in the same viewing window. What effect does changing σ have on the shape of the curve?

Additional Integration Topics

INTRODUCTION

This chapter contains additional topics on integration. Since they are essentially independent of one another, they may be taken up in any order, and certain sections may be omitted if desired.

Section 7-1

Area Between Curves

❑ AREA BETWEEN A CURVE AND THE X AXIS
❑ AREA BETWEEN TWO CURVES
❑ APPLICATION: INCOME DISTRIBUTION

In Chapter 6 we found that the definite integral $\int_a^b f(x)\,dx$ represents the sum of the signed areas between the graph of $y = f(x)$ and the x axis from $x = a$ to $x = b$, where the areas above the x axis are counted positively and the areas below the x axis are counted negatively (see Fig. 1). In this section we are interested in using the definite integral to find the actual area between a curve and the x axis or the actual area between two curves. These areas are always nonnegative quantities—**area measure is never negative.**

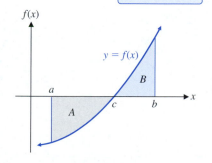

FIGURE 1
$$\int_a^b f(x)\,dx = -A + B$$

❑ AREA BETWEEN A CURVE AND THE X AXIS

In Figure 1, A represents the area between $y = f(x)$ and the x axis from $x = a$ to $x = c$, and B represents the area between $y = f(x)$ and the x axis from $x = c$ to $x = b$. Both A and B are positive quantities. Since $f(x) \geqslant 0$ on the interval $[c, b]$,

$$\int_c^b f(x)\,dx = B$$

433

and since $f(x) \le 0$ on the interval $[a, c]$,

$$\int_a^c f(x)\, dx = -A$$

or

$$A = -\int_a^c f(x)\, dx = \int_a^c [-f(x)]\, dx$$

Thus:

The area between the graph of a negative function and the x axis is equal to the definite integral of the negative of the function.

Explore–Discuss 1

Sketch a graph of a function f such that $f(x) \le 0$ over the interval $[1, 5]$. (No equation is necessary.) Sketch a graph of $y = -f(x)$ over the same interval in the same coordinate system. Explain how these figures relate to the discussion above.

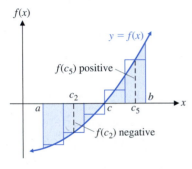

FIGURE 2
Area and Riemann sums

All the above interpretation of the definite integral relative to area, as we saw in Chapter 6, is based on the definition of the definite integral as the limit of a Riemann sum:

$$\int_a^b f(x)\, dx = \lim_{n \to \infty} \sum_{k=1}^n f(c_k)\, \Delta x_k$$

Figure 2 shows a particular Riemann sum, the midpoint sum M_5, over the interval $[a, b]$. The product $f(c_k)\, \Delta x_k$ is negative for any rectangle on $[a, c]$ and positive for any rectangle on $[c, b]$. Thus, $f(c_k)\, \Delta x_k$ represents the negative of the area of a rectangle on $[a, c]$, and $-f(c_k)\, \Delta x_k$ represents the actual area of the rectangle. Consequently, for the interval $[a, c]$, where $f(x) \le 0$,

$$\text{area} = \lim_{n \to \infty} \sum_{k=1}^n [-f(c_k)]\, \Delta x_k = \int_a^c [-f(x)]\, dx$$

But for the interval $[c, b]$, where $f(x) \ge 0$,

$$\text{area} = \lim_{n \to \infty} \sum_{k=1}^n f(c_k)\, \Delta x_k = \int_c^b f(x)\, dx$$

In summary:

Area Between a Curve and the x Axis

For a function f that is continuous over $[a, b]$, the area between $y = f(x)$ and the x axis from $x = a$ to $x = b$ can be found using definite integrals as follows:

For $f(x) \ge 0$ over $[a, b]$: $\quad$ Area $= \displaystyle\int_a^b f(x)\, dx$

For $f(x) \le 0$ over $[a, b]$: $\quad$ Area $= \displaystyle\int_a^b [-f(x)]\, dx$

If $f(x)$ is positive for some values of x and negative for others on an interval (as in Fig. 1), the area between the graph of f and the x axis can be obtained by dividing the interval into subintervals over which f is always positive or always negative, finding the area over each subinterval, and then summing these areas.

Example 1 ⇋ **Area Between a Curve and the x Axis**   Find the area bounded by $f(x) = 6x - x^2$ and $y = 0$ for $1 \leqslant x \leqslant 4$.

SOLUTION   We sketch a graph of the region first (Fig. 3). (The solution of every area problem should begin with a sketch.) Since $f(x) \geqslant 0$ on $[1, 4]$,

$$A = \int_1^4 (6x - x^2)\, dx = \left(3x^2 - \frac{x^3}{3}\right)\Big|_1^4$$

$$= \left[3(4)^2 - \frac{(4)^3}{3}\right] - \left[3(1)^2 - \frac{(1)^3}{3}\right]$$

$$= 48 - \tfrac{64}{3} - 3 + \tfrac{1}{3}$$

$$= 48 - 21 - 3$$

$$= 24$$

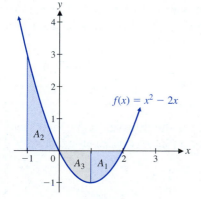

FIGURE 3

$y = f(x) = 6x - x^2$

A

Matched Problem 1 ⇋ Find the area bounded by $f(x) = x^2 + 1$ and $y = 0$ for $-1 \leqslant x \leqslant 3$.

Example 2 ⇋ **Area Between a Curve and the x Axis**   Find the area between the graph of $f(x) = x^2 - 2x$ and the x axis over the indicated intervals:

(A) $[1, 2]$ (B) $[-1, 1]$

SOLUTION   We begin by sketching the graph of f, as shown in Figure 4.

(A) From the graph, we see that $f(x) \leqslant 0$ for $1 \leqslant x \leqslant 2$, so we integrate $-f(x)$:

$$A_1 = \int_1^2 [-f(x)]\, dx$$

$$= \int_1^2 (2x - x^2)\, dx$$

$$= \left(x^2 - \frac{x^3}{3}\right)\Big|_1^2$$

$$= \left[(2)^2 - \frac{(2)^3}{3}\right] - \left[(1)^2 - \frac{(1)^3}{3}\right]$$

$$= 4 - \tfrac{8}{3} - 1 + \tfrac{1}{3} = \tfrac{2}{3} \approx 0.667$$

$f(x) = x^2 - 2x$

A_2 A_3 A_1

FIGURE 4

(B) Since the graph shows that $f(x) \geqslant 0$ on $[-1, 0]$ and $f(x) \leqslant 0$ on $[0, 1]$, the computation of this area will require two integrals:

$$A = A_2 + A_3$$

$$= \int_{-1}^{0} f(x)\, dx + \int_{0}^{1} [-f(x)]\, dx$$

$$= \int_{-1}^{0} (x^2 - 2x)\, dx + \int_{0}^{1} (2x - x^2)\, dx$$

$$= \left(\frac{x^3}{3} - x^2 \right) \Big|_{-1}^{0} + \left(x^2 - \frac{x^3}{3} \right) \Big|_{0}^{1}$$

$$= \tfrac{4}{3} + \tfrac{2}{3} = 2$$

Matched Problem 2   Find the area between the graph of $f(x) = x^2 - 9$ and the x axis over the indicated intervals:

(A) $[0, 2]$ (B) $[2, 4]$

❑   ## AREA BETWEEN TWO CURVES

FIGURE 5

Now we want to consider the area bounded by $y = f(x)$ and $y = g(x)$, where $f(x) \geqslant g(x) \geqslant 0$, for $a \leqslant x \leqslant b$, as indicated in Figure 5.

$$\left(\begin{matrix} \text{Area } A \text{ between} \\ f(x) \text{ and } g(x) \end{matrix} \right) = \left(\begin{matrix} \text{area} \\ \text{under } f(x) \end{matrix} \right) - \left(\begin{matrix} \text{area} \\ \text{under } g(x) \end{matrix} \right)$$ Areas are from $x = a$ to $x = b$ above the x axis.

$$= \int_{a}^{b} f(x)\, dx - \int_{a}^{b} g(x)\, dx$$ Use definite integral property 4 (Section 6-5).

$$= \int_{a}^{b} [f(x) - g(x)]\, dx$$

It can be shown that the preceding result does not require $f(x)$ or $g(x)$ to remain positive over the interval $[a, b]$. A more general result is stated in the box:

Area Between Two Curves

If f and g are continuous and $f(x) \geqslant g(x)$ over the interval $[a, b]$, then the area bounded by $y = f(x)$ and $y = g(x)$ for $a \leqslant x \leqslant b$ is given exactly by

$$A = \int_{a}^{b} [f(x) - g(x)]\, dx$$

Explore–Discuss 2

A Riemann sum for the integral representing the area between the graphs of $y = f(x)$ and $y = g(x)$ has the form

$$\sum_{k=1}^{n} [f(c_k) - g(c_k)]\, \Delta x_k$$

If $f(x) \geqslant g(x)$, each term in this sum represents the area of a rectangle with height $f(c_k) - g(c_k)$ and width Δx. Discuss the relationship between these rectangles and the area between the graphs of $y = f(x)$ and $y = g(x)$.

Example 3 ➭ **Area Between Two Curves**   Find the area bounded by the graphs of $f(x) = \frac{1}{2}x + 3$, $g(x) = -x^2 + 1$, $x = -2$, and $x = 1$.

SOLUTION   We first sketch the area (Fig. 6), and then set up and evaluate an appropriate definite integral. We observe from the graph that $f(x) \geqslant g(x)$ for $-2 \leqslant x \leqslant 1$, so

$$A = \int_{-2}^{1} [f(x) - g(x)]\, dx = \int_{-2}^{1} \left[\left(\frac{x}{2} + 3 \right) - (-x^2 + 1) \right] dx$$

$$= \int_{-2}^{1} \left(x^2 + \frac{x}{2} + 2 \right) dx$$

$$= \left(\frac{x^3}{3} + \frac{x^2}{4} + 2x \right) \Big|_{-2}^{1}$$

$$= \left(\frac{1}{3} + \frac{1}{4} + 2 \right) - \left(\frac{-8}{3} + \frac{4}{4} - 4 \right) = \frac{33}{4} = 8.25 \qquad ■$$

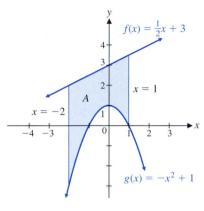

FIGURE 6

Matched Problem 3 ➭ Find the area bounded by $f(x) = x^2 - 1$, $g(x) = -\frac{1}{2}x - 3$, $x = -1$, and $x = 2$.   ■

Example 4 ➭ **Area Between Two Curves**   Find the area bounded by $f(x) = 5 - x^2$ and $g(x) = 2 - 2x$.

SOLUTION   First, graph f and g on the same coordinate system, as shown in Figure 7. Since the statement of the problem does not include any limits on the values of x, we must determine the appropriate values from the graph. The graph of f is a parabola and the graph of g is a line, as shown. The area bounded by these two graphs extends from the intersection point on the left to the intersection point on the right. To find these intersection points, we solve the equation $f(x) = g(x)$ for x:

$$f(x) = g(x)$$
$$5 - x^2 = 2 - 2x$$
$$x^2 - 2x - 3 = 0$$
$$x = -1, 3$$

You should check these values in the original equations. (Note that the area between the graphs for $x < -1$ is unbounded on the left, and the area between the graphs for $x > 3$ is unbounded on the right.) Figure 7 shows that $f(x) \geqslant g(x)$ over the interval $[-1, 3]$, so we have

FIGURE 7

$$A = \int_{-1}^{3} [f(x) - g(x)]\, dx = \int_{-1}^{3} [5 - x^2 - (2 - 2x)]\, dx$$

$$= \int_{-1}^{3} (3 + 2x - x^2)\, dx$$

$$= \left(3x + x^2 - \frac{x^3}{3} \right)\Big|_{-1}^{3}$$

$$= \left[3(3) + (3)^2 - \frac{(3)^3}{3} \right] - \left[3(-1) + (-1)^2 - \frac{(-1)^3}{3} \right] = \frac{32}{3} \approx 10.667$$

Matched Problem 4 ⮑ Find the area bounded by $f(x) = 6 - x^2$ and $g(x) = x$.

Example 5 ⮑ **Area Between Two Curves** Find the area bounded by $f(x) = x^2 - x$ and $g(x) = 2x$ for $-2 \le x \le 3$.

SOLUTION The graphs of f and g are shown in Figure 8. Examining the graph, we see that $f(x) \ge g(x)$ on the interval $[-2, 0]$, but $g(x) \ge f(x)$ on the interval $[0, 3]$. Thus, two integrals are required to compute this area:

$$A_1 = \int_{-2}^{0} [f(x) - g(x)]\, dx \qquad f(x) \ge g(x) \text{ on } [-2, 0]$$

$$= \int_{-2}^{0} [x^2 - x - 2x]\, dx$$

$$= \int_{-2}^{0} (x^2 - 3x)\, dx$$

$$= \left(\frac{x^3}{3} - \frac{3}{2} x^2 \right)\Big|_{-2}^{0}$$

FIGURE 8

$$= (0) - \left[\frac{(-2)^3}{3} - \frac{3}{2}(-2)^2 \right] = \frac{26}{3} \approx 8.667$$

$$A_2 = \int_{0}^{3} [g(x) - f(x)]\, dx \qquad g(x) \ge f(x) \text{ on } [0, 3]$$

$$= \int_{0}^{3} [2x - (x^2 - x)]\, dx$$

$$= \int_{0}^{3} (3x - x^2)\, dx$$

$$= \left(\frac{3}{2} x^2 - \frac{x^3}{3} \right)\Big|_{0}^{3}$$

$$= \left[\frac{3}{2}(3)^2 - \frac{(3)^3}{3} \right] - (0) = \frac{9}{2} = 4.5$$

The total area between the two graphs is

$$A = A_1 + A_2 = \tfrac{26}{3} + \tfrac{9}{2} = \tfrac{79}{6} \approx 13.167$$

Matched Problem 5 ⮑ Find the area bounded by $f(x) = 2x^2$ and $g(x) = 4 - 2x$ for $-2 \le x \le 2$.

Example 6 ⮑ **Computing Areas Using a Numerical Integration Routine** Find the area (to three decimal places) bounded by $f(x) = e^{-x^2}$ and $g(x) = x^2 - 1$.

SOLUTION First, we use a graphing utility to graph the functions f and g and find their intersection points (see Fig. 9A). We see that the graph of f is bell-shaped and the graph of g is a parabola, and note that $f(x) \geqslant g(x)$ on the interval $[-1.131, 1.131]$. Now we compute the area A by a numerical integration routine (see Fig. 9B):

$$A = \int_{-1.131}^{1.131} [e^{-x^2} - (x^2 - 1)]\, dx = 2.876$$

(A)   (B)

FIGURE 9

Matched Problem 6   Find the area (to three decimal places) bounded by the graphs of $f(x) = x^2 \ln x$ and $g(x) = 3x - 3$.

❑   APPLICATION: INCOME DISTRIBUTION

The U.S. Bureau of the Census compiles and analyzes a great deal of data having to do with the distribution of income among families in the United States. For 1997 the Bureau reported that the lowest 20% of families received 4% of all family income, and the top 20% received 47%. Table 1 and Figure 10 give a detailed picture of the distribution of family income in 1997.

TABLE 1

FAMILY INCOME DISTRIBUTION IN THE UNITED STATES, 1997

INCOME LEVEL	x	y
Under $21,000	0.20	0.04
Under $36,000	0.40	0.14
Under $54,000	0.60	0.30
Under $80,000	0.80	0.53

Source: U.S. Bureau of the Census

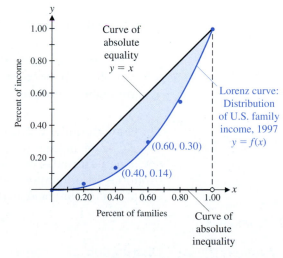

FIGURE 10 Lorenz chart

The graph of $y = f(x)$ in Figure 10 is called a **Lorenz curve** and is generally found using *regression analysis,* a technique of fitting a particular elementary function to a data set over a given interval. The variable x **represents the**

cumulative percentage of families at or below a given income level and **y represents the cumulative percentage of total family income received.** For example, data point $(0.40, 0.14)$ in Table 1 indicates that the bottom 40% of families (those with incomes under \$36,000) receive 14% of the total income for all families; data point $(0.60, 0.30)$ indicates that the bottom 60% of families receive 30% of the total income for all families; and so on.

Absolute equality of income would occur if the area between the Lorenz curve and $y = x$ were 0. In this case, the Lorenz curve would be $y = x$ and all families would receive equal shares of the total income. That is, 5% of the families would receive 5% of the income, 20% of the families would receive 20% of the income, 65% of the families would receive 65% of the income, and so on. The maximum possible area between a Lorenz curve and $y = x$ is $\frac{1}{2}$, the area of the triangle below $y = x$. In this case, we would have **absolute inequality**—all the income would be in the hands of one family and the rest would have none. In actuality, Lorenz curves lie between these two extremes. But as the shaded area increases, the greater the inequality of income distribution.

The ratio of the area bounded by $y = x$ and the Lorenz curve $y = f(x)$ to the area of the triangle under the line $y = x$ from $x = 0$ to $x = 1$ is called the **index of income concentration.** The area bounded by $y = x$ and $y = f(x)$ is given by $\int_0^1 [x - f(x)]\, dx$, and the area of the triangle below $y = x$ is $\frac{1}{2}$. Thus, we have the following:

Index of Income Concentration

If $y = f(x)$ is the equation of a Lorenz curve, then

$$\textbf{index of income concentration} = 2 \int_0^1 [x - f(x)]\, dx$$

The index of income concentration is always a number between 0 and 1:

A measure of 0 indicates absolute equality—all people share equally in the income. A measure of 1 indicates absolute inequality—one person has all the income and the rest have none.

The closer the index is to 0, the closer the income is to being equally distributed. The closer the index is to 1, the closer the income is to being concentrated in a few hands. The index of income concentration is used to compare income distributions at various points in time, between different groups of people, before and after taxes are paid, between different countries, and so on.

Example 7 ⇨ **Distribution of Income** The Lorenz curve for the distribution of income in a certain country in 1990 is given by $f(x) = x^{2.6}$. Economists predict that the Lorenz curve for the country in the year 2010 will be given by $g(x) = x^{1.8}$. Find the index of income concentration for each curve, and interpret the results.

SOLUTION   The Lorenz curves are shown in Figure 11.

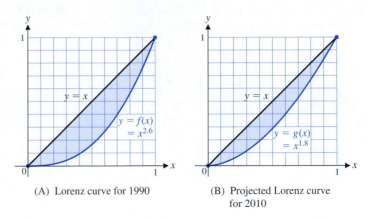

(A) Lorenz curve for 1990   (B) Projected Lorenz curve
for 2010

FIGURE 11

The index of income concentration in 1990 is (see Fig. 11A)

$$2 \int_0^1 [x - f(x)] \, dx = 2 \int_0^1 [x - x^{2.6}] \, dx = 2 \left(\frac{1}{2} x^2 - \frac{1}{3.6} x^{3.6} \right) \bigg|_0^1$$

$$= 2 \left(\frac{1}{2} - \frac{1}{3.6} \right) \approx 0.444$$

The projected index of income concentration in 2010 is (see Fig. 11B)

$$2 \int_0^1 [x - g(x)] \, dx = 2 \int_0^1 [x - x^{1.8}] \, dx = 2 \left(\frac{1}{2} x^2 - \frac{1}{2.8} x^{2.8} \right) \bigg|_0^1$$

$$= 2 \left(\frac{1}{2} - \frac{1}{2.8} \right) \approx 0.286$$

If this projection is correct, the index of income concentration will decrease, and income will be more equally distributed in the year 2010 than in 1990. ■

Matched Problem 7 ✏   Repeat Example 7 if the projected Lorenz curve in the year 2010 is given by $g(x) = x^{3.8}$. ■

Answers to Matched Problems   **1.** $A = \int_{-1}^3 (x^2 + 1) \, dx = \frac{40}{3} \approx 13.333$

2. (A) $A = \int_0^2 (9 - x^2) \, dx = \frac{46}{3} \approx 15.333$

     (B) $A = \int_2^3 (9 - x^2) \, dx + \int_3^4 (x^2 - 9) \, dx = 6$

3. $A = \int_{-1}^2 \left[(x^2 - 1) - \left(-\frac{x}{2} - 3 \right) \right] dx = \frac{39}{4} = 9.75$

4. $A = \int_{-3}^2 [(6 - x^2) - x] \, dx = \frac{125}{6} \approx 20.833$

5. $A = \int_{-2}^1 [(4 - 2x) - 2x^2] \, dx + \int_1^2 [2x^2 - (4 - 2x)] \, dx = \frac{38}{3} \approx 12.667$

6. 0.443

7. Index of income concentration ≈ 0.583; income will be less equally distributed in 2010.

Exercise 7-1

A *Problems 1–6 refer to Figures A–D. Set up definite integrals in Problems 1–4 that represent the indicated shaded area.*

(A)   (B)

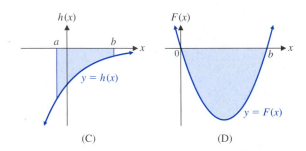

(C)   (D)

1. Shaded area in Figure B

2. Shaded area in Figure A

3. Shaded area in Figure C

4. Shaded area in Figure D

5. Explain why $\int_a^b h(x)\, dx$ does not represent the area between the graph of $y = h(x)$ and the x axis from $x = a$ to $x = b$ in Figure C.

6. Explain why $\int_a^b [-h(x)]\, dx$ represents the area between the graph of $y = h(x)$ and the x axis from $x = a$ to $x = b$ in Figure C.

In Problems 7–16, find the area bounded by the graphs of the indicated equations over the given intervals. Compute answers to three decimal places.

7. $y = -2x - 1;\ y = 0,\ 0 \le x \le 4$

8. $y = 2x - 4;\ y = 0,\ -2 \le x \le 1$

9. $y = x^2 + 2;\ y = 0,\ -1 \le x \le 0$

10. $y = 3x^2 + 1;\ y = 0,\ -2 \le x \le 0$

11. $y = x^2 - 4;\ y = 0,\ -1 \le x \le 2$

12. $y = 3x^2 - 12;\ y = 0,\ -2 \le x \le 1$

13. $y = e^x;\ y = 0,\ -1 \le x \le 2$

14. $y = e^{-x};\ y = 0,\ -2 \le x \le 1$

15. $y = -1/t;\ y = 0,\ 0.5 \le t \le 1$

16. $y = -1/t;\ y = 0,\ 0.1 \le t \le 1$

B *Problems 17–26 refer to Figures A and B. Set up definite integrals in Problems 17–24 that represent the indicated shaded areas over the given intervals.*

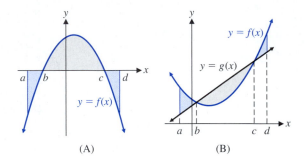

(A)   (B)

17. Over interval $[a, b]$ in Figure A

18. Over interval $[c, d]$ in Figure A

19. Over interval $[b, d]$ in Figure A

20. Over interval $[a, c]$ in Figure A

21. Over interval $[c, d]$ in Figure B

22. Over interval $[a, b]$ in Figure B

23. Over interval $[a, c]$ in Figure B

24. Over interval $[b, d]$ in Figure B

25. Referring to Figure B, explain how you would use definite integrals and the functions f and g to find the area bounded by the two functions from $x = a$ to $x = d$.

26. Referring to Figure A, explain how you would use definite integrals to find the area between the graph of $y = f(x)$ and the x axis from $x = a$ to $x = d$.

In Problems 27–42, find the area bounded by the graphs of the indicated equations over the given intervals (when stated). Compute answers to three decimal places.

27. $y = -x;\ y = 0,\ -2 \le x \le 1$

28. $y = -x + 1;\ y = 0,\ -1 \le x \le 2$

29. $y = x^2 - 4;\ y = 0,\ 0 \le x \le 3$

30. $y = 4 - x^2;\ y = 0,\ 0 \le x \le 4$

31. $y = 4 - x^2;\ y = 0,\ -3 \le x \le 4$

32. $y = x^2 - 4;\ y = 0,\ -4 \le x \le 3$

33. $y = -2x + 8;\ y = 12;\ -1 \le x \le 2$

34. $y = 2x + 6;\ y = 3;\ -1 \le x \le 2$

35. $y = 3x^2;\ y = 12$

36. $y = x^2;\ y = 9$

37. $y = 4 - x^2;\ y = -5$

38. $y = x^2 - 1;\ y = 3$

39. $y = x^2 + 1;\ y = 2x - 2;\ -1 \le x \le 2$

40. $y = x^2 - 1; y = x - 2; -2 \leq x \leq 1$

41. $y = e^{0.5x}; y = -\dfrac{1}{x}; 1 \leq x \leq 2$

42. $y = \dfrac{1}{x}; y = -e^x; 0.5 \leq x \leq 1$

 In Problems 43–46, use a graphing utility to graph the equations and find relevant intersection points. Then find the area bounded by the curves. Compute answers to three decimal places.

43. $y = 3 - 5x - 2x^2; y = 2x^2 + 3x - 2$

44. $y = 3 - 2x^2; y = 2x^2 - 4x$

45. $y = -0.5x + 2.25; y = \dfrac{1}{x}$

46. $y = x - 4.25; y = -\dfrac{1}{x}$

C *In Problems 47–54, find the area bounded by the graphs of the indicated equations over the given intervals (when stated). Compute answers to three decimal places.*

47. $y = 10 - 2x; y = 4 + 2x; 0 \leq x \leq 4$

48. $y = 3x; y = x + 5; 0 \leq x \leq 5$

49. $y = x^3; y = 4x$

50. $y = x^3 + 1; y = x + 1$

51. $y = x^3 - 3x^2 - 9x + 12; y = x + 12$

52. $y = x^3 - 6x^2 + 9x; y = x$

53. $y = x^4 - 4x^2 + 1; y = x^2 - 3$

54. $y = x^4 - 6x^2; y = 4x^2 - 9$

 In Problems 55–60, use a graphing utility to graph the equations and find relevant intersection points. Then find the area bounded by the curves. Compute answers to three decimal places.

55. $y = x^3 - x^2 + 2; y = -x^3 + 8x - 2$

56. $y = 2x^3 + 2x^2 - x; y = -2x^3 - 2x^2 + 2x$

57. $y = e^{-x}; y = 3 - 2x$

58. $y = 2 - (x + 1)^2; y = e^{x+1}$

59. $y = e^x; y = 5x - x^3$

60. $y = 2 - e^x; y = x^3 + 3x^2$

 In Problems 61–64, use a numerical integration routine on a graphing utility to find the area bounded by the graphs of the indicated equations over the given interval (when stated). Compute answers to three decimal places.

61. $y = e^{-x}; y = \sqrt{\ln x}; 2 \leq x \leq 5$

62. $y = x^2 + 3x + 1; y = e^{e^x}; -3 \leq x \leq 0$

63. $y = e^{x^2}; y = x + 2$

64. $y = \ln(\ln x); y = 0.01x$

Applications

In the following applications it is helpful to sketch graphs to get a clearer understanding of each problem and to interpret results. A graphing utility will prove useful if you have one, but it is not necessary.

Business & Economics

65. *Oil production.* Using data from the first 3 years of production as well as geological studies, the management of an oil company estimates that oil will be pumped from a producing field at a rate given by

$$R(t) = \frac{100}{t + 10} + 10 \qquad 0 \leq t \leq 15$$

where $R(t)$ is the rate of production (in thousands of barrels per year) t years after pumping begins. Find the area between the graph of R and the t axis over the interval $[5, 10]$ and interpret the results.

66. *Oil production.* In Problem 65, if the rate is found to be

$$R(t) = \frac{100t}{t^2 + 25} + 4 \qquad 0 \leq t \leq 25$$

find the area between the graph of R and the t axis over the interval $[5, 15]$ and interpret the results.

67. *Useful life.* An amusement company maintains records for each video game it installs in an arcade. Suppose that $C(t)$ and $R(t)$ represent the total accumulated costs and revenues (in thousands of dollars), respectively, t years after a particular game has been installed. If

$$C'(t) = 2 \qquad \text{and} \qquad R'(t) = 9e^{-0.3t}$$

find the area between the graphs of C' and R' over the interval on the t axis from 0 to the useful life of the game and interpret the results.

68. *Useful life.* Repeat Problem 67 if

$$C'(t) = 2t \qquad \text{and} \qquad R'(t) = 5te^{-0.1t^2}$$

69. *Income distribution.* As part of a study of the effects of World War II on the economy of the United States, an economist used data from the U.S. Bureau of the Census to produce the following Lorenz curves for distribution of income in the United States in 1935 and in 1947:

$$f(x) = x^{2.4} \qquad \text{Lorenz curve for 1935}$$
$$g(x) = x^{1.6} \qquad \text{Lorenz curve for 1947}$$

Find the index of income concentration for each Lorenz curve and interpret the results.

70. *Income distribution.* Using data from the U.S. Bureau of the Census, an economist produced the following Lorenz curves for distribution of income in the United States in 1962 and in 1972:

$$f(x) = \tfrac{3}{10}x + \tfrac{7}{10}x^2 \qquad \text{Lorenz curve for 1962}$$
$$g(x) = \tfrac{1}{2}x + \tfrac{1}{2}x^2 \qquad \text{Lorenz curve for 1972}$$

Find the index of income concentration for each Lorenz curve and interpret the results.

71. *Distribution of wealth.* Lorenz curves also can be
WWW used to provide a relative measure of the distribution of the total assets of a country. Using data in a report by the U.S. Congressional Joint Economic Committee, an economist produced the following Lorenz curves for the distribution of total assets in the United States in 1963 and in 1983:

$$f(x) = x^{10} \qquad \text{Lorenz curve for 1963}$$
$$g(x) = x^{12} \qquad \text{Lorenz curve for 1983}$$

Find the index of income concentration for each Lorenz curve and interpret the results.

72. *Income distribution.* The government of a small country is planning sweeping changes in the tax structure in order to provide a more equitable distribution of income. The Lorenz curves for the current income distribution and for the projected income distribution after enactment of the tax changes are given below. Find the index of income concentration for each Lorenz curve. Will the proposed changes provide a more equitable income distribution? Explain.

$$f(x) = x^{2.3} \qquad \text{Current Lorenz curve}$$
$$g(x) = 0.4x + 0.6x^2 \qquad \text{Projected Lorenz curve after changes in tax laws}$$

73. *Distribution of wealth.* The data in the table describe the distribution of wealth in a country:

x	0	0.20	0.40	0.60	0.80	1
y	0	0.12	0.31	0.54	0.78	1

(A) Use quadratic regression to find the equation of a Lorenz curve for the data.

(B) Use the regression equation and a numerical integration routine to approximate the index of income concentration.

74. *Distribution of wealth.* Refer to Problem 73.

(A) Use cubic regression to find the equation of a Lorenz curve for the data.

(B) Use the cubic regression equation and a numerical integration routine to approximate the index of income concentration.

Life Sciences

75. *Biology.* A yeast culture is growing at a rate of $W'(t) = 0.3e^{0.1t}$ grams per hour. Find the area between the graph of W' and the t axis over the interval $[0, 10]$ and interpret the results.

76. *Natural resource depletion.* The instantaneous rate
WWW of change of the demand for lumber in the United States since 1970 ($t = 0$) in billions of cubic feet per year is estimated to be given by

$$Q'(t) = 12 + 0.006t^2 \qquad 0 \leq t \leq 50$$

Find the area between the graph of Q' and the t axis over the interval $[15, 20]$ and interpret the results.

Social Sciences

77. *Learning.* A beginning high school language class was chosen for an experiment on learning. Using a list of 50 words, the experiment involved measuring the rate of vocabulary memorization at different times during a continuous 5-hour study session. It was found that the average rate of learning for the entire class was inversely proportional to the time spent studying and was given approximately by

$$V'(t) = \frac{15}{t} \qquad 1 \leq t \leq 5$$

Find the area between the graph of V' and the t axis over the interval $[2, 4]$ and interpret the results.

78. *Learning.* Repeat Problem 77 if $V'(t) = 13/t^{1/2}$ and the interval is changed to $[1, 4]$.

Section 7-2	# Applications in Business and Economics

- ❑ PROBABILITY DENSITY FUNCTIONS
- ❑ CONTINUOUS INCOME STREAM
- ❑ FUTURE VALUE OF A CONTINUOUS INCOME STREAM
- ❑ CONSUMERS' AND PRODUCERS' SURPLUS

This section contains a number of important applications of the definite integral from business and economics. Included are three independent topics: probability density functions, continuous income streams, and consumers' and producers' surplus. Any of the three may be covered as time and interests dictate, and in any order.

❑ PROBABILITY DENSITY FUNCTIONS

We now take a brief look at the use of the definite integral to determine probabilities. Our approach will be intuitive and informal. A more formal treatment of the subject requires use of the special "improper" integral form $\int_{-\infty}^{\infty} f(x)\, dx$, which we have not defined or discussed at this point.

Suppose that an experiment is designed in such a way that any real number x on the interval $[c, d]$ is a possible outcome. For example, x may represent an IQ score, the height of a person in inches, or the life of a light bulb in hours. Technically, we refer to x as a *continuous random variable.*

In certain situations it is possible to find a function f with x as an independent variable such that the function f can be used to determine the probability that the outcome x of an experiment will be in the interval $[c, d]$. Such a function, called a **probability density function,** must satisfy the following three conditions (see Fig. 1):

1. $f(x) \geq 0$ for all real x
2. The area under the graph of $f(x)$ over the interval $(-\infty, \infty)$ is exactly 1.
3. If $[c, d]$ is a subinterval of $(-\infty, \infty)$, then

$$\text{Probability } (c \leq x \leq d) = \int_{c}^{d} f(x)\, dx$$

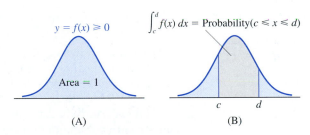

FIGURE 1 Probability density function

Let: $f(x) = \begin{cases} \frac{1}{4}e^{-x/4} & \text{if } x \geq 0 \\ 0 & \text{otherwise} \end{cases}$

(A) Explain why $f(x) \geq 0$ over the interval $(-\infty, \infty)$.

(B) Find $\int_{0}^{10} f(x)\, dx$, $\int_{0}^{20} f(x)\, dx$, and $\int_{0}^{30} f(x)\, dx$.

(C) On the basis of part (B), what would you conjecture the area under the graph of $f(x)$ over the interval $(-\infty, \infty)$ to be equal to?

f(t)

FIGURE 2

Example 1 ⇔ **Duration of Telephone Calls** Suppose that the length of telephone calls (in minutes) in a public telephone booth is a continuous random variable with probability density function shown in Figure 2.

$$f(t) = \begin{cases} \frac{1}{4}e^{-t/4} & \text{if } t \geq 0 \\ 0 & \text{otherwise} \end{cases}$$

(A) Determine the probability that a call selected at random will last between 2 and 3 minutes.

(B) Find b (to two decimal places) so that the probability of a call selected at random lasting between 2 and b minutes is .5.

Solution (A) Probability $(2 \leq t \leq 3) = \int_2^3 \frac{1}{4}e^{-t/4}\, dt$

$$= (-e^{-t/4})|_2^3$$
$$= -e^{-3/4} + e^{-1/2} \approx .13$$

(B) We want to find b such that Probability $(2 \leq t \leq b) = .5$.

$$\int_2^b \frac{1}{4}e^{-t/4}\, dt = .5$$
$$-e^{-b/4} + e^{-1/2} = .5 \qquad \textit{Solve for b.}$$
$$e^{-b/4} = e^{-.5} - .5$$
$$-\frac{b}{4} = \ln(e^{-.5} - .5)$$
$$b = 8.96 \text{ minutes}$$

Thus, the probability of a call selected at random lasting from 2 to 8.96 minutes is .5.

■

Matched Problem 1 ⇔ (A) In Example 1, find the probability that a call selected at random will last 4 minutes or less.

(B) Find b (to two decimal places) so that the probability of a call selected at random lasting b minutes or less is .9

■

$f(x) = \dfrac{1}{\sigma\sqrt{2\pi}}\, e^{-(x-\mu)^2/2\sigma^2}$

μ

FIGURE 3 Normal curve

Group Activity 2 at the end of Chapter 6 investigated one of the most important probability density functions, the **normal probability density function** defined below and graphed in Figure 3.

$$f(x) = \frac{1}{\sigma\sqrt{2\pi}}e^{-(x-\mu)^2/2\sigma^2} \qquad \begin{array}{l} \mu \text{ is the mean.} \\ \sigma \text{ is the standard deviation.} \end{array}$$

It can be shown (but not easily) that the area under the normal curve in Figure 3 over the interval $(-\infty, \infty)$ is exactly 1. Since $\int e^{-x^2}\, dx$ is noninte-grable in terms of elementary functions (that is, the antiderivative cannot be expressed as a finite combination of simple functions), probabilities such as

$$\text{Probability}(c \leq x \leq d) = \frac{1}{\sigma\sqrt{2\pi}}\int_c^d e^{-(x-\mu)^2/2\sigma^2}\, dx$$

are generally determined by making an appropriate substitution in the integrand and then using a table of areas under the standard normal curve (that is, the normal curve with $\mu = 0$ and $\sigma = 1$). Such tables are readily available in most mathematical handbooks. A table can be constructed by using a rectangle rule, as discussed in Section 6-5; however, computers that employ refined techniques are generally used for this purpose. Some calculators have the capability of computing normal curve areas directly.

❑ CONTINUOUS INCOME STREAM

We start with a simple example having an obvious solution and generalize the concept to examples having less obvious solutions.

Suppose an aunt has established a trust that pays you $2,000 a year for 10 years. What is the total amount you will receive from the trust by the end of the 10th year? Since there are 10 payments of $2,000 each, you will receive

$$10 \times \$2,000 = \$20,000$$

We now look at the same problem from a different point of view, a point of view that will be useful in more complex problems. Let us assume that the income stream is continuous at a rate of $2,000 per year. In Figure 4 the area under the graph of $f(t) = 2,000$ from 0 to t represents the income accumulated t years after the start. For example, for $t = \frac{1}{4}$ year, the income would be $\frac{1}{4}(2,000) = \$500$; for $t = \frac{1}{2}$ year, the income would be $\frac{1}{2}(2,000) = \$1,000$; for $t = 1$ year, the income would be $1(2,000) = \$2,000$; for $t = 5.3$ years, the income would be $5.3(2,000) = \$10,600$; and for $t = 10$ years, the income would be $10(2,000) = \$20,000$. The total income over a 10-year period—that is, the area under the graph of $f(t) = 2,000$ from 0 to 10—is also given by the definite integral

$$\int_0^{10} 2,000 \, dt = 2,000t \Big|_0^{10} = 2,000(10) - 2,000(0) = \$20,000$$

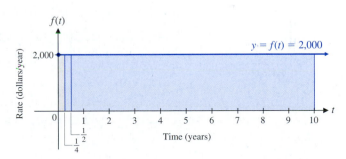

FIGURE 4 Continuous income stream

We now apply the idea of a continuous income stream to a less obvious problem.

Example 2 ↩   **Continuous Income Stream**   The rate of change of the income produced by a vending machine located at an airport is given by

$$f(t) = 5,000e^{0.04t}$$

where t is time in years since the installation of the machine. Find the total income produced by the machine during the first 5 years of operation.

SOLUTION The area under the graph of the rate of change function from 0 to 5 represents the total change in income over the first 5 years (Fig. 5), and hence is given by a definite integral:

f(t)

$$\text{Total income} = \int_0^5 5{,}000e^{0.04t}\, dt$$

$$= 125{,}000e^{0.04t}\Big|_0^5$$

$$= 125{,}000e^{0.04(5)} - 125{,}000e^{0.04(0)}$$

$$= 152{,}675 - 125{,}000$$

$$= \$27{,}675 \quad \textit{Rounded to the nearest dollar}$$

FIGURE 5 Continuous income stream

Thus, the vending machine produces a total income of \$27,675 during the first 5 years of operation.

Matched Problem 2 ⤶ Referring to Example 2, find the total income produced (to the nearest dollar) during the second 5 years of operation.

In reality, income from a vending machine is not usually received as a single payment at the end of each year, even though the rate is given as a yearly rate. Income is usually collected on a daily or weekly basis. In problems of this type it is convenient to assume that income is actually received in a **continuous stream;** that is, we assume that income is a continuous function of time and the rate of change is an instantaneous rate. The rate of change is called the **rate of flow** of the continuous income stream. In general, we have the following:

Total Income for a Continuous Income Stream

If $f(t)$ is the rate of flow of a continuous income stream, the **total income** produced during the time period from $t = a$ to $t = b$ is

$$\text{total income} = \int_a^b f(t)\, dt$$

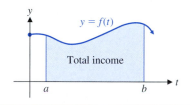

□ FUTURE VALUE OF A CONTINUOUS INCOME STREAM

In Section 5-1 we discussed the continuous compound interest formula

$$A = Pe^{rt}$$

where P is the principal (or present value), A is the amount (or future value), r is the annual rate of continuous compounding (expressed as a decimal), and t is time in years. For example, if money is worth 12% compounded continuously, then the future value of a \$10,000 investment in 5 years is (to the nearest dollar)

$$A = 10{,}000e^{0.12(5)} = \$18{,}221$$

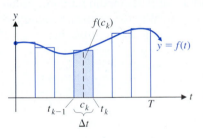

FIGURE 6

Now we want to apply the future value concept to the income produced by a continuous income stream. Suppose that $f(t)$ is the rate of flow of a continuous income stream, and the income produced by this continuous income stream is invested as soon as it is received at a rate r, compounded continuously. We already know how to find the total income produced after T years, but how can we find the total of the income produced and the interest earned by this income? Since the income is received in a continuous flow, we cannot just use the formula $A = Pe^{rt}$. This formula is valid only for a single deposit P, not for a continuous flow of income. Instead, we use a Riemann sum approach that will allow us to apply the formula $A = Pe^{rt}$ repeatedly. To begin, we divide the time interval $[0, T]$ into n equal subintervals of length Δt and choose an arbitrary point c_k in each subinterval, as illustrated in Figure 6.

The total income produced during the time period from $t = t_{k-1}$ to $t = t_k$ is equal to the area under the graph of $f(t)$ over this subinterval and is approximately equal to $f(c_k) \, \Delta t$, the area of the shaded rectangle in Figure 6. The income received during this time period will earn interest for approximately $T - c_k$ years. Thus, using the future value formula $A = Pe^{rt}$ with $P = f(c_k) \, \Delta t$ and $t = T - c_k$, the future value of the income produced during the time period from $t = t_{k-1}$ to $t = t_k$ is approximately equal to

$$f(c_k) \, \Delta t \, e^{(T-c_k)r}$$

The total of these approximate future values over n subintervals is then

$$f(c_1) \, \Delta t \, e^{(T-c_1)r} + f(c_2) \, \Delta t \, e^{(T-c_2)r} + \cdots + f(c_n) \, \Delta t \, e^{(T-c_n)r} = \sum_{k=1}^{n} f(c_k) e^{r(T-c_k)} \, \Delta t$$

This has the form of a Riemann sum, and the limit of this sum is a definite integral. (See the definition of definite integral in Section 6-5). Thus, the *future value, FV,* of the income produced by the continuous income stream is given by

$$FV = \int_0^T f(t) e^{r(T-t)} \, dt$$

Since r and T are constants, we also can write

$$FV = \int_0^T f(t) e^{rT} e^{-rt} \, dt = e^{rT} \int_0^T f(t) e^{-rt} \, dt \qquad (1)$$

This last form is preferable, since the integral is usually easier to evaluate than the first form.

Future Value of a Continuous Income Stream

If $f(t)$ is the rate of flow of a continuous income stream, $0 \le t \le T$, and if the income is continuously invested at a rate r compounded continuously, then the **future value, FV,** at the end of T years is given by

$$FV = \int_0^T f(t) e^{r(T-t)} \, dt = e^{rT} \int_0^T f(t) e^{-rt} \, dt$$

The future value of a continuous income stream is the total value of all money produced by the continuous income stream (income and interest) at the end of T years.

We return to the trust set up for you by your aunt. Suppose that the $2,000 per year you receive from the trust is invested as soon as it is received at 8% compounded continuously. We consider the trust income a continuous income stream with a flow rate of $2,000 per year. What is its future value (to the nearest dollar) by the end of the 10th year? Using the definite integral for future value from the box, we have

$$FV = e^{rT} \int_0^T f(t)e^{-rt}\, dt$$

$$FV = e^{0.08(10)} \int_0^{10} 2,000e^{-0.08t}\, dt \qquad r = 0.08,\, T = 10,\, f(t) = 2,000$$

$$= 2,000e^{0.8} \int_0^{10} e^{-0.08t}\, dt$$

$$= 2,000e^{0.8} \left[\frac{e^{-0.08t}}{-0.08} \right]\Big|_0^{10}$$

$$= -25,000e^{0.8}\left[e^{-0.08(10)} - e^{-0.08(0)}\right] = \$30,639$$

Thus, at the end of 10 years you will have received $30,639, including interest. How much is interest? Since you received $20,000 in income from the trust, the interest is the difference between the future value and income. Thus,

$$\$30,639 - \$20,000 = \$10,639$$

is the interest earned by the income received from the trust over the 10-year period.

Explore–Discuss 2

Suppose that a trust fund is set up to pay you $2,000 per year for 20 years with the continuous income stream invested at 8% compounded continuously. When will its future value be equal to $50,000?

We now apply the same analysis to Example 2, the slightly more involved vending machine problem.

Example 3 **Future Value of a Continuous Income Stream** Using the continuous income rate of flow for the vending machine in Example 2,

$$f(t) = 5,000e^{0.04t}$$

find the future value of this income stream at 12% compounded continuously for 5 years, and find the total interest earned. Compute answers to the nearest dollar.

SOLUTION Using the formula

$$FV = e^{rT} \int_0^T f(t)e^{-rt}\, dt$$

with $r = 0.12$, $T = 5$, and $f(t) = 5,000e^{0.04t}$, we have

$$FV = e^{0.12(5)} \int_0^5 5,000e^{0.04t}e^{-0.12t} \, dt$$

$$= 5,000e^{0.6} \int_0^5 e^{-0.08t} \, dt$$

$$= 5,000e^{0.6} \left(\frac{e^{-0.08t}}{-0.08} \right) \Big|_0^5$$

$$= 5,000e^{0.6}(-12.5e^{-0.4} + 12.5)$$

$$= \$37,545 \qquad \text{Rounded to the nearest dollar}$$

Thus, the future value of the income stream at 12% compounded continuously at the end of 5 years is $37,545.

In Example 2, we saw that the total income produced by this vending machine over a 5-year period was $27,675. The difference between the future value and income is interest. Thus,

$$\$37,545 - \$27,675 = \$9,870$$

is the interest earned by the income produced by the vending machine during the 5-year period.

Matched Problem 3   Repeat Example 3 if the interest rate is 9% compounded continuously.

□ CONSUMERS' AND PRODUCERS' SURPLUS

Let $p = D(x)$ be the price–demand equation for a product, where x is the number of units of the product that consumers will purchase at a price of p per unit. Suppose that $\bar{p}$ is the current price and $\bar{x}$ is the number of units that can be sold at that price. The price–demand curve in Figure 7 shows that if the price is higher than $\bar{p}$, the demand x is less than $\bar{x}$, but some consumers are still willing to pay the higher price. Consumers who are willing to pay more than $\bar{p}$ but are still able to buy the product at $\bar{p}$ have saved money. We want to determine the total amount saved by all the consumers who are willing to pay a price higher than $\bar{p}$ for this product.

To do this, consider the interval $[c_k, c_k + \Delta x]$, where $c_k + \Delta x < \bar{x}$. If the price remained constant over this interval, the savings on each unit would be the difference between $D(c_k)$, the price consumers are willing to pay, and $\bar{p}$, the price they actually pay. Since Δx represents the number of units purchased by consumers over the interval, the total savings to consumers over this interval is approximately equal to

$$[D(c_k) - \bar{p}] \, \Delta x \qquad \text{(savings per unit)} \times \text{(number of units)}$$

which is the area of the shaded rectangle shown in Figure 7. If we divide the interval $[0, \bar{x}]$ into n equal subintervals, the total savings to consumers is approximately equal to

$$[D(c_1) - \bar{p}] \, \Delta x + [D(c_2) - \bar{p}] \, \Delta x + \cdots + [D(c_n) - \bar{p}] \, \Delta x = \sum_{k=1}^{n} [D(c_k) - \bar{p}] \, \Delta x$$

which we recognize as a Riemann sum for the following integral:

$$\int_0^{\bar{x}} [D(x) - \bar{p}] \, dx$$

FIGURE 7

Thus, we define the *consumers' surplus* to be this integral.

Consumers' Surplus

If $(\bar{x}, \bar{p})$ is a point on the graph of the price–demand equation $p = D(x)$ for a particular product, then the **consumers' surplus, CS,** at a price level of $\bar{p}$ is

$$CS = \int_0^{\bar{x}} [D(x) - \bar{p}] \, dx$$

which is the area between $p = \bar{p}$ and $p = D(x)$ from $x = 0$ to $x = \bar{x}$, as shown in the margin.

The consumers' surplus represents the total savings to consumers who are willing to pay more than $\bar{p}$ for the product but are still able to buy the product for $\bar{p}$.

Example 4 **Consumers' Surplus** Find the consumers' surplus at a price level of \$8 for the price–demand equation

$$p = D(x) = 20 - 0.05x$$

SOLUTION **Step 1.** Find $\bar{x}$, the demand when the price is $\bar{p} = 8$:

$$\bar{p} = 20 - 0.05\bar{x}$$
$$8 = 20 - 0.05\bar{x}$$
$$0.05\bar{x} = 12$$
$$\bar{x} = 240$$

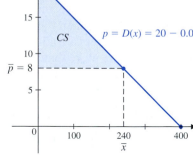

FIGURE 8

Step 2. Sketch a graph, as shown in Figure 8.

Step 3. Find the consumers' surplus (the shaded area in the graph):

$$CS = \int_0^{\bar{x}} [D(x) - \bar{p}] \, dx$$
$$= \int_0^{240} (20 - 0.05x - 8) \, dx$$
$$= \int_0^{240} (12 - 0.05x) \, dx$$
$$= (12x - 0.025x^2)\big|_0^{240}$$
$$= 2{,}880 - 1{,}440 = \$1{,}440$$

Thus, the total savings to consumers who are willing to pay a higher price for the product is \$1,440. ∎

Matched Problem 4 Repeat Example 4 for a price level of \$4. ∎

If $p = S(x)$ is the price–supply equation for a product, $\bar{p}$ is the current price, and $\bar{x}$ is the current supply, some suppliers are still willing to supply some units at a lower price than $\bar{p}$. The additional money that these suppliers gain from the higher price is called the *producers' surplus* and can be expressed in terms of a definite integral (proceeding as we did for the consumers' surplus).

Producers' Surplus

If $(\bar{x}, \bar{p})$ is a point on the graph of the price–supply equation $p = S(x)$, then the **producers' surplus, PS,** at a price level of $\bar{p}$ is

$$PS = \int_0^{\bar{x}} [\bar{p} - S(x)]\, dx$$

which is the area between $p = \bar{p}$ and $p = S(x)$ from $x = 0$ to $x = \bar{x}$, as shown in the margin.

The producers' surplus represents the total gain to producers who are willing to supply units at a lower price than $\bar{p}$ but are still able to supply units at $\bar{p}$.

Example 5   **Producers' Surplus**   Find the producers' surplus at a price level of $20 for the price–supply equation

$$p = S(x) = 2 + 0.0002x^2$$

SOLUTION   **Step 1.**   Find $\bar{x}$, the supply when the price is $\bar{p} = 20$:

$$\bar{p} = 2 + 0.0002\bar{x}^2$$
$$20 = 2 + 0.0002\bar{x}^2$$
$$0.0002\bar{x}^2 = 18$$
$$\bar{x}^2 = 90{,}000$$
$$\bar{x} = 300 \qquad \text{\textcolor{gray}{There is only one solution since } } \bar{x} \geqslant 0.$$

Step 2.   Sketch a graph, as shown in Figure 9.

Step 3.   Find the producers' surplus (the shaded area in the graph):

$$PS = \int_0^{\bar{x}} [\bar{p} - S(x)]\, dx = \int_0^{300} [20 - (2 + 0.0002x^2)]\, dx$$

$$= \int_0^{300} (18 - 0.0002x^2)\, dx = \left. \left(18x - 0.0002\frac{x^3}{3}\right) \right|_0^{300}$$

$$= 5{,}400 - 1{,}800 = \$3{,}600$$

Thus, the total gain to producers who are willing to supply units at a lower price is $3,600.

FIGURE 9

Matched Problem 5   Repeat Example 5 for a price level of $4.

In a free competitive market, the price of a product is determined by the relationship between supply and demand. If $p = D(x)$ and $p = S(x)$ are the price–demand and price–supply equations, respectively, for a product and if $(\bar{x}, \bar{p})$ is the point of intersection of these equations, $\bar{p}$ is called the **equilibrium price** and $\bar{x}$ is called the **equilibrium quantity.** If the price stabilizes at the equilibrium price $\bar{p}$, this is the price level that will determine both the consumers' surplus and the producers' surplus.

Example 6   **Equilibrium Price and Consumers' and Producers' Surplus**   Find the equilibrium price and then find the consumers' surplus and producers' surplus at the equilibrium price level if

$$p = D(x) = 20 - 0.05x \qquad \text{and} \qquad p = S(x) = 2 + 0.0002x^2$$

SOLUTION **Step 1.** Find the equilibrium point. Set $D(x)$ equal to $S(x)$ and solve:

$$D(x) = S(x)$$
$$20 - 0.05x = 2 + 0.0002x^2$$
$$0.0002x^2 + 0.05x - 18 = 0$$
$$x^2 + 250x - 90{,}000 = 0$$
$$x = 200, -450$$

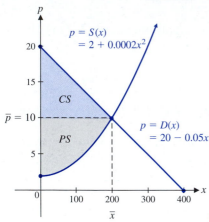

FIGURE 10

Since x cannot be negative, the only solution is $x = 200$. The equilibrium price can be determined by using $D(x)$ or $S(x)$. We will use both to check our work:

$$\bar{p} = D(200) \qquad\qquad \bar{p} = S(200)$$
$$= 20 - 0.05(200) = 10 \qquad = 2 + 0.0002(200)^2 = 10$$

Thus, the equilibrium price is $\bar{p} = 10$, and the equilibrium quantity is $\bar{x} = 200$.

Step 2. Sketch a graph, as shown in Figure 10.

Step 3. Find the consumers' surplus:

$$CS = \int_0^{\bar{x}} [D(x) - \bar{p}]\, dx$$
$$= \int_0^{200} (20 - 0.05x - 10)\, dx$$
$$= \int_0^{200} (10 - 0.05x)\, dx$$
$$= (10x - 0.025x^2)\big|_0^{200}$$
$$= 2{,}000 - 1{,}000 = \$1{,}000$$

Step 4. Find the producers' surplus:

$$PS = \int_0^{\bar{x}} [\bar{p} - S(x)]\, dx$$
$$= \int_0^{200} [10 - (2 + 0.0002x^2)]\, dx$$
$$= \int_0^{200} (8 - 0.0002x^2)\, dx$$
$$= \left(8x - 0.0002\frac{x^3}{3}\right)\bigg|_0^{200}$$
$$= 1{,}600 - \tfrac{1{,}600}{3} \approx \$1{,}067 \qquad \textit{Rounded to the nearest dollar}$$

 A graphing utility offers an alternative approach to finding the equilibrium point for Example 6 (Fig. 11A). A numerical integration routine can then be used to find the consumers' and producers' surplus (Fig. 11B).

(A)   (B)

FIGURE 11

Matched Problem 6 ✏ Repeat Example 6 for

$$p = D(x) = 25 - 0.001x^2 \quad \text{and} \quad p = S(x) = 5 + 0.1x$$

Answers to Matched Problems   **1.** (A) .63   (B) 9.21 min   **2.** $33,803
3. $FV = \$34,691$; interest $= \$7,016$ **4.** $2,560   **5.** $133
6. $\bar{p} = 15$; $CS = \$667$; $PS = \$500$

Exercise 7-2

A *In Problems 1–4, evaluate each definite integral to two decimal places.*

1. $\int_0^5 e^{-0.08t}\, dt$

2. $\int_0^5 e^{0.08(5-t)}\, dt$

3. $\int_0^{30} e^{0.06t}\, e^{0.12(30-t)}\, dt$

4. $\int_0^{20} 1{,}000e^{0.03t}\, e^{0.15(20-t)}\, dt$

5. (A) $\int_0^8 e^{0.07(8-t)}\, dt$ (B) $\int_0^8 (e^{0.56} - e^{0.07t})\, dt$

(C) $e^{0.56} \int_0^8 e^{-0.07t}\, dt$

6. (A) $\int_0^{10} 2{,}000e^{0.05t}e^{0.12(10-t)}\, dt$

(B) $2{,}000e^{1.2} \int_0^{10} e^{-0.07t}\, dt$

(C) $2{,}000e^{0.05} \int_0^{10} e^{0.12(10-t)}\, dt$

B *In Problems 5 and 6, explain which of (A), (B), and (C) are equal before evaluating the expressions. Then evaluate each expression to two decimal places.*

Applications

Business & Economics

Unless stated to the contrary, compute all monetary answers to the nearest dollar.

7. The life expectancy (in years) of a certain brand of clock radio is a continuous random variable with probability density function

$$f(x) = \begin{cases} 2/(x+2)^2 & \text{if } x \geqslant 0 \\ 0 & \text{otherwise} \end{cases}$$

(A) Find the probability that a randomly selected clock radio lasts at most 6 years.

(B) Find the probability that a randomly selected clock radio lasts from 6 to 12 years.

(C) Graph $y = f(x)$ for $[0, 12]$ and show the shaded region for part (A).

8. The shelf life (in years) of a certain brand of flashlight batteries is a continuous random variable with probability density function

$$f(x) = \begin{cases} 1/(x+1)^2 & \text{if } x \geqslant 0 \\ 0 & \text{otherwise} \end{cases}$$

(A) Find the probability that a randomly selected battery has a shelf life of 3 years or less.

(B) Find the probability that a randomly selected battery has a shelf life of from 3 to 9 years.

(C) Graph $y = f(x)$ for $[0, 10]$ and show the shaded region for part (A).

9. In Problem 7, find d so that the probability of a randomly selected clock radio lasting d years or less is .8.

10. In Problem 8, find d so that the probability of a randomly selected battery lasting d years or less is .5.

11. A manufacturer guarantees a product for 1 year. The time to failure of the product after it is sold is given by the probability density function

$$f(t) = \begin{cases} .01e^{-.01t} & \text{if } t \geqslant 0 \\ 0 & \text{otherwise} \end{cases}$$

where t is time in months. What is the probability that a buyer chosen at random will have a product failure:

(A) During the warranty period?

(B) During the second year after purchase?

12. In a certain city, the daily use of water (in hundreds of gallons) per household is a continuous random variable with probability density function

$$f(x) = \begin{cases} .15e^{-.15x} & \text{if } x \geqslant 0 \\ 0 & \text{otherwise} \end{cases}$$

Find the probability that a household chosen at random will use:

(A) At most 400 gallons of water per day

(B) Between 300 and 600 gallons of water per day

13. In Problem 11, what is the probability that the product will last at least 1 year? [*Hint:* Recall that the total area under the probability density function curve is 1.]

14. In Problem 12, what is the probability that a household will use more than 400 gallons of water per day? [See the hint in Problem 13.]

15. Find the total income produced by a continuous income stream in the first 5 years if the rate of flow is $f(t) = 2,500$.

16. Find the total income produced by a continuous income stream in the first 10 years if the rate of flow is $f(t) = 3,000$.

17. Interpret the results in Problem 15 with both a graph and a verbal description of the graph.

18. Interpret the results in Problem 16 with both a graph and a verbal description of the graph.

19. Find the total income produced by a continuous income stream in the first 3 years if the rate of flow is $f(t) = 400e^{0.05t}$.

20. Find the total income produced by a continuous income stream in the first 2 years if the rate of flow is $f(t) = 600e^{0.06t}$.

21. Interpret the results in Problem 19 with both a graph and a verbal description of the graph.

22. Interpret the results in Problem 20 with both a graph and a verbal description of the graph.

23. Starting at age 25, you deposit $2,000 a year into an **WWW** IRA account for retirement. Treat the yearly deposits into the account as a continuous income stream. If money in the account earns 5% compounded continuously, how much will be in the account 40 years later when you retire at age 65? How much of the final amount is interest?

24. Suppose in Problem 23 that you start the IRA deposits at age 30, but the account earns 6% compounded continuously. Treat the yearly deposits into the account as a continuous income stream. How much will be in the account 35 years later when you retire at age 65? How much of the final amount is interest?

25. Find the future value at 6.25% interest compounded continuously for 4 years for the continuous income stream with rate of flow $f(t) = 1,650e^{-0.02t}$.

26. Find the future value at 5.75% interest compounded continuously for 6 years for the continuous income stream with rate of flow $f(t) = 2,000e^{0.06t}$.

27. Compute the interest earned in Problem 25.

28. Compute the interest earned in Problem 26.

29. An investor is presented with a choice of two investments, an established clothing store and a new computer store. Each choice requires the same initial investment and each produces a continuous income stream of 10% compounded continuously. The rate of flow of income from the clothing store is $f(t) = 12,000$, and the rate of flow of income from the computer store is expected to be $g(t) = 10,000e^{0.05t}$. Compare the future values of these investments to determine which is the better choice over the next 5 years.

30. Refer to Problem 29. Which investment is the better choice over the next 10 years?

31. An investor has $10,000 to invest in either a bond that matures in 5 years or a business that will produce a continuous stream of income over the next 5 years with rate of flow $f(t) = 2,000$. If both the bond and the continuous income stream earn 8% compounded continuously, which is the better investment?

32. Refer to Problem 31. Which is the better investment if the rate of the income from the business is $f(t) = 3,000$?

33. A business is planning to purchase a piece of equipment that will produce a continuous stream of income for 8 years with rate of flow $f(t) = 9,000$. If the continuous income stream earns 6.95% compounded continuously, what single deposit into an account earning the same interest rate will produce the same future value as the continuous income stream? (This deposit is called the **present value** of the continuous income stream.)

34. Refer to Problem 33. Find the present value of a continuous income stream at 7.65% compounded continuously for 12 years if the rate of flow is $f(t) = 1,000e^{0.03t}$.

35. Find the future value at a rate r compounded continuously for T years for a continuous income stream with rate of flow $f(t) = k$, where k is a constant.

36. Find the future value at a rate r compounded continuously for T years for a continuous income stream with rate of flow $f(t) = ke^{ct}$, where c and k are constants, $c \neq r$.

37. Find the consumers' surplus at a price level of $\bar{p} = \$150$ for the price–demand equation

$$p = D(x) = 400 - 0.05x$$

38. Find the consumers' surplus at a price level of $\bar{p} = \$120$ for the price–demand equation

$$p = D(x) = 200 - 0.02x$$

39. Interpret the results in Problem 37 with both a graph and a verbal description of the graph.

40. Interpret the results in Problem 38 with both a graph and a verbal description of the graph.

41. Find the producers' surplus at a price level of $\bar{p} = \$67$ for the price–supply equation

$$p = S(x) = 10 + 0.1x + 0.0003x^2$$

42. Find the producers' surplus at a price level of $\bar{p} = \$55$ for the price–supply equation

$$p = S(x) = 15 + 0.1x + 0.003x^2$$

43. Interpret the results in Problem 41 with both a graph and a verbal description of the graph.

44. Interpret the results in Problem 42 with both a graph and a verbal description of the graph.

In Problems 45–52, find the consumers' surplus and the producers' surplus at the equilibrium price level for the given price–demand and price–supply equations. Include a graph that identifies the consumers' surplus and the producers' surplus. Round all values to the nearest integer.

45. $p = D(x) = 50 - 0.1x; p = S(x) = 11 + 0.05x$

46. $p = D(x) = 25 - 0.004x^2; p = S(x) = 5 + 0.004x^2$

47. $p = D(x) = 80e^{-0.001x}; p = S(x) = 30e^{0.001x}$

48. $p = D(x) = 185e^{-0.005x}; p = S(x) = 25e^{0.005x}$

49. $p = D(x) = 80 - 0.04x; p = S(x) = 30e^{0.001x}$

50. $p = D(x) = 190 - 0.2x; p = S(x) = 25e^{0.005x}$

51. $p = D(x) = 80e^{-0.001x}; p = S(x) = 15 + 0.0001x^2$

52. $p = D(x) = 185e^{-0.005x}; p = S(x) = 20 + 0.002x^2$

53. The tables give price–demand and price–supply data for the sale of soybeans at a grain market, where x is the number of bushels of soybeans (in thousands of bushels) and p is the price per bushel (in dollars):

Price–Demand		Price–Supply	
x	$p = D(x)$	x	$p = S(x)$
0	6.70	0	6.43
10	6.59	10	6.45
20	6.52	20	6.48
30	6.47	30	6.53
40	6.45	40	6.62

Use quadratic regression to model the price–demand data and linear regression to model the price–supply data.

(A) Find the equilibrium quantity (to three decimal places) and equilibrium price (to the nearest cent).

(B) Use a numerical integration routine to find the consumers' surplus and producers' surplus at the equilibrium price level.

54. Repeat Problem 53 using quadratic regression to model both sets of data.

Section 7-3   # Integration by Parts

In Section 6-1 we said we would return later to the indefinite integral

$$\int \ln x \, dx$$

since none of the integration techniques considered up to that time could be used to find an antiderivative for $\ln x$. We now develop a very useful technique, called *integration by parts,* that will enable us to find not only the integral above, but also many others, including integrals such as

$$\int x \ln x \, dx \quad \text{and} \quad \int xe^x \, dx$$

The technique of integration by parts is also used to derive many integration formulas that are tabulated in mathematical handbooks. Some of these handbook formulas are discussed in the next section.

The method of integration by parts is based on the product formula for derivatives. If f and g are differentiable functions, then

$$\frac{d}{dx}[f(x)g(x)] = f(x)g'(x) + g(x)f'(x)$$

which can be written in the equivalent form

$$f(x)g'(x) = \frac{d}{dx}[f(x)g(x)] - g(x)f'(x)$$

Integrating both sides, we obtain

$$\int f(x)g'(x)\,dx = \int \frac{d}{dx}[f(x)g(x)]\,dx - \int g(x)f'(x)\,dx$$

The first integral to the right of the equal sign is $f(x)g(x) + C$. (Why?) We will leave out the constant of integration for now, since we can add it after integrating the second integral to the right of the equal sign. So we have

$$\int f(x)g'(x)\,dx = f(x)g(x) - \int g(x)f'(x)\,dx$$

This equation can be transformed into a more convenient form by letting $u = f(x)$ and $v = g(x)$; then $du = f'(x)\,dx$ and $dv = g'(x)\,dx$. Making these substitutions, we obtain the **integration by parts formula:**

Integration by Parts Formula

$$\int u\,dv = uv - \int v\,du$$

This formula can be very useful when the integral on the left is difficult or impossible to integrate using standard formulas. If u and dv are chosen with care—this is the crucial part of the process—then the integral on the right side may be easier to integrate than the one on the left. The formula provides us with another tool that is helpful in many, but not all, cases. We are able to easily check the results by differentiating to get the original integrand, a good habit to develop. Several examples will demonstrate the use of the formula.

Example 1 ✏ **Integration by Parts** Find $\int xe^x\,dx$ using integration by parts and check the result.

SOLUTION First, write the integration by parts formula:

$$\int u\,dv = uv - \int v\,du \tag{1}$$

Now try to identify u and dv in $\int xe^x\,dx$ so that the integral $\int v\,du$ on the right side of (1) is easier to integrate than $\int u\,dv = \int xe^x\,dx$ on the left side. There are essentially two reasonable choices in selecting u and dv in $\int xe^x\,dx$:

$$
\begin{array}{cc}
\text{Choice 1} & \text{Choice 2} \\
u \quad dv & u \quad dv \\
\int x\ e^x\,dx & \int e^x\ x\,dx
\end{array}
$$

We pursue choice 1 and leave choice 2 for you to explore (see Explore–Discuss 1 following this example).

From choice 1, $u = x$ and $dv = e^x dx$. Looking at formula (1), we need du and v to complete the right side. It is convenient to proceed with the following arrangement: Let

$$u = x \qquad dv = e^x \, dx$$

Then

$$du = dx \qquad \int dv = \int e^x \, dx$$
$$v = e^x$$

Any constant may be added to v, but we will always choose 0 for simplicity. The general arbitrary constant of integration will be added at the end of the process.

Substituting these results in formula (1), we obtain

$$\int u \, dv = uv - \int v \, du$$
$$\int xe^x \, dx = xe^x - \int e^x \, dx \qquad \text{The right integral is easy to integrate.}$$
$$= xe^x - e^x + C \qquad \text{Now add the arbitrary constant } C.$$

Check: $\dfrac{d}{dx}(xe^x - e^x + C) = xe^x + e^x - e^x = xe^x$

■

Explore–Discuss 1

Pursue choice 2 in Example 1 using the integration by parts formula, and explain why this choice does not work out.

Matched Problem 1   ⇨   Find: $\displaystyle\int xe^{2x} \, dx$

■

Example 2   ⇨   **Integration by Parts**   Find: $\displaystyle\int x \ln x \, dx$

SOLUTION   As before, we have essentially two choices in choosing u and dv:

Choice 1   Choice 2

$u \quad dv$ $u \quad dv$

$\displaystyle\int x \, \ln x \, dx$ $\displaystyle\int \ln x \, x \, dx$

Choice 1 is rejected, since we do not yet know how to find an antiderivative of $\ln x$. So we move to choice 2 and choose $u = \ln x$ and $dv = x \, dx$; then proceed as in Example 1. Let

$$u = \ln x \qquad dv = x \, dx$$

Then

$$du = \frac{1}{x}dx \qquad \int dv = \int x\, dx$$

$$v = \frac{x^2}{2}$$

Substitute these results into the integration by parts formula:

$$\int u\, dv = uv - \int v\, du$$

$$\int x \ln x\, dx = (\ln x)\left(\frac{x^2}{2}\right) - \int \left(\frac{x^2}{2}\right)\left(\frac{1}{x}\right) dx$$

$$= \frac{x^2}{2} \ln x - \int \frac{x}{2}\, dx \qquad \text{\textcolor{blue}{An easy integral to evaluate}}$$

$$= \frac{x^2}{2} \ln x - \frac{x^2}{4} + C$$

Check: $\quad \dfrac{d}{dx}\left(\dfrac{x^2}{2} \ln x - \dfrac{x^2}{4} + C\right) = x \ln x + \left(\dfrac{x^2}{2} \cdot \dfrac{1}{x}\right) - \dfrac{x}{2} = x \ln x$

Matched Problem 2 ✐ Find: $\int x \ln 2x\, dx$

As you should have discovered in Explore–Discuss 1, some choices for u and dv will lead to integrals that are more complicated than the original integral. This does not mean that there is an error in the calculations or the integration by parts formula. It simply means that the particular choice of u and dv does not change the problem into one we can solve. When this happens, we must look for a different choice of u and dv. In some problems, it is possible that no choice will work. These observations and some guidelines for selecting u and dv are summarized in the box.

Integration by Parts: Selection of u and dv

For $\int u\, dv = uv - \int v\, du$:

1. The product $u\, dv$ must equal the original integrand.
2. It must be possible to integrate dv (preferably by using standard formulas or simple substitutions).
3. The new integral $\int v\, du$ should not be any more involved than the original integral $\int u\, dv$.
4. For integrals involving $x^p e^{ax}$, try

$$u = x^p \qquad \text{and} \qquad dv = e^{ax}\, dx$$

5. For integrals involving $x^p (\ln x)^q$, try

$$u = (\ln x)^q \qquad \text{and} \qquad dv = x^p\, dx$$

In some cases, repeated use of the integration by parts formula will lead to the evaluation of the original integral. The next example provides an illustration of such a case.

Example 3 ✏ **Repeated Use of Integration by Parts**   Find: $\int x^2 e^{-x}\, dx$

SOLUTION   Following suggestion 4 in the box, we choose

$$u = x^2 \qquad dv = e^{-x}\, dx$$

Then,

$$du = 2x\, dx \qquad v = -e^{-x}$$

and

$$x^2 e^{-x}\, dx = x^2(-e^{-x}) - \int (-e^{-x})2x\, dx$$

$$= -x^2 e^{-x} + 2\int xe^{-x}\, dx \tag{2}$$

The new integral is not one we can evaluate by standard formulas, but it is simpler than the original integral. Applying the integration by parts formula to it will produce an even simpler integral. For the integral $\int xe^{-x}\, dx$, we choose

$$u = x \qquad dv = e^{-x}\, dx$$

Then,

$$du = dx \qquad v = -e^{-x}$$

and

$$\int xe^{-x}\, dx = x(-e^{-x}) - \int (-e^{-x})\, dx$$

$$= -xe^{-x} + \int e^{-x}\, dx$$

$$= -xe^{-x} - e^{-x} \qquad \text{Choose 0 for the constant.} \tag{3}$$

Substituting equation (3) into equation (2), we have

$$\int x^2 e^{-x}\, dx = -x^2 e^{-x} + 2(-xe^{-x} - e^{-x}) + C \qquad \text{Add an arbitrary constant here.}$$

$$= -x^2 e^{-x} - 2xe^{-x} - 2e^{-x} + C$$

Check: $\dfrac{d}{dx}(-x^2 e^{-x} - 2xe^{-x} - 2e^{-x} + C) = x^2 e^{-x} - 2xe^{-x} + 2xe^{-x} - 2e^{-x} + 2e^{-x}$

$$= x^2 e^{-x}$$

■

Matched Problem 3 ✏ Find: $\int x^2 e^{2x}\, dx$

■

Example 4 ✏ **Using Integration by Parts**   Find $\int_1^e \ln x\, dx$ and interpret geometrically.

SOLUTION   First, we will find $\int \ln x\, dx$, and then return to the definite integral. Following suggestion 5 in the box (with $p = 0$), we choose

$$u = \ln x \qquad dv = dx$$

Then,

$$du = \frac{1}{x}dx \qquad v = x$$

Hence,

$$\int \ln x \, dx = (\ln x)(x) - \int (x)\frac{1}{x}dx$$
$$= x \ln x - x + C$$

Note that this is the important result we mentioned at the beginning of this section. Now, we have

$$\int_1^e \ln x \, dx = (x \ln x - x)\Big|_1^e$$
$$= (e \ln e - e) - (1 \ln 1 - 1)$$
$$= (e - e) - (0 - 1)$$
$$= 1$$

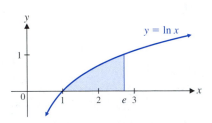

FIGURE 1

The integral represents the area under the curve $y = \ln x$ from $x = 1$ to $x = e$, as shown in Figure 1. ∎

Matched Problem 4 ⇨   Find: $\int_1^2 \ln 3x \, dx$ ∎

Explore–Discuss 2

Try using the integration by parts formula for $\int e^{x^2} \, dx$ and explain why it does not solve the problem.

Answer to Matched Problems

1. $\dfrac{x}{2}e^{2x} - \dfrac{1}{4}e^{2x} + C$ **2.** $\dfrac{x^2}{2}\ln 2x - \dfrac{x^2}{4} + C$

3. $\dfrac{x^2}{2}e^{2x} - \dfrac{x}{2}e^{2x} + \dfrac{1}{4}e^{2x} + C$ **4.** $2 \ln 6 - \ln 3 - 1 \approx 1.4849$

Exercise 7-3

A *In Problems 1–4, integrate using the method of integration by parts. Assume that $x > 0$ whenever the natural logarithm function is involved.*

1. $\displaystyle\int xe^{3x} \, dx$ **2.** $\displaystyle\int xe^{4x} \, dx$

3. $\displaystyle\int x^2 \ln x \, dx$ **4.** $\displaystyle\int x^3 \ln x \, dx$

B

5. If you want to use integration by parts to find $\int (x + 1)^5(x + 2) \, dx$, which is the better choice for u: $u = (x + 1)^5$ or $u = x + 2$? Explain your choice, and then integrate.

6. If you want to use integration by parts to find $\int (5x - 7)(x - 1)^4 \, dx$, which is the better choice for u: $u = 5x - 7$ or $u = (x - 1)^4$? Explain your choice, and then integrate.

Problems 7–20 are mixed—some require integration by parts and others can be solved using techniques we have considered earlier. Integrate as indicated, assuming $x > 0$ whenever the natural logarithm function is involved.

7. $\int xe^{-x} \, dx$ **8.** $\int (x - 1)e^{-x} \, dx$

9. $\int xe^{x^2} \, dx$ **10.** $\int xe^{-x^2} \, dx$

11. $\int_0^1 (x - 3)e^x \, dx$ **12.** $\int_0^1 (x + 1)e^x \, dx$

13. $\int_1^3 \ln 2x \, dx$ **14.** $\int_1^2 \ln\left(\dfrac{x}{2}\right) dx$

15. $\int \dfrac{2x}{x^2 + 1} \, dx$ **16.** $\int \dfrac{x^2}{x^3 + 5} \, dx$

17. $\int \dfrac{\ln x}{x} \, dx$ **18.** $\int \dfrac{e^x}{e^x + 1} \, dx$

19. $\int \sqrt{x} \ln x \, dx$ **20.** $\int \dfrac{\ln x}{\sqrt{x}} \, dx$

In Problems 21–24, illustrate each integral graphically and describe what the integral represents in terms of areas.

21. Problem 11   **22.** Problem 12

23. Problem 13   **24.** Problem 14

Problems 25–42 are mixed—some may require use of the integration by parts formula along with techniques we have considered earlier; others may require repeated use of the integration by parts formula. Assume that $g(x) > 0$ whenever $\ln g(x)$ is involved.

25. $\int x^2 e^x \, dx$ **26.** $\int x^3 e^x \, dx$

27. $\int xe^{ax} \, dx, \, a \neq 0$ **28.** $\int \ln(ax) \, dx, \, a > 0$

29. $\int_1^e \dfrac{\ln x}{x^2} \, dx$ **30.** $\int_1^2 x^3 e^{x^2} \, dx$

31. $\int_0^2 \ln(x + 4) \, dx$ **32.** $\int_0^2 \ln(4 - x) \, dx$

33. $\int xe^{x-2} \, dx$ **34.** $\int xe^{x+1} \, dx$

35. $\int x \ln(1 + x^2) \, dx$ **36.** $\int x \ln(1 + x) \, dx$

37. $\int e^x \ln(1 + e^x) \, dx$ **38.** $\int \dfrac{\ln(1 + \sqrt{x})}{\sqrt{x}} \, dx$

39. $\int (\ln x)^2 \, dx$ **40.** $\int x(\ln x)^2 \, dx$

41. $\int (\ln x)^3 \, dx$ **42.** $\int x(\ln x)^3 \, dx$

 In Problems 43–46, use a graphing utility to graph each equation over the indicated interval, and find the area between the curve and the x axis over that interval. Find answers to two decimal places.

43. $y = x - 2 - \ln x$; $1 \leqslant x \leqslant 4$

44. $y = 6 - x^2 - \ln x$; $1 \leqslant x \leqslant 4$

45. $y = 5 - xe^x$; $0 \leqslant x \leqslant 3$

46. $y = xe^x + x - 6$; $0 \leqslant x \leqslant 3$

Applications

Business & Economics

47. *Profit.* If the marginal profit (in millions of dollars per year) is given by

$$P'(t) = 2t - te^{-t}$$

find the total profit earned over the first 5 years of operation (to the nearest million dollars) by use of an appropriate definite integral.

48. *Production.* An oil field is estimated to produce oil at a rate of $R(t)$ thousand barrels per month t months from now, as given by

$$R(t) = 10te^{-0.1t}$$

Find the total production in the first year of operation (to the nearest thousand barrels) by use of an appropriate definite integral.

49. *Profit.* Interpret the results in Problem 47 with both a graph and a verbal description of the graph.

50. *Production.* Interpret the results in Problem 48 with both a graph and a verbal description of the graph.

51. *Continuous income stream.* Find the future value at 8% compounded continuously for 5 years of a continuous income stream with a rate of flow of

$$f(t) = 1,000 - 200t$$

52. *Continuous income stream.* Find the interest earned at 10% compounded continuously for 4 years for a continuous income stream with a rate of flow of

$$f(t) = 1,000 - 250t$$

53. *Income distribution.* Find the index of income concentration for the Lorenz curve with equation

$$y = xe^{x-1}$$

54. *Income distribution.* Find the index of income concentration for the Lorenz curve with equation

$$y = x^2 e^{x-1}$$

55. *Income distribution.* Interpret the results in Problem 53 with both a graph and a verbal description of the graph.

56. *Income distribution.* Interpret the results in Problem 54 with both a graph and a verbal description of the graph.

57. *Sales analysis.* The monthly sales of a particular personal computer are expected to decline at the rate of

$$S'(t) = -4te^{0.1t}$$

computers per month, where t is time in months and $S(t)$ is the number of computers sold each month. The company plans to stop manufacturing this computer when the monthly sales reach 800 computers. If the monthly sales now ($t = 0$) are 2,000 computers, find $S(t)$. How long, to the nearest month, will the company continue to manufacture this computer?

58. *Sales analysis.* The rate of change of the monthly sales of a new home video game cartridge is given by

$$S'(t) = 350 \ln(t + 1) \qquad S(0) = 0$$

where t is the number of months since the game was released and $S(t)$ is the number of cartridges sold each month. Find $S(t)$. When, to the nearest month, will the monthly sales reach 15,000 cartridges?

59. *Consumers' surplus.* Find the consumers' surplus (to the nearest dollar) at a price level of $\bar{p} = \$2.089$ for the price–demand equation

$$p = D(x) = 9 - \ln(x + 4)$$

Use $\bar{x}$ computed to the nearest higher unit.

60. *Producers' surplus.* Find the producers' surplus (to the nearest dollar) at a price level of $\bar{p} = \$26$ for the price–supply equation

$$p = S(x) = 5 \ln(x + 1)$$

Use $\bar{x}$ computed to the nearest higher unit.

61. *Consumers' surplus.* Interpret the results in Problem 59 with both a graph and a verbal description of the graph.

62. *Producers' surplus.* Interpret the results in Problem 60 with both a graph and a verbal description of the graph.

Life Sciences

63. *Pollution.* The concentration of particulate matter (in parts per million) t hours after a factory ceases operation for the day is given by

$$C(t) = \frac{20 \ln(t + 1)}{(t + 1)^2}$$

Find the average concentration for the time period from $t = 0$ to $t = 5$.

64. *Medicine.* After a person takes a pill, the drug contained in the pill is assimilated into the bloodstream. The rate of assimilation t minutes after taking the pill is

$$R(t) = te^{-0.2t}$$

Find the total amount of the drug that is assimilated into the bloodstream during the first 10 minutes after the pill is taken.

Social Sciences

65. *Learning.* In a particular business college, it was found that an average student enrolled in an advanced typing class progressed at a rate of

$$N'(t) = (t + 6)e^{-0.25t}$$

words per minute per week, t weeks after enrolling in a 15-week course. If at the beginning of the course a student could type 40 words per minute, how many words per minute, $N(t)$, would the student be expected to type t weeks into the course? How long, to the nearest week, should it take the student to achieve the 70-word per minute level? How many words per minute should the student be able to type by the end of the course?

66. *Learning.* In the same business college, it was also found that an average student enrolled in a beginning shorthand class progressed at a rate of

$$N'(t) = (t + 10)e^{-0.1t}$$

words per minute per week, t weeks after enrolling in a 15-week course. If at the beginning of the course a student had no knowledge of shorthand (that is, could take dictation in shorthand at 0 words per minute), how many words per minute, $N(t)$, would the student be expected to handle t weeks into the course? How long, to the nearest week, should it take the student to achieve 90 words per minute? How many words per minute should the student be able to handle by the end of the course?

67. *Politics.* The number of voters (in thousands) in a certain city is given by

$$N(t) = 20 + 4t - 5te^{-0.1t}$$

where t is time in years. Find the average number of voters during the time period from $t = 0$ to $t = 5$.

<table>
<tr><td>Section 7-4</td><td>

Integration Using Tables

</td></tr>
</table>

- ❑ USING A TABLE OF INTEGRALS
- ❑ SUBSTITUTION AND INTEGRAL TABLES
- ❑ REDUCTION FORMULAS
- ❑ APPLICATION

A **table of integrals** is a list of integration formulas used to evaluate integrals. People who frequently evaluate complex integrals may refer to tables that contain hundreds of formulas. Tables of this type are included in mathematical handbooks available in most college bookstores. Table II of Appendix C contains a short list of integral formulas illustrating the types found in more extensive tables. Some of these formulas can be derived using the integration techniques discussed earlier, while others require techniques we have not considered. However, it is possible to verify each formula by differentiating the right side.

❑ USING A TABLE OF INTEGRALS

The formulas in Table II (and in larger integral tables) are organized by categories, such as "Integrals Involving $a + bu$," "Integrals Involving $\sqrt{u^2 - a^2}$," and so on. The variable u is the variable of integration. All other symbols represent constants. To use a table to evaluate an integral, you must first find the category that most closely agrees with the form of the integrand and then find a formula in that category that can be made to match the integrand exactly by assigning values to the constants in the formula. The following examples illustrate this process.

Example 1 ➭ **Integration Using Tables** Use Table II to find: $\displaystyle\int \frac{x}{(5 + 2x)(4 - 3x)}\, dx$

SOLUTION Since the integrand

$$f(x) = \frac{x}{(5 + 2x)(4 - 3x)}$$

is a rational function involving terms of the form $a + bu$ and $c + du$, we examine formulas 15 to 20 in Table II to see if any of the integrands in these formulas can be made to match $f(x)$ exactly. Comparing the integrand in formula 16 with $f(x)$, we see that this integrand will match $f(x)$ if we let $a = 5, b = 2, c = 4$, and $d = -3$. Letting $u = x$ and substituting for a, b, c, and d in formula 16, we have

$$\int \frac{u}{(a + bu)(c + du)}\, du = \frac{1}{ad - bc}\left(\frac{a}{b}\ln|a + bu| - \frac{c}{d}\ln|c + du|\right) \qquad \text{Formula 16}$$

$$\int \frac{x}{(5 + 2x)(4 - 3x)}\, dx = \frac{1}{5\cdot(-3) - 2\cdot 4}\left(\frac{5}{2}\ln|5 + 2x| - \frac{4}{-3}\ln|4 - 3x|\right) + C$$

$$\qquad\quad a\quad\ b\quad c\quad\ d \qquad\qquad\qquad a\cdot d - b\cdot c = 5\cdot(-3) - 2\cdot 4 = -23$$

$$= -\tfrac{5}{46}\ln|5 + 2x| - \tfrac{4}{69}\ln|4 - 3x| + C$$

Notice that the constant of integration C is not included in any of the formulas in Table II. However, you must still include C in all antiderivatives. ∎

Matched Problem 1 ⇔ Use Table II to find: $\displaystyle\int \frac{1}{(5 + 3x)^2(1 + x)}\,dx$

Example 2 ⇔ **Integration Using Tables** Evaluate: $\displaystyle\int_3^4 \frac{1}{x\sqrt{25 - x^2}}\,dx$

SOLUTION First, we use Table II to find

$$\int \frac{1}{x\sqrt{25 - x^2}}\,dx$$

Since the integrand involves the expression $\sqrt{25 - x^2}$, we examine formulas 29 to 31 and select formula 29 with $a^2 = 25$ and $a = 5$:

$$\int \frac{1}{u\sqrt{a^2 - u^2}}\,du = -\frac{1}{a}\ln\left|\frac{a + \sqrt{a^2 - u^2}}{u}\right| \qquad \text{Formula 29}$$

$$\int \frac{1}{x\sqrt{25 - x^2}}\,dx = -\frac{1}{5}\ln\left|\frac{5 + \sqrt{25 - x^2}}{x}\right| + C$$

Thus,

$$\int_3^4 \frac{1}{x\sqrt{25 - x^2}}\,dx = -\frac{1}{5}\ln\left|\frac{5 + \sqrt{25 - x^2}}{x}\right|\Big|_3^4$$

$$= -\frac{1}{5}\ln\left|\frac{5 + 3}{4}\right| + \frac{1}{5}\ln\left|\frac{5 + 4}{3}\right|$$

$$= -\tfrac{1}{5}\ln 2 + \tfrac{1}{5}\ln 3 = \tfrac{1}{5}\ln 1.5 \approx 0.0811$$

Matched Problem 2 ⇔ Evaluate: $\displaystyle\int_6^8 \frac{1}{x^2\sqrt{100 - x^2}}\,dx$

❑ SUBSTITUTION AND INTEGRAL TABLES

As Examples 1 and 2 illustrate, if the integral we want to evaluate can be made to match one in the table exactly, evaluating the indefinite integral consists of simply substituting the correct values of the constants into the formula. What happens if we cannot match an integral with one of the formulas in the table? In many cases, a substitution will change the given integral into one that corresponds to a table entry. The following examples illustrate several frequently used substitutions.

Example 3 ⇔ **Integration Using Substitution and Tables** Find: $\displaystyle\int \frac{x^2}{\sqrt{16x^2 - 25}}\,dx$

SOLUTION   In order to relate this integral to one of the formulas involving $\sqrt{u^2 - a^2}$ (formulas 40 to 45), we observe that if $u = 4x$, then

$$u^2 = 16x^2 \quad \text{and} \quad \sqrt{16x^2 - 25} = \sqrt{u^2 - 25}$$

Thus, we will use the substitution $u = 4x$ to change this integral into one that appears in the table:

$$\int \frac{x^2}{\sqrt{16x^2 - 25}}\, dx = \frac{1}{4} \int \frac{\frac{1}{16} u^2}{\sqrt{u^2 - 25}}\, du \qquad \begin{array}{l} \text{Substitution:} \\ u = 4x,\ du = 4\, dx,\ x = \tfrac{1}{4}u \end{array}$$

$$= \frac{1}{64} \int \frac{u^2}{\sqrt{u^2 - 25}}\, du$$

This last integral can be evaluated by using formula 44 with $a = 5$:

$$\int \frac{u^2}{\sqrt{u^2 - a^2}}\, du = \frac{1}{2}\left(u\sqrt{u^2 - a^2} + a^2 \ln|u + \sqrt{u^2 - a^2}|\right) \qquad \text{Formula 44}$$

$$\int \frac{x^2}{\sqrt{16x^2 - 25}}\, dx = \frac{1}{64} \int \frac{u^2}{\sqrt{u^2 - 25}}\, du \qquad \text{Use formula 44 with } a = 5.$$

$$= \tfrac{1}{128}\left(u\sqrt{u^2 - 25} + 25\ln|u + \sqrt{u^2 - 25}|\right) + C \qquad \text{Substitute } u = 4x.$$

$$= \tfrac{1}{128}\left(4x\sqrt{16x^2 - 25} + 25\ln|4x + \sqrt{16x^2 - 25}|\right) + C$$

Matched Problem 3 ↪ Find: $\int \sqrt{9x^2 - 16}\, dx$

Example 4 ↪ **Integration Using Substitution and Tables**   Find: $\int \frac{x}{\sqrt{x^4 + 1}}\, dx$

SOLUTION   None of the formulas in the table involve fourth powers; however, if we let $u = x^2$, then

$$\sqrt{x^4 + 1} = \sqrt{u^2 + 1}$$

and this form does appear in formulas 32 to 39. Thus, we substitute $u = x^2$:

$$\int \frac{1}{\sqrt{x^4 + 1}} x\, dx = \frac{1}{2} \int \frac{1}{\sqrt{u^2 + 1}}\, du \qquad \begin{array}{l} \text{Substitution:} \\ u = x^2,\ du = 2x\, dx \end{array}$$

We recognize the last integral as formula 36 with $a = 1$:

$$\int \frac{1}{\sqrt{u^2 + a^2}}\, du = \ln|u + \sqrt{u^2 + a^2}| \qquad \text{Formula 36}$$

$$\int \frac{x}{\sqrt{x^4 + 1}}\, dx = \frac{1}{2} \int \frac{1}{\sqrt{u^2 + 1}}\, du \qquad \text{Use formula 36 with } a = 1.$$

$$= \tfrac{1}{2}\ln|u + \sqrt{u^2 + 1}| + C \qquad \text{Substitute } u = x^2.$$

$$= \tfrac{1}{2}\ln|x^2 + \sqrt{x^4 + 1}| + C$$

Matched Problem 4 ⇔ Find: $\displaystyle\int x\sqrt{x^4 + 1}\, dx$

❑ REDUCTION FORMULAS

Example 5 ⇔ **Using Reduction Formulas**   Use Table II to find: $\displaystyle\int x^2 e^{3x}\, dx$

SOLUTION   Since the integrand involves the function e^{3x}, we examine formulas 46–48 and conclude that formula 47 can be used for this problem. Letting $u = x, n = 2$, and $a = 3$ in formula 47, we have

$$\int u^n e^{au}\, du = \frac{u^n e^{au}}{a} - \frac{n}{a}\int u^{n-1} e^{au}\, du \qquad \text{Formula 47}$$

$$\int x^2 e^{3x}\, dx = \frac{x^2 e^{3x}}{3} - \frac{2}{3}\int x e^{3x}\, dx$$

Notice that the expression on the right still contains an integral, but the exponent of x has been reduced by 1. Formulas of this type are called **reduction formulas** and are designed to be applied repeatedly until an integral that can be evaluated is obtained. Applying formula 47 to $\int x e^{3x}\, dx$ with $n = 1$, we have

$$\int x^2 e^{3x}\, dx = \frac{x^2 e^{3x}}{3} - \frac{2}{3}\left(\frac{x e^{3x}}{3} - \frac{1}{3}\int e^{3x}\, dx\right)$$

$$= \frac{x^2 e^{3x}}{3} - \frac{2x e^{3x}}{9} + \frac{2}{9}\int e^{3x}\, dx$$

This last expression contains an integral that is easy to evaluate:

$$\int e^{3x}\, dx = \tfrac{1}{3}e^{3x}$$

After making a final substitution and adding a constant of integration, we have

$$\int x^2 e^{3x}\, dx = \frac{x^2 e^{3x}}{3} - \frac{2x e^{3x}}{9} + \frac{2}{27}e^{3x} + C$$

Matched Problem 5 ⇔ Use Table II to find: $\displaystyle\int (\ln x)^2\, dx$

❑ APPLICATION

Example 6 ⇔ **Producers' Surplus**   Find the producers' surplus at a price level of $20 for the price–supply equation

$$p = S(x) = \frac{5x}{500 - x}$$

SOLUTION   **Step 1.**   Find $\bar{x}$, the supply when the price is $\bar{p} = 20$:

$$\bar{p} = \frac{5\bar{x}}{500 - \bar{x}}$$

$$20 = \frac{5\bar{x}}{500 - \bar{x}}$$

$$10{,}000 - 20\bar{x} = 5\bar{x}$$

$$10{,}000 = 25\bar{x}$$

$$\bar{x} = 400$$

$$p = S(x) = \frac{5x}{500 - x}$$

$$\bar{p} = 20$$

PS

FIGURE 1

Step 2.   Sketch a graph, as shown in Figure 1.

Step 3.   Find the producers' surplus (the shaded area in the graph):

$$PS = \int_0^{\bar{x}} [\bar{p} - S(x)]\,dx$$

$$= \int_0^{400} \left(20 - \frac{5x}{500 - x}\right) dx$$

$$= \int_0^{400} \frac{10{,}000 - 25x}{500 - x}\,dx$$

Use formula 20 with $a = 10{,}000$, $b = -25$, $c = 500$, and $d = -1$:

$$\int \frac{a + bu}{c + du}\,du = \frac{bu}{d} + \frac{ad - bc}{d^2}\ln|c + du| \qquad \text{Formula 20}$$

$$PS = \left(25x + 2{,}500\ln|500 - x|\right)\Big|_0^{400}$$

$$= 10{,}000 + 2{,}500\ln|100| - 2{,}500\ln|500|$$

$$\approx \$5{,}976$$

Matched Problem 6   Find the consumers' surplus at a price level of \$10 for the price–demand equation

$$p = D(x) = \frac{20x - 8{,}000}{x - 500}$$

Explore–Discuss 1

Use algebraic manipulation, including algebraic long division, on the integrand in Example 6 to show that

$$\frac{5\bar{x}}{500 - \bar{x}} = \frac{-5\bar{x}}{\bar{x} - 500} = -5 - \frac{2{,}500}{\bar{x} - 500}$$

Use this result to find the indefinite integral in Example 6 without resorting to table formulas.

Answers to Matched Problems   **1.** $\frac{1}{2}\left(\dfrac{1}{5 + 3x}\right) + \frac{1}{4}\ln\left|\dfrac{1 + x}{5 + 3x}\right| + C$ **2.** $\frac{7}{1{,}200} \approx 0.0058$

3. $\frac{1}{6}(3x\sqrt{9x^2 - 16} - 16\ln|3x + \sqrt{9x^2 - 16}|) + C$

4. $\frac{1}{4}(x^2\sqrt{x^4 + 1} + \ln|x^2 + \sqrt{x^4 + 1}|) + C$ **5.** $x(\ln x)^2 - 2x\ln x + 2x + C$

6. $3{,}000 + 2{,}000\ln 200 - 2{,}000\ln 500 \approx \$1{,}167$

Exercise 7-4

A *Use Table II to find each indefinite integral in Problems 1–14.*

1. $\int \dfrac{1}{x(1 + x)}\, dx$

2. $\int \dfrac{1}{x^2(1 + x)}\, dx$

3. $\int \dfrac{1}{(3 + x)^2(5 + 2x)}\, dx$

4. $\int \dfrac{x}{(5 + 2x)^2(2 + x)}\, dx$

5. $\int \dfrac{x}{\sqrt{16 + x}}\, dx$

6. $\int \dfrac{1}{x\sqrt{16 + x}}\, dx$

7. $\int \dfrac{1}{x\sqrt{1 - x^2}}\, dx$

8. $\int \dfrac{\sqrt{9 - x^2}}{x}\, dx$

9. $\int \dfrac{1}{x\sqrt{x^2 + 4}}\, dx$

10. $\int \dfrac{1}{x^2\sqrt{x^2 - 16}}\, dx$

11. $\int x^2 \ln x\, dx$

12. $\int x^3 \ln x\, dx$

13. $\int \dfrac{1}{1 + e^x}\, dx$

14. $\int \dfrac{1}{5 + 2e^{3x}}\, dx$

Evaluate each definite integral in Problems 15–20. Use Table II to find the antiderivative.

15. $\displaystyle\int_1^3 \dfrac{x^2}{3 + x}\, dx$

16. $\displaystyle\int_2^6 \dfrac{x}{(6 + x)^2}\, dx$

17. $\displaystyle\int_0^7 \dfrac{1}{(3 + x)(1 + x)}\, dx$

18. $\displaystyle\int_0^7 \dfrac{x}{(3 + x)(1 + x)}\, dx$

19. $\displaystyle\int_0^4 \dfrac{1}{\sqrt{x^2 + 9}}\, dx$

20. $\displaystyle\int_4^5 \sqrt{x^2 - 16}\, dx$

B *In Problems 21–32, use substitution techniques and Table II to find each indefinite integral.*

21. $\int \dfrac{\sqrt{4x^2 + 1}}{x^2}\, dx$

22. $\int x^2\sqrt{9x^2 - 1}\, dx$

23. $\int \dfrac{x}{\sqrt{x^4 - 16}}\, dx$

24. $\int x\sqrt{x^4 - 16}\, dx$

25. $\int x^2\sqrt{x^6 + 4}\, dx$

26. $\int \dfrac{x^2}{\sqrt{x^6 + 4}}\, dx$

27. $\int \dfrac{1}{x^3\sqrt{4 - x^4}}\, dx$

28. $\int \dfrac{\sqrt{x^4 + 4}}{x}\, dx$

29. $\int \dfrac{e^x}{(2 + e^x)(3 + 4e^x)}\, dx$

30. $\int \dfrac{e^x}{(4 + e^x)^2(2 + e^x)}\, dx$

31. $\int \dfrac{\ln x}{x\sqrt{4 + \ln x}}\, dx$

32. $\int \dfrac{1}{x \ln x\sqrt{4 + \ln x}}\, dx$

C *In Problems 33–38, use Table II to find each indefinite integral.*

33. $\int x^2 e^{5x}\, dx$

34. $\int x^2 e^{-4x}\, dx$

35. $\int x^3 e^{-x}\, dx$

36. $\int x^3 e^{2x}\, dx$

37. $\int (\ln x)^3\, dx$

38. $\int (\ln x)^4\, dx$

Problems 39–46 are mixed—some require the use of Table II and others can be solved using techniques we considered earlier.

39. $\displaystyle\int_3^5 x\sqrt{x^2 - 9}\, dx$

40. $\displaystyle\int_3^5 x^2\sqrt{x^2 - 9}\, dx$

41. $\displaystyle\int_2^4 \dfrac{1}{x^2 - 1}\, dx$

42. $\displaystyle\int_2^4 \dfrac{x}{(x^2 - 1)^2}\, dx$

43. $\int \dfrac{\ln x}{x^2}\, dx$

44. $\int \dfrac{(\ln x)^2}{x}\, dx$

45. $\int \dfrac{x}{\sqrt{x^2 - 1}}\, dx$

46. $\int \dfrac{x^2}{\sqrt{x^2 - 1}}\, dx$

In Problems 47–50, find the area bounded by the graphs of $y = f(x)$ and $y = g(x)$ to two decimal places. Use a graphing utility to approximate intersection points to two decimal places.

47. $f(x) = \dfrac{10}{\sqrt{x^2 + 1}};\ g(x) = x^2 + 3x$

48. $f(x) = \sqrt{1 + x^2};\ g(x) = 5x - x^2$

49. $f(x) = x\sqrt{4 + x};\ g(x) = 1 + x$

50. $f(x) = \dfrac{x}{\sqrt{x + 4}};\ g(x) = x - 2$

Applications

Use Table II to evaluate all integrals involved in any solutions of Problems 51–74.

Business & Economics

51. *Consumers' surplus.* Find the consumers' surplus at a price level of $\bar{p} = \$15$ for the price–demand equation

$$p = D(x) = \frac{7,500 - 30x}{300 - x}$$

52. *Producers' surplus.* Find the producers' surplus at a price level of $\bar{p} = \$20$ for the price–supply equation

$$p = S(x) = \frac{10x}{300 - x}$$

53. *Consumers' surplus.* For Problem 51, graph the price–demand equation and the price level equation $\bar{p} = 15$ in the same coordinate system. What region represents the consumers' surplus?

54. *Producers' surplus.* For Problem 52, graph the price–supply equation and the price level equation $\bar{p} = 20$ in the same coordinate system. What region represents the producers' surplus?

55. *Cost.* A company manufactures downhill skis. It has fixed costs of $25,000 and a marginal cost given by

$$C'(x) = \frac{250 + 10x}{1 + 0.05x}$$

where $C(x)$ is the total cost at an output of x pairs of skis. Find the cost function $C(x)$ and determine the production level (to the nearest unit) that produces a cost of $150,000. What is the cost (to the nearest dollar) for a production level of 850 pairs of skis?

56. *Cost.* A company manufactures a portable CD player. It has fixed costs of $11,000 per week and a marginal cost given by

$$C'(x) = \frac{65 + 20x}{1 + 0.4x}$$

where $C(x)$ is the total cost per week at an output of x players per week. Find the cost function $C(x)$ and determine the production level (to the nearest unit) that produces a cost of $52,000 per week. What is the cost (to the nearest dollar) for a production level of 700 players per week?

57. *Continuous income stream.* Find the future value at 10% compounded continuously for 10 years for the continuous income stream with rate of flow $f(t) = 50t^2$.

58. *Continuous income stream.* Find the interest earned at 8% compounded continuously for 5 years for the continuous income stream with rate of flow $f(t) = 200t$.

59. *Income distribution.* Find the index of income concentration for the Lorenz curve with equation

$$y = \tfrac{1}{2}x\sqrt{1 + 3x}$$

60. *Income distribution.* Find the index of income concentration for the Lorenz curve with equation

$$y = \tfrac{1}{2}x^2\sqrt{1 + 3x}$$

61. *Income distribution.* For Problem 59, graph $y = x$ and the Lorenz curve over the interval $[0, 1]$. Discuss the effect of the area bounded by $y = x$ and the Lorenz curve getting smaller relative to the equitable distribution of income.

62. *Income distribution.* For Problem 60, graph $y = x$ and the Lorenz curve over the interval $[0, 1]$. Discuss the effect of the area bounded by $y = x$ and the Lorenz curve getting larger relative to the equitable distribution of income.

63. *Marketing.* After test marketing a new high-fiber cereal, the market research department of a major food producer estimates that monthly sales (in millions of dollars) will grow at the monthly rate of

$$S'(t) = \frac{t^2}{(1 + t)^2}$$

t months after the cereal is introduced. If we assume 0 sales at the time the cereal is introduced, find the total sales, $S(t)$, t months after the cereal is introduced. Find the total sales during the first 2 years this cereal is on the market.

64. *Average price.* At a discount department store, the price–demand equation for premium motor oil is given by

$$p = D(x) = \frac{50}{\sqrt{100 + 6x}}$$

where x is the number of cans of oil that can be sold at a price of $\$p$. Find the average price over the demand interval $[50, 250]$.

65. *Marketing*. In Problem 63, show the sales over the first 2 years geometrically, and verbally describe the geometric representation.

66. *Price–demand*. In Problem 64, graph the price–demand equation and the line representing the average price in the same coordinate system over the interval $[50, 250]$. Describe how the areas under the two curves over the interval $[50, 250]$ are related.

67. *Profit*. The marginal profit for a small car agency that sells x cars per week is given by

$$P'(x) = x\sqrt{2 + 3x}$$

where $P(x)$ is the profit in dollars. The agency's profit on the sale of only 1 car per week is $-\$2,000$. Find the profit function and the number of cars that must be sold (to the nearest unit) to produce a profit of $\$13,000$ per week. How much weekly profit (to the nearest dollar) will the agency have if 80 cars are sold per week?

68. *Revenue*. The marginal revenue for a company that manufactures and sells x solar-powered calculators per week is given by

$$R'(x) = \frac{x}{\sqrt{1 + 2x}} \qquad R(0) = 0$$

where $R(x)$ is the revenue in dollars. Find the revenue function and the number of calculators that must be sold (to the nearest unit) to produce $\$10,000$ in revenue per week. How much weekly revenue (to the nearest dollar) will the company have if 1,000 calculators are sold per week?

Life Sciences

69. *Pollution*. An oil tanker aground on a reef is losing oil and producing an oil slick that is radiating outward at a rate given approximately by

$$\frac{dR}{dt} = \frac{100}{\sqrt{t^2 + 9}} \qquad t \geq 0$$

where R is the radius (in feet) of the circular slick after t minutes. Find the radius of the slick after 4 minutes if the radius is 0 when $t = 0$.

70. *Pollution*. The concentration of particulate matter (in parts per million) during a 24-hour period is given approximately by

$$C(t) = t\sqrt{24 - t} \qquad 0 \leq t \leq 24$$

where t is time in hours. Find the average concentration during the time period from $t = 0$ to $t = 24$.

Social Sciences

71. *Learning*. A person learns N items at a rate given approximately by

$$N'(t) = \frac{60}{\sqrt{t^2 + 25}} \qquad t \geq 0$$

where t is the number of hours of continuous study. Determine the total number of items learned in the first 12 hours of continuous study.

72. *Politics*. The number of voters (in thousands) in a metropolitan area is given approximately by

$$f(t) = \frac{500}{2 + 3e^{-t}} \qquad t \geq 0$$

where t is time in years. Find the average number of voters during the time period from $t = 0$ to $t = 10$.

73. *Learning*. Interpret Problem 71 geometrically. Verbally describe the geometric interpretation.

74. *Politics*. In Problem 72, graph $y = f(t)$ and the line representing the average number of voters over the interval $[0, 10]$ in the same coordinate system. Describe how the areas under the two curves over the interval $[0, 10]$ are related.

Review Exercise

Work through all the problems in this chapter review and check your answers in the back of the book. Answers to all review problems are there along with section numbers in italics to indicate where each type of problem is discussed. Where weaknesses show up, review appropriate sections in the text.

Compute all numerical answers to three decimal places unless directed otherwise.

A *In Problems 1–3, set up definite integrals that represent the shaded areas in the figure over the indicated intervals.*

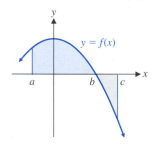

1. Interval $[a, b]$ **2.** Interval $[b, c]$ **3.** Interval $[a, c]$

4. Sketch a graph of the area between the graphs of $y = \ln x$ and $y = 0$ over the interval $[0.5, e]$, and find the area.

In Problems 5–10, evaluate each integral.

5. $\int xe^{4x}\, dx$

6. $\int x \ln x\, dx$

7. $\int \dfrac{\ln x}{x}\, dx$

8. $\int \dfrac{x}{1 + x^2}\, dx$

9. $\int \dfrac{1}{x(1 + x)^2}\, dx$

10. $\int \dfrac{1}{x^2\sqrt{1 + x}}\, dx$

B *In Problems 11–14, set up definite integrals that represent the shaded areas in the figure over the indicated intervals.*

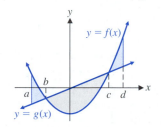

11. Interval $[a, b]$ **12.** Interval $[b, c]$

13. Interval $[b, d]$ **14.** Interval $[a, d]$

15. Sketch a graph of the area bounded by the graphs of $y = x^2 - 6x + 9$ and $y = 9 - x$, and find the area.

In Problems 16–21, evaluate each integral.

16. $\int_0^1 xe^x\, dx$

17. $\int_0^3 \dfrac{x^2}{\sqrt{x^2 + 16}}\, dx$

18. $\int \sqrt{9x^2 - 49}\, dx$

19. $\int te^{-0.5t}\, dt$

20. $\int x^2 \ln x\, dx$

21. $\int \dfrac{1}{1 + 2e^x}\, dx$

22. Sketch a graph of the area bounded by the indicated graphs and find the area. In part (B), approximate intersection points and area to two decimal places.

(A) $y = x^3 - 6x^2 + 9x;\ y = x$

(B) $y = x^3 - 6x^2 + 9x;\ y = x + 1$

C *In Problems 23–30, evaluate each integral.*

23. $\int \dfrac{(\ln x)^2}{x}\, dx$

24. $\int x(\ln x)^2\, dx$

25. $\int \dfrac{x}{\sqrt{x^2 - 36}}\, dx$

26. $\int \dfrac{x}{\sqrt{x^4 - 36}}\, dx$

27. $\int_0^4 x \ln(10 - x)\, dx$

28. $\int (\ln x)^2\, dx$

29. $\int xe^{-2x^2}\, dx$

30. $\int x^2 e^{-2x}\, dx$

31. Use a numerical integration routine on a graphing utility to find the area (to three decimal places) in the first quadrant that is below the graph of

$$y = \dfrac{6}{2 + 5e^{-x}}$$

and above the graph of $y = 0.2x + 1.6$.

Applications

Business & Economics

32. *Product warranty.* A manufacturer warrants a product for parts and labor for 1 year, and for parts only for a second year. The time to a failure of the product after it is sold is given by the probability density function

$$f(t) = \begin{cases} 0.21e^{-0.21t} & \text{if } \geq 0 \\ 0 & \text{otherwise} \end{cases}$$

What is the probability that a buyer chosen at random will have a product failure:

(A) During the first year of warranty?

(B) During the second year of warranty?

33. *Product warranty.* Graph the probability density function for Problem 32 over the interval $[0, 3]$, interpret part (B) of Problem 32 geometrically, and verbally describe the geometric representation.

34. *Revenue function.* The weekly marginal revenue from the sale of x hair dryers is given by

$$R'(x) = 65 - 6\ln(x + 1) \qquad R(0) = 0$$

where $R(x)$ is the revenue in dollars. Find the revenue function and the production level (to the nearest unit) for a revenue of $20,000 per week. What is the weekly revenue (to the nearest dollar) at a production level of 1,000 hair dryers per week?

35. *Continuous income stream.* The rate of flow (in dollars per year) of a continuous income stream for a 5-year period is given by

$$f(t) = 2,500e^{0.05t} \qquad 0 \le t \le 5$$

(A) Graph $y = f(t)$ over $[0, 5]$ and shade in the area that represents the total income received from the end of the first year to the end of the fourth year.

(B) Find the total income received, to the nearest dollar, from the end of the first year to the end of the fourth year.

36. *Future value of a continuous income stream.* The continuous income stream in Problem 35 is invested as it is received at 15% compounded continuously.

(A) Find the future value (to the nearest dollar) at the end of the 5-year period.

(B) Find the interest earned (to the nearest dollar) during this 5-year period.

37. *Income distribution.* An economist produced the following Lorenz curves for the current income distribution and the projected income distribution 10 years from now in a certain country:

$$f(x) = 0.1x + 0.9x^2 \qquad \text{Current Lorenz curve}$$
$$g(x) = x^{1.5} \qquad \text{Projected Lorenz curve}$$

(A) Graph $y = x$ and the current Lorenz curve on one set of coordinate axes for $[0, 1]$, and graph $y = x$ and the projected Lorenz curve on another set of coordinate axes over the same interval.

(B) Looking at the areas bounded by the Lorenz curves and $y = x$, can you say that the income will be more or less equitably distributed 10 years from now?

(C) Compute the index of income concentration (to one decimal place) for the current and projected curves. Now what can you say about the distribution of income 10 years from now? More equitable or less?

38. *Consumers' and producers' surplus.* Find the consumers' surplus and the producers' surplus at the equilibrium price level for each pair of price–demand and price–supply equations. Include a graph that identifies the consumers' surplus and the producers' surplus. Round all values to the nearest integer.

(A) $p = D(x) = 70 - 0.2x$;
$\quad p = S(x) = 13 + 0.0012x^2$

(B) $p = D(x) = 70 - 0.2x; p = S(x) = 13e^{0.006x}$

39. *Producers' surplus.* The table below gives price–supply data for the sale of hogs at a livestock market, where x is the number of pounds (in thousands) and p is the price per pound (in cents).

Price–Supply	
x	$p = S(x)$
0	43.50
10	46.74
20	50.05
30	54.72
40	59.18

(A) Using quadratic regression to model the data, find the demand at a price of 52.50 cents per pound.

(B) Use a numerical integration routine to find the producers' surplus (to the nearest dollar) at a price level of 52.50 cents per pound.

Life Sciences

40. *Drug assimilation.* The rate at which the body eliminates a drug (in milliliters per hour) is given by

$$R(t) = \frac{60t}{(t + 1)^2(t + 2)}$$

where t is the number of hours since the drug was administered. How much of the drug is eliminated during the first hour after it was administered? During the fourth hour?

41. With the aid of a graphing utility, illustrate Problem 40 geometrically.

42. *Medicine.* For a particular doctor, the length of time (in hours) spent with a patient per office visit has the probability density function

$$f(t) = \begin{cases} \dfrac{\frac{4}{3}}{(t + 1)^2} & \text{if } 0 \le t \le 3 \\ 0 & \text{otherwise} \end{cases}$$

(A) What is the probability that this doctor will spend less than 1 hour with a randomly selected patient?

(B) What is the probability that this doctor will spend more than 1 hour with a randomly selected patient?

43. *Medicine.* Illustrate part (B) in Problem 42 geometrically. Describe the geometric interpretation verbally.

Social Sciences

44. *Politics.* The rate of change of the voting population of a city with respect to time t (in years) is estimated to be

$$N'(t) = \frac{100t}{(1 + t^2)^2}$$

where $N(t)$ is in thousands. If $N(0)$ is the current voting population, how much will this population increase during the next 3 years?

45. *Psychology.* Rats were trained to go through a maze by rewarding them with a food pellet upon successful completion. After the seventh successful run, it was found that the probability density function for length of time (in minutes) until success on the eighth trial is given by

$$f(t) = \begin{cases} .5e^{-.5t} & \text{if } t \geq 0 \\ 0 & \text{otherwise} \end{cases}$$

What is the probability that a rat selected at random after seven successful runs will take 2 or more minutes to complete the eighth run successfully? [Recall that the area under a probability density function curve from $-\infty$ to ∞ is 1.]

Group Activity 1   ***Analysis of Income Concentration from Raw Data***

This group activity may be done without the use of a graphing utility, but additional insight into mathematical modeling will be gained if one is available.

We start with raw data on income distribution supplied in table form by the U.S. Bureau of the Census (Table 1). From the raw data in the table, we wish to compare income distribution among whites and income distribution among blacks in the United States in 1997. The approach will be numeric, geometric, and symbolic. We will first organize in table form the data that correspond to data points for a Lorenz curve. We will then find Lorenz curves of the form $f(x) = x^p$ for each set of data points. We will interpret the income distribution geometrically by graphing the Lorenz curves and $y = x$ for both sets of data. Finally, we compute the index of income concentration for blacks and for whites, and interpret the results. (See the discussion of Lorenz curves in Section 7-1 for relevant background material.)

www

TABLE 1

INCOME DISTRIBUTION BY POPULATION FIFTHS

FAMILIES, 1997

UPPER LIMIT OF EACH FIFTH*

RACE	Lowest	Second	Third	Fourth	Top 5%[†]
TOTAL	$20,586	$36,000	$53,616	$80,000	$137,080
WHITE	22,576	38,258	55,783	82,442	142,400
BLACK	11,396	21,875	36,052	57,000	95,684

PERCENTAGE DISTRIBUTION OF TOTAL INCOME

FAMILIES, 1997 RACE	Lowest fifth	Second fifth	Third fifth	Fourth fifth	Highest fifth	Top 5%
TOTAL	4.2	9.9	15.7	23.0	47.2	20.7
WHITE	4.6	10.2	15.7	22.8	46.8	20.7
BLACK	3.4	9.1	15.6	25.1	46.8	17.6

*The highest fifth does not have an upper limit.
[†]Lower limit for top 5%.
Source: U.S. Bureau of the Census, U.S. Department of Commerce

Part 1. *Numeric analysis.* Complete Tables 2 and 3 using the data in Table 1. Round income levels to the nearest thousand dollars and represent percents in decimal form to two decimal places. Remember that x represents the cumulative percentage of families in a given category and y represents the

TABLE 2

BLACK FAMILY INCOME
DISTRIBUTION, 1997

INCOME LEVEL	x	y
Under $11,000	0.20	0.03
Under		0.13
Under		
Under $57,000	0.80	

TABLE 3

WHITE FAMILY INCOME
DISTRIBUTION, 1997

INCOME LEVEL	x	y
Under $23,000	0.20	0.05
Under		0.15
Under		
Under $82,000	0.80	

corresponding cumulative percentage of income received by these families. Verbally describe the meaning of the last two lines in Tables 2 and 3 after they are completed.

Part 2. *Geometric analysis*

(A) Sketch separate graphs on suitable graph paper for each table by plotting the data points from the x and y columns in Tables 2 and 3. Also, graph the line $y = x$ over the interval $[0, 1]$ in each graph.

(B) Find p (to one decimal place) so that the graph of the function $f(x) = x^p$ goes through the point $(0.20, 0.03)$ in Table 2. Plot this curve on the corresponding graph, $0 \leqslant x \leqslant 1$. Also, find q (to one decimal place) so that the function $g(x) = x^q$ goes through the point $(0.20, 0.05)$ in Table 3. Plot this curve on the corresponding graph, $0 \leqslant x \leqslant 1$. Repeat this process for each of the remaining points in each table. Compare all the graphs based on Table 2 and select the value of p that best fits the data (by eye). Do the same for all the graphs based on Table 3.

(C) From the final graphs chosen in part (B) can you draw any conclusions about whether income is distributed more equitably among blacks or whites? Verbally support your conclusions.

Part 3. *Symbolic analysis.* Use the values of p and q you determined in part 2B to compute the index of income concentration for blacks and for whites, and interpret.

Group Activity 2 **Grain Exchange**

The tables below give price–demand and price–supply data for the sale of wheat at a grain market, where x is the number of bushels of wheat (in thousands) and p is the price per bushel (in cents).

Price–Demand		Price–Supply	
x	$p = D(x)$	x	$p = S(x)$
20	345	20	311
25	336	25	312
30	323	30	321
35	320	35	323
40	318	40	326
45	307	45	338

(A) Find quadratic, logarithmic, and exponential regression models for each set of data.

(B) Use the sum of squares of the residuals (see Group Activity 2 in Chapter 2) to compare the models from part (A).

(C) Using the models that best fit the data, approximate the supply and demand supported by a price of $3.50 per bushel; by a price of $3.25 per bushel; by a price of $3.00 per bushel.

(D) Using the models that best fit the data, find the equilibrium quantity and equilibrium price.

(E) Use a numerical integration routine to find the consumers' surplus and the producers' surplus at the equilibrium price level.

Multivariable Calculus

Section 8-1

Functions of Several Variables

- ❑ FUNCTIONS OF TWO OR MORE INDEPENDENT VARIABLES
- ❑ EXAMPLES OF FUNCTIONS OF SEVERAL VARIABLES
- ❑ THREE-DIMENSIONAL COORDINATE SYSTEMS

❑ FUNCTIONS OF TWO OR MORE INDEPENDENT VARIABLES

In Section 1-1 we introduced the concept of a function with one independent variable. Now we broaden the concept to include functions with more than one independent variable. We start with an example.

A small manufacturing company produces a standard type of surfboard and no other products. If fixed costs are $500 per week and variable costs are $70 per board produced, the weekly cost function is given by

$$C(x) = 500 + 70x \tag{1}$$

where x is the number of boards produced per week. The cost function is a function of a single independent variable x. For each value of x from the domain of C there exists exactly one value of $C(x)$ in the range of C.

Now, suppose that the company decides to add a high-performance competition board to its line. If the fixed costs for the competition board are $200 per week and the variable costs are $100 per board, the cost function (1) must be modified to

$$C(x, y) = 700 + 70x + 100y \tag{2}$$

where $C(x, y)$ is the cost for weekly output of x standard boards and y competition boards. Equation (2) is an example of a function with two independent

477

variables, x and y. Of course, as the company expands its product line even further, its weekly cost function must be modified to include more and more independent variables, one for each new product produced.

In general, an equation of the form

$$z = f(x, y)$$

describes a **function of two independent variables** if for each permissible ordered pair (x, y), there is one and only one value of z determined by $f(x, y)$. The variables x and y are **independent variables,** and the variable z is a **dependent variable.** The set of all ordered pairs of permissible values of x and y is the **domain** of the function, and the set of all corresponding values $f(x, y)$ is the **range** of the function. Unless otherwise stated, we will assume that the domain of a function specified by an equation of the form $z = f(x, y)$ is the set of all ordered pairs of real numbers (x, y) such that $f(x, y)$ is also a real number. It should be noted, however, that certain conditions in practical problems often lead to further restrictions of the domain of a function.

We can similarly define functions of three independent variables, $w = f(x, y, z)$; of four independent variables, $u = f(w, x, y, z)$; and so on. In this chapter we concern ourselves primarily with functions of two independent variables.

Example 1 ⇔ **Evaluating a Function of Two Independent Variables** For the cost function $C(x, y) = 700 + 70x + 100y$ described earlier, find $C(10, 5)$.

Solution
$$C(10, 5) = 700 + 70(10) + 100(5)$$
$$= \$1,900$$

Matched Problem 1 ⇔ Find $C(20, 10)$ for the cost function in Example 1.

Example 2 ⇔ **Evaluating a Function of Three Independent Variables** For the function $f(x, y, z) = 2x^2 - 3xy + 3z + 1$, find $f(3, 0, -1)$.

Solution
$$f(3, 0, -1) = 2(3)^2 - 3(3)(0) + 3(-1) + 1$$
$$= 18 - 0 - 3 + 1 = 16$$

Matched Problem 2 ⇔ Find $f(-2, 2, 3)$ for f in Example 2.

Example 3 ⇔ **Revenue, Cost, and Profit Functions** The surfboard company discussed at the beginning of this section has determined that the demand equations for the two types of boards they produce are given by

$$p = 210 - 4x + y$$
$$q = 300 + x - 12y$$

where p is the price of the standard board, q is the price of the competition board, x is the weekly demand for standard boards, and y is the weekly demand for competition boards.

(A) Find the weekly revenue function $R(x, y)$, and evaluate $R(20, 10)$.

(B) If the weekly cost function is

$$C(x, y) = 700 + 70x + 100y$$

find the weekly profit function $P(x, y)$, and evaluate $P(20, 10)$.

SOLUTION   (A)

$$\text{Revenue} = \begin{pmatrix} \text{demand for} \\ \text{standard} \\ \text{boards} \end{pmatrix} \times \begin{pmatrix} \text{price of a} \\ \text{standard} \\ \text{board} \end{pmatrix} + \begin{pmatrix} \text{demand for} \\ \text{competition} \\ \text{boards} \end{pmatrix} \times \begin{pmatrix} \text{price of a} \\ \text{competition} \\ \text{board} \end{pmatrix}$$

$$
\begin{aligned}
R(x, y) &= xp + yq \\
&= x(210 - 4x + y) + y(300 + x - 12y) \\
&= 210x + 300y - 4x^2 + 2xy - 12y^2 \\
R(20, 10) &= 210(20) + 300(10) - 4(20)^2 + 2(20)(10) - 12(10)^2 \\
&= \$4{,}800
\end{aligned}
$$

(B)   Profit = revenue − cost

$$
\begin{aligned}
P(x, y) &= R(x, y) - C(x, y) \\
&= 210x + 300y - 4x^2 + 2xy - 12y^2 - 700 - 70x - 100y \\
&= 140x + 200y - 4x^2 + 2xy - 12y^2 - 700 \\
P(20, 10) &= 140(20) + 200(10) - 4(20)^2 + 2(20)(10) - 12(10)^2 - 700 \\
&= \$1{,}700
\end{aligned}
$$
■

Matched Problem 3 ⇌ Repeat Example 3 if the demand and cost equations are given by

$$
\begin{aligned}
p &= 220 - 6x + y \\
q &= 300 + 3x - 10y \\
C(x, y) &= 40x + 80y + 1{,}000
\end{aligned}
$$
■

❑ EXAMPLES OF FUNCTIONS OF SEVERAL VARIABLES

A number of concepts we have already considered can be thought of in terms of functions of two or more variables. We list a few of these below.

Area of a rectangle $A(x, y) = xy$

Volume of a box $V(x, y, z) = xyz$

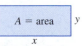

Volume of a right circular cylinder $V(r, h) = \pi r^2 h$

Simple interest $A(P, r, t) = P(1 + rt)$

$A = $ amount
$P = $ principal
$r = $ annual rate
$t = $ time in years

Compound interest $\quad A(P, r, t, n) = P\left(1 + \dfrac{r}{n}\right)^{nt}$

A = amount
P = principal
r = annual rate
t = time in years
n = compound periods per year

IQ $\qquad\qquad Q(M, C) = \dfrac{M}{C}(100)$

Q = IQ = intelligence quotient
M = MA = mental age
C = CA = chronological age

Resistance for blood flow in a vessel (Poiseuille's law) $\qquad R(L, r) = k\dfrac{L}{r^4}$

R = resistance
L = length of vessel
r = radius of vessel
k = constant

Example 4 **Package Design** A company uses a box with a square base and an open top for one of its products (see the figure). If x is the length (in inches) of each side of the base and y is the height (in inches), find the total amount of material $M(x, y)$ required to construct one of these boxes, and evaluate $M(5, 10)$.

Solution

$$\text{Area of base} = x^2$$
$$\text{Area of one side} = xy$$
$$\text{Total material} = (\text{area of base}) + 4(\text{area of one side})$$
$$M(x, y) = x^2 + 4xy$$
$$M(5, 10) = (5)^2 + 4(5)(10)$$
$$= 225 \text{ square inches}$$

Matched Problem 4 For the box in Example 4, find the volume $V(x, y)$, and evaluate $V(5, 10)$.

The next example concerns the **Cobb–Douglas production function,**

$$f(x, y) = kx^m y^n$$

where k, m, and n are positive constants with $m + n = 1$. Economists use this function to describe the number of units $f(x, y)$ produced from the utilization of x units of labor and y units of capital (for equipment such as tools, machinery, buildings, and so on). Cobb–Douglas production functions are also used to describe the productivity of a single industry, of a group of industries producing the same product, or even of an entire country.

Example 5 **Productivity** The productivity of a steel manufacturing company is given approximately by the function

$$f(x, y) = 10x^{0.2}y^{0.8}$$

with the utilization of x units of labor and y units of capital. If the company uses 3,000 units of labor and 1,000 units of capital, how many units of steel will be produced?

SOLUTION   The number of units of steel produced is given by

$$f(3,000, 1,000) = 10(3,000)^{0.2}(1,000)^{0.8} \quad \textit{Use a calculator.}$$
$$\approx 12,457 \text{ units}$$

Matched Problem 5 ⇨ Refer to Example 5. Find the steel production if the company uses 1,000 units of labor and 2,000 units of capital.

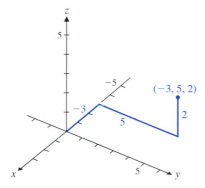

FIGURE 1 Rectangular coordinate system

❑   THREE-DIMENSIONAL COORDINATE SYSTEMS

We now take a brief look at some graphs of functions of two independent variables. Since functions of the form $z = f(x, y)$ involve two independent variables, x and y, and one dependent variable, z, we need a *three-dimensional coordinate system* for their graphs. A **three-dimensional coordinate system** is formed by three mutually perpendicular number lines intersecting at their origins (see Fig. 1). In such a system, every ordered **triplet of numbers** (x, y, z) can be associated with a unique point, and conversely.

Example 6 ⇨ **Three-Dimensional Coordinates**   Locate $(-3, 5, 2)$ in a rectangular coordinate system.

SOLUTION

Matched Problem 6 ⇨ Find the coordinates of the corners A, C, G, and D of the rectangular box shown in the figure.

Imagine that you are facing the front of a classroom whose rectangular walls meet at right angles. Suppose that the point of intersection of the floor, front wall, and left side wall is the origin of a three-dimensional coordinate system in which every point in the room has nonnegative coordinates. Then the plane $z = 0$ (or, equivalently, the xy plane) can be described as "the floor," and the plane $z = 2$ can be described as "the plane parallel to, but 2 units above, the floor." Give similar descriptions of the following planes:

(A) $x = 0$ (B) $x = 3$ (C) $y = 0$
(D) $y = 4$ (E) $x = -1$

What does the graph of $z = x^2 + y^2$ look like? If we let $x = 0$ and graph $z = 0^2 + y^2 = y^2$ in the yz plane, we obtain a parabola; if we let $y = 0$ and graph $z = x^2 + 0^2 = x^2$ in the xz plane, we obtain another parabola. It can be shown that the graph of $z = x^2 + y^2$ is either one of these parabolas rotated around the z axis (see Fig. 2). This cup-shaped figure is a *surface* and is called a **paraboloid.**

In general, the graph of any function of the form $z = f(x, y)$ is called a **surface.** The graph of such a function is the graph of all ordered triplets of numbers (x, y, z) that satisfy the equation. Graphing functions of two independent variables is often a very difficult task, and the general process will not be dealt with in this book. We present only a few simple graphs to suggest extensions of earlier geometric interpretations of the derivative and local maxima and minima to functions of two variables. Note that $z = f(x, y) = x^2 + y^2$ appears (see Fig. 2) to have a local minimum at $(x, y) = (0, 0)$. Figure 3 shows a local maximum at $(x, y) = (0, 0)$.

FIGURE 2 Paraboloid

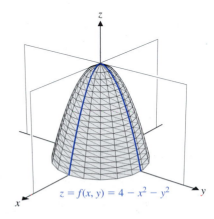

FIGURE 3 Local maximum: $f(0, 0) = 4$

Figure 4 shows a point at $(x, y) = (0, 0)$, called a **saddle point,** which is neither a local minimum nor a local maximum. Note that if $x = 0$, the saddle point is a local minimum, and if $y = 0$, the saddle point is a local maximum. More will be said about local maxima and minima in Section 8-3.

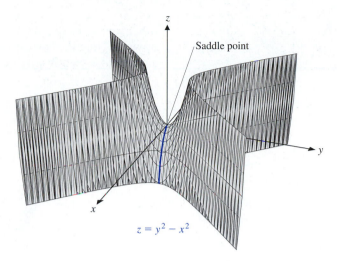

FIGURE 4 Saddle point at $(0, 0, 0)$

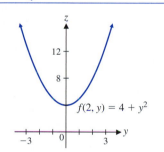

FIGURE 5

Explore–Discuss 2

(A) Let $f(x, y) = x^2 + y^2$. The cross section of the surface $z = f(x, y)$ by the plane $x = 2$ is the graph of $z = f(2, y) = 4 + y^2$, which is a parabola (see Fig. 5). Explain why each cross section of $z = f(x, y)$ by a plane parallel to the yz plane is a parabola that opens upward. Explain why each cross section of $z = f(x, y)$ by a plane parallel to the xz plane is a parabola that opens upward.

(B) Let $g(x, y) = y^2 - x^2$. Explain why each cross section of $z = g(x, y)$ by a plane parallel to the yz plane is a parabola that opens upward (see Fig. 4). Explain why each cross section of $z = f(x, y)$ by a plane parallel to the xz plane is a parabola that opens downward.

 Some graphing utilities are designed to draw graphs (like those of Figs. 2, 3, and 4) of functions of two independent variables. Others, like the graphing calculator used for the displays in this book, are designed to draw graphs of functions of just a single independent variable. When using the latter type of calculator, we can graph cross sections by planes parallel to the xz plane or yz plane to gain insight into the graph of a function of two independent variables.

Example 7 ⇔ **Graphing Cross Sections**

(A) Describe the cross sections of $f(x, y) = 2x^2 + y^2$ in the planes $y = 0$, $y = 1, y = 2, y = 3$, and $y = 4$.

(B) Describe the cross sections of $f(x, y) = 2x^2 + y^2$ in the planes $x = 0$, $x = 1, x = 2, x = 3$, and $x = 4$.

SOLUTION

(A) The cross section of $f(x, y) = 2x^2 + y^2$ by the plane $y = 0$ is the graph of the function $f(x, 0) = 2x^2$ in this plane. We can examine the shape of this cross section by graphing $y_1 = 2x^2$ on a graphing utility (Fig. 6). Similarly, the graphs of $y_2 = f(x, 1) = 2x^2 + 1$, $y_3 = f(x, 2) = 2x^2 + 4$, $y_4 = f(x, 3) = 2x^2 + 9$, and $y_5 = f(x, 4) = 2x^2 + 16$ show the shapes of the other four cross sections (see Fig. 6). Each of these is a parabola that opens upward. Note the correspondence between the graphs in Figure 6 and the actual cross sections of $f(x, y) = 2x^2 + y^2$ shown in Figure 7.

FIGURE 6

$y_1 = 2x^2$ $y_4 = 2x^2 + 9$
$y_2 = 2x^2 + 1$ $y_5 = 2x^2 + 16$
$y_3 = 2x^2 + 4$

FIGURE 7

FIGURE 8

$y_1 = x^2$ $y_4 = 18 + x^2$
$y_2 = 2 + x^2$ $y_5 = 32 + x^2$
$y_3 = 8 + x^2$

(B) The five cross sections are represented by the graphs of the functions $f(0, y) = y^2$, $f(1, y) = 2 + y^2$, $f(2, y) = 8 + y^2$, $f(3, y) = 18 + y^2$, and $f(4, y) = 32 + y^2$. These five functions are graphed in Figure 8. (Note that changing the name of the independent variable from y to x for graphing purposes does not affect the graph displayed.) Each of the five cross sections is a parabola that opens upward. ∎

Matched Problem 7

(A) Describe the cross sections of $g(x, y) = y^2 - x^2$ in the planes $y = 0, y = 1, y = 2, y = 3$, and $y = 4$.

(B) Describe the cross sections of $g(x, y) = y^2 - x^2$ in the planes $x = 0, x = 1, x = 2, x = 3$, and $x = 4$. ∎

Answers to Matched Problems

1. $3,100   **2.** 30

3. (A) $R(x, y) = 220x + 300y - 6x^2 + 4xy - 10y^2$; $R(20, 10) = \$4,800$
    (B) $P(x, y) = 180x + 220y - 6x^2 + 4xy - 10y^2 - 1,000$; $P(20, 10) = \$2,200$

4. $V(x, y) = x^2y$; $V(5, 10) = 250$ in.^{3} **5.** 17,411 units

6. $A(0, 0, 0)$; $C(2, 4, 0)$; $G(2, 4, 3)$; $D(0, 4, 0)$

7. (A) Each cross section is a parabola that opens downward.
    (B) Each cross section is a parabola that opens upward.

Exercise 8-1

A *In Problems 1–10, find the indicated values of the functions $f(x, y) = 7xy - 8$ and $g(x, y) = 4x - y^2 + 12$.*

1. $f(0, 0)$ **2.** $f(2, 1)$ **3.** $f(-3, 1)$

4. $f(2, -7)$ **5.** $g(0, 0)$ **6.** $g(0, -1)$

7. $g(2, -1)$ **8.** $g(-1, 2)$ **9.** $g(3, 4) - 6f(3, 4)$

10. $f(-2, 5) + 9g(-2, 5)$

B *Find the indicated value of the given function in Problems 11–24.*

11. $A(2, 3)$ for $A(x, y) = xy$

12. $V(2, 4, 3)$ for $V(x, y, z) = xyz$

13. $Q(12, 8)$ for $Q(M, C) = \dfrac{M}{C}(100)$

14. $T(50, 17)$ for $T(V, x) = \dfrac{33V}{x + 33}$

15. $V(2, 4)$ for $V(r, h) = \pi r^2 h$

16. $S(4, 2)$ for $S(x, y) = 5x^2 y^3$

17. $R(1, 2)$ for
$R(x, y) = -5x^2 + 6xy - 4y^2 + 200x + 300y$

18. $P(2, 2)$ for
$P(x, y) = -x^2 + 2xy - 2y^2 - 4x + 12y + 5$

19. $R(6, 0.5)$ for $R(L, r) = 0.002\dfrac{L}{r^4}$

20. $L(2,000, 50)$ for $L(w, v) = (1.25 \times 10^{-5})wv^2$

21. $A(100, 0.06, 3)$ for $A(P, r, t) = P + Prt$

22. $A(10, 0.04, 3, 2)$ for $A(P, r, t, n) = P\left(1 + \dfrac{r}{n}\right)^{tn}$

23. $P(0.05, 12)$ for $P(r, T) = \displaystyle\int_0^T 4{,}000e^{-rt}\, dt$

24. $F(0.07, 10)$ for $F(r, T) = \displaystyle\int_0^T 4{,}000e^{r(T-t)}\, dt$

25. Let $F(x, y) = x^2 + e^x y - y^2$. Find all values of x such that $F(x, 2) = 0$.

26. Let $G(a, b, c) = a^3 + b^3 + c^3 - (ab + ac + bc) - 6$. Find all values of b such that $G(2, b, 1) = 0$.

C

27. For the function $f(x, y) = x^2 + 2y^2$, find:
$$\frac{f(x + h, y) - f(x, y)}{h}$$

28. For the function $f(x, y) = x^2 + 2y^2$, find:
$$\frac{f(x, y + k) - f(x, y)}{k}$$

29. For the function $f(x, y) = 2xy^2$, find:
$$\frac{f(x + h, y) - f(x, y)}{h}$$

30. For the function $f(x, y) = 2xy^2$, find:
$$\frac{f(x, y + k) - f(x, y)}{k}$$

31. Find the coordinates of E and F in the figure for Matched Problem 6 (in the text).

32. Find the coordinates of B and H in the figure for Matched Problem 6 (in the text).

In Problems 33–38, use a graphing utility as necessary to explore the graphs of the indicated cross sections.

33. Let $f(x, y) = x^2$.

(A) Explain why the cross sections of the surface $z = f(x, y)$ by planes parallel to $y = 0$ are parabolas.

(B) Describe the cross sections of the surface by the planes $x = 0, x = 1$, and $x = 2$.

(C) Describe the surface $z = f(x, y)$.

34. Let $f(x, y) = \sqrt{4 - y^2}$.

(A) Explain why the cross sections of the surface $z = f(x, y)$ by planes parallel to $x = 0$ are semicircles of radius 2.

(B) Describe the cross sections of the surface by the planes $y = 0, y = 2$, and $y = 3$.

(C) Describe the surface $z = f(x, y)$.

35. Let $f(x, y) = \sqrt{36 - x^2 - y^2}$.

(A) Describe the cross sections of the surface $z = f(x, y)$ by the planes $y = 1, y = 2, y = 3, y = 4$, and $y = 5$.

(B) Describe the cross sections of the surface by the planes $x = 0, x = 1, x = 2, x = 3, x = 4$, and $x = 5$.

(C) Describe the surface $z = f(x, y)$.

36. Let $f(x, y) = 100 + 10x + 25y - x^2 - 5y^2$.

(A) Describe the cross sections of the surface $z = f(x, y)$ by the planes $y = 0, y = 1, y = 2$, and $y = 3$.

(B) Describe the cross sections of the surface by the planes $x = 0, x = 1, x = 2$, and $x = 3$.

(C) Describe the surface $z = f(x, y)$.

37. Let $f(x, y) = e^{-(x^2 + y^2)}$.

(A) Explain why $f(a, b) = f(c, d)$ whenever (a, b) and (c, d) are points on the same circle centered at the origin in the xy plane.

(B) Describe the cross sections of the surface $z = f(x, y)$ by the planes $x = 0, y = 0$, and $x = y$.

(C) Describe the surface $z = f(x, y)$.

38. Let $f(x, y) = 4 - \sqrt{x^2 + y^2}$.

    (A) Explain why $f(a, b) = f(c, d)$ whenever (a, b) and (c, d) are points on the same circle with center at the origin in the xy plane.

(B) Describe the cross sections of the surface $z = f(x, y)$ by the planes $x = 0, y = 0$, and $x = y$.

(C) Describe the surface $z = f(x, y)$.

Applications

Business & Economics

39. *Cost function.* A small manufacturing company produces two models of a surfboard: a standard model and a competition model. If the standard model is produced at a variable cost of $70 each, the competition model at a variable cost of $100 each, and the total fixed costs per month are $2,000, the monthly cost function is given by

$$C(x, y) = 2,000 + 70x + 100y$$

    where x and y are the numbers of standard and competition models produced per month, respectively. Find $C(20, 10)$, $C(50, 5)$, and $C(30, 30)$.

40. *Advertising and sales.* A company spends $x thousand per week on newspaper advertising and $y thousand per week on television advertising. Its weekly sales are found to be given by

$$S(x, y) = 5x^2y^3$$

    Find $S(3, 2)$ and $S(2, 3)$.

41. *Revenue function.* A supermarket sells two brands of coffee: brand A at $p per pound and brand B at $q per pound. The daily demand equations for brands A and B are, respectively

$$x = 200 - 5p + 4q$$
$$y = 300 + 2p - 4q$$

    (both in pounds). Find the daily revenue function $R(p, q)$. Evaluate $R(2, 3)$ and $R(3, 2)$.

42. *Revenue, cost, and profit functions.* A company manufactures ten- and three-speed bicycles. The weekly demand and cost equations are

$$p = 230 - 9x + y$$
$$q = 130 + x - 4y$$
$$C(x, y) = 200 + 80x + 30y$$

where $p is the price of a ten-speed bicycle, $q is the price of a three-speed bicycle, x is the weekly demand for ten-speed bicycles, y is the weekly demand for three-speed bicycles, and $C(x, y)$ is the cost function. Find the weekly revenue function $R(x, y)$ and the weekly profit function $P(x, y)$. Evaluate $R(10, 15)$ and $P(10, 15)$.

43. *Productivity.* The Cobb–Douglas production function for a petroleum company is given by

$$f(x, y) = 20x^{0.4}y^{0.6}$$

    where x is the utilization of labor and y is the utilization of capital. If the company uses 1,250 units of labor and 1,700 units of capital, how many units of petroleum will be produced?

44. *Productivity.* The petroleum company in Problem 43 is taken over by another company that decides to double both the units of labor and the units of capital utilized in the production of petroleum. Use the Cobb–Douglas production function given in Problem 43 to find the amount of petroleum that will be produced by this increased utilization of labor and capital. What is the effect on productivity of doubling both the units of labor and the units of capital?

45. *Future value.* At the end of each year, $2,000 is invested into an IRA earning 9% compounded annually.

    (A) How much will be in the account at the end of 30 years? Use the annuity formula

$$F(P, i, n) = P\frac{(1 + i)^n - 1}{i}$$

    where

$$P = \text{periodic payment}$$
$$i = \text{rate per period}$$
$$n = \text{number of payments (periods)}$$
$$F = FV = \text{future value}$$

    (B) Use graphical approximation methods to determine the rate of interest that would produce $500,000 in the account at the end of 30 years.

46. *Package design.* The packaging department in a company has been asked to design a rectangular box with no top and a partition down the middle (see the figure). Let x, y, and z be the dimensions (in inches).

Figure for 46

(A) Find the total amount of material $M(x, y, z)$ used in constructing one of these boxes, and evaluate $M(10, 12, 6)$.

(B) Suppose that the box is to have a square base and a volume of 720 cubic inches. Use graphical approximation methods to determine the dimensions that require the least amount of material.

Life Sciences

47. *Marine biology.* In using scuba diving gear, a marine
WWW biologist estimates the time of a dive according to the equation

$$T(V, x) = \frac{33V}{x + 33}$$

where

  T = time of dive in minutes

  V = volume of air, at sea level pressure, compressed into tanks

  x = depth of dive in feet

Find $T(70, 47)$ and $T(60, 27)$.

48. *Blood flow.* Poiseuille's law states that the resistance, R, for blood flowing in a blood vessel varies directly as the length of the vessel, L, and inversely as the fourth power of its radius, r. Stated as an equation,

$$R(L, r) = k\frac{L}{r^4} \qquad k \text{ a constant}$$

Find $R(8, 1)$ and $R(4, 0.2)$.

49. *Physical anthropology.* Anthropologists, in their
WWW study of race and human genetic groupings, often use an index called the *cephalic index*. The cephalic index, C, varies directly as the width W, of the head, and inversely as the length, L, of the head (both viewed from the top). In terms of an equation,

$$C(W, L) = 100\frac{W}{L}$$

where

  W = width in inches

  L = length in inches

Find $C(6, 8)$ and $C(8.1, 9)$.

Social Sciences

50. *Safety research.* Under ideal conditions, if a person driving a car slams on the brakes and skids to a stop, the length of the skid marks (in feet) is given by the formula

$$L(w, v) = kwv^2$$

where

  k = constant

  w = weight of car in pounds

  v = speed of car in miles per hour

For $k = 0.000\ 013\ 3$, find $L(2,000, 40)$ and $L(3,000, 60)$.

51. *Psychology.* The intelligence quotient (IQ) is defined to be the ratio of mental age (MA), as determined by certain tests, and chronological age (CA), multiplied by 100. Stated as an equation,

$$Q(M, C) = \frac{M}{C} \cdot 100$$

where

  Q = IQ M = MA C = CA

Find $Q(12, 10)$ and $Q(10, 12)$.

Partial Derivatives

- ❑ PARTIAL DERIVATIVES
- ❑ SECOND-ORDER PARTIAL DERIVATIVES

❑ PARTIAL DERIVATIVES

We know how to differentiate many kinds of functions of one independent variable and how to interpret the results. What about functions with two or more independent variables? Let us return to the surfboard example considered at the beginning of the chapter.

For the company producing only the standard board, the cost function was

$$C(x) = 500 + 70x$$

Differentiating with respect to x, we obtain the marginal cost function

$$C'(x) = 70$$

Since the marginal cost is constant, $70 is the change in cost for a 1-unit increase in production at any output level.

For the company producing two types of boards, a standard model and a competition model, the cost function was

$$C(x, y) = 700 + 70x + 100y$$

Now suppose that we differentiate with respect to x, holding y fixed, and denote this by $C_x(x, y)$; or suppose we differentiate with respect to y, holding x fixed, and denote this by $C_y(x, y)$. Differentiating in this way, we obtain

$$C_x(x, y) = 70 \qquad C_y(x, y) = 100$$

Each of these is called a **partial derivative,** and, in this example, each represents marginal cost. The first is the change in cost due to a 1-unit increase in production of the standard board with the production of the competition model held fixed. The second is the change in cost due to a 1-unit increase in production of the competition board with the production of the standard board held fixed.

In general, if $z = f(x, y)$, then the **partial derivative of f with respect to x,** denoted by $\partial z / \partial x, f_x$, or $f_x(x, y)$, is defined by

$$\frac{\partial z}{\partial x} = \lim_{h \to 0} \frac{f(x + h, y) - f(x, y)}{h}$$

provided that the limit exists. We recognize this as the ordinary derivative of f with respect to x, holding y constant. Thus, we are able to continue to use all the derivative rules and properties discussed in Chapters 3 to 5 for partial derivatives.

Similarly, the **partial derivative of f with respect to y,** denoted by $\partial z / \partial y, f_y$, or $f_y(x, y)$, is defined by

$$\frac{\partial z}{\partial y} = \lim_{k \to 0} \frac{f(x, y + k) - f(x, y)}{k}$$

which is the ordinary derivative with respect to y, holding x constant.

Parallel definitions and interpretations hold for functions with three or more independent variables.

Example 1 ➭ **Partial Derivatives** For $z = f(x, y) = 2x^2 - 3x^2y + 5y + 1$, find:

(A) $\partial z / \partial x$ (B) $f_x(2, 3)$

Solution   (A) $z = 2x^2 - 3x^2y + 5y + 1$

Differentiating with respect to x, holding y constant (that is, treating y as a constant), we obtain

$$\frac{\partial z}{\partial x} = 4x - 6xy$$

(B) $f(x, y) = 2x^2 - 3x^2y + 5y + 1$

First, differentiate with respect to x. From part (A) we have

$$f_x(x, y) = 4x - 6xy$$

Then evaluate at $(2, 3)$:

$$f_x(2, 3) = 4(2) - 6(2)(3) = -28$$

 In Example 1B, an alternative approach would be to substitute $y = 3$ into $f(x, y)$ and graph the function $f(x, 3) = -7x^2 + 16$, which represents the cross section of the surface $z = f(x, y)$ by the plane $y = 3$. Then determine the slope of the tangent line when $x = 2$. Again, we conclude that $f_x(2, 3) = -28$ (see Fig. 1).

FIGURE 1 $y_1 = -7x^2 + 16$

Matched Problem 1   For f in Example 1, find:

(A) $\partial z / \partial y$ (B) $f_y(2, 3)$

Example 2   **Partial Derivatives Using the Chain Rule**   For $z = f(x, y) = e^{x^2 + y^2}$, find:

(A) $\partial z / \partial x$ (B) $f_y(2, 1)$

Solution   (A) Using the chain rule [thinking of $z = e^u$, $u = u(x)$; y is held constant], we obtain

$$\frac{\partial z}{\partial x} = e^{x^2 + y^2} \frac{\partial(x^2 + y^2)}{\partial x}$$

$$= 2xe^{x^2 + y^2}$$

(B) $f_y(x, y) = e^{x^2 + y^2} \dfrac{\partial(x^2 + y^2)}{\partial y} = 2ye^{x^2 + y^2}$

$$f_y(2, 1) = 2(1)e^{(2)^2 + (1)^2}$$
$$= 2e^5$$

Matched Problem 2   For $z = f(x, y) = (x^2 + 2xy)^5$, find:

(A) $\partial z / \partial y$ (B) $f_x(1, 0)$

Example 3 **Profit** The profit function for the surfboard company in Example 3 in Section 8-1 was

$$P(x, y) = 140x + 200y - 4x^2 + 2xy - 12y^2 - 700$$

Find $P_x(15, 10)$ and $P_x(30, 10)$, and interpret the results.

SOLUTION
$$P_x(x, y) = 140 - 8x + 2y$$
$$P_x(15, 10) = 140 - 8(15) + 2(10) = 40$$
$$P_x(30, 10) = 140 - 8(30) + 2(10) = -80$$

At a production level of 15 standard and 10 competition boards per week, increasing the production of standard boards by 1 unit and holding the production of competition boards fixed at 10 will increase profit by approximately \$40. At a production level of 30 standard and 10 competition boards per week, increasing the production of standard boards by 1 unit and holding the production of competition boards fixed at 10 will decrease profit by approximately \$80. ∎

Matched Problem 3 For the profit function in Example 3, find $P_y(25, 10)$ and $P_y(25, 15)$, and interpret the results. ∎

Explore–Discuss 1

Let $P(x, y)$ be the profit function of Example 3 and Matched Problem 3.

(A) Assume that the production of competition boards remains fixed at 10. Which production level of standard boards will yield a maximum profit? Calculate that maximum profit.

(B) Assume that the production of standard boards remains fixed at 25. Which production level of competition boards will yield a maximum profit? Calculate that maximum profit.

Example 4 **Productivity** The productivity of a major computer manufacturer is given approximately by the Cobb–Douglas production function

$$f(x, y) = 15x^{0.4}y^{0.6}$$

with the utilization of x units of labor and y units of capital. The partial derivative $f_x(x, y)$ represents the rate of change of productivity with respect to labor and is called the **marginal productivity of labor.** The partial derivative $f_y(x, y)$ represents the rate of change of productivity with respect to capital and is called the **marginal productivity of capital.** If the company is currently utilizing 4,000 units of labor and 2,500 units of capital, find the marginal productivity of labor and the marginal productivity of capital. For the greatest increase in productivity, should the management of the company encourage increased use of labor or increased use of capital?

SOLUTION
$$f_x(x, y) = 6x^{-0.6}y^{0.6}$$
$$f_x(4,000, 2,500) = 6(4,000)^{-0.6}(2,500)^{0.6}$$
$$\approx 4.53 \qquad \text{Marginal productivity of labor}$$

$$f_y(x, y) = 9x^{0.4}y^{-0.4}$$
$$f_y(4,000, 2,500) = 9(4,000)^{0.4}(2,500)^{-0.4}$$
$$\approx 10.86 \qquad \text{Marginal productivity of capital}$$

At the current level of utilization of 4,000 units of labor and 2,500 units of capital, each 1-unit increase in labor utilization (keeping capital utilization fixed at 2,500 units) will increase production by approximately 4.53 units, and each 1-unit increase in capital utilization (keeping labor utilization fixed at 4,000 units) will increase production by approximately 10.86 units. Thus, the management of the company should encourage increased use of capital. ∎

Matched Problem 4 ⇨ The productivity of an airplane manufacturing company is given approximately by the Cobb–Douglas production function

$$f(x, y) = 40x^{0.3}y^{0.7}$$

(A) Find $f_x(x, y)$ and $f_y(x, y)$.

(B) If the company is currently using 1,500 units of labor and 4,500 units of capital, find the marginal productivity of labor and the marginal productivity of capital.

(C) For the greatest increase in productivity, should the management of the company encourage increased use of labor or increased use of capital? ∎

Partial derivatives have simple geometric interpretations, as indicated in Figure 2. If we hold x fixed, say, $x = a$, then $f_y(a, y)$ is the slope of the curve obtained by intersecting the plane $x = a$ with the surface $z = f(x, y)$. A similar interpretation is given to $f_x(x, b)$.

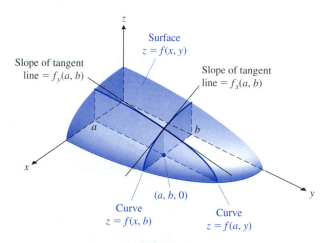

FIGURE 2

Explore–Discuss 2 Let $f(x, y) = x^2y - 2xy^2 + 3$ and consider the surface $z = f(x, y)$.

(A) The cross section of the surface by the plane $x = 2$ is the graph of $f(2, y) = 4y - 4y^2 + 3$, which is a parabola. Use an ordinary derivative to find the slope of the tangent line to this parabola when $y = \frac{1}{2}$.

(B) Use partial derivatives to confirm your answer to part (A).

(C) Explain why the cross section of the surface by the plane $y = 1$ is also a parabola. Use an ordinary derivative to find the slope of the tangent line to this parabola when $x = 3$.

(D) Use partial derivatives to confirm your answer to part (C).

□ SECOND-ORDER PARTIAL DERIVATIVES

The function

$$z = f(x, y) = x^4 y^7$$

has two **first-order partial derivatives,**

$$\frac{\partial z}{\partial x} = f_x = f_x(x, y) = 4x^3 y^7 \quad \text{and} \quad \frac{\partial z}{\partial y} = f_y = f_y(x, y) = 7x^4 y^6$$

Each of these partial derivatives, in turn, has two partial derivatives called **second-order partial derivatives** of $z = f(x, y)$. Generalizing the various notations we have for first-order partial derivatives, the four second-order partial derivatives of $z = f(x, y) = x^4 y^7$ are written as

Equivalent notations

$$f_{xx} = f_{xx}(x, y) = \frac{\partial^2 z}{\partial x^2} = \frac{\partial}{\partial x}\left(\frac{\partial z}{\partial x}\right) = \frac{\partial}{\partial x}(4x^3 y^7) = 12x^2 y^7$$

$$f_{xy} = f_{xy}(x, y) = \frac{\partial^2 z}{\partial y\, \partial x} = \frac{\partial}{\partial y}\left(\frac{\partial z}{\partial x}\right) = \frac{\partial}{\partial y}(4x^3 y^7) = 28x^3 y^6$$

$$f_{yx} = f_{yx}(x, y) = \frac{\partial^2 z}{\partial x\, \partial y} = \frac{\partial}{\partial x}\left(\frac{\partial z}{\partial y}\right) = \frac{\partial}{\partial x}(7x^4 y^6) = 28x^3 y^6$$

$$f_{yy} = f_{yy}(x, y) = \frac{\partial^2 z}{\partial y^2} = \frac{\partial}{\partial y}\left(\frac{\partial z}{\partial y}\right) = \frac{\partial}{\partial y}(7x^4 y^6) = 42x^4 y^5$$

In the mixed partial derivative $\partial^2 z / \partial y\, \partial x = f_{xy}$, we started with $z = f(x, y)$ and first differentiated with respect to x (holding y constant). Then we differentiated with respect to y (holding x constant). In the other mixed partial derivative, $\partial^2 z / \partial x\, \partial y = f_{yx}$, the order of differentiation was reversed; however, the final result was the same—that is, $f_{xy} = f_{yx}$. Although it is possible to find functions for which $f_{xy} \neq f_{yx}$, such functions rarely occur in applications involving partial derivatives. Thus, for all the functions in this book, we will assume that $f_{xy} = f_{yx}$.

In general, we have the following definitions:

Second-Order Partial Derivatives

If $z = f(x, y)$, then

$$f_{xx} = f_{xx}(x, y) = \frac{\partial^2 z}{\partial x^2} = \frac{\partial}{\partial x}\left(\frac{\partial z}{\partial x}\right)$$

$$f_{xy} = f_{xy}(x, y) = \frac{\partial^2 z}{\partial y\, \partial x} = \frac{\partial}{\partial y}\left(\frac{\partial z}{\partial x}\right)$$

$$f_{yx} = f_{yx}(x, y) = \frac{\partial^2 z}{\partial x\, \partial y} = \frac{\partial}{\partial x}\left(\frac{\partial z}{\partial y}\right)$$

$$f_{yy} = f_{yy}(x, y) = \frac{\partial^2 z}{\partial y^2} = \frac{\partial}{\partial y}\left(\frac{\partial z}{\partial y}\right)$$

Example 5 ⇔ **Second-Order Partial Derivatives**   For $z = f(x, y) = 3x^2 - 2xy^3 + 1$, find:

(A) $\dfrac{\partial^2 z}{\partial x\, \partial y}, \dfrac{\partial^2 z}{\partial y\, \partial x}$ (B) $\dfrac{\partial^2 z}{\partial x^2}$ (C) $f_{yx}(2, 1)$

SOLUTION   (A) First differentiate with respect to y and then with respect to x:

$$\frac{\partial z}{\partial y} = -6xy^2 \qquad \frac{\partial^2 z}{\partial x\, \partial y} = \frac{\partial}{\partial x}\!\left(\frac{\partial z}{\partial y}\right) = \frac{\partial}{\partial x}(-6xy^2) = -6y^2$$

First differentiate with respect to x and then with respect to y:

$$\frac{\partial z}{\partial x} = 6x - 2y^3 \qquad \frac{\partial^2 z}{\partial y\, \partial x} = \frac{\partial}{\partial y}\!\left(\frac{\partial z}{\partial x}\right) = \frac{\partial}{\partial y}(6x - 2y^3) = -6y^2$$

(B) Differentiate with respect to x twice:

$$\frac{\partial z}{\partial x} = 6x - 2y^3 \qquad \frac{\partial^2 z}{\partial x^2} = \frac{\partial}{\partial x}\!\left(\frac{\partial z}{\partial x}\right) = 6$$

(C) First find $f_{yx}(x, y)$; then evaluate at $(2, 1)$. Again, remember that f_{yx} means to differentiate with respect to y first and then with respect to x. Thus,

$$f_y(x, y) = -6xy^2 \qquad f_{yx}(x, y) = -6y^2$$

and

$$f_{yx}(2, 1) = -6(1)^2 = -6$$

■

Matched Problem 5 ⇔   For $z = f(x, y) = x^3 y - 2y^4 + 3$, find:

(A) $\dfrac{\partial^2 z}{\partial y\, \partial x}$ (B) $\dfrac{\partial^2 z}{\partial y^2}$

(C) $f_{xy}(2, 3)$ (D) $f_{yx}(2, 3)$

■

Answers to Matched Problems   **1.** (A) $\partial z / \partial y = -3x^2 + 5$ (B) $f_y(2, 3) = -7$

2. (A) $10x(x^2 + 2xy)^4$ (B) 10

3. $P_y(25, 10) = 10$: at a production level of $x = 25$ and $y = 10$, increasing y by 1 unit and holding x fixed at 25 will increase profit by approx. \$10; $P_y(25, 15) = -110$: at a production level of $x = 25$ and $y = 15$, increasing y by 1 unit and holding x fixed at 25 will decrease profit by approx. \$110

4. (A) $f_x(x, y) = 12x^{-0.7}y^{0.7}$; $f_y(x, y) = 28x^{0.3}y^{-0.3}$

(B) Marginal productivity of labor ≈ 25.89;
marginal productivity of capital ≈ 20.14

(C) Labor

5. (A) $3x^2$ (B) $-24y^2$

(C) 12 (D) 12

Exercise 8-2

A *In Problems 1–4, let z = f(x, y) = 7 − 2x + 5y and find the indicated function or value.*

1. $\partial z / \partial x$ **2.** $\partial z / \partial y$

3. $f_y(1, 2)$ **4.** $f_x(1, 2)$

In Problems 5–10, let z = f(x, y) = 6x²y + 3y³ and find the indicated function or value.

5. $\partial z / \partial y$ **6.** $\partial z / \partial x$

7. $\dfrac{\partial^2 z}{\partial y^2}$ **8.** $\dfrac{\partial^2 z}{\partial x^2}$

9. $f_x(2, 3)$ **10.** $f_y(2, 3)$

In Problems 11–18, let C(x, y) = 5x² − 6xy + 3y² − 4x + 10y − 15 and find the indicated function or value.

11. $C_x(x, y)$ **12.** $C_y(x, y)$

13. $C_x(2, 2)$ **14.** $C_y(2, 2)$

15. $C_{xy}(x, y)$ **16.** $C_{yx}(x, y)$

17. $C_{xx}(x, y)$ **18.** $C_{yy}(x, y)$

B *In Problems 19–24, let S(x, y) = 4x³y + y ln x and find the indicated function or value.*

19. $S_x(x, y)$ **20.** $S_y(x, y)$

21. $S_y(2, 1)$ **22.** $S_x(2, 1)$

23. $S_{xy}(x, y)$ **24.** $S_{yx}(x, y)$

In Problems 25–32, let z = f(x, y) = e^{4x²+5y} and find the indicated function or value.

25. $\dfrac{\partial z}{\partial x}$ **26.** $\dfrac{\partial z}{\partial y}$

27. $\dfrac{\partial^2 z}{\partial x \, \partial y}$ **28.** $\dfrac{\partial^2 z}{\partial y \, \partial x}$

29. $f_{xy}(1, 0)$ **30.** $f_{yx}(0, 1)$

31. $f_{xx}(0, 1)$ **32.** $f_{yy}(1, 0)$

In Problems 33–42, find $f_x(x, y)$ and $f_y(x, y)$ for each function f.

33. $f(x, y) = (x^2 - y^3)^3$ **34.** $f(x, y) = \sqrt{2x - y^2}$

35. $f(x, y) = (3x^2y - 1)^4$ **36.** $f(x, y) = (3 + 2xy^2)^3$

37. $f(x, y) = \ln(x^2 + y^2)$ **38.** $f(x, y) = \ln(2x - 3y)$

39. $f(x, y) = y^2 e^{xy^2}$ **40.** $f(x, y) = x^3 e^{x^2y}$

41. $f(x, y) = \dfrac{x^2 - y^2}{x^2 + y^2}$ **42.** $f(x, y) = \dfrac{2x^2y}{x^2 + y^2}$

43. (A) Let $f(x, y) = y^3 + 4y^2 - 5y + 3$. Show that $\partial f / \partial x = 0$.

(B) Explain why there are an infinite number of functions $g(x, y)$ such that $\partial g / \partial x = 0$.

44. (A) Find an example of a function $f(x, y)$ such that $\partial f / \partial x = 3$ and $\partial f / \partial y = 2$.

(B) How many such functions are there? Explain.

In Problems 45–50, find $f_{xx}(x, y), f_{xy}(x, y), f_{yx}(x, y),$ and $f_{yy}(x, y)$ for each function f.

45. $f(x, y) = x^2y^2 + x^3 + y$

46. $f(x, y) = x^3y^3 + x + y^2$

47. $f(x, y) = \dfrac{x}{y} - \dfrac{y}{x}$ **48.** $f(x, y) = \dfrac{x^2}{y} - \dfrac{y^2}{x}$

49. $f(x, y) = xe^{xy}$ **50.** $f(x, y) = x \ln(xy)$

C

51. For
$$P(x, y) = -x^2 + 2xy - 2y^2 - 4x + 12y - 5$$
find all values of x and y such that
$$P_x(x, y) = 0 \quad \text{and} \quad P_y(x, y) = 0$$
simultaneously.

52. For
$$C(x, y) = 2x^2 + 2xy + 3y^2 - 16x - 18y + 54$$
find all values of x and y such that
$$C_x(x, y) = 0 \quad \text{and} \quad C_y(x, y) = 0$$
simultaneously.

53. For
$$F(x, y) = x^3 - 2x^2y^2 - 2x - 4y + 10$$
find all values of x and y such that
$$F_x(x, y) = 0 \quad \text{and} \quad F_y(x, y) = 0$$
simultaneously.

54. For
$$G(x, y) = x^2 \ln y - 3x - 2y + 1$$
find all values of x and y such that
$$G_x(x, y) = 0 \quad \text{and} \quad G_y(x, y) = 0$$
simultaneously.

55. Let $f(x, y) = 3x^2 + y^2 - 4x - 6y + 2$.

(A) Find the minimum value of $f(x, y)$ when $y = 1$.

(B) Explain why the answer to part (A) is not the minimum value of the function $f(x, y)$.

56. Let $f(x, y) = 5 - 2x + 4y - 3x^2 - y^2$.

(A) Find the maximum value of $f(x, y)$ when $x = 2$.

(B) Explain why the answer to part (A) is not the maximum value of the function $f(x, y)$.

57. Let $f(x, y) = 4 - x^4y + 3xy^2 + y^5$.

(A) Use graphical approximation methods to find c (to three decimal places) such that $f(c, 2)$ is the maximum value of $f(x, y)$ when $y = 2$.

(B) Find $f_x(c, 2)$ and $f_y(c, 2)$.

58. Let $f(x, y) = e^x + 2e^y + 3xy^2 + 1$.

(A) Use graphical approximation methods to find d (to three decimal places) such that $f(1, d)$ is the minimum value of $f(x, y)$ when $x = 1$.

(B) Find $f_x(1, d)$ and $f_y(1, d)$.

In Problems 59 and 60, show that the function f satisfies $f_{xx}(x, y) + f_{yy}(x, y) = 0.$

59. $f(x, y) = \ln(x^2 + y^2)$ **60.** $f(x, y) = x^3 - 3xy^2$

61. For $f(x, y) = x^2 + 2y^2$, find:

(A) $\lim_{h \to 0} \dfrac{f(x + h, y) - f(x, y)}{h}$

(B) $\lim_{k \to 0} \dfrac{f(x, y + k) - f(x, y)}{k}$

62. For $f(x, y) = 2xy^2$, find:

(A) $\lim_{h \to 0} \dfrac{f(x + h, y) - f(x, y)}{h}$

(B) $\lim_{k \to 0} \dfrac{f(x, y + k) - f(x, y)}{k}$

Applications

Business & Economics

63. *Profit function.* A firm produces two types of calculators, x of type A and y of type B each week. The weekly revenue and cost functions (in dollars) are

$$R(x, y) = 80x + 90y + 0.04xy - 0.05x^2 - 0.05y^2$$
$$C(x, y) = 8x + 6y + 20,000$$

Find $P_x(1,200, 1,800)$ and $P_y(1,200, 1,800)$, and interpret the results.

64. *Advertising and sales.* A company spends $\$x$ per week on newspaper advertising and $\$y$ per week on television advertising. Its weekly sales were found to be given by

$$S(x, y) = 10x^{0.4}y^{0.8}$$

Find $S_x(3,000, 2,000)$ and $S_y(3,000, 2,000)$, and interpret the results.

65. *Demand equations.* A supermarket sells two brands of coffee, brand A at $\$p$ per pound and brand B at $\$q$ per pound. The daily demand equations for brands A and B are, respectively,

$$x = 200 - 5p + 4q$$
$$y = 300 + 2p - 4q$$

Find $\partial x/\partial p$ and $\partial y/\partial p$, and interpret the results.

66. *Revenue and profit functions.* A company manufactures ten- and three-speed bicycles. The weekly demand and cost functions are

$$p = 230 - 9x + y$$
$$q = 130 + x - 4y$$
$$C(x, y) = 200 + 80x + 30y$$

where $\$p$ is the price of a ten-speed bicycle, $\$q$ is the price of a three-speed bicycle, x is the weekly demand for ten-speed bicycles, y is the weekly demand for three-speed bicycles, and $C(x, y)$ is the cost function. Find $R_x(10, 5)$ and $P_x(10, 5)$, and interpret the results.

67. *Productivity.* The productivity of a certain third-world country is given approximately by the function

$$f(x, y) = 10x^{0.75}y^{0.25}$$

with the utilization of x units of labor and y units of capital.

(A) Find $f_x(x, y)$ and $f_y(x, y)$.

(B) If the country is now using 600 units of labor and 100 units of capital, find the marginal productivity of labor and the marginal productivity of capital.

(C) For the greatest increase in the country's productivity, should the government encourage increased use of labor or increased use of capital?

68. *Productivity.* The productivity of an automobile manufacturing company is given approximately by the function

$$f(x, y) = 50\sqrt{xy} = 50x^{0.5}y^{0.5}$$

with the utilization of x units of labor and y units of capital.

(A) Find $f_x(x, y)$ and $f_y(x, y)$.

(B) If the company is now using 250 units of labor and 125 units of capital, find the marginal productivity of labor and the marginal productivity of capital.

(C) For the greatest increase in the company's productivity, should the management encourage increased use of labor or increased use of capital?

Problems 69–72 refer to the following: If a decrease in demand for one product results in an increase in demand for another product, the two products are said to be **competitive, or substitute, products.** *(Real whipping cream and imitation whipping cream are examples of competitive, or substitute, products.) If a decrease in demand for one product results in a decrease in demand for another product, the two products are said to be* **complementary products.** *(Fishing boats and outboard motors are examples of complementary products.) Partial derivatives can be used to test whether two products are competitive, complementary, or neither. We start with demand functions for two products where the demand for either depends on the prices for both:*

$$x = f(p, q) \qquad \text{Demand function for product A}$$
$$y = g(p, q) \qquad \text{Demand function for product B}$$

The variables x and y represent the number of units demanded of products A and B, respectively, at a price p for 1 unit of product A and a price q for 1 unit of product B. Normally, if the price of A increases while the price of B is held constant, then the demand for A will decrease; that is, $f_p(p, q) < 0$. Then, if A and B are competitive products, the demand for B will increase; that is, $g_p(p, q) > 0$. Similarly, if the price of B increases while the price of A is held constant, the demand for B will decrease; that is, $g_q(p, q) < 0$. And if A and B are competitive products, the demand for A will increase; that is, $f_q(p, q) > 0$. Reasoning similarly for complementary products, we arrive at the following test:

Test for Competitive and Complementary Products

PARTIAL DERIVATIVES		PRODUCTS A AND B
$f_q(p, q) > 0$ and	$g_p(p, q) > 0$	Competitive (substitute)
$f_q(p, q) < 0$ and	$g_p(p, q) < 0$	Complementary
$f_q(p, q) \geq 0$ and	$g_p(p, q) \leq 0$	Neither
$f_q(p, q) \leq 0$ and	$g_p(p, q) \geq 0$	Neither

Use this test in Problems 69–72 to determine whether the indicated products are competitive, complementary, or neither.

69. *Product demand.* The weekly demand equations for the sale of butter and margarine in a supermarket are

$$x = f(p, q) = 8{,}000 - 0.09p^2 + 0.08q^2 \quad \text{Butter}$$
$$y = g(p, q) = 15{,}000 + 0.04p^2 - 0.3q^2 \quad \text{Margarine}$$

70. *Product demand.* The daily demand equations for the sale of brand A coffee and brand B coffee in a supermarket are

$$x = f(p, q) = 200 - 5p + 4q \quad \text{Brand A coffee}$$
$$y = g(p, q) = 300 + 2p - 4q \quad \text{Brand B coffee}$$

71. *Product demand.* The monthly demand equations for the sale of skis and ski boots in a sporting goods store are

$$x = f(p, q) = 800 - 0.004p^2 - 0.003q^2 \quad \text{Skis}$$
$$y = g(p, q) = 600 - 0.003p^2 - 0.002q^2 \quad \text{Ski boots}$$

72. *Product demand.* The monthly demand equations for the sale of tennis rackets and tennis balls in a sporting goods store are

$$x = f(p, q) = 500 - 0.5p - q^2 \quad \text{Tennis rackets}$$
$$y = g(p, q) = 10{,}000 - 8p - 100q^2 \quad \text{Tennis balls (cans)}$$

Life Sciences

73. *Medicine.* The following empirical formula relates the surface area A (in square inches) of an average human body to its weight w (in pounds) and its height h (in inches):

$$A = f(w, h) = 15.64w^{0.425}h^{0.725}$$

Knowing the surface area of a human body is useful, for example, in studies pertaining to hypothermia (heat loss due to exposure).

(A) Find $f_w(w, h)$ and $f_h(w, h)$.

(B) For a 65-pound child who is 57 inches tall, find $f_w(65, 57)$ and $f_h(65, 57)$, and interpret the results.

74. *Blood flow.* Poiseuille's law states that the resistance, R, for blood flowing in a blood vessel varies directly as the length of the vessel, L, and inversely as the fourth power of its radius, r. Stated as an equation,

$$R(L, r) = k\frac{L}{r^4} \qquad k \text{ a constant}$$

Find $R_L(4, 0.2)$ and $R_r(4, 0.2)$, and interpret the results.

Social Sciences

75. *Physical anthropology.* Anthropologists, in their study of race and human genetic groupings, often use the cephalic index, C, which varies directly as the width, W, of the head, and inversely as the length, L, of the head (both viewed from the top). In terms of an equation,

$$C(W, L) = 100 \frac{W}{L}$$

where

W = width in inches L = length in inches

Find $C_W(6, 8)$ and $C_L(6, 8)$, and interpret the results.

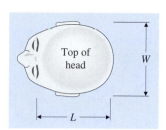

76. *Safety research.* Under ideal conditions, if a person driving a car slams on the brakes and skids to a stop, the length of the skid marks (in feet) is given by the formula

$$L(w, v) = kwv^2$$

where

k = constant

w = weight of car in pounds

v = speed of car in miles per hour

For $k = 0.000\ 013\ 3$,   find $L_w(2,500, 60)$ and $L_v(2,500, 60)$, and interpret the results.

| Section 8-3 | ## Maxima and Minima |

We are now ready to undertake a brief but useful analysis of local maxima and minima for functions of the type $z = f(x, y)$. Basically, we are going to extend the second-derivative test developed for functions of a single independent variable. To start, we assume that all second-order partial derivatives exist for the function f in some circular region in the xy plane. This guarantees that the surface $z = f(x, y)$ has no sharp points, breaks, or ruptures. In other words, we are dealing only with smooth surfaces with no edges (like the edge of a box); or breaks (like an earthquake fault); or sharp points (like the bottom point of a golf tee). See Figure 1.

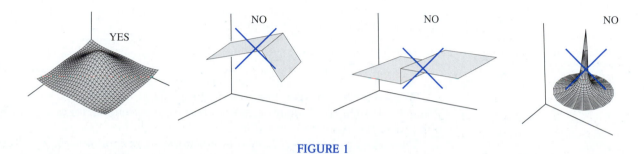

FIGURE 1

In addition, we will not concern ourselves with boundary points or absolute maxima–minima theory. Despite these restrictions, the procedure we are now going to describe will help us solve a large number of useful problems.

What does it mean for $f(a, b)$ to be a local maximum or a local minimum? We say that $f(a, b)$ **is a local maximum** if there exists a circular region in the domain of f with (a, b) as the center, such that

$$f(a, b) \geq f(x, y)$$

for all (x, y) in the region. Similarly, we say that $f(a, b)$ **is a local minimum** if there exists a circular region in the domain of f with (a, b) as the center, such that

$$f(a, b) \leq f(x, y)$$

for all (x, y) in the region. Figure 2A illustrates a local maximum, Figure 2B a local minimum, and Figure 2C a **saddle point,** which is neither a local maximum nor a local minimum.

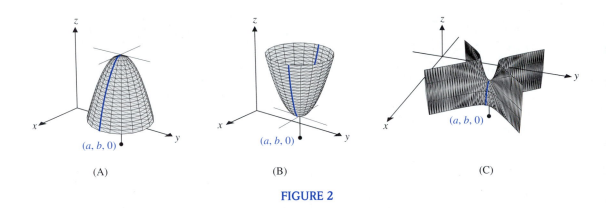

(A)   (B)   (C)

FIGURE 2

What happens to $f_x(a, b)$ and $f_y(a, b)$ if $f(a, b)$ is a local minimum or a local maximum and the partial derivatives of f exist in a circular region containing (a, b)? Figure 2 suggests that $f_x(a, b) = 0$ and $f_y(a, b) = 0$, since the tangent lines to the given curves are horizontal. Theorem 1 indicates that our intuitive reasoning is correct.

THEOREM 1   Local Extrema and Partial Derivatives

Let $f(a, b)$ be a local extremum (a local maximum or a local minimum) for the function f. If both f_x and f_y exist at (a, b) then

$$f_x(a, b) = 0 \quad \text{and} \quad f_y(a, b) = 0 \tag{1}$$

The converse of this theorem is false. That is, if $f_x(a, b) = 0$ and $f_y(a, b) = 0$, then $f(a, b)$ may or may not be a local extremum; for example, the point $(a, b, f(a, b))$ may be a saddle point (see Fig. 2C).

Theorem 1 gives us *necessary* (but not *sufficient*) conditions for $f(a, b)$ to be a local extremum. We thus find all points (a, b) such that $f_x(a, b) = 0$ and $f_y(a, b) = 0$ and test these further to determine whether $f(a, b)$ is a local extremum or a saddle point. Points (a, b) such that conditions (1) hold are called **critical points.**

(A) Let $f(x, y) = y^2 + 1$. Explain why $f(x, y)$ has a local minimum at every point on the x axis. Verify that every point on the x axis is a critical point. Explain why the graph of $z = f(x, y)$ could be described as a "trough."

(B) Let $g(x, y) = x^3$. Show that every point on the y axis is a critical point. Explain why no point on the y axis is a local extremum. Explain why the graph of $z = g(x, y)$ could be described as a "slide."

The next theorem, using second-derivative tests, gives us *sufficient* conditions for a critical point to produce a local extremum or a saddle point. (As was the case with Theorem 1, we state this theorem without proof.)

THEOREM 2   Second-Derivative Test for Local Extrema

Given:

1. $z = f(x, y)$
2. $f_x(a, b) = 0$ and $f_y(a, b) = 0$ $[(a, b)$ is a critical point]
3. All second-order partial derivatives of f exist in some circular region containing (a, b) as a center.
4. $A = f_{xx}(a, b), \quad B = f_{xy}(a, b), \quad C = f_{yy}(a, b)$

Then:

Case 1.   If $AC - B^2 > 0$ and $A < 0$, then $f(a, b)$ is a local maximum.
Case 2.   If $AC - B^2 > 0$ and $A > 0$, then $f(a, b)$ is a local minimum.
Case 3.   If $AC - B^2 < 0$, then f has a saddle point at (a, b).
Case 4.   If $AC - B^2 = 0$, the test fails.

FIGURE 3

To illustrate the use of Theorem 2, we will first find the local extremum for a very simple function whose solution is almost obvious: $z = f(x, y) = x^2 + y^2 + 2$. From the function f itself and its graph (Fig. 3), it is clear that a local minimum is found at $(0, 0)$. Let us see how Theorem 2 confirms this observation.

Step 1.   Find critical points: Find (x, y) such that $f_x(x, y) = 0$ and $f_y(x, y) = 0$ simultaneously:

$$f_x(x, y) = 2x = 0 \qquad f_y(x, y) = 2y = 0$$
$$x = 0 \qquad\qquad\qquad y = 0$$

The only critical point is $(a, b) = (0, 0)$.

Step 2.   Compute $A = f_{xx}(0, 0), B = f_{xy}(0, 0)$, and $C = f_{yy}(0, 0)$:

$$
\begin{aligned}
f_{xx}(x, y) &= 2 &\quad \text{thus} \quad& A = f_{xx}(0, 0) = 2 \\
f_{xy}(x, y) &= 0 &\quad \text{thus} \quad& B = f_{xy}(0, 0) = 0 \\
f_{yy}(x, y) &= 2 &\quad \text{thus} \quad& C = f_{yy}(0, 0) = 2
\end{aligned}
$$

Step 3.   Evaluate $AC - B^2$ and try to classify the critical point $(0, 0)$ using Theorem 2:

$$AC - B^2 = (2)(2) - (0)^2 = 4 > 0 \qquad \text{and} \qquad A = 2 > 0$$

Therefore, case 2 in Theorem 2 holds. That is, $f(0, 0) = 2$ is a local minimum.

We will now use Theorem 2 in the following examples to analyze extrema without the aid of graphs.

Example 1   **Finding Local Extrema**   Use Theorem 2 to find local extrema for

$$f(x, y) = -x^2 - y^2 + 6x + 8y - 21$$

SOLUTION   **Step 1.**   Find critical points: Find (x, y) such that $f_x(x, y) = 0$ and $f_y(x, y) = 0$ simultaneously:

$$f_x(x, y) = -2x + 6 = 0 \qquad f_y(x, y) = -2y + 8 = 0$$
$$x = 3 \qquad\qquad\qquad y = 4$$

The only critical point is $(a, b) = (3, 4)$.

Step 2.   Compute $A = f_{xx}(3, 4)$, $B = f_{xy}(3, 4)$, and $C = f_{yy}(3, 4)$:

$$
\begin{array}{lll}
f_{xx}(x, y) = -2 & \text{thus} & A = f_{xx}(3, 4) = -2 \\
f_{xy}(x, y) = 0 & \text{thus} & B = f_{xy}(3, 4) = 0 \\
f_{yy}(x, y) = -2 & \text{thus} & C = f_{yy}(3, 4) = -2
\end{array}
$$

Step 3.   Evaluate $AC - B^2$ and try to classify the critical point $(3, 4)$ using Theorem 2:

$$AC - B^2 = (-2)(-2) - (0)^2 = 4 > 0 \qquad \text{and} \qquad A = -2 < 0$$

Therefore, case 1 in Theorem 2 holds. That is, $f(3, 4) = 4$ is a local maximum.

■

Matched Problem 1   Use Theorem 2 to find local extrema for:

$$f(x, y) = x^2 + y^2 - 10x - 2y + 36$$

■

Example 2   **Finding Local Extrema: Multiple Critical Points**   Use Theorem 2 to find local extrema for:

$$f(x, y) = x^3 + y^3 - 6xy$$

SOLUTION   **Step 1.**   Find critical points for $f(x, y) = x^3 + y^3 - 6xy$:

$$
\begin{aligned}
f_x(x, y) = 3x^2 - 6y &= 0 && \textit{Solve for y.} \\
6y &= 3x^2 \\
y &= \tfrac{1}{2}x^2 && \text{(2)}
\end{aligned}
$$

$$
\begin{aligned}
f_y(x, y) = 3y^2 - 6x &= 0 \\
3y^2 &= 6x && \textit{Use equation (2) to eliminate y.} \\
3(\tfrac{1}{2}x^2)^2 &= 6x \\
\tfrac{3}{4}x^4 &= 6x && \textit{Solve for x.} \\
3x^4 - 24x &= 0 \\
3x(x^3 - 8) &= 0
\end{aligned}
$$

$$
\begin{array}{lll}
x = 0 & \text{or} & x = 2 \\
y = 0 & & y = \tfrac{1}{2}(2)^2 = 2
\end{array}
$$

The critical points are $(0, 0)$ and $(2, 2)$.

Since there are two critical points, steps 2 and 3 must be performed twice.

Test (0, 0)　**Step 2.**　Compute $A = f_{xx}(0, 0)$, $B = f_{xy}(0, 0)$, and $C = f_{yy}(0, 0)$:

$$f_{xx}(x, y) = 6x \qquad \text{thus} \qquad A = f_{xx}(0, 0) = 0$$
$$f_{xy}(x, y) = -6 \qquad \text{thus} \qquad B = f_{xy}(0, 0) = -6$$
$$f_{yy}(x, y) = 6y \qquad \text{thus} \qquad C = f_{yy}(0, 0) = 0$$

Step 3.　Evaluate $AC - B^2$ and try to classify the critical point $(0, 0)$ using Theorem 2:

$$AC - B^2 = (0)(0) - (-6)^2 = -36 < 0$$

Therefore, case 3 in Theorem 2 applies. That is, f has a saddle point at $(0, 0)$.

Now we will consider the second critical point, $(2, 2)$.

Test (2, 2)　**Step 2.**　Compute $A = f_{xx}(2, 2)$, $B = f_{xy}(2, 2)$, and $C = f_{yy}(2, 2)$:

$$f_{xx}(x, y) = 6x \qquad \text{thus} \qquad A = f_{xx}(2, 2) = 12$$
$$f_{xy}(x, y) = -6 \qquad \text{thus} \qquad B = f_{xy}(2, 2) = -6$$
$$f_{yy}(x, y) = 6y \qquad \text{thus} \qquad C = f_{yy}(2, 2) = 12$$

Step 3.　Evaluate $AC - B^2$ and try to classify the critical point $(2, 2)$ using Theorem 2:

$$AC - B^2 = (12)(12) - (-6)^2 = 108 > 0 \qquad \text{and} \qquad A = 12 > 0$$

Thus, case 2 in Theorem 2 applies, and $f(2, 2) = -8$ is a local minimum.　■

　Our conclusions in Example 2 may be confirmed geometrically by graphing cross sections of the function f. The cross sections of f by the planes $y = 0, x = 0, y = x$, and $y = -x$ [each of these planes contains $(0, 0)$] are represented by the graphs of the functions $f(x, 0) = x^3, f(0, y) = y^3, f(x, x) = 2x^3 - 6x^2$, and $f(x, -x) = 6x^2$, respectively, as shown in Figure 4A (note that the first two functions have the same graph). The cross sections of f by the planes $y = 2, x = 2, y = x$, and $y = 4 - x$ [each of these planes contains $(2, 2)$] are represented by the graphs of the functions $f(x, 2) = x^3 - 12x + 8$, $f(2, y) = y^3 - 12y + 8$, $\qquad f(x, x) = 2x^3 - 6x^2$, $\qquad$ and $\qquad f(x, 4 - x) = x^3 + (4 - x)^3 + 6x^2 - 24x$, respectively, as shown in Figure 4B (the first two functions have the same graph). Figure 4B illustrates the fact that since f has a local minimum at $(2, 2)$, each of the cross sections of f through $(2, 2)$ has a local minimum of -8 at $(2, 2)$. Figure 4A, on the other hand, indicates that some cross sections of f through $(0, 0)$ have a local minimum, some a local maximum, and some neither one, at $(0, 0)$.

(A) $y_1 = x^3$
$\quad y_2 = 2x^3 - 6x^2$
$\quad y_3 = 6x^2$

(B) $y_1 = x^3 - 12x + 8$
$\quad y_2 = 2x^3 - 6x^2$
$\quad y_3 = x^3 + (4 - x)^3 + 6x^2 - 24x$

FIGURE 4

Matched Problem 2 ⇔ Use Theorem 2 to find local extrema for: $f(x, y) = x^3 + y^2 - 6xy$ ∎

Explore–Discuss 2

Let $f(x, y) = x^4 + y^2 + 3$, $g(x, y) = 10 - x^2 - y^4$, and $h(x, y) = x^3 + y^2$.

(A) Show that each function has only $(0, 0)$ as a critical point.

(B) Explain why $f(x, y)$ has a minimum at $(0, 0)$, $g(x, y)$ has a maximum at $(0, 0)$, and $h(x, y)$ has neither a minimum nor a maximum at $(0, 0)$.

(C) Are the results of part (B) consequences of Theorem 2? Explain.

Example 3 ⇔ **Profit** Suppose that the surfboard company discussed earlier has developed the yearly profit equation

$$P(x, y) = -22x^2 + 22xy - 11y^2 + 110x - 44y - 23$$

where x is the number (in thousands) of standard surfboards produced per year, y is the number (in thousands) of competition surfboards produced per year, and P is profit (in thousands of dollars). How many of each type of board should be produced per year to realize a maximum profit? What is the maximum profit?

Solution **Step 1.** Find critical points:

$$P_x(x, y) = -44x + 22y + 110 = 0$$
$$P_y(x, y) = 22x - 22y - 44 = 0$$

Solving this system, we obtain $(3, 1)$ as the only critical point.

Step 2. Compute $A = P_{xx}(3, 1)$, $B = P_{xy}(3, 1)$, and $C = P_{yy}(3, 1)$:

$$P_{xx}(x, y) = -44 \qquad \text{thus} \qquad A = P_{xx}(3, 1) = -44$$
$$P_{xy}(x, y) = 22 \qquad \text{thus} \qquad B = P_{xy}(3, 1) = 22$$
$$P_{yy}(x, y) = -22 \qquad \text{thus} \qquad C = P_{yy}(3, 1) = -22$$

Step 3. Evaluate $AC - B^2$ and try to classify the critical point $(3, 1)$ using Theorem 2:

$$AC - B^2 = (-44)(-22) - 22^2 = 484 > 0 \qquad \text{and} \qquad A = -44 < 0$$

Therefore, case 1 in Theorem 2 applies. That is, $P(3, 1) = 120$ is a local maximum. A maximum profit of $\$120,000$ is obtained by producing and selling 3,000 standard boards and 1,000 competition boards per year. ∎

Matched Problem 3 ⇔ Repeat Example 3 with:

$$P(x, y) = -66x^2 + 132xy - 99y^2 + 132x - 66y - 19$$

∎

Example 4 ⇔ **Package Design** The packaging department in a company has been asked to design a rectangular box with no top and a partition down the middle. The box must have a volume of 48 cubic inches. Find the dimensions that will minimize the amount of material used to construct the box.

Solution Refer to Figure 5. The amount of material used in constructing this box is

$$\overset{\text{Base}}{M} = \overset{\substack{\text{Front,}\\\text{back}}}{xy} + \overset{\substack{\text{Sides,}\\\text{partition}}}{2xz} + 3yz \tag{3}$$

FIGURE 5

The volume of the box is

$$V = xyz = 48 \tag{4}$$

Since Theorem 2 applies only to functions with two independent variables, we must use equation (4) to eliminate one of the variables in equation (3):

$$M = xy + 2xz + 3yz \qquad \text{\color{blue}Substitute } z = 48/xy.$$

$$= xy + 2x\left(\frac{48}{xy}\right) + 3y\left(\frac{48}{xy}\right)$$

$$= xy + \frac{96}{y} + \frac{144}{x}$$

Thus, we must find the minimum value of

$$M(x, y) = xy + \frac{96}{y} + \frac{144}{x} \qquad x > 0 \qquad \text{and} \qquad y > 0$$

Step 1. Find critical points:

$$M_x(x, y) = y - \frac{144}{x^2} = 0$$

$$y = \frac{144}{x^2} \tag{5}$$

$$M_y(x, y) = x - \frac{96}{y^2} = 0$$

$$x = \frac{96}{y^2} \qquad \text{\color{blue}Solve for } y^2.$$

$$y^2 = \frac{96}{x} \qquad \text{\color{blue}Use equation (5) to eliminate } y$$
$$\qquad\qquad \text{\color{blue}and solve for } x.$$

$$\left(\frac{144}{x^2}\right)^2 = \frac{96}{x}$$

$$\frac{20{,}736}{x^4} = \frac{96}{x} \qquad \text{\color{blue}Multiply both sides by } x^4/96$$
$$\qquad\qquad \text{\color{blue}(recall, } x > 0\text{).}$$

$$x^3 = \frac{20{,}736}{96} = 216$$

$$x = 6 \qquad \text{\color{blue}Use equation (5) to find } y.$$

$$y = \frac{144}{36} = 4$$

Thus, $(6, 4)$ is the only critical point.

Step 2. Compute $A = M_{xx}(6, 4)$, $B = M_{xy}(6, 4)$, and $C = M_{yy}(6, 4)$:

$$M_{xx}(x, y) = \frac{288}{x^3} \qquad \text{thus} \qquad A = M_{xx}(6, 4) = \tfrac{288}{216} = \tfrac{4}{3}$$

$$M_{xy}(x, y) = 1 \qquad \text{thus} \qquad B = M_{xy}(6, 4) = 1$$

$$M_{yy}(x, y) = \frac{192}{y^3} \qquad \text{thus} \qquad C = M_{yy}(6, 4) = \tfrac{192}{64} = 3$$

Step 3. Evaluate $AC - B^2$ and try to classify the critical point $(6, 4)$ using Theorem 2:

$$AC - B^2 = (\tfrac{4}{3})(3) - (1)^2 = 3 > 0 \qquad \text{and} \qquad A = \tfrac{4}{3} > 0$$

Therefore, case 2 in Theorem 2 applies; $M(x, y)$ has a local minimum at $(6, 4)$. If $x = 6$ and $y = 4$, then

$$z = \frac{48}{xy} = \frac{48}{(6)(4)} = 2$$

Thus, the dimensions that will require the minimum amount of material are 6 inches by 4 inches by 2 inches. ■

Matched Problem 4 If the box in Example 4 must have a volume of 384 cubic inches, find the dimensions that will require the least amount of material. ■

Answers to Matched Problems

1. $f(5, 1) = 10$ is a local minimum

2. f has a saddle point at $(0, 0)$; $f(6, 18) = -108$ is a local minimum

3. Local maximum for $x = 2$ and $y = 1$; $P(2, 1) = 80$; a maximum profit of $80,000 is obtained by producing and selling 2,000 standard boards and 1,000 competition boards

4. 12 in. by 8 in. by 4 in.

Exercise 8-3

A *In Problems 1–4, find $f_x(x, y)$ and $f_y(x, y)$ and explain, using Theorem 1, why $f(x, y)$ has no local extrema.*

1. $f(x, y) = 4x + 5y - 6$

2. $f(x, y) = 10 - 2x - 3y + x^2$

3. $f(x, y) = 3.7 - 1.2x + 6.8y + 0.2y^3 + x^4$

4. $f(x, y) = x^3 - y^2 + 7x + 3y + 1$

Find local extrema in Problems 5–24 using Theorem 2.

5. $f(x, y) = 6 - x^2 - 4x - y^2$

6. $f(x, y) = 3 - x^2 - y^2 + 6y$

7. $f(x, y) = x^2 + y^2 + 2x - 6y + 14$

8. $f(x, y) = x^2 + y^2 - 4x + 6y + 23$

B

9. $f(x, y) = xy + 2x - 3y - 2$

10. $f(x, y) = x^2 - y^2 + 2x + 6y - 4$

11. $f(x, y) = -3x^2 + 2xy - 2y^2 + 14x + 2y + 10$

12. $f(x, y) = -x^2 + xy - 2y^2 + x + 10y - 5$

13. $f(x, y) = 2x^2 - 2xy + 3y^2 - 4x - 8y + 20$

14. $f(x, y) = 2x^2 - xy + y^2 - x - 5y + 8$

C

15. $f(x, y) = e^{xy}$

16. $f(x, y) = x^2y - xy^2$

17. $f(x, y) = x^3 + y^3 - 3xy$

18. $f(x, y) = 2y^3 - 6xy - x^2$

19. $f(x, y) = 2x^4 + y^2 - 12xy$

20. $f(x, y) = 16xy - x^4 - 2y^2$

21. $f(x, y) = x^3 - 3xy^2 + 6y^2$

22. $f(x, y) = 2x^2 - 2x^2y + 6y^3$

23. $f(x, y) = y^3 + 2x^2y^2 - 3x - 2y + 8$

24. $f(x, y) = x \ln y + x^2 - 4x - 5y + 3$

25. Explain why $f(x, y) = x^2$ has an infinite number of local extrema.

26. (A) Find the local extrema of the functions $f(x, y) = x + y, g(x, y) = x^2 + y^2$, and $h(x, y) = x^3 + y^3$.

(B) Discuss the local extrema of the function $k(x, y) = x^n + y^n$, where n is a positive integer.

27. (A) Show that $(0, 0)$ is a critical point for the function $f(x, y) = x^4e^y + x^2y^4 + 1$, but that the second-derivative test for local extrema fails.

(B) Use cross-sections, as in Example 2, to decide whether f has a local maximum, a local minimum, or a saddle point at $(0, 0)$.

28. (A) Show that $(0, 0)$ is a critical point for the function $g(x, y) = e^{xy^2} + x^2y^3 + 2$, but that the second-derivative test for local extrema fails.

(B) Use cross sections, as in Example 2, to decide whether g has a local maximum, a local minimum, or a saddle point at $(0, 0)$.

Applications

Business & Economics

29. *Product mix for maximum profit.* A firm produces two types of calculators, x thousand of type A and y thousand of type B per year. If the revenue and cost equations for the year are (in millions of dollars)

$$R(x, y) = 2x + 3y$$
$$C(x, y) = x^2 - 2xy + 2y^2 + 6x - 9y + 5$$

determine how many of each type of calculator should be produced per year to maximize profit. What is the maximum profit?

30. *Automation–labor mix for minimum cost.* The annual labor and automated equipment cost (in millions of dollars) for a company's production of television sets is given by

$$C(x, y) = 2x^2 + 2xy + 3y^2 - 16x - 18y + 54$$

where x is the amount spent per year on labor and y is the amount spent per year on automated equipment (both in millions of dollars). Determine how much should be spent on each per year to minimize this cost. What is the minimum cost?

31. *Maximizing profit.* A department store sells two brands of inexpensive calculators. The store pays $6 for each brand A calculator and $8 for each brand B calculator. The research department has estimated the following weekly demand equations for these two competitive products:

$x = 116 - 30p + 20q$ *Demand equation for brand A*
$y = 144 + 16p - 24q$ *Demand equation for brand B*

where p is the selling price for brand A and q is the selling price for brand B.

(A) Determine the demands x and y when $p = \$10$ and $q = \$12$; when $p = \$11$ and $q = \$11$.

(B) How should the store price each calculator to maximize weekly profits? What is the maximum weekly profit? [*Hint:* $C = 6x + 8y$, $R = px + qy$, and $P = R - C$.]

32. *Maximizing profit.* A store sells two brands of color print film. The store pays $2 for each roll of brand A film and $3 for each roll of brand B film. A consulting firm has estimated the following daily demand equations for these two competitive products:

$x = 75 - 40p + 25q$ *Demand equation for brand A*
$y = 80 + 20p - 30q$ *Demand equation for brand B*

where p is the selling price for brand A and q is the selling price for brand B.

(A) Determine the demands x and y when $p = \$4$ and $q = \$5$; when $p = \$4$ and $q = \$4$.

(B) How should the store price each brand of film to maximize daily profits? What is the maximum daily profit? [*Hint:* $C = 2x + 3y$, $R = px + qy$, and $P = R - C$.]

33. *Minimizing cost.* A satellite television reception station is to be located at $P(x, y)$ so that the sum of the squares of the distances from P to the three towns A, B, and C is minimum (see the figure). Find the coordinates of P. This location will minimize the cost of providing satellite cable television for all three towns.

WWW

Figure for 33

34. *Minimizing cost.* Repeat Problem 33, replacing the coordinates of B with $B(6, 9)$ and the coordinates of C with $C(9, 0)$.

35. *Minimum material.* A rectangular box with no top and two parallel partitions (see the figure) is to be made to hold a volume of 64 cubic inches. Find the dimensions that will require the least amount of material.

Figure for 35

36. *Minimum material.* A rectangular box with no top and two intersecting partitions (see the figure) is to be made to hold a volume of 72 cubic inches. What should its dimensions be in order to use the least amount of material in its construction?

Figure for 36

37. *Maximum volume.* A mailing service states that a
rectangular package shall have the sum of the length
and girth not to exceed 120 inches (see the figure).
What are the dimensions of the largest (in volume)
mailing carton that can be constructed meeting these
restrictions?

Length

Girth

Figure for 37

38. *Maximum shipping volume.* A shipping box is to be
reinforced with steel bands in all three directions, as
indicated in the figure. A total of 150 inches of steel
tape is to be used, with 6 inches of waste because of
a 2-inch overlap in each direction. Find the dimensions of the box with maximum volume that can be
taped as indicated.

Figure for 38

Section 8-4

Maxima and Minima Using Lagrange Multipliers

❑ FUNCTIONS OF TWO INDEPENDENT VARIABLES
❑ FUNCTIONS OF THREE INDEPENDENT VARIABLES

❑ FUNCTIONS OF TWO INDEPENDENT VARIABLES

We will now consider a particularly powerful method of solving a certain class
of maxima–minima problems. The method is due to Joseph Louis Lagrange
(1736–1813), an eminent eighteenth-century French mathematician, and it is
called the **method of Lagrange multipliers.** We introduce the method through
an example; then we formalize the discussion in the form of a theorem.

A rancher wants to construct two feeding pens of the same size along an existing fence (see Fig. 1). If the rancher has 720 feet of fencing materials available, how long should x and y be in order to obtain the maximum total area?
What is the maximum area?

The total area is given by

$$f(x, y) = xy$$

which can be made as large as we like, provided that there are no restrictions
on x and y. But there are restrictions on x and y, since we have only 720 feet of
fencing. That is, x and y must be chosen so that

$$3x + y = 720$$

This restriction on x and y, called a **constraint,** leads to the following maxima–
minima problem:

Maximize $f(x, y) = xy$ (1)
subject to $3x + y = 720$ or $3x + y - 720 = 0$ (2)

Existing
fence

FIGURE 1

This problem is a special case of a general class of problems of the form

$$\text{Maximize (or minimize)} \quad z = f(x, y) \tag{3}$$
$$\text{subject to} \qquad\qquad g(x, y) = 0 \tag{4}$$

Of course, we could try to solve equation (4) for y in terms of x, or for x in terms of y, then substitute the result into equation (3). and use methods developed in Section 4-4 for functions of a single variable. But what if equation (4) were more complicated than equation (2), and solving for one variable in terms of the other was either very difficult or impossible? In the method of Lagrange multipliers, we will work with $g(x, y)$ directly and avoid having to solve equation (4) for one variable in terms of the other. In addition, the method generalizes to functions of arbitrarily many variables subject to one or more constraints.

Now, to the method. We form a new function F, using functions f and g in equations (3) and (4), as follows:

$$F(x, y, \lambda) = f(x, y) + \lambda g(x, y) \tag{5}$$

where λ (the Greek lowercase letter lambda) is called a **Lagrange multiplier.** Theorem 1 gives the basis for the method.

THEOREM 1   **Method of Lagrange Multipliers for Functions of Two Variables**

Any local maxima or minima of the function $z = f(x, y)$ subject to the constraint $g(x, y) = 0$ will be among those points (x_0, y_0) for which (x_0, y_0, λ_0) is a solution to the system

$$F_x(x, y, \lambda) = 0$$
$$F_y(x, y, \lambda) = 0$$
$$F_\lambda(x, y, \lambda) = 0$$

where $F(x, y, \lambda) = f(x, y) + \lambda g(x, y)$, provided that all the partial derivatives exist.

We now solve the fence problem using the method of Lagrange multipliers.

Step 1.   Formulate the problem in the form of equations (3) and (4):

$$\text{Maximize} \quad f(x, y) = xy$$
$$\text{subject to} \quad g(x, y) = 3x + y - 720 = 0$$

Step 2.   Form the function F, introducing the Lagrange multiplier λ:

$$F(x, y, \lambda) = f(x, y) + \lambda g(x, y)$$
$$= xy + \lambda(3x + y - 720)$$

Step 3.   Solve the system $F_x = 0$, $F_y = 0$, $F_\lambda = 0$. (The solutions are called **critical points** for F.)

$$F_x = y + 3\lambda = 0$$
$$F_y = x + \lambda = 0$$
$$F_\lambda = 3x + y - 720 = 0$$

From the first two equations, we see that

$$y = -3\lambda$$
$$x = -\lambda$$

Substitute these values for x and y into the third equation and solve for λ:

$$-3\lambda - 3\lambda = 720$$
$$-6\lambda = 720$$
$$\lambda = -120$$

Thus,

$$y = -3(-120) = 360 \text{ feet}$$
$$x = -(-120) = 120 \text{ feet}$$

and $(x_0, y_0, \lambda_0) = (120, 360, -120)$ is the only critical point for F.

Step 4. According to Theorem 1, if the function $f(x, y)$, subject to the constraint $g(x, y) = 0$, has a local maximum or minimum, it must occur at $x = 120, y = 360$. Although it is possible to develop a test similar to Theorem 2 in Section 8-3 to determine the nature of this local extremum, we will not do so. [Note that Theorem 2 cannot be applied to $f(x, y)$ at $(120, 360)$, since this point is not a critical point of the unconstrained function $f(x, y)$.] We simply assume that the maximum value of $f(x, y)$ must occur for $x = 120, y = 360$. Thus,

$$\text{Max } f(x, y) = f(120, 360)$$
$$= (120)(360) = 43{,}200 \text{ square feet}$$

The key steps in applying the method of Lagrange multipliers are listed in the following box:

Method of Lagrange Multipliers: Key Steps

Step 1. Formulate the problem in the form

Maximize (or minimize) $z = f(x, y)$
subject to $g(x, y) = 0$

Step 2. Form the function F:

$F(x, y, \lambda) = f(x, y) + \lambda g(x, y)$

Step 3. Find the critical points for F; that is, solve the system

$F_x(x, y, \lambda) = 0$
$F_y(x, y, \lambda) = 0$
$F_\lambda(x, y, \lambda) = 0$

Step 4. If (x_0, y_0, λ_0) is the only critical point of F, we assume that (x_0, y_0) will always produce the solution to the problems we consider. If F has more than one critical point, we evaluate $z = f(x, y)$ at (x_0, y_0) for each critical point (x_0, y_0, λ_0) of F. For the problems we consider, we assume that the largest of these values is the maximum value of $f(x, y)$, subject to the constraint $g(x, y) = 0$, and the smallest is the minimum value of $f(x, y)$, subject to the constraint $g(x, y) = 0$.

Example 1 ➭ **Minimization Subject to a Constraint** Minimize $f(x, y) = x^2 + y^2$ subject to $x + y = 10$.

SOLUTION **Step 1.** Minimize $f(x, y) = x^2 + y^2$
subject to $g(x, y) = x + y - 10 = 0$

Step 2. $F(x, y, \lambda) = x^2 + y^2 + \lambda(x + y - 10)$
Step 3. $F_x = 2x + \lambda = 0$
$\qquad\quad F_y = 2y + \lambda = 0$
$\qquad\quad F_\lambda = x + y - 10 = 0$

From the first two equations, $x = -\lambda/2$ and $y = -\lambda/2$. Substituting these into the third equation, we obtain

$$-\frac{\lambda}{2} - \frac{\lambda}{2} = 10$$

$$-\lambda = 10$$

$$\lambda = -10$$

The only critical point is $(x_0, y_0, \lambda_0) = (5, 5, -10)$.

Step 4.   Since $(5, 5, -10)$ is the only critical point for F, we conclude that (see step 4 in the box)

$$\text{Min } f(x, y) = f(5, 5) = (5)^2 + (5)^2 = 50$$

FIGURE 2 $h(x) = x^2 + (10 - x)^2$

Since $g(x, y)$ in Example 1 has a relatively simple form, an alternative to the method of Lagrange multipliers is to solve $g(x, y) = 0$ for y, and then substitute into $f(x, y)$ to obtain the function $h(x) = f(x, 10 - x) = x^2 + (10 - x)^2$ in the single variable x. Then minimize h (see Fig. 2). Again, we conclude that Min $f(x, y) = f(5, 5) = 50$. This technique depends on being able to solve the constraint for one of the two variables, and thus is not always available as an alternative to the method of Lagrange multipliers.

Matched Problem 1   Maximize $f(x, y) = 25 - x^2 - y^2$ subject to $x + y = 4$.

Figures 3 and 4 illustrate the results obtained in Example 1 and Matched Problem 1, respectively.

FIGURE 3

FIGURE 4

Explore–Discuss 1

Consider the problem of minimizing $f(x, y) = 3x^2 + 5y^2$ subject to the constraint $g(x, y) = 2x + 3y - 6 = 0$.

(A) Compute the value of $f(x, y)$ when x and y are integers, $0 \le x \le 3, 0 \le y \le 2$. Record your answers next to the points (x, y) in Figure 5.

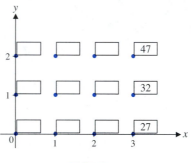

FIGURE 5

(B) Graph the constraint $g(x, y) = 0$.

(C) Estimate the minimum value of f on the basis of your graph and the computations from part (A).

(D) Use the method of Lagrange multipliers to solve the minimization problem.

Example 2

Productivity The Cobb–Douglas production function for a new product is given by

$$N(x, y) = 16x^{0.25}y^{0.75}$$

where x is the number of units of labor and y is the number of units of capital required to produce $N(x, y)$ units of the product. Each unit of labor costs \$50 and each unit of capital costs \$100. If \$500,000 has been budgeted for the production of this product, how should this amount be allocated between labor and capital in order to maximize production? What is the maximum number of units that can be produced?

SOLUTION The total cost of using x units of labor and y units of capital is $50x + 100y$. Thus, the constraint imposed by the \$500,000 budget is

$$50x + 100y = 500,000$$

Step 1. Maximize $N(x, y) = 16x^{0.25}y^{0.75}$
subject to $g(x, y) = 50x + 100y - 500,000 = 0$

Step 2. $F(x, y, \lambda) = 16x^{0.25}y^{0.75} + \lambda(50x + 100y - 500,000)$

Step 3. $F_x = 4x^{-0.75}y^{0.75} + 50\lambda = 0$
$F_y = 12x^{0.25}y^{-0.25} + 100\lambda = 0$
$F_\lambda = 50x + 100y - 500,000 = 0$

From the first two equations,

$$\lambda = -\tfrac{2}{25}x^{-0.75}y^{0.75} \qquad \text{and} \qquad \lambda = -\tfrac{3}{25}x^{0.25}y^{-0.25}$$

Thus,

$$-\tfrac{2}{25}x^{-0.75}y^{0.75} = -\tfrac{3}{25}x^{0.25}y^{-0.25} \qquad \text{Multiply both sides by } x^{0.75}y^{0.25}.$$
$$-\tfrac{2}{25}y = -\tfrac{3}{25}x \qquad \text{(We can assume that } x \neq 0 \text{ and } y \neq 0.)$$
$$y = \tfrac{3}{2}x$$

Now, substitute for y in the third equation and solve for x:

$$50x + 100(\tfrac{3}{2}x) - 500{,}000 = 0$$
$$200x = 500{,}000$$
$$x = 2{,}500$$

Thus,

$$y = \tfrac{3}{2}(2{,}500) = 3{,}750$$

and

$$\lambda = -\tfrac{2}{25}(2{,}500)^{-0.75}(3{,}750)^{0.75} \approx -0.1084$$

The only critical point of F is $(2{,}500, 3{,}750, -0.1084)$.

Step 4.   Since F has only one critical point, we conclude that maximum productivity occurs when 2,500 units of labor and 3,750 units of capital are used (see step 4 in the method of Lagrange multipliers). Thus,

$$\text{Max } N(x, y) = N(2{,}500, 3{,}750)$$
$$= 16(2{,}500)^{0.25}(3{,}750)^{0.75}$$
$$\approx 54{,}216 \text{ units}$$

The negative of the value of the Lagrange multiplier found in step 3 is called the **marginal productivity of money** and gives the approximate increase in production for each additional dollar spent on production. In Example 2, increasing the production budget from $500,000 to $600,000 would result in an approximate increase in production of

$$0.1084(100{,}000) = 10{,}840 \text{ units}$$

Note that simplifying the constraint equation

$$50x + 100y - 500{,}000 = 0$$

to

$$x + 2y - 10{,}000 = 0$$

before forming the function $F(x, y, \lambda)$ would make it difficult to interpret $-\lambda$ correctly. Thus, **in marginal productivity problems, the constraint equation should not be simplified.**

Matched Problem 2   The Cobb–Douglas production function for a new product is given by

$$N(x, y) = 20x^{0.5}y^{0.5}$$

where x is the number of units of labor and y is the number of units of capital required to produce $N(x, y)$ units of the product. Each unit of labor costs $40 and each unit of capital costs $120.

(A) If $300,000 has been budgeted for the production of this product, how should this amount be allocated in order to maximize production? What is the maximum production?

(B) Find the marginal productivity of money in this case, and estimate the increase in production if an additional $40,000 is budgeted for production.

■

FIGURE 6

Explore–Discuss 2

Consider the problem of maximizing $f(x, y) = 4 - x^2 - y^2$ subject to the constraint $g(x, y) = y - x^2 + 1 = 0$.

(A) Explain why $f(x, y) = 3$ whenever (x, y) is a point on the circle of radius 1 centered at the origin. What is the value of $f(x, y)$ when (x, y) is a point on the circle of radius 2 centered at the origin? On the circle of radius 3 centered at the origin? (See Fig. 6.)

(B) Explain why some points on the parabola $y - x^2 + 1 = 0$ lie inside the circle $x^2 + y^2 = 1$.

(C) In light of part (B), would you guess that the maximum value of $f(x, y)$ subject to the constraint is greater than 3? Explain.

(D) Use Lagrange multipliers to solve the maximization problem.

❑ FUNCTIONS OF THREE INDEPENDENT VARIABLES

We have indicated that the method of Lagrange multipliers can be extended to functions with arbitrarily many independent variables with one or more constraints. We now state a theorem for functions with three independent variables and one constraint, and consider an example that will demonstrate the advantage of the method of Lagrange multipliers over the method used in Section 8-3.

THEOREM 2   **Method of Lagrange Multipliers for Functions of Three Variables**

Any local maxima or minima of the function $w = f(x, y, z)$ subject to the constraint $g(x, y, z) = 0$ will be among the set of points (x_0, y_0, z_0) for which $(x_0, y_0, z_0, \lambda_0)$ is a solution to the system

$$F_x(x, y, z, \lambda) = 0$$
$$F_y(x, y, z, \lambda) = 0$$
$$F_z(x, y, z, \lambda) = 0$$
$$F_\lambda(x, y, z, \lambda) = 0$$

where $F(x, y, z, \lambda) = f(x, y, z) + \lambda g(x, y, z)$, provided that all the partial derivatives exist.

Example 3   **Package Design**   A rectangular box with an open top and one partition is to be constructed from 162 square inches of cardboard (Fig. 7). Find the dimensions that will result in a box with the largest possible volume.

SOLUTION   We must maximize

$$V(x, y, z) = xyz$$

subject to the constraint that the amount of material used is 162 square inches. Thus, x, y, and z must satisfy

$$xy + 2xz + 3yz = 162$$

Step 1.   Maximize $V(x, y, z) = xyz$

subject to $g(x, y, z) = xy + 2xz + 3yz - 162 = 0$

FIGURE 7

Step 2. $F(x, y, z, \lambda) = xyz + \lambda(xy + 2xz + 3yz - 162)$

Step 3. $F_x = yz + \lambda(y + 2z) = 0$
$F_y = xz + \lambda(x + 3z) = 0$
$F_z = xy + \lambda(2x + 3y) = 0$
$F_\lambda = xy + 2xz + 3yz - 162 = 0$

From the first two equations, we can write

$$\lambda = \frac{-yz}{y + 2z} \qquad \lambda = \frac{-xz}{x + 3z}$$

Eliminating λ, we have

$$\frac{-yz}{y + 2z} = \frac{-xz}{x + 3z}$$
$$-xyz - 3yz^2 = -xyz - 2xz^2$$
$$3yz^2 = 2xz^2 \qquad \text{We can assume that } z \neq 0.$$
$$3y = 2x$$
$$x = \tfrac{3}{2}y$$

From the second and third equations,

$$\lambda = \frac{-xz}{x + 3z} \qquad \lambda = \frac{-xy}{2x + 3y}$$

Eliminating λ, we have

$$\frac{-xz}{x + 3z} = \frac{-xy}{2x + 3y}$$
$$-2x^2z - 3xyz = -x^2y - 3xyz$$
$$2x^2z = x^2y \qquad \text{We can assume that } x \neq 0.$$
$$2z = y$$
$$z = \tfrac{1}{2}y$$

Substituting $x = \tfrac{3}{2}y$ and $z = \tfrac{1}{2}y$ in the fourth equation, we have

$$(\tfrac{3}{2}y)y + 2(\tfrac{3}{2}y)(\tfrac{1}{2}y) + 3y(\tfrac{1}{2}y) - 162 = 0$$
$$\tfrac{3}{2}y^2 + \tfrac{3}{2}y^2 + \tfrac{3}{2}y^2 = 162$$
$$y^2 = 36 \qquad \text{We can assume that } y > 0.$$
$$y = 6$$
$$x = \tfrac{3}{2}(6) = 9 \quad \text{Using } x = \tfrac{3}{2}y$$
$$z = \tfrac{1}{2}(6) = 3 \quad \text{Using } z = \tfrac{1}{2}y$$

and finally,

$$\lambda = \frac{-(6)(3)}{6 + 2(3)} = -\frac{3}{2} \quad \text{Using } \lambda = \frac{-yz}{y + 2z}$$

Thus, the only critical point of F with x, y, and z all positive is $(9, 6, 3, -\tfrac{3}{2})$.

Step 4. The box with maximum volume has dimensions 9 inches by 6 inches by 3 inches.

■

Matched Problem 3   A box of the same type as described in Example 3 is to be constructed from 288 square inches of cardboard. Find the dimensions that will result in a box with the largest possible volume.

■

Suppose that we had decided to solve Example 3 by the method used in Section 8-3. First, we would have to solve the material constraint for one of the variables, say, z:

$$z = \frac{162 - xy}{2x + 3y}$$

Then we would eliminate z in the volume function and maximize

$$V(x, y) = xy\frac{162 - xy}{2x + 3y}$$

Using the method of Lagrange multipliers allows us to avoid the formidable task of finding the partial derivatives of V.

Answers to Matched Problems   **1.** Max $f(x, y) = f(2, 2) = 17$ (see Fig. 4)

2. (A) 3,750 units of labor and 1,250 units of capital;
Max $N(x, y) = N(3,750, 1,250) \approx 43,301$ units
(B) Marginal productivity of money ≈ 0.1443;
increase in production $\approx 5,774$ units

3. 12 in. by 8 in. by 4 in.

Exercise 8-4

A *Use the method of Lagrange multipliers in Problems 1–4.*

1. Maximize $f(x, y) = 2xy$
subject to $x + y = 6$

2. Minimize $f(x, y) = 6xy$
subject to $y - x = 6$

3. Minimize $f(x, y) = x^2 + y^2$
subject to $3x + 4y = 25$

4. Maximize $f(x, y) = 25 - x^2 - y^2$
subject to $2x + y = 10$

B *In Problems 5 and 6, use Theorem 1 to explain why no maxima or minima exist.*

5. Minimize $f(x, y) = 4y - 3x$
subject to $2x + 5y = 3$

6. Maximize $f(x, y) = 6x + 5y + 24$
subject to $3x + 2y = 4$

Use the method of Lagrange multipliers in Problems 7–16.

7. Find the maximum and minimum of $f(x, y) = 2xy$ subject to $x^2 + y^2 = 18$.

8. Find the maximum and minimum of $f(x, y) = x^2 - y^2$ subject to $x^2 + y^2 = 25$.

9. Maximize the product of two numbers if their sum must be 10.

10. Minimize the product of two numbers if their difference must be 10.

C

11. Minimize $f(x, y, z) = x^2 + y^2 + z^2$
subject to $2x - y + 3z = -28$

12. Maximize $f(x, y, z) = xyz$
subject to $2x + y + 2z = 120$

13. Maximize and minimize $f(x, y, z) = x + y + z$
subject to $x^2 + y^2 + z^2 = 12$

14. Maximize and minimize $f(x, y, z) = 2x + 4y + 4z$
subject to $x^2 + y^2 + z^2 = 9$

15. Maximize $f(x, y) = y + xy^2$
subject to $x + y^2 = 1$

16. Maximize and minimize $f(x, y) = x + e^y$
subject to $x^2 + y^2 = 1$

In Problems 17 and 18, use Theorem 1 to explain why no maxima or minima exist.

17. Maximize $f(x, y) = e^x + 3e^y$
subject to $x - 2y = 6$

18. Minimize $f(x, y) = x^3 + 2y^3$
subject to $6x - 2y = 1$

19. Consider the problem of maximizing $f(x, y)$ subject to $g(x, y) = 0$, where $g(x, y) = y - 5$. Explain how the maximization problem can be solved without using the method of Lagrange multipliers.

20. Consider the problem of minimizing $f(x, y)$ subject to $g(x, y) = 0$, where $g(x, y) = 4x - y + 3$. Explain how the minimization problem can be solved without using the method of Lagrange multipliers.

21. Consider the problem of maximizing $f(x, y) = e^{-(x^2+y^2)}$ subject to the constraint $g(x, y) = x^2 + y - 1 = 0$.

(A) Solve the constraint equation for y, and then substitute into $f(x, y)$ to obtain a function $h(x)$ of the single variable x. Graph h and solve the original maximization problem by maximizing h (round answers to three decimal places).

(B) Confirm your answer using the method of Lagrange multipliers.

22. Consider the problem of minimizing

$$f(x, y) = x^2 + 2y^2$$

subject to the constraint $g(x, y) = ye^{x^2} - 1 = 0$.

(A) Solve the constraint equation for y, and then substitute into $f(x, y)$ to obtain a function $h(x)$ of the single variable x. Graph h and solve the original minimization problem by minimizing h (round answers to three decimal places).

(B) Confirm your answer using the method of Lagrange multipliers.

Applications

Business & Economics

23. *Budgeting for least cost.* A manufacturing company produces two models of a television set, x units of model A and y units of model B per week, at a cost (in dollars) of

$$C(x, y) = 6x^2 + 12y^2$$

If it is necessary (because of shipping considerations) that

$$x + y = 90$$

how many of each type of set should be manufactured per week to minimize cost? What is the minimum cost?

24. *Budgeting for maximum production.* A manufacturing firm has budgeted $60,000 per month for labor and materials. If $x thousand is spent on labor and $y thousand is spent on materials, and if the monthly output (in units) is given by

$$N(x, y) = 4xy - 8x$$

how should the $60,000 be allocated to labor and materials in order to maximize N? What is the maximum N?

25. *Productivity.* A consulting firm for a manufacturing company arrived at the following Cobb–Douglas production function for a particular product:

$$N(x, y) = 50x^{0.8}y^{0.2}$$

where x is the number of units of labor and y is the number of units of capital required to produce $N(x, y)$ units of the product. Each unit of labor costs $40 and each unit of capital costs $80.

(A) If $400,000 is budgeted for production of the product, determine how this amount should be allocated to maximize production, and find the maximum production.

(B) Find the marginal productivity of money in this case, and estimate the increase in production if an additional $50,000 is budgeted for the production of this product.

26. *Productivity.* The research department for a manufacturing company arrived at the following Cobb–Douglas production function for a particular product:

$$N(x, y) = 10x^{0.6}y^{0.4}$$

where x is the number of units of labor and y is the number of units of capital required to produce $N(x, y)$ units of the product. Each unit of labor costs $30 and each unit of capital costs $60.

(A) If $300,000 is budgeted for production of the product, determine how this amount should be allocated to maximize production, and find the maximum production.

(B) Find the marginal productivity of money in this case, and estimate the increase in production if an additional $80,000 is budgeted for the production of this product.

27. *Maximum volume.* A rectangular box with no top and two intersecting partitions is to be constructed from 192 square inches of cardboard (see the figure). Find the dimensions that will maximize the volume.

Figure for 27

28. *Maximum volume.* A mailing service states that a rectangular package shall have the sum of the length and girth not to exceed 120 inches (see the figure). What are the dimensions of the largest (in volume) mailing carton that can be constructed meeting these restrictions?

WWW

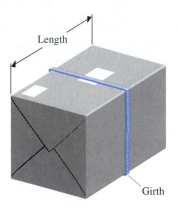

Length

Girth

Figure for 28

Life Sciences

29. *Agriculture.* Three pens of the same size are to be built along an existing fence (see the figure). If 400 feet of fencing is available, what length should x and y be to produce the maximum total area? What is the maximum area?

Existing fence

Figure for 29

30. *Diet and minimum cost.* A group of guinea pigs is to receive 25,600 calories per week. Two available foods produce $200xy$ calories for a mixture of x kilograms of type M food and y kilograms of type N food. If type M costs \$1 per kilogram and type N costs \$2 per kilogram, how much of each type of food should be used to minimize weekly food costs? What is the minimum cost?

WWW

[*Note:* $x \geqslant 0, y \geqslant 0$]

Section 8-5

Method of Least Squares

❑ LEAST SQUARES APPROXIMATION

❑ APPLICATIONS

TABLE 1

NUMBER OF UNITS x (hundreds)	COST y (thousand \$)
2	4
5	6
6	7
9	8

FIGURE 1

❑ **LEAST SQUARES APPROXIMATION**

Regression analysis is the process of fitting an elementary function to a set of data points using the **method of least squares.** The mechanics of using regression techniques were introduced in Chapter 2. Now, using the optimization techniques of Section 8-3, we are able to develop and explain the mathematical foundation of the method of least squares. We begin with **linear regression,** the process of finding the equation of the line that is the "best" approximation to a set of data points.

A manufacturer wants to approximate the cost function for a product. The value of the cost function has been determined for certain levels of production, as listed in Table 1. Although these points do not all lie on a line (see Fig. 1), they are very close to being linear. The manufacturer would like to approximate the cost function by a linear function, that is, determine values a and b so that the line

$$y = ax + b$$

is, in some sense, the "best" approximation to the cost function.

What do we mean by "best"? Since the line $y = ax + b$ will not go through all four points, it is reasonable to examine the differences between the y coordinates of the points listed in the table and the y coordinates of the corresponding points on the line. Each of these differences is called the **residual** at that point (see Fig. 2). For example, at $x = 2$, the point from Table 1 is $(2, 4)$ and the point on the line is $(2, 2a + b)$, so the residual is

$$4 - (2a + b) = 4 - 2a - b$$

All the residuals are listed in Table 2.

TABLE 2			
x	y	$ax + b$	RESIDUAL
2	4	$2a + b$	$4 - 2a - b$
5	6	$5a + b$	$6 - 5a - b$
6	7	$6a + b$	$7 - 6a - b$
9	8	$9a + b$	$8 - 9a - b$

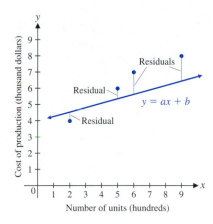

FIGURE 2

Our criterion for the "best" approximation is the following: Determine the values of a and b that *minimize the sum of the squares* of the residuals. The resulting line is called the **least squares line,** or the **regression line.** To this end, we minimize

$$F(a, b) = (4 - 2a - b)^2 + (6 - 5a - b)^2 + (7 - 6a - b)^2 + (8 - 9a - b)^2$$

Step 1. Find critical points:

$$\begin{aligned} F_a(a, b) &= 2(4 - 2a - b)(-2) + 2(6 - 5a - b)(-5) \\ &\quad + 2(7 - 6a - b)(-6) + 2(8 - 9a - b)(-9) \\ &= -304 + 292a + 44b = 0 \\ F_b(a, b) &= 2(4 - 2a - b)(-1) + 2(6 - 5a - b)(-1) \\ &\quad + 2(7 - 6a - b)(-1) + 2(8 - 9a - b)(-1) \\ &= -50 + 44a + 8b = 0 \end{aligned}$$

After dividing each equation by 2, we solve the system

$$\begin{aligned} 146a + 22b &= 152 \\ 22a + 4b &= 25 \end{aligned}$$

obtaining $(a, b) = (0.58, 3.06)$ as the only critical point.

Step 2. Compute $A = F_{aa}(a, b), B = F_{ab}(a, b),$ and $C = F_{bb}(a, b)$:

$$\begin{aligned} F_{aa}(a, b) &= 292 & \text{thus} \quad & A = F_{aa}(0.58, 3.06) = 292 \\ F_{ab}(a, b) &= 44 & \text{thus} \quad & B = F_{ab}(0.58, 3.06) = 44 \\ F_{bb}(a, b) &= 8 & \text{thus} \quad & C = F_{bb}(0.58, 3.06) = 8 \end{aligned}$$

Step 3. Evaluate $AC - B^2$ and try to classify the critical point (a, b) using Theorem 2 in Section 8-3:

$$AC - B^2 = (292)(8) - (44)^2 = 400 > 0 \quad \text{and} \quad A = 292 > 0$$

Cost of production (thousand dollars)

$y = 0.58x + 3.06$

Number of units (hundreds)

FIGURE 3

Therefore, case 2 in Theorem 2 applies, and $F(a, b)$ has a local minimum at the critical point $(0.58, 3.06)$.

Thus, the least squares line for the given data is

$$y = 0.58x + 3.06 \qquad \text{\textit{Least squares line}}$$

The sum of the squares of the residuals is minimized for this choice of a and b (see Fig. 3).

This linear function can now be used by the manufacturer to estimate any of the quantities normally associated with the cost function—such as costs, marginal costs, average costs, and so on. For example, the cost of producing 2,000 units is approximately

$$y = (0.58)(20) + 3.06 = 14.66 \qquad \text{or} \qquad \$14,660$$

The marginal cost function is

$$\frac{dy}{dx} = 0.58$$

The average cost function is

$$\bar{y} = \frac{0.58x + 3.06}{x}$$

In general, if we are given a set of n points $(x_1, y_1), (x_2, y_2), \ldots, (x_n, y_n)$, we want to determine the line $y = ax + b$ for which the sum of the squares of the residuals is minimized. Using summation notation, the sum of the squares of the residuals is given by

$$F(a, b) = \sum_{k=1}^{n} (y_k - ax_k - b)^2$$

Note that in this expression the variables are a and b, and the x_k and y_k are all known values. To minimize $F(a, b)$, we thus compute the partial derivatives with respect to a and b and set them equal to 0:

$$F_b(a, b) = \sum_{k=1}^{n} 2(y_k - ax_k - b)(-x_k) = 0$$

$$F_b(a, b) = \sum_{k=1}^{n} 2(y_k - ax_k - b)(-1) = 0$$

Dividing each equation by 2 and simplifying, we see that the coefficients a and b of the least squares line $y = ax + b$ must satisfy the following system of *normal equations:*

$$\left(\sum_{k=1}^{n} x_k^2 \right)a + \left(\sum_{k=1}^{n} x_k \right)b = \sum_{k=1}^{n} x_k y_k$$

$$\left(\sum_{k=1}^{n} x_k \right)a + nb = \sum_{k=1}^{n} y_k$$

Solving this system for a and b produces the formulas given in the box.

Least Squares Approximation

For a set of n points $(x_1, y_1), (x_2, y_2), \ldots, (x_n, y_n)$, the coefficients of the least squares line $y = ax + b$ are the solutions of the system of **normal equations**

$$\left(\sum_{k=1}^{n} x_k^2\right)a + \left(\sum_{k=1}^{n} x_k\right)b = \sum_{k=1}^{n} x_k y_k \tag{1}$$

$$\left(\sum_{k=1}^{n} x_k\right)a + nb = \sum_{k=1}^{n} y_k$$

and are given by the formulas

$$a = \frac{n\left(\sum_{k=1}^{n} x_k y_k\right) - \left(\sum_{k=1}^{n} x_k\right)\left(\sum_{k=1}^{n} y_k\right)}{n\left(\sum_{k=1}^{n} x_k^2\right) - \left(\sum_{k=1}^{n} x_k\right)^2} \tag{2}$$

$$b = \frac{\sum_{k=1}^{n} y_k - a\left(\sum_{k=1}^{n} x_k\right)}{n} \tag{3}$$

TABLE 3

	x_k	y_k	$x_k y_k$	x_k^2
	2	4	8	4
	5	6	30	25
	6	7	42	36
	9	8	72	81
Totals	22	25	152	146

Now we return to the data in Table 1 and tabulate the sums required for the normal equations and their solution in Table 3.

The normal equations (1) are then

$$146a + 22b = 152$$
$$22a + 4b = 25$$

The solution to the normal equations given by equations (2) and (3) is

$$a = \frac{4(152) - (22)(25)}{4(146) - (22)^2} = 0.58$$

$$b = \frac{25 - 0.58(22)}{4} = 3.06$$

Compare these results with step 1 on page 517. Note that Table 3 provides a convenient format for the computation of step 1 above.

 Many graphing utilities have a linear regression feature that solves the system of normal equations obtained by setting the partial derivatives of the sum of squares of the residuals equal to 0. Therefore, in practice, we simply enter the given data points and use the linear regression feature to determine the line $y = ax + b$ that best fits the data (see Fig. 4). There is no need to compute partial derivatives, or even to tabulate sums (as in Table 3).

(A)

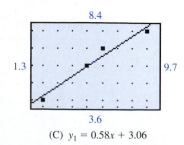

(B)   (C) $y_1 = 0.58x + 3.06$

FIGURE 4

(A) Plot the four points $(0, 0)$, $(0, 1)$, $(10, 0)$, and $(10, 1)$. Which line would you guess "best" fits these four points? Use formulas (2) and (3) to test your conjecture.

(B) Plot the four points $(0, 0)$, $(0, 10)$, $(1, 0)$ and $(1, 10)$. Which line would you guess "best" fits these four points? Use formulas (2) and (3) to test your conjecture.

(C) If either of your conjectures was wrong, explain how your reasoning was mistaken.

The method of least squares can also be applied to find the quadratic equation $y = ax^2 + bx + c$ that best fits a set of data points. In this case, the sum of the squares of the residuals is a function of three variables:

$$F(a, b, c) = \sum_{k=1}^{n} (y_k - ax_k^2 - bx_k - c)^2$$

There are now three partial derivatives to compute and set equal to 0:

$$F_a(a, b, c) = \sum_{k=1}^{n} 2(y_k - ax_k^2 - bx_k - c)(-x_k^2) = 0$$

$$F_b(a, b, c) = \sum_{k=1}^{n} 2(y_k - ax_k^2 - bx_k - c)(-x_k) = 0$$

$$F_c(a, b, c) = \sum_{k=1}^{n} 2(y_k - ax_k^2 - bx_k - c)(-1) = 0$$

The resulting set of three linear equations in the three variables a, b, and c is called the *set of normal equations for quadratic regression*.

 A quadratic regression feature on a calculator is designed to solve such normal equations after the given set of points has been entered. Figure 5 illustrates the computation for the data of Table 1.

(A)   (B)   (C) $y_1 = -0.0417x^2 + 1.0383x + 2.06$

FIGURE 5

(A) Use the graphs in Figures 4 and 5 to predict which technique, linear regression or quadratic regression, yields the smaller sum of squares of the residuals for the data of Table 1. Explain.

 (B) Confirm your prediction by computing the sum of squares of the residuals in each case.

The method of least squares can also be applied to other regression equations—for example, cubic, quartic, logarithmic, exponential, and power regression models. Details are explored in some of the exercises at the end of this section.

Example 1 ⇔   **Educational Testing**   Table 4 lists the midterm and final examination scores for 10 students in a calculus course.

TABLE 4			
MIDTERM	FINAL	MIDTERM	FINAL
49	61	78	77
53	47	83	81
67	72	85	79
71	76	91	93
74	68	99	99

(A) Use formulas (1), (2), and (3) to find the normal equations and the least squares line for the data given in Table 4.

(B) Use the linear regression feature on a graphing utility to find and graph the least squares line.

(C) Use the least squares line to predict the final examination score for a student who scored 95 on the midterm examination.

SOLUTION   (A) Table 5 shows a convenient way to compute all the sums in the formulas for a and b.

TABLE 5			
	x_k	y_k	$x_k y_k$	x_k^2
---	---	---	---	---
	49	61	2,989	2,401
	53	47	2,491	2,809
	67	72	4,824	4,489
	71	76	5,396	5,041
	74	68	5,032	5,476
	78	77	6,006	6,084
	83	81	6,723	6,889
	85	79	6,715	7,225
	91	93	8,463	8,281
	99	99	9,801	9,801
Totals	750	753	58,440	58,496

From the last line in Table 5, we have

$$\sum_{k=1}^{10} x_k = 750 \qquad \sum_{k=1}^{10} y_k = 753 \qquad \sum_{k=1}^{10} x_k y_k = 58,440 \qquad \sum_{k=1}^{10} x_k^2 = 58,496$$

and the normal equations are

$$58{,}496a + 750b = 58{,}440$$
$$750a + 10b = 753$$

Using formulas (2) and (3),

$$a = \frac{10(58{,}440) - (750)(753)}{10(58{,}496) - (750)^2} = \frac{19{,}650}{22{,}460} \approx 0.875$$

$$b = \frac{753 - 0.875(750)}{10} = 9.675$$

The least squares line is given (approximately) by

$$y = 0.875x + 9.675$$

(B) We enter the data and use the linear regression feature, as shown in Figure 6. (The discrepancy between values of a and b above and those in Figure 6B is due to rounding in part (A).)

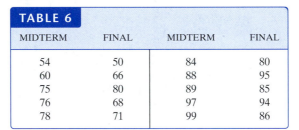

| (A) | (B) | (C) |

FIGURE 6

(C) If $x = 95$, then $y = 0.875(95) + 9.675 \approx 92.8$ is the predicted score on the final exam. This is also indicated in Figure 6C. If we assume that the exam score must be an integer, we would predict a score of 93. ∎

Matched Problem 1   Repeat Example 1 for the scores listed in Table 6.

TABLE 6

MIDTERM	FINAL	MIDTERM	FINAL
54	50	84	80
60	66	88	95
75	80	89	85
76	68	97	94
78	71	99	86

∎

Example 2   **Energy Consumption** The use of fuel oil for home heating in the United States has declined steadily for several decades. Table 7 lists the percentage of occupied housing units in the United States that were heated by fuel oil for various years between 1960 and 1991. Use the data in the table and linear regression to predict the percentage of occupied housing units in the United States that will be heated by fuel oil in the year 2010.

TABLE 7

OCCUPIED HOUSING UNITS HEATED BY FUEL OIL			
YEAR	PERCENT	YEAR	PERCENT
1960	32.4	1985	14.1
1970	26.0	1989	13.3
1975	22.5	1993	12.9
1980	18.1	1997	10.9

SOLUTION   We enter the data, with $x = 0$ representing 1960, $x = 10$ representing 1970, and so on, and use the linear regression feature, as shown in Figure 7.

(A)

(B)

(C)

FIGURE 7

Figure 7 indicates that the least squares line is $y = -0.60x + 31.43$. The estimated percentage of occupied housing units heated by fuel oil in the year 2010 (corresponding to $x = 50$) is thus $-0.60(50) + 31.43 = 1.43\%$. ■

Matched Problem 2   In 1950, coal was still a major source of fuel for home energy consumption, and the percentage of occupied housing units heated by fuel oil was only 22.1. Add the data for 1950 to the data for Example 2, and compute the new least squares line and the new estimate for the percentage of occupied housing units heated by fuel oil in the year 2010. Discuss the discrepancy between the two estimates. ■

Explore–Discuss 3

The data in Table 8 give the amount A (in milligrams) of the radioactive isotope gallium-67, used in the diagnosis of malignant tumors, at time t (in hours).

TABLE 8

t(hours)	A(milligrams)
0	50.0
5	46.4
10	43.1
15	39.9
20	37.1

(A) Since the data describe radioactive decay, we would expect the relationship between A and t to be exponential—that is, $A = ae^{bt}$ for some constants a and b. Show that if $A = ae^{bt}$, the relationship between $\ln A$ and t is linear.

(B) Compute $\ln A$ for each data point and find the line that "best" fits the data $(t, \ln A)$.

(C) Determine the "best" values of a and b.

(D) Confirm your results by using the exponential regression feature on a graphing utility.

Answers to Matched Problems   **1.** (A) $y = 0.85x + 9.47$

(B)

(C) 90.3

2. $y = -0.38x + 25.87; 6.87\%$

Exercise 8-5

A *In Problems 1–6, find the least squares line. Graph the data and the least squares line.*

1.

x	y
1	1
2	3
3	4
4	3

2.

x	y
1	-2
2	-1
3	3
4	5

3.

x	y
1	8
2	5
3	4
4	0

4.

x	y
1	20
2	14
3	11
4	3

5.

x	y
1	3
2	4
3	5
4	6

6.

x	y
1	2
2	3
3	3
4	2

B *In Problems 7–14, find the least squares line and use it to estimate y for the indicated value of x. Round answers to two decimal places.*

7.

x	y
1	3
2	1
2	2
3	0

Estimate y when x = 2.5.

8.

x	y
1	0
3	1
3	6
3	4

Estimate y when x = 3.

9.

x	y
0	10
5	22
10	31
15	46
20	51

Estimate y when x = 25.

10.

x	y
-5	60
0	50
5	30
10	20
15	15

Estimate y when x = 20.

11.

x	y
-1	14
1	12
3	8
5	6
7	5

Estimate y when x = 2.

12.

x	y
2	-4
6	0
10	8
14	12
18	14

Estimate y when x = 15.

13.

x	y	x	y
0.5	25	9.5	12
2	22	11	11
3.5	21	12.5	8
5	21	14	5
6.5	18	15.5	1

Estimate y when x = 8.

14.

x	y	x	y
0	-15	12	11
2	-9	14	13
4	-7	16	19
6	-7	18	25
8	-1	20	33

Estimate y when x = 10.

C

15. To find the coefficients of the parabola

$$y = ax^2 + bx + c$$

that is the "best" fit for the points $(1, 2), (2, 1), (3, 1),$ and $(4, 3)$, minimize the sum of the squares of the residuals

$$\begin{aligned} F(a, b, c) = (a + b + c - 2)^2 \\ + (4a + 2b + c - 1)^2 \\ + (9a + 3b + c - 1)^2 \\ + (16a + 4b + c - 3)^2 \end{aligned}$$

by solving the system of normal equations

$$F_a(a, b, c) = 0 \qquad F_b(a, b, c) = 0 \qquad F_c(a, b, c) = 0$$

for a, b, and c. Graph the points and the parabola.

16. Repeat Problem 15 for the points $(-1, -2)$, $(0, 1), (1, 2),$ and $(2, 0)$.

Problems 17 and 18 refer to the system of normal equations and the formulas for a and b given in the text.

17. Verify formulas (2) and (3) by solving the system of normal equations (1) for a and b.

18. If

$$\bar{x} = \frac{1}{n} \sum_{k=1}^{n} x_k \qquad \text{and} \qquad \bar{y} = \frac{1}{n} \sum_{k=1}^{n} y_k$$

are the averages of the x and y coordinates, respectively, show that the point $(\bar{x}, \bar{y})$ satisfies the equation of the least squares line $y = ax + b$.

19. (A) Suppose that $n = 5$ and that the x coordinates of the data points $(x_1, y_1), (x_2, y_2), \ldots, (x_n, y_n)$ are $-2, -1, 0, 1, 2$. Show that system (1) implies that

$$a = \frac{\sum x_k y_k}{\sum x_k^2}$$

and that b is equal to the average of the values of y_k.

(B) Show that the conclusion of part (A) holds whenever the average of the x coordinates of the data points is 0.

20. (A) Give an example of a set of six data points such that half of the points lie above the least squares line and half lie below.

(B) Give an example of a set of six data points such that just one of the points lies above the least squares line and five lie below.

21. (A) Find the linear and quadratic functions that best fit the data points $(0, 1.3)$, $(1, 0.6)$, $(2, 1.5)$, $(3, 3.6)$ and $(4, 7.4)$. (Round coefficients to two decimal places.)

(B) Which of the two functions best fits the data? Explain.

22. (A) Find the linear, quadratic, and logarithmic functions that best fit the data points $(1, 3.2)$, $(2, 4.2)$, $(3, 4.7)$, $(4, 5.0)$, and $(5, 5.3)$. (Round coefficients to two decimal places.)

(B) Which of the three functions best fits the data? Explain.

23. Describe the normal equations for cubic regression. How many equations are there? What are the variables? What techniques could be used to solve them?

24. Describe the normal equations for quartic regression. How many equations are there? What are the variables? What techniques could be used to solve them?

Applications

Business & Economics

25. *Motor vehicle production.* Data for motor vehicle production in Canada for the years 1990 to 1999 is given in the table.

www

Motor Vehicle Production in Canada			
YEAR	THOUSANDS	YEAR	THOUSANDS
1990	1,928	1995	2,408
1991	1,888	1996	2,397
1992	1,961	1997	2,571
1993	2,246	1998	2,568
1994	2,321	1999	3,026

(A) Find the least squares line for the data using $x = 0$ for 1990.

(B) Use the least squares line to estimate the annual production of motor vehicles in Canada in 2010.

26. *Beef production.* Data for U.S. production of beef for the years 1990 to 1999 is given in the table.

www

U.S. Beef Production			
YEAR	MILLION POUNDS	YEAR	MILLION POUNDS
1990	22,743	1995	25,222
1991	22,917	1996	25,525
1992	23,086	1997	25,490
1993	23,049	1998	25,760
1994	24,386	1999	26,493

(A) Find the least squares line for the data using $x = 0$ for 1990.

(B) Use the least squares line to estimate the annual production of beef in the United States in 2012.

27. *Maximizing profit.* The market research department for a drugstore chain chose two summer resort areas to test market a new sunscreen lotion packaged in 4-ounce plastic bottles. After a summer of varying the selling price and recording the monthly demand, the research department arrived at the demand table given below, where y is the number of bottles purchased per month (in thousands) at x dollars per bottle.

x	y
5.0	2.0
5.5	1.8
6.0	1.4
6.5	1.2
7.0	1.1

(A) Find a demand equation using the method of least squares.

(B) If each bottle of sunscreen costs the drugstore chain $4, how should it be priced to achieve a maximum monthly profit? [*Hint*: Use the result of part (A), with $C = 4y$, $R = xy$, and $P = R - C$.]

28. *Maximizing profit.* A market research consultant for a supermarket chain chose a large city to test market a new brand of mixed nuts packaged in 8-ounce cans. After a year of varying the selling price and recording the monthly demand, the consultant arrived at the demand table given below, where y is the number of cans purchased per month (in thousands) at x dollars per can.

x	y
4.0	4.2
4.5	3.5
5.0	2.7
5.5	1.5
6.0	0.7

(A) Find a demand equation using the method of least squares.

(B) If each can of nuts costs the supermarket chain $3, how should it be priced to achieve a maximum monthly profit?

Life Sciences

29. *Medicine.* If a person dives into cold water, a neural reflex response automatically shuts off blood circulation to the skin and muscles and reduces the pulse rate. A medical research team conducted an experiment using a group of ten 2-year-olds. A child's face was placed momentarily in cold water, and the corresponding reduction in pulse rate was recorded. The data for the average reduction in heart rate for each temperature is summarized in the table.

WATER TEMPERATURE (°F)	PULSE RATE REDUCTION
50	15
55	13
60	10
65	6
70	2

(A) If T is water temperature (in degrees Fahrenheit) and P is pulse rate reduction (in beats per minute), use the method of least squares to find a linear equation relating T and P.

(B) Use the equation found in part (A) to find P when $T = 57$.

30. *Biology.* In biology there is an approximate rule, called the *bioclimatic rule for temperate climates,* that has been known for a couple of hundred years. This rule states that in spring and early summer, periodic phenomena such as blossoming of flowers, appearance of insects, and ripening of fruit usually come about 4 days later for each 500 feet of altitude. Stated as a formula,

$$d = 8h \qquad 0 \leqslant h \leqslant 4$$

where d is the change in days and h is the altitude (in thousands of feet). To test this rule, an experiment was set up to record the difference in blossoming time of the same type of apple tree at different altitudes. A summary of the results is given in the table.

h	d
0	0
1	7
2	18
3	28
4	33

(A) Use the method of least squares to find a linear equation relating h and d. Does the bioclimatic rule, $d = 8h$, appear to be approximately correct?

(B) How much longer will it take this type of apple tree to blossom at 3.5 thousand feet than at sea level? [Use the linear equation found in part (A).]

 31. *Global warming.* Average global temperatures from 1885 to 1995 are given in the table.

 www

Average Global Temperatures			
YEAR	°F	YEAR	°F
1885	56.65	1945	57.13
1895	56.64	1955	57.06
1905	56.52	1965	57.05
1915	56.57	1975	57.04
1925	56.74	1985	57.36
1935	57.00	1995	57.64

(A) Find the least squares line for the data using $x = 0$ for 1885.

(B) Use the least squares line to estimate the average global temperature in 2085.

 32. *Air pollution.* Data for emissions of air pollutants in the United States for the years 1989 to 1998 is given in the table.

www

Emissions of Air Pollutants in the United States			
YEAR	MILLION SHORT TONS	YEAR	MILLION SHORT TONS
1989	179.6	1994	174.6
1990	170.5	1995	161.4
1991	172.5	1996	160.9
1992	169.0	1997	160.6
1993	169.6	1998	154.3

(A) Find the least squares line for the data using $x = 0$ for 1989.

(B) Use the least squares line to estimate the emissions of air pollutants in the United States in 2015.

Social Sciences

33. *Political science.* Association of economic class and party affiliation did not start with Roosevelt's New Deal; it goes back to the time of Andrew Jackson (1767–1845). Paul Lazarsfeld of Columbia University published an article in the November 1950 issue of *Scientific American* in which he discusses statistical investigations of the relationships between economic class and party affiliation. The data in the table is taken from this article.

Political Affiliations, 1836

WARD	AVERAGE ASSESSED VALUE PER PERSON (HUNDRED $)	DEMOCRATIC VOTES (%)
12	1.7	51
3	2.1	49
1	2.3	53
5	2.4	36
2	3.6	65
11	3.7	35
10	4.7	29
4	6.2	40
6	7.1	34
9	7.4	29
8	8.7	20
7	11.9	23

(A) If A represents the average assessed value per person in a given ward in 1836 and D represents the percentage of people in that ward voting Democratic in 1836, use the method of least squares to find a linear equation relating A and D.

(B) If the average assessed value per person in a ward had been $300, what is the predicted percentage of people in that ward that would have voted Democratic?

34. *Education.* The table lists the high school grade-point averages (GPAs) of 10 students, along with their grade-point averages after one semester of college.

HIGH SCHOOL GPA	COLLEGE GPA	HIGH SCHOOL GPA	COLLEGE GPA
2.0	1.5	3.0	2.3
2.2	1.5	3.1	2.5
2.4	1.6	3.3	2.9
2.7	1.8	3.4	3.2
2.9	2.1	3.7	3.5

(A) Find the least squares line for the data.

(B) Estimate the college GPA for a student with a high school GPA of 3.5.

(C) Estimate the high school GPA necessary for a college GPA of 2.7.

35. *Olympic Games.* The table gives the winning heights in the pole vault in the Olympic Games from 1896 to 2000.

Olympic Pole Vault Winning Height

YEAR	HEIGHT (FT)	YEAR	HEIGHT (FT)
1896	10.81	1956	14.96
1900	10.82	1960	15.43
1904	11.50	1964	16.73
1906	11.60	1968	17.71
1908	12.17	1972	18.04
1912	12.96	1976	18.04
1920	13.46	1980	18.96
1924	12.96	1984	18.85
1928	13.78	1988	19.35
1932	14.16	1992	19.02
1936	14.27	1996	19.42
1948	14.10	2000	19.35
1952	14.93		

(A) Use a graphing utility to find the least squares line for the data using $x = 0$ for 1896.

(B) Estimate the winning height in the pole vault in the Olympic Games of 2016.

Section 8-6

Double Integrals over Rectangular Regions

- ❑ INTRODUCTION
- ❑ DEFINITION OF THE DOUBLE INTEGRAL
- ❑ AVERAGE VALUE OVER RECTANGULAR REGIONS
- ❑ VOLUME AND DOUBLE INTEGRALS

❑   INTRODUCTION

We have generalized the concept of differentiation to functions with two or more independent variables. How can we do the same with integration, and how can we interpret the results? Let us first look at the operation of antidifferentiation. We can antidifferentiate a function of two or more variables with respect to one of the variables by treating all the other variables as though they were constants. Thus, this operation is the reverse operation of partial differentiation, just as ordinary antidifferentiation is the reverse operation of ordinary differentiation. We write $\int f(x, y) \, dx$ to indicate that we are to antidifferentiate $f(x, y)$ with respect to x, holding y fixed; we write $\int f(x, y) \, dy$ to indicate that we are to antidifferentiate $f(x, y)$ with respect to y, holding x fixed.

Example 1 ⇨ **Partial Antidifferentiation**   Evaluate:

(A) $\int (6xy^2 + 3x^2) \, dy$ (B) $\int (6xy^2 + 3x^2) \, dx$

SOLUTION   (A) Treating x as a constant and using the properties of antidifferentiation from Section 6-1, we have

$$\int (6xy^2 + 3x^2) \, dy = \int 6xy^2 \, dy + \int 3x^2 \, dy$$

$$= 6x \int y^2 \, dy + 3x^2 \int dy$$

$$= 6x \left(\frac{y^3}{3} \right) + 3x^2(y) + C(x)$$

$$= 2xy^3 + 3x^2y + C(x)$$

The dy tells us we are looking for the anti-derivative of $6xy^2 + 3x^2$ with respect to y only, holding x constant.

Notice that the constant of integration actually can be *any function of x alone,* since for any such function

$$\frac{\partial}{\partial y} C(x) = 0$$

Check: We can verify that our answer is correct by using partial differentiation:

$$\frac{\partial}{\partial y} [2xy^3 + 3x^2y + C(x)] = 6xy^2 + 3x^2 + 0$$

$$= 6xy^2 + 3x^2$$

(B) Now we treat y as a constant:

$$\int (6xy^2 + 3x^2) \, dx = \int 6xy^2 \, dx + \int 3x^2 \, dx$$

$$= 6y^2 \int x \, dx + 3 \int x^2 \, dx$$

$$= 6y^2 \left(\frac{x^2}{2} \right) + 3 \left(\frac{x^3}{3} \right) + E(y)$$

$$= 3x^2y^2 + x^3 + E(y)$$

This time, the antiderivative contains an arbitrary function $E(y)$ of y alone.

Check: $\dfrac{\partial}{\partial x}[3x^2y^2 + x^3 + E(y)] = 6xy^2 + 3x^2 + 0$

$$= 6xy^2 + 3x^2$$

Matched Problem 1 ⇨ Evaluate:

(A) $\int (4xy + 12x^2y^3) \, dy$ (B) $\int (4xy + 12x^2y^3) \, dx$

Now that we have extended the concept of antidifferentiation to functions with two variables, we also can evaluate definite integrals of the form

$$\int_a^b f(x, y)\, dx \qquad \text{or} \qquad \int_c^d f(x, y)\, dy$$

Example 2 ⇨ **Evaluating a Partial Antiderivative**   Evaluate, substituting the limits of integration in y if dy is used and in x if dx is used:

(A) $\int_0^2 (6xy^2 + 3x^2)\, dy$ (B) $\int_0^1 (6xy^2 + 3x^2)\, dx$

SOLUTION   (A) From Example 1A, we know that

$$\int (6xy^2 + 3x^2)\, dy = 2xy^3 + 3x^2y + C(x)$$

According to properties of the definite integral for a function of one variable, we can use any antiderivative to evaluate the definite integral. Thus, choosing $C(x) = 0$, we have

$$\int_0^2 (6xy^2 + 3x^2)\, dy = (2xy^3 + 3x^2y)\Big|_{y=0}^{y=2}$$
$$= [2x(2)^3 + 3x^2(2)] - [2x(0)^3 + 3x^2(0)]$$
$$= 16x + 6x^2$$

(B) From Example 1B, we know that

$$\int (6xy^2 + 3x^2)\, dx = 3x^2y^2 + x^3 + E(y)$$

Thus, choosing $E(y) = 0$, we have

$$\int_0^1 (6xy^2 + 3x^2)\, dx = (3x^2y^2 + x^3)\Big|_{x=0}^{x=1}$$
$$= [3y^2(1)^2 + (1)^3] - [3y^2(0)^2 + (0)^3]$$
$$= 3y^2 + 1$$

■

Matched Problem 2 ⇨ Evaluate:

(A) $\int_0^1 (4xy + 12x^2y^3)\, dy$ (B) $\int_0^3 (4xy + 12x^2y^3)\, dx$

■

Notice that integrating and evaluating a definite integral with integrand $f(x, y)$ with respect to y produces a function of x alone (or a constant). Likewise, integrating and evaluating a definite integral with integrand $f(x, y)$ with respect to x produces a function of y alone (or a constant). Each of these results, involving at most one variable, can now be used as an integrand in a second definite integral.

Example 3 ⇨ **Evaluating Iterated Integrals**   Evaluate:

(A) $\int_0^1 \left[\int_0^2 (6xy^2 + 3x^2)\, dy \right] dx$ (B) $\int_0^2 \left[\int_0^1 (6xy^2 + 3x^2)\, dx \right] dy$

SOLUTION (A) Example 2A showed that

$$\int_0^2 (6xy^2 + 3x^2)\, dy = 16x + 6x^2$$

Thus,

$$\int_0^1 \left[\int_0^2 (6xy^2 + 3x^2)\, dy \right] dx = \int_0^1 (16x + 6x^2)\, dx$$

$$= (8x^2 + 2x^3)\Big|_{x=0}^{x=1}$$

$$= [8(1)^2 + 2(1)^3] - [8(0)^2 + 2(0)^3] = 10$$

(B) Example 2B showed that

$$\int_0^1 (6xy^2 + 3x^2)\, dx = 3y^2 + 1$$

Thus,

$$\int_0^2 \left[\int_0^1 (6xy^2 + 3x^2)\, dx \right] dy = \int_0^2 (3y^2 + 1)\, dy$$

$$= (y^3 + y)\Big|_{y=0}^{y=2}$$

$$= [(2)^3 + 2] - [(0)^3 + 0] = 10 \qquad \blacksquare$$

FIGURE 1

A numerical integration routine can be used as an alternative to the fundamental theorem of calculus to evaluate the last integrals in Examples 3A and 3B, $\int_0^1 (16x + 6x^2)\, dx$ and $\int_0^2 (3y^2 + 1)\, dy$, since the integrand in each case is a function of a single variable (see Fig. 1).

Matched Problem 3 ⇔ Evaluate:

(A) $\displaystyle \int_0^3 \left[\int_0^1 (4xy + 12x^2y^3)\, dy \right] dx$

(B) $\displaystyle \int_0^1 \left[\int_0^3 (4xy + 12x^2y^3)\, dx \right] dy$

❑ DEFINITION OF THE DOUBLE INTEGRAL

Notice that the answers in Examples 3A and 3B are identical. This is not an accident. In fact, it is this property that enables us to define the *double integral,* as follows:

Double Integral

The **double integral** of a function $f(x, y)$ over a rectangle

$$R = \{(x, y) | a \leqslant x \leqslant b, \quad c \leqslant y \leqslant d\}$$

is

$$\iint_R f(x, y)\, dA = \int_a^b \left[\int_c^d f(x, y)\, dy \right] dx$$

$$= \int_c^d \left[\int_a^b f(x, y)\, dx \right] dy$$

In the double integral $\iint_R f(x, y)\, dA$, $f(x, y)$ is called the **integrand** and R is called the **region of integration.** The expression dA indicates that this is an integral over a two-dimensional region. The integrals

$$\int_a^b \left[\int_c^d f(x, y)\, dy \right] dx \qquad \text{and} \qquad \int_c^d \left[\int_a^b f(x, y)\, dx \right] dy$$

are referred to as **iterated integrals** (the brackets are often omitted), and the order in which dx and dy are written indicates the order of integration. This is not the most general definition of the double integral over a rectangular region; however, it is equivalent to the general definition for all the functions we will consider.

Example 4 ✏ **Evaluating a Double Integral**   Evaluate:

$$\iint_R (x + y)\, dA \qquad \text{over} \qquad R = \{(x, y) | 1 \leqslant x \leqslant 3, \ -1 \leqslant y \leqslant 2\}$$

Solution   Region R is illustrated in Figure 2. We can choose either order of iteration. As a check, we will evaluate the integral both ways:

FIGURE 2

$$\iint_R (x + y)\, dA = \int_1^3 \int_{-1}^2 (x + y)\, dy\, dx$$

$$= \int_1^3 \left[\left(xy + \frac{y^2}{2} \right) \Big|_{y=-1}^{y=2} \right] dx$$

$$= \int_1^3 \left[(2x + 2) - (-x + \tfrac{1}{2}) \right] dx$$

$$= \int_1^3 (3x + \tfrac{3}{2})\, dx$$

$$= (\tfrac{3}{2}x^2 + \tfrac{3}{2}x) \Big|_{x=1}^{x=3}$$

$$= (\tfrac{27}{2} + \tfrac{9}{2}) - (\tfrac{3}{2} + \tfrac{3}{2}) = 18 - 3 = 15$$

$$\iint_R (x + y)\, dA = \int_{-1}^2 \int_1^3 (x + y)\, dx\, dy$$

$$= \int_{-1}^2 \left[\left(\frac{x^2}{2} + xy \right) \Big|_{x=1}^{x=3} \right] dy$$

$$= \int_{-1}^2 \left[(\tfrac{9}{2} + 3y) - (\tfrac{1}{2} + y) \right] dy$$

$$= \int_{-1}^2 (4 + 2y)\, dy$$

$$= (4y + y^2) \Big|_{y=-1}^{y=2}$$

$$= (8 + 4) - (-4 + 1) = 12 - (-3) = 15$$

∎

Matched Problem 4 ✏ Evaluate both ways:

$$\iint_R (2x - y)\, dA \qquad \text{over} \qquad R = \{(x, y) | -1 \leqslant x \leqslant 5, \ 2 \leqslant y \leqslant 4\}$$

Example 5 ➭ **Double Integral of an Exponential Function** Evaluate:

$$\iint_R 2xe^{x^2+y}\, dA \qquad \text{over} \qquad R = \{(x, y)|0 \leqslant x \leqslant 1, \ -1 \leqslant y \leqslant 1\}$$

SOLUTION Region R is illustrated in Figure 3.

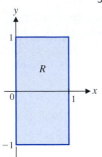

FIGURE 3

$$\iint_R 2xe^{x^2+y}\, dA = \int_{-1}^{1} \int_{0}^{1} 2xe^{x^2+y}\, dx\, dy$$

$$= \int_{-1}^{1} \left[(e^{x^2+y}) \Big|_{x=0}^{x=1} \right] dy$$

$$= \int_{-1}^{1} (e^{1+y} - e^y)\, dy$$

$$= (e^{1+y} - e^y) \Big|_{y=-1}^{y=1}$$

$$= (e^2 - e) - (e^0 - e^{-1})$$

$$= e^2 - e - 1 + e^{-1}$$

Matched Problem 5 ➭ Evaluate: $\displaystyle\iint_R \frac{x}{y^2} e^{x/y}\, dA \qquad$ over $\qquad R = \{(x, y)|0 \leqslant x \leqslant 1, \ 1 \leqslant y \leqslant 2\}$

❑ AVERAGE VALUE OVER RECTANGULAR REGIONS

In Section 6-5, the average value of a function $f(x)$ over an interval $[a, b]$ was defined as

$$\frac{1}{b - a} \int_a^b f(x)\, dx$$

This definition is easily extended to functions of two variables over rectangular regions, as shown in the box. Notice that the denominator in the expression given in the box, $(b - a)(d - c)$, is simply the area of the rectangle R.

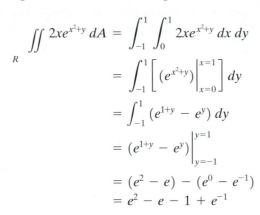

> *Average Value over Rectangular Regions*
>
> The **average value** of the function $f(x, y)$ over the rectangle
>
> $$R = \{(x, y)|a \leqslant x \leqslant b, \ c \leqslant y \leqslant d\}$$
>
> is
>
> $$\frac{1}{(b - a)(d - c)} \iint_R f(x, y)\, dA$$

Example 6 ➭ **Average Value** Find the average value of $f(x, y) = 4 - \frac{1}{2}x - \frac{1}{2}y$ over the rectangle $R = \{(x, y)|0 \leqslant x \leqslant 2, \ 0 \leqslant y \leqslant 2\}$.

SOLUTION   Region R is illustrated in Figure 4.

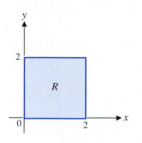

FIGURE 4

$$\frac{1}{(b-a)(d-c)} \iint\limits_{R} f(x,y)\, dA = \frac{1}{(2-0)(2-0)} \iint\limits_{R} \left(4 - \frac{1}{2}x - \frac{1}{2}y\right) dA$$

$$= \tfrac{1}{4} \int_{0}^{2}\int_{0}^{2} (4 - \tfrac{1}{2}x - \tfrac{1}{2}y)\, dy\, dx$$

$$= \tfrac{1}{4} \int_{0}^{2} \left[(4y - \tfrac{1}{2}xy - \tfrac{1}{4}y^2) \Big|_{y=0}^{y=2} \right] dx$$

$$= \tfrac{1}{4} \int_{0}^{2} (7 - x)\, dx$$

$$= \tfrac{1}{4}(7x - \tfrac{1}{2}x^2) \Big|_{x=0}^{x=2}$$

$$= \tfrac{1}{4}(12) = 3$$

Figure 5 illustrates the surface $z = f(x, y)$, and our calculations show that 3 is the average of the z values over the region R.

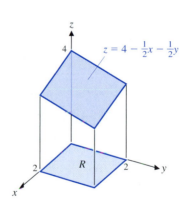

FIGURE 5

Matched Problem 6   Find the average value of $f(x, y) = x + 2y$ over the rectangle

$$R = \{(x, y) | 0 \leq x \leq 2, \quad 0 \leq y \leq 1\}$$

Explore–Discuss 1

(A) Which of the two functions, $f(x, y) = 4 - x^2 - y^2$ or $g(x, y) = 4 - x - y$, would you guess has the greater average value over the rectangle $R = \{(x, y) | 0 \leq x \leq 1, \quad 0 \leq y \leq 1\}$? Explain.

(B) Use double integrals to check the correctness of your guess in part (A).

❑   VOLUME AND DOUBLE INTEGRALS

One application of the definite integral of a function with one variable is the calculation of areas, so it is not surprising that the definite integral of a function of two variables can be used to calculate volumes of solids.

Volume Under a Surface

If $f(x, y) \geq 0$ over a rectangle $R = \{(x, y)|a \leq x \leq b, \quad c \leq y \leq d\}$, then the volume of the solid formed by graphing f over the rectangle R is given by

$$V = \iint\limits_{R} f(x, y) \, dA$$

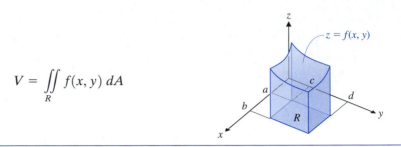

A proof of the statement in the box is left to a more advanced text.

Example 7 ➭ **Volume** Find the volume of the solid under the graph of $f(x, y) = 1 + x^2 + y^2$ over the rectangle $R = \{(x, y)|0 \leq x \leq 1, \quad 0 \leq y \leq 1\}$.

SOLUTION Figure 6 shows the region R, and Figure 7 illustrates the volume under consideration.

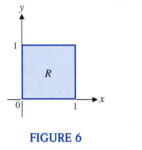

FIGURE 6

$$V = \iint\limits_{R} (1 + x^2 + y^2) \, dA$$

$$= \int_0^1 \int_0^1 (1 + x^2 + y^2) \, dx \, dy$$

$$= \int_0^1 \left[(x + \tfrac{1}{3}x^3 + xy^2) \Big|_{x=0}^{x=1} \right] dy$$

$$= \int_0^1 (\tfrac{4}{3} + y^2) \, dy$$

$$= (\tfrac{4}{3}y + \tfrac{1}{3}y^3) \Big|_{y=0}^{y=1} = \tfrac{5}{3} \text{ cubic units}$$

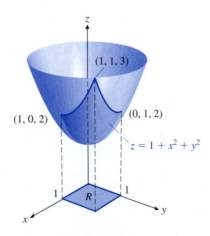

FIGURE 7

Matched Problem 7 ➭ Find the volume of the solid under the graph of $f(x, y) = 1 + x + y$ over the rectangle $R = \{(x, y)|0 \leq x \leq 1, \quad 0 \leq y \leq 2\}$.

Explore–Discuss 2	Consider the solid under the graph of $f(x, y) = 4 - y^2$ and above the rectangle $R = \{(x, y)|0 \leqslant x \leqslant 3, \ 0 \leqslant y \leqslant 2\}$.

(A) Explain why each cross section of the solid by a plane parallel to the yz plane has the same area, and compute that area.

(B) Compute the areas of the cross sections of the solid by the planes $y = 0, y = \frac{1}{2}$, and $y = 1$.

(C) Compute the volume of the solid in two different ways.

Answers to Matched Problems   **1.** (A) $2xy^2 + 3x^2y^4 + C(x)$ (B) $2x^2y + 4x^3y^3 + E(y)$

2. (A) $2x + 3x^2$ (B) $18y + 108y^3$ **3.** (A) 36   (B) 36

4. 12   **5.** $e - 2e^{1/2} + 1$ **6.** 2   **7.** 5 cubic units

Exercise 8-6

A *In Problems 1–8, find each antiderivative. Then use the antiderivative to evaluate the definite integral.*

1. (A) $\displaystyle\int 12x^2y^3 \, dy$ (B) $\displaystyle\int_0^1 12x^2y^3 \, dy$

2. (A) $\displaystyle\int 12x^2y^3 \, dx$ (B) $\displaystyle\int_{-1}^2 12x^2y^3 \, dx$

3. (A) $\displaystyle\int (4x + 6y + 5) \, dx$

  (B) $\displaystyle\int_{-2}^3 (4x + 6y + 5) \, dx$

4. (A) $\displaystyle\int (4x + 6y + 5) \, dy$

  (B) $\displaystyle\int_1^4 (4x + 6y + 5) \, dy$

5. (A) $\displaystyle\int \frac{x}{\sqrt{y + x^2}} \, dx$ (B) $\displaystyle\int_0^2 \frac{x}{\sqrt{y + x^2}} \, dx$

6. (A) $\displaystyle\int \frac{x}{\sqrt{y + x^2}} \, dy$ (B) $\displaystyle\int_1^5 \frac{x}{\sqrt{y + x^2}} \, dy$

7. (A) $\displaystyle\int \frac{\ln x}{xy} \, dy$ (B) $\displaystyle\int_1^{e^2} \frac{\ln x}{xy} \, dy$

8. (A) $\displaystyle\int \frac{\ln x}{xy} \, dx$ (B) $\displaystyle\int_1^e \frac{\ln x}{xy} \, dx$

B *In Problems 9–16, evaluate each iterated integral. (See the indicated problem for the evaluation of the inner integral.)*

9. $\displaystyle\int_{-1}^2 \int_0^1 12x^2y^3 \, dy \, dx$ **10.** $\displaystyle\int_0^1 \int_{-1}^2 12x^2y^3 \, dx \, dy$

  (See Problem 1.)   (See Problem 2.)

11. $\displaystyle\int_1^4 \int_{-2}^3 (4x + 6y + 5) \, dx \, dy$

  (See Problem 3.)

12. $\displaystyle\int_{-2}^3 \int_1^4 (4x + 6y + 5) \, dy \, dx$

  (See Problem 4.)

13. $\displaystyle\int_1^5 \int_0^2 \frac{x}{\sqrt{y + x^2}} \, dx \, dy$

  (See Problem 5.)

14. $\displaystyle\int_0^2 \int_1^5 \frac{x}{\sqrt{y + x^2}} \, dy \, dx$

  (See Problem 6.)

15. $\displaystyle\int_1^e \int_1^{e^2} \frac{\ln x}{xy} \, dy \, dx$

  (See Problem 7.)

16. $\displaystyle\int_1^{e^2} \int_1^e \frac{\ln x}{xy} \, dx \, dy$

  (See Problem 8.)

Use both orders of iteration to evaluate each double integral in Problems 17–20.

17. $\displaystyle\iint_R xy \, dA; R = \{(x, y)|0 \leqslant x \leqslant 2, \ 0 \leqslant y \leqslant 4\}$

18. $\displaystyle\iint_R \sqrt{xy} \, dA; R = \{(x, y)|1 \leqslant x \leqslant 4, \ 1 \leqslant y \leqslant 9\}$

19. $\displaystyle\iint_R (x + y)^5 \, dA;$

  $R = \{(x, y)|-1 \leqslant x \leqslant 1, \ 1 \leqslant y \leqslant 2\}$

20. $\displaystyle\iint_R xe^y \, dA; R = \{(x, y)|-2 \leqslant x \leqslant 3, \ 0 \leqslant y \leqslant 2\}$

In Problems 21–24, find the average value of each function over the given rectangle.

21. $f(x, y) = (x + y)^2$;

  $R = \{(x, y)|1 \leqslant x \leqslant 5, \ -1 \leqslant y \leqslant 1\}$

22. $f(x, y) = x^2 + y^2$;

  $R = \{(x, y)|-1 \leqslant x \leqslant 2, \ 1 \leqslant y \leqslant 4\}$

23. $f(x, y) = x/y; R = \{(x, y)|1 \le x \le 4, \quad 2 \le y \le 7\}$

24. $f(x, y) = x^2 y^3;$
$R = \{(x, y)|-1 \le x \le 1, \quad 0 \le y \le 2\}$

In Problems 25–28, find the volume of the solid under the graph of each function over the given rectangle.

25. $f(x, y) = 2 - x^2 - y^2;$
$R = \{(x, y)|0 \le x \le 1, \quad 0 \le y \le 1\}$

26. $f(x, y) = 5 - x;$
$R = \{(x, y)|0 \le x \le 5, \quad 0 \le y \le 5\}$

27. $f(x, y) = 4 - y^2;$
$R = \{(x, y)|0 \le x \le 2, \quad 0 \le y \le 2\}$

28. $f(x, y) = e^{-x-y};$
$R = \{(x, y)|0 \le x \le 1, \quad 0 \le y \le 1\}$

C *Evaluate each double integral in Problems 29–32. Select the order of integration carefully—each problem is easy to do one way and difficult the other.*

29. $\iint\limits_R xe^{xy}\, dA; R = \{(x, y)|0 \le x \le 1, \quad 1 \le y \le 2\}$

30. $\iint\limits_R xye^{x^2y}\, dA; R = \{(x, y)|0 \le x \le 1, \quad 1 \le y \le 2\}$

31. $\iint\limits_R \frac{2y + 3xy^2}{1 + x^2}\, dA;$
$R = \{(x, y)|0 \le x \le 1, \quad -1 \le y \le 1\}$

32. $\iint\limits_R \frac{2x + 2y}{1 + 4y + y^2}\, dA;$
$R = \{(x, y)|1 \le x \le 3, \quad 0 \le y \le 1\}$

33. Show that $\int_0^2 \int_0^2 (1 - y)\, dx\, dy = 0$. Does the double integral represent the volume of a solid? Explain.

34. (A) Find the average values of the functions $f(x, y) = x + y, g(x, y) = x^2 + y^2$, and $h(x, y) = x^3 + y^3$ over the rectangle

$$R = \{(x, y)|0 \le x \le 1, \quad 0 \le y \le 1\}$$

(B) Does the average value of $k(x, y) = x^n + y^n$ over the rectangle

$$R_1 = \{(x, y)|0 \le x \le 1, \quad 0 \le y \le 1\}$$

increase or decrease as n increases? Explain.

(C) Does the average value of $k(x, y) = x^n + y^n$ over the rectangle

$$R_2 = \{(x, y)|0 \le x \le 2, \quad 0 \le y \le 2\}$$

increase or decrease as n increases? Explain.

35. Let $f(x, y) = x^3 + y^2 - e^{-x} - 1$.

(A) Find the average value of $f(x, y)$ over the rectangle $R = \{(x, y)|-2 \le x \le 2, \quad -2 \le y \le 2\}$.

(B) Graph the set of all points (x, y) in R for which $f(x, y) = 0$.

(C) For which points (x, y) in R is $f(x, y)$ greater than 0? Less than 0? Explain.

36. Find the dimensions of the square S centered at the origin for which the average value of $f(x, y) = x^2 e^y$ over S is equal to 100.

Applications

Business & Economics

37. *Multiplier principle.* Suppose that Congress enacts a one-time-only 10% tax rebate that is expected to infuse $\$y$ billion, $5 \le y \le 7$, into the economy. If every person and every corporation is expected to spend a proportion x, $0.6 \le x \le 0.8$, of each dollar received, then by the **multiplier principle** in economics, the total amount of spending S (in billions of dollars) generated by this tax rebate is given by

$$S(x, y) = \frac{y}{1 - x}$$

What is the average total amount of spending for the indicated ranges of the values of x and y? Set up a double integral and evaluate.

38. *Multiplier principle.* Repeat Problem 37 if $6 \le y \le 10$ and $0.7 \le x \le 0.9$.

39. *Cobb–Douglas production function.* If an industry invests x thousand labor-hours, $10 \le x \le 20$, and $\$y$ million, $1 \le y \le 2$, in the production of N thousand units of a certain item, then N is given by

$$N(x, y) = x^{0.75} y^{0.25}$$

What is the average number of units produced for the indicated ranges of x and y? Set up a double integral and evaluate.

40. *Cobb–Douglas production function.* Repeat Problem 39 for

$$N(x, y) = x^{0.5} y^{0.5}$$

where $10 \le x \le 30$ and $1 \le y \le 3$.

Life Sciences

41. *Population distribution.* In order to study the
www population distribution of a certain species of
insects, a biologist has constructed an artificial
habitat in the shape of a rectangle 16 feet long and
12 feet wide. The only food available to the insects
in this habitat is located at its center. The biologist
has determined that the concentration C of insects
per square foot at a point d units from the food
supply (see the figure) is given approximately by

$$C = 10 - \tfrac{1}{10}d^2$$

What is the average concentration of insects throughout the habitat? Express C as a function of x and y, set
up a double integral, and evaluate.

Figure for 41

42. *Population distribution.* Repeat Problem 41 for a
square habitat that measures 12 feet on each side,
where the insect concentration is given by

$$C = 8 - \tfrac{1}{10}d^2$$

43. *Pollution.* A heavy industrial plant located in the cen-
www ter of a small town emits particulate matter into the
atmosphere. Suppose that the concentration of particulate matter (in parts per million) at a point d miles
from the plant (see the figure) is given by

$$C = 100 - 15d^2$$

If the boundaries of the town form a rectangle 4 miles
long and 2 miles wide, what is the average concentration of particulate matter throughout the city? Express C as a function of x and y, set up a double
integral, and evaluate.

Figure for 43

44. *Pollution.* Repeat Problem 43 if the boundaries of
the town form a rectangle 8 miles long and 4 miles
wide and the concentration of particulate matter is
given by

$$C = 100 - 3d^2$$

Social Sciences

45. *Safety research.* Under ideal conditions, if a person
driving a car slams on the brakes and skids to a
stop, the length of the skid marks (in feet) is given
by the formula

$$L = 0.000\ 013\ 3xy^2$$

where x is the weight of the car (in pounds) and y is
the speed of the car (in miles per hour). What is the
average length of the skid marks for cars weighing
between 2,000 and 3,000 pounds and traveling at
speeds between 50 and 60 miles per hour? Set up a
double integral and evaluate.

46. *Safety research.* Repeat Problem 45 for cars weighing between 2,000 and 2,500 pounds and traveling at
speeds between 40 and 50 miles per hour.

47. *Psychology.* The intelligence quotient Q for a person
with mental age x and chronological age y is given by

$$Q(x, y) = 100\frac{x}{y}$$

In a group of sixth-graders, the mental age varies between 8 and 16 years and the chronological age varies
between 10 and 12 years. What is the average intelligence quotient for this group? Set up a double integral and evaluate.

48. *Psychology.* Repeat Problem 47 for a group with
mental ages between 6 and 14 years and chronological ages between 8 and 10 years.

Important Terms and Symbols

8-1 *Functions of Several Variables.* Functions of two
independent variables; functions of several independent variables; Cobb–Douglas production
function; three-dimensional coordinate system; triplet

of numbers (x, y, z); paraboloid; surface; saddle
point

$$z = f(x, y); \qquad w = f(x, y, z)$$

8-2 *Partial Derivatives.* Partial derivative of f with respect to x; partial derivative of f with respect to y; marginal productivity of labor; marginal productivity of capital; second-order partial derivatives

$$\frac{\partial z}{\partial x}; \quad \frac{\partial z}{\partial y}; \quad f_x(x, y); \quad f_y(x, y);$$

$$\frac{\partial^2 z}{\partial x^2} = f_{xx}(x, y); \quad \frac{\partial^2 z}{\partial x \, \partial y} = f_{yx}(x, y);$$

$$\frac{\partial^2 z}{\partial y \, \partial x} = f_{xy}(x, y); \quad \frac{\partial^2 z}{\partial y^2} = f_{yy}(x, y)$$

8-3 *Maxima and Minima.* Local maximum; local minimum; saddle point; critical point; second-derivative test

8-4 *Maxima and Minima Using Lagrange Multipliers.* Constraint; Lagrange multiplier; critical point; method of Lagrange multipliers for functions of two variables; marginal productivity of money; method of Lagrange multipliers for functions of three variables

8-5 *Method of Least Squares.* Least squares approximation; regression analysis; method of least squares; linear regression; residual; least squares line; regression line; normal equations

8-6 *Double Integrals over Rectangular Regions.* Double integral; integrand; region of integration; iterated integral; average value over rectangular region; volume under a surface

$$\iint\limits_{R} f(x, y) \, dA = \int_a^b \left[\int_c^d f(x, y) \, dy \right] dx$$

$$= \int_c^d \left[\int_a^b f(x, y) \, dx \right] dy;$$

$$\frac{1}{(b - a)(d - c)} \iint\limits_{R} f(x, y) \, dA;$$

$$V = \iint\limits_{R} f(x, y) \, dA$$

Review Exercise

Work through all the problems in this chapter review and check your answers in the back of the book. Answers to all review problems are there along with section numbers in italics to indicate where each type of problem is discussed. Where weaknesses show up, review appropriate sections in the text.

A

1. For $f(x, y) = 2,000 + 40x + 70y$, find $f(5, 10)$, $f_x(x, y)$, and $f_y(x, y)$.

2. For $z = x^3 y^2$, find $\partial^2 z / \partial x^2$ and $\partial^2 z / \partial x \, \partial y$.

3. Evaluate: $\int (6xy^2 + 4y) \, dy$

4. Evaluate: $\int (6xy^2 + 4y) \, dx$

5. Evaluate: $\int_0^1 \int_0^1 4xy \, dy \, dx$

6. For $f(x, y) = 6 + 5x - 2y + 3x^2 + x^3$, find $f_x(x, y)$ and $f_y(x, y)$, and explain why $f(x, y)$ has no local extrema.

B

7. For $f(x, y) = 3x^2 - 2xy + y^2 - 2x + 3y - 7$, find $f(2, 3), f_x(x, y)$, and $f_y(2, 3)$.

8. For $f(x, y) = -4x^2 + 4xy - 3y^2 + 4x + 10y + 81$, find $[f_{xx}(2, 3)][f_{yy}(2, 3)] - [f_{xy}(2, 3)]^2$

9. If $f(x, y) = x + 3y$ and $g(x, y) = x^2 + y^2 - 10$, find the critical points of $F(x, y, \lambda) = f(x, y) + \lambda g(x, y)$.

10. Use the least squares line for the data in the table to estimate y when $x = 10$.

x	y
2	12
4	10
6	7
8	3

11. For $R = \{(x, y) | -1 \le x \le 1, \ 1 \le y \le 2\}$, evaluate the following in two ways:

$$\iint\limits_{R} (4x + 6y) \, dA$$

C

12. For $f(x, y) = e^{x^2 + 2y}$, find f_x, f_y, and f_{xy}.

13. For $f(x, y) = (x^2 + y^2)^5$, find f_x and f_{xy}.

14. Find all critical points and test for extrema for

$$f(x, y) = x^3 - 12x + y^2 - 6y$$

15. Use Lagrange multipliers to maximize $f(x, y) = xy$ subject to $2x + 3y = 24$.

16. Use Lagrange multipliers to minimize $f(x, y, z) = x^2 + y^2 + z^2$ subject to $2x + y + 2z = 9$.

17. Find the least squares line for the data in the table.

x	y	x	y
10	50	60	80
20	45	70	85
30	50	80	90
40	55	90	90
50	65	100	110

18. Find the average value of $f(x, y) = x^{2/3}y^{1/3}$ over the rectangle

$$R = \{(x, y)|-8 \leq x \leq 8, \quad 0 \leq y \leq 27\}$$

19. Find the volume of the solid under the graph of $z = 3x^2 + 3y^2$ over the rectangle

$$R = \{(x, y)|0 \leq x \leq 1, \quad -1 \leq y \leq 1\}$$

20. Without doing any computation, predict the average value of $f(x, y) = x + y$ over the rectangle $R = \{(x, y)|-10 \leq x \leq 10, \quad -10 \leq y \leq 10\}$. Then check the correctness of your prediction by evaluating a double integral.

21. (A) Find the dimensions of the square S centered at the origin such that the average value of

$$f(x, y) = \frac{e^x}{y + 10}$$

over S is equal to 5.

(B) Is there a square centered at the origin over which

$$f(x, y) = \frac{e^x}{y + 10}$$

has average value 0.05? Explain.

22. Explain why the function $f(x, y) = 4x^3 - 5y^3$, subject to the constraint $3x + 2y = 7$, has no maxima or minima.

Applications

Business & Economics

23. *Maximizing profit.* A company produces x units of product A and y units of product B (both in hundreds per month). The monthly profit equation (in thousands of dollars) is found to be

$$P(x, y) = -4x^2 + 4xy - 3y^2 + 4x + 10y + 81$$

(A) Find $P_x(1, 3)$ and interpret the results.

(B) How many of each product should be produced each month to maximize profit? What is the maximum profit?

24. *Minimizing material.* A rectangular box with no top and six compartments (see the figure) is to have a volume of 96 cubic inches. Find the dimensions that will require the least amount of material.

Figure for 24

25. *Profit.* A company's annual profits (in millions of dollars) over a 5-year period are given in the table. Use the least squares line to estimate the profit for the sixth year.

YEAR	PROFIT
1	2
2	2.5
3	3.1
4	4.2
5	4.3

26. *Productivity.* The Cobb–Douglas production function for a product is

$$N(x, y) = 10x^{0.8}y^{0.2}$$

where x is the number of units of labor and y is the number of units of capital required to produce N units of the product.

(A) Find the marginal productivity of labor and the marginal productivity of capital at $x = 40$ and $y = 50$. For the greatest increase in productivity, should management encourage increased use of labor or increased use of capital?

(B) If each unit of labor costs $100, each unit of capital costs $50, and $10,000 is budgeted for production of this product, use the method of Lagrange multipliers to determine the allocations of labor and capital that will maximize the number of units produced and find the maximum production. Find the marginal productivity of money and approximate the increase in production that would result from an increase of $2,000 in the amount budgeted for production.

(C) If $50 \leqslant x \leqslant 100$ and $20 \leqslant y \leqslant 40$, find the average number of units produced. Set up a definite integral and evaluate.

Life Sciences

27. *Marine biology.* The function used for timing dives with scuba gear is

$$T(V, x) = \frac{33V}{x + 33}$$

where T is the time of the dive in minutes, V is the volume of air (in cubic feet, at sea level pressure) compressed into tanks, and x is the depth of the dive in feet. Find $T_x(70, 17)$ and interpret the results.

28. *Pollution.* A heavy industrial plant located in the center of a small town emits particulate matter into the atmosphere. Suppose that the concentration of particulate matter (in parts per million) at a point d miles from the plant is given by

$$C = 100 - 24d^2$$

If the boundaries of the town form a square 4 miles long and 4 miles wide, what is the average concentration of particulate matter throughout the town? Express C as a function of x and y, set up a double integral, and evaluate.

Social Sciences

29. *Sociology.* Joseph Cavanaugh, a sociologist, found that the number of long-distance telephone calls, n, between two cities in a given period of time varied (approximately) jointly as the populations P_1 and P_2 of the two cities, and varied inversely as the distance, d, between the two cities. In terms of an equation for a time period of 1 week,

$$n(P_1, P_2, d) = 0.001 \frac{P_1 P_2}{d}$$

Find $n(100,000, 50,000, 100)$.

30. *Education.* At the beginning of the semester, students in a foreign language course are given a proficiency exam. The same exam is given at the end of the

semester. The results for 5 students are given in the table. Use the least squares line to estimate the score on the second exam for a student who scored 40 on the first exam.

FIRST EXAM	SECOND EXAM
30	60
50	75
60	80
70	85
90	90

31. *Population density.* The table gives the population per square mile in the United States for the years 1900–2000.

U.S. Population Density			
YEAR	POPULATION (PER SQUARE MILE)	YEAR	POPULATION (PER SQUARE MILE)
1900	25.6	1960	50.6
1910	31.0	1970	57.4
1920	35.6	1980	64.0
1930	41.2	1990	70.4
1940	44.2	2000	77.8
1950	50.7		

(A) Find the least squares line for the data using $x = 0$ for 1900.

(B) Use linear regression to estimate the population density in the United States in the year 2020.

(C) Use quadratic regression and exponential regression to make the estimate of part (B).

32. *Life expectancy.* The table gives life expectancies for males and females in a sample of Central and South American countries.

Life Expectancies for Central and South American Countries			
MALES	FEMALES	MALES	FEMALES
62.30	67.50	70.15	74.10
68.05	75.05	62.93	66.58
72.40	77.04	68.43	74.88
63.39	67.59	66.68	72.80
55.11	59.43		

(A) Find the least squares line for the data.

(B) Use linear regression to estimate the life expectancy of a female in a Central or South American country in which the life expectancy for males is 60 years.

(C) Use quadratic and logarithmic regression to make the estimate of part (B).

33. *Women in the workforce.* It is reasonable to conjecture from the data given in the table that many Japanese women tend to leave the workforce to marry and have children, but then reenter the workforce when the children are grown.

WWW

(A) Explain why you might expect cubic regression to provide a better fit to the data than linear or quadratic regression.

(B) Investigate your expectation by plotting the data points and graphing the curve of best fit using linear, quadratic, and cubic regression.

Women in the Workforce in Japan, 1997

AGE	PERCENTAGE OF WOMEN EMPLOYED	AGE	PERCENTAGE OF WOMEN EMPLOYED
22	73	47	72
27	65	52	68
32	56	57	59
37	63	62	42
42	71		

Group Activity 1   **City Planning**

A city planning commission is seeking to identify prime locations for a new zoo and a new hospital. The city's economy is heavily dependent on two industrial plants located relatively near the city center. Both emit particulate matter into the atmosphere, and the resulting air pollution is of concern to the commission and will influence its decisions. The consensus of the commission is that the new zoo should be built in the least polluted area within the city limits, and the new hospital should be built in the least polluted location within 2 miles of the city center.

The boundaries of the city form a rectangle 10 miles from east to west and 6 miles from north to south. When a coordinate system is chosen with the origin at the center of the rectangle (the city center), industrial plant 1 has coordinates $(-1, 1)$ and industrial plant 2 has coordinates $(1, 0)$. At a point (x, y), the concentration of particulate matter (in parts per million) due to emissions from plant 1 is given by $C_1 = 200 - 3(d_1)^2$, where d_1 is the distance from (x, y) to plant 1. Similarly, the concentration due to emissions from plant 2 is given by $C_2 = 200 - 3(d_2)^2$, where d_2 is the distance from (x, y) to plant 2.

(A) Find the point within the city limits that has the greatest concentration of particulate matter.

(B) Find the points on the city boundaries that have the greatest and least concentrations of particulate matter.

(C) Find the average concentration of particulate matter throughout the city.

(D) Find the points on the circle of radius 2 miles, centered at the origin, that have the greatest and least concentrations of particulate matter.

(E) Determine the optimal locations for the city's new zoo and new hospital.

Group Activity 2   ***Numerical Integration of Multivariable Functions***

A definite integral $\int_a^b f(x)\,dx$ is a limit of Riemann sums of the form

$$\sum_{k=1}^{n} f(c_k)\,\Delta x_k$$

(see Section 6-5). Analogously, a double integral $\iint_R f(x, y)\,dx\,dy$ over a rectangle $R = \{(x, y)|a \le x \le b, \ c \le x \le d\}$ is a limit of sums of the form

$$\sum_{j=1}^{m} \sum_{k=1}^{n} f(c_j, d_k)\,\Delta x_j\,\Delta y_k$$

where $\quad a = x_0 < x_1 < x_2 < \cdots < x_m = b, c = y_0 < y_1 < y_2 < \cdots < y_n = d,$
$x_{j-1} \le c_j \le x_j$ for $j = 1, 2, \ldots, m$, and $y_{k-1} \le d_k \le y_k$ for $k = 1, 2, \ldots, n$.

It follows that numerical integration methods such as Simpson's rule (see Group Activity 1 in Chapter 6) can also be used to approximate double integrals. We illustrate with an example.

Example 1 ⇔   **Using Simpson's Rule to Approximate Double Integrals**   Use Simpson's rule to approximate

$$\int_0^4 \int_0^6 (x^2 y + 2y^2 + 3x + 1)\,dy\,dx$$

by partitioning $[0, 4]$ into four equal subintervals and $[0, 6]$ into six equal subintervals.

SOLUTION   We evaluate the function $f(x, y) = x^2 y + 2y^2 + 3x + 1$ at each of the 35 intersection points of the grid determined by the two partitions (see Table 1).

TABLE 1

	$y = 0$	1	2	3	4	5	6	R_j
$x = 0$	1	3	9	19	33	51	73	150
1	4	7	14	25	40	59	82	186
2	7	13	23	37	55	77	103	258
3	10	21	36	55	78	105	136	366
4	13	31	53	79	109	143	181	510

Then we apply Simpson's rule for six equal subdivisions of an interval to each row of the table to compute the entries in the column labeled R_j. For example, letting $g(y) = f(x_0, y)$, we have

$$R_0 = [g(y_0) + 4g(y_1) + 2g(y_2) + 4g(y_3) + 2g(y_4) + 4g(y_5) + g(y_6)]\frac{\Delta y}{3}$$

$$= [(1) + 4(3) + 2(9) + 4(19) + 2(33) + 4(51) + 1(73)]\tfrac{1}{3}$$
$$= 150$$

Finally, we apply Simpson's rule for four equal subdivisions of an interval to the values R_0, R_1, R_2, R_3, R_4:

$$S = [R_0 + 4R_1 + 2R_2 + 4R_3 + R_4]\frac{\Delta x}{3}$$

$$= [150 + 4(186) + 2(258) + 4(366) + 510]\tfrac{1}{3}$$
$$= 1{,}128$$

Therefore, the value of the double integral $\int_0^4 \int_0^6 (x^2y + 2y^2 + 3x + 1)\, dy\, dx$ is approximately 1,128.   ■

In Example 1, the double integral can be evaluated as an iterated integral. The exact value of the double integral is also 1,128. This is not surprising since Simpson's rule gives exact results for polynomials of degree 3 or less. But the technique illustrated above can also be applied when it is impossible to find antiderivatives, or when we have only a table of values (and not a formula) for $f(x, y)$.

(A) Evaluate the double integral of Example 1 as an iterated integral.

(B) Show that the sum S in Example 1 may also be found by applying Simpson's rule to each column of Table 1, obtaining $C_0, C_1, \ldots, C_6$ and then applying Simpson's rule to $C_0, C_1, \ldots, C_6$.

(C) Approximate

$$\int_0^3 \int_0^3 e^{-(x^2+y^2)}\, dx\, dy$$

by partitioning each of the intervals $[0, 3]$ into six equal subintervals, evaluating the function $f(x, y) = e^{-(x^2+y^2)}$ (to four decimal places) at each of the 49 intersection points of the grid determined by the two partitions, and applying Simpson's rule.

(D) Repeat part (C), but use a numerical integration routine on a graphing utility, rather than Simpson's rule, to approximate the R_j.

(E) It can be shown that the volume under the graph of $f(x, y) = e^{-(x^2+y^2)}$ and above the entire first quadrant is $\pi/4$. Compare your answers to parts (C) and (D) with this value and explain any discrepancy.

Differential Equations

INTRODUCTION

Mathematical models for many processes in nature and society are best developed by understanding the *rate* at which the process occurs. For example, the rate of growth of a population may be proportional to the size of the population, or proportional to the difference between the size and a fixed limit, or proportional to both the size and the difference between the size and a fixed limit. These descriptions of the rate may be expressed succinctly by *differential equations* such as

$$\frac{dy}{dt} = ky, \qquad \frac{dy}{dt} = k(M - y), \qquad \frac{dy}{dt} = ky(M - y)$$

Solutions to these differential equations provide models for several types of exponential growth—unlimited, limited, and logistic (see Table 1, page 385). Unfortunately, there is no single method that will solve all the differential equations that may be encountered—even in very simple applications. In this chapter, after discussing some basic concepts in the first section, we will consider methods for solving several types of differential equations that have significant applications.

Section 9-1

Basic Concepts

❑ SOLUTIONS OF DIFFERENTIAL EQUATIONS
❑ IMPLICIT SOLUTIONS
❑ APPLICATION

In this section we review the basic concepts introduced in Section 6-3 and discuss in more detail what is meant by a solution to a differential equation, including explicit and implicit representations. Specific techniques for solving differential equations will be discussed in the next two sections.

❑ SOLUTIONS OF DIFFERENTIAL EQUATIONS

A **differential equation** is an equation involving an unknown function, usually denoted by y, and one or more of its derivatives. For example,

$$y' = 0.2xy \tag{1}$$

is a differential equation. Since only the first derivative of the unknown function y appears in this equation, it is called a **first-order** differential equation. In general, the **order** of a differential equation is the highest derivative of the unknown function present in the equation.

Notice that we used y' rather than dy/dx to represent the derivative of y with respect to x in equation (1). This is customary practice in the study of differential equations. Unless indicated otherwise, we will assume that y is a function of the independent variable x and that y' refers to the derivative of y with respect to x.

Explore–Discuss 1

Consider the differential equation

$$y' = 2x \tag{2}$$

(A) Which of the following functions satisfy this equation?

$$y = x^2 \qquad y = 2x^2 \qquad y = x^2 + 2$$

(B) Can you find other functions that satisfy equation (2)? What form do all these functions have?

(C) Discuss possible solution methods for any differential equation of the form

$$y' = f(x)$$

Figure 1A shows a *slope field* for equation (1). Recall from Section 6-3 that the **slope field** for a differential equation is obtained by drawing tangent line segments determined by the equation at each point in a grid. This provides a geometric interpretation of the equation that indicates the general shape of solutions to the equation. Due to the large number of line segments, slope fields are usually generated by a graphing utility. Figure 1B shows the slope field for $y' = 0.2xy$ on a TI-83 Plus.

Now consider the function

$$y = 2e^{0.1x^2}$$

whose derivative is

$$y' = 2e^{0.1x^2}(0.2x) = 0.4xe^{0.1x^2}$$

Substituting for y and y' in equation (1) gives

$$y' = 0.2xy \tag{1}$$
$$0.4xe^{0.1x^2} = 0.2x(2e^{0.1x^2})$$
$$0.4xe^{0.1x^2} = 0.4xe^{0.1x^2}$$

which is certainly true for all values of x. This shows that the function $y = 2e^{0.1x^2}$ is a **solution** of equation (1). But this function is not the only solution. In fact, if C is any constant, substituting $y = Ce^{0.1x^2}$ and $y' = 0.2xCe^{0.1x^2}$ in equation (1)

(A)

(B)

FIGURE 1 Slope field for $y' = 0.2xy$

yields the identity

$$y' = 0.2xy$$
$$0.2xCe^{0.1x^2} = 0.2xCe^{0.1x^2}$$

It turns out that all solutions of $y' = 0.2xy$ can be obtained from $y = Ce^{0.1x^2}$ by assigning C appropriate values; hence, $y = Ce^{0.1x^2}$ is called the **general solution** of equation (1). The collection of all functions of the form $y = Ce^{0.1x^2}$ is called the **family of solutions** of equation (1). The function $y = 2e^{0.1x^2}$, obtained by letting $C = 2$ in the general solution, is called a **particular solution** of equation (1). Other particular solutions are

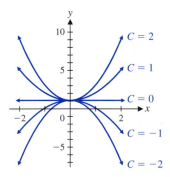

$$
\begin{array}{ll}
y = -2e^{0.1x^2} & C = -2 \\
y = -e^{0.1x^2} & C = -1 \\
y = 0 & C = 0 \\
y = e^{0.1x^2} & C = 1
\end{array}
$$

FIGURE 2 Particular solutions of $y' = 0.2xy$

These particular solutions are graphed in Figure 2, along with the slope field for equation (1). Notice how the particular solutions conform to the shapes indicated by the slope field.

REMARK: RECOGNIZING GENERAL SOLUTIONS

Most, but not all, first-order differential equations have general solutions that involve one arbitrary constant. Whenever we find a function, such as $y = Ce^{0.1x^2}$, that satisfies a given differential equation for any value of the constant C, we will assume this function is the general solution.

Example 1 ➱ **Verifying the Solution of a Differential Equation**   Show that

$$y = Cx^2 + 1$$

is the general solution of the differential equation

$$xy' = 2y - 2$$

On the same set of axes, graph the particular solutions obtained by letting $C = -2, -1, 0, 1,$ and 2.

FIGURE 3 Particular solutions of $xy' = 2y - 2$

SOLUTION
Substituting $y = Cx^2 + 1$ and $y' = 2Cx$ in the differential equation, we have

$$xy' = 2y - 2$$
$$x(2Cx) = 2(Cx^2 + 1) - 2$$
$$2Cx^2 = 2Cx^2 + 2 - 2$$
$$2Cx^2 = 2Cx^2$$

which shows that $y = Cx^2 + 1$ is the general solution. The particular solutions corresponding to $C = -2, -1, 0, 1,$ and 2 are graphed in Figure 3. ■

Matched Problem 1 ➱ Show that

$$y = Cx + 1$$

is the general solution of the differential equation

$$xy' = y - 1$$

On the same set of axes, graph the particular solutions obtained by letting $C = -2, -1, 0, 1,$ and 2. ■

In many applications, we will be interested in finding a particular solution $y(x)$ that satisfies an **initial condition** of the form $y(x_0) = y_0$. The value of the constant C in the general solution must then be selected so that this initial condition is satisfied; that is, so that the solution curve (from the family of solution curves) passes through (x_0, y_0).

Example 2 ⇔ **Finding a Particular Solution** Using the general solution in Example 1, find a particular solution of the differential equation $xy' = 2y - 2$ that satisfies the indicated initial condition, if such a solution exists.

(A) $y(1) = 3$ (B) $y(0) = 3$ (C) $y(0) = 1$

SOLUTION (A) *Initial condition* $y(1) = 3$. From Example 1, the general solution of the differential equation is

$$y = Cx^2 + 1$$

Substituting $x = 1$ and $y = 3$ in this general solution yields

$$3 = C(1)^2 + 1$$
$$= C + 1$$
$$C = 2$$

Thus, the particular solution satisfying the initial condition $y(1) = 3$ is

$$y = 2x^2 + 1$$

See the graph labeled $C = 2$ in Figure 3.

(B) *Initial condition* $y(0) = 3$. Substituting $x = 0$ and $y = 3$ in the general solution yields

$$3 = C(0)^2 + 1$$

No matter what value of C we select, $C(0)^2 = 0$ and this equation reduces to

$$3 = 1$$

which is *never* valid. Thus, we must conclude that there is no particular solution of this differential equation that will satisfy the initial condition $y(0) = 3$.

(C) *Initial condition* $y(0) = 1$. Substituting $x = 0$ and $y = 1$ in the general solution, we have

$$1 = C(0)^2 + 1$$
$$1 = 1$$

This equation is valid for *all* values of C. Thus, all solutions of this differential equation satisfy the initial condition $y(0) = 1$ (see Fig. 3). ■

REMARK: NATURE OF PARTICULAR SOLUTIONS

The situation illustrated in Example 2A is the most common. Most first-order differential equations you will encounter will have exactly one particular solution that satisfies a given initial condition. However, we must always be aware of the possibility that a differential equation may not have any particular solutions that satisfy a given initial condition, or the possibility that it may have many particular solutions satisfying a given initial condition.

Matched Problem 2 ⇔ Using the general solution in Matched Problem 1, find a particular solution of the differential equation $xy' = y - 1$ that satisfies the indicated initial condition, if such a solution exists.

(A) $y(1) = 2$ (B) $y(0) = 2$ (C) $y(0) = 1$

❑ IMPLICIT SOLUTIONS

The solution method we will discuss in the next section always produces an implicitly defined solution of a differential equation. In most of the differential equations we will consider, an explicit form of the solution can then be found by solving this implicit equation for the dependent variable in terms of the independent variable.

Explore–Discuss 2

Let y be a function defined implicitly by the equation

$$xy = C \qquad \text{where } C \text{ is a constant} \tag{3}$$

(A) Differentiate equation (3) implicitly to obtain a differential equation involving y and y'.

(B) Solve equation (3) for y, and verify that the resulting function satisfies the differential equation you found in part (A).

Example 3 ⇔ **Verifying an Implicit Solution**   If y is defined implicitly by the equation

$$y^3 + e^y - x^4 = C \tag{4}$$

show that y satisfies the differential equation

$$(3y^2 + e^y)y' = 4x^3$$

SOLUTION   We use implicit differentiation to show that y satisfies the given differential equation:

$$y^3 + e^y - x^4 = C$$

$$\frac{d}{dx}(y^3 + e^y - x^4) = \frac{d}{dx}C \qquad \text{Remember, } \frac{d}{dx}x^n = nx^{n-1} \text{ and } \frac{d}{dx}e^x = e^x,$$

$$\frac{d}{dx}y^3 + \frac{d}{dx}e^y - \frac{d}{dx}x^4 = 0 \qquad \text{but } \frac{d}{dx}y^n = ny^{n-1}y' \text{ and } \frac{d}{dx}e^y = e^y y'$$

$$3y^2y' + e^y y' - 4x^3 = 0 \qquad \text{since } y \text{ is a function of } x.$$

$$(3y^2 + e^y)y' = 4x^3$$

Since the last equation is the given differential equation, our calculations show that any function y defined implicitly by equation (4) is a solution of this differential equation. We cannot find an explicit formula for y, since none exists in terms of finite combinations of elementary functions.

Matched Problem 3 ⇔ If y is defined implicitly by the equation

$$y + e^{y^2} - x^2 = C$$

show that y satisfies the differential equation

$$(1 + 2ye^{y^2})y' = 2x$$

Example 4 ⌐ **Finding an Explicit Solution** If y is defined implicitly by the equation

$$y^2 - x^2 = C$$

show that y satisfies the differential equation

$$yy' = x$$

Find an explicit expression for the particular solution that satisfies the initial condition $y(0) = 2$.

SOLUTION Using implicit differentiation, we have

$$y^2 - x^2 = C$$
$$\frac{d}{dx}(y^2 - x^2) = \frac{d}{dx}C$$
$$2yy' - 2x = 0$$
$$2yy' = 2x$$
$$yy' = x$$

which shows that y satisfies the given differential equation. Substituting $x = 0$ and $y = 2$ in $y^2 - x^2 = C$, we have

$$y^2 - x^2 = C$$
$$(2)^2 - (0)^2 = C$$
$$4 = C$$

Thus, the particular solution satisfying $y(0) = 2$ is a solution of the equation

$$y^2 - x^2 = 4$$

or

$$y^2 = 4 + x^2$$

Solving the last equation for y yields two explicit solutions,

$$y_1(x) = \sqrt{4 + x^2} \qquad \text{and} \qquad y_2(x) = -\sqrt{4 + x^2}$$

The first of these two solutions satisfies

$$y_1(0) = \sqrt{4 + (0)^2} = \sqrt{4} = 2$$

while the second satisfies

$$y_2(0) = -\sqrt{4 + (0)^2} = -\sqrt{4} = -2$$

Thus, the particular solution of the differential equation $yy' = x$ that satisfies the initial condition $y(0) = 2$ is

$$y(x) = \sqrt{4 + x^2}$$

CHECK
$$y'(x) = \frac{1}{2}(4 + x^2)^{-1/2}2x \qquad\qquad yy' = x$$

$$= \frac{x}{\sqrt{4 + x^2}} \qquad\qquad \sqrt{4 + x^2}\left(\frac{x}{\sqrt{4 + x^2}}\right) \overset{?}{=} x$$

$$x \overset{?}{=} x$$

Matched Problem 4   ⇨   If y is defined implicitly by the equation

$$y^2 - x = C$$

show that y satisfies the differential equation

$$2yy' = 1$$

Find an explicit expression for the particular solution that satisfies the initial condition $y(0) = 3$.

■

□   **APPLICATION**

www   In economics, the price of a product is often studied over a period of time, and so it is natural to view price as a function of time. Let $p(t)$ be the price of a particular product at time t. If $p(t)$ approaches a limiting value $\bar{p}$ as t approaches infinity, then the price for this product is said to be **dynamically stable** and $\bar{p}$ is referred to as the **equilibrium price.** (Later in this chapter, this definition of equilibrium price will be related to the one given in Section 7-2.) In order to study the behavior of price as a function of time, economists often assume that the price satisfies a differential equation. This approach is illustrated in the next example.

Example 5   ⇨   **Dynamic Price Stability**   The price $p(t)$ of a product is assumed to satisfy the differential equation

$$\frac{dp}{dt} = 10 - 0.5p \qquad t \geqslant 0$$

(A) Show that

$$p(t) = 20 - Ce^{-0.5t}$$

is the general solution of this differential equation, and evaluate

$$\bar{p} = \lim_{t \to \infty} p(t)$$

(B) On the same set of axes, graph the three particular solutions that satisfy the initial conditions $p(0) = 40, p(0) = 10$, and $p(0) = 20$.

(C) Discuss the long-term behavior of the price of this product.

SOLUTION   (A) $$p(t) = 20 - Ce^{-0.5t}$$
$$p'(t) = 0.5Ce^{-0.5t}$$

Substituting in the given differential equation, we have

$$\frac{dp}{dt} = 10 - 0.5p$$
$$0.5Ce^{-0.5t} = 10 - 0.5(20 - Ce^{-0.5t})$$
$$= 10 - 10 + 0.5Ce^{-0.5t}$$
$$= 0.5Ce^{-0.5t}$$

which shows that $p(t) = 20 - Ce^{-0.5t}$ is the general solution of this differential equation. To find the equilibrium price, we must evaluate $\lim_{t \to \infty} p(t)$.

$$\bar{p} = \lim_{t \to \infty} p(t)$$
$$= \lim_{t \to \infty} (20 - Ce^{-0.5t})$$
$$= 20 - C \lim_{t \to \infty} e^{-0.5t} \qquad \lim_{t \to \infty} e^{-0.5t} = 0$$
$$= 20 - C \cdot 0$$
$$= 20$$

$p(t) = 20 + 20e^{-0.5t}$

$p(t) = 20 = \bar{p}$

$p(t) = 20 - 10e^{-0.5t}$

FIGURE 4 Particular solutions of $\dfrac{dp}{dt} = 10 - 0.5p$

(B) Before we can graph the three particular solutions, we must evaluate the constant C for each of the indicated initial conditions. In each case, we will make use of the equation

$$p(0) = 20 - Ce^0 = 20 - C$$

$p(0) = 40$	$p(0) = 10$	$p(0) = 20$
$20 - C = 40$	$20 - C = 10$	$20 - C = 20$
$C = -20$	$C = 10$	$C = 0$
$p(t) = 20 + 20e^{-0.5t}$	$p(t) = 20 - 10e^{-0.5t}$	$p(t) = 20$

The graphs are shown in Figure 4.

(C) The equilibrium price $\bar{p}$ does not depend on the constant C and consequently does not depend on the initial value of the price function $p(0)$ (see Fig. 4). If $p(0) > \bar{p}$, then the price decreases and approaches $\bar{p}$ as a limit. If $p(0) < \bar{p}$, then the price increases and approaches $\bar{p}$ as a limit. If $p(0) = \bar{p}$, then the price remains constant for all t.   ∎

Matched Problem 5

(A) Show that the price function

$$p(t) = 25 - Ce^{-0.2t}$$

is the general solution of the differential equation

$$\frac{dp}{dt} = 5 - 0.2p \qquad t \geq 0$$

Find the equilibrium price $\bar{p}$.

(B) On the same set of axes, graph the three particular solutions that satisfy the initial conditions $p(0) = 40, p(0) = 5$, and $p(0) = 25$.

(C) Discuss the long-term behavior of the price of this product.   ∎

Answers to Matched Problems

1.

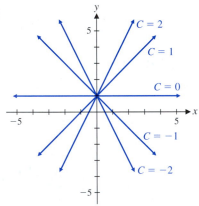

2. (A) $y = x + 1$
   (B) No particular solution exists.
   (C) $y = Cx + 1$ for any C

3. $\dfrac{d}{dx}y + \dfrac{d}{dx}e^{y^2} - \dfrac{d}{dx}x^2 = \dfrac{d}{dx}C$

   $y' + e^{y^2}2yy' - 2x = 0$

   $(1 + 2ye^{y^2})y' = 2x$

4. $y(x) = \sqrt{9 + x}$

5. (A) $\bar{p} = 25$
   (B)

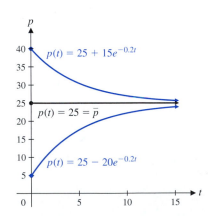

   (C) If $p(0) > \bar{p}$, then the price decreases and approaches $\bar{p}$ as a limit. If $p(0) < \bar{p}$, then the price increases and approaches $\bar{p}$ as a limit. If $p(0) = \bar{p}$, then the price remains constant for all t.

Exercise 9-1

A *In Problems 1–6, show that the given function is the general solution of the indicated differential equation.*

1. $y = Cx^2; xy' = 2y$

2. $y = C(x - 1)^2; (x - 1)y' = 2y$

3. $y = \dfrac{C}{x}; xy' = -y$

4. $y = \dfrac{C}{x + 1}; (x + 1)y' = -y$

5. $y = 4 + Ce^t; y' = y - 4$

6. $y = 1.5 + Ce^{-2t}; y' = 3 - 2y$

In Problems 7–10, determine which of the slope fields (A)–(D) shown below is associated with the indicated differential equation. Briefly justify your answer.

(A)

(B)

(C)

(D)

7. $xy' = 2y$ (From Problem 1)

8. $(x - 1)y' = 2y$ (From Problem 2)

9. $xy' = -y$ (From Problem 3)

10. $(x + 1)y' = -y$ (From Problem 4)

In Problems 11–14, use the appropriate slope field — (A), (B), (C), or (D), given for Problems 7–10 (or a copy) — to graph the particular solutions obtained by letting C = -2, -1, 0, 1, and 2.

11. $y = Cx^2$ (From Problem 1)

12. $y = C(x - 1)^2$ (From Problem 2)

13. $y = \dfrac{C}{x}$ (From Problem 3)

14. $y = \dfrac{C}{x + 1}$ (From Problem 4)

B *In Problems 15–20, show that the given function y is the general solution of the indicated differential equation. Find a particular solution satisfying the given initial condition.*

15. $y = 3 + 3x + Ce^x; y' = y - 3x; y(0) = 4$

16. $y = -4 - 4x + Ce^x; y' = y + 4x; y(0) = 1$

17. $y = xe^{-x} + Ce^{-x}; y' = -y + e^{-x}; y(0) = 0$

18. $y = xe^{2x} + Ce^{2x}; y' = 2y + e^{2x}; y(0) = 0$

19. $y = 2x + \dfrac{C}{x}; xy' = 4x - y; y(1) = 1$

20. $y = -3x + \dfrac{C}{x}; xy' + y = -6x; y(1) = 2$

If y is defined implicitly by the given equation, use implicit differentiation in Problems 21–24 to show that y satisfies the indicated differential equation.

21. $y^3 + xy - x^3 = C; (3y^2 + x)y' = 3x^2 - y$

22. $y + x^3y^3 - 2x = C; (1 + 3x^3y^2)y' = 2 - 3x^2y^3$

23. $xy + e^{y^2} - x^2 = C; (x + 2ye^{y^2})y' = 2x - y$

24. $y + e^{xy} - x = C; (1 + xe^{xy})y' = 1 - ye^{xy}$

If y is defined implicitly by the given equation, use implicit differentiation in Problems 25–28 to show that y satisfies the indicated differential equation. Find an explicit expression for the particular solution that satisfies the given initial condition.

25. $y^2 + x^2 = C; yy' = -x; y(0) = 3$

26. $y^2 + 2x^2 = C; yy' = -2x; y(0) = -4$

27. $\ln(2 - y) = x + C; y' = y - 2; y(0) = 1$

28. $\ln(5 - y) = 2x + C; y' = 2(y - 5); y(0) = 2$

In Problems 29–32, use the general solution y of the differential equation to find a particular solution that satisfies the indicated initial condition. Graph the particular solutions for x ≥ 0.

29. $y = 2 + Ce^{-x}; y' = 2 - y$

   (A) $y(0) = 1$ (B) $y(0) = 2$ (C) $y(0) = 3$

30. $y = 4 + Ce^{-x}; y' = 4 - y$

   (A) $y(0) = 2$ (B) $y(0) = 4$ (C) $y(0) = 5$

31. $y = 2 + Ce^x; y' = y - 2$

   (A) $y(0) = 1$ (B) $y(0) = 2$ (C) $y(0) = 3$

32. $y = 4 + Ce^x; y' = y - 4$

   (A) $y(0) = 2$ (B) $y(0) = 4$ (C) $y(0) = 5$

In Problems 33–36, write a differential equation that expresses the given description of a rate. Explain what each symbol in your differential equation represents.

33. The number of employees of a corporation increases at a rate of 650 employees per year.

34. The blood pressure in the aorta decreases between heartbeats at a rate proportional to the blood pressure.

35. When a pizza is taken out of a hot oven, it cools off at a rate proportional to the difference between the temperature of the pizza and the room temperature of 72°F.

36. When a gallon of milk is taken out of a refrigerator, it warms up at a rate proportional to the difference between the room temperature of 75°F and the temperature of the milk.

C *In Problems 37 and 38, use the general solution y of the differential equation to find a particular solution that satisfies the indicated initial condition. Graph the particular solutions for x ⩾ 0.*

37. $y = \dfrac{10}{1 + Ce^{-x}}; y' = 0.1y(10 - y)$

(A) $y(0) = 1$ (B) $y(0) = 10$ (C) $y(0) = 20$
[*Hint:* The particular solution in part (A) has an inflection point at $x = \ln 9$.]

38. $y = \dfrac{5}{1 + Ce^{-x}}; y' = 0.2y(5 - y)$

(A) $y(0) = 1$ (B) $y(0) = 5$ (C) $y(0) = 10$
[*Hint:* The particular solution in part (A) has an inflection point at $x = \ln 4$.]

In Problems 39 and 40, use the general solution y of the differential equation to find a particular solution that satisfies the indicated initial condition.

39. $y = Cx^3 + 2; xy' = 3y - 6$

(A) $y(0) = 2$ (B) $y(0) = 0$ (C) $y(1) = 1$

40. $y = Cx^4 + 1; xy' = 4y - 4$

(A) $y(0) = 1$ (B) $y(0) = 0$ (C) $y(1) = 2$

In Problems 41 and 42, use window dimensions Xmin = −5, Xmax = 5, Ymin = −5, and Ymax = 5.

41. Given that $y = x + Ce^{-x}$ is the general solution of the differential equation $y' + y = 1 + x$:

(A) In the same viewing window, graph the particular solutions obtained by letting $C = 0, 1, 2,$ and 3.

(B) What do the graphs of the solutions for $C > 0$ have in common?

(C) In the same viewing window, graph the particular solutions obtained by letting $C = 0, -1, -2,$ and -3.

(D) What do the graphs of the solutions for $C < 0$ have in common?

42. Repeat Problem 41, given that $y = x + (C/x)$ is the general solution of the differential equation $xy' = 2x - y$.

Applications

Business & Economics

43. *Price stability.* The price $p(t)$ of a product is assumed to satisfy the differential equation

$$\frac{dp}{dt} = 0.5 - 0.1p$$

(A) Show that

$$p(t) = 5 - Ce^{-0.1t}$$

is the general solution of this differential equation, and evaluate

$$\bar{p} = \lim_{t \to \infty} p(t)$$

(B) Graph the particular solutions that satisfy the initial conditions $p(0) = 1$ and $p(0) = 10$.

(C) Discuss the long-term behavior of the price of this product.

44. *Price stability.* The price $p(t)$ of a product is assumed to satisfy the differential equation

$$\frac{dp}{dt} = 0.8 - 0.2p$$

(A) Show that

$$p(t) = 4 - Ce^{-0.2t}$$

is the general solution of this differential equation, and evaluate

$$\bar{p} = \lim_{t \to \infty} p(t)$$

(B) Graph the particular solutions that satisfy the initial conditions $p(0) = 2$ and $p(0) = 8$.

(C) Discuss the long-term behavior of the price of this product.

45. *Continuous compound interest.* If money is deposited at the continuous rate of $200 per year into an account earning 8% compounded continuously, then the amount A in the account after t years satisfies the differential equation

$$\frac{dA}{dt} = 0.08A + 200$$

(A) Show that

$$A = Ce^{0.08t} - 2{,}500$$

is the general solution of this differential equation.

(B) Graph the particular solutions satisfying $A(0) = 0$ and $A(0) = 1{,}000$.

(C) Compare the long-term behavior of the two particular solutions in part (B). Discuss the effect of the value of $A(0)$ on the amount in the account.

46. *Continuous compound interest.* If money is withdrawn at the continuous rate of $500 per year from an account earning 10% compounded continuously, then the amount A in the account after t years satisfies the differential equation

$$\frac{dA}{dt} = 0.1A - 500$$

(A) Show that

$$A = 5{,}000 + Ce^{0.1t}$$

is the general solution of this differential equation.

(B) Graph the particular solutions satisfying $A(0) = 4{,}000$ and $A(0) = 6{,}000$.

(C) Compare the long-term behavior of the two particular solutions in part (B). Discuss the effect of the value of $A(0)$ on the amount in the account.

Life Sciences

47. *Population growth—Verhulst growth law.* The number $N(t)$ of bacteria in a culture at time t is
WWW assumed to satisfy the Verhulst growth law

$$\frac{dN}{dt} = 100 - 0.5N$$

(A) Show that

$$N(t) = 200 - Ce^{-0.5t}$$

is the general solution of this differential equation, and evaluate

$$\overline{N} = \lim_{t \to \infty} N(t)$$

where $\overline{N}$ is the equilibrium size of the population.

(B) Graph the particular solutions that satisfy $N(0) = 50$ and $N(0) = 300$.

(C) Discuss the long-term behavior of this population.

48. *Population growth—logistic growth.* The population $N(t)$ of a certain species of animal in a controlled habitat at time t is assumed to satisfy the logistic growth law

$$\frac{dN}{dt} = \frac{1}{500}N(1{,}000 - N)$$

(A) Show that

$$N(t) = \frac{1{,}000}{1 + Ce^{-2t}}$$

is the general solution of this differential equation, and evaluate $\overline{N} = \lim_{t \to \infty} N(t)$.

(B) Graph the particular solutions that satisfy $N(0) = 200$ and $N(0) = 2{,}000$.

(C) Discuss the long-term behavior of this population.

Social Sciences

49. *Rumor spread—Gompertz growth law.* The rate of
WWW propagation of a rumor is assumed to satisfy the Gompertz growth law

$$\frac{dN}{dt} = Ne^{-0.5t}$$

where $N(t)$ is the number of individuals who have heard the rumor at time t.

(A) Show that $N(t) = Ce^{-2e^{-0.5t}}$ is the general solution of this differential equation, and evaluate $\overline{N} = \lim_{t \to \infty} N(t)$.

(B) Graph the particular solutions that satisfy $N(0) = 100$ and $N(0) = 200$.

(C) Discuss the effect of the value of $N(0)$ on the long-term propagation of this rumor.

❑ SEPARATION OF VARIABLES

In this section we will develop a technique called *separation of variables,* which can be used to solve differential equations that can be expressed in the form

$$f(y)y' = g(x) \tag{1}$$

We have already used this technique informally in Section 6-3 to derive some important exponential growth laws. We will now present a more formal development of this method.

Explore–Discuss 1

Use algebraic manipulation to express each of the following equations in the form $f(y)y' = g(x)$, or explain why it is not possible to do so.

(A) $y' = x^2y$ (B) $y' = x^2 + xy$ (C) $y' = x^2y + xy$

The solution of differential equations by the method of separating the variables is based on the substitution formula for indefinite integrals. If $y = y(x)$ is a differentiable function of x, then

$$\int f(y)\, dy = \int f[y(x)]y'(x)\, dx \qquad \text{Substitution formula} \atop \text{for indefinite integrals} \tag{2}$$

If $y(x)$ is also the solution of (1), then $y(x)$ and $y'(x)$ must satisfy (1). That is,

$$f[y(x)]y'(x) = g(x)$$

Substituting $g(x)$ for $f[y(x)]y'(x)$ in (2), we have

$$\int f(y)\, dy = \int g(x)\, dx$$

Thus, the solution $y(x)$ of (1) is given implicitly by this equation. If both indefinite integrals can be evaluated, and if the resulting equation can be solved for y in terms of x, then we have an explicit solution of the differential equation (1).

This discussion is summarized in Theorem 1:

THEOREM 1   Separation of Variables

The solution of the differential equation

$$f(y)y' = g(x)$$

is given implicitly by the equation

$$\int f(y)\, dy = \int g(x)\, dx$$

Several examples will demonstrate the use of this method.

Example 1   ➯   **Using Separation of Variables**   Solve: $y' = 2xy^2$.

SOLUTION

$$y' = 2xy^2$$ Multiply by $1/y^2$ to separate the variables.

$$\frac{1}{y^2}y' = 2x$$ Convert to an equation involving indefinite integrals.

$$\int \frac{1}{y^2}\,dy = \int 2x\,dx$$ Evaluate each indefinite integral.

$$-\frac{1}{y} + C_1 = x^2 + C_2$$ Combine the two constants of integration into a single arbitrary constant.

$$-\frac{1}{y} = x^2 + C$$ Solve for y.

$$y = -\frac{1}{x^2 + C}$$ General solution of $y' = 2xy^2$

CHECK

$$y' = \frac{d}{dx}\left(-\frac{1}{x^2 + C}\right) = \frac{2x}{(x^2 + C)^2}$$

$$y' = 2xy^2$$ Substitute $y' = 2x/(x^2 + C)^2$ and

$$\frac{2x}{(x^2 + C)^2} \overset{?}{=} 2x\left(-\frac{1}{x^2 + C}\right)^2$$ $y = -1/(x^2 + C)$.

$$\frac{2x}{(x^2 + C)^2} \overset{\checkmark}{=} \frac{2x}{(x^2 + C)^2}$$

This verifies that our general solution is correct. (You should develop the habit of checking the solution of each differential equation you solve, as we have done here. From now on, we leave it to you to check most of the examples worked in the text.)   ■

Matched Problem 1   ➯   Solve: $y' = 4x^3y^2$.   ■

In some cases the general solution obtained by the technique of separation of variables does not include all the solutions to a differential equation. For example, the constant function $y = 0$ also satisfies the differential equation in Example 1. (Verify this.) Yet the solution $y = 0$ cannot be obtained from the expression

$$y = -\frac{1}{x^2 + C}$$

for any choice of the constant C. Solutions of this type are referred to as **singular solutions** and are usually discussed in more advanced courses. We will not attempt to find the singular solutions of any of the differential equations we consider.

Example 2   ➯   **Using Separation of Variables**   Find the general solution of

$$(1 + x^2)y' = 2x(y - 1)$$

Then find the particular solution that satisfies the initial condition $y(0) = 3$.

SOLUTION First, we find the general solution:

$$(1 + x^2)y' = 2x(y - 1)$$

Multiply by $1/(1 + x^2)$ and $1/(y - 1)$ to separate the variables.

$$\frac{y'}{y - 1} = \frac{2x}{1 + x^2}$$

Convert to an equation involving indefinite integrals.

$$\int \frac{dy}{y - 1} = \int \frac{2x\, dx}{1 + x^2}$$

Evaluate each indefinite integral.

$$\ln|y - 1| = \ln(1 + x^2) + C$$

where C is an arbitrary constant. Notice that the two constants of integration always can be combined to form a single arbitrary constant. Also note that we can use $1 + x^2$ instead of $|1 + x^2|$, since $1 + x^2$ is always positive. In order to solve for y, we convert this last equation to exponential form:

$$
\begin{aligned}
|y - 1| &= e^{\ln(1+x^2)+C} \\
&= e^C e^{\ln(1+x^2)} \qquad \text{Use the property } e^{\ln r} = r. \\
&= e^C(1 + x^2)
\end{aligned}
$$

It can be shown that if we replace e^C with an arbitrary constant K, then we can omit the absolute value signs on the left side of the last equation.* The resulting equation can then be used to find the general solution to the original differential equation. Thus,

$$
\begin{aligned}
y - 1 &= K(1 + x^2) \\
y &= 1 + K(1 + x^2) \qquad \text{General solution}
\end{aligned}
$$

To find the particular solution that satisfies $y(0) = 3$, we substitute $x = 0$ and $y = 3$ in the general solution and solve for K:

$$
\begin{aligned}
3 &= 1 + K(1 + 0) \\
K &= 2 \\
y &= 1 + 2(1 + x^2) = 3 + 2x^2 \qquad \text{Particular solution}
\end{aligned}
$$

∎

Matched Problem 2 Find the general solution of

$$(2 + x^4)y' = 4x^3(y - 3)$$

Then find the particular solution that satisfies the initial condition $y(0) = 5$.

∎

Explore–Discuss 2 Let y be a positive quantity that is changing with respect to time t, such as the price of a product, the amount of a drug in the bloodstream, or the population of a country. If y satisfies the differential equation

---

*In any problem involving separation of variables, you may assume that the equations

$$e^{\ln|f(y)|} = e^C g(x) \qquad \text{and} \qquad f(y) = Kg(x)$$

are equivalent (both C and K are arbitrary constants). Justification of this assumption involves properties of the absolute value and exponential functions.

$$\frac{dy}{dt} = ky \qquad k \text{ a constant}$$

then we say that the rate of change of y with respect to time t is proportional to y. If $k > 0$, y is increasing (growing); if $k = 0$, y is constant; and if $k < 0$, y is decreasing (declining or decaying). For each of the following differential equations, write a brief verbal description of the rate of change of y and discuss the possible increasing/decreasing properties of y.

(A) $\dfrac{dy}{dt} = k$ (B) $\dfrac{dy}{dt} = k(100 - y)$ (C) $\dfrac{dy}{dt} = ky(100 - y)$

❑ EXPONENTIAL GROWTH

WWW   We now return to the study of exponential growth laws first begun in Section 6-3. This time we will place more emphasis on determining the relevant growth law for a particular application and on using separation of variables to solve the corresponding differential equation in each problem. We begin with the familiar exponential growth law.

Exponential Growth Law

If the rate of change with respect to time t of a quantity y is proportional to the amount present, then y satisfies the differential equation

$$\frac{dy}{dt} = ky$$

Exponential growth includes both the case where y is increasing and the case where y is decreasing (or decaying).

Example 3 **Product Analysis**   Mothballs of a certain brand evaporate at a rate proportional to their volume, losing half their volume every 4 weeks. If the volume of each mothball is initially 15 cubic centimeters and a mothball becomes ineffective when its volume reaches 1 cubic centimeter, how long will these mothballs be effective?

SOLUTION   The volume of each mothball is decaying at a rate proportional to its volume. If V is the volume of a mothball after t weeks, then

$$\frac{dV}{dt} = kV$$

Since the initial volume is 15 cubic centimeters, we know that $V(0) = 15$. After 4 weeks, the volume will be half the original volume, so $V(4) = 7.5$. Summarizing these requirements, we have the following exponential decay model:

$$\frac{dV}{dt} = kV$$

$$V(0) = 15 \qquad V(4) = 7.5$$

We want to determine the value of t that satisfies the equation $V(t) = 1$. First, we use separation of variables to find the general solution of the differential equation:

$$\frac{dV}{dt} = kV$$

$$\frac{1}{V}\frac{dV}{dt} = k$$

$$\int \frac{dV}{V} = \int k\,dt$$

$$\ln V = kt + C \qquad \text{\textit{We can write} ln V \textit{in place of} ln|V|, \textit{since} V > 0.}$$

$$V = e^{kt+C} = e^{C}e^{kt} = Ae^{kt} \quad \textit{General solution}$$

where $A = e^{C}$ is a positive constant. Now we use the initial condition to determine the value of the constant A:

$$V(0) = Ae^{0} = A = 15$$

$$V(t) = 15e^{kt}$$

Next, we apply the condition $V(4) = 7.5$ to determine the constant k:

$$V(4) = 15e^{4k} = 7.5$$

$$e^{4k} = \frac{7.5}{15} = 0.5$$

$$4k = \ln 0.5$$

$$k = \frac{\ln 0.5}{4} \approx -0.1733$$

$$V(t) = 15e^{(t/4)\ln 0.5} \qquad \textit{Particular solution}$$

The graph of $V(t)$ is shown in Figure 1. To determine how long the mothballs will be effective, we find t when $V = 1$:

$$V(t) = 1$$

$$15e^{(t/4)\ln 0.5} = 1$$

$$e^{(t/4)\ln 0.5} = \frac{1}{15}$$

$$\frac{t}{4}\ln 0.5 = \ln\frac{1}{15}$$

$$t = \frac{4\ln\frac{1}{15}}{\ln 0.5} \approx 15.6 \text{ weeks}$$

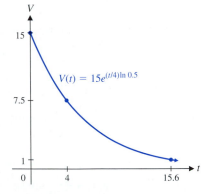

$V(t) = 15e^{(t/4)\ln 0.5}$

FIGURE 1 Exponential decay

Matched Problem 3 Repeat Example 3 if the mothballs lose half their volume every 5 weeks.

Refer to the solution of Example 3. The intersection routine on a graphing utility can be used to find both the constant k (Fig. 2) and the solution of the equation $V(t) = 1$ (Fig. 3).

FIGURE 2 $y_1 = 15e^{4x}$; $y_2 = 7.5$
$k \approx -0.1733$

FIGURE 3 $y_1 = 15e^{-0.1733x}$; $y_2 = 1$
$t \approx 15.6$ weeks

❑   LIMITED GROWTH

In certain situations there is an upper limit (or a lower limit), say M, on the values a variable can assume. This limiting value leads to the *limited growth law*.

> ### Limited Growth Law
>
> If the rate of change with respect to time t of a quantity y is proportional to the difference between y and a limiting value M, then y satisfies the differential equation
>
> $$\frac{dy}{dt} = k(M - y)$$

When we speak of limited growth, we will include both the case where y increases and approaches M from below and the case where y decreases and approaches M from above.

Example 4 ⇨   **Sales Growth**   The annual sales of a new company are expected to grow at a rate proportional to the difference between the sales and an upper limit of $20 million. The sales are 0 initially and $4 million for the second year of operation.

(A)   What should the company expect the sales to be during the 10th year?

(B)   In what year should the sales be expected to reach $15 million?

SOLUTION   If S is the annual sales (in millions of dollars) during year t, then the model for this problem is

$$\frac{dS}{dt} = k(20 - S) \tag{3}$$

$$S(0) = 0 \qquad S(2) = 4$$

This is a limited growth model. For part (A) we want to find $S(10)$, and for part (B) we want to solve $S(t) = 15$ for t. First, separating the variables in (3) and integrating both sides, we obtain

$$\int \frac{dS}{20 - S} = \int k \, dt$$

$$-\ln(20 - S) = kt + C \qquad \text{\color{blue}We can write } -\ln(20 - S) \text{ in place}$$
$$\ln(20 - S) = -kt - C \qquad \text{\color{blue}of } -\ln|20 - S|, \text{ since } 0 < S < 20.$$
$$20 - S = e^{-kt-C} = e^{-C}e^{-kt} = Ae^{-kt} \qquad \text{\color{blue}} A = e^{-C}$$
$$S = 20 - Ae^{-kt} \qquad \text{\color{blue}General solution}$$

Now we use the conditions $S(0) = 0$ and $S(2) = 4$ to determine the constants A and k:

$$S(0) = 20 - Ae^0$$
$$= 20 - A = 0$$
$$A = 20$$
$$S(t) = 20 - 20e^{-kt}$$
$$S(2) = 20 - 20e^{-2k} = 4$$
$$20e^{-2k} = 16$$
$$e^{-2k} = \frac{16}{20} = 0.8$$
$$-2k = \ln 0.8$$
$$k = -\frac{\ln 0.8}{2} \approx 0.1116$$
$$S(t) = 20 - 20e^{(t/2)\ln 0.8} \qquad \textit{Particular solution}$$

The graph of $S(t)$ is shown in Figure 4. Notice that the upper limit of $20 million is a horizontal asymptote.

(A) $S(10) = 20 - 20e^{5\ln 0.8} \approx \$13.45$ million

(B) $S(t) = 20 - 20e^{(t/2)\ln 0.8} = 15$
$$20e^{(t/2)\ln 0.8} = 5$$
$$e^{(t/2)\ln 0.8} = \frac{5}{20} = 0.25$$
$$\frac{t}{2} \ln 0.8 = \ln 0.25$$
$$t = \frac{2 \ln 0.25}{\ln 0.8} \approx 12.43 \text{ years}$$

The annual sales will exceed $15 million in the 13th year.

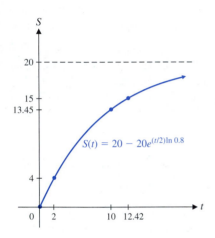

FIGURE 4 Limited growth

Matched Problem 4 Repeat Example 4 if the sales during the 2nd year are $3 million.

❑ LOGISTIC GROWTH

www If a quantity first begins to grow exponentially but then starts to approach a limiting value, it is said to exhibit *logistic growth*. More formally, we have

---

Logistic Growth Law

If the rate of change with respect to time t of a quantity y is proportional to both the amount present and the difference between y and a limiting value M, then y satisfies the differential equation

$$\frac{dy}{dt} = ky(M - y)$$

Theoretically, functions satisfying logistic growth laws can be increasing or decreasing, just as was the case for the exponential and limited growth laws. However, decreasing functions are seldom encountered in actual practice.

Example 5   ⇔   **Population Growth**   In a study of ciliate protozoans, it has been shown that the rate of growth of the number of *Paramecium caudatum* in a medium with fixed volume is proportional to the product of the number present and the difference between an upper limit of 375 and the number present. Suppose the medium initially contains 25 paramecia. After 1 hour there are 125 paramecia. How many paramecia are present after 2 hours?

SOLUTION   If P is the number of paramecia in the medium at time t, then the model for this problem is the following logistic growth model:

$$\frac{dP}{dt} = kP(375 - P)$$

$$P(0) = 25 \qquad P(1) = 125$$

We want to find $P(2)$. First, we separate the variables and convert to an equation involving indefinite integrals:

$$\frac{1}{P(375 - P)} \frac{dP}{dt} = k$$

$$\int \frac{1}{P(375 - P)} dP = \int k \, dt \tag{4}$$

The integral on the left side of (4) can be evaluated either by using an algebraic identity* or by using formula 9 in Table II. We will use formula 9 with $u = P$, $a = 375$, and $b = -1$:

$$\int \frac{1}{u(a + bu)} du = \frac{1}{a} \ln \left| \frac{u}{a + bu} \right| \qquad \text{Formula 9}$$

$$\int \frac{1}{P(375 - P)} dP = \frac{1}{375} \ln \left| \frac{P}{375 - P} \right| \qquad \begin{array}{l}\text{Since } 0 < P < 375, \text{ the ab-}\\ \text{solute value signs can be omitted.}\end{array}$$

$$= \frac{1}{375} \ln \left(\frac{P}{375 - P} \right)$$

Returning to equation (4), we have

$$\frac{1}{375} \ln \left(\frac{P}{375 - P} \right) = \int k \, dt = kt + D \qquad \text{D is a constant.}$$

$$\ln \left(\frac{P}{375 - P} \right) = 375kt + 375D$$

$$\frac{P}{375 - P} = e^{375kt + 375D} = e^{Bt} e^C \qquad B = 375k, \ C = 375D$$

$$P = 375 e^{Bt} e^C - P e^{Bt} e^C$$

$$P(e^{Bt} e^C + 1) = 375 e^{Bt} e^C$$

$$P = \frac{375 e^{Bt} e^C}{e^{Bt} e^C + 1} \qquad \begin{array}{l}\text{Multiply numerator and denomi-}\\ \text{nator by } e^{-Bt} e^{-C} \text{ and let } A = e^{-C}.\end{array}$$

$$= \frac{375}{1 + A e^{-Bt}} \qquad \text{General solution}$$

---

*If you do not wish to use Table II to evaluate integrals, you can use the algebraic identity

$$\frac{1}{x(a - x)} = \frac{1}{a} \left[\frac{(a - x) + x}{x(a - x)} \right] = \frac{1}{a} \left(\frac{1}{x} + \frac{1}{a - x} \right)$$

to evaluate this integral and in all logistic growth problems in Exercise 9-2.

Now we use the conditions $P(0) = 25$ and $P(1) = 125$ to evaluate the constants A and B:

$$P(0) = \frac{375}{1 + A} = 25$$

$$375 = 25 + 25A$$

$$A = 14$$

$$P(t) = \frac{375}{1 + 14e^{-Bt}}$$

$$P(1) = \frac{375}{1 + 14e^{-B}} = 125$$

$$375 = 125 + 1{,}750e^{-B}$$

$$e^{-B} = \frac{250}{1{,}750} = \frac{1}{7}$$

$$-B = \ln\frac{1}{7} = -\ln 7$$

$$B = \ln 7$$

$$P(t) = \frac{375}{1 + 14e^{-t\ln 7}} \qquad \textcolor{blue}{\text{Particular solution}}$$

To determine the population after 2 hours, we evaluate $P(2)$:

$$P(2) = \frac{375}{1 + 14e^{-2\ln 7}} \approx 292 \text{ paramecia}$$

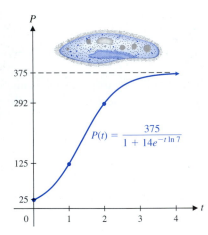

P

375

292

125

25

$P(t) = \dfrac{375}{1 + 14e^{-t\ln 7}}$

0   1   2   3   4 t

FIGURE 5 Logistic growth

The graph of $P(t)$ is shown in Figure 5. This S-shaped curve is typical of logistic growth functions. Notice that the upper limit of 375 is a horizontal asymptote. ■

Figures 6, 7, and 8 illustrate an alternative, graphical approach to finding the constants A and B for Example 5, as well as the population $P(2)$ after 2 hours.

FIGURE 6

$y_1 = \dfrac{375}{1 + x}; y_2 = 25$

$A = 14$

FIGURE 7

$y_1 = \dfrac{375}{1 + 14e^{-x}}; y_2 = 125$

$B \approx 1.9459$

FIGURE 8

$y_1 = \dfrac{375}{1 + 14e^{-1.9459x}};$

$y_2 = 375$

$P(2) \approx 292$

Logistic growth

Matched Problem 5   Repeat Example 5 for 15 paramecia initially and 150 paramecia after 1 hour.

Answers to Matched Problems   **1.** $y = -1/(x^4 + C)$

2. General solution: $y = 3 + K(2 + x^4)$; particular solution: $y = 5 + x^4$

3. 19.5 weeks

4. (A) \$11.13 million   (B) 17th year ($t \approx 17.06$ yr)

5. 343 paramecia

Exercise 9-2

A *In Problems 1–4, write a differential equation that describes the rate of change of the indicated quantity.*

1. The annual sales y of a company are increasing at the rate of \$100,000 per year.

2. The annual sales y of a company are increasing at a rate proportional to the annual sales.

3. The fish population y in a lake is growing at a rate proportional to the difference between the population and an upper limit of 10,000 fish.

4. In a community of 100,000, the number of people y who have contracted an infectious disease is growing at a rate proportional to the product of the number of people who have contracted the disease and the number who have not.

In Problems 5–8, write a verbal description of the rate of change of the given quantity satisfying the indicated differential equation, and discuss the increasing/decreasing properties of the quantity.

5. The annual sales y (in millions of dollars) of a company satisfy $dy/dt = 0.1y$ and $y(0) = 2$.

6. The annual sales y (in millions of dollars) of a company satisfy $dy/dt = 0.1(8 - y)$ and $y(0) = 2$.

7. The number of people y who have heard a rumor satisfies $dy/dt = 0.2y(5,000 - y)$ and $y(0) = 1$.

8. The amount of a drug y (in milliliters) in a patient's bloodstream satisfies $dy/dt = -0.2y$ and $y(0) = 5$.

In Problems 9–12, show that the technique of separation of variables is applicable by writing each differential equation in the form $f(y)y' = g(x)$.

9. $xy' = y^2$ **10.** $y' + xy = 2x$

11. $xy' + xy = 3y$ **12.** $y' + \dfrac{x}{y} = \dfrac{2}{y}$

In Problems 13–22, find the general solution for each differential equation. Then find the particular solution satisfying the initial condition.

13. $y' = 3x^2$; $y(0) = -1$ **14.** $y' = 2e^{2x}$; $y(0) = 5$

15. $y' = \dfrac{2}{x}$; $y(1) = 2$ **16.** $y' = \dfrac{2}{\sqrt[3]{x}}$; $y(1) = 2$

17. $y' = y$; $y(0) = 10$ **18.** $y' = y - 10$; $y(0) = 15$

19. $y' = 25 - y$; $y(0) = 5$

20. $y' = 3x^2y$; $y(0) = \dfrac{1}{2}$

21. $y' = \dfrac{y}{x}$; $y(1) = 5$; $x > 0$

22. $y' = \dfrac{y}{x^2}$; $y(-1) = 2e$

B *In Problems 23–32, find the general solution for each differential equation. Then find the particular solution satisfying the initial condition.*

23. $y' = \dfrac{1}{y^2}$; $y(1) = 3$ **24.** $y' = \dfrac{x^2}{y^2}$; $y(0) = 2$

25. $y' = ye^x$; $y(0) = 3e$ **26.** $y' = -y^2e^x$; $y(0) = \dfrac{1}{2}$

27. $y' = \dfrac{e^x}{e^y}$; $y(0) = \ln 2$

28. $y' = y^2(2x + 1)$; $y(0) = -\dfrac{1}{5}$

29. $y' = xy + x$; $y(0) = 2$

30. $y' = (2x + 4)(y - 3)$; $y(0) = 2$

31. $y' = (2 - y)^2e^x$; $y(0) = 1$

32. $y' = \dfrac{x^{1/2}}{(y - 5)^2}$; $y(1) = 7$

In Problems 33–38, find the general solution for each differential equation. Do not attempt to find an explicit expression for the solution.

33. $y' = \dfrac{1 + x^2}{1 + y^2}$ **34.** $y' = \dfrac{6x - 9x^2}{2y + 4}$

35. $xyy' = (1 + x^2)(1 + y^2)$

36. $(xy^2 - x) y' = x^2 y - x^2$

37. $x^2 e^y y' = x^3 + x^3 e^y$ **38.** $y' = \dfrac{xe^x}{\ln y}$

C *In Problems 39–44, find an explicit expression for the particular solution for each differential equation.*

39. $xyy' = \ln x$; $y(1) = 1$

40. $y' = \dfrac{xe^{x^2}}{y}$; $y(0) = 2$

41. $xy' = x\sqrt{y} + 2\sqrt{y}$; $y(1) = 4$

42. $y' = x(x - 1)^{1/2}(y - 1)^{1/2}$; $y(1) = 1$

43. $yy' = xe^{-y^2}$; $y(0) = 1$

44. $yy' = x(1 + y^2)$; $y(0) = 1$

In Problems 45–48, discuss why the technique of separation of variables does not produce an explicit solution $y = h(x)$ of the differential equation.

45. $y' + xy = 2$ **46.** $y' - 5x^2 y = e^x$

47. $3yy' = e^{x^2} + y$ **48.** $y' = \dfrac{9x^2 + 2}{5y^4 - 6}$

49. In many applications involving limited growth, the limiting value M is not known in advance and must be determined from data. Suppose that the rate of change with respect to time t of a quantity y is proportional to the difference between y and an unknown upper limit M, and that $y = 0$ when $t = 0$. Using the limited growth law, it follows that

$$y = M(1 - e^{-kt}) \qquad M > 0, \ k > 0$$

(A) Using the data given in the table, find two equations that M and k must satisfy, and solve each equation for M.

t	1	1.5
y	3	4

(B) Use a graphing utility and the equations from part (A) to approximate M to one decimal place.

50. Refer to Problem 49. Suppose that the rate of change with respect to time of a quantity y is proportional to the difference between y and an unknown lower limit M, and that $y = 10$ when $t = 0$. Using the limited growth law, it follows that

$$y = M + (10 - M)e^{-kt} \qquad M > 0, k > 0$$

(A) Using the data given in the table, find two equations that M and k must satisfy, and solve each equation for M.

t	1	5
y	9	7

(B) Use a graphing utility and the equations from part (A) to approximate M to one decimal place.

Applications

Business & Economics

51. *Continuous compound interest.* Provident Bank offers a CD (certificate of deposit) that earns 5.5% compounded continuously. How much will a $5,000 investment be worth in 5 years?

52. *Continuous compound interest.* Equitable Federal offers a CD that earns 4.9% compounded continuously. How much will a $10,000 investment be worth in 4 years?

53. *Advertising.* A company is using radio advertising to introduce a new product to a community of 100,000 people. Suppose the rate at which people learn about the new product is proportional to the number who have not yet heard of it.

(A) If no one is aware of the product at the start of the advertising campaign and after 7 days 20,000 people are aware of the product, how long will it take for 50,000 people to become aware of the product?

(B) Suppose the company is dissatisfied with the result in part (A) and wants to decrease to 14 days the amount of time it takes for 50,000 people to become aware of their product. In order for this to happen, how many people must become aware of the product during the first 7 days?

54. *Advertising.* Prior to the beginning of an advertising campaign, 10% of the potential users of a certain brand are aware of the brand name. After the first week of the campaign, 20% of the consumers are aware of the brand name. Assume that the percentage of informed consumers is growing at a rate proportional to the product of the percentage of informed consumers and the percentage of uninformed consumers.

(A) What percentage of consumers will be aware of the brand name after 5 weeks of advertising?

(B) Suppose the company is dissatisfied with the result in part (A) and increases the intensity of the advertising campaign so that 25% of the consumers are aware of the brand name after the first week. What effect does this have on the percentage of consumers that will be aware of the brand name after 5 weeks of advertising?

55. *Product analysis.* A company wishes to analyze a new room deodorizer. The active ingredient evaporates at a rate proportional to the amount present. Half of the ingredient evaporates in the first 30 days after the deodorizer is installed. If the deodorizer becomes ineffective after 90% of the active ingredient has evaporated, how long will one of these deodorizers remain effective?

56. *Corporate profits.* According to the U.S. Bureau of
www Economic Analysis, total corporate profits in domestic industries were $576.8 billion in 1995 and $946.2 billion in 2000. If corporate profits are increasing at a rate proportional to total corporate profits, what will the total corporate profits be in the year 2010?

57. *Sales growth.* The annual sales of a new company are expected to grow at a rate proportional to the difference between the sales and an upper limit of $5 million. If the sales are 0 initially and $1 million during the fourth year of operation, when will the sales reach $4 million?

58. *Sales analysis.* The annual sales of a company have declined from $8 million 2 years ago to $6 million today. If the annual sales continue to decline at a rate proportional to the difference between the annual sales and a lower limit of $3 million, find the annual sales 3 years from now.

59. *Sales analysis.* A new company has 0 sales initially, sales of $2 million during the first year, and sales of $5 million during the third year. If the annual sales S are assumed to be growing at a rate proportional to the difference between the sales and an unknown upper limit M, then by the limited growth law,
$$S(t) = M(1 - e^{-kt}) \qquad M > 0, k > 0$$
where t is time (in years) and $S(t)$ represents sales (in millions of dollars). Use approximation techniques to find k to one decimal place and M to the nearest million. (See Problem 49.)

60. *Sales analysis.* Refer to Problem 59. Approximate k to one decimal place and M to the nearest million if the sales during the third year are $4 million and all other information is unchanged.

www *Newton's law of cooling states that the rate of change of the temperature of an object is proportional to the difference between the temperature of the object and the temperature of the surrounding medium. Use this law to formulate models for Problems 61–64, and then solve using the techniques discussed in this section.*

61. *Manufacturing.* As part of a manufacturing process, a metal bar is to be heated in an oven until its temperature reaches 500°F. The oven is maintained at a constant temperature of 800°F. Before it is placed in the oven, the temperature of the bar is 80°F. After 2 minutes in the oven, the temperature of the bar is 200°F. How long should the bar be left in the oven?

62. *Manufacturing.* The next step in the manufacturing process described in Problem 61 calls for the heated bar to be cooled in a vat of water until its temperature reaches 100°F. The water in the vat is maintained at a constant temperature of 50°F. If the temperature

of the bar is 500°F when it is first placed in the water and the bar has cooled to 400°F after 5 minutes in the water, how long should the bar be left in the water?

Life Sciences

63. *Food preparation.* A pie is removed from an oven where the temperature is 325°F and placed in a freezer with a constant temperature of 25°F. After 1 hour in the freezer, the temperature of the pie is 225°F. What is the temperature of the pie after 4 hours in the freezer?

64. *Food preparation.* A roast is taken from a freezer where the temperature is 20°F and placed in an oven with a constant temperature of 350°F. After 1 hour in the oven, the temperature of the roast is 185°F. What is the temperature of the roast after 3 hours in the oven?

65. *Population growth.* A culture of bacteria is growing at a rate proportional to the number present. The culture initially contains 100 bacteria. After 1 hour there are 140 bacteria in the culture.

(A) How many bacteria will be present after 5 hours?
(B) When will the culture contain 1,000 bacteria?

66. *Population growth.* A culture of bacteria is growing in a medium that can support a maximum of 1,100 bacteria. The rate of change of the number of bacteria is proportional to the product of the number present and the difference between 1,100 and the number present. The culture initially contains 100 bacteria. After 1 hour there are 140 bacteria.

(A) How many bacteria are present after 5 hours?
(B) When will the culture contain 1,000 bacteria?

67. *Simple epidemic.* An influenza epidemic has spread throughout a community of 50,000 people at a rate proportional to the product of the number of people who have been infected and the number who have not been infected. If 100 individuals were infected initially and 500 were infected 10 days later:

(A) How many people will be infected after 20 days?
(B) When will half the community be infected?

68. *Ecology.* A fish population in a large lake is declining at a rate proportional to the difference between the population and a lower limit of 5,000 fish.

(A) If the population has declined from 15,000 fish 3 years ago to 10,000 today, find the population 6 years from now.
(B) Suppose the fish population 6 years from now turns out to be 8,000, indicating that the lake might be able to support more than the original lower limit of 5,000 fish. Assuming that the rate of change of the population is still proportional to the difference between the population and an unknown lower limit M, approximate M to the nearest hundred.

Social Sciences

69. *Sensory perception.* A person is subjected to a physical stimulus that has a measurable magnitude, but the intensity of the resulting sensation is difficult to measure. If s is the magnitude of the stimulus and $I(s)$ is the intensity of sensation, experimental evidence suggests that

$$\frac{dI}{ds} = k\frac{I}{s}$$

for some constant k. Express I as a function of s.

70. *Learning.* The number of words per minute, N, a person can type increases with practice. Suppose the rate of change of N is proportional to the difference between N and an upper limit of 140. It is reasonable to assume that a beginner cannot type at all. Thus,

$N = 0$ when $t = 0$. If a person can type 35 words per minute after 10 hours of practice:

(A) How many words per minute can that individual type after 20 hours of practice?

(B) How many hours must that individual practice to be able to type 105 words per minute?

71. *Rumor spread.* A rumor spreads through a population of 1,000 people at a rate proportional to the product of the number who have heard it and the number who have not heard it. If 5 people initiated a rumor and 10 people had heard it after 1 day:

(A) How many people will have heard the rumor after 7 days?

(B) How long will it take for 850 people to hear the rumor?

Section 9-3

First-Order Linear Differential Equations

- ❑ The Product Rule Revisited
- ❑ Solution of First-Order Linear Differential Equations
- ❑ Applications

❑ The Product Rule Revisited

The solution method we will discuss in this section is based in part on the product rule. Recall that if F and S are two differentiable functions, then

$$(FS)' = FS' + F'S$$

In Chapter 3, we started with F and S and used the product rule to find $(FS)'$. Now we want to use the rule in the reverse direction. That is, we want to be able to recognize when a sum of two terms is the derivative of a product.

Explore–Discuss 1

Match each sum on the left with the derivative of a product on the right, where y is an unknown differentiable function of x.

(A) $x^2 y' + 2xy$ (1) $(e^{x^2}y)'$

(B) $e^{-2x}y' - 2e^{-2x}y$ (2) $\left(\dfrac{1}{x^2}y\right)'$

(C) $\dfrac{1}{x^2}y' - \dfrac{2}{x^3}y$ (3) $(e^{-2x}y)'$

(D) $e^{x^2}y' + 2xe^{x^2}y$ (4) $(x^2 y)'$

In general, if $m(x)y' + n(x)y$ is the derivative of a product, how must $m(x)$ and $n(x)$ be related?

❑ Solution of First-Order Linear differential Equations

A differential equation that can be expressed in the form

$$y' + f(x)y = g(x)$$

is called a **first-order linear differential equation.**

For example,

$$y' + \frac{2}{x}y = x \tag{1}$$

is a first-order linear differential equation with $f(x) = 2/x$ and $g(x) = x$. This equation cannot be solved by the method of separation of variables. (Try to separate the variables to convince yourself that this is true.) Instead, we will change the form of the equation by multiplying both sides by x^2:

$$x^2y' + 2xy = x^3 \tag{2}$$

How was x^2 chosen? We will discuss that below. Let us first see how this choice leads to a solution of the problem.

Since $(x^2)' = 2x$, the left side of (2) is the derivative of a product:

$$x^2y' + 2xy = (x^2y)'$$

Thus, we can write equation (2) as

$$(x^2y)' = x^3$$

Now we can integrate both sides:

$$\int (x^2y)'\, dx = \int x^3\, dx$$

$$x^2y = \frac{x^4}{4} + C$$

Solving for y, we obtain the general solution

$$y = \frac{x^2}{4} + \frac{C}{x^2}$$

The function x^2, which we used to transform the original equation into one we could solve as illustrated, is called an *integrating factor*. It turns out that there is a specific formula for determining the integrating factor for any first-order linear differential equation. Furthermore, this integrating factor can then be used to find the solution of the differential equation, just as we used x^2 to find the solution to (1). The formula for the integrating factor and a step-by-step summary of the solution process are given in the box.

---

www

> ## *Solving First-Order Linear Differential Equations*
>
> **Step 1.** Write the equation in the **standard form:**
>
> $$y' + f(x)y = g(x)$$
>
> **Step 2.** Compute the **integrating factor:**
>
> $$I(x) = e^{\int f(x)\, dx}$$
>
> [When evaluating $\int f(x)\, dx$, choose 0 for the constant of integration.]
>
> **Step 3.** Multiply both sides of the standard form by the integrating factor $I(x)$. The left side should now be in the form $[I(x)y]'$:
>
> $$[I(x)y]' = I(x)g(x)$$
>
> **Step 4.** Integrate both sides:
>
> $$I(x)y = \int I(x)g(x)\, dx$$

[When evaluating $\int I(x)g(x)\,dx$, include an arbitrary constant of integration.]

Step 5. Solve for y to obtain the **general solution:**

$$y = \frac{1}{I(x)} \int I(x)g(x)\,dx$$

Example 1 ⇔ **Using an Integrating Factor** Solve: $2xy' + y = 10x^2$.

SOLUTION **Step 1.** Multiply both sides by $1/(2x)$ to obtain the standard form:

$$y' + \frac{1}{2x}y = 5x \qquad f(x) = \frac{1}{2x} \text{ and } g(x) = 5x$$

Step 2. Find the integrating factor:

$$
\begin{aligned}
I(x) &= e^{\int f(x)\,dx} \\
&= e^{\int [1/(2x)]\,dx} \qquad && \text{Assume } x > 0 \text{ and choose } 0 \text{ for the constant of} \\
& && \text{integration.} \\
&= e^{(1/2)\ln x} \qquad && r\ln t = \ln t^r \\
&= e^{\ln x^{1/2}} \qquad && e^{\ln r} = r, r > 0 \\
&= x^{1/2} \qquad && \text{Integrating factor}
\end{aligned}
$$

Step 3. Multiply both sides of the standard form by the integrating factor:

$$x^{1/2}\left(y' + \frac{1}{2x}y\right) = x^{1/2}(5x)$$

$$x^{1/2}y' + \tfrac{1}{2}x^{-1/2}y = 5x^{3/2} \qquad \text{The left side should have the form } [I(x)y]'.$$

$$(x^{1/2}y)' = 5x^{3/2}$$

Step 4. Integrate both sides:

$$\int (x^{1/2}y)'\,dx = \int 5x^{3/2}\,dx \qquad \begin{array}{l}\text{Include an arbitrary constant of integration} \\ \text{on the right side.}\end{array}$$

$$x^{1/2}y = 2x^{5/2} + C$$

Step 5. Solve for y:

$$y = \frac{1}{x^{1/2}}(2x^{5/2} + C)$$

$$= 2x^2 + \frac{C}{x^{1/2}} \qquad \text{General solution}$$

Matched Problem 1 ⇔ Solve: $xy' + 3y = 4x$.

In Example 1, notice that we assumed $x > 0$ to avoid introducing absolute value signs in the integrating factor. Many of the problems in this section will require the evaluation of expressions of the form $e^{\int [h'(x)/h(x)]\,dx}$. In order to avoid the complications caused by the introduction of absolute value signs, we will assume that the domain of $h(x)$ has been restricted so that $h(x) > 0$. This will simplify the solution process. We state the following familiar formulas for convenient reference (in the first formula the constant of integration is 0, in accordance with step 2):

> ### Basic Formulas Involving the Natural Logarithm Function
>
> If the domain of $h(x)$ is restricted so that $h(x) > 0$, then
>
> $$\int \frac{h'(x)}{h(x)}\, dx = \ln h(x) \qquad \text{and} \qquad e^{\ln h(x)} = h(x)$$

If a first-order linear differential equation is written in standard form, then multiplying both sides of the equation by its integrating factor will always convert the left side of the equation into the derivative of $I(x)y$. Thus, it is possible to omit steps 3 and 4, and proceed directly to step 5. This approach is illustrated in the next example. You decide which is easier to use—the step-by-step procedure or the formula in step 5.

Example 2 ⟅ **Using an Integrating Factor**   Find the particular solution of the equation

$$y' + 2xy = 4x$$

satisfying the initial condition $y(0) = 5$.

SOLUTION   Since the equation is already in standard form, we begin by finding the integrating factor:

$$I(x) = e^{\int 2x\, dx} = e^{x^2} \qquad f(x) = 2x$$

Proceeding directly to step 5, we have

$$y = \frac{1}{I(x)} \int I(x)g(x)\, dx \qquad g(x) = 4x$$

$$= \frac{1}{e^{x^2}} \int e^{x^2}(4x)\, dx$$

$$= e^{-x^2} \int 4xe^{x^2}\, dx \qquad \text{Let } u = x^2,\ du = 2x\, dx.$$

$$= e^{-x^2}(2e^{x^2} + C)$$

$$= 2 + Ce^{-x^2} \qquad \text{General solution}$$

Substituting $x = 0$ and $y = 5$ in the general solution, we have

$$5 = 2 + C$$

$$C = 3$$

$$y = 2 + 3e^{-x^2} \qquad \text{Particular solution}$$ ■

Matched Problem 2 ⟅ Find the particular solution of $y' + 3x^2y = 9x^2$ satisfying the initial condition $y(0) = 7$.   ■

CAUTION

  1. When integrating both sides of an equation such as

$$(x^{1/2}y)' = 5x^{3/2}$$

remember that

$$x^{1/2}y = \int 5x^{3/2}\,dx \neq 2x^{5/2} \qquad \text{"}+C\text{" is missing in the antiderivative.}$$

Remember: A constant of integration must be included when evaluating

$$\int I(x)g(x)\,dx.$$

If you omit this constant, you will not be able to find the general solution of the differential equation. See step 4 of Example 1 for the correct procedure.

2. $$\frac{1}{e^{x^2}} \int 4xe^{x^2}\,dx \neq \frac{1}{e^{x^2}} \int 4x e^{x^2}\,dx = \int 4x\,dx$$

Just as a variable factor cannot be moved across the integral sign, **a variable factor outside the integral sign cannot be used to cancel a factor inside the integral sign.** See Example 2 for the correct procedure.

□ APPLICATIONS

If P is the initial amount deposited into an account earning $100r\%$ compounded continuously, and A is the amount in the account after t years, then we saw in Section 6-3 that A satisfies the exponential growth equation

$$\frac{dA}{dt} = rA \qquad A(0) = P$$

Now suppose that money is continuously withdrawn from this account at a rate of $\$m$ per year. Then the amount A in the account at time t must satisfy

$$\begin{pmatrix} \text{Rate of change} \\ \text{of amount } A \end{pmatrix} = \begin{pmatrix} \text{Rate of growth} \\ \text{from continuous} \\ \text{compounding} \end{pmatrix} - \begin{pmatrix} \text{Rate of} \\ \text{withdrawal} \end{pmatrix}$$

$$\frac{dA}{dt} \qquad = \qquad rA \qquad - \qquad m$$

or

$$\frac{dA}{dt} - rA = -m \tag{3}$$

Explore–Discuss 2

How does equation (3) change if instead of making continuous withdrawals, money is continuously deposited into the account at the rate of $\$m$ per year?

Example 3 ⇔   **Continuous Compound Interest**   An initial deposit of $\$10,000$ is made into an account earning 8% compounded continuously. Money is then continuously withdrawn at a constant rate of $\$1,000$ a year until the account is depleted. Find the amount in the account at any time t. When will the amount be 0? What is the total amount withdrawn from this account?

SOLUTION   The amount A in the account at any time t must satisfy [see equation (3)]

$$\frac{dA}{dt} - 0.08A = -1,000 \qquad A(0) = 10,000$$

The integrating factor for this equation is

$$I(t) = e^{\int -0.08\,dt} = e^{-0.08t} \qquad f(t) = -0.08$$

Multiplying both sides of the differential equation by $I(t)$ and following the step-by-step procedure, we have

$$e^{-0.08t}\frac{dA}{dt} - 0.08e^{-0.08t}A = -1{,}000e^{-0.08t}$$

$$(e^{-0.08t}A)' = -1{,}000e^{-0.08t}$$

$$e^{-0.08t}A = \int -1{,}000e^{-0.08t}\,dt$$

$$= 12{,}500e^{-0.08t} + C$$

$$A = 12{,}500 + Ce^{0.08t} \qquad \textit{General solution}$$

Applying the initial condition $A(0) = 10{,}000$ yields

$$A(0) = 12{,}500 + C = 10{,}000$$

$$C = -2{,}500$$

$$A(t) = 1{,}2500 - 2{,}500e^{0.08t} \qquad \textit{Amount in the account at any time } t$$

To determine when the amount in the account is 0, we solve $A(t) = 0$ for t:

$$A(t) = 0$$

$$12{,}500 - 2{,}500e^{0.08t} = 0$$

$$12{,}500 = 2{,}500e^{0.08t}$$

$$5 = e^{0.08t}$$

$$t = \frac{\ln 5}{0.08} \approx 20.118 \text{ years}$$

10,000

0   25

Root
X=20.117974 Y=0

−3,000

FIGURE 1 $y_1 = 12{,}500 - 2{,}500e^{0.08x}$

Thus, the account is depleted after 20.118 years (see Fig. 1, which illustrates an alternative approach using the root approximation routine on a graphing utility). Since money is being withdrawn at the rate of $1,000 per year, the total amount withdrawn is

$$1{,}000(20.118) = \$20{,}118$$

■

Matched Problem 3   Repeat Example 3 if the account earns 5% compounded continuously.   ■

Example 4   **Equilibrium Price**   In economics, the supply S and the demand D for a commodity often can be considered as functions of both the price, $p(t)$, and the rate of change of the price, $p'(t)$. (Thus, S and D are ultimately functions of time t.) The **equilibrium price at time t** is the solution of the equation $S = D$. If $p(t)$ is the solution of this equation, then the **long-range equilibrium price** is

$$\bar{p} = \lim_{t \to \infty} p(t)$$

For example, if

$$D = 50 - 2p(t) + 2p'(t) \qquad S = 20 + 4p(t) + 5p'(t)$$

and $p(0) = 15$, then the equilibrium price at time t is the solution of the equation

$$50 - 2p(t) + 2p'(t) = 20 + 4p(t) + 5p'(t)$$

This simplifies to

$$p'(t) + 2p(t) = 10 \qquad \textit{f(t) = 2 and g(t) = 10}$$

which is a first-order linear equation with integrating factor

$$I(t) = e^{\int 2\,dt} = e^{2t}$$

Proceeding directly to step 5, we have

$$p(t) = \frac{1}{e^{2t}} \int 10e^{2t}\,dt$$

$$= e^{-2t}(5e^{2t} + C)$$

$$p(t) = 5 + Ce^{-2t} \qquad \textcolor{blue}{\textit{General solution}}$$

$$p(0) = 5 + C = 15$$

$$C = 10$$

$$p(t) = 5 + 10e^{-2t} \qquad \textcolor{blue}{\textit{Equilibrium price at time t (see Fig. 2)}}$$

$$\bar{p} = \lim_{t \to \infty}(5 + 10e^{-2t}) = 5 \qquad \textcolor{blue}{\textit{Long-range equilibrium price}}$$

p(t)

$p(t) = 5 + 10e^{-2t}$

$\bar{p} = 5$

FIGURE 2

Matched Problem 4   If $D = 70 + 2p(t) + 2p'(t)$, $S = 30 + 6p(t) + 3p'(t)$, and $p(0) = 25$, find the equilibrium price at time t, and find the long-range equilibrium price.

Example 5   **Pollution Control**   A company has a 1,000 gallon holding tank which is used to control the release of pollutants into a sewage system. Initially, the tank contains 500 gallons of water. Each gallon of water contains 2 pounds of pollutants. Additional polluted water containing 5 pounds of pollutants per gallon is pumped into the tank at the rate of 100 gallons per hour and is thoroughly mixed with the water already present in the tank. At the same time, the uniformly mixed water in the tank is released into the sewage system at a rate of 50 gallons per hour. This process continues for 5 hours. At the end of this 5 hour period, determine

(A) The total amount of pollutants in the tank

(B) The rate (in pounds per gallon) at which pollutants are being released into the sewage system

SOLUTION   Let $p(t)$ be the total amount (in pounds) of pollutants in the tank t hours after this process begins. Since the tank initially contains 500 gallons of water and each gallon of water contains 2 pounds of pollutants,

$$p(0) = 2 \cdot 500 = 1{,}000 \qquad \textcolor{orange}{\textit{Initial amount of pollutants in the tank}}$$

Since polluted water is entering and leaving the tank at different rates and with different concentrations of pollutants, the rate of change of the amount of pollutants in the tank will depend on the rate at which pollutants enter the tank and the rate at which they leave the tank:

$$\left(\begin{array}{c}\text{Rate of change}\\\text{of pollutants}\end{array}\right) = \left(\begin{array}{c}\text{Rate pollutants}\\\text{enter the tank}\end{array}\right) - \left(\begin{array}{c}\text{Rate pollutants}\\\text{leave the tank}\end{array}\right)$$

Finding expressions for the two rates on the right side of this equation will produce a differential equation involving $p'(t)$.

Since water containing 5 pounds of pollutants per gallon is entering the tank at a constant rate of 100 gallons per hour, pollutants are entering the tank at a constant rate of

$$\left(\begin{array}{c}5\text{ pounds}\\\text{per gallon}\end{array}\right) \times \left(\begin{array}{c}100\text{ gallons}\\\text{per hour}\end{array}\right) = 500\text{ pounds per hour}$$

Water enters tank at 100 gallons per hour

Water level is rising

Water leaves tank at 50 gallons per hour

FIGURE 3

How fast are the pollutants leaving the tank? Since the total amount of pollutants in the tank is increasing, the rate at which the pollutants leave the tank will depend on the amount of pollutants in the tank at time t and the amount of water in the tank at time t. Since 100 gallons of water enter the tank each hour and only 50 gallons leave each hour, the amount of water in the tank increases at the rate of 50 gallons per hour. Thus, the amount of water in the tank at time t is

$$\left(\begin{array}{c}\text{Initial amount}\\\text{of water}\end{array}\right) + \left(\begin{array}{c}\text{Gallons}\\\text{per hour}\end{array}\right) \times \left(\begin{array}{c}\text{Number}\\\text{of hours}\end{array}\right)$$

$$500 \qquad + \qquad 50t$$

The amount of pollutants in each gallon of water at any time t is

$$\frac{p(t)}{500 + 50t} \quad \frac{\left(\begin{array}{c}\text{Total amount}\\\text{of pollutants}\end{array}\right)}{\left(\begin{array}{c}\text{Total amount}\\\text{of water}\end{array}\right)} = \text{Pollutants per gallon}$$

Since water is leaving the tank at the rate of 50 gallons per hour, the rate at which the pollutants are leaving the tank is

$$\frac{50p(t)}{500 + 50t} = \frac{p(t)}{10 + t}$$

Thus, the rate of change of $p(t)$ must satisfy

$$\left(\begin{array}{c}\text{Rate of change}\\\text{of pollutants}\end{array}\right) = \left(\begin{array}{c}\text{Rate pollutants}\\\text{enter the tank}\end{array}\right) - \left(\begin{array}{c}\text{Rate pollutants}\\\text{leave the tank}\end{array}\right)$$

$$p'(t) \qquad = \qquad 500 \qquad - \qquad \frac{p(t)}{10 + t}$$

This gives the following model for this problem:

$$p'(t) = 500 - \frac{p(t)}{10 + t} \qquad p(0) = 1{,}000$$

or

$$p'(t) + \frac{1}{10 + t}\,p(t) = 500 \qquad \text{\textcolor{blue}{First-order linear equation with}}$$
$$\text{\textcolor{blue}{$f(t) = 1/(10 + t)$ and $g(t) = 500$}}$$

$$I(t) = e^{\int dt/(10+t)} = e^{\ln(10+t)} = 10 + t \quad \text{\textcolor{blue}{Integrating factor}}$$

$$p(t) = \frac{1}{10 + t}\int 500(10 + t)\,dt \qquad \text{\textcolor{blue}{Proceeding directly to step 5}}$$

$$= \frac{1}{10 + t}\left[250(10 + t)^2 + C\right]$$

$$p(t) = 250(10 + t) + \frac{C}{10 + t} \qquad \text{\textcolor{blue}{General solution}}$$

$$1{,}000 = 250(10) + \frac{C}{10} \qquad \text{\textcolor{blue}{Initial condition: $p(0) = 1{,}000$}}$$

$$\frac{C}{10} = -1{,}500$$

$$C = -15{,}000$$

$$p(t) = 250(10 + t) - \frac{15{,}000}{10 + t} \qquad \text{\textcolor{blue}{Particular solution (see Fig. 4)}}$$

$$p(t) = 250(10 + t) - \frac{15{,}000}{10 + t}$$

FIGURE 4

(A) To find the total amount of pollutants after 5 hours, we evaluate $p(5)$:

$$p(5) = 250(15) - \frac{15,000}{15} = 2,750 \text{ pounds}$$

(B) After 5 hours, the tank contains 750 gallons of water. The rate at which pollutants are being released into the sewage system is

$$\frac{2,750}{750} \approx 3.67 \text{ pounds per gallon}$$

Matched Problem 5 ⤳   Repeat Example 5 if water is released from the tank at the rate of 75 gallons per hour.

Answers to Matched Problems

1. $I(x) = x^3; y = x + (C/x^3)$

2. $I(x) = e^{x^3}; y = 3 + 4e^{-x^3}$

3. $A(t) = 20,000 - 10,000e^{0.05t}; A = 0$ when $t = (\ln 2)/0.05 \approx 13.863$ yr; total withdrawals $= \$13,863$

4. $p(t) = 10 + 15e^{-4t}; \bar{p} = 10$

5. $p(t) = 125(20 + t) - \dfrac{12,000,000}{(20 + t)^3}$ (A) 2,357 lb   (B) Approx. 3.77 lb/gal

Exercise 9-3

In all problems, assume $h(x) > 0$ whenever $\ln h(x)$ is involved.

A *In Problems 1–4, is the differential equation first-order linear? If so, find $f(x)$ and $g(x)$ so that the equation can be expressed in the form $y' + f(x)y = g(x)$.*

1. $y' = ky$

2. $y' = k(M - y)$

3. $y' = ky(M - y)$

4. $y' = kx^2(M - y)$

*In Problems 5–12, replace each **?** with a function of x that will make the integrand equal to the derivative of a product, and then find the antiderivative. Choose 0 for the constant of integration.*

5. $\int (x^3 y' + ?y)\, dx$

6. $\int (e^{2x} y' + ?y)\, dx$

7. $\int (e^{-3x} y' + ?y)\, dx$

8. $\int (x^{1/2} y' + ?y)\, dx$

9. $\int (?y' + 4x^3 y)\, dx$

10. $\int (?y' + 0.2e^{0.2x} y)\, dx$

11. $\int (?y' - 0.5e^{-0.5x} y)\, dx$

12. $\int \left(?y' - \frac{1}{x^2} y\right) dx$

In Problems 13–24, find the integrating factor, the general solution, and the particular solution satisfying the given initial condition.

13. $y' + 2y = 4; y(0) = 1$

14. $y' - 3y = 3; y(0) = -1$

15. $y' + y = e^{-2x}; y(0) = 3$

16. $y' - 2y = e^{3x}; y(0) = 2$

17. $y' - y = 2e^x; y(0) = -4$

18. $y' + 4y = 3e^{-4x}; y(0) = 5$

19. $y' + y = 9x^2 e^{-x}; y(0) = 2$

20. $y' - 3y = 6\sqrt{x}\, e^{3x}; y(0) = -2$

21. $xy' + y = 2x; y(1) = 1$

22. $xy' - 3y = 4x; y(1) = 1$

23. $xy' + 2y = 10x^3; y(2) = 8$

24. $xy' - y = \dfrac{9}{x^2}; y(2) = 3$

B *In Problems 25–34, find the integrating factor $I(x)$ for each equation, and then find the general solution.*

25. $y' + xy = 5x$

26. $y' - 2xy = 6x$

27. $y' - 2y = 4x$

28. $y' + y = x^2$

29. $xy' + y = xe^x$

30. $xy' + 2y = xe^{3x}$

31. $xy' + y = x \ln x$

32. $xy' - y = x \ln x$

33. $2xy' + 3y = 20x$

34. $2xy' - y = 6x^2$

In Problems 35–38, suppose that the indicated "solutions" were given to you by a student whom you are tutoring in this class.

(A) How would you have the student check each solution?

(B) Is the solution right or wrong? If the solution is wrong, explain what is wrong and how it can be corrected.

(C) Show a correct solution for each incorrect solution, and check the result.

35. Equation: $y' + \dfrac{1}{x} y = (x + 1)^2$

   Solution: $y = \dfrac{1}{x} \displaystyle\int x(x + 1)^2 dx$

   $= \dfrac{1}{3}(x + 1)^3 + C$

36. Equation: $y' + \dfrac{2}{x} y = x - 1$

   Solution: $y = \dfrac{1}{x^2} \displaystyle\int x^2(x - 1)\, dx$

   $= \dfrac{1}{2}(x - 1)^2 + C$

37. Equation: $y' + 3y = e^{-x}$

   Solution: $y = \dfrac{1}{e^{3x}} \displaystyle\int e^{2x} dx$

   $= \dfrac{1}{2} e^{-x}$

38. Equation: $y' - 2y = e^x$

   Solution: $y = e^{2x} \displaystyle\int e^{-x} dx$

   $= -e^x$

C *In Problems 39–44, find the general solution two ways. First, use an integrating factor and then use separation of variables.*

39. $y' = \dfrac{1 - y}{x}$ **40.** $y' = \dfrac{y + 2}{x + 1}$

41. $y' = \dfrac{2x + 2xy}{1 + x^2}$ **42.** $y' = \dfrac{4x + 2xy}{4 + x^2}$

43. $y' = 2x(y + 1)$ **44.** $y' = 3x^2(y + 2)$

45. Use an integrating factor to find the general solution of the unlimited growth model,

$$\frac{dy}{dt} = ky$$

[*Hint:* Remember, the antiderivative of the constant function 0 is an arbitrary constant C.]

46. Use an integrating factor to find the general solution of the limited growth model,

$$\frac{dy}{dt} = k(L - y)$$

47. Discuss how the solution process for

$$y' + f(x)y = g(x)$$

simplifies if $f(x)$ and $g(x)$ are constants $a \neq 0$ and b, respectively, and find the general solution.

48. Discuss how the solution process for

$$y' + f(x)y = g(x)$$

simplifies if $g(x) = 0$ for all x, and find the general solution.

Applications

Business & Economics

49. *Continuous compound interest.* An initial deposit of $20,000 is made into an account that earns 4.8% compounded continuously. Money is then withdrawn at a constant rate of $3,000 a year until the amount in the account is 0. Find the amount in the account at any time t. When is the amount 0? What is the total amount withdrawn from the account?

50. *Continuous compound interest.* An initial deposit of $50,000 is made into an account that earns 4.5% compounded continuously. Money is then withdrawn at a constant rate of $5,400 a year until the amount in the

account is 0. Find the amount in the account at any time t. When is the amount 0? What is the total amount withdrawn from the account?

51. *Continuous compound interest.* An initial deposit of $P is made into an account that earns 4.75% compounded continuously. Money is then withdrawn at a constant rate of $1,900 a year. After 10 years of continuous withdrawals, the amount in the account is 0. Find the initial deposit P.

52. *Continuous compound interest.* An initial deposit of $P is made into an account that earns 4.25% compounded continuously. Money is then withdrawn at a constant rate of $3,400 a year. After 5 years of continuous withdrawals, the amount in the account is 0. Find the initial deposit P.

53. *Continuous compound interest.* An initial deposit of $5,000 is made into an account earning 7% compounded continuously. Thereafter, money is deposited into the account at a constant rate of $2,100 per year. Find the amount in this account at any time t. How much is in this account after 5 years?

54. *Continuous compound interest.* An initial deposit of $10,000 is made into an account earning 6.25% compounded continuously. Thereafter, money is deposited into the account at a constant rate of $5,000 per year. Find the amount in this account at any time t. How much is in this account after 10 years?

55. *Continuous compound interest.* An initial investment of $10,000 is made in a mutual fund. Additional investments are made at the rate of $1,000 per year. After 10 years the fund has a value of $30,000. What continuous compound rate of interest has this fund earned over this 10 year period? Express the answer as a percentage, correct to two decimal places.

56. *Continuous compound interest.* A trust fund is established by depositing $100,000 into an account. Withdrawals are made at the rate of $20,000 per year. After 5 years, the trust fund contains $15,000. What continuous compound rate of interest has this fund earned over this 5 year period? Express the answer as a percentage, correct to two decimal places.

57. *Supply–demand.* The supply S and demand D for a certain commodity satisfy the equations

$$S = 35 - 2p(t) + 3p'(t) \quad \text{and}$$
$$D = 95 - 5p(t) + 2p'(t)$$

If $p(0) = 30$, find the equilibrium price at time t and the long-range equilibrium price.

58. *Supply–demand.* The supply S and demand D for a certain commodity satisfy the equations

$$S = 70 - 3p(t) + 2p'(t) \quad \text{and}$$
$$D = 100 - 5p(t) + p'(t)$$

If $p(0) = 5$, find the equilibrium price at time t and the long-range equilibrium price.

Life Sciences

59. *Pollution.* A 1,000-gallon holding tank contains 200 gallons of water. Initially, each gallon of water in the tank contains 2 pounds of pollutants. Water containing 3 pounds of pollutants per gallon enters the tank at a rate of 75 gallons per hour, and the uniformly mixed water is released from the tank at a rate of 50 gallons per hour. How many pounds of pollutants are in the tank after 2 hours? At what rate (in pounds per gallon) are the pollutants being released after 2 hours?

Water enters tank at 75 gallons per hour

Water leaves tank at 50 gallons per hour

60. *Pollution.* Rework Problem 59 if water is entering the tank at the rate of 100 gallons per hour.

61. *Pollution.* Rework Problem 59 if water is entering the tank at the rate of 50 gallons per hour.

62. *Pollution.* Rework Problem 59 if water is entering the tank at the rate of 150 gallons per hour.

63. *Pollution.* Refer to Problem 59. When will the tank contain 1,000 pounds of pollutants? Round answer to one decimal place.

64. *Pollution.* Refer to Problem 62. When will the tank contain 1,500 pounds of pollutants? Round answer to one decimal place.

In an article in the College Mathematics Journal *(Jan. 1987, 18:1), Arthur Segal proposed the following model for weight loss or gain:*

$$\frac{dw}{dt} + 0.005w = \frac{1}{3,500}C$$

where $w(t)$ is a person's weight (in pounds) after t days of consuming exactly C calories per day. Use this model to solve Problems 65–68.

65. *Weight loss.* A person weighing 160 pounds goes on a diet of 2,100 calories per day. How much will this person weigh after 30 days on this diet? How long will it take this person to lose 10 pounds? Find $\lim\limits_{t \to \infty} w(t)$ and interpret the results.

66. *Weight loss.* A person weighing 200 pounds goes on a diet of 2,800 calories per day. How much will this person weigh after 90 days on this diet? How long will it take this person to lose 25 pounds? Find $\lim\limits_{t \to \infty} w(t)$ and interpret the results.

67. *Weight loss.* A person weighing 130 pounds would like to lose 5 pounds during a 30 day period. How many calories per day should this person consume to reach this goal?

68. *Weight loss.* A person weighing 175 pounds would like to lose 10 pounds during a 45 day period. How many calories per day should this person consume to reach this goal?

Social Sciences

www *In 1960, William K. Estes proposed the following model for measuring a student's performance in the classroom:*

$$\frac{dk}{dt} + \ell k = \lambda \ell$$

where $k(t)$ is the student's knowledge after t weeks (expressed as a percentage and measured by performance on examinations), ℓ is a constant called the coefficient of learning and representing the student's ability to learn (expressed as a percentage and determined by IQ or some similar general intelligence predictor), and λ is a constant representing the fraction of available time the student spends performing helpful acts that should increase knowledge of the subject (studying, going to class, and so on). Use this model to solve Problems 69 and 70.

69. *Learning theory.* Students enrolled in a beginning Spanish class are given a pretest the first day of class in order to determine their initial knowledge of the subject. The results of the pretest, the coefficient of learning, and the fraction of time spent performing helpful acts for two students in the class are given in the table. Use the Estes model to predict the knowledge of each student after 6 weeks in the class.

	SCORE ON PRETEST	COEFFICIENT OF LEARNING	FRACTION OF HELPFUL ACTS
STUDENT A	0.1 (10%)	0.8	0.9
STUDENT B	0.4 (40%)	0.8	0.7

70. *Learning theory.* Refer to Problem 69. When will both students have the same level of knowledge? Round answer to one decimal place.

Important Terms and Symbols

9-1 *Basic Concepts.* Differential equation; first-order differential equation; order; slope field; solution; general solution; family of solutions; particular solution; initial condition; implicit solution; dynamically stable price; equilibrium price

9-2 *Separation of Variables.* Separation of variables; singular solution; exponential growth law; limited growth law; logistic growth law

$$\int f(y)\,dy = \int g(x)\,dx; \quad \frac{dy}{dt} = ky;$$

$$\frac{dy}{dt} = k(M - y); \quad \frac{dy}{dt} = ky(M - y)$$

9-3 *First-Order Linear Differential Equations.* First-order linear differential equation; standard form; integrating factor; general solution; equilibrium price at time t; long-range equilibrium price

$$y' + f(x)y = g(x); \quad I(x) = e^{\int f(x)\,dx};$$
$$y = \frac{1}{I(x)} \int I(x)g(x)\,dx$$

Review Exercise

Work through all the problems in this chapter review and check your answers in the back of the book. Answers to all review problems are there along with section numbers in italics to indicate where each type of problem is discussed. Where weaknesses show up, review appropriate sections in the text.

A *In Problems 1 and 2, show that the given function is the general solution of the indicated differential equation.*

1. $y = C\sqrt{x}; \; 2xy' = y$

2. $y = 1 + Ce^{-x/3}; \; 3y' + y = 1$

In Problems 3 and 4, determine which of the following slope fields is associated with the indicated differential equation. Briefly justify your answer.

(A)   (B)

3. $2xy' = y$ (From Problem 1)

4. $3y' + y = 1$ (From Problem 2)

In Problems 5 and 6, use the appropriate slope field—(A) or (B), given for Problems 3 and 4 (or a copy)—to graph the particular solutions obtained by letting C = −2, −1, 0, 1, and 2.

5. $y = C\sqrt{x}$ (From Problem 1)

6. $y = 1 + Ce^{-x/3}$ (From Problem 2)

In Problems 7 and 8, write a differential equation that describes the rate of change of the indicated quantity.

7. The price y of a product is decreasing at a rate proportional to the difference between the price and a lower limit of $5.

8. The price y of a product is increasing at a rate proportional to the price.

In Problems 9 and 10, write a verbal description of the rate of change of the given quantity satisfying the indicated differential equation and discuss the increasing/decreasing properties of the quantity.

9. The number of people y who have contracted an infectious disease satisfies $dy/dt = 0.2y(50{,}000 - y)$ and $y(0) = 1$.

10. The amount of radioactive material y (in grams) at the site of a nuclear accident satisfies $dy/dt = -0.001y$ and $y(0) = 100$.

In Problems 11–14, determine whether the differential equation can be written in the separation of variables form $f(y)y' = g(x)$, the first-order linear form $y' + f(x)y = g(x)$, both forms, or neither form.

11. $y' + 3x^2y + 9 = e^x(1 - x + y)$

12. $y' + 10y + 25 = 5xy^2$

13. $y' + 3y = xy + 2x - 6$ **14.** $y^2y' = 4xe^{x-y}$

In Problems 15–20, find the general solution.

15. $y' = -\dfrac{4y}{x}$ **16.** $y' = -\dfrac{4y}{x} + x$

17. $y' = 3x^2y^2$ **18.** $y' = 2y - e^x$

19. $y' = \dfrac{5}{x}y + x^6$ **20.** $y' = \dfrac{3 + y}{2 + x}$, $x > -2$

B *In Problems 21–28, find the particular solution that satisfies the given condition.*

21. $y' = 10 - y$; $y(0) = 0$

22. $y' + y = x$; $y(0) = 0$ **23.** $y' = 2ye^{-x}$; $y(0) = 1$

24. $y' = \dfrac{2x - y}{x + 4}$; $y(0) = 1$

25. $y' = \dfrac{x}{y + 4}$; $y(0) = 0$

26. $y' + \dfrac{2}{x}y = \ln x$; $y(1) = 2$

27. $yy' = \dfrac{x(1 + y^2)}{1 + x^2}$; $y(0) = 1$

28. $y' + 2xy = 2e^{-x^2}$; $y(0) = 1$

C **29.** Solve the following differential equation two ways. First, use an integrating factor, and then use separation of variables.

$$xy' - 4y = 8$$

30. Give an example of an equation that can be solved using an integrating factor but cannot be solved using separation of variables. Solve the equation.

31. Give an example of an equation that can be solved using separation of variables but cannot be solved using an integrating factor. Solve the equation.

32. Find the general solution of the following differential equation and then find the particular solutions satisfying the indicated initial conditions.

$$xy' - 5y = -10$$

(A) $y(0) = 2$ (B) $y(0) = 0$ (C) $y(1) = 1$

33. Find an explicit expression for the particular solution of the differential equation $yy' = x$ satisfying the initial condition $y(0) = 4$.

34. Given that $y = x^3 + Cx$ is the general solution of the differential equation $xy' - y = 2x^3$:

(A) In the same viewing window, graph the particular solutions obtained by letting $C = 0, 1, 2,$ and 3.

(B) What do the graphs of the solutions for $C > 0$ have in common?

(C) In the same viewing window, graph the particular solutions obtained by letting $C = 0, -1, -2,$ and -3.

(D) What do the graphs of the solutions for $C < 0$ have in common?

35. The rate of change with respect to time t of a quantity y is proportional to the difference between y and an unknown upper limit M, and $y = 0$ when $t = 0$. Using the limited growth law, it follows that

$$y = M(1 - e^{-kt}) \qquad M > 0, k > 0$$

(A) Using the data given in the table, find two equations that M and k must satisfy, and solve each equation for M.

t	2	5
y	4	7

(B) Use a graphing utility and the equations from part (A) to approximate M to one decimal place.

Applications

Business & Economics

36. Depreciation. A refrigerator costs $500 when it is new.
www The value of the refrigerator depreciates to $25 over a 20 year period. If the rate of change of the value is proportional to the value, find the value of the refrigerator 5 years after it was purchased.

37. Sales growth. The rate of growth of the annual sales of a new company is proportional to the difference between the annual sales S and an upper limit of $200,000. Assume that $S(0) = 0$.

(A) If the sales are $50,000 for the first year, how long will it take for the annual sales to reach $150,000? Round answer to the nearest year.

(B) Suppose the company is dissatisfied with the results in part (A) and wants the annual sales to reach $150,000 after just 3 years. In order for this to happen, what must the sales be for the first year? Round answer to the nearest thousand dollars.

38. Price stability. The supply S and the demand D for a certain commodity satisfy the equations

$$S = 100 + p + p' \quad \text{and} \quad D = 200 - p - p'$$

(A) Find the equilibrium price p at any time t.

(B) Find the long-range equilibrium price $\bar{p}$.

(C) Graph the particular solutions that satisfy the initial conditions $p(0) = 75$ and $p(0) = 25$.

(D) Discuss the long-term behavior of the price of this commodity.

39. Continuous compound interest. An initial deposit of $60,000 is made into an account that earns 5% compounded continuously. Money is then withdrawn at a constant rate of $5,000 per year until the amount in the account is 0. Find the amount in the account at any time t. When is the amount 0? What is the total amount withdrawn from the account?

40. Continuous compound interest. An initial investment of $15,000 is made in a mutual fund. Additional investments are made at the rate of $2,000 per year. After 10 years the fund has a value of $50,000. What continuous compound rate of interest has this fund earned over this 10 year period? Express the answer as a percentage, correct to two decimal places.

Life Sciences

41. Crop yield. The yield $y(t)$ (in bushels per acre) of a corn crop satisfies the equation

$$\frac{dy}{dt} = 100 + e^{-t} - y$$

If $y(0) = 0$, find y at any time t.

42. Pollution. A 1,000-gallon holding tank contains 100 gallons of unpolluted water. Water containing 2 pounds of pollutant per gallon is pumped into the tank at the rate of 75 gallons per hour. The uniformly mixed water is released from the tank at 50 gallons per hour.

(A) Find the total amount of pollutants in the tank after 2 hours. Round answer to one decimal place.

(B) When will the tank contain 700 pounds of pollutants? Round answer to one decimal place.

43. Ecology. A bird population on an island is declining at a rate proportional to the difference between the population and a lower limit of 200 birds.

(A) If the population has declined from 1,000 birds 5 years ago to 500 today, find the population 4 years from now.

(B) Suppose the bird population 4 years from now turns out to be 400, indicating that the island might be able to support more than the original lower limit of 200 birds. Assuming that the rate of change of the population is still proportional to the difference between the population and an unknown lower limit M, approximate M to the nearest integer.

Social Sciences

44. Rumor spread. A single individual starts a rumor in a community of 200 people. The rumor spreads at a rate proportional to the number of people who have not yet heard the rumor. After 2 days, 10 people have heard the rumor.

(A) How many people will have heard the rumor after 5 days?

(B) How long will it take for the rumor to spread to 100 people?

Group Activity 1   *Torricelli's Law*

www   Imagine water flowing out of a hole in the side of a container (see Fig. 1 on the next page). Let h denote the height of water above the hole. When h is large, h is decreasing rapidly since water is flowing rapidly out of the hole. When h is small, h is decreasing slowly since water is flowing slowly out of

FIGURE 1

the hole. One might guess that dh/dt is proportional to h—that is, that h satisfies the differential equation

$$\frac{dh}{dt} = kh \tag{1}$$

where k is a constant. But the differential equations

$$\frac{dh}{dt} = kh^2 \tag{2}$$

and

$$\frac{dh}{dt} = k\sqrt{h} \tag{3}$$

also imply that water flows more slowly as h decreases. Which provides the best model for the water flow? We can answer this question by conducting a simple experiment.

Drill a small hole approximately 4 millimeters in diameter near the bottom of the cylindrical portion of a clear plastic container. (It is not necessary for the entire container to be a perfect cylinder—a 2-liter soft drink bottle or a gallon container for windshield washer fluid will work.) Mark a scale on a piece of tape and affix it to the cylindrical portion of the bottle so the height h can be measured to the nearest millimeter (see Fig. 1). Fill the bottle with water while holding a finger over the hole, and then let the water flow out of the hole into a bucket. Start a stopwatch when the height falls to a convenient mark on the tape (11.0 centimeters, for example). Record the height of the water at intervals of 10 or 15 seconds as the water level drops (choose a time interval to allow approximately ten readings before the jet of water becomes a trickle).

(A) Construct a table in the format of Table 1 and enter your recorded data in the first two columns. (Change the 15 second time intervals and the number of rows, as necessary, depending on your apparatus.)

(B) Find the particular solutions h_1, h_2, and h_3 to the differential equations (1), (2), and (3), respectively, that agree with both the first and last rows of your recorded data (see Table 1).

(C) Compute values of h_1, h_2, and h_3 to complete the last three columns of your table.

TABLE 1

Sample of Recorded Data

TIME (SEC)	OBSERVED HEIGHT (CM)	h_1	h_2	h_3
0	11.0	11.0	11.0	11.0
15	9.8			
30	8.9			
45	7.9			
60	7.1			
75	6.2			
90	5.4			
105	4.6			
120	4.0			
135	3.4			
150	2.8			
165	2.4			
180	1.9	1.9	1.9	1.9

(D) Graph the functions h_1, h_2, h_3, and the observed heights on the same co-ordinate axes.

(E) Use parts (C) and (D) to determine which of $h_1, h_2,$ or h_3 provides the best model of water flow.

Torricelli's law implies that the rate dh/dt is proportional to the square root of h (for containers with constant cross-sectional area). The law is due to Evangelista Torricelli (1608–1647), the inventor of the mercury barometer.

(F) Discuss the extent to which your work confirms Torricelli's law.

Group Activity 2   *Euler's Method*

In this chapter we developed techniques for finding the solutions to certain types of differential equations. But many differential equations do not have solutions that can be expressed in terms of a finite combination of familiar functions. In these cases it is necessary to approximate the solutions. Here, we will consider a method of approximating the solution to any **initial-value problem** of the form

$$y' = f(x, y) \qquad y(a) = C \qquad a \leq x \leq b \tag{1}$$

To avoid confusion between the actual solution of an initial-value problem and an approximation to that solution, we will call $y = y(x)$ the **exact solution** of (1) if

$$y'(x) = f(x, y(x)) \qquad \text{for } a \leq x \leq b \qquad \text{and} \qquad y(a) = C$$

Note that an exact solution is always a particular solution. General solutions involving arbitrary constants cannot be approximated numerically.

The technique for approximating solutions to initial-value problems that we will discuss is called **Euler's method.** This method is based on the following observation about the slope of a line: If m is the slope of the line segment through the points (x_0, y_0) and (x_1, y_1), and $\Delta x = x_1 - x_0$ is the change in x, then $y_1 = y_0 + m\Delta x$ (see Fig. 1).

FIGURE 1

To introduce Euler's method, we consider the following initial-value problem:

$$y' = y \qquad y(0) = 1 \tag{2}$$

Of course, this is a very familiar equation whose exact solution is $y = e^x$. Studying this simple equation will enable us to compare the approximate solution with the exact solution. We want to approximate the solution to (2) for $0 \leq x \leq 1$. To do this we divide the interval $[0, 1]$ into five equal subintervals, each of length $\Delta x = 0.2$, and find approximate values of the solution at the end points of each of these intervals. Recall from our earlier discussion of slope fields that at a given point in the plane, the equation $y' = y$ gives the slope of the tangent line to a solution through that point (see Fig. 2).

FIGURE 2 Slope field for $y' = y$ FIGURE 3 Approximating $y(0.2)$

The slope at the given initial point $(0, y(0)) = (0, 1)$ is $y' = 1$. We draw a line with slope 1 starting at $(0, 1)$ and ending at $(0.2, y_1)$ (see Fig. 3). From our earlier observation about slope, we conclude that

$$y_1 = 1 + 0.2(1) = 1.2 \approx y(0.2)$$

(A) Find the slope at $(0.2, 1.2)$, draw a line segment with this slope starting at $(0.2, 1.2)$ and ending at $(0.4, y_2)$, and use y_2 to approximate $y(0.4)$.

(B) Repeat part (A) to find approximate values for $y(0.6)$, $y(0.8)$, and $y(1)$.

Figure 4 shows the line segments connecting the approximate values of $y(x)$, and Figure 5 compares this graph with the graph of the exact solution.

FIGURE 4   FIGURE 5

The ideas introduced above are summarized in the following box.

Euler's Method

If $y(x)$ is the (exact) solution to the initial-value problem

$$y' = f(x, y) \qquad y(x_0) = y_0$$

$x_0, x_1, \ldots, x_n$ is the sequence of values defined by

$$x_k = x_{k-1} + \Delta x \qquad k = 1, 2, \ldots, n$$

and $y_1, y_2, \ldots, y_n$ is the sequence of values defined by

$$y_k = y_{k-1} + f(x_{k-1}, y_{k-1})\Delta x \qquad k = 1, 2, \ldots, n \tag{3}$$

then

$$y(x_k) \approx y_k \qquad k = 1, 2, \ldots, n$$

(C) Formula (3) in Euler's method is a **recursion formula;** that is, each approximate value is calculated in terms of the preceding one. Verify that this formula gives the same approximate values you found in parts (A) and (B).

(D) Discuss the relationship between formula (3) and the slope of a line segment as illustrated in Figure 1.

(E) The distance between successive x values, Δx, is called the **step size.** We have used equal step sizes. Will this procedure work if the step sizes are not equal?

Now let's consider the following initial value problem:

$$y' = y + 7x - 3 \qquad y(0) = 1 \tag{4}$$

(F) Find the exact solution of (4).

(G) Approximate the solution for $0 \leqslant x \leqslant 1$ using Euler's method with a step size of $\Delta x = 0.2$. Construct a table comparing the approximate values with the exact values.

(H) Repeat part (G) with a step size of $\Delta x = 0.1$.

(I) To compare the exact solution from part (F) and the approximate solution from part (G) on a graphing utility, enter the values of x_k and the approximate values of y_k as a data set in the statistics editor. Then enter the exact solution in the function editor, and graph the data set and the function in the same viewing window. Repeat for the approximate values from part (H). Computing and graphing approximate values is easily automated by writing a short program. Table 1 on the next page lists programs for several graphing calculators, and Figure 6 on the next page illustrates the use of one of these programs for the data set from part (G). In Figure 6A, note that the function $f(x, y)$ was entered in capital letters preceded by quotation marks. The TI-85/TI-86 program also requires the quotation marks, but the function must be entered in lowercase letters.

(J) The distance between the exact and approximate solutions represents the error in the approximation. Based on the tables and graphs you have constructed, what happens to this error each time you compute the next approximate value? What happens to this error if you decrease the step size?

TABLE 1

GRAPHING CALCULATOR PROGRAMS FOR EULER'S METHOD

TI-82/TI-83	TI-85/TI-86*
$\{0\} \to L_1$	$\{0\} \to$ xStat
$\{0\} \to L_2$	$\{0\} \to$ yStat
Input "X0? ",X	Input "X0: ",x
Input "Y0? ",Y	Input "Y0: ",y
Input "DELTA X? ",D	Input "DELTA x: ",D
Input "NBR OF STEPS? ",N	Input "NUMBER OF STEPS: ",N
Input "F(X,Y)? ",Y0	Input "f(x,y): ",Y
FnOff 0	ST▶Eq(Y, y9)
$0 \to I$	FnOff 9
Lb1 A	$0 \to I$
$X \to L_1(I+1)$	Lb1 A
$Y \to L_2(I+1)$	$x \to x$ Stat(I+1)
$X+Y_0*D \to Y$	$y \to$ yStat(I+1)
$X+D \to X$	$y+y9*D \to y$
$I+1 \to I$	$X+D \to x$
If I>N	$I+1 \to I$
Stop	If I>N
Goto A	Stop
	Goto A

*The TI-86 has a built-in routine named "Euler," so you must give this program a different name, such as "Eulr."

(A) Input to program   (B) Output from program   (C) Graph of exact and
approximate solutions

FIGURE 6 Using the TI-82/TI-83 program for Euler's method

So far we have considered only equations whose exact solution is known. Now let's consider an equation whose exact solution cannot be found by any of the methods we have discussed:

$$y' = \frac{1}{x} + \frac{1}{y} \qquad y(1) = 1$$

(K) Use Euler's method with a step size of 0.2 to approximate $y(2)$.

(L) Repeat part (K) with smaller step sizes until you believe you have approximated $y(2)$ correctly to one decimal place.

Taylor Polynomials and Infinite Series

10

CHAPTER

INTRODUCTION

The circuits inside a calculator are capable only of performing the basic operations of addition, subtraction, multiplication, and division. Yet, many calculators have keys that allow you to evaluate functions such as e^x, $\ln x$, and $\sin x$. How is this done? In most cases, the values of these functions are *approximated* by using a carefully selected polynomial, and, of course, polynomials can be evaluated by using the basic arithmetic operations. Thus, the approximating polynomials give the calculator the capability of evaluating nonpolynomial functions. In this chapter we will study one type of approximating polynomial, called a *Taylor polynomial* after the English mathematician Brook Taylor (1685–1731). For a given function f, we will determine the coefficients of the Taylor polynomial, the values of x where the Taylor polynomial approximates $f(x)$, and the accuracy of this approximation. We will also consider various applications involving Taylor polynomials, including the approximation of definite integrals.

Section 10-1

Taylor Polynomials

❑ HIGHER-ORDER DERIVATIVES
❑ APPROXIMATING e^x WITH POLYNOMIALS
❑ TAYLOR POLYNOMIALS AT 0
❑ TAYLOR POLYNOMIALS AT a
❑ APPLICATION

❑   HIGHER-ORDER DERIVATIVES

Up to this point, we have considered only the first and second derivatives of a function. In the work that follows, we will need to find higher-order derivatives. For example, if we start with the function f defined by

$$f(x) = x^{-1}$$

then the first derivative of f is

$$f'(x) = -x^{-2}$$

and the second derivative of f is

$$f''(x) = \frac{d}{dx}f'(x) = \frac{d}{dx}(-x^{-2}) = 2x^{-3}$$

Additional higher-order derivatives are found by successive differentiation. Thus, the third derivative is

$$f^{(3)}(x) = \frac{d}{dx}f''(x) = \frac{d}{dx}(2x^{-3}) = -6x^{-4}$$

the fourth derivative is

$$f^{(4)}(x) = \frac{d}{dx}f^{(3)}(x) = \frac{d}{dx}(-6x^{-4}) = 24x^{-5}$$

and so on.

In general, the symbol $f^{(n)}$ is used to represent the **nth derivative of the function f.** When stating formulas involving a function and its higher-order derivatives, it is convenient to let $f^{(0)}$ represent the function f (that is, the zeroth derivative of a function is just the function itself).

The order of the derivative must be enclosed in parentheses, because in most contexts, $f^n(x)$ is interpreted to mean the nth power of f, not the nth derivative. Thus, $f^2(x) = [f(x)]^2 = f(x)f(x)$, while $f^{(2)}(x) = f''(x)$.

Finding a particular higher-order derivative is a routine calculation. However, finding a formula for the nth derivative for arbitrary n requires careful observation of the patterns that develop as each successive derivative is found. Study the next example carefully. Many problems in this chapter will involve similar concepts.

Example 1 ⇔   **Finding nth Derivatives**   Find the nth derivative of

$$f(x) = \frac{1}{1 + x} = (1 + x)^{-1}$$

SOLUTION   We begin by finding the first four derivatives of f:

$$f(x) = (1 + x)^{-1}$$
$$f'(x) = (-1)(1 + x)^{-2}$$
$$f''(x) = (-1)(-2)(1 + x)^{-3}$$
$$f^{(3)}(x) = (-1)(-2)(-3)(1 + x)^{-4}$$
$$f^{(4)}(x) = (-1)(-2)(-3)(-4)(1 + x)^{-5}$$

Notice that we did not multiply out the coefficient of $(1 + x)^{-k}$ in each derivative. Our objective is to look for a pattern in the form of each derivative. Multiplying out these coefficients would tend to obscure any pattern that is developing. Instead, we observe that each coefficient is a product of successive negative integers that can be written as follows:

$$(-1) = (-1)^1(1)$$
$$(-1)(-2) = (-1)^2(1)(2)$$
$$(-1)(-2)(-3) = (-1)^3(1)(2)(3)$$
$$(-1)(-2)(-3)(-4) = (-1)^4(1)(2)(3)(4)$$

Next we note that the product of natural numbers in each expression on the right can be written in terms of a factorial:*

$$(-1) = (-1)^1 1! \qquad 1! = 1$$
$$(-1)(-2) = (-1)^2 2! \qquad 2! = 1 \cdot 2$$
$$(-1)(-2)(-3) = (-1)^3 3! \qquad 3! = 1 \cdot 2 \cdot 3$$
$$(-1)(-2)(-3)(-4) = (-1)^4 4! \qquad 4! = 1 \cdot 2 \cdot 3 \cdot 4$$

Substituting these last expressions in the derivatives of f, we have

$$f'(x) = (-1)^1 1!(1 + x)^{-2} \qquad n = 1$$
$$f''(x) = (-1)^2 2!(1 + x)^{-3} \qquad n = 2$$
$$f^{(3)}(x) = (-1)^3 3!(1 + x)^{-4} \qquad n = 3$$
$$f^{(4)}(x) = (-1)^4 4!(1 + x)^{-5} \qquad n = 4$$

This suggests that

$$f^{(n)}(x) = (-1)^n n!(1 + x)^{-(n+1)} \qquad \text{Arbitrary } n$$

Matched Problem 1   Find the nth derivative of $f(x) = \ln x$.

Explore–Discuss 1

(A) Compute the first six derivatives of the polynomial function $p(x) = 7x^3 - 9x^2 + 4x - 15$.

(B) Let $q(x)$ be a polynomial of degree n. For which orders are the higher-order derivatives of $q(x)$ equal to the constant function $f(x) = 0$?

❑ APPROXIMATING e^x WITH POLYNOMIALS

We have already seen that the irrational number e and the exponential function e^x play important roles in many applications, including continuous compound interest, population growth, and exponential decay, to name a few. Now, given the function

$$f(x) = e^x$$

we would like to construct a polynomial function p whose values are close to the values of f, at least for some values of x. If we are successful, then we can use the values of p (which are easily computed) to approximate the values of f. We begin by trying to approximate f for values of x near 0 with a first-degree polynomial of the form

$$p_1(x) = a_0 + a_1 x \tag{1}$$

We want to place conditions on p_1 that will enable us to determine the unknown coefficients a_0 and a_1. Since we want to approximate f for values near 0, it is

---

*For n a natural number, $n! = n(n - 1)! = n(n - 1)(n - 2) \cdots \cdot 2 \cdot 1$ and $0! = 1$.

reasonable to require that f and p_1 agree at 0. Thus,

$$a_0 = p_1(0) = f(0) = e^0 = 1$$

This determines the value of a_0. To determine the value of a_1, we require that both functions have the same slope at 0. Since $p_1'(x) = a_1$ and $f'(x) = e^x$, this implies that

$$a_1 = p_1'(0) = f'(0) = e^0 = 1$$

Thus, after substituting $a_0 = 1$ and $a_1 = 1$ into (1), we obtain

$$p_1(x) = 1 + x$$

which is a first-degree polynomial satisfying

$$p_1(0) = f(0) \quad \text{and} \quad p_1'(0) = f'(0)$$

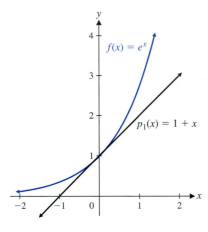

FIGURE 1

How well does $1 + x$ approximate e^x? Examining the graph in Figure 1, it appears that $1 + x$ is a good approximation to e^x for x very close to 0. However, as x moves away from 0, in either direction, the distance between the values of $1 + x$ and e^x increases and the accuracy of the approximation decreases.

Now we will try to improve this approximation by using a second-degree polynomial of the form

$$p_2(x) = a_0 + a_1 x + a_2 x^2 \tag{2}$$

We need three conditions to determine the coefficients a_0, a_1, and a_2. We still require that $p_2(0) = f(0)$ and $p_2'(0) = f'(0)$, and add the condition that $p_2''(0) = f''(0)$. This ensures that the graphs of p_2 and f have the same concavity at $x = 0$. Proceeding as before, we compute the first and second derivatives of p_2 and f and apply these conditions:

$$
\begin{aligned}
p_2(x) &= a_0 + a_1 x + a_2 x^2 & f(x) &= e^x \\
p_2'(x) &= \quad\quad\; a_1 + 2a_2 x & f'(x) &= e^x \\
p_2''(x) &= \quad\quad\quad\quad\; 2a_2 & f''(x) &= e^x
\end{aligned}
$$

Thus,

$$
\begin{array}{llll}
a_0 = p_2(0) = f(0) = e^0 = 1 & \text{implies} & a_0 = 1 \\
a_1 = p_2'(0) = f'(0) = e^0 = 1 & \text{implies} & a_1 = 1 \\
2a_2 = p_2''(0) = f''(0) = e^0 = 1 & \text{implies} & a_2 = \tfrac{1}{2}
\end{array}
$$

and, substituting in (2), we obtain

$$p_2(x) = 1 + x + \tfrac{1}{2}x^2$$

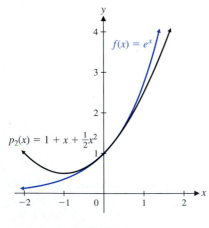

FIGURE 2

The graph is shown in Figure 2.

Comparing Figures 1 and 2, we see that for values of x near 0, the graph of $p_2(x) = 1 + x + \tfrac{1}{2}x^2$ is closer to the graph of $f(x) = e^x$ than was the graph of $p_1(x) = 1 + x$. Thus, the polynomial $1 + x + \tfrac{1}{2}x^2$ can be used to approximate e^x over a larger range of x values than the polynomial $1 + x$.

It seems reasonable to assume that a third-degree polynomial would yield a still better approximation. Using the notation for higher-order derivatives discussed earlier in this section, we can state the required condition as

Find

$$p_3(x) = a_0 + a_1 x + a_2 x^2 + a_3 x^3 \tag{3}$$

Satisfying

$$p_3^{(k)}(0) = f^{(k)}(0) \qquad k = 0, 1, 2, 3$$

As before, we obtain the value of the additional coefficient a_3 by adding the requirement that

$$p_3^{(3)}(0) = f^{(3)}(0)$$

Thus,

$$
\begin{array}{ll}
p_3(x) = a_0 + a_1 x + a_2 x^2 + a_3 x^3 & f(x) = e^x \\
p_3'(x) = a_1 + 2a_2 x + 3a_3 x^2 & f'(x) = e^x \\
p_3''(x) = 2a_2 + 6a_3 x & f''(x) = e^x \\
p_3^{(3)}(x) = 6a_3 & f^{(3)}(x) = e^x
\end{array}
$$

Applying the conditions $p_3^{(k)}(0) = f^{(k)}(0)$, we have

$$
\begin{array}{lllll}
a_0 = p_3(0) & = f(0) & = e^0 = 1 & \text{implies} & a_0 = 1 \\
a_1 = p_3'(0) & = f'(0) & = e^0 = 1 & \text{implies} & a_1 = 1 \\
2a_2 = p_3''(0) & = f''(0) & = e^0 = 1 & \text{implies} & a_2 = \frac{1}{2} \\
6a_3 = p_3^{(3)}(0) & = f^{(3)}(0) & = e^0 = 1 & \text{implies} & a_3 = \frac{1}{6}
\end{array}
$$

Substituting in (3), we obtain

$$p_3(x) = 1 + x + \tfrac{1}{2}x^2 + \tfrac{1}{6}x^3$$

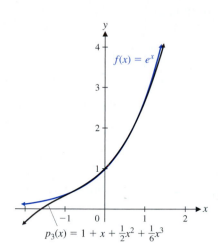

$p_3(x) = 1 + x + \frac{1}{2}x^2 + \frac{1}{6}x^3$

FIGURE 3

Figure 3 indicates that the approximations provided by p_3 are an improvement over those provided by p_2 and p_1.

In order to compare all three approximations, Table 1 lists the values of p_1, p_2, p_3, and e^x at selected values of x, and Table 2 lists the absolute value of the difference between these polynomials and e^x for the same x values. The values of e^x were obtained by using a calculator and are rounded to six decimal places. Comparing the columns in Table 2, we see that increasing the degree of the approximating polynomial decreases the difference between the polynomial and e^x. Thus, p_2 provides a better approximation than p_1, and p_3 provides a better approximation than p_2. We will have more to say about the accuracy of these approximations later in this chapter.

TABLE 1

$p_1(x) = 1 + x, p_2(x) = 1 + x + \frac{1}{2}x^2, p_3(x) = 1 + x + \frac{1}{2}x^2 + \frac{1}{6}x^3$

x	$p_1(x)$	$p_2(x)$	$p_3(x)$	e^x
−0.2	0.8	0.820	0.818 667	0.818 731
−0.1	0.9	0.905	0.904 833	0.904 837
0	1	1	1	1
0.1	1.1	1.105	1.105 167	1.105 171
0.2	1.2	1.220	1.221 333	1.221 403

TABLE 2

x	$\lvert p_1(x) - e^x \rvert$	$\lvert p_2(x) - e^x \rvert$	$\lvert p_3(x) - e^x \rvert$
−0.2	0.018 731	0.001 269	0.000 064
−0.1	0.004 837	0.000 163	0.000 004
0	0	0	0
0.1	0.005 171	0.000 171	0.000 004
0.2	0.021 403	0.001 403	0.000 069

(A) Let $p(x)$ be a polynomial of degree $n \geq 1$. Explain why each of the following is true:

(1) The graph of $p(x)$ intersects any vertical line in exactly one point.
(2) The graph of $p(x)$ intersects any horizontal line in at most n points.
(3) The graph of $p(x)$ has neither horizontal nor vertical asymptotes.

(B) Discuss the graphs of

$$f(x) = e^x \qquad g(x) = \frac{5x^2 - 3x + 1}{2x^2 + 3} \qquad h(x) = \frac{1}{x - 2} \qquad k(x) = e^{-x^2}$$

Use part (A) to explain why none of the graphs of f, g, h, or k is the graph of a polynomial function.

❑ TAYLOR POLYNOMIALS AT 0

The process we have used to determine p_1, p_2, and p_3 can be continued. Given any positive integer n, we define

$$p_n(x) = a_0 + a_1 x + a_2 x^2 + \cdots + a_n x^n$$

and require that

$$p_n^{(k)}(0) = f^{(k)}(0) \qquad k = 0, 1, 2, \ldots, n$$

The polynomial p_n is called a *Taylor polynomial*. Before determining p_n for $f(x) = e^x$, it will be convenient to make some general statements concerning the relationship between a_k, $p_n^{(k)}(0)$, and $f^{(k)}(0)$ for an arbitrary function f. First, $p_n(x)$ is differentiated n times to obtain the following relationships:

$$p_n(x) = a_0 + a_1 x + a_2 x^2 + a_3 x^3 + \cdots + a_n x^n$$
$$p_n'(x) = a_1 + 2a_2 x + 3a_3 x^2 + \cdots + na_n x^{n-1}$$
$$p_n''(x) = 2a_2 + (3)(2)a_3 x + \cdots + n(n-1)a_n x^{n-2}$$
$$p_n^{(3)}(x) = (3)(2)a_3 + \cdots + n(n-1)(n-2)a_n x^{n-3}$$
$$\vdots$$
$$p_n^{(n)}(x) = n(n-1)(n-2) \cdots (1)a_n = n!a_n$$

Evaluating each derivative at 0 and applying the requirement that $p_n^{(k)}(0) = f^{(k)}(0)$ leads to the following equations:

$$a_0 = p_n(0) = f(0)$$
$$a_1 = p_n'(0) = f'(0)$$
$$2a_2 = p_n''(0) = f''(0)$$
$$(3)(2)a_3 = p_n^{(3)}(0) = f^{(3)}(0)$$
$$\vdots \qquad \vdots$$
$$n!a_n = p_n^{(n)}(0) = f^{(n)}(0)$$

Solving each equation for a_k, we have

$$a_k = \frac{p_n^{(k)}(0)}{k!} = \frac{f^{(k)}(0)}{k!}$$

This relationship enables us to state the general definition of a Taylor polynomial.

Taylor Polynomial at 0*

If f has n derivatives at 0, then the **nth-degree Taylor polynomial for f at 0** is

$$p_n(x) = a_0 + a_1 x + a_2 x^2 + \cdots + a_n x^n = \sum_{k=0}^{n} a_k x^k$$

where

$$p_n^{(k)}(0) = f^{(k)}(0) \quad \text{and} \quad a_k = \frac{f^{(k)}(0)}{k!} \qquad k = 0, 1, 2, \ldots, n$$

---

*Taylor polynomials at 0 are also often referred to as *Maclaurin polynomials,* but we will not use this terminology in this book.

This result can be stated in a form that is more readily remembered as follows:

Taylor Polynomial at 0: *Concise Form*

The nth-degree Taylor polynomial for f at 0 is

$$p_n(x) = f(0) + f'(0)x + \frac{f''(0)}{2!}x^2 + \cdots + \frac{f^{(n)}(0)}{n!}x^n = \sum_{k=0}^{n} \frac{f^{(k)}(0)}{k!}x^k$$

provided f has n derivatives at 0.

In each of the preceding definitions, notice that we have used both the **expanded notation**

$$p_n(x) = a_0 + a_1 x + a_2 x^2 + \cdots + a_n x^n$$
$$= f(0) + f'(0)x + \frac{f''(0)}{2!}x^2 + \cdots + \frac{f^{(n)}(0)}{n!}x^n$$

and the more compact **summation notation**

$$p_n(x) = \sum_{k=0}^{n} a_k x^k = \sum_{k=0}^{n} \frac{f^{(k)}(0)}{k!}x^k$$

to represent a finite sum. Recall that we used summation notation earlier in Chapters 6 and 7. In this chapter we will place more emphasis on the expanded notation, since it is usually easier to visualize a polynomial written in this notation, but we will continue to include the summation notation where appropriate.

The nth-degree Taylor polynomial for f at 0 is always a polynomial, but it does not necessarily have degree n, since it is possible that $f^{(n)}(0) = 0$. For example, if $g(x) = x^3 + 2x + 1$, then $g'(x) = 3x^2 + 2$ and $g''(x) = 6x$, so $g(0) = 1, g'(0) = 2$, and $g''(0) = 0$. Therefore, for this function g at 0, the second-degree Taylor polynomial is $p_2(x) = 1 + 2x$, which actually has degree 1, not degree 2.

Returning to our original function $f(x) = e^x$, it is now an easy matter to find the nth-degree Taylor polynomial for this function. Since $(d/dx)e^x = e^x$, it follows that

$$f^{(k)}(x) = e^x \qquad f^{(k)}(0) = e^0 = 1 \qquad a_k = \frac{f^{(k)}(0)}{k!} = \frac{1}{k!}$$

for all values of k. Thus, for any n, the nth-degree Taylor polynomial for e^x is

$$p_n(x) = 1 + x + \frac{1}{2!}x^2 + \frac{1}{3!}x^3 + \cdots + \frac{1}{n!}x^n = \sum_{k=0}^{n} \frac{1}{k!}x^k$$

Example 2 ⇔ **Approximation Using a Taylor Polynomial**   Find the third-degree Taylor polynomial at 0 for $f(x) = \sqrt{x + 4}$. Use p_3 to approximate $\sqrt{5}$.

SOLUTION   **Step 1.**   *Find the derivatives:*

$$f(x) = (x + 4)^{1/2}$$
$$f'(x) = \tfrac{1}{2}(x + 4)^{-1/2}$$
$$f''(x) = -\tfrac{1}{4}(x + 4)^{-3/2}$$
$$f^{(3)}(x) = \tfrac{3}{8}(x + 4)^{-5/2}$$

Step 2.   *Evaluate the derivatives at 0:*

$$f(0) = 4^{1/2} = 2$$
$$f'(0) = \tfrac{1}{2}(4^{-1/2}) = \tfrac{1}{4}$$
$$f''(0) = -\tfrac{1}{4}(4^{-3/2}) = -\tfrac{1}{32}$$
$$f^{(3)}(0) = \tfrac{3}{8}(4^{-5/2}) = \tfrac{3}{256}$$

Step 3.   *Find the coefficients of the Taylor polynomial:*

$$a_0 = \frac{f(0)}{0!} = f(0) = 2$$

$$a_1 = \frac{f'(0)}{1!} = f'(0) = \frac{1}{4}$$

$$a_2 = \frac{f''(0)}{2!} = \frac{-\frac{1}{32}}{2} = -\frac{1}{64}$$

$$a_3 = \frac{f^{(3)}(0)}{3!} = \frac{\frac{3}{256}}{6} = \frac{1}{512}$$

Step 4.   *Write down the Taylor polynomial:*

$$p_3(x) = 2 + \tfrac{1}{4}x - \tfrac{1}{64}x^2 + \tfrac{1}{512}x^3 \qquad \text{Taylor polynomial}$$

To use p_3 to approximate $\sqrt{5}$, we must first determine the appropriate value of x:

$$f(x) = \sqrt{x + 4} = \sqrt{5} \qquad \text{Square both sides.}$$
$$x + 4 = 5 \qquad \text{Solve for } x.$$
$$x = 1$$

Thus,

$$\sqrt{5} = f(1) \approx p_3(1) = 2 + \tfrac{1}{4} - \tfrac{1}{64} + \tfrac{1}{512} \approx 2.236\ 328\ 1$$

[*Note:* The value of $\sqrt{5}$ obtained by using a calculator is, to six decimal places, 2.236 068.] ∎

FIGURE 4 Graphs of $f(x) = \sqrt{x + 4}$ and $p_3(x)$

Compare the evaluation of $p_3(1)$ shown in Figure 4 to the value obtained in Example 2. The figure illustrates graphically that $p_3(x)$ is a good approximation to $f(x)$ for values of x near 0.

Matched Problem 2 ➭ Find the second-degree Taylor polynomial at 0 for $f(x) = \sqrt{x + 9}$. Use $p_2(x)$ to approximate $\sqrt{10}$.

■

Many of the problems in this section involve approximating the values of functions that also can be evaluated with the use of a calculator. In such problems you should compare the Taylor polynomial approximation with the calculator value (which is also an approximation). This will give you some indication of the accuracy of the Taylor polynomial approximation. Later in this chapter we will discuss methods for determining the accuracy of any Taylor polynomial approximation.

Example 3 ➭ **Taylor Polynomials at 0**   Find the nth-degree Taylor polynomial at 0 for $f(x) = e^{2x}$.

SOLUTION   **Step 1**   **Step 2**   **Step 3**

$$f(x) = e^{2x} \qquad f(0) = 1 \qquad a_0 = f(0) \quad = 1$$
$$f'(x) = 2e^{2x} \qquad f'(0) = 2 \qquad a_1 = f'(0) \quad = 2$$

$$f''(x) = 2^2 e^{2x} \qquad f''(0) = 2^2 \qquad a_2 = \frac{f''(0)}{2!} = \frac{2^2}{2!}$$

$$f^{(3)}(x) = 2^3 e^{2x} \qquad f^{(3)}(0) = 2^3 \qquad a_3 = \frac{f^{(3)}(0)}{3!} = \frac{2^3}{3!}$$

$$\vdots \qquad\qquad \vdots \qquad\qquad \vdots \qquad \vdots$$

$$f^{(n)}(x) = 2^n e^{2x} \qquad f^{(n)}(0) = 2^n \qquad a_n = \frac{f^{(n)}(0)}{n!} = \frac{2^n}{n!}$$

Step 4.   *Write down the Taylor polynomial:*

$$p_n(x) = 1 + 2x + \frac{2^2}{2!}x^2 + \frac{2^3}{3!}x^3 + \cdots + \frac{2^n}{n!}x^n = \sum_{k=0}^{n} \frac{2^k}{k!}x^k$$

■

Matched Problem 3 ➭ Find the nth-degree Taylor polynomial at 0 for $f(x) = e^{x/3}$.

■

❑   TAYLOR POLYNOMIALS AT a

Suppose we want to approximate the function $f(x) = \sqrt[4]{x}$ with a polynomial. Since $f^{(k)}(0)$ is undefined for $k = 1, 2, \ldots$, this function does not have a Taylor polynomial at 0. However, all the derivatives of f exist at any $x > 0$. How can we generalize the definition of the Taylor polynomial in order to approximate functions such as this one? Before answering this question, we need to review a basic property of polynomials.

We are used to expressing polynomials in powers of x. However, it is also possible to express any polynomial in powers of $x - a$ for an arbitrary number a. For example, the following three expressions all represent the same polynomial:

$$p(x) = 3 + 2x \qquad + x^2 \qquad \text{Powers of } x$$
$$= 6 + 4(x - 1) + (x - 1)^2 \quad \text{Powers of } x - 1$$
$$= 3 - 2(x + 2) + (x + 2)^2 \quad \text{Powers of } x + 2$$

To verify this statement, we expand the second and third expressions:

$$6 + 4(x - 1) + (x - 1)^2 = 6 + 4x - 4 + x^2 - 2x + 1$$
$$= 3 + 2x + x^2$$
$$3 - 2(x + 2) + (x + 2)^2 = 3 - 2x - 4 + x^2 + 4x + 4$$
$$= 3 + 2x + x^2$$

Now we return to the problem of generalizing the definition of the Taylor polynomial. Proceeding as we did before, given a function f with n derivatives at a number a, we want to find an nth-degree polynomial p_n with the property that

$$p^{(k)}(a) = f^{(k)}(a) \qquad k = 0, 1, \ldots, n$$

That is, we require that p_n and its first n derivatives agree with f and its first n derivatives at the number a. It turns out that it is much easier to find p_n when it is expressed in powers of $x - a$. The general expression for an nth-degree polynomial in powers of $x - a$ and its first n derivatives are as follows:

$$
\begin{aligned}
p_n(x) &= a_0 + a_1(x - a) + a_2(x - a)^2 + &&a_3(x - a)^3 + \cdots + a_n(x - a)^n \\
p_n'(x) &= a_1 + 2a_2(x - a) + &&3a_3(x - a)^2 + \cdots + na_n(x - a)^{n-1} \\
p_n''(x) &= 2a_2 + &&(3)(2)a_3(x - a) + \cdots + n(n - 1)a_n(x - a)^{n-2} \\
p_n^{(3)}(x) &= &&(3)(2)(1)a_3 + \cdots + n(n - 1)(n - 2)a_n(x - a)^{n-3} \\
&\ \ \vdots
\end{aligned}
$$

$$p_n^{(n)}(x) = n(n - 1) \cdot \cdots \cdot (1)a_n = n!a_n$$

Now we evaluate each function at a and apply the appropriate condition:

$$
\begin{aligned}
a_0 &= p_n(a) &&= f(a) &&k = 0 \\
a_1 &= p_n'(a) &&= f'(a) &&k = 1 \\
2a_2 &= p_n''(a) &&= f''(a) &&k = 2 \\
(3)(2)(1)a_3 &= p_n^{(3)}(a) &&= f^{(3)}(a) &&k = 3 \\
&\ \ \vdots &&\ \ \vdots &&\ \ \vdots \\
n!a_n &= p_n^{(n)}(a) &&= f^{(n)}(a) &&k = n
\end{aligned}
$$

Thus, each coefficient of p_n satisfies

$$a_k = \frac{f^{(k)}(a)}{k!} \qquad k = 0, 1, 2, \ldots, n$$

Taylor Polynomial at a

The **nth-degree Taylor polynomial at a** for a function f is

$$p_n(x) = f(a) + f'(a)(x - a) + \frac{f''(a)}{2!}(x - a)^2 + \cdots + \frac{f^{(n)}(a)}{n!}(x - a)^n$$

$$= \sum_{k=0}^{n} \frac{f^{(k)}(a)}{k!}(x - a)^k$$

provided f has n derivatives at a.

Theoretically, a function f has an nth-degree Taylor polynomial at any value a where it has n derivatives. In practice, we usually choose a so that f and its derivatives are easy to evaluate at $x = a$. For example, if $f(x) = \sqrt{x}$, then good choices for a are 1, 4, 9, 16, and so on.

Example 4 ⟿ **Taylor Polynomials at a**   Find the third-degree Taylor polynomial at $a = 1$ for $f(x) = \sqrt[4]{x}$. Use $p_3(x)$ to approximate $\sqrt[4]{2}$.

SOLUTION   **Step 1.**   *Find the derivatives:*

$$f(x) = x^{1/4}$$
$$f'(x) = \tfrac{1}{4}x^{-3/4}$$
$$f''(x) = -\tfrac{3}{16}x^{-7/4}$$
$$f^{(3)}(x) = \tfrac{21}{64}x^{-11/4}$$

Step 2.   *Evaluate the derivatives at $a = 1$:*

$$f(1) = 1$$
$$f'(1) = \tfrac{1}{4}$$
$$f''(1) = -\tfrac{3}{16}$$
$$f^{(3)}(1) = \tfrac{21}{64}$$

Step 3.   *Find the coefficients of the Taylor polynomial:*

$$a_0 = f(1) = 1$$
$$a_1 = f'(1) = \tfrac{1}{4}$$
$$a_2 = \frac{f''(1)}{2!} = \frac{-\tfrac{3}{16}}{2} = -\frac{3}{32}$$
$$a_3 = \frac{f^{(3)}(1)}{3!} = \frac{\tfrac{21}{64}}{6} = \frac{7}{128}$$

Step 4.   *Write down the Taylor polynomial:*

$$p_3(x) = 1 + \tfrac{1}{4}(x-1) - \tfrac{3}{32}(x-1)^2 + \tfrac{7}{128}(x-1)^3$$

Now we use the Taylor polynomial to approximate $\sqrt[4]{2}$:

$$\sqrt[4]{2} = f(2) \approx p_3(2) = 1 + \tfrac{1}{4} - \tfrac{3}{32} + \tfrac{7}{128} = 1.210\ 937\ 5$$

[*Note:* The value obtained by using a calculator is $\sqrt[4]{2} \approx 1.189\ 207\ 1$.] ▪

Matched Problem 4 ⇨   Find the second-degree Taylor polynomial at $a = 8$ for $f(x) = \sqrt[3]{x}$. Use $p_2(x)$ to approximate $\sqrt[3]{9}$. ▪

❑   APPLICATION

Example 5 ⇨   **Average Price**   Given the demand function

$$p = D(x) = \sqrt{2{,}500 - x^2}$$

use the second-degree Taylor polynomial at 0 to approximate the average price (in dollars) over the demand interval $[10, 40]$.

SOLUTION   The average price over the demand interval $[10, 40]$ is given by

$$\text{Average price} = \frac{1}{30} \int_{10}^{40} \sqrt{2{,}500 - x^2}\, dx$$

This integral cannot be evaluated by any of the techniques we have discussed. However, we can use a Taylor polynomial to approximate the value of the

FIGURE 5

integral. Omitting the details (which you should supply), the second-degree Taylor polynomial at 0 for $D(x) = \sqrt{2{,}500 - x^2}$ is

$$p_2(x) = 50 - \tfrac{1}{100}x^2$$

Assuming that $D(x) \approx p_2(x)$ for $10 \le x \le 40$ (see Fig. 5), we have

$$\text{Average price} \approx \frac{1}{30} \int_{10}^{40}\left(50 - \frac{1}{100}x^2\right) dx$$

$$= \frac{1}{30}\left(50x - \frac{1}{300}x^3\right)\Big|_{10}^{40}$$

$$= \$43$$

Matched Problem 5   Given the demand function

$$p = D(x) = \sqrt{400 - x^2}$$

use the second-degree Taylor polynomial at 0 to approximate the average price (in dollars) over the demand interval $[5, 15]$.

Example 5 illustrates the convenience of using Taylor polynomials to approximate more complicated functions, but it also raises several important questions. First, how can we determine whether it is reasonable to use the Taylor polynomial to approximate a function on a given interval? Second, is it possible to find the Taylor polynomial without resorting to successive differentiation of the function? Finally, can we determine the accuracy of an approximation such as the approximate average price found in Example 5? These questions will be discussed in the next three sections.

Answers to Matched Problems

1. $f^{(n)}(x) = (-1)^{n-1}(n-1)!x^{-n}$

2. $p_2(x) = 3 + \tfrac{1}{6}x - \tfrac{1}{216}x^2$; $\sqrt{10} \approx 3.162\,037$

3. $p_n(x) = 1 + \dfrac{1}{3}x + \dfrac{1}{3^2}\dfrac{1}{2!}x^2 + \dfrac{1}{3^3}\dfrac{1}{3!}x^3 + \cdots + \dfrac{1}{3^n}\dfrac{1}{n!}x^n = \displaystyle\sum_{k=0}^{n}\dfrac{1}{3^k}\dfrac{1}{k!}x^k$

4. $p_2(x) = 2 + \tfrac{1}{12}(x - 8) - \tfrac{1}{288}(x - 8)^2$; $\sqrt[3]{9} \approx 2.079\,861\,1$

5. $\dfrac{1}{10}\displaystyle\int_{5}^{15}\left(20 - \dfrac{x^2}{40}\right) dx = \dfrac{415}{24} \approx \$17.29$

Exercise 10-1

A   *In Problems 1–4, find $f^{(4)}(x)$.*

1. $f(x) = \dfrac{1}{2 - x}$ **2.** $f(x) = \ln(3 + x)$

3. $f(x) = e^{-2x}$ **4.** $f(x) = \sqrt{x}$

In Problems 5–12, find the indicated Taylor polynomial at 0.

5. $f(x) = e^{-x}$; $p_4(x)$ **6.** $f(x) = e^{4x}$; $p_3(x)$

7. $f(x) = (x + 1)^3$; $p_4(x)$

8. $f(x) = (1 - x)^4$; $p_3(x)$

9. $f(x) = \ln(1 + 2x)$; $p_3(x)$

10. $f(x) = \ln(1 + \tfrac{1}{2}x)$; $p_4(x)$

11. $f(x) = \sqrt[3]{x + 1}$; $p_3(x)$

12. $f(x) = \sqrt[4]{x + 16}$; $p_2(x)$

13. (A) Find the third-degree Taylor polynomial $p_3(x)$ for $f(x) = x^4 - 1$ at 0. For which values of x is $|p_3(x) - f(x)| < 0.1$?

(B) Find the fourth-degree Taylor polynomial $p_4(x)$ for $f(x) = x^4 - 1$ at 0. For which values of x is $|p_4(x) - f(x)| < 0.1$?

14. (A) Find the fourth-degree Taylor polynomial $p_4(x)$ for $f(x) = x^5$ at 0. For which values of x is $|p_4(x) - f(x)| < 0.01$?

(B) Find the fifth-degree Taylor polynomial $p_5(x)$ for $f(x) = x^5$ at 0. For which values of x is $|p_5(x) - f(x)| < 0.01$?

In Problems 15–20, find the indicated Taylor polynomial at the given value of a.

15. $f(x) = e^{x-1}; p_4(x)$ at 1

16. $f(x) = e^{2x}; p_3(x)$ at $\frac{1}{2}$

17. $f(x) = x^3; p_3(x)$ at 1

18. $f(x) = x^2 - 6x + 10; p_2(x)$ at $x = 3$

19. $f(x) = \ln(3x); p_3(x)$ at $\frac{1}{3}$

20. $f(x) = \ln(2 - x); p_4(x)$ at 1

21. Use the third-degree Taylor polynomial at 0 for $f(x) = e^{-2x}$ and $x = 0.25$ to approximate $e^{-0.5}$.

22. Use the fourth-degree Taylor polynomial at 0 for $f(x) = \ln(1 - x)$ and $x = 0.1$ to approximate $\ln 0.9$.

23. Use the second-degree Taylor polynomial at 0 for $f(x) = \sqrt{x + 16}$ and $x = 1$ to approximate $\sqrt{17}$.

24. Use the third-degree Taylor polynomial at 0 for for $f(x) = \sqrt{(x + 4)^3}$ and $x = 1$ to approximate $\sqrt{125}$.

25. Use the fourth-degree Taylor polynomial at 1 for $f(x) = \sqrt{x}$ and $x = 1.2$ to approximate $\sqrt{1.2}$.

26. Use the third-degree Taylor polynomial at 4 for $f(x) = \sqrt{x}$ and $x = 3.95$ to approximate $\sqrt{3.95}$.

B *In Problems 27–32, find $f^{(n)}(x)$.*

27. $f(x) = \dfrac{1}{4 - x}$ **28.** $f(x) = \dfrac{4}{1 + x}$

29. $f(x) = e^{3x}$

30. $f(x) = \ln(2x + 1)$

31. $f(x) = \ln(6 - x)$

32. $f(x) = e^{x/2}$

In Problems 33–38, find the nth-degree Taylor polynomial at 0. Write the answer in expanded notation.

33. $f(x) = \dfrac{1}{4 - x}$ **34.** $f(x) = \dfrac{4}{1 + x}$

35. $f(x) = e^{3x}$

36. $f(x) = \ln(2x + 1)$

37. $f(x) = \ln(6 - x)$ **38.** $f(x) = e^{x/2}$

In Problems 39–46, find the nth-degree Taylor polynomial at the indicated value of a. Write the answer in expanded notation.

39. $f(x) = \dfrac{1}{x}$; at -1 **40.** $f(x) = \dfrac{2}{x}$; at 1

41. $f(x) = \ln x$; at 1   **42.** $f(x) = e^x$; at -2

43. $f(x) = e^{-x}$; at 2   **44.** $f(x) = \ln(2 - x)$; at -1

45. $f(x) = x \ln x$; at 3   **46.** $f(x) = xe^x$; at 1

C **47.** Find the first three Taylor polynomials at 0 for $f(x) = \ln(1 + x)$, and use these polynomials to complete the following two tables. Round all table entries to six decimal places.

x	$p_1(x)$	$p_2(x)$	$p_3(x)$	$f(x)$
-0.2				
-0.1				
0				
0.1				
0.2				

x	$\|p_1(x) - f(x)\|$	$\|p_2(x) - f(x)\|$	$\|p_3(x) - f(x)\|$
-0.2			
-0.1			
0			
0.1			
0.2			

48. Repeat Problem 47 for $f(x) = \sqrt{1 + x}$.

49. Refer to Problem 47. Graph $f(x) = \ln(1 + x)$ and its first three Taylor polynomials in the same viewing window. Use $[-3, 3]$ for both the x range and the y range.

50. Refer to Problem 48. Graph $f(x) = \sqrt{1 + x}$ and its first three Taylor polynomials in the same viewing window. Use $[-3, 3]$ for both the x range and the y range.

51. Consider $f(x) = e^x$ and its third-degree Taylor polynomial $p_3(x)$ at 0. Use graphical approximation techniques to find all values of x such that $|p_3(x) - e^x| < 0.1$.

52. Consider $f(x) = \ln(1 + x)$ and its third-degree Taylor polynomial $p_3(x)$ at 0. Use graphical approximation techniques to find all values of x such that

$$|p_3(x) - \ln(1 + x)| < 0.1$$

In Problems 53–56, write the answer in both expanded notation and summation notation.

53. Find the nth-degree Taylor polynomial for $f(x) = e^x$ at any point a.

54. Find the nth-degree Taylor polynomial for $f(x) = \ln x$ at any $a > 0$.

55. Find the nth-degree Taylor polynomial for $f(x) = 1/x$ at any $a \neq 0$.

56. Find the nth-degree Taylor polynomial at 0 for $f(x) = (x + c)^n$, where c is a constant.

57. Let $f(x)$ be a polynomial. For which values of n is the nth-degree Taylor polynomial for f at 0 equal to $f(x)$? Explain.

58. Explain how any polynomial expressed in powers of x may be rewritten in powers of $x - 1$ using Taylor polynomials. Illustrate your method for $f(x) = 1 + x^2 + x^4$.

Applications

Business & Economics

59. *Average price.* Given the demand equation

$$p = D(x) = \tfrac{1}{10}\sqrt{10{,}000 - x^2}$$

use the second-degree Taylor polynomial at 0 to approximate the average price (in dollars) over the demand interval $[0, 30]$.

60. *Average price.* Given the demand equation

$$p = D(x) = \tfrac{1}{5}\sqrt{1{,}600 - x^2}$$

use the second-degree Taylor polynomial at 0 to approximate the average price (in dollars) over the demand interval $[0, 15]$.

61. *Average price.* Refer to Problem 59. Use the second-degree Taylor polynomial at $a = 60$ to approximate the average price over the demand interval $[60, 80]$.

62. *Average price.* Refer to Problem 60. Use the second-degree Taylor polynomial at $a = 24$ to approximate the average price over the demand interval $[24, 32]$.

63. *Production.* The rate of production of a mine (in **www** millions of dollars per year) is given by

$$R(t) = 2 + 8e^{-0.1t^2}$$

Use the second-degree Taylor polynomial at 0 to approximate the total production during the first 2 years of operation of the mine.

64. *Production.* The rate of production of an oil well (in millions of dollars per year) is given by

$$R(t) = 5 + 10e^{-0.05t^2}$$

Use the second-degree Taylor polynomial at 0 to approximate the total production during the first 3 years of operation of the well.

Life Sciences

65. *Medicine.* The rate of healing for a skin wound (in square centimeters per day) is given by

$$A'(t) = \frac{-75}{t^2 + 25}$$

The initial wound has an area of 12 square centimeters. Use the second-degree Taylor polynomial at 0 for $A'(t)$ to approximate the area of the wound after 2 days.

66. *Medicine.* Rework Problem 65 for $A'(t) = \dfrac{-60}{t^2 + 20}$.

67. *Pollution.* On an average summer day in a particular **www** large city the air pollution level (in parts per million) is given by

$$P(x) = 20\sqrt{3x^2 + 25} - 80$$

where x is the number of hours elapsed since 8:00 AM. Use the second-degree Taylor polynomial for $P(x)$ at $a = 5$ to approximate the average pollution level during the 10-hour period from 8:00 AM to 6:00 PM.

68. *Pollution.* On an average summer day in another large city the air pollution level (in parts per million) is given by

$$P(x) = 10\sqrt{4x^2 + 36} - 50$$

where x is the number of hours elapsed since 8:00 AM. Use the second-degree Taylor polynomial for $P(x)$ at $a = 4$ to approximate the average pollution level during the 8-hour period from 8:00 AM to 4:00 PM.

Social Sciences

69. *Learning.* In a particular business college, it was found that an average student enrolled in an advanced typing class progresses at a rate of $N'(t) = 6e^{-0.01t^2}$ words per minute per week, t weeks after enrolling in a 15-week course. At the beginning of the course an average student could type 40 words per minute. Use the second-degree Taylor polynomial at 0 for $N'(t)$ to approximate the improvement in typing after 5 weeks in the course.

70. *Learning.* In the same business college, it was also found that an average student enrolled in a beginning shorthand class progressed at a rate of $N'(t) = 12e^{-0.005t^2}$ words per minute per week, t weeks after enrolling in a 15 week course. At the beginning of the course none of the students could take any dictation by shorthand. Use the second-degree Taylor polynomial at 0 for $N'(t)$ to approximate the improvement after 5 weeks in the course.

In Problems 71–78, use a graphing utility to graph the given function and its second-degree Taylor polynomial at a in the indicated viewing window.

Business & Economics

71. *Average price.* From Problem 59,

$$p = D(x) = \tfrac{1}{10}\sqrt{10,000 - x^2}$$
$a = 0$
x range:　$[0, 100]$
p range:　$[0, 10]$

72. *Average price.* From Problem 60,

$$p = D(x) = \tfrac{1}{5}\sqrt{1,600 - x^2}$$
$a = 0$
x range:　$[0, 40]$
p range:　$[0, 8]$

73. *Production.* From Problem 63,

$$y = R(t) = 2 + 8e^{-0.1t^2}$$
$a = 0$
t range:　$[0, 5]$
y range:　$[0, 10]$

74. *Production.* From Problem 64,

$$y = R(t) = 5 + 10e^{-0.05t^2}$$

$a = 0$
t range:　$[0, 5]$
y range:　$[0, 15]$

Life Sciences

75. *Pollution.* From Problem 67,

$$y = P(x) = 20\sqrt{3x^2 + 25} - 80$$
$a = 5$
x range:　$[0, 10]$
y range:　$[0, 300]$

76. *Pollution.* From Problem 68,

$$y = P(x) = 10\sqrt{4x^2 + 36} - 50$$
$a = 4$
x range:　$[0, 10]$
y range:　$[0, 200]$

Social Sciences

77. *Learning.* From Problem 69,

$$y = N'(t) = 6e^{-0.01t^2}$$
$a = 0$
t range:　$[0, 10]$
y range:　$[0, 6]$

78. *Learning.* From Problem 70,

$$y = N'(t) = 12e^{-0.005t^2}$$
$a = 0$
t range:　$[0, 15]$
y range:　$[0, 12]$

Section 10-2　# Taylor Series

- ❑　INTRODUCTION
- ❑　TAYLOR SERIES
- ❑　REPRESENTATION OF FUNCTIONS BY TAYLOR SERIES

❑　## INTRODUCTION

www　If f is a function with derivatives of all order at a point a, then we can construct the Taylor polynomial p_n at a for any integer n. Now we are interested in the relationship between the original function f and the corresponding Taylor polynomial p_n as n assumes larger and larger values. If the Taylor polynomial is to be a useful tool for approximating functions, then the accuracy of the approximation should improve as we increase the size of n. Indeed, Figures 1, 2, and 3 and Tables 1 and 2 in the preceding section indicate that this is the case for the function e^x.

For a given value of x, we would like to know whether we can make $p_n(x)$ arbitrarily close to $f(x)$ by making n sufficiently large. In other words, we want to know whether

$$\lim_{n \to \infty} p_n(x) = f(x) \tag{1}$$

It turns out that for most functions with derivatives of all order at a point, there is a set of values of x for which equation (1) is valid. We will begin by considering a specific example to illustrate this.

Let

$$f(x) = \frac{1}{1-x} \quad \text{and} \quad a = 0$$

First, we find p_n:

Step 1

$$f(x) = (1-x)^{-1}$$
$$f'(x) = (-1)(1-x)^{-2}(-1) = (1-x)^{-2}$$
$$f''(x) = (-2)(1-x)^{-3}(-1) = 2(1-x)^{-3}$$
$$f^{(3)}(x) = (-3)(2)(1-x)^{-4}(-1) = 3!(1-x)^{-4}$$
$$\vdots$$
$$f^{(n)}(x) = n!(1-x)^{-n-1}$$

Step 2

$$f(0) = 1$$
$$f'(0) = 1$$
$$f''(0) = 2$$
$$f^{(3)}(0) = 3!$$
$$\vdots$$
$$f^{(n)}(0) = n!$$

Step 3

$$a_0 = f(0) = 1$$
$$a_1 = f'(0) = 1$$
$$a_2 = \frac{f''(0)}{2!} = \frac{2}{2!} = 1$$
$$a_3 = \frac{f^{(3)}(0)}{3!} = \frac{3!}{3!} = 1$$
$$\vdots$$
$$a_n = \frac{f^{(n)}(0)}{n!} = \frac{n!}{n!} = 1$$

Step 4. *Write down the Taylor polynomial:*

$$p_n(x) = 1 + x + x^2 + x^3 + \cdots + x^n \qquad (2)$$

$$= \sum_{k=0}^{n} x^k$$

Now we want to evaluate

$$\lim_{n \to \infty} p_n(x) = \lim_{n \to \infty} \sum_{k=0}^{n} x^k$$

As we saw in Chapter 6, it is not possible to evaluate limits written in this form. The difficulty lies in the fact that the number of terms is increasing as n increases. First, we must find a closed form for the summation that does not involve a sum of n terms. [You may recognize that $p_n(x)$ is a finite geometric series with common ratio x. See Appendix B-2.]

If we multiply both sides of (2) by x and subtract this new equation from the original, we obtain an equation that we can solve for $p_n(x)$. Notice how many terms drop out when the subtraction is performed.

$$p_n(x) = 1 + x + x^2 + \cdots + x^{n-1} + x^n$$
$$xp_n(x) = \quad\ \ x + x^2 + \cdots + x^{n-1} + x^n + x^{n+1}$$
$$\overline{}$$
$$p_n(x) - xp_n(x) = 1 \qquad\qquad\qquad\qquad\qquad -x^{n+1}$$
$$p_n(x)(1-x) = 1 - x^{n+1}$$
$$p_n(x) = \frac{1 - x^{n+1}}{1 - x} \qquad x \neq 1$$

Since we had to exclude the value $x = 1$ to avoid division by 0, we must consider $x = 1$ as a special case:

$$\overbrace{p_n(1) = 1 + 1 + 1^2 + \cdots + 1^n}^{n + 1 \text{ terms}}$$
$$= n + 1$$

Thus,

$$p_n(x) = \begin{cases} \dfrac{1 - x^{n+1}}{1 - x} & \text{if } x \neq 1 \\ n + 1 & \text{if } x = 1 \end{cases}$$

If $x = 1$, it is clear that

$$\lim_{n \to \infty} p_n(1) = \lim_{n \to \infty} (n + 1) = \infty$$

That is, $\lim_{n \to \infty} p_n(1)$ does not exist. What happens to this limit for other values of x? It can be shown that

$$\lim_{n \to \infty} x^{n+1} = \begin{cases} 0 & \text{if } -1 < x < 1 \\ 1 & \text{if } x = 1 \\ \text{Does not exist} & \text{if } x \leqslant -1 \text{ or } x > 1 \end{cases}$$

Experimenting with a calculator for various values of x and large values of n should convince you that this statement is valid.

Thus,

$$\lim_{n \to \infty} p_n(x) = \begin{cases} \dfrac{1}{1 - x} & \text{if } -1 < x < 1 & \lim_{n \to \infty} x^{n+1} = 0 \\ \text{Does not exist} & \text{if } x = 1 & \lim_{n \to \infty}(n+1) \text{ does not exist} \\ \text{Does not exist} & \text{if } x \leqslant -1 \text{ or } x > 1 & \lim_{n \to \infty} x^{n+1} \text{ does not exist} \end{cases}$$

Explore–Discuss 1

(A) Construct a table with 8 rows labeled with values of n from 1 to 8, and 7 columns labeled with the following values of x: $-1.5, -1, -0.5, 0, 0.5, 1$, and 1.5. Complete the table by computing each entry $p_n(x)$.

(B) Explain how the table from part (A) provides supporting evidence for the statement that $\lim_{n \to \infty} p_n(x)$ equals $1/(1 - x)$ if $-1 < x < 1$, but does not exist if $x \leqslant -1$ or $x \geqslant 1$.

We can now conclude that

$$\begin{aligned} f(x) &= \frac{1}{1 - x} \\ &= \lim_{n \to \infty} p_n(x) \quad -1 < x < 1 \\ &= \lim_{n \to \infty} \sum_{k=0}^{n} x^k \end{aligned}$$

and that the limit does not exist for any other values of x. This is often abbreviated by writing

$$\begin{aligned} \frac{1}{1 - x} &= 1 + x + x^2 + \cdots + x^n + \cdots \quad -1 < x < 1 \\ &= \sum_{k=0}^{\infty} x^k \end{aligned}$$

The expression $\sum_{k=0}^{\infty} x^k$ is called the *Taylor series for f at* 0, and the expanded notation

$$1 + x + x^2 + \cdots + x^n + \cdots$$

is just another way to write this series. The Taylor series is said to *converge at* x if $\lim_{n \to \infty} p_n(x)$ exists and to *diverge at* x if this limit fails to exist. Thus,

(A) $f(x) = \dfrac{1}{1 - x}$

Defined for $x \neq 1$

(B) $1 + x + x^2 + \cdots + x^n + \cdots$

Defined for $-1 < x < 1$

FIGURE 1

$\sum_{k=0}^{\infty} x^k$ converges for $-1 < x < 1$ and diverges for $x \leqslant -1$ or $x \geqslant 1$. The set of values $\{x \mid -1 < x < 1\} = (-1, 1)$ where this Taylor series converges is called the *interval of convergence*. Since $f(x)$ is equal to the Taylor series for any x in the interval of convergence, we say that f is represented by its Taylor series throughout this interval. That is,

$$f(x) = \frac{1}{1 - x}$$
$$= 1 + x + x^2 + \cdots + x^n + \cdots \qquad -1 < x < 1$$

There can be no relationship between f and its Taylor series outside the interval of convergence since the series is not defined there (see Fig. 1).

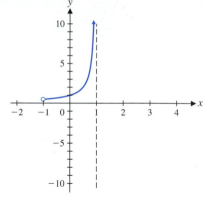

FIGURE 2

(A) The six functions $p_n(x) = 1 + x + \cdots + x^n$, $n = 1, 2, \ldots, 6$, are graphed in Figure 2. Which graph belongs to which function?

(B) Based solely on the graphs, would you conjecture that the interval of convergence of the Taylor series representation

$$\frac{1}{1 - x} = 1 + x + x^2 + \cdots + x^n + \cdots$$

is $-1 < x < 1$? Explain.

❑   TAYLOR SERIES

We now generalize the concepts introduced in the preceding discussion to arbitrary functions.

Taylor Series

If f is a function with derivatives of all order at a point a, $p_n(x)$ is the nth-degree Taylor polynomial at a for f, and $a_n = f^{(n)}(a)/n!$, $n = 0, 1, 2, \ldots$, then

$$\sum_{k=0}^{\infty} a_k(x - a)^k = a_0 + a_1(x - a) + a_2(x - a)^2 +$$
$$\cdots + a_n(x - a)^n + \cdots$$

is the **Taylor series at a for f**. The Taylor series **converges at x** if

$$\lim_{n\to\infty} p_n(x) = \lim_{n\to\infty} \sum_{k=0}^{n} a_k(x-a)^k$$

exists and **diverges at x** if this limit does not exist. The set of values of x for which this limit exists is called the **interval of convergence.**

We must emphasize that we are not really adding up an infinite number of terms in a Taylor series, nor is a Taylor series an "infinite" polynomial. Rather, for x in the interval of convergence,

$$\sum_{k=0}^{\infty} a_k(x-a)^k = \lim_{n\to\infty} \sum_{k=0}^{n} a_k(x-a)^k$$

Thus, when we write

$$\frac{1}{1-x} = 1 + x + x^2 + \cdots + x^n + \cdots \qquad -1 < x < 1$$

we mean

$$\frac{1}{1-x} = \lim_{n\to\infty} (1 + x + x^2 + \cdots + x^n) \qquad -1 < x < 1$$

In general, it is very difficult to find the interval of convergence by directly evaluating

$$\lim_{n\to\infty} \sum_{k=0}^{n} a_k(x-a)^k$$

For many Taylor series, the following theorem can be used to find the interval of convergence.

THEOREM 1   Interval of Convergence

Let f be a function with derivatives of all order at a point a, let $a_n = f^{(n)}(a)/n!$, $n = 0, 1, 2, \ldots$, and let

$$\sum_{k=0}^{\infty} a_k(x-a)^k = a_0 + a_1(x-a) + a_2(x-a)^2 + \cdots + a_n(x-a)^n + \cdots$$

be the Taylor series at a for f. If $a_n \neq 0$ for $n \geqslant n_0$, then:

Case 1. If

$$\lim_{n\to\infty} \left| \frac{a_{n+1}}{a_n} \right| = L > 0 \qquad \text{and} \qquad R = \frac{1}{L}$$

then the series converges for $|x-a| < R$ and diverges for $|x-a| > R$.

Case 2. If

$$\lim_{n\to\infty} \left| \frac{a_{n+1}}{a_n} \right| = 0$$

then the series converges for all values of x.

Case 3. If

$$\lim_{n\to\infty} \left| \frac{a_{n+1}}{a_n} \right| = \infty$$

then the series converges only at $x = a$.

The various possibilities for the interval of convergence are illustrated in Figure 3. In case 1, the series may or may not converge at the end points $x = a - R$ and $x = a + R$. (Determination of the behavior of a series at the end points of the interval of convergence requires techniques that we will not discuss.)

Case 1. $\lim\limits_{n \to \infty} \left| \dfrac{a_{n+1}}{a_n} \right| = L > 0$

$R = 1/L$

Case 2. $\lim\limits_{n \to \infty} \left| \dfrac{a_{n+1}}{a_n} \right| = 0$

Case 3. $\lim\limits_{n \to \infty} \left| \dfrac{a_{n+1}}{a_n} \right| = \infty$

FIGURE 3 Intervals of convergence

Notice that the set of values where a series converges can be expressed in interval notation as $(a - R, a + R)$ in case 1 and as $(-\infty, \infty)$ in case 2. We were anticipating this result when we used the term *interval of convergence* to describe the set of points where a Taylor series converges.

Finally, the requirement that all the coefficients be nonzero from some point on is necessary to ensure that the ratio a_{n+1}/a_n is well defined.

REMARK

It is possible that

$$\lim_{n \to \infty} \left| \frac{a_{n+1}}{a_n} \right|$$

fails to exist so that none of the cases in Theorem 1 hold. In that event, the interval of convergence will still have one of the forms illustrated by Figure 3, but other techniques are required to determine which one.

The next two examples illustrate the application of Theorem 1. The details of finding the Taylor series are omitted. In the next section we will discuss methods that greatly simplify the process of finding Taylor series. For now, we want to concentrate on understanding the significance of the interval of convergence.

Example 1 ⇔ **Finding the Interval of Convergence** The Taylor series at 0 for $f(x) = e^{2x}$ is

$$1 + 2x + \frac{2^2}{2!}x^2 + \cdots + \frac{2^n}{n!}x^n + \cdots$$

Find the interval of convergence.

SOLUTION

$$a_n = \frac{2^n}{n!} \qquad \textit{Notice that } a_n > 0 \textit{ for } n \geqslant 0.$$

$$a_{n+1} = \frac{2^{n+1}}{(n+1)!}$$

$$\frac{a_{n+1}}{a_n} = \frac{\dfrac{2^{n+1}}{(n+1)!}}{\dfrac{2^n}{n!}} \qquad \textit{Form the ratio } a_{n-1}/a_n \textit{ and simplify.}$$

$$= \frac{2^{n+1}}{(n+1)!} \cdot \frac{n!}{2^n} \qquad \frac{n!}{(n+1)!} = \frac{n!}{(n+1)n!} = \frac{1}{n+1}$$

$$= \frac{2}{n+1}$$

$$\lim_{n \to \infty} \left| \frac{a_{n+1}}{a_n} \right| = \lim_{n \to \infty} \left| \frac{2}{n+1} \right| \qquad \textit{Absolute value signs can be dropped since } a_{n+1}/a_n > 0.$$

$$= \lim_{n \to \infty} \frac{2}{n+1} = 0^*$$

Case 2 in Theorem 1 applies, and the series converges for all values of x. ∎

Matched Problem 1 ⇒ The Taylor series at 0 for $f(x) = e^{x/3}$ is

$$1 + \frac{1}{3}x + \frac{1}{3^2 2!}x^2 + \cdots + \frac{1}{3^n n!}x^n + \cdots$$

Find the interval of convergence.

Example 2 ⇒ **Finding the Interval of Convergence**   The Taylor series at 0 for $f(x) = 1/(1 + 5x)^2$ is

$$1 - 2 \cdot 5x + 3 \cdot 5^2 x^2 - \cdots + (-1)^n (n+1) 5^n x^n + \cdots$$

Find the interval of convergence.

SOLUTION

$$a_n = (-1)^n (n+1) 5^n \qquad \textit{Notice that } a_n \neq 0 \textit{ for } n \geqslant 0.$$

$$a_{n+1} = (-1)^{n+1} (n+2) 5^{n+1}$$

$$\frac{a_{n+1}}{a_n} = \frac{(-1)^{n+1}(n+2)5^{n+1}}{(-1)^n(n+1)5^n}$$

$$= \frac{-5(n+2)}{n+1}$$

$$\lim_{n \to \infty} \left| \frac{a_{n+1}}{a_n} \right| = \lim_{n \to \infty} \left| \frac{-5(n+2)}{n+1} \right|$$

$$= \lim_{n \to \infty} \frac{5(n+2)}{n+1}$$

$$= \lim_{n \to \infty} \frac{5n+10}{n+1} \qquad \textit{Divide numerator and denominator by } n.$$

$$= \lim_{n \to \infty} \frac{5 + \dfrac{10}{n}}{1 + \dfrac{1}{n}} = 5$$

---

*The limit expressions $\lim_{x \to \infty} g(x)$ and $\lim_{n \to \infty} g(n)$ are very similar. If $\lim_{x \to \infty} g(x) = L$ [that is, $g(x)$ can be made arbitrarily close to L by making the *real number* x sufficiently large], then $\lim_{n \to \infty} g(n) = L$ [that is, $g(n)$ can be made arbitrarily close to L by making the *integer n* sufficiently large]. In particular,

$$\lim_{n \to \infty} \frac{2}{n+1} = \lim_{x \to \infty} \frac{2}{x+1} = 0$$

Thus, the results of Section 4-3 on limits at infinity for rational functions can be helpful in evaluating limits of the form $\lim_{n \to \infty} g(n)$.

Applying case 1 in Theorem 1, we have $L = 5, R = \frac{1}{5}$, and the series converges for $-\frac{1}{5} < x < \frac{1}{5}$.

∎

Matched Problem 2 ⇝ The Taylor series at 0 for $f(x) = 4/(2 + x)^2$ is

$$1 - \frac{2}{2}x + \frac{3}{2^2}x^2 - \cdots + \frac{(-1)^n(n + 1)}{2^n}x^n + \cdots$$

Find the interval of convergence.

∎

❑ REPRESENTATION OF FUNCTIONS BY TAYLOR SERIES

Referring to the function $f(x) = 1/(1 - x)$ discussed earlier in this section, our calculations showed that the Taylor series at 0 for f is

$$\lim_{n\to\infty} p_n(x) = 1 + x + x^2 + \cdots + x^n + \cdots$$

with interval of convergence $(-1, 1)$. Furthermore, we showed that f is represented by its Taylor series throughout the interval of convergence. That is,

$$f(x) = \frac{1}{1 - x} = \lim_{n\to\infty} p_n(x)$$

$$= 1 + x + x^2 + \cdots + x^n + \cdots \qquad -1 < x < 1$$

A similar statement can be made for most functions that have derivatives of all order at a point a. But not for all. For example, the absolute value function $g(x) = |x|$ has Taylor series (verify this)

$$1 + (x - 1) = x$$

at the point $a = 1$. Clearly, this very simple Taylor series converges for all x, but $g(x) = x$ only for $x \geq 0$. So g is not represented by its Taylor series throughout the interval of convergence of the series. The problem with g is that it is not differentiable at 0. Even if we require that a function f has derivatives of all order at every point in the interval of convergence of its Taylor series, it is still possible that f is not represented by its Taylor series. Since we will not consider such functions, we will make the following assumption.

Assumption on the Representation of Functions by Taylor Series

If f is a function with derivatives of all order throughout the interval of convergence of its Taylor series at a, then f **is represented by its Taylor series throughout the interval of convergence.** Thus, if $p_n(x)$ is the nth-degree Taylor polynomial at a for f and $a_n = f^{(n)}(a)/n!, n = 0, 1, 2, \ldots$, then

$$f(x) = \lim_{n\to\infty} p_n(x) = \sum_{k=0}^{\infty} a_k(x - a)^k$$

for any x in the interval of convergence of the Taylor series at a for f.

Consequently, the interval of convergence determines the values of x for which the Taylor polynomials $p_n(x)$ can be used to approximate the values of the function f.

Referring to Examples 1 and 2, we can now write

$$e^{2x} = 1 + 2x + \frac{2^2}{2!}x^2 + \cdots + \frac{2^n}{n!}x^n + \cdots \qquad -\infty < x < \infty$$

and

$$\frac{1}{(1+5x)^2} = 1 - 2 \cdot 5x + 3 \cdot 5^2 x^2 - \cdots + (-1)^n(n+1)5^n x^n + \cdots \qquad -\tfrac{1}{5} < x < \tfrac{1}{5}$$

Example 3   ➦   **Representation of a Function by Its Taylor Series**   Let $f(x) = \ln x$.

(A) Find the nth-degree Taylor polynomial at $a = 1$ for f.
(B) Find the Taylor series at $a = 1$ for f.
(C) Determine the values of x for which $f(x) = \lim\limits_{n\to\infty} p_n(x)$.

SOLUTION   (A) We use the four-step process from the preceding section to find the nth-degree Taylor polynomial:

Step 1

$$f(x) = \ln x$$

$$f'(x) = \frac{1}{x} = x^{-1}$$

$$f''(x) = (-1)x^{-2}$$

$$f^{(3)}(x) = (-1)(-2)x^{-3}$$
$$= (-1)^2 2! x^{-3}$$

$$f^{(4)}(x) = (-1)(-2)(-3)x^{-4}$$
$$= (-1)^3 3! x^{-4}$$
$$\vdots$$

$$f^{(n)}(x) = (-1)^{n-1}(n-1)! x^{-n}$$

Step 2

$$f(1) = \ln 1 = 0$$

$$f'(1) = 1$$

$$f''(1) = -1$$

$$f^{(3)}(1) = (-1)^2 2!$$

$$f^{(4)}(1) = (-1)^3 3!$$
$$\vdots$$

$$f^{(n)}(1) = (-1)^{n-1}(n-1)!$$

Step 3

$$a_0 = f(1) \quad = 0$$

$$a_1 = f'(1) \quad = 1$$

$$a_2 = \frac{f''(1)}{2!} \quad = -\frac{1}{2}$$

$$a_3 = \frac{f^{(3)}(1)}{3!} = \frac{(-1)^2 2!}{3!} = \frac{1}{3}$$

$$a_4 = \frac{f^{(4)}(1)}{4!} = \frac{(-1)^3 3!}{4!} = \frac{-1}{4}$$
$$\vdots$$

$$a_n = \frac{f^{(n)}(1)}{n!} = \frac{(-1)^{n-1}(n-1)!}{n!} = \frac{(-1)^{n-1}}{n}$$

Step 4. The nth-degree Taylor polynomial at $a = 1$ for $f(x) = \ln x$ is

$$p_n(x) = (x-1) - \frac{1}{2}(x-1)^2 + \frac{1}{3}(x-1)^3 - \cdots + \frac{(-1)^{n-1}}{n}(x-1)^n$$

(B) Once the nth-degree Taylor polynomial for a function has been determined, finding the Taylor series is simply a matter of using the notation correctly. Using the Taylor polynomial from part (A), the Taylor series at $a = 1$ for $f(x) = \ln x$ can be written as

$$\lim_{n\to\infty} p_n(x) = (x-1) - \frac{1}{2}(x-1)^2 + \frac{1}{3}(x-1)^3 - \cdots + \frac{(-1)^{n-1}}{n}(x-1)^n + \cdots$$

or, using summation notation,* as

$$\lim_{n\to\infty} p_n(x) = \sum_{k=0}^{\infty} \frac{(-1)^{k-1}}{k}(x-1)^k$$

---

*From this point on, we will use the expanded notation exclusively when writing Taylor series.

(C) To determine the values for which $f(x) = \lim_{n \to \infty} p_n(x)$, we use Theorem 1 to find the interval of convergence of the Taylor series in part (B):

$$a_n = \frac{(-1)^{n-1}}{n}$$

$$a_{n+1} = \frac{(-1)^n}{n+1}$$

$$\frac{a_{n+1}}{a_n} = \frac{\dfrac{(-1)^n}{n+1}}{\dfrac{(-1)^{n-1}}{n}} = \frac{(-1)^n}{n+1} \cdot \frac{n}{(-1)^{n-1}} = \frac{-n}{n+1}$$

$$\lim_{n \to \infty} \left| \frac{a_{n+1}}{a_n} \right| = \lim_{n \to \infty} \left| \frac{-n}{n+1} \right|$$

$$= \lim_{n \to \infty} \frac{n}{n+1} \qquad \text{Divide numerator and denominator by } n.$$

$$= \lim_{n \to \infty} \frac{1}{1 + \dfrac{1}{n}} = 1$$

From case 1 of Theorem 1, $L = 1$, $R = 1/L = 1$, and the series converges for $|x - 1| < 1$. Converting this to double inequalities, we have

$$|x - 1| < 1 \qquad |y| < c \text{ is equivalent to } -c < y < c.$$
$$-1 < x - 1 < 1 \qquad \text{Add 1 to each side.}$$
$$0 < \quad x \quad < 2 \qquad \text{Interval of convergence}$$

Thus, $f(x) = \lim_{n \to \infty} p_n(x)$ for $0 < x < 2$ or, equivalently,

$$\ln x = (x - 1) - \frac{1}{2}(x-1)^2 + \frac{1}{3}(x-1)^3 - \cdots + \frac{(-1)^{n-1}}{n}(x-1)^n + \cdots \qquad 0 < x < 2$$

Matched Problem 3   ⇔   Repeat Example 3 for $f(x) = 1/x$.

Table 1 displays the values of the Taylor polynomials for $f(x) = \ln x$ at $x = 1.5$ and at $x = 3$ for $n = 1, 2, \dots, 10$ (see Example 3). Notice that $x = 1.5$

TABLE 1

$f(x) = \ln x$

$$p_n(x) = (x - 1) - \frac{1}{2}(x-1)^2 + \frac{1}{3}(x-1)^3 - \cdots + \frac{(-1)^{n-1}}{n}(x-1)^n$$

n	$p_n(1.5)$	$p_n(3)$
1	0.5	2
2	0.375	0
3	0.416 667	2.666 667
4	0.401 042	−1.333 333
5	0.407 292	5.066 667
6	0.404 688	−5.6
7	0.405 804	12.685 714
8	0.405 315	−19.314 286
9	0.405 532	37.574 603
10	0.405 435	−64.825 397
	$\ln 1.5 = 0.405\ 465$	$\ln 3 = 1.098\ 612$

is in the interval of convergence of the Taylor series at 1 for $f(x)$ and that the values of $p_n(1.5)$ are approaching ln 1.5 as n increases. On the other hand, $x = 3$ is *not* in the interval of convergence and the values of $p_n(3)$ do not approach ln 3. These calculations illustrate the importance of the interval of convergence.

The values of x must be in the interval of convergence in order to use Taylor polynomials to approximate the values of a function.

Answers to Matched Problems   **1.** $-\infty < x < \infty$

2. $-2 < x < 2$

3. (A) $1 - (x - 1) + (x - 1)^2 - \cdots + (-1)^n(x - 1)^n$
(B) $1 - (x - 1) + (x - 1)^2 - \cdots + (-1)^n(x - 1)^n + \cdots$
(C) $0 < x < 2$

Exercise 10-2

A   *In Problems 1–12, find the interval of convergence of the given Taylor series representation.*

1. $\dfrac{2}{1 - x} = 2 + 2x + 2x^2 + \cdots + 2x^n + \cdots$

2. $\dfrac{1}{1 + x} = 1 - x + x^2 - \cdots + (-1)^n x^n + \cdots$

3. $\dfrac{1}{1 - 2x} = 1 + 2x + 2^2 x^2 + \cdots + 2^n x^n + \cdots$

4. $\dfrac{1}{1 - 3x} = 1 + 3x + 3^2 x^2 + \cdots + 3^n x^n + \cdots$

5. $\ln(1 + x) = x - \dfrac{1}{2}x^2 + \dfrac{1}{3}x^3 - \cdots$
$\qquad + \dfrac{(-1)^{n-1}}{n}x^n + \cdots$

6. $\ln(1 - x) = -x - \dfrac{1}{2}x^2 - \dfrac{1}{3}x^3 - \cdots$
$\qquad - \dfrac{1}{n}x^n - \cdots$

7. $\dfrac{1}{5 - x} = \dfrac{1}{5} + \dfrac{1}{5^2}x + \dfrac{1}{5^3}x^2 + \cdots$
$\qquad + \dfrac{1}{5^{n+1}}x^n + \cdots$

8. $\dfrac{3}{4 + x} = \dfrac{3}{4} - \dfrac{3}{4^2}x + \dfrac{3}{4^3}x^2 - \cdots$
$\qquad + \dfrac{3(-1)^n}{4^{n+1}}x^n + \cdots$

9. $\dfrac{1}{(1 + x)^2} = 1 - 2x + 3x^2 - \cdots$
$\qquad + (-1)^n(n + 1)x^n + \cdots$

10. $\dfrac{x}{(1 - x)^2} = x + 2x^2 + 3x^3 + \cdots + nx^n + \cdots$

11. $e^x = 1 + x + \dfrac{1}{2!}x^2 + \cdots + \dfrac{1}{n!}x^n + \cdots$

12. $e^{-x} = 1 - x + \dfrac{1}{2!}x^2 - \cdots + \dfrac{(-1)^n}{n!}x^n + \cdots$

13. The Taylor series for e^x at 0 is

$$e^x = 1 + x + \dfrac{1}{2!}x^2 + \dfrac{1}{3!}x^3 + \cdots$$
$$+ \dfrac{1}{n!}x^n + \cdots$$

(A) Find the smallest value of n such that
$$|p_n(2) - e^2| < 0.1.$$

(B) Does there exist a value of n such that
$$|p_n(100) - e^{100}| < 0.1? \text{ Explain.}$$

14. Suppose the function f has the following Taylor series representation at 0:

$$f(x) = 1 + x + 2!x^2 + 3!x^3 + \cdots$$
$$+ n!x^n + \cdots$$

(A) Find the interval of convergence.
(B) Compute $p_n(0.2)$ for $n = 1, 2, 3, \dots, 10$, where $p_n(x)$ is the nth-degree Taylor polynomial for f at 0. Explain why your computations are consistent with your answer to part (A).

B   *In Problems 15–22, find the interval of convergence of the given Taylor series representation.*

15. $\ln(4 + x) = (x + 3) - \dfrac{1}{2}(x + 3)^2$
$\qquad + \dfrac{1}{3}(x + 3)^3 - \cdots + \dfrac{(-1)^{n+1}}{n}(x + 3)^n + \cdots$

16. $\ln(9 - 2x) = -2(x - 4) - \dfrac{2^2}{2}(x - 4)^2$
$\qquad - \dfrac{2^3}{3}(x - 4)^3 - \cdots - \dfrac{2^n}{n}(x - 4)^n - \cdots$

17. $e^{2x} = 1 + 2x + \dfrac{2^2}{2!}x^2 + \cdots + \dfrac{2^n}{n!}x^n + \cdots$

18. $e^{-x/3} = 1 - \dfrac{1}{3}x + \dfrac{1}{2!3^2}x^2 - \cdots$

$\qquad + \dfrac{(-1)^n}{n!3^n}x^n + \cdots$

19. $\dfrac{1}{6 - x} = \dfrac{1}{4} + \dfrac{1}{4^2}(x - 2) + \dfrac{1}{4^3}(x - 2)^2 + \cdots$

$\qquad + \dfrac{1}{4^{n+1}}(x - 2)^n + \cdots$

20. $\dfrac{3}{2 + 3x} = -3 - 3^2(x + 1) - 3^3(x + 1)^2 - \cdots$

$\qquad - 3^{n+1}(x + 1)^n - \cdots$

21. $\dfrac{1}{6 - x} = -\dfrac{1}{2} + \dfrac{1}{2^2}(x - 8) - \dfrac{1}{2^3}(x - 8)^2$

$\qquad + \cdots + \dfrac{(-1)^n}{2^{n+1}}(x - 8)^n + \cdots$

22. $\dfrac{3}{2 + 3x} = \dfrac{3}{5} - \dfrac{3^2}{5^2}(x - 1) + \dfrac{3^3}{5^3}(x - 1)^2 - \cdots$

$\qquad + \dfrac{(-1)^n 3^{n+1}}{5^{n+1}}(x - 1)^n + \cdots$

23. (A) Graph the nth-degree Taylor polynomials at 0 for $f(x) = e^x, n = 1, 2, 3, 4, 5$. Use the viewing window Xmin = −3, Xmax = 3, Ymin = −5, and Ymax = 15. Based solely on the graphs, what would you conjecture the interval of convergence to be? Explain.

(B) What is the actual interval of convergence?

24. (A) Graph the nth-degree Taylor polynomials at 0 for $f(x) = \ln(1 + x), n = 1, 2, 3, 4, 5$. Use the viewing window Xmin = −3, Xmax = 3, Ymin = −15, and Ymax = 15. Based solely on the graphs, what would you conjecture the interval of convergence to be? Explain.

(B) What is the actual interval of convergence?

C *In Problems 25–30, find the nth-degree Taylor polynomial at 0 for f, find the Taylor series at 0 for f, and determine the values of x for which $f(x) = \lim\limits_{n \to \infty} p_n(x)$.*

25. $f(x) = e^{4x}$ **26.** $f(x) = e^{-x/2}$

27. $f(x) = \ln(1 + 2x)$ **28.** $f(x) = \ln(1 - 4x)$

29. $f(x) = \dfrac{2}{2 - x}$ **30.** $f(x) = \dfrac{3}{3 + x}$

In Problems 31–34, find the nth-degree Taylor polynomial at the indicated value of a for f, find the Taylor series at a for f, and determine the values of x for which $f(x) = \lim\limits_{n \to \infty} p_n(x)$.

31. $f(x) = \ln(2 - x); a = 1$

32. $f(x) = \ln(2 + x); a = -1$

33. $f(x) = \dfrac{1}{1 - x}; a = 2$

34. $f(x) = \dfrac{1}{4 - x}; a = 3$

35. (A) Find the interval of convergence of the Taylor series representation

$$\dfrac{1}{4 - x} = \dfrac{1}{4} + \dfrac{1}{4^2}x + \dfrac{1}{4^3}x^2 + \cdots$$

$$+ \dfrac{1}{4^{n+1}}x^n + \cdots$$

(B) Explain why Theorem 1 is not directly applicable to the Taylor series representation

$$\dfrac{1}{4 - x^2} = \dfrac{1}{4} + \dfrac{1}{4^2}x^2 + \dfrac{1}{4^3}x^4 + \cdots$$

$$+ \dfrac{1}{4^{n+1}}x^{2n} + \cdots$$

(C) Use part (A) to find the interval of convergence of the Taylor series of part (B).

36. (A) Explain why Theorem 1 is not directly applicable to the Taylor series representation

$$e^{x^2} = 1 + x^2 + \dfrac{1}{2!}x^4 + \dfrac{1}{3!}x^6 + \cdots$$

$$+ \dfrac{1}{n!}x^{2n} + \cdots$$

(B) Use the Taylor series for e^x at 0 to find the interval of convergence of the Taylor series of part (A).

37. (A) Explain why the interval of convergence of the Taylor series

$$\dfrac{1}{1 - x} + \dfrac{1}{1 - x^2} = 2 + x + 2x^2 + x^3$$

$$+ 2x^4 + x^5 + \cdots$$

cannot be determined by applying Theorem 1.

(B) If two Taylor series have the same interval of convergence, then so does their term-by-term sum. Use this fact to find the interval of convergence of the Taylor series of part (A).

38. (A) Explain why the interval of convergence of the Taylor series

$$e^x - e^{x^2} = x + \left(\dfrac{1}{2!} - 1\right)x^2 + \dfrac{1}{3!}x^3$$

$$+ \left(\dfrac{1}{4!} - \dfrac{1}{2!}\right)x^4 + \dfrac{1}{5!}x^5 + \cdots$$

cannot be determined by applying Theorem 1.

(B) If two Taylor series have the same interval of convergence, then so does their term-by-term difference. Use this fact to find the interval of convergence of the Taylor series of part (A).

Section 10-3 # Operations on Taylor Series

- ❑ BASIC TAYLOR SERIES
- ❑ ADDITION AND MULTIPLICATION
- ❑ DIFFERENTIATION AND INTEGRATION
- ❑ SUBSTITUTION

❑ BASIC TAYLOR SERIES

In the preceding sections, we used repeated differentiation and the formula $a_n = f^{(n)}(a)/n!$ to find Taylor polynomials and Taylor series for a variety of simple functions. In this section we will see how to use the series for these simple functions and some basic properties of Taylor series to find the Taylor series for more complicated functions. Table 1 lists the series we will use in this process for convenient reference.

www

TABLE 1

BASIC TAYLOR SERIES

FUNCTION $f(x)$	TAYLOR SERIES AT 0	INTERVAL OF CONVERGENCE
$\dfrac{1}{1-x}$	$= 1 + x + x^2 + \cdots + x^n + \cdots$	$-1 < x < 1$
$\dfrac{1}{1+x}$	$= 1 - x + x^2 - \cdots + (-1)^n x^n + \cdots$	$-1 < x < 1$
$\ln(1-x)$	$= -x - \dfrac{1}{2}x^2 - \dfrac{1}{3}x^3 - \cdots - \dfrac{1}{n}x^n - \cdots$	$-1 < x < 1$
$\ln(1+x)$	$= x - \dfrac{1}{2}x^2 + \dfrac{1}{3}x^3 - \cdots + \dfrac{(-1)^{n-1}}{n}x^n + \cdots$	$-1 < x < 1$
e^x	$= 1 + x + \dfrac{1}{2!}x^2 + \cdots + \dfrac{1}{n!}x^n + \cdots$	$-\infty < x < \infty$
e^{-x}	$= 1 - x + \dfrac{1}{2!}x^2 - \cdots + \dfrac{(-1)^n}{n!}x^n + \cdots$	$-\infty < x < \infty$

Most of the discussion in this section involves Taylor series at 0. At the end of the section we will see that Taylor series at points other than 0 can be obtained from Taylor series at 0 by using a simple substitution process.

❑ ADDITION AND MULTIPLICATION

PROPERTY 1   Addition

Two Taylor series at 0 can be added term by term:

If
$$f(x) = a_0 + a_1 x + a_2 x^2 + \cdots + a_n x^n + \cdots$$
and
$$g(x) = b_0 + b_1 x + b_2 x^2 + \cdots + b_n x^n + \cdots$$
then
$$f(x) + g(x) = (a_0 + b_0) + (a_1 + b_1)x + (a_2 + b_2)x^2 + \cdots$$
$$+ (a_n + b_n)x^n + \cdots$$

This operation is valid in the intersection of the intervals of convergence of the series for f and g.

Example 1 ➾ **Addition of Taylor Series** Find the Taylor series at 0 for $f(x) = e^x + \ln(1 + x)$, and find the interval of convergence.

SOLUTION From Table 1,

$$e^x = 1 + x + \frac{1}{2!}x^2 + \frac{1}{3!}x^3 + \cdots + \frac{1}{n!}x^n + \cdots \qquad -\infty < x < \infty$$

and

$$\ln(1 + x) = x - \frac{1}{2}x^2 + \frac{1}{3}x^3 - \cdots + \frac{(-1)^{n-1}}{n}x^n + \cdots \qquad -1 < x < 1$$

Adding the coefficients of corresponding powers of x, we have

$$f(x) = e^x + \ln(1 + x)$$

$$= 1 + 2x + \left(\frac{1}{2!} - \frac{1}{2}\right)x^2 + \left(\frac{1}{3!} + \frac{1}{3}\right)x^3 + \cdots$$

$$+ \left[\frac{1}{n!} + \frac{(-1)^{n-1}}{n}\right]x^n + \cdots$$

The series for e^x converges for all x; however, the series for $\ln(1 + x)$ converges only for $-1 < x < 1$. Thus, the combined series converges for $-1 < x < 1$, the intersection of these two intervals of convergence. ∎

Matched Problem 1 ➾ Find the Taylor series at 0 for $f(x) = e^x + 1/(1 - x)$, and find the interval of convergence. ∎

PROPERTY 2 **Multiplication**

A Taylor series at 0 can be multiplied term by term by an expression of the form cx^r, where c is a nonzero constant and r is a nonnegative integer:

If

$$f(x) = a_0 + a_1x + a_2x^2 + \cdots + a_nx^n + \cdots$$

then

$$cx^r f(x) = ca_0x^r + ca_1x^{r+1} + ca_2x^{r+2} + \cdots + ca_nx^{r+n} + \cdots$$

The Taylor series for $cx^r f(x)$ has the same interval of convergence as the Taylor series for f. ∎

Example 2 ➾ **Multiplication of Taylor Series** Find the Taylor series at 0 for $f(x) = 2x^2 e^{-x}$, and find the interval of convergence.

SOLUTION From Table 1,

$$e^{-x} = 1 - x + \frac{1}{2!}x^2 - \cdots + \frac{(-1)^n}{n!}x^n + \cdots \qquad -\infty < x < \infty$$

Multiplying each term of this series by $2x^2$, the series for $f(x)$ is

$$f(x) = 2x^2 e^{-x}$$

$$= 2x^2\left[1 - x + \frac{1}{2!}x^2 - \cdots + \frac{(-1)^n}{n!}x^n + \cdots\right]$$

$$= 2x^2 - 2x^3 + \frac{2}{2!}x^4 - \cdots + \frac{2(-1)^n}{n!}x^{n+2} + \cdots$$

Since the series for e^{-x} converges for all x, the series for $f(x)$ also converges for $-\infty < x < \infty$.

∎

Matched Problem 2 ➭ Find the Taylor series at 0 for $f(x) = 3x^3 \ln(1-x)$, and find the interval of convergence.

∎

❑   DIFFERENTIATION AND INTEGRATION

PROPERTY 3   Differentiation

A Taylor series at 0 can be differentiated term by term:

If

$$f(x) = a_0 + a_1x + a_2x^2 + a_3x^3 + \cdots + a_nx^n + \cdots$$

then

$$f'(x) = a_1 + 2a_2x + 3a_3x^2 + \cdots + na_nx^{n-1} + \cdots$$

The Taylor series for f' has the same interval of convergence as the Taylor series for f.

━━━━━━━━━ ∎

Example 3 ➭ **Differentiation of Taylor Series**   Find the Taylor series at 0 for $f(x) = 1/(1-x)^2$, and find the interval of convergence.

SOLUTION   We want to relate f to the derivative of one of the functions in Table 1, so that we can apply property 3. Since

$$\frac{d}{dx}\left(\frac{1}{1-x}\right) = \frac{d}{dx}(1-x)^{-1}$$

$$= -(1-x)^{-2}(-1)$$

$$= \frac{1}{(1-x)^2}$$

$$= f(x)$$

we can apply property 3 to the series for $1/(1-x)$. From Table 1,

$$\frac{1}{1-x} = 1 + x + x^2 + x^3 + \cdots + x^n + \cdots \qquad -1 < x < 1$$

Differentiating both sides of this equation,

$$\frac{d}{dx}\left(\frac{1}{1-x}\right) = \frac{d}{dx}(1 + x + x^2 + x^3 + \cdots + x^n + \cdots)$$

$$\frac{1}{(1-x)^2} = \frac{d}{dx}1 + \frac{d}{dx}x + \frac{d}{dx}x^2 + \frac{d}{dx}x^3 + \cdots + \frac{d}{dx}x^n + \cdots$$

$$f(x) = 1 + 2x + 3x^2 + \cdots + nx^{n-1} + \cdots$$

Since the series for $1/(1-x)$ converges for $-1 < x < 1$, the series for $f(x)$ also converges for $-1 < x < 1$. [Verify this statement by applying Theorem 1 in the preceding section to the series for $f(x)$.]

∎

Matched Problem 3 ➭ Find the Taylor series at 0 for $f(x) = 1/(1 + x)^2$, and find the interval of convergence.

■

Explore–Discuss 1

The Taylor series representation for a function f at 0 is

$$f(x) = x - \frac{1}{3!}x^3 + \frac{1}{5!}x^5 - \frac{1}{7!}x^7 + \cdots$$

$$+ \frac{(-1)^n}{(2n + 1)!}x^{2n+1} + \cdots \qquad -\infty < x < \infty$$

(A) Find the Taylor series for $f'(x)$.

(B) Find the Taylor series for $f''(x)$. How are f and f'' related?

PROPERTY 4 **Integration**

A Taylor series at 0 can be integrated term by term:

If

$$f(x) = a_0 + a_1x + a_2x^2 + \cdots + a_nx^n + \cdots$$

then

$$\int f(x)\, dx = C + a_0x + \frac{1}{2}a_1x^2 + \frac{1}{3}a_2x^3 + \cdots + \frac{1}{n + 1}a_nx^{n+1} + \cdots$$

where C is the constant of integration. The Taylor series for $\int f(x)\, dx$ has the same interval of convergence as the Taylor series for f.

■

Example 4 ➭ **Integration of Taylor Series** If f is a function that satisfies $f'(x) = x^2e^x$ and $f(0) = 2$, find the Taylor series at 0 for f. Find the interval of convergence.

SOLUTION Using the series for e^x from Table 1 and property 2, the series for f' is

$$f'(x) = x^2e^x$$

$$= x^2\left(1 + x + \frac{1}{2!}x^2 + \cdots + \frac{1}{n!}x^n + \cdots\right) \qquad -\infty < x < \infty$$

$$= x^2 + x^3 + \frac{1}{2!}x^4 + \cdots + \frac{1}{n!}x^{n+2} + \cdots$$

Integrating term by term produces a series for f:

$$f(x) = \int f'(x)\, dx$$

$$= \int\left(x^2 + x^3 + \frac{1}{2!}x^4 + \cdots + \frac{1}{n!}x^{n+2} + \cdots\right)dx$$

$$= \int x^2\, dx + \int x^3\, dx + \frac{1}{2!}\int x^4\, dx + \cdots + \frac{1}{n!}\int x^{n+2}\, dx + \cdots$$

$$= C + \frac{1}{3}x^3 + \frac{1}{4}x^4 + \frac{1}{(5)2!}x^5 + \cdots + \frac{1}{(n + 3)n!}x^{n+3} + \cdots$$

Now we use the condition $f(0) = 2$ to evaluate the constant of integration C:

$$2 = f(0)$$
$$= C + 0 + 0 + \cdots + 0 + \cdots$$
$$= C$$

Thus,

$$f(x) = 2 + \frac{1}{3}x^3 + \frac{1}{4}x^4 + \cdots + \frac{1}{(n+3)n!}x^{n+3} + \cdots$$

Since the series for f' converges for $-\infty < x < \infty$, the series for f converges for $-\infty < x < \infty$.

■

Matched Problem 4 ✏   If f is a function that satisfies $f'(x) = x \ln(1 + x)$ and $f(0) = 4$, find the Taylor series at 0 for f. Find the interval of convergence.

■

❑   ␣ SUBSTITUTION

Example 5 ✏   **Using Substitution to Find Taylor Series**   Find the Taylor series at 0 for $f(x) = e^{-x^2}$, and find the interval of convergence.

SOLUTION   Suppose we try to solve this problem by finding the general form of the nth derivative of f:

$$f(x) = e^{-x^2}$$
$$f'(x) = -2xe^{-x^2}$$
$$f''(x) = -2e^{-x^2} + 4x^2e^{-x^2}$$
$$f^{(3)}(x) = 4xe^{-x^2} + 8xe^{-x^2} - 8x^3e^{-x^2}$$
$$= 12xe^{-x^2} - 8x^3e^{-x^2}$$
$$f^{(4)}(x) = 12e^{-x^2} - 24x^2e^{-x^2} - 24x^2e^{-x^2} + 16x^4e^{-x^2}$$
$$= 12e^{-x^2} - 48x^2e^{-x^2} + 16x^4e^{-x^2}$$

Since the higher-order derivatives are becoming very complicated and no general pattern is emerging, we will try another approach. How can we relate f to one of the functions in Table 1? If we let $g(x) = e^{-x}$, then f and g are related by

$$f(x) = e^{-x^2} = g(x^2)$$

From Table 1,

$$g(x) = e^{-x}$$

$$= 1 - x + \frac{1}{2!}x^2 - \frac{1}{3!}x^3 + \cdots + \frac{(-1)^n}{n!}x^n + \cdots \qquad -\infty < x < \infty$$

Substituting x^2 for x in the series for g, we have

$$f(x) = g(x^2)$$

$$= 1 - x^2 + \frac{1}{2!}(x^2)^2 - \frac{1}{3!}(x^2)^3 + \cdots + \frac{(-1)^n}{n!}(x^2)^n + \cdots$$

$$= 1 - x^2 + \frac{1}{2!}x^4 - \frac{1}{3!}x^6 + \cdots + \frac{(-1)^n}{n!}x^{2n} + \cdots$$

Since the series for g converges for all values of x, the series for f must also converge for all values of x.

■

Explore–Discuss 2

FIGURE 1

(A) The function $f(x) = e^{-x^2}$ and the Taylor polynomials

$$1 - x^2 + \frac{1}{2!}x^4 - \cdots + \frac{(-1)^n}{n!}x^{2n}$$

are graphed for $n = 1, 2, 3, 4, 5$ in Figure 1. Which curve belongs to which function?

(B) Does Figure 1 provide supporting evidence for the statement that the Taylor series for f at 0 converges for all values of x? Explain.

Matched Problem 5   Find the Taylor series at 0 for $f(x) = e^{x^3}$, and find the interval of convergence.

Making a substitution in a known Taylor series in order to obtain a new series is a very useful technique. Since there are many different substitutions that can be used, it is difficult to make a general statement concerning the effect of a substitution on the interval of convergence. The following examples illustrate some of the possibilities that may occur.

Example 6   **Using Substitution to Find Taylor Series**   Find the Taylor series at 0 for $f(x) = 1/(4 - x)$, and find the interval of convergence.

SOLUTION   If we factor a 4 out of the denominator of f, we can establish a relationship between $f(x)$ and $g(x) = 1/(1 - x)$. Thus,

$$f(x) = \frac{1}{4 - x}$$

$$= \left(\frac{1}{4}\right)\frac{1}{1 - (x/4)} \quad g(x) = \frac{1}{1 - x}$$

$$= \frac{1}{4}g\left(\frac{x}{4}\right)$$

From Table 1,

$$g(x) = \frac{1}{1 - x}$$

$$= 1 + x + x^2 + \cdots + x^n + \cdots \qquad -1 < x < 1$$

Substituting $x/4$ for x in this series and multiplying by $\frac{1}{4}$, we have

$$f(x) = \frac{1}{4}g\left(\frac{x}{4}\right)$$

$$= \frac{1}{4}\left[1 + \left(\frac{x}{4}\right) + \left(\frac{x}{4}\right)^2 + \cdots + \left(\frac{x}{4}\right)^n + \cdots\right]$$

$$= \frac{1}{4} + \frac{1}{4^2}x + \frac{1}{4^3}x^2 + \cdots + \frac{1}{4^{n+1}}x^n + \cdots$$

Since the original series for g converges for $-1 < x < 1$ and we substituted $x/4$ for x in that series, the series for f converges for

$$-1 < \frac{x}{4} < 1 \quad \text{Multiply each member by 4.}$$

$$-4 < x < 4$$

Matched Problem 6 ➭ Find the Taylor series at 0 for $f(x) = 1/(3 + x)$, and find the interval of convergence.

Example 7 ➭ **Using Substitution to Find Taylor Series**   Find the Taylor series at 0 for $f(x) = \ln(1 + 4x^2)$, and find the interval of convergence.

SOLUTION   If we let $g(x) = \ln(1 + x)$, then

$$f(x) = \ln(1 + 4x^2)$$
$$= g(4x^2)$$

From Table 1, the series for g is

$$g(x) = x - \frac{1}{2}x^2 + \frac{1}{3}x^3 - \cdots + \frac{(-1)^{n-1}}{n}x^n + \cdots \qquad -1 < x < 1$$

(A)

Substituting $4x^2$ for x in this series, we have

$$f(x) = g(4x^2)$$
$$= 4x^2 - \frac{1}{2}(4x^2)^2 + \frac{1}{3}(4x^2)^3 - \cdots + \frac{(-1)^{n-1}}{n}(4x^2)^n + \cdots$$
$$= 4x^2 - \frac{4^2}{2}x^4 + \frac{4^3}{3}x^6 - \cdots + \frac{(-1)^{n-1}4^n}{n}x^{2n} + \cdots$$

The series for g converges for $-1 < x < 1$. Since we substituted $4x^2$ for x, the series for f converges for

$$-1 < 4x^2 < 1$$

Inequalities of this type are easier to solve if we use absolute value notation:

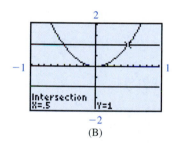
(B)

FIGURE 2 Graphical solution of $-1 < 4x^2 < 1$

$$-1 < 4x^2 < 1 \qquad \textit{Change to absolute value notation.}$$
$$|4x^2| < 1 \qquad \textit{Multiply by } \tfrac{1}{4}.$$
$$|x^2| < \tfrac{1}{4}$$
$$|x|^2 < \tfrac{1}{4}$$
$$|x| < \tfrac{1}{2} \qquad \textit{Convert to double inequalities.}$$
$$-\tfrac{1}{2} < x < \tfrac{1}{2} \qquad \textit{See Figure 2 for a graphical solution.}$$

Thus, the series for f converges for $-\frac{1}{2} < x < \frac{1}{2}$.

Matched Problem 7 ➭ Find the Taylor series at 0 for $f(x) = 1/(1 + 9x^2)$, and find the interval of convergence.

Up to this point in this section we have restricted our attention to Taylor series at 0. How can we use the techniques discussed to find a Taylor series at a point $a \neq 0$? Properties 1–4 could be stated in terms of Taylor series at an arbitrary point a; however, there is an easier way to proceed. The method of substitution allows us to use Taylor series at 0 to find Taylor series at other points. The following example illustrates this technique.

Example 8 ➭ **Using Substitution to Find Taylor Series**   Find the Taylor series at 1 for $f(x) = 1/(2 - x)$, and find the interval of convergence.

SOLUTION   In order to find a Taylor series for f in powers of $x - 1$, we will use the substitution $t = x - 1$ to express f as a function of t. If we find the Taylor series at

0 for this new function and then replace t with $x - 1$, we will have obtained the Taylor series at 1 for f.

$$t = x - 1 \qquad \text{Solve for x.}$$
$$x = t + 1 \qquad \text{Substitute for x in } f(x) = 1/(2 - x).$$

$$\frac{1}{2 - x} = \frac{1}{2 - (t + 1)}$$

$$= \frac{1}{1 - t} \qquad \text{Find the Taylor series at 0 for this function of t.}$$

$$= 1 + t + t^2 + \cdots + t^n + \cdots \qquad \text{Substitute x } - 1 \text{ for t.}$$
$$-1 < t < 1$$
$$= 1 + (x - 1) + (x - 1)^2 + \cdots + (x - 1)^n + \cdots$$

Since $t = x - 1$ and the series in powers of t converges for $-1 < t < 1$, the series in powers of $x - 1$ converges for

$$-1 < x - 1 < 1 \qquad \text{Add 1 to each member.}$$
$$0 < \quad x \quad < 2$$

■

Matched Problem 8 ⮑ Find the Taylor series at 1 for $f(x) = 1/x$, and find the interval of convergence.

■

Answers to Matched Problems

1. $2 + 2x + \left(\dfrac{1}{2!} + 1\right)x^2 + \cdots + \left(\dfrac{1}{n!} + 1\right)x^n + \cdots, -1 < x < 1$

2. $-3x^4 - \dfrac{3}{2}x^5 - x^6 - \cdots - \dfrac{3}{n}x^{n+3} - \cdots, -1 < x < 1$

3. $1 - 2x + 3x^2 - \cdots + (-1)^{n+1}nx^{n-1} + \cdots, -1 < x < 1$

4. $4 + \dfrac{1}{3}x^3 - \dfrac{1}{8}x^4 + \dfrac{1}{15}x^5 - \cdots + \dfrac{(-1)^{n-1}}{n(n + 2)}x^{n+2} + \cdots, -1 < x < 1$

5. $1 + x^3 + \dfrac{1}{2!}x^6 + \cdots + \dfrac{1}{n!}x^{3n} + \cdots, -\infty < x < \infty$

6. $\dfrac{1}{3} - \dfrac{1}{3^2}x + \dfrac{1}{3^3}x^2 - \cdots + \dfrac{(-1)^n}{3^{n+1}}x^n + \cdots, -3 < x < 3$

7. $1 - 9x^2 + 9^2x^4 - \cdots + (-1)^n 9^n x^{2n} + \cdots, -\tfrac{1}{3} < x < \tfrac{1}{3}$

8. $1 - (x - 1) + (x - 1)^2 - \cdots + (-1)^n(x - 1)^n + \cdots, 0 < x < 2$

Exercise 10-3

Solve all the problems in this exercise by performing operations on the Taylor series in Table 1. State the interval of convergence for each series you find.

A *In Problems 1–10, find the Taylor series at 0.*

1. $f(x) = \dfrac{1}{1 - x} + \ln(1 + x)$

2. $f(x) = \dfrac{1}{1 - x} + \ln(1 - x)$

3. $f(x) = \dfrac{x^3}{1 - x}$

4. $f(x) = x^3 e^x$

5. $f(x) = xe^{x^2}$

6. $f(x) = xe^{-2x^2}$

7. $f(x) = \dfrac{1}{1 - x^3}$

8. $f(x) = \ln(1 + x^3)$

9. $f(x) = \dfrac{1}{1 + x^2}$

10. $f(x) = \dfrac{3}{1 + \dfrac{x}{5}}$

B *In Problems 11–18, find the Taylor series at 0.*

11. $f(x) = \dfrac{1}{2 - x}$

12. $f(x) = \dfrac{1}{5 + x}$

13. $f(x) = \dfrac{1}{1 - 8x^3}$

14. $f(x) = \ln(1 + 16x^2)$

15. $f(x) = 10^x$

16. $f(x) = \log(1 + x)$

17. $f(x) = \log_2(1 - x)$ **18.** $f(x) = 3^x$

19. $f(x) = \dfrac{1}{4 + x^2}$ **20.** $f(x) = \dfrac{1}{9 - x^2}$

21. Substituting $\sqrt{x}$ for x in the Taylor series at 0 for e^x gives the formula

$$e^{\sqrt{x}} = 1 + x^{1/2} + \frac{1}{2!}x + \frac{1}{3!}x^{3/2} + \cdots + \frac{1}{n!}x^{n/2} + \cdots$$

(A) For which values of x is the formula valid?

(B) Is the formula a Taylor series at 0 for $e^{\sqrt{x}}$? Explain.

22. Substituting $1/x$ for x in the Taylor series at 0 for $1/(1 - x)$ gives the formula

$$\frac{1}{1 - \dfrac{1}{x}} = 1 + x^{-1} + x^{-2} + \cdots + x^{-n} + \cdots$$

(A) For which values of x is the formula valid?

(B) Explain why the formula is not a Taylor series at 0 for

$$f(x) = \frac{1}{1 - \dfrac{1}{x}}$$

(C) Find a Taylor series at 0 that equals

$$\frac{1}{1 - \dfrac{1}{x}}$$

for all $x \neq 0$ in the interval $-1 < x < 1$.

23. Find the Taylor series at 0 for

(A) $f(x) = \dfrac{1}{1 - x^2}$

(B) $g(x) = \dfrac{2x}{(1 - x^2)^2}$

[*Hint*: Compare $f'(x)$ and $g(x)$.]

24. Find the Taylor series at 0 for

(A) $f(x) = \dfrac{x}{1 - x^2}$

(B) $g(x) = \dfrac{1 + x^2}{(1 - x^2)^2}$

[*Hint*: Compare $f'(x)$ and $g(x)$.]

25. If $f(x)$ satisfies $f'(x) = 1/(1 + x^2)$ and $f(0) = 0$, find the Taylor series at 0 for $f(x)$.

26. If $f(x)$ satisfies $f'(x) = \ln(1 + x^2)$ and $f(0) = 1$, find the Taylor series at 0 for $f(x)$.

27. If $f(x)$ satisfies $f'(x) = x^2 \ln(1 - x)$ and $f(0) = 5$, find the Taylor series at 0 for $f(x)$.

28. If $f(x)$ satisfies $f'(x) = xe^x$ and $f(0) = -3$, find the Taylor series at 0 for $f(x)$.

29. Most graphing calculators include the hyperbolic sine function, $f(x) = \sinh x$, as a built-in function. The Taylor series for the hyperbolic sine at 0 is

$$\sinh x = x + \frac{1}{3!}x^3 + \frac{1}{5!}x^5 + \cdots$$
$$+ \frac{1}{(2n + 1)!}x^{2n+1} + \cdots \qquad -\infty < x < \infty$$

(A) Graph f and its Taylor polynomials $p_1(x), p_3(x)$, and $p_5(x)$ in the viewing window $-5 \leqslant x \leqslant 5$, $-30 \leqslant y \leqslant 30$.

(B) Explain why $\dfrac{d^2}{dx^2}(\sinh x) = \sinh x$.

30. Most graphing calculators include the hyperbolic cosine function, $g(x) = \cosh x$, as a built-in function. The Taylor series for the hyperbolic cosine at 0 is

$$\cosh x = 1 + \frac{1}{2!}x^2 + \frac{1}{4!}x^4 + \frac{1}{6!}x^6 + \cdots$$
$$+ \frac{1}{(2n)!}x^{2n} + \cdots \qquad -\infty < x < \infty$$

(A) Graph g and its Taylor polynomials $p_2(x)$, $p_4(x)$, and $p_6(x)$ in the viewing window $-5 \leqslant x \leqslant 5, 0 \leqslant y \leqslant 60$.

(B) Explain why $\dfrac{d^2}{dx^2}(\cosh x) = \cosh x$.

In Problems 31–36, use the substitution $t = x - a$ to find the Taylor series at the indicated value of a.

31. $f(x) = \dfrac{1}{4 - x}$; at 3   **32.** $f(x) = \dfrac{1}{3 + x}$; at -2

33. $f(x) = \ln x$; at 1   **34.** $f(x) = \ln(3 - x)$; at 2

35. $f(x) = \dfrac{1}{4 - 3x}$; at 1

36. $f(x) = \dfrac{1}{5 - 2x}$; at 2

C **37.** Use the Taylor series at 0 for $1/(1 - x)$ and repeated applications of property 3 to find the Taylor series at 0 for

$$f(x) = \frac{1}{(1 - x)^3}$$

38. Use the Taylor series at 0 for $1/(1 + x)$ and repeated applications of property 3 to find the Taylor series at 0 for

$$f(x) = \frac{1}{(1 + x)^4}$$

39. Suppose that the Taylor series for f at 0 is

$$f(x) = x + \frac{1}{3}x^3 + \frac{1}{5}x^5 + \cdots$$
$$+ \frac{1}{2n + 1}x^{2n+1} + \cdots \qquad -1 < x < 1$$

(A) Find the Taylor series for f'.

(B) Explain why $f'(x) = \dfrac{1}{1 - x^2}$ for $-1 < x < 1$.

40. Suppose that the Taylor series for g at 0 is

$$g(x) = x - \frac{1}{3}x^3 + \frac{1}{5}x^5 - \cdots$$
$$+ \frac{(-1)^n}{2n + 1}x^{2n+1} + \cdots \qquad -1 < x < 1$$

(A) Find the Taylor series for g'.

(B) Explain why $g'(x) = \dfrac{1}{1 + x^2}$ for $-1 < x < 1$.

41. Find the Taylor series at 0 for $f(x) = \dfrac{1 + x}{1 - x}$.

$$\left[Note: \quad \frac{1 + x}{1 - x} = \frac{1}{1 - x} + \frac{x}{1 - x} \right]$$

42. Find the Taylor series at 0 for $f(x) = \dfrac{1 - 2x}{1 + x}$.

$$\left[Note: \quad \frac{1 - 2x}{1 + x} = \frac{1}{1 + x} - \frac{2x}{1 + x} \right]$$

43. Find the Taylor series at 0 for $f(x) = \dfrac{1}{2}\ln\!\left(\dfrac{1 + x}{1 - x}\right)$.

$$\left[Note: \quad \ln\!\left(\frac{1 + x}{1 - x}\right) = \ln(1 + x) - \ln(1 - x) \right]$$

44. Find the Taylor series at 0 for $f(x) = \ln(1 - 2x + x^2)$.

45. Find the Taylor series at 0 for $f(x) = \dfrac{e^x + e^{-x}}{2}$.

46. Find the Taylor series at 0 for $f(x) = \dfrac{e^x - e^{-x}}{2}$.

47. If $f(x)$ satisfies $f''(x) = \ln(1 + x)$, $f'(0) = 3$, and $f(0) = -2$, find the Taylor series at 0 for $f(x)$.

48. If $f(x)$ satisfies $f''(x) = \dfrac{1}{1 - x^2}$, $f'(0) = 4$, and $f(0) = 5$, find the Taylor series at 0 for $f(x)$.

49. Find the Taylor series at any $a > 0$ for $f(x) = \ln x$.

50. Find the Taylor series at any a for $f(x) = e^x$.

51. If a and b are constants $(b \neq 0, a \neq b^{-1})$, find the Taylor series at a for

$$f(x) = \frac{1}{1 - bx}$$

52. If a and b are constants $(a \neq b)$, find the Taylor series at a for

$$f(x) = \frac{1}{b - x}$$

Section 10-4   ## Approximations Using Taylor Series

- ❑ THE REMAINDER
- ❑ TAYLOR'S FORMULA FOR THE REMAINDER
- ❑ TAYLOR SERIES WITH ALTERNATING TERMS
- ❑ APPROXIMATING DEFINITE INTEGRALS

❑ **THE REMAINDER**

Now that we can find Taylor series for a variety of functions, we return to our original goal: approximating the values of a function.

　　If x is in the interval of convergence of the Taylor series for a function f and $p_n(x)$ is the nth-degree Taylor polynomial, then

$$f(x) = \lim_{n \to \infty} p_n(x)$$

and $p_n(x)$ can be used to approximate $f(x)$. We want to consider two questions:

1. If we select a particular value of n, how accurate is the approximation $f(x) \approx p_n(x)$?
2. If we want the approximation $f(x) \approx p_n(x)$ to have a specified accuracy, how do we select the proper value of n?

It turns out that both of these questions can be answered by examining the difference between $f(x)$ and $p_n(x)$. This difference is called the *remainder* and is defined in the following box.

The Remainder of a Taylor Series

If p_n is the nth-degree Taylor polynomial for f, then the **remainder** is

$$R_n(x) = f(x) - p_n(x)$$

The **error** in the approximation $f(x) \approx p_n(x)$ is

$$|f(x) - p_n(x)| = |R_n(x)|$$

If the Taylor series at 0 for f is

$$f(x) = \overbrace{a_0 + a_1x + a_2x^2 + \cdots + a_nx^n}^{p_n(x)} + \overbrace{a_{n+1}x^{n-1} + \cdots}^{R_n(x)}$$

then

$$p_n(x) = a_0 + a_1x + a_2x^2 + \cdots + a_nx^n$$

and

$$R_n(x) = a_{n+1}x^{n+1} + a_{n+2}x^{n+2} + \cdots$$

Similar statements can be made for Taylor series at a.

It follows from our basic assumption (see page 608) for the functions we consider that

$$\lim_{n\to\infty} R_n(x) = \lim_{n\to\infty}[f(x) - p_n(x)] = f(x) - f(x) = 0$$

if and only if x is in the interval of convergence of f.*

In general, it is difficult to find the exact value of $R_n(x)$. In fact, since $f(x) = p_n(x) + R_n(x)$, this is equivalent to finding the exact value of $f(x)$. Instead, we will discuss two methods for *estimating* the value of $R_n(x)$. The first method works in all cases, but can be difficult to apply, while the second method is easy to apply, but does not work in all cases.

Explore–Discuss 1

Let $p_1(x), p_2(x)$, and $p_3(x)$ be Taylor polynomials for $f(x) = e^x$ at 0. The corresponding remainder functions $R_1(x), R_2(x)$, and $R_3(x)$ are graphed in Figure 1.

(A) Which curve belongs to which function?

(B) Use graphical approximation techniques to estimate those values of x for which the error in each of the approximations $f(x) \approx p_n(x)$ is less than 0.01.

FIGURE 1

❑ TAYLOR'S FORMULA FOR THE REMAINDER

The first method for estimating the remainder of a Taylor series is based on *Taylor's formula for the remainder*.

---

*In more advanced texts, the statement $\lim_{n\to\infty} R_n(x) = 0$ for x in the interval of convergence is actually proved for functions such as e^x, making our basic assumption unnecessary.

> **Taylor's Formula for the Remainder**
>
> If f has derivatives of all order at 0, then
>
> $$R_n(x) = \frac{f^{(n+1)}(t)x^{n+1}}{(n+1)!}$$
>
> for some number t between 0 and x.

A similar formula can be stated for the remainder of a Taylor series at an arbitrary point a. (We will not use this formula in this text.)

In most applications of the remainder formula, it is not possible to find the value of t. However, if we can find a number M satisfying

$$|f^{(n+1)}(t)| < M \qquad \text{for all } t \text{ between 0 and } x$$

then we can estimate $R_n(x)$ as follows:

$$|R_n(x)| = \left|\frac{f^{(n+1)}(t)x^{n+1}}{(n+1)!}\right| < \frac{M|x|^{n+1}}{(n+1)!}$$

This technique is illustrated in the next example.

Example 1 ✏ **Taylor's Formula for the Remainder** Use the second-degree Taylor polynomial at 0 for $f(x) = e^x$ to approximate $e^{0.1}$. Use Taylor's formula for the remainder to estimate the error in this approximation.

SOLUTION Since $f^{(n)}(x) = e^x$ for any n, we can write

$$f(x) = e^x = p_2(x) + R_2(x)$$

$$= 1 + x + \frac{1}{2!}x^2 + \overbrace{\frac{e^t x^3}{3!}}^{R_2(x)} \quad f^{(3)}(t) = e^t$$

for some number t between 0 and x. Thus,

$$
\begin{aligned}
f(0.1) &= e^{0.1} \\
&= p_2(0.1) + R_2(0.1) \\
&\approx p_2(0.1) \\
&= 1 + 0.1 + \tfrac{1}{2}(0.1)^2 \\
&= 1.105
\end{aligned}
$$

To estimate the error in this approximation, we must estimate

$$|R_2(0.1)| = \left|\frac{e^t(0.1)^3}{3!}\right| = \frac{e^t}{6{,}000}$$

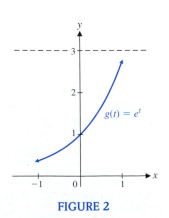

FIGURE 2

where $0 \le t \le 0.1$. In order to estimate $|R_2(0.1)|$, we must estimate $g(t) = e^t$ for $0 \le t \le 0.1$. Since e^t is always increasing, $e^t \le e^{0.1}$ for $0 \le t \le 0.1$. However, $e^{0.1}$ is the number we are trying to approximate. We do not want to use this number in our estimate of the error. (This situation occurs frequently in approximation problems involving the exponential function.) Instead, we will use the following rough estimate for e^t (see Fig. 2):

$$\text{If} \qquad t \le 1, \qquad \text{then} \qquad e^t \le 3. \tag{1}$$

Since (1) holds for $t \le 1$, it certainly holds for $0 \le t \le 0.1$. Thus,

$$|R_2(0.1)| = \frac{e^t}{6{,}000} \le \frac{3}{6{,}000} = 0.0005$$

and we can conclude that the approximate value 1.105 is within ± 0.0005 of the exact value of $e^{0.1}$.

■

Matched Problem 1 ➯ Use the second-degree Taylor polynomial at 0 for $f(x) = e^x$ to approximate $e^{0.2}$. Use Taylor's formula for the remainder to estimate the error in the approximation.

■

❑   TAYLOR SERIES WITH ALTERNATING TERMS

Taylor's formula for the remainder can be difficult to apply for large values of n. For most functions, the formula for the nth derivative becomes very complicated as n increases. For a certain type of problem, there is another method that does not require estimation of the nth derivative. If the series of numbers that is formed by evaluating a Taylor series at a given number x_0 *alternates in sign and decreases in absolute value,* then the remainder can be estimated by simply examining the numbers in the series. Series of numbers whose terms alternate in sign are called **alternating series.** The estimate for the remainder in an alternating series is given in Theorem 1.

THEOREM 1   Error Estimation for Alternating Series

If x_0 is a number in the interval of convergence for

$$f(x) = a_0 + a_1 x + a_2 x^2 + \cdots + a_k x^k + \cdots$$

and the terms in the series

$$f(x_0) = a_0 + a_1 x_0 + a_2 x_0^2 + \cdots + a_k x_0^k + \cdots$$

are alternating in sign and decreasing in absolute value, then the error in the approximation

$$f(x_0) \approx a_0 + a_1 x_0 + a_2 x_0^2 + \cdots + a_n x_0^n$$

is strictly less than the absolute value of the next term. That is,

$$|R_n(x_0)| < |a_{n+1} x_0^{n+1}|$$

■

Example 2 ➯ **Estimating the Remainder for Alternating Series**   Use the Taylor series at 0 for $f(x) = e^{-x}$ to approximate $e^{-0.3}$ with an error of no more than 0.0005.

SOLUTION   From Table 1 in Section 10-3, the Taylor series at 0 for $f(x) = e^{-x}$ is

$$e^{-x} = 1 - x + \frac{1}{2!}x^2 - \frac{1}{3!}x^3 + \cdots + \frac{(-1)^k}{k!}x^k + \cdots \qquad -\infty < x < \infty$$

If we substitute $x = 0.3$ in this series, we obtain

$$
\begin{aligned}
f(0.3) &= e^{-0.3} \\
&= 1 - 0.3 + \tfrac{1}{2}(0.3)^2 - \tfrac{1}{6}(0.3)^3 + \tfrac{1}{24}(0.3)^4 - \cdots \\
&= 1 - 0.3 + 0.045 - 0.0045 + 0.000\ 337\ 5 - \cdots
\end{aligned}
$$

Since the terms in this series are alternating in sign and decreasing in absolute value, Theorem 1 applies. If we use the first four terms in this series to approximate $e^{-0.3}$, then the error in this approximation is less than the absolute value of the fifth term. That is,

$$|R_3(0.3)| < 0.000\ 337\ 5 < 0.0005$$

Thus,

$$e^{-0.3} \approx 1 - 0.3 + 0.045 - 0.0045 = 0.7405$$

and the error in this approximation is less than the specified accuracy of 0.0005. ∎

Matched Problem 2 ⤳ Use the Taylor series at 0 for $f(x) = e^{-x}$ to approximate $e^{-0.1}$ with an error of no more than 0.0005. ∎

As Example 2 illustrates, Theorem 1 is much easier to use than Taylor's formula for the remainder. Notice that we did not have to find an estimate for $f^{(n+1)}(t)$. However, it is very important to understand that Theorem 1 can be applied only if the terms in the series *alternate in sign after they have been evaluated at* x_0. For example, if we try to use the series for e^{-x} to approximate $e^{0.3}$ by substituting $x_0 = -0.3$, we have

$$e^{0.3} = 1 - (-0.3) + \tfrac{1}{2}(-0.3)^2 - \tfrac{1}{6}(-0.3)^3 + \cdots$$
$$= 1 + 0.3 + \tfrac{1}{2}(0.3)^2 + \tfrac{1}{6}(0.3)^3 + \cdots$$

Since these numbers do not alternate in sign, Theorem 1 does not apply. Taylor's formula for the remainder would have to be used to estimate the error in approximations obtained from this series.

❑ APPROXIMATING DEFINITE INTEGRALS

In order to find the exact value of a definite intergral $\int_a^b f(x)\,dx$, we must first find an antiderivative of the function f. But suppose we cannot find an antiderivative of f (it may not even exist in a convenient form). In Section 6-5 we saw that Riemann sums can be used to approximate definite integrals. Taylor series techniques provide an alternative method for approximating definite integrals that is often more efficient, and, in the case of alternating series, automatically determines the accuracy of the approximation.

Example 3 ⤳ **Using Taylor Series to Approximate Definite Integrals** Approximate $\int_0^1 e^{-x^2}\,dx$ with a maximum error of 0.005.

SOLUTION If $F(x)$ is an antiderivative of e^{-x^2}, then

$$\int_0^1 e^{-x^2}\,dx = F(1) - F(0)$$

It is not possible to express $F(x)$ as a finite combination of simple functions; however, it is possible to find a Taylor series for F. This series can be used to approximate the values of F and, consequently, of the definite integral.

Step 1. *Find a Taylor series for the integrand:* From Table 1 in Section 10-3,

$$e^x = 1 + x + \frac{1}{2!}x^2 + \cdots + \frac{1}{n!}x^n + \cdots \qquad -\infty < x < \infty$$

Thus,

$$e^{-x^2} = 1 + (-x^2) + \frac{1}{2!}(-x^2)^2 + \cdots + \frac{1}{n!}(-x^2)^n + \cdots$$

$$= 1 - x^2 + \frac{1}{2!}x^4 - \cdots + \frac{(-1)^n}{n!}x^{2n} + \cdots \qquad -\infty < x < \infty$$

Step 2. *Find the Taylor series for the antiderivative:* Integrating term by term, we have

$$F(x) = \int e^{-x^2} dx$$

$$= \int \left[1 - x^2 + \frac{1}{2!}x^4 - \cdots + \frac{(-1)^n}{n!}x^{2n} + \cdots \right] dx$$

$$= C + x - \frac{1}{3}x^3 + \frac{1}{10}x^5 - \cdots + \frac{(-1)^n}{n!(2n+1)}x^{2n+1} + \cdots \qquad -\infty < x < \infty$$

Step 3. *Approximate the definite integral:* If we choose $C = 0$, then $F(0) = 0$ and

$$\int_0^1 e^{-x^2}\, dx = F(1) - F(0)$$

$$= \left[x - \frac{1}{3}x^3 + \frac{1}{10}x^5 - \cdots + \frac{(-1)^n}{n!(2n+1)}x^{2n+1} + \cdots \right]\Big|_0^1$$

$$= \left[1 - \frac{1}{3} + \frac{1}{10} - \cdots + \frac{(-1)^n}{n!(2n+1)} + \cdots \right] - 0$$

$$= 1 - \frac{1}{3} + \frac{1}{10} - \frac{1}{42} + \frac{1}{216} - \cdots$$

$$= 1 - 0.333\ 333 + 0.1 - 0.023\ 809 + 0.004\ 630 - \cdots$$

Since the Taylor series for $F(x)$ converges for all values of x, we can use this last series of numbers to approximate $F(1)$. Notice that the terms in this series are alternating in sign and decreasing in absolute value. Applying Theorem 1, we conclude that the error introduced by approximating $F(1)$ with the first four terms of this series will be no more than 0.004 630, the absolute value of the fifth term. Since this is less than the specified error of 0.005, we have

$$\int_0^1 e^{-x^2}\, dx = F(1) - F(0)$$

$$\approx 1 - 0.333\ 333 + 0.1 - 0.023\ 809$$

$$\approx 0.743 \quad \textcolor{blue}{\textit{Rounded to three decimal places}}$$

Matched Problem 3   Approximate $\int_0^{0.5} e^{-x^2}$ with a maximum error of 0.005.

Explore–Discuss 2   Suppose you wish to use a Taylor series for

$$f(x) = \frac{1}{1-x^2} \qquad \text{to approximate} \qquad \int_2^3 \frac{dx}{1-x^2}$$

(A) Explain why you would not use the Taylor series for f at 0.

(B) If you use the Taylor series for f at a, which value of a would you expect to yield the best approximation to the integral? Explain.

Example 4 ⇔   **Income Distribution**   Approximate the index of income concentration for the Lorenz curve given by

$$f(x) = \frac{11x^4}{10 + x^2}$$

with an error of no more than 0.005.

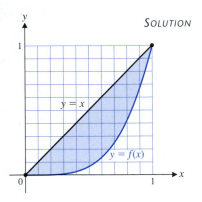

FIGURE 3

SOLUTION   Referring to Section 7-1, the index of income concentration for a Lorenz curve is twice the area between the graph of the Lorenz curve and the graph of the line $y = x$ (see Fig. 3). Thus, we must evaluate the integral

$$2\int_0^1 [x - f(x)]\, dx = 2\int_0^1 x\, dx - 2\int_0^1 f(x)\, dx$$

$$= \int_0^1 2x\, dx - \int_0^1 \frac{22x^4}{10 + x^2}\, dx \qquad (2)$$

The first integral in (2) is easy to evaluate:

$$\int_0^1 2x\, dx = x^2 \Big|_0^1 = 1 - 0 = 1$$

Since the second integral in (2) cannot be evaluated by any of the techniques we have discussed, we will use a Taylor series to approximate this integral.

Step 1.   *Find a Taylor series for the integrand:*

$$\frac{22x^4}{10 + x^2} = \frac{22x^4}{10}\left[\frac{1}{1 + (x^2/10)}\right] \qquad \text{Substitute } x^2/10 \text{ for } x \text{ in the series for } 1/(1 + x).$$

$$= 2.2x^4\left[1 - \frac{x^2}{10} + \left(\frac{x^2}{10}\right)^2 - \cdots + (-1)^n\left(\frac{x^2}{10}\right)^n + \cdots\right] \quad -1 < \frac{x^2}{10} < 1$$

$$= 2.2x^4 - \frac{2.2}{10}x^6 + \frac{2.2}{10^2}x^8 - \cdots + \frac{2.2(-1)^n}{10^n}x^{2n+4} + \cdots$$

To find the interval of convergence, we solve $-1 < x^2/10 < 1$ for x:

$$-1 < \frac{x^2}{10} < 1 \qquad \text{Change to absolute value notation.}$$

$$\left|\frac{x^2}{10}\right| < 1 \qquad \text{Multiply by 10.}$$

$$|x^2| < 10 \qquad \text{Take the square root of both sides.}$$

$$|x| < \sqrt{10} \qquad \text{Convert to double inequalities.}$$

$$-\sqrt{10} < x < \sqrt{10} \qquad \text{Interval of convergence}$$

Step 2.   *Find the Taylor series for the antiderivative:*   Using property 4,

$$\int \frac{22x^4}{10 + x^2}\, dx = \int\left[2.2x^4 - \frac{2.2}{10}x^6 + \frac{2.2}{10^2}x^8 - \cdots + \frac{2.2(-1)^n}{10^n}x^{2n+4} + \cdots\right] dx$$

$$= C + \frac{2.2}{5}x^5 - \frac{2.2}{7\cdot 10}x^7 + \frac{2.2}{9\cdot 10^2}x^9 - \cdots$$

$$+ \frac{2.2(-1)^n}{(2n+5)10^n}x^{2n+5} + \cdots$$

This series also converges for $-\sqrt{10} < x < \sqrt{10}$.

Step 3. *Approximate the definite integral:*   Choosing $C = 0$ in the antiderivative, we have

$$\int_0^1 \frac{22x^4}{10 + x^2}\, dx = \left[\frac{2.2}{5} x^5 - \frac{2.2}{7 \cdot 10} x^7 + \frac{2.2}{9 \cdot 10^2} x^9 - \cdots \right.$$

$$\left. + \frac{2.2(-1)^n}{(2n + 5)10^n} x^{2n+5} + \cdots \right]\Bigg|_0^1$$

$$= \left[\frac{2.2}{5} - \frac{2.2}{7 \cdot 10} + \frac{2.2}{9 \cdot 10^2} - \cdots + \frac{2.2(-1)^n}{(2n + 5)10^n} + \cdots \right] - 0$$

$$= 0.44 - 0.031\ 429 + 0.002\ 444 - \cdots$$

Since the limits of integration, 0 and 1, are within the interval of convergence of the Taylor series for the antiderivative, we can use this series to approximate the value of the definite integral. Notice that the numbers in this series are alternating in sign and decreasing in absolute value. Since the absolute value of the third term is less than 0.005, Theorem 1 implies that we can use the first two terms of this series to approximate the definite integral to the specified accuracy. Thus,

$$\int_0^1 \frac{22x^4}{10 + x^2}\, dx \approx 0.44 - 0.031\ 429$$

$$\approx 0.409 \quad \text{Rounded to three decimal places}$$

Returning to (2), we have

$$2\int_0^1 [x - f(x)]\, dx = 2\int_0^1 x\, dx - 2\int_0^1 f(x)\, dx$$

$$\approx 1 - 0.409$$

$$= 0.591 \quad \text{Index of income concentration}$$

Matched Problem 4 ➭   Repeat Example 4 for $f(x) = \dfrac{11x^5}{10 + x^2}$.

Answers to Matched Problems

1. $e^{0.2} \approx 1 + 0.2 + \frac{1}{2}(0.2)^2 = 1.22; |R_2(0.2)| < 0.004$

2. $e^{-0.1} \approx 1 - 0.1 + 0.005 = 0.905$ within $\pm 0.000\ 167$

3. $\displaystyle\int_0^{0.5} e^{-x^2} dx \approx 0.5 - \frac{1}{3}(0.5)^3 \approx 0.458$ within $\pm 0.003\ 125$

4. $\displaystyle\int_0^1 \frac{22x^5}{10 + x^2} dx \approx \frac{2.2}{6} - \frac{2.2}{80} \approx 0.339$ within ± 0.0022;

index of income concentration ≈ 0.661

Exercise 10-4

In Problems 1–30, use Theorem 1 to perform the indicated error estimations.

A *Evaluating the Taylor series at 0 for $f(x) = e^{-x}$ at $x = 0.6$ produces the following series:*

$$e^{-0.6} = 1 - 0.6 + 0.18 - 0.036 + 0.0054$$
$$- 0.000\ 648 + \cdots$$

In Problems 1–4, use the indicated number of terms in this series to approximate $e^{-0.6}$, and then estimate the error in this approximation.

1. Two terms

2. Three terms

3. Four terms

4. Five terms

Evaluating the Taylor series at 0 for $f(x) = \ln(1 + x)$ at $x = 0.3$ produces the following series:

$$\ln 1.3 = 0.3 - 0.045 + 0.009 - 0.002\ 025 + 0.000\ 486$$
$$- 0.000\ 121\ 5 + \cdots$$

In Problems 5–8, use the indicated number of terms in this series to approximate $\ln 1.3$, and then estimate the error in this approximation.

5. Two terms   **6.** Three terms

7. Four terms   **8.** Five terms

In Problems 9–12, use the third-degree Taylor polynomial at 0 for $f(x) = e^{-x}$ to approximate each expression, and then estimate the error in the approximation.

9. $e^{-0.2}$ **10.** $e^{-0.5}$

11. $e^{-0.03}$ **12.** $e^{-0.06}$

In Problems 13–16, use the third-degree Taylor polynomial at 0 for $f(x) = \ln(1 + x)$ to approximate each expression, and then estimate the error in the approximation.

13. $\ln 1.6$ **14.** $\ln 1.8$

15. $\ln 1.06$ **16.** $\ln 1.08$

B *In Problems 17–24, use a Taylor polynomial at 0 to approximate each expression with an error of no more than 0.0005. Select the polynomial of lowest degree that can be used to obtain this accuracy and state the degree of this polynomial.*

17. $e^{-0.4}$ **18.** $e^{-0.8}$ **19.** $e^{-0.04}$

20. $e^{-0.08}$ **21.** $\ln 1.1$ **22.** $\ln 1.2$

23. $\ln 1.01$ **24.** $\ln 1.02$

In Problems 25–30, use a Taylor series at 0 to approximate each integral with an error of no more than 0.0005.

25. $\displaystyle\int_0^{0.2} \frac{1}{1 + x^2}\, dx$ **26.** $\displaystyle\int_0^{0.5} \frac{x}{1 + x^4}\, dx$

27. $\displaystyle\int_0^{0.6} \ln(1 + x^2)\, dx$ **28.** $\displaystyle\int_0^{0.7} x\ln(1 + x^4)\, dx$

29. $\displaystyle\int_0^{0.4} x^2 e^{-x^2}\, dx$ **30.** $\displaystyle\int_0^{0.8} x^4 e^{-x^2}\, dx$

In Problems 31–34, assume that $f(x)$ is a function such that $|f^{(n)}(x)| \le 1$ for all n and all x [the trigonometric functions $f(x) = \sin x$ and $f(x) = \cos x$, for example, satisfy that property] and let $p_n(x)$ be the nth-degree Taylor polynomial for f at 0.

31. Use Taylor's formula for the remainder to find the smallest value of n such that the error in the approximation of $f(2)$ by $p_n(2)$ is guaranteed to be less than 0.001.

32. Use Taylor's formula for the remainder to find the smallest value of n such that the error in the approximation of $f(10)$ by $p_n(10)$ is guaranteed to be less than 0.001.

33. Use Taylor's formula for the remainder to determine the values of x such that the error in the approximation of $f(x)$ by $p_5(x)$ is guaranteed to be less than 0.05.

34. Use Taylor's formula for the remainder to determine the values of x such that the error in the approximation of $f(x)$ by $p_{10}(x)$ is guaranteed to be less than 0.05.

35. Let $f(x) = xe^x$ and consider the third-degree Taylor polynomial $p_3(x)$ for f at 0. Use graphical approximation techniques to determine those values of x for which the error in the approximation $f(x) \approx p_3(x)$ is less than 0.005.

36. Let $f(x) = 1/(4 - x^2)$ and consider the fourth-degree Taylor polynomial $p_4(x)$ for f at 0. Use graphical approximation techniques to determine those values of x for which the error in the approximation $f(x) \approx p_4(x)$ is less than 0.0001.

C *In Problems 37–40, use the second-degree Taylor polynomial at 0 for $f(x) = e^x$ to approximate the given number. Use Taylor's formula for the remainder to estimate the error in each approximation.*

37. $e^{-0.3}$ **38.** $e^{-0.4}$ **39.** $e^{0.05}$ **40.** $e^{0.01}$

41. The Taylor series at 0 for $f(x) = \sqrt{16 + x}$ converges for $-16 < x < 16$. Use the second-degree Taylor polynomial at 0 to approximate $\sqrt{17}$. Use Taylor's formula for the remainder to estimate the error in this approximation.

42. The Taylor series at 0 for $f(x) = \sqrt{(4 + x)^3}$ converges for $-4 < x < 4$. Use the third-degree Taylor polynomial at 0 to approximate $\sqrt{125}$. Use Taylor's formula for the remainder to estimate the error in this approximation.

43. To estimate

$$\int_{-2}^{0} \frac{1}{1 - x}\, dx$$

a student takes the first five terms of the Taylor series for $f(x) = 1/(1 - x)$ at 0 and integrates term by term. He obtains the estimate 5.067. A second student estimates the integral by noting that $F(x) = -\ln(1 - x)$ is an antiderivative for $f(x)$. She evaluates $F(x)$ between -2 and 0 to obtain the estimate 1.099. Is either computation correct? How do you account for the large discrepancy between the two estimates?

44. To estimate

$$\int_0^{1.5} \frac{1}{1 + x^2}\, dx$$

a student takes the first five nonzero terms of the Taylor series for $f(x) = 1/(1 + x^2)$ at 0 and integrates term by term. He obtains the estimate 3.724. A second student doubts the estimate. She claims that since $1/(1 + x^2) \le 1$ for $0 \le x \le 1.5$, the value of the integral must be less than 1.5. Is either student correct? How do you account for the large discrepancy between their estimates?

There are different ways to approximate a function f by polynomials. If, for example, $f(a), f'(a),$ and $f''(a)$ are known, then we can construct the second-degree Taylor polynomial $p_2(x)$ at a for $f(x)$; $p_2(x)$ and $f(x)$ will have the same value at a and the same first and second derivatives at a. If, on the other hand, $f(x_1), f(x_2),$ and $f(x_3)$ are known, then we can compute the quadratic regression polynomial $q_2(x)$ for the points $(x_1, f(x_1))$, $(x_2, f(x_2)), (x_3, f(x_3))$; $q_2(x)$ and $f(x)$ will have the same values at x_1, x_2, x_3. Problems 45 and 46 concern these contrasting methods of approximation by polynomials.

45. **(A)** Find the second-degree Taylor polynomial $p_2(x)$ at 0 for $f(x) = e^x$, and use a graphing utility to compute the quadratic regression polynomial $q_2(x)$ for the points $(-0.1, e^{-0.1})$, $(0, e^0)$, and $(0.1, e^{0.1})$.

(B) Use graphical approximation techniques to find the maximum error for $-0.1 \le x \le 0.1$ in approximating $f(x) = e^x$ by $p_2(x)$ and by $q_2(x)$.

(C) Which polynomial, $p_2(x)$ or $q_2(x)$, gives the better approximation to

$$\int_{-0.1}^{0.1} e^x \, dx?$$

46. **(A)** Find the fourth-degree Taylor polynomial $p_4(x)$ at 0 for $f(x) = \ln(1 + x)$, and use a graphing utility to compute the quartic regression polynomial $q_4(x)$ for the points $(0, \ln 1)$,

$$\left(-\tfrac{1}{2}, \ln(1 - \tfrac{1}{2})\right), \left(-\tfrac{3}{4}, \ln(1 - \tfrac{3}{4})\right),$$
$$\left(-\tfrac{7}{8}, \ln(1 - \tfrac{7}{8})\right), \left(-\tfrac{15}{16}, \ln(1 - \tfrac{15}{16})\right).$$

(B) Use graphical approximation techniques to find the maximum error for $-\tfrac{15}{16} \le x \le 0$ in approximating $f(x) = \ln(1 + x)$ by $p_4(x)$ and by $q_4(x)$.

(C) Which polynomial, $p_4(x)$ or $q_4(x)$, gives the better approximation to

$$\int_{-15/16}^{0} \ln(1 + x) \, dx?$$

Applications

In Problems 47–58, use Theorem 1 to perform the indicated error estimations.

Business & Economics

47. *Income distribution.* The income distribution for a
www certain country is represented by the Lorenz curve with the equation

$$f(x) = \frac{5x^6}{4 + x^2}$$

Approximate the index of income concentration to within ±0.005.

48. *Income distribution.* Repeat Problem 47 for

$$f(x) = \frac{10x^4}{9 + x^2}$$

49. *Marketing.* A soft drink manufacturer is ready to introduce a new diet soda by a national sales campaign. After test marketing the soda in a carefully selected city, the market research department estimates that sales (in millions of dollars) will increase at the monthly rate of

$$S'(t) = 10 - 10e^{-0.01t^2} \qquad 0 \le t \le 12$$

t months after the national campaign is started. Use the fourth-degree Taylor polynomial at 0 for $S'(t)$ to approximate the total sales during the first 4 months

of the campaign, and estimate the error in this approximation.

50. *Marketing.* Repeat Problem 49 if the monthly rate of increase in sales is given by

$$S'(t) = 10 - 10e^{-0.005t^2} \qquad 0 \le t \le 12$$

51. *Useful life.* A computer store rents time on desktop publishing systems. The total accumulated costs $C(t)$ and revenues $R(t)$ (in thousands of dollars) from a particular system satisfy

$$C'(t) = 4 \qquad \text{and} \qquad R'(t) = \frac{80}{16 + t^2}$$

where t is the time in years that the system has been available for rental. Find the useful life of the system, and approximate the total profit during the useful life to within ±0.005.

52. *Average price.* Given the demand equation

$$p = D(x) = 10 - 20 \ln\left(1 + \frac{x^2}{2,500}\right) \quad 0 \le x \le 40$$

approximate the average price (in dollars) over the demand interval [0, 20] to within ±0.005.

Life Sciences

53. *Temperature.* The temperature (in degrees Celsius) in an artificial habitat is made to change according to the equation

$$C(t) = 20 + 800 \ln\left(1 + \frac{t^2}{100}\right) \quad 0 \le t \le 2$$

Use a Taylor series at 0 to approximate the average temperature over the time interval [0, 2] to within ±0.005.

54. *Temperature.* Repeat Problem 53 for

$$C(t) = 10 + 200 \ln\left(1 + \frac{t^2}{50}\right) \quad 0 \le t \le 2$$

55. *Medicine.* The rate of healing for a skin wound (in
WWW square centimeters per day) is given by

$$A'(t) = \frac{-75}{t^2 + 25}$$

The initial wound has an area of 12 square centimeters. Use the second-degree Taylor polynomial at 0 for $A'(t)$ to approximate the area of the wound after 2 days, and estimate the error in this approximation.

56. *Medicine.* Repeat Problem 55 for

$$A'(t) = \frac{-60}{t^2 + 20}$$

Social Sciences

57. *Learning.* In a particular business college, it was found that an average student enrolled in an advanced typing class progresses at a rate of $N'(t) = 6e^{-0.01t^2}$ words per minute per week, t weeks after enrolling in a 15 week course. At the beginning of the course an average student could type 40 words per minute. Use the second-degree Taylor polynomial at 0 for $N'(t)$ to approximate the improvement in typing after 5 weeks in the course, and estimate the error in this approximation.

58. *Learning.* In the same business college, it was found that an average student in a beginning shorthand class progressed at a rate of $N'(t) = 12e^{-0.005t^2}$ words per minute per week, t weeks after enrolling in a 15 week course. At the beginning of the course none of the students could take any dictation by shorthand. Use the second-degree Taylor polynomial at 0 for $N'(t)$ to approximate the improvement after 5 weeks in the course, and estimate the error in this approximation.

Important Terms and Symbols

10-1 *Taylor Polynomials.* Higher-order derivatives; nth derivative; Taylor polynomial at 0; expanded notation; summation notation; Taylor polynomial at a

$$p_n(x) = \sum_{k=0}^{n} \frac{f^{(k)}(0)}{k!} x^k$$

$$= f(0) + f'(0)x + \frac{f''(0)}{2!} x^2 + \cdots + \frac{f^{(n)}(0)}{n!} x^n$$

$$p_n(x) = \sum_{k=0}^{n} \frac{f^{(k)}(a)}{k!} (x - a)^k$$

$$= f(a) + f'(a)(x - a) + \frac{f''(a)}{2!} (x - a)^2 + \cdots$$
$$+ \frac{f^{(n)}(a)}{n!} (x - a)^n$$

10-2 *Taylor Series.* Taylor series at a; converge; diverge; interval of convergence; using $\lim_{n\to\infty} |a_{n+1}/a_n|$ to find the interval of convergence; assumption on the representation of functions by Taylor series

$$f(x) = \sum_{k=0}^{\infty} \frac{f^{(k)}(a)}{k!} (x - a)^k$$
$$\text{for } x \text{ in the interval of convergence}$$

10-3 *Operations on Taylor Series.* Basic Taylor series; addition; multiplication; differentiation; integration; substitution

10-4 *Approximations Using Taylor Series.* Remainder; error; Taylor's formula for the remainder; alternating series; error estimation; approximating definite integrals

$$R_n(x) = f(x) - p_n(x); \quad R_n(x) = \frac{f^{(n+1)}(t)x^{n+1}}{(n + 1)!}$$

Work through all the problems in this chapter review and check your answers in the back of the book. Answers to all review problems are there along with section numbers in italics to indicate where each type of problem is discussed. Where weaknesses show up, review appropriate sections in the text.

Unless directed otherwise, use Theorem 1 of Section 10-4 in all problems involving error estimation.

A 1. Find $f^{(4)}(x)$ for $f(x) = \ln(x + 5)$.

2. Use the third-degree Taylor polynomial at 0 for $f(x) = \sqrt[3]{1 + x}$ and $x = 0.01$ to approximate $\sqrt[3]{1.01}$.

3. Use the third-degree Taylor polynomial at $a = 3$ for $f(x) = \sqrt{1 + x}$ and $x = 2.9$ to approximate $\sqrt{3.9}$.

4. Use the second-degree Taylor polynomial at 0 for $f(x) = \sqrt{9 + x^2}$ and $x = 0.1$ to approximate $\sqrt{9.01}$.

Use Theorem 1 of Section 10-2 to find the interval of convergence of each Taylor series representation given in Problems 5–8.

5. $\dfrac{1}{1 - 4x} = 1 + 4x + 4^2x^2 + \cdots + 4^nx^n + \cdots$

6. $\dfrac{5}{x - 1} = 1 - \dfrac{1}{5}(x - 6) + \dfrac{1}{5^2}(x - 6)^2 - \cdots$
$$+ \dfrac{(-1)^n}{5^n}(x - 6)^n + \cdots$$

7. $\dfrac{2x}{(1 - x)^3} = 1 \cdot 2x + 2 \cdot 3x^2 + 3 \cdot 4x^3 + \cdots$
$$+ n(n + 1)x^n + \cdots$$

8. $e^{10x} = 1 + 10x + \dfrac{10^2}{2!}x^2 + \cdots + \dfrac{10^n}{n!}x^n + \cdots$

9. Find the nth derivative of $f(x) = e^{-9x}$.

B *In Problems 10 and 11, use the formula $a_n = f^{(n)}(a)/n!$ to find the Taylor series at the indicated value of a. Use Theorem 1 of Section 10-2 to find the interval of convergence.*

10. $f(x) = \dfrac{1}{7 - x}$; at 0   11. $f(x) = \ln x$; at 2

In Problems 12–16, use Table 1 of Section 10-3 and the properties of Taylor series to find the Taylor series of each function at the indicated value of a. Find the interval of convergence.

12. $f(x) = \dfrac{1}{10 + x}$; at 0   13. $f(x) = \dfrac{x^2}{4 - x^2}$; at 0

14. $f(x) = x^2e^{3x}$; at 0   15. $f(x) = x \ln(e + x)$; at 0

16. $f(x) = \dfrac{1}{4 - x}$; at 2

17. (A) Explain why Theorem 1 of Section 10-2 is not directly applicable to the Taylor series representation
$$\ln(1 - 5x^2) = -5x^2 - \dfrac{5^2}{2}x^4 - \dfrac{5^3}{3}x^6 - \cdots - \dfrac{5^n}{n}x^{2n} - \cdots$$
(B) Use another method to find the interval of convergence of the Taylor series in part (A).

18. Substituting $\sqrt{x}$ for x in the Taylor series at 0 for $1/(1 + x)$ gives the formula
$$\dfrac{1}{1 + \sqrt{x}} = 1 - x^{1/2} + x - x^{3/2} + \cdots + (-1)^nx^{n/2} + \cdots$$
(A) For which values of x is the formula valid?
(B) Is the formula a Taylor series at 0 for $1/(1 + \sqrt{x})$? Explain.

In Problems 19 and 20, find the Taylor series at 0 for $f(x)$ and use the relationship $g(x) = f'(x)$ to find the Taylor series at 0 for $g(x)$. Find the interval of convergence for both series.

19. $f(x) = \dfrac{1}{2 - x}$; $g(x) = \dfrac{1}{(2 - x)^2}$

20. $f(x) = \dfrac{x^2}{1 + x^2}$; $g(x) = \dfrac{2x}{(1 + x^2)^2}$

In Problems 21 and 22, find the Taylor series at 0, and find the interval of convergence.

21. $f(x) = \displaystyle\int_0^x \dfrac{t^2}{9 + t^2}\, dt$ 22. $f(x) = \displaystyle\int_0^x \dfrac{t^4}{16 - t^2}\, dt$

23. (A) Compute the Taylor series for $f(x) = x^3 - 3x^2 + 4$ at $a = 1$ and at $a = -1$.
(B) Explain how the two series are related.

24. (A) Explain why $f(x) = |x|$ does not have any Taylor polynomials at 0.
(B) If $a \neq 0$, find the Taylor polynomials for $f(x) = |x|$ at a.

In Problems 25 and 26, use the second-degree Taylor polynomial at 0 for $f(x) = e^x$ to approximate the indicated quantity. Use Taylor's formula for the remainder to estimate the error in the approximation.

25. $e^{0.6}$ 26. $e^{0.06}$

In Problems 27 and 28, use a Taylor polynomial at 0 for $f(x) = \ln(1 + x)$ to estimate the indicated quantity to within ± 0.0005. Give the degree of the Taylor polynomial of lowest degree that will provide this accuracy.

27. $\ln 1.3$

28. $\ln 1.03$

29. If $f(x)$ satisfies $f'(x) = x \ln(1 - x)$ and $f(0) = 5$, find the Taylor series at 0 for $f(x)$.

30. If $f(x)$ satisfies $f''(x) = xe^{-x}$, $f'(0) = -4$, and $f(0) = 3$, find the Taylor series at 0 for $f(x)$.

C *In Problems 31 and 32, approximate the integral to within ± 0.0005.*

31. $\int_0^1 \frac{1}{16 + x^2}\, dx$

32. $\int_0^1 x^2 e^{-0.1x^2}\, dx$

In Problems 33 and 34, assume that $f(x)$ is a function such that $|f^{(n)}(x)| \leq 10$ for all n and all x, and let $p_n(x)$ be the nth-degree Taylor polynomial for f at 0.

33. Use Taylor's formula for the remainder to find the smallest value of n such that the error in the approximation of $f(0.5)$ by $p_n(0.5)$ is guaranteed to be less than 10^{-6}.

34. Use Taylor's formula for the remainder to determine the values of x such that the error in the approximation of $f(x)$ by $p_8(x)$ is guaranteed to be less than 10^{-6}.

35. Let $f(x) = e^{x^2}$ and consider the fourth-degree Taylor polynomial $p_4(x)$ for f at 0. Use graphical approximation techniques to determine those values of x for which the error in the approximation $f(x) \approx p_4(x)$ is less than 0.01.

36. Let $f(x) = \ln(1 - x^2)$ and consider the sixth-degree Taylor polynomial $p_6(x)$ for f at 0. Use graphical approximation techniques to determine those values of x for which the error in the approximation $f(x) \approx p_6(x)$ is less than 0.001.

Applications

Business & Economics

37. *Average price.* Given the demand equation

$$p = D(x) = \tfrac{1}{10}\sqrt{2{,}500 - x^2}$$

use the second-degree Taylor polynomial at 0 to approximate the average price (in dollars) over the demand interval [0, 15].

38. *Production.* The rate of production of an oil well (in thousands of dollars per year) is given by

$$R(t) = 6 + 3e^{-0.01t^2}$$

Use the second-degree Taylor polynomial at 0 to approximate the total production during the first 10 years of operation of the well.

39. *Income distribution.* The income distribution for a certain country is represented by the Lorenz curve with the equation

$$f(x) = \frac{9x^3}{8 + x^2}$$

Approximate the index of income concentration to within ± 0.005.

40. *Marketing.* A cereal manufacturer is ready to introduce a new high-fiber cereal by a national sales campaign. After test-marketing the cereal in a carefully selected city, the market research department estimates that sales (in millions of dollars) will increase at the monthly rate of

$$S'(t) = 20 - 20e^{-0.001t^2} \qquad 0 \leq t \leq 12$$

t months after the national campaign is started. Use the fourth-degree Taylor polynomial at 0 for $S'(t)$ to approximate the total sales during the first 8 months of the campaign, and estimate the error in this approximation.

Life Sciences

41. *Medicine.* The rate of healing for a skin wound (in square centimeters per day) is given by

$$A'(t) = \frac{-100}{t^2 + 40}$$

The initial wound has an area of 15 square centimeters. Use the second-degree Taylor polynomial at 0 for $A'(t)$ to approximate the area of the wound after 2 days, and estimate the error in this approximation.

42. *Medicine.* A large injection of insulin is administered to a patient. The level of insulin in the blood-stream t minutes after the injection is given approximately by

$$L(t) = \frac{5{,}000t^2}{10{,}000 + t^4}$$

Express the average insulin level over the time interval [0, 5] as a definite integral, and use a Taylor series at 0 to approximate this integral to within ± 0.005.

Social Sciences

43. *Politics.* In a newly incorporated city, the number of voters (in thousands) t years after incorporation is given by

$$N(t) = 10 + 2t - 5e^{-0.01t^2} \qquad 0 \leq t \leq 5$$

Express the average number of voters over the time interval [0, 5] as a definite integral, and use a Taylor series at 0 to approximate this integral to within ± 0.05.

Group Activity 1   *L'Hôpital's Rule*

The expressions

$$\lim_{x \to 2} \frac{x^2 - 4}{x - 2} \quad \text{and} \quad \lim_{x \to 1} \frac{e^x - e}{x^2 + 3x - 4}$$

are examples of $0/0$ indeterminate forms. The numerator and denominator both have limit 0, and the quotient property for limits (see Section 3-1) does not apply. But the first limit expression can be evaluated by performing some algebraic simplifications:

$$\lim_{x \to 2} \frac{x^2 - 4}{x - 2} = \lim_{x \to 2} \frac{(x - 2)(x + 2)}{x - 2} = \lim_{x \to 2}(x + 2) = 4$$

To evaluate the second limit expression, we can write the numerator $f(x) = e^x - e$ and the denominator $g(x) = x^2 + 3x - 4$ as Taylor series at $a = 1$:

$$\lim_{x \to 1} \frac{f(x)}{g(x)} = \lim_{x \to 1} \frac{e^x - e}{x^2 + 3x - 4}$$

$$= \lim_{x \to 1} \frac{e(x - 1) + \dfrac{e}{2!}(x - 1)^2 + \dfrac{e}{3!}(x - 1)^3 + \cdots}{5(x - 1) + (x - 1)^2}$$

$$= \lim_{x \to 1} \frac{(x - 1)\left[e + \dfrac{e}{2!}(x - 1) + \dfrac{e}{3!}(x - 1)^2 + \cdots\right]}{(x - 1)[5 + (x - 1)]}$$

$$= \lim_{x \to 1} \frac{\left[e + \dfrac{e}{2!}(x - 1) + \dfrac{e}{3!}(x - 1)^2 + \cdots\right]}{[5 + (x - 1)]} = \frac{e}{5}$$

The value of the limit, $e/5$, is thus the ratio of the coefficients of $x - 1$ in the Taylor series for f and g. But the coefficient of $x - 1$ in the Taylor series for f is $f'(1)$, and the coefficient of $x - 1$ in the Taylor series for g is $g'(1)$, so the limit is $f'(1)/g'(1)$.

Our reasoning suggests a simple method for computing each of the above limits. We compute the limit of the derivative of the numerator divided by the derivative of the denominator (without calculating Taylor series) as follows:

$$\lim_{x \to 2} \frac{x^2 - 4}{x - 2} = \lim_{x \to 2} \frac{\dfrac{d}{dx}(x^2 - 4)}{\dfrac{d}{dx}(x - 2)} = \lim_{x \to 2} \frac{2x}{1} = 4$$

$$\lim_{x \to 1} \frac{e^x - e}{x^2 + 3x - 4} = \lim_{x \to 1} \frac{\dfrac{d}{dx}(e^x - e)}{\dfrac{d}{dx}(x^2 + 3x - 4)} = \lim_{x \to 1} \frac{e^x}{2x + 3} = \frac{e}{5}$$

Note that the derivative of the numerator and the derivative of the denominator are computed separately (we are not using the quotient rule!).

The simple computations above are illustrations of a powerful technique for evaluating limits of quotients called *L'Hôpital's rule*. The rule is named after the French mathematician Marquis de L'Hôpital (1661–1704), who is generally credited with writing the first calculus textbook. (In Appendix B-5, L'Hôpital's rule is discussed without reference to Taylor series.)

L'Hôpital's Rule for $0/0$ Indeterminate Forms

For c a real number, if $\lim_{x \to c} f(x) = 0$ and $\lim_{x \to c} g(x) = 0$, then

$$\lim_{x \to c} \frac{f(x)}{g(x)} = \lim_{x \to c} \frac{f'(x)}{g'(x)}$$

provided the second limit exists or is $+\infty$ or $-\infty$. The rule remains valid if the symbol $x \to c$ is replaced everywhere it occurs with one of the following symbols:

$$x \to c^+ \qquad x \to c^- \qquad x \to \infty \qquad x \to -\infty$$

---

1. Find each limit in two ways: by using algebraic simplification and by using L'Hôpital's rule.

   (A) $\lim\limits_{x \to 2} \dfrac{x^4 - 16}{x^3 - 8}$

   (B) $\lim\limits_{x \to 1} \dfrac{x^6 - 1}{x^5 - 1}$

   (C) $\lim\limits_{x \to 2} \dfrac{x^2 + x - 6}{x^2 + 6x - 16}$

   (D) $\lim\limits_{x \to 4} \dfrac{x^2 - 8x + 16}{x^2 - 5x + 4}$

2. Find each limit in two ways: by using Taylor series expansions and by using L'Hôpital's rule.

   (A) $\lim\limits_{x \to 0} \dfrac{e^x - 1}{x}$

   (B) $\lim\limits_{x \to 1} \dfrac{\ln x}{x - 1}$

   (C) $\lim\limits_{x \to 0} \dfrac{\ln(1 + 4x)}{x}$

   (D) $\lim\limits_{x \to 0} \dfrac{e^{2x} - 1}{x}$

Group Activity 2   **Taylor Series Solutions to Differential Equations**

Since Taylor series at 0 can be differentiated term by term, they can be used to solve differential equations. The idea is simple: Substitute

$$f(x) = a_0 + a_1 x + a_2 x^2 + \cdots + a_n x^n + \cdots \tag{1}$$

in the differential equation and solve for the coefficients $a_0, a_1, a_2, \ldots$. The following examples will illustrate the technique.

Example 1 ⟴ **Taylor Series Solution**   Find the particular solution of $y' = y$ that satisfies $y(0) = 1$.

SOLUTION   Substituting $f(x)$ from (1) for y in the differential equation and $f'(x)$ for y', gives

$$a_1 + 2a_2x + 3a_3x^2 + 4a_4x^3 + \cdots = a_0 + a_1x + a_2x^2 + a_3x^3 + \cdots$$

By equating the constant terms, then the coefficients of x, then the coefficients of x^2, and so on, we obtain the following sequence of equations:

$$a_1 = a_0$$
$$2a_2 = a_1$$
$$3a_3 = a_2$$
$$4a_4 = a_3$$
$$\vdots$$

Since $f(0) = a_0$, the initial condition $y(0) = 1$ implies that $a_0 = 1$. Now the sequence of equations can be used to solve successively for a_1, a_2, a_3, a_4:

$$a_1 = 1, \qquad a_2 = \tfrac{1}{2}, \qquad a_3 = \tfrac{1}{6}, \qquad a_4 = \tfrac{1}{24}$$

Therefore,

$$f(x) = 1 + x + \tfrac{1}{2}x^2 + \tfrac{1}{6}x^3 + \tfrac{1}{24}x^4 + \cdots$$

■

(A) Identify a well-known function whose Taylor series at 0 agrees with these first five terms of $f(x)$, and show that your function satisfies the differential equation and initial condition of Example 1.

In general, we will not recognize the Taylor series solution to a differential equation as a known function, but we will be able to compute as many terms of the Taylor series solution as desired.

Example 2 ⇨   **Taylor Series Solution**   Find the first six nonzero terms of the Taylor series solution to $y'' + xy' = 3x^2 - 2y$, $y(0) = 5$, $y'(0) = -3$.

SOLUTION   Substituting $f(x)$ from (1) for y, $f'(x)$ for y', and $f''(x)$ for y'', gives

$$(2a_2 + 6a_3x + 12a_4x^2 + 20a_5x^3 + \cdots) + x(a_1 + 2a_2x + 3a_3x^2 + 4a_4x^3 + \cdots) =$$
$$3x^2 - 2(a_0 + a_1x + a_2x^2 + a_3x^3 + \cdots)$$

By equating the constant terms, then the coefficients of x, then the coefficients of x^2, and so on, we obtain the following sequence of equations:

$$2a_2 = -2a_0$$
$$6a_3 + a_1 = -2a_1$$
$$12a_4 + 2a_2 = 3 - 2a_2$$
$$20a_5 + 3a_3 = -2a_3$$

Since $f(0) = a_0$, the initial condition $y(0) = 5$ implies that $a_0 = 5$. Since $f'(0) = a_1$, the initial condition $y'(0) = -3$ implies that $a_1 = -3$. Now the sequence of equations can be used to solve successively for a_2, a_3, a_4, and a_5:

$$a_2 = -5, \qquad a_3 = \tfrac{3}{2}, \qquad a_4 = \tfrac{23}{12}, \qquad a_5 = \tfrac{-3}{8}$$

Therefore,

$$f(x) = 5 - 3x - 5x^2 + \tfrac{3}{2}x^3 + \tfrac{23}{12}x^4 - \tfrac{3}{8}x^5 + \cdots$$

■

6

−1   1

0

FIGURE 1

The first-, second-, third-, fourth-, and fifth-degree Taylor polynomials for $f(x)$ in Example 2 are graphed in Figure 1. Note from the graphs that each polynomial satisfies the initial conditions of the differential equation.

(B) Find the coefficients of x^6 and x^7 in the Taylor series solution in Example 2.

(C) Find the first four nonzero terms of the Taylor series solution to $y' + 4xy = 6x$, $y(0) = 1$.

(D) Find the first four nonzero terms of the Taylor series solution to $y'' = -y$, $y(0) = 1$, $y'(0) = 0$.

(E) It is known that the function

$$g(x) = \tfrac{3}{2} - \tfrac{1}{2}e^{-2x^2}$$

is a solution to the first-order linear differential equation and initial condition of part (C). Use a graphing utility to graph $g(x)$ and the sum of the first four nonzero terms of the Taylor series of part (C). Discuss how the graphs differ.

(F) It is known that the function $h(x) = \cos x$ is a solution to the differential equation and initial condition of part (D). Use a graphing utility to graph $h(x)$ and the sum of the first four nonzero terms of the Taylor series of part (D). Discuss how the graphs differ.

Probability and Calculus

11 CHAPTER

INTRODUCTION

In Section 7-2 we briefly touched on applications of calculus to probability. In this chapter, we discuss this topic in much more detail. In the first section we introduce improper integrals, a new integral form that is fundamental to the work that follows. After discussing basic concepts in the next two sections, we consider several commonly used probability distributions, culminating with the very important normal distribution.

Section 11-1 — Improper Integrals

- ❑ IMPROPER INTEGRALS
- ❑ APPLICATION: CAPITAL VALUE

❑ IMPROPER INTEGRALS

We are now going to consider an integral form that has wide application in probability studies as well as other areas. Earlier, when we introduced the idea of a definite integral,

$$\int_a^b f(x)\, dx \tag{1}$$

we required f to be continuous over a closed interval $[a, b]$. Now we are going to extend the meaning of (1) so that the interval $[a, b]$ may become infinite in length.

Explore–Discuss 1

For each of the following functions, find the area under the graph from $x = 1$ to $x = b, b > 1$ (see Fig. 1). Discuss the behavior of this area as b assumes larger and larger values.

(A) $f(x) = \dfrac{1}{x^{5/4}} = x^{-5/4}$ (B) $g(x) = \dfrac{1}{x^{4/5}} = x^{-4/5}$

FIGURE 1

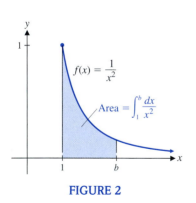

FIGURE 2

Let us investigate a particular example that will motivate several general definitions. What would be a reasonable interpretation for the following expression?

$$\int_1^\infty \frac{dx}{x^2}$$

Sketching a graph of $f(x) = 1/x^2, x \geq 1$ (see Fig. 2), we note that for any fixed $b > 1$, $\int_1^b f(x)\, dx$ is the area between the graph of $y = 1/x^2$, the x axis, $x = 1$, and $x = b$.

Let us see what happens when we let $b \to \infty$; that is, when we compute the following limit:

$$\lim_{b \to \infty} \int_1^b \frac{dx}{x^2} = \lim_{b \to \infty} \left[(-x^{-1}) \Big|_1^b \right]$$

$$= \lim_{b \to \infty} \left(-\frac{1}{b} + 1 \right) = 1$$

Did you expect this result? No matter how large b is taken, the area under the graph from $x = 1$ to $x = b$ never exceeds 1, and in the limit it is 1 (see Fig. 3). This suggests that we write

$$\int_1^\infty \frac{dx}{x^2} = \lim_{b \to \infty} \int_1^b \frac{dx}{x^2} = 1$$

This integral is an example of an *improper integral*. In general, the forms

$$\int_{-\infty}^b f(x)\, dx \qquad \int_a^\infty f(x)\, dx \qquad \int_{-\infty}^\infty f(x)\, dx$$

where f is continuous over the indicated interval, are called **improper integrals.** (These integrals are "improper" because the interval of integration is unbounded, as indicated by the use of ∞ for one or both limits of integration. There are other types of improper integrals involving certain types of

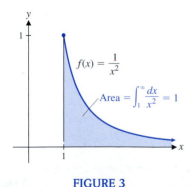

FIGURE 3

points of discontinuity within the interval of integration, but these will not be considered here.) Each type of improper integral above is formally defined in the following box:

Improper Integrals

If f is continuous over the indicated interval and the limit exists, then

1. $\displaystyle \int_a^\infty f(x)\,dx = \lim_{b \to \infty} \int_a^b f(x)\,dx$

2. $\displaystyle \int_{-\infty}^b f(x)\,dx = \lim_{a \to -\infty} \int_a^b f(x)\,dx$

3. $\displaystyle \int_{-\infty}^\infty f(x)\,dx = \int_{-\infty}^c f(x)\,dx + \int_c^\infty f(x)\,dx$

where c is any point on $(-\infty, \infty)$, provided *both* improper integrals on the right exist.

If the indicated limit exists, then the improper integral is said to exist or to **converge;** if the limit does not exist, then the improper integral is said not to exist or to **diverge** (and no value is assigned to it).

Example 1 ➤ **Evaluating an Improper Integral** Evaluate the following, if it converges:

$$\int_2^\infty \frac{dx}{x}$$

SOLUTION $\displaystyle \int_2^\infty \frac{dx}{x} = \lim_{b \to \infty} \int_2^b \frac{dx}{x}$

$$= \lim_{b \to \infty}\left[(\ln x)\Big|_2^b \right]$$

$$= \lim_{b \to \infty}(\ln b - \ln 2)$$

Since $\ln b \to \infty$ as $b \to \infty$, the limit does not exist. Hence, the improper integral diverges. ∎

Matched Problem 1 ➤ Evaluate the following, if it converges: $\displaystyle \int_3^\infty \frac{dx}{(x-1)^2}$. ∎

Example 2 ➤ **Evaluating an Improper Integral** Evaluate the following, if it converges:

$$\int_{-\infty}^2 e^x\,dx$$

SOLUTION $\displaystyle \int_{-\infty}^2 e^x\,dx = \lim_{a \to -\infty} \int_a^2 e^x\,dx$

$$= \lim_{a \to -\infty}\left(e^x\Big|_a^2 \right)$$

$$= \lim_{a \to -\infty}(e^2 - e^a) = e^2 - 0 = e^2 \qquad \text{The integral converges.}$$

∎

Matched Problem 2 ➭ Evaluate the following, if it converges: $\int_{-\infty}^{-1} x^{-2}\, dx$.

■

Example 3 ➭ **Evaluating an Improper Integral** Evaluate the following, if it converges:

$$\int_{-\infty}^{\infty} \frac{2x}{(1 + x^2)^2}\, dx$$

SOLUTION

$$\int_{-\infty}^{\infty} \frac{2x}{(1 + x^2)^2}\, dx = \int_{-\infty}^{0} (1 + x^2)^{-2} 2x\, dx + \int_{0}^{\infty} (1 + x^2)^{-2} 2x\, dx$$

$$= \lim_{a \to -\infty} \int_{a}^{0} (1 + x^2)^{-2} 2x\, dx$$

$$+ \lim_{b \to \infty} \int_{0}^{b} (1 + x^2)^{-2} 2x\, dx$$

$$= \lim_{a \to -\infty} \left[\frac{(1 + x^2)^{-1}}{-1} \Big|_{a}^{0} \right] + \lim_{b \to \infty} \left[\frac{(1 + x^2)^{-1}}{-1} \Big|_{0}^{b} \right]$$

$$= \lim_{a \to -\infty} \left[-1 + \frac{1}{1 + a^2} \right] + \lim_{b \to \infty} \left[-\frac{1}{1 + b^2} + 1 \right]$$

$$= -1 + 1 = 0 \qquad \text{\textit{The integral converges.}}$$

■

Matched Problem 3 ➭ Evaluate the following, if it converges: $\int_{-\infty}^{\infty} \frac{dx}{e^x}$.

■

Explore–Discuss 2

Let

$$f(x) = \begin{cases} 2 - x & \text{if } 0 \leq x \leq 1 \\ 0 & \text{otherwise} \end{cases}$$

(A) Graph $y = f(x)$ and explain why f is not continuous at $x = 0$ and at $x = 1$.

(B) Find $\int_{0}^{1} f(x)\, dx$.

(C) Discuss possible interpretations of the improper integral $\int_{-\infty}^{\infty} f(x)\, dx$.

In probability studies, we frequently encounter functions that are 0 over one or more intervals on the real axis. An improper integral involving such a function can be simplified by deleting the intervals where the function is 0 and considering the integral over the remaining intervals. The next example illustrates this process.

Example 4 ➭ **Evaluating an Improper Integral** Evaluate $\int_{-\infty}^{\infty} f(x)\, dx$ for

$$f(x) = \begin{cases} \dfrac{4}{(x + 2)^2} & \text{if } x \geq 0 \\ 0 & \text{otherwise} \end{cases}$$

SOLUTION   The function f is discontinuous at $x = 0$ (see Fig. 4). However, since $f(x) = 0$ for $x < 0$, we can proceed as follows:

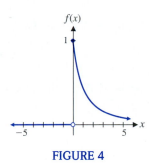

FIGURE 4

$$\int_{-\infty}^{\infty} f(x)\,dx = \int_{0}^{\infty} f(x)\,dx \qquad f(x) = 0 \text{ for } x < 0.$$

$$= \int_{0}^{\infty} \frac{4}{(x + 2)^2}\,dx$$

$$= \lim_{b \to \infty} \int_{0}^{b} \frac{4}{(x + 2)^2}\,dx$$

$$= \lim_{b \to \infty} \left[-4(x + 2)^{-1} \Big|_{0}^{b} \right]$$

$$= \lim_{b \to \infty} \left(\frac{-4}{b + 2} + 2 \right)$$

$$= 2 \qquad\qquad \textit{The integral converges.}$$

Matched Problem 4   Evaluate $\displaystyle\int_{-\infty}^{\infty} f(x)\,dx$ for $f(x) = \begin{cases} 4x - x^2 & \text{if } 0 \le x \le 4 \\ 0 & \text{otherwise.} \end{cases}$

Example 5   **Oil Production**   It is estimated that an oil well will produce oil at a rate of $R(t)$ million barrels per year t years from now, as given by

$$R(t) = 8e^{-0.05t} - 8e^{-0.1t}$$

Estimate the total amount of oil that will be produced by this well.

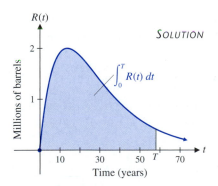

FIGURE 5

SOLUTION   The total amount of oil produced in T years of operation is $\int_{0}^{T} R(t)\,dt$ (Fig. 5). At some point in the future, the annual production rate will become so low that it will no longer be economically feasible to operate the well. However, since we do not know when this will occur, it is convenient to assume that the well is operated indefinitely so we can use an improper integral. Thus, the total amount of oil produced is approximately

$$\int_{0}^{\infty} R(t)\,dt = \lim_{T \to \infty} \int_{0}^{T} R(t)\,dt$$

$$= \lim_{T \to \infty} \int_{0}^{T} (8e^{-0.05t} - 8e^{-0.1t})\,dt$$

$$= \lim_{T \to \infty} \left[(-160e^{-0.05t} + 80e^{-0.1t}) \Big|_{0}^{T} \right]$$

$$= \lim_{T \to \infty} (-160e^{-0.05T} + 80e^{-0.1T} + 80)$$

$$= 80 \quad \text{or} \quad 80,000,000 \text{ barrels}$$

X	Y1
50	67.405
60	72.232
70	75.241
80	77.096
90	78.232
100	78.926
110	79.347

Y1■-160e^(-.05X...

FIGURE 6 Total production over $[0, x]$

The total production over any finite time period $[0, T]$ is less than 80,000,000 barrels, and the larger T is, the closer the total production will be to 80,000,000 barrels. Figure 6 displays the total production over $[0, T]$ for values of $T(= X)$ from 50 to 110 years.

Matched Problem 5   The annual production rate (in millions of barrels) for an oil well is given by

$$R(t) = 4e^{-0.1t} - e^{-0.2t}$$

Assuming that the well is operated indefinitely, find the total production.   ■

❑   APPLICATION: CAPITAL VALUE

Recall that if money is invested at a rate r compounded continuously for T years, then the present value, PV, and the future value, FV, are related by the continuous compound interest formula (see Section 5-1)

$$FV = PVe^{rT}$$

which also can be written as

$$PV = e^{-rT}FV$$

This relationship is valid for both single deposits and amounts generated by continuous income streams. Thus, if $f(t)$ is a continuous income stream, the future value is given by (see Section 7-2)

$$FV = e^{rT} \int_0^T f(t)e^{-rt}\, dt \qquad \text{Future value}$$

and, using the compound interest formula, the present value is given by

$$PV = e^{-rT}FV = e^{-rT}e^{rT} \int_0^T f(t)e^{-rt}\, dt$$

Since $e^{-rT}e^{rT} = 1$, we have

$$PV = \int_0^T f(t)e^{-rt}\, dt \qquad \text{Present value}$$

If we let T approach ∞ in this formula for present value, we obtain an improper integral that represents the *capital value* of the income stream.

> ### Capital Value of a Perpetual Income Stream
>
> A continuous income stream is called **perpetual** if it never stops producing income. The **capital value, CV,** of a perpetual income stream $f(t)$ at a rate r compounded continuously is the present value over the time interval $[0, \infty)$. That is,
>
> $$CV = \int_0^\infty f(t)e^{-rt}\, dt$$

Capital value provides a method for expressing the worth (in terms of today's dollars) of an investment that will produce income for an indefinite period of time.

Example 6   **Capital Value**   A family has leased the oil rights of a property to a petroleum company in return for a perpetual annual payment of $1,200. Find the capital value of this lease at 5% compounded continuously.

SOLUTION   The annual payments from the oil company produce a continuous income stream with rate of flow $f(t) = 1,200$ that continues indefinitely. (It is common practice to treat a sequence of equal periodic payments as a continuous income stream with a constant rate of flow, even if the income is received only at the end of each period.) Thus, the capital value is

$$CV = \int_0^\infty f(t)e^{-rt}\, dt$$

$$CV = \int_0^\infty 1{,}200e^{-0.05t}\, dt$$

$$= \lim_{T \to \infty} \int_0^T 1{,}200e^{-0.05t}\, dt$$

$$= \lim_{T \to \infty} \left[-24{,}000e^{-0.05t} \Big|_0^T \right]$$

$$= \lim_{T \to \infty} (-24{,}000e^{-0.05T} + 24{,}000) = \$24{,}000$$

Matched Problem 6   Repeat Example 6 if the interest rate is 6% compounded continuously.

Answers to Matched Problems   **1.** $\frac{1}{2}$ **2.** 1   **3.** Diverges   **4.** $\frac{32}{3}$ **5.** 35,000,000 barrels   **6.** \$20,000

Exercise 11-1

A *In Problems 1–16, find the value of each improper integral that converges.*

1. $\displaystyle\int_1^\infty \frac{dx}{x^4}$

2. $\displaystyle\int_1^\infty \frac{dx}{x^3}$

3. $\displaystyle\int_0^\infty 10^{-6}x\, dx$

4. $\displaystyle\int_1^\infty \frac{10^8}{x^2}\, dx$

5. $\displaystyle\int_0^\infty e^{-x/2}\, dx$

6. $\displaystyle\int_0^\infty e^{-x}\, dx$

B   **7.** $\displaystyle\int_1^\infty \frac{dx}{\sqrt{x}}$

8. $\displaystyle\int_1^\infty \frac{dx}{\sqrt[3]{x}}$

9. $\displaystyle\int_0^\infty \frac{dx}{(x+1)^2}$

10. $\displaystyle\int_0^\infty \frac{dx}{(x+1)^3}$

11. $\displaystyle\int_0^\infty \frac{dx}{(x+1)^{2/3}}$

12. $\displaystyle\int_0^\infty \frac{dx}{\sqrt{x+1}}$

13. $\displaystyle\int_1^\infty \frac{dx}{x^{0.99}}$

14. $\displaystyle\int_1^\infty \frac{dx}{x^{1.01}}$

15. $0.3 \displaystyle\int_0^\infty e^{-0.3x}\, dx$

16. $0.01 \displaystyle\int_0^\infty e^{-0.1x}\, dx$

In Problems 17–22, graph $y = f(x)$ *and find the value of* $\int_{-\infty}^\infty f(x)\, dx$ *if it converges.*

17. $f(x) = \begin{cases} 1 + x^2 & \text{if } 0 \le x \le 2 \\ 0 & \text{otherwise} \end{cases}$

18. $f(x) = \begin{cases} 2 + x & \text{if } 0 \le x \le 3 \\ 0 & \text{otherwise} \end{cases}$

19. $f(x) = \begin{cases} e^{-0.1x} & \text{if } x \ge 0 \\ 0 & \text{otherwise} \end{cases}$

20. $f(x) = \begin{cases} 2 - e^{-0.2x} & \text{if } x \ge 0 \\ 0 & \text{otherwise} \end{cases}$

21. $f(x) = \begin{cases} 4/(x+2) & \text{if } x \ge 2 \\ 0 & \text{otherwise} \end{cases}$

22. $f(x) = \begin{cases} 4/(x+1)^2 & \text{if } x \ge 1 \\ 0 & \text{otherwise} \end{cases}$

In Problems 23–26, discuss the validity of each statement. If the statement is always true, explain why. If not, give a counterexample.

23. If f is a continuous increasing function on $[0, \infty)$ such that $f(x) > 0$ for all x, then $\int_0^\infty f(x)\, dx$ diverges.

24. If f is a continuous decreasing function on $[0, \infty)$ such that $f(x) > 0$ for all x, then $\int_0^\infty f(x)\, dx$ converges.

25. If f is a continuous function on $[1, \infty)$ such that $f(x) > 0$ for all x and $f(x) \ge 10$ for $1 \le x \le 1{,}000$, then $\int_1^\infty f(x)\, dx$ diverges.

26. If f is a continuous decreasing function on $[1, \infty)$ such that $0 < f(x) \le 0.001$ for all x, then $\int_1^\infty f(x)\, dx$ converges.

In Problems 27–36, find F(b), use a graphing utility to graph F(b), and use the graph to estimate $\lim_{b \to \infty} F(b)$.

27. From Problem 1, $F(b) = \int_1^b \dfrac{dx}{x^4}$

28. From Problem 2, $F(b) = \int_1^b \dfrac{dx}{x^3}$

29. From Problem 3, $F(b) = \int_0^b 10^{-6}x \, dx$

30. From Problem 4, $F(b) = \int_1^b \dfrac{10^8}{x^2} \, dx$

31. From Problem 5, $F(b) = \int_0^b e^{-x/2} \, dx$

32. From Problem 6, $F(b) = \int_0^b e^{-x} \, dx$

33. From Problem 7, $F(b) = \int_1^b \dfrac{dx}{\sqrt{x}}$

34. From Problem 8, $F(b) = \int_1^b \dfrac{dx}{\sqrt[3]{x}}$

35. From Problem 9, $F(b) = \int_0^b \dfrac{dx}{(x+1)^2}$

36. From Problem 10, $F(b) = \int_0^b \dfrac{dx}{(x+1)^3}$

37. If f is continuous on $[0, \infty)$ and $\int_0^\infty f(x) \, dx$ converges, does $\int_1^\infty f(x) \, dx$ converge? Explain.

38. If f is continuous on $[0, \infty)$ and $\int_0^\infty f(x) \, dx$ diverges, does $\int_1^\infty f(x) \, dx$ diverge? Explain.

C *In Problems 39–48, find the value of each improper integral that converges.*

39. $\int_0^\infty \dfrac{1}{k} e^{-x/k} \, dx, k > 0$ **40.** $\int_1^\infty \dfrac{1}{x^p} \, dx, p > 1$

41. $\int_{-\infty}^\infty \dfrac{x}{1+x^2} \, dx$ **42.** $\int_{-\infty}^\infty xe^{-x^2} \, dx$

43. $\int_0^\infty (e^{-x} - e^{-2x}) \, dx$ **44.** $\int_0^\infty (e^{-x} + 2e^x) \, dx$

45. $\int_{-\infty}^0 \dfrac{1}{\sqrt{1-x}} \, dx$ **46.** $\int_{-\infty}^0 \dfrac{1}{\sqrt[3]{(1-x)^4}} \, dx$

47. $\int_1^\infty \dfrac{\ln x}{x} \, dx$ **48.** $\int_0^\infty \dfrac{e^{-x}}{1+e^{-x}} \, dx$

Applications

Business & Economics

49. *Capital value.* The perpetual annual rent for a property is $6,000. Find the capital value at 5% compounded continuously.

50. *Capital value.* The perpetual annual rent for a property is $10,000. Find the capital value at 4% compounded continuously.

51. *Capital value.* A trust fund produces a perpetual stream of income with rate of flow

$$f(t) = 1,500e^{0.04t}$$

Find the capital value at 7% compounded continuously.

52. *Capital value.* A trust fund produces a perpetual stream of income with rate of flow

$$f(t) = 1,000e^{0.02t}$$

Find the capital value at 6% compounded continuously.

53. *Capital value.* Refer to Problem 49. Discuss the effect on the capital value if the interest rate is increased to 6%. Decreased to 4%. Interpret these results.

54. *Capital value.* Refer to Problem 50. Discuss the effect on the capital value if the interest rate is increased to 5%. Decreased to 2%. Interpret these results.

55. *Production.* The rate of production of a natural gas well (in billions of cubic feet per year) is given by (see the figure)

$$R(t) = 3e^{-0.2t} - 3e^{-0.4t}$$

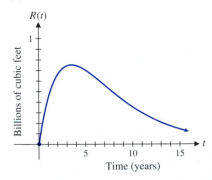

(A) Assuming that the well is operated indefinitely, find the total production.

(B) When will the output from the well reach 50% of the total production? Round answer to two decimal places.

56. *Production.* The rate of production of an oil well (in millions of barrels per year) is given by

$$R(t) = \frac{1,000t}{(50 + t^2)^2}$$

(A) Assuming that the well is operated indefinitely, find the total production.

(B) When will the output from the well reach 50% of the total production? Round answer to two decimal places.

Life Sciences

57. *Pollution.* It has been estimated that the rate of
www seepage of toxic chemicals from a waste dump is $R(t)$ gallons per year t years from now, where

$$R(t) = \frac{500}{(1 + t)^2}$$

Assuming that this seepage continues indefinitely, find the total amount of toxic chemicals that seep from the dump.

58. *Drug assimilation.* When a person takes a drug, the body does not assimilate all of the drug. One way to determine the amount of the drug assimilated is to measure the rate at which the drug is eliminated from the body. If the rate of elimination of the drug (in milliliters per minute) is given by

$$R(t) = 3e^{-0.1t} - 3e^{-0.3t}$$

where t is the time in minutes since the drug was administered, how much of the drug is eliminated from the body?

Social Sciences

59. *Immigration.* In order to control rising population,
www a country is planning to institute a new policy for immigration. Analysts estimate that the rate of immigration into the country t years after the policy is instituted (in millions of immigrants per year) will be given by

$$R(t) = \frac{400}{(5 + t)^3}$$

Find the total number of immigrants that will enter the country under this policy, assuming that the policy is applied indefinitely.

Continuous Random Variables

- CONTINUOUS RANDOM VARIABLES
- PROBABILITY DENSITY FUNCTIONS
- CUMULATIVE DISTRIBUTION FUNCTIONS

CONTINUOUS RANDOM VARIABLES

Probability theory is concerned with determining the long-run frequency of the outcomes of an experiment when the outcome each time the experiment is performed is uncertain. For example, when a fair coin is tossed, it is uncertain whether it will turn up heads or tails. However, probability theory enables us to predict that the coin will turn up heads approximately 50% of the time.

Assigning a value to each outcome of an experiment defines a function called a *random variable.* For example, consider the experiment of tossing 3 coins and the function X whose values are the number of heads. The possible outcomes of the experiment and the corresponding values of X are shown in Table 1. Since the set of possible outcomes for this experiment is a finite set, X is called a **discrete random variable.**

Things are more complicated if an experiment has an infinite number of outcomes. For example, consider the experiment of turning on light bulbs and letting them remain lit until they burn out. One bulb may be defective and not light at all. Another may still be lit 50 or 100 years from now. Let X be the function whose values (in hours) are the life expectancies of the light bulbs. Theoretically, there is no reason to exclude any of the values in the interval $[0, \infty)$ as a possibility for the life expectancy of the bulb. Thus,

TABLE 1

NUMBER OF HEADS IN THE TOSS OF 3 COINS

SAMPLE SPACE S	NUMBER OF HEADS $X(e_j)$
e_1: TTT	0
e_2: TTH	1
e_3: THT	1
e_4: HTT	1
e_5: THH	2
e_6: HTH	2
e_7: HHT	2
e_8: HHH	3

we assume that the range of X is $[0, \infty)$, and we call X a *continuous random variable.*

Continuous Random Variable

A **continuous random variable** X is a function that assigns a numeric value to each outcome of an experiment. The set of possible values of X is an interval of real numbers. This interval may be open or closed, and it may be bounded or unbounded.

The term *continuous* is not used in the same sense here as it was used in Section 3-2. In the present case, it refers to the fact that the values of the random variable form a continuous set of real numbers, such as $[0, \infty)$, rather than a discrete set, such as $\{0, 1, 2, 3\}$ or $\{2, 4, 8, \dots\}$.

Explore–Discuss 1

For each of the following experiments, determine whether the indicated random variable is discrete or continuous.

(A) Two dice are rolled, and X is the sum of the dots on the upturned faces of the dice.

(B) A dart is thrown at a circular dart board, and X is the distance from the dart to the center of the board.

(C) The drive-up window of a bank is open 10 hours a day, and X is the number of customers served in one day.

(D) The drive-up window of a bank is open 10 hours a day, and X is the time (in minutes) each customer must wait for service.

❑ PROBABILITY DENSITY FUNCTIONS

If X is a discrete random variable, then the probability that X lies in a given interval can be computed by addition. For example, consider the experiment of tossing 3 coins, and let X represent the number of heads. The *probability distribution* for X is shown in Table 2 and illustrated in the histogram in Figure 1. We denote the probability that X is between 0 and 2, inclusively, by $P(0 \le X \le 2)$ and compute it by addition:

$$P(0 \le X \le 2) = P(X = 0) + P(X = 1) + P(X = 2)$$
$$= \tfrac{1}{8} + \tfrac{3}{8} + \tfrac{3}{8} = \tfrac{7}{8}$$

If X is a continuous random variable, the same approach will not work. Since X can now assume any real number in the interval $[0, 2]$, it is impossible to write $P(0 \le X \le 2)$ as a finite or even an infinite sum. (What would be the second term in such a sum? That is, what is the "next" real number after 0? Think about this.) Instead, we introduce a new type of function, called a *probability density function,* and use integrals involving this function to compute the probability that a continuous random variable X lies in a given interval. For example, we might use a probability density function to find the probability that the actual amount of soda in a 12 ounce can is between 11.9 and 12.1 ounces, or that the speed of a car involved in an accident was between 60 and 65 miles per hour.

For convenience in stating definitions and formulas, we will assume that the value of a continuous random variable can be any real number; that is, the range is $(-\infty, \infty)$.

TABLE 2

PROBABILITY DISTRIBUTION

NUMBER OF HEADS X	0	1	2	3
PROBABILITY $p(x)$	$\tfrac{1}{8}$	$\tfrac{3}{8}$	$\tfrac{3}{8}$	$\tfrac{1}{8}$

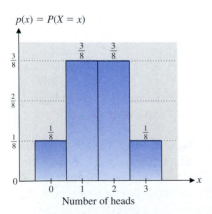

FIGURE 1 Histogram for a probability distribution

Probability Density Function

The function $f(x)$ is a **probability density function** for a continuous random variable X if

1. $f(x) \geqslant 0$ for all $x \in (-\infty, \infty)$

2. $\displaystyle\int_{-\infty}^{\infty} f(x)\, dx = 1$

3. The probability that X lies in the interval $[c, d]$ is given by

$$P(c \leqslant X \leqslant d) = \int_{c}^{d} f(x)\, dx$$

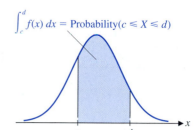

Range of $X = (-\infty, \infty) = $ Domain of f

Example 1 ⇨   **Using a Probability Density Function**   Let

$$f(x) = \begin{cases} 12x^2 - 12x^3 & \text{if } 0 \leqslant x \leqslant 1 \\ 0 & \text{otherwise} \end{cases}$$

Graph f and verify that f satisfies the first two conditions for a probability density function. Then find each of the following probabilities and illustrate graphically:

(A) $P(.4 \leqslant X \leqslant .7)$ (B) $P(X \leqslant .5)$
(C) $P(X \geqslant .6)$ (D) $P(X = .3)$

SOLUTION   The graph of f (Fig. 2) shows that $f(x) \geqslant 0$ for all x. Also,

$$\int_{-\infty}^{\infty} f(x)\, dx = \int_{0}^{1} (12x^2 - 12x^3)\, dx = (4x^3 - 3x^4)\Big|_{0}^{1} = 1$$

(A) $\displaystyle P(.4 \leqslant X \leqslant .7) = \int_{.4}^{.7} f(x)\, dx$

$$= \int_{.4}^{.7} (12x^2 - 12x^3)\, dx = (4x^3 - 3x^4)\Big|_{.4}^{.7} = .4725$$

FIGURE 2

(B) $P(X \le .5) = \int_0^{.5} f(x)\,dx$ Note that $f(x) = 0$ for $x < 0$.

$$= \int_0^{.5} (12x^2 - 12x^3)\,dx = (4x^3 - 3x^4)\Big|_0^{.5} = .3125$$

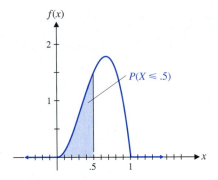

(C) $P(X \ge .6) = \int_{.6}^1 f(x)\,dx$ Note that $f(x) = 0$ for $x > 1$.

$$= \int_{.6}^1 (12x^2 - 12x^3)\,dx = (4x^3 - 3x^4)\Big|_{.6}^1 = .5248$$

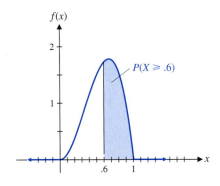

(D) $P(X = .3) = \int_{.3}^{.3} f(x)\,dx = 0$ Definite integral property 1 (Section 6-5)

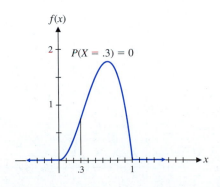

Matched Problem 1 ⇔ Let

$$f(x) = \begin{cases} 6x - 6x^2 & \text{if } 0 \leqslant x \leqslant 1 \\ 0 & \text{otherwise} \end{cases}$$

Graph f and verify that f satisfies the first two conditions for a probability density function. Then find each of the following probabilities and illustrate graphically:

(A) $P(.3 \leqslant X \leqslant .8)$ (B) $P(X \leqslant .7)$

(C) $P(X \geqslant .4)$ (D) $P(X = .5)$

■

The probability in Example 1D illustrates a fundamental difference between discrete and continuous random variables. In the discrete case, there is a *probability distribution* $P(x)$ that gives the probability of each possible value of the random variable. Thus, if c is one of the values of the random variable, then $P(X = c) = P(c)$. In the continuous case, the *integral of the probability density function* $f(x)$ gives the probability that the outcome lies in a certain interval. If c is any real number, then the probability that the outcome is *exactly* c is

$$P(X = c) = P(c \leqslant X \leqslant c) = \int_c^c f(x)\, dx = 0$$

Thus, $P(X = c) = 0$ for any number c.

The fact that $P(X = c) = 0$ also implies that excluding either end point from an interval does not change the probability that the random variable lies in that interval; that is,

$$P(a < X < b) = P(a < X \leqslant b) = P(a \leqslant X < b) = P(a \leqslant X \leqslant b)$$

$$= \int_a^b f(x)\, dx$$

Example 2 ⇔ **Using a Probability Density Function**   Use the probability density function in Example 1 to compute $P(.1 < X \leqslant .2)$ and $P(X > .9)$.

SOLUTION

$$P(.1 < X \leqslant .2) = \int_{.1}^{.2} f(x)\, dx \qquad f(x) = \begin{cases} 12x^2 - 12x^3 & \text{if } 0 \leqslant x \leqslant 1 \\ 0 & \text{otherwise} \end{cases}$$

$$= \int_{.1}^{.2} (12x^2 - 12x^3)\, dx$$

$$= (4x^3 - 3x^4)\Big|_{.1}^{.2}$$

$$= .0272 - .0037 = .0235$$

$$P(X > .9) = \int_{.9}^{\infty} f(x)\, dx$$

$$= \int_{.9}^{1} (12x^2 - 12x^3)\, dx$$

$$= (4x^3 - 3x^4)\Big|_{.9}^{1}$$

$$= 1 - .9477 = .0523$$

■

Matched Problem 2 ⟿ Use the probability density function in Matched Problem 1 to compute the probabilities $P(.2 \leq X < .4)$ and $P(X < .8)$. ■

Example 3 ⟿ **Shelf-Life**   The shelf-life (in months) of a certain drug is a continuous random variable with probability density function (see Fig. 3)

$$f(x) = \begin{cases} 50/(x + 50)^2 & \text{if } x \geq 0 \\ 0 & \text{otherwise} \end{cases}$$

Find the probability that the drug has a shelf-life of

(A) Between 10 and 20 months

(B) At most 30 months

(C) Over 25 months

FIGURE 3

SOLUTION   (A) $P(10 \leq X \leq 20) = \displaystyle\int_{10}^{20} f(x)\, dx = \int_{10}^{20} \dfrac{50}{(x + 50)^2}\, dx = \dfrac{-50}{x + 50}\bigg|_{10}^{20}$

$$= \left(-\tfrac{50}{70}\right) - \left(-\tfrac{50}{60}\right) = \tfrac{5}{42}$$

(B) $P(X \leq 30) = \displaystyle\int_{-\infty}^{30} f(x)\, dx = \int_{0}^{30} \dfrac{50}{(x + 50)^2}\, dx = \dfrac{-50}{x + 50}\bigg|_{0}^{30}$

$$= \left(-\tfrac{50}{80}\right) - (-1) = \tfrac{3}{8}$$

(C) $P(X > 25) = \displaystyle\int_{25}^{\infty} f(x)\, dx = \int_{25}^{\infty} \dfrac{50}{(x + 50)^2}\, dx$

This improper integral can be evaluated directly using the techniques discussed in Section 11-1. However, there is another method that does not involve evaluation of any improper integrals. Since f is a probability density function, we can write

$$1 = \int_{-\infty}^{\infty} f(x)\, dx$$

$$= \int_{0}^{\infty} \dfrac{50}{(x + 50)^2}\, dx \qquad \int_{a}^{b} f(x)\, dx = \int_{a}^{c} f(x)\, dx + \int_{c}^{b} f(x)\, dx$$

$$= \int_{0}^{25} \dfrac{50}{(x + 50)^2}\, dx + \int_{25}^{\infty} \dfrac{50}{(x + 50)^2}\, dx$$

Solving this last equation for $\displaystyle\int_{25}^{\infty} [50/(x + 50)^2]\, dx$, we have

$$\int_{25}^{\infty} \dfrac{50}{(x + 50)^2}\, dx = 1 - \int_{0}^{25} \dfrac{50}{(x + 50)^2}\, dx$$

$$= 1 - \dfrac{-50}{x + 50}\bigg|_{0}^{25}$$

$$= 1 - \left(\tfrac{-50}{75}\right) - 1 = \tfrac{2}{3}$$

Thus, $P(X > 25) = \tfrac{2}{3}$. ■

Matched Problem 3   In Example 3 find the probability that the drug has a shelf-life of

(A) Between 50 and 100 months

(B) At most 20 months

(C) Over 10 months

❑   CUMULATIVE DISTRIBUTION FUNCTIONS

Explore–Discuss 2   Let

$$f(x) = \begin{cases} \frac{1}{2} & \text{if } 0 \leq x \leq 2 \\ 0 & \text{otherwise} \end{cases} \quad \text{and} \quad F(x) = \begin{cases} 0 & \text{if } x < 0 \\ \frac{1}{2}x & \text{if } 0 \leq x \leq 2 \\ 1 & \text{if } x > 2 \end{cases}$$

(A) Graph each function.

(B) Is f a probability density function? Is F?

(C) Is there any relationship between f and F?

(D) Evaluate $\displaystyle\int_{.5}^{1.5} f(x)\, dx$.

(E) Can you use F to evaluate the integral in part (D)?

Each time we compute the probability for a continuous random variable, we must find the antiderivative of the probability density function. This antiderivative is used so often that it is convenient to give it a name.

Cumulative Distribution Function

If f is a probability density function, then the associated **cumulative distribution function F** is defined by

$$F(x) = P(X \leq x) = \int_{-\infty}^{x} f(t)\, dt$$

Furthermore,

$$P(c \leq X \leq d) = F(d) - F(c)$$

Figure 4 gives a geometric interpretation of these ideas.

(A) Cumulative distribution function

(B) Probability density function

(C)

FIGURE 4

Notice that $F(x) = \int_{-\infty}^{x} f(t)\, dt$ is a function of x, the upper limit of integration, not t, the variable in the integrand. We state some important properties of cumulative distribution functions in the next box. These properties follow directly from the fact that $F(x)$ can be interpreted geometrically as the area under the graph of $y = f(t)$ from $-\infty$ to x (see Fig. 4B on preceding page).

Properties of Cumulative Distribution Functions

If f is a probability density function and

$$F(x) = \int_{-\infty}^{x} f(t)\, dt$$

is the associated cumulative distribution function, then

1. $F'(x) = f(x)$ wherever f is continuous
2. $0 \le F(x) \le 1, \quad -\infty < x < \infty$
3. $F(x)$ is nondecreasing on $(-\infty, \infty)$*

---

*A function $F(x)$ is nondecreasing on (a, b) if $F(x_1) \le F(x_2)$ for $a < x_1 < x_2 < b$.

Example 4 ⇨ **Using a Cumulative Distribution Function** Find the cumulative distribution function for the probability density function in Example 1, and use it to compute $P(.1 \le X \le .9)$.

SOLUTION If $x < 0$, then

$$F(x) = \int_{-\infty}^{x} f(t)\, dt \qquad f(x) = \begin{cases} 12x^2 - 12x^3 & \text{if } 0 \le x \le 1 \\ 0 & \text{otherwise} \end{cases}$$

$$= \int_{-\infty}^{x} 0\, dt = 0$$

If $0 \le x \le 1$, then

$$F(x) = \int_{-\infty}^{x} f(t)\, dt = \int_{-\infty}^{0} f(t)\, dt + \int_{0}^{x} f(t)\, dt$$

$$= 0 + \int_{0}^{x} (12t^2 - 12t^3)\, dt = (4t^3 - 3t^4)\Big|_{0}^{x}$$

$$= 4x^3 - 3x^4$$

If $x > 1$, then

$$F(x) = \int_{-\infty}^{x} f(t)\, dt = \int_{-\infty}^{0} f(t)\, dt + \int_{0}^{1} f(t)\, dt + \int_{1}^{x} f(t)\, dt$$

$$= 0 + 1 + 0 = 1$$

Thus,

$$F(x) = \begin{cases} 0 & \text{if } x < 0 \\ 4x^3 - 3x^4 & \text{if } 0 \le x \le 1 \\ 1 & \text{if } x > 1 \end{cases}$$

And

$$P(.1 \leqslant X \leqslant .9) = F(.9) - F(.1) = .9477 - .0037 = .944$$

See Figure 5.

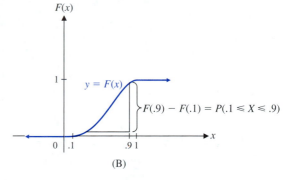

(A)   (B)

FIGURE 5

Matched Problem 4 ⇨ Find the cumulative distribution function for the probability density function in Matched Problem 1, and use it to compute $P(.3 \leqslant X \leqslant .7)$.

Example 5 ⇨ **Shelf-Life**   Returning to the discussion of the shelf-life of a drug in Example 3, suppose a pharmacist wants to be 95% certain that the drug is still good when it is sold. How long is it safe to leave the drug on the shelf?

SOLUTION   Let x be the number of months the drug has been on the shelf when it is sold. The probability that the shelf-life of the drug is less than the number of months it has been sitting on the shelf is $P(0 \leqslant X \leqslant x)$. The pharmacist wants this probability to be .05. Thus, we must solve the equation $P(0 \leqslant X \leqslant x) = .05$ for x. First, we will find the cumulative distribution function F. For $x < 0$, we see that $F(x) = 0$. For $x \geqslant 0$,

$$F(x) = \int_0^x \frac{50}{(50 + t)^2}\, dt = \left.\frac{-50}{50 + t}\right|_0^x = \frac{-50}{50 + x} - (-1) = 1 - \frac{50}{50 + x}$$

$$= \frac{x}{50 + x}$$

Thus,

$$F(x) = \begin{cases} 0 & \text{if } x < 0 \\ x/(50 + x) & \text{if } x \geqslant 0 \end{cases}$$

Now, to solve the equation $P(0 \leqslant X \leqslant x) = .05$, we solve

$$F(x) - F(0) = .05 \qquad F(0) = 0$$

$$\frac{x}{50 + x} = .05$$

$$x = 2.5 + .05x$$

$$.95x = 2.5$$

$$x \approx 2.6$$

If the drug is sold during the first 2.6 months it is on the shelf, then the probability that it is still good is .95.

Figure 6 shows an alternative method for solving Example 5 on a graphing utility.

FIGURE 6

Matched Problem 5 ⇔ Repeat Example 5 if the pharmacist wants the probability that the drug is still good to be .99.

Answers to Matched Problems

1.

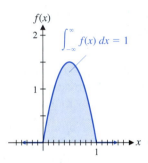

$$\int_{-\infty}^{\infty} f(x)\,dx = 1$$

(A) .68

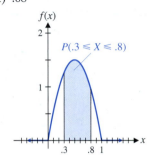

$P(.3 \le X \le .8)$

(B) .784

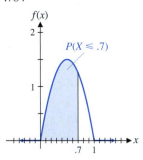

$P(X \le .7)$

(C) .648

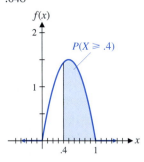

$P(X \ge .4)$

(D) 0

$P(X = .5) = 0$

2. .248; .896　　**3.** (A) $\frac{1}{6}$　(B) $\frac{2}{7}$　(C) $\frac{5}{6}$

4. $F(x) = \begin{cases} 0 & \text{if } x < 0 \\ 3x^2 - 2x^3 & \text{if } 0 \leq x \leq 1; \\ 1 & \text{if } x > 1 \end{cases}$　$P(.3 \leq X \leq .7) = .568$

5. Approx. $\frac{1}{2}$ month, or 15 days

Exercise 11-2

A *In Problems 1 and 2, graph f, and show that f satisfies the first two conditions for a probability density function.*

1. $f(x) = \begin{cases} \frac{1}{8}x & \text{if } 0 \leq x \leq 4 \\ 0 & \text{otherwise} \end{cases}$

2. $f(x) = \begin{cases} \frac{1}{9}x^2 & \text{if } 0 \leq x \leq 3 \\ 0 & \text{otherwise} \end{cases}$

In Problems 3–6, is f a probability density function? Explain.

3. $f(x) = \begin{cases} \frac{1}{8}x & \text{if } -4 \leq x \leq 0 \\ 0 & \text{otherwise} \end{cases}$

4. $f(x) = \begin{cases} \frac{1}{9}x^2 & \text{if } 0 \leq x \leq 4 \\ 0 & \text{otherwise} \end{cases}$

5. $f(x) = \begin{cases} 1 & \text{if } x = 1 \\ 0 & \text{otherwise} \end{cases}$

6. $f(x) = \begin{cases} 1/x & \text{if } x \geq 1 \\ 0 & \text{otherwise} \end{cases}$

7. Use the function in Problem 1 to find the indicated probabilities. Illustrate each probability with a graph.
(A) $P(1 \leq X \leq 3)$
(B) $P(X \leq 2)$
(C) $P(X > 3)$

8. Use the function in Problem 2 to find the indicated probabilities. Illustrate each probability with a graph.
(A) $P(1 \leq X \leq 2)$
(B) $P(X \geq 1)$
(C) $P(X < 2)$

9. Use the function in Problem 1 to find the indicated probabilities.
(A) $P(X = 1)$
(B) $P(X > 5)$
(C) $P(X < 5)$

10. Use the function in Problem 2 to find the indicated probabilities.
(A) $P(X = 2)$
(B) $P(X > 4)$
(C) $P(X < 4)$

11. Find and graph the cumulative distribution function associated with the function in Problem 1.

12. Find and graph the cumulative distribution function associated with the function in Problem 2.

13. Use the cumulative distribution function from Problem 11 to find the indicated probabilities.
(A) $P(2 \leq X \leq 4)$
(B) $P(0 < X < 2)$

14. Use the cumulative distribution function from Problem 12 to find the indicated probabilities.
(A) $P(0 \leq X \leq 1)$
(B) $P(2 < X < 3)$

15. Use the cumulative distribution function from Problem 11 to find the value of x that satisfies each equation.
(A) $P(0 \leq X \leq x) = \frac{1}{4}$
(B) $P(0 \leq X \leq x) = \frac{1}{9}$

16. Use the cumulative distribution function from Problem 12 to find the value of x that satisfies each equation.
(A) $P(0 \leq X \leq x) = \frac{1}{8}$
(B) $P(0 \leq X \leq x) = \frac{1}{64}$

B *In Problems 17 and 18, graph f, and show that f satisfies the first two conditions for a probability density function.*

17. $f(x) = \begin{cases} 2/(1 + x)^3 & \text{if } x \geq 0 \\ 0 & \text{otherwise} \end{cases}$

18. $f(x) = \begin{cases} 2/(2 + x)^2 & \text{if } x \geq 0 \\ 0 & \text{otherwise} \end{cases}$

19. Use the function in Problem 17 to find the indicated probabilities.
(A) $P(1 \leq X \leq 4)$
(B) $P(X > 3)$
(C) $P(X \leq 2)$

20. Use the function in Problem 18 to find the indicated probabilities.
(A) $P(2 \leq X \leq 8)$
(B) $P(X \geq 3)$
(C) $P(X < 1)$

21. Find and graph the cumulative distribution function associated with the function in Problem 17.

22. Find and graph the cumulative distribution function associated with the function in Problem 18.

23. Use the cumulative distribution function from Problem 21 to find the value of x that satisfies each equation.

(A) $P(0 \leqslant X \leqslant x) = \frac{3}{4}$

(B) $P(X \geqslant x) = \frac{1}{16}$

24. Use the cumulative distribution function from Problem 22 to find the value of x that satisfies each equation.

(A) $P(0 \leqslant X \leqslant x) = \frac{3}{4}$

(B) $P(X > x) = \frac{1}{3}$

In Problems 25–28, find the associated cumulative distribution function. Graph both functions (on separate sets of axes).

25. $f(x) = \begin{cases} \frac{3}{2}x - \frac{3}{4}x^2 & \text{if } 0 \leqslant x \leqslant 2 \\ 0 & \text{otherwise} \end{cases}$

26. $f(x) = \begin{cases} \frac{3}{4} - \frac{3}{4}x^2 & \text{if } -1 \leqslant x \leqslant 1 \\ 0 & \text{otherwise} \end{cases}$

27. $f(x) = \begin{cases} \frac{1}{2} + \frac{1}{2}x^3 & \text{if } -1 \leqslant x \leqslant 1 \\ 0 & \text{otherwise} \end{cases}$

28. $f(x) = \begin{cases} \frac{3}{4} - \frac{3}{8}\sqrt{x} & \text{if } 0 \leqslant x \leqslant 4 \\ 0 & \text{otherwise} \end{cases}$

 In Problems 29–32, use a graphing utility to approximate (to two decimal places) the value of x that satisfies the given equation for the indicated cumulative distribution function $F(x)$.

29. $P(0 \leqslant X \leqslant x) = .2$ for $F(x)$ from Problem 25

30. $P(-1 \leqslant X \leqslant x) = .4$ for $F(x)$ from Problem 26

31. $P(-1 \leqslant X \leqslant x) = .6$ for $F(x)$ from Problem 27

32. $P(0 \leqslant X \leqslant x) = .7$ for $F(x)$ from Problem 28

For each function f in Problems 33–36, find a constant k so that kf is a probability density function, or explain why no such constant exists.

33. $f(x) = \begin{cases} 1/x & \text{if } x \geqslant 1 \\ 0 & \text{otherwise} \end{cases}$

34. $f(x) = \begin{cases} 1/x^4 & \text{if } x \geqslant 1 \\ 0 & \text{otherwise} \end{cases}$

35. $f(x) = \begin{cases} 1/x^3 & \text{if } x \geqslant 1 \\ 0 & \text{otherwise} \end{cases}$

36. $f(x) = \begin{cases} 1/\sqrt{x} & \text{if } x \geqslant 1 \\ 0 & \text{otherwise} \end{cases}$

37. Can the function

$$F(x) = \begin{cases} 2 - 2x & \text{if } 0 \leqslant x \leqslant 1 \\ 0 & \text{otherwise} \end{cases}$$

be a cumulative distribution function? Explain.

38. Can the function

$$F(x) = \begin{cases} 0.5 - 0.5x^3 & \text{if } -1 \leqslant x \leqslant 1 \\ 0 & \text{otherwise} \end{cases}$$

be a cumulative distribution function? Explain.

C *In Problems 39 and 40, use a graphing utility to graph f and compute the integrals $\int_{-N}^{N} f(x)\,dx$ for $N = 1, 2,$ and 3. Discuss the evidence for concluding that f is a probability density function.*

39. $f(x) = \dfrac{1}{\sqrt{2\pi}} e^{-0.5x^2}$ **40.** $f(x) = \dfrac{2}{\sqrt{2\pi}} e^{-2(x-1.5)^2}$

In Problems 41–44, $F(x)$ is the cumulative distribution function for a continuous random variable X. Find the probability density function $f(x)$ associated with each $F(x)$.

41. $F(x) = \begin{cases} 0 & \text{if } x < 0 \\ x^2 & \text{if } 0 \leqslant x \leqslant 1 \\ 1 & \text{if } x > 1 \end{cases}$

42. $F(x) = \begin{cases} 0 & \text{if } x < 1 \\ \frac{1}{2}x - \frac{1}{2} & \text{if } 1 \leqslant x \leqslant 3 \\ 1 & \text{if } x > 3 \end{cases}$

43. $F(x) = \begin{cases} 0 & \text{if } x < 0 \\ 6x^2 - 8x^3 + 3x^4 & \text{if } 0 \leqslant x \leqslant 1 \\ 1 & \text{if } x > 1 \end{cases}$

44. $F(x) = \begin{cases} 1 - (1/x^3) & \text{if } x \geqslant 1 \\ 0 & \text{otherwise} \end{cases}$

In Problems 45 and 46, find the associated cumulative distribution function.

45. $f(x) = \begin{cases} x & \text{if } 0 \leqslant x \leqslant 1 \\ 2 - x & \text{if } 1 < x \leqslant 2 \\ 0 & \text{otherwise} \end{cases}$

46. $f(x) = \begin{cases} \frac{1}{4} & \text{if } 0 \leqslant x \leqslant 1 \\ \frac{1}{2} & \text{if } 1 < x \leqslant 2 \\ \frac{1}{4} & \text{if } 2 < x \leqslant 3 \\ 0 & \text{otherwise} \end{cases}$

Applications

Business & Economics

47. *Electricity consumption.* The daily demand for electricity (in millions of kilowatt-hours) in a large city is a continuous random variable with probability density function

$$f(x) = \begin{cases} .2 - .02x & \text{if } 0 \leq x \leq 10 \\ 0 & \text{otherwise} \end{cases}$$

(A) Evaluate $\int_2^6 f(x)\,dx$ and interpret the results.

(B) What is the probability that the daily demand for electricity is less than 8 million kilowatt-hours?

(C) What is the probability that 5 million kilowatt-hours will not be sufficient to meet the daily demand?

48. *Gasoline consumption.* The daily demand for gasoline (in millions of gallons) in a large city is a continuous random variable with probability density function

$$f(x) = \begin{cases} .4 - .08x & \text{if } 0 \leq x \leq 5 \\ 0 & \text{otherwise} \end{cases}$$

(A) Evaluate $\int_1^4 f(x)\,dx$ and interpret the results.

(B) What is the probability that the daily demand is less than 2 million gallons?

(C) What is the probability that 3 million gallons will not be sufficient to meet the daily demand?

49. *Time-sharing.* In a computer time-sharing network, the time it takes (in seconds) to respond to a user's request is a continuous random variable with probability density function given by

$$f(x) = \begin{cases} \frac{1}{10}e^{-x/10} & \text{if } x \geq 0 \\ 0 & \text{otherwise} \end{cases}$$

(A) Evaluate $\int_5^{10} f(x)\,dx$ and interpret the results.

(B) What is the probability that the computer responds within 1 second?

(C) What is the probability that a user must wait over 4 seconds for a response?

50. *Waiting time.* The time (in minutes) a customer must wait in line at a bank is a continuous random variable with probability density function given by

$$f(x) = \begin{cases} \frac{1}{2}e^{-x/2} & \text{if } x \geq 0 \\ 0 & \text{otherwise} \end{cases}$$

(A) Evaluate $\int_2^6 f(x)\,dx$ and interpret the results.

(B) What is the probability that a customer waits less than 3 minutes?

(C) What is the probability that a customer waits more than 5 minutes?

51. *Demand.* The weekly demand for hamburger (in thousands of pounds) for a chain of supermarkets is

a continuous random variable with probability density function given by

$$f(x) = \begin{cases} .003x\sqrt{100 - x^2} & \text{if } 0 \leq x \leq 10 \\ 0 & \text{otherwise} \end{cases}$$

(A) What is the probability that more than 4,000 pounds of hamburger are demanded?

(B) The manager of the meat department orders 8,000 pounds of hamburger. What is the probability that the demand will not exceed this amount?

(C) The manager wants the probability that the demand does not exceed the amount ordered to be .9. How much hamburger meat should be ordered?

52. *Demand.* The demand for a weekly sports magazine (in thousands of copies) in a certain city is a continuous random variable with probability density function given by

$$f(x) = \begin{cases} \frac{3}{125}x\sqrt{25 - x^2} & \text{if } 0 \leq x \leq 5 \\ 0 & \text{otherwise} \end{cases}$$

(A) The magazine's distributor in this city orders 3,000 copies of the magazine. What is the probability that the demand exceeds this number?

(B) What is the probability that the demand does not exceed 4,000 copies?

(C) If the distributor wants to be 95% certain that the demand does not exceed the number ordered, how many copies should be ordered?

Life Sciences

53. *Life expectancy.* The life expectancy (in minutes) of a certain microscopic organism is a continuous random variable with probability density function given by

$$f(x) = \begin{cases} \frac{1}{5,000}(10x^3 - x^4) & \text{if } 0 \leq x \leq 10 \\ 0 & \text{otherwise} \end{cases}$$

(A) What is the probability that an organism lives for at least 7 minutes?

(B) What is the probability that an organism lives for at most 5 minutes?

54. *Life expectancy.* The life expectancy (in months) of plants of a certain species is a continuous random variable with probability density function given by

$$f(x) = \begin{cases} \frac{1}{36}(6x - x^2) & \text{if } 0 \leq x \leq 6 \\ 0 & \text{otherwise} \end{cases}$$

(A) What is the probability that one of these plants survives for at least 4 months?

(B) What is the probability that one of these plants survives for at most 5 months?

55. *Shelf-life.* The shelf-life (in days) of a perishable drug is a continuous random variable with

probability density function given by

$$f(x) = \begin{cases} 800x/(400 + x^2)^2 & \text{if } x \geqslant 0 \\ 0 & \text{otherwise} \end{cases}$$

(A) What is the probability that the drug has a shelf-life of at most 20 days?

(B) What is the probability that the shelf-life exceeds 15 days?

(C) If the user wants the probability that the drug is still good to be .8, when is the last time it should be used?

56. *Shelf-life.* Repeat Problem 55 if

$$f(x) = \begin{cases} 200x/(100 + x^2)^2 & \text{if } x \geqslant 0 \\ 0 & \text{otherwise} \end{cases}$$

Social Sciences

57. *Learning.* The number of words per minute a beginner can type after 1 week of practice is a continuous random variable with probability density function given by

$$f(x) = \begin{cases} \frac{1}{20}e^{-x/20} & \text{if } x \geqslant 0 \\ 0 & \text{otherwise} \end{cases}$$

(A) What is the probability that a beginner can type at least 30 words per minute after 1 week of practice?

(B) What is the probability that a beginner can type at least 80 words per minute after 1 week of practice?

58. *Learning.* The number of hours it takes a chimpanzee **www** to learn a new task is a continuous random variable with probability density function given by

$$f(x) = \begin{cases} \frac{4}{9}x^2 - \frac{4}{27}x^3 & \text{if } 0 \leqslant x \leqslant 3 \\ 0 & \text{otherwise} \end{cases}$$

(A) What is the probability that the chimpanzee learns the task in the first hour?

(B) What is the probability that the chimpanzee does not learn the task in the first 2 hours?

Section 11-3   # Expected Value, Standard Deviation, and Median

- ❑ EXPECTED VALUE (OR MEAN) AND STANDARD DEVIATION
- ❑ ALTERNATIVE FORMULA FOR VARIANCE
- ❑ MEDIAN

❑ EXPECTED VALUE (OR MEAN) AND STANDARD DEVIATION

Statisticians use a measure of central tendency called the *mean* and a measure of dispersion called the *standard deviation* to help describe sets of data. These same terms are used to describe both discrete and continuous random variables. For example, if X is the discrete random variable representing the number of heads when 3 coins are tossed (Table 1), then the mean and standard deviation are computed as follows:

TABLE 1

x	0	1	2	3
$p(x)$	$\frac{1}{8}$	$\frac{3}{8}$	$\frac{3}{8}$	$\frac{1}{8}$

$$\text{Mean} = 0 \cdot \tfrac{1}{8} + 1 \cdot \tfrac{3}{8} + 2 \cdot \tfrac{3}{8} + 3 \cdot \tfrac{1}{8} = \tfrac{3}{2}$$

$$\text{Variance} = (0 - \tfrac{3}{2})^2 \cdot \tfrac{1}{8} + (1 - \tfrac{3}{2})^2 \cdot \tfrac{3}{8} + (2 - \tfrac{3}{2})^2 \cdot \tfrac{3}{8} + (3 - \tfrac{3}{2})^2 \cdot \tfrac{1}{8}$$

$$= \tfrac{3}{4}$$

$$\text{Standard deviation} = \sqrt{\text{Variance}} = \sqrt{\tfrac{3}{4}} \approx .8660$$

Notice that these quantities were computed using finite sums and the probabilities in Table 1.

The extension of these ideas to continuous random variables makes use of integration and the probability density function.

(A) $\sigma = 1$

(B) $\sigma = 4$

FIGURE 1

Expected Value (or Mean) and Standard Deviation for a Continuous Random Variable

Let $f(x)$ be the probability density function for a continuous random variable X. The **expected value, or mean, of X** is

$$\mu = E(X) = \int_{-\infty}^{\infty} x f(x)\, dx$$

The **variance** is

$$V(X) = \int_{-\infty}^{\infty} (x - \mu)^2 f(x)\, dx$$

and the **standard deviation** is

$$\sigma = \sqrt{V(X)}$$

Just as in the discrete case, the mean of a continuous random variable is a measure of central tendency for the variable, and the standard deviation is a measure of the dispersion of the variable about the mean. This is illustrated in Figure 1. The probability density function in Figure 1A has a standard deviation of 1. Most of the area under the curve is near the mean. In Figure 1B, the standard deviation is four times as large, and the area under the graph is much more spread out.

Example 1 ➭ **Computing Expected Value (Mean), Variance, and Standard Deviation**
Find the expected value (mean), variance, and standard deviation for

$$f(x) = \begin{cases} 12x^2 - 12x^3 & \text{if } 0 \leqslant x \leqslant 1 \\ 0 & \text{otherwise} \end{cases}$$

SOLUTION

$$\mu = E(X) = \int_{-\infty}^{\infty} x f(x)\, dx = \int_0^1 x(12x^2 - 12x^3)\, dx = \int_0^1 (12x^3 - 12x^4)\, dx$$

$$= \left(3x^4 - \tfrac{12}{5}x^5\right)\Big|_0^1 = \tfrac{3}{5} \qquad \text{Expected value (mean)}$$

$$V(X) = \int_{-\infty}^{\infty} (x - \mu)^2 f(x)\, dx = \int_0^1 \left(x - \tfrac{3}{5}\right)^2 (12x^2 - 12x^3)\, dx$$

$$= \int_0^1 \left(x^2 - \tfrac{6}{5}x + \tfrac{9}{25}\right)(12x^2 - 12x^3)\, dx$$

$$= \int_0^1 \left(\tfrac{108}{25}x^2 - \tfrac{468}{25}x^3 + \tfrac{132}{5}x^4 - 12x^5\right) dx$$

$$= \left(\tfrac{36}{25}x^3 - \tfrac{117}{25}x^4 + \tfrac{132}{25}x^5 - 2x^6\right)\Big|_0^1 = \tfrac{1}{25} \qquad \text{Variance}$$

$$\sigma = \sqrt{V(X)} = \sqrt{\tfrac{1}{25}} = \tfrac{1}{5} \qquad \text{Standard deviation}$$

■

Matched Problem 1 ➭ Find the expected value (mean), variance, and standard deviation for

$$f(x) = \begin{cases} 6x - 6x^2 & \text{if } 0 \leqslant x \leqslant 1 \\ 0 & \text{otherwise} \end{cases}$$

■

$\mu = .6$

(B)

FIGURE 2

Consider the probability density function

$$f(x) = \begin{cases} \frac{1}{2}x & \text{if } 0 \leqslant x \leqslant 2 \\ 0 & \text{otherwise} \end{cases}$$

(A) Find the mean μ.

(B) The vertical line $x = \mu$ divides the area under the graph of $y = f(x)$ into two regions. Do these regions have equal areas? If not, find a number m with the property that the vertical line $x = m$ divides the area under the graph of $y = f(x)$ into two regions with equal areas.

The graph of the probability density function considered in Example 1 is shown in Figure 2A, with the areas to the left and right of the mean indicated (computations omitted). As you discovered in the preceding Explore–Discuss 1, the vertical line through the mean does not divide the region under the graph of f into two regions with equal areas. The value of x that accomplishes this is the *median,* which we will discuss later in this section. However, the mean does have a geometric interpretation. If the shaded region in Figure 2A were drawn on a piece of wood of uniform thickness and the wood cut around the outside of the region, then the resulting object would balance on a wedge placed at the mean $\mu = .6$ (see Fig. 2B). For this reason, the mean μ is often referred to as a *measure of central tendency.*

Example 2 **Life Expectancy** The life expectancy (in hours) for a particular brand of light bulbs is a continuous random variable with probability density function

$$f(x) = \begin{cases} \frac{1}{100} - \frac{1}{20,000}x & \text{if } 0 \leqslant x \leqslant 200 \\ 0 & \text{otherwise} \end{cases}$$

(A) What is the average life expectancy of one of these light bulbs?

(B) What is the probability that a bulb will last longer than this average?

Solution (A) Since the value of this random variable is the number of hours a bulb lasts, the average life expectancy is just the expected value of the random variable. Thus,

$$E(X) = \int_{-\infty}^{\infty} xf(x)\,dx = \int_{0}^{200} x\left(\tfrac{1}{100} - \tfrac{1}{20,000}x\right) dx$$

$$= \int_{0}^{200} \left(\tfrac{1}{100}x - \tfrac{1}{20,000}x^2\right) dx = \left(\tfrac{1}{200}x^2 - \tfrac{1}{60,000}x^3\right)\Big|_{0}^{200}$$

$$= \tfrac{200}{3} \approx 66.67 \text{ hours}$$

(B) The probability that a bulb lasts longer than $\frac{200}{3}$ hours is

$$P\left(X > \tfrac{200}{3}\right) = \int_{200/3}^{\infty} f(x)\,dx = \int_{200/3}^{200} \left(\tfrac{1}{100} - \tfrac{1}{20,000}x\right) dx$$

$$= \left(\tfrac{1}{100}x - \tfrac{1}{40,000}x^2\right)\Big|_{200/3}^{200}$$

$$= 1 - \tfrac{5}{9} = \tfrac{4}{9} \approx 0.44$$

Matched Problem 2 Repeat Example 2 if the probability density function is

$$f(x) = \begin{cases} \frac{1}{200} - \frac{1}{90{,}000}x & \text{if } 0 \leqslant x \leqslant 300 \\ 0 & \text{otherwise} \end{cases}$$

❑   ALTERNATIVE FORMULA FOR VARIANCE

The term $(x - \mu)^2$ in the formula for $V(X)$ introduces some complicated algebraic manipulations in the evaluation of the integral. We can use the properties of the definite integral to simplify this formula. Thus,

$$V(X) = \int_{-\infty}^{\infty} (x - \mu)^2 f(x)\, dx$$

$$= \int_{-\infty}^{\infty} (x^2 - 2x\mu + \mu^2) f(x)\, dx \qquad\qquad \text{Algebra}$$

$$= \int_{-\infty}^{\infty} [x^2 f(x) - 2x\mu f(x) + \mu^2 f(x)]\, dx \qquad\qquad \text{Algebra}$$

$$= \int_{-\infty}^{\infty} x^2 f(x)\, dx - \int_{-\infty}^{\infty} 2x\mu f(x)\, dx + \int_{-\infty}^{\infty} \mu^2 f(x)\, dx \qquad \text{Definite integral property}$$

$$= \int_{-\infty}^{\infty} x^2 f(x)\, dx - 2\mu \int_{-\infty}^{\infty} x f(x)\, dx + \mu^2 \int_{-\infty}^{\infty} f(x)\, dx \qquad \text{Definite integral property}$$

$$= \int_{-\infty}^{\infty} x^2 f(x)\, dx - 2\mu(\mu) + \mu^2(1) \qquad\qquad \int_{-\infty}^{\infty} xf(x)\, dx = \mu, \quad \int_{-\infty}^{\infty} f(x)\, dx = 1$$

$$= \int_{-\infty}^{\infty} x^2 f(x)\, dx - \mu^2$$

In general, it will be easier to evaluate $\int_{-\infty}^{\infty} x^2 f(x)\, dx$ than to evaluate $\int_{-\infty}^{\infty} (x - \mu)^2 f(x)\, dx$.

THEOREM 1   Alternative Formula for Variance

$$V(X) = \int_{-\infty}^{\infty} x^2 f(x)\, dx - \mu^2$$

Example 3 **Computing Variance**   Use the alternative formula for variance (Theorem 1) to compute the variance in Example 1.

SOLUTION   From Example 1, we have $\mu = \int_{-\infty}^{\infty} xf(x)\, dx = \frac{3}{5}$. Thus,

$$\int_{-\infty}^{\infty} x^2 f(x)\, dx = \int_0^1 x^2 (12x^2 - 12x^3)\, dx \quad f(x) = \begin{cases} 12x^2 - 12x^3 & \text{if } 0 \leqslant x \leqslant 1 \\ 0 & \text{otherwise} \end{cases}$$

$$= \int_0^1 (12x^4 - 12x^5)\, dx$$

$$= \left. \left(\tfrac{12}{5}x^5 - \tfrac{12}{6}x^6 \right) \right|_0^1 = \tfrac{2}{5} = .4$$

$$V(X) = \int_{-\infty}^{\infty} x^2 f(x)\, dx - \mu^2 = \tfrac{2}{5} - \left(\tfrac{3}{5}\right)^2 = \tfrac{1}{25} = .04$$

Matched Problem 3 Use the alternative formula for variance (Theorem 1) to compute the variance in Matched Problem 1.

Example 4   ✏   **Computing Expected Value (Mean), Variance, and Standard Deviation**
Find the expected value (mean), variance, and standard deviation for

$$f(x) = \begin{cases} 3/x^4 & \text{if } x \geqslant 1 \\ 0 & \text{otherwise} \end{cases}$$

SOLUTION

$$\mu = \int_{-\infty}^{\infty} x f(x)\, dx = \int_{1}^{\infty} x \frac{3}{x^4}\, dx = \lim_{R \to \infty} \int_{1}^{R} \frac{3}{x^3}\, dx$$

$$= \lim_{R \to \infty}\left[-\frac{3}{2}\left(\frac{1}{x^2}\right)\right]\Big|_{1}^{R} = \lim_{R \to \infty}\left[-\frac{3}{2}\left(\frac{1}{R^2}\right) + \frac{3}{2}\right] = \frac{3}{2} = 1.5$$

$$\int_{-\infty}^{\infty} x^2 f(x)\, dx = \int_{1}^{\infty} x^2 \frac{3}{x^4}\, dx = \lim_{R \to \infty} \int_{1}^{R} \frac{3}{x^2}\, dx = \lim_{R \to \infty}\left(-\frac{3}{x}\right)\Big|_{1}^{R}$$

$$= \lim_{R \to \infty}\left(-\frac{3}{R} + 3\right) = 3$$

$$V(X) = \int_{-\infty}^{\infty} x^2 f(x)\, dx - \mu^2 = 3 - \left(\frac{3}{2}\right)^2 = \frac{3}{4} = .75$$

$$\sigma = \sqrt{V(X)} = \sqrt{\frac{3}{4}} = \frac{\sqrt{3}}{2} \approx .8660$$

∎

Matched Problem 4   ✏   Find the expected value (mean), variance, and standard deviation for

$$f(x) = \begin{cases} 24/x^4 & \text{if } x \geqslant 2 \\ 0 & \text{otherwise} \end{cases}$$

∎

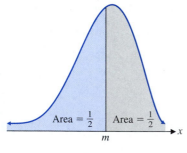

Area = $\frac{1}{2}$ Area = $\frac{1}{2}$

m

FIGURE 3

❑ MEDIAN

Another measurement often used to describe the properties of a random variable is the median. A **median** is a value of the random variable that divides the area under the graph of the probability density function into two equal parts (see Fig. 3). If m is a median, then m must satisfy

$$P(X \leqslant m) = \tfrac{1}{2}$$

Generally, this equation is solved by first finding the cumulative distribution function.

Example 5   ✏   **Finding the Median**   Find the median of the continuous random variable with probability density function

$$f(x) = \begin{cases} 3/x^4 & \text{if } x \geqslant 1 \\ 0 & \text{otherwise} \end{cases}$$

SOLUTION   **Step 1.**   *Find the cumulative distribution function:*   For $x < 1$, we have $F(x) = 0$. If $x \geqslant 1$, then

$$F(x) = \int_{-\infty}^{x} f(t)\, dt = \int_{1}^{x} \frac{3}{t^4}\, dt = -\frac{1}{t^3}\Big|_{1}^{x} = -\frac{1}{x^3} + 1 = 1 - \frac{1}{x^3}$$

Step 2.   *Solve the equation $P(X \leqslant m) = \tfrac{1}{2}$ for m:*

$$F(m) = P(X \leqslant m)$$

$$1 - \frac{1}{m^3} = \frac{1}{2}$$

$$\frac{1}{2} = \frac{1}{m^3}$$

$$m^3 = 2$$

$$m = \sqrt[3]{2}$$

Thus, the median is $\sqrt[3]{2} \approx 1.26$.

 Figure 4 shows an alternative method for solving Example 5 on a graphing utility.

FIGURE 4

Matched Problem 5 ➭ Find the median of the continuous random variable with probability density function

$$f(x) = \begin{cases} 24/x^4 & \text{if } x \geqslant 2 \\ 0 & \text{otherwise} \end{cases}$$

Explore–Discuss 2   Consider the probability density function f defined by the graph in Figure 5. Does a median exist? Is it unique? Explain.

FIGURE 5

Example 6 ➭ **Life Expectancy** In Example 2, find the median life expectancy of a light bulb.

SOLUTION   **Step 1.** *Find the cumulative distribution function:* If $x < 0$, we have $F(x) = 0$. If $0 \leqslant x \leqslant 200$, then

$$F(x) = \int_{-\infty}^{x} f(t) \, dt \qquad\qquad f(x) = \begin{cases} \frac{1}{100} - \frac{1}{20{,}000}x & \text{if } 0 \leqslant x \leqslant 200 \\ 0 & \text{otherwise} \end{cases}$$

$$= \int_{0}^{x} \left(\tfrac{1}{100} - \tfrac{1}{20{,}000} t \right) dt$$

$$= \left. \left(\tfrac{1}{100} t - \tfrac{1}{40{,}000} t^2 \right) \right|_{0}^{x}$$

$$= \tfrac{1}{100} x - \tfrac{1}{40{,}000} x^2$$

If $x > 200$, then

$$F(x) = \int_{-\infty}^{x} f(t)\, dt = \int_{-\infty}^{0} f(t)\, dt + \int_{0}^{200} f(t)\, dt + \int_{200}^{x} f(t)\, dt$$

$$= 0 + 1 + 0 = 1$$

Thus,

$$F(x) = \begin{cases} 0 & \text{if } x < 0 \\ \frac{1}{100}x - \frac{1}{40,000}x^2 & \text{if } 0 \leq x \leq 200 \\ 1 & \text{if } x > 200 \end{cases}$$

Step 2. *Solve the equation* $P(X \leq m) = \frac{1}{2}$ *for* m:

$$F(m) = P(X \leq m) = \frac{1}{2}$$

$$\frac{1}{100}m - \frac{1}{40,000}m^2 = \frac{1}{2}$$

$$m^2 - 400m + 20,000 = 0 \qquad \textcolor{blue}{\textit{The solution must occur for } 0 \leq m \leq 200.}$$

This quadratic equation has two solutions, $200 + 100\sqrt{2}$ and $200 - 100\sqrt{2}$. Since m must lie in the interval $[0, 200]$, the second solution is the correct answer.

Thus, the median life expectancy is $200 - 100\sqrt{2} \approx 58.58$ hours.

Matched Problem 6 In Matched Problem 2, find the median life expectancy of a light bulb.

Figure 6 shows the mean and the median for two of the probability density functions discussed in this section (see Examples 2, 4, 5, and 6). As we noted earlier, the mean and the median generally are not equal.

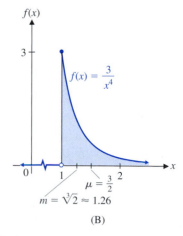

$f(x) = \frac{1}{100} - \frac{1}{20,000}x$

$\mu = \frac{200}{3}$

$m = 200 - 100\sqrt{2} \approx 58.58$

(A)

$f(x) = \frac{3}{x^4}$

$\mu = \frac{3}{2}$

$m = \sqrt[3]{2} \approx 1.26$

(B)

FIGURE 6

Answers to Matched Problems

1. $\mu = \frac{1}{2}$; $V(X) = \frac{1}{20}$; $\sigma \approx .2236$

2. (A) 125 hr   (B) $\frac{133}{288}$ **3.** $\frac{1}{20}$

4. $\mu = 3$; $V(X) = 3$; $\sigma = \sqrt{3} \approx 1.732$

5. $m = \sqrt[3]{16} = 2\sqrt[3]{2} \approx 2.52$

6. $m = 450 - 150\sqrt{5} \approx 114.59$ hr

Exercise 11-3

A *In Problems 1–6, find the mean, variance, and standard deviation.*

1. $f(x) = \begin{cases} \frac{1}{2}x & \text{if } 0 \leqslant x \leqslant 2 \\ 0 & \text{otherwise} \end{cases}$

2. $f(x) = \begin{cases} 3x^2 & \text{if } 0 \leqslant x \leqslant 1 \\ 0 & \text{otherwise} \end{cases}$

3. $f(x) = \begin{cases} \frac{1}{3} & \text{if } 2 \leqslant x \leqslant 5 \\ 0 & \text{otherwise} \end{cases}$

4. $f(x) = \begin{cases} \frac{1}{2} & \text{if } 1 \leqslant x \leqslant 3 \\ 0 & \text{otherwise} \end{cases}$

5. $f(x) = \begin{cases} 4 - 2x & \text{if } 1 \leqslant x \leqslant 2 \\ 0 & \text{otherwise} \end{cases}$

6. $f(x) = \begin{cases} 2 - \frac{1}{2}x & \text{if } 2 \leqslant x \leqslant 4 \\ 0 & \text{otherwise} \end{cases}$

In Problems 7–12, find the median.

7. $f(x) = \begin{cases} 2x & \text{if } 0 \leqslant x \leqslant 1 \\ 0 & \text{otherwise} \end{cases}$

8. $f(x) = \begin{cases} \frac{1}{8}x & \text{if } 0 \leqslant x \leqslant 4 \\ 0 & \text{otherwise} \end{cases}$

9. $f(x) = \begin{cases} \frac{1}{6}x & \text{if } 2 \leqslant x \leqslant 4 \\ 0 & \text{otherwise} \end{cases}$

10. $f(x) = \begin{cases} \frac{1}{4}x & \text{if } 1 \leqslant x \leqslant 3 \\ 0 & \text{otherwise} \end{cases}$

11. $f(x) = \begin{cases} \frac{1}{2} - \frac{1}{8}x & \text{if } 0 \leqslant x \leqslant 4 \\ 0 & \text{otherwise} \end{cases}$

12. $f(x) = \begin{cases} 2 - 2x & \text{if } 0 \leqslant x \leqslant 1 \\ 0 & \text{otherwise} \end{cases}$

B *Problems 13 and 14 refer to random variables X_1 and X_2 with probability density functions f_1 and f_2, respectively, where*

$$f_1(x) = \begin{cases} 0.5 & \text{if } 0 \leqslant x \leqslant 2 \\ 0 & \text{otherwise} \end{cases}$$

$$f_2(x) = \begin{cases} 0.25 & \text{if } -2 \leqslant x \leqslant 2 \\ 0 & \text{otherwise} \end{cases}$$

13. Explain how you can predict from the graphs of f_1 and f_2 which random variable, X_1 or X_2, has the greater mean. Check your prediction by computing the mean of each.

14. Explain how you can predict from the graphs of f_1 and f_2 which random variable, X_1 or X_2, has the greater variance. Check your prediction by computing the variance of each.

Problems 15 and 16 refer to random variables X_1 and X_2 with probability density functions g_1 and g_2, respectively, where

$$g_1(x) = \begin{cases} 0.5 & \text{if } 0 \leqslant x < 1 \\ 0.25 & \text{if } 1 \leqslant x \leqslant 3 \\ 0 & \text{otherwise} \end{cases}$$

$$g_2(x) = \begin{cases} 0.5 & \text{if } 0 \leqslant x < 1 \\ 0.125 & \text{if } 1 \leqslant x \leqslant 5 \\ 0 & \text{otherwise} \end{cases}$$

15. Explain how you can predict from the graphs of g_1 and g_2 which random variable, X_1 or X_2, has the greater median. Check your prediction by computing the median of each.

16. Explain how you can predict from the graphs of g_1 and g_2 which random variable, X_1 or X_2, has the greater mean. Check your prediction by computing the mean of each.

In Problems 17–20, find the mean, variance, and standard deviation.

17. $f(x) = \begin{cases} 4/x^5 & \text{if } x \geqslant 1 \\ 0 & \text{otherwise} \end{cases}$

18. $f(x) = \begin{cases} 5/x^6 & \text{if } x \geqslant 1 \\ 0 & \text{otherwise} \end{cases}$

19. $f(x) = \begin{cases} 64/x^5 & \text{if } x \geqslant 2 \\ 0 & \text{otherwise} \end{cases}$

20. $f(x) = \begin{cases} 81/x^4 & \text{if } x \geqslant 3 \\ 0 & \text{otherwise} \end{cases}$

 In Problems 21 and 22, use a graphing utility to approximate the mean, variance, and standard deviation by replacing integrals over $(-\infty, \infty)$ by integrals over $[-10, 10]$.

21. $f(x) = \dfrac{2}{\sqrt{2\pi}} e^{-2(x-1)^2}$

22. $f(x) = \dfrac{1}{1.4\sqrt{2\pi}} e^{-(x+2.1)^2/3.92}$

In Problems 23–30, find the median.

23. $f(x) = \begin{cases} 1/x & \text{if } 1 \leqslant x \leqslant e \\ 0 & \text{otherwise} \end{cases}$

24. $f(x) = \begin{cases} 1/(2x) & \text{if } 1 \leqslant x \leqslant e^2 \\ 0 & \text{otherwise} \end{cases}$

25. $f(x) = \begin{cases} 4/(2 + x)^2 & \text{if } 0 \leqslant x \leqslant 2 \\ 0 & \text{otherwise} \end{cases}$

26. $f(x) = \begin{cases} 2/(1 + x)^2 & \text{if } 0 \leqslant x \leqslant 1 \\ 0 & \text{otherwise} \end{cases}$

27. $f(x) = \begin{cases} 1/(1 + x)^2 & \text{if } x \geqslant 0 \\ 0 & \text{otherwise} \end{cases}$

28. $f(x) = \begin{cases} 3/(3 + x)^2 & \text{if } x \geqslant 0 \\ 0 & \text{otherwise} \end{cases}$

29. $f(x) = \begin{cases} 2e^{-2x} & \text{if } x \geqslant 0 \\ 0 & \text{otherwise} \end{cases}$

30. $f(x) = \begin{cases} e^{-x} & \text{if } x \geqslant 0 \\ 0 & \text{otherwise} \end{cases}$

In Problems 31 and 32, use the graph of the cumulative distribution function F to find the median. (Assume each portion of the graph of F is a straight line segment.) Is the median unique? Explain.

31.

32.

In Problems 33 and 34, use the graph of the probability density function f to find the cumulative distribution function F. (Assume each portion of the graph of f is a straight line segment.) Find the median. Is the median unique? Explain.

33.

34.

C *In Problems 35 and 36, f(x) is a continuous probability density function with mean μ and standard deviation σ;*

a and b are constants. Evaluate each integral, expressing the result in terms of a, b, μ, and σ.

35. $\displaystyle\int_{-\infty}^{\infty} (ax + b)f(x)\, dx$ **36.** $\displaystyle\int_{-\infty}^{\infty} (x - a)^2 f(x)\, dx$

*The **quartile points** for a probability density function are the values x_1, x_2, x_3 that divide the area under the graph of the function into four equal parts. Find the quartile points for the probability density functions in Problems 37–40.*

37. $f(x) = \begin{cases} \frac{1}{2}x & \text{if } 0 \leqslant x \leqslant 2 \\ 0 & \text{otherwise} \end{cases}$

38. $f(x) = \begin{cases} 3x^2 & \text{if } 0 \leqslant x \leqslant 1 \\ 0 & \text{otherwise} \end{cases}$

39. $f(x) = \begin{cases} 3/(3 + x)^2 & \text{if } x \geqslant 0 \\ 0 & \text{otherwise} \end{cases}$

40. $f(x) = \begin{cases} 1/(1 + x)^2 & \text{if } x \geqslant 0 \\ 0 & \text{otherwise} \end{cases}$

 In Problems 41–44, use a graphing utility to approximate the median of the indicated probability density function f (to two decimal places).

41. $f(x) = \begin{cases} 4x - 4x^3 & \text{if } 0 \leqslant x \leqslant 1 \\ 0 & \text{otherwise} \end{cases}$

42. $f(x) = \begin{cases} 3x - 3x^5 & \text{if } 0 \leqslant x \leqslant 1 \\ 0 & \text{otherwise} \end{cases}$

43. $f(x) = \begin{cases} \dfrac{1}{2x^2} + \dfrac{3}{2x^4} & \text{if } x \geqslant 1 \\ 0 & \text{otherwise} \end{cases}$

44. $f(x) = \begin{cases} \dfrac{2}{3x^3} + \dfrac{2}{x^4} & \text{if } x \geqslant 1 \\ 0 & \text{otherwise} \end{cases}$

Applications

Business & Economics

45. *Profit.* A building contractor's profit (in thousands of dollars) on each unit in a subdivision is a continuous random variable with probability density function given by

$$f(x) = \begin{cases} 0.08(20 - x) & \text{if } 15 \leqslant x \leqslant 20 \\ 0 & \text{otherwise} \end{cases}$$

(A) Find the contractor's expected profit.

(B) Find the median profit.

46. *Electricity consumption.* The daily consumption of
www electricity (in millions of kilowatt-hours) in a large
city is a continuous random variable with probability
density function

$$f(x) = \begin{cases} .2 - .02x & \text{if } 0 \leq x \leq 10 \\ 0 & \text{otherwise} \end{cases}$$

(A) Find the expected daily consumption of
electricity.

(B) Find the median daily consumption of electricity.

47. *Waiting time.* The time (in minutes) a customer must
wait in line at a bank is a continuous random variable
with probability density function given by

$$f(x) = \begin{cases} \frac{1}{3}e^{-x/3} & \text{if } x \geq 0 \\ 0 & \text{otherwise} \end{cases}$$

Find the median waiting time.

48. *Product life.* The life expectancy (in years) of an automobile battery is a continuous random variable with
probability density function given by

$$f(x) = \begin{cases} \frac{1}{2}e^{-x/2} & \text{if } x \geq 0 \\ 0 & \text{otherwise} \end{cases}$$

Find the median life expectancy.

49. *Water consumption.* The daily consumption of water (in
millions of gallons) in a small city is a continuous random variable with probability density function given by

$$f(x) = \begin{cases} 1/(1 + x^2)^{3/2} & \text{if } x \geq 0 \\ 0 & \text{otherwise} \end{cases}$$

Find the expected daily consumption.

50. *Gasoline consumption.* The daily consumption of gasoline (in millions of gallons) in a large city is a continuous random variable with probability density function

$$f(x) = \begin{cases} 4/(4 + x^2)^{3/2} & \text{if } x \geq 0 \\ 0 & \text{otherwise} \end{cases}$$

Find the expected daily consumption of gasoline.

Life Sciences

51. *Life expectancy.* The life expectancy of a certain
microscopic organism (in minutes) is a continuous
random variable with probability density function

given by

$$f(x) = \begin{cases} \frac{1}{5,000}(10x^3 - x^4) & \text{if } 0 \leq x \leq 10 \\ 0 & \text{otherwise} \end{cases}$$

Find the mean life expectancy of one of these
organisms.

52. *Life expectancy.* The life expectancy (in months) of
plants of a certain species is a continuous random variable with probability density function given by

$$f(x) = \begin{cases} \frac{1}{36}(6x - x^2) & \text{if } 0 \leq x \leq 6 \\ 0 & \text{otherwise} \end{cases}$$

Find the mean life expectancy of one of these plants.

53. *Shelf-life.* The shelf-life (in days) of a perishable drug
www is a continuous random variable with probability density function given by

$$f(x) = \begin{cases} 800x/(400 + x^2)^2 & \text{if } x \geq 0 \\ 0 & \text{otherwise} \end{cases}$$

Find the median shelf-life.

54. *Shelf-life.* Repeat Problem 53 if

$$f(x) = \begin{cases} 200x/(100 + x^2)^2 & \text{if } x \geq 0 \\ 0 & \text{otherwise} \end{cases}$$

Social Sciences

55. *Learning.* The number of hours it takes a chimpanzee
to learn a new task is a continuous random variable
with probability density function given by

$$f(x) = \begin{cases} \frac{4}{9}x^2 - \frac{4}{27}x^3 & \text{if } 0 \leq x \leq 3 \\ 0 & \text{otherwise} \end{cases}$$

What is the expected number of hours it will take a
chimpanzee to learn the task?

56. *Voter turnout.* The number of registered voters (in
thousands) who vote in an off-year election in a small
town is a continuous random variable with probability density function given by

$$f(x) = \begin{cases} 1 - \frac{1}{8}x & \text{if } 4 \leq x \leq 8 \\ 0 & \text{otherwise} \end{cases}$$

(A) Find the expected voter turnout.

(B) Find the median voter turnout.

Section 11-4 **Special Probability Distributions**

- Uniform Distributions
- Exponential Distributions
- Normal Distributions

Now that we have developed some general properties of probability density
functions, we can examine some specific functions that are used extensively
in applications.

❑ UNIFORM DISTRIBUTIONS

Explore–Discuss 1

Consider a random variable X with probability density function

$$f(x) = \begin{cases} \frac{1}{10} & \text{if } 0 \leq x \leq 10 \\ 0 & \text{otherwise} \end{cases}$$

Graph f and use the graph to find the indicated probabilities.

(A) Find $P(0 \leq X \leq 1)$, and $P(2 \leq X \leq 3)$, $P(6 \leq X \leq 7)$, $P(9 \leq X \leq 10)$.

(B) If $a \geq 0$, $n \geq 0$, and $a + n \leq 10$, find $P(a \leq X \leq a + n)$.

(C) If I is an interval of length n that lies in the interval $[0, 10]$, what is the probability that the random variable X lies in I? Does this probability depend on the location of the interval I? Explain.

Simple probability density functions like the example given in the preceding Explore–Discuss 1 turn out to be quite useful. The next example will lead to some general observations about functions of this form.

Example 1 ⇨

Waiting Time A bus arrives every 30 minutes at a particular bus stop. If an individual arrives at the bus stop at a random time (that is, with no knowledge of the bus schedule), then the random variable X, representing the time this individual must spend waiting for the next bus, is said to be *uniformly distributed* on the interval $[0, 30]$. This means that the probability that X lies in a small interval of fixed length is independent of the location of that interval within $[0, 30]$. Thus, the probability of waiting between 0 and 5 minutes is the same as the probability of waiting between 5 and 10 minutes or the probability of waiting between 18 and 23 minutes. It can be shown that the probability density function for this uniformly distributed random variable is

$$f(x) = \begin{cases} \frac{1}{30} & \text{if } 0 \leq x \leq 30 \\ 0 & \text{otherwise} \end{cases}$$

Thus,

FIGURE 1

$$P(0 \leq X \leq 5) = \int_0^5 \frac{1}{30}\,dx = \frac{x}{30}\Big|_0^5 = \frac{5}{30} - \frac{0}{30} = \frac{1}{6}$$

$$P(5 \leq X \leq 10) = \int_5^{10} \frac{1}{30}\,dx = \frac{x}{30}\Big|_5^{10} = \frac{10}{30} - \frac{5}{30} = \frac{1}{6}$$

$$P(18 \leq X \leq 23) = \int_{18}^{23} \frac{1}{30}\,dx = \frac{x}{30}\Big|_{18}^{23} = \frac{23}{30} - \frac{18}{30} = \frac{1}{6}$$

Each of these probabilities is represented as an area under the graph of f in Figure 1. ∎

Matched Problem 1 ⇨

Use the probability density function given in Example 1 to find the probability that the individual waits:

(A) Between 0 and 10 minutes

(B) Between 10 and 20 minutes

(C) Between 17 and 27 minutes

In general, if the outcomes of an experiment lie in an interval $[a, b]$ and if the probability of the outcome lying in a small interval of fixed length is independent of the location of this small interval within $[a, b]$, then we say that the continuous random variable for this experiment is **uniformly distributed** on the interval $[a, b]$. The **uniform probability density function** is

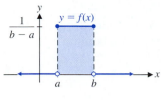

$$f(x) = \begin{cases} \dfrac{1}{b - a} & \text{if } a \leqslant x \leqslant b \\ 0 & \text{otherwise} \end{cases}$$

FIGURE 2

See Figure 2. Since $f(x) \geqslant 0$ and

$$\int_{-\infty}^{\infty} f(x)\, dx = \int_{a}^{b} \frac{1}{b - a}\, dx = \frac{x}{b - a}\Big|_{a}^{b} = \frac{b}{b - a} - \frac{a}{b - a} = 1$$

f satisfies the necessary conditions for a probability density function.

If F is the associated cumulative distribution function, then for $x < a$, $F(x) = 0$. For $a \leqslant x \leqslant b$, we have

$$F(x) = \int_{-\infty}^{x} f(t)\, dt = \int_{a}^{x} \frac{1}{b - a}\, dt = \frac{t}{b - a}\Big|_{a}^{x}$$

$$= \frac{x}{b - a} - \frac{a}{b - a} = \frac{x - a}{b - a}$$

For $x > b$, $F(x) = 1$.

Using the techniques discussed in the preceding section, it can be shown that (see Problems 49–52 in Exercise 11-4)

$$\mu = m = \frac{1}{2}(a + b) \qquad \text{and} \qquad \sigma = \frac{1}{\sqrt{12}}(b - a)$$

These properties are summarized in the next box.

Uniform Probability Density Function

$$f(x) = \begin{cases} \dfrac{1}{b - a} & \text{if } a \leqslant x \leqslant b \\ 0 & \text{otherwise} \end{cases} \qquad F(x) = \begin{cases} 0 & \text{if } x < a \\ \dfrac{x - a}{b - a} & \text{if } a \leqslant x \leqslant b \\ 1 & \text{if } x > b \end{cases}$$

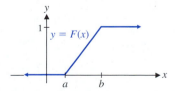

Mean $\mu = \frac{1}{2}(a + b)$ **Median** $m = \frac{1}{2}(a + b)$

Standard deviation $\sigma = \dfrac{1}{\sqrt{12}}(b - a)$

Example 2 ⇔

Electrical Current Standard electrical current is uniformly distributed between 110 and 120 volts. What is the probability that the current is between 113 and 118 volts?

SOLUTION Since we are told that the current is uniformly distributed on the interval $[110, 120]$, we choose the uniform probability density function

$$f(x) = \begin{cases} \frac{1}{10} & \text{if } 110 \leq x \leq 120 \\ 0 & \text{otherwise} \end{cases}$$

Then

$$P(113 \leq X \leq 118) = \int_{113}^{118} \frac{1}{10}\, dx = \left.\frac{x}{10}\right|_{113}^{118} = \frac{118}{10} - \frac{113}{10} = \frac{1}{2}$$

■

Matched Problem 2 ⇔

In Example 2, what is the probability that the current is at least 116 volts?

■

❑ EXPONENTIAL DISTRIBUTIONS

Explore–Discuss 2

Determine a relationship between the positive constants a and b that will make the following function a probability density function.

$$f(x) = \begin{cases} ae^{-bx} & \text{if } x \geq 0 \\ 0 & \text{otherwise} \end{cases}$$

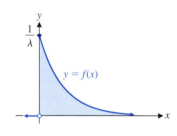

A continuous random variable has an **exponential distribution** and is referred to as an **exponential random variable** if its probability density function is the **exponential probability density function,**

$$f(x) = \begin{cases} (1/\lambda)e^{-x/\lambda} & \text{if } x \geq 0 \\ 0 & \text{otherwise} \end{cases}$$

where λ is a positive constant. Exponential random variables are used in a variety of applications, including studies of the length of telephone conversations, the time customers spend waiting in line at a bank, and the life expectancy of machine parts.

Since $f(x) \geq 0$ and

$$\int_{-\infty}^{\infty} f(x)\, dx = \int_{0}^{\infty} \frac{1}{\lambda} e^{-x/\lambda}\, dx$$

$$= \lim_{R \to \infty} \int_{0}^{R} \frac{1}{\lambda} e^{-x/\lambda}\, dx$$

$$= \lim_{R \to \infty} \left. (-e^{-x/\lambda}) \right|_{0}^{R}$$

$$= \lim_{R \to \infty} (-e^{-R/\lambda} + 1) = 1$$

f satisfies the conditions for a probability density function. If F is the cumulative distribution function, we see that $F(x) = 0$ for $x < 0$. For $x \geq 0$, we have

$$F(x) = \int_{-\infty}^{x} f(t)\, dt = \int_{0}^{x} \frac{1}{\lambda} e^{-t/\lambda}\, dt$$

$$= \left. -e^{-t/\lambda} \right|_{0}^{x} = 1 - e^{-x/\lambda}$$

To find the median, we solve

$$F(m) = P(X \leq m) = \frac{1}{2}$$

$$1 - e^{-m/\lambda} = \frac{1}{2}$$

$$\frac{1}{2} = e^{-m/2}$$

$$\ln \frac{1}{2} = -\frac{m}{\lambda}$$

$$m = -\lambda \ln \frac{1}{2} = \lambda \ln 2 \quad \text{Note: } \ln \frac{1}{2} = -\ln 2$$

Integration by parts can be used to show that $\mu = \lambda$ and $\sigma = \lambda$. The calculations are not included here. These results are summarized in the next box.

Exponential Probability Density Function

$$f(x) = \begin{cases} (1/\lambda)e^{-x/\lambda} & \text{if } x \geq 0 \\ 0 & \text{otherwise} \end{cases} \qquad F(x) = \begin{cases} 1 - e^{-x/\lambda} & \text{if } x \geq 0 \\ 0 & \text{otherwise} \end{cases}$$

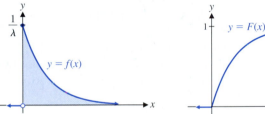

Mean $\mu = \lambda$ **Median** $m = \lambda \ln 2$

Standard deviation $\sigma = \lambda$

Example 3   **Arrival Rates**   The length of time between calls received by the switchboard in a large legal firm is an exponential random variable. The average length of time between calls is 20 seconds. If a call has just been received, what is the probability that no calls are received in the next 30 seconds?

SOLUTION   Let X be the random variable that represents the length of time between calls received (in seconds). Since the average length of time is 20 seconds, we have $\mu = \lambda = 20$. Thus, the probability density function for X is

$$f(x) = \begin{cases} \frac{1}{20}e^{-x/20} & \text{if } x \geq 0 \\ 0 & \text{otherwise} \end{cases}$$

and

$$P(X \geq 30) = \int_{30}^{\infty} \tfrac{1}{20}e^{-x/20}\, dx \qquad \int_{0}^{30} f(x)\,dx + \int_{30}^{\infty} f(x)\,dx = 1$$

$$= 1 - \int_{0}^{30} \tfrac{1}{20}e^{-x/20}\, dx$$

$$= 1 - (-e^{-x/20})\Big|_{0}^{30} = e^{-1.5} \approx .223$$

Matched Problem 3   In Example 3, if a call has just been received, what is the probability that no calls are received in the next 10 seconds?

❑ NORMAL DISTRIBUTIONS

We now consider the most important of all the probability density functions, the *normal probability density function*. This function is at the heart of a great deal of statistical theory, and it is also a useful tool in its own right for solving problems.

A continuous random variable X has a **normal distribution** and is referred to as a **normal random variable** if its probability density function is the **normal probability density function**

$$f(x) = \frac{1}{\sigma\sqrt{2\pi}} e^{-(x-\mu)^2/(2\sigma^2)}$$

where μ is any constant and σ is any positive constant. It can be shown, but not easily, that

$$\int_{-\infty}^{\infty} f(x)\,dx = 1$$

and

$$E(X) = \int_{-\infty}^{\infty} xf(x)\,dx = \mu \qquad V(X) = \int_{-\infty}^{\infty} (x-\mu)^2 f(x)\,dx = \sigma^2$$

Thus, μ is the mean of the normal probability density function and σ is the standard deviation. The graph of $f(x)$ is always a bell-shaped curve called a **normal curve.** Figure 3 illustrates three normal curves for different values of μ and σ.

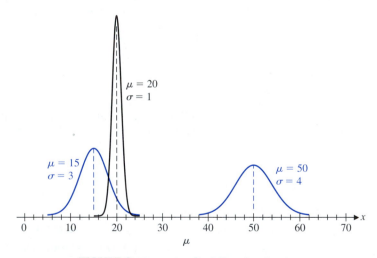

FIGURE 3 Normal probability distributions

The standard deviation measures the dispersion of the normal probability density function about the mean. A small standard deviation indicates a tight clustering about the mean and thus a tall, narrow curve; a large standard deviation indicates a large deviation from the mean and thus a broad, flat curve. Notice that each of the normal curves in Figure 3 is symmetric about a vertical line through the mean. This is true for any normal curve. Thus, the line $x = \mu$ divides the region under a normal curve into two regions with equal areas. Since the total area under a normal curve is always 1, the area of each of these regions is .5. This implies that the median of a normal random variable is always equal to the mean (see Fig. 4).

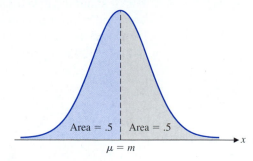

FIGURE 4 The mean and median of a normal random variable

The properties of the normal probability density function are summarized in the next box for ease of reference.

Normal Probability Density Function

$$f(x) = \frac{1}{\sigma\sqrt{2\pi}}\, e^{-(x-\mu)^2/(2\sigma^2)} \qquad \sigma > 0$$

Mean μ
Median μ
Standard deviation σ

The graph of $f(x)$ is symmetric with respect to the line $x = \mu$.

The cumulative distribution function for a normal random variable is given formally by

$$F(x) = \frac{1}{\sigma\sqrt{2\pi}} \int_{-\infty}^{x} e^{-(t-\mu)^2/(2\sigma^2)}\, dt$$

It is not possible to express $F(x)$ as a finite combination of the functions with which we are familiar. Furthermore, we cannot use antidifferentiation to evaluate probabilities such as

$$P(c \leq X \leq d) = \frac{1}{\sigma\sqrt{2\pi}} \int_{c}^{d} e^{-(x-\mu)^2/(2\sigma^2)}\, dx$$

However, numerical integration can be used to approximate probabilities of this type, as illustrated in Figure 5, where we use a graphing utility to approximate $P(0 \leq X \leq 1.5)$ for the normal probability density function with $\mu = 0$ and $\sigma = 1$.

It is also possible to use a table to approximate probabilities involving the normal probability density function. Remarkably, the area under a normal curve between a mean μ and a given number of standard deviations to the right (or left) of μ is the same, regardless of the shape of the normal curve. For example, the area under the normal curve with $\mu = 3, \sigma = 5$ from $\mu = 3$ to $\mu + 1.5\sigma = 10.5$ is equal to the area under the normal curve with $\mu = 15, \sigma = 2$ from $\mu = 15$ to $\mu = 1.5\sigma = 18$ (see Fig. 6 on the next page, noting that the shaded regions have the same areas, or equivalently, the same numbers of pixels). Therefore, such areas for any normal curve can be easily

FIGURE 5

determined from the areas for the **standard normal curve,** that is, the normal curve with mean 0 and standard deviation 1. In fact, if z represents the number of standard deviations that a measurement x is from a mean μ, then the area under a normal curve from μ to $\mu + z\sigma$ equals the area under the standard normal curve from 0 to z (see Fig. 7). Table III in Appendix C lists those areas for the standard normal curve.

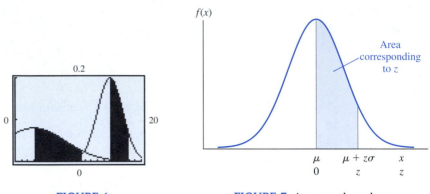

FIGURE 6   **FIGURE 7** Areas and z values

In the discussion that follows we will concentrate on using Table III in Appendix C to approximate areas under the normal curve. If desired, numerical integration on a graphing utility can be used to obtain the same results.

Example 4   ⤳   **Finding Probabilities for a Normal Distribution**   A manufacturing process produces light bulbs with life expectancies that are normally distributed with a mean of 500 hours and a standard deviation of 100 hours. What percentage of the light bulbs can be expected to last between 500 and 670 hours?

SOLUTION   To answer this question, we first determine how many standard deviations 670 is from 500, the mean. This is easily done by dividing the distance between 500 and 670 by 100, the standard deviation. Thus,

$$z = \frac{670 - 500}{100} = \frac{170}{100} = 1.70$$

That is, 670 is 1.7 standard deviations from 500, the mean. Referring to Table III in Appendix C, we see that .4554 corresponds to $z = 1.70$. And since the total area under a normal curve is 1, we conclude that 45.54% of the light bulbs produced will last between 500 and 670 hours (see Fig. 8).

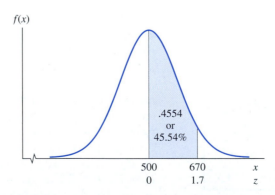

FIGURE 8 Light bulb life expectancy—positive z

Matched Problem 4 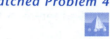   In Example 4, what percentage of the light bulbs can be expected to last between 500 and 750 hours?

In general, to find how many standard deviations a measurement x is from a mean μ, first determine the distance between x and μ, and then divide by σ:

$$z = \frac{\text{Distance between } x \text{ and } \mu}{\text{Standard deviation}} = \frac{x - \mu}{\sigma}$$

Example 5   **Finding Probabilities for a Normal Distribution**   From all light bulbs produced (see Example 4), what is the probability of a light bulb chosen at random lasting between 380 and 500 hours?

SOLUTION   To answer this, we first find z:

$$z = \frac{x - \mu}{\sigma} = \frac{380 - 500}{100} = -1.20$$

It is usually a good idea to draw a rough sketch of a normal curve and insert relevant data (see Fig. 9).

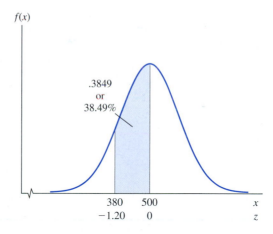

FIGURE 9 Light bulb life expectancy—negative z

Table III in Appendix C does not include negative values for z, but because normal curves are symmetrical with respect to a vertical line through the mean, we simply use the absolute value (positive value) of z for the table. Thus, the area corresponding to $z = -1.20$ is the same as the area corresponding to $z = 1.20$, which is .3849. And since the area under the whole normal curve is 1, we conclude that the probability of a light bulb chosen at random lasting between 380 and 500 hours is .3849.

Matched Problem 5   In Example 5, what is the probability of a light bulb chosen at random lasting between 400 and 500 hours?

Answers to Matched Problems   **1.** (A) $\frac{1}{3}$ (B) $\frac{1}{3}$ (C) $\frac{1}{3}$ **2.** $\frac{2}{5}$ **3.** $e^{-0.5} \approx .607$ **4.** 49.38%   **5.** .3413

Exercise 11-4

A *In Problems 1–4, find the probability density function f and the associated cumulative distribution function F for the continuous random variable X if*

1. X is uniformly distributed on $[0, 2]$.

2. X is uniformly distributed on $[3, 6]$.

3. X is an exponential random variable with $\lambda = \frac{1}{2}$.

4. X is an exponential random variable with $\lambda = \frac{1}{4}$.

In Problems 5–8, find the mean, median, and standard deviation of the continuous random variable X if

5. X is uniformly distributed on $[1, 5]$.

6. X is uniformly distributed on $[2, 8]$.

7. X is an exponential random variable with $\lambda = 5$.

8. X is an exponential random variable with $\lambda = 3$.

In Problems 9–16, use Table III to find the area under the standard normal curve from 0 to the indicated value of z.

9. 1   **10.** 2   **11.** −3   **12.** −1

13. .9   **14.** −1.7   **15.** 2.47   **16.** −1.96

Given a normal distribution with mean 50 and standard deviation 10, use Table III to find the area under this normal curve from the mean to the indicated measurement in Problems 17–24.

17. 65   **18.** 75   **19.** 83   **20.** 79

21. 45   **22.** 38   **23.** 42   **24.** 26

B *Problems 25 and 26 refer to random variables X_1 and X_2 with probability density functions f_1 and f_2, respectively, where*

$$f_1(x) = \frac{1}{\sqrt{2\pi}} e^{-x^2/2}$$

$$f_2(x) = \begin{cases} 0.4 & \text{if } -1.25 \leq x \leq 1.25 \\ 0 & \text{otherwise} \end{cases}$$

25. Explain how you can predict from the graphs of f_1 and f_2 which random variable, X_1 or X_2, has the greater standard deviation. Check your prediction by finding the standard deviation of each.

26. Explain how you can predict from the graphs of f_1 and f_2 which random variable, X_1 or X_2, has the greater median. Check your prediction by finding the median of each.

Problems 27 and 28 refer to random variables X_1 and X_2 with probability density functions g_1 and g_2, respectively, where

$$g_1(x) = \frac{1}{\sqrt{2\pi}} e^{-(x-1)^2/2}$$

$$g_2(x) = \begin{cases} e^{-x} & \text{if } x \geq 0 \\ 0 & \text{otherwise} \end{cases}$$

27. Explain how you can predict from the graphs of g_1 and g_2 which random variable, X_1 or X_2, has the greater median. Check your prediction by finding the median of each.

28. Explain how you can predict from the graphs of g_1 and g_2 which random variable, X_1 or X_2, has the greater mean. Check your prediction by finding the mean of each.

In Problems 29–32, X is a continuous random variable with mean μ. Find μ, and then find $P(X \leq \mu)$ if

29. X is uniformly distributed on $[0, 4]$.

30. X is uniformly distributed on $[0, 10]$.

31. X is an exponential random variable with $\lambda = 1$.

32. X is an exponential random variable with $\lambda = 2$.

In Problems 33–36, X is a continuous random variable with mean μ and standard deviation σ. Find μ and σ, and then find $P(\mu - \sigma \leq X \leq \mu + \sigma)$ if

33. X is uniformly distributed on $[-5, 5]$.

34. X is uniformly distributed on $[-2, 2]$.

35. X is an exponential random variable with $m = 6 \ln 2$.

36. X is an exponential random variable with $m = 4 \ln 2$.

Given a normal random variable X with mean 70 and standard deviation 8, find the indicated probabilities in Problems 37–44.

37. $P(60 \leq X \leq 80)$ **38.** $P(50 \leq X \leq 90)$

39. $P(62 \leq X \leq 74)$ **40.** $P(66 \leq X \leq 78)$

41. $P(X \geq 88)$ **42.** $P(X \geq 90)$

43. $P(X \leq 60)$ **44.** $P(X \leq 56)$

 Problems 45–48 refer to the normal random variable X with mean μ and standard deviation σ. Use a graphing utility with a built-in numerical integration routine to compute the indicated probabilities.

45. Approximate $P(\mu - \sigma \leq X \leq \mu + \sigma)$ to four decimal places for $\mu = 0$ and

(A) $\sigma = .5$ (B) $\sigma = 1$ (C) $\sigma = 2$

Write a brief verbal explanation of the result indicated by these calculations.

46. Approximate $P(\mu - 2\sigma \leq X \leq \mu + 2\sigma)$ to four decimal places for $\mu = 0$ and

(A) $\sigma = .5$ (B) $\sigma = 1$ (C) $\sigma = 2$

Write a brief verbal explanation of the result indicated by these calculations.

47. Approximate $P(\mu - 3\sigma \leq X \leq \mu + 3\sigma)$ to four decimal places for $\mu = 0$ and

(A) $\sigma = .5$ (B) $\sigma = 1$ (C) $\sigma = 2$

Write a brief verbal explanation of the result indicated by these calculations.

48. In Problems 45–47, if μ is changed from 0 to 5, will any of the probabilities change? Explain.

C *Problems 49–52 refer to the uniformly distributed random variable X with probability density function*

$$f(x) = \begin{cases} \dfrac{1}{b-a} & \text{if } a \leq x \leq b \\ 0 & \text{otherwise} \end{cases}$$

49. Show that $\mu = (a+b)/2$.

50. Show that $m = (a+b)/2$.

51. Show that $\int_{-\infty}^{\infty} x^2 f(x)\, dx = (b^2 + ab + a^2)/3$.

52. Show that $V(X) = (b-a)^2/12$.

53. Use numerical integration on a graphing utility to approximate the mean, median, and standard deviation of the probability density function

$$f(x) = \begin{cases} 5e^{-5x} & \text{if } x \geq 0 \\ 0 & \text{otherwise} \end{cases}$$

Explain why it is reasonable to replace integrals over the interval $(-\infty, \infty)$ by integrals over the interval $[0, 10]$. Verify that your answers agree with the results expected for exponential probability density functions.

54. Use numerical integration on a graphing utility to approximate the mean, median, and standard deviation of the probability density function

$$f(x) = \begin{cases} 0.36e^{-0.36x} & \text{if } x \geq 0 \\ 0 & \text{otherwise} \end{cases}$$

Explain why it is reasonable to replace integrals over the interval $(-\infty, \infty)$ by integrals over the interval $[0, 100]$. Verify that your answers agree with the results expected for exponential probability density functions.

Problems 55–58 refer to a random variable X with probability density function

$$f(x) = \begin{cases} p/x^{p+1} & \text{if } x \geq 1 \\ 0 & \text{otherwise} \end{cases}$$

*where p is a positive constant. This random variable is often referred to as a **Pareto*** random variable*

55. Verify that f satisfies the necessary conditions for a probability density function and find the mean, if it exists. What restrictions must you place on p to ensure that the mean exists?

56. Find the standard deviation, if it exists. What restrictions must you place on p to ensure that the standard deviation exists?

57. Find the median.

58. The mean and the median for a uniformly distributed random variable are equal, as are the mean and standard deviation for an exponentially distributed random variable. Use a graph to illustrate the relationship between the mean, median, and standard deviation for a Pareto random variable. Approximate abscissas of intersection points to two decimal places.

Applications

Business & Economics

59. *Waiting time.* The time (in minutes) applicants must wait to receive a driver's examination is uniformly distributed on the interval $[0, 40]$. What is the probability that an applicant must wait more than 25 minutes?

60. *Waiting time.* The time (in minutes) passengers must wait for a commuter plane in a large airport is uniformly distributed on the interval $[0, 60]$. What is the probability that a passenger waits less than 20 minutes?

61. *Communications.* The length of time for telephone conversations (in minutes) is exponentially distributed. The average (mean) length of a conversation is 3 minutes. What is the probability that a conversation lasts less than 2 minutes?

62. *Waiting time.* The waiting time (in minutes) for customers at a drive-in bank is an exponential random variable. The average (mean) time a customer waits is 4 minutes. What is the probability that a customer waits more than 5 minutes?

63. *Service time.* The time between failures of a photo copier is an exponential random variable. Half the copiers require service during the first 2 years of operation. What is the probability that a copier requires service during the first year of operation?

64. *Component failure.* The life expectancy (in years) of a component in a computer is an exponential random variable. Half the components fail in the first 3 years. The company that manufactures the component offers a 1 year warranty. What is the probability that a component will fail during the warranty period?

65. *Sales.* The annual sales for salespeople at a business machine company are normally distributed. The average (mean) annual sales are $200,000 with a standard deviation of $20,000. What percentage of the

*Named after Vilfredo Pareto (1848–1923), an Italian sociologist and economist. Although best known for his theories on political behavior, Pareto also developed new applications of mathematics to economics.

salespeople would be expected to make annual sales of $240,000 or more?

66. *Guarantees.* The life expectancy of a car battery is normally distributed. The average (mean) lifetime is 170 weeks with a standard deviation of 10 weeks. If the company guarantees the battery for 3 years, what percentage of the batteries sold would be expected to be returned before the end of the warranty period?

67. *Quality control.* A manufacturing process produces a
www precision part whose length is a normal random variable with mean 100 millimeters and standard deviation 2 millimeters. All parts deviating by more than 5 millimeters from the mean must be rejected. What percentage of the parts must be rejected, on the average?

68. *Quality control.* An automated manufacturing process produces a component whose width is a normal random variable with mean 7.55 centimeters and standard deviation 0.02 centimeter. All components deviating by more than 0.05 centimeter from the mean must be rejected. What percentage of the parts must be rejected, on the average?

Life Sciences

69. *Survival time.* The time of death (in years) after pa-
www tients have contracted a certain disease is exponentially distributed. The probability that a patient dies within 1 year is .3.

(A) What is the expected time of death?

(B) What is the probability that a patient survives longer than the expected time of death?

70. *Survival time.* Repeat Problem 69 if the probability that a patient dies within 1 year is .5.

71. *Medicine.* The healing time of a certain type of incision is normally distributed with an average

healing time of 240 hours and a standard deviation of 20 hours. What percentage of the people having this incision would heal in 8 days or less?

72. *Agriculture.* The height of a hay crop is normally distributed with an average height of 38 inches and a standard deviation of 1.5 inches. What percentage of the crop will be 40 inches or more?

Social Sciences

73. *Learning.* The time (in minutes) it takes an adult to memorize a sequence of random digits is an exponential random variable. The average time is 2 minutes. What is the probability that it takes an adult over 5 minutes to memorize the digits?

74. *Psychology.* The time (in seconds) it takes rats to find their way through a maze is exponentially distributed. The average time is 30 seconds. What is the probability that it takes a rat over 1 minute to find a path through the maze?

75. *Grading on a curve.* An instructor grades on a curve by assuming the grades on a test are normally distributed. If the average grade is 70 and the standard deviation is 8, find the test scores for each grade interval if the instructor wishes to assign grades as follows: 10% A's, 20% B's, 40% C's, 20% D's, and 10% F's. Round answers to one decimal place.

76. *Psychology.* A test devised to measure aggressive–passive personalities was standardized on a large group of people. The scores were normally distributed with a mean of 50 and a standard deviation of 10. If we want to designate the highest 10% as aggressive, the next 20% as moderately aggressive, the middle 40% as average, the next 20% as moderately passive, and the lowest 10% as passive, what range of scores will be covered by these five designations? Round answers to one decimal place.

Important Terms And Symbols

11-1 *Improper Integrals.* Improper integral; converge; diverge; perpetual income stream; capital value

$$\int_a^\infty f(x)\, dx = \lim_{b\to\infty} \int_a^b f(x)\, dx$$

$$\int_{-\infty}^b f(x)\, dx = \lim_{a\to-\infty} \int_a^b f(x)\, dx$$

$$\int_{-\infty}^\infty f(x)\, dx = \int_{-\infty}^c f(x)\, dx + \int_c^\infty f(x)\, dx$$

$$CV = \int_0^\infty f(t)e^{-rt}\, dt$$

11-2 *Continuous Random Variables.* Discrete random variable; continuous random variable; probability density

function; cumulative distribution function

$$P(c \le X \le d) = \int_c^d f(x)\, dx$$

$$F(x) = P(X \le x) = \int_{-\infty}^x f(t)\, dt$$

11-3 *Expected Value, Standard Deviation, and Median.* Expected value; mean; variance; standard deviation; alternative formula for variance; median

Expected value (Mean): $\mu = E(X) = \int_{-\infty}^\infty x f(x)\, dx$

Variance: $V(X) = \int_{-\infty}^{\infty} (x - \mu)^2 f(x)\, dx$

$$= \int_{-\infty}^{\infty} x^2 f(x)\, dx - \mu^2$$

Standard deviation: $\sigma = \sqrt{V(X)}$

Median: Solve $P(X \le m) = \frac{1}{2}$ for m.

11-4 *Special Probability Distributions.* Uniformly distributed; uniform probability density function; exponential distributions; exponential random variable; exponential probability density function; normal distribution; normal random variable; normal probability density function; normal curve; standard normal curve.

Uniform probability density function:

$$f(x) = \begin{cases} \dfrac{1}{b - a} & \text{if } a \le x \le b \\ 0 & \text{otherwise} \end{cases}$$

Exponential probability density function:

$$f(x) = \begin{cases} (1/\lambda)e^{-x/\lambda} & \text{if } x \ge 0 \\ 0 & \text{otherwise} \end{cases}$$

Normal probability density function:

$$f(x) = \frac{1}{\sigma\sqrt{2\pi}} e^{-(x-\mu)^2/(2\sigma^2)} \quad \sigma > 0$$

Review Exercise

Work through all the problems in this chapter review and check your answers in the back of the book. Answers to all review problems are there along with section numbers in italics to indicate where each type of problem is discussed. Where weaknesses show up, review appropriate sections in the text.

A *In Problems 1–3, evaluate each improper integral, if it converges.*

1. $\displaystyle\int_0^{\infty} e^{-2x}\, dx$

2. $\displaystyle\int_0^{\infty} \frac{1}{x + 1}\, dx$

3. $\displaystyle\int_1^{\infty} \frac{16}{x^3}\, dx$

Problems 4–7 refer to the continuous random variable X with probability density function

$$f(x) = \begin{cases} 1 - \frac{1}{2}x & \text{if } 0 \le x \le 2 \\ 0 & \text{otherwise} \end{cases}$$

4. Find $P(0 \le X \le 1)$ and illustrate with a graph.

5. Find the mean, variance, and standard deviation.

6. Find and graph the associated cumulative distribution function.

7. Find the median.

8. If X is a normal random variable with a mean of 100 and a standard deviation of 10, find $P(100 \le X \le 118)$.

In Problems 9 and 10, find the probability density function f and the associated cumulative distribution function F for the continuous random variable X if

9. X is uniformly distributed on $[5, 15]$.

10. X is an exponential random variable with $\lambda = \frac{1}{5}$.

In Problems 11 and 12, is f a probability density function? Explain.

11. $f(x) = e^{-x^2}$

12. $f(x) = \begin{cases} 0.75(x^2 - 4x + 3) & \text{if } 0 \le x \le 4 \\ 0 & \text{otherwise} \end{cases}$

B *Problems 13–16 refer to the continuous random variable X with probability density function*

$$f(x) = \begin{cases} \frac{5}{2}x^{-7/2} & \text{if } x \ge 1 \\ 0 & \text{otherwise} \end{cases}$$

13. Find $P(1 \le X \le 4)$ and illustrate with a graph.

14. Find the mean, variance, and standard deviation.

15. Find and graph the associated cumulative distribution function.

16. Find the median.

Problems 17–20 refer to an exponentially distributed random variable X.

17. If $P(4 \le X) = e^{-2}$, find the probability density function.

18. Find $P(0 \le X \le 2)$.

19. Find the associated cumulative distribution function.

20. Find the mean, standard deviation, and median.

21. Given a normal distribution with mean 50 and standard deviation 6, find the area under the normal curve:

(A) Between 41 and 62   (B) From 59 on

22. Given a normal random variable X with mean 82 and standard deviation 8, find:

(A) $P(84 \le X \le 94)$ (B) $P(X \ge 60)$

In Problems 23 and 24, evaluate each improper integral, if it converges.

23. $\displaystyle\int_{-\infty}^{0} e^x\, dx$ **24.** $\displaystyle\int_0^{\infty} \frac{1}{(x + 3)^2}\, dx$

25. If f is continuous on $[-1, \infty)$ and $\int_{-1}^{\infty} f(x)\, dx$ converges, does $\int_{1}^{\infty} f(x)\, dx$ converge? Explain.

In Problems 26 and 27, find a constant k so that kf is a probability density function, or explain why no such k exists.

26. $f(x) = \begin{cases} e^{-10x} & \text{if } x \geqslant 0 \\ 0 & \text{otherwise} \end{cases}$

27. $f(x) = \begin{cases} e^{10x} & \text{if } x \geqslant 0 \\ 0 & \text{otherwise} \end{cases}$

28. If X is an exponentially distributed random variable with median $m = 3 \ln 2$ and mean μ, find $P(X \leqslant \mu)$.

C **29.** Find the mean and the median of the continuous random variable with probability density function

$$f(x) = \begin{cases} 50/(x + 5)^3 & \text{if } x \geqslant 0 \\ 0 & \text{otherwise} \end{cases}$$

30. Use a graphing utility to approximate (to two decimal places) the median of the continuous random variable with probability density function

$$f(x) = \begin{cases} \dfrac{.8}{x^2} + \dfrac{.8}{x^5} & \text{if } x \geqslant 1 \\ 0 & \text{otherwise} \end{cases}$$

31. Find $\displaystyle\int_{-\infty}^{\infty} \frac{e^x}{(1 + e^x)^2}\, dx$, if it converges.

32. If f is a continuous probability density function with mean μ and standard deviation σ and a, b, and c are constants, evaluate the following integral. Express the result in terms of μ, σ, a, b, and c.

$$\int_{-\infty}^{\infty} (ax^2 + bx + c)f(x)\, dx$$

Problems 33–36 refer to random variables X_1 and X_2 with probability density functions f_1 and f_2, respectively, where

$$f_1(x) = \begin{cases} 0.25xe^{-x/2} & \text{if } x \geqslant 0 \\ 0 & \text{otherwise} \end{cases}$$

$$f_2(x) = \begin{cases} 0.0625x^2e^{-x/2} & \text{if } x \geqslant 0 \\ 0 & \text{otherwise} \end{cases}$$

33. Explain how you can predict from the graphs of f_1 and f_2 which random variable, X_1 or X_2, has the greater mean.

34. Explain how you can predict from the graphs of f_1 and f_2 which random variable, X_1 or X_2, has the greater variance.

35. Check your prediction in Problem 33 by using numerical integration to approximate the means of X_1 and X_2.

36. Check your prediction in Problem 34 by using numerical integration to approximate the variances of X_1 and X_2.

Applications

Business & Economics

37. *Production.* The rate of production of an oil well (in millions of barrels per year) is given by (see the figure)

$$R(t) = 12e^{-0.3t} - 12e^{-0.6t}$$

Time (years)

(A) Assuming that the well is operated indefinitely, find the total production.

(B) When will the output from the well reach 50% of the total production? Round answer to two decimal places.

38. *Demand.* The manager of a movie theater has determined that the weekly demand for popcorn (in pounds) is a continuous random variable with probability density function

$$f(x) = \begin{cases} .02(1 - .01x) & \text{if } 0 \leqslant x \leqslant 100 \\ 0 & \text{otherwise} \end{cases}$$

(A) Evaluate $\int_{40}^{100} f(x)\,dx$ and interpret.

(B) If the manager has 50 pounds of popcorn on hand at the beginning of the week, what is the probability that this will be enough to meet the weekly demand?

(C) If the manager wants the probability that the supply on hand exceeds the weekly demand to be .96, how much popcorn must be on hand at the beginning of the week?

39. *Capital value.* The perpetual annual rent for a property is $2,400. Find the capital value at 6% compounded continuously.

40. *Credit applications.* The percentage of applications (expressed as a decimal between 0 and 1) for a national credit card processed on the same day they are received is a continuous random variable with probability density function

$$f(x) = \begin{cases} 6x(1-x) & \text{if } 0 \le x \le 1 \\ 0 & \text{otherwise} \end{cases}$$

(A) What is the probability that at least 20% of the applications received are processed the same day they arrive?

(B) What is the expected percentage of applications processed the same day they arrive?

(C) What is the median percentage of applications processed the same day they arrive?

41. *Computer failure.* A computer manufacturer has determined that the time between failures for its computer is an exponentially distributed random variable with a mean failure time of 4,000 hours. Suppose a particular computer has just been repaired.

(A) What is the probability that the computer operates for the next 4,000 hours without failure?

(B) What is the probability that the computer fails in the next 1,000 hours?

42. *Radial tire failure.* The life expectancy (in miles) of a certain brand of radial tires is a normal random variable with a mean of 35,000 miles and a standard deviation of 5,000 miles. What is the probability that a tire fails during the first 25,000 miles of use?

43. *Personnel screening.* The scores on a screening test for new technicians are normally distributed with mean 100 and standard deviation 10. Find the approximate percentage of applicants taking the test who score

(A) Between 92 and 108   (B) 115 or higher

Life Sciences

44. *Medicine.* The shelf-life (in months) of a certain drug is a continuous random variable with probability density function

$$f(x) = \begin{cases} 10/(x+10)^2 & \text{if } x \ge 0 \\ 0 & \text{otherwise} \end{cases}$$

(A) Evaluate $\int_2^8 f(x)\,dx$ and interpret.

(B) What is the probability that the drug is still usable after 5 months?

(C) What is the median shelf-life?

45. *Life expectancy.* The life expectancy (in months) after dogs have contracted a certain disease is an exponentially distributed random variable. The probability of surviving more than 1 month is e^{-2}. After contracting the disease,

(A) What is the probability of the dog surviving more than 2 months?

(B) What is the mean life expectancy?

46. *Drug assimilation.* The rate at which the body eliminates a drug (in milliliters per hour) is given by

$$R(t) = 15e^{-0.2t} - 15e^{-0.3t}$$

where t is the number of hours since the drug was administered.

(A) What is the total amount of the drug that is eliminated by the body?

(B) How long will it take for 50% of the total amount to be eliminated? Round answer to two decimal places.

Social Sciences

47. *Testing.* The IQ scores for 6-year-old children in a certain area are normally distributed with a mean of 108 and a standard deviation of 12. What percentage of the children can be expected to have IQ scores of 135 or more?

48. *Politics.* The rate of change of the voting population of a city with respect to time t (in years) is estimated to be

$$N'(t) = \frac{100t}{(1+t^2)^2}$$

where $N(t)$ is in thousands. If $N(0)$ is the current voting population, how much will this population increase during the next 3 years? If the population continues to grow at this rate indefinitely, what is the total increase in the voting population?

Group Activity 1   *Beta Distributions*

WWW

An area of applied statistics called *Bayesian inference,* named after the Presbyterian minister Thomas Bayes (1702–1763), has received a great deal of attention in recent years. Effective use of this method requires experience with a variety of random variables with values that can be expressed as fractions or percentages. One particular random variable that has been used extensively in this area of statistics is the *beta random variable.* Typical applications of beta random variables include the percentage of fast-food restaurants that make a profit during their first year of operation, the percentage of time each year a manufacturing plant is shut down, the consumption of natural gas as a percentage of capacity, and the percentage of job application forms that contain errors.

A continuous random variable has a **beta distribution*** and is referred to as a **beta random variable** if its probability density function is the **beta probability density function**

$$f(x) = \begin{cases} (\beta + 1)(\beta + 2)x^{\beta}(1 - x) & \text{if } 0 \leqslant x \leqslant 1 \\ 0 & \text{otherwise} \end{cases}$$

where β is a constant, $\beta \geqslant 0$. The value of β is usually determined by examining the results of a particular experiment. The values of a beta random variable can be expressed as fractions or percentages; however, percentages should be converted to fractions before performing calculations involving a beta random variable.

(A) Show that f satisfies the requirements for a probability density function.

(B) Graph f for $\beta = 0, 1, 2,$ and 3. Write a general description of the graph of f for any $\beta > 1$, including the location of all local extrema and inflection points in the interval $(0, 1)$.

(C) Find the mean and standard deviation.

(D) Find the cumulative distribution function.

(E) Find the median for $\beta = 0$ and $\beta = 1$.

(F) Approximate the median (to two decimal places) for $\beta = 2$ and $\beta = 3$.

Group Activity 2   *Chi-Square Distributions*

A continuous random variable has a **chi-square distribution** with $2m$ **degrees of freedom** (where m is a positive integer) if its probability density function is

$$f(x) = \begin{cases} \dfrac{1}{(m - 1)!2^{m}} x^{m-1}e^{-x/2} & \text{if } 0 \leqslant x < \infty \\ 0 & \text{otherwise} \end{cases}$$

For $m = 1$, but for no other values of m, $f(x)$ is an exponential probability density function.

Chi-square distributions have natural connections to normal distributions. They also have many applications in statistical analysis. For example, chi-square distributions can be used to test whether a survey reveals a significant difference between the opinions of men and women in their approval ratings of the president.

(A) Use a graphing utility to show that the function f satisfies the require-

---

*There is a more general definition of a beta distribution, but we will not consider it here.

ments for a probability density function for $m = 1, 2, 3, 4$, and 5. Discuss why it is reasonable to replace integrals over the interval $(-\infty, \infty)$ by integrals over the interval $[0, 50]$.

(B) Graph f for $m = 1, 2, 3, 4$, and 5. Describe how the graphs change as m increases.

(C) Use numerical integration to approximate the mean of the chi-square distribution for $m = 1, 2, 3, 4$, and 5. Use your answers to conjecture a formula for the mean of a chi-square distribution with $2m$ degrees of freedom.

(D) Use numerical integration to approximate the variance of the chi-square distribution for $m = 1, 2, 3, 4$, and 5. Use your answers to conjecture a formula for the variance of a chi-square distribution with $2m$ degrees of freedom.

(E) Use numerical integration to approximate the median of the chi-square distribution for $m = 1, 2, 3, 4$, and 5. Describe how the median changes as m increases.

Trigonometric Functions

12 C H A P T E R

INTRODUCTION

Until now we have restricted our attention to algebraic, logarithmic, and exponential functions. These functions were used to model many real-life situations from business, economics, and the life and social sciences. Now we turn our attention to another important class of functions, called the *trigonometric functions*. These functions are particularly useful in describing periodic phenomena—that is, phenomena that repeat in cycles. Consider the sunrise times for a 2-year period starting January 1, as pictured in the figure. We see that the cycle repeats after 1 year. Business cycles, blood pressure in the aorta, seasonal growth, water waves, and amounts of pollution in the atmosphere are often periodic and can be modeled with similar types of graphs.

www

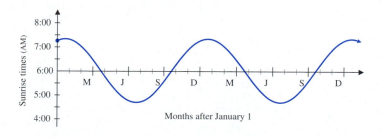

We assume that the reader has had a course in trigonometry. In Section 12-1 we provide a brief review of the topics that are most important for our purposes.

❑   ANGLES: DEGREE–RADIAN MEASURE

We start our discussion of trigonometry with the concept of *angle.* An **angle** is formed by rotating a half-line, called a **ray,** around its endpoint. One ray, *m,* called the **initial side** of the angle, remains fixed; a second ray, *n,* called the **terminal side** of the angle, starts in the initial side position and rotates around the common endpoint *P* in a plane until it reaches its terminal position. The common endpoint *P* is called the **vertex** (see Fig. 1).

There is no restriction on the amount or direction of rotation in a given plane. When the terminal side is rotated counterclockwise, the angle formed is **positive** (see Fig. 2A); when it is rotated clockwise, the angle formed is **negative** (see Fig. 2B). Two different angles may have the same initial and terminal sides, as shown in Figure 2C. Such angles are said to be **coterminal.**

FIGURE 1 Angle θ

(A) Positive angle α:
     Counterclockwise rotation

(B) Negative angle β:
     Clockwise rotation

(C) α and β are coterminal

FIGURE 2

There are two widely used measures of angles—the *degree* and the *radian.* A central angle of a circle subtended by an arc $\frac{1}{360}$ of the circumference of the circle is said to have **degree measure 1,** written **1°** (see Fig. 3A). It follows that a central angle subtended by an arc $\frac{1}{4}$ the circumference has degree measure 90; $\frac{1}{2}$ the circumference has degree measure 180; and the whole circumference of a circle has degree measure 360.

$\frac{1}{360}$ circumference

(A) 1°

(B) 1 radian

FIGURE 3 Degree and radian measure

The other measure of angles, which we use extensively in the next two sections, is radian measure. A central angle subtended by an arc of length equal to

FIGURE 4

the radius (R) of the circle is said to have **radian measure 1,** written **1 radian** or **1 rad** (see Fig. 3B). In general, a central angle subtended by an arc of length s has radian measure determined as follows:

$$\theta_{\text{rad}} = \text{radian measure of } \theta = \frac{\text{arc length}}{\text{radius}} = \frac{s}{R}$$

See Figure 4. [*Note:* If $R = 1$, then $\theta_{\text{rad}} = s$.]

What is the radian measure of an angle of 180°? A central angle of 180° is subtended by an arc of $\frac{1}{2}$ the circumference of a circle. Thus,

$$s = \frac{C}{2} = \frac{2\pi R}{2} = \pi R \quad \text{and} \quad \theta_{\text{rad}} = \frac{s}{R} = \frac{\pi R}{R} = \pi \text{ rad}$$

The following proportion can be used to convert degree measure to radian measure, and vice versa:

Degree–Radian Conversion Formula

$$\frac{\theta_{\text{deg}}}{180°} = \frac{\theta_{\text{rad}}}{\pi \text{ rad}}$$

Example 1 ⇨ **From Degrees to Radians**   Find the radian measure of 1°.

SOLUTION
$$\frac{1°}{180°} = \frac{\theta_{\text{rad}}}{\pi \text{ rad}}$$

$$\theta_{\text{rad}} = \frac{\pi}{180} \text{ rad} \approx 0.0175 \text{ rad}$$

■

Matched Problem 1 ⇨   Find the degree measure of 1 rad.

■

A comparison of degree and radian measure for a few important angles is given in the following table:

RADIAN	0	$\pi/6$	$\pi/4$	$\pi/3$	$\pi/2$	π	2π
DEGREE	0	30°	45°	60°	90°	180°	360°

Before defining trigonometric functions, we introduce the idea of an angle in *standard position.* Starting with a rectangular coordinate system and two half-lines coinciding with the nonnegative x axis, the initial side of the angle remains fixed and the terminal side rotates until it reaches its terminal position. Remember, when the terminal side is rotated counterclockwise, the angle formed is considered positive (see Fig. 5A). When it is rotated clockwise, the angle formed is considered negative (see Fig. 5B). Angles located in a coordinate system in this manner are said to be in a **standard position.**

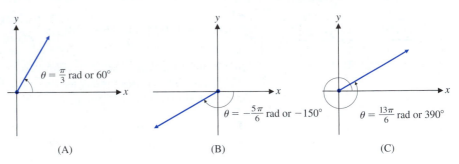

FIGURE 5 Generalized angles

□ TRIGONOMETRIC FUNCTIONS

FIGURE 6

www Let us locate a unit circle (radius 1) in a coordinate system with center at the origin (Fig. 6). The terminal side of any angle in standard position will pass through this circle at some point P. The abscissa of this point P is called the **cosine of θ** (abbreviated **cos θ**), and the ordinate of the point is the **sine of θ** (abbreviated **sin θ**). Thus, the set of all ordered pairs of the form $(\theta, \cos \theta)$, and the set of all ordered pairs of the form $(\theta, \sin \theta)$ constitute, respectively, the **cosine** and **sine functions.** The **domain** of these two functions is the set of all angles, positive or negative, with measure either in degrees or radians. The **range** is a subset of the set of real numbers.

It is desirable, and necessary for our work in calculus, to define these two trigonometric functions in terms of real number domains. This is easily done as follows:

Sine and Cosine Functions with Real Number Domains

For any real number x,

$$\sin x = \sin(x \text{ radians}) \qquad \cos x = \cos(x \text{ radians})$$

Example 2 ⤶ **Evaluating Sine and Cosine Functions** Referring to Figure 6, find:

(A) $\cos 90°$ (B) $\sin(-\pi/2 \text{ rad})$ (C) $\cos \pi$

SOLUTION (A) The terminal side of an angle of degree measure 90 passes through $(0, 1)$ on the unit circle. This point has abscissa 0. Thus,

$$\cos 90° = 0$$

(B) The terminal side of an angle of radian measure $-\pi/2$ $(-90°)$ passes through $(0, -1)$ on the unit circle. This point has ordinate -1. Thus,

$$\sin\left(-\frac{\pi}{2} \text{rad}\right) = -1$$

(C) $\cos \pi = \cos(\pi \text{ rad}) = -1$, since the terminal side of an angle of radian measure π $(180°)$ passes through $(-1, 0)$ on the unit circle and this point has abscissa -1. ∎

Matched Problem 2 ⇨ Referring to Figure 6, find:

(A) $\sin 180°$ (B) $\cos(2\pi \, \text{rad})$ (C) $\sin(-\pi)$

Explore–Discuss 1

(A) For the sine and cosine functions with angle domains, discuss the range for each function by referring to Figure 6.

(B) For the sine and cosine functions with real number domains, discuss the range for each function by referring to Figure 6.

To find the value of either the sine or the cosine function for any angle or any real number by direct use of the definition is not easy. Calculators with $\boxed{\sin}$ and $\boxed{\cos}$ keys are used. Calculators generally have degree and radian options, so we can use a calculator to evaluate these functions for most of the real numbers in which we might have an interest. The following table includes a few values produced by a calculator in the radian mode.

x	1	-7	35.26	-105.9
$\sin x$	0.8415	-0.6570	-0.6461	0.7920
$\cos x$	0.5403	0.7539	-0.7632	0.6105

Explore–Discuss 2

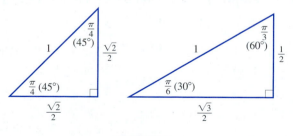

Many errors in trigonometry can be traced to having the calculator in the wrong mode, radian instead of degree, or vice versa, when performing calculations. The window display from a calculator shown here gives two different values for cos 30. Experiment with your calculator and explain the discrepancy.

Exact values of the sine and cosine functions can be obtained for multiples of the special angles shown in the triangles in Figure 7, because these triangles can be used to find the coordinate of the intersection of the terminal side of each angle with the unit circle.

FIGURE 7

REMARK

We now drop the word *radian* after $\pi/4$ and interpret $\pi/4$ as the radian measure of an angle or simply as a real number, depending on the context.

Example 3 ⮂ **Finding Exact Values for Special "Angles"**   Find the exact value of each of the following using Figure 7 on the preceding page:

(A) $\cos\dfrac{\pi}{4}$ (B) $\sin\dfrac{\pi}{6}$ (C) $\sin\left(-\dfrac{\pi}{6}\right)$

SOLUTION   (A) $\cos\dfrac{\pi}{4} = \dfrac{\sqrt{2}}{2}$ (B) $\sin\dfrac{\pi}{6} = \dfrac{1}{2}$

(C) $\sin\left(-\dfrac{\pi}{6}\right) = -\dfrac{1}{2}$

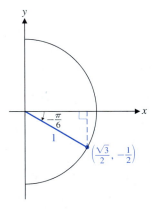

Matched Problem 3 ⮂ Find the exact value of each of the following using Figure 7 on the preceding page:

(A) $\sin\dfrac{\pi}{4}$

(B) $\cos\dfrac{\pi}{3}$

(C) $\cos\left(-\dfrac{\pi}{3}\right)$

www ❏ GRAPHS OF THE SINE AND COSINE FUNCTIONS

To graph $y = \sin x$ or $y = \cos x$ for x a real number, we could use a calculator to produce a table, and then plot the ordered pairs from the table in a coordinate system. However, we can speed up the process by returning to basic defi-

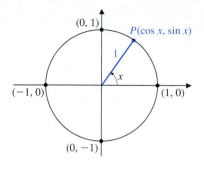

FIGURE 8

nitions. Referring to Figure 8, since cos x and sin x are the coordinates of a point on the unit circle, we see that

$$-1 \leq \sin x \leq 1 \quad \text{and} \quad -1 \leq \cos x \leq 1$$

for all real numbers x. Furthermore, as x increases and P moves around the unit circle in a counterclockwise (positive) direction, both sin x and cos x behave in uniform ways, as indicated in the table:

As x increases from	$y = \sin x$	$y = \cos x$
0 to $\pi/2$	Increases from 0 to 1	Decreases from 1 to 0
$\pi/2$ to π	Decreases from 1 to 0	Decreases from 0 to −1
π to $3\pi/2$	Decreases from 0 to −1	Increases from −1 to 0
$3\pi/2$ to 2π	Increases from −1 to 0	Increases from 0 to 1

Note that P has completed one revolution and is back at its starting place. If we let x continue to increase, the second and third columns in the table will be repeated every 2π units. In general, it can be shown that

$$\sin(x + 2\pi) = \sin x \qquad \cos(x + 2\pi) = \cos x$$

for all real numbers x. Functions such that

$$f(x + p) = f(x)$$

for some positive constant p and all real numbers x for which the functions are defined are said to be **periodic.** The smallest such value of p is called the **period** of the function. Thus, both the sine and cosine functions are periodic (a very important property) with period 2π.

Putting all this information together, and, perhaps adding a few values obtained from a calculator or Figure 7, we obtain the graphs of the sine and cosine functions illustrated in Figure 9. Notice that these curves are continuous. It can be shown that **the sine and cosine functions are continuous for all real numbers.**

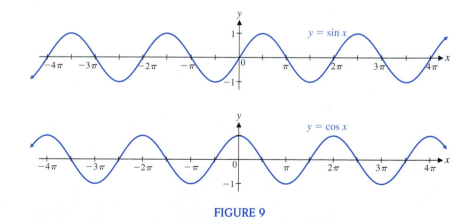

FIGURE 9

❑ FOUR OTHER TRIGONOMETRIC FUNCTIONS

The sine and cosine functions are only two of six trigonometric functions. They are, however, the most important of the six for many applications. We define the other four trigonometric functions in the box. Problems involving these functions may be found in the exercise set that follows.

Four Other Trigonometric Functions

$$\tan x = \frac{\sin x}{\cos x} \qquad \cos x \neq 0 \qquad \sec x = \frac{1}{\cos x} \qquad \cos x \neq 0$$

$$\cot x = \frac{\cos x}{\sin x} \qquad \sin x \neq 0 \qquad \csc x = \frac{1}{\sin x} \qquad \sin x \neq 0$$

Answers to Matched Problems
1. $180/\pi \approx 57.3°$ 2. (A) 0   (B) 1   (C) 0
3. (A) $\sqrt{2}/2$ (B) $\frac{1}{2}$ (C) $\frac{1}{2}$

Exercise 12-1

A *Recall that 180° corresponds to π radians. Mentally convert each degree measure given in Problems 1–6 to radian measure in terms of π.*

1. 60°   2. 90°   3. 45°
4. 360°   5. 30°   6. 120°

In Problems 7–12, indicate the quadrant in which the terminal side of each angle lies.

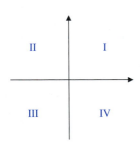

7. 150°   8. −190°   9. −π/3 rad
10. 7π/6 rad   11. 400°   12. −250°

Use Figure 6 to find the exact value of each expression in Problems 13–18.

13. $\cos 0°$ 14. $\sin 90°$ 15. $\sin \pi$
16. $\cos \dfrac{\pi}{2}$ 17. $\cos(-\pi)$ 18. $\sin \dfrac{3\pi}{2}$

B *Recall that π rad corresponds to 180° Mentally convert each radian measure given in Problems 19–24 to degree measure.*

19. $\pi/3$ rad   20. 2π rad   21. $\pi/4$ rad
22. $\pi/2$ rad   23. $\pi/6$ rad   24. $5\pi/6$ rad

Use Figure 7 to find the exact value of each expression in Problems 25–30.

25. $\cos 30°$ 26. $\sin(-45°)$ 27. $\sin \dfrac{\pi}{6}$
28. $\sin\left(-\dfrac{\pi}{3}\right)$ 29. $\cos \dfrac{5\pi}{6}$ 30. $\cos(-120°)$

Use a calculator (set in radian mode) to find the value (to four decimal places) of each expression in Problems 31–36.

31. $\sin 3$ 32. $\cos 13$ 33. $\cos 33.74$
34. $\sin 325.9$ 35. $\sin(-43.06)$ 36. $\cos(-502.3)$

C *In Problems 37 and 38, convert to radian measure.*

37. 27°   38. 18°

In Problems 39 and 40, convert to degree measure.

39. $\pi/12$ rad   40. $\pi/60$ rad

Use Figure 7 to find the exact value of each expression in Problems 41–46.

41. $\tan 45°$ 42. $\cot 45°$
43. $\sec \dfrac{\pi}{3}$ 44. $\csc \dfrac{\pi}{6}$
45. $\cot \dfrac{\pi}{3}$ 46. $\tan \dfrac{\pi}{6}$

47. Refer to Figure 6 and use the Pythagorean theorem to show that

$$(\sin x)^2 + (\cos x)^2 = 1$$

for all x.

48. Use the results of Problem 47 and basic definitions to show that

(A) $(\tan x)^2 + 1 = (\sec x)^2$

(B) $1 + (\cot x)^2 = (\csc x)^2$

 In Problems 49–52, graph each trigonometric form using a graphing utility set in radian mode.

49. $y = 2 \sin \pi x; 0 \leq x \leq 2, -2 \leq y \leq 2$

50. $y = -0.5 \cos 2x; 0 \leq x \leq 2\pi, -0.5 \leq y \leq 0.5$

51. $y = 4 - 4 \cos \dfrac{\pi x}{2}; 0 \leq x \leq 8, 0 \leq y \leq 8$

52. $y = 6 + 6 \sin \dfrac{\pi x}{26}; 0 \leq x \leq 104, 0 \leq y \leq 12$

Applications

Business & Economics

53. *Seasonal business cycle.* Suppose that profits on the sale of swimming suits in a department store over a 2-year period are given approximately by

$$P(t) = 5 - 5 \cos \frac{\pi t}{26} \qquad 0 \leq t \leq 104$$

where P is profit (in hundreds of dollars) for a week of sales t weeks after January 1. The graph of the profit function is shown in the figure.

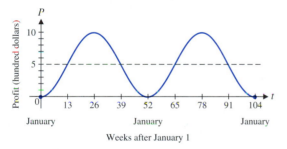

Figure for 53

(A) Find the exact values of $P(13), P(26), P(39)$, and $P(52)$ without using a calculator.

(B) Use a calculator to find $P(30)$ and $P(100)$. Interpret the results verbally.

(C) Use a graphing utility to confirm the graph shown here for $y = P(t)$.

54. *Seasonal business cycle.* A soft drink company has revenues from sales over a 2-year period as given approximately by

$$R(t) = 4 - 3 \cos \frac{\pi t}{6} \qquad 0 \leq t \leq 24$$

where $R(t)$ is revenue (in millions of dollars) for a month of sales t months after February 1. The graph of the revenue function is shown in the figure.

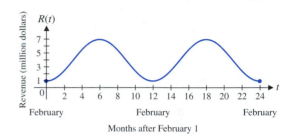

Figure for 54

(A) Find the exact values of $R(0), R(2), R(3)$, and $R(18)$ without using a calculator.

(B) Use a calculator to find $R(5)$ and $R(23)$. Interpret the results verbally.

(C) Use a graphing utility to confirm the graph shown here for $y = R(t)$.

Life Sciences

55. *Physiology.* A normal seated adult breathes in and exhales about 0.8 liter of air every 4 seconds. The volume of air $V(t)$ in the lungs t seconds after exhaling is given approximately by

$$V(t) = 0.45 - 0.35 \cos \frac{\pi t}{2} \qquad 0 \leq t \leq 8$$

The graph for two complete respirations is shown in the figure.

Figure for 55

(A) Find the exact value of $V(0)$, $V(1)$, $V(2)$, $V(3)$, and $V(7)$ without using a calculator.

(B) Use a calculator to find $V(3.5)$ and $V(5.7)$. Interpret the results verbally.

(C) Use a graphing utility to confirm the graph shown here for $y = V(t)$.

56. *Pollution.* In a large city, the amount of sulfur dioxide pollutant released into the atmosphere due to the burning of coal and oil for heating purposes varies seasonally. Suppose that the number of tons of pollutant released into the atmosphere during the nth week after January 1 is given approximately by

$$P(n) = 1 + \cos\frac{\pi n}{26} \qquad 0 \leq n \leq 104$$

The graph of the pollution function is shown in the figure.

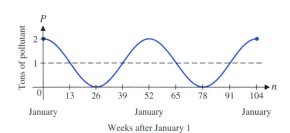

Figure for 56

(A) Find the exact values of $P(0)$, $P(39)$, $P(52)$, and $P(65)$. without using a calculator.

(B) Use a calculator to find $P(10)$ and $P(95)$. Interpret the results verbally.

(C) Use a graphing utility to confirm the graph shown here for $y = P(n)$.

Social Sciences

57. *Psychology: perception.* An important area of study in psychology is perception. Individuals perceive objects differently in different settings. Consider the well-known illusions shown in Figure A. Lines that appear parallel in one setting may appear to be curved in another (the two vertical lines are actually parallel). Lines of the same length may appear to be of different lengths in two different settings (the two horizontal lines are actually the same length). An interesting experiment in visual perception was conducted by psychologists Berliner and Berliner (*American Journal of Psychology*, 1952, 65:271–277). They reported that when subjects were presented with a large tilted field of parallel lines and were asked to estimate the position of a horizontal line in the field, most of the subjects were consistently off. They found that the difference in degrees, d, between the estimates and the actual horizontal could be approximated by the equation

$$d = a + b \sin 4\theta$$

where a and b are constants associated with a particular person and θ is the angle of tilt of the visual field (in degrees). Suppose that for a given person, $a = -2.1$ and $b = -4$. Find d if:

(A) $\theta = 30°$ (B) $\theta = 10°$

(A)   (B)

5

Section 12-2 **Derivatives of Trigonometric Functions**

❑ DERIVATIVE FORMULAS
❑ APPLICATION

❑ DERIVATIVE FORMULAS

In this section we discuss derivative formulas for the sine and cosine functions. Once we have these formulas, we automatically have integral formulas for the same functions, which we discuss in the next section.

From the definition of the derivative (Section 3-3),

$$\frac{d}{dx}\sin x = \lim_{h \to 0} \frac{\sin(x + h) - \sin x}{h}$$

Using trigonometric identities and some special trigonometric limits, it can be shown that the limit on the right is cos x. Similarly, it can be shown that

$$\frac{d}{dx}\cos x = -\sin x$$

We now add the following important derivative formulas to our list of derivative formulas:

Derivative Formulas for Sine and Cosine

Basic Form

$$\frac{d}{dx}\sin x = \cos x \qquad \frac{d}{dx}\cos x = -\sin x$$

Generalized Form

For $u = u(x)$:

$$\frac{d}{dx}\sin u = \cos u \frac{du}{dx} \qquad \frac{d}{dx}\cos u = -\sin u \frac{du}{dx}$$

Example 1   ➩   **Derivatives Involving Sine and Cosine**

www

(A) $\dfrac{d}{dx}\sin x^2 = (\cos x^2)\dfrac{d}{dx}x^2 = (\cos x^2)2x = 2x \cos x^2$

(B) $\dfrac{d}{dx}\cos(2x - 5) = -\sin(2x - 5)\dfrac{d}{dx}(2x - 5) = -2 \sin(2x - 5)$

(C) $\dfrac{d}{dx}(3x^2 - x)\cos x = (3x^2 - x)\dfrac{d}{dx}\cos x + (\cos x)\dfrac{d}{dx}(3x^2 - x)$

$\qquad\qquad = -(3x^2 - x)\sin x + (6x - 1)\cos x$

$\qquad\qquad = (x - 3x^2)\sin x + (6x - 1)\cos x$ ∎

Matched Problem 1   ➩   Find each of the following derivatives:

(A) $\dfrac{d}{dx}\cos x^3$ (B) $\dfrac{d}{dx}\sin(5 - 3x)$ (C) $\dfrac{d}{dx}\dfrac{\sin x}{x}$ ∎

Example 2 ➭ **Slope**   Find the slope of the graph of $f(x) = \sin x$ at $(\pi/2, 1)$, and sketch in
www   the tangent line to the graph at this point.

SOLUTION   Slope at $(\pi/2, 1) = f'(\pi/2) = \cos(\pi/2) = 0$.

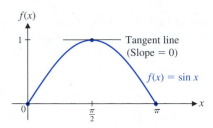

Matched Problem 2 ➭ Find the slope of the graph of $f(x) = \cos x$ at $(\pi/6, \sqrt{3}/2)$.

Explore–Discuss 1

From the graph of $y = f'(x)$ shown in Figure 1, describe the shape of the graph of $y = f(x)$ relative to increasing, decreasing, concavity, and local maxima and minima. Make a sketch of a possible graph of $y = f(x), 0 \leqslant x \leqslant 2\pi$, given that it has x intercepts at $(0, 0)$, $(\pi, 0)$, and $(2\pi, 0)$. Can you identify $f(x)$ and $f'(x)$ in terms of sine or cosine functions?

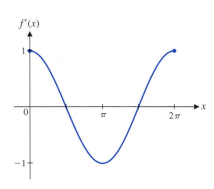

FIGURE 1

Example 3 ➭ **Derivative of Secant**   Find: $\dfrac{d}{dx} \sec x$

SOLUTION
$$\frac{d}{dx} \sec x = \frac{d}{dx} \frac{1}{\cos x} = \frac{d}{dx} (\cos x)^{-1} \qquad \text{Since sec x } = \frac{1}{\cos x}$$

$$= -(\cos x)^{-2} \frac{d}{dx} \cos x$$

$$= -(\cos x)^{-2}(-\sin x)$$

$$= \frac{\sin x}{(\cos x)^2} = \left(\frac{\sin x}{\cos x} \right)\left(\frac{1}{\cos x} \right)$$

$$= \tan x \sec x \qquad \text{Since tan x } = \frac{\sin x}{\cos x}$$

Matched Problem 3 ➭ Find: $\dfrac{d}{dx} \csc x$

□ APPLICATION

Example 4 ⇔   **Revenue**   A sporting goods store has revenues from the sale of ski jackets that are given approximately by

$$R(t) = 1.55 + 1.45 \cos\frac{\pi t}{26} \qquad 0 \le t \le 104$$

where $R(t)$ is revenue (in thousands of dollars) for a week of sales t weeks after January 1.

(A) What is the rate of change of revenue t weeks after the first of the year?

(B) What is the rate of change of revenue 10 weeks after the first of the year? 26 weeks after the first of the year? 40 weeks after the first of the year?

(C) Find all local maxima and minima for $0 < t < 104$.

(D) Find the absolute maximum and minimum for $0 \le t \le 104$.

(E) Illustrate the results from parts (A)–(D) by sketching a graph of $y = R(t)$ with the aid of a graphing utility.

SOLUTION   (A) $R'(t) = -\dfrac{1.45\pi}{26}\sin\dfrac{\pi t}{26} \qquad 0 \le t \le 104$

(B) $R'(10) \approx -\$0.164$ thousand   or $-\$164$ per week
$R'(26) = \$0$ per week
$R'(40) \approx \$0.174$ thousand   or $\$174$ per week

(C) Find the critical points:

$$R'(t) = -\frac{1.45\pi}{26}\sin\frac{\pi t}{26} = 0 \qquad 0 < t < 104$$

$$\sin\frac{\pi t}{26} = 0$$

$$\frac{\pi t}{26} = \pi, 2\pi, 3\pi \quad \textit{Note: } 0 < t < 104 \text{ implies that } 0 < \frac{\pi t}{26} < 4\pi.$$

$$t = 26, 52, 78$$

Use the second-derivative test to get the results shown in Table 1.

$$R''(t) = -\frac{1.45\pi^2}{26^2}\cos\frac{\pi t}{26}$$

TABLE 1		
t	$R''(t)$	GRAPH OF R
26	+	Local minimum
52	−	Local maximum
78	+	Local minimum

(D) Evaluate $R(t)$ at endpoints $t = 0$ and $t = 104$ and at the critical points found in part (C), as listed in Table 2.

TABLE 2	
t	$R(t)$
0	\$3,000 Absolute maximum
26	\$100 Absolute minimum
52	\$3,000 Absolute maximum
78	\$100 Absolute minimum
104	\$3,000 Absolute maximum

(E) The results from parts (A)–(D) can be visualized as shown in the graph of $y = R(t)$ in Figure 2.

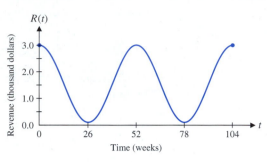

FIGURE 2

Matched Problem 4 Suppose that in Example 4 revenues from the sale of ski jackets are given approximately by

$$R(t) = 6.2 + 5.8 \cos\frac{\pi t}{6} \qquad 0 \leqslant t \leqslant 24$$

where $R(t)$ is revenue (in thousands of dollars) for a month of sales t months after January 1.

(A) What is the rate of change of revenue t months after the first of the year?

(B) What is the rate of change of revenue 2 months after the first of the year? 12 months after the first of the year? 23 months after the first of the year?

(C) Find all local maxima and minima for $0 < t < 24$.

(D) Find the absolute maximum and minimum for $0 \leqslant t \leqslant 24$.

(E) Illustrate the results from parts (A) to (D) by sketching a graph of $y = R(t)$ with the aid of a graphing utility.

Answers to Matched Problems

1. (A) $-3x^2 \sin x^3$

(B) $-3 \cos(5 - 3x)$

(C) $\dfrac{x \cos x - \sin x}{x^2}$

2. $-\frac{1}{2}$ **3.** $-\cot x \csc x$

4. (A) $R'(t) = -\dfrac{5.8\pi}{6} \sin\dfrac{\pi t}{6}, 0 < t < 24$

(B) $R'(2) \approx -\$2.630$ thousand or $-\$2,630$/month; $R'(12) = \$0$/month; $R'(23) \approx \$1.518$ thousand or $\$1,518$/month

(C) Local minima at $t = 6$ and $t = 18$; local maximum at $t = 12$

(D)

	t	$R(t)$	
Endpoint	0	$12,000	Absolute maximum
	6	$400	Absolute minimum
	12	$12,000	Absolute maximum
	18	$400	Absolute minimum
Endpoint	24	$12,000	Absolute maximum

(E) R(t)

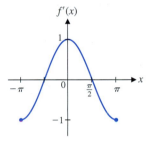

Exercise 12-2

Find the indicated derivatives in Problems 1–14.

A

1. $\dfrac{d}{dt}\cos t$

2. $\dfrac{d}{dw}\sin w$

3. $\dfrac{d}{dx}\sin x^3$

4. $\dfrac{d}{dx}\cos(x^2 - 1)$

B

5. $\dfrac{d}{dt}t\sin t$

6. $\dfrac{d}{du}u\cos u$

7. $\dfrac{d}{dx}\sin x \cos x$

8. $\dfrac{d}{dx}\dfrac{\sin x}{\cos x}$

9. $\dfrac{d}{dx}(\sin x)^5$

10. $\dfrac{d}{dx}(\cos x)^8$

11. $\dfrac{d}{dx}\sqrt{\sin x}$

12. $\dfrac{d}{dx}\sqrt{\cos x}$

13. $\dfrac{d}{dx}\cos\sqrt{x}$

14. $\dfrac{d}{dx}\sin\sqrt{x}$

15. Find the slope of the graph of $f(x) = \sin x$ at $x = \pi/6$.

16. Find the slope of the graph of $f(x) = \cos x$ at $x = \pi/4$.

17. From the graph of $y = f'(x)$ shown here, describe the shape of the graph of $y = f(x)$ relative to increasing, decreasing, concavity, and local maxima and minima. Make a sketch of a possible graph of $y = f(x), -\pi \le x \le \pi$, given it has x intercepts at $(-\pi/2, 0)$ and $(\pi/2, 0)$. Identify $f(x)$ and $f'(x)$ as particular trigonometric functions.

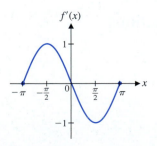

Figure for 17

18. From the graph of $y = f'(x)$ shown here, describe the shape of the graph of $y = f(x)$ relative to increasing, decreasing, concavity, and local maxima and minima. Make a sketch of a possible graph of $y = f(x), -\pi \le x \le \pi$, given it has x intercepts at $(-\pi/2, 0)$ and $(\pi/2, 0)$. Identify $f(x)$ and $f'(x)$ as particular trigonometric functions.

f'(x)

Figure for 18

C *Find the indicated derivatives in Problems 19–22.*

19. $\dfrac{d}{dx}\tan x$

20. $\dfrac{d}{dx}\cot x$

21. $\dfrac{d}{dx}\sin\sqrt{x^2 - 1}$

22. $\dfrac{d}{dx}\cos\sqrt{x^4 - 1}$

In Problems 23 and 24, find $f''(x)$.

23. $f(x) = e^x \sin x$

24. $f(x) = e^x \cos x$

 In Problems 25–30, graph each function on a graphing utility.

25. $y = x \sin \pi x; 0 \le x \le 9, -9 \le y \le 9$

26. $y = -x \cos \pi x; 0 \le x \le 9, -9 \le y \le 9$

27. $y = \dfrac{\cos \pi x}{x}; 0 \le x \le 8, -2 \le y \le 3$

28. $y = \dfrac{\sin \pi x}{0.5x}; 0 \le x \le 8, -2 \le y \le 3$

29. $y = e^{-0.3x}\sin \pi x; 0 \le x \le 10, -1 \le y \le 1$

30. $y = e^{-0.2x}\cos \pi x; 0 \le x \le 10, -1 \le y \le 1$

Applications

Business & Economics

31. *Profit.* Suppose that profits on the sale of swimming suits in a department store are given approximately by

$$P(t) = 5 - 5 \cos \frac{\pi t}{26} \qquad 0 \leq t \leq 104$$

where $P(t)$ is profit (in hundreds of dollars) for a week of sales t weeks after January 1.

(A) What is the rate of change of profit t weeks after the first of the year?

(B) What is the rate of change of profit 8 weeks after the first of the year? 26 weeks after the first of the year? 50 weeks after the first of the year?

(C) Find all local maxima and minima for $0 < t < 104$.

(D) Find the absolute maximum and minimum for $0 \leq t \leq 104$.

(E) Repeat part (C) using a graphing utility.

32. *Revenue.* A soft drink company has revenues from sales over a 2-year period as given approximately by

$$R(t) = 4 - 3 \cos \frac{\pi t}{6} \qquad 0 \leq t \leq 24$$

where $R(t)$ is revenue (in millions of dollars) for a month of sales t months after February 1.

(A) What is the rate of change of revenue t months after February 1?

(B) What is the rate of change of revenue 1 month after February 1? 6 months after February 1? 11 months after February 1?

(C) Find all local maxima and minima for $0 < t < 24$.

(D) Find the absolute maximum and minimum for $0 \leq t \leq 24$.

(E) Repeat part (C) using a graphing utility.

Life Sciences

33. *Physiology.* A normal seated adult breathes in and exhales about 0.8 liter of air every 4 seconds. The volume of air $V(t)$ in the lungs t seconds after exhaling is given approximately by

$$V(t) = 0.45 - 0.35 \cos \frac{\pi t}{2} \qquad 0 \leq t \leq 8$$

(A) What is the rate of flow of air t seconds after exhaling?

(B) What is the rate of flow of air 3 seconds after exhaling? 4 seconds after exhaling? 5 seconds after exhaling?

(C) Find all local maxima and minima for $0 < t < 8$.

(D) Find the absolute maximum and minimum for $0 \leq t \leq 8$.

(E) Repeat part (C) using a graphing utility.

34. *Pollution.* In a large city, the amount of sulfur dioxide pollutant released into the atmosphere due to the burning of coal and oil for heating purposes varies seasonally. Suppose that the number of tons of pollutant released into the atmosphere during the nth week after January 1 is given approximately by

$$P(n) = 1 + \cos \frac{\pi n}{26} \qquad 0 \leq n \leq 104$$

(A) What is the rate of change of pollutant n weeks after the first of the year?

(B) What is the rate of change of pollutant 13 weeks after the first of the year? 26 weeks after the first of the year? 30 weeks after the first of the year?

(C) Find all local maxima and minima for $0 < t < 104$.

(D) Find the absolute maximum and minimum for $0 \leq t \leq 104$.

(E) Repeat part (C) using a graphing utility.

Section 12-3 | ## Integration of Trigonometric Functions

❑ INTEGRAL FORMULAS
❑ APPLICATION

❑ **INTEGRAL FORMULAS**

Now that we know the derivative formulas

$$\frac{d}{dx} \sin x = \cos x \qquad \text{and} \qquad \frac{d}{dx} \cos x = -\sin x$$

from the definition of the indefinite integral of a function (Section 6-1), we automatically have the two integral formulas

$$\int \cos x \, dx = \sin x + C \qquad \text{and} \qquad \int \sin x \, dx = -\cos x + C$$

Example 1   **Area Under a Sine Curve**   Find the area under the sine curve $y = \sin x$
WWW   from 0 to π.

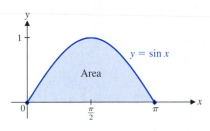

SOLUTION
$$\text{Area} = \int_0^{\pi} \sin x \, dx = -\cos x \Big|_0^{\pi}$$
$$= (-\cos \pi) - (-\cos 0)$$
$$= [-(-1)] - [-(1)] = 2$$

■

Matched Problem 1   Find the area under the cosine curve $y = \cos x$ from 0 to $\pi/2$.

■

Explore–Discuss 1

From the graph of $y = \sin x$, $0 \le x \le 2\pi$, guess the value for $\int_0^{2\pi} \sin x \, dx$ and explain the reasoning behind your guess. Confirm the guess by direct evaluation of the definite integral.

From the general derivative formulas

$$\frac{d}{dx} \sin u = \cos u \, \frac{du}{dx} \qquad \text{and} \qquad \frac{d}{dx} \cos u = -\sin u \, \frac{du}{dx}$$

we obtain the general integral formulas below.

Integral Formulas for Sine and Cosine

For $u = u(x)$,

$$\int \sin u \, du = -\cos u + C \qquad \text{and} \qquad \int \cos u \, du = \sin u + C$$

Example 2 ⇔ **Indefinite Integrals and Trigonometric Functions** Find: $\int x \sin x^2 \, dx$

SOLUTION
www

$$\int x \sin x^2 \, dx = \tfrac{1}{2} \int 2x \sin x^2 \, dx$$

$$= \tfrac{1}{2} \int (\sin x^2) 2x \, dx \qquad \text{Let } u = x^2; \text{ then } du = 2x \, dx.$$

$$= \tfrac{1}{2} \int \sin u \, du$$

$$= -\tfrac{1}{2} \cos u + C$$
$$= -\tfrac{1}{2} \cos x^2 + C \qquad \text{Since } u = x^2$$

Check: To check, we differentiate the result to obtain the original integrand:

$$\frac{d}{dx} \left(-\tfrac{1}{2} \cos x^2 \right) = -\tfrac{1}{2} \frac{d}{dx} \cos x^2$$

$$= -\tfrac{1}{2} (-\sin x^2) \frac{d}{dx} x^2$$

$$= -\tfrac{1}{2} (-\sin x^2)(2x)$$

$$= x \sin x^2$$

Matched Problem 2 ⇔ Find: $\int \cos 20\pi t \, dt$

Example 3 ⇔ **Indefinite Integrals and Trigonometric Functions**

Find: $\int (\sin x)^5 \cos x \, dx$

SOLUTION This is of the form $\int u^p \, du$, where $u = \sin x$ and $du = \cos x \, dx$. Thus,

$$\int (\sin x)^5 \cos x \, dx = \frac{(\sin x)^6}{6} + C$$

Matched Problem 3 ⇔ Find: $\int \sqrt{\sin x} \cos x \, dx$

Example 4 ⇔ **Definite Integrals and Trigonometric Functions** Evaluate: $\int_2^{3.5} \cos x \, dx$

SOLUTION

$$\int_2^{3.5} \cos x \, dx = \sin x \Big|_2^{3.5}$$

$$= \sin 3.5 - \sin 2 \qquad \text{Use a calculator in radian mode.}$$
$$= -0.3508 - 0.9093$$
$$= -1.2601$$

Matched Problem 4 ⇔ Use a calculator to evaluate: $\int_1^{1.5} \sin x \, dx$

❑ APPLICATION

Example 5 ⇔ **Total Revenue** In Example 4 (Section 12-2), we were given the following revenue equation from the sale of ski jackets:

$$R(t) = 1.55 + 1.45 \cos \frac{\pi t}{26} \qquad 0 \le t \le 104$$

where $R(t)$ is revenue (in thousands of dollars) for a week of sales t weeks after January 1.

(A) Find the total revenue taken in over the 2-year period—that is, from $t = 0$ to $t = 104$.

(B) Find the total revenue taken in from $t = 39$ to $t = 65$.

SOLUTION   (A) The area under the graph of the revenue equation for the 2-year period approximates the total revenue taken in for that period:

This area (and total revenue) is given by the following definite integral:

$$\text{total revenue} \approx \int_0^{104} \left(1.55 + 1.45 \cos \frac{\pi t}{26} \right) dt$$

$$= \left[1.55t + 1.45 \left(\frac{26}{\pi} \right) \sin \frac{\pi t}{26} \right] \Bigg|_0^{104}$$

$$= \$161.200 \text{ thousand} \quad \text{or} \quad \$161,200$$

(B) The total revenue from $t = 39$ to $t = 65$ is approximated by the area under the curve from $t = 39$ to $t = 65$:

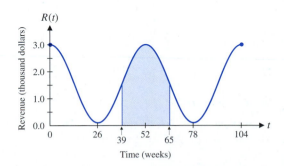

$$\text{Total revenue} \approx \int_{39}^{65} \left(1.55 + 1.45 \cos \frac{\pi t}{26} \right) dt$$

$$= \left[1.55t + 1.45 \left(\frac{26}{\pi} \right) \sin \frac{\pi t}{26} \right] \Bigg|_{39}^{65}$$

$$= \$64.301 \text{ thousand} \quad \text{or} \quad \$64,301$$

Matched Problem 5 ✏️ Suppose that in Example 5 revenues from the sale of ski jackets are given approximately by

$$R(t) = 6.2 + 5.8 \cos\frac{\pi t}{6} \qquad 0 \le t \le 24$$

where $R(t)$ is revenue (in thousands of dollars) for a month of sales t months after January 1.

(A) Find the total revenue taken in over the 2 year period—that is, from $t = 0$ to $t = 24$.

(B) Find the total revenue taken in from $t = 4$ to $t = 8$.

■

Answers to Matched Problems   **1.** 1   **2.** $\dfrac{1}{20\pi}\sin 20\pi t + C$ **3.** $\frac{2}{3}(\sin x)^{3/2} + C$ **4.** 0.4696

5. (A) $148.8 thousand or $148,800   (B) $5.614 thousand or $5,614

Exercise 12-3

Find each of the indefinite integrals in Problems 1–10.

A

1. $\displaystyle\int \sin t \, dt$

2. $\displaystyle\int \cos w \, dw$

3. $\displaystyle\int \cos 3x \, dx$

4. $\displaystyle\int \sin 2x \, dx$

5. $\displaystyle\int (\sin x)^{12} \cos x \, dx$

6. $\displaystyle\int \sin x \cos x \, dx$

B

7. $\displaystyle\int \sqrt[3]{\cos x}\,\sin x \, dx$

8. $\displaystyle\int \frac{\cos x}{\sqrt{\sin x}}\, dx$

9. $\displaystyle\int x^2 \cos x^3 \, dx$

10. $\displaystyle\int (x + 1)\sin(x^2 + 2x)\, dx$

Evaluate each of the definite integrals in Problems 11–14.

11. $\displaystyle\int_0^{\pi/2} \cos x \, dx$

12. $\displaystyle\int_0^{\pi/4} \cos x \, dx$

13. $\displaystyle\int_{\pi/2}^{\pi} \sin x \, dx$

14. $\displaystyle\int_{\pi/6}^{\pi/3} \sin x \, dx$

15. Find the shaded area under the cosine curve in the figure:

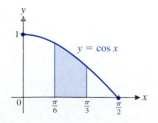

16. Find the shaded area under the sine curve in the figure:

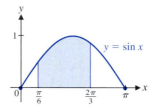

Use a calculator to evaluate the definite integrals in Problems 17–20 after performing the indefinite integration. (Remember that the limits are real numbers, so the radian mode must be used on the calculator.)

17. $\displaystyle\int_0^2 \sin x \, dx$

18. $\displaystyle\int_0^{0.5} \cos x \, dx$

19. $\displaystyle\int_1^2 \cos x \, dx$

20. $\displaystyle\int_1^3 \sin x \, dx$

C   *Find each of the indefinite integrals in Problems 21–26.*

21. $\displaystyle\int e^{\sin x} \cos x \, dx$

22. $\displaystyle\int e^{\cos x} \sin x \, dx$

23. $\displaystyle\int \frac{\cos x}{\sin x}\, dx$

24. $\displaystyle\int \frac{\sin x}{\cos x}\, dx$

25. $\displaystyle\int \tan x \, dx$

26. $\displaystyle\int \cot x \, dx$

27. Given the definite integral

$$I = \int_0^3 e^{-x} \sin x \, dx$$

(A) Graph the integrand $f(x) = e^{-x} \sin x$ over $[0, 3]$.

(B) Approximate I using the midpoint sum M_6 (see Section 6-5).

(C) Show that 2 is an upper bound for $|f''(x)|$ on the interval $[0, 3]$ using a graphing utility.

(D) Use the result from part (C) to find an error estimate for part (B).

28. Given the definite integral

$$I = \int_0^3 e^{-x} \cos x \, dx$$

(A) Graph the integrand $f(x) = e^{-x} \cos x$ over $[0, 3]$.

(B) Approximate I using the midpoint sum M_6 (see Section 6–5).

(C) Show that 0.7 is an upper bound for $|f''(x)|$ on the interval $[0, 3]$ using a graphing utility.

(D) Use the result from part (C) to find an error estimate for part (B).

Applications

Business & Economics

29. *Seasonal business cycle.* Suppose that profits on the sale of swimming suits in a department store are given approximately by

$$P(t) = 5 - 5 \cos \frac{\pi t}{26} \qquad 0 \leq t \leq 104$$

where $P(t)$ is profit (in hundreds of dollars) for a week of sales t weeks after January 1. Use definite integrals to approximate:

(A) The total profit earned during the 2-year period

(B) The total profit earned from $t = 13$ to $t = 26$

(C) Illustrate part (B) graphically with an appropriate shaded region representing the total profit earned.

30. *Seasonal business cycle.* A soft drink company has revenues from sales over a 2-year period as given approximately by

$$R(t) = 4 - 3 \cos \frac{\pi t}{6} \qquad 0 \leq t \leq 24$$

where $R(t)$ is revenue (in millions of dollars) for a month of sales t months after February 1. Use definite integrals to approximate:

(A) Total revenues taken in over the 2-year period

(B) Total revenues taken in from $t = 8$ to $t = 14$

(C) Illustrate part (B) graphically with an appropriate shaded region representing the total revenues taken in.

Life Sciences

31. *Pollution.* In a large city, the amount of sulfur dioxide pollutant released into the atmosphere due to the burning of coal and oil for heating purposes is given approximately by

$$P(n) = 1 + \cos \frac{\pi n}{26} \qquad 0 \leq n \leq 104$$

where $P(n)$ is the amount of sulfur dioxide (in tons) released during the nth week after January 1.

(A) How many tons of pollutants were emitted into the atmosphere over the 2-year period?

(B) How many tons of pollutants were emitted into the atmosphere from $n = 13$ to $n = 52$?

(C) Illustrate part (B) graphically with an appropriate shaded region representing the total tons of pollutants emitted into the atmosphere.

Important Terms and Symbols

12-1 *Trigonometric Functions Review.* Angle; ray; initial side; terminal side; vertex; positive angle; negative angle; coterminal angles; degree measure; radian measure; conversion formula; standard position; sine and cosine functions; graphs of sine and cosine functions; periodic function; period of a function; tangent, cotangent, secant, and cosecant functions

$$\sin x; \quad \cos x; \quad \tan x; \quad \cot x; \quad \sec x; \quad \csc x$$

12-2 *Derivatives of Trigonometric Functions*

$$\frac{d}{dx} \sin x = \cos x; \quad \frac{d}{dx} \cos x = -\sin x;$$

$$\frac{d}{dx} \sin u = \cos u \frac{du}{dx}; \quad \frac{d}{dx} \cos u = -\sin u \frac{du}{dx}$$

12-3 *Integration of Trigonometric Functions*

$$\int \cos u \, du = \sin u + C; \qquad \int \sin u \, du = -\cos u + C$$

Review Exercise

Work through all the problems in this chapter review and check your answers in the back of the book. Answers to all review problems are there along with section numbers in italics to indicate where each type of problem is discussed. Where weaknesses show up, review appropriate sections in the text.

A

1. Convert to radian measure in terms of π:

    (A) 30°   (B) 45°   (C) 60°   (D) 90°

2. Evaluate without using a calculator:

    (A) $\cos \pi$ (B) $\sin 0$ (C) $\sin \dfrac{\pi}{2}$

In Problems 3–6, find each derivative or integral.

3. $\dfrac{d}{dm} \cos m$ **4.** $\dfrac{d}{du} \sin u$ **5.** $\dfrac{d}{dx} \sin(x^2 - 2x + 1)$

6. $\displaystyle\int \sin 3t\, dt$

B

7. Convert to degree measure:

    (A) $\pi/6$ (B) $\pi/4$ (C) $\pi/3$ (D) $\pi/2$

8. Evaluate without using a calculator:

    (A) $\sin \dfrac{\pi}{6}$ (B) $\cos \dfrac{\pi}{4}$ (C) $\sin \dfrac{\pi}{3}$

9. Evaluate using a calculator:

    (A) $\cos 33.7$ (B) $\sin (-118.4)$

In Problems 10–16, find each derivative or integral.

10. $\dfrac{d}{dx} (x^2 - 1)\sin x$ **11.** $\dfrac{d}{dx}(\sin x)^6$

12. $\dfrac{d}{dx} \sqrt[3]{\sin x}$ **13.** $\displaystyle\int t \cos(t^2 - 1)dt$

14. $\displaystyle\int_0^\pi \sin u\, du$ **15.** $\displaystyle\int_0^{\pi/3} \cos x\, dx$

16. $\displaystyle\int_1^{2.5} \cos x\, dx$

17. Find the slope of the cosine curve $y = \cos x$ at $x = \pi/4$.

18. Find the area under the sine curve $y = \sin x$ from $x = \pi/4$ to $x = 3\pi/4$.

19. Given the definite integral

$$I = \int_1^5 \frac{\sin x}{x}\, dx$$

    (A) Graph the integrand

$$f(x) = \frac{\sin x}{x}$$

    over $[1, 5]$.

    (B) Approximate I using the midpoint sum M_4.

    (C) Show that 0.3 is an upper bound for $|f''(x)|$ on the interval $[1, 5]$ using a graphing utility.

    (D) Use the result from part (C) to find an error estimate for part (B).

C

20. Convert 15° to radian measure.

21. Evaluate without using a calculator:

    (A) $\sin \dfrac{3\pi}{2}$ (B) $\cos \dfrac{5\pi}{6}$ (C) $\sin\left(\dfrac{-\pi}{6}\right)$

In Problems 22–26, find each derivative or integral.

22. $\dfrac{d}{du} \tan u$ **23.** $\dfrac{d}{dx} e^{\cos x^2}$

24. $\displaystyle\int e^{\sin x} \cos x\, dx$ **25.** $\displaystyle\int \tan x\, dx$

26. $\displaystyle\int_2^5 (5 + 2 \cos 2x)\, dx$

 In Problems 27–29, graph each function on a graphing utility set in radian mode.

27. $y = \dfrac{\sin \pi x}{0.2x}; 1 \le x \le 8, -4 \le y \le 4$

28. $y = 0.5x \cos \pi x; 0 \le x \le 8, -5 \le y \le 5$

29. $y = 3 - 2 \cos \pi x; 0 \le x \le 6, 0 \le y \le 5$

Applications

Business & Economics

Problems 30–32 refer to the following: Revenues from sweater sales in a sportswear chain are given approximately by

$$R(t) = 3 + 2 \cos\frac{\pi t}{6} \qquad 0 \le t \le 24$$

where $R(t)$ is the revenue (in thousands of dollars) for a month of sales t months after January 1.

30. (A) Find the exact values of $R(0)$, $R(2)$, $R(3)$, and $R(6)$ without using a calculator.

    (B) Use a calculator to find $R(1)$ and $R(22)$. Interpret the results verbally.

31. (A) What is the rate of change of revenue t months after January 1?

(B) What is the rate of change of revenue 3 months after January 1? 10 months after January 1? 18 months after January 1?

(C) Find all local maxima and minima for $0 < t < 24$.

(D) Find the absolute maximum and minimum for $0 \leq t \leq 24$.

(E) Repeat part (C) using a graphing utility.

32. (A) Find the total revenues taken in over the 2-year period.

(B) Find the total revenues taken in from $t = 5$ to $t = 9$.

(C) Illustrate part (B) graphically with an appropriate shaded region representing the total revenue taken in.

Group Activity 1   *Seasonal Business Cycles*

A large soft drink company's profits are seasonal, with the greatest profits occurring during the warm part of the year and the least profits occurring during the cold part of the year. Table 1 gives the profit for a month of sales (profit per month) t months after February 1 over a 2-year period.

TABLE 1

Profit t Months After February 1 (Million Dollars)

t	$P(t)$	t	$P(t)$	t	$P(t)$	t	$P(t)$
0	1.551						
1	1.082	7	10.447	13	0.932	19	9.578
2	1.804	8	10.353	14	1.785	20	9.116
3	4.255	9	8.662	15	3.461	21	8.533
4	5.692	10	5.960	16	7.011	22	5.506
5	8.137	11	2.847	17	8.245	23	3.725
6	9.907	12	1.158	18	9.137	24	1.879

(A) Plot the data points from Table 1 on graph paper.

(B) Estimate the total profit from $t = 1$ to $t = 13$ using the average of the left and right sums (see Section 6-4) and estimate the error.
[*Hint:* Break the problem into two parts where $P(t)$ is monotonic in each.]

(C) Sketch a graph of

$$P(t) = 5.682 - 4.425 \cos\left(\frac{\pi t}{6} - 0.43\right)$$

on top of the plot from part (A) over the interval $[1, 13]$. [The equation $y = P(t)$ provides a reasonable model for the data in Table 1.]

(D) With an appropriate use of a definite integral involving the profit function in part (C), estimate the total profit earned from $t = 1$ to $t = 13$.

Group Activity 2   ***Heating Degree Days***

www

Estimating the energy demand for heating and cooling homes is an important activity for government agencies, local utilities, oil companies, and other energy suppliers. **Heating degree days** are used to estimate the amount of energy required for residential space heating. If the average temperature T on a given day is less than 65°F, the number of heating degree days for that day is $65 - T$. If $T \geq 65$, the number of heating degree days is 0. The sum of the heating degree days over the entire year is used to predict the annual heating demand.

Table 1 gives the average monthly temperature $T(x)$ for Tulsa, Oklahoma, where x is days after January 1.

TABLE 1

AVERAGE MONTHLY TEMPERATURES FOR TULSA (°F)

MONTH	Jan.	Feb.	Mar.	Apr.	May	June	July	Aug.	Sept.	Oct.	Nov.	Dec.
x	15	45	75	105	135	165	195	225	255	285	315	345
$T(x)$	35	40	51	61	69	78	83	82	73	62	50	39

(A) To construct a model that uses monthly average temperatures to approximate the annual heating demand in Tulsa, we assume that each month has exactly 30 days. Enter x, the midpoint of each month, and the corresponding temperature $T(x)$ in the statistics editor of a graphing utility. If your graphing utility supports sinusoidal regression, find a model of the form $T(x) = a \sin(bx + c) + d$. Otherwise, find a fourth-degree polynomial model for $T(x)$.

(B) Graph $T(x)$ and the line $y = 65$ for $0 \leq x \leq 360$, and describe the region whose area would approximate the annual heating demand.

(C) Express the area from part (B) in terms of one or more definite integrals, and use numerical integration to approximate this area.

In a similar manner, **cooling degree days** are defined to be the difference between the average daily temperature T and 65 if $T > 65$, and 0 otherwise.

(D) Describe the region on the graph in part (B) that would approximate the annual cooling demand.

(E) Express the area from part (D) in terms of one or more definite integrals, and use numerical integration to approximate this area.

Appendix A

Basic Algebra Review

INTRODUCTION

Appendix A reviews some important basic algebra concepts usually studied in earlier courses. The material may be studied systematically before beginning the rest of the book or reviewed as needed. The Self-Test on Basic Algebra that precedes Section A-1 may be taken to locate areas of weakness. All the answers to the self-test are in the answer section in the back of the book and are keyed to the sections in Appendix A where the related topics are discussed.

Self-Test on Basic Algebra

Work through all the problems in this self-test and check your answers in the back of the book. All answers are there and are keyed to relevant sections in Appendix A. Where weaknesses show up, review appropriate sections in the appendix.

1. Indicate true (T) or false (F):

(A) $7 \notin \{4, 6, 8\}$ (B) $\{8\} \subset \{4, 6, 8\}$

(C) $\varnothing \notin \{4, 6, 8\}$ (D) $\varnothing \subset \{4, 6, 8\}$

2. Replace each question mark with an appropriate expression that will illustrate the use of the indicated real number property:

(A) Commutative ($\cdot$): $x(y + z) = ?$

(B) Associative (+): $2 + (x + y) = ?$

(C) Distributive: $(2 + 3)x = ?$

Problems 3–7 refer to the following polynomials:

(A) $3x - 4$
(B) $x + 2$
(C) $3x^2 + x - 8$
(D) $x^3 + 8$

3. Add all four.

4. Subtract the sum of (A) and (C) from the sum of (B) and (D).

5. Multiply (C) and (D).

6. What is the degree of (D)?

7. What is the coefficient of the second term in (C)?

In Problems 8–13, perform the indicated operations and simplify.

8. $5x^2 - 3x[4 - 3(x - 2)]$

9. $(2x + y)(3x - 4y)$

10. $(2a - 3b)^2$

11. $(2x - y)(2x + y) - (2x - y)^2$

12. $(m^2 + 2mn - n^2)(m^2 - 2mn - n^2)$

13. $(x - 2y)^3$

14. Write in scientific notation:

(A) 4,065,000,000,000   (B) 0.0073

15. Write in standard decimal form:

(A) 2.55×10^8 (B) 4.06×10^{-4}

16. If $U = \{2, 4, 5, 6, 8\}$, $M = \{2, 4, 5\}$, and $N = \{5, 6\}$, find:

(A) $M \cup N$ (B) $M \cap N$
(C) $(M \cup N)'$ (D) $M \cap N'$

17. Given the Venn diagram shown in the figure, how many elements are in each of the following sets:

(A) $A \cup B$ (B) $A \cap B$
(C) $(A \cup B)'$ (D) $A \cap B'$

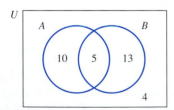

18. Indicate true (T) or false (F):

(A) A natural number is a rational number.
(B) A number with a repeating decimal expansion is an irrational number.

19. Give an example of an integer that is not a natural number.

Simplify Problems 20–28 and write answers using positive exponents only. All variables represent positive real numbers.

20. $6(xy^3)^5$

21. $\dfrac{9u^8v^6}{3u^4v^8}$

22. $(2 \times 10^5)(3 \times 10^{-3})$

23. $(x^{-3}y^2)^{-2}$

24. $u^{5/3}u^{2/3}$

25. $(9a^4b^{-2})^{1/2}$

26. $\dfrac{5^0}{3^2} + \dfrac{3^{-2}}{2^{-2}}$

27. $(x^{1/2} + y^{1/2})^2$

28. $(3x^{1/2} - y^{1/2})(2x^{1/2} + 3y^{1/2})$

Write Problems 29–34 in completely factored form relative to the integers. If a polynomial cannot be factored further relative to the integers, say so.

29. $12x^2 + 5x - 3$

30. $8x^2 - 18xy + 9y^2$

31. $t^2 - 4t - 6$

32. $6n^3 - 9n^2 - 15n$

33. $(4x - y)^2 - 9x^2$

34. $2x^2 + 4xy - 5y^2$

In Problems 35–40, perform the indicated operations and reduce to lowest terms. Represent all compound fractions as simple fractions reduced to lowest terms.

35. $\dfrac{2}{5b} - \dfrac{4}{3a^3} - \dfrac{1}{6a^2b^2}$

36. $\dfrac{3x}{3x^2 - 12x} + \dfrac{1}{6x}$

37. $\dfrac{x}{x^2 - 16} - \dfrac{x + 4}{x^2 - 4x}$

38. $\dfrac{y - 2}{y^2 - 4y + 4} \div \dfrac{y^2 + 2y}{y^2 + 4y + 4}$

39. $\dfrac{\dfrac{1}{7 + h} - \dfrac{1}{7}}{h}$

40. $\dfrac{x^{-1} + y^{-1}}{x^{-2} - y^{-2}}$

41. Each statement illustrates the use of one of the following real number properties or definitions. Indicate which one.

Commutative $(+, \cdot)$ Associative $(+, \cdot)$ Distributive
Identity $(+, \cdot)$ Inverse $(+, \cdot)$ Subtraction
Division   Negatives   Zero

(A) $(-7) - (-5) = (-7) + [-(-5)]$
(B) $5u + (3v + 2) = (3v + 2) + 5u$
(C) $(5m - 2)(2m + 3) =$
$(5m - 2)2m + (5m - 2)3$
(D) $9 \cdot (4y) = (9 \cdot 4)y$
(E) $\dfrac{u}{-(v - w)} = -\dfrac{u}{v - w}$
(F) $(x - y) + 0 = (x - y)$

42. In a freshman class of 100 students, 70 are taking English, 45 are taking math, and 25 are taking both English and math.

(A) How many students are taking either English or math?

(B) How many students are taking English and not math?

43. Change to rational exponent form:

$$6\sqrt[5]{x^2} - 7\sqrt[4]{(x-1)^3}$$

44. Change to radical form: $2x^{1/2} - 3x^{2/3}$

45. Write in the form $ax^p + bx^q$, where a and b are real numbers and p and q are rational numbers:

$$\frac{4\sqrt{x} - 3}{2\sqrt{x}}$$

In Problems 46 and 47, rationalize the denominator.

46. $\dfrac{3x}{\sqrt{3x}}$ **47.** $\dfrac{x-5}{\sqrt{x} - \sqrt{5}}$

In Problems 48 and 49, rationalize the numerator.

48. $\dfrac{\sqrt{x} - 5}{x - 5}$ **49.** $\dfrac{\sqrt{u+h} - \sqrt{u}}{h}$

Solve Problems 50–54 for x.

50. $\dfrac{x}{12} - \dfrac{x-3}{3} = \dfrac{1}{2}$ **51.** $x^2 = 5x$

52. $3x^2 - 21 = 0$ **53.** $x^2 - x - 20 = 0$

54. $2x = 3 + \dfrac{1}{x}$

In Problems 55–57, solve and graph on a real number line.

55. $2(x + 4) > 5x - 4$

56. $1 - \dfrac{x-3}{3} \le \dfrac{1}{2}$

57. $-2 \le \dfrac{x}{2} - 3 < 3$

In Problems 58 and 59, solve for y in terms of x.

58. $2x - 3y = 6$

59. $xy - y = 3$

60. If $A \cap B = A$, is it always true that $A \subset B$? Explain.

Applications

61. *Economics.* If the gross domestic product (GDP) was $8,511,000,000,000 for the United States in 1998 and the population was 270,300,000, determine the GDP per person using scientific notation. Express the answer in scientific notation and in standard decimal form to the nearest dollar.

62. *Investment.* An investor has $60,000 to invest. If part is invested at 8% and the rest at 14%, how much

should be invested at each rate to yield 12% on the total amount?

63. *Break-even analysis.* A producer of educational videos is producing an instructional video. The producer estimates that it will cost $72,000 to shoot the video and $12 per unit to copy and distribute the tape. If the wholesale price of the tape is $30, how many tapes must be sold for the producer to break even?

Section A-1

Sets

- ❑ SET PROPERTIES AND SET NOTATION
- ❑ SET OPERATIONS
- ❑ APPLICATION

In this section we review a few key ideas from set theory. Set concepts and notation not only help us talk about certain mathematical ideas with greater clarity and precision, but are indispensable to a clear understanding of probability.

❑ SET PROPERTIES AND SET NOTATION

We can think of a **set** as any collection of objects specified in such a way that we can tell whether any given object is or is not in the collection. Capital letters, such as A, B, and C, are often used to designate particular sets. Each object in a set is called a **member,** or **element,** of the set. Symbolically:

$a \in A$	means	"a is an element of set A"
$a \notin A$	means	"a is not an element of set A"

A set without any elements is called the **empty,** or **null, set.** For example, the set of all people over 20 feet tall is an empty set. Symbolically:

$\varnothing$	represents	"the empty, or null, set"

A set is usually described either by listing all its elements between braces { } or by enclosing a rule within braces that determines the elements of the set. Thus, if $P(x)$ is a statement about x, then

$S = \{x \mid P(x)\}$	means	"S is the set of all x such that $P(x)$ is true"

Recall that the vertical bar within the braces is read "such that." The following example illustrates the rule and listing methods of representing sets.

Example 1 ⇨ **Representing Sets**

Rule   Listing

$$\{x \mid x \text{ is a weekend day}\} = \{\text{Saturday, Sunday}\}$$
$$\{x \mid x^2 = 4\} = \{-2, 2\}$$
$$\{x \mid x \text{ is an odd positive counting number}\} = \{1, 3, 5, \ldots\}$$

■

The three dots (. . .) in the last set given in Example 1 indicate that the pattern established by the first three entries continues indefinitely. The first two sets in Example 1 are **finite sets** (we intuitively know that the elements can be counted, and there is an end); the last set is an **infinite set** (we intuitively know that there is no end in counting the elements). When listing the elements in a set, we do not list an element more than once, and the order in which the elements are listed does not matter.

Matched Problem 1 ⇨ Let G be the set of all numbers such that $x^2 = 9$.

(A) Denote G by the rule method. (B) Denote G by the listing method.
(C) Indicate whether the following are true or false: $3 \in G, 9 \notin G$.

■

If each element of a set A is also an element of set B, we say that A is a **subset** of B. For example, the set of all women students in a class is a subset of the whole class. Note that the definition allows a set to be a subset of itself. If set A and set B have exactly the same elements, then the two sets are said to be **equal.** Symbolically:

$A \subset B$	means	"A is a subset of B"
$A = B$	means	"A and B have exactly the same elements"
$A \not\subset B$	means	"A is not a subset of B"
$A \neq B$	means	"A and B do not have exactly the same elements"

It can be proved that:

⊘ **is a subset of every set.**

Example 2 ⇨ **Set Notation** If $A = \{-3, -1, 1, 3\}$, $B = \{3, -3, 1, -1\}$, and $C = \{-3, -2, -1, 0, 1, 2, 3\}$, each of the following statements is true:

$$A = B \qquad A \subset C \qquad A \subset B$$
$$C \neq A \qquad C \not\subset A \qquad B \subset A$$
$$\varnothing \subset A \qquad \varnothing \subset C \qquad \varnothing \notin A$$

∎

Matched Problem 2 ⇨ Given $A = \{0, 2, 4, 6\}$, $B = \{0, 1, 2, 3, 4, 5, 6\}$, and $C = \{2, 6, 0, 4\}$, indicate whether the following relationships are true (T) or false (F):

(A) $A \subset B$ (B) $A \subset C$ (C) $A = C$
(D) $C \subset B$ (E) $B \not\subset A$ (F) $\varnothing \subset B$

∎

Example 3 ⇨ **Subsets** List all the subsets of the set $\{a, b, c\}$.

SOLUTION $\{a, b, c\}, \quad \{a, b\}, \quad \{a, c\}, \quad \{b, c\}, \quad \{a\}, \quad \{b\}, \quad \{c\}, \quad \varnothing$ ∎

Matched Problem 3 ⇨ List all the subsets of the set $\{1, 2\}$.

∎

Explore–Discuss 1

Which of the following statements are true?

(A) $\varnothing \subset \varnothing$ (B) $\varnothing \in \varnothing$ (C) $\varnothing = \{0\}$ (D) $\varnothing \subset \{0\}$

❑ SET OPERATIONS

The **union** of sets A and B, denoted by $A \cup B$, is the set of all elements formed by combining all the elements of A and all the elements of B into one set. Symbolically:

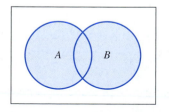

FIGURE 1 $A \cup B$ is the shaded region.

Union

$$A \cup B = \{x \mid x \in A \text{ **or** } x \in B\}$$

Here we use the word **or** in the way it is always used in mathematics; that is, x may be an element of set A or set B or both.

Venn diagrams are useful in visualizing set relationships. The union of two sets can be illustrated as shown in Figure 1. Note that

$$A \subset A \cup B \qquad \text{and} \qquad B \subset A \cup B$$

The **intersection** of sets A and B, denoted by $A \cap B$, is the set of elements in set A that are also in set B. Symbolically:

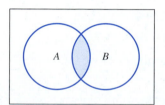

FIGURE 2 $A \cap B$ is the shaded region.

Intersection

$$A \cap B = \{x \mid x \in A \text{ **and** } x \in B\}$$

This relationship is easily visualized in the Venn diagram shown in Figure 2. Note that

$$A \cap B \subset A \qquad \text{and} \qquad A \cap B \subset B$$

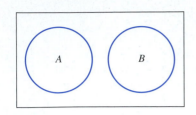

FIGURE 3 $A \cap B = \emptyset$; A and B are disjoint.

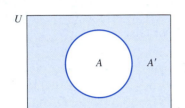

FIGURE 4 The complement of A is A'.

If $A \cap B = \emptyset$, the sets A and B are said to be **disjoint;** this is illustrated in Figure 3.

The set of all elements under consideration is called the **universal set U.** Once the universal set is determined for a particular case, all other sets under discussion must be subsets of U.

We now define one more operation on sets, called the *complement.* The **complement** of A (relative to U), denoted by A', is the set of elements in U that are not in A (see Fig. 4). Symbolically:

> *Complement*
>
> $$A' = \{x \in U \mid x \notin A\}$$

Example 4 ➭ **Union, Intersection, and Complement** If $A = \{3, 6, 9\}$, $B = \{3, 4, 5, 6, 7\}$, $C = \{4, 5, 7\}$, and $U = \{1, 2, 3, 4, 5, 6, 7, 8, 9\}$, then

$$A \cup B = \{3, 4, 5, 6, 7, 9,\}$$
$$A \cap B = \{3, 6\}$$
$$A \cap C = \emptyset \qquad \textit{A and C are disjoint.}$$
$$B' = \{1, 2, 8, 9\}$$

Matched Problem 4 ➭ If $R = \{1, 2, 3, 4\}$, $S = \{1, 3, 5, 7\}$, $T = \{2, 4\}$, and $U = \{1, 2, 3, 4, 5, 6, 7, 8, 9\}$, find:

(A) $R \cup S$ (B) $R \cap S$ (C) $S \cap T$ (D) S'

❑ APPLICATION

Example 5 ➭ **Marketing Survey**   In a survey of 100 randomly chosen students, a marketing questionnaire included the following three questions:

1. Do you own a TV?

2. Do you own a car?

3. Do you own a TV and a car?

Seventy-five answered yes to 1, 45 answered yes to 2, and 35 answered yes to 3.

(A) How many students owned either a car or a TV?

(B) How many students did not own either a car or a TV?

SOLUTION   Venn diagrams are very useful for this type of problem. If we let

$$U = \text{set of students in sample (100)}$$
$$T = \text{set of students who own TV sets (75)}$$
$$C = \text{set of students who own cars (45)}$$
$$T \cap C = \text{set of students who own cars and TV sets (35)}$$

then:

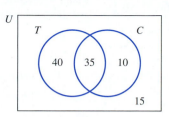

Place the number in the intersection first; then work outward:
40 = 75 − 35
10 = 45 − 35
15 = 100 − (40 + 35 + 10)

(A) The number of students who own either a car or a TV is the number of students in the set $T \cup C$. You might be tempted to say that this is just the number of students in T plus the number of students in C, $75 + 45 = 120$, but this sum is larger than the sample we started with! What is wrong? We have actually counted the number in the intersection (35) twice. The correct answer, as seen in the Venn diagram, is

$$40 + 35 + 10 = 85$$

(B) The number of students who do not own either a car or a TV is the number of students in the set $(T \cup C)'$, that is, 15.

Matched Problem 5 ⇔ Referring to Example 5:

(A) How many students owned a car but not a TV set?

(B) How many students did not own both a car and a TV set?

Note in Example 5 and Matched Problem 5 that the word **and** is associated with intersection and the word **or** is associated with union.

Explore–Discuss 2

In Example 5, find the number of students in the set $(T \cup C) \cap C'$. Describe this set verbally and with a Venn diagram.

Answers to Matched Problems   **1.** (A) $\{x \mid x^2 = 9\}$ (B) $\{-3, 3\}$ (C) True; true

2. All are true   **3.** $\{1, 2\}, \{1\}, \{2\}, \varnothing$

4. (A) $\{1, 2, 3, 4, 5, 7\}$ (B) $\{1, 3\}$ (C) $\varnothing$ (D) $\{2, 4, 6, 8, 9\}$

5. (A) 10, the number in $T' \cap C$ (B) 65, the number in $(T \cap C)'$

Exercise A-1

A *Indicate true (T) or false (F) in Problems 1–8.*

1. $4 \in \{2, 3, 4\}$ **2.** $6 \notin \{2, 3, 4\}$

3. $\{2, 3\} \subset \{2, 3, 4\}$ **4.** $\{3, 2, 4\} = \{2, 3, 4\}$

5. $\{3, 2, 4\} \subset \{2, 3, 4\}$ **6.** $\{3, 2, 4\} \in \{2, 3, 4\}$

7. $\varnothing \subset \{2, 3, 4\}$ **8.** $\varnothing = \{0\}$

In Problems 9–20, write the resulting set using the listing method.

9. $\{1, 3, 5\} \cup \{2, 3, 4\}$ **10.** $\{3, 4, 6, 7\} \cup \{3, 4, 5\}$

11. $\{1, 3, 4\} \cap \{2, 3, 4\}$ **12.** $\{3, 4, 6, 7\} \cap \{3, 4, 5\}$

13. $\{1, 5, 9\} \cap \{3, 4, 6, 8\}$

14. $\{6, 8, 9, 11\} \cap \{3, 4, 5, 7\}$

B

15. $\{x \mid x - 2 = 0\}$ **16.** $\{x \mid x + 7 = 0\}$

17. $\{x \mid x^2 = 49\}$ **18.** $\{x \mid x^2 = 100\}$

19. $\{x \mid x$ is an odd number between 1 and 9, inclusive$\}$

20. $\{x \mid x$ is a month starting with $M\}$

21. For $U = \{1, 2, 3, 4, 5\}$ and $A = \{2, 3, 4\}$, find A'.

22. For $U = \{7, 8, 9, 10, 11\}$ and $A = \{7, 11\}$, find A'.

Problems 23–34 refer to the Venn diagram below. How many elements are in each of the indicated sets?

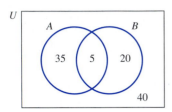

23. A **24.** U **25.** A'

26. B' **27.** $A \cup B$ **28.** $A \cap B$

29. $A' \cap B$ **30.** $A \cap B'$ **31.** $(A \cap B)'$

32. $(A \cup B)'$ **33.** $A' \cap B'$ **34.** U'

35. If $R = \{1, 2, 3, 4\}$ and $T = \{2, 4, 6\}$, find:

    (A) $\{x \,|\, x \in R \text{ or } x \in T\}$ (B) $R \cup T$

36. If $R = \{1, 3, 4\}$ and $T = \{2, 4, 6\}$, find:

    (A) $\{x \,|\, x \in R \text{ and } x \in T\}$ (B) $R \cap T$

37. For $P = \{1, 2, 3, 4\}$, $Q = \{2, 4, 6\}$, and $R = \{3, 4, 5, 6\}$, find $P \cup (Q \cap R)$.

38. For P, Q, and R in Problem 37, find $P \cap (Q \cup R)$.

In Problems 39–48, discuss the validity of each statement. Venn diagrams may be of help. If the statement is true, explain why. If not, give a counterexample.

39. If $A \subset B$, then $A \cap B = A$.

40. If $A \subset B$, then $A \cup B = A$.

41. If $A \cup B = A$, then $A \subset B$.

42. If $A \cap B = A$, then $A \subset B$.

43. If $A \cap B = \varnothing$, then $A = \varnothing$.

44. If $A = \varnothing$, then $A \cap B = \varnothing$.

45. If $A \subset B$, then $A' \subset B'$.

46. If $A \subset B$, then $B' \subset A'$.

47. The empty set is an element of every set.

48. The empty set is a subset of the empty set.

49. How many subsets does each of the following sets contain?

    (A) $\{a\}$
    (B) $\{a, b\}$
    (C) $\{a, b, c\}$
    (D) $\{a, b, c, d\}$

50. Let A be a set that contains exactly n elements. Find a formula in terms of n for the number of subsets of A.

51. Are the sets $\varnothing$ and $\{\varnothing\}$ equal? Explain.

52. Are the sets $\{\varnothing\}$ and $\{\varnothing, \{\varnothing\}\}$ equal? Explain.

Applications

Business & Economics

Marketing survey. Problems 53–64 refer to the following survey: A marketing survey of 1,000 car commuters found that 600 answered yes to listening to the news, 500 answered yes to listening to music, and 300 answered yes to listening to both. Let

N = *set of commuters in the sample who listen to news*

M = *set of commuters in the sample who listen to music*

Following the procedures in Example 5, find the number of commuters in each set described below.

53. $N \cup M$ **54.** $N \cap M$ **55.** $(N \cup M)'$

56. $(N \cap M)'$ **57.** $N' \cap M$ **58.** $N \cap M'$

59. Set of commuters who listen to either news or music

60. Set of commuters who listen to both news and music

61. Set of commuters who do not listen to either news or music

62. Set of commuters who do not listen to both news and music

63. Set of commuters who listen to music but not news

64. Set of commuters who listen to news but not music

65. *Committee selection.* The management of a company, a president and three vice-presidents, denoted by the set $\{P, V_1, V_2, V_3\}$, wish to select a committee of 2 people from among themselves. How many ways can this committee be formed? That is, how many 2-person subsets can be formed from a set of 4 people?

66. *Voting coalition.* The management of the company in Problem 65 decides for or against certain measures as follows: The president has 2 votes and each vice-president has 1 vote. Three favorable votes are needed to pass a measure. List all minimal winning coalitions; that is, list all subsets of $\{P, V_1, V_2, V_3\}$ that represent exactly 3 votes.

Life Sciences

Blood types. *When receiving a blood transfusion, a recipient must have all the antigens of the donor. A person may have one or more of the three antigens A, B, and Rh, or none at all. Eight blood types are possible, as indicated in the following Venn diagram, where U is the set of all people under consideration:*

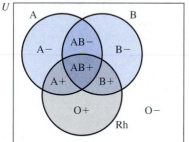

An A– person has A antigens but no B or Rh; an O+ person has Rh but neither A nor B; an AB– person has A and B antigens but no Rh; and so on.

Using the Venn diagram, indicate which of the eight blood types are included in the sets in Problems 67–74.

67. $A \cap Rh$ **68.** $A \cap B$ **69.** $A \cup Rh$

70. $A \cup B$ **71.** $(A \cup B)'$ **72.** $(A \cup B \cup Rh)'$

73. $A' \cap B$ **74.** $Rh' \cap A$

Social Sciences

Group structures.   *R. D. Luce and A. D. Perry, in a study on group structure (Psychometrika, 1949, 14:95–116), used the idea of sets to formally define the notion of a clique within a group. Let G be the set of all persons in the group and let C ⊂ G. Then C is a clique provided that:*

1. *C contains at least 3 elements.*
2. *For every a, b ∈ C, a **R** b and b **R** a.*
3. *For every a ∉ C, there is at least one b ∈ C such that a **R** b or b **R** a or both.*

*[Note: Interpret "a **R** b" to mean "a relates to b," "a likes b," "a is as wealthy as b," and so on. Of course, "a **R̶** b" means "a does not relate to b," and so on.]*

75. Translate statement 2 into ordinary English.

76. Translate statement 3 into ordinary English.

Section A-2

Algebra and Real Numbers

- ❑   SET OF REAL NUMBERS
- ❑   REAL NUMBER LINE
- ❑   BASIC REAL NUMBER PROPERTIES
- ❑   FURTHER PROPERTIES
- ❑   FRACTION PROPERTIES

The rules for manipulating and reasoning with symbols in algebra depend, in large measure, on properties of the real numbers. In this section we look at some of the important properties of this number system. To make our discussions here and elsewhere in the book clearer and more precise, we occasionally make use of simple *set* concepts and notation. Refer to Section A-1 if you are not yet familiar with the basic ideas concerning sets.

❑   SET OF REAL NUMBERS

What number system have you been using most of your life? The *real number system.* Informally, a **real number** is any number that has a decimal representation. Table 1 describes the set of real numbers and some of its important subsets. Figure 1 illustrates how these sets of numbers are related.

TABLE 1

SET OF REAL NUMBERS

SYMBOL	NAME	DESCRIPTION	EXAMPLES
N	Natural numbers	Counting numbers (also called positive integers)	$1, 2, 3, \ldots$
Z	Integers	Natural numbers, their negatives, and 0	$\ldots, -2, -1, 0, 1, 2, \ldots$
Q	Rational numbers	Numbers that can be represented as a/b, where a and b are integers and $b \neq 0$; decimal representations are repeating or terminating	$-4, 0, 1, 25, \frac{-3}{5}, \frac{2}{3}, 3.67, -0.33\overline{3}, 5.272\ \overline{727}*$
I	Irrational numbers	Numbers that can be represented as nonrepeating and nonterminating decimal numbers	$\sqrt{2}, \pi, \sqrt[3]{7}, 1.414\ 213 \ldots, 2.718\ 281\ 82 \ldots$
R	Real numbers	Rational and irrational numbers	

*The overbar indicates that the number (or block of numbers) repeats indefinitely. The space after every third digit is used to help keep track of the number of decimal places.

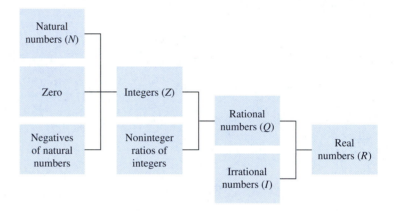

FIGURE 1 Real numbers and important subsets

The set of integers contains all the natural numbers and something else—their negatives and 0. The set of rational numbers contains all the integers and something else—noninteger ratios of integers. And the set of real numbers contains all the rational numbers and something else—irrational numbers.

❑ REAL NUMBER LINE

A one-to-one correspondence exists between the set of real numbers and the set of points on a line. That is, each real number corresponds to exactly one point, and each point corresponds to exactly one real number. A line with a real number associated with each point, and vice versa, as shown in Figure 2, is called a **real number line,** or simply a **real line.** Each number associated with a point is called the **coordinate** of the point.

The point with coordinate 0 is called the **origin.** The arrow on the right end of the line indicates a positive direction. The coordinates of all points to the right of the origin are called **positive real numbers,** and those to the left of the origin are called **negative real numbers.** The real number 0 is neither positive nor negative.

FIGURE 2 Real number line

❑ BASIC REAL NUMBER PROPERTIES

We now take a look at some of the basic properties of the real number system that enable us to convert algebraic expressions into *equivalent forms.* These assumed properties become operational rules in the algebra of real numbers.

Basic Properties of the Set of Real Numbers

Let a, b, and c be arbitrary elements in the set of real numbers R.

Addition Properties

Associative: $(a + b) + c = a + (b + c)$

Commutative: $a + b = b + a$

Identity: 0 is the additive identity; that is, $0 + a = a + 0 = a$ for all a in R, and 0 is the only element in R with this property.

Inverse: For each a in R, $-a$ is its unique additive inverse; that is, $a + (-a) = (-a) + a = 0$, and $-a$ is the only element in R relative to a with this property.

Multiplication Properties

Associative: $(ab)c = a(bc)$

Commutative: $ab = ba$

Identity: 1 is the multiplicative identity; that is, $(1)a = a(1) = a$ for all a in R, and 1 is the only element in R with this property.

Inverse: For each a in R, $a \neq 0$, $1/a$ is its unique multiplicative inverse; that is, $a(1/a) = (1/a)a = 1$, and $1/a$ is the only element in R relative to a with this property.

Distributive Properties

$a(b + c) = ab + ac \qquad (a + b)c = ac + bc$

Do not be intimidated by the names of these properties. Most of the ideas presented here are quite simple. In fact, you have been using many of these properties in arithmetic for a long time.

You are already familiar with the **commutative properties** for addition and multiplication. They indicate that the order in which the addition or multiplication of two numbers is performed does not matter. For example,

$$7 + 2 = 2 + 7 \qquad \text{and} \qquad 3 \cdot 5 = 5 \cdot 3$$

Is there a commutative property relative to subtraction or division? That is, does $a - b = b - a$ or does $a \div b = b \div a$ for all real numbers a and b (division by 0 excluded)? The answer is no, since, for example,

$$8 - 6 \neq 6 - 8 \qquad \text{and} \qquad 10 \div 5 \neq 5 \div 10$$

When computing

$$3 + 2 + 6 \qquad \text{or} \qquad 3 \cdot 2 \cdot 6$$

why do we not need parentheses to indicate which two numbers are to be added or multiplied first? The answer is to be found in the **associative properties.** These properties allow us to write

$$(3 + 2) + 6 = 3 + (2 + 6) \qquad \text{and} \qquad (3 \cdot 2) \cdot 6 = 3 \cdot (2 \cdot 6)$$

so it does not matter how we group numbers relative to either operation. Is there an associative property for subtraction or division? The answer is no, since, for example,

$$(12 - 6) - 2 \neq 12 - (6 - 2) \qquad \text{and} \qquad (12 \div 6) \div 2 \neq 12 \div (6 \div 2)$$

Evaluate each side of each equation to see why.

> **Relative to addition, commutativity and associativity permit us to change the order of addition at will and insert or remove parentheses as we please. The same is true for multiplication, but not for subtraction and division.**

What number added to a given number will give that number back again? What number times a given number will give that number back again? The answers are 0 and 1, respectively. Because of this, 0 and 1 are called the **identity elements** for the real numbers. Hence, for any real numbers a and b,

$$0 + 5 = 5 \qquad \text{and} \qquad (a + b) + 0 = a + b$$
$$1 \cdot 4 = 4 \qquad \text{and} \qquad (a + b) \cdot 1 = a + b$$

We now consider **inverses.** For each real number a, there is a unique real number $-a$ such that $a + (-a) = 0$. The number $-a$ is called the **additive inverse** of a, or the **negative** of a. For example, the additive inverse (or negative) of 7 is -7, since $7 + (-7) = 0$. The additive inverse (or negative) of -7 is $-(-7) = 7$, since $-7 + [-(-7)] = 0$. It is important to remember that:

> **$-a$ is not necessarily a negative number; it is positive if a is negative and negative if a is positive.**

For each nonzero real number a, there is a unique real number $1/a$ such that $a(1/a) = 1$. The number $1/a$ is called the **multiplicative inverse** of a, or the **reciprocal** of a. For example, the multiplicative inverse (or reciprocal) of 4 is $\frac{1}{4}$, since $4(\frac{1}{4}) = 1$. (Also note that 4 is the multiplicative inverse of $\frac{1}{4}$.) The number 0 has no multiplicative inverse.

We now turn to the **distributive properties,** which involve both multiplication and addition. Consider the following two computations:

$$5(3 + 4) = 5 \cdot 7 = 35 \qquad 5 \cdot 3 + 5 \cdot 4 = 15 + 20 = 35$$

Thus,

$$5(3 + 4) = 5 \cdot 3 + 5 \cdot 4$$

and we say that multiplication by 5 *distributes* over the sum $(3 + 4)$. In general, **multiplication distributes over addition** in the real number system. Two more illustrations are

$$9(m + n) = 9m + 9n \qquad (7 + 2)u = 7u + 2u$$

Example 1 ➯ **Real Number Properties** State the real number property that justifies the indicated statement.

STATEMENT	PROPERTY ILLUSTRATED
(A) $x(y + z) = (y + z)x$	Commutative ($\cdot$)
(B) $5(2y) = (5 \cdot 2)y$	Associative ($\cdot$)
(C) $2 + (y + 7) = 2 + (7 + y)$	Commutative (+)
(D) $4z + 6z = (4 + 6)z$	Distributive
(E) If $m + n = 0$, then $n = -m$.	Inverse (+)

Matched Problem 1   State the real number property that justifies the indicated statement.

(A) $8 + (3 + y) = (8 + 3) + y$
(B) $(x + y) + z = z + (x + y)$
(C) $(a + b)(x + y) = a(x + y) + b(x + y)$
(D) $5xy + 0 = 5xy$
(E) If $xy = 1, x \neq 0$, then $y = 1/x$.

☐ **FURTHER PROPERTIES**

Subtraction and *division* can be defined in terms of addition and multiplication, respectively:

Subtraction and Division

For all real numbers a and b:

Subtraction: $a - b = a + (-b)$ $7 - (-5) = 7 + [-(-5)]$
$$= 7 + 5 = 12$$

Division: $a \div b = a\left(\dfrac{1}{b}\right), b \neq 0$ $9 \div 4 = 9\left(\dfrac{1}{4}\right) = \dfrac{9}{4}$

Thus, to subtract b from a, add the negative (the additive inverse) of b to a. To divide a by b, multiply a by the reciprocal (the multiplicative inverse) of b. Note that division by 0 is not defined, since 0 does not have a reciprocal. Thus:

0 can never be used as a divisor!

The following properties of negatives can be proved using the preceding assumed properties and definitions.

Properties of Negatives

For all real numbers a and b:

1. $-(-a) = a$

2. $(-a)b = -(ab)$
$= a(-b) = -ab$

3. $(-a)(-b) = ab$

4. $(-1)a = -a$

5. $\dfrac{-a}{b} = -\dfrac{a}{b} = \dfrac{a}{-b}, b \neq 0$

6. $\dfrac{-a}{-b} = -\dfrac{-a}{b} = -\dfrac{a}{-b} = \dfrac{a}{b}, b \neq 0$

We now state two important properties involving 0:

Zero Properties

For all real numbers a and b:

1. $a \cdot 0 = 0$ $0 \cdot 0 = 0$ $(-35)(0) = 0$

2. $ab = 0$ if and only if $a = 0$ or $b = 0$
If $(3x + 2)(x - 7) = 0$, then either $3x + 2 = 0$ or $x - 7 = 0$.

Explore–Discuss 1

In general, a set of numbers is closed under an operation if performing the operation on numbers in the set always produces another number in the set. For example, the real numbers R are closed under addition, multiplication, subtraction, and division, excluding division by 0. Replace each ? in the following tables with T (true) or F (false), and illustrate each false statement with an example. (See Table 1 for the definitions of the sets N, Z, Q, I, and R.)

	CLOSED UNDER ADDITION	CLOSED UNDER MULTIPLICATION		CLOSED UNDER SUBTRACTION	CLOSED UNDER DIVISION*
N	?	?	N	?	?
Z	?	?	Z	?	?
Q	?	?	Q	?	?
I	?	?	I	?	?
R	T	T	R	T	T

*Excluding division by 0.

□ **FRACTION PROPERTIES**

Recall that the quotient $a \div b\,(b \neq 0)$ written in the form a/b is called a **fraction.** The quantity a is called the **numerator,** and the quantity b is called the **denominator.**

Fraction Properties

For all real numbers a, b, c, d, and k (division by 0 excluded):

1. $\dfrac{a}{b} = \dfrac{c}{d}$ if and only if $ad = bc$ $\dfrac{4}{6} = \dfrac{6}{9}$ $since\ 4 \cdot 9 = 6 \cdot 6$

2. $\dfrac{ka}{kb} = \dfrac{a}{b}$

$\dfrac{7 \cdot 3}{7 \cdot 5} = \dfrac{3}{5}$

3. $\dfrac{a}{b} \cdot \dfrac{c}{d} = \dfrac{ac}{bd}$

$\dfrac{3}{5} \cdot \dfrac{7}{8} = \dfrac{3 \cdot 7}{5 \cdot 8}$

4. $\dfrac{a}{b} \div \dfrac{c}{d} = \dfrac{a}{b} \cdot \dfrac{d}{c}$

$\dfrac{2}{3} \div \dfrac{5}{7} = \dfrac{2}{3} \cdot \dfrac{7}{5}$

5. $\dfrac{a}{b} + \dfrac{c}{b} = \dfrac{a + c}{b}$

$\dfrac{3}{6} + \dfrac{5}{6} = \dfrac{3 + 5}{6}$

6. $\dfrac{a}{b} - \dfrac{c}{b} = \dfrac{a - c}{b}$

$\dfrac{7}{8} - \dfrac{3}{8} = \dfrac{7 - 3}{8}$

7. $\dfrac{a}{b} + \dfrac{c}{d} = \dfrac{ad + bc}{bd}$

$\dfrac{2}{3} + \dfrac{3}{5} = \dfrac{2 \cdot 5 + 3 \cdot 3}{3 \cdot 5}$

Answers to Matched Problems **1.** (A) Associative (+)   (B) Commutative (+)   (C) Distributive
(D) Identity (+)   (E) Inverse (·)

Exercise A-2

All variables represent real numbers.

A

In Problems 1–6, replace each question mark with an appropriate expression that will illustrate the use of the indicated real number property.

1. Commutative property (·): $uv = ?$

2. Commutative property (+): $x + 7 = ?$

3. Associative property (+): $3 + (7 + y) = ?$

4. Associative property (·): $x(yz) = ?$

5. Identity property (•): $1(u + v) = ?$

6. Identity property (+): $0 + 9m = ?$

In Problems 7–26, indicate true (T) or false (F).

7. $5(8m) = (5 \cdot 8)m$

8. $a + cb = a + bc$

9. $5x + 7x = (5 + 7)x$

10. $uv(w + x) = uvw + uvx$

11. $7 - 11 = 7 + (-11)$

12. $8 \div (-5) = 8\left(\frac{1}{-5}\right)$

13. $(x + 3) + 2x = 2x + (x + 3)$

14. $(4x + 3) + (x + 2) = 4x + [3 + (x + 2)]$

15. $\dfrac{2x}{-(x + 3)} = -\dfrac{2x}{x + 3}$

16. $-\dfrac{2x}{-(x - 3)} = \dfrac{2x}{x - 3}$

17. $(-3)\left(\frac{1}{-3}\right) = 1$

18. $(-0.5) + (0.5) = 0$

19. $-x^2y^2 = (-1)x^2y^2$

20. $[-(x + 2)](-x) = (x + 2)x$

21. $\dfrac{a}{b} + \dfrac{c}{d} = \dfrac{a + c}{b + d}$

22. $\dfrac{k}{k + b} = \dfrac{1}{1 + b}$

23. $(x + 8)(x + 6) = (x + 8)x + (x + 8)6$

24. $u(u - 2v) + v(u - 2v) = (u + v)(u - 2v)$

25. If $(x - 2)(2x + 3) = 0$, then either $x - 2 = 0$ or $2x + 3 = 0$.

26. If either $x - 2 = 0$ or $2x + 3 = 0$, then $(x - 2)(2x + 3) = 0$.

B

27. If $uv = 1$, does either u or v have to be 1? Explain.

28. If $uv = 0$, does either u or v have to be 0? Explain.

29. Indicate whether the following are true (T) or false (F):

(A) All integers are natural numbers.

(B) All rational numbers are real numbers.

(C) All natural numbers are rational numbers.

30. Indicate whether the following are true (T) or false (F):

(A) All natural numbers are integers.

(B) All real numbers are irrational.

(C) All rational numbers are real numbers.

31. Give an example of a real number that is not a rational number.

32. Give an example of a rational number that is not an integer.

33. Given the sets of numbers N (natural numbers), Z (integers), Q (rational numbers), and R (real numbers), indicate to which set(s) each of the following numbers belongs:

(A) 8   (B) $\sqrt{2}$ (C) -1.414 (D) $\frac{-5}{2}$

34. Given the sets of numbers N, Z, Q, and R (see Problem 33), indicate to which set(s) each of the following numbers belongs:

(A) -3 (B) 3.14   (C) π (D) $\frac{2}{3}$

35. Indicate true (T) or false (F), and for each false statement find real number replacements for a, b, and c that will provide a counterexample. For all real numbers a, b, and c:

(A) $(a + b) + c = a + (b + c)$

(B) $(a - b) - c = a - (b - c)$

(C) $a(bc) = (ab)c$

(D) $(a \div b) \div c = a \div (b \div c)$

36. Indicate true (T) or false (F), and for each false statement find real number replacements for a and b that will provide a counterexample. For all real numbers a and b:

(A) $a + b = b + a$

(B) $a - b = b - a$

(C) $ab = ba$

(D) $a \div b = b \div a$

C

37. If $c = 0.151\ 515\ldots$, then $100c = 15.151\ 5\ldots$ and

$$100c - c = 15.151\ 5\ldots - 0.151\ 515\ldots$$
$$99c = 15$$
$$c = \tfrac{15}{99} = \tfrac{5}{33}$$

Proceeding similarly, convert the repeating decimal $0.090\ 909\ldots$ into a fraction. (All repeating decimals are rational numbers, and all rational numbers have repeating decimal representations.)

38. Repeat Problem 37 for $0.181\ 818\ldots$.

Use a calculator to express each number in Problems 39 and 40 as a decimal to the capacity of your calculator. Observe the repeating decimal representation of the rational numbers and the nonrepeating decimal representation of the irrational numbers.

39. (A) $\frac{13}{6}$ (B) $\sqrt{21}$ (C) $\frac{7}{16}$ (D) $\frac{29}{111}$

40. (A) $\frac{8}{9}$ (B) $\frac{3}{11}$ (C) $\sqrt{5}$ (D) $\frac{11}{8}$

Operations on Polynomials

- ❑ NATURAL NUMBER EXPONENTS
- ❑ POLYNOMIALS
- ❑ COMBINING LIKE TERMS
- ❑ ADDITION AND SUBTRACTION
- ❑ MULTIPLICATION
- ❑ COMBINED OPERATIONS

This section covers basic operations on *polynomials,* a mathematical form that is encountered frequently. Our discussion starts with a brief review of natural number exponents. Integer and rational exponents and their properties will be discussed in detail in subsequent sections. (Natural numbers, integers, and rational numbers are important parts of the real number system; see Table 1 and Figure 1 in Appendix A-2.)

❑ NATURAL NUMBER EXPONENTS

We define a **natural number exponent** as follows:

Natural Number Exponent

For n a natural number and b any real number,

$$b^n = b \cdot b \cdot \cdots \cdot b \qquad n \text{ factors of } b$$
$$3^5 = 3 \cdot 3 \cdot 3 \cdot 3 \cdot 3 \qquad 5 \text{ factors of } 3$$

where n is called the **exponent** and b is called the **base.**

Along with this definition, we state the **first property of exponents:**

THEOREM 1 **First Property of Exponents**

For any natural numbers m and n, and any real number b:

$$b^m b^n = b^{m+n} \qquad (2t^4)(5t^3) = 2 \cdot 5 t^{4+3} = 10t^7$$

❑ POLYNOMIALS

Algebraic expressions are formed by using constants and variables and the algebraic operations of addition, subtraction, multiplication, division, raising to powers, and taking roots. Special types of algebraic expressions are called *polynomials.* A **polynomial in one variable** x is constructed by adding or subtracting constants and terms of the form ax^n, where a is a real number and n is a natural number. A **polynomial in two variables** x and y is constructed by adding and subtracting constants and terms of the form $ax^m y^n$, where a is a real number and m and n are natural numbers. Polynomials in three and more variables are defined in a similar manner.

Polynomials		**Not Polynomials**	
8	0	$\dfrac{1}{x}$	$\dfrac{x-y}{x^2+y^2}$
$3x^3 - 6x + 7$	$6x + 3$		
$2x^2 - 7xy - 8y^2$	$9y^3 + 4y^2 - y + 4$	$\sqrt{x^3 - 2x}$	$2x^{-2} - 3x^{-1}$
$2x - 3y + 2$	$u^5 - 3u^3v^2 + 2uv^4 - v^4$		

Polynomial forms are encountered frequently in mathematics. For the efficient study of polynomials it is useful to classify them according to their *degree.* If a term in a polynomial has only one variable as a factor, then the **degree of the term** is the power of the variable. If two or more variables are present in a term as factors, then the **degree of the term** is the sum of the powers of the variables. The **degree of a polynomial** is the degree of the nonzero term with the highest degree in the polynomial. Any nonzero constant is defined to be a **polynomial of degree 0.** The number 0 is also a polynomial but is not assigned a degree.

Example 1 ⇨ **Degree**

(A) The degree of the first term in $5x^3 + \sqrt{3}x - \frac{1}{2}$ is 3, the degree of the second term is 1, the degree of the third term is 0, and the degree of the whole polynomial is 3 (the same as the degree of the term with the highest degree).

(B) The degree of the first term in $8u^3v^2 - \sqrt{7}uv^2$ is 5, the degree of the second term is 3, and the degree of the whole polynomial is 5. ∎

Matched Problem 1 ⇨ (A) Given the polynomial $6x^5 + 7x^3 - 2$, what is the degree of the first term? The second term? The third term? The whole polynomial?

(B) Given the polynomial $2u^4v^2 - 5uv^3$, what is the degree of the first term? The second term? The whole polynomial? ∎

In addition to classifying polynomials by degree, we also call a single-term polynomial a **monomial,** a two-term polynomial a **binomial,** and a three-term polynomial a **trinomial.**

❑   COMBINING LIKE TERMS

The concept of *coefficient* plays a central role in the process of combining *like terms.* A constant in a term of a polynomial, including the sign that precedes it, is called the **numerical coefficient,** or simply, the **coefficient,** of the term. If a constant does not appear, or only a + sign appears, the coefficient is understood to be 1. If only a − sign appears, the coefficient is understood to be − 1. Thus, given the polynomial

$$5x^4 - x^3 - 3x^2 + x - 7 \boxed{= 5x^4 + (-1)x^3 + (-3)x^2 + 1x + (-7)}$$

the coefficient of the first term is 5, the coefficient of the second term is −1, the coefficient of the third term is −3, the coefficient of the fourth term is 1, and the coefficient of the fifth term is −7.

The following distributive properties are fundamental to the process of combining *like terms.*

> *Distributive Properties of Real Numbers*
>
> **1.** $a(b + c) = (b + c)a = ab + ac$
> **2.** $a(b - c) = (b - c)a = ab - ac$
> **3.** $a(b + c + \cdots + f) = ab + ac + \cdots + af$

Two terms in a polynomial are called **like terms** if they have exactly the same variable factors to the same powers. The numerical coefficients may or may not be the same. Since constant terms involve no variables, all constant terms are like terms. If a polynomial contains two or more like terms, these terms can be combined into a single term by making use of distributive properties. The following example illustrates the reasoning behind the process:

$$
\begin{aligned}
3x^2y - 5xy^2 + x^2y - 2x^2y &= 3x^2y + x^2y - 2x^2y - 5xy^2 \\
&= (3x^2y + 1x^2y - 2x^2y) - 5xy^2 \\
&= (3 + 1 - 2)x^2y - 5xy^2 \\
&= 2x^2y - 5xy^2
\end{aligned}
$$

Note the use of distributive properties.

It should be clear that free use is made of the real number properties discussed in Appendix A-2. The steps shown in the dashed box are usually done mentally, and the process is quickly mechanized as follows:

Like terms in a polynomial are combined by adding their numerical coefficients.

How can we simplify expressions such as $4(x - 2y) - 3(2x - 7y)$? We clear the expression of parentheses using distributive properties, and combine like terms:

$$
\begin{aligned}
4(x - 2y) - 3(2x - 7y) &= 4x - 8y - 6x + 21y \\
&= -2x + 13y
\end{aligned}
$$

Example 2 ⇔ **Removing Parentheses** Remove parentheses and simplify:

(A) $2(3x^2 - 2x + 5) + (x^2 + 3x - 7)$
$$
\begin{aligned}
&= 2(3x^2 - 2x + 5) + 1(x^2 + 3x - 7) \\
&= 6x^2 - 4x + 10 + x^2 + 3x - 7 \\
&= 7x^2 - x + 3
\end{aligned}
$$

(B) $(x^3 - 2x - 6) - (2x^3 - x^2 + 2x - 3)$
$$
\begin{aligned}
&= 1(x^3 - 2x - 6) + (-1)(2x^3 - x^2 + 2x - 3) \\
&= x^3 - 2x - 6 - 2x^3 + x^2 - 2x + 3 \\
&= -x^3 + x^2 - 4x - 3
\end{aligned}
$$
Be careful with the sign here.

(C) $[3x^2 - (2x + 1)] - (x^2 - 1) = [3x^2 - 2x - 1] - (x^2 - 1)$
$$
\begin{aligned}
&= 3x^2 - 2x - 1 - x^2 + 1 \\
&= 2x^2 - 2x
\end{aligned}
$$
Remove inner parentheses first.

Matched Problem 2 ⇔ Remove parentheses and simplify:

(A) $3(u^2 - 2v^2) + (u^2 + 5v^2)$
(B) $(m^3 - 3m^2 + m - 1) - (2m^3 - m + 3)$
(C) $(x^3 - 2) - [2x^3 - (3x + 4)]$

❑   ADDITION AND SUBTRACTION

Addition and subtraction of polynomials can be thought of in terms of removing parentheses and combining like terms, as illustrated in Example 2. Horizontal and vertical arrangements are illustrated in the next two examples. You should be able to work either way, letting the situation dictate your choice.

Example 3   ➩   **Adding Polynomials**   Add horizontally and vertically:

$$x^4 - 3x^3 + x^2, \quad -x^3 - 2x^2 + 3x, \quad \text{and} \quad 3x^2 - 4x - 5$$

SOLUTION   Add horizontally:

$$(x^4 - 3x^3 + x^2) + (-x^3 - 2x^2 + 3x) + (3x^2 - 4x - 5)$$
$$= x^4 - 3x^3 + x^2 - x^3 - 2x^2 + 3x + 3x^2 - 4x - 5$$
$$= x^4 - 4x^3 + 2x^2 - x - 5$$

Or vertically, by lining up like terms and adding their coefficients:

$$
\begin{array}{r}
x^4 - 3x^3 + x^2 \\
- x^3 - 2x^2 + 3x \\
3x^2 - 4x - 5 \\
\hline
x^4 - 4x^3 + 2x^2 - x - 5
\end{array}
$$

Matched Problem 3   ➩   Add horizontally and vertically:

$$3x^4 - 2x^3 - 4x^2, \quad x^3 - 2x^2 - 5x, \quad \text{and} \quad x^2 + 7x - 2$$

Example 4   ➩   **Subtracting Polynomials**   Subtract $4x^2 - 3x + 5$ from $x^2 - 8$, both horizontally and vertically.

SOLUTION

$$(x^2 - 8) - (4x^2 - 3x + 5) \qquad \text{or} \qquad
\begin{array}{r}
x^2 \qquad - 8 \\
-4x^2 + 3x - 5 \\
\hline
-3x^2 + 3x - 13
\end{array}
\quad \leftarrow \text{Change signs}$$
$$= x^2 - 8 - 4x^2 + 3x - 5 \qquad\qquad\qquad\qquad\qquad\qquad \text{and add.}$$
$$= -3x^2 + 3x - 13$$

Matched Problem 4   ➩   Subtract $2x^2 - 5x + 4$ from $5x^2 - 6$, both horizontally and vertically.

❑   MULTIPLICATION

Multiplication of algebraic expressions involves the extensive use of distributive properties for real numbers, as well as other real number properties.

Example 5   ➩ $(2x - 3)(3x^2 - 2x + 3) \quad = 2x(3x^2 - 2x + 3) - 3(3x^2 - 2x + 3)$

SOLUTION

$$= 6x^3 - 4x^2 + 6x - 9x^2 + 6x - 9$$
$$= 6x^3 - 13x^2 + 12x - 9$$

Or, using a vertical arrangement,

$$
\begin{array}{r}
3x^2 - 2x + 3 \\
2x - 3 \\
\hline
6x^3 - 4x^2 + 6x \\
- 9x^2 + 6x - 9 \\
\hline
6x^3 - 13x^3 + 12x - 9
\end{array}
$$

Matched Problem 5 ⇨   Multiply: $(2x - 3)(2x^2 + 3x - 2)$ ■

Thus, to multiply two polynomials, multiply each term of one by each term of the other, and combine like terms.

Products of binomial factors occur frequently, so it is useful to develop procedures that will enable us to write down their products by inspection. To find the product $(2x - 1)(3x + 2)$, we proceed as follows:

$$(2x - 1)(3x + 2) \; = 6x^2 + 4x - 3x - 2$$

The inner and outer products are like terms, so combine into a single term.

$$= 6x^2 + x - 2$$

To speed the process, we do the step in the dashed box mentally.

Products of certain binomial factors occur so frequently that it is useful to learn formulas for their products. The following formulas are easily verified by multiplying the factors on the left:

> ### Special Products
>
> **1.** $(a - b)(a + b) = a^2 - b^2$
> **2.** $(a + b)^2 = a^2 + 2ab + b^2$
> **3.** $(a - b)^2 = a^2 - 2ab + b^2$

Explore–Discuss 1

(A) Explain the relationship between special product formula 1 and the areas of the rectangles in the figure.

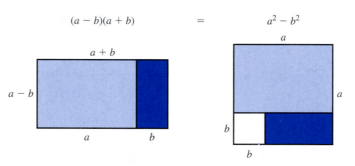

$(a - b)(a + b)$ $\qquad = \qquad$ $a^2 - b^2$

(B) Construct similar figures to provide geometric interpretations for special product formulas 2 and 3.

Example 6 ⇨   **Special Products**   Multiply mentally where possible:

(A) $(2x - 3y)(5x + 2y)$ (B) $(3a - 2b)(3a + 2b)$
(C) $(5x - 3)^2$ (D) $(m + 2n)^3$

SOLUTION   (A) $(2x - 3y)(5x + 2y) \; = 10x^2 + 4xy - 15xy - 6y^2$

$$= 10x^2 - 11xy - 6y^2$$

(B) $(3a - 2b)(3a + 2b) \; = (3a)^2 - (2b)^2$

$$= 9a^2 - 4b^2$$

(C) $(5x - 3)^2 \boxed{= (5x)^2 - 2(5x)(3) + 3^2}$

$$= 25x^2 - 30x + 9$$

(D) $(m + 2n)^3 = (m + 2n)^2(m + 2n)$

$$= (m^2 + 4mn + 4n^2)(m + 2n)$$
$$= m^2(m + 2n) + 4mn(m + 2n) + 4n^2(m + 2n)$$
$$= m^3 + 2m^2n + 4m^2n + 8mn^2 + 4mn^2 + 8n^3$$
$$= m^3 + 6m^2n + 12mn^2 + 8n^3$$

Matched Problem 6 Multiply mentally where possible:

(A) $(4u - 3v)(2u + v)$ (B) $(2xy + 3)(2xy - 3)$
(C) $(m + 4n)(m - 4n)$ (D) $(2u - 3v)^2$
(E) $(2x - y)^3$

❑ COMBINED OPERATIONS

We complete this section by considering several examples that use all the operations just discussed. Note that in simplifying, we usually remove grouping symbols starting from the inside. That is, we remove parentheses ( ) first, then brackets [ ], and finally braces { }, if present. Also:

Multiplication and division precede addition and subtraction, and taking powers precedes multiplication and division.

Example 7 **Combined Operations** Perform the indicated operations and simplify:

(A) $3x - \{5 - 3[x - x(3 - x)]\} = 3x - \{5 - 3[x - 3x + x^2]\}$
$$= 3x - \{5 - 3x + 9x - 3x^2\}$$
$$= 3x - 5 + 3x - 9x + 3x^2$$
$$= 3x^2 - 3x - 5$$

(B) $(x - 2y)(2x + 3y) - (2x + y)^2 = 2x^2 - xy - 6y^2 - (4x^2 + 4xy + y^2)$
$$= 2x^2 - xy - 6y^2 - 4x^2 - 4xy - y^2$$
$$= -2x^2 - 5xy - 7y^2$$

Matched Problem 7 Perform the indicated operations and simplify:

(A) $2t - \{7 - 2[t - t(4 + t)]\}$
(B) $(u - 3v)^2 - (2u - v)(2u + v)$

Answers to Matched Problems
1. (A) $5, 3, 0, 5$ (B) $6, 4, 6$
2. (A) $4u^2 - v^2$ (B) $-m^3 - 3m^2 + 2m - 4$ (C) $-x^3 + 3x + 2$
3. $3x^4 - x^3 - 5x^2 + 2x - 2$ **4.** $3x^2 + 5x - 10$ **5.** $4x^3 - 13x + 6$
6. (A) $8u^2 - 2uv - 3v^2$ (B) $4x^2y^2 - 9$ (C) $m^2 - 16n^2$
    (D) $4u^2 - 12uv + 9v^2$ (E) $8x^3 - 12x^2y + 6xy^2 - y^3$
7. (A) $-2t^2 - 4t - 7$ (B) $-3u^2 - 6uv + 10v^2$

Exercise A-3

A *Problems 1–8 refer to the following polynomials:*

(A) $2x - 3$ (B) $2x^2 - x + 2$ (C) $x^3 + 2x^2 - x + 3$

1. What is the degree of (C)?

2. What is the degree of (A)?

3. Add (B) and (C).   4. Add (A) and (B).

5. Subtract (B) from (C).   6. Subtract (A) from (B).

7. Multiply (B) and (C).   8. Multiply (A) and (C).

In Problems 9–30, perform the indicated operations and simplify.

9. $2(u - 1) - (3u + 2) - 2(2u - 3)$

10. $2(x - 1) + 3(2x - 3) - (4x - 5)$

11. $4a - 2a[5 - 3(a + 2)]$

12. $2y - 3y[4 - 2(y - 1)]$

13. $(a + b)(a - b)$ 14. $(m - n)(m + n)$

15. $(3x - 5)(2x + 1)$ 16. $(4t - 3)(t - 2)$

17. $(2x - 3y)(x + 2y)$ 18. $(3x + 2y)(x - 3y)$

19. $(3y + 2)(3y - 2)$ 20. $(2m - 7)(2m + 7)$

21. $(3m + 7n)(2m - 5n)$ 22. $(6x - 4y)(5x + 3y)$

23. $(4m + 3n)(4m - 3n)$ 24. $(3x - 2y)(3x + 2y)$

25. $(3u + 4v)^2$ 26. $(4x - y)^2$

27. $(a - b)(a^2 + ab + b^2)$

28. $(a + b)(a^2 - ab + b^2)$

29. $(4x + 3y)^2$ 30. $(3x + 2)^2$

B *In Problems 31–44, perform the indicated operations and simplify.*

31. $m - \{m - [m - (m - 1)]\}$

32. $2x - 3\{x + 2[x - (x + 5)] + 1\}$

33. $(x^2 - 2xy + y^2)(x^2 + 2xy + y^2)$

34. $(2x^2 + x - 2)(x^2 - 3x + 5)$

35. $(3a - b)(3a + b) - (2a - 3b)^2$

36. $(2x - 1)^2 - (3x + 2)(3x - 2)$

37. $(m - 2)^2 - (m - 2)(m + 2)$

38. $(x - 3)(x + 3) - (x - 3)^2$

39. $(x - 2y)(2x + y) - (x + 2y)(2x - y)$

40. $(3m + n)(m - 3n) - (m + 3n)(3m - n)$

41. $(u + v)^3$ 42. $(x - y)^3$

43. $(x - 2y)^3$ 44. $(2m - n)^3$

45. Subtract the sum of the last two polynomials from the sum of the first two: $2x^2 - 4xy + y^2$, $3xy - y^2$, $x^2 - 2xy - y^2$, $-x^2 + 3xy - 2y^2$

46. Subtract the sum of the first two polynomials from the sum of the last two: $3m^2 - 2m + 5$, $4m^2 - m$, $3m^2 - 3m - 2$, $m^3 + m^2 + 2$

C *In Problems 47–50, perform the indicated operations and simplify.*

47. $(2x - 1)^3 - 2(2x - 1)^2 + 3(2x - 1) + 7$

48. $2(x - 2)^3 - (x - 2)^2 - 3(x - 2) - 4$

49. $2\{(x - 3)(x^2 - 2x + 1) - x[3 - x(x - 2)]\}$

50. $-3x\{x[x - x(2 - x)] - (x + 2)(x^2 - 3)\}$

51. If you are given two polynomials, one of degree m and the other of degree n, where m is greater than n, what is the degree of their product?

52. What is the degree of the sum of the two polynomials in Problem 51?

53. How does the answer to Problem 51 change if the two polynomials can have the same degree?

54. How does the answer to Problem 52 change if the two polynomials can have the same degree?

55. Show by example that, in general, $(a + b)^2 \neq a^2 + b^2$. Discuss possible conditions on a and b that would make this a valid equation.

56. Show by example that, in general, $(a - b)^2 \neq a^2 - b^2$. Discuss possible conditions on a and b that would make this a valid equation.

Applications

Business & Economics

57. *Investment.* You have $10,000 to invest, part at 9% and the rest at 12%. If x is the amount invested at 9%, write an algebraic expression that represents the total annual income from both investments. Simplify the expression.

58. *Investment.* A person has $100,000 to invest. If $x are invested in a money market account yielding 7% and twice that amount in certificates of deposit yielding 9%, and if the rest is invested in high-grade bonds yielding 11%, write an algebraic expression that represents the total annual income from all three investments. Simplify the expression.

59. *Gross receipts.* Four thousand tickets are to be sold for a musical show. If x tickets are to be sold for $10 each and three times that number for $30 each, and if

the rest are sold for $50 each, write an algebraic expression that represents the gross receipts from ticket sales, assuming all tickets are sold. Simplify the expression.

60. *Gross receipts.* Six thousand tickets are to be sold for a concert, some for $9 each and the rest for $15 each. If x is the number of $9 tickets sold, write an algebraic expression that represents the gross receipts from ticket sales, assuming all tickets are sold. Simplify the expression.

Life Sciences

61. *Nutrition.* Food mix A contains 2% fat, and food mix B contains 6% fat. A 10 kilogram diet mix of foods A

and B is formed. If x kilograms of food A are used, write an algebraic expression that represents the total number of kilograms of fat in the final food mix. Simplify the expression.

62. *Nutrition.* Each ounce of food M contains 8 units of calcium, and each ounce of food N contains 5 units of calcium. A 160-ounce diet mix is formed using foods M and N. If x is the number of ounces of food M used, write an algebraic expression that represents the total number of units of calcium in the diet mix. Simplify the expression.

Section A-4

Factoring Polynomials

- ❑ COMMON FACTORS
- ❑ FACTORING BY GROUPING
- ❑ FACTORING SECOND-DEGREE POLYNOMIALS
- ❑ SPECIAL FACTORING FORMULAS
- ❑ COMBINED FACTORING TECHNIQUES

A polynomial is written in factored form if it is written as the product of two or more polynomials. The following polynomials are written in factored form:

$$4x^2y - 6xy^2 = 2xy(2x - 3y) \qquad 2x^3 - 8x = 2x(x - 2)(x + 2)$$
$$x^2 - x - 6 = (x - 3)(x + 2) \qquad 5m^2 + 20 = 5(m^2 + 4)$$

Unless stated to the contrary, we will limit our discussion of factoring of polynomials to polynomials with integer coefficients.

A polynomial with integer coefficients is said to be **factored completely** if each factor cannot be expressed as the product of two or more polynomials with integer coefficients, other than itself or 1. All the polynomials above, as we will see by the conclusion of this section, are factored completely.

Writing polynomials in completely factored form is often a difficult task. But accomplishing it can lead to the simplification of certain algebraic expressions and to the solution of certain types of equations and inequalities. The distributive properties for real numbers are central to the factoring process.

❑ **COMMON FACTORS**

Generally, a first step in any factoring procedure is to factor out all factors common to all terms.

Example 1 ⇔ **Common Factors**   Factor out all factors common to all terms:

(A) $3x^3y - 6x^2y^2 - 3xy^3$ (B) $3y(2y + 5) + 2(2y + 5)$

SOLUTION (A) $3x^3y - 6x^2y^2 - 3xy^3 = (3xy)x^2 - (3xy)2xy - (3xy)y^2$

$$= 3xy(x^2 - 2xy - y^2)$$

(B) $3y(2y + 5) + 2(2y + 5) = 3y(2y + 5) + 2(2y + 5)$

$$= (3y + 2)(2y + 5)$$

Matched Problem 1 Factor out all factors common to all terms:

(A) $2x^3y - 8x^2y^2 - 6xy^3$ (B) $2x(3x - 2) - 7(3x - 2)$

❑ FACTORING BY GROUPING

Occasionally, polynomials can be factored by grouping terms in such a way that we obtain results that look like Example 1B. We can then complete the factoring following the steps used in that example. This process will prove useful in the next subsection, where an efficient method is developed for factoring a second-degree polynomial as the product of two first-degree polynomials, if such factors exist.

Example 2 **Factoring by Grouping** Factor by grouping:

(A) $3x^2 - 3x - x + 1$ (B) $4x^2 - 2xy - 6xy + 3y^2$
(C) $y^2 + xz + xy + yz$

SOLUTION (A) $3x^2 - 3x - x + 1$

$$= (3x^2 - 3x) - (x - 1) \qquad \text{Group the first two and the last two terms.}$$
$$= 3x(x - 1) - (x - 1) \qquad \text{Factor out any common factors from each}$$
$$= (x - 1)(3x - 1) \qquad \text{group. The common factor } (x - 1) \text{ can be}$$
$$\qquad\qquad\qquad\qquad\qquad \text{taken out, and the factoring is complete.}$$

(B) $4x^2 - 2xy - 6xy + 3y^2 = (4x^2 - 2xy) - (6xy - 3y^2)$

$$= 2x(2x - y) - 3y(2x - y)$$
$$= (2x - y)(2x - 3y)$$

(C) If, as in parts (A) and (B), we group the first two terms and the last two terms of $y^2 + xz + xy + yz$, no common factor can be taken out of each group to complete the factoring. However, if the two middle terms are reversed, we can proceed as before:

$$y^2 + xz + xy + yz = y^2 + xy + xz + yz$$
$$= (y^2 + xy) + (xz + yz)$$
$$= y(y + x) + z(x + y)$$
$$= y(x + y) + z(x + y)$$
$$= (x + y)(y + z)$$

Matched Problem 2 Factor by grouping:

(A) $6x^2 + 2x + 9x + 3$ (B) $2u^2 + 6uv - 3uv - 9v^2$
(C) $ac + bd + bc + ad$

❑ FACTORING SECOND-DEGREE POLYNOMIALS

We now turn our attention to factoring second-degree polynomials of the form

$$2x^2 - 5x - 3 \qquad \text{and} \qquad 2x^2 + 3xy - 2y^2$$

into the product of two first-degree polynomials with integer coefficients. Since many second-degree polynomials with integer coefficients cannot be factored in this way, it would be useful to know ahead of time that the factors we are seeking actually exist. The factoring approach we use, involving the *ac test*, determines at the beginning whether first-degree factors with integer coefficients do exist. Then, if they exist, the test provides a simple method for finding them.

---

ac Test for Factorability

If in polynomials of the form

$$ax^2 + bx + c \qquad \text{or} \qquad ax^2 + bxy + cy^2 \qquad (1)$$

the product ac has two integer factors p and q whose sum is the coefficient b of the middle term; that is, if integers p and q exist so that

$$pq = ac \qquad \text{and} \qquad p + q = b \qquad (2)$$

then the polynomials have first-degree factors with integer coefficients. If no integers p and q exist that satisfy equations (2), then the polynomials in equations (1) will not have first-degree factors with integer coefficients.

If integers p and q exist that satisfy equations (2) in the *ac* test, the factoring always can be completed as follows: Using $b = p + q$, split the middle terms in equations (1) to obtain

$$ax^2 + bx + c = ax^2 + px + qx + c$$
$$ax^2 + bxy + cy^2 = ax^2 + pxy + qxy + cy^2$$

Complete the factoring by grouping the first two terms and the last two terms as in Example 2. This process always works, and it does not matter if the two middle terms on the right are interchanged.

Several examples should make the process clear. After a little practice, you will perform many of the steps mentally and will find the process fast and efficient.

Example 3 ⮑   **Factoring Second-Degree Polynomials**   Factor, if possible, using integer coefficients:

(A) $4x^2 - 4x - 3$ (B) $2x^2 - 3x - 4$ (C) $6x^2 - 25xy + 4y^2$

SOLUTION   (A) $4x^2 - 4x - 3$

Step 1.   Use the *ac* test to test for factorability. Comparing $4x^2 - 4x - 3$ with $ax^2 + bx + c$, we see that $a = 4$, $b = -4$, and $c = -3$. Multiply a and c to obtain

$$ac = (4)(-3) = -12$$

pq
$(1)(-12)$
$(-1)(12)$
$(2)(-6)$
$(-2)(6)$
$(3)(-4)$
$(-3)(4)$

All factor pairs of $-12 = ac$

List all pairs of integers whose product is -12, as shown in the margin. These are called **factor pairs** of -12. Then try to find a factor pair that sums to $b = -4$, the coefficient of the middle term in $4x^2 - 4x - 3$. (In practice, this part of step 1 is often done mentally and can be done rather quickly.) Notice that the factor pair 2 and -6 sums to -4. Thus, by the *ac* test, $4x^2 - 4x - 3$ has first-degree factors with integer coefficients.

Step 2. Split the middle term, using $b = p + q$, and complete the factoring by grouping. Using $-4 = 2 + (-6)$, we split the middle term in $4x^2 - 4x - 3$ and complete the factoring by grouping:

$$
\begin{aligned}
4x^2 - 4x - 3 &= 4x^2 + 2x - 6x - 3 \\
&= (4x^2 + 2x) - (6x + 3) \\
&= 2x\mathbf{(2x + 1)} - 3\mathbf{(2x + 1)} \\
&= \mathbf{(2x + 1)}(2x - 3)
\end{aligned}
$$

The result can be checked by multiplying the two factors to obtain the original polynomial.

(B) $2x^2 - 3x - 4$

Step 1. Use the ac test to test for factorability:

$$ac = (2)(-4) = -8$$

Does -8 have a factor pair whose sum is -3? None of the factor pairs listed in the margin sums to $-3 = b$, the coefficient of the middle term in $2x^2 - 3x - 4$. According to the ac test, we can conclude that $2x^2 - 3x - 4$ does not have first-degree factors with integer coefficients, and we say that the polynomial is **not factorable.**

pq
$(-1)(8)$
$(1)(-8)$
$(-2)(4)$
$(2)(-4)$

All factor pairs of $-8 = ac$

(C) $6x^2 - 25xy + 4y^2$

Step 1. Use the ac test to test for factorability:

$$ac = (6)(4) = 24$$

Mentally checking through the factor pairs of 24, keeping in mind that their sum must be $-25 = b$, we see that if $p = -1$ and $q = -24$, then

$$pq = (-1)(-24) = 24 = ac$$

and

$$p + q = (-1) + (-24) = -25 = b$$

Thus, the polynomial is factorable.

Step 2. Split the middle term, using $b = p + q$, and complete the factoring by grouping. Using $-25 = (-1) + (-24)$, we split the middle term in $6x^2 - 25xy + 4y^2$ and complete the factoring by grouping:

$$
\begin{aligned}
6x^2 - 25xy + 4y^2 &= 6x^2 - xy - 24xy + 4y^2 \\
&= (6x^2 - xy) - (24xy - 4y^2) \\
&= x(6x - y) - 4y(6x - y) \\
&= (6x - y)(x - 4y)
\end{aligned}
$$

The check is left to the reader.

Matched Problem 3 ⇔ Factor, if possible, using integer coefficients:

(A) $2x^2 + 11x - 6$ (B) $4x^2 + 11x - 6$ (C) $6x^2 + 5xy - 4y^2$

❑ SPECIAL FACTORING FORMULAS

The factoring formulas listed in the following box will enable us to factor certain polynomial forms that occur frequently. These formulas can be established by multiplying the factors on the right.

> **Special Factoring Formulas**
>
> | | |
> |---|---|
> | **Perfect square:** | 1. $u^2 + 2uv + v^2 = (u + v)^2$ |
> | **Perfect square:** | 2. $u^2 - 2uv + v^2 = (u - v)^2$ |
> | **Difference of squares:** | 3. $u^2 - v^2 = (u - v)(u + v)$ |
> | **Difference of cubes:** | 4. $u^3 - v^3 = (u - v)(u^2 + uv + v^2)$ |
> | **Sum of cubes:** | 5. $u^3 + v^3 = (u + v)(u^2 - uv + v^2)$ |

CAUTION

Notice that $u^2 + v^2$ is not included in the list of special factoring formulas. In general,

$$u^2 + v^2 \neq (au + bv)(cu + dv)$$

for any choice of real number coefficients a, b, c, and d.

Example 4 ➰ **Factoring**   Factor completely:

(A) $4m^2 - 12mn + 9n^2$ (B) $x^2 - 16y^2$ (C) $z^3 - 1$
(D) $m^3 + n^3$ (E) $a^2 - 4(b + 2)^2$

SOLUTION   (A) $4m^2 - 12mn + 9n^2 = (2m - 3n)^2$

(B) $x^2 - 16y^2 \;\boxed{= x^2 - (4y)^2}\; = (x - 4y)(x + 4y)$

(C) $z^3 - 1 = (z - 1)(z^2 + z + 1)$
(D) $m^3 + n^3 = (m + n)(m^2 - mn + n^2)$
(E) $a^2 - 4(b + 2)^2 = [a - 2(b + 2)][a + 2(b + 2)]$ ∎

Matched Problem 4 ➰ Factor completely:

(A) $x^2 + 6xy + 9y^2$ (B) $9x^2 - 4y^2$ (C) $8m^3 - 1$
(D) $x^3 + y^3z^3$ (E) $9(m - 3)^2 - 4n^2$

Explore–Discuss 1

(A) Verify the following factoring formulas for $u^4 - v^4$:

$$u^4 - v^4 = (u - v)(u + v)(u^2 + v^2)$$
$$= (u - v)(u^3 + u^2v + uv^2 + v^3)$$

(B) Discuss the pattern in the following formulas:

$$u^2 - v^2 = (u - v)(u + v)$$
$$u^3 - v^3 = (u - v)(u^2 + uv + v^2)$$
$$u^4 - v^4 = (u - v)(u^3 + u^2v + uv^2 + v^3)$$

(C) Use the pattern you discovered in part (B) to write similar formulas for $u^5 - v^5$ and $u^6 - v^6$. Verify your formulas by multiplication.

❑ COMBINED FACTORING TECHNIQUES

We complete this section by considering several factoring problems that involve combinations of the preceding techniques. Generally speaking: **when factoring a polynomial, we first take out all factors common to all terms, if they are**

present. Then we continue, using techniques discussed previously, until the polynomial is in a completely factored form.

Example 5 ➯ **Combined Factoring Techniques**   Factor completely:

(A) $3x^3 - 48x$ (B) $3u^4 - 3u^3v - 9u^2v^2$ (C) $3m^4 - 24mn^3$
(D) $3x^4 - 5x^2 + 2$

SOLUTION   (A) $3x^3 - 48x = 3x(x^2 - 16) = 3x(x - 4)(x + 4)$
(B) $3u^4 - 3u^3v - 9u^2v^2 = 3u^2(u^2 - uv - 3v^2)$
(C) $3m^4 - 24mn^3 = 3m(m^3 - 8n^3) = 3m(m - 2n)(m^2 + 2mn + 4n^2)$
(D) $3x^4 - 5x^2 + 2 = (3x^2 - 2)(x^2 - 1) = (3x^2 - 2)(x - 1)(x + 1)$ ■

Matched Problem 5 ➯ Factor completely:

(A) $18x^3 - 8x$ (B) $4m^3n - 2m^2n^2 + 2mn^3$ (C) $2t^4 - 16t$
(D) $2y^4 - 5y^2 - 12$

Answers to Matched Problems   **1.** (A) $2xy(x^2 - 4xy - 3y^2)$ (B) $(2x - 7)(3x - 2)$
2. (A) $(3x + 1)(2x + 3)$ (B) $(u + 3v)(2u - 3v)$ (C) $(a + b)(c + d)$
3. (A) $(2x - 1)(x + 6)$ (B) Not factorable   (C) $(3x + 4y)(2x - y)$
4. (A) $(x + 3y)^2$ (B) $(3x - 2y)(3x + 2y)$
   (C) $(2m - 1)(4m^2 + 2m + 1)$ (D) $(x + yz)(x^2 - xyz + y^2z^2)$
   (E) $[3(m - 3) - 2n][3(m - 3) + 2n]$
5. (A) $2x(3x - 2)(3x + 2)$ (B) $2mn(2m^2 - mn + n^2)$
   (C) $2t(t - 2)(t^2 + 2t + 4)$ (D) $(2y^2 + 3)(y - 2)(y + 2)$

Exercise A-4

A *In Problems 1–8, factor out all factors common to all terms.*

1. $6m^4 - 9m^3 - 3m^2$

2. $6x^4 - 8x^3 - 2x^2$

3. $8u^3v - 6u^2v^2 + 4uv^3$

4. $10x^3y + 20x^2y^2 - 15xy^3$

5. $7m(2m - 3) + 5(2m - 3)$

6. $5x(x + 1) - 3(x + 1)$

7. $a(3c + d) - 4b(3c + d)$

8. $2w(y - 2z) - x(y - 2z)$

In Problems 9–18, factor by grouping.

9. $2x^2 - x + 4x - 2$ **10.** $x^2 - 3x + 2x - 6$

11. $3y^2 - 3y + 2y - 2$ **12.** $2x^2 - x + 6x - 3$

13. $2x^2 + 8x - x - 4$ **14.** $6x^2 + 9x - 2x - 3$

15. $wy - wz + xy - xz$ **16.** $ac + ad + bc + bd$

17. $am - bn - bm + an$ **18.** $ab + 6 + 2a + 3b$

B *In Problems 19–56, factor completely. If a polynomial cannot be factored, say so.*

19. $3y^2 - y - 2$ **20.** $2x^2 + 5x - 3$

21. $u^2 - 2uv - 15v^2$ **22.** $x^2 - 4xy - 12y^2$

23. $m^2 - 6m - 3$ **24.** $x^2 + x - 4$

25. $w^2x^2 - y^2$ **26.** $25m^2 - 16n^2$

27. $9m^2 - 6mn + n^2$ **28.** $x^2 + 10xy + 25y^2$

29. $y^2 + 16$ **30.** $u^2 + 81$

31. $4z^2 - 28z + 48$ **32.** $6x^2 + 48x + 72$

33. $2x^4 - 24x^3 + 40x^2$ **34.** $2y^3 - 22y^2 + 48y$

35. $4xy^2 - 12xy + 9x$ **36.** $16x^2y - 8xy + y$

37. $6m^2 - mn - 12n^2$ **38.** $6s^2 + 7st - 3t^2$

39. $4u^3v - uv^3$ **40.** $x^3y - 9xy^3$

41. $2x^3 - 2x^2 + 8x$ **42.** $3m^3 - 6m^2 + 15m$

43. $r^3 - t^3$ **44.** $m^3 + n^3$

45. $a^3 + 1$ **46.** $c^3 - 1$

C

47. $(x + 2)^2 - 9y^2$

48. $(a - b)^2 - 4(c - d)^2$

49. $5u^2 + 4uv - 2v^2$

50. $3x^2 - 2xy - 4y^2$

51. $6(x - y)^2 + 23(x - y) - 4$

52. $4(A + B)^2 - 5(A + B) - 6$

53. $y^4 - 3y^2 - 4$ **54.** $m^4 - n^4$

55. $27a^2 + a^5b^3$ **56.** $s^4t^4 - 8st$

In Problems 57–60, discuss the validity of each statement. If the statement is true, explain why. If not, give a counterexample.

57. If n is a positive integer greater than 1, then $u^n - v^n$ can be factored.

58. If m and n are positive integers and $m \neq n$, then $u^m - v^n$ is not factorable.

59. If n is a positive integer greater than 1, then $u^n + v^n$ can be factored.

60. If k is a positive integer, then $u^{2k+1} + v^{2k+1}$ can be factored.

Section A-5

Operations on Rational Expressions

❑   REDUCING TO LOWEST TERMS

❑   MULTIPLICATION AND DIVISION

❑   ADDITION AND SUBTRACTION

❑   COMPOUND FRACTIONS

We now turn our attention to fractional forms. A quotient of two algebraic expressions (division by 0 excluded) is called a **fractional expression.** If both the numerator and the denominator are polynomials, the fractional expression is called a **rational expression.** Some examples of rational expressions are

$$\frac{1}{x^3 + 2x} \qquad \frac{5}{x} \qquad \frac{x + 7}{3x^2 - 5x + 1} \qquad \frac{x^2 - 2x + 4}{1}$$

In this section we discuss basic operations on rational expressions, including multiplication, division, addition, and subtraction.

Since variables represent real numbers in the rational expressions we will consider, the properties of real number fractions summarized in Appendix A-2 will play a central role in much of the work that we will do.

Even though not always explicitly stated, we always assume that variables are restricted so that division by 0 is excluded.

For example, given the rational expression

$$\frac{2x + 5}{x(x + 2)(x - 3)}$$

the variable x is understood to be restricted from being $0, -2,$ or 3, since these values would cause the denominator to be 0.

❑   **REDUCING TO LOWEST TERMS**

Central to the process of reducing rational expressions to *lowest terms* is the *fundamental property of fractions,* which we restate here for convenient reference:

> **Fundamental Property of Fractions**
>
> If a, b, and k are real numbers with $b, k \neq 0$, then
>
> $$\frac{ka}{kb} = \frac{a}{b} \qquad \frac{5 \cdot 2}{5 \cdot 7} = \frac{2}{7} \qquad \frac{x(x+4)}{2(x+4)} = \frac{x}{2}, \quad x \neq -4$$

Using this property from left to right to eliminate all common factors from the numerator and the denominator of a given fraction is referred to as **reducing a fraction to lowest terms.** We are actually dividing the numerator and denominator by the same nonzero common factor.

Using the property from right to left—that is, multiplying the numerator and denominator by the same nonzero factor—is referred to as **raising a fraction to higher terms.** We will use the property in both directions in the material that follows.

Example 1 ➾ **Reducing to Lowest Terms**   Reduce each rational expression to lowest terms.

(A) $\dfrac{6x^2 + x - 1}{2x^2 - x - 1} = \dfrac{(2x+1)(3x-1)}{(2x+1)(x-1)}$ *Factor numerator and denominator completely.*

$= \dfrac{3x - 1}{x - 1}$ *Divide numerator and denominator by the common factor $(2x + 1)$.*

(B) $\dfrac{x^4 - 8x}{3x^3 - 2x^2 - 8x} = \dfrac{x(x-2)(x^2 + 2x + 4)}{x(x-2)(3x+4)}$

$= \dfrac{x^2 + 2x + 4}{3x + 4}$ ∎

Matched Problem 1 ➾ Reduce each rational expression to lowest terms.

(A) $\dfrac{x^2 - 6x + 9}{x^2 - 9}$ (B) $\dfrac{x^3 - 1}{x^2 - 1}$ ∎

❑   MULTIPLICATION AND DIVISION

Since we are restricting variable replacements to real numbers, multiplication and division of rational expressions follow the rules for multiplying and dividing real number fractions summarized in Appendix A-2.

> **Multiplication and Division**
>
> If a, b, c, and d are real numbers, then:
>
> **1.** $\dfrac{a}{b} \cdot \dfrac{c}{d} = \dfrac{ac}{bd}$, $\quad b, d \neq 0$ $\dfrac{3}{5} \cdot \dfrac{x}{x+5} = \dfrac{3x}{5(x+5)}$
>
> **2.** $\dfrac{a}{b} \div \dfrac{c}{d} = \dfrac{a}{b} \cdot \dfrac{d}{c}$, $\quad b, c, d \neq 0$ $\dfrac{3}{5} \div \dfrac{x}{x+5} = \dfrac{3}{5} \cdot \dfrac{x+5}{x}$

Explore–Discuss 1   Write a verbal description of the process of multiplying two rational expressions. Do the same for the quotient of two rational expressions.

Example 2   **Multiplication and Division**   Perform the indicated operations and reduce to lowest terms.

(A) $\dfrac{10x^3y}{3xy + 9y} \cdot \dfrac{x^2 - 9}{4x^2 - 12x}$

Factor numerators and denominators. Then divide any numerator and any denominator with a like common factor.

$$= \dfrac{\overset{5x^2}{\cancel{10x^3y}}}{\underset{3 \cdot 1}{\cancel{3y}(x + 3)}} \cdot \dfrac{\overset{1 \cdot 1}{\cancel{(x - 3)}\cancel{(x + 3)}}}{\underset{2 \cdot 1}{\cancel{4x}\cancel{(x - 3)}}}$$

$$= \dfrac{5x^2}{6}$$

(B) $\dfrac{4 - 2x}{4} \div (x - 2) = \dfrac{\overset{1}{\cancel{2}}(2 - x)}{\underset{2}{\cancel{4}}} \cdot \dfrac{1}{x - 2}$ $x - 2 = \dfrac{x - 2}{1}$

$$= \dfrac{2 - x}{2(x - 2)} = \dfrac{\overset{-1}{\cancel{-(x - 2)}}}{\underset{1}{2\cancel{(x - 2)}}}$$ $b - a = -(a - b)$, *a useful change in some problems*

$$= -\dfrac{1}{2}$$

Matched Problem 2   Perform the indicated operations and reduce to lowest terms.

(A) $\dfrac{12x^2y^3}{2xy^2 + 6xy} \cdot \dfrac{y^2 + 6y + 9}{3y^3 + 9y^2}$ (B) $(4 - x) \div \dfrac{x^2 - 16}{5}$

❑   ADDITION AND SUBTRACTION

Again, because we are restricting variable replacements to real numbers, addition and subtraction of rational expressions follow the rules for adding and subtracting real number fractions.

Addition and Subtraction

For a, b, and c real numbers:

1. $\dfrac{a}{b} + \dfrac{c}{b} = \dfrac{a + c}{b}$, $b \neq 0$ $\dfrac{x}{x + 5} + \dfrac{8}{x + 5} = \dfrac{x + 8}{x + 5}$

2. $\dfrac{a}{b} - \dfrac{c}{b} = \dfrac{a - c}{b}$, $b \neq 0$ $\dfrac{x}{3x^2y^2} - \dfrac{x + 7}{3x^2y^2} = \dfrac{x - (x + 7)}{3x^2y^2}$

Thus, we add rational expressions with the same denominators by adding or subtracting their numerators and placing the result over the common denominator. If the denominators are not the same, we raise the fractions to higher terms, using the fundamental property of fractions to obtain common denominators, and then proceed as described.

Even though any common denominator will do, our work will be simplified if the *least common denominator (LCD)* is used. Often, the LCD is obvious, but if it is not, the steps in the next box describe how to find it.

Least Common Denominator

The least common denominator (LCD) of two or more rational expressions is found as follows:

1. Factor each denominator completely, including integer factors.
2. Identify each different factor from all the denominators.
3. Form a product using each different factor to the highest power that occurs in any one denominator. This product is the LCD.

Example 3 ↩ **Addition and Subtraction** Combine into a single fraction and reduce to lowest terms.

(A) $\dfrac{3}{10} + \dfrac{5}{6} - \dfrac{11}{45}$ (B) $\dfrac{4}{9x} - \dfrac{5x}{6y^2} + 1$ (C) $\dfrac{1}{x-1} - \dfrac{1}{x} - \dfrac{2}{x^2-1}$

SOLUTION (A) To find the LCD, factor each denominator completely:

$$\left.\begin{array}{l} 10 = 2 \cdot 5 \\ 6 = 2 \cdot 3 \\ 45 = 3^2 \cdot 5 \end{array}\right\} \quad \text{LCD} = 2 \cdot 3^2 \cdot 5 = 90$$

Now use the fundamental property of fractions to make each denominator 90:

$$\frac{3}{10} + \frac{5}{6} - \frac{11}{45} = \frac{9 \cdot 3}{9 \cdot 10} + \frac{15 \cdot 5}{15 \cdot 6} - \frac{2 \cdot 11}{2 \cdot 45}$$

$$= \frac{27}{90} + \frac{75}{90} - \frac{22}{90}$$

$$= \frac{27 + 75 - 22}{90} = \frac{80}{90} = \frac{8}{9}$$

(B) $\left.\begin{array}{l} 9x = 3^2 x \\ 6y^2 = 2 \cdot 3y^2 \end{array}\right\} \quad \text{LCD} = 2 \cdot 3^2 xy^2 = 18xy^2$

$$\frac{4}{9x} - \frac{5x}{6y^2} + 1 = \frac{2y^2 \cdot 4}{2y^2 \cdot 9x} - \frac{3x \cdot 5x}{3x \cdot 6y^2} + \frac{18xy^2}{18xy^2}$$

$$= \frac{8y^2 - 15x^2 + 18xy^2}{18xy^2}$$

(C) $\dfrac{1}{x-1} - \dfrac{1}{x} - \dfrac{2}{x^2-1}$

$$= \frac{1}{x-1} - \frac{1}{x} - \frac{2}{(x-1)(x+1)} \qquad \text{LCD} = x(x-1)(x+1)$$

$$= \frac{x(x+1) - (x-1)(x+1) - 2x}{x(x-1)(x+1)}$$

$$= \frac{x^2 + x - x^2 + 1 - 2x}{x(x-1)(x+1)}$$

$$= \frac{1-x}{x(x-1)(x+1)}$$

$$= \frac{\overset{-1}{-\cancel{(x-1)}}}{\underset{1}{x\cancel{(x-1)}}(x+1)} = \frac{-1}{x(x+1)}$$

■

Matched Problem 3 ⇔ Combine into a single fraction and reduce to lowest terms.

(A) $\dfrac{5}{28} - \dfrac{1}{10} + \dfrac{6}{35}$ (B) $\dfrac{1}{4x^2} - \dfrac{2x+1}{3x^3} + \dfrac{3}{12x}$

(C) $\dfrac{2}{x^2 - 4x + 4} + \dfrac{1}{x} - \dfrac{1}{x-2}$

■

Explore–Discuss 2

What is the value of $\dfrac{\frac{16}{4}}{2}$?

What is the result of entering $16 \div 4 \div 2$ on a calculator?

What is the difference between $16 \div (4 \div 2)$ and $(16 \div 4) \div 2$?

How could you use fraction bars to distinguish between these two cases when writing $\dfrac{\frac{16}{4}}{2}$?

□   ## COMPOUND FRACTIONS

A fractional expression with fractions in its numerator, denominator, or both is called a **compound fraction.** It is often necessary to represent a compound fraction as a **simple fraction**—that is (in all cases we will consider), as the quotient of two polynomials. The process does not involve any new concepts. It is a matter of applying old concepts and processes in the correct sequence.

Example 4 ⇔ **Simplifying Compound Fractions**   Express as a simple fraction reduced to lowest terms:

(A) $\dfrac{\dfrac{1}{5+h} - \dfrac{1}{5}}{h}$ (B) $\dfrac{\dfrac{y}{x^2} - \dfrac{x}{y^2}}{\dfrac{y}{x} - \dfrac{x}{y}}$

SOLUTION   We will simplify the expressions in parts (A) and (B) using two different methods—each is suited to the particular type of problem.

(A) We simplify this expression by combining the numerator into a single fraction and using division of rational forms.

$$\frac{\dfrac{1}{5+h} - \dfrac{1}{5}}{h} = \left[\frac{1}{5+h} - \frac{1}{5} \right] \div \frac{h}{1}$$

$$= \frac{5 - 5 - h}{5(5+h)} \cdot \frac{1}{h}$$

$$= \frac{-h}{5(5+h)h} = \frac{-1}{5(5+h)}$$

(B) The method used here makes effective use of the fundamental property of fractions in the form

$$\frac{a}{b} = \frac{ka}{kb} \qquad b, k \neq 0$$

Multiply the numerator and denominator by the LCD of all fractions in the numerator and denominator—in this case, x^2y^2:

$$\frac{x^2y^2 \left(\dfrac{y}{x^2} - \dfrac{x}{y^2} \right)}{x^2y^2 \left(\dfrac{y}{x} - \dfrac{x}{y} \right)} = \frac{x^2y^2\dfrac{y}{x^2} - x^2y^2\dfrac{x}{y^2}}{x^2y^2\dfrac{y}{x} - x^2y^2\dfrac{x}{y}} = \frac{y^3 - x^3}{xy^3 - x^3y}$$

$$= \frac{\overset{1}{\cancel{(y - x)}}(y^2 + xy + x^2)}{xy\underset{1}{\cancel{(y - x)}}(y + x)}$$

$$= \frac{y^2 + xy + x^2}{xy(y + x)} \qquad \text{or} \qquad \frac{x^2 + xy + y^2}{xy(x + y)}$$

Matched Problem 4 ⟿   Express as a simple fraction reduced to lowest terms:

(A) $\dfrac{\dfrac{1}{2 + h} - \dfrac{1}{2}}{h}$ (B) $\dfrac{\dfrac{a}{b} - \dfrac{b}{a}}{\dfrac{a}{b} + 2 + \dfrac{b}{a}}$

Answers to Matched Problems   **1.** (A) $\dfrac{x - 3}{x + 3}$ (B) $\dfrac{x^2 + x + 1}{x + 1}$ **2.** (A) $2x$ (B) $\dfrac{-5}{x + 4}$

3. (A) $\dfrac{1}{4}$ (B) $\dfrac{3x^2 - 5x - 4}{12x^3}$ (C) $\dfrac{4}{x(x - 2)^2}$

4. (A) $\dfrac{-1}{2(2 + h)}$ (B) $\dfrac{a - b}{a + b}$

Exercise A-5

A *In Problems 1–18, perform the indicated operations and reduce answers to lowest terms.*

1. $\dfrac{d^5}{3a} \div \left(\dfrac{d^2}{6a^2} \cdot \dfrac{a}{4d^3} \right)$

2. $\left(\dfrac{d^5}{3a} \div \dfrac{d^2}{6a^2} \right) \cdot \dfrac{a}{4d^3}$

3. $\dfrac{x^2}{12} + \dfrac{x}{18} - \dfrac{1}{30}$

4. $\dfrac{2y}{18} - \dfrac{-1}{28} - \dfrac{y}{42}$

5. $\dfrac{4m - 3}{18m^3} + \dfrac{3}{4m} - \dfrac{2m - 1}{6m^2}$

6. $\dfrac{3x + 8}{4x^2} - \dfrac{2x - 1}{x^3} - \dfrac{5}{8x}$

7. $\dfrac{x^2 - 9}{x^2 - 3x} \div (x^2 - x - 12)$

8. $\dfrac{2x^2 + 7x + 3}{4x^2 - 1} \div (x + 3)$

9. $\dfrac{2}{x} - \dfrac{1}{x - 3}$

10. $\dfrac{3}{m} - \dfrac{2}{m + 4}$

11. $\dfrac{3}{x^2 - 1} - \dfrac{2}{x^2 - 2x + 1}$

12. $\dfrac{1}{a^2 - b^2} + \dfrac{1}{a^2 + 2ab + b^2}$

13. $\dfrac{x + 1}{x - 1} - 1$

14. $m - 3 - \dfrac{m - 1}{m - 2}$

15. $\dfrac{3}{a - 1} - \dfrac{2}{1 - a}$

16. $\dfrac{5}{x - 3} - \dfrac{2}{3 - x}$

17. $\dfrac{2x}{x^2 - 16} - \dfrac{x - 4}{x^2 + 4x}$

18. $\dfrac{m + 2}{m^2 - 2m} - \dfrac{m}{m^2 - 4}$

B *In Problems 19–30, perform the indicated operations and reduce answers to lowest terms. Represent any compound fractions as simple fractions reduced to lowest terms.*

19. $\dfrac{x^2}{x^2 + 2x + 1} + \dfrac{x - 1}{3x + 3} - \dfrac{1}{6}$

20. $\dfrac{y}{y^2 - y - 2} - \dfrac{1}{y^2 + 5y - 14} - \dfrac{2}{y^2 + 8y + 7}$

21. $\dfrac{2 - x}{2x + x^2} \cdot \dfrac{x^2 + 4x + 4}{x^2 - 4}$

22. $\dfrac{9 - m^2}{m^2 + 5m + 6} \cdot \dfrac{m + 2}{m - 3}$

23. $\dfrac{c + 2}{5c - 5} - \dfrac{c - 2}{3c - 3} + \dfrac{c}{1 - c}$

24. $\dfrac{x + 7}{ax - bx} + \dfrac{y + 9}{by - ay}$

25. $\dfrac{1 + \dfrac{3}{x}}{x - \dfrac{9}{x}}$

26. $\dfrac{1 - \dfrac{y^2}{x^2}}{1 - \dfrac{y}{x}}$

27. $\dfrac{\dfrac{1}{2(x + h)} - \dfrac{1}{2x}}{h}$

28. $\dfrac{\dfrac{1}{x + h} - \dfrac{1}{x}}{h}$

29. $\dfrac{\dfrac{x}{y} - 2 + \dfrac{y}{x}}{\dfrac{x}{y} - \dfrac{y}{x}}$

30. $\dfrac{1 + \dfrac{2}{x} - \dfrac{15}{x^2}}{1 + \dfrac{4}{x} - \dfrac{5}{x^2}}$

In Problems 31–38, imagine that the indicated "solutions" were given to you by a student whom you were tutoring in this class.

(A) *Is the solution correct? If the solution is incorrect, explain what is wrong and how it can be corrected.*

(B) *Show a correct solution for each incorrect solution.*

31. $\dfrac{x^2 + 4x + 3}{x + 3} = \dfrac{x^2 + 4x}{x} = x + 4$

32. $\dfrac{x^2 - 3x - 4}{x - 4} = \dfrac{x^2 - 3x}{x} = x - 3$

33. $\dfrac{(x + h)^2 - x^2}{h} = (x + 1)^2 - x^2 = 2x + 1$

34. $\dfrac{(x + h)^3 - x^3}{h} = (x + 1)^3 - x^3 = 3x^2 + 3x + 1$

35. $\dfrac{x^2 - 3x}{x^2 - 2x - 3} + x - 3 = \dfrac{x^2 - 3x + x - 3}{x^2 - 2x - 3} = 1$

36. $\dfrac{2}{x - 1} - \dfrac{x + 3}{x^2 - 1} = \dfrac{2x + 2 - x - 3}{x^2 - 1} = \dfrac{1}{x + 1}$

37. $\dfrac{2x^2}{x^2 - 4} - \dfrac{x}{x - 2} = \dfrac{2x^2 - x^2 - 2x}{x^2 - 4} = \dfrac{x}{x + 2}$

38. $x + \dfrac{x - 2}{x^2 - 3x + 2} = \dfrac{x + x - 2}{x^2 - 3x + 2} = \dfrac{2}{x - 2}$

C *Represent the compound fractions in Problems 39–42 as simple fractions reduced to lowest terms.*

39. $\dfrac{\dfrac{1}{3(x + h)^2} - \dfrac{1}{3x^2}}{h}$

40. $\dfrac{\dfrac{1}{(x + h)^2} - \dfrac{1}{x^2}}{h}$

41. $1 - \dfrac{1}{1 - \dfrac{1}{1 - \dfrac{1}{x}}}$

42. $2 - \dfrac{1}{1 - \dfrac{2}{a + 2}}$

Section A-6

Integer Exponents and Scientific Notation

❑ INTEGER EXPONENTS
❑ SCIENTIFIC NOTATION

We now review basic operations on integer exponents and scientific notation and its use.

❑   **INTEGER EXPONENTS**

Definitions for **integer exponents** are listed below.

Definition of a^n

For n an integer and a a real number:

1. For n a positive integer,

$$a^n = a \cdot a \cdot \cdots \cdot a \qquad n \text{ factors of } a \qquad 5^4 = 5 \cdot 5 \cdot 5 \cdot 5$$

2. For $n = 0$,

$$a^0 = 1 \qquad a \neq 0 \qquad 12^0 = 1$$

0^0 is not defined.

3. For n a negative integer,

$$a^n = \frac{1}{a^{-n}} \qquad a \neq 0 \qquad a^{-3} = \frac{1}{a^{-(-3)}} = \frac{1}{a^3}$$

[If n is negative, then $(-n)$ is positive.]
Note: It can be shown that for *all* integers n,

$$a^{-n} = \frac{1}{a^n} \qquad \text{and} \qquad a^n = \frac{1}{a^{-n}} \qquad a \neq 0 \qquad a^5 = \frac{1}{a^{-5}}, \quad a^{-5} = \frac{1}{a^5}$$

The following integer exponent properties are very useful in manipulating integer exponent forms.

THEOREM 1   **Exponent Properties**

For n and m integers and a and b real numbers:

1. $a^m a^n = a^{m+n}$ $\qquad\qquad\qquad\qquad$ $a^8 a^{-3} = a^{8+(-3)} = a^5$

2. $(a^n)^m = a^{mn}$ $\qquad\qquad\qquad\qquad$ $(a^{-2})^3 = a^{3(-2)} = a^{-6}$

3. $(ab)^n = a^m b^m$ $\qquad\qquad\qquad\qquad$ $(ab)^{-2} = a^{-2}b^{-2}$

4. $\left(\dfrac{a}{b}\right)^m = \dfrac{a^m}{b^m} \qquad b \neq 0$ $\qquad\qquad$ $\left(\dfrac{a}{b}\right)^5 = \dfrac{a^5}{b^5}$

5. $\dfrac{a^m}{a^n} = a^{m-n} = \dfrac{1}{a^{n-m}} \qquad a \neq 0$ $\qquad$ $\dfrac{a^{-3}}{a^7} = \dfrac{1}{a^{7-(-3)}} = \dfrac{1}{a^{10}}$

Explore–Discuss 1

Property 1 in Theorem 1 can be expressed verbally as follows:

To find the product of two exponential forms with the same base, add the exponents and use the same base.

Express the other properties in Theorem 1 verbally. Decide which you find easier to remember—a formula or a verbal description.

Exponent forms are frequently encountered in algebraic applications. You should sharpen your skills in using these forms by reviewing the above basic definitions and properties and the examples that follow.

Example 1   ⇔   **Simplifying Exponent Forms**   Simplify, and express the answers using positive exponents only.

(A) $(2x^3)(3x^5) = 2 \cdot 3x^{3+5} = 6x^8$ $\qquad$ (B) $x^5 x^{-9} = x^{-4} = \dfrac{1}{x^4}$

(C) $\dfrac{x^5}{x^7} = x^{5-7} = x^{-2} = \dfrac{1}{x^2}$ $\qquad\qquad$ (D) $\dfrac{x^{-3}}{y^{-4}} = \dfrac{y^4}{x^3}$

$\qquad$ or $\dfrac{x^5}{x^7} = \dfrac{1}{x^{7-5}} = \dfrac{1}{x^2}$

(E) $(u^{-3}v^2)^{-2} = (u^{-3})^{-2}(v^2)^{-2} = u^6v^{-4} = \dfrac{u^6}{v^4}$

(F) $\left(\dfrac{y^{-5}}{y^{-2}}\right)^{-2} = \dfrac{(y^{-5})^{-2}}{(y^{-2})^{-2}} = \dfrac{y^{10}}{y^4} = y^6$

(G) $\dfrac{4m^{-3}n^{-5}}{6m^{-4}n^3} = \dfrac{2m^{-3-(-4)}}{3n^{3-(-5)}} = \dfrac{2m}{3n^8}$ ■

Matched Problem 1 ✐ Simplify, and express the answers using positive exponents only.

(A) $(3y^4)(2y^3)$ (B) m^2m^{-6} (C) $(u^3v^{-2})^{-2}$

(D) $\left(\dfrac{y^{-6}}{y^{-2}}\right)^{-1}$ (E) $\dfrac{8x^{-2}y^{-4}}{6x^{-5}y^2}$ ■

Example 2 ✐ **Converting to a Simple Fraction**   Write as a simple fraction with positive exponents:

$$\dfrac{1-x}{x^{-1}-1}$$

Solution   First note that

$$\dfrac{1-x}{x^{-1}-1} \neq \dfrac{x(1-x)}{-1}\qquad \textcolor{blue}{\textit{A common error}}$$

The original expression is a complex fraction, and we proceed to simplify it as follows:

$$\dfrac{1-x}{x^{-1}-1} = \dfrac{1-x}{\dfrac{1}{x}-1}\qquad \textcolor{blue}{\textit{Multiply numerator and denominator by x to clear internal fractions.}}$$

$$= \dfrac{x(1-x)}{x\left(\dfrac{1}{x}-1\right)}$$

$$= \dfrac{x(1-x)}{1-x} = x$$ ■

Matched Problem 2 ✐ Write as a simple fraction with positive exponents: $\dfrac{1+x^{-1}}{1-x^{-2}}$ ■

❑ ## Scientific Notation

In the real world, one often encounters very large numbers. For example:

The public debt in the United States in 1998, to the nearest billion dollars, was

$5,526,000,000,000

The world population in the year 2025, to the nearest million, is projected to be

7,896,000,000

Very small numbers are also encountered:

The sound intensity of a normal conversation is

0.000 000 000 316 watt per square centimeter*

It is generally troublesome to write and work with numbers of this type in standard decimal form. The first and last example cannot even be entered into many calculators as they are written. But with exponents defined for all integers, we can now express any finite decimal form as the product of a number between 1 and 10 and an integer power of 10, that is, in the form

$$a \times 10^n \qquad 1 \le a < 10, \quad a \text{ in decimal form}, \quad n \text{ an integer}$$

A number expressed in this form is said to be in **scientific notation.** The following are some examples of numbers in standard decimal notation and in scientific notation:

DECIMAL AND SCIENTIFIC NOTATION

$7 = 7 \times 10^0$	$0.5 = 5 \times 10^{-1}$
$67 = 6.7 \times 10$	$0.45 = 4.5 \times 10^{-1}$
$580 = 5.8 \times 10^2$	$0.0032 = 3.2 \times 10^{-3}$
$43,000 = 4.3 \times 10^4$	$0.000\ 045 = 4.5 \times 10^{-5}$
$73,400,000 = 7.34 \times 10^7$	$0.000\ 000\ 391 = 3.91 \times 10^{-7}$

Note that the power of 10 used corresponds to the number of places we move the decimal to form a number between 1 and 10. The power is positive if the decimal is moved to the left and negative if it is moved to the right. Positive exponents are associated with numbers greater than or equal to 10; negative exponents are associated with positive numbers less than 1; and a zero exponent is associated with a number that is 1 or greater, but less than 10.

Example 3   **Scientific Notation**

(A) Write each number in scientific notation:

7,320,000   and   0.000 000 54

(B) Write each number in standard decimal form:

4.32×10^6 and 4.32×10^{-5}

SOLUTION   (A) $7,320,000 = 7.320\ 000. \times 10^6 = 7.32 \times 10^6$

6 places left
Positive exponent

$0.000\ 000\ 54 = 0.000\ 000\ 5.4 \times 10^{-7} = 5.4 \times 10^{-7}$

7 places right
Negative exponent

*We write 0.000 000 000 316 in place of 0.000000000316, because it is then easier to keep track of the number of decimal places. We follow this convention when there are more than five decimal places to the right of the decimal.

(B) $4.32 \times 10^6 = 4{,}320{,}000$ $4.32 \times 10^{-5} = \dfrac{4.32}{10^5} = 0.000\ 043\ 2$

6 places right   5 places left

Positive exponent 6   Negative exponent −5

Matched Problem 3   (A) Write each number in scientific notation: 47,100;   2,443,000,000;   1.45
(B) Write each number in standard decimal form: 3.07×10^8; 5.98×10^{-6}

Explore–Discuss 3   Scientific and graphing calculators can calculate in either standard decimal mode or scientific notation mode. If the result of a computation in decimal mode is either too large or too small to be displayed, most calculators will automatically display the answer in scientific notation. Read the manual for your calculator and experiment with some operations on very large and very small numbers in both decimal mode and scientific notation mode. For example, show that

$$\frac{216{,}700{,}000{,}000}{0.000\ 000\ 000\ 000\ 078\ 8} = 2.75 \times 10^{24}$$

Answers to Matched Problems   **1.** (A) $6y^7$ (B) $\dfrac{1}{m^4}$ (C) $\dfrac{v^4}{u^6}$ (D) y^4 (E) $\dfrac{4x^3}{3y^6}$ **2.** $\dfrac{x}{x-1}$
3. (A) 4.7×10^4; 2.443×10^9; 1.45×10^0 (B) 307,000,000; 0.000 005 98

Exercise A-6

A *In Problems 1–14, simplify and express answers using positive exponents only. Variables are restricted to avoid division by 0.*

1. $2x^{-9}$ **2.** $3y^{-5}$ **3.** $\dfrac{3}{2w^{-7}}$

4. $\dfrac{5}{4x^{-9}}$ **5.** $2x^{-8}x^5$ **6.** $3c^{-9}c^4$

7. $\dfrac{w^{-8}}{w^{-3}}$ **8.** $\dfrac{m^{-11}}{m^{-5}}$ **9.** $5v^8v^{-8}$

10. $7d^{-4}d^4$ **11.** $(a^{-3})^2$ **12.** $(b^4)^{-3}$

13. $(x^6y^{-3})^{-2}$ **14.** $(a^{-3}b^4)^{-3}$

Write each number in Problems 15–20 in scientific notation.

15. 82,300,000,000   **16.** 5,380,000

17. 0.783   **18.** 0.019

19. 0.000 034   **20.** 0.000 000 007 832

Write each number in Problems 21–28 in standard decimal notation.

21. 4×10^4 **22.** 9×10^6

23. 7×10^{-3} **24.** 2×10^{-5}

25. 6.171×10^7 **26.** 3.044×10^3

27. 8.08×10^{-4} **28.** 1.13×10^{-2}

B *In Problems 29–38, simplify and express answers using positive exponents only.*

29. $(22 + 31)^0$ **30.** $(2x^3y^4)^0$

31. $\dfrac{10^{-3} \cdot 10^4}{10^{-11} \cdot 10^{-2}}$ **32.** $\dfrac{10^{-17} \cdot 10^{-5}}{10^{-3} \cdot 10^{-14}}$

33. $(5x^2y^{-3})^{-2}$ **34.** $(2m^{-3}n^2)^{-3}$

35. $\dfrac{8 \times 10^{-3}}{2 \times 10^{-5}}$ **36.** $\dfrac{18 \times 10^{12}}{6 \times 10^{-4}}$

37. $\dfrac{8x^{-3}y^{-1}}{6x^2y^{-4}}$ **38.** $\dfrac{9m^{-4}n^3}{12m^{-1}n^{-1}}$

In Problems 39–42, write each expression in the form $ax^p + bx^q$ or $ax^p + bx^q + cx^r$, where a, b, and c are real numbers and p, q, and r are integers. For example,

$$\frac{2x^4 - 3x^2 + 1}{2x^3} = \frac{2x^4}{2x^3} - \frac{3x^2}{2x^3} + \frac{1}{2x^3} = x - \frac{3}{2}x^{-1} + \frac{1}{2}x^{-3}$$

39. $\dfrac{7x^5 - x^2}{4x^5}$ **40.** $\dfrac{5x^3 - 2}{3x^2}$

41. $\dfrac{3x^4 - 4x^2 - 1}{4x^3}$ **42.** $\dfrac{2x^3 - 3x^2 + x}{2x^2}$

Write each expression in Problems 43–46 with positive exponents only, and as a single fraction reduced to lowest terms.

43. $\dfrac{3x^2(x - 1)^2 - 2x^3(x - 1)}{(x - 1)^4}$

44. $\dfrac{5x^4(x + 3)^2 - 2x^5(x + 3)}{(x + 3)^4}$

45. $2x^{-2}(x - 1) - 2x^{-3}(x - 1)^2$

46. $2x(x + 3)^{-1} - x^2(x + 3)^{-2}$

In Problems 47–50, convert each number to scientific notation and simplify. Express the answer in both scientific notation and in standard decimal form.

47. $\dfrac{9{,}600{,}000{,}000}{(1{,}600{,}000)(0.000\ 000\ 25)}$

48. $\dfrac{(60{,}000)(0.000\ 003)}{(0.0004)(1{,}500{,}000)}$

49. $\dfrac{(1{,}250{,}000)(0.000\ 38)}{0.0152}$

50. $\dfrac{(0.000\ 000\ 82)(230{,}000)}{(625{,}000)(0.0082)}$

51. What is the result of entering 2^{3^2} on a calculator?

52. Refer to Problem 51. What is the difference between $2^{(3^2)}$ and $(2^3)^2$? Which agrees with the value of 2^{3^2} obtained with a calculator?

53. If $n = 0$, then property 1 in Theorem 1 implies that $a^m a^0 = a^{m+0} = a^m$. Explain how this helps motivate the definition of a^0.

54. If $m = -n$, then property 1 in Theorem 1 implies that $a^{-n} a^n = a^0 = 1$. Explain how this helps motivate the definition of a^{-n}.

C *Write the fractions in Problems 55–58 as simple fractions reduced to lowest terms.*

55. $\dfrac{u + v}{u^{-1} + v^{-1}}$ **56.** $\dfrac{x^{-1} - y^{-1}}{x - y}$

57. $\dfrac{b^{-2} - c^{-2}}{b^{-3} - c^{-3}}$ **58.** $\dfrac{xy^{-2} - yx^{-2}}{y^{-1} - x^{-1}}$

Applications

Business & Economics

Problems 59 and 60 refer to Table 1.

TABLE 1

ASSETS OF THE FIVE LARGEST U.S. COMMERCIAL BANKS, 1998

BANK	ASSETS ($)
Citigroup, New York	667,400,000,000
BankAmerica, San Francisco	617,679,000,000
Chase Manhattan, New York	365,875,000,000
Bank One, Chicago	261,496,000,000
J.P. Morgan, New York	261,067,000,000

59. *Financial assets*

(A) Write Citigroup's assets in scientific notation.

(B) After converting to scientific notation, determine the ratio of the assets of Citigroup to the assets of Chase Manhattan. Write the answer in standard decimal form to four decimal places.

(C) Repeat part (B) with the banks reversed.

60. *Financial assets*

(A) Write BankAmerica's assets in scientific notation.

(B) After converting to scientific notation, determine the ratio of the assets of BankAmerica to the assets of J. P. Morgan. Write the answer in standard decimal form to four decimal places.

(C) Repeat part (B) with the banks reversed.

Problems 61 and 62 refer to Table 2.

TABLE 2

U.S. PUBLIC DEBT, INTEREST ON DEBT, AND POPULATION

YEAR	PUBLIC DEBT ($)	INTEREST ON DEBT ($)	POPULATION
1990	3,233,300,000,000	264,800,000,000	248,765,170
1998	5,526,200,000,000	363,800,000,000	270,299,000

61. *Public debt.* Carry out the following computations using scientific notation, and write final answers in standard decimal form.

(A) What was the per capita debt in 1998 (to the nearest dollar)?

(B) What was the per capita interest paid on the debt in 1998 (to the nearest dollar)?

(C) What was the percentage interest paid on the debt in 1998 (to two decimal places)?

62. *Public debt.* Carry out the following computations using scientific notation, and write final answers in standard decimal form.

  (A) What was the per capita debt in 1990 (to the nearest dollar)?

  (B) What was the per capita interest paid on the debt in 1990 (to the nearest dollar)?

  (C) What was the percentage interest paid on the debt in 1990 (to two decimal places)?

Life Sciences

Air pollution. *Air quality standards establish maximum amounts of pollutants considered acceptable in the air. The amounts are frequently given in parts per million (ppm). A standard of 30 ppm also can be expressed as follows:*

$$30 \text{ ppm} = \frac{30}{1,000,000} = \frac{3 \times 10}{10^6}$$

$$= 3 \times 10^{-5} = 0.000\ 03 = 0.003\%$$

In Problems 63 and 64, express the given standard:

(A) In scientific notation

(B) In standard decimal notation

(C) As a percent

63. 9 ppm, the standard for carbon monoxide, when averaged over a period of 8 hours

64. 0.03 ppm, the standard for sulfur oxides, when averaged over a year

Social Sciences

65. *Crime.* In 1998, the United States had a violent crime rate of 566.4 per 100,000 people and a population of 270.3 million people. How many violent crimes occurred that year? Compute the answer using scientific notation and convert the answer to standard decimal form (to the nearest thousand).

66. *Population density*. The United States had a 1998 population of 270.3 million people and a land area of 3,539,000 square miles. What was the population density? Compute the answer using scientific notation and convert the answer to standard decimal form (to one decimal place).

Section A-7

Rational Exponents and Radicals

❑ *n*TH ROOTS OF REAL NUMBERS
❑ RATIONAL EXPONENTS AND RADICALS
❑ PROPERTIES OF RADICALS

Square roots may now be generalized to *nth roots,* and the meaning of exponent may be generalized to include all rational numbers.

❑   *n*TH ROOTS OF REAL NUMBERS

Consider a square of side r with area 36 square inches. We can write

$$r^2 = 36$$

and conclude that side r is a number whose square is 36. We say that r is a **square root** of b if $r^2 = b$. Similarly, we say that r is a **cube root** of b if $r^3 = b$. And, in general:

---

nth Root

For any natural number n,

  r is an **nth root** of b if $r^n = b$

Thus, 4 is a square root of 16, since $4^2 = 16$, and -2 is a cube root of -8, since $(-2)^3 = -8$. Since $(-4)^2 = 16$, we see that -4 is also a square root of 16. It can be shown that any positive number has two real square roots, two real 4th roots, and, in general, two real *n*th roots if n is even. Negative numbers have no real square roots, no real 4th roots, and, in general, no real *n*th roots if *n* is even. The reason is that no real number raised to an even power can be negative. For odd roots the situation is simpler. Every real number has exactly one real cube root, one real 5th root, and, in general, one real *n*th root if *n* is odd.

Additional roots can be considered in the *complex number system.* But in this book we restrict our interest to *real roots of real numbers,* and "root" will always be interpreted to mean "real root."

❑ RATIONAL EXPONENTS AND RADICALS

We now turn to the question of what symbols to use to represent *n*th roots. For *n* a natural number greater than 1, we use

$$b^{1/n} \quad \text{or} \quad \sqrt[n]{b}$$

to represent a **real *n*th root of *b*.** The exponent form is motivated by the fact that $(b^{1/n})^n = b$ if exponent laws are to continue to hold for rational exponents. The other form is called an ***n*th root radical.** In the expression below, the symbol $\sqrt{}$ is called a **radical,** *n* is the **index** of the radical, and *a* is the **radicand:**

$$\text{Index} \longrightarrow \sqrt[n]{a} \longleftarrow \text{Radical}$$
$$\underset{\uparrow}{} \text{Radicand}$$

When the index is 2, it is usually omitted. That is, when dealing with square roots, we simply use $\sqrt{b}$ rather than $\sqrt[2]{b}$. If there are two real *n*th roots, both $b^{1/n}$ and $\sqrt[n]{b}$ denote the positive root, called the **principal *n*th root.**

Example 1 ➭ **Finding *n*th Roots** Evaluate each of the following:

(A) $4^{1/2}$ and $\sqrt{4}$ (B) $-4^{1/2}$ and $-\sqrt{4}$ (C) $(-4)^{1/2}$ and $\sqrt{-4}$
(D) $8^{1/3}$ and $\sqrt[3]{8}$ (E) $(-8)^{1/3}$ and $\sqrt[3]{-8}$ (F) $-8^{1/3}$ and $-\sqrt[3]{8}$

SOLUTION (A) $4^{1/2} = \sqrt{4} = 2$ $(\sqrt{4} \neq \pm 2)$ (B) $-4^{1/2} = -\sqrt{4} = -2$
(C) $(-4)^{1/2}$ and $\sqrt{-4}$ are not real numbers
(D) $8^{1/3} = \sqrt[3]{8} = 2$ (E) $(-8)^{1/3} = \sqrt[3]{-8} = -2$
(F) $-8^{1/3} = -\sqrt[3]{8} = -2$ ◼

Matched Problem 1 ➭ Evaluate each of the following:

(A) $16^{1/2}$ (B) $-\sqrt{16}$ (C) $\sqrt[3]{-27}$ (D) $(-9)^{1/2}$ (E) $(\sqrt[4]{81})^3$ ◼

COMMON ERROR

The symbol $\sqrt{4}$ represents the single number 2, not ± 2. Do not confuse $\sqrt{4}$ with the solutions of the equation $x^2 = 4$, which are usually written in the form $x = \pm\sqrt{4} = \pm 2$.

We now define b^r for any rational number $r = m/n$.

Rational Exponents

If m and n are natural numbers without common prime factors, b is a real number, and b is nonnegative when n is even, then

$$b^{m/n} = \begin{cases} (b^{1/n})^m = (\sqrt[n]{b})^m \\ (b^m)^{1/n} = \sqrt[n]{b^m} \end{cases}$$

$8^{2/3} = (8^{1/3})^2 = (\sqrt[3]{8})^2 = 2^2 = 4$

$8^{2/3} = (8^2)^{1/3} = \sqrt[3]{8^2} = \sqrt[3]{64} = 4$

and

$$b^{-m/n} = \frac{1}{b^{m/n}} \quad b \neq 0 \qquad 8^{-2/3} = \frac{1}{8^{2/3}} = \frac{1}{4}$$

Note that the two definitions of $b^{m/n}$ are equivalent under the indicated restrictions on $m, n,$ and b.

All the properties listed for integer exponents in Theorem 1 in Section A-6 also hold for rational exponents, provided that b is nonnegative when n is even. Unless stated to the contrary, all variables in the rest of the discussion represent positive real numbers.

Example 2   **From Rational Exponent Form to Radical Form and Vice Versa**   Change rational exponent form to radical form.

(A) $x^{1/7} = \sqrt[7]{x}$

(B) $(3u^2v^3)^{3/5} = \sqrt[5]{(3u^2v^3)^3}$ or $(\sqrt[5]{3u^2v^3})^3$ *The first is usually preferred.*

(C) $y^{-2/3} = \dfrac{1}{y^{2/3}} = \dfrac{1}{\sqrt[3]{y^2}}$ or $\sqrt[3]{y^{-2}}$ or $\sqrt[3]{\dfrac{1}{y^2}}$

Change radical form to rational exponent form.

(D) $\sqrt[5]{6} = 6^{1/5}$ (E) $-\sqrt[3]{x^2} = -x^{2/3}$

(F) $\sqrt{x^2 + y^2} = (x^2 + y^2)^{1/2}$ *Note that $(x^2 + y^2)^{1/2} \neq x + y$. Why?*   ∎

Matched Problem 2   Convert to radical form.

(A) $u^{1/5}$ (B) $(6x^2y^5)^{2/9}$ (C) $(3xy)^{-3/5}$

Convert to rational exponent form.

(D) $\sqrt[4]{9u}$ (E) $-\sqrt[7]{(2x)^4}$ (F) $\sqrt[3]{x^3 + y^3}$ ∎

Example 3   **Working with Rational Exponents**   Simplify each and express answers using positive exponents only. If rational exponents appear in final answers, convert to radical form.

(A) $(3x^{1/3})(2x^{1/2}) = 6x^{1/3+1/2} = 6x^{5/6} = 6\sqrt[6]{x^5}$

(B) $(-8)^{5/3} = [(-8)^{1/3}]^5 = (-2)^5 = -32$

(C) $(2x^{1/3}y^{-2/3})^3 = 8xy^{-2} = \dfrac{8x}{y^2}$

(D) $\left(\dfrac{4x^{1/3}}{x^{1/2}}\right)^{1/2} = \dfrac{4^{1/2}x^{1/6}}{x^{1/4}} = \dfrac{2}{x^{1/4-1/6}} = \dfrac{2}{x^{1/12}} = \dfrac{2}{\sqrt[12]{x}}$ ∎

Matched Problem 3 ⇔ Simplify each and express answers using positive exponents only. If rational exponents appear in final answers, convert to radical form.

(A) $9^{3/2}$ (B) $(-27)^{4/3}$ (C) $(5y^{1/4})(2y^{1/3})$ (D) $(2x^{-3/4}y^{1/4})^4$

(E) $\left(\dfrac{8x^{1/2}}{x^{2/3}}\right)^{1/3}$

■

Explore–Discuss 1

In each of the following, evaluate both radical forms:

$$16^{3/2} = \sqrt{16^3} = (\sqrt{16})^3$$
$$27^{2/3} = \sqrt[3]{27^2} = (\sqrt[3]{27})^2$$

Which radical conversion form is easier to use if you are performing the calculations by hand?

Example 4 ⇔ **Working with Rational Exponents**   Multiply, and express answers using positive exponents only.

(A) $3y^{2/3}(2y^{1/3} - y^2)$ (B) $(2u^{1/2} + v^{1/2})(u^{1/2} - 3v^{1/2})$

SOLUTION   (A) $3y^{2/3}(2y^{1/3} - y^2) \;\vert\; = 6y^{2/3+1/3} - 3y^{2/3+2}$

$$= 6y - 3y^{8/3}$$

(B) $(2u^{1/2} + v^{1/2})(u^{1/2} - 3v^{1/2}) = 2u - 5u^{1/2}v^{1/2} - 3v$

■

Matched Problem 4 ⇔ Multiply, and express answers using positive exponents only.

(A) $2c^{1/4}(5c^3 - c^{3/4})$ (B) $(7x^{1/2} - y^{1/2})(2x^{1/2} + 3y^{1/2})$

■

Example 5 ⇔ **Working with Rational Exponents**   Write the following expression in the form $ax^p + bx^q$, where a and b are real numbers and p and q are rational numbers:

$$\frac{2\sqrt{x} - 3\sqrt[3]{x^2}}{2\sqrt[3]{x}}$$

SOLUTION $\dfrac{2\sqrt{x} - 3\sqrt[3]{x^2}}{2\sqrt[3]{x}} = \dfrac{2x^{1/2} - 3x^{2/3}}{2x^{1/3}}$ *Change to rational exponent form.*

$$= \frac{2x^{1/2}}{2x^{1/3}} - \frac{3x^{2/3}}{2x^{1/3}}$$ *Separate into two fractions.*

$$= x^{1/6} - 1.5x^{1/3}$$

■

Matched Problem 5 ⇔ Write the following expression in the form $ax^p + bx^q$, where a and b are real numbers and p and q are rational numbers:

$$\frac{5\sqrt[3]{x} - 4\sqrt{x}}{2\sqrt{x^3}}$$

■

❑ PROPERTIES OF RADICALS

Changing or simplifying radical expressions is aided by several properties of radicals that follow directly from the properties of exponents considered earlier.

> ### Properties of Radicals
>
> If c, n, and m are natural numbers greater than or equal to 2, and if x and y are positive real numbers, then:
>
> **1.** $\sqrt[n]{x^n} = x$ $\qquad \sqrt[3]{x^3} = x$ $\qquad$ **3.** $\sqrt[n]{\dfrac{x}{y}} = \dfrac{\sqrt[n]{x}}{\sqrt[n]{y}}$ $\qquad \sqrt[4]{\dfrac{x}{y}} = \dfrac{\sqrt[4]{x}}{\sqrt[4]{y}}$
>
> **2.** $\sqrt[n]{xy} = \sqrt[n]{x}\sqrt[n]{y}$ $\qquad \sqrt[5]{xy} = \sqrt[5]{x}\,\sqrt[5]{y}$

Example 6 ⇔ **Applying Properties of Radicals** Simplify using properties of radicals:

$\quad$ (A) $\sqrt[4]{(3x^4y^3)^4}$ $\qquad$ (B) $\sqrt[4]{8}\,\sqrt[4]{2}$ $\qquad$ (C) $\sqrt[3]{\dfrac{xy}{27}}$

Solution $\quad$ (A) $\sqrt[4]{(3x^4y^3)^4} = 3x^4y^3$ $\qquad\qquad\qquad$ Property 1

$\qquad$ (B) $\sqrt[4]{8}\,\sqrt[4]{2} = \sqrt[4]{16} = \sqrt[4]{2^4} = 2$ $\qquad$ Properties 2 and 1

$\qquad$ (C) $\sqrt[3]{\dfrac{xy}{27}} = \dfrac{\sqrt[3]{xy}}{\sqrt[3]{27}} = \dfrac{\sqrt[3]{xy}}{3}$ $\quad$ or $\quad \dfrac{1}{3}\sqrt[3]{xy}$ $\quad$ Properties 3 and 1

Matched Problem 6 ⇔ Simplify using properties of radicals:

$\quad$ (A) $\sqrt[7]{(x^3 + y^3)^7}$ $\qquad$ (B) $\sqrt[3]{8y^3}$ $\qquad$ (C) $\dfrac{\sqrt[3]{16x^4y}}{\sqrt[3]{2xy}}$

Explore–Discuss 2

Multiply:

$$(\sqrt{a} - \sqrt{b})(\sqrt{a} + \sqrt{b})$$

How can this product be used to simplify radical forms?

$\qquad$ A question arises regarding the best form in which a radical expression should be left. There are many answers, depending on what use we wish to make of the expression. In deriving certain formulas, it is sometimes useful to clear either a denominator or a numerator of radicals. The process is referred to as **rationalizing** the denominator or numerator. Examples 7 and 8 illustrate the rationalizing process.

Example 7 ⇔ **Rationalizing Denominators** Rationalize each denominator:

$\quad$ (A) $\dfrac{6x}{\sqrt{2x}}$ $\qquad$ (B) $\dfrac{6}{\sqrt{7} - \sqrt{5}}$ $\qquad$ (C) $\dfrac{x - 4}{\sqrt{x} + 2}$

Solution $\quad$ (A) $\dfrac{6x}{\sqrt{2x}} = \dfrac{6x}{\sqrt{2x}} \cdot \dfrac{\sqrt{2x}}{\sqrt{2x}} = \dfrac{6x\sqrt{2x}}{2x} = 3\sqrt{2x}$

$\qquad$ (B) $\dfrac{6}{\sqrt{7} - \sqrt{5}} = \dfrac{6}{\sqrt{7} - \sqrt{5}} \cdot \dfrac{\sqrt{7} + \sqrt{5}}{\sqrt{7} + \sqrt{5}}$

$\qquad\qquad = \dfrac{6(\sqrt{7} + \sqrt{5})}{2} = 3(\sqrt{7} + \sqrt{5})$

(C) $\dfrac{x-4}{\sqrt{x}+2} = \dfrac{x-4}{\sqrt{x}+2} \cdot \dfrac{\sqrt{x}-2}{\sqrt{x}-2}$

$= \dfrac{(x-4)(\sqrt{x}-2)}{x-4} = \sqrt{x}-2$

Matched Problem 7 ➭ Rationalize each denominator:

(A) $\dfrac{12ab^2}{\sqrt{3ab}}$ (B) $\dfrac{9}{\sqrt{6}+\sqrt{3}}$ (C) $\dfrac{x^2-y^2}{\sqrt{x}-\sqrt{y}}$

Example 8 ➭ **Rationalizing Numerators**   Rationalize each numerator:

(A) $\dfrac{\sqrt{2}}{2\sqrt{3}}$ (B) $\dfrac{3+\sqrt{m}}{9-m}$ (C) $\dfrac{\sqrt{2+h}-\sqrt{2}}{h}$

Solution   (A) $\dfrac{\sqrt{2}}{2\sqrt{3}} = \dfrac{\sqrt{2}}{2\sqrt{3}} \cdot \dfrac{\sqrt{2}}{\sqrt{2}} = \dfrac{2}{2\sqrt{6}} = \dfrac{1}{\sqrt{6}}$

(B) $\dfrac{3+\sqrt{m}}{9-m} = \dfrac{3+\sqrt{m}}{9-m} \cdot \dfrac{3-\sqrt{m}}{3-\sqrt{m}} = \dfrac{9-m}{(9-m)(3-\sqrt{m})} = \dfrac{1}{3-\sqrt{m}}$

(C) $\dfrac{\sqrt{2+h}-\sqrt{2}}{h} = \dfrac{\sqrt{2+h}-\sqrt{2}}{h} \cdot \dfrac{\sqrt{2+h}+\sqrt{2}}{\sqrt{2+h}+\sqrt{2}}$

$= \dfrac{h}{h(\sqrt{2+h}+\sqrt{2})} = \dfrac{1}{\sqrt{2+h}+\sqrt{2}}$

Matched Problem 8 ➭ Rationalize each numerator:

(A) $\dfrac{\sqrt{3}}{3\sqrt{2}}$ (B) $\dfrac{2-\sqrt{n}}{4-n}$ (C) $\dfrac{\sqrt{3+h}-\sqrt{3}}{h}$

Answers to Matched Problems   **1.** (A) 4   (B) -4 (C) -3 (D) Not a real number   (E) 27
2. (A) $\sqrt[5]{u}$ (B) $\sqrt[9]{(6x^2y^5)^2}$ or $(\sqrt[9]{6x^2y^5})^2$ (C) $1/\sqrt[5]{(3xy)^3}$
   (D) $(9u)^{1/4}$ (E) $-(2x)^{4/7}$ (F) $(x^3+y^3)^{1/3}$ (not $x+y$)
3. (A) 27   (B) 81   (C) $10y^{7/12} = 10\sqrt[12]{y^7}$ (D) $16y/x^3$
   (E) $2/x^{1/18} = 2/\sqrt[18]{x}$
4. (A) $10c^{13/4} - 2c$ (B) $14x + 19x^{1/2}y^{1/2} - 3y$ **5.** $2.5x^{-7/6} - 2x^{-1}$
6. (A) x^3+y^3 (B) $2y$ (C) $2x$
7. (A) $4b\sqrt{3ab}$ (B) $3(\sqrt{6}-\sqrt{3})$ (C) $(x+y)(\sqrt{x}+\sqrt{y})$
8. (A) $\dfrac{1}{\sqrt{6}}$ (B) $\dfrac{1}{2+\sqrt{n}}$ (C) $\dfrac{1}{\sqrt{3+h}+\sqrt{3}}$

Exercise A-7

A *Change each expression in Problems 1–6 to radical form.
Do not simplify.*

*Change each expression in Problems 7–12 to rational
exponent form. Do not simplify.*

1. $6x^{3/5}$ **2.** $7y^{2/5}$ **3.** $(4xy^3)^{2/5}$

7. $5\sqrt[4]{x^3}$ **8.** $7m\sqrt[5]{n^2}$ **9.** $\sqrt[9]{(2x^2y)^3}$

4. $(7x^2y)^{5/7}$ **5.** $(x^2+y^2)^{1/2}$ **6.** $x^{1/2}+y^{1/2}$

10. $\sqrt[9]{(3m^4n)^2}$ **11.** $\sqrt[3]{x}+\sqrt[3]{y}$ **12.** $\sqrt[3]{x^2+y^3}$

In Problems 13–24, find rational number representations for each, if they exist.

13. $25^{1/2}$ **14.** $64^{1/3}$ **15.** $16^{3/2}$

16. $16^{3/4}$ **17.** $-36^{1/2}$ **18.** $-32^{3/5}$

19. $(-36)^{1/2}$ **20.** $(-32)^{3/5}$ **21.** $\left(\frac{4}{25}\right)^{3/2}$

22. $\left(\frac{8}{27}\right)^{2/3}$ **23.** $9^{-3/2}$ **24.** $8^{-2/3}$

In Problems 25–34, simplify each expression and write answers using positive exponents only. All variables represent positive real numbers.

25. $x^{4/5}x^{-2/5}$ **26.** $y^{-3/7}y^{4/7}$ **27.** $\dfrac{m^{2/3}}{m^{-1/3}}$

28. $\dfrac{x^{1/4}}{x^{3/4}}$ **29.** $(8x^3y^{-6})^{1/3}$ **30.** $(4u^{-2}v^4)^{1/2}$

31. $\left(\dfrac{4x^{-2}}{y^4}\right)^{-1/2}$ **32.** $\left(\dfrac{w^4}{9x^{-2}}\right)^{-1/2}$ **33.** $\dfrac{8x^{-1/3}}{12x^{1/4}}$

34. $\dfrac{6a^{3/4}}{15a^{-1/3}}$

Simplify each expression in Problems 35–40 using properties of radicals. All variables represent positive real numbers.

35. $\sqrt[5]{(2x+3)^5}$ **36.** $\sqrt[3]{(7+2y)^3}$ **37.** $\sqrt{18x^3}\sqrt{2x^3}$

38. $\sqrt{2a^3}\sqrt{32a^5}$ **39.** $\dfrac{\sqrt{6x}\sqrt{10}}{\sqrt{15x}}$ **40.** $\dfrac{\sqrt{8}\sqrt{12y}}{\sqrt{6y}}$

B   In Problems 41–48, multiply, and express answers using positive exponents only.

41. $3x^{3/4}(4x^{1/4} - 2x^8)$

42. $2m^{1/3}(3m^{2/3} - m^6)$

43. $(3u^{1/2} - v^{1/2})(u^{1/2} - 4v^{1/2})$

44. $(a^{1/2} + 2b^{1/2})(a^{1/2} - 3b^{1/2})$

45. $(5m^{1/2} + n^{1/2})(5m^{1/2} - n^{1/2})$

46. $(2x^{1/2} - 3y^{1/2})(2x^{1/2} + 3y^{1/2})$

47. $(3x^{1/2} - y^{1/2})^2$

48. $(x^{1/2} + 2y^{1/2})^2$

Write each expression in Problems 49–54 in the form $ax^p + bx^q$, where a and b are real numbers and p and q are rational numbers.

49. $\dfrac{\sqrt[3]{x^2} + 2}{2\sqrt[3]{x}}$ **50.** $\dfrac{12\sqrt{x} - 3}{4\sqrt{x}}$

51. $\dfrac{2\sqrt[4]{x^3} + \sqrt[3]{x}}{3x}$ **52.** $\dfrac{3\sqrt[3]{x^2} + \sqrt{x}}{5x}$

53. $\dfrac{2\sqrt[3]{x} - \sqrt{x}}{4\sqrt{x}}$ **54.** $\dfrac{x^2 - 4\sqrt{x}}{2\sqrt[3]{x}}$

Rationalize the denominators in Problems 55–60.

55. $\dfrac{12mn^2}{\sqrt{3mn}}$ **56.** $\dfrac{14x^2}{\sqrt{7x}}$ **57.** $\dfrac{2}{\sqrt{x} - 2}$

58. $\dfrac{3}{\sqrt{x+4}}$ **59.** $\dfrac{7(x-y)^2}{\sqrt{x} - \sqrt{y}}$ **60.** $\dfrac{3a - 3b}{\sqrt{a} + \sqrt{b}}$

Rationalize the numerators in Problems 61–66.

61. $\dfrac{\sqrt{5xy}}{5x^2y^2}$ **62.** $\dfrac{\sqrt{3mn}}{3mn}$

63. $\dfrac{\sqrt{x+h} - \sqrt{x}}{h}$ **64.** $\dfrac{\sqrt{2(a+h)} - \sqrt{2a}}{h}$

65. $\dfrac{\sqrt{t} - \sqrt{x}}{t - x}$ **66.** $\dfrac{\sqrt{x} - \sqrt{y}}{\sqrt{x} + \sqrt{y}}$

Problems 67–70 illustrate common errors involving rational exponents. In each case, find numerical examples that show that the left side is not always equal to the right side.

67. $(x + y)^{1/2} \neq x^{1/2} + y^{1/2}$

68. $(x^3 + y^3)^{1/3} \neq x + y$

69. $(x + y)^{1/3} \neq \dfrac{1}{(x+y)^3}$

70. $(x + y)^{-1/2} \neq \dfrac{1}{(x+y)^2}$

C   In Problems 71–78, discuss the validity of each statement. If the statement is true, explain why. If not, give a counterexample.

71. If $r < 0$, then r has no cube roots.

72. If $r < 0$, then r has no square roots.

73. If $r > 0$, then r has two square roots.

74. If $r > 0$, then r has three cube roots.

75. The fourth roots of 100 are $\sqrt{10}$ and $-\sqrt{10}$.

76. The square roots of $2\sqrt{6} - 5$ are $\sqrt{3} - \sqrt{2}$ and $\sqrt{2} - \sqrt{3}$.

77. $\sqrt{355 - 60\sqrt{35}} = 5\sqrt{7} - 6\sqrt{5}$

78. $\sqrt[3]{7 - 5\sqrt{2}} = 1 - \sqrt{2}$

In Problems 79–84, simplify by writing each expression as a simple or single fraction reduced to lowest terms and without negative exponents.

79. $-\frac{1}{2}(x - 2)(x + 3)^{-3/2} + (x + 3)^{-1/2}$

80. $2(x - 2)^{-1/2} - \frac{1}{2}(2x + 3)(x - 2)^{-3/2}$

81. $\dfrac{(x - 1)^{1/2} - x(\frac{1}{2})(x - 1)^{-1/2}}{x - 1}$

82. $\dfrac{(2x - 1)^{1/2} - (x + 2)(\frac{1}{2})(2x - 1)^{-1/2}(2)}{2x - 1}$

83. $\dfrac{(x + 2)^{2/3} - x(\frac{2}{3})(x + 2)^{-1/3}}{(x + 2)^{4/3}}$

84. $\dfrac{2(3x - 1)^{1/3} - (2x + 1)(\frac{1}{3})(3x - 1)^{-2/3}(3)}{(3x - 1)^{2/3}}$

In Problems 85–90, evaluate using a calculator. (Refer to the instruction book for your calculator to see how exponential forms are evaluated.)

85. $22^{3/2}$ **86.** $15^{5/4}$ **87.** $827^{-3/8}$

88. $103^{-3/4}$ **89.** $37.09^{7/3}$ **90.** $2.876^{8/5}$

In Problems 91 and 92, evaluate each expression on a calculator and determine which pairs have the same value. Verify these results algebraically.

91. (A) $\sqrt{3} + \sqrt{5}$

(B) $\sqrt{2 + \sqrt{3}} + \sqrt{2 - \sqrt{3}}$

(C) $1 + \sqrt{3}$

(D) $\sqrt[3]{10 + 6\sqrt{3}}$

(E) $\sqrt{8 + \sqrt{60}}$

(F) $\sqrt{6}$

92. (A) $2\sqrt[3]{2} + \sqrt{5}$

(B) $\sqrt{8}$

(C) $\sqrt{3} + \sqrt{7}$

(D) $\sqrt{3 + \sqrt{8}} + \sqrt{3 - \sqrt{8}}$

(E) $\sqrt{10 + \sqrt{84}}$

(F) $1 + \sqrt{5}$

Section A-8	# Linear Equations and Inequalities in One Variable

- ❑ LINEAR EQUATIONS
- ❑ LINEAR INEQUALITIES
- ❑ APPLICATIONS

The equation

$$3 - 2(x + 3) = \frac{x}{3} - 5$$

and the inequality

$$\frac{x}{2} + 2(3x - 1) \geq 5$$

are both first degree in one variable. In general, a **first-degree, or linear, equation** in one variable is any equation that can be written in the form

Standard form: $ax + b = 0$ $a \neq 0$ (1)

If the equality symbol, $=$, in (1) is replaced by $<$, $>$, $\leq$, or $\geq$, the resulting expression is called a **first-degree, or linear, inequality.**

A **solution** of an equation (or inequality) involving a single variable is a number that when substituted for the variable makes the equation (or inequality) true. The set of all solutions is called the **solution set.** When we say that we **solve an equation** (or inequality), we mean that we find its solution set.

Knowing what is meant by the solution set is one thing; finding it is another. We start by recalling the idea of equivalent equations and equivalent inequalities. If we perform an operation on an equation (or inequality) that produces another equation (or inequality) with the same solution set, then the two equations (or inequalities) are said to be **equivalent.** The basic idea in solving equations and inequalities is to perform operations on these forms that produce simpler equivalent forms, and to continue the process until we obtain an equation or inequality with an obvious solution.

❑   ### LINEAR EQUATIONS

Linear equations are generally solved using the following equality properties:

> ### *Equality Properties*
>
> An equivalent equation will result if:
>
> **1.** The same quantity is added to or subtracted from each side of a given equation.
>
> **2.** Each side of a given equation is multiplied by or divided by the same nonzero quantity.

Several examples should remind you of the process of solving equations.

Example 1 ➯   **Solving a Linear Equation**   Solve and check:

$$8x - 3(x - 4) = 3(x - 4) + 6$$

SOLUTION

$$8x - 3(x - 4) = 3(x - 4) + 6$$
$$8x - 3x + 12 = 3x - 12 + 6$$
$$5x + 12 = 3x - 6$$
$$2x = -18$$
$$x = -9$$

Check

$$8x - 3(x - 4) = 3(x - 4) + 6$$
$$8(-9) - 3[(-9) - 4] \overset{?}{=} 3[(-9) - 4] + 6$$
$$-72 - 3(-13) \overset{?}{=} 3(-13) + 6$$
$$-33 \overset{✓}{=} -33$$

Matched Problem 1 ➯   Solve and check: $3x - 2(2x - 5) = 2(x + 3) - 8$

> **Explore–Discuss 1**
>
> According to equality property 2, multiplying both sides of an equation by a nonzero number always produces an equivalent equation. What is the smallest positive number that you could use to multiply both sides of the following equation to produce an equivalent equation without fractions?
>
> $$\frac{x + 1}{3} - \frac{x}{4} = \frac{1}{2}$$

Example 2 ➯   **Solving a Linear Equation**   Solve and check: $\dfrac{x + 2}{2} - \dfrac{x}{3} = 5$

SOLUTION   What operations can we perform on

$$\frac{x + 2}{2} - \frac{x}{3} = 5$$

to eliminate the denominators? If we can find a number that is exactly divisible by each denominator, we can use the multiplication property of equality to clear the denominators. The LCD (least common denominator) of the fractions, 6, is exactly what we are looking for! Actually, any common denominator will do, but the LCD results in a simpler equivalent equation. Thus, we multiply both sides of the equation by 6:

$$6\left(\frac{x+2}{2} - \frac{x}{3}\right) = 6 \cdot 5$$

$$\overset{3}{\cancel{6}} \cdot \frac{(x+2)}{\underset{1}{\cancel{2}}} - \overset{2}{\cancel{6}} \cdot \frac{x}{\underset{1}{\cancel{3}}} = 30$$

$$3(x+2) - 2x = 30$$
$$3x + 6 - 2x = 30$$
$$x = 24$$

Check
$$\frac{x+2}{2} - \frac{x}{3} = 5$$

$$\frac{24+2}{2} - \frac{24}{3} \overset{?}{=} 5$$

$$13 - 8 \overset{?}{=} 5$$

$$5 \overset{\checkmark}{=} 5$$

■

Matched Problem 2 ➭ Solve and check: $\dfrac{x+1}{3} - \dfrac{x}{4} = \dfrac{1}{2}$

■

In many applications of algebra, formulas or equations must be changed to alternative equivalent forms. The following examples are typical.

Example 3 ➭ **Solving a Formula for a Particular Variable** Solve the amount formula for simple interest, $A = P + Prt$, for:

(A) r in terms of the other variables
(B) P in terms of the other variables

SOLUTION (A) $\qquad A = P + Prt$ Reverse equation.
$\qquad\qquad P + Prt = A$ Now isolate r on the left side.
$\qquad\qquad\quad Prt = A - P$ Divide both members by Pt.
$\qquad\qquad\qquad r = \dfrac{A-P}{Pt}$

(B) $\qquad\quad A = P + Prt$ Reverse equation.
$\qquad\qquad P + Prt = A$ Factor out P (note the use of the distributive property).
$\qquad\qquad P(1 + rt) = A$ Divide by (1 + rt).
$\qquad\qquad\qquad P = \dfrac{A}{1+rt}$

■

Matched Problem 3 ➭ Solve $M = Nt + Nr$ for:

(A) t (B) N

■

❑ **LINEAR INEQUALITIES**

Before we start solving linear inequalities, let us recall what we mean by $<$ (less than) and $>$ (greater than). If a and b are real numbers, we write

$a < b$ *a is less than b*

if there exists a positive number p such that $a + p = b$. Certainly, we would expect that if a positive number was added to any real number, the sum would be larger than the original. That is essentially what the definition states. If $a < b$, we may also write

$b > a$ *b is greater than a.*

Example 4 ⇔ **Inequalities**

(A) $3 < 5$ *Since $3 + 2 = 5$*
(B) $-6 < -2$ *Since $-6 + 4 = -2$*
(C) $0 > -10$ *Since $-10 < 0$*

Matched Problem 4 ⇔ Replace each question mark with either $<$ or $>$.

(A) $2 \, ? \, 8$ (B) $-20 \, ? \, 0$ (C) $-3 \, ? \, -30$

The inequality symbols have a very clear geometric interpretation on the real number line. If $a < b$, then a is to the left of b on the number line; if $c > d$, then c is to the right of d (Fig. 1). Check this geometric property with the inequalities in Example 4.

FIGURE 1 $a < b, c > d$

Explore–Discuss 2

Replace **?** with $<$ or $>$ in each of the following:

(A) $-1 \, ? \, 3$ and $2(-1) \, ? \, 2(3)$
(B) $-1 \, ? \, 3$ and $-2(-1) \, ? \, -2(3)$

(C) $12 \, ? \, -8$ and $\dfrac{12}{4} \, ? \, \dfrac{-8}{4}$

(D) $12 \, ? \, -8$ and $\dfrac{12}{-4} \, ? \, \dfrac{-8}{-4}$

Based on these examples, describe verbally the effect of multiplying both sides of an inequality by a number.

Now let us turn to the problem of solving linear inequalities in one variable. Recall that a **solution** of an inequality involving one variable is a number that, when substituted for the variable, makes the inequality true. The set of all solutions is called the **solution set.** When we say that we **solve an inequality,** we mean that we find its solution set. The procedures used to solve linear inequalities in one variable are almost the same as those used to solve linear equations in one variable but with one important exception, as noted in property 3 below.

Inequality Properties

An equivalent inequality will result and the **sense will remain the same** if each side of the original inequality:

1. Has the same real number added to or subtracted from it.
2. Is multiplied or divided by the same positive number.

An equivalent inequality will result and the **sense will reverse** if each side of the original inequality:

3. Is multiplied or divided by the same negative number.

Note: Multiplication by 0 and division by 0 are not permitted.

Thus, we can perform essentially the same operations on inequalities that we perform on equations, with the exception that **the sense of the inequality reverses if we multiply or divide both sides by a negative number.** Otherwise, the sense of the inequality does not change. For example, if we start with the true statement

$$-3 > -7$$

and multiply both sides by 2, we obtain

$$-6 > -14$$

and the sense of the inequality stays the same. But if we multiply both sides of $-3 > -7$ by -2, the left side becomes 6 and the right side becomes 14, so we must write

$$6 < 14$$

to have a true statement. Thus, the sense of the inequality reverses.

If $a < b$, the double inequality $a < x < b$ means that $x > a$ **and** $x < b$; that is, x is between a and b. Other variations, as well as a useful interval notation, are given in Table 1. Note that an endpoint on a line graph has a square bracket through it if it is included in the inequality and a parenthesis through it if it is not.

TABLE 1		
INTERVAL NOTATION	**INEQUALITY NOTATION**	**LINE GRAPH**
$[a, b]$	$a \leq x \leq b$	
$[a, b)$	$a \leq x < b$	
$(a, b]$	$a < x \leq b$	
(a, b)	$a < x < b$	
$(-\infty, a]$	$x \leq a$	
$(-\infty, a)$	$x < a$	
$[b, \infty)*$	$x \geq b$	
(b, ∞)	$x > b$	

*The symbol ∞ (read "infinity") is not a number. When we write $[b, \infty)$, we are simply referring to the interval starting at b and continuing indefinitely to the right. We would never write $[b, \infty]$.

The following terminology is useful in connection with the interval notation of Table 1: If $a < b$, an interval of the form $[a, b]$, containing both endpoints, is a *closed interval*; an interval of the form (a, b), containing neither endpoint, is an *open interval*; and intervals of the form $(a, b]$ or $[a, b)$ are *half-open* and *half-closed*.

Example 5 ⇨ **Interval and Inequality Notation, and Line Graphs**

(A) Write $[-2, 3)$ as a double inequality and graph.
(B) Write $x \geq -5$ in interval notation and graph.

SOLUTION   (A) $[-2, 3)$ is equivalent to $-2 \leq x < 3$.

(B) $x \geq -5$ is equivalent to $[-5, \infty)$.

Matched Problem 5 ⇨ (A) Write $(-7, 4]$ as a double inequality and graph.
(B) Write $x < 3$ in interval notation and graph.

Explore–Discuss 3

The solution to Example 5B shows the graph of the inequality $x \geq -5$. What is the graph of $x < -5$? What is the corresponding interval? Describe the relationship between these sets.

Example 6 ⇨ **Solving a Linear Inequality**   Solve and graph:

$$2(2x + 3) < 6(x - 2) + 10$$

SOLUTION
$$2(2x + 3) < 6(x - 2) + 10$$
$$4x + 6 < 6x - 12 + 10$$
$$4x + 6 < 6x - 2$$
$$-2x + 6 < -2$$
$$-2x < -8$$
$$x > 4 \quad \text{or} \quad (4, \infty)$$

Notice that the sense of the inequality reverses when we divide both sides by -2.

Notice that in the graph of $x > 4$, we use a parenthesis through 4, since the point 4 is not included in the graph.

Matched Problem 6 ⇨ Solve and graph: $3(x - 1) \leq 5(x + 2) - 5$

Example 7 ⇨ **Solving a Double Inequality**   Solve and graph: $-3 < 2x + 3 \leq 9$

SOLUTION We are looking for all numbers x such that $2x + 3$ is between -3 and 9, including 9 but not -3. We proceed as before except that we try to isolate x in the middle:

$$-3 < 2x + 3 \leq 9$$

$$-3 - \mathbf{3} < 2x + 3 - \mathbf{3} \leq 9 - \mathbf{3}$$

$$-6 < 2x \leq 6$$

$$\frac{-6}{\mathbf{2}} < \frac{2x}{\mathbf{2}} \leq \frac{6}{\mathbf{2}}$$

$$-3 < x \leq 3 \quad \text{or} \quad (-3, 3]$$

Matched Problem 7 ⇨ Solve and graph: $-8 \leq 3x - 5 < 7$

Note that a linear equation usually has exactly one solution, while a linear inequality usually has infinitely many solutions.

  □ APPLICATIONS

To realize the full potential of algebra, we must be able to translate real-world problems into mathematical forms. In short, we must be able to do word problems.

The first example below involves the important concept of **break-even analysis,** which is encountered in several places in this text. Any manufacturing company has **costs, C,** and **revenues, R.** The company will have a **loss** if $R < C$, will **break even** if $R = C$, and will have a **profit** if $R > C$. Costs involve **fixed costs,** such as plant overhead, product design, setup, and promotion; and **variable costs,** which are dependent on the number of items produced at a certain cost per item.

Example 8 ⇨ **Break-Even Analysis** A recording company produces compact disks (CDs). One-time fixed costs for a particular CD are $24,000, which includes costs such as recording, album design, and promotion. Variable costs amount to $6.20 per CD and include the manufacturing, distribution, and royalty costs for each disk actually manufactured and sold to a retailer. The CD is sold to retail outlets at $8.70 each. How many CDs must be manufactured and sold for the company to break even?

SOLUTION Let x = number of CDs manufactured and sold

C = cost of producing x CDs

R = revenue (return) on sales of x CDs

The company breaks even if $R = C$, with

C = fixed costs + variable costs

= $24,000 + $6.20x$

$R = \$8.70x$

Find x such that $R = C$; that is, such that

$$8.7x = 24,000 + 6.2x$$

$$2.5x = 24,000$$
$$x = 9,600 \text{ CDs}$$

For $x = 9,600$,

$$
\begin{aligned}
C &= 24,000 + 6.2x & R &= 8.7x \\
&= 24,000 + 6.2(9,600) & &= 8.7(9,600) \\
&= \$83,520 & &= \$83,520
\end{aligned}
$$

Matched Problem 8 ➭ What is the break-even point in Example 8 if fixed costs are $18,000, variable costs are $5.20 per CD, and the CDs are sold to retailers for $7.60 each?

Algebra has many different types of applications—so many, in fact, that no single approach applies to all. However, the following suggestions may help you get started:

Suggestions for Solving Word Problems

1. Read the problem very carefully.
2. Write down important facts and relationships.
3. Identify unknown quantities in terms of a single letter, if possible.
4. Write an equation (or inequality) relating the unknown quantities and the facts in the problem.
5. Solve the equation (or inequality).
6. Write all solutions requested in the original problem.
7. Check the solution(s) in the original problem.

Example 9 ➭   **Consumer Price Index**   The Consumer Price Index (CPI) is a measure of the average change in prices over time from a designated reference period, which equals 100. The index is based on prices of basic consumer goods and services, and is published at regular intervals by the Bureau of Labor Statistics. Table 2 lists the CPI for several years from 1960 to 1998. What net annual salary in 1998 would have the same purchasing power as a net annual salary of $13,000 in 1960? Compute the answer to the nearest dollar.

TABLE 2

CPI (1982–1984 = 100)	
YEAR	INDEX
1960	29.6
1970	38.8
1980	82.4
1990	130.7
1998	163.0

SOLUTION   To have the same purchasing power, the ratio of a salary in 1998 to a salary in 1960 would have to be the same as the ratio of the CPI in 1998 to the CPI in 1960. Thus, if x is the net annual salary in 1998, we solve the equation

$$\frac{x}{13,000} = \frac{163.0}{29.6}$$

$$x = 13,000 \cdot \frac{163.0}{29.6}$$

$$= \$71,588 \text{ per year}$$

Matched Problem 9 ➭   What net annual salary in 1970 would have had the same purchasing power as a net annual salary of $75,000 in 1998? Compute the answer to the nearest dollar.

Answers to Matched Problems

1. $x = 4$ **2.** $x = 2$ **3.** (A) $t = \dfrac{M - Nr}{N}$ (B) $N = \dfrac{M}{t + r}$

4. (A) $<$ (B) $<$ (C) $>$

5. (A) $-7 < x \le 4$;   (B) $(-\infty, 3)$

6. $x \ge -4$ or $[-4, \infty)$

7. $-1 \le x < 4$ or $[-1, 4)$

8. 7,500 CD's   **9.** $17,853

Exercise A-8

A *Solve Problems 1–6.*

1. $2m + 9 = 5m - 6$

2. $3y - 4 = 6y - 19$

3. $x + 5 < -4$

4. $x - 3 > -2$

5. $-3x \ge -12$

6. $-4x \le 8$

Solve Problems 7–10 and graph.

7. $-4x - 7 > 5$

8. $-2x + 8 < 4$

9. $2 \le x + 3 \le 5$

10. $-3 < y - 5 < 8$

Solve Problems 11–26.

11. $\dfrac{y}{7} - 1 = \dfrac{1}{7}$

12. $\dfrac{m}{5} - 2 = \dfrac{3}{5}$

13. $\dfrac{x}{3} > -2$

14. $\dfrac{y}{-2} \le -1$

15. $\dfrac{y}{3} = 4 - \dfrac{y}{6}$

16. $\dfrac{x}{4} = 9 - \dfrac{x}{2}$

B

17. $10x + 25(x - 3) = 275$

18. $-3(4 - x) = 5 - (x + 1)$

19. $3 - y \le 4(y - 3)$

20. $x - 2 \ge 2(x - 5)$

21. $\dfrac{x}{5} - \dfrac{x}{6} = \dfrac{6}{5}$

22. $\dfrac{y}{4} - \dfrac{y}{3} = \dfrac{1}{2}$

23. $\dfrac{m}{5} - 3 < \dfrac{3}{5} - m$

24. $u - \dfrac{2}{3} > \dfrac{u}{3} + 2$

25. $0.1(x - 7) + 0.05x = 0.8$

26. $0.4(u + 5) - 0.3u = 17$

Solve Problems 27–30 and graph.

27. $2 \le 3x - 7 < 14$

28. $-4 \le 5x + 6 < 21$

29. $-4 \le \dfrac{9}{5}C + 32 \le 68$ **30.** $-1 \le \dfrac{2}{3}t + 5 \le 11$

C *Solve Problems 31–38 for the indicated variable.*

31. $3x - 4y = 12$;   for y

32. $y = -\dfrac{2}{3}x + 8$;   for x

33. $Ax + By = C$;   for y $(B \ne 0)$

34. $y = mx + b$;   for m

35. $F = \dfrac{9}{5}C + 32$;   for C

36. $C = \dfrac{5}{9}(F - 32)$;   for F

37. $A = Bm - Bn$;   for B

38. $U = 3C - 2CD$;   for C

Solve Problems 39 and 40 and graph.

39. $-3 \le 4 - 7x < 18$

40. $-1 < 9 - 2u \le 5$

41. What can be said about the signs of the numbers a and b in each case?

   (A) $ab > 0$ (B) $ab < 0$

   (C) $\dfrac{a}{b} > 0$ (D) $\dfrac{a}{b} < 0$

42. What can be said about the signs of the numbers a, b, and c in each case?

   (A) $abc > 0$ (B) $\dfrac{ab}{c} < 0$

   (C) $\dfrac{a}{bc} > 0$ (D) $\dfrac{a^2}{bc} < 0$

43. Replace each question mark with $<$ or $>$, as appropriate:

   (A) If $a - b = 2$, then a **?** b.

   (B) If $c - d = -1$, then c **?** d.

44. For what c and d is $c + d < c - d$?

45. If both a and b are positive numbers and b/a is greater than 1, then is $a - b$ positive or negative?

46. If both a and b are negative numbers and b/a is greater than 1, then is $a - b$ positive or negative?

In Problems 47–52, discuss the validity of each statement. If the statement is true, explain why. If not, give a counterexample.

47. If the intersection of two open intervals is nonempty, then their intersection is an open interval.

48. If the intersection of two closed intervals is nonempty, then their intersection is a closed interval.

49. The union of any two open intervals is an open interval.

50. The union of any two closed intervals is a closed interval.

51. If the intersection of two open intervals is nonempty, then their union is an open interval.

52. If the intersection of two closed intervals is nonempty, then their union is a closed interval.

Applications

Business & Economics

53. *Puzzle.* A jazz concert brought in $165,000 on the sale of 8,000 tickets. If the tickets sold for $15 and $25 each, how many of each type of ticket were sold?

54. *Puzzle.* An all-day parking meter takes only dimes and quarters. If it contains 100 coins with a total value of $14.50, how many of each type of coin are in the meter?

55. *Investing.* You have $12,000 to invest. If part is invested at 10% and the rest at 15%, how much should be invested at each rate to yield 12% on the total amount?

56. *Investing.* An investor has $20,000 to invest. If part is invested at 8% and the rest at 12%, how much should be invested at each rate to yield 11% on the total amount?

57. *Inflation.* If the price change of cars parallels the change in the CPI (see Table 2 in Example 9), what would a car sell for (to the nearest dollar) in 1998 if a comparable model sold for $5,000 in 1970?

58. *Inflation.* If the price change in houses parallels the CPI (see Table 2 in Example 9), what would a house valued at $200,000 in 1998 be valued at (to the nearest dollar) in 1960?

59. *Break-even analysis.* A publisher for a promising new novel figures fixed costs (overhead, advances, promotion, copy editing, typesetting, and so on) at $55,000, and variable costs (printing, paper, binding, shipping) at $1.60 for each book produced. If the book is sold to distributors for $11 each, what is the break-even point for the publisher?

60. *Break-even analysis.* The publisher of a new book called *Muscle-Powered Sports* figures fixed costs at $92,000 and variable costs at $2.10 for each book produced. If the book is sold to distributors for $15 each, how many must be sold for the publisher to break even?

61. *Break-even analysis.* The publisher in Problem 59 finds that rising prices for paper increase the variable costs to $2.10 per book.

(A) Discuss possible strategies the company might use to deal with this increase in costs.

(B) If the company continues to sell the books for $11, how many books must they sell now to make a profit?

(C) If the company wants to start making a profit at the same production level as before the cost increase, how much should they sell the book for now?

62. *Break-even analysis.* The publisher in Problem 60 finds that rising prices for paper increase the variable costs to $2.70 per book.

(A) Discuss possible strategies the company might use to deal with this increase in costs.

(B) If the company continues to sell the books for $15, how many books must they sell now to make a profit?

(C) If the company wants to start making a profit at the same production level as before the cost increase, how much should they sell the book for now?

Life Sciences

63. *Wildlife management.* A naturalist for a fish and game department estimated the total number of rainbow trout in a certain lake using the popular capture–mark–recapture technique. He netted, marked, and released 200 rainbow trout. A week later, allowing for thorough mixing, he again netted 200 trout and found 8 marked ones among them. Assuming that the proportion of marked fish in the second sample was the same as the proportion of all marked fish in the total population, estimate the number of rainbow trout in the lake.

64. *Ecology.* If the temperature for a 24 hour period at an Antarctic station ranged between $-49°F$ and $14°F$ (that is, $-49 \le F \le 14$), what was the range in degrees Celsius? [*Note*: $F = \frac{9}{5}C + 32$.]

Social Sciences

65. *Psychology.* The IQ (intelligence quotient) is found by dividing the mental age (MA), as indicated on standard tests, by the chronological age (CA) and multiplying by 100. For example, if a child has a mental age of 12 and a chronological age of 8, the calculated IQ is 150. If a 9-year-old girl has an IQ of 140, compute her mental age.

66. *Anthropology.* In their study of genetic groupings, anthropologists use a ratio called the *cephalic index.* This is the ratio of the width of the head to its length (looking down from above) expressed as a percentage. Symbolically,

$$C = \frac{100W}{L}$$

where C is the cephalic index, W is the width, and L is the length. If an Indian tribe in Baja California (Mexico) had an average cephalic index of 66 and the average width of their heads was 6.6 inches, what was the average length of their heads?

Figure for 66

Section A-9

Quadratic Equations

❑ SOLUTION BY SQUARE ROOT
❑ SOLUTION BY FACTORING
❑ QUADRATIC FORMULA
❑ QUADRATIC FORMULA AND FACTORING
❑ APPLICATION: SUPPLY AND DEMAND

A **quadratic equation** in one variable is any equation that can be written in the form

$$ax^2 + bx + c = 0 \qquad a \ne 0$$

where x is a variable and a, b, and c are constants. We will refer to this form as the **standard form.** The equations

$$5x^2 - 3x + 7 = 0 \qquad \text{and} \qquad 18 = 32t^2 - 12t$$

are both quadratic equations, since they are either in the standard form or can be transformed into this form.

We restrict our review to finding real solutions to quadratic equations.

❑ SOLUTION BY SQUARE ROOT

The easiest type of quadratic equation to solve is the special form where the first-degree term is missing:

$$ax^2 + c = 0 \qquad a \ne 0$$

The method of solution of this special form makes direct use of the square root property:

Square Root Property

If $a^2 = b$, then $a = \pm\sqrt{b}$.

Determine whether each of the following pairs of equations are equivalent. Explain.

(A) $x^2 = 4$ and $x = 2$
(B) $x^2 = 4$ and $x = -2$
(C) $x = \sqrt{4}$ and $x = 2$
(D) $x = \sqrt{4}$ and $x = -2$
(E) $x = -\sqrt{4}$ and $x = -2$

Example 1 ⤙ **Square Root Method**   Use the square root property to solve each equation.

(A) $x^2 - 7 = 0$ (B) $2x^2 - 10 = 0$ (C) $3x^2 + 27 = 0$
(D) $(x - 8)^2 = 9$

SOLUTION   (A) $x^2 - 7 = 0$

$$x^2 = 7 \qquad \text{What real number squared is 7?}$$
$$x = \pm\sqrt{7} \qquad \text{Short for } \sqrt{7} \text{ and } -\sqrt{7}$$

(B) $2x^2 - 10 = 0$

$$2x^2 = 10$$
$$x^2 = 5 \qquad \text{What real number squared is 5?}$$
$$x = \pm\sqrt{5}$$

(C) $3x^2 + 27 = 0$

$$3x^2 = -27$$
$$x^2 = -9 \qquad \text{What real number squared is } -9?$$

No real solution, since no real number squared is negative.

(D) $(x - 8)^2 = 9$

$$x - 8 = \pm\sqrt{9}$$
$$x - 8 = \pm 3$$
$$x = 8 \pm 3 = 5 \quad \text{or} \quad 11$$

■

Matched Problem 1 ⤙ Use the square root property to solve each equation.

(A) $x^2 - 6 = 0$ (B) $3x^2 - 12 = 0$ (C) $x^2 + 4 = 0$
(D) $(x + 5)^2 = 1$

■

❑ SOLUTION BY FACTORING

If the left side of a quadratic equation when written in standard form can be factored, the equation can be solved very quickly. The method of solution by factoring rests on the following important property of real numbers (see Appendix A-2):

If a and b are real numbers, then $ab = 0$ if and only if $a = 0$ or $b = 0$ (or both).

Example 2 ⤙ **Factoring Method**   Solve by factoring using integer coefficients, if possible.

(A) $3x^2 - 6x - 24 = 0$ (B) $3y^2 = 2y$ (C) $x^2 - 2x - 1 = 0$

SOLUTION (A) $3x^2 - 6x - 24 = 0$ *Divide both sides by 3, since 3 is a factor of each coefficient.*

$$x^2 - 2x - 8 = 0$$ *Factor the left side, if possible.*
$$(x - 4)(x + 2) = 0$$
$$x - 4 = 0 \quad \text{or} \quad x + 2 = 0$$
$$x = 4 \quad \text{or} \quad x = -2$$

(B) $3y^2 = 2y$

$$3y^2 - 2y = 0$$ *We lose the solution y = 0 if both sides are divided by y*
$$y(3y - 2) = 0$$ *($3y^2 = 2y$ and $3y = 2$ are not equivalent).*
$$y = 0 \quad \text{or} \quad 3y - 2 = 0$$
$$3y = 2$$
$$y = \tfrac{2}{3}$$

(C) $x^2 - 2x - 1 = 0$

This equation cannot be factored using integer coefficients. We will solve this type of equation by another method, considered below. ■

Matched Problem 2 ➭ Solve by factoring using integer coefficients, if possible.

(A) $2x^2 + 4x - 30 = 0$ (B) $2x^2 = 3x$ (C) $2x^2 - 8x + 3 = 0$ ■

Note that an equation such as $x^2 = 25$ can be solved by either the square root or the factoring method, and the results are the same (as they should be). Solve this equation both ways and compare.

Also, note that the factoring method can be extended to higher-degree polynomial equations. Consider the following:

$$x^3 - x = 0$$
$$x(x^2 - 1) = 0$$
$$x(x - 1)(x + 1) = 0$$
$$x = 0 \quad \text{or} \quad x - 1 = 0 \quad \text{or} \quad x + 1 = 0$$
Solution: $x = 0, 1, -1$

Check these solutions in the original equation.

The factoring and square root methods are fast and easy to use when they apply. However, there are quadratic equations that look simple but cannot be solved by either method. For example, as was noted in Example 2C, the polynomial in

$$x^2 - 2x - 1 = 0$$

cannot be factored using integer coefficients. This brings us to the well-known and widely used *quadratic formula.*

❑ QUADRATIC FORMULA

There is a method called *completing the square* that will work for all quadratic equations. After briefly reviewing this method, we will then use it to develop the famous quadratic formula—a formula that will enable us to solve any quadratic equation quite mechanically.

Replace ? in each of the following with a number that makes the equation valid.

(A) $(x + 1)^2 = x^2 + 2x +$ **?**   (B) $(x + 2)^2 = x^2 + 4x +$ **?**
(C) $(x + 3)^2 = x^2 + 6x +$ **?**   (D) $(x + 4)^2 = x^2 + 8x +$ **?**

Replace ? in each of the following with a number that makes the trinomial a perfect square.

(E) $x^2 + 10x +$ **?**   (F) $x^2 + 12x +$ **?**   (G) $x^2 + bx +$ **?**

The method of **completing the square** is based on the process of transforming a quadratic equation in standard form,

$$ax^2 + bx + c = 0$$

into the form

$$(x + A)^2 = B$$

where A and B are constants. Then, this last equation can be solved easily (if it has a real solution) by the square root method discussed above.

Consider the equation from Example 2C:

$$x^2 - 2x - 1 = 0 \tag{1}$$

Since the left side does not factor using integer coefficients, we add 1 to each side to remove the constant term from the left side:

$$x^2 - 2x = 1 \tag{2}$$

Now we try to find a number that we can add to each side to make the left side a square of a first-degree polynomial. Note the following square of a binomial:

$$(x + m)^2 = x^2 + 2mx + m^2$$

We see that the third term on the right is the square of one-half the coefficient of x in the second term on the right. To complete the square in equation (2), we add the square of one-half the coefficient of x, $(-\frac{2}{2})^2 = 1$, to each side. (This rule works only when the coefficient of x^2 is 1, that is, $a = 1$.) Thus,

$$x^2 - 2x + \mathbf{1} = 1 + \mathbf{1}$$

The left side is the square of $x - 1$, and we write

$$(x - 1)^2 = 2$$

What number squared is 2?

$$x - 1 = \pm\sqrt{2}$$
$$x = 1 \pm \sqrt{2}$$

And equation (1) is solved!

Let us try the method on the general quadratic equation

$$ax^2 + bx + c = 0 \qquad a \neq 0 \tag{3}$$

and solve it once and for all for x in terms of the coefficients a, b, and c. We start by multiplying both sides of equation (3) by $1/a$ to obtain

$$x^2 + \frac{b}{a}x + \frac{c}{a} = 0$$

Add $-c/a$ to both sides:

$$x^2 + \frac{b}{a}x = -\frac{c}{a}$$

Now we complete the square on the left side by adding the square of one-half the coefficient of x, that is, $(b/2a)^2 = b^2/4a^2$, to each side:

$$x^2 + \frac{b}{a}x + \frac{b^2}{4a^2} = \frac{b^2}{4a^2} - \frac{c}{a}$$

Writing the left side as a square and combining the right side into a single fraction, we obtain

$$\left(x + \frac{b}{2a}\right)^2 = \frac{b^2 - 4ac}{4a^2}$$

Now we solve by the square root method:

$$x + \frac{b}{2a} = \pm\sqrt{\frac{b^2 - 4ac}{4a^2}}$$

$$x = -\frac{b}{2a} \pm \frac{\sqrt{b^2 - 4ac}}{2a} \qquad \text{Since } \pm\sqrt{4a^2} = \pm 2a \text{ for any real number } a$$

When this is written as a single fraction, it becomes the **quadratic formula:**

Quadratic Formula

If $ax^2 + bx + c = 0$, $a \neq 0$, then

$$x = \frac{-b \pm \sqrt{b^2 - 4ac}}{2a}$$

TABLE 1	
$b^2 - 4ac$	$ax^2 + bx + c = 0$
Positive	Two real solutions
Zero	One real solution
Negative	No real solutions

This formula is generally used to solve quadratic equations when the square root or factoring methods do not work. The quantity $b^2 - 4ac$ under the radical is called the **discriminant,** and it gives us the useful information about solutions listed in Table 1.

Example 3 ➦ **Quadratic Formula Method**   Solve $x^2 - 2x - 1 = 0$ using the quadratic formula.

SOLUTION

$$x^2 - 2x - 1 = 0$$

$$x = \frac{-b \pm \sqrt{b^2 - 4ac}}{2a} \qquad a = 1, b = -2, c = -1$$

$$= \frac{-(-2) \pm \sqrt{(-2)^2 - 4(1)(-1)}}{2(1)}$$

$$= \frac{2 \pm \sqrt{8}}{2} = \frac{2 \pm 2\sqrt{2}}{2} = 1 \pm \sqrt{2} \approx -0.414 \quad \text{or} \quad 2.414$$

Check

$$x^2 - 2x - 1 = 0$$

When $x = 1 + \sqrt{2}$,

$$(1 + \sqrt{2})^2 - 2(1 + \sqrt{2}) - 1 = 1 + 2\sqrt{2} + 2 - 2 - 2\sqrt{2} - 1 = 0$$

When $x = 1 - \sqrt{2}$,

$$(1 - \sqrt{2})^2 - 2(1 - \sqrt{2}) - 1 = 1 - 2\sqrt{2} + 2 - 2 + 2\sqrt{2} - 1 = 0$$

Matched Problem 3 ⇔ Solve $2x^2 - 4x - 3 = 0$ using the quadratic formula.

If we try to solve $x^2 - 6x + 11 = 0$ using the quadratic formula, we obtain

$$x = \frac{6 \pm \sqrt{-8}}{2}$$

which is not a real number. (Why?)

❑   QUADRATIC FORMULA AND FACTORING

As in Section A-4, we restrict our interest in factoring to polynomials with integer coefficients. If a polynomial cannot be factored as a product of lower-degree polynomials with integer coefficients, we say that the polynomial is **not factorable in the integers.**

Suppose that you were asked to factor

$$x^2 - 19x - 372 \tag{4}$$

The larger the coefficients, the more difficult the process of applying the *ac* test discussed in Section A-4. The quadratic formula provides a simple and efficient method of factoring a second-degree polynomial with integer coefficients as the product of two first-degree polynomials with integer coefficients, if the factors exist. We illustrate the method using equation (4), and generalize the process from this experience.

We start by solving the corresponding quadratic equation using the quadratic formula:

$$x^2 - 19x - 372 = 0$$

$$x = \frac{-(-19) \pm \sqrt{(-19)^2 - 4(1)(-372)}}{2}$$

$$= \frac{19 \pm \sqrt{1,849}}{2}$$

$$= \frac{19 \pm 43}{2} = -12 \quad \text{or} \quad 31$$

Now, we write

$$x^2 - 19x - 372 = [x - (-12)](x - 31) = (x + 12)(x - 31)$$

Multiplying the two factors on the right produces the second-degree polynomial on the left.

What is behind this procedure? The following two theorems justify and generalize the process:

THEOREM 1   Factorability Theorem

A second-degree polynomial, $ax^2 + bx + c$, with integer coefficients can be expressed as the product of two first-degree polynomials with integer coefficients if and only if $\sqrt{b^2 - 4ac}$ is an integer.

THEOREM 2   Factor Theorem

If r_1 and r_2 are solutions to the second-degree polynomial $ax^2 + bx + c = 0$, then

$$ax^2 + bx + c = a(x - r_1)(x - r_2)$$

Example 4 ⇦ **Factoring with the Aid of the Discriminant**   Factor, if possible, using integer coefficients:

(A) $4x^2 - 65x + 264$ (B) $2x^2 - 33x - 306$

SOLUTION   (A) $4x^2 - 65x + 264$

Step 1.   Test for factorability:

$$\sqrt{b^2 - 4ac} = \sqrt{(-65)^2 - 4(4)(264)} = 1$$

Since the result is an integer, the polynomial has first-degree factors with integer coefficients.

Step 2.   Factor, using the factor theorem. Find the solutions to the corresponding quadratic equation using the quadratic formula:

$$4x^2 - 65x + 264 = 0 \quad \text{From step 1}$$

$$x = \frac{-(-65) \pm 1}{2 \cdot 4} = \frac{33}{4} \quad \text{or} \quad 8$$

Thus,

$$4x^2 - 65x + 264 = 4\left(x - \frac{33}{4}\right)(x - 8)$$
$$= (4x - 33)(x - 8)$$

(B) $2x^2 - 33x - 306$

Step 1.   Test for factorability:

$$\sqrt{b^2 - 4ac} = \sqrt{(-33)^2 - 4(2)(-306)} = \sqrt{3{,}537}$$

Since $\sqrt{3{,}537}$ is not an integer, the polynomial is not factorable in the integers. ■

Matched Problem 4 ⇦ Factor, if possible, using integer coefficients:

(A) $3x^2 - 28x - 464$ (B) $9x^2 + 320x - 144$

□   APPLICATION: SUPPLY AND DEMAND

Supply and demand analysis is a very important part of business and economics. In general, producers are willing to supply more of an item as the price of an item increases, and less of an item as the price decreases. Similarly, buyers are willing to buy less of an item as the price increases, and more of an item as the price decreases. Thus, we have a dynamic situation where the price, supply, and demand fluctuate until a price is reached at which the supply is equal to the demand. In economic theory, this point is called the **equilibrium point**—if the price increases from this point, the supply will increase and the demand will decrease; if the price decreases from this point, the supply will decrease and the demand will increase.

Example 5 ➭

Supply and Demand   At a large beach resort in the summer, the weekly supply and demand equations for folding beach chairs are

$$p = \frac{x}{140} + \frac{3}{4}$$ *Supply equation*

$$p = \frac{5,670}{x}$$ *Demand equation*

The supply equation indicates that the supplier is willing to sell x units at a price of p dollars per unit. The demand equation indicates that consumers are willing to buy x units at a price of p dollars per unit. How many units are required for supply to equal demand? At what price will supply equal demand?

SOLUTION   Set the right side of the supply equation equal to the right side of the demand equation and solve for x:

$$\frac{x}{140} + \frac{3}{4} = \frac{5,670}{x}$$ *Multiply by 140x, the LCD.*

$$x^2 + 105x = 793,800$$ *Write in standard form.*

$$x^2 + 105x - 793,800 = 0$$ *Use the quadratic formula.*

$$x = \frac{-105 \pm \sqrt{105^2 - 4(1)(-793,800)}}{2}$$

$$x = 840 \text{ units}$$

The negative root is discarded, since a negative number of units cannot be produced or sold. Substitute $x = 840$ back into either the supply equation or the demand equation to find the equilibrium price (we use the demand equation):

$$p = \frac{5,670}{x} = \frac{5,670}{840} = \$6.75$$

Thus, at a price of \$6.75 the supplier is willing to supply 840 chairs and consumers are willing to buy 840 chairs during a week.   ■

Matched Problem 5 ➭

Repeat Example 5 if near the end of summer the supply and demand equations are

$$p = \frac{x}{80} - \frac{1}{20}$$ *Supply equation*

$$p = \frac{1,264}{x}$$ *Demand equation*   ■

Answers to Matched Problems   **1.** (A) $\pm\sqrt{6}$ (B) ± 2 (C) No real solution   (D) $-6, -4$
2. (A) $-5, 3$ (B) $0, \frac{3}{2}$ (C) Cannot be factored using integer coefficients
3. $(2 \pm \sqrt{10})/2$
4. (A) Cannot be factored using integer coefficients   (B) $(9x - 4)(x + 36)$
5. 320 chairs at \$3.95 each

Exercise A-9

Find only real solutions in the problems below. If there are no real solutions, say so.

A *Solve Problems 1–4 by the square root method.*

1. $2x^2 - 22 = 0$ **2.** $3m^2 - 21 = 0$

3. $(x - 1)^2 = 4$ **4.** $(x + 2)^2 = 9$

Solve Problems 5–8 by factoring.

5. $2u^2 - 8u - 24 = 0$ **6.** $3x^2 - 18x + 15 = 0$

7. $x^2 = 2x$ **8.** $n^2 = 3n$

Solve Problems 9–12 by using the quadratic formula.

9. $x^2 - 6x - 3 = 0$ **10.** $m^2 + 8m + 3 = 0$

11. $3u^2 + 12u + 6 = 0$ **12.** $2x^2 - 20x - 6 = 0$

B *Solve Problems 13–30 by using any method.*

13. $2x^2 = 4x$ **14.** $2x^2 = -3x$

15. $4u^2 - 9 = 0$ **16.** $9y^2 - 25 = 0$

17. $8x^2 + 20x = 12$ **18.** $9x^2 - 6 = 15x$

19. $x^2 = 1 - x$ **20.** $m^2 = 1 - 3m$

21. $2x^2 = 6x - 3$ **22.** $2x^2 = 4x - 1$

23. $y^2 - 4y = -8$ **24.** $x^2 - 2x = -3$

25. $(x + 4)^2 = 11$ **26.** $(y - 5)^2 = 7$

27. $\dfrac{3}{p} = p$ **28.** $x - \dfrac{7}{x} = 0$

29. $2 - \dfrac{2}{m^2} = \dfrac{3}{m}$ **30.** $2 + \dfrac{5}{u} = \dfrac{3}{u^2}$

In Problems 31–38, factor, if possible, as the product of two first-degree polynomials with integer coefficients. Use the quadratic formula and the factor theorem.

31. $x^2 + 40x - 84$ **32.** $x^2 - 28x - 128$

33. $x^2 - 32x + 144$ **34.** $x^2 + 52x + 208$

35. $2x^2 + 15x - 108$ **36.** $3x^2 - 32x - 140$

37. $4x^2 + 241x - 434$ **38.** $6x^2 - 427x - 360$

C

39. Solve $A = P(1 + r)^2$ for r in terms of A and P; that is, isolate r on the left side of the equation (with coefficient 1) and end up with an algebraic expression on the right side involving A and P but not r. Write the answer using positive square roots only.

40. Solve $x^2 + mx + n = 0$ for x in terms of m and n.

41. Consider the quadratic equation

$$x^2 + 4x + c = 0$$

where c is a real number. Discuss the relationship between the values of c and the three types of roots listed in Table 1.

42. Consider the quadratic equation

$$x^2 - 2x + c = 0$$

where c is a real number. Discuss the relationship between the values of c and the three types of roots listed in Table 1.

Applications

Business & Economics

43. *Supply and demand.* A company wholesales a certain brand of shampoo in a particular city. Their marketing research department established the following weekly supply and demand equations:

$$p = \frac{x}{450} + \frac{1}{2} \qquad \text{Supply equation}$$

$$p = \frac{6{,}300}{x} \qquad \text{Demand equation}$$

How many units are required for supply to equal demand? At what price per bottle will supply equal demand?

44. *Supply and demand.* An importer sells a certain brand of automatic camera to outlets in a large metropolitan area. During the summer, the weekly supply and demand equations are

$$p = \frac{x}{6} + 9 \qquad \text{Supply equation}$$

$$p = \frac{24{,}840}{x} \qquad \text{Demand equation}$$

How many units are required for supply to equal demand? At what price will supply equal demand?

45. *Interest rate.* If P dollars are invested at $100r$ percent compounded annually, at the end of 2 years it will grow to $A = P(1 + r)^2$. At what interest rate will $100 grow to $144 in 2 years? [*Note:* If $A = 144$ and $P = 100$, find r.]

46. *Interest rate.* Using the formula in Problem 45, determine the interest rate that will make $1,000 grow to $1,210 in 2 years.

Life Sciences

47. *Ecology.* An important element in the erosive force of moving water is its velocity. To measure the velocity v (in feet per second) of a stream, we position a hollow L-shaped tube with one end under the water pointing upstream and the other end pointing straight up a couple of feet out of the water. The water will then be pushed up the tube a certain distance h (in feet) above the surface of the stream. Physicists have shown that $v^2 = 64h$. Approximately how fast is a stream flowing if $h = 1$ foot? If $h = 0.5$ foot?

Social Sciences

48. *Safety research.* It is of considerable importance to know the least number of feet d in which a car can be stopped, including reaction time of the driver, at various speeds v (in miles per hour). Safety research has produced the formula $d = 0.044v^2 + 1.1v$. If it took a car 550 feet to stop, estimate the car's speed at the moment the stopping process was started.

Appendix B

Special Topics

Section B-1   Sequences, Series, and Summation Notation

- ❑   SEQUENCES
- ❑   SERIES AND SUMMATION NOTATION

If someone asked you to list all natural numbers that are perfect squares, you might begin by writing

$$1, 4, 9, 16, 25, 36$$

But you would soon realize that it is impossible to actually list all the perfect squares, since there are an infinite number of them. However, you could represent this collection of numbers in several different ways. One common method is to write

$$1, 4, 9, \ldots, n^2, \ldots \qquad n \in N$$

where N is the set of natural numbers. A list of numbers such as this is generally called a *sequence.* Sequences and related topics form the subject matter of this section.

❑   SEQUENCES

Consider the function f given by

$$f(n) = 2n + 1 \tag{1}$$

where the domain of f is the set of natural numbers N. Note that

$$f(1) = 3, \qquad f(2) = 5, \qquad f(3) = 7, \ldots$$

The function f is an example of a sequence. In general, a **sequence** is a function with domain a set of successive integers. Instead of the standard function notation used in equation (1), sequences are usually defined in terms of a special notation.

The range value $f(n)$ is usually symbolized more compactly with a symbol such as a_n. Thus, in place of equation (1), we write

$$a_n = 2n + 1$$

779

and the domain is understood to be the set of natural numbers unless something is said to the contrary or the context indicates otherwise. The elements in the range are called **terms of the sequence;** a_1 is the first term, a_2 is the second term, and a_n is the **nth term,** or **general term.**

$$a_1 = 2(1) + 1 = 3 \qquad \text{First term}$$
$$a_2 = 2(2) + 1 = 5 \qquad \text{Second term}$$
$$a_3 = 2(3) + 1 = 7 \qquad \text{Third term}$$
$$\vdots$$
$$a_n = 2n + 1 \qquad\qquad \text{General term}$$

The ordered list of elements

$$3, 5, 7, \dots, 2n + 1, \dots$$

obtained by writing the terms of the sequence in their natural order with respect to the domain values is often informally referred to as a sequence. A sequence also may be represented in the abbreviated form $\{a_n\}$, where a symbol for the nth term is written within braces. For example, we could refer to the sequence $3, 5, 7, \dots, 2n + 1, \dots$ as the sequence $\{2n + 1\}$.

If the domain of a sequence is a finite set of successive integers, then the sequence is called a **finite sequence.** If the domain is an infinite set of successive integers, then the sequence is called an **infinite sequence.** The sequence $\{2n + 1\}$ discussed above is an infinite sequence.

Example 1 ⇨ **Writing the Terms of a Sequence** Write the first four terms of each sequence:

(A) $a_n = 3n - 2$ (B) $\left\{ \dfrac{(-1)^n}{n} \right\}$

SOLUTION (A) $1, 4, 7, 10$ (B) $-1, \dfrac{1}{2}, \dfrac{-1}{3}, \dfrac{1}{4}$ ∎

Matched Problem 1 ⇨ Write the first four terms of each sequence:

(A) $a_n = -n + 3$ (B) $\left\{ \dfrac{(-1)^n}{2^n} \right\}$ ∎

Explore–Discuss 1

(A) A multiple-choice test question asked for the next term in the sequence

$$2, 4, 8, \dots$$

and gave the following choices:

(1) 16 (2) 14 (3) $\frac{25}{2}$

Which is the correct answer?

(B) Compare the first four terms of the following sequences:

(1) $a_n = 2^2$ (2) $b_n = n^2 - n + 2$ (3) $c_n = 5n + \dfrac{6}{n} - 9$

Now, which of the choices in part (A) appears to be correct?

Now that we have seen how to use the general term to find the first few terms in a sequence, we consider the reverse problem. That is, can a sequence be defined just by listing the first three or four terms of the sequence? And can we then use these initial terms to find a formula for the *n*th term? In general, without other information, the answer to the first question is no. As Explore–Discuss 1 illustrates, many different sequences may start off with the same terms. Simply listing the first three terms (or any other finite number of terms) does not specify a particular sequence. In fact, it can be shown that given any list of *m* numbers, there are an infinite number of sequences whose first *m* terms agree with these given numbers.

What about the second question? That is, given a few terms, can we find the general formula for at least one sequence whose first few terms agree with the given terms? The answer to this question is a qualified yes. If we can observe a simple pattern in the given terms, we usually can construct a general term that will produce that pattern. The next example illustrates this approach.

Example 2 ➯ **Finding the General Term of a Sequence**   Find the general term of a sequence whose first four terms are:

(A) $3, 4, 5, 6, \ldots$ (B) $5, -25, 125, -625, \ldots$

SOLUTION   (A) Since these terms are consecutive integers, one solution is $a_n = n, n \geqslant 3$. If we want the domain of the sequence to be all natural numbers, another solution is $b_n = n + 2$.

(B) Each of these terms can be written as the product of a power of 5 and a power of -1:

$$5 = (-1)^0 5^1 = a_1$$
$$-25 = (-1)^1 5^2 = a_2$$
$$125 = (-1)^2 5^3 = a_3$$
$$-625 = (-1)^3 5^4 = a_4$$

If we choose the domain to be all natural numbers, a solution is

$$a_n = (-1)^{n-1} 5^n$$

Matched Problem 2 ➯ Find the general term of a sequence whose first four terms are:

(A) $3, 6, 9, 12, \ldots$ (B) $1, -2, 4, -8, \ldots$

In general, there is usually more than one way of representing the *n*th term of a given sequence (see the solution of Example 2A). However, unless something is stated to the contrary, we assume that the domain of the sequence is the set of natural numbers *N*.

❑   SERIES AND SUMMATION NOTATION

If $a_1, a_2, a_3, \ldots, a_n, \ldots$ is a sequence, the expression

$$a_1 + a_2 + a_3 + \cdots + a_n + \cdots$$

is called a **series.** If the sequence is finite, the corresponding series is a **finite series.** If the sequence is infinite, the corresponding series is an **infinite series.** We consider only finite series in this section. For example,

$$1, 3, 5, 7, 9 \qquad \text{Finite sequence}$$
$$1 + 3 + 5 + 7 + 9 \qquad \text{Finite series}$$

Notice that we can easily evaluate this series by adding the five terms:

$$1 + 3 + 5 + 7 + 9 = 25$$

Series are often represented in a compact form called **summation notation.** Consider the following examples:

$$\sum_{k=3}^{6} k^2 = 3^2 + 4^2 + 5^2 + 6^2$$
$$= 9 + 16 + 25 + 36 = 86$$

$$\sum_{k=0}^{2} (4k + 1) = (4 \cdot 0 + 1) + (4 \cdot 1 + 1) + (4 \cdot 2 + 1)$$
$$= 1 + 5 + 9 = 15$$

In each case, the terms of the series on the right are obtained from the expression on the left by successively replacing the **summing index k** with integers, starting with the number indicated below the **summation sign** Σ and ending with the number that appears above Σ. The summing index may be represented by letters other than k and may start at any integer and end at any integer greater than or equal to the starting integer. Thus, if we are given the finite sequence

$$\frac{1}{2}, \frac{1}{4}, \frac{1}{8}, \dots, \frac{1}{2^n}$$

the corresponding series is

$$\frac{1}{2} + \frac{1}{4} + \frac{1}{8} + \cdots + \frac{1}{2^n} = \sum_{j=1}^{n} \frac{1}{2^j}$$

where we have used j for the summing index.

Example 3 ➭ **Summation Notation**   Write

$$\sum_{k=1}^{5} \frac{k}{k^2 + 1}$$

without summation notation. Do not evaluate the sum.

SOLUTION
$$\sum_{k=1}^{5} \frac{k}{k^2 + 1} = \frac{1}{1^2 + 1} + \frac{2}{2^2 + 1} + \frac{3}{3^2 + 1} + \frac{4}{4^2 + 1} + \frac{5}{5^2 + 1}$$
$$= \frac{1}{2} + \frac{2}{5} + \frac{3}{10} + \frac{4}{17} + \frac{5}{26}$$

Matched Problem 3 ➭ Write

$$\sum_{k=1}^{5} \frac{k + 1}{k}$$

without summation notation. Do not evaluate the sum.

(A) Find the smallest value of n for which the value of the series

$$\sum_{k=1}^{n} \frac{k}{k^2 + 1}$$

is greater than 3.

(B) Find the smallest value of n for which the value of the series

$$\sum_{j=1}^{n} \frac{1}{2^j}$$

is greater than 0.99. Greater than 0.999.

If the terms of a series are alternately positive and negative, we call the series an **alternating series.** The next example deals with the representation of such a series.

Example 4   **Summation Notation**   Write the alternating series

$$\frac{1}{2} - \frac{1}{4} + \frac{1}{6} - \frac{1}{8} + \frac{1}{10} - \frac{1}{12}$$

using summation notation with:

(A) The summing index k starting at 1

(B) The summing index j starting at 0

SOLUTION   (A) $(-1)^{k+1}$ provides the alternation of sign, and $1/(2k)$ provides the other part of each term. Thus, we can write

$$\frac{1}{2} - \frac{1}{4} + \frac{1}{6} - \frac{1}{8} + \frac{1}{10} - \frac{1}{12} = \sum_{k=1}^{6} \frac{(-1)^{k+1}}{2k}$$

(B) $(-1)^j$ provides the alternation of sign, and $1/[2(j + 1)]$ provides the other part of each term. Thus, we can write

$$\frac{1}{2} - \frac{1}{4} + \frac{1}{6} - \frac{1}{8} + \frac{1}{10} - \frac{1}{12} = \sum_{j=0}^{5} \frac{(-1)^j}{2(j + 1)}$$

Matched Problem 4   Write the alternating series

$$1 - \frac{1}{3} + \frac{1}{9} - \frac{1}{27} + \frac{1}{81}$$

using summation notation with:

(A) The summing index k starting at 1

(B) The summing index j starting at 0

Summation notation provides a compact notation for the sum of any list of numbers, even if the numbers are not generated by a formula. For example, suppose that the results of an examination taken by a class of 10 students are given in the following list:

87, 77, 95, 83, 86, 73, 95, 68, 75, 86

If we let $a_1, a_2, a_3, \ldots, a_{10}$ represent these 10 scores, the average test score is given by

$$\frac{1}{10} \sum_{k=1}^{10} a_k = \frac{1}{10} (87 + 77 + 95 + 83 + 86 + 73 + 95 + 68 + 75 + 86)$$

$$= \frac{1}{10} (825) = 82.5$$

More generally, in statistics, the **arithmetic mean** $\bar{a}$ of a list of n numbers $a_1, a_2, \ldots, a_n$ is defined as

$$\bar{a} = \frac{1}{n} \sum_{k=1}^{n} a_k$$

Example 5 ➭ **Arithmetic Mean**   Find the arithmetic mean of 3, 5, 4, 7, 4, 2, 3, and 6.

SOLUTION $\bar{a} = \frac{1}{8} \sum_{k=1}^{8} a_k = \frac{1}{8} (3 + 5 + 4 + 7 + 4 + 2 + 3 + 6) = \frac{1}{8} (34) = 4.25$ ■

Matched Problem 5 ➭ Find the arithmetic mean of 9, 3, 8, 4, 3, and 6.   ■

Answers to Matched Problems
1. (A) $2, 1, 0, -1$ (B) $\frac{-1}{2}, \frac{1}{4}, \frac{-1}{8}, \frac{1}{16}$
2. (A) $a_n = 3n$ (B) $a_n = (-2)^{n-1}$ **3.** $2 + \frac{3}{2} + \frac{4}{3} + \frac{5}{4} + \frac{6}{5}$
4. (A) $\sum_{k=1}^{5} \frac{(-1)^{k-1}}{3^{k-1}}$ (B) $\sum_{j=0}^{4} \frac{(-1)^j}{3^j}$ **5.** 5.5

Exercise B-1

A Write the first four terms for each sequence in Problems 1–6.

1. $a_n = 2n + 3$ **2.** $a_n = 4n - 3$

3. $a_n = \dfrac{n + 2}{n + 1}$ **4.** $a_n = \dfrac{2n + 1}{2n}$

5. $a_n = (-3)^{n+1}$ **6.** $a_n = \left(-\frac{1}{4}\right)^{n-1}$

7. Write the 10th term of the sequence in Problem 1.

8. Write the 15th term of the sequence in Problem 2.

9. Write the 99th term of the sequence in Problem 3.

10. Write the 200th term of the sequence in Problem 4.

In Problems 11–16, write each series in expanded form without summation notation, and evaluate.

11. $\sum_{k=1}^{6} k$ **12.** $\sum_{k=1}^{5} k^2$ **13.** $\sum_{k=4}^{7} (2k - 3)$

14. $\sum_{k=0}^{4} (-2)^k$ **15.** $\sum_{k=0}^{3} \frac{1}{10^k}$ **16.** $\sum_{k=1}^{4} \frac{1}{2^k}$

Find the arithmetic mean of each list of numbers in Problems 17–20.

17. 5, 4, 2, 1, and 6   **18.** 7, 9, 9, 2, and 4

19. 96, 65, 82, 74, 91, 88, 87, 91, 77, and 74

20. 100, 62, 95, 91, 82, 87, 70, 75, 87, and 82

B Write the first five terms of each sequence in Problems 21–26.

21. $a_n = \dfrac{(-1)^{n+1}}{2^n}$ **22.** $a_n = (-1)^n (n - 1)^2$

23. $a_n = n[1 + (-1)^n]$ **24.** $a_n = \dfrac{1 - (-1)^n}{n}$

25. $a_n = \left(-\dfrac{3}{2}\right)^{n-1}$ **26.** $a_n = \left(-\dfrac{1}{2}\right)^{n+1}$

In Problems 27–42, find the general term of a sequence whose first four terms agree with the given terms.

27. $-2, -1, 0, 1, \ldots$ **28.** $4, 5, 6, 7, \ldots$
29. $4, 8, 12, 16, \ldots$ **30.** $-3, -6, -9, -12, \ldots$
31. $\frac{1}{2}, \frac{3}{4}, \frac{5}{6}, \frac{7}{8}, \ldots$ **32.** $\frac{1}{2}, \frac{2}{3}, \frac{3}{4}, \frac{4}{5}, \ldots$
33. $1, -2, 3, -4, \ldots$ **34.** $-2, 4, -8, 16, \ldots$
35. $1, -3, 5, -7, \ldots$ **36.** $3, -6, 9, -12, \ldots$
37. $1, \frac{2}{5}, \frac{4}{25}, \frac{8}{125}, \ldots$ **38.** $\frac{4}{3}, \frac{16}{9}, \frac{64}{27}, \frac{256}{81}, \ldots$
39. $x, x^2, x^3, x^4, \ldots$ **40.** $1, 2x, 3x^2, 4x^3, \ldots$

41. $x, -x^3, x^5, -x^7, \ldots$

42. $x, \dfrac{x^2}{2}, \dfrac{x^3}{3}, \dfrac{x^4}{4}, \ldots$

Write each series in Problems 43–50 in expanded form without summation notation. Do not evaluate.

43. $\displaystyle\sum_{k=1}^{5} (-1)^{k+1}(2k-1)^2$

44. $\displaystyle\sum_{k=1}^{4} \dfrac{(-2)^{k+1}}{2k+1}$

45. $\displaystyle\sum_{k=2}^{5} \dfrac{2^k}{2k+3}$

46. $\displaystyle\sum_{k=3}^{7} \dfrac{(-1)^k}{k^2-k}$

47. $\displaystyle\sum_{k=1}^{5} x^{k-1}$

48. $\displaystyle\sum_{k=1}^{3} \dfrac{1}{k}x^{k+1}$

49. $\displaystyle\sum_{k=0}^{4} \dfrac{(-1)^k x^{2k+1}}{2k+1}$

50. $\displaystyle\sum_{k=0}^{4} \dfrac{(-1)^k x^{2k}}{2k+2}$

Write each series in Problems 51–54 using summation notation with:

(A) The summing index k starting at k = 1

(B) The summing index j starting at j = 0

51. $2 + 3 + 4 + 5 + 6$

52. $1^2 + 2^2 + 3^2 + 4^2$

53. $1 - \frac{1}{2} + \frac{1}{3} - \frac{1}{4}$

54. $1 - \frac{1}{3} + \frac{1}{5} - \frac{1}{7} + \frac{1}{9}$

Write each series in Problems 55–58 using summation notation with the summing index k starting at k = 1.

55. $2 + \dfrac{3}{2} + \dfrac{4}{3} + \cdots + \dfrac{n+1}{n}$

56. $1 + \dfrac{1}{2^2} + \dfrac{1}{3^2} + \cdots + \dfrac{1}{n^2}$

57. $\dfrac{1}{2} - \dfrac{1}{4} + \dfrac{1}{8} - \cdots + \dfrac{(-1)^{n+1}}{2^n}$

58. $1 - 4 + 9 - \cdots + (-1)^{n+1}n^2$

C *In Problems 59–62, discuss the validity of each statement. If the statement is true, explain why. If not, give a counterexample.*

59. For each positive integer n, the sum of the series
$1 + \dfrac{1}{2} + \dfrac{1}{3} + \cdots + \dfrac{1}{n}$ is less than 4.

60. For each positive integer n, the sum of the series
$\dfrac{1}{2} + \dfrac{1}{4} + \dfrac{1}{8} + \cdots + \dfrac{1}{2^n}$ is less than 1.

61. For each positive integer n, the sum of the series
$\dfrac{1}{2} - \dfrac{1}{4} + \dfrac{1}{8} - \cdots + \dfrac{(-1)^{n+1}}{2^n}$ is greater than or equal
to $\dfrac{1}{4}$.

62. For each positive integer n, the sum of the series
$1 - \dfrac{1}{2} + \dfrac{1}{3} - \dfrac{1}{4} + \cdots + \dfrac{(-1)^{n+1}}{n}$ is greater than or
equal to $\dfrac{1}{2}$.

*Some sequences are defined by a **recursion formula**— that is, a formula that defines each term of the sequence in terms of one or more of the preceding terms. For example, if $\{a_n\}$ is defined by*

$$a_1 = 1 \quad and \quad a_n = 2a_{n-1} + 1 \quad for\ n \geq 2$$

then

$$a_2 = 2a_1 + 1 = 2 \cdot 1 + 1 = 3$$
$$a_3 = 2a_2 + 1 = 2 \cdot 3 + 1 = 7$$
$$a_4 = 2a_3 + 1 = 2 \cdot 7 + 1 = 15$$

and so on. In Problems 63–66, write the first five terms of each sequence.

63. $a_1 = 2$ and $a_n = 3a_{n-1} + 2$ for $n \geq 2$

64. $a_1 = 3$ and $a_n = 2a_{n-1} - 2$ for $n \geq 2$

65. $a_1 = 1$ and $a_n = 2a_{n-1}$ for $n \geq 2$

66. $a_1 = 1$ and $a_n = -\frac{1}{3}a_{n-1}$ for $n \geq 2$

If A is a positive real number, the terms of the sequence defined by

$$a_1 = \dfrac{A}{2} \quad and \quad a_n = \dfrac{1}{2}\left(a_{n-1} + \dfrac{A}{a_{n-1}}\right) \quad for\ n \geq 2$$

can be used to approximate $\sqrt{A}$ to any decimal place accuracy desired. In Problems 67 and 68, compute the first four terms of this sequence for the indicated value of A, and compare the fourth term with the value of $\sqrt{A}$ obtained from a calculator.

67. $A = 2$ **68.** $A = 6$

69. The sequence defined recursively by $a_1 = 1$, $a_2 = 1$, $a_n = a_{n-1} + a_{n-2}$ for $n \geq 3$ is called the *Fibonacci sequence*. Find the first ten terms of the Fibonacci sequence.

70. The sequence defined by $b_n = \dfrac{\sqrt{5}}{5}\left(\dfrac{1+\sqrt{5}}{2}\right)^n$ is related to the Fibonacci sequence. Find the first ten terms (to three decimal places) of the sequence $\{b_n\}$ and describe the relationship.

For most sequences it is difficult to sum an arbitrary number of terms of the sequence without adding term by term. But particular types of sequences—*arithmetic sequences* and *geometric sequences*—have certain properties that lead to convenient and useful formulas for the sums of the corresponding *arithmetic series* and *geometric series*.

❑ ARITHMETIC AND GEOMETRIC SEQUENCES

The sequence $5, 7, 9, 11, 13, \ldots, 5 + 2(n - 1), \ldots$, where each term after the first is obtained by adding 2 to the preceding term, is an example of an arithmetic sequence. The sequence $5, 10, 20, 40, 80, \ldots, 5(2)^{n-1}, \ldots$, where each term after the first is obtained by multiplying the preceding term by 2, is an example of a geometric sequence.

Arithmetic Sequence

A sequence of numbers

$$a_1, a_2, a_3, \ldots, a_n, \ldots$$

is called an **arithmetic sequence** if there is a constant d, called the **common difference,** such that

$$a_n - a_{n-1} = d$$

That is,

$$a_n = a_{n-1} + d \qquad \text{for every } n > 1$$

Geometric Sequence

A sequence of numbers

$$a_1, a_2, a_3, \ldots, a_n, \ldots$$

is called a **geometric sequence** if there exists a nonzero constant r, called a **common ratio,** such that

$$\frac{a_n}{a_{n-1}} = r$$

That is,

$$a_n = ra_{n-1} \qquad \text{for every } n > 1$$

Explore–Discuss 1

(A) Describe verbally all arithmetic sequences with common difference 2.

(B) Describe verbally all geometric sequences with common ratio 2.

Example 1 ➾ **Recognizing Arithmetic and Geometric Sequences**   Which of the following can be the first four terms of an arithmetic sequence? Of a geometric sequence?

(A) $1, 2, 3, 5, \ldots$ (B) $-1, 3, -9, 27, \ldots$

(C) $3, 3, 3, 3, \ldots$ (D) $10, 8.5, 7, 5.5, \ldots$

Solution   (A) Since $2 - 1 \neq 5 - 3$, there is no common difference, so the sequence is not an arithmetic sequence. Since $2/1 \neq 3/2$, there is no common ratio, so the sequence is not geometric either.

(B) The sequence is geometric with common ratio -3. It is not arithmetic.

(C) The sequence is arithmetic with common difference 0, and is also geometric with common ratio 1.

(D) The sequence is arithmetic with common difference -1.5. It is not geometric.

  ■

Matched Problem 1 ➾ Which of the following can be the first four terms of an arithmetic sequence? Of a geometric sequence?

(A) $8, 2, 0.5, 0.125, \ldots$ (B) $-7, -2, 3, 8, \ldots$

(C) $1, 5, 25, 100, \ldots$

  ■

❑ nTH-TERM FORMULAS

If $\{a_n\}$ is an arithmetic sequence with common difference d, then

$$a_2 = a_1 + d$$
$$a_3 = a_2 + d = a_1 + 2d$$
$$a_4 = a_3 + d = a_1 + 3d$$

This suggests that:

nth Term of an Arithmetic Sequence

$$a_n = a_1 + (n - 1)d \qquad \text{for all } n > 1 \tag{1}$$

Similarly, if $\{a_n\}$ is a geometric sequence with common ratio r, then

$$a_2 = a_1 r$$
$$a_3 = a_2 r = a_1 r^2$$
$$a_4 = a_3 r = a_1 r^3$$

This suggests that:

nth Term of a Geometric Sequence

$$a_n = a_1 r^{n-1} \qquad \text{for all } n > 1 \tag{2}$$

Example 2 ➷ **Finding Terms in Arithmetic and Geometric Sequences**

(A) If the 1st and 10th terms of an arithmetic sequence are 3 and 30, respectively, find the 40th term of the sequence.

(B) If the 1st and 10th terms of a geometric sequence are 3 and 30, find the 40th term to three decimal places.

Solution (A) First use formula (1) with $a_1 = 3$ and $a_{10} = 30$ to find d:

$$a_n = a_1 + (n - 1)d$$
$$a_{10} = a_1 + (10 - 1)d$$
$$30 = 3 + 9d$$
$$d = 3$$

Now find a_{40}:

$$a_{40} = 3 + 39 \cdot 3 = 120$$

(B) First use formula (2) with $a_1 = 3$ and $a_{10} = 30$ to find r:

$$a_n = a_1 r^{n-1}$$
$$a_{10} = a_1 r^{10-1}$$
$$30 = 3r^9$$
$$r^9 = 10$$
$$r = 10^{1/9}$$

Now find a_{40}:

$$a_{40} = 3(10^{1/9})^{39} = 3(10^{39/9}) = 64{,}633.041$$
■

Matched Problem 2 ➷ (A) If the 1st and 15th terms of an arithmetic sequence are -5 and 23, respectively, find the 73rd term of the sequence.

(B) Find the 8th term of the geometric sequence

$$\frac{1}{64}, \frac{-1}{32}, \frac{1}{16}, \ldots$$
■

❑ **SUM FORMULAS FOR FINITE ARITHMETIC SERIES**

If $a_1, a_2, a_3, \ldots, a_n$ is a finite arithmetic sequence, then the corresponding series $a_1 + a_2 + a_3 + \cdots + a_n$ is called a *finite arithmetic series*. We will derive two simple and very useful formulas for the sum of a finite arithmetic series. Let d be the common difference of the arithmetic sequence $a_1, a_2, a_3, \ldots, a_n$ and let S_n denote the sum of the series $a_1 + a_2 + a_3 + \cdots + a_n$. Then

$$S_n = a_1 + (a_1 + d) + \cdots + [a_1 + (n - 2)d] + [a_1 + (n - 1)d]$$

Reversing the order of the sum, we obtain

$$S_n = [a_1 + (n - 1)d] + [a_1 + (n - 2)d] + \cdots + (a_1 + d) + a_1$$

Something interesting happens if we combine these last two equations by addition (adding corresponding terms on the right sides):

$$2S_n = [2a_1 + (n - 1)d] + [2a_1 + (n - 1)d] + \cdots + [2a_1 + (n - 1)d] + [2a_1 + (n - 1)d]$$

All the terms on the right side are the same, and there are n of them. Thus,

$$2S_n = n[2a_1 + (n - 1)d]$$

and we have the following general formula:

> **Sum of a Finite Arithmetic Series: First Form**
>
> $$S_n = \frac{n}{2}[2a_1 + (n-1)d] \tag{3}$$

Replacing

$$[a_1 + (n-1)d] \quad \text{in} \quad \frac{n}{2}[a_1 + a_1 + (n-1)d]$$

by a_n from equation (1), we obtain a second useful formula for the sum:

> **Sum of a Finite Arithmetic Series: Second Form**
>
> $$S_n = \frac{n}{2}(a_1 + a_n) \tag{4}$$

Example 3   **Finding a Sum**   Find the sum of the first 30 terms in the arithmetic sequence:
3, 8, 13, 18, ...

Solution   Use formula (3) with $n = 30$, $a_1 = 3$, and $d = 5$:

$$S_{30} = \frac{30}{2}[2 \cdot 3 + (30-1)5] = 2{,}265$$

Matched Problem 3   Find the sum of the first 40 terms in the arithmetic sequence:
15, 13, 11, 9, ...

Example 4   **Finding a Sum**   Find the sum of all the even numbers between 31 and 87.

Solution   First, find n using equation (1):

$$a_n = a_1 + (n-1)d$$
$$86 = 32 + (n-1)2$$
$$n = 28$$

Now find S_{28} using formula (4):

$$S_n = \frac{n}{2}(a_1 + a_n)$$

$$S_{28} = \frac{28}{2}(32 + 86) = 1{,}652$$

Matched Problem 4   Find the sum of all the odd numbers between 24 and 208.

❑ SUM FORMULAS FOR FINITE GEOMETRIC SERIES

If $a_1, a_2, a_3, \ldots, a_n$ is a finite geometric sequence, then the corresponding series $a_1 + a_2 + a_3 + \cdots + a_n$ is called a *finite geometric series*. As with arithmetic series, we can derive two simple and very useful formulas for the sum of a finite geometric series. Let r be the common ratio of the geometric sequence $a_1, a_2, a_3, \ldots, a_n$ and let S_n denote the sum of the series $a_1 + a_2 + a_3 + \cdots + a_n$. Then

$$S_n = a_1 + a_1 r + a_1 r^2 + \cdots + a_1 r^{n-2} + a_1 r^{n-1}$$

If we multiply both sides by r, we obtain

$$r S_n = a_1 r + a_1 r^2 + a_1 r^3 + \cdots + a_1 r^{n-1} + a_1 r^n$$

Now combine these last two equations by subtraction to obtain

$$r S_n - S_n = (a_1 r + a_1 r^2 + a_1 r^3 + \cdots + a_1 r^{n-1} + a_1 r^n) - (a_1 + a_1 r + a_1 r^2 + \cdots + a_1 r^{n-2} + a_1 r^{n-1})$$
$$(r - 1) S_n = a_1 r^n - a_1$$

Notice how many terms drop out on the right side. Solving for S_n, we have:

Sum of a Finite Geometric Series: First Form

$$S_n = \frac{a_1(r^n - 1)}{r - 1} \qquad r \neq 1 \tag{5}$$

Since $a_n = a_1 r^{n-1}$, or $r a_n = a_1 r^n$, formula (5) also can be written in the form:

Sum of a Finite Geometric Series: Second Form

$$S_n = \frac{r a_n - a_1}{r - 1} \qquad r \neq 1 \tag{6}$$

Example 5 ✎ **Finding a Sum** Find the sum of the first ten terms of the geometric sequence:

$$1, 1.05, 1.05^2, \ldots$$

SOLUTION Use formula (5) with $a_1 = 1$, $r = 1.05$, and $n = 10$:

$$S_n = \frac{a_1(r^n - 1)}{r - 1}$$

$$S_{10} = \frac{1(1.05^{10} - 1)}{1.05 - 1}$$

$$\approx \frac{0.6289}{0.05} \approx 12.58$$

∎

Matched Problem 5 ✎ Find the sum of the first eight terms of the geometric sequence:

$$100, 100(1.08), 100(1.08)^2, \ldots$$

∎

❑ SUM FORMULA FOR INFINITE GEOMETRIC SERIES

Explore–Discuss 2 (A) For any n, the sum of a finite geometric series with $a_1 = 5$ and $r = \frac{1}{2}$ is given by [see formula (5)]

$$S_n = \frac{5[(\frac{1}{2})^n - 1]}{\frac{1}{2} - 1} = 10 - 10\left(\frac{1}{2}\right)^n$$

Discuss the behavior of S_n as n increases.

(B) Repeat part (A) if $a_1 = 5$ and $r = 2$.

Given a geometric series, what happens to the sum S_n of the first n terms as n increases without stopping? To answer this question, let us write formula (5) in the form

$$S_n = \frac{a_1 r^n}{r - 1} - \frac{a_1}{r - 1}$$

It is possible to show that if $-1 < r < 1$, then r^n will approach 0 as n increases. Thus, the first term above will approach 0 and S_n can be made as close as we please to the second term, $-a_1/(r - 1)$ [which can be written as $a_1/(1 - r)$], by taking n sufficiently large. Thus, if the common ratio r is between -1 and 1, we define the sum of an infinite geometric series to be:

Sum of an Infinite Geometric Series

$$S_\infty = \frac{a_1}{1 - r} \qquad -1 < r < 1 \tag{7}$$

If $r \leqslant -1$ or $r \geqslant 1$, then an infinite geometric series has no sum.

   ❑   APPLICATIONS

Example 6 ⇔ **Loan Repayment**   A person borrows \$3,600 and agrees to repay the loan in monthly installments over a period of 3 years. The agreement is to pay 1% of the unpaid balance each month for using the money and \$100 each month to reduce the loan. What is the total cost of the loan over the 3 years?

SOLUTION   Let us look at the problem relative to a time line:

The total cost of the loan is

$$1 + 2 + \cdots + 34 + 35 + 36$$

The terms form a finite arithmetic series with $n = 36$, $a_1 = 1$, and $a_{36} = 36$, so we can use formula (4):

$$S_n = \frac{n}{2}(a_1 + a_n)$$

$$S_{36} = \frac{36}{2}(1 + 36) = \$666$$

We conclude that the total cost of the loan over the period of 3 years is \$666. ▪

Matched Problem 6 ⇔ Repeat Example 6 with a loan of \$6,000 over a period of 5 years. ▪

Example 7 **Economy Stimulation** The government has decided on a tax rebate program to stimulate the economy. Suppose that you receive $600 and you spend 80% of this, and each of the people who receive what you spend also spend 80% of what they receive, and this process continues without end. According to the **multiplier principle** in economics, the effect of your $600 tax rebate on the economy is multiplied many times. What is the total amount spent if the process continues as indicated?

SOLUTION We need to find the sum of an infinite geometric series with the first amount spent being $a_1 = (0.8)(\$600) = \$480$ and $r = 0.8$. Using formula (7), we obtain

$$S_\infty = \frac{a_1}{1-r}$$

$$= \frac{\$480}{1-0.8} = \$2,400$$

Thus, assuming the process continues as indicated, we would expect the $600 tax rebate to result in about $2,400 of spending.

Matched Problem 7 Repeat Example 7 with a tax rebate of $1,000.

Answers to Matched Problems
1. (A) The sequence is geometric with $r = \frac{1}{4}$. It is not arithmetic.
   (B) The sequence is arithmetic with $d = 5$. It is not geometric.
   (C) The sequence is neither arithmetic nor geometric.
2. (A) 139   (B) -2 3. -960 4. 10,672
5. 1,063.66   6. $1,830   7. $4,000

Exercise B-2

A *In Problems 1 and 2, determine whether the indicated sequence can be the first three terms of an arithmetic or geometric sequence, and, if so, find the common difference or common ratio and the next two terms of the sequence.*

1. (A) $-11, -16, -21, \ldots$ (B) $2, -4, 8, \ldots$
   (C) $1, 4, 9, \ldots$ (D) $\frac{1}{2}, \frac{1}{6}, \frac{1}{18}, \ldots$

2. (A) $5, 20, 100, \ldots$ (B) $-5, -5, -5, \ldots$
   (C) $7, 6.5, 6, \ldots$ (D) $512, 256, 128, \ldots$

B

In Problem 3–8, determine whether the finite series is arithmetic, geometric, both, or neither. If the series is arithmetic or geometric, find its sum.

3. $\displaystyle\sum_{k=1}^{101} (-1)^{k+1}$

4. $\displaystyle\sum_{k=1}^{200} 3$

5. $1 + \dfrac{1}{2} + \dfrac{1}{3} + \cdots + \dfrac{1}{50}$

6. $3 - 9 + 27 - \cdots - 3^{20}$

7. $5 + 4.9 + 4.8 + \cdots + 0.1$

8. $1 - \dfrac{1}{4} + \dfrac{1}{9} - \cdots - \dfrac{1}{100^2}$

Let $a_1, a_2, a_3, \ldots, a_n, \ldots$ be an arithmetic sequence. In Problems 9–14, find the indicated quantities.

9. $a_1 = 7; d = 4; a_2 = ?; a_3 = ?$
10. $a_1 = -2; d = -3; a_2 = ?; a_3 = ?$
11. $a_1 = 2; d = 4; a_{21} = ?; S_{31} = ?$
12. $a_1 = 8; d = -10; a_{15} = ?; S_{23} = ?$
13. $a_1 = 18; a_{20} = 75; S_{20} = ?$
14. $a_1 = 203; a_{30} = 261; S_{30} = ?$

Let $a_1, a_2, a_3, \ldots, a_n, \ldots$ be a geometric sequence. In Problems 15–24, find the indicated quantities.

15. $a_1 = 3; r = -2; a_2 = ?; a_3 = ?; a_4 = ?$
16. $a_1 = 32; r = -\frac{1}{2}; a_2 = ?; a_3 = ?; a_4 = ?$
17. $a_1 = 1; a_7 = 729; r = -3; S_7 = ?$

18. $a_1 = 3$; $a_7 = 2,187$; $r = 3$; $S_7 = $?

19. $a_1 = 100$; $r = 1.08$; $a_{10} = $?

20. $a_1 = 240$; $r = 1.06$; $a_{12} = $?

21. $a_1 = 100$; $a_9 = 200$; $r = $?

22. $a_1 = 100$; $a_{10} = 300$; $r = $?

23. $a_1 = 500$; $r = 0.6$; $S_{10} = $?; $S_\infty = $?

24. $a_1 = 8,000$; $r = 0.4$; $S_{10} = $?; $S_\infty = $?

25. $S_{41} = \sum\limits_{k=1}^{41} (3k + 3) = $?

26. $S_{50} = \sum\limits_{k=1}^{50} (2k - 3) = $?

27. $S_8 = \sum\limits_{k=1}^{8} (-2)^{k-1} = $?

28. $S_8 = \sum\limits_{k=1}^{8} 2^k = $?

29. Find the sum of all the odd integers between 12 and 68.

30. Find the sum of all the even integers between 23 and 97.

31. Find the sum of each infinite geometric sequence (if it exists).

(A) $2, 4, 8, \ldots$ (B) $2, -\frac{1}{2}, \frac{1}{8}, \ldots$

32. Repeat Problem 31 for:

(A) $16, 4, 1, \ldots$ (B) $1, -3, 9, \ldots$

C

33. Find $f(1) + f(2) + f(3) + \cdots + f(50)$ if $f(x) = 2x - 3$.

34. Find $g(1) + g(2) + g(3) + \cdots + g(100)$ if $g(t) = 18 - 3t$.

35. Find $f(1) + f(2) + \cdots + f(10)$ if $f(x) = \left(\frac{1}{2}\right)^x$.

36. Find $g(1) + g(2) + \cdots + g(10)$ if $g(x) = 2^x$.

37. Show that the sum of the first n odd positive integers is n^2, using appropriate formulas from this section.

38. Show that the sum of the first n even positive integers is $n + n^2$, using formulas in this section.

39. If $r = 1$, neither the first form nor the second form for the sum of a finite geometric series is valid. Find a formula for the sum of a finite geometric series if $r = 1$.

40. If all of the terms of an infinite geometric series are less than 1, could the sum be greater tham 1000? Explain.

41. Does there exist a finite arithmetic series with $a_1 = 1$ and $a_n = 1.1$ that has sum equal to 100? Explain.

42. Does there exist a finite arithmetic series with $a_1 = 1$ and $a_n = 1.1$ that has sum equal to 105? Explain.

43. Does there exist an infinite geometric series with $a_1 = 10$ that has sum equal to 6? Explain.

44. Does there exist an infinite geometric series with $a_1 = 10$ that has sum equal to 5? Explain.

Applications

Business & Economics

45. *Loan repayment.* If you borrow $4,800 and repay the loan by paying $200 per month to reduce the loan and 1% of the unpaid balance each month for the use of the money, what is the total cost of the loan over 24 months?

46. *Loan repayment.* If you borrow $5,400 and repay the loan by paying $300 per month to reduce the loan and 1.5% of the unpaid balance each month for the use of the money, what is the total cost of the loan over 18 months?

47. *Economy stimulation.* The government, through a subsidy program, distributes $5,000,000. If we assume that each person or agency spends 70% of what is received, and 70% of this is spent, and so on, how much total increase in spending results from this government action? (Let $a_1 = $ $3,500,000$.)

48. *Economy stimulation.* Due to reduced taxes, a person has an extra $1,200 in spendable income. If we as-sume that the person spends 65% of this on consumer goods, and the producers of these goods in turn spend 65% on consumer goods, and that this process continues indefinitely, what is the total amount spent (to the nearest dollar) on consumer goods?

49. *Compound interest.* If $1,000 is invested at 5% compounded annually, the amount A present after n years forms a geometric sequence with common ratio $1 + 0.05 = 1.05$. Use a geometric sequence formula to find the amount A in the account (to the nearest cent) after 10 years. After 20 years. [*Hint:* Use a time line.]

50. *Compound interest.* If P is invested at $100r$% compounded annually, the amount A present after n years forms a geometric sequence with common ratio $1 + r$. Write a formula for the amount present after n years. [*Hint:* Use a time line.]

Section B-3 # Binomial Theorem

❑ FACTORIAL
❑ DEVELOPMENT OF THE BINOMIAL THEOREM

The binomial form

$$(a + b)^n$$

where n is a natural number, appears more frequently than you might expect. The coefficients in the expansion play an important role in probability studies. The *binomial formula*, which we will derive informally, enables us to expand $(a + b)^n$ directly for n any natural number. Since the formula involves *factorials*, we digress for a moment here to introduce this important concept.

❑ FACTORIAL

For n a natural number, **n factorial**, denoted by **$n!$**, is the product of the first n natural numbers. **Zero factorial** is defined to be 1. That is:

$$n! = n \cdot (n - 1) \cdot \cdots \cdot 2 \cdot 1$$
$$1! = 1$$
$$0! = 1$$

It is also useful to note that:

$$n! = n \cdot (n - 1)! \qquad n \geq 1$$

Example 1 ⇔ **Factorial Forms** Evaluate each:

(A) $5! = 5 \cdot 4 \cdot 3 \cdot 2 \cdot 1 = 120$ (B) $\dfrac{8!}{7!} = \dfrac{8 \cdot \cancel{7!}}{\cancel{7!}} = 8$

(C) $\dfrac{10!}{7!} = \dfrac{10 \cdot 9 \cdot 8 \cdot \cancel{7!}}{\cancel{7!}} = 720$

Matched Problem 1 ⇔ Evaluate each:

(A) $4!$ (B) $\dfrac{7!}{6!}$ (C) $\dfrac{8!}{5!}$

The following important formula involving factorials has applications in many areas of mathematics and statistics. We will use this formula to provide a more concise form for the expressions encountered later in this discussion.

For n and r integers satisfying $0 \leq r \leq n$,

$$C_{n,r} = \frac{n!}{r!(n - r)!}$$

Example 2 ➩ **Evaluating $C_{n,r}$**

(A) $C_{9,2} = \dfrac{9!}{2!(9-2)!} = \dfrac{9!}{2!7!} = \dfrac{9\cdot 8\cdot 7!}{2\cdot 7!} = 36$

(B) $C_{5,5} = \dfrac{5!}{5!(5-5)!} = \dfrac{5!}{5!0!} = \dfrac{5!}{5!} = 1$

Matched Problem 2 ➩ Find:

(A) $C_{5,2}$ (B) $C_{6,0}$

❑ **DEVELOPMENT OF THE BINOMIAL THEOREM**

Let us expand $(a+b)^n$ for several values of n to see if we can observe a pattern that leads to a general formula for the expansion for any natural number n:

$(a+b)^1 = a + b$
$(a+b)^2 = a^2 + 2ab + b^2$
$(a+b)^3 = a^3 + 3a^2b + 3ab^2 + b^3$
$(a+b)^4 = a^4 + 4a^3b + 6a^2b^2 + 4ab^3 + b^4$
$(a+b)^5 = a^5 + 5a^4b + 10a^3b^2 + 10a^2b^3 + 5ab^4 + b^5$

OBSERVATIONS

1. The expansion of $(a+b)^n$ has $(n+1)$ terms.
2. The power of a decreases by 1 for each term as we move from left to right.
3. The power of b increases by 1 for each term as we move from left to right.
4. In each term, the sum of the powers of a and b always equals n.
5. Starting with a given term, we can get the coefficient of the next term by multiplying the coefficient of the given term by the exponent of a and dividing by the number that represents the position of the term in the series of terms. For example, in the expansion of $(a+b)^4$ above, the coefficient of the third term is found from the second term by multiplying 4 and 3, and then dividing by 2 [that is, the coefficient of the third term $= (4\cdot 3)/2 = 6$].

We now postulate these same properties for the general case:

$(a+b)^n = a^n + \dfrac{n}{1}a^{n-1}b + \dfrac{n(n-1)}{1\cdot 2}a^{n-2}b^2 + \dfrac{n(n-1)(n-2)}{1\cdot 2\cdot 3}a^{n-3}b^3 + \cdots + b^n$

$= \dfrac{n!}{0!(n-0)!}a^n + \dfrac{n!}{1!(n-1)!}a^{n-1}b + \dfrac{n!}{2!(n-2)!}a^{n-2}b^2 + \dfrac{n!}{3!(n-3)!}a^{n-3}b^3 + \cdots + \dfrac{n!}{n!(n-n)!}b^n$

$= C_{n,0}a^n + C_{n,1}a^{n-1}b + C_{n,2}a^{n-2}b^2 + C_{n,3}a^{n-3}b^3 + \cdots + C_{n,n}b^n$

And we are led to the formula in the binomial theorem (a formal proof requires mathematical induction, which is beyond the scope of this book):

Binomial Theorem

For all natural numbers n,

$(a+b)^n = C_{n,0}a^n + C_{n,1}a^{n-1}b + C_{n,2}a^{n-2}b^2 + C_{n,3}a^{n-3}b^3 + \cdots + C_{n,n}b^n$

Example 3 ➾ **Using the Binomial Theorem** Use the binomial formula to expand $(u + v)^6$.

SOLUTION

$$(u + v)^6 = C_{6,0}u^6 + C_{6,1}u^5v + C_{6,2}u^4v^2 + C_{6,3}u^3v^3 + C_{6,4}u^2v^4 + C_{6,5}uv^5 + C_{6,6}v^6$$
$$= u^6 + 6u^5v + 15u^4v^2 + 20u^3v^3 + 15u^2v^4 + 6uv^5 + v^6$$

■

Matched Problem 3 ➾ Use the binomial formula to expand $(x + 2)^5$.

■

Example 4 ➾ **Using the Binomial Theorem** Use the binomial formula to find the sixth term in the expansion of $(x - 1)^{18}$.

SOLUTION Sixth term $= C_{18,5}x^{13}(-1)^5 = \dfrac{18!}{5!(18 - 5)!}x^{13}(-1)$

$$= -8{,}568x^{13}$$

■

Matched Problem 4 ➾ Use the binomial formula to find the fourth term in the expansion of $(x - 2)^{20}$.

■

Explore–Discuss 1

(A) Use the formula for $C_{n,r}$ to find

$$C_{6,0} + C_{6,1} + C_{6,2} + C_{6,3} + C_{6,4} + C_{6,5} + C_{6,6}$$

(B) Write the binomial theorem for $n = 6$, $a = 1$, and $b = 1$, and compare with the results of part (A).

(C) For any natural number n, find

$$C_{n,0} + C_{n,1} + C_{n,2} + \cdots + C_{n,n}$$

Answers to Matched Problems **1.** (A) 24 (B) 7 (C) 336 **2.** (A) 10 (B) 1
3. $x^5 + 10x^4 + 40x^3 + 80x^2 + 80x + 32$
4. $-9{,}120x^{17}$

Exercise B-3

A *In Problems 1–20, evaluate each expression.*

1. $6!$ **2.** $7!$ **3.** $\dfrac{10!}{9!}$

4. $\dfrac{20!}{19!}$ **5.** $\dfrac{12!}{9!}$ **6.** $\dfrac{10!}{6!}$

7. $\dfrac{5!}{2!3!}$ **8.** $\dfrac{7!}{3!4!}$ **9.** $\dfrac{6!}{5!(6 - 5)!}$

10. $\dfrac{7!}{4!(7 - 4)!}$ **11.** $\dfrac{20!}{3!17!}$ **12.** $\dfrac{52!}{50!2!}$

B

13. $C_{5,3}$ **14.** $C_{7,3}$ **15.** $C_{6,5}$ **16.** $C_{7,4}$

17. $C_{5,0}$ **18.** $C_{5,5}$ **19.** $C_{18,15}$ **20.** $C_{18,3}$

Expand each expression in Problems 21–26 using the binomial formula.

21. $(a + b)^4$ **22.** $(m + n)^5$ **23.** $(x - 1)^6$

24. $(u - 2)^5$ **25.** $(2a - b)^5$ **26.** $(x - 2y)^5$

Find the indicated term in each expansion in Problems 27–32.

27. $(x - 1)^{18}$; 5th term   **28.** $(x - 3)^{20}$; 3rd term

29. $(p + q)^{15}$; 7th term   **30.** $(p + q)^{15}$; 13th term

31. $(2x + y)^{12}$; 11th term   **32.** $(2x + y)^{12}$; 3rd term

C

33. Show that $C_{n,0} = C_{n,n}$ for $n \geqslant 0$.

34. Show that $C_{n,r} = C_{n,n-r}$ for $n \geqslant r \geqslant 0$.

35. The triangle below is called **Pascal's triangle.** Can you guess what the next two rows at the bottom are? Compare these numbers with the coefficients of binomial expansions.

```
            1
          1   1
        1   2   1
      1   3   3   1
    1   4   6   4   1
```

36. Explain why the sum of the entries in each row of Pascal's triangle is a power of 2. [*Hint*: Let $a = b = 1$ in the binomial theorem.]

37. Explain why the alternating sum of the entries (e.g., $1 - 4 + 6 - 4 + 1$) in each row of Pascal's triangle is equal to 0.

38. Show that $C_{n,r} = \frac{n - r + 1}{r} C_{n,r-1}$ for $n \geqslant r \geqslant 1$.

39. Show that $C_{n,r-1} + C_{n,r} = C_{n+1,r}$ for $n \geqslant r \geqslant 1$.

Section B-4

Increments and Differentials

❑ INCREMENTS
❑ DIFFERENTIALS
❑ APPROXIMATIONS USING DIFFERENTIALS

Differential notation is used in Chapters 6 and 7 as a bookkeeping device to aid in the process of finding indefinite integrals. Now we discuss the definition of *differentials* in greater depth, introduce the concept of *increments,* and show how differentials can be used to approximate increments.

❑ INCREMENTS

Given $y = f(x)$, we are sometimes interested in how y is affected by a small change in x. For example, if y represents profit and x represents price per item, we might be interested in the change in profit resulting from a small change in price.

Given $y = f(x) = x^3$, if x changes from 2 to 2.1, then y will change from $y = f(2) = 2^3 = 8$ to $y = f(2.1) = 2.1^3 = 9.261$. Mathematically, the change in x and the corresponding change in y, called *increments in x and y,* respectively, are denoted by Δx and Δy (read "delta x" and "delta y," since Δ is the Greek letter delta). In terms of the given example, we write

$$\Delta x = 2.1 - 2 = 0.1 \qquad \text{Change in x}$$
$$\Delta y = f(2.1) - f(2)$$
$$= 2.1^3 - 2^3 = 9.261 - 8 = 1.261 \qquad \text{Corresponding change in y}$$

It is important to remember that Δx and Δy do not represent products—each is a variable with replacements from the real numbers. The general definitions are given in the following box:

Increments

For $y = f(x)$,

$$\Delta x = x_2 - x_1 \qquad \Delta y = y_2 - y_1$$
$$x_2 = x_1 + \Delta x \qquad\quad = f(x_2) - f(x_1)$$
$$= f(x_1 + \Delta x) - f(x_1)$$

Δy represents the change in y corresponding to a Δx change in x.

Δx can be either positive or negative.

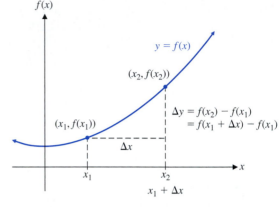

[*Note:* Δy depends on the function f, the input x, and the increment Δx.]

Example 1 ➩ **Increments**   Given the function $y = f(x) = \dfrac{x^2}{2}$,

(A) Find Δx, Δy, and $\Delta y / \Delta x$ for $x_1 = 1$ and $x_2 = 2$.

(B) Find $\dfrac{f(x_1 + \Delta x) - f(x_1)}{\Delta x}$ for $x_1 = 1$ and $\Delta x = 2$.

Solution   (A) $\Delta x = x_2 - x_1 = 2 - 1 = 1$

$\Delta y = f(x_2) - f(x_1)$

$= f(2) - f(1) = \dfrac{4}{2} - \dfrac{1}{2} = \dfrac{3}{2}$

$\dfrac{\Delta y}{\Delta x} = \dfrac{f(x_2) - f(x_1)}{x_2 - x_1} = \dfrac{\frac{3}{2}}{1} = \dfrac{3}{2}$

(B) $\dfrac{f(x_1 + \Delta x) - f(x_1)}{\Delta x} = \dfrac{f(1 + 2) - f(1)}{2}$

$= \dfrac{f(3) - f(1)}{2} = \dfrac{\frac{9}{2} - \frac{1}{2}}{2} = \dfrac{4}{2} = 2$

Matched Problem 1 ➩ Given the function $y = f(x) = x^2 + 1$,

(A) Find Δx, Δy, and $\Delta y / \Delta x$ for $x_1 = 2$ and $x_2 = 3$.

(B) Find $\dfrac{f(x_1 + \Delta x) - f(x_1)}{\Delta x}$ for $x_1 = 1$ and $\Delta x = 2$.

In Example 1, we observe another notation for the familiar difference quotient

$$\frac{f(x + h) - f(x)}{h} \tag{1}$$

It is common to refer to h, the change in x, as Δx. Then, the difference quotient (1) takes on the form

$$\frac{f(x + \Delta x) - f(x)}{\Delta x} \qquad \text{or} \qquad \frac{\Delta y}{\Delta x} \qquad \Delta y = f(x + \Delta x) - f(x)$$

and the derivative is defined by

$$f'(x) = \lim_{\Delta x \to 0} \frac{f(x + \Delta x) - f(x)}{\Delta x}$$

or

$$f'(x) = \lim_{\Delta x \to 0} \frac{\Delta y}{\Delta x} \tag{2}$$

if the limit exists.

Explore–Discuss 1

Suppose that $y = f(x)$ defines a function whose domain is the set of all real numbers. If every increment Δy is equal to 0, what is the range of f?

❏ DIFFERENTIALS

Assume that the limit in equation (2) exists. Then, for small Δx, the difference quotient $\Delta y / \Delta x$ provides a good approximation for $f'(x)$. Also, $f'(x)$ provides a good approximation for $\Delta y / \Delta x$. We indicate the latter by writing

$$\frac{\Delta y}{\Delta x} \approx f'(x) \qquad \Delta x \text{ is small but} \neq 0 \tag{3}$$

Multiplying both sides of (3) by Δx gives us

$$\Delta y \approx f'(x) \, \Delta x \qquad \Delta x \text{ is small but} \neq 0 \tag{4}$$

From equation (4) we see that $f'(x) \, \Delta x$ provides a good approximation for Δy when Δx is small.

Because of the practical and theoretical importance of $f'(x) \, \Delta x$, we give it the special name **differential** and represent it by special symbols, **dy** or **df**:

$$\boldsymbol{dy = f'(x) \, \Delta x} \qquad \text{or} \qquad \boldsymbol{df = f'(x) \, \Delta x}$$

For example,

$$d(2x^3) = (2x^3)' \, \Delta x = 6x^2 \, \Delta x$$

$$d(x) = (x)' \, \Delta x = 1 \, \Delta x = \Delta x$$

In the second example, we usually drop the parentheses in $d(x)$ and simply write

$$\boldsymbol{dx = \Delta x}$$

In summary, we have the following:

> ### *Differentials*
>
> If $y = f(x)$ defines a differential function, then the **differential dy, or df,** is defined as the product of $f'(x)$ and dx, where $dx = \Delta x$. Symbolically,
>
> $$dy = f'(x)\, dx \qquad \text{or} \qquad df = f'(x)\, dx$$
>
> where
>
> $$dx = \Delta x$$
>
> [*Note*: The differential dy (or df) is actually a function involving two independent variables, x and dx; a change in either one or both will affect dy (or df).]

Example 2 ↝ **Differentials**   Find dy for $f(x) = x^2 + 3x$. Evaluate dy for

(A) $x = 2$ and $dx = 0.1$ (B) $x = 3$ and $dx = 0.1$

(C) $x = 1$ and $dx = 0.02$

Solution
$$
\begin{aligned}
dy &= f'(x)\, dx \\
&= (2x + 3)\, dx
\end{aligned}
$$

(A) When $x = 2$ and $dx = 0.1$,   (B) When $x = 3$ and $dx = 0.1$,

$$dy = [2(2) + 3]0.1 = 0.7 \qquad\qquad dy = [2(3) + 3]0.1 = 0.9$$

(C) When $x = 1$ and $dx = 0.02$,

$$dy = [2(1) + 3]0.02 = 0.1$$

∎

Matched Problem 2 ↝ Find dy for $f(x) = \sqrt{x} + 3$. Evaluate dy for

(A) $x = 4$ and $dx = 0.1$
(B) $x = 9$ and $dx = 0.12$
(C) $x = 1$ and $dx = 0.01$

∎

We now have two interpretations of the symbol dy/dx. Referring to the function $y = f(x) = x^2 + 3x$ in Example 2 with $x = 2$ and $dx = 0.1$, we have

$$\frac{dy}{dx} = f'(2) = 7 \qquad \text{Derivative}$$

and

$$\frac{dy}{dx} = \frac{0.7}{0.1} = 7 \qquad \text{Ratio of differentials}$$

❑ APPROXIMATIONS USING DIFFERENTIALS

Earlier, we noted that for small Δx,

$$\frac{\Delta y}{\Delta x} \approx f'(x) \qquad \text{and} \qquad \Delta y \approx f'(x)\, \Delta x$$

Also, since

$$dy = f'(x)\, dx$$

it follows that

$$\Delta y \approx dy$$

and dy can be used to approximate Δy.

To interpret this result geometrically, we need to recall a basic property of slope. The vertical change in a line is equal to the product of the slope and the horizontal change, as shown in Figure 1.

Now consider the line tangent to the graph of $y = f(x)$, as shown in Figure 2. Since $f'(x)$ is the slope of the tangent line and dx is the horizontal change in the tangent line, it follows that the vertical change in the tangent line is given by $dy = f'(x)\, dx$, as indicated in Figure 2.

Slope = m

$m = \dfrac{b}{a}$ or $b = ma$

FIGURE 1

FIGURE 2

Example 3   **Graphs and Tables of Increments and Differentials**

(A) Graph the functions Δy and dy (of the independent variable $\Delta x = dx$) for $f(x) = 6x - x^2$ when $x = 2$.

(B) Find Δy and dy for $f(x) = 6x - x^2$ when $x = 2$ and $\Delta x = dx = 0.1, 0.2$, and 0.3.

SOLUTION (A) $\Delta y = f(2 + \Delta x) - f(2) = 6(2 + \Delta x) - (2 + \Delta x)^2 - 8$

Since $f'(x) = 6 - 2x$, then $dy = (6 - 2 \cdot 2)\, dx = 2\, dx$. A graphing utility can be used to graph each function, as shown in Figure 3. (Note that the independent variable $\Delta x = dx$ has been represented by X, the independent variable of the graphing utility.) From Figure 3 we can see that the differential provides a linear approximation of the increment, and that the approximation is better for values of $\Delta x = dx$ close to 0.

FIGURE 3

(B) The values can be computed from the graphs, as illustrated by Figure 3 ($\Delta y = 0.19$ when $\Delta x = 0.1$), or they may be displayed in table form, as shown in Figure 4. For $\Delta x = dx = 0.1, 0.2$, and 0.3, the corresponding values of the increment are $\Delta y = 0.19, 0.36$, and 0.51, and the corresponding values of the differential are $dy = 0.2, 0.4$, and 0.6.

X	Y₁	Y₂
.1	.19	.2
.2	.36	.4
.3	.51	.6

Y₁=.19

FIGURE 4

Matched Problem 3 ⇨ Repeat Example 3 for $x = 4$ and $\Delta x = dx = -0.1, -0.2$, and -0.3.

Explore–Discuss 2   Suppose that $y = f(x)$ defines a differentiable function. Explain why the graph of the function dy (of the independent variable $\Delta x = dx$) is a straight line through the origin for a fixed value of x.

Example 4 ⇨   **Cost–Revenue**   A company manufactures and sells x transistor radios per week. If the weekly cost and revenue equations are

$$C(x) = 5{,}000 + 2x \qquad R(x) = 10x - \frac{x^2}{1{,}000} \qquad 0 \le x \le 8{,}000$$

find the approximate changes in revenue and profit if production is increased from 2,000 to 2,010 units per week.

SOLUTION   We will approximate ΔR and ΔP with dR and dP, respectively, using $x = 2{,}000$ and $dx = 2{,}010 - 2{,}000 = 10$.

$$R(x) = 10x - \frac{x^2}{1{,}000} \qquad\qquad P(x) = R(x) - C(x) = 10x - \frac{x^2}{1{,}000} - 5{,}000 - 2x$$

$$dR = R'(x)\,dx \qquad\qquad\qquad\qquad = 8x - \frac{x^2}{1{,}000} - 5{,}000$$

$$= \left(10 - \frac{x}{500}\right) dx \qquad\qquad dP = P'(x)\,dx$$

$$= \left(10 - \frac{2{,}000}{500}\right) 10 \qquad\qquad = \left(8 - \frac{x}{500}\right) dx$$

$$= \$60 \text{ per week} \qquad\qquad = \left(8 - \frac{2{,}000}{500}\right) 10$$

$$\qquad\qquad\qquad\qquad\qquad = \$40 \text{ per week}$$

Matched Problem 4 ⇨   Repeat Example 4 with production increasing from 6,000 to 6,010.

Comparing the results in Example 4 and Matched Problem 4, we see that an increase in production results in a revenue and profit increase at the 2,000 production level, but a revenue and profit loss at the 6,000 production level.

Before closing, we should mention that even though differentials can be used to approximate certain quantities, the error can be substantial in certain cases. Generally, the smaller Δx is, the better the approximation $\Delta y \approx dy$.

Answers to Matched Problems

1. (A) $\Delta x = 1, \Delta y = 5, \Delta y / \Delta x = 5$ (B) 4

2. $dy = \dfrac{1}{2\sqrt{x}} dx$ (A) 0.025   (B) 0.02   (C) 0.005

3. (A)

(B)

4. $dR = -\$20/\text{wk}; dP = -\$40/\text{wk}$

Exercise B-4

A *In Problems 1–6, find the indicated quantities for* $y = f(x) = 3x^2$.

1. $\Delta x, \Delta y,$ and $\Delta y / \Delta x$; given $x_1 = 1$ and $x_2 = 4$

2. $\Delta x, \Delta y,$ and $\Delta y / \Delta x$; given $x_1 = 2$ and $x_2 = 5$

3. $\dfrac{f(x_1 + \Delta x) - f(x_1)}{\Delta x}$; given $x_1 = 1$ and $\Delta x = 2$

4. $\dfrac{f(x_1 + \Delta x) - f(x_1)}{\Delta x}$; given $x_1 = 2$ and $\Delta x = 1$

5. $\Delta y / \Delta x$; given $x_1 = 1$ and $x_2 = 3$

6. $\Delta y / \Delta x$; given $x_1 = 2$ and $x_2 = 3$

In Problems 7–12, find dy for each function.

7. $y = 30 + 12x^2 - x^3$ **8.** $y = 200x - \dfrac{x^2}{30}$

9. $y = x^2\left(1 - \dfrac{x}{9}\right)$ **10.** $y = x^3(60 - x)$

11. $y = \dfrac{590}{\sqrt{x}}$ **12.** $y = 52\sqrt{x}$

B *In Problems 13 and 14, find the indicated quantities for* $y = f(x) = 3x^2$.

13. (A) $\dfrac{f(2 + \Delta x) - f(2)}{\Delta x}$ (Simplify)

(B) What does the quantity in part (A) approach as Δx approaches 0?

14. (A) $\dfrac{f(3 + \Delta x) - f(3)}{\Delta x}$ (Simplify)

(B) What does the quantity in part (A) approach as Δx approaches 0?

In Problems 15–18, find dy for each function.

15. $y = (2x + 1)^3$

16. $y = (3x + 5)^5$

17. $y = \dfrac{x}{x^2 + 9}$

18. $y = \dfrac{x^2}{(x + 1)^2}$

In Problems 19–22, evaluate dy and Δy for each function for the indicated values.

19. $y = f(x) = x^2 - 3x + 2; x = 5, dx = \Delta x = 0.2$

20. $y = f(x) = 30 + 12x^2 - x^3; x = 2, dx = \Delta x = 0.1$

21. $y = f(x) = 75\left(1 - \dfrac{2}{x}\right); x = 5, dx = \Delta x = -0.5$

22. $y = f(x) = 100\left(x - \dfrac{4}{x^2}\right); x = 2, dx = \Delta x = -0.1$

23. A cube with sides 10 inches long is covered with a coat of fiberglass 0.2 inch thick. Use differentials to estimate the volume of the fiberglass shell.

24. A sphere with a radius of 5 centimeters is coated with ice 0.1 centimeter thick. Use differentials to estimate the volume of the ice. [Recall that $V = \frac{4}{3}\pi r^3$.]

C *In Problems 25–28,*

(A) *Graph the functions Δy and dy (of the independent variable $\Delta x = dx$) for the function $f(x)$ at the indicated value of x.*

(B) *Find Δy and dy for the function $f(x)$ at the indicated values of x and $\Delta x = dx$.*

25. $f(x) = x^2 + 2x + 3$;
$x = -0.5$, $\Delta x = dx = 0.1, 0.2, 0.3$

26. $f(x) = x^2 + 2x + 3$;
$x = -2$, $\Delta x = dx = -0.1, -0.2, -0.3$

27. $f(x) = x^3 - 2x^2$; $x = 1$, $\Delta x = dx = 0.05, 0.10, 0.15$

28. $f(x) = x^3 - 2x^2$;
$x = 2$, $\Delta x = dx = -0.05, -0.10, -0.15$

In Problems 29–32, discuss the validity of each statement. If the statement is always true, explain why. If not, give a counterexample.

29. If the graph of the function $y = f(x)$ is a line, then the functions Δy and dy (of the independent variable $\Delta x = dx$) for $f(x)$ at $x = 3$ are identical functions.

30. If the graph of the function $y = f(x)$ is a parabola, then the functions Δy and dy (of the independent

variable $\Delta x = dx$) for $f(x)$ at $x = 0$ are identical functions.

31. Suppose that $y = f(x)$ defines a differentiable function whose domain is the set of all real numbers. If every differential dy at $x = 2$ is equal to 0, then $f(x)$ is a constant function.

32. Suppose that $y = f(x)$ defines a function whose domain is the set of all real numbers. If every increment at $x = 2$ is equal to 0, then $f(x)$ is a constant function.

33. Find dy if $y = \sqrt[3]{3x^2 - 2x + 1}$.

34. Find dy if $y = (2x^2 - 4)\sqrt{x + 2}$.

35. Find dy and Δy for $y = 52\sqrt{x}$, $x = 4$, and $\Delta x = dx = 0.3$.

36. Find dy and Δy for $y = 590/\sqrt{x}$, $x = 64$, and $\Delta x = dx = 1$.

Applications

Use differential approximations in the following problems.

Business & Economics

37. *Advertising.* Using past records, it is estimated that a company will sell N units of a product after spending $\$x$ thousand in advertising, as given by

$$N = 60x - x^2 \qquad 5 \leqslant x \leqslant 30$$

Approximately what increase in sales will result by increasing the advertising budget from $\$10,000$ to $\$11,000$? From $\$20,000$ to $\$21,000$?

38. *Price–demand.* Suppose, in a grocery chain, the daily demand (in pounds) for chocolate candy at $\$x$ per pound is given by

$$D = 1,000 - 40x^2 \qquad 1 \leqslant x \leqslant 5$$

If the price is increased from $\$3.00$ per pound to $\$3.20$ per pound, what is the approximate change in demand?

39. *Average cost.* For a company that manufactures tennis rackets, the average cost per racket, $\overline{C}$, is found to be

$$\overline{C} = \frac{400}{x} + 5 + \frac{1}{2}x \qquad x \geqslant 1$$

where x is the number of rackets produced per hour. What will the approximate change in average cost per racket be if production is increased from 20 per hour to 25 per hour? From 40 per hour to 45 per hour?

40. *Revenue and profit.* A company manufactures and sells x televisions per month. If the cost and revenue

equations are

$$C(x) = 72,000 + 60x$$

$$R(x) = 200x - \frac{x^2}{30} \qquad 0 \leqslant x \leqslant 6,000$$

what will the approximate changes in revenue and profit be if production is increased from 1,500 to 1,510? From 4,500 to 4,510?

Life Sciences

41. *Pulse rate.* The average pulse rate y (in beats per minute) of a healthy person x inches tall is given approximately by

$$y = \frac{590}{\sqrt{x}} \qquad 30 \leqslant x \leqslant 75$$

Approximately how will the pulse rate change for a height change from 36 to 37 inches? From 64 to 65 inches?

42. *Measurement.* An egg of a particular bird is very nearly spherical. If the radius to the inside of the shell is 5 millimeters and the radius to the outside of the shell is 5.3 millimeters, approximately what is the volume of the shell? [Remember that $V = \frac{4}{3}\pi r^3$.]

43. *Medicine.* A drug is given to a patient to dilate her arteries. If the radius of an artery is increased from 2 to 2.1 millimeters, approximately how much is the cross-sectional area increased? [Assume the cross section of the artery is circular; $A = \pi r^2$.]

44. *Drug sensitivity.* One hour after x milligrams of a particular drug are given to a person, the change in body temperature T (in degrees Fahrenheit) is given by

$$T = x^2\left(1 - \frac{x}{9}\right) \qquad 0 \leq x \leq 6$$

Approximate the changes in body temperature produced by the following changes in drug dosages:

(A) From 2 to 2.1 milligrams

(B) From 3 to 3.1 milligrams

(C) From 4 to 4.1 milligrams

Social Sciences

45. *Learning.* A particular person learning to type has an achievement record given approximately by

$$N = 75\left(1 - \frac{2}{t}\right) \qquad 3 \leq t \leq 20$$

where N is the number of words per minute typed after t weeks of practice. What is the approximate improvement from 5 to 5.5 weeks of practice?

46. *Learning.* If a person learns y items in x hours, as given approximately by

$$y = 52\sqrt{x} \qquad 0 \leq x \leq 9$$

what is the approximate increase in the number of items learned when x changes from 1 to 1.1 hours? From 4 to 4.1 hours?

47. *Politics.* In a newly incorporated city, it is estimated that the voting population (in thousands) will increase according to

$$N(t) = 30 + 12t^2 - t^3 \qquad 0 \leq t \leq 8$$

where t is time in years. Find the approximate change in votes for the following time changes:

(A) From 1 to 1.1 years

(B) From 4 to 4.1 years

(C) From 7 to 7.1 years

Section B-5

L'Hôpital's Rule

- ❑ Introduction
- ❑ L'Hôpital's Rule and the Indeterminate Form $0/0$
- ❑ One-Sided Limits and Limits at ∞
- ❑ L'Hôpital's Rule and the Indeterminate Form ∞/∞

❑ Introduction

The ability to evaluate a wide variety of different types of limits is one of the skills necessary to apply the techniques of calculus successfully. Limits play a fundamental role in the development of the derivative and are an important graphing tool. Important applications involving limits are discussed in Chapters 7 and 11. In order to deal effectively with some of these applications, we need to develop some additional methods for evaluating limits.

In this section we discuss a powerful technique for evaluating limits of quotients called *L'Hôpital's rule*. This rule is named after the French mathematician Marquis de L'Hôpital (1661–1704), who is generally credited with writing the first calculus textbook. To use L'Hôpital's rule, it is necessary to be very familiar with the limit properties of some basic functions. Figure 1 reviews some limits involving powers of x that we have discussed earlier.

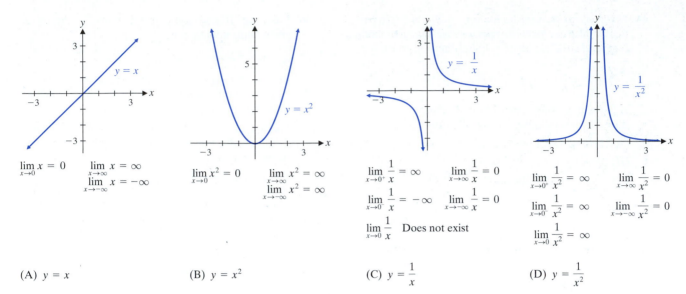

FIGURE 1 Limits involving powers of x

The limits in Figure 1 are easily extended to functions of the form $f(x) = (x - c)^n$ and $g(x) = 1/(x - c)^n$. In general, if n is an odd integer, limits involving $(x - c)^n$ or $1/(x - c)^n$ as x approaches c (or $\pm\infty$) behave like the limits of x and $1/x$ as x approaches 0 (or $\pm\infty$). If n is an even integer, limits involving these expressions behave like the limits of x^2 and $1/x^2$ as x approaches 0 (or $\pm\infty$).

Example 1   ➭   **Limits Involving Powers of $x - c$**

(A) $\lim\limits_{x \to 2} \dfrac{5}{(x - 2)^4} = \infty$ *Compare with* $\lim\limits_{x \to 0} \dfrac{1}{x^2}$ *in Figure 1.*

(B) $\lim\limits_{x \to -1^-} \dfrac{4}{(x + 1)^3} = -\infty$ *Compare with* $\lim\limits_{x \to 0^-} \dfrac{1}{x}$ *in Figure 1.*

(C) $\lim\limits_{x \to \infty} \dfrac{4}{(x - 9)^6} = 0$ *Compare with* $\lim\limits_{x \to \infty} \dfrac{1}{x^2}$ *in Figure 1.*

(D) $\lim\limits_{x \to -\infty} 3x^3 = -\infty$ *Compare with* $\lim\limits_{x \to -\infty} x$ *in Figure 1.*

Matched Problem 1   ➭   Evaluate each limit.

(A) $\lim\limits_{x \to 3^+} \dfrac{7}{(x - 3)^5}$ (B) $\lim\limits_{x \to -4} \dfrac{6}{(x + 4)^6}$

(C) $\lim\limits_{x \to -\infty} \dfrac{3}{(x + 2)^3}$ (D) $\lim\limits_{x \to \infty} 5x^4$

Figure 2 reviews some limits of exponential and logarithmic functions.

$$\lim_{x \to 0} e^x = 1 \qquad \lim_{x \to \infty} e^x = \infty$$

$$\lim_{x \to -\infty} e^x = 0$$

$$\lim_{x \to 0} e^{-x} = 1 \qquad \lim_{x \to \infty} e^{-x} = 0$$

$$\lim_{x \to -\infty} e^{-x} = \infty$$

$$\lim_{x \to 0^+} \ln x = -\infty \qquad \lim_{x \to \infty} \ln x = \infty$$

(A) $y = e^x$ (B) $y = e^{-x}$ (C) $y = \ln x$

FIGURE 2 Limits involving exponential and logarithmic functions

The limits in Figure 2 also generalize to other simple exponential and logarithmic forms.

Example 2 ➭ **Limits Involving Exponential and Logarithmic Forms**

(A) $\displaystyle\lim_{x \to \infty} 2e^{3x} = \infty$ *Compare with $\displaystyle\lim_{x \to \infty} e^x$ in Figure 2.*

(B) $\displaystyle\lim_{x \to \infty} 4e^{-5x} = 0$ *Compare with $\displaystyle\lim_{x \to \infty} e^{-x}$ in Figure 2.*

(C) $\displaystyle\lim_{x \to \infty} \ln(x + 4) = \infty$ *Compare with $\displaystyle\lim_{x \to \infty} \ln x$ in Figure 2.*

(D) $\displaystyle\lim_{x \to 2^+} \ln(x - 2) = -\infty$ *Compare with $\displaystyle\lim_{x \to 0^+} \ln x$ in Figure 2.* ∎

Matched Problem 2 ➭ Evaluate each limit.

(A) $\displaystyle\lim_{x \to -\infty} 2e^{-6x}$ (B) $\displaystyle\lim_{x \to -\infty} 3e^{2x}$

(C) $\displaystyle\lim_{x \to -4^+} \ln(x + 4)$ (D) $\displaystyle\lim_{x \to \infty} \ln(x - 10)$ ∎

Now that we have reviewed the limit properties of some basic functions, we are ready to consider the main topic of this section, L'Hôpital's rule—a powerful tool for evaluating certain types of limits.

Explore–Discuss 1

It follows from Figure 2 that

$$\lim_{x \to \infty} e^{-x} = 0 \qquad \text{and} \qquad \lim_{x \to \infty} \ln(1 + 3e^{-x}) = 0$$

Does

$$\lim_{x \to \infty} \frac{\ln(1 + 3e^{-x})}{e^{-x}}$$

exist? If so, what is its value? Use graphical methods to support your answers.

❏ L'Hôpital's Rule and the Indeterminate Form 0/0

Recall that the limit

$$\lim_{x \to c} \frac{f(x)}{g(x)}$$

is a 0/0 indeterminate form (Section 3-1) if

$$\lim_{x \to c} f(x) = 0 \quad \text{and} \quad \lim_{x \to c} g(x) = 0$$

The quotient property for limits in Section 3-1 does not apply since $\lim_{x \to c} g(x) = 0$.

If we are dealing with a 0/0 indeterminate form, the limit may or may not exist, and we cannot tell which is true without further investigation.

Each of the following is a 0/0 indeterminate form:

$$\lim_{x \to 2} \frac{x^2 - 4}{x - 2} \quad \text{and} \quad \lim_{x \to 1} \frac{e^x - e}{x - 1}$$

The first limit can be evaluated by performing some algebraic simplifications, such as

$$\lim_{x \to 2} \frac{x^2 - 4}{x - 2} = \lim_{x \to 2} \frac{(x - 2)(x + 2)}{x - 2} = \lim_{x \to 2} (x + 2) = 4$$

But the second cannot. Instead, we turn to the very powerful **L'Hôpital's rule,** which we now state without proof. This rule can be used whenever a limit is a 0/0 indeterminate form.

THEOREM 1   **L'Hôpital's Rule for 0/0 Indeterminate Forms: Version 1**

For c a real number,
If $\lim_{x \to c} f(x) = 0$ and $\lim_{x \to c} g(x) = 0$, then

$$\lim_{x \to c} \frac{f(x)}{g(x)} = \lim_{x \to c} \frac{f'(x)}{g'(x)}$$

provided the second limit exists or is $+\infty$ or $-\infty$.

The use of L'Hôpital's rule is best illustrated through examples.

Example 3 ⇔ **L'Hôpital's Rule**   Evaluate: $\lim_{x \to 1} \dfrac{e^x - e}{x - 1}$

SOLUTION   **Step 1.**   *Check to see if L'Hôpital's rule applies:*

$$\lim_{x \to 1} (e^x - e) = e^1 - e = 0 \quad \text{and} \quad \lim_{x \to 1} (x - 1) = 1 - 1 = 0$$

L'Hôpital's rule does apply.

Step 2.   *Apply L'Hôpital's rule:*

$$\underset{0/0 \text{ form}}{\lim_{x \to 1} \frac{e^x - e}{x - 1}} = \lim_{x \to 1} \frac{\dfrac{d}{dx}(e^x - e)}{\dfrac{d}{dx}(x - 1)}$$

$$= \lim_{x \to 1} \frac{e^x}{1} \qquad \text{L'Hôpital's rule}$$

$$= \frac{e^1}{1} = e \qquad \text{Limit property}$$

■

■

Matched Problem 3 ⤳ Evaluate: $\displaystyle \lim_{x \to 4} \frac{e^x - e^4}{x - 4}$

REMARK

In L'Hôpital's rule, the symbol $f'(x)/g'(x)$ represents the derivative of $f(x)$ divided by the derivative of $g(x)$, not the derivative of the quotient $f(x)/g(x)$.

When applying L'Hôpital's rule to a 0/0 indeterminate form, be certain that you differentiate the numerator and denominator separately.

 The functions

$$y_1 = \frac{e^x - e}{x - 1} \qquad \text{and} \qquad y_2 = \frac{e^x}{1}$$

of Example 3 are different functions (see Fig. 3), but both functions have the same limit e as x approaches 1. Although y_1 is undefined at $x = 1$, the graph of y_1 provides a check of the answer to Example 3.

FIGURE 3

Example 4 ⤳ **L'Hôpital's Rule**   Evaluate: $\displaystyle \lim_{x \to 0} \frac{\ln(1 + x^2)}{x^4}$

SOLUTION   **Step 1.** *Check to see if L'Hôpital's rule applies:*

$$\lim_{x \to 0} \ln(1 + x^2) = \ln 1 = 0 \qquad \text{and} \qquad \lim_{x \to 0} x^4 = 0$$

L'Hôpital's rule does apply.

Step 2. *Apply L'Hôpital's rule:*

$$\lim_{x \to 0} \frac{\overset{0/0\,\text{form}}{\ln(1 + x^2)}}{x^4} = \lim_{x \to 0} \frac{\dfrac{d}{dx}\ln(1 + x^2)}{\dfrac{d}{dx}x^4}$$

$$\lim_{x \to 0} \frac{\ln(1 + x^2)}{x^4} = \lim_{x \to 0} \frac{2x/(1 + x^2)}{4x^3} \qquad \text{\textcolor{blue}{L'Hôpital's rule}}$$

$$= \lim_{x \to 0} \frac{1}{2x^2(1 + x^2)} \qquad \text{\textcolor{blue}{Algebra}}$$

$$= \infty \qquad \text{\textcolor{blue}{Vertical asymptote at } x = 0}$$

FIGURE 4

The last step follows since $x = 0$ is a vertical asymptote for

$$f(x) = \frac{1}{2x^2(1 + x^2)}$$

and values of $f(x)$ become very large positive numbers as x approaches 0 from the left or right (see Fig. 4).

Matched Problem 4 ➯ Evaluate: $\displaystyle\lim_{x \to 1} \frac{\ln x}{(x - 1)^3}$

Example 5 ➯ **L'Hôpital's Rule May Not Be Applicable** Evaluate: $\displaystyle\lim_{x \to 1} \frac{\ln x}{x}$

SOLUTION **Step 1.** *Check to see if L'Hôpital's rule applies:*

$$\lim_{x \to 1} \ln x = \ln 1 = 0 \qquad \text{but} \qquad \lim_{x \to 1} x = 1 \neq 0$$

L'Hôpital's rule does not apply.

Step 2. *Evaluate by another method.* The quotient property for limits from Section 3-1 does apply, and we have

$$\lim_{x \to 1} \frac{\ln x}{x} = \frac{\displaystyle\lim_{x \to 1} \ln x}{\displaystyle\lim_{x \to 1} x} = \frac{\ln 1}{1} = \frac{0}{1} = 0$$

Note that applying L'Hôpital's rule would give us an incorrect result:

$$\lim_{x \to 1} \frac{\ln x}{x} \neq \lim_{x \to 1} \frac{\dfrac{d}{dx} \ln x}{\dfrac{d}{dx} x} = \lim_{x \to 1} \frac{1/x}{1} = 1$$

Matched Problem 5 ➯ Evaluate: $\displaystyle\lim_{x \to 0} \frac{x}{e^x}$

REMARK

As Example 5 illustrates, all limits involving quotients are not $0/0$ indeterminate forms.

You must always check to see if L'Hôpital's rule applies before you use it.

Example 6 ➯ **Repeated Application of L'Hôpital's Rule** Evaluate: $\displaystyle\lim_{x \to 0} \frac{x^2}{e^x - 1 - x}$

SOLUTION **Step 1.** *Check to see if L'Hôpital's rule applies:*

$$\lim_{x \to 0} x^2 = 0 \qquad \text{and} \qquad \lim_{x \to 0}(e^x - 1 - x) = 0$$

L'Hôpital's rule does apply.

Step 2. *Apply L'Hôpital's rule:*

$$\lim_{x \to 0} \frac{x^2}{e^x - 1 - x} \overset{0/0\,\text{form}}{=} \lim_{x \to 0} \frac{\dfrac{d}{dx}x^2}{\dfrac{d}{dx}(e^x - 1 - x)} = \lim_{x \to 0} \frac{2x}{e^x - 1}$$

Since $\lim_{x \to 0} 2x = 0$ and $\lim_{x \to 0}(e^x - 1) = 0$, the new limit obtained is also a 0/0 indeterminate form, and L'Hôpital's rule can be applied again.

Step 3. *Apply L'Hôpital's rule again:*

$$\lim_{x \to 0} \frac{2x}{e^x - 1} \overset{0/0\,\text{form}}{=} \lim_{x \to 0} \frac{\dfrac{d}{dx}2x}{\dfrac{d}{dx}(e^x - 1)} = \lim_{x \to 0} \frac{2}{e^x} = \frac{2}{e^0} = 2$$

Thus,

$$\lim_{x \to 0} \frac{x^2}{e^x - 1 - x} = \lim_{x \to 0} \frac{2x}{e^x - 1} = \lim_{x \to 0} \frac{2}{e^x} = 2$$

 A check that the limit is 2 as x approaches 0 is provided in Figure 5. The figure shows the graph of

$$y_1 = \frac{x^2}{e^x - 1 - x}$$

and a table of selected values of y_1.

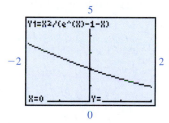

FIGURE 5

Matched Problem 6   ↪   Evaluate: $\displaystyle \lim_{x \to 0} \frac{e^{2x} - 1 - 2x}{x^2}$

❑   ONE-SIDED LIMITS AND LIMITS AT ∞

In addition to the limit as x approaches c, we have discussed one-sided limits and limits at ∞ in Chapters 3 and 4. L'Hôpital's rule is valid in these cases also.

THEOREM 2   L'Hôpital's Rule for 0/0 Indeterminate Forms: Version 2
(For One-Sided Limits and Limits at Infinity)

The first version of L'Hôpital's rule (Theorem 1) remains valid if the symbol $x \to c$ is replaced everywhere it occurs with one of the following symbols:

$$x \to c^+ \qquad x \to c^- \qquad x \to \infty \qquad x \to -\infty$$

For example, if $\lim\limits_{x \to \infty} f(x) = 0$ and $\lim\limits_{x \to \infty} g(x) = 0$, then

$$\lim_{x \to \infty} \frac{f(x)}{g(x)} = \lim_{x \to \infty} \frac{f'(x)}{g'(x)}$$

provided the second limit exists or is $+\infty$ or $-\infty$. Similar rules can be written for $x \to c^{+}$, $x \to c^{-}$, and $x \to -\infty$.

Example 7 ⟳ **L'Hôpital's Rule for One-Sided Limits**   Evaluate: $\lim\limits_{x \to 1^{+}} \dfrac{\ln x}{(x-1)^2}$

SOLUTION   **Step 1.**   *Check to see if L'Hôpital's rule applies:*

$$\lim_{x \to 1^{+}} \ln x = 0 \qquad \text{and} \qquad \lim_{x \to 1^{+}} (x-1)^2 = 0$$

L'Hôpital's rule does apply.

Step 2.   *Apply L'Hôpital's rule:*

$$\underset{0/0\ \text{form}}{\lim_{x \to 1^{+}} \frac{\ln x}{(x-1)^2}} = \lim_{x \to 1^{+}} \frac{\dfrac{d}{dx}(\ln x)}{\dfrac{d}{dx}(x-1)^2}$$

$$= \lim_{x \to 1^{+}} \frac{1/x}{2(x-1)} \qquad \text{L'Hôpital's rule}$$

$$= \lim_{x \to 1^{+}} \frac{1}{2x} \cdot \frac{1}{x-1} \qquad \text{Algebra}$$

$$= \infty \qquad \text{Limit property}$$

The last step follows, since

$$\lim_{x \to 1^{+}} \frac{1}{2x} = \frac{1}{2} \qquad \text{and} \qquad \lim_{x \to 1^{+}} \frac{1}{x-1} = \infty$$

Matched Problem 7 ⟳   Evaluate: $\lim\limits_{x \to 1^{-}} \dfrac{\ln x}{(x-1)^2}$

Example 8 ⟳ **L'Hôpital's Rule for Limits at Infinity**   Evaluate: $\lim\limits_{x \to \infty} \dfrac{\ln(1 + e^{-x})}{e^{-x}}$

SOLUTION   **Step 1.**   *Check to see if L'Hôpital's rule applies:*

$$\lim_{x \to \infty} \ln(1 + e^{-x}) = \ln(1 + 0) = \ln 1 = 0 \qquad \text{and} \qquad \lim_{x \to \infty} e^{-x} = 0$$

L'Hôpital's rule does apply.

Step 2.   *Apply L'Hôpital's rule:*

$$\underset{0/0\ \text{form}}{\lim_{x \to \infty} \frac{\ln(1 + e^{-x})}{e^{-x}}} = \lim_{x \to \infty} \frac{\dfrac{d}{dx}[\ln(1 + e^{-x})]}{\dfrac{d}{dx} e^{-x}}$$

$$= \lim_{x \to \infty} \frac{-e^{-x}/(1 + e^{-x})}{-e^{-x}} \qquad \text{L'Hôpital's rule}$$

$$\lim_{x \to \infty} \frac{\ln(1 + e^{-x})}{e^{-x}} = \lim_{x \to \infty} \frac{1}{1 + e^{-x}} \qquad \text{Algebra}$$

$$= \frac{1}{1 + 0} = 1 \qquad \text{Limit property}$$

■

Matched Problem 8 ⇨   Evaluate: $\displaystyle \lim_{x \to -\infty} \frac{\ln(1 + 2e^x)}{e^x}$

■

❑   **L'Hôpital's Rule and the Indeterminate Form ∞/∞**

In Section 4-3 we discussed techniques for evaluating limits of rational functions such as

$$\lim_{x \to \infty} \frac{2x^2}{x^3 + 3} \qquad \lim_{x \to \infty} \frac{4x^3}{2x^2 + 5} \qquad \lim_{x \to \infty} \frac{3x^3}{5x^3 + 6}$$

Each of these limits is an ∞/∞ *indeterminate form.* In general, if $\lim_{x \to c} f(x) = \pm\infty$ and $\lim_{x \to c} g(x) = \pm\infty$, then

$$\lim_{x \to c} \frac{f(x)}{g(x)}$$

is called an **∞/∞ indeterminate form.** Furthermore, $x \to c$ can be replaced in all three limits above with $x \to c^+, x \to c^-, x \to \infty$, or $x \to -\infty$. It can be shown that L'Hôpital's rule also applies to these ∞/∞ indeterminate forms.

THEOREM 3   **L'Hôpital's Rule for the Indeterminate Form ∞/∞: Version 3**

Versions 1 and 2 of L'Hôpital's rule for the indeterminate form 0/0 are also valid if the limit of *f* and the limit of *g* are both infinite; that is, both $+\infty$ and $-\infty$ are permissible for either limit.

▬▬▬▬▬

For example, if $\lim_{x \to c^+} f(x) = \infty$ and $\lim_{x \to c^+} g(x) = -\infty$, then L'Hôpital's rule can be applied to $\lim_{x \to c^+} [f(x)/g(x)]$.

Example 9 ⇨   **L'Hôpital's Rule for the Indeterminate Form ∞/∞**

Evaluate: $\displaystyle \lim_{x \to \infty} \frac{\ln x}{x^2}$

SOLUTION   **Step 1.**   *Check to see if L'Hôpital's rule applies:*

$$\lim_{x \to \infty} \ln x = \infty \qquad \text{and} \qquad \lim_{x \to \infty} x^2 = \infty$$

L'Hôpital's rule does apply.

Step 2.   *Apply L'Hôpital's rule:*

∞/∞ form
$$\lim_{x \to \infty} \frac{\ln x}{x^2} = \lim_{x \to \infty} \frac{\dfrac{d}{dx}(\ln x)}{\dfrac{d}{dx} x^2}$$

$$= \lim_{x \to \infty} \frac{1/x}{2x} \qquad \text{L'Hôpital's rule}$$

$$\lim_{x \to \infty} \frac{\ln x}{x^2} = \lim_{x \to \infty} \frac{1}{2x^2} \qquad \textcolor{blue}{\textit{Algebra}}$$

$$= 0 \qquad \textcolor{blue}{\textit{Limit property}}$$

Matched Problem 9 ➭ Evaluate: $\lim_{x \to \infty} \dfrac{\ln x}{x}$

Example 10 ➭ **L'Hôpital's Rule for the Indeterminate Form** ∞/∞

Evaluate: $\lim_{x \to \infty} \dfrac{e^x}{x^2}$

SOLUTION **Step 1.** *Check to see if L'Hôpital's rule applies:*

$$\lim_{x \to \infty} e^x = \infty \qquad \text{and} \qquad \lim_{x \to \infty} x^2 = \infty$$

L'Hôpital's rule does apply.

Step 2. *Apply L'Hôpital's rule:*

∞/∞ form

$$\lim_{x \to \infty} \frac{e^x}{x^2} = \boxed{\lim_{x \to \infty} \frac{\dfrac{d}{dx} e^x}{\dfrac{d}{dx} x^2} = \lim_{x \to \infty} \frac{e^x}{2x}}$$

Since $\lim_{x \to \infty} e^x = \infty$ and $\lim_{x \to \infty} 2x = \infty$, this limit is an ∞/∞ indeterminate form and L'Hôpital's rule can be applied again.

Step 3. *Apply L'Hôpital's rule again:*

∞/∞ form

$$\lim_{x \to \infty} \frac{e^x}{2x} = \boxed{\lim_{x \to \infty} \frac{\dfrac{d}{dx} e^x}{\dfrac{d}{dx} 2x} = \lim_{x \to \infty} \frac{e^x}{2} = \infty}$$

Thus,

$$\lim_{x \to \infty} \frac{e^x}{x^2} = \lim_{x \to \infty} \frac{e^x}{2x} = \lim_{x \to \infty} \frac{e^x}{2} = \infty$$

Matched Problem 10 ➭ Evaluate: $\lim_{x \to \infty} \dfrac{e^{2x}}{x^2}$

Explore–Discuss 2

Let n be a positive integer. Explain how L'Hôpital's rule can be used to show that

$$\lim_{x \to \infty} \frac{e^x}{x^n} = \infty$$

Does this imply that e^x increases more rapidly than any power of x?

Answers to Matched Problems
1. (A) ∞ (B) ∞ (C) 0 (D) ∞
2. (A) ∞ (B) 0 (C) $-\infty$ (D) ∞
3. e^4 **4.** ∞ **5.** 0 **6.** 2 **7.** $-\infty$ **8.** 2 **9.** 0 **10.** ∞

Exercise B-5

A *Find each limit in Problems 1–14 using L'Hôpital's rule.*

1. $\displaystyle\lim_{x\to 2}\frac{x^4-16}{x^3-8}$

2. $\displaystyle\lim_{x\to 1}\frac{x^6-1}{x^5-1}$

3. $\displaystyle\lim_{x\to 2}\frac{x^2+x-6}{x^2+6x-16}$

4. $\displaystyle\lim_{x\to 4}\frac{x^2-8x+16}{x^2-5x+4}$

5. $\displaystyle\lim_{x\to 0}\frac{e^x-1}{x}$

6. $\displaystyle\lim_{x\to 1}\frac{\ln x}{x-1}$

7. $\displaystyle\lim_{x\to 0}\frac{\ln(1+4x)}{x}$

8. $\displaystyle\lim_{x\to 0}\frac{e^{2x}-1}{x}$

9. $\displaystyle\lim_{x\to\infty}\frac{2x^2+7}{5x^3+9}$

10. $\displaystyle\lim_{x\to\infty}\frac{3x^4+6}{2x^2+5}$

11. $\displaystyle\lim_{x\to\infty}\frac{e^{3x}}{x}$

12. $\displaystyle\lim_{x\to\infty}\frac{x}{e^{4x}}$

13. $\displaystyle\lim_{x\to\infty}\frac{x^2}{\ln x}$

14. $\displaystyle\lim_{x\to\infty}\frac{\ln x}{x^4}$

In Problems 15–18, explain why L'Hôpital's rule does not apply. If the limit exists, find it by other means.

15. $\displaystyle\lim_{x\to 1}\frac{x^2+5x+4}{x^3+1}$

16. $\displaystyle\lim_{x\to\infty}\frac{e^{-x}}{\ln x}$

17. $\displaystyle\lim_{x\to 2}\frac{x+2}{(x-2)^4}$

18. $\displaystyle\lim_{x\to -3}\frac{x^2}{(x+3)^5}$

B *Find each limit in Problems 19–42. Note that L'Hôpital's rule does not apply to every problem, and some problems will require more than one application of L'Hôpital's rule.*

 Check each limit by graphing the quotient.

19. $\displaystyle\lim_{x\to 0}\frac{e^{4x}-1-4x}{x^2}$

20. $\displaystyle\lim_{x\to 0}\frac{3x+1-e^{3x}}{x^2}$

21. $\displaystyle\lim_{x\to 2}\frac{\ln(x-1)}{x-1}$

22. $\displaystyle\lim_{x\to -1}\frac{\ln(x+2)}{x+2}$

23. $\displaystyle\lim_{x\to 0^+}\frac{\ln(1+x^2)}{x^3}$

24. $\displaystyle\lim_{x\to 0^-}\frac{\ln(1+2x)}{x^2}$

25. $\displaystyle\lim_{x\to 0^+}\frac{\ln(1+\sqrt{x})}{x}$

26. $\displaystyle\lim_{x\to 0^+}\frac{\ln(1+x)}{\sqrt{x}}$

27. $\displaystyle\lim_{x\to -2}\frac{x^2+2x+1}{x^2+x+1}$

28. $\displaystyle\lim_{x\to 1}\frac{2x^3-3x^2+1}{x^3-3x+2}$

29. $\displaystyle\lim_{x\to -1}\frac{x^3+x^2-x-1}{x^3+4x^2+5x+2}$

30. $\displaystyle\lim_{x\to 3}\frac{x^3+3x^2-x-3}{x^2+6x+9}$

31. $\displaystyle\lim_{x\to 2^-}\frac{x^3-12x+16}{x^3-6x^2+12x-8}$

32. $\displaystyle\lim_{x\to 1^+}\frac{x^3-x^2-x+1}{x^3-3x^2+3x-1}$

33. $\displaystyle\lim_{x\to\infty}\frac{3x^2+5x}{4x^3+7}$

34. $\displaystyle\lim_{x\to\infty}\frac{4x^2+9x}{5x^2+8}$

35. $\displaystyle\lim_{x\to\infty}\frac{x^2}{e^{2x}}$

36. $\displaystyle\lim_{x\to\infty}\frac{e^{3x}}{x^3}$

37. $\displaystyle\lim_{x\to\infty}\frac{1+e^{-x}}{1+x^2}$

38. $\displaystyle\lim_{x\to -\infty}\frac{1+e^{-x}}{1+x^2}$

39. $\displaystyle\lim_{x\to\infty}\frac{e^{-x}}{\ln(1+4e^{-x})}$

40. $\displaystyle\lim_{x\to\infty}\frac{\ln(1+2e^{-x})}{\ln(1+e^{-x})}$

41. $\displaystyle\lim_{x\to 0}\frac{e^x-e^{-x}-2x}{x^3}$

42. $\displaystyle\lim_{x\to 0}\frac{e^{2x}-1-2x-2x^2}{x^3}$

C **43.** Find: $\displaystyle\lim_{x\to 0^+}(x\ln x)$

[*Hint*:   Write $x\ln x=(\ln x)/x^{-1}$.]

44. Find: $\displaystyle\lim_{x\to 0^+}(\sqrt{x}\ln x)$

[*Hint*:   Write $\sqrt{x}\ln x=(\ln x)/x^{-1/2}$.]

In Problems 45–48, n is a positive integer. Find each limit.

45. $\displaystyle\lim_{x\to\infty}\frac{\ln x}{x^n}$

46. $\displaystyle\lim_{x\to\infty}\frac{x^n}{\ln x}$

47. $\displaystyle\lim_{x\to\infty}\frac{e^x}{x^n}$

48. $\displaystyle\lim_{x\to\infty}\frac{x^n}{e^x}$

In Problems 49–52, show that repeated application of L'Hôpital's rule does not lead to a solution.

 Use a graphing utility to graph each function and estimate the limit from the graph.

49. $\displaystyle\lim_{x\to\infty}\frac{\sqrt{1+x^2}}{x}$

50. $\displaystyle\lim_{x\to -\infty}\frac{x}{\sqrt{4+x^2}}$

51. $\displaystyle\lim_{x\to -\infty}\frac{\sqrt[3]{x^3+1}}{x}$

52. $\displaystyle\lim_{x\to\infty}\frac{x^2}{\sqrt[3]{(x^3+1)^2}}$

FIGURE 1

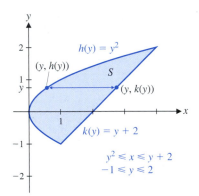

FIGURE 2

Section B-6

Double Integrals Over More General Regions

- ❑ REGULAR REGIONS
- ❑ DOUBLE INTEGRALS OVER REGULAR REGIONS
- ❑ REVERSING THE ORDER OF INTEGRATION
- ❑ VOLUME AND DOUBLE INTEGRALS

In this section we extend the concept of double integration discussed in Section 8-6 to nonrectangular regions. We begin with an example and some new terminology.

❑ REGULAR REGIONS

Let R be the region graphed in Figure 1. We can describe R with the following inequalities:

$$R = \{(x, y) | x \leq y \leq 6x - x^2, \quad 0 \leq x \leq 5\}$$

The region R can be viewed as a union of vertical line segments. For each x in the interval $[0, 5]$, the line segment from the point $(x, g(x))$ to the point $(x, f(x))$ lies in the region R. Any region that can be covered by vertical line segments in this manner is called a *regular x region*.

Now consider the region S in Figure 2. It can be described with the following inequalities:

$$S = \{(x, y) | y^2 \leq x \leq y + 2, \quad -1 \leq y \leq 2\}$$

The region S can be viewed as a union of horizontal line segments going from the graph of $h(y) = y^2$ to the graph of $k(y) = y + 2$ on the interval $[-1, 2]$. Regions that can be described in this manner are called *regular y regions*.

In general, *regular regions* are defined as follows:

Regular Regions

A region R in the xy plane is a **regular x region** if there exist functions $f(x)$ and $g(x)$ and numbers a and b so that

$$R = \{(x, y) | g(x) \leq y \leq f(x), \quad a \leq x \leq b\}$$

A region R is a **regular y region** if there exist functions $h(y)$ and $k(y)$ and numbers c and d so that

$$R = \{(x, y) | h(y) \leq x \leq k(y), \quad c \leq y \leq d\}$$

See Figure 3 (at the top of the next page) for a geometric interpretation.

REMARK

A region is *not* a regular y region if some horizontal line has a nonempty intersection with the region that is neither a closed interval nor a point. Similarly, a region is *not* a regular x region if some vertical line has a nonempty intersection with the region that is neither a closed interval nor a point (see Fig. 3).

FIGURE 3

Example 1 ⇔   **Describing a Regular *x* Region**   The region *R* is bounded by the graphs of $y = 4 - x^2$ and $y = x - 2$, $x \geq 0$, and the *y* axis. Graph *R* and use set notation and double inequalities to describe *R* as a regular *x* region.

SOLUTION   As the solid line in the figure indicates, *R* can be covered by vertical line segments that go from the graph of $y = x - 2$ to the graph of $y = 4 - x^2$. Thus, *R* is a regular *x* region. In terms of set notation and double inequalities, we can write

$$R = \{(x, y) \mid x - 2 \leq y \leq 4 - x^2, \quad 0 \leq x \leq 2\}$$

_____

REMARK

The region *R* of Example 1 is also a regular *y* region, since $R = \{(x, y) \mid 0 \leq x \leq k(y), \quad -2 \leq y \leq 4\}$, where

$$k(y) = \begin{cases} 2 + y & \text{if } -2 \leq y \leq 0 \\ \sqrt{4 - y} & \text{if } 0 \leq y \leq 4 \end{cases}$$

But since $k(y)$ is piecewise-defined, this description is more complicated than the description of *R* in Example 1 as a regular *x* region.

Matched Problem 1   Describe the region R bounded by the graphs of $x = 6 - y$, $x = y^2$, $y \geqslant 0$, and the x axis as a regular y region.

Example 2   **Describing Regular Regions**   The region R is bounded by the graphs of $x + y^2 = 9$ and $x + 3y = 9$. Graph R and describe R as a regular x region, a regular y region, both, or neither. Represent R using set notation and double inequalities.

SOLUTION

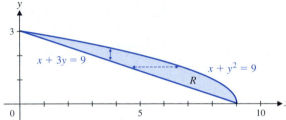

TEST FOR X REGION   Region R can be covered by vertical line segments that go from the graph of $x + 3y = 9$ to the graph of $x + y^2 = 9$. Thus, R is a regular x region. In order to describe R with inequalities, we must solve each equation for y in terms of x:

$$x + 3y = 9 \qquad\qquad x + y^2 = 9$$
$$3y = 9 - x \qquad\qquad y^2 = 9 - x$$
$$y = 3 - \tfrac{1}{3}x \qquad\qquad y = \sqrt{9 - x}$$

We use the positive square root since the graph is in the first quadrant.

Thus,

$$R = \{(x, y) | 3 - \tfrac{1}{3}x \leqslant y \leqslant \sqrt{9 - x}, \;\; 0 \leqslant x \leqslant 9\}$$

TEST FOR Y REGION   Since region R also can be covered by horizontal line segments (see the dashed line in the figure) that go from the graph of $x + 3y = 9$ to the graph of $x + y^2 = 9$, it is a regular y region. Now we must solve each equation for x in terms of y:

$$x + 3y = 9 \qquad\qquad x + y^2 = 9$$
$$x = 9 - 3y \qquad\qquad x = 9 - y^2$$

Thus,

$$R = \{(x, y) | 9 - 3y \leqslant x \leqslant 9 - y^2, \;\; 0 \leqslant y \leqslant 3\}$$

Matched Problem 2 ⇔ Repeat Example 2 for the region bounded by the graphs of $2y - x = 4$ and $y^2 - x = 4$, as shown in the following figure:

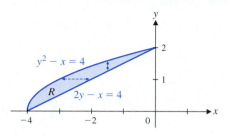

Explore–Discuss 1

Consider the vowels A, E, I, O, U, written in block letters as shown in the margin, to be regions of the plane. One of the vowels is a regular x region but not a regular y region; one is a regular y region but not a regular x region; one is both; two are neither. Explain.

A
E
I
O
U

❑ **DOUBLE INTEGRALS OVER REGULAR REGIONS**

Now we want to extend the definition of double integration to include regular x regions and regular y regions. Notice in the box that the order of integration now depends on the nature of the region R. If R is a regular x region, we integrate with respect to y first, while if R is a regular y region, we integrate with respect to x first.

It is also important to note that the variable limits of integration (when present) are always on the inner integral, and the constant limits of integration are always on the outer integral.

Double Integration Over Regular Regions

Regular x Region

If $R = \{(x, y) \mid g(x) \le y \le f(x), \quad a \le x \le b\}$, then

$$\iint\limits_{R} F(x, y)\, dA = \int_{a}^{b} \left[\int_{g(x)}^{f(x)} F(x, y)\, dy \right] dx$$

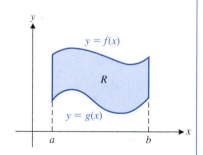

Regular y Region

If $R = \{(x, y) \mid h(y) \le x \le k(y), \quad c \le y \le d\}$, then

$$\iint\limits_{R} F(x, y)\, dA = \int_{c}^{d} \left[\int_{h(y)}^{k(y)} F(x, y)\, dx \right] dy$$

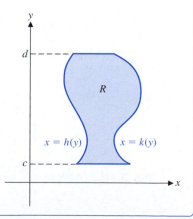

Example 3   ➭   **Evaluating a Double Integral**   Evaluate $\iint\limits_{R} 2xy\, dA$,   where R is the region bounded by the graphs of $y = -x$ and $y = x^2, x \geq 0$, and the graph of $x = 1$.

SOLUTION   From the graph we can see that R is a regular x region described by

$$R = \{(x, y)|-x \leq y \leq x^2, \quad 0 \leq x \leq 1\}$$

Thus,

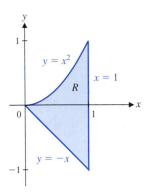

$$\iint\limits_{R} 2xy\, dA = \int_{0}^{1}\left[\int_{-x}^{x^2} 2xy\, dy\right] dx$$

$$= \int_{0}^{1}\left[xy^2\Big|_{y=-x}^{|y=x^2}\right] dx$$

$$= \int_{0}^{1}[x(x^2)^2 - x(-x)^2]\, dx$$

$$= \int_{0}^{1}(x^5 - x^3)\, dx$$

$$= \left(\frac{x^6}{6} - \frac{x^4}{4}\right)\Big|_{x=0}^{x=1}$$

$$= (\tfrac{1}{6} - \tfrac{1}{4}) - (0 - 0) = -\tfrac{1}{12}$$

Matched Problem 3   ➭   Evaluate $\iint\limits_{R} 3xy^2\, dA$, where R is the region in Example 3.

Example 4   ➭   **Evaluating a Double Integral**   Evaluate $\iint\limits_{R}(2x + y)\, dA$, where R is the region bounded by the graphs of $y = \sqrt{x}, x + y = 2$, and $y = 0$.

SOLUTION   From the graph we can see that R is a regular y region. After solving each equation for x, we can write

$$R = \{(x, y)|y^2 \leq x \leq 2 - y, \quad 0 \leq y \leq 1\}$$

Thus,

$$\iint\limits_{R}(2x + y)\, dA = \int_{0}^{1}\left[\int_{y^2}^{2-y}(2x + y)\, dx\right] dy$$

$$= \int_{0}^{1}\left[(x^2 + yx)\Big|_{x=y^2}^{|x=2-y}\right] dy$$

$$= \int_{0}^{1}\{[(2 - y)^2 + y(2 - y)] - [(y^2)^2 + y(y^2)]\}\, dy$$

$$= \int_{0}^{1}(4 - 2y - y^3 - y^4)\, dy$$

$$= (4y - y^2 - \tfrac{1}{4}y^4 - \tfrac{1}{5}y^5)\Big|_{y=0}^{|y=1}$$

$$= (4 - 1 - \tfrac{1}{4} - \tfrac{1}{5}) - 0 = \tfrac{51}{20}$$

Matched Problem 4   ➭   Evaluate $\iint\limits_{R}(y - 4x)\, dA$, where R is the region in Example 4.

Example 5   ⇨   **Evaluating a Double Integral**   The region R is bounded by the graphs of $y = \sqrt{x}$ and $y = \frac{1}{2}x$. Evaluate $\iint\limits_R 4xy^3\, dA$ two different ways.

SOLUTION   Region R is both a regular x region and a regular y region:

$$R = \{(x, y)\,|\,\tfrac{1}{2}x \leqslant y \leqslant \sqrt{x},\ \ 0 \leqslant x \leqslant 4\} \quad \text{Regular x region}$$
$$R = \{(x, y)\,|\,y^2 \leqslant x \leqslant 2y,\ \ 0 \leqslant y \leqslant 2\} \quad \text{Regular y region}$$

Using the first representation (a regular x region), we obtain

$$\iint\limits_R 4xy^3\, dA = \int_0^4 \left[\int_{x/2}^{\sqrt{x}} 4xy^3\, dy\right] dx$$

$$= \int_0^4 \left[xy^4 \Big|_{y=x/2}^{y=\sqrt{x}}\right] dx$$

$$= \int_0^4 \left[x(\sqrt{x})^4 - x(\tfrac{1}{2}x)^4\right] dx$$

$$= \int_0^4 (x^3 - \tfrac{1}{16}x^5)\, dx$$

$$= (\tfrac{1}{4}x^4 - \tfrac{1}{96}x^6)\Big|_{x=0}^{x=4}$$

$$= (64 - \tfrac{128}{3}) - 0 = \tfrac{64}{3}$$

Using the second representation (a regular y region), we obtain

$$\iint\limits_R 4xy^3\, dA = \int_0^2 \left[\int_{y^2}^{2y} 4xy^3\, dx\right] dy$$

$$= \int_0^2 \left[2x^2y^3 \Big|_{x=y^2}^{x=2y}\right] dy$$

$$= \int_0^2 [2(2y)^2y^3 - 2(y^2)^2y^3]\, dy$$

$$= \int_0^2 (8y^5 - 2y^7)\, dy$$

$$= (\tfrac{4}{3}y^6 - \tfrac{1}{4}y^8)\Big|_{y=0}^{y=2}$$

$$= (\tfrac{256}{3} - 64) - 0 = \tfrac{64}{3} \quad ■$$

Matched Problem 5   ⇨   The region R is bounded by the graphs of $y = x$ and $y = \frac{1}{2}x^2$. Evaluate $\iint\limits_R 4x^3y\, dA$ two different ways.

■

❑   REVERSING THE ORDER OF INTEGRATION

Example 5 shows that

$$\iint\limits_R 4xy^3\, dA = \int_0^4 \left[\int_{x/2}^{\sqrt{x}} 4xy^3\, dy\right] dx = \int_0^2 \left[\int_{y^2}^{2y} 4xy^3\, dx\right] dy$$

In general, if R is both a regular x region and a regular y region, the two iterated integrals are equal. In rectangular regions, reversing the order of integration in an iterated integral was a simple matter. As Example 5 illustrates, the process is more complicated in nonrectangular regions. The next example illustrates how to start with an iterated integral and reverse the order of integration. Since we are interested in the reversal process and not in the value of either integral, the integrand will not be specified.

Example 6 ⇋ **Reversing the Order of Integration** Reverse the order of integration in

$$\int_1^3 \left[\int_0^{x-1} f(x, y) \, dy \right] dx$$

SOLUTION The order of integration indicates that the region of integration is a regular x region:

$$R = \{(x, y) | 0 \le y \le x - 1, \quad 1 \le x \le 3\}$$

Graph region R to determine whether it is also a regular y region. The graph shows that R is also a regular y region, and we can write

$$R = \{(x, y) | y + 1 \le x \le 3, \quad 0 \le y \le 2\}$$

Thus,

$$\int_1^3 \left[\int_0^{x-1} f(x, y) \, dy \right] dx = \int_0^2 \left[\int_{y+1}^3 f(x, y) \, dx \right] dy$$

Matched Problem 6 ⇋ Reverse the order of integration in $\int_2^4 \left[\int_0^{4-x} f(x, y) \, dy \right] dx$.

Explore–Discuss 2 Explain the difficulty in evaluating $\int_0^2 \int_{x^2}^4 x e^{y^2} dy \, dx$ and how it can be overcome by reversing the order of integration.

❑ VOLUME AND DOUBLE INTEGRALS

In Section 8-6 we used the double integral to calculate the volume of a solid with a rectangular base. In general, if a solid can be described by the graph of a positive function $f(x, y)$ over a regular region R (not necessarily a rectangle), then the double integral of the function f over the region R still represents the volume of the corresponding solid.

Example 7 ⇋ **Volume** The region R is bounded by the graphs of $x + y = 1$, $y = 0$, and $x = 0$. Find the volume of the solid under the graph of $z = 1 - x - y$ over the region R.

SOLUTION

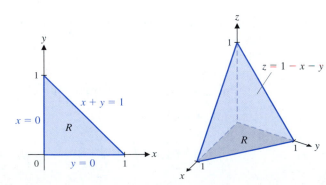

The graph of R indicates that R is both a regular x region and a regular y region. We choose to use the regular x region:

$$R = \{(x, y)|0 \leqslant y \leqslant 1 - x, \quad 0 \leqslant x \leqslant 1\}$$

Thus, the volume of the solid is

$$V = \iint\limits_{R} (1 - x - y)\, dA = \int_0^1 \left[\int_0^{1-x} (1 - x - y)\, dy \right] dx$$

$$= \int_0^1 \left[(y - xy - \tfrac{1}{2}y^2) \Big|_{y=0}^{y=1-x} \right] dx$$

$$= \int_0^1 [(1 - x) - x(1 - x) - \tfrac{1}{2}(1 - x)^2]\, dx$$

$$= \int_0^1 (\tfrac{1}{2} - x + \tfrac{1}{2}x^2)\, dx$$

$$= (\tfrac{1}{2}x - \tfrac{1}{2}x^2 + \tfrac{1}{6}x^3) \Big|_{x=0}^{x=1}$$

$$= (\tfrac{1}{2} - \tfrac{1}{2} + \tfrac{1}{6}) - 0 = \tfrac{1}{6}$$

Matched Problem 7   The region R is bounded by the graphs of $y + 2x = 2$, $y = 0$, and $x = 0$. Find the volume of the solid under the graph of $z = 2 - 2x - y$ over the region R. [*Hint*: Sketch the region first—the solid does not have to be sketched.]

Answers to Matched Problems
   1. $R = \{(x, y)|y^2 \leqslant x \leqslant 6 - y, \quad 0 \leqslant y \leqslant 2\}$
   2. R is both a regular x region and a regular y region:

$$R = \{(x, y)|\tfrac{1}{2}x + 2 \leqslant y \leqslant \sqrt{x + 4}, \quad -4 \leqslant x \leqslant 0\}$$

$$R = \{(x, y)|y^2 - 4 \leqslant x \leqslant 2y - 4, \quad 0 \leqslant y \leqslant 2\}$$

   3. $\frac{13}{40}$ **4.** $-\frac{77}{20}$ **5.** $\frac{16}{3}$ **6.** $\int_0^2 \int_2^{4-y} f(x, y)\, dx\, dy$ **7.** $\frac{2}{3}$

Exercise B-6

A *In Problems 1–6, graph the region R bounded by the graphs of the equations. Use set notation and double inequalities to describe R as a regular x region and as a regular y region in Problems 1 and 2, and as a regular x region or a regular y region, whichever is simpler, in Problems 3–6.*

   1. $y = 4 - x^2, y = 0, 0 \leqslant x \leqslant 2$

   2. $y = x^2, y = 9, 0 \leqslant x \leqslant 3$

   3. $y = x^3, y = 12 - 2x, x = 0$

   4. $y = 5 - x, y = 1 + x, y = 0$

   5. $y^2 = 2x, y = x - 4$

   6. $y = 4 + 3x - x^2, x + y = 4$

Evaluate each integral in Problems 7–10.

   7. $\int_0^1 \int_0^x (x + y)\, dy\, dx$

   8. $\int_0^2 \int_0^y xy\, dx\, dy$

   9. $\int_0^1 \int_{y^3}^{\sqrt{y}} (2x + y)\, dx\, dy$

   10. $\int_1^4 \int_x^{x^2} (x^2 + 2y)\, dy\, dx$

B *In Problems 11–14, give a verbal description of the region R and determine whether R is a regular x region, a regular y region, both, or neither.*

   11. $R = \{(x, y)||x| \leqslant 2, \quad |y| \leqslant 3\}$

   12. $R = \{(x, y)|1 \leqslant x^2 + y^2 \leqslant 4\}$

13. $R = \{(x, y)|x^2 + y^2 \geqslant 1, \quad |x| \leqslant 2, \quad 0 \leqslant y \leqslant 2\}$

14. $R = \{(x, y)||x| + |y| \leqslant 1\}$

In Problems 15–20, use the description of the region R to evaluate the indicated integral.

15. $\displaystyle\iint\limits_{R} (x^2 + y^2)\, dA;$

$R = \{(x, y)|0 \leqslant y \leqslant 2x, \quad 0 \leqslant x \leqslant 2\}$

16. $\displaystyle\iint\limits_{R} 2x^2 y\, dA;$

$R = \{(x, y)|0 \leqslant y \leqslant 9 - x^2, \quad -3 \leqslant x \leqslant 3\}$

17. $\displaystyle\iint\limits_{R} (x + y - 2)^3\, dA;$

$R = \{(x, y)|0 \leqslant x \leqslant y + 2, \quad 0 \leqslant y \leqslant 1\}$

18. $\displaystyle\iint\limits_{R} (2x + 3y)\, dA;$

$R = \{(x, y)|y^2 - 4 \leqslant x \leqslant 4 - 2y, \quad 0 \leqslant y \leqslant 2\}$

19. $\displaystyle\iint\limits_{R} e^{x+y}\, dA;$

$R = \{(x, y)| - x \leqslant y \leqslant x, \quad 0 \leqslant x \leqslant 2\}$

20. $\displaystyle\iint\limits_{R} \frac{x}{\sqrt{x^2 + y^2}}\, dA;$

$R = \{(x, y)|0 \leqslant x \leqslant \sqrt{4y - y^2}, \quad 0 \leqslant y \leqslant 2\}$

In Problems 21–26, graph the region R bounded by the graphs of the indicated equations. Describe R in set notation with double inequalities, and evaluate the indicated integral.

21. $y = x + 1, \quad y = 0, \quad x = 0, \quad x = 1;$

$\displaystyle\iint\limits_{R} \sqrt{1 + x + y}\, dA$

22. $y = x^2, \quad y = \sqrt{x}; \quad \displaystyle\iint\limits_{R} 12xy\, dA$

23. $y = 4x - x^2, \quad y = 0; \quad \displaystyle\iint\limits_{R} \sqrt{y + x^2}\, dA$

24. $x = 1 + 3y, \quad x = 1 - y, \quad y = 1;$

$\displaystyle\iint\limits_{R} (x + y + 1)^3\, dA$

25. $y = 1 - \sqrt{x}, \quad y = 1 + \sqrt{x}, \quad x = 4;$

$\displaystyle\iint\limits_{R} x(y - 1)^2\, dA$

26. $y = \frac{1}{2}x, \quad y = 6 - x, \quad y = 1; \quad \displaystyle\iint\limits_{R} \frac{1}{x + y}\, dA$

In Problems 27–32, evaluate each integral. Graph the region of integration, reverse the order of integration, and then evaluate the integral with the order reversed.

27. $\displaystyle\int_0^3 \int_0^{3-x} (x + 2y)\, dy\, dx$

28. $\displaystyle\int_0^2 \int_0^y (y - x)^4\, dx\, dy$

29. $\displaystyle\int_0^1 \int_0^{1-x^2} x\sqrt{y}\, dy\, dx$

30. $\displaystyle\int_0^2 \int_{x^3}^{4x} (1 + 2y)\, dy\, dx$

31. $\displaystyle\int_0^4 \int_{x/4}^{\sqrt{x/2}} x\, dy\, dx$

32. $\displaystyle\int_0^4 \int_{y^2/4}^{2\sqrt{y}} (1 + 2xy)\, dx\, dy$

In Problems 33–36, find the volume of the solid under the graph of $f(x, y)$ over the region R bounded by the graphs of the indicated equations. Sketch the region R— the solid does not have to be sketched.

33. $f(x, y) = 4 - x - y;$ R is bounded by the graphs of $x + y = 4, y = 0, x = 0$

34. $f(x, y) = (x - y)^2;$ R is the region bounded by the graphs of $y = x, y = 2, x = 0$

35. $f(x, y) = 4;$ R is the region bounded by the graphs of $y = 1 - x^2$ and $y = 0$ for $0 \leqslant x \leqslant 1$

36. $f(x, y) = 4xy;$ R is the region bounded by the graphs of $y = \sqrt{1 - x^2}$ and $y = 0$ for $0 \leqslant x \leqslant 1$

C   *In Problems 37–40, reverse the order of integration for each integral. Evaluate the integral with the order reversed. Do not attempt to evaluate the integral in the original form.*

37. $\displaystyle\int_0^2 \int_{x^2}^4 \frac{4x}{1 + y^2}\, dy\, dx$

38. $\displaystyle\int_0^1 \int_y^1 \sqrt{1 - x^2}\, dx\, dy$

39. $\displaystyle\int_0^1 \int_{y^2}^1 4ye^{x^2}\, dx\, dy$

40. $\displaystyle\int_0^4 \int_{\sqrt{x}}^2 \sqrt{3x + y^2}\, dy\, dx$

 In Problems 41–46, use a graphing utility to graph the region R bounded by the graphs of the indicated equations. Use approximation techniques to find intersection points correct to two decimal places. Describe R in set notation with double inequalities, and evaluate the indicated integral correct to two decimal places.

41. $y = 1 + \sqrt{x}, \quad y = x^2, \quad x = 0; \quad \displaystyle\iint\limits_{R} x\, dA$

42. $y = 1 + \sqrt[3]{x}, \quad y = x, \quad x = 0; \quad \displaystyle\iint\limits_{R} x\, dA$

43. $y = \sqrt[3]{x}, \quad y = 1 - x, \quad y = 0; \quad \iint\limits_{R} 24xy\, dA$

44. $y = x^3, \quad y = 1 - x, \quad y = 0; \quad \iint\limits_{R} 48xy\, dA$

45. $y = e^{-x}, \quad y = 3 - x; \quad \iint\limits_{R} 4y\, dA$

46. $y = e^x, \quad y = 2 + x; \quad \iint\limits_{R} 8y\, dA$

 In Problems 47–50, evaluate each integral to three decimal places using numerical integration on a graphing utility where appropriate.

47. $\displaystyle\int_0^2 \int_0^{9-x^2} ye^{x^2}\, dy\, dx$

48. $\displaystyle\int_1^3 \int_y^{y^2} \frac{\ln x}{xy^3}\, dx\, dy$

49. $\displaystyle\int_1^{10} \int_{1+y}^{\ln y} \frac{\sqrt{x}}{1 + y^2}\, dx\, dy$

50. $\displaystyle\int_1^5 \int_{\ln x}^{e^x} e^{x-y}\, dy\, dx$

Section B-7

Interpolating Polynomials and Divided Differences

- ❑ INTRODUCTION
- ❑ THE INTERPOLATING POLYNOMIAL
- ❑ DIVIDED DIFFERENCE TABLES
- ❑ APPLICATION

Given two points in the plane with distinct x coordinates, we can use the point-slope form of the equation of a line to find a polynomial whose graph passes through these two points. If we are given a set of three, four, or more points with distinct x coordinates, is there a polynomial whose graph will pass through all the given points? In this section, we will see that the answer to this question is yes, and we will discuss several methods for finding this polynomial, called the *interpolating polynomial*. The principal use of interpolating polynomials is to approximate y coordinates for points not in the given set. For example, a retail sales firm may have obtained a table of prices at various demands by examining past sales records. Prices for demands not in the table can be approximated by an interpolating polynomial. Interpolating polynomials also have applications to computer graphics. If a computer user selects a set of points on a drawing, the interpolating polynomial can be used to produce a smooth curve that passes through the selected points.

❑ INTRODUCTION

We usually write polynomials in standard form using either increasing or decreasing powers of x. Both of the following polynomials are written in standard form:

$$p(x) = 1 + x^2 - 2x^3 \qquad q(x) = -3x^5 + 2x^4 - 5x$$

In this section, we will find it convenient to write polynomials in a different form. The following activity will give you some experience with this new form.

Consider the points in Table 1.

TABLE 1

x	1	2	3	4
y	4	7	4	1

(A) Let $p_1(x) = a_0 + a_1(x - 1)$. Determine a_0 and a_1 so that the graph of $y = p_1(x)$ passes through the first two points in Table 1.

(B) Let $p_2(x) = a_0 + a_1(x - 1) + a_2(x - 1)(x - 2)$. Determine a_0, a_1, and a_2 so that the graph of $y = p_2(x)$ passes through the first three points in Table 1.

(C) Let
$$p_3(x) = a_0 + a_1(x - 1) + a_2(x - 1)(x - 2) + a_3(x - 1)(x - 2)(x - 3).$$

Determine a_0, a_1, a_2, and a_3 so that the graph of $y = p_3(x)$ passes through all four points in Table 1.

The following example will illustrate basic concepts.

Example 1 ✏ **Approximating Revenue** A manufacturing company has defined the revenue function for one of its products by examining past records and listing the revenue (in thousands of dollars) for certain levels of production (in thousands of units). Use the revenue function defined by Table 2 to estimate the revenue if 3,000 units are produced and if 7,000 units are produced.

SOLUTION   One way to approximate values of a function defined by a table is to use a **piecewise linear approximation.** To form the piecewise linear approximation for Table 2, we simply use the point-slope formula to find the equation of the line joining each successive pair of points in the table (see Fig. 1).

TABLE 2

REVENUE R DEFINED AS A FUNCTION OF PRODUCTION x BY A TABLE

x	1	4	6	8
$R(x)$	65	80	40	16

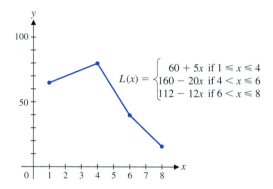

$$L(x) = \begin{cases} 60 + 5x & \text{if } 1 \le x \le 4 \\ 160 - 20x & \text{if } 4 < x \le 6 \\ 112 - 12x & \text{if } 6 < x \le 8 \end{cases}$$

FIGURE 1 $L(x)$ is the piecewise linear approximation for $R(x)$.

This type of approximation is very useful in certain applications, but it has several disadvantages. First, the piecewise linear approximation usually has a sharp corner at each point in the table and thus is not differentiable at these points. Second, the piecewise linear approximation requires the use of a different formula between each successive pair of points in the table (see Fig. 1). Instead of using the piecewise linear approximation, we will outline a method that will produce a polynomial whose values agree with $R(x)$ at each point in Table 2. This will provide us with a differentiable function given by a single formula that can be used to approximate $R(x)$ for any value of x between 1 and 8.

Suppose $p(x)$ is a polynomial whose values agree with the values of $R(x)$ at the four x values given in Table 2. Instead of expressing $p(x)$ in terms of powers of x, the standard method for writing polynomial forms, we use the first three x values in the table to write

$$p(x) = a_0 + a_1(x - 1) + a_2(x - 1)(x - 4) + a_3(x - 1)(x - 4)(x - 6)$$

As we will see, writing $p(x)$ in this special form will greatly simplify our work.

Since $p(x)$ is to agree with $R(x)$ at each x value in Table 2, we can write the following equations involving the coefficients a_0, a_1, a_2, and a_3:

$$65 = R(1) = p(1) = a_0 \tag{1}$$
$$80 = R(4) = p(4) = a_0 + 3a_1 \tag{2}$$
$$40 = R(6) = p(6) = a_0 + 5a_1 + 10a_2 \tag{3}$$
$$16 = R(8) = p(8) = a_0 + 7a_1 + 28a_2 + 56a_3 \tag{4}$$

From equation (1), we see that $a_0 = 65$. Solving equation (2) for a_1 and substituting for a_0, we have

$$a_1 = \frac{1}{3}(80 - a_0) = \frac{1}{3}(80 - 65) = 5$$

Proceeding the same way with equations (3) and (4), we have

$$a_2 = \frac{1}{10}(40 - a_0 - 5a_1) = \frac{1}{10}(40 - 65 - 25) = -5$$

$$a_3 = \frac{1}{56}(16 - a_0 - 7a_1 - 28a_2) = \frac{1}{56}(16 - 65 - 35 + 140) = 1$$

Thus,

$$p(x) = 65 + 5(x - 1) - 5(x - 1)(x - 4) + (x - 1)(x - 4)(x - 6)$$

The polynomial $p(x)$ agrees with $R(x)$ at each x value in Table 2 (verify this) and can be used to approximate $R(x)$ for values of x between 1 and 8 (see Fig. 2).

TABLE 2				
x	1	4	6	8
$R(x)$	65	80	40	16

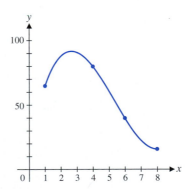

FIGURE 2 $p(x) = 65 + 5(x - 1) - 5(x - 1)(x - 4) + (x - 1)(x - 4)(x - 6)$

If 3,000 units are produced, then the revenue can be approximated by evaluating $p(3)$:

$$R(3) \approx p(3) = 65 + 5(2) - 5(2)(-1) + (2)(-1)(-3)$$
$$= 91 \text{ or } \$91{,}000$$

If 7,000 units are produced, then

$$R(7) \approx p(7) = 65 + 5(6) - 5(6)(3) + (6)(3)(1)$$
$$= 23 \text{ or } \$23,000$$

■

Matched Problem 1 ↝   Refer to Example 1. Approximate the revenue if 2,000 units are produced and if 5,000 units are produced.

■

Since the revenue function in Example 1 was defined by a table, we have no information about this function for any value of x other than those listed in the table. Thus, we cannot say anything about the accuracy of the approximations obtained by using $p(x)$. As we mentioned earlier, the piecewise linear approximation might provide a better approximation in some cases. The primary advantage of using $p(x)$ is that we have a differentiable function that is defined by a single equation and agrees with the revenue function at every value of x in the table.

❑   THE INTERPOLATING POLYNOMIAL

The procedure we used to find a polynomial approximation for the revenue function in Example 1 can be applied to any function that is defined by a table. The polynomial that is obtained in this way is referred to as the *interpolating polynomial*. The basic concepts are summarized in the next box.

The Interpolating Polynomial

If $f(x)$ is the function defined by the following table of $n + 1$ points,

x	x_0	x_1	...	x_n
$f(x)$	y_0	y_1	...	y_n

then the **interpolating polynomial** for $f(x)$ is the polynomial $p(x)$ of degree less than or equal to n that satisfies

$$p(x_0) = y_0 = f(x_0)$$
$$p(x_1) = y_1 = f(x_1)$$
$$\vdots \qquad \vdots$$
$$p(x_n) = y_n = f(x_n)$$

Newton's form for the interpolating polynomial is

$$p(x) = a_0 + a_1(x - x_0) + a_2(x - x_0)(x - x_1) + \cdots$$
$$+ a_n(x - x_0)(x - x_1) \cdot \cdots \cdot (x - x_{n-1})$$

Notice that if we graph the points in the defining table and the interpolating polynomial $p(x)$ on the same set of axes, then the graph of $p(x)$ will pass through every point given in the table (see Fig. 2). Is it possible to find a polynomial that is different from $p(x)$ and also has a graph that passes through all the points in the table? In more advanced texts, it is shown that

The interpolating polynomial is the only polynomial of degree less than or equal to n whose graph will pass through every point in the table.

Any other polynomial whose graph goes through all these points must be of degree greater than n. The steps we used in finding the interpolating polynomial are summarized in the following box.

Steps for Finding the Interpolating Polynomial

Step 1. Write Newton's form for $p(x)$.

Step 2. Use the conditions $p(x_i) = y_i$, $i = 0, 1, \ldots, n$ to write $n + 1$ linear equations for the coefficients. Do not change the order of these equations. This system of equations is called a **lower triangular system.**

Step 3. Starting with the first and proceeding down the list, solve each equation for the coefficient with the largest subscript and substitute all the previously determined coefficients. This method of solving for the coefficients is called **forward substitution.**

Example 2 ⇔   **Finding the Interpolating Polynomial**   Find the interpolating polynomial for the function defined by the following table:

x	0	1	2	3
$f(x)$	5	4	−3	−4

SOLUTION   **Step 1.** Newton's form for $p(x)$ is

$$p(x) = a_0 + a_1 x + a_2 x(x - 1) + a_3 x(x - 1)(x - 2)$$

Step 2. The lower triangular system is

$$
\begin{aligned}
a_0 &= 5 & p(0) = f(0) &= 5 \\
a_0 + a_1 &= 4 & p(1) = f(1) &= 4 \\
a_0 + 2a_1 + 2a_2 &= -3 & p(2) = f(2) &= -3 \\
a_0 + 3a_1 + 6a_2 + 6a_3 &= -4 & p(3) = f(3) &= -4
\end{aligned}
$$

Step 3. Solving this system by forward substitution, we have

$$a_0 = 5$$
$$a_1 = 4 - a_0 = 4 - 5 = -1$$
$$a_2 = \frac{1}{2}(-3 - a_0 - 2a_1) = \frac{1}{2}(-3 - 5 + 2) = -3$$
$$a_3 = \frac{1}{6}(-4 - a_0 - 3a_1 - 6a_2) = \frac{1}{6}(-4 - 5 + 3 + 18) = 2$$

Thus, Newton's form for this interpolating polynomial is

$$p(x) = 5 - x - 3x(x - 1) + 2x(x - 1)(x - 2)$$

This form of $p(x)$ is suitable for evaluating the polynomial. For other operations, such as differentiation, integration, or graphing, it may be preferable to perform the indicated operations, collect like terms, and express $p(x)$ in the standard polynomial form

$$p(x) = 2x^3 - 9x^2 + 6x + 5$$

The graph of $p(x)$ is shown in Figure 3.   ▪

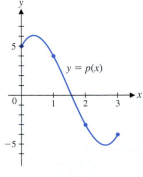

FIGURE 3

$p(x) = 2x^3 - 9x^2 + 6x + 5$

Matched Problem 2 ✏ Find the interpolating polynomial for the function defined by the following table:

x	−1	0	1	2
$f(x)$	5	3	3	11

■

Explore–Discuss 2

Given the following polynomial and table,

$$p(x) = x^3 - 10x^2 + 29x - 17$$

x	1	3	6
y	3	7	13

(A) Show that the graph of $p(x)$ passes through each point in the table.

(B) Is $p(x)$ the interpolating polynomial for this table? If not, what is the interpolating polynomial for this table?

(C) Discuss the relationship between the number of points in a table and the degree of the interpolating polynomial for that table.

❑ DIVIDED DIFFERENCE TABLES

We now present a simple computational procedure for finding the coefficients $a_0, a_1, \ldots, a_n$ in Newton's form for an interpolating polynomial. To introduce this method, we return to Table 2 in Example 1, which we restate here.

TABLE 2

x	1	4	6	8
$f(x)$	65	80	40	16

The coefficients in Newton's form for the interpolating polynomial for this table were $a_0 = 65$, $a_1 = 5$, $a_2 = -5$, and $a_3 = 1$. We will now construct a table, called a *divided difference table,* which will produce these coefficients with a minimum of computation. To begin, we place the x and y values in the first two columns of a new table. Then we compute the ratio of the change in y to the change in x for each successive pair of points in the table, and place the result on the line between the two points (see Table 3). These ratios are called the **first divided differences.**

To form the next column in the table, we repeat this process, using the change in the first divided differences in the numerator and the change in *two* successive values of x in the denominator. These ratios are called the **second divided differences** and are placed on the line between the corresponding first divided differences (see Table 4).

TABLE 3

FIRST DIVIDED DIFFERENCES

x_k	y_k	FIRST DIVIDED DIFFERENCE
1	65	
		$\dfrac{80-65}{4-1}=\dfrac{15}{3}=5$
4	80	
		$\dfrac{40-80}{6-4}=\dfrac{-40}{2}=-20$
6	40	
		$\dfrac{16-40}{8-6}=\dfrac{-24}{2}=-12$
8	16	

TABLE 4

SECOND DIVIDED DIFFERENCES

x_k	y_k	FIRST DIVIDED DIFFERENCE	SECOND DIVIDED DIFFERENCE
1	65		
		5	
4	80		$\dfrac{-20-5}{6-1}=\dfrac{-25}{5}=-5$
		−20	
6	40		$\dfrac{-12-(-20)}{8-4}=\dfrac{8}{4}=2$
		−12	
8	16		

To form the next column of the table, we form the ratio of the change in the second divided differences to the change in *three* successive values of x. These ratios are called the **third divided differences** and are placed on the line between the corresponding second divided differences (see Table 5). Since our table has only two second divided differences, there is only one third divided difference and this process is now complete.

TABLE 5

Third Divided Differences

x_k	y_k	FIRST DIVIDED DIFFERENCE	SECOND DIVIDED DIFFERENCE	THIRD DIVIDED DIFFERENCE
1	65			
		5		
4	80		-5	
		-20		$\dfrac{2-(-5)}{8-1}=\dfrac{7}{7}=1$
6	40		2	
		-12		
8	16			

We have presented each step in constructing the divided difference table here in a separate table to clearly illustrate this process. In applications of this technique, these steps are combined into a single table. With a little practice, you should be able to proceed quickly from the defining table for the function (Table 2) to the final form of the divided difference table (Table 6).

TABLE 6

Divided Difference Table—Final Form

x_k	y_k	FIRST DIVIDED DIFFERENCE	SECOND DIVIDED DIFFERENCE	THIRD DIVIDED DIFFERENCE
1	**65**			
		$\dfrac{80-65}{4-1}=$ **5**		
4	80		$\dfrac{-20-5}{6-1}=$ **-5**	
		$\dfrac{40-80}{6-4}=-20$		$\dfrac{2-(-5)}{8-1}=$ **1**
6	40		$\dfrac{-12-(-20)}{8-4}=2$	
		$\dfrac{16-40}{8-6}=-12$		
8	16			

Now that we have computed the divided difference table, how do we use it? If we write the first number from each column of the divided difference table, beginning with the second column:

65   5   −5   1

we see that these numbers are the coefficients of the interpolating polynomial for Table 1 (see Example 1). Thus, Table 6 contains all the information we need to write the interpolating polynomial:

$$p(x) = 65 + 5(x - 1) - 5(x - 1)(x - 4) + (x - 1)(x - 4)(x - 6)$$

The divided difference table provides an alternate method for finding interpolating polynomials that generally requires fewer computations and can be implemented easily on a computer. The ideas introduced in the preceding discussion are summarized in the following box.

Divided Difference Tables and Interpolating Polynomials

Given the defining table for a function $f(x)$ with $n + 1$ points,

x	x_0	x_1	$\cdots$	x_n
$f(x)$	y_0	y_1	$\cdots$	y_n

where $x_0 < x_1 < \cdots < x_n$, then the **divided difference table** is computed as follows:

Column 1: x values from the defining table

Column 2: y values from the defining table

Column 3:   First divided differences computed using columns 1 and 2

Column 4:   Second divided differences computed using columns 1 and 3

$\vdots$

Column $n + 2$: nth divided differences computed using columns 1 and $n + 1$

The coefficients in Newton's form for the interpolating polynomial,

$$p(x) = a_0 + a_1(x - x_0) + a_2(x - x_0)(x - x_1) + \cdots \\ + a_n(x - x_0)(x - x_1) \cdot \cdots \cdot (x - x_{n-1})$$

are the first numbers in each column of the divided difference table, beginning with column 2.

REMARKS

1. The points in the defining table must be arranged with increasing x values before computing the divided difference table. If the x values are out of order, then the divided difference table will not contain the coefficients of Newton's form for the interpolating polynomial.

2. Since each column in the divided difference table uses all the values in the preceding column, it is necessary to compute all the numbers in every column, even though we are interested only in the first number in each column.

3. Other methods can be used to find interpolating polynomials. Referring to Table 1, we could write $p(x)$ in standard polynomial notation

$$p(x) = b_3 x^3 + b_2 x^2 + b_1 x + b_0$$

and use the points in the table to write the following system of linear equations:

$$
\begin{aligned}
p(1) &= & b_3 + & b_2 + & b_1 + & b_0 &= 65 \\
p(4) &= & 64b_3 + & 16b_2 + & 4b_1 + & b_0 &= 80 \\
p(6) &= & 216b_3 + & 36b_2 + & 6b_1 + & b_0 &= 40 \\
p(8) &= & 512b_3 + & 64b_2 + & 8b_1 + & b_0 &= 16
\end{aligned}
$$

The computations required to solve this system of equations are far more complicated than those involved in finding the divided difference table.

Example 3   ➭   **Using a Divided Difference Table**

TABLE 7					
x	0	1	2	3	4
$f(x)$	35	25	19	-7	-29

(A) Find the divided difference table for the points in Table 7.

(B) Use the divided difference table to find the interpolating polynomial.

SOLUTION   (A) The divided difference table is as follows.

x_k	y_k	FIRST DIVIDED DIFFERENCE	SECOND DIVIDED DIFFERENCE	THIRD DIVIDED DIFFERENCE	FOURTH DIVIDED DIFFERENCE
0	35				
		$\dfrac{25-35}{1-0}=-10$			
1	25		$\dfrac{-6-(-10)}{2-0}=2$		
		$\dfrac{19-25}{2-1}=-6$		$\dfrac{-10-2}{3-0}=-4$	
2	19		$\dfrac{-26-(-6)}{3-1}=-10$		$\dfrac{4-(-4)}{4-0}=2$
		$\dfrac{-7-19}{3-2}=-26$		$\dfrac{2-(-10)}{4-1}=4$	
3	-7		$\dfrac{-22-(-26)}{4-2}=2$		
		$\dfrac{-29-(-7)}{4-3}=-22$			
4	-29				

(B) Newton's form for the interpolating polynomial is

$$
\begin{aligned}
p(x) = a_0 &+ a_1 x + a_2 x(x-1) + a_3 x(x-1)(x-2) \\
&+ a_4 x(x-1)(x-2)(x-3)
\end{aligned}
$$

Substituting the values from the divided difference table for the coefficients in Newton's form, we have

FIGURE 4

$$p(x) = 35 - 10x + 2x(x - 1) - 4x(x - 1)(x - 2)$$
$$+ 2x(x - 1)(x - 2)(x - 3)$$
$$= 35 - 32x + 36x^2 - 16x^3 + 2x^4 \quad \text{Standard form}$$

Multiplication details for the standard form are omitted. Figure 4 verifies that the values of the interpolating polynomial agree with the values in Table 7.

Matched Problem 3

TABLE 8					
x	0	1	2	3	4
$f(x)$	5	1	−1	−7	1

(A) Find the divided difference table for the points in Table 8.
(B) Use the divided difference table to find the interpolating polynomial.

Explore–Discuss 3

A graphing utility can be used to calculate a divided difference table. Figure 5A shows a program on a TI-83 that calculates divided difference tables and Figure 5B shows the input and output generated when we use this program to solve Example 3.

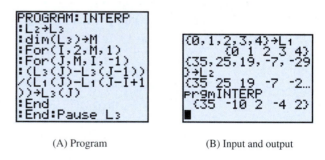

(A) Program   (B) Input and output

FIGURE 5

Enter this program into your TI-83 and use it to solve Matched Problem 3. If you have a different graphing utility, consult the manual for the correct format for each statement in the program.

❑ APPLICATION

Example 4

Inventory A store orders 8,000 units of a new product. The inventory I on hand t weeks after the order arrived is given in the following table:

INVENTORY					
t	0	2	4	6	8
$I(t)$	8,000	5,952	3,744	1,568	0

Use the interpolating polynomial to approximate the inventory after 5 weeks and the average inventory during the first 5 weeks after the order arrived.

SOLUTION   The divided difference table is as follows:

t_k	y_k	FIRST DIVIDED DIFFERENCE	SECOND DIVIDED DIFFERENCE	THIRD DIVIDED DIFFERENCE	FOURTH DIVIDED DIFFERENCE
0	8,000				
		−1,024			
2	5,952		−20		
		−1,104		4	
4	3,744		4		1
		−1,088		12	
6	1,568		76		
		−784			
8	0				

The interpolating polynomial is

$$p(t) = 8{,}000 - 1{,}024t - 20t(t - 2) + 4t(t - 2)(t - 4) \\ + t(t - 2)(t - 4)(t - 6)$$

or, after simplifying,

$$p(t) = t^4 - 8t^3 - 1{,}000t + 8{,}000$$

The inventory after 5 weeks is given approximately by

$$p(5) = 5^4 - 8(5^3) - 1{,}000(5) + 8{,}000 = 2{,}625 \text{ units}$$

The average inventory during the first 5 weeks is given approximately by

$$\frac{1}{5}\int_0^5 p(t)\,dt = \frac{1}{5}\int_0^5 (t^4 - 8t^3 - 1{,}000t + 8{,}000)\,dt$$

$$= \frac{1}{5}\left(\frac{1}{5}t^5 - 2t^4 - 500t^2 + 8{,}000t\right)\Big|_0^5$$

$$= \frac{1}{5}(625 - 1{,}250 - 12{,}500 + 40{,}000) - \frac{1}{5}(0)$$

$$= 5{,}375 \text{ units}$$

Matched Problem 4 ↩   Refer to Example 4. Approximate the inventory after 7 weeks and the average inventory during the first 7 weeks.

Answers to Matched Problems   **1.** $p(2) = 88$ or \$88,000; $p(5) = 61$ or \$61,000
2. $p(x) = 5 - 2(x + 1) + (x + 1)x + (x + 1)x(x - 1)$

3. (A)

x_k	y_k	FIRST DIVIDED DIFFERENCE	SECOND DIVIDED DIFFERENCE	THIRD DIVIDED DIFFERENCE	FOURTH DIVIDED DIFFERENCE
0	5				
		-4			
1	1		1		
		-2		-1	
2	-1		-2		1
		-6		3	
3	-7		7		
		8			
4	1				

$$
\begin{aligned}
\text{(B)}\quad p(x) &= 5 - 4x + x(x-1) - x(x-1)(x-2) \\
&\quad + x(x-1)(x-2)(x-3) \\
&= 5 - 13x + 15x^2 - 7x^3 + x^4
\end{aligned}
$$

4. 657 units; 4,294.2 units

Exercise B-7

A *In Problems 1–4,*

  (A) Write Newton's form for the interpolating polynomial.

  (B) Write the associated lower triangular system for the coefficients.

  (C) Use forward substitution to find the interpolating polynomial.

1.

x	1	3	4
$f(x)$	2	6	11

2.

x	-1	1	2
$f(x)$	1	3	7

3.

x	-1	0	2	4
$f(x)$	6	5	15	-39

4.

x	-1	0	2	3
$f(x)$	5	1	5	1

In Problems 5–10, find the divided difference table and then find the interpolating polynomial.

5.

x	1	2	3
$f(x)$	4	8	14

6.

x	1	2	3
$f(x)$	1	3	7

7.

x	-1	0	1	2
$f(x)$	-3	1	3	9

8.

x	-1	0	1	2
$f(x)$	5	6	3	2

9.

x	-2	1	2	4
$f(x)$	25	10	17	13

10.

x	-1	0	3	5
$f(x)$	17	10	25	5

B **11.** Can a table with three points have a linear inter- polating polynomial? A quadratic interpolating polynomial? A cubic interpolating polynomial? Explain.

12. Can a table with four points have a linear interpolat- ing polynomial? A quadratic interpolating polynomi- al? A cubic interpolating polynomial? A quartic interpolating polynomial? Explain.

In Problems 13–20, use the interpolating polynomial to approximate the value of the function defined by the table at the indicated values of x.

13.

x	−4	0	4	8
$f(x)$	−64	32	0	224

(A) $f(2) \approx ?$ (B) $f(6) \approx ?$

14.

x	−5	0	5	10
$f(x)$	250	50	100	−350

(A) $f(-3) \approx ?$ (B) $f(8) \approx ?$

15.

x	−1	0	1	4
$f(x)$	0	0	0	15

(A) $f(2) \approx ?$ (B) $f(3) \approx ?$

16.

x	−2	0	2	6
$f(x)$	0	0	0	−96

(A) $f(1) \approx ?$ (B) $f(4) \approx ?$

17.

x	−4	−2	0	2	4
$f(x)$	24	2	0	−6	8

(A) $f(-3) \approx ?$ (B) $f(1) \approx ?$

18.

x	−6	−2	0	2	6
$f(x)$	19	3	10	3	19

(A) $f(1) \approx ?$ (B) $f(5) \approx ?$

19.

x	−3	−2	−1	1	2	3
$f(x)$	−24	−6	0	0	6	24

(A) $f(-0.5) \approx ?$ (B) $f(2.5) \approx ?$

20.

x	−3	−2	−1	0	1	2	3
$f(x)$	40	0	0	4	0	0	40

(A) $f(-2.5) \approx ?$ (B) $f(1.5) \approx ?$

In Problems 21–30, find the interpolating polynomial. Graph the interpolating polynomial and the points in the given table on the same set of axes.

21.

x	−2	0	2
$f(x)$	2	0	2

22.

x	−2	0	2
$f(x)$	2	0	−2

23.

x	0	1	2
$f(x)$	−4	−2	0

24.

x	0	1	2
$f(x)$	−4	−3	0

25.

x	−1	0	2	3
$f(x)$	0	2	0	−4

26.

x	−3	−1	0	1
$f(x)$	0	4	3	0

27.

x	−2	−1	0	1	2
$f(x)$	1	5	3	1	5

28.

x	−2	−1	0	1	2
$f(x)$	−8	0	2	4	12

29.

x	−2	−1	0	1	2
$f(x)$	−3	0	5	0	−3

30.

x	−1	0	1	2	3
$f(x)$	6	2	0	−6	2

Use the quartic regression routine on a graphing utility to fit a fourth degree polynomial to the tables in the indicated problems. Compare this polynomial with the interpolating polynomial.

31. Problem 27   **32.** Problem 28

33. Problem 29   **34.** Problem 30

C 35. The following table was obtained from the function $f(x) = \sqrt{x}$:

x	1	4	9
$f(x)$	1	2	3

Find the interpolating polynomial for this table. Compare the values of the interpolating polynomial $p(x)$ and the original function $f(x) = \sqrt{x}$ by completing the table below. Use a calculator to evaluate $\sqrt{x}$ and round each value to one decimal place.

x	1	2	3	4	5	6	7	8	9
$p(x)$	1			2					3
$\sqrt{x}$	1			2					3

36. The following table was obtained from the function $f(x) = 6/\sqrt{x}$:

x	1	4	9
$f(x)$	6	3	2

Find the interpolating polynomial for this table. Compare the values of the interpolating polynomial $p(x)$

and the original function $f(x) = 6/\sqrt{x}$ by completing the table below. Use a calculator to evaluate $6/\sqrt{x}$ and round each value to one decimal place.

x	1	2	3	4	5	6	7	8	9
$p(x)$	6			3					2
$6/\sqrt{x}$	6			3					2

37. The following table was obtained from the function $f(x) = 10x/(1 + x^2)$:

x	−2	−1	0	1	2
$f(x)$	−4	−5	0	5	4

Find the interpolating polynomial $p(x)$ for this table. Graph $p(x)$ and $f(x)$ on the same set of axes.

38. The following table was obtained from the function $f(x) = (9 - x^2)/(1 + x^2)$:

x	−2	−1	0	1	2
$f(x)$	1	4	9	4	1

Find the interpolating polynomial $p(x)$ for this table. Graph $p(x)$ and $f(x)$ on the same set of axes.

39. Find the equation of the parabola whose graph passes through the points $(-x_1, y_1)$, $(0, y_2)$, and (x_1, y_1), where $x_1 > 0$ and $y_1 \neq y_2$.

40. Find the equation of the parabola whose graph passes through the points $(0, 0)$, (x_1, y_1), and $(2x_1, 0)$, where $x_1 > 0$ and $y_1 \neq 0$.

Applications

Business & Economics

41. *Cash reserves.* Suppose the cash reserves C (in thousands of dollars) for a small business are given by the following table, where t is the number of months after the first of the year.

t	0	4	8	12
$C(t)$	2	32	38	20

(A) Find the interpolating polynomial for this table.

(B) Use the interpolating polynomial to approximate (to the nearest thousand dollars) the cash reserves after 6 months.

(C) Use the interpolating polynomial to approximate (to the nearest hundred dollars) the average cash reserves for the first quarter.

42. *Inventory.* A hardware store orders 147 lawnmowers. The inventory I of lawnmowers on hand t months after the order arrived is given in the table.

t	0	1	2	3
$I(t)$	147	66	19	0

(A) Find the interpolating polynomial for this table.

(B) Use the interpolating polynomial to approximate (to the nearest integer) the average number of lawnmowers on hand for this three-month period.

43. *Income distribution.* The income distribution for the United States in 1999 is represented by the Lorenz curve $y = f(x)$, where $f(x)$ is given in the table.

x	0	0.2	0.8	1
$f(x)$	0	0.04	0.52	1

(A) Find the interpolating polynomial for this table.

(B) Use the interpolating polynomial to approximate (to four decimal places) the index of income concentration.

44. *Income distribution.* Refer to Problem 43. After making a series of adjustments for things like taxes, fringe benefits, and returns on home equity, the income distribution for the United States in 1999 is represented by the Lorenz curve $y = g(x)$, where $g(x)$ is given in the table.

x	0	0.2	0.8	1
$g(x)$	0	0.06	0.54	1

(A) Find the interpolating polynomial for this table.

(B) Use the interpolating polynomial to approximate (to four decimal places) the index of income concentration.

45. *Maximum revenue.* The revenue R (in thousands of dollars) from the sale of x thousand table lamps is given in the table.

x	2	4	6
$R(x)$	24.4	36	34.8

(A) Find the interpolating polynomial for this table.

(B) Use the interpolating polynomial to approximate (to the nearest thousand dollars) the revenue if 5,000 table lamps are produced.

(C) Use the interpolating polynomial to approximate (to the nearest integer) the production level that will maximize the revenue.

46. *Minimum average cost.* The cost C (in thousands of dollars) of producing x thousand microwave ovens is given in the table.

x	1	3	5
$C(x)$	215	535	1,055

(A) Find the interpolating polynomial for this table.

(B) Use the interpolating polynomial to approximate (to the nearest thousand dollars) the cost of producing 4,000 ovens.

(C) Use the interpolating polynomial to approximate (to the nearest integer) the production level that will minimize the average cost.

Life Sciences

47. *Temperature.* The temperature C (in degrees celsius) in an artificial habitat after t hours is given in the table.

t	0	1	2	3	4
$C(t)$	14	13	16	17	10

(A) Find the interpolating polynomial for this table.

(B) Use the interpolating polynomial to approximate (to the nearest tenth of a degree) the average temperature over this 4-hour period.

48. *Drug concentration.* The concentration C (in milligrams per cubic centimeter) of a particular drug in a patient's bloodstream t hours after the drug is taken is given in the table.

t	0	1	2	3	4
$C(t)$	0	0.032	0.036	0.024	0.008

(A) Find the interpolating polynomial for this table.

(B) Use the interpolating polynomial to approximate (to two decimal places) the number of hours it will take for the drug concentration to reach its maximum level.

49. *Bacteria control.* A lake that is used for recreational swimming is treated periodically to control harmful bacteria growth. The concentration C (in bacteria per cubic centimeter) t days after a treatment is given in the table.

t	0	2	4	6
$C(t)$	450	190	90	150

(A) Find the interpolating polynomial for this table.

(B) Use the interpolating polynomial to approximate (to two decimal places) the number of days it will take for the bacteria concentration to reach its minimum level.

50. *Medicine-respiration.* Physiologists use a machine called a pneumotachograph to produce a graph of the rate of flow R on air into the lungs (inspiration) and out (expiration). The figure gives the graph of the inspiration phase of the breathing cycle of an individual at rest.

(A) Use the values given by the graph at $t = 0, 1, 2$, and 3 to find the interpolating polynomial for R.

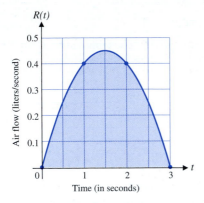

R(t)

Time (in seconds)

Air flow (liters/second)

(B) Use the interpolating polynomial to approximate (to one decimal place) the total volume of air inhaled.

Social Sciences

51. *Voter registration.* The number N of registered voters in a precinct over a 30-year period is given in the table.

t	0	10	20	30
$N(t)$	10,000	13,500	20,000	23,500

(A) Find the interpolating polynomial for this table.

(B) Use the interpolating polynomial to approximate (to the nearest thousand) the average number of voters over the first 20 years of this period.

52. *Voter registration.* The number N of registered voters in a precinct over a 10-year period is given in the table.

t	0	4	6	10
$N(t)$	15,000	18,800	22,200	26,000

(A) Find the interpolating polynomial for this table.

(B) Use the interpolating polynomial to approximate (to the nearest integer) the year t in which the rate of increase in the number of voters is most rapid.

Appendix C

Tables

TABLE I

BASIC GEOMETRIC FORMULAS

1. Similar Triangles
   (A) Two triangles are similar if two angles of one triangle have the same measure as two angles of the other.
   (B) If two triangles are similar, their corresponding sides are proportional:

$$\frac{a}{a'} = \frac{b}{b'} = \frac{c}{c'}$$

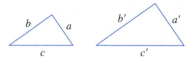

2. Pythagorean Theorem

$$c^2 = a^2 + b^2$$

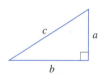

3. Rectangle

$$A = ab \qquad \text{Area}$$
$$P = 2a + 2b \qquad \text{Perimeter}$$

4. Parallelogram

$$h = \text{height}$$
$$A = ah = ab \sin \theta \qquad \text{Area}$$
$$P = 2a + 2b \qquad \text{Perimeter}$$

5. Triangle

$$h = \text{height}$$
$$A = \tfrac{1}{2}hc \qquad \text{Area}$$
$$P = a + b + c \qquad \text{Perimeter}$$
$$s = \tfrac{1}{2}(a + b + c) \qquad \text{Semiperimeter}$$
$$A = \sqrt{s(s - a)(s - b)(s - c)} \qquad \text{Area: Heron's formula}$$

6. Trapezoid
   Base a is parallel to base b.

$$h = \text{height}$$
$$A = \tfrac{1}{2}(a + b)h \qquad \text{Area}$$

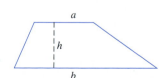

(continued)

TABLE I

(continued)

7. Circle

R = radius
D = diameter
$D = 2R$
$A = \pi R^2 = \frac{1}{4}\pi D^2$ Area
$C = 2\pi R = \pi D$ Circumference
$\dfrac{C}{D} = \pi$ For all circles
$\pi \approx 3.141\,59$

8. Rectangular Solid

$V = abc$ Volume
$T = 2ab + 2ac + 2bc$ Total surface area

9. Right Circular Cylinder

R = radius of base
h = height
$V = \pi R^2 h$ Volume
$S = 2\pi Rh$ Lateral surface area
$T = 2\pi R(R + h)$ Total surface area

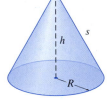

10. Right Circular Cone

R = radius of base
h = height
s = slant height
$V = \frac{1}{3}\pi R^2 h$ Volume
$S = \pi Rs = \pi R\sqrt{R^2 + h^2}$ Lateral surface area
$T = \pi R(R + s) = \pi R(R + \sqrt{R^2 + h^2})$ Total surface area

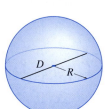

11. Sphere

R = radius
D = diameter
$D = 2R$
$V = \frac{4}{3}\pi R^3 = \frac{1}{6}\pi D^3$ Volume
$S = 4\pi R^2 = \pi D^2$ Surface area

TABLE II

INTEGRATION FORMULAS

[*Note:* The constant of integration is omitted for each integral, but must be included in any particular application of a formula. The variable u is the variable of integration; all other symbols represent constants.]

Integrals Involving u^n

1. $\int u^n \, du = \dfrac{u^{n+1}}{n+1}, \quad n \ne -1$

2. $\int u^{-1} \, du = \int \dfrac{1}{u} \, du = \ln|u|$

Integrals Involving $a + bu$, $a \ne 0$ and $b \ne 0$

3. $\int \dfrac{1}{a+bu} \, du = \dfrac{1}{b} \ln|a+bu|$

4. $\int \dfrac{u}{a+bu} \, du = \dfrac{u}{b} - \dfrac{a}{b^2} \ln|a+bu|$

5. $\int \dfrac{u^2}{a+bu} \, du = \dfrac{(a+bu)^2}{2b^3} - \dfrac{2a(a+bu)}{b^3} + \dfrac{a^2}{b^3} \ln|a+bu|$

6. $\int \dfrac{u}{(a+bu)^2} \, du = \dfrac{1}{b^2}\left(\ln|a+bu| + \dfrac{a}{a+bu} \right)$

7. $\int \dfrac{u^2}{(a+bu)^2} \, du = \dfrac{(a+bu)}{b^3} - \dfrac{a^2}{b^3(a+bu)} - \dfrac{2a}{b^3} \ln|a+bu|$

8. $\int u(a+bu)^n \, du = \dfrac{(a+bu)^{n+2}}{(n+2)b^2} - \dfrac{a(a+bu)^{n+1}}{(n+1)b^2}, \quad n \ne -1, -2$

9. $\int \dfrac{1}{u(a+bu)} \, du = \dfrac{1}{a} \ln\left| \dfrac{u}{a+bu} \right|$

10. $\int \dfrac{1}{u^2(a+bu)} \, du = -\dfrac{1}{au} + \dfrac{b}{a^2} \ln\left| \dfrac{a+bu}{u} \right|$

11. $\int \dfrac{1}{u(a+bu)^2} \, du = \dfrac{1}{a(a+bu)} + \dfrac{1}{a^2} \ln\left| \dfrac{u}{a+bu} \right|$

12. $\int \dfrac{1}{u^2(a+bu)^2} \, du = -\dfrac{a+2bu}{a^2u(a+bu)} + \dfrac{2b}{a^3} \ln\left| \dfrac{a+bu}{u} \right|$

Integrals Involving $a^2 - u^2$, $a > 0$

13. $\int \dfrac{1}{u^2-a^2} \, du = \dfrac{1}{2a} \ln\left| \dfrac{u-a}{u+a} \right|$

14. $\int \dfrac{1}{a^2-u^2} \, du = \dfrac{1}{2a} \ln\left| \dfrac{u+a}{u-a} \right|$

Integrals Involving $(a + bu)$ and $(c + du)$, $b \ne 0$, $d \ne 0$, and $ad - bc \ne 0$

15. $\int \dfrac{1}{(a+bu)(c+du)} \, du = \dfrac{1}{ad-bc} \ln\left| \dfrac{c+du}{a+bu} \right|$

16. $\int \dfrac{u}{(a+bu)(c+du)} \, du = \dfrac{1}{ad-bc}\left(\dfrac{a}{b} \ln|a+bu| - \dfrac{c}{d} \ln|c+du| \right)$

17. $\int \dfrac{u^2}{(a+bu)(c+du)} \, du = \dfrac{1}{bd} u - \dfrac{1}{ad-bc}\left(\dfrac{a^2}{b^2} \ln|a+bu| - \dfrac{c^2}{d^2} \ln|c+du| \right)$

18. $\int \dfrac{1}{(a+bu)^2(c+du)} \, du = \dfrac{1}{ad-bc} \dfrac{1}{a+bu} + \dfrac{d}{(ad-bc)^2} \ln\left| \dfrac{c+du}{a+bu} \right|$

19. $\int \dfrac{u}{(a+bu)^2(c+du)} \, du = -\dfrac{a}{b(ad-bc)} \dfrac{1}{a+bu} - \dfrac{c}{(ad-bc)^2} \ln\left| \dfrac{c+du}{a+bu} \right|$

20. $\int \dfrac{a+bu}{c+du} \, du = \dfrac{bu}{d} + \dfrac{ad-bc}{d^2} \ln|c+du|$

(continued)

TABLE II

(continued)

Integrals Involving $\sqrt{a + bu}$, $a \neq 0$ and $b \neq 0$

21. $\displaystyle\int \sqrt{a + bu}\, du = \frac{2\sqrt{(a + bu)^3}}{3b}$

22. $\displaystyle\int u\sqrt{a + bu}\, du = \frac{2(3bu - 2a)}{15b^2}\sqrt{(a + bu)^3}$

23. $\displaystyle\int u^2\sqrt{a + bu}\, du = \frac{2(15b^2u^2 - 12abu + 8a^2)}{105b^3}\sqrt{(a + bu)^3}$

24. $\displaystyle\int \frac{1}{\sqrt{a + bu}}\, du = \frac{2\sqrt{a + bu}}{b}$

25. $\displaystyle\int \frac{u}{\sqrt{a + bu}}\, du = \frac{2(bu - 2a)}{3b^2}\sqrt{a + bu}$

26. $\displaystyle\int \frac{u^2}{\sqrt{a + bu}}\, du = \frac{2(3b^2u^2 - 4abu + 8a^2)}{15b^3}\sqrt{a + bu}$

27. $\displaystyle\int \frac{1}{u\sqrt{a + bu}}\, du = \frac{1}{\sqrt{a}}\ln\left|\frac{\sqrt{a + bu} - \sqrt{a}}{\sqrt{a + bu} + \sqrt{a}}\right|, \quad a > 0$

28. $\displaystyle\int \frac{1}{u^2\sqrt{a + bu}}\, du = -\frac{\sqrt{a + bu}}{au} - \frac{b}{2a\sqrt{a}}\ln\left|\frac{\sqrt{a + bu} - \sqrt{a}}{\sqrt{a + bu} + \sqrt{a}}\right|, \quad a > 0$

Integrals Involving $\sqrt{a^2 - u^2}$, $a > 0$

29. $\displaystyle\int \frac{1}{u\sqrt{a^2 - u^2}}\, du = -\frac{1}{a}\ln\left|\frac{a + \sqrt{a^2 - u^2}}{u}\right|$

30. $\displaystyle\int \frac{1}{u^2\sqrt{a^2 - u^2}}\, du = -\frac{\sqrt{a^2 - u^2}}{a^2u}$

31. $\displaystyle\int \frac{\sqrt{a^2 - u^2}}{u}\, du = \sqrt{a^2 - u^2} - a\ln\left|\frac{a + \sqrt{a^2 - u^2}}{u}\right|$

Integrals Involving $\sqrt{u^2 + a^2}$, $a > 0$

32. $\displaystyle\int \sqrt{u^2 + a^2}\, du = \frac{1}{2}(u\sqrt{u^2 + a^2} + a^2\ln|u + \sqrt{u^2 + a^2}|)$

33. $\displaystyle\int u^2\sqrt{u^2 + a^2}\, du = \frac{1}{8}[u(2u^2 + a^2)\sqrt{u^2 + a^2} - a^4\ln|u + \sqrt{u^2 + a^2}|]$

34. $\displaystyle\int \frac{\sqrt{u^2 + a^2}}{u}\, du = \sqrt{u^2 + a^2} - a\ln\left|\frac{a + \sqrt{u^2 + a^2}}{u}\right|$

35. $\displaystyle\int \frac{\sqrt{u^2 + a^2}}{u^2}\, du = -\frac{\sqrt{u^2 + a^2}}{u} + \ln|u + \sqrt{u^2 + a^2}|$

36. $\displaystyle\int \frac{1}{\sqrt{u^2 + a^2}}\, du = \ln|u + \sqrt{u^2 + a^2}|$

37. $\displaystyle\int \frac{1}{u\sqrt{u^2 + a^2}}\, du = \frac{1}{a}\ln\left|\frac{u}{a + \sqrt{u^2 + a^2}}\right|$

38. $\displaystyle\int \frac{u^2}{\sqrt{u^2 + a^2}}\, du = \frac{1}{2}(u\sqrt{u^2 + a^2} - a^2\ln|u + \sqrt{u^2 + a^2}|)$

39. $\displaystyle\int \frac{1}{u^2\sqrt{u^2 + a^2}}\, du = -\frac{\sqrt{u^2 + a^2}}{a^2u}$

Integrals Involving $\sqrt{u^2 - a^2}$, $a > 0$

40. $\displaystyle\int \sqrt{u^2 - a^2}\, du = \frac{1}{2}(u\sqrt{u^2 - a^2} - a^2\ln|u + \sqrt{u^2 - a^2}|)$

41. $\displaystyle\int u^2\sqrt{u^2 - a^2}\, du = \frac{1}{8}[u(2u^2 - a^2)\sqrt{u^2 - a^2} - a^4\ln|u + \sqrt{u^2 - a^2}|]$

(continued)

TABLE II

(continued)

42. $\displaystyle\int \frac{\sqrt{u^2 - a^2}}{u^2}\, du = -\frac{\sqrt{u^2 - a^2}}{u} + \ln|u + \sqrt{u^2 - a^2}|$

43. $\displaystyle\int \frac{1}{\sqrt{u^2 - a^2}}\, du = \ln|u + \sqrt{u^2 - a^2}|$

44. $\displaystyle\int \frac{u^2}{\sqrt{u^2 - a^2}}\, du = \frac{1}{2}(u\sqrt{u^2 - a^2} + a^2 \ln|u + \sqrt{u^2 - a^2}|)$

45. $\displaystyle\int \frac{1}{u^2\sqrt{u^2 - a^2}}\, du = \frac{\sqrt{u^2 - a^2}}{a^2 u}$

Integrals Involving e^{au}, $a \neq 0$

46. $\displaystyle\int e^{au}\, du = \frac{e^{au}}{a}$

47. $\displaystyle\int u^n e^{au}\, du = \frac{u^n e^{au}}{a} - \frac{n}{a}\int u^{n-1} e^{au}\, du$

48. $\displaystyle\int \frac{1}{c + de^{au}}\, du = \frac{u}{c} - \frac{1}{ac}\ln|c + de^{au}|, \quad c \neq 0$

Integrals Involving ln u

49. $\displaystyle\int \ln u\, du = u \ln u - u$

50. $\displaystyle\int \frac{\ln u}{u}\, du = \frac{1}{2}(\ln u)^2$

51. $\displaystyle\int u^n \ln u \cdot du = \frac{u^{n+1}}{n + 1}\ln u - \frac{u^{n+1}}{(n + 1)^2}, \quad n \neq -1$

52. $\displaystyle\int (\ln u)^n\, du = u(\ln u)^n - n\int (\ln u)^{n-1}\, du$

Integrals Involving Trigonometric Functions of au, $a \neq 0$

53. $\displaystyle\int \sin au\, du = -\frac{1}{a}\cos au$

54. $\displaystyle\int \cos au\, du = \frac{1}{a}\sin au$

55. $\displaystyle\int \tan au\, du = -\frac{1}{a}\ln|\cos au|$

56. $\displaystyle\int \cot au\, du = \frac{1}{a}\ln|\sin au|$

57. $\displaystyle\int \sec au\, du = \frac{1}{a}\ln|\sec au + \tan au|$

58. $\displaystyle\int \csc au\, du = \frac{1}{a}\ln|\csc au - \cot au|$

59. $\displaystyle\int (\sin au)^2\, du = \frac{u}{2} - \frac{1}{4a}\sin 2au$

60. $\displaystyle\int (\cos au)^2\, du = \frac{u}{2} + \frac{1}{4a}\sin 2au$

61. $\displaystyle\int (\sin au)^n\, du = -\frac{1}{an}(\sin au)^{n-1}\cos au + \frac{n-1}{n}\int (\sin au)^{n-2}\, du, \quad n \neq 0$

62. $\displaystyle\int (\cos au)^n\, du = \frac{1}{an}\sin au(\cos au)^{n-1} + \frac{n-1}{n}\int (\cos au)^{n-2}\, du, \quad n \neq 0$

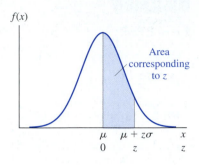

f(x)

Area corresponding to z

TABLE III

AREA UNDER THE STANDARD NORMAL CURVE

(TABLE ENTRIES REPRESENT THE AREA UNDER THE STANDARD NORMAL CURVE FROM 0 TO z, z ≥ 0)

z	.00	.01	.02	.03	.04	.05	.06	.07	.08	.09
0.0	0.0000	0.0040	0.0080	0.0120	0.0160	0.0199	0.0239	0.0279	0.0319	0.0359
0.1	0.0398	0.0438	0.0478	0.0517	0.0557	0.0596	0.0636	0.0675	0.0714	0.0753
0.2	0.0793	0.0832	0.0871	0.0910	0.0948	0.0987	0.1026	0.1064	0.1103	0.1141
0.3	0.1179	0.1217	0.1255	0.1293	0.1331	0.1368	0.1406	0.1443	0.1480	0.1517
0.4	0.1554	0.1591	0.1628	0.1664	0.1700	0.1736	0.1772	0.1808	0.1844	0.1879
0.5	0.1915	0.1950	0.1985	0.2019	0.2054	0.2088	0.2123	0.2157	0.2190	0.2224
0.6	0.2257	0.2291	0.2324	0.2357	0.2389	0.2422	0.2454	0.2486	0.2517	0.2549
0.7	0.2580	0.2611	0.2642	0.2673	0.2704	0.2734	0.2764	0.2794	0.2823	0.2852
0.8	0.2881	0.2910	0.2939	0.2967	0.2995	0.3023	0.3051	0.3078	0.3106	0.3133
0.9	0.3159	0.3186	0.3212	0.3238	0.3264	0.3289	0.3315	0.3340	0.3365	0.3389
1.0	0.3413	0.3438	0.3461	0.3485	0.3508	0.3531	0.3554	0.3577	0.3599	0.3621
1.1	0.3643	0.3665	0.3686	0.3708	0.3729	0.3749	0.3770	0.3790	0.3810	0.3830
1.2	0.3849	0.3869	0.3888	0.3907	0.3925	0.3944	0.3962	0.3980	0.3997	0.4015
1.3	0.4032	0.4049	0.4066	0.4082	0.4099	0.4115	0.4131	0.4147	0.4162	0.4177
1.4	0.4192	0.4207	0.4222	0.4236	0.4251	0.4265	0.4279	0.4292	0.4306	0.4319
1.5	0.4332	0.4345	0.4357	0.4370	0.4382	0.4394	0.4406	0.4418	0.4429	0.4441
1.6	0.4452	0.4463	0.4474	0.4484	0.4495	0.4505	0.4515	0.4525	0.4535	0.4545
1.7	0.4554	0.4564	0.4573	0.4582	0.4591	0.4599	0.4608	0.4616	0.4625	0.4633
1.8	0.4641	0.4649	0.4656	0.4664	0.4671	0.4678	0.4686	0.4693	0.4699	0.4706
1.9	0.4713	0.4719	0.4726	0.4732	0.4738	0.4744	0.4750	0.4756	0.4761	0.4767
2.0	0.4772	0.4778	0.4783	0.4788	0.4793	0.4798	0.4803	0.4808	0.4812	0.4817
2.1	0.4821	0.4826	0.4830	0.4834	0.4838	0.4842	0.4846	0.4850	0.4854	0.4857
2.2	0.4861	0.4864	0.4868	0.4871	0.4875	0.4878	0.4881	0.4884	0.4887	0.4890
2.3	0.4893	0.4896	0.4898	0.4901	0.4904	0.4906	0.4909	0.4911	0.4913	0.4916
2.4	0.4918	0.4920	0.4922	0.4925	0.4927	0.4929	0.4931	0.4932	0.4934	0.4936
2.5	0.4938	0.4940	0.4941	0.4943	0.4945	0.4946	0.4948	0.4949	0.4951	0.4952
2.6	0.4953	0.4955	0.4956	0.4957	0.4959	0.4960	0.4961	0.4962	0.4963	0.4964
2.7	0.4965	0.4966	0.4967	0.4968	0.4969	0.4970	0.4971	0.4972	0.4973	0.4974
2.8	0.4974	0.4975	0.4976	0.4977	0.4977	0.4978	0.4979	0.4979	0.4980	0.4981
2.9	0.4981	0.4982	0.4982	0.4983	0.4984	0.4984	0.4985	0.4985	0.4986	0.4986
3.0	0.4987	0.4987	0.4987	0.4988	0.4988	0.4989	0.4989	0.4989	0.4990	0.4990
3.1	0.4990	0.4991	0.4991	0.4991	0.4992	0.4992	0.4992	0.4992	0.4993	0.4993
3.2	0.4993	0.4993	0.4994	0.4994	0.4994	0.4994	0.4994	0.4995	0.4995	0.4995
3.3	0.4995	0.4995	0.4995	0.4996	0.4996	0.4996	0.4996	0.4996	0.4996	0.4997
3.4	0.4997	0.4997	0.4997	0.4997	0.4997	0.4997	0.4997	0.4997	0.4997	0.4998
3.5	0.4998	0.4998	0.4998	0.4998	0.4998	0.4998	0.4998	0.4998	0.4998	0.4998
3.6	0.4998	0.4998	0.4999	0.4999	0.4999	0.4999	0.4999	0.4999	0.4999	0.4999
3.7	0.4999	0.4999	0.4999	0.4999	0.4999	0.4999	0.4999	0.4999	0.4999	0.4999
3.8	0.4999	0.4999	0.4999	0.4999	0.4999	0.4999	0.4999	0.4999	0.4999	0.4999
3.9	0.5000	0.5000	0.5000	0.5000	0.5000	0.5000	0.5000	0.5000	0.5000	0.5000

CHAPTER 1

Exercise 1-1

1. Function   **3.** Not a function   **5.** Function   **7.** Function   **9.** Not a function   **11.** Function
13. 1   **15.** -5 **17.** 15   **19.** -3 **21.** 3   **23.** 7   **25.** 7   **27.** -45 **29.** 0   **31.** $y = 0$ **33.** $y = 4$
35. $x = -5, 0, 4$ **37.** $x = -6$ **39.** All real numbers   **41.** All real numbers except -4
43. All real numbers except -4 and 1   **45.** $x \leq 7$ **47.** $x < 7$
49. $f(2) = 0$, and 0 is a number; therefore, $f(2)$ exists. On the other hand, $f(3)$ is not defined, since the denominator would be 0; therefore, we say that $f(3)$ does not exist.
51. $g(x) = 2x^3 - 5$ **53.** $G(x) = 2\sqrt{x} - x^2$
55. Function f multiplies the domain element by 2 and subtracts 3 from the result.
57 Function F multiplies the cube of the domain element by 3 and subtracts twice the square root of the domain element from the result.
59. A function with domain R **61.** A function with domain R **63.** Not a function; for example, when $x = 1, y = \pm3$
65. A function with domain all real numbers except $x = 4$ **67.** Not a function; for example, when $x = 4, y = \pm3$
69. 4   **71.** $h - 1$ **73.** 4   **75.** $8a + 4h - 7$ **77.** $3a^2 + 3ah + h^2$ **79.** $\dfrac{1}{\sqrt{a + h} + \sqrt{a}}$

81. $P(w) = 2w + \dfrac{50}{w}, \quad w > 0$ **83.** $A(l) = l(50 - l), \quad 0 \leq l \leq 50$

85. \$54, \$42   **87.** (A) $R(x) = (75 - 3x)x, \quad 1 \leq x \leq 20$ (B)

(B)

x	$R(x)$
1	72
4	252
8	408
12	468
16	432
20	300

(C)
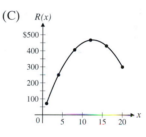

89. (A) $P(x) = 59x - 3x^2 - 125, \quad 1 \leq x \leq 20$ (B)

x	$P(x)$
1	-69
4	63
8	155
12	151
16	51
20	-145

(C)
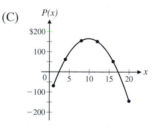

91. (A) $V(x) = x(8 - 2x)(12 - 2x)$ (B) $0 \leq x \leq 4$ (C)

x	$V(x)$
1	60
2	64
3	36

(D)

93. (A) The graph indicates that there is a value of x near 2 that will produce a volume of 65. The table shows $x = 1.9$ to one decimal place:

x	1.8	1.9	2
$V(x)$	66.5	65.4	64

(B) $x = 1.93$ to two decimal places

95. $v = \dfrac{75 - w}{15 + w}$; 1.9032 cm/sec

Exercise 1-2

1. Domain: all real numbers; range: all real numbers
3. Domain: $[0, \infty)$; range: $(-\infty, 0]$
5. Domain: all real numbers; range: $[0, \infty)$
7. Domain: all real numbers; range: all real numbers

9.

11.

13.

15.

17.

19.

21. The graph of $g(x) = -|x + 3|$ is the graph of $y = |x|$ reflected in the x axis and shifted 3 units to the left.

23. The graph of $f(x) = (x - 4)^2 - 3$ is the graph of $y = x^2$ shifted 4 units to the right and 3 units down.

25. The graph of $f(x) = 7 - \sqrt{x}$ is the graph of $y = \sqrt{x}$ reflected in the x axis and shifted 7 units up.

27. The graph of $h(x) = -3|x|$ is the graph of $y = |x|$ reflected in the x axis and vertically expanded by a factor of 3.

29. The graph of the basic function $y = x^2$ is shifted 2 units to the left and 3 units down. Equation: $y = (x + 2)^2 - 3$.
31. The graph of the basic function $y = x^2$ is reflected in the x axis and shifted 3 units to the right and 2 units up. Equation: $y = 2 - (x - 3)^2$.
33. The graph of the basic function $y = \sqrt{x}$ is reflected in the x axis and shifted 4 units up. Equation: $y = 4 - \sqrt{x}$.
35. The graph of the basic function $y = x^3$ is shifted 2 units to the left and 1 unit down. Equation: $y = (x + 2)^3 - 1$.

37. $g(x) = \sqrt{x-2} - 3$ **39.** $g(x) = -|x+3|$ **41.** $g(x) = -(x-2)^3 - 1$ **43.**

45.

47.

49. The graph of the basic function $y = |x|$ is reflected in the x axis and vertically contracted by a factor of 0.5.
Equation: $y = -0.5|x|$.

51. The graph of the basic function $y = x^2$ is reflected in the x axis and vertically expanded by a factor of 2.
Equation: $y = -2x^2$.

53. The graph of the basic function $y = \sqrt[3]{x}$ is reflected in the x axis and vertically expanded by a factor of 3.
Equation: $y = -3\sqrt[3]{x}$.

55. Reversing the order does not change the result.

57. Reversing the order can change the result.

59. Reversing the order can change the result.

61. (A) The graph of the basic function $y = \sqrt{x}$ is reflected in the x axis, vertically expanded by a factor of 4, and shifted up 115 units.

(B)

63. (A) The graph of the basic function $y = x^3$ is vertically contracted by a factor of 0.000 48 and shifted right 500 units and up 60,000 units.

(B)

65. (A) $S(x) = \begin{cases} 8.5 + 0.065x & \text{if } 0 \le x \le 700 \\ -9 + 0.09x & \text{if } x > 700 \end{cases}$

(B)

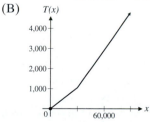

67. (A) $T(x) = \begin{cases} 0.035x & \text{if } 0 \le x \le 30{,}000 \\ 0.0625x - 825 & \text{if } 30{,}000 < x \le 60{,}000 \\ 0.0645x - 945 & \text{if } x > 60{,}000 \end{cases}$

(B)

(C) \$1,675; \$3,570

69. (A) The graph of the basic function $y = x$ is vertically expanded by a factor of 5.5 and shifted down 220 units.

(B)

71. (A) The graph of the basic function $y = \sqrt{x}$ is vertically expanded by a factor of 7.08.

(B)

Exercise 1-3

1. (D)   **3.** (C); slope is 0   **5.**

7.

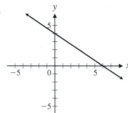

9. Slope $= 3$; y intercept $= 1$

11. Slope $= -\frac{3}{7}$; y intercept $= -6$ **13.** $y = -2x + 3$ **15.** $y = \frac{4}{3}x - 4$

17.

19.

21.

23. -4 **25.** $-\frac{3}{5}$

27. (A)

(B) x intercept: 3.5;
y intercept: -4.2

(C)

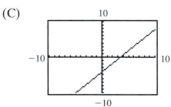

(D) x intercept: 3.5;
y intercept: -4.2
(E) $x > 3.5$ or $(3.5, \infty)$

29. $x = 4, y = -3$ **31.** $x = -1.5, y = -3.5$ **33.** $y = -4x + 5$ **35.** $y = \frac{3}{2}x + 1$ **37.** $y = 4.6$ **39.** $\frac{2}{3}$
41. $-\frac{5}{4}$ **43.** Not defined   **45.** 0   **47.** $2x - 3y = -11$; linear function   **49.** $5x + 4y = -14$; linear function
51. $x = 5$; neither   **53.** $y = 5$; constant function   **55.** The graphs have the same y intercept, $(0, 2)$.
57. (A)

(B) Varying C produces a family of parallel lines. This is verified by observing that varying C does not change the slope of the lines but changes the intercepts.

59. The graph of g is the same as the graph of f for x satisfying $mx + b \geqslant 0$ and the reflection of the graph of f in the x axis for x satisfying $mx + b < 0$. The function g is never linear.

61. (A) $130; $220
   (B)
   (C) The slope is 6. The amount in the account is growing at the rate of $6 per year.

63. (A) $C(x) = 180x + 200$
   (B) $2,360
   (C)

65. (A)
   (B) $p(x) = -\frac{1}{60}x + 200$
   (C) $p(3,000) = 150
   (D) The slope is $-\frac{1}{60} \approx -0.02$. The price decreases $0.02 for each unit increase in demand.

67. (A) Supply: $p = 0.4x - 0.9$; demand: $p = -0.4x + 6.42$ (B) 9.15 million bushels at $2.76 per bushel

69. (A)

x	0	1	2	3	4
SALES	16.7	19.8	23.6	26.9	32.7
$f(x)$	16.1	20.0	23.9	27.9	31.8

(B)

   (C) $55.22 billion, $74.77 billion

71. (A) $f(x) = -1.25x + 97$ (B) 63.25%, 30.75%   (C) Mid-1997

73. $0.2x + 0.1y = 20$

75. (A) 64 g; 35 g   (B)   (C) $-\frac{1}{5}$

Exercise 1-4

1. $(x - 2)^2 - 1$ **3.** $-(x - 3)^2 + 5$ **5.** The graph of $f(x)$ is the graph of $y = x^2$ shifted right 2 units and down 1 unit.

7. The graph of $m(x)$ is the graph of $y = x^2$ reflected in the x axis, then shifted right 3 units and up 5 units.

9. (A) m (B) g (C) f (D) n

11. (A) x intercepts: 1, 3; y intercept: -3 (B) Vertex: $(2, 1)$ (C) Maximum: 1   (D) Range: $y \le 1$ or $(-\infty, 1]$
   (E) Increasing interval: $x \le 2$ or $(-\infty, 2]$ (F) Decreasing interval: $x \ge 2$ or $[2, \infty)$

13. (A) x intercepts: $-3, -1$; y intercept: 3   (B) Vertex: $(-2, -1)$ (C) Minimum: -1 (D) Range: $y \ge -1$ or $[-1, \infty)$
   (E) Increasing interval: $x \ge -2$ or $[-2, \infty)$ (F) Decreasing interval: $x \le -2$ or $(-\infty, -2]$

15. (A) x intercepts: $3 \pm \sqrt{2}$, y intercept: -7 (B) Vertex: $(3, 2)$ (C) Maximum: 2   (D) Range: $y \le 2$ or $(-\infty, 2]$

17. (A) x intercepts: $-1 \pm \sqrt{2}$, y intercept: -1 (B) Vertex: $(-1, -2)$ (C) Minimum: -2 (D) Range: $y \ge -2$ or $[-2, \infty)$

19. $y = -[x - (-2)]^2 + 5$ or $y = -(x + 2)^2 + 5$ **21.** $y = (x - 1)^2 - 3$

23. Standard form: $(x - 4)^2 - 4$ (A) x intercepts: 2 and 6, y intercept: 12   (B) Vertex: $(4, -4)$ (C) Minimum: -4
   (D) Range: $y \ge -4$ or $[-4, \infty)$

25. Standard form: $-4(x - 2)^2 + 1$ (A) x intercepts: 1.5 and 2.5, y intercept: -15 (B) Vertex: $(2, 1)$ (C) Maximum: 1
(D) Range: $y \leqslant 1$ or $(-\infty, 1]$
27. Standard form: $0.5(x - 2)^2 + 3$ (A) x intercepts: none, y intercept: 5   (B) Vertex: $(2, 3)$ (C) Minimum: 3
(D) Range: $y \geqslant 3$ or $[3, \infty)$ **29.** (A) $-4.87, 8.21$ (B) $-3.44, 6.78$ (C) No solution
31. The vertex of the parabola is on the x axis.
33. $g(x) = 0.25(x - 3)^2 - 9.25$ (A) x intercepts: $-3.08, 9.08$; y intercept: -7 (B) Vertex: $(3, -9.25)$
(C) Minimum: -9.25 (D) Range: $y \geqslant -9.25$ or $[-9.25, \infty)$
35. $f(x) = -0.12(x - 4)^2 + 3.12$ (A) x intercepts: $-1.1, 9.1$; y intercept: 1.2   (B) Vertex: $(4, 3.12)$ (C) Maximum: 3.12
(D) Range: $y \leqslant 3.12$ or $(-\infty, 3.12]$ **37.** $x = -5.37, 0.37$
39. $-1.37 < x < 2.16$ **41.** $x \leqslant -0.74$ or $x \geqslant 4.19$ **43.** Axis: $x = 2$; vertex: $(2, 4)$; range: $y \geqslant 4$ or $[4, \infty)$; no x intercepts
45. (A)   (B) 1.64, 7.61
(C) $1.64 < x < 7.61$
(D) $0 \leqslant x < 1.64$
or $7.61 < x \leqslant 10$

47. (A)   (B) 1.10, 5.57
(C) $1.10 < x < 5.57$
(D) $0 \leqslant x < 1.10$
or $5.57 < x \leqslant 8$

49. $f(x) = x^2 + 1$ and $g(x) = -(x - 4)^2 - 1$ are two examples.
The graphs do not cross the x axis.

51. (A)

x	28	30	32	34	36
MILEAGE	45	52	55	51	47
$f(x)$	45.3	51.8	54.2	52.4	46.5

(B)

(C) $f(31) = 53.50$ thousand miles; $f(35) = 49.95$ thousand miles

53. (A)   (B) 12,500,000 chips; $468,750,000
(C) $37.50

55. (A)   (B) 2,415,000 chips and 17,251,000 chips
(C) Loss: $1 \leqslant x < 2.415$ or $17.251 < x \leqslant 20$; profit: $2.415 < x < 17.251$

57. (A) $P(x) = 59x - 3x^2 - 125$

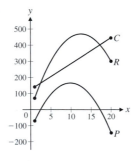

(C) Intercepts and break-even points: 2,415,000 chips and 17,251,000 chips
(E) Maximum profit is $165,083,000 at a production level of 9,833,000 chips. This is much smaller than the maximum revenue of $468,750,000.

59. $x = 0.14$ cm

Review Exercise

1. *(1-1)*

2. (A) Not a function   (B) A function   (C) A function   (D) Not a function *(1-1)*
3. (A) -2 (B) -8 (C) 0   (D) Not defined *(1-1)*
4. (A) $y = 4$ (B) $x = 0$ (C) $y = 1$ (D) $x = -1$ or 1   (E) $y = -2$
(F) $x = -5$ or 5 *(1-1)*

5. (A)   (B)   (C)   (D)

(1-2)

6. (A) n (B) g (C) m; slope is zero (D) f; slope is not defined *(1-3)* **7.** $y = -\frac{2}{3}x + 6$ *(1-3)*

8. Vertical line: $x = -6$; horizontal line: $y = 5$ *(1-3)*

9. x intercept $= 9$; y intercept $= -6$; slope $= \frac{2}{3}$ *(1-3)*

10. $f(x) = -(x - 2)^2 + 4$. The graph of $f(x)$ is the graph of $y = x^2$ reflected in the x axis, then shifted right 2 units and up 4 units.

11. (A) g (B) m (C) n (D) f *(1-2, 1-4)*

12. (A) x intercepts: $-4, 0$; y intercept: 0
    (B) Vertex: $(-2, -4)$ (C) Minimum: -4
    (D) Range: $y \geqslant -4$ or $[-4, \infty)$
    (E) Increasing on $[-2, \infty)$
    (F) Decreasing on $(-\infty, -2]$ *(1-4)*

13. Linear functions: (A), (C), (E), (F);
    constant function: (D) *(1-3)*

14. (A) All real numbers except $x = -2$ and 3
    (B) $x < 5$ *(1-1)*

15. Function g multiplies a domain element by 2 and then subtracts 3 times the square root of the domain element from the result. *(1-1)*

16. Standard form: $4(x + \frac{1}{2})^2 - 4$
    (A) x intercepts: $-\frac{3}{2}$ and $\frac{1}{2}$; y intercept: -3
    (B) Vertex: $(-\frac{1}{2}, -4)$ (C) Minimum: -4
    (D) Range: $y \geqslant -4$ or $[-4, \infty)$

17. The graph of $x = -3$ is a vertical line with x intercept -3, and the graph of $y = 2$ is a horizontal line with y intercept 2. *(1-3)*

18. (A)

    (B) The equation $f(x) = 3$ has one solution; the equation $f(x) = 2$ has two solutions; and the equation $f(x) = 1$ has three solutions.
    (C) $f(x) = 3$: $x = -4.28$; $f(x) = 2$: $x = -4.19, 1.19$; $f(x) = 1$: $x = -4.10, 0.35, 1.75$ *(1-1)*

19. -2 *(1-1)*   **20.** $2a + h - 3$ *(1-1)*

21. The graph of function m is the graph of $y = |x|$ reflected in the x axis and shifted to the right 4 units. *(1-2)*

22. The graph of function g is the graph of $y = x^3$ vertically contracted by a factor of 0.3 and shifted up 3 units. *(1-2)*

23. The graph of $y = x^2$ is vertically expanded by a factor of 2, reflected in the x axis, and shifted to the left 3 units.
   Equation: $y = -2(x + 3)^2$ *(1-2)*

24. $f(x) = 2\sqrt{x + 3} - 1$ *(1-2)*   **25.**

(1-2)

26. (A) $y = -\frac{2}{3}x$ (B) $y = 3$ *(1-3)*   **27.** (A) $3x + 2y = 1$ (B) $y = 5$ (C) $x = -2$ *(1-3)*

28. $y = -(x - 4)^2 + 3$ *(1-2, 1-4)*

29. $f(x) = -0.4(x - 4)^2 + 7.6$ (A) x intercepts: $-0.4, 8.4$; y intercept: 1.2 (B) Vertex: $(4.0, 7.6)$
   (C) Maximum: 7.6 (D) Range: $x \leq 7.6$ or $(-\infty, 7.6]$ *(1-4)*

30.

   (A) x intercepts: $-0.4, 8.4$; y intercept: 1.2
   (B) Vertex: $(4.0, 7.6)$
   (C) Maximum: 7.6
   (D) Range: $x \leq 7.6$ or $(-\infty, 7.6]$ *(1-4)*

31. The graph of $y = \sqrt[3]{x}$ is vertically expanded by a factor of 2, reflected in the x axis, and shifted 1 unit left and 1 unit down.
   Equation: $y = -2\sqrt[3]{x + 1} - 1$. *(1-2)*

32. The graphs appear to be perpendicular to each other. (It can be shown that if the slopes of two slant lines are the negative reciprocals of each other, then the two lines are perpendicular.) *(1-3)*

33. (A) $\dfrac{1}{\sqrt{x + h} + \sqrt{x}}$ (B) $\dfrac{-1}{x(x + h)}$ *(1-2)*

34. $G(x) = 0.3(x + 2)^2 - 8.1$ (A) x intercepts: $-7.2, 3.2$; y intercept: -6.9 (B) Vertex: $(-2, -8.1)$ (C) Minimum: -8.1
   (D) Range: $x \geq -8.1$ or $[-8.1, \infty)$ (E) Decreasing: $(-\infty, -2]$; increasing: $[-2, \infty)$ *(1-4)*

35.

   (A) x intercepts: $-7.2, 3.2$;
      y intercept: -6.9
   (B) Vertex: $(-2, -8.1)$
   (C) Minimum: -8.1
   (D) Range: $x \geq -8.1$ or $[-8.1, \infty)$
   (E) Decreasing: $(-\infty, -2]$;
      increasing: $[-2, \infty)$ *(1-4)*

36. (A) $V(t) = -1,250t + 12,000, \quad 0 \leq t \leq 8$

   (B) $V(5) = \$5,750$ *(1-3)*

37. (A)

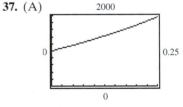

   (B) $r = 0.1447$ or 14.47%
      compounded annually
      (1-1, 1-2)

38. (A) $R = 1.6C$ (B) \$192   (C) \$110   (D) 1.6; the slope indicates the change in retail price per unit change in cost. *(1-3)*

39. (A)

x	0	5	10	15	20
CONSUMPTION	271	255	233	236	256
$f(x)$	274	248.5	237	239.5	256

(B)

(C) $286.5, 331$ *(1-4)*

40. (A) $S(x) = \begin{cases} 3 & \text{if } 0 \leq x \leq 20 \\ 0.057x + 1.86 & \text{if } 20 < x \leq 200 \\ 0.0346x + 6.34 & \text{if } 200 < x \leq 1{,}000 \\ 0.0217x + 19.24 & \text{if } x > 1{,}000 \end{cases}$

(B)

(1-2)

41. (A) Supply $p = 0.04x - 14.5$; demand: $p = -0.02x + 11$
   (B) 425 million bushels at \$2.50 per bushel *(1-3)*

42. (A) $C = 84{,}000 + 15x$; $R = 50x$

(B) $R = C$ at $x = 2{,}400$ units; $R < C$ for $0 \leq x < 2{,}400$; $R > C$ for $x > 2{,}400$
(C) Same as (B) *(1-3)*

43. (A)

(B) $R = C$ for $x = 4.686$ thousand units (4,686 units) and for $x = 27.314$ thousand units (27,314 units); $R < C$ for $1 \leq x < 4.686$ or $27.314 < x \leq 40$; $R > C$ for $4.686 < x < 27.314$.
(C) Maximum revenue is 500 thousand dollars (\$500,000). This occurs at an output of 20 thousand units (20,000 units). At this output, the wholesale price is $p(20) = \$25$. *(1-3, 1-4)*

44. (A) $P(x) = R(x) - C(x) = x(50 - 1.25x) - (160 + 10x)$

(B) $P = 0$ for $x = 4.686$ thousand units (4,686 units) and for $x = 27.314$ thousand units (27,314 units); $P < 0$ for $1 \leq x < 4.686$ or $27.314 < x \leq 40$; $P > 0$ for $4.686 < x < 27.314$.
(C) Maximum profit is 160 thousand dollars (\$160,000). This occurs at an output of 16 thousand units (16,000 units). At this output, the wholesale price is $p(16) = \$30$. *(1-4)*

45. (A) $A(x) = -\frac{3}{2}x^2 + 420x$
   (B) Domain: $0 \leq x \leq 280$
   (C)

   (D) There are two solutions to the equation $A(x) = 25{,}000$, one near 90 and another near 190.
   (E) 86 ft; 194 ft
   (F) Maximum combined area is 29,400 ft^2. This occurs for $x = 140$ ft and $y = 105$ ft. *(1-4)*

46. (A) $P(x) = 15x + 20$
   (B) 95
   (C)

   (D) 15 *(1-4)*

47. (A) 1 lb; 3 lb
   (B)

   (C) $\frac{1}{30}$ *(1-4)*

CHAPTER 2

Exercise 2-1

1. (A) 2 (B) 1 (C) 2 (D) 0 (E) 1 (F) 1 **3.** (A) 5 (B) 4 (C) 5 (D) 1 (E) 1 (F) 1
5. (A) 6 (B) 5 (C) 6 (D) 0 (E) 1 (F) 1 **7.** (A) 3 (B) 4 (C) Negative **9.** (A) 4 (B) 5 (C) Negative
11. (A) 0 (B) 1 (C) Negative **13.** (A) 5 (B) 6 (C) Positive
15. (A) x intercept: -2; y intercept: -1 (B) Domain: all real numbers except 2
  (C) Vertical asymptote: $x = 2$; horizontal asymptote: $y = 1$ (D)   (E)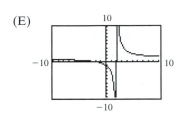

17. (A) x intercept: 0; y intercept: 0 (B) Domain: all real numbers except -2
  (C) Vertical asymptote: $x = -2$; horizontal asymptote: $y = 3$ (D)   (E)

19. (A) x intercept: 2; y intercept: -1 (B) Domain: all real numbers except 4
  (C) Vertical asymptote: $x = 4$; horizontal asymptote: $y = -2$ (D)   (E)

21. The graph will look more and more like the graph of $y = 2x^4$.
23. The graph will look more and more like the graph of $y = -x^5$.
25. (A)   (B)

$$y = 2x^4 \qquad y = 2x^4 - 5x^2 + x + 2 \qquad y = 2x^4 \qquad y = 2x^4 - 5x^2 + x + 2$$

27. (A)   (B)

$$y = -x^5 \qquad y = -x^5 + 4x^3 - 4x + 1 \qquad y = -x^5 \qquad y = -x^5 + 4x^3 - 4x + 1$$

29. $[-9/2, 9/2], -1.84, 0.42, 1.92$
31. $[-5, 5], -2.50, -1.22, 0.22, 1.50$
33. $[-16, 16], -0.92, 1.30, 11.38$
37. (A) x intercept: 0; y intercept: 0   (B) Vertical asymptotes: $x = -2, x = 3$; horizontal asymptote: $y = 2$
   (C)   (D)

39. (A) x intercept: $\pm\sqrt{3}$; y intercept: $-\frac{2}{3}$ (B) Vertical asymptotes: $x = -3, x = 3$; horizontal asymptote: $y = -2$
   (C)   (D)

41. (A) x intercept: 0; y intercept: 0   (B) Vertical asymptotes: $x = -3, x = 2$; horizontal asymptote: $y = 0$
   (C)   (D)   **43.** $f(x) = x^2 - x - 2$
                                                45. $f(x) = 4x - x^3$

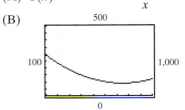

47. (A) $C(x) = 180x + 200$
   (B) $\overline{C}(x) = \dfrac{180x + 200}{x}$
   (C)   (D) $180 per board

49. (A) $\overline{C}(n) = \dfrac{2{,}500 + 175n + 25n^2}{n}$ (D) 10 yr; $675.00 per year
   (B)

   (C) 10 yr; $675.00 per year

51. (A) $\overline{C}(x) = \dfrac{0.00048(x - 500)^3 + 60{,}000}{x}$
   (B)   (C) 750 cases per
                                    month; $90 per
                                    case

53. (A)
```
LinReg
y=ax+b
a=-.0506666667
b=19.96666667
```
```
QuadReg
y=ax²+bx+c
a=2.4444444E-4
b=-.0065555556
c=2.086111111
```
   (B) $\overline{x} = 195; \overline{p} = \$10.09$

55. (A)   (B) $2,215.4 billion

57. (A) 0.06 cm/sec
(B)

59. (A)   (B) 7.0, 5.8

Exercise 2-2

1. (A) k (B) g (C) h (D) f

3.

5.

7.

9.

11.

13.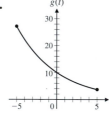

15. 4^{6xy} **17.** e **19.** $8e^{3.6t}$

21. The graph of g is a reflection of the graph f in the x axis.

23. The graph of g is the graph of f shifted 1 unit to the left.

25. The graph of g is the graph of f shifted 1 unit up.

27. The graph of g is the graph of f vertically expanded by a factor of 2 and shifted to the left 2 units.

29. (A)   (B)

(C)   (D)

31.

33.

35.

37.

39.

41. $a = 1, -1$

43. $x = 1$ **45.** $x = -1, 6$ **47.** $x = 3$ **49.** $x = 3$ **51.** $x = -3, 0$

53.

55.

57. $x = 1.40$ **59.** $x = -0.73$

61. (A) \$2,633.56   **63.** (A) \$11,871.65
   (B) \$7,079.54   (B) \$20,427.93

65. \$10,706
67. (A) \$10,715.44   (B) \$10,720.72   (C) \$10,703.65
69. \$32,201.82
71. N approaches 2 as t increases without bound.   **73.** (A) 1998: \$1,492,000; 2010: \$26,781,000

```
ExpReg
y=a*b^x
a=217.5431432
b=1.272080123
```

75. (A) 10%   (B) 1%
77. (A) $N = 47.3e^{0.116t}$ million cases
   (B) 2005: 106,500,000; 2010: 190,300,000
   (C)

79. (A) $P = 5.7e^{0.0114t}$
   (B) 6.8 billion; 8.5 billion
   (C)

81. (A) 22,033,000,000

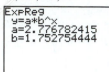

Exercise 2-3

1. $27 = 3^3$ **3.** $10^0 = 1$ **5.** $8 = 4^{3/2}$ **7.** $\log_7 49 = 2$ **9.** $\log_4 8 = \frac{3}{2}$ **11.** $\log_b A = u$ **13.** 0 **15.** 1
17. 1 **19.** 3 **21.** −3 **23.** 3 **25.** $\log_b P - \log_b Q$ **27.** $5 \log_b L$ **29.** $\log_b p - \log_b q - \log_b r - \log_b s$
31. $x = 9$ **33.** $y = 2$ **35.** $b = 10$ **37.** $x = 2$ **39.** $y = -2$ **41.** $b = 100$ **43.** $5 \log_b x - 3 \log_b y$
45. $\frac{1}{3} \log_b N$ **47.** $2 \log_b x + \frac{1}{3} \log_b y$ **49.** $\log_b 50 - 0.2t \log_b 2$ **51.** $\log_b P + t \log_b(1 + r)$ **53.** $\log_e 100 - 0.01t$
55. $x = 2$ **57.** $x = 8$ **59.** $x = 7$ **61.** No solution
63.

65. The graph of $y = \log_2(x - 2)$ is the graph of $y = \log_2 x$ shifted to the right 2 units.
67. Domain: $(-1, \infty)$; range: all real numbers
69. (A) 3.547 43 (B) −2.160 32 (C) 5.626 29 (D) −3.197 04
71. (A) 13.4431 (B) 0.0089 (C) 16.0595 (D) 0.1514
73. 1.0792 **75.** 1.4595 **77.** 30.6589 **79.** 18.3559

81. Increasing: $(0, \infty)$ **83.** Decreasing: $(0,1]$ **85.** Increasing: $(-2, \infty)$ **87.** Increasing: $(0, \infty)$
Increasing: $[1, \infty)$

89. Because $b^0 = 1$ for any permissible base b ($b > 0, b \neq 1$). **91.** $y = c \cdot 10^{0.8x}$ **93.** $x > \sqrt{x} > \ln x$ for $1 < x \leq 16$
95. 4 yr **97.** 9.87 yr; 9.80 yr **99.** 6.758%
101. (A) 5,373 (B) 7,220 **105.** 147.5 bushels/acre **107.** 901 yr

Review Exercise

1. $v = \ln u$ *(2-3)* **2.** $y = \log x$ *(2-3)* **3.** $M = e^N$ *(2-3)* **4.** $u = 10^v$ *(2-3)* **5.** 5^{2x} *(2-2)* **6.** e^{2u^2} *(2-2)*
7. $x = 9$ *(2-3)* **8.** $x = 6$ *(2-3)* **9.** $x = 4$ *(2-3)* **10.** $x = 2.157$ *(2-3)* **11.** $x = 13.128$ *(2-3)*
12. $x = 1,273.503$ *(2-3)* **13.** $x = 0.318$ *(2-3)* **14.** (A) 3 (B) 2 (C) 3 (D) 1 (E) 1 (F) 1 *(2-1)*
15. (A) 4 (B) 3 (C) 4 (D) 0 (E) 1 (F) 1 *(2-1)* **16.** (A) 2 (B) 3 (C) Positive *(2-1)*

17. (A) 3   (B) 4   (C) Negative *(2-1)*
18. (A) x intercept: -4; y intercept: -2 (B) All real numbers, except $x = 2$ (C) Vertical asymptote: $x = 2$;
horizontal asymptote: $y = 1$

(D)   (E)   *(2-1)*

19. (A) x intercept: $\frac{4}{3}$; y intercept: -2 (B) All real numbers, except $x = -2$
(C) Vertical asymptote: $x = -2$; horizontal asymptote: $y = 3$

(D)   (E)   *(2-1)*

20. $x = 8$ *(2-3)*   **21.** $x = 3$ *(2-3)*   **22.** $x = 3$ *(2-2)*   **23.** $x = -1, 3$ *(2-2)*   **24.** $x = 0, \frac{3}{2}$ *(2-2)*
25. $x = -2$ *(2-3)*   **26.** $x = \frac{1}{2}$ *(2-3)*   **27.** $x = 27$ *(2-3)*   **28.** $x = 13.3113$ *(2-3)*   **29.** $x = 158.7552$ *(2-3)*
30. $x = 0.0097$ *(2-3)*   **31.** $x = 1.4359$ *(2-3)*   **32.** $x = 1.4650$ *(2-3)*   **33.** $x = 92.1034$ *(2-3)*   **34.** $x = 9.0065$ *(2-3)*
35. $x = 2.1081$ *(2-3)*   **36.** $x = 2.8074$ *(2-3)*   **37.** $x = -1.0387$ *(2-3)*   **38.** They look very much alike. *(2-1)*
39. (A)   (B)   *(2-1)*

40. $x = -1.14, 6.78$ *(2-1)*   **41.** $(-1.54, -0.79); (0.69, 0.99)$ *(2-2, 2-3)*   **42.** $1 + 2e^x - e^{-x}$ *(2-2)*   **43.** $2e^{-2x} - 2$ *(2-2)*
44. Increasing: $[-2, 4]$ *(2-2)*   **45.** Decreasing: $[0, \infty)$ *(2-2)*   **46.** Increasing: $(-1, 10]$ *(2-3)*

47. $\log 10^{\pi} = \pi$ and $10^{\log \sqrt{2}} = \sqrt{2}$; $\ln e^{\pi} = \pi$ and $e^{\ln \sqrt{2}} = \sqrt{2}$ *(2-3)*   **48.** $x = 2$ *(2-3)*   **49.** $x = 2$ *(2-3)*
50. $x = 1$ *(2-3)*   **51.** $x = 300$ *(2-3)*   **52.** $y = ce^{-5t}$ *(2-3)*
53. If $\log_1 x = y$, then $1^y = x$; that is, $1 = x$ for all positive real numbers x, which is not possible. *(2-3)*
54. \$6,951.36 *(2-2)*   **55.** \$6,947.68 *(2-2)*
56. 16.7 yr *(2-3)*   **57.** 10.6 yr *(2-3)*

58. (A) $C(x) = 40x + 300$; $\overline{C}(x) = \dfrac{40x + 300}{x}$

(B)

(C) $y = 40$ (D) $40 per pair *(2-1)*

59. (A) $\overline{C}(x) = \dfrac{20x^3 - 360x^2 + 2{,}300x - 1{,}000}{x}$

(B)

(C) 8.667 thousand cases (8,667)

(D) $567 per case *(2-1)*

60. (A) 2,833 sets (B) 4,836

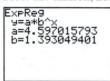

```
QuadReg
y=ax²+bx+c
a=5.9477212ε-6
b=-.1024018814
c=422.3467853
```

```
LinReg
y=ax+b
a=.0387421907
b=-7.364689544
```

(D) Equilibrium price: $131.59; equilibrium quantity: 3,587 cookware sets *(2-1)*

61. (A) 2000: 127 million; 2010: 3,482 million *(2-2)*

```
ExpReg
y=a*b^x
a=4.597015793
b=1.393049401
```

62. (A) $N = 2^{2t}$ or $N = 4^t$ (B) 15 days *(2-2, 2-3)* **63.** $k = 0.009\,42$; 489 ft *(2-2, 2-3)*

64. (A) 1996: 1,724 million bushels; 2010: 2,426 million bushels. *(2-3)*

```
LnReg
y=a+blnx
a=-21796.9294
b=5153.244133
```

65. 23.4 yr *(2-2, 2-3)* **66.** 23.1 yr *(2-2, 2-3)* **67.** (A) 2010: $886 billion

```
ExpReg
y=a*b^x
a=39.23474084
b=1.109502846
```

(B) Midway through 2004 *(2-2)*

CHAPTER 3

Exercise 3-1

1. 12 **3.** 1 **5.** −4 **7.** −1.5 **9.** 3

11. −4 **13.** 36 **15.** $\frac{5}{9}$ **17.** $\sqrt{5}$ **19.** 1.4

21.

23.

25. (A) 2 (B) 2 (C) 2 (D) 2

27. (A) 1 (B) 2 (C) Does not exist (D) 2 (E) No

29. (A) 1 (B) 2 (C) Does not exist (D) Does not exist (E) No

31. (A) 1 (B) 1 (C) 1 (D) 3 (E) Yes, define $g(3) = 1$.

33. −6 **35.** 1 **37.** −1 **39.** Does not exist **41.** 1

43. 0.5 **45.** −1 **47.** −5 **49.** $\frac{2}{3}$ **51.** 0 **53.** 3

55. 4 **57.** $1/(2\sqrt{2})$ **59.** Does not exist

61. (A) The limit does not exist. The values of $1/x$ are large negative numbers when x is close to 0 on the left.
(B) The limit does not exist. The values of $1/x$ are large positive numbers when x is close to 0 on the right.

63. (A) $\lim_{x \to 1^-} f(x) = 2$ (B) $\lim_{x \to 1^-} f(x) = 3$ (C) $m = 1.5$ (D) The graph in (A) is broken
$\lim_{x \to 1^+} f(x) = 3$ $\lim_{x \to 1^+} f(x) = 2$ when it jumps from $(1, 2)$
up to $(1, 3)$. The graph in
(B) is also broken when it
jumps down from $(1, 3)$ to
$(1, 2)$. The graph in (C) is
one continuous piece, with
no breaks or jumps.

65. $2a$ **67.** $1/(2\sqrt{a})$
69. 10   **71.** 0.693   **73.** 2.718   **75.** Typical values of n are 95 on a TI-81, 94 on a TI-82 and TI-83, and 126 on a TI-85.
77. $\lim_{x \to -2^-} f(x) = \lim_{x \to -2^+} f(x) = 2$ **79.** $\lim_{x \to -3^-} f(x) = -3, \lim_{x \to -3^+} f(x) = 3$
$\lim_{x \to 2^-} f(x) = \lim_{x \to 2^+} f(x) = 2$ $\lim_{x \to 3^-} f(x) = -3, \lim_{x \to 3^+} f(x) = 3$

   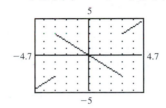

81. (A) $F(x) = \begin{cases} 0.99 & \text{if } 0 < x \le 20 \\ 0.07x - 0.41 & \text{if } x \ge 20 \end{cases}$

85. (A) $D(x) = \begin{cases} x & \text{if } 0 \le x < 300 \\ 0.97x & \text{if } 300 \le x < 1{,}000 \\ 0.95x & \text{if } 1{,}000 \le x < 3{,}000 \\ 0.93x & \text{if } 3{,}000 \le x < 5{,}000 \\ 0.9x & \text{if } 0 \ge 5{,}000 \end{cases}$

(B)
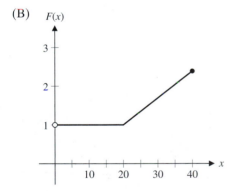

(C) All 3 limits are 0.99.

(B) $\lim_{x \to 1{,}000} D(x)$ does not exist because
$\lim_{x \to 1{,}000^-} D(x) = 970$ and $\lim_{x \to 1{,}000^+} D(x) = 950$;
$\lim_{x \to 3{,}000} D(x)$ does not exist because
$\lim_{x \to 3{,}000^-} D(x) = 2{,}850$ and $\lim_{x \to 3{,}000^+} D(x) = 2{,}790$

87. (A) $F(x) = \begin{cases} 20x & \text{if } 0 < x \le 4{,}000 \\ 80{,}000 & \text{if } x \ge 4{,}000 \end{cases}$
(B) $\lim_{x \to 4{,}000} F(x) = 80{,}000$; $\lim_{x \to 8{,}000} F(x) = 80{,}000$

89. $\lim_{x \to 5} f(x)$ does not exist; $\lim_{x \to 10} f(x) = 0$;
$\lim_{x \to 5} g(x) = 0$; $\lim_{x \to 10} g(x) = 1$

Exercise 3-2

1. f is continuous at $x = 1$,   **3.** f is discontinuous at $x = 1$,   **5.** f is discontinuous at $x = 1$,
since $\lim_{x \to 1} f(x) = f(1)$.   since $\lim_{x \to 1} f(x) \ne f(1)$.   since $\lim_{x \to 1} f(x)$ does not exist.

7.

9.

11. All x **13.** All x, except $x = -2$

15. All x, except $x = -4$ and $x = 1$

17. (A) $\lim_{x \to 1^-} f(x) = 2$; $\lim_{x \to 1^+} f(x) = 1$; $\lim_{x \to 1} f(x)$ does not exist; $f(1) = 1$

(B) No, because $\lim_{x \to 1} f(x)$ does not exist.

19. (A) $\lim_{x \to -2^-} f(x) = 1$; $\lim_{x \to -2^+} f(x) = 1$; $\lim_{x \to -2} f(x) = 1$; $f(-2) = 3$

(B) No, because $\lim_{x \to -2} f(x) \neq f(-2)$.

21. (A)

(B) 1
(C) 2
(D) No
(E) All integers

$-2 > x > 0$

23. f is discontinuous at $x = -3$; $\lim_{x \to -3^-} f(x) = -\infty$ and $\lim_{x \to -3^+} f(x) = \infty$; the line $x = -3$ is a vertical asymptote

25. h is discontinuous at $x = -2$; $\lim_{x \to -2^-} h(x) = \infty$ and $\lim_{x \to -2^+} h(x) = -\infty$; the line $x = -2$ is a vertical asymptote; h is discontinuous at $x = 2$; $\lim_{x \to 2^-} h(x) = -\infty$ and $\lim_{x \to 2^+} h(x) = \infty$; the line $x = 2$ is a vertical asymptote

27. F is continuous for all real numbers x and has no vertical asymptotes.

29. H is discontinuous at $x = 1$; $\lim_{x \to 1^-} H(x) = -\infty$ and $\lim_{x \to 1^+} H(x) = \infty$; the line $x = 1$ is a vertical asymptote; H is discontinuous at $x = 3$; $\lim_{x \to 3} H(x) = 2$, but $H(3)$ does not exist; there is no vertical asymptote at $x = 3$

31. T is discontinuous at $x = 0$; $\lim_{x \to 0^-} T(x) = -\infty$ and $\lim_{x \to 0^+} T(x) = -\infty$; the line $x = 0$ (the y axis) is a vertical asymptote; T is discontinuous at $x = 4$; $\lim_{x \to 4^-} T(x) = \infty$ and $\lim_{x \to 4^+} T(x) = \infty$; the line $x = 4$ is a vertical asymptote

33. $-3 < x < 4$; $(-3, 4)$ **35.** $x < 3$ or $x > 7$; $(-\infty, 3) \cup (7, \infty)$ **37.** $-5 < x < 0$ or $x > 3$; $(-5, 0) \cup (3, \infty)$

39. (A) $(-1.33, 1.20) \cup (3.13, \infty)$ (B) $(-\infty, -1.33) \cup (1.20, 3.13)$

41. (A) $(-\infty, -2.53) \cup (-0.72, \infty)$ (B) $(-2.53, -0.72)$

43. (A) $(-2.15, -0.52) \cup (1, 2.67)$ (B) $(-\infty, -2.15) \cup (-0.52, 1) \cup (2.76, \infty)$

45. $[6, \infty)$ **47.** $(-\infty, \infty)$ **49.** $(-\infty, -3] \cup [3, \infty)$ **51.** $(-\infty, \infty)$

53. Since $\lim_{x \to 1^-} f(x) = 2$ and $\lim_{x \to 1^+} f(x) = 4$, $\lim_{x \to 1} f(x)$ does not exist and f is not continuous at $x = 1$.

55. This function is continuous for all x.

57. Since $\lim_{x \to 0} f(x) = 0$ and $f(0) = 1$, $\lim_{x \to 0} f(x) \neq f(0)$ and f is not continuous at $x = 0$.

59. Since $\lim_{x \to 2^-} f(x) = 0$ and $\lim_{x \to 2^+} f(x) = 4$, $\lim_{x \to 2} f(x)$ does not exist. Furthermore, $f(2)$ is not defined. Thus, f is not continuous at $x = 2$.

61. Since $f(-1)$ and $f(1)$ are not defined, f is not continuous at $x = -1$ and $x = 1$, even though $\lim_{x \to -1} f(x) = 2$ and $\lim_{x \to 1} f(x) = 2$.

63. (A) Yes   (B) Yes   (C) Yes   (D) Yes   **65.** (A) Yes   (B) No   (C) Yes   (D) No   (E) Yes

67. x intercepts: $x = -5, 2$ **69.** x intercepts: $x = -6, -1, 4$ **71.** No, but this does not contradict Theorem 2, since f is discontinuous at $x = 1$.

73. The following sketches illustrate that either condition is possible. Theorem 2 implies that one of these two conditions must occur.

75. (A) $P(x) = \begin{cases} 0.34 & \text{if } 0 < x \leqslant 1 \\ 0.55 & \text{if } 1 < x \leqslant 2 \\ 0.76 & \text{if } 2 < x \leqslant 3 \\ 0.97 & \text{if } 3 < x \leqslant 4 \\ 1.18 & \text{if } 4 < x \leqslant 5 \\ \vdots \end{cases}$

(B)

(C) Yes; no

79. (A) $S(x) = \begin{cases} 5 + 0.63x & \text{if } 0 \leqslant x \leqslant 50 \\ 14 + 0.45x & \text{if } 50 < x \end{cases}$ **(B)**   **(C)** Yes

81. (A)   **(B)** $\lim_{s \to 10,000} E(s) = \$1,000$; $E(10,000) = \$1,000$
(C) $\lim_{s \to 20,000} E(s)$ does not exist; $E(20,000) = \$2,000$
(D) Yes; no

83. (A) t_2, t_3, t_4, t_6, t_7
(B) $\lim_{t \to t_5} N(t) = 7$; $N(t_5) = 7$
(C) $\lim_{t \to t_3} N(t)$ does not exist; $N(t_3) = 4$

Exercise 3-3

1. (A) -3; slope of the secant line through $(1, f(1))$ and $(2, f(2))$
(B) $-2 - h$; slope of the secant line through $(1, f(1))$ and $(1 + h, f(1 + h))$
(C) -2; slope of the tangent line at $(1, f(1))$
3. (A) 15   **(B)** 15   **(C)** $6 + 3h$ **(D)** 6   **(E)** 6   **(F)** 6   **(G)** $y = 6x - 3$
5. $f'(x) = 0; f'(1) = 0, f'(2) = 0, f'(3) = 0$ **7.** $f'(x) = 2; f'(1) = 2, f'(2) = 2, f'(3) = 2$
9. $f'(x) = -2x; f'(1) = -2, f'(2) = -4, f'(3) = -6$ **11.** $f'(x) = -\dfrac{2}{x^2}; f'(1) = -2; f'(2) = -\dfrac{1}{2}; f'(3) = -\dfrac{2}{9}$
13. (A) 5   **(B)** $3 + h$ **(C)** 3   **(D)** $y = 3x - 1$
15. (A) 5 m/s   **(B)** $3 + h$ m/s   **(C)** 3 m/s
17. $f'(x) = 5 - 4x; f'(1) = 1, f'(2) = -3, f'(3) = -7$
19. $f'(x) = -1/\sqrt{x}; f'(1) = -1, f'(2) = -1/\sqrt{2}, f'(3) = -1/\sqrt{3}$
21. $f'(x) = -1/(x + 3)^2; f'(1) = -\frac{1}{16}, f'(2) = -\frac{1}{25}, f'(3) = -\frac{1}{36}$ **23.** Yes   **25.** No   **27.** No   **29.** Yes
31. (A) $f'(x) = 2x - 4$ **(B)** $-4, 0, 4$ **33.** $v = f'(x) = 8x - 2$; 6 ft/s, 22 ft/s, 38 ft/s   **35.** 0.69   **37.** 0.71
(C)

39. (A) The graphs of g and h are vertical translations of the graph of f. All three functions should have the same derivative.
    (B) $2x$
41. (A) The slope of the graph of f is 0 at any point on the graph.
43. f is nondifferentiable at $x = 1$ **45.** f is differentiable for all real numbers   **47.** No   **49.** No

51. $f'(0) = 0$ **53.** 6 s; 192 ft/s
55. (A) \$8.75 (B) $R'(x) = 60 - 0.05x$
    (C) $R(1,000) = 35,000$; $R'(1,000) = 10$; At a production level of 1,000 car seats, the revenue is \$35,000 and is increasing
    at the rate of \$10 per seat.
57. (A) $S'(t) = 1/\sqrt{t + 10}$
    (B) $S(15) = 10$; $S'(15) = 0.2$. After 15 months, the total sales are \$10 million and are increasing at the rate of
    \$0.2 million, or \$200,000, per month.
    (C) The estimated total sales are \$10.2 million after 16 months and \$10.4 million after 17 months.
59. (A) $p'(t) = 28t - 6.6$
    (B) $p(15) = 3,653.4$; $p'(15) = 413.4$; In 2010, 3,653.4 thousand tons of zinc are produced and this quantity is increasing at
    the rate of 413.4 thousand tons per year.

61. (A)
```
QuadReg
y=ax²+bx+c
a=-.0583731517
b=-1.509819783
c=50.23597577
```

    (B) In 2010, retail sales will be 1,546.8 billion kilowatthours and will be increasing at the rate of 41.9 billion kilowatthours
    per year.
63. (A) $P'(t) = 12 - 2t$
    (B) $P(3) = 107$; $P'(3) = 6$. After 3 hours, the ozone level is 107 ppb and is increasing at the rate of 6 ppb per hour.
65. (A) $f'(t) = 0.022t - 1$
    (B) $f(40) = 7.4$; $f'(40) = -0.12$; In 2000, the number of male infant deaths per 100,000 births was 7.4 and was decreasing
    at the rate of -0.12 death per 100,000 births per year.

Exercise 3-4

1. 0   **3.** $9x^8$ **5.** $3x^2$ **7.** $-4x^{-5}$ **9.** $\dfrac{8}{3}x^{5/3}$

11. $-\dfrac{10}{x^{11}}$ **13.** $10x$ **15.** $2.8x^6$ **17.** $\dfrac{x^2}{6}$

19. 12   **21.** 2   **23.** 9   **25.** $-10x^{-6} = -10/x^6$ **27.** $-16x^{-5} = -16/x^5$ **29.** $x^{-3} = 1/x^3$

31. $-x^{-2/3} = -1/x^{2/3}$ **33.** $4.8x - 3.5$ **35.** $15x^4 - 6x^2$ **37.** $-12x^{-5} - 4x^{-3} = \dfrac{-12}{x^5} - \dfrac{4}{x^3}$

39. $-\dfrac{1}{2}x^{-2} + 2x^{-4} = \dfrac{-1}{2x^2} + \dfrac{2}{x^4}$ **41.** $2x^{-1/3} - \dfrac{5}{3}x^{-2/3} = \dfrac{2}{x^{1/3}} - \dfrac{5}{3x^{2/3}}$ **43.** $-\dfrac{9}{5}x^{-8/5} + 3x^{-3/2} = \dfrac{-9}{5x^{8/5}} + \dfrac{3}{x^{3/2}}$

45. $-\dfrac{1}{3}x^{-4/3} = \dfrac{-1}{3x^{4/3}}$ **47.** $-0.6x^{-3/2} + 6.4x^{-3} + 1 = -\dfrac{0.6}{x^{3/2}} + \dfrac{6.4}{x^3} + 1$

49. (A) $f'(x) = 6 - 2x$ (B) $f'(2) = 2$; $f'(4) = -2$ (C) $y = 2x + 4$; $y = -2x + 16$ (D) $x = 3$
51. (A) $f'(x) = 12x^3 - 12x$ (B) $f'(2) = 72$; $f'(4) = 720$ (C) $y = 72x - 127$; $y = 720x - 2,215$ (D) $x = -1, 0, 1$
53. (A) $v = f'(x) = 176 - 32x$ (B) $f'(0) = 176$ ft/s; $f'(3) = 80$ ft/s (C) 5.5 s

55. (A) $v = f'(x) = 3x^2 - 18x + 15$ (B) $f'(0) = 15$ ft/s; $f'(3) = -12$ ft/s   (C) $x = 1$ s, $x = 5$ s

57. $f'(x) = 2x - 3 - 2x^{-1/2} = 2x - 3 - \dfrac{2}{x^{1/2}}$; $x = 2.18$ **59.** $f'(x) = 4\sqrt[3]{x} - 3x - 3$; $x = -2.90$

61. $f'(x) = 0.2x^3 + 0.3x^2 - 3x - 1.6$; $x = -4.46, -0.52, 3.48$ **63.** $f'(x) = 0.8x^3 - 9.36x^2 + 32.5x - 28.25$; $x = 1.30$

65. $x = -b/(2a)$ **67.** (A) $x^3 + x$ (B) x^3 (C) $x^3 - x$ **69.** $-20x^{-2} = -20/x^2$

71. $2x - 3 - 10x^{-3} = 2x - 3 - (10/x^3)$

75. The domain of $f'(x)$ is all real numbers except $x = 0$. At $x = 0$, the graph of $y = f(x)$ is smooth, but it has a vertical tangent.

77. (A) $C'(x) = 60 - (x/2)$
   (B) $C'(60) = \$30$ per racket. At a production level of 60 rackets, the rate of change of total cost relative to production is $30 per racket; thus, the cost of producing 1 more racket at this level of production is approximately $30.
   (C) $29.75; the marginal cost of $30 per racket found in part (B) is a close approximation to this value.
   (D) $C'(80) = \$20$ per racket. At a production level of 80 rackets, the rate of change of total cost relative to production is $20 per racket; thus, the cost of producing 1 more racket at this level of production is approximately $20.

79. (A)
```
CubicReg
y=ax³+bx²+cx+d
a=7.445341406
b=-218.3357137
c=1657.552951
d=2545.706745
```
   (B) In 1992, 3,900 limousines were produced and limousine production was decreasing at the rate of 400 limousines per year.
   (C) In 1998, 3,900 limousines were produced and limousine production was increasing at the rate of 600 limousines per year.

81. The approximate cost of producing the 101st oven is greater than that of the 401st oven. Since these marginal costs are decreasing, the manufacturing process is becoming more efficient.

83. (A) $N'(x) = 3,780/x^2$
   (B) $N'(10) = 37.8$. At the $10,000 level of advertising, sales are increasing at the rate of 37.8 boats per $1,000 spent on advertising.
   $N'(20) = 9.45$. At the $20,000 level of advertising, sales are increasing at the rate of 9.45 boats per $1,000 spent on advertising.

85. (A) -1.37 beats/min   (B) -0.58 beat/min   **87.** (A) 25 items/h   (B) 8.33 items/h

Exercise 3-5

1. $2x^3(2x) + (x^2 - 2)(6x^2) = 10x^4 - 12x^2$ **3.** $(x - 3)(2) + (2x - 1)(1) = 4x - 7$

5. $\dfrac{(x - 3)(1) - x(1)}{(x - 3)^2} = \dfrac{-3}{(x - 3)^2}$ **7.** $\dfrac{(x - 2)(2) - (2x + 3)(1)}{(x - 2)^2} = \dfrac{-7}{(x - 2)^2}$

9. $(x^2 + 1)(2) + (2x - 3)(2x) = 6x^2 - 6x + 2$ **11.** $(0.4x + 2)(0.5) + (0.5x - 5)(0.4) = 0.4x - 1$

13. $\dfrac{(2x - 3)(2x) - (x^2 + 1)(2)}{(2x - 3)^2} = \dfrac{2x^2 - 6x - 2}{(2x - 3)^2}$ **15.** $(x^2 + 2)2x + (x^2 - 3)2x = 4x^3 - 2x$

17. $\dfrac{(x^2 - 3)2x - (x^2 + 2)2x}{(x^2 - 3)^2} = \dfrac{-10x}{(x^2 - 3)^2}$ **19.** 1   **21.** $-\frac{19}{9}$ **23.** $-\frac{2}{9}$

25. $(2x + 1)(2x - 3) + (x^2 - 3x)(2) = 6x^2 - 10x - 3$

27. $(2.5x - x^2)(4) + (4x + 1.4)(2.5 - 2x) = -12x^2 + 17.2x + 3.5$

29. $\dfrac{(x^2 + 2x)(5) - (5x - 3)(2x + 2)}{(x^2 + 2x)^2} = \dfrac{-5x^2 + 6x + 6}{(x^2 + 2x)^2}$

31. $\dfrac{(x^2 - 1)(2x - 3) - (x^2 - 3x + 1)(2x)}{(x^2 - 1)^2} = \dfrac{3x^2 - 4x + 3}{(x^2 - 1)^2}$

33. $f'(x) = (1 + 3x)(-2) + (5 - 2x)(3)$; $y = -11x + 29$ **35.** $f'(x) = \dfrac{(3x - 4)(1) - (x - 8)(3)}{(3x - 4)^2}$; $y = 5x - 13$

37. $f'(x) = (2x - 15)(2x) + (x^2 + 18)(2) = 6(x - 2)(x - 3)$; $x = 2, x = 3$

39. $f'(x) = \dfrac{(x^2 + 1)(1) - x(2x)}{(x^2 + 1)^2} = \dfrac{1 - x^2}{(x^2 + 1)^2}$; $x = -1, x = 1$ **41.** $7x^6 - 3x^2$ **43.** $-27x^{-4} = -\dfrac{27}{x^4}$

45. $(2x^4 - 3x^3 + x)(2x - 1) + (x^2 - x + 5)(8x^3 - 9x^2 + 1) = 12x^5 - 25x^4 + 52x^3 - 42x^2 - 2x + 5$

47. $\dfrac{(4x^2 + 5x - 1)(6x - 2) - (3x^2 - 2x + 3)(8x + 5)}{(4x^2 + 5x - 1)^2} = \dfrac{23x^2 - 30x - 13}{(4x^2 + 5x - 1)^2}$

49. $9x^{1/3}(3x^2) + (x^3 + 5)(3x^{-2/3}) = \dfrac{30x^3 + 15}{x^{2/3}}$ **51.** $\dfrac{(x^2 - 3)(2x^{-2/3}) - 6x^{1/3}(2x)}{(x^2 - 3)^2} = \dfrac{-10x^2 - 6}{(x^2 - 3)^2 x^{2/3}}$

53. $x^{-2/3}(3x^2 - 4x) + (x^3 - 2x^2)(-\tfrac{2}{3}x^{-5/3}) = -\tfrac{8}{3}x^{1/3} + \tfrac{7}{3}x^{4/3}$

55. $\dfrac{(x^2 + 1)[(2x^2 - 1)(2x) + (x^2 + 3)(4x)] - (2x^2 - 1)(x^2 + 3)(2x)}{(x^2 + 1)^2} = \dfrac{4x^5 + 8x^3 + 16x}{(x^2 + 1)^2}$

59. $f'(x) = n[u(x)]^{n-1}u'(x)$ **61.** $x = 1.21$ **63.** $x = -3.72, x = -1, x = 1.36, x = 3.36$

65. (A) $S'(t) = \dfrac{(t^2 + 50)(180t) - 90t^2(2t)}{(t^2 + 50)^2} = \dfrac{9{,}000t}{(t^2 + 50)^2}$

  (B) $S(10) = 60; S'(10) = 4$. After 10 months, the total sales are 60,000 CDs, and sales are increasing at the rate of 4,000 CDs per month.

  (C) Approximately 64,000 CDs

67. (A) $\dfrac{dx}{dp} = \dfrac{(0.1p + 1)(0) - 4{,}000(0.1)}{(0.1p + 1)^2} = \dfrac{-400}{(0.1p + 1)^2}$

  (B) $x = 800; dx/dp = -16$. At a price level of $40, the demand is 800 CD players per week, and demand is decreasing at the rate of 16 players per dollar.

  (C) Approximately 784 CD players

69. (A) $C'(t) = \dfrac{(t^2 + 1)(0.14) - 0.14t(2t)}{(t^2 + 1)^2} = \dfrac{0.14 - 0.14t^2}{(t^2 + 1)^2}$

  (B) $C'(0.5) = 0.0672$. After 0.5 h, concentration is increasing at the rate of 0.0672 mg/cm³ per hour.
  $C'(3) = -0.0112$. After 3 h, concentration is decreasing at the rate of 0.0112 mg/cm³ per hour.

71. (A) $N'(x) = \dfrac{(x + 32)(100) - (100x + 200)}{(x + 32)^2} = \dfrac{3{,}000}{(x + 32)^2}$ (B) $N'(4) = 2.31; N'(68) = 0.30$

Exercise 3-6

1. 3   **3.** $(-4x)$ **5.** $(2 + 6x)$ **7.** $6(2x + 5)^2$ **9.** $-8(5 - 2x)^3$ **11.** $5(4 + 0.2x)^4(0.2) = (4 + 0.2x)^4$

13. $30x(3x^2 + 5)^4$ **15.** $8(x^3 - 2x^2 + 2)^7(3x^2 - 4x)$ **17.** $(2x - 5)^{-1/2} = \dfrac{1}{(2x - 5)^{1/2}}$

19. $-8x^3(x^4 + 1)^{-3} = \dfrac{-8x^3}{(x^4 + 1)^3}$ **21.** $f'(x) = 6(2x - 1)^2; y = 6x - 5; x = \tfrac{1}{2}$

23. $f'(x) = 2(4x - 3)^{-1/2} = \dfrac{2}{(4x - 3)^{1/2}}; y = \dfrac{2}{3}x + 1;$ none   **25.** $24x(x^2 - 2)^3$

27. $-6(x^2 + 3x)^{-4}(2x + 3) = \dfrac{-6(2x + 3)}{(x^2 + 3x)^4}$ **29.** $x(x^2 + 8)^{-1/2} = \dfrac{x}{(x^2 + 8)^{1/2}}$ **31.** $(3x + 4)^{-2/3} = \dfrac{1}{(3x + 4)^{2/3}}$

33. $\dfrac{1}{4}(0.8x + 3.6)^{-3/4}(0.8) = \dfrac{0.2}{(0.8x + 3.6)^{3/4}}$ **35.** $(x^2 - 4x + 2)^{-1/2}(x - 2) = \dfrac{x - 2}{(x^2 - 4x + 2)^{1/2}}$

37. $-2(2x + 4)^{-2} = \dfrac{-2}{(2x + 4)^2}$ **39.** $-15x^2(x^3 + 4)^{-6} = \dfrac{-15x^2}{(x^3 + 4)^6}$ **41.** $(-8x + 4)(4x^2 - 4x + 1)^{-2} = \dfrac{-4}{(2x - 1)^3}$

43. $-2(2x - 3)(x^2 - 3x)^{-3/2} = \dfrac{-2(2x - 3)}{(x^2 - 3x)^{3/2}}$

45. $f'(x) = (4 - x)^3 - 3x(4 - x)^2 = 4(4 - x)^2(1 - x); y = -16x + 48$

47. $f'(x) = \dfrac{(2x - 5)^3 - 6x(2x - 5)^2}{(2x - 5)^6} = \dfrac{-4x - 5}{(2x - 5)^4}; y = -17x + 54$

49. $f'(x) = (2x + 2)^{1/2} + x(2x + 2)^{-1/2} = \dfrac{3x + 2}{(2x + 2)^{1/2}}; y = \tfrac{5}{2}x - \tfrac{1}{2}$

51. $f'(x) = 2x(x - 5)^3 + 3x^2(x - 5)^2 = 5x(x - 5)^2(x - 2); x = 0, 2, 5$

53. $f'(x) = \dfrac{(2x + 5)^2 - 4x(2x + 5)}{(2x + 5)^4} = \dfrac{5 - 2x}{(2x + 5)^3}; x = \tfrac{5}{2}$

55. $f'(x) = (x^2 - 8x + 20)^{-1/2}(x - 4) = \dfrac{x - 4}{(x^2 - 8x + 20)^{1/2}}; x = 4$

57. $x = -1.49, 0.48, 1.77$ **59.** $x = 0, 1.55$ **61.** $x = -1.77, 0.08, 1.69$

63. $18x^2(x^2 + 1)^2 + 3(x^2 + 1)^3 = 3(x^2 + 1)^2(7x^2 + 1)$

65. $\dfrac{24x^5(x^3 - 7)^3 - (x^3 - 7)^46x^2}{4x^6} = \dfrac{3(x^3 - 7)^3(3x^3 + 7)}{2x^4}$

67. $(2x - 3)^2[12x(2x^2 + 1)^2] + (2x^2 + 1)^3[4(2x - 3)] = 4(2x^2 + 1)^2(2x - 3)(8x^2 - 9x + 1)$

69. $4x^3(x^2 - 1)^{-1/2} + 8x(x^2 - 1)^{1/2} = \dfrac{12x^3 - 8x}{(x^2 - 1)^{1/2}}$ **71.** $\dfrac{(x - 3)^{1/2}(2) - x(x - 3)^{-1/2}}{x - 3} = \dfrac{x - 6}{(x - 3)^{3/2}}$

73. $\frac{1}{2}[(2x - 1)^3(x^2 + 3)^4]^{-1/2}[8x(2x - 1)^3(x^2 + 3)^3 + 6(x^2 + 3)^4(2x - 1)^2] = (2x - 1)^{1/2}(x^2 + 3)(11x^2 - 4x + 9)$

75. (A) $C'(x) = (2x + 16)^{-1/2} = \dfrac{1}{(2x + 16)^{1/2}}$

  (B) $C'(24) = \frac{1}{8}$, or 12.50. At a production level of 24 calculators, total cost is increasing at the rate of $12.50 per calculator; also, the cost of producing the 25th calculator is approximately $12.50.
  $C'(42) = \frac{1}{10}$, or 10.00. At a production level of 42 calculators, total cost is increasing at the rate of $10.00 per calculator; also, the cost of producing the 43rd calculator is approximately $10.00.

77. (A) $\dfrac{dx}{dp} = 40(p + 25)^{-1/2} = \dfrac{40}{(p + 25)^{1/2}}$

  (B) $x = 400$ and $dx/dp = 4$. At a price of $75, the supply is 400 speakers per week, and supply is increasing at the rate of 4 speakers per dollar.

79. $4,000(1 + \frac{1}{12}r)^{47}$ **81.** $\dfrac{(4 \times 10^6)x}{(x^2 - 1)^{5/3}}$

83. (A) $f'(n) = n(n - 2)^{-1/2} + 2(n - 2)^{1/2}$

  (B) $f'(11) = \frac{29}{3} = 9.67$. When the list contains 11 items, the learning time is increasing at the rate of 9.67 min per item.
  $f'(27) = \frac{77}{5} = 15.4$. When the list contains 27 items, the learning time is increasing at the rate of 15.4 min per item.

Exercise 3-7

1. (A) $29.50 (B) $30
3. (A) $420
  (B) $\overline{C}'(500) = -0.24$. At a production level of 500 frames, average cost is decreasing at the rate of 24¢ per frame.
  (C) Approximately $419.76
5. (A) $14.70 (B) $15
7. (A) $P'(450) = 0.5$. At a production level of 450 cassettes, profit is increasing at the rate of 50¢ per cassette.
  (B) $P'(750) = -2.5$. At a production level of 750 cassettes, profit is decreasing at the rate of $2.50 per cassette.
9. (A) $13.50
  (B) $\overline{P}'(50) = 0.27$. At a production level of 50 mowers, the average profit per mower is increasing at the rate of $0.27 per mower.
  (C) Approximately $13.77
11. (A) $p = 100 - 0.025x$, domain: $0 \leqslant x \leqslant 4,000$
  (B) $R(x) = 100x - 0.025x^2$, domain: $0 \leqslant x \leqslant 4,000$
  (C) $R'(1,600) = 20$. At a production level of 1,600 radios, revenue is increasing at the rate of $20 per radio.
  (D) $R'(2,500) = -25$. At a production level of 2,500 radios, revenue is decreasing at the rate of $25 per radio.
13. (A) $p = 200 - \frac{1}{30}x$, domain: $0 \leqslant x \leqslant 6,000$
  (B) $C'(x) = 60$ (C) $R(x) = 200x - (x^2/30)$, domain: $0 \leqslant x \leqslant 6,000$ (D) $R'(x) = 200 - (x/15)$
  (E) $R'(1,500) = 100$. At a production level of 1,500 saws, revenue is increasing at the rate of $100 per saw.
  $R'(4,500) = -100$. At a production level of 4,500 saws, revenue is decreasing at the rate of $100 per saw.
  (F) Break-even points; $(600, 108,000)$ and $(3,600, 288,000)$ (G) $P(x) = -(x^2/30) + 140x - 72,000$
  (H) $P'(x) = -(x/15) + 140$
  (I) $P'(1,500) = 40$. At a production level of 1,500 saws, profit is increasing at the rate of $40 per saw.
  $P'(3,000) = -60$. At a production level of 3,000 saws, profit is decreasing at the rate of $60 per saw.

15. (A) $p = 20 - 0.02x$, domain: $0 \leqslant x \leqslant 1,000$ (B) $R(x) = 20x - 0.02x^2$, domain: $0 \leqslant x \leqslant 1,000$
(C) $C(x) = 4x + 1,400$ (D) Break-even points: $(100, 1,800)$ and $(700, 4,200)$ (E) $P(x) = 16x - 0.02x^2 - 1,400$

(F) $P'(250) = 6$. At a production level of 250 toasters, profit is increasing at the rate of \$6 per toaster. $P'(475) = -3$. At a production level of 475 toasters, profit is decreasing at the rate of \$3 per toaster.

17. (A) $x = 500$ (B) $P(x) = 176x - 0.2x^2 - 21,900$
(C) $x = 440$
(D) Break-even points: $(150, 25,500)$ and $(730, 39,420)$; x intercepts for $P(x)$: $x = 150$ and $x = 730$

19. (A) $R(x) = 20x - x^{3/2}$
(B) Break-even points: $(44, 588)$, $(258, 1,016)$

21. (A)

(B) Fixed costs $\approx \$721,680$ (C) $(713, 807,703)$, $(5,423, 1,376,227)$
Variable costs $\approx \$121$ (D) $\$254 \leqslant p \leqslant \$1,133$

```
LinReg
y=ax+b
a=120.7047281
b=721680.1282
r=.9934384133
```

Review Exercise

1. (A) 16   (B) 8   (C) 8   (D) 4   (E) 4   (F) 4 *(3-3)*   **2.** $f'(x) = -3$ *(3-3)*

3. (A) 22   (B) 8   (C) 2 *(3-1)*   **4.** (A) 1   (B) 1   (C) 1   (D) 1 *(3-1)*

5. (A) 2   (B) 3   (C) Does not exist   (D) 3 *(3-1)*   **6.** (A) 4   (B) 4   (C) 4   (D) Does not exist *(3-1)*

7. (A) Does not exist   (B) 3   (C) No *(3-2)*   **8.** (A) 2   (B) Not defined   (C) No *(3-2)*

9. (A) 1   (B) 1   (C) Yes *(3-2)*   **10.** $f'(x) = 10x$ *(3-3)*

11. (A) -11 (B) -14 (C) $\frac{5}{2}$ (D) -8 *(3-4, 3-5, 3-6)*   **12.** $(6x + 4)$ *(3-6)*   **13.** $12x^3 - 4x$ *(3-4)*

14. $x^{-1/2} - 3 = \dfrac{1}{x^{1/2}} - 3$ *(3-4)*   **15.** 0 *(3-4)*   **16.** $-x^{-3} + x = \dfrac{-1}{x^3} + x$ *(3-4)*   **17.** $-2x^{-5} + x^3 = \dfrac{-2}{x^5} + x^3$ *(3-4)*

18. $(2x - 1)(3) + (3x + 2)(2) = 12x + 1$ *(3-5)*   **19.** $(x^2 - 1)(3x^2) + (x^3 - 3)(2x) = 5x^4 - 3x^2 - 6x$ *(3-5)*

20. $(0.2x - 1.5)(0.5) + (0.5x + 0.4)(0.2) = 0.2x - 0.67$ *(3-5)*   **21.** $\dfrac{(x^2 + 2)2 - 2x(2x)}{(x^2 + 2)^2} = \dfrac{4 - 2x^2}{(x^2 + 2)^2}$ *(3-5)*

22. $(-1)(3x + 2)^{-2}(3) = \dfrac{-3}{(3x + 2)^2}$ *(3-6)*   **23.** $3(2x - 3)^2(2) = 6(2x - 3)^2$ *(3-6)*

24. $-2(x^2 + 2)^{-3}(2x) = \dfrac{-4x}{(x^2 + 2)^3}$ *(3-6)*   **25.** (A) 4   (B) 6   (C) Does not exist   (D) 6   (E) No *(3-2)*

26. (A) 3   (B) 3   (C) 3   (D) 3   (E) Yes *(3-2)*   **27.** $(-3, 4)$ *(3-2)*   **28.** $(-3, 0) \cup (5, \infty)$ *(3-2)*

29. $(-2.34, -0.47) \cup (1.81, \infty)$ *(3-2)*   **30.** (A) 3   (B) $2 + 0.5h$ (C) 2 *(3-3)*   **31.** $12x^3 + 6x^{-4} = 12x^3 + \dfrac{6}{x^4}$ *(3-4)*

32. $(2x^2 - 3x + 2)(2x + 2) + (x^2 + 2x - 1)(4x - 3) = 8x^3 + 3x^2 - 12x + 7$ *(3-5)*

33. $\dfrac{(x - 1)^2(2) - (2x - 3)(2)(x - 1)}{(x - 1)^4} = \dfrac{4 - 2x}{(x - 1)^3}$ *(3-5)*   **34.** $x^{-1/2} - 2x^{-3/2} = \dfrac{1}{x^{1/2}} - \dfrac{2}{x^{3/2}}$ *(3-4)*

35. $0.6x^{-2/3} - 0.3x^{-4/3} = \dfrac{0.6}{x^{2/3}} - \dfrac{0.3}{x^{4/3}}$ *(3-4)*

36. $(x^2 - 1)[2(2x + 1)(2)] + (2x + 1)^2(2x) = 2(2x + 1)(4x^2 + x - 2)$ *(3-5, 3-6)*

37. $\dfrac{1}{3}(x^3 - 5)^{-2/3}(3x^2) = \dfrac{x^2}{(x^3 - 5)^{2/3}}$ *(3-6)*   **38.** $-8x^{-3} = \dfrac{-8}{x^3}$ *(3-4)*

39. $\dfrac{(2x - 3)(4)(x^2 + 2)^3(2x) - (x^2 + 2)^4(2)}{(2x - 3)^2} = \dfrac{2(x^2 + 2)^3(7x^2 - 12x - 2)}{(2x - 3)^2}$ *(3-5, 3-6)*

40. (A) $m = f'(1) = 2$ (B) $y = 2x + 3$ *(3-3, 3-4)*   **41.** (A) $m = f'(1) = 16$ (B) $y = 16x - 12$ *(3-3, 3-5)*

42. $x = 5$ *(3-4)*   **43.** $x = -5, x = 3$ *(3-5)*   **44.** $x = -2, x = 2$ *(3-5)*   **45.** $x = 0, x = 3, x = \frac{15}{2}$ *(3-5)*

46. $x = -1.34, 0.58, 2.26$ *(3-4)*   **47.** $x = -0.41, 1.80$ *(3-5)*   **48.** $x = -0.89, 0, 0.89$ *(3-5, 3-6)*

49. (A) $v = f'(x) = 32x - 4$ (B) $f'(3) = 92$ ft/s *(3-4)*   **50.** (A) $v = f'(x) = 96 - 32x$ (B) $x = 3$ s *(3-4)*

51. (A) The graph of g is the graph of f shifted 4 units to the right, and the graph of h is the graph of f shifted 3 units to the left:

(B) The graph of g' is the graph of f' shifted 4 units to the right, and the graph of h' is the graph of f' shifted 3 units to the left:

(3-3, 3-6)

52. (A) The graph of g is a horizontal translation of the graph of f, and the graph of g' is a horizontal translation of the graph of f'.
(B) The graph of g is a vertical translation of the graph of f, and the graph of g' is the same as the graph of f'. *(3-3)*

53. $(-\infty, \infty)$ *(3-2)*   **54.** $(-\infty, 2) \cup (2, \infty)$ *(3-2)*   **55.** $(-\infty, -4) \cup (-4, 1) \cup (1, \infty)$ *(3-2)*   **56.** $(-\infty, \infty)$ *(3-2)*

57. $[-2, 2]$ *(3-2)*   **58.** ∞ *(3-2)*   **59.** $-\infty$ *(3-2)*   **60.** Does not exist *(3-2)*   **61.** ∞ *(3-2)*

62. $\frac{3}{8}$ *(3-1)*   **63.** 16 *(3-1)*   **64.** -1 *(3-1)*   **65.** $\frac{1}{6}$ *(3-1)*   **66.** -1 *(3-1)*   **67.** 1 *(3-1)*   **68.** Does not exist *(3-1)*

69. 4 *(3-1)*   **70.** 4 *(3-1)*   **71.** $\dfrac{-1}{(x + 2)^2}$ *(3-1)*

72. (A) $\lim_{x \to -2^-} f(x) = -6$; $\lim_{x \to -2^+} f(x) = 6$; $\lim_{x \to -2} f(x)$ does not exist   (B) $\lim_{x \to 0} f(x) = 4$
(C) $\lim_{x \to 2^-} f(x) = 2$; $\lim_{x \to 2^+} f(x) = -2$; $\lim_{x \to 2} f(x)$ does not exist *(3-1)*

73. $2x - 1$ *(3-3)*   **74.** $1/(2\sqrt{x})$ *(3-3)*   **75.** No *(3-3)*   **76.** No *(3-3)*   **77.** No *(3-3)*   **78.** Yes *(3-3)*

79. $(x - 4)^4(3)(x + 3)^2 + (x + 3)^3(4)(x - 4)^3 = 7x(x - 4)^3(x + 3)^2$ *(3-5, 3-6)*

80. $\dfrac{(2x + 1)^4(5x^4) - x^5(4)(2x + 1)^3(2)}{(2x + 1)^8} = \dfrac{x^4(2x + 5)}{(2x + 1)^5}$ *(3-5, 3-6)*

81. $\dfrac{x(\frac{1}{2})(x^2 - 1)^{-1/2}(2x) - (x^2 - 1)^{1/2}}{x^2} = \dfrac{1}{x^2(x^2 - 1)^{1/2}}$ *(3-5, 3-6)*

82. $\dfrac{(x^2 + 4)^{1/2} - x(\frac{1}{2})(x^2 + 4)^{-1/2}(2x)}{x^2 + 4} = \dfrac{4}{(x^2 + 4)^{3/2}}$ *(3-5, 3-6)*

83. The domain of $f'(x)$ is all real numbers except $x = 0$. At $x = 0$, the graph of $y = f(x)$ is smooth, but it has a vertical tangent. *(3-3)*

84. (A) $\lim_{x \to 1^-} f(x) = 1$; $\lim_{x \to 1^+} f(x) = -1$ (B) $\lim_{x \to 1^-} f(x) = -1$; $\lim_{x \to 1^+} f(x) = 1$ (C) $m = 1$

(D) The graphs in (A) and (B) have discontinuities at $x = 1$; the graph in (C) does not. *(3-2)*

85. (A) 1 (B) -1 (C) Does not exist   (D) No *(3-3)*

86. (A) $S(x) = \begin{cases} 7.47 & + 0.4x & \text{if } 0 \le x \le 90 \\ 24.786 + 0.2076x & \text{if } 90 < x \end{cases}$ (B)

(C) Yes *(3-2)*

87. (A) $179.90   (B) $180 *(3-7)*

88. (A) $C(100) = 9,500$; $C'(100) = 50$. At a production level of 100 bicycles, the total cost is $9,500, and cost is increasing at the rate of $50 per bicycle.

(B) $\overline{C}(100) = 95$; $\overline{C}'(100) = -0.45$. At a production level of 100 bicycles, the average cost is $95, and average cost is decreasing at a rate of $0.45 per bicycle. *(3-7)*

89. The approximate cost of producing the 201st printer is greater than that of the 601st printer. Since these marginal costs are decreasing, the manufacturing process is becoming more efficient. *(3-7)*

90. (A) $C'(x) = 2$; $\overline{C}(x) = 2 + \dfrac{9,000}{x}$; $\overline{C}'(x) = \dfrac{-9,000}{x^2}$

(B) $R(x) = xp = 25x - 0.01x^2$; $R'(x) = 25 - 0.02x$; $\overline{R}(x) = 25 - 0.01x$; $\overline{R}'(x) = -0.01$

(C) $P(x) = R(x) - C(x) = 23x - 0.01x^2 - 9,000$; $P'(x) = 23 - 0.02x$; $\overline{P}(x) = 23 - 0.01x - \dfrac{9,000}{x}$;

$\overline{P}'(x) = -0.01 + \dfrac{9,000}{x^2}$

(D) $(500, 10,000)$ and $(1,800, 12,600)$

(E) $P'(1,000) = 3$. Profit is increasing at the rate of $3 per umbrella.

$P'(1,150) = 0$. Profit is flat.

$P'(1,400) = -5$. Profit is decreasing at the rate of $5 per umbrella.

(F)

(3-7)

91. (A) 2 components/day   (B) 3.2 components/day *(3-5)*

92. $N(5) = 15$; $N'(5) = 3.833$. After 5 months, the total sales are 15,000 pools, and sales are increasing at the rate of 3,833 pools per month. *(3-6)*

93. (A)
```
CubicReg
y=ax³+bx²+cx+d
a=.001225
b=-.0819285714
c=1.564642857
d=12.08428571
```
(B) $N(50) = 38.6$; $N'(50) = 2.6$. In 2010, natural gas consumption is 38.6 trillion cubic feet and is increasing at the rate of 2.6 trillion cubic feet per year *(3-3)*

94. (A)
```
LinReg
y=ax+b
a=-.0384180791
b=13.59887006
r=-.9897782666
```
(B) Fixed costs: $484.21; variable costs per kringle: $2.11
```
LinReg
y=ax+b
a=2.107344633
b=484.2090395
r=.9939318704
```
(C) $(51, 591.15)$, $(248, 1,007.62)$
(D) $4.07 < p < 11.64$ *(3-7)*

95. $C'(9) = -1$ ppm/m; $C'(99) = -0.001$ ppm/m *(3-6)*

96. $F(3) = 100$; $F'(3) = -0.25$. After 3 h, the body temperature is 100°F, and the temperature is decreasing at the rate of 0.25°F/h. *(3-6)*

97. (A) 10 items/h   (B) 5 items/h *(3-4)*

CHAPTER 4

Exercise 4-1

1. $(a, b); (d, f); (g, h)$ **3.** $(b, c); (c, d); (f, g)$ **5.** c, d, f **7.** b, f
9. Local maximum at $x = a$; local minimum at $x = c$; no local extrema at $x = b$ and $x = d$
11. Decreasing on $(-\infty, 4)$; increasing on $(4, \infty)$; local minimum at $x = 4$
13. Increasing on $(-\infty, -4)$; decreasing on $(-4, \infty)$; local maximum at $x = -4$ **15.** Increasing for all x; no local extrema
17. Decreasing on $(-\infty, 0)$ and $(6, \infty)$; increasing on $(0, 6)$; local minimum at $x = 0$; local maximum at $x = 6$
19. Increasing on $(-\infty, -2)$ and $(3, \infty)$; decreasing on $(-2, 3)$; local maximum at $x = -2$, local minimum at $x = 3$
21. Decreasing on $(-\infty, -3)$ and $(0, 3)$; increasing on $(-3, 0)$ and $(3, \infty)$; local minima at $x = -3$ and $x = 3$; local maximum at $x = 0$
23. Increasing on $(-\infty, 2)$; decreasing on $(2, \infty)$; local maximum at $x = 2$
25. Decreasing on $(-\infty, -0.39)$; increasing on $(-0.39, \infty)$; local minimum at $x = -0.39$
27. Decreasing on $(-\infty, -0.77)$ and $(1.08, 2.69)$; increasing on $(-0.77, 1.08)$ and $(2.69, \infty)$; local minima at $x = -0.77$ and $x = 2.69$; local maximum at $x = 1.08$
29. Decreasing on $(-\infty, -1.22)$ and $(0.35, 2.38)$; increasing on $(-1.22, 0.35)$ and $(2.38, \infty)$; local minima at $x = -1.22$ and $x = 2.38$; local maximum at $x = 0.35$
31. Increasing on $(-\infty, 4)$
Decreasing on $(4, \infty)$
Horizontal tangent at $x = 4$

33. Increasing on $(-\infty, -1), (1, \infty)$
Decreasing on $(-1, 1)$
Horizontal tangents at $x = -1, 1$

35. Decreasing for all x
Horizontal tangent at $x = 2$

37. **39.** **41.** **43.**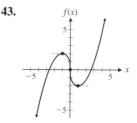

45. g_4 **47.** g_6 **49.** g_2
51. Increasing on $(-1, 2)$; decreasing on $(-\infty, -1)$ and $(2, \infty)$; local minimum at $x = -1$; local maximum at $x = 2$

53. Increasing on $(-1, 2)$ and $(2, \infty)$; decreasing on $(-\infty, -1)$; local minimum at $x = -1$

55. Increasing on $(-2, 0)$ and $(3, \infty)$; decreasing on $(-\infty, -2)$ and $(0, 3)$; local minima at $x = -2$ and $x = 3$; local maximum at $x = 0$

57. $f'(x) > 0$ on $(-\infty, -1)$ and $(3, \infty)$; $f'(x) < 0$ on $(-1, 3)$; $f'(x) = 0$ at $x = -1$ and $x = 3$

59. $f'(x) > 0$ on $(-2, 1)$ and $(3, \infty)$; $f'(x) < 0$ on $(-\infty, -2)$ and $(1, 3)$; $f'(x) = 0$ at $x = -2, x = 1$, and $x = 3$

61. Critical values: $x = -2, x = 2$; increasing on $(-\infty, -2)$ and $(2, \infty)$; decreasing on $(-2, 0)$ and $(0, 2)$; local maximum at $x = -2$; local minimum at $x = 2$

63. Critical value: $x = -2$; increasing on $(-2, 0)$; decreasing on $(-\infty, -2)$ and $(0, \infty)$; local minimum at $x = -2$

65. Critical values: $x = 0, x = 4$; increasing on $(-\infty, 0)$ and $(4, \infty)$; decreasing on $(0, 2)$ and $(2, 4)$; local maximum at $x = 0$; local minimum at $x = 4$

67. Critical values: $x = 0, x = 4, x = 6$; increasing on $(0, 4)$ and $(6, \infty)$; decreasing on $(-\infty, 0)$ and $(4, 6)$; local maximum at $x = 4$; local minima at $x = 0$ and $x = 6$

69. Critical value: $x = 2$; increasing on $(2, \infty)$; decreasing on $(-\infty, 2)$; local minimum at $x = 2$

71. Critical value: $x = 1$; increasing on $(0, 1)$; decreasing on $(1, \infty)$; local maximum at $x = 1$

73. (A) There are no critical values and no local extrema. The function is increasing for all x.
    (B) There are two critical values, $x = \pm\sqrt{-k/3}$. The function increases on $(-\infty, -\sqrt{-k/3})$ to a local maximum at $x = -\sqrt{-k/3}$, decreases on $(-\sqrt{-k/3}, \sqrt{-k/3})$ to a local minimum at $x = \sqrt{-k/3}$, and increases on $(\sqrt{-k/3}, \infty)$.
    (C) The only critical value is $x = 0$. There are no local extrema. The function is increasing for all x.

75. (A) The marginal profit is positive on $(0, 600)$, 0 at $x = 600$, and negative on $(600, 1{,}000)$.

77. (A) The price decreases for the first 15 months to a local minimum, increases for the next 40 months to a local maximum, and then decreases for the remaining 15 months.

    (B) $P'(x)$

    (B) $B(t)$

79. (A) $\overline{C}(x) = 0.05x + 20 + \dfrac{320}{x}$
    (B) Critical value: $x = 80$; decreasing for $0 < x < 80$; increasing for $80 < x < 150$; local minimum at $x = 80$

81. $P(x)$ is increasing over (a, b) if $P'(x) = R'(x) - C'(x) > 0$ over (a, b); that is, if $R'(x) > C'(x)$ over (a, b).

83. Critical value: $t = 1$; increasing for $0 < t < 1$; decreasing for $1 < t < 24$; local maximum at $t = 1$

85. Critical value: $t = 7$; increasing for $0 < t < 7$; decreasing for $7 < t < 24$; local maximum at $t = 7$

Exercise 4-2

1. $(a, c), (c, d), (e, g)$ **3.** $(d, e), (g, h)$ **5.** $(a, c), (c, d), (e, g)$ **7.** d, e, g **9.** (C)   **11.** (D)

13. $12x - 8$ **15.** $4x^{-3} - 18x^{-4}$ **17.** $2 + \dfrac{9}{2}x^{-3/2}$ **19.** $8(x^2 + 9)^3 + 48x^2(x^2 + 9)^2 = 8(x^2 + 9)^2(7x^2 + 9)$

21. $f(2) = -2$ is a local minimum   **23.** $f(-1) = 2$ is a local maximum; $f(2) = -25$ is a local minimum

25. No local extrema   **27.** $f(-2) = -6$ is a local minimum; $f(0) = 10$ is a local maximum; $f(2) = -6$ is a local minimum

29. $f(0) = 2$ is a local minimum   **31.** $f(-4) = -8$ is a local maximum; $f(4) = 8$ is a local minimum.

33. Concave upward for all x; no inflection points

35. Concave downward on $(-\infty, \frac{4}{3})$; concave upward on $(\frac{4}{3}, \infty)$; inflection point at $x = \frac{4}{3}$

37. Concave downward on $(-\infty, 0)$ and $(6, \infty)$; concave upward on $(0, 6)$; inflection points at $x = 0$ and $x = 6$

39. Local minimum   **41.** Unable to determine   **43.** Neither

45.

47.

49.

51.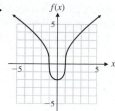

53. Local maximum at $x = 0$
Local minimum at $x = 4$
Inflection point at $x = 2$

55. Inflection point at $x = 0$

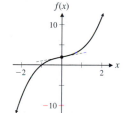

57. Inflection point at $x = 2$

59. Local maximum at $x = -2$
Local minimum at $x = 2$
Inflection point at $x = 0$

61.

x	$f'(x)$	$f(x)$
$-\infty < x < -1$	Positive and decreasing	Increasing and concave downward
$x = -1$	x intercept	Local maximum
$-1 < x < 0$	Negative and decreasing	Decreasing and concave downward
$x = 0$	Local minimum	Inflection point
$0 < x < 2$	Negative and increasing	Decreasing and concave upward
$x = 2$	Local maximum	Inflection point
$2 < x < \infty$	Negative and decreasing	Decreasing and concave downward

63.

x	$f'(x)$	$f(x)$
$-\infty < x < -2$	Negative and increasing	Decreasing and concave upward
$x = -2$	Local maximum	Inflection point
$-2 < x < 0$	Negative and decreasing	Decreasing and concave downward
$x = 0$	Local minimum	Inflection point
$0 < x < 2$	Negative and increasing	Decreasing and concave upward
$x = 2$	Local maximum	Inflection point
$2 < x < \infty$	Negative and decreasing	Decreasing and concave downward

65. Inflection points at $x = 0.27$ and $x = 3.73$; concave upward on $(-\infty, 0.27)$ and $(3.73, \infty)$; concave downward on $(0.27, 3.73)$

67. Inflection point at $x = -1.40$; concave upward on $(-1.40, \infty)$; concave downward on $(-\infty, -1.40)$

69. Inflection points at $x = -0.61, x = 0.66$, and $x = 1.74$; concave upward on $(-0.61, 0.66)$ and $(1.74, \infty)$; concave downward on $(-\infty, -0.61)$ and $(0.66, 1.74)$

71. If $f'(x)$ has a local extremum at $x = c$, then $f'(x)$ must change from increasing to decreasing or from decreasing to increasing at $x = c$. Thus, the graph of $y = f(x)$ must change concavity at $x = c$, and there must be an inflection point at $x = c$.

73. If there is an inflection point on the graph of $y = f(x)$ at $x = c$, then $f(x)$ must change concavity at $x = c$. Consequently, $f'(x)$ must change from increasing to decreasing or from decreasing to increasing at $x = c$, and $x = c$ is a local extremum for $f'(x)$.

75. Inflection points at $x = -2$ and $x = 2$ **77.** Inflection points at $x = -6, x = 0$, and $x = 6$

79. The graph of the CPI is concave upward.

81. The graph of $y = C'(x)$ is positive and decreasing. Since marginal costs are decreasing, the production process is becoming more efficient as production increases.

83. (A) Local maximum at $x = 60$
(B) Concave downward on the whole interval $(0, 80)$

85. (A) Increasing on $(10, 25)$; decreasing on $(25, 40)$
(B) Inflection point at $x = 25$
(C)

(D) Max $N'(x) = N'(25) = 2{,}025$

87. (A)

(B) 32 ads to sell 574 cars per month

89. (A) Increasing on $(0, 10)$; decreasing on $(10, 20)$
(B) Inflection point at $t = 10$
(C)

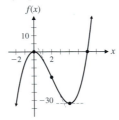

(D) $N'(10) = 300$

91. (A) Increasing on $(5, \infty)$; decreasing on $(0, 5)$
(B) Inflection point at $n = 5$

(C) $T'(5) = 0$

Exercise 4-3

1. $(-\infty, b), (0, e), (e, g)$ **3.** $(b, d), (d, 0), (g, \infty)$ **5.** $x = 0$ **7.** $(-\infty, a), (d, e), (h, \infty)$ **9.** $(a, d), (e, h)$
11. $x = a, x = h$ **13.** $x = d, x = e$ **15.** $-\infty$ **17.** ∞ **19.** $-\frac{3}{2}$ **21.** ∞ **23.** 0
25. Horizontal asymptote: $y = 3$; vertical asymptote: $x = 0$
27. Horizontal asymptote: $y = 1$; no vertical asymptotes
29. Horizontal asymptote: $y = 0$; vertical asymptotes: $x = -1, x = 1$
31. No horizontal asymptote; vertical asymptotes: $x = -4, x = 4$
33. Horizontal asymptote $y = 2$; vertical asymptote: $x = 1$
35. Horizontal asymptote: $y = \frac{3}{2}$; vertical asymptotes: $x = -4, x = \frac{5}{2}$
37. Domain: All real numbers
y intercept: 0; x intercepts: 0, 6
Increasing on $(-\infty, 0)$ and $(4, \infty)$
Decreasing on $(0, 4)$
Local maximum at $x = 0$
Local minimum at $x = 4$
Concave upward on $(2, \infty)$
Concave downward on $(-\infty, 2)$
Inflection points at $x = 2$

39. Domain: All real numbers
y intercept: 16; x intercepts: $-4, 2$
Increasing on $(-\infty, -2)$ and $(2, \infty)$
Decreasing on $(-2, 2)$
Local maximum at $x = -2$
Local minimum at $x = 2$
Concave upward on $(0, \infty)$
Concave downward on $(-\infty, 0)$
Inflection point at $x = 0$

41. Domain: All real numbers
y intercept: 0; x intercepts: 0, 4
Increasing on $(-\infty, 3)$
Decreasing on $(3, \infty)$
Local maximum at $x = 3$
Concave upward on $(0, 2)$
Concave downward on $(-\infty, 0)$ and $(2, \infty)$
Inflection points at $x = 0$ and $x = 2$

43. Domain: All real numbers, except 3
y intercept: -1; x intercept: -3
Horizontal asymptote: $y = 1$
Vertical asymptote: $x = 3$
Decreasing on $(-\infty, 3)$ and $(3, \infty)$
Concave upward on $(3, \infty)$
Concave downward on $(-\infty, 3)$

45. Domain: All real numbers, except 2
y intercept: 0; x intercept: 0
Horizontal asymptote: $y = 1$
Vertical asymptote: $x = 2$
Decreasing on $(-\infty, 2)$ and $(2, \infty)$
Concave downward on $(-\infty, 2)$
Concave upward on $(2, \infty)$

47.

49.

51.

53.

55. For any $n \geq 1$, the limit will be ∞ if $a > 0$ and $-\infty$ if $a < 0$.

57. (A) $\lim_{x \to \infty} p'(x) = \infty$; $\lim_{x \to \infty} p''(x) = \infty$. The graph of $y = p(x)$ is increasing and concave upward for large positive values of x.

    (B) $\lim_{x \to -\infty} p'(x) = \infty$; $\lim_{x \to -\infty} p''(x) = -\infty$. The graph of $y = p(x)$ is increasing and concave downward for large negative values of x.

59. Domain: All real numbers, except 0
Vertical asymptote: $x = 0$
Increasing on $(-\infty, -1)$ and $(1, \infty)$
Decreasing on $(-1, 0)$ and $(0, 1)$
Local maximum at $x = -1$
Local minimum at $x = 1$
Concave upward on $(0, \infty)$
Concave downward on $(-\infty, 0)$

61. Domain: All real numbers
y intercept: 0; x intercepts: $-1, 0, 1$
Increasing on $(-\infty, -\sqrt{3}/3)$ and $(\sqrt{3}/3, \infty)$
Decreasing on $(-\sqrt{3}/3, \sqrt{3}/3)$
Local maximum at $x = -\sqrt{3}/3$
Local minimum at $x = \sqrt{3}/3$
Concave downward on $(-\infty, 0)$
Concave upward on $(0, \infty)$
Inflection point at $x = 0$

63. Domain: All real numbers
y intercept: 27; x intercepts: $-3, 3$
Increasing on $(-\infty, -\sqrt{3})$ and $(0, \sqrt{3})$
Decreasing on $(-\sqrt{3}, 0)$ and $(\sqrt{3}, \infty)$
Local maxima at $x = -\sqrt{3}$ and $x = \sqrt{3}$
Local minimum at $x = 0$
Concave upward on $(-1, 1)$
Concave downward on $(-\infty, -1)$ and $(1, \infty)$
Inflection points at $x = -1$ and $x = 1$

65. Domain: All real numbers
y intercept: 16; x intercepts: $-2, 2$
Decreasing on $(-\infty, -2)$ and $(0, 2)$
Increasing on $(-2, 0)$ and $(2, \infty)$
Local minima at $x = -2$ and $x = 2$
Local maximum at $x = 0$
Concave upward on $(-\infty, -2\sqrt{3}/3)$ and $(2\sqrt{3}/3, \infty)$
Concave downward on $(-2\sqrt{3}/3, 2\sqrt{3}/3)$
Inflection points at $x = -2\sqrt{3}/3$ and $x = 2\sqrt{3}/3$

67. Domain: All real numbers
y intercept: 0; x intercepts: 0, 1.5
Decreasing on $(-\infty, 0)$ and $(0, 1.25)$
Increasing on $(1.25, \infty)$
Local minimum at $x = 1.25$
Concave upward on $(-\infty, 0)$ and $(1, \infty)$
Concave downward on $(0, 1)$
Inflection points at $x = 0$ and $x = 1$

69. Domain: All real numbers, except ± 2
y intercept: 0; x intercept: 0
Horizontal asymptote: $y = 0$
Vertical asymptotes: $x = -2, x = 2$
Decreasing on $(-\infty, -2), (-2, 2)$, and $(2, \infty)$
Concave upward on $(-2, 0)$ and $(2, \infty)$
Concave downward on $(-\infty, -2)$ and $(0, 2)$
Inflection point at $x = 0$

71. Domain: All real numbers
y intercept: 1
Horizontal asymptote: $y = 0$
Increasing on $(-\infty, 0)$
Decreasing on $(0, \infty)$
Local maximum at $x = 0$
Concave upward on $(-\infty, -\sqrt{3}/3)$ and $(\sqrt{3}/3, \infty)$
Concave downward on $(-\sqrt{3}/3, \sqrt{3}/3)$
Inflection points at $x = -\sqrt{3}/3$ and $x = \sqrt{3}/3$

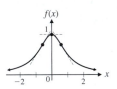

73. Domain: All real numbers
x intercepts: $-1.18, 0.61, 1.87, 3.71$
y intercept: -5
Decreasing on $(-\infty, -0.53)$ and $(1.24, 3.04)$
Increasing on $(-0.53, 1.24)$ and $(3.04, \infty)$
Local minima at $x = -0.53$ and $x = 3.04$
Local maximum at $x = 1.24$
Concave upward on $(-\infty, 0.22)$ and $(2.28, \infty)$
Concave downward on $(0.22, 2.28)$
Inflection points at $x = 0.22$ and $x = 2.28$

75. Domain: All real numbers
y intercept: 100; x intercepts: 8.01, 13.36
Increasing on $(-0.10, 4.57)$ and $(11.28, \infty)$
Decreasing on $(-\infty, -0.10)$ and $(4.57, 11.28)$
Local maximum at $x = 4.57$
Local minima at $x = -0.10$ and $x = 11.28$
Concave upward on $(-\infty, 1.95)$ and $(8.55, \infty)$
Concave downward on $(1.95, 8.55)$
Inflection points at $x = 1.95$ and $x = 8.55$

77. Domain: All real numbers
x intercepts: $-2.40, 1.16$; y intercept: 3
Increasing on $(-\infty, -1.58)$
Decreasing on $(-1.58, \infty)$
Local maximum at $x = -1.58$
Concave downward on $(-\infty, -0.88)$ and $(0.38, \infty)$
Concave upward on $(-0.88, 0.38)$
Inflection points at $x = -0.88$ and $x = 0.38$

79. Domain: All real numbers
x intercepts: $-6.68, -3.64, -0.72$; y intercept: 30
Decreasing on $(-5.59, -2.27)$ and $(1.65, 3.82)$
Increasing on $(-\infty, -5.59), (-2.27, 1.65)$, and $(3.82, \infty)$
Local minima at $x = -2.27$ and $x = 3.82$
Local maxima at $x = -5.59$ and $x = 1.65$
Concave upward on $(-4.31, -0.40)$ and $(2.91, \infty)$
Concave downward on $(-\infty, -4.31)$ and $(-0.40, 2.91)$
Inflection points at $x = -4.31, x = -0.40$, and $x = 2.91$

81.

83. (A) Increasing on $(0, 1)$
(B) Concave upward on $(0, 1)$
(C) $x = 1$ is a vertical asymptote
(D) The origin is both an x and a y intercept
(E)

85. (A) $\overline{C}(n) = \dfrac{3,200}{n} + 250 + 50n$

(B)

(C) 8 yr

87. (A)

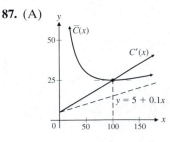

(B) 25 at $x = 100$

89. (A)

(B) Minimum average cost is $4.35 when 177 pizzas are produced daily.

91.

93.

Exercise 4-4

1. Min $f(x) = f(0) = 0$; Max $f(x) = f(10) = 14$ **3.** Min $f(x) = f(0) = 0$; Max $f(x) = f(3) = 9$
5. Min $f(x) = f(1) = f(7) = 5$; Max $f(x) = f(10) = 14$
7. Min $f(x) = f(1) = f(7) = 5$; Max $f(x) = f(3) = f(9) = 9$
9. Min $f(x) = f(5) = 7$; Max $f(x) = f(3) = 9$
11. Min $f(x) = f(1) = 2$; no maximum
13. Max $f(x) = f(-3) = 18$; no minimum
15. No absolute extrema   **17.** Max $f(x) = f(2) = 8$
19. Min $f(x) = f(4) = 22$ **21.** Max $f(x) = f(3.87) = 46.29$
23. (A) Max $f(x) = f(5) = 14$; Min $f(x) = f(-1) = -22$ (B) Max $f(x) = f(1) = -2$; Min $f(x) = f(-1) = -22$
(C) Max $f(x) = f(5) = 14$; Min $f(x) = f(3) = -6$
25. (A) Max $f(x) = f(0) = 126$; Min $f(x) = f(2) = -26$ (B) Max $f(x) = f(7) = 49$; Min $f(x) = f(2) = -26$
(C) Max $f(x) = f(6) = 6$; Min $f(x) = f(3) = -15$
27. Exactly in half   **29.** 15 and -15 **31.** A square of side 25 cm; maximum area $= 625$ cm^2
33. If x and y are the dimensions of the rectangle and A is the fixed area, the model is: Minimize $C = 2Bx + 2AB/x, x > 0$. This mathematical problem always has a solution. This agrees with our economic intuition that there should be a cheapest way to build the fence.
35. If x and y are the dimensions of the rectangle and C is the fixed amount to be spent, the model is: Maximize $A = x(C - 2Bx)/(2B), 0 \leq x \leq C/(2B)$. This mathematical problem always has a solution. This agrees with our economic intuition that there should be a largest area that can be enclosed with a fixed amount of fencing.
37. (A) Max $R(x) = R(3,000) = \$300,000$
(B) Maximum profit is $75,000 when 2,100 sets are manufactured and sold for $130 each.
(C) Maximum profit is $64,687.50 when 2,025 sets are manufactured and sold for $132.50 each.
39. (A)   (B)   (C) The maximum profit is $118,996 when the price per saw is $195.

41. $35; $6,125   **43.** 40 trees; 1,600 lb   **45.** $(10 - 2\sqrt{7})/3 = 1.57$ in. squares
47. 20 ft by 40 ft (with the expensive side being one of the short sides)   **49.** 10,000 books in 5 printings
51. (A) $x = 5.1$ mi   (B) $x = 10$ mi   **53.** 4 days; 20 bacteria/cm^3 **55.** 50 mice per order   **57.** 1 month; 2 ft
59. 4 yr from now

Review Exercise

1. $(a, c_1), (c_3, c_6)$ (4-1, 4-2)   **2.** $(c_1, c_3), (c_6, b)$ (4-1, 4-2)   **3.** $(a, c_2), (c_4, c_5), (c_7, b)$ (4-1, 4-2)   **4.** c_3 (4-1)
5. c_1, c_6 (4-1)   **6.** c_1, c_3, c_5 (4-1)   **7.** c_4, c_6 (4-1)   **8.** c_2, c_4, c_5, c_7 (4-2)
9. (4-3)   **10.** (4-3)

11. $f''(x) = 12x^2 + 30x$ (4-2)   **12.** $y'' = 8/x^3$ (4-2)

13. (A) All real numbers (B) y intercept: 0; x intercepts: 0, 9
(C) No horizontal or vertical asymptotes *(4-3)*

14. (A) 3, 9 (B) 3, 9
(C) Increasing on $(-\infty, 3)$ and $(9, \infty)$; decreasing on $(3, 9)$
(D) Local maximum at $x = 3$; local minimum at $x = 9$ *(4-3)*

15. (A) Concave downward on $(-\infty, 6)$; concave upward on $(6, \infty)$
(B) Inflection point at $x = 6$ *(4-3)*

16. *(4-3)*

17. (A) All real numbers, except -2 (B) y intercept: 0; x intercept: 0
(C) Horizontal asymptote: $y = 3$; vertical asymptote: $x = -2$ *(4-3)*

18. (A) None (B) -2 (C) Increasing on $(-\infty, -2)$ and $(-2, \infty)$ (D) None *(4-3)*

19. (A) Concave upward on $(-\infty, -2)$; concave downward on $(-2, \infty)$
(B) No inflection points *(4-3)*

20. *(4-3)*

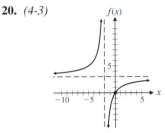

21.

x	$f'(x)$	$f(x)$
$-\infty < x < -2$	Negative and increasing	Decreasing and concave upward
$x = -2$	x intercept	Local minimum
$-2 < x < -1$	Positive and increasing	Increasing and concave upward
$x = -1$	Local maximum	Inflection point
$-1 < x < 1$	Positive and decreasing	Increasing and concave downward
$x = 1$	Local minimum	Inflection point
$1 < x < \infty$	Positive and increasing	Increasing and concave upward

(4-2)

22. (C) *(4-2)* **23.** Local maximum at $x = -1$; local minimum at $x = 5$ *(4-2)*

24. Min $f(x) = f(2) = -4$; Max $f(x) = f(5) = 77$ *(4-4)* **25.** Min $f(x) = f(2) = 8$ *(4-4)*

26. $-\infty$ *(4-3)* **27.** $-\infty$ *(4-3)* **28.** ∞ *(4-3)*

29. 2 *(4-3)* **30.** ∞ *(4-3)* **31.** 0 *(4-3)*

32. Horizontal asymptote: $y = 0$; no vertical asymptotes *(4-3)*

33. No horizontal asymptote; vertical asymptotes: $x = -2, x = 2$ *(4-3)*

34. Yes. Since f is continuous on $[a, b]$, f has an absolute maximum on $[a, b]$. But each endpoint is a local minimum; hence, the absolute maximum must occur between a and b. *(4-4)*

35. No, increasing/decreasing properties apply to intervals in the domain of f. It is correct to say that $f(x)$ is decreasing on $(-\infty, 0)$ and $(0, \infty)$. *(4-1)*

36. A critical value for $f(x)$ is a partition number for $f'(x)$ that is also in the domain of f. For example, if $f(x) = x^{-1}$, then 0 is a partition number for $f'(x) = -x^{-2}$, but 0 is not a critical value for $f(x)$ since 0 is not in the domain of f. *(4-1)*

37. Max $f'(x) = f'(2) = 12$ *(4-4)* **38.** Each number is 20; minimum sum is 40 *(4-4)*

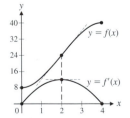

39. Domain: all real numbers
y intercept: -3; x intercepts: $-3, 1$
No vertical or horizontal asymptotes
Increasing on $(-2, \infty)$
Decreasing on $(-\infty, -2)$
Local minimum at $x = -2$
Concave upward on $(-\infty, -1)$ and $(1, \infty)$
Concave downward on $(-1, 1)$
Inflection points at $x = -1$ and $x = 1$ *(4-3)*

40. Domain: All real numbers
x intercepts: 0.79, 1.64; y intercept: 4
Increasing on $(-1.68, -0.35)$ and $(1.28, \infty)$
Decreasing on $(-\infty, -1.68)$ and $(-0.35, 1.28)$
Local minima at $x = -1.68$ and $x = 1.28$
Local maximum at $x = -0.35$
Concave downward on $(-1.10, 0.60)$
Concave upward on $(-\infty, -1.10)$ and $(0.60, \infty)$
Inflection points at $x = -1.10$ and $x = 0.60$ *(4-3)*

41. Domain: All real numbers
x intercepts; 0, 11.10; y intercept: 0
Increasing on $(1.87, 4.19)$ and $(8.94, \infty)$
Decreasing on $(-\infty, 1.87)$ and $(4.19, 8.94)$
Local maximum at $x = 4.19$
Local minima at $x = 1.87$ and $x = 8.94$
Concave upward on $(-\infty, 2.92)$ and $(7.08, \infty)$
Concave downward on $(2.92, 7.08)$
Inflection points at $x = 2.92$ and $x = 7.08$ *(4-3)*

42. (A) For the first 15 months, the graph of the price is increasing and concave downward, with a local maximum at $t = 15$. For the next 15 months, the graph of the price is decreasing and concave downward, with an inflection point at $t = 30$. For the next 15 months, the graph of the price is decreasing and concave upward, with a local minimum at $t = 45$. For the remaining 15 months, the graph of the price is increasing and concave upward.

(B) *(4-2)*

43. (A) Max $R(x) = R(10,000) = \$2,500,000$
(B) Maximum profit is \$175,000 when 3,000 stoves are manufactured and sold for \$425 each.
(C) Maximum profit is \$119,000 when 2,600 stoves are manufactured and sold for \$435 each. *(4-4)*

44. (A) The expensive side is 50 ft; the other side is 100 ft. (B) The expensive side is 75 ft; the other side is 150 ft. *(4-4)*

45. \$49; \$6,724 *(4-4)*   **46.** 12 orders/yr *(4-4)*

47. Min $\overline{C}(x) = \overline{C}(200) = 50$

 (4-3)

48. (A)
```
QuadReg
y=ax²+bx+c
a=.0061285714
b=.1224285714
c=102.2
```
(B) Min $\overline{C}(x) = \overline{C}(129) = \$1.71$ *(4-3)*

49. (A)
```
CubicReg
y=ax³+bx²+cx+d
a=-.01
b=.83
c=-2.3
d=221
```
(B) 28 ads to sell 588 refrigerators per month *(4-3)*

50. 3 days *(4-1)*   **51.** 2 yr from now *(4-1)*

CHAPTER 5

Exercise 5-1

1. \$1,221.40; \$1,648.72; \$2,225.54

3.

5. 11.55
7. 10.99
9. 0.14

11.

n	$[1 + (1/n)]^n$
10	2.593 74
100	2.704 81
1,000	2.716 92
10,000	2.718 15
100,000	2.718 27
1,000,000	2.718 28
10,000,000	2.718 28
$\downarrow$	$\downarrow$
∞	$e = 2.718\ 281\ 828\ 459\ldots$

13. $\lim_{n \to \infty}(1 + n)^{1/n} = 1$

15.
$$y_3 = \left(1 + \frac{1}{n}\right)^{n+1}$$
$$y_2$$
$$y_1 = \left(1 + \frac{1}{n}\right)^n$$

17. (A) \$17,349.87 (B) 7.36 yr
19. \$11,890.41
21. 8.11%

23. (A)

(B) $\lim_{t \to \infty} 10,000e^{-0.08t} = 0$

25. 9.90 yr
27. 8.66%
29. 7.3 yr

31. (A) $A = Pe^{rt}$ **(B)**

$2P = Pe^{rt}$

$2 = e^{rt}$

$rt = \ln 2$

$t = \dfrac{\ln 2}{r}$

Although r could be any positive number, the restrictions on r are reasonable in the sense that most investments would be expected to earn a return of between 2% and 30%.

(C) The doubling times (in years) are 13.86, 6.93, 4.62, 3.47, 2.77, and 2.31, respectively.

33. $t = -(\ln 0.5)/0.000\ 433\ 2 \approx 1{,}600$ yr   **35.** $r = (\ln 0.5)/30 \approx -0.0231$ **37.** 53.3 yr   **39.** 1.39%

Exercise 5-2

1. $f'(x) = 2e^x + \dfrac{3}{x}$ **3.** $\dfrac{dy}{dx} = 4e^x - 3ex^{e-1}$ **5.** $y' = \dfrac{3}{x}$ **7.** $\dfrac{6(\ln x)^5}{x}$ **9.** $x^3 + 4x^3 \ln x = x^3(1 + 4\ln x)$

11. $x^3 e^x + 3x^2 e^x = x^2 e^x(x + 3)$ **13.** $\dfrac{(x^2 + 9)e^x - 2xe^x}{(x^2 + 9)^2} = \dfrac{e^x(x^2 - 2x + 9)}{(x^2 + 9)^2}$ **15.** $\dfrac{x^3 - 4x^3 \ln x}{x^8} = \dfrac{1 - 4\ln x}{x^5}$

17. $3(x + 2)^2 \ln x + \dfrac{(x + 2)^3}{x} = (x + 2)^2 \left(3 \ln x + \dfrac{x + 2}{x}\right)$ **19.** $(x + 1)^3 e^x + 3(x + 1)^2 e^x = (x + 1)^2 e^x(x + 4)$

21. $\dfrac{2xe^x - (x^2 + 1)e^x}{(e^x)^2} = \dfrac{2x - x^2 - 1}{e^x}$ **23.** $(\ln x)^3 + 3(\ln x)^2 = (\ln x)^2(\ln x + 3)$

25. $3(4 - 5e^x)^2(-5e^x) = -15e^x(4 - 5e^x)^2$ **27.** $\dfrac{1}{2}(1 + \ln x)^{-1/2}\left(\dfrac{1}{x}\right) = \dfrac{1}{2x(1 + \ln x)^{1/2}}$ **29.** $xe^x + e^x - e^x = xe^x$

31. $2x^2\left(\dfrac{1}{x}\right) + 4x \ln x - 2x = 4x \ln x$ **33.** $y = ex$ **35.** $y = \dfrac{1}{e}x$

37. Yes, she is correct. In fact, for any real number c, the tangent line to $y = e^x$ at the point (c, e^c) has equation $y - e^c = e^c(x - c)$, and thus the tangent line passes through the point $(c - 1, 0)$.

39. $\dfrac{d}{dx} e^{kx} = ke^{kx}$ for any constant k **41.** Max $f(x) = f(e^{1.5}) = 2e^{1.5} \approx 8.963$

43. Min $f(x) = f(2) = \dfrac{e^2}{4} \approx 1.847$ **45.** Max $f(x) = f(e^{2.5}) = \dfrac{e^5}{2} \approx 74.207$

47. Domain: All real numbers
y intercept: 0; x intercept: 0
Horizontal asymptote: $y = 1$
Decreasing on $(-\infty, \infty)$
Concave downward on $(-\infty, \infty)$

49. Domain: $(0, \infty)$
Vertical asymptote: $x = 0$
Increasing on $(1, \infty)$
Decreasing on $(0, 1)$
Local minimum at $x = 1$
Concave upward on $(0, \infty)$

51. Domain: All real numbers
y intercept: 3; x intercept: 3
Horizontal asymptote: $y = 0$
Increasing on $(-\infty, 2)$
Decreasing on $(2, \infty)$
Local maximum at $x = 2$
Concave upward on $(-\infty, 1)$
Concave downward on $(1, \infty)$
Inflection point at $x = 1$

53. Domain: $(0, \infty)$
x intercept: 1
Increasing on $(e^{-1/2}, \infty)$
Decreasing on $(0, e^{-1/2})$
Local minimum at $x = e^{-1/2}$
Concave upward on $(e^{-3/2}, \infty)$
Concave downward on $(0, e^{-3/2})$
Inflection point at $x = e^{-3/2}$

55. Critical values: $x = 0.36$, $x = 2.15$
Increasing on $(-\infty, 0.36)$ and $(2.15, \infty)$
Decreasing on $(0.36, 2.15)$
Local maximum at $x = 0.36$
Local minimum at $x = 2.15$

57. Critical value: $x = 2.21$
Increasing on $(0, 2.21)$
Decreasing on $(2.21, \infty)$
Local maximum at $x = 2.21$

59. $(-0.82, 0.44), (1.43, 4.18), (8.61, 5503.66)$ **61.** $(3.65, 1.30), (332,105.11, 12.71)$
63. $p = \$2$ **65.** $p = \$27.57$ **67.** Min $\overline{C}(x) = \overline{C}(e^7) \approx \$99.91$

69. (A) At \$3.68 each, the maximum revenue will be \$3,680/wk (in the test city).
(B)

71. At the 40-lb weight level, blood pressure would increase at the rate of 0.44 mm of mercury per pound of weight gain. At the 90-lb weight level, blood pressure would increase at the rate of 0.19 mm of mercury per pound of weight gain.

73. (A) After 1 hr, the concentration is decreasing at the rate of 1.60 mg/mL per hour; after 4 hr, the concentration is decreasing at the rate of 0.08 mg/mL per hour.

(B)

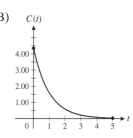

75. $dR/dS = k/S$

Exercise 5-3

1. $(3x^2 + 2)^3$ **3.** $\ln(2x^2 + 5)$ **5.** e^{-x^2} **7.** $y = u^4; u = 3x^2 - x + 5$ **9.** $y = \ln u; u = 3x^2 - 2x + 9$
11. $y = e^u; u = 1 + x + x^2$ **13.** $y = (2 + e^x)^2; dy/dx = 2e^x(2 + e^x)$ **15.** $y = e^{2-x^4}; dy/dx = -4x^3 e^{2-x^4}$
17. $y = \ln(4x^5 - 7); \dfrac{dy}{dx} = \dfrac{20x^4}{4x^5 - 7}$ **19.** $\dfrac{1}{x - 3}$ **21.** $\dfrac{-2}{3 - 2t}$ **23.** $6e^{2x}$ **25.** $-8e^{-4t}$
27. $-3e^{-0.03x}$ **29.** $\dfrac{4}{x + 1}$ **31.** $4e^{2x} - 3e^x$ **33.** $(6x - 2)e^{3x^2 - 2x}$ **35.** $\dfrac{2t + 3}{t^2 + 3t}$ **37.** $\dfrac{x}{x^2 + 1}$
39. $\dfrac{4[\ln(t^2 + 1)]^3(2t)}{t^2 + 1} = \dfrac{8t[\ln(t^2 + 1)]^3}{t^2 + 1}$ **41.** $4(e^{2x} - 1)^3(2e^{2x}) = 8e^{2x}(e^{2x} - 1)^3$
43. $\dfrac{(x^2 + 1)(2e^{2x}) - e^{2x}(2x)}{(x^2 + 1)^2} = \dfrac{2e^{2x}(x^2 - x + 1)}{(x^2 + 1)^2}$ **45.** $(x^2 + 1)(-e^{-x}) + e^{-x}(2x) = e^{-x}(2x - x^2 - 1)$
47. $\dfrac{e^{-x}}{x} - e^{-x} \ln x = \dfrac{e^{-x}(1 - x \ln x)}{x}$ **49.** $\dfrac{-2x}{(1 + x^2)[\ln(1 + x^2)]^2}$ **51.** $\dfrac{-2x}{3(1 - x^2)[\ln(1 - x^2)]^{2/3}}$
53. The graph of $y = \log_b x$ is a vertical expansion of the graph of $y = \ln x$ if $1 < b < e$, and a vertical contraction if $b > e$.
55. Domain: $(-\infty, \infty)$
y intercept: 0; x intercept: 0
Horizontal asymptote: $y = 1$
Increasing on $(-\infty, \infty)$
Concave downward on $(-\infty, \infty)$

57. Domain: $(-\infty, 1)$
y intercept: 0; x intercept: 0
Vertical asymptote: $x = 1$
Decreasing on $(-\infty, 1)$
Concave downward on $(-\infty, 1)$

59. Domain: $(-\infty, \infty)$
y intercept: 1
Horizontal asymptote: $y = 0$
Increasing on $(-\infty, 0)$
Decreasing on $(0, \infty)$
Local maximum at $x = 0$
Concave upward on $(-\infty, -1)$ and $(1, \infty)$
Concave downward on $(-1, 1)$
Inflection points at $x = -1$ and $x = 1$

61. $y = 1 + [\ln(2 + e^x)]^2; \dfrac{dy}{dx} = \dfrac{2e^x \ln(2 + e^x)}{2 + e^x}$

63. $\dfrac{1}{\ln 2}\left(\dfrac{6x}{3x^2 - 1}\right)$

65. $(2x + 1)(10^{x^2+x})(\ln 10)$

67. $\dfrac{12x^2 + 5}{(4x^3 + 5x + 7)\ln 3}$

69. $2^{x^3-x^2+4x+1}(3x^2 - 2x + 4)\ln 2$

73. (A) $g(x)$ is not negative when $f(x)$ is decreasing

(B) $f'(x) = \dfrac{2x}{x^2 + 1}$

75. $f'(x) = g'(x) = \dfrac{8x}{x^2 + 3}$

77. A maximum revenue of $735.80 is realized at a production level of 20 units at $36.79 each.

79. A maximum profit of $224.61 is realized at a production level of 17 units at $42.74 each.

81. $159.68

83. $-$27,145/yr; $-$18,196/yr; $-$11,036/yr

85. (A) 23 days; $26,685; about 50%

(B)

87. 2.27 mm of mercury/yr; 0.81 mm of mercury/yr; 0.41 mm of mercury/yr

89. $A'(t) = 2(\ln 2)5,000e^{2t \ln 2} = 10,000(\ln 2)2^{2t}; A'(1) = 27,726$ bacteria/hr (rate of change at the end of the first hour); $A'(5) = 7,097,827$ bacteria/hr (rate of change at the end of the fifth hour)

91.

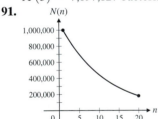

Exercise 5-4

1. $y' = -\dfrac{3}{5}$ **3.** $y' = \dfrac{3x}{2}$ **5.** $y' = 10x; 10$ **7.** $y' = \dfrac{2x}{3y^2}; \dfrac{4}{3}$ **9.** $y' = -\dfrac{3}{2y + 2}; -\dfrac{3}{4}$ **11.** $y' = -\dfrac{y}{x}; -\dfrac{3}{2}$

13. $y' = -\dfrac{2y}{2x + 1}; 4$ **15.** $y' = \dfrac{6 - 2y}{x}; -1$ **17.** $y' = \dfrac{2x}{e^y - 2y}; 2$ **19.** $y' = \dfrac{3x^2y}{y + 1}; \dfrac{3}{2}$

21. $y' = \dfrac{6x^2y - y \ln y}{x + 2y}; 2$ **23.** $x' = \dfrac{2tx - 3t^2}{2x - t^2}; 8$ **25.** $y'\big|_{(1.6, 1.8)} = -\dfrac{3}{4}; y'\big|_{(1.6, 0.2)} = \dfrac{3}{4}$ **27.** $y = -x + 5$

29. $y = \dfrac{2}{5}x - \dfrac{12}{5}; y = \dfrac{3}{5}x + \dfrac{12}{5}$ **31.** $y' = -\dfrac{1}{x}$ **33.** $y' = \dfrac{1}{3(1 + y)^2 + 1}; \dfrac{1}{13}$ **35.** $y' = \dfrac{3(x - 2y)^2}{6(x - 2y)^2 + 4y}; \dfrac{3}{10}$

37. $y' = \dfrac{3x^2(7 + y^2)^{1/2}}{y}; 16$ **39.** $y' = \dfrac{y}{2xy^2 - x}; 1$ **41.** $y = 0.63x + 1.04$ **43.** $p' = \dfrac{1}{2p - 2}$

45. $p' = -\dfrac{\sqrt{10,000 - p^2}}{p}$ **47.** $\dfrac{dL}{dV} = \dfrac{-(L + m)}{V + n}$

Exercise 5-5

1. 30   **3.** $-\dfrac{16}{3}$ **5.** $-\dfrac{16}{7}$ **7.** Decreasing at 9 units/sec   **9.** Approx. -3.03 ft/sec   **11.** $dA/dt \approx 126$ ft²/sec

13. 3,768 cm³/min   **15.** 6 lb/in.²/hr   **17.** $-\dfrac{9}{4}$ ft/sec   **19.** $\dfrac{20}{3}$ ft/sec

21. 0.0214 ft/sec; 0.0135 ft/sec; yes, at $t = 0.000\ 19$ sec   **23.** 3.835 units/sec
25. (A) $dC/dt = \$15{,}000/\text{wk}$ (B) $dR/dt = -\$50{,}000/\text{wk}$ (C) $dP/dt = -\$65{,}000/\text{wk}$ **27.** $ds/dt = \$2{,}207/\text{wk}$
29. (A) $dx/dt = -12.73$ units/month   (B) $dp/dt = \$1.53/\text{month}$ **31.** Approximately 100 ft^3/min

Review Exercise

1. \$3,136.62; \$4,919.21; \$12,099.29 *(5-1)*   **2.** $\dfrac{2}{x} + 3e^x$ *(5-2)*   **3.** $2e^{2x-3}$ *(5-2)*   **4.** $\dfrac{2}{2x+7}$ *(5-2)*

5. (A) $y = \ln(3 + e^x)$ (B) $\dfrac{dy}{dx} = \dfrac{e^x}{3+e^x}$ *(5-3)*   **6.** $y' = \dfrac{9x^2}{4y}; \dfrac{9}{8}$ *(5-4)*   **7.** $dy/dt = 216$ *(5-4)*

8. Domain: All real numbers
y intercept: 100
Horizontal asymptote: $y = 0$
Decreasing on $(-\infty, \infty)$
Concave upward on $(-\infty, \infty)$ *(5-2)*

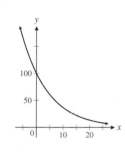

9. $\displaystyle\lim_{n\to\infty}\left(1 + \dfrac{2}{n}\right)^n = e^2 \approx 7.389\ 06$ *(5-1)*

10. $\dfrac{7[(\ln z)^6 + 1]}{z}$ *(5-3)*   **11.** $x^5(1 + 6\ln x)$ *(5-2)*

12. $\dfrac{e^x(x-6)}{x^7}$ *(5-2)*   **13.** $\dfrac{6x^2 - 3}{2x^3 - 3x}$ *(5-3)*

14. $(3x^2 - 2x)e^{x^3-x^2}$ *(5-3)*

15. $\dfrac{1 - 2x\ln 5x}{xe^{2x}}$ *(5-3)*

16. $y = -x + 2;\ y = -ex + 1$ *(5-2)*

17. $y' = \dfrac{3y - 2x}{8y - 3x}; \dfrac{8}{19}$ *(5-4)*   **18.** $x' = \dfrac{4tx}{3x^2 - 2t^2}; -4$ *(5-4)*   **19.** $y' = \dfrac{1}{e^y + 2y}; 1$ *(5-4)*

20. $y' = \dfrac{2xy}{1 + 2y^2}; \dfrac{2}{3}$ *(5-4)*   **21.** $dy/dt = -2$ units/sec *(5-5)*   **22.** 0.27 ft/sec *(5-5)*

23. $dR/dt = 1/\pi \approx 0.318$ in./min *(5-5)*
24. Max $f(x) = f(e^{4.5}) = 2e^{4.5} \approx 180.03$ *(5-2)*   **25.** Max $f(x) = f(0.5) = 5e^{-1} \approx 1.84$ *(5-3)*
26. Max $f(x) = f(1.373) = 2.487$ *(5-2)*   **27.** Max $f(x) = f(1.763) = 0.097$ *(5-2)*

28. Domain: All real numbers
y intercept: 0; x intercept: 0
Horizontal asymptote: $y = 5$
Increasing on $(-\infty, \infty)$
Concave downward on $(-\infty, \infty)$

(5-2)

29. Domain: $(0, \infty)$
x intercept: 1
Increasing on $(e^{-1/3}, \infty)$
Decreasing on $(0, e^{-1/3})$
Local minimum at $x = e^{-1/3}$
Concave upward on $(e^{-5/6}, \infty)$
Concave downward on $(0, e^{-5/6})$
Inflection point at $x = e^{-5/6}$

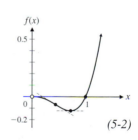

(5-2)

30. (A) $y = [\ln(4 - e^x)]^3$ (B) $\dfrac{dy}{dx} = \dfrac{-3e^x[\ln(4 - e^x)]^2}{4 - e^x}$ *(5-3)*   **31.** $2x(5^{x^2-1})(\ln 5)$ *(5-3)*   **32.** $\left(\dfrac{1}{\ln 5}\right)\dfrac{2x - 1}{x^2 - x}$ *(5-3)*

33. $\dfrac{2x + 1}{2(x^2 + x)\sqrt{\ln(x^2 + x)}}$ *(5-3)*   **34.** $y' = \dfrac{2x - e^{xy}y}{xe^{xy} - 1}; 0$ *(5-4)*

35. The rate of increase of area is proportional to the radius R, so it is smallest when $R = 0$, and has no largest value. *(5-5)*
36. Yes, for $-\sqrt{3}/3 < x < \sqrt{3}/3$ *(5-5)*   **37.** (A) 15 yr   (B) 13.9 yr *(5-1)*
38. $A'(t) = 10e^{0.1t}; A'(1) = \$11.05/\text{yr}; A'(10) = \$27.18/\text{yr}$ *(5-1)*   **39.** $R'(x) = (1{,}000 - 20x)e^{-0.02x}$ *(5-3)*
40. A maximum revenue of \$18,394 is realized at a production level of 50 units at \$367.88 each. *(5-3)*

41. *(5-3)*

R(x)

20,000
16,000
12,000
8,000
4,000

0 20 40 60 80 100   x

42. \$549.15; \$9,864 *(5-3)*   **43.** \$1.52 *(5-2)*

44. Min $\overline{C}(x) = \overline{C}(e^5) \approx \$49.66$ *(5-2)*   **45.** $p' = \dfrac{-(5{,}000 - 2p^3)^{1/2}}{3p^2}$ *(5-4)*

46. $dR/dt = \$110/\text{day}$ *(5-5)*
47. -1.111 mg/mL per hour; -0.335 mg/mL per hour *(5-3)*
48. $dR/dt = -3/(2\pi)$; approx. 0.477 mm/day *(5-5)*

49. (A) Increasing at the rate of 2.68 units/day at the end of 1 day   **50.** $dT/dt = -1/27 \approx -0.037$ min/operation hour *(5-5)*
of training; increasing at the rate of 0.54 unit/day after
5 days of training

(B)

(5-3)

CHAPTER 6

Exercise 6-1

1. $2.7x + C$ **3.** $(x^9/9) + C$ **5.** $-(t^{-3}/3) + C$ **7.** $2\pi t + C$ **9.** $(3u^7/7) + C$
11. $u + u^3 + C$ **13.** $5e^t + C$ **15.** $3 \ln|z| + C$ **17.** $(x^2/2) + (x^3/3) + C$

19.

21.

23. $y = 40x^5 + C$ **25.** $P = 24x - 3x^2 + C$ **27.** $y = \frac{1}{3}u^6 - u^3 - u + C$
29. $y = e^x + 3x + C$ **31.** $x = 5 \ln|t| + t + C$ **33.** (A) False (B) True
35. No, since one graph cannot be obtained from another by a vertical translation.
37. Yes, since one graph can be obtained from another by a vertical translation.
39. $15x^{4/3} + C$ **41.** $-3t^{-3} + C$ **43.** $2\sqrt{u} + C$ **45.** $-(x^{-2}/8) + C$ **47.** $4 \ln|u| + u + C$
49. $x^3 + 2x^{-1} + C$ **51.** $2x^5 + 2x^{-4} - 2x + C$ **53.** $2x^{3/2} + 4x^{1/2} + C$ **55.** $\frac{3}{5}x^{5/3} + 2x^{-2} + C$
57. $(e^x/4) - (3x^2/8) + C$ **59.** $-4z^{-3} - \frac{5}{2}z^{-2} - 3 \ln|z| + C$

61.

63.

65. $y = x^2 - 3x + 5$ **67.** $C(x) = 2x^3 - 2x^2 + 3,000$ **69.** $x = 40\sqrt{t}$ **71.** $y = -2x^{-1} + 3 \ln|x| - x + 3$
73. $x = 4e^t - 2t - 3$ **75.** $y = 2x^2 - 3x + 1$ **77.** $x^2 + x^{-1} + C$ **79.** $\frac{1}{2}x^2 + x^{-2} + C$ **81.** $e^x - 2 \ln|x| + C$
83. $M = t + t^{-1} + \frac{3}{4}$ **85.** $y = 3x^{5/3} + 3x^{2/3} - 6$ **87.** $p(x) = 10x^{-1} + 10$ **89.** x^3 **91.** $x^4 + 3x^2 + C$

93. $\overline{C}(x) = 15 + \dfrac{1,000}{x}$; $C(x) = 15x + 1,000$; $C(0) = \$1,000$

95. (A) The cost function increases from 0 to 8, is concave downward from 0 to 4, and is concave upward from 4 to 8. There is an inflection point at $x = 4$.

(B) $C(x) = x^3 - 12x^2 + 53x + 30$; $C(4) = \$114,000$; $C(8) = \$198,000$

(C)

(D) Manufacturing plants are often inefficient at low and high levels of production.

97. $S(t) = 2,000 - 15t^{5/3}$; $80^{3/5} \approx 14$ mo   **99.** $S(t) = 2,000 - 15t^{5/3} - 70t$; $t \approx 8.92$ mo

101. $L(x) = 4,800x^{1/2}$; $L(25) = 24,000$ labor-hours   **103.** $W(h) = 0.0005h^3$; $W(70) = 171.5$ lb   **105.** 19,400

Exercise 6-2

1. $\frac{1}{3}(4x - 3)^3 + C$ **3.** $\frac{1}{5}(x^2 + 1)^4 + C$ **5.** $e^{2x} + C$ **7.** $\ln(1 + x^4) + C$ **9.** $\frac{1}{11}(x + 3)^{11} + C$

11. $-\frac{1}{6}(6t - 7)^{-1} + C$ **13.** $\frac{1}{12}(t^2 + 1)^6 + C$ **15.** $\frac{1}{2}e^{x^2} + C$ **17.** $\frac{1}{5}\ln|5x + 4| + C$ **19.** $-e^{1-t} + C$

21. $-\frac{1}{18}(3t^2 + 1)^{-3} + C$ **23.** $\frac{1}{3}(4 - x^3)^{-1} + C$ **25.** $\frac{2}{5}(x + 4)^{5/2} - \frac{8}{3}(x + 4)^{3/2} + C$

27. $\frac{2}{3}(x - 3)^{3/2} + 6(x - 3)^{1/2} + C$ **29.** $\frac{1}{11}(x - 4)^{11} + \frac{2}{5}(x - 4)^{10} + C$ **31.** $\frac{1}{8}(1 + e^{2x})^4 + C$

33. $\frac{1}{2}\ln|4 + 2x + x^2| + C$ **35.** $e^{x^2+x+1} + C$ **37.** $\frac{1}{4}(e^x - 2x)^4 + C$ **39.** $-\frac{1}{12}(x^4 + 2x^2 + 1)^{-3} + C$

41. (A) Differentiate the right side to get the integrand on the left side.

(B) Wrong, since $\dfrac{d}{dx}[\ln|2x - 3| + C] = \dfrac{2}{2x - 3} \neq \dfrac{1}{2x - 3}$. If $u = 2x - 3$, then $du = 2\,dx$. The integrand was not adjusted for the missing constant factor 2.

(C) $\displaystyle\int \frac{1}{2x - 3}\,dx = \frac{1}{2}\int \frac{2}{2x - 3}\,dx = \frac{1}{2}\ln|2x - 3| + C$ Check: $\dfrac{d}{dx}\left[\dfrac{1}{2}\ln|2x - 3| + C\right] = \dfrac{1}{2x - 3}$

43. (A) Differentiate the right side to get the integrand on the left side.

(B) Wrong, since $\dfrac{d}{dx}[e^{x^4} + C] = 4x^3 e^{x^4} \neq x^3 e^{x^4}$. If $u = x^4$, then $du = 4x^3\,dx$. The integrand was not adjusted for the missing constant factor 4.

(C) $\displaystyle\int x^3 e^{x^4}\,dx = \frac{1}{4}\int 4x^3 e^{x^4}\,dx = \frac{1}{4}e^{x^4} + C$ Check: $\dfrac{d}{dx}\left[\dfrac{1}{4}e^{x^4} + C\right] = x^3 e^{x^4}$

45. (A) Differentiate the right side to get the integrand on the left side.

(B) Wrong, since $\dfrac{d}{dx}\left[\dfrac{(x^2 - 2)^2}{3x} + C\right] = \dfrac{3x^4 - 4x^2 - 4}{3x^2} \neq 2(x^2 - 2)^2$. If $u = x^2 - 2$, then $du = 2x\,du$. It appears that the student moved a variable factor across the integral sign as follows (which is *not* valid):

$$\int 2(x^2 - 2)^2\,dx = \frac{1}{x}\int 2x(x^2 - 2)^2\,dx.$$

(C) $\displaystyle\int 2(x^2 - 2)^2\,dx = \int (2x^4 - 8x^2 + 8)\,dx = \frac{2}{5}x^5 - \frac{8}{3}x^3 + 8x + C$

Check: $\dfrac{d}{dx}\left[\dfrac{2}{5}x^5 - \dfrac{8}{3}x^3 + 8x + C\right] = 2x^4 - 8x^2 + 8 = 2(x^2 - 2)^2$

47. $\frac{1}{9}(3x^2 + 7)^{3/2} + C$ **49.** $\frac{1}{8}x^8 + \frac{4}{5}x^5 + 2x^2 + C$ **51.** $\frac{1}{9}(x^3 + 2)^3 + C$ **53.** $\frac{1}{4}(2x^4 + 3)^{1/2} + C$ **55.** $\frac{1}{4}(\ln x)^4 + C$

57. $e^{-1/x} + C$ **59.** $x = \frac{1}{3}(t^3 + 5)^7 + C$ **61.** $y = 3(t^2 - 4)^{1/2} + C$ **63.** $p = -(e^x - e^{-x})^{-1} + C$

67. $p(x) = 2,000/(3x + 50)$; 250 bottles   **69.** $C(x) = 12x + 500\ln(x + 1) + 2,000$; $\overline{C}(1,000) = \$17.45$

71. (A) $S(t) = 10t + 100e^{-0.1t} - 100$, $0 \le t \le 24$ (B) $S(12) \approx \$50$ million   (C) 18.41 mo

73. $Q(t) = 100\ln(t + 1) + 5t$, $0 \le t \le 20$; $Q(9) \approx 275$ thousand barrels   **75.** $W(t) = 2e^{0.1t}$; $W(8) \approx 4.45$ g

77. (A) $-1,000$ bacteria/mL per day   (B) $N(t) = 5,000 - 1,000\ln(1 + t^2)$; 385 bacteria/mL   (C) 7.32 days

79. $N(t) = 100 - 60e^{-0.1t}$, $0 \le t \le 15$; $N(15) \approx 87$ words/min

81. $E(t) = 12,000 - 10,000(t + 1)^{-1/2}$; $E(15) = 9,500$ students

Exercise 6-3

1. $y = 2.08x^2 + C$ **3.** $y = 2e^{0.5x} + C$ **5.** $y = \dfrac{x^3}{3} - \dfrac{x^2}{2}$ **7.** $y = e^{-x^2} + 2$

9. Figure B. When $x = 1$, the slope $dy/dx = 1 - 1 = 0$ for any y. When $x = 0$, the slope $dy/dx = 0 - 1 = -1$ for any y. Both are consistent with the slope field shown in Figure B.

11. $y = \dfrac{x^2}{2} - x + C; y = \dfrac{x^2}{2} - x - 2$

13.

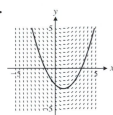

15. $y = Ce^{2t}$
17. $y = 100e^{-0.5x}$
19. $x = Ce^{-8t}$

21. Figure A. When $y = 1$, the slope $dy/dx = 1 - 1 = 0$ for any x. When $y = 2$, the slope $dy/dx = 1 - 2 = -1$ for any x. Both are consistent with the slope field shown in Figure A.

23. $y = 1 - e^{-x}$ **25.**

27.

29.

31.

33.

35.

37. Apply the second-derivative test to $f(y) = ky(M - y)$.   **39.** 1999   **41.** $A = 1{,}000e^{0.08t}$ **43.** $A = 8{,}000e^{0.06t}$

45. (A) $p(x) = 100e^{-0.05x}$ **47.** (A) $N = L(1 - e^{-0.051t})$ **49.** $I = I_0e^{-0.00942x}; x \approx 74$ ft
    (B) \$60.65 per unit   (B) 22.5%
    (C)   (C) 32 days

    (D)

51. (A) $Q = 3e^{-0.04t}$ **53.** 0.023 117   **59.** (A) 7 people; 353 people
    (B) $Q(10) = 2.01$ mL   **55.** Approx. 24,200 yr   (B) 400
    (C) 27.47 hr   **57.** 104 times; 67 times   (C)
    (D)

Exercise 6-4

1.

3. Figure A: $L_3 = 13$; $R_3 = 20$, $A_3 = 16.5$; Figure B: $L_3 = 14$, $R_3 = 7$, $A_3 = 10.5$

5. $L_3 \leq \int_1^4 f(x)\, dx \leq R_3$; $R_3 \leq \int_1^4 g(x)\, dx \leq L_3$; since $f(x)$ is increasing, L_3 underestimates the area and R_3 overestimates the area; since $g(x)$ is decreasing, the reverse is true.

7. Both figures: Error bound for L_3 and R_3 is 7; error bound for A_3 is 3.5.

9. The exact area under the graph of $y = f(x)$ is within 3.5 units (either way) of the average of the left sum and right sum estimates, $A_3 = 16.5$.

11. (A) $L_4 = 320$, $R_4 = 192$, $A_4 = 256$; error bound for L_4 and R_4 is 128; error bound for A_4 is 64
   (B) The height of each rectangle represents an instantaneous rate, and the base of each rectangle represents a time interval; rate times time is distance.
   (C) $n > 256$

13. (A) False   (B) True   **15.** $A_{10} = 311{,}100$ ft^2; error bound is 25,000 ft^2; $n \geq 100$

17. (A) $A_7 = 298$ ft; error bound $= 55$ ft   (B) $n > 77$

19. (A) $P = 2$

(B)

(C) To the left of $P = 2$, the left rectangles underestimate the true area and the right rectangles overestimate the true area. To the right of $P = 2$, the left rectangles overestimate the true area and the right rectangles underestimate the true area.
(D) $26 \leq \int_0^5 f(x)\, dx \leq 39$

21. $L_6 = -3.53$, $R_6 = -0.91$, $A_6 = -2.22$; error bound for L_6 and R_6 is 2.63; error bound for A_6 is 1.31. Geometrically, the definite integral over the interval $[2, 5]$ is the sum of the areas between the curve and the x axis from $x = 2$ to $x = 5$, with the areas below the x axis counted negatively and those above the x axis counted positively.

23. $(-\infty, 0], [0, \infty)$ **25.** $(-\infty, -1], [-1, 0], [0, 1], [1, \infty)$ **27.** $n \geq 22$ **29.** $n \geq 104$ **31.** $n \geq 30$

33. If we let I represent the definite integral on the left side, then we know that $|I - L_n| \leq |f(b) - f(a)| \dfrac{b - a}{n}$. Since the limit of the right side is 0 as $n \to \infty$, the result follows.

35. $A_2 = \$180{,}000$; error bound $= \$30{,}000$ **37.** $L_4 = \$4{,}251$, $R_4 = \$4{,}605$; $\$4{,}251 \leq \int_2^6 800e^{0.08t}\, dt \leq \$4{,}605$

39. First 60 days: $A_3 = 3{,}240$ units; error bound $= 660$ units; second 60 days: $A_3 = 4{,}920$ units; error bound $= 100$ units

41. (A) $\int_{100}^{200} R'(x)\, dx$ represents the area under the marginal revenue curve from $x = 100$ to $x = 200$; it also represents the total change in revenue going from sales of 100 six-packs per day to sales of 200 six-packs per day.
   (B) $A_4 = \$200$; error bound $= \$50$
   (C) $\$200$; both $R(200) - R(100)$ and $\int_{100}^{200} R'(x)\, dx$ represent total change in revenue going from sales of 100 six-packs per day to sales of 200 six-packs per day. This suggests that $\int_{100}^{200} R'(x)\, dx = R(200) - R(100)$.

43. (A) $L_5 = 3.72$ cm^2, $R_5 = 3.37$ cm^2 (B) $3.37 \leq \int_0^5 A'(t)\, dt \leq 3.72$

45. $A_3 = 108$ code symbols; error bound $= 6$ code symbols

Exercise 6-5

1. -2.33 **3.** 8.34 **5.** 1.067 **7.** 5 **9.** 5 **11.** 2 **13.** $-\frac{7}{3} \approx -2.333$ **15.** 2

17. $\frac{1}{2}(e^2 - 1) \approx 3.195$ **19.** $2 \ln 3.5 \approx 2.506$ **21.** -10.67 **23.** -5.04 **25.** 240 ft   **27.** -2

29. 14   **31.** $5^6 = 15,625$ **33.** $\ln 4 \approx 1.386$ **35.** $20(e^{0.25} - e^{-0.5}) \approx 13.550$
37. $\frac{1}{2}$ **39.** $\frac{8}{3} \approx 2.667$ **41.** $\frac{1}{2}(1 - e^{-1}) \approx 0.316$ **43.** $-\frac{3}{2} - \ln 2 \approx -2.193$

45. (A) Average $f(x) = 250$

(B)

47. (A) Average $f(t) = 2$

(B)

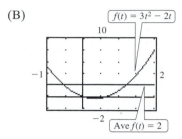

49. (A) Average $f(x) = \frac{45}{28} \approx 1.61$

(B)

51. (A) Average $f(x) = 2(1 - e^{-2}) \approx 1.73$

(B)

53. $I = 10 \pm 0.67$ **55.** $10.67; |I - M_4| = 0.67$; yes   **57.** $n \geqslant 47$
59. $\int_2^5 (1 - x^2)\, dx = -36$ **61.** $\int_2^{12} (3x^2 - 2x + 3)\, dx = 1,610$
63. $\frac{1}{6}(15^{3/2} - 5^{3/2}) \approx 7.819$ **65.** $\frac{1}{2}(\ln 2 - \ln 3) \approx -0.203$ **67.** 0
69. (A) $\ln 2 = 0.6919 \pm 0.0033$ (B) $\ln 2 = 0.6931$ (C) 0.0012; yes
71. $n \geqslant 13$ **73.** 4.566   **75.** 2.214
77. $\int_{300}^{900} \left(500 - \dfrac{x}{3}\right) dx = \$180,000$
79. $\int_0^5 500(t - 12)\, dt = -\$23,750; \int_5^{10} 500(t - 12)\, dt = -\$11,250$

81. (A)

(B) 6,505   **83.** Useful life $= \sqrt{\ln 55} \approx 2$ yr;
total profit $= \frac{51}{22} - \frac{5}{2}e^{-4} \approx 2.272$
or \$2,272
85. (A) \$420 (B) \$135,000

87. (A)

(B) \$100,505   **89.** $50e^{0.6} - 50e^{0.4} - 10 \approx \$6.51$
91. 4,800 labor-hours
93. (A) $I = -200t + 600$
(B) $\frac{1}{3}\int_0^3 (-200t + 600)\, dt = 300$

95. $100 \ln 11 + 50 \approx 290$ thousand barrels; $100 \ln 21 - 100 \ln 11 + 50 \approx 115$ thousand barrels   **97.** \$10,000
99. $3,120,000$ ft^2 **101.** $2e^{0.8} - 2 \approx 2.45$ g; $2e^{1.6} - 2e^{0.8} \approx 5.45$ g   **103.** 10°C   **105.** 1.1 liters
107. $0.6 \ln 2 + 0.1 \approx 0.516$; $(4.2 \ln 625 + 2.4 - 4.2 \ln 49)/24 \approx 0.546$

Review Exercise

1. $3x^2 + 3x + C$ *(6-1)*
2. 50 *(6-5)*
3. -207 *(6-5)*
4. $-\frac{1}{8}(1 - t^2)^4 + C$ *(6-2)*
5. $\ln |u| + \frac{1}{4}u^4 + C$ *(6-1)*
6. 0.216 *(6-5)*
7. e^{-x^2} *(6-1)*   **8.** $\sqrt{4 + 5x} + C$ *(6-1)*   **9.** $y = f(x) = x^3 - 2x + 4$ *(6-3)*
10. Increasing on $[0, 2]$; decreasing on $[2, 4]$; concave downward on $[0, 4]$; local maximum at $x = 2$. Antiderivative graphs differ by a vertical translation. *(6-1)*

11. *(6-1)*

12. (A) $2x^4 - 2x^2 - x + C$ (B) $e^t - 4\ln|t| + C$ *(6-1)*

13. $\int_1^5 (x^2 + 1)\,dx = 44 \pm 1.333$ *(6-5)*

14. $45.333; |I - M_2| = 1.333$ *(6-5)*   **15.** 30.8 *(6-5)*   **16.** 7 *(6-5)*

17. Width $= 2 - (-1) = 3$; height $=$ average $f(x) = 7$ *(6-5)*   **18.** -10 *(6-4, 6-5)*

19. 0.4 *(6-4, 6-5)*   **20.** 1.4 *(6-4, 6-5)*   **21.** 0 *(6-4, 6-5)*   **22.** 0.4 *(6-4, 6-5)*

23. 2 *(6-4, 6-5)*   **24.** -2 *(6-4, 6-5)*   **25.** -0.4 *(6-4, 6-5)*

26. Increasing on $[0, 1]$ and $[3, 4]$; decreasing on $[1, 3]$; concave downward on $[0, 2]$; concave upward on $[2, 4]$; local maximum at $x = 1$; local minimum at $x = 3$; inflection point at $x = 2$; graphs of antiderivatives differ by a vertical translation. *(6-1)*

27. *(6-1)*

28. (A) $1; 1$ (B) $4; 4$ *(6-3)*

29. $dy/dx = (2y)/x$; the slopes computed in Problem 28A are compatible with the slope field shown. *(6-3)*

31. $y = \frac{1}{4}x^2$; $y = -\frac{1}{4}x^2$ *(6-3)*

32. *(6-3)*

33. *(6-3)*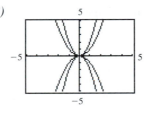

34. $\frac{2}{3}(2)^{3/2} \approx 1.886$ *(6-5)*   **35.** $\frac{1}{6} \approx 0.167$ *(6-5)*   **36.** $-5e^{-t} + C$ *(6-1)*   **37.** $\frac{1}{2}(1 + e^2)$ *(6-1)*

38. $\frac{1}{6}e^{3x^2} + C$ *(6-2)*   **39.** $2(\sqrt{5} - 1) \approx 2.472$ *(6-5)*   **40.** $\frac{1}{2}\ln 10 \approx 1.151$ *(6-5)*   **41.** 0.45 *(6-5)*

42. $\frac{1}{48}(2x^4 + 5)^6 + C$ *(6-2)*   **43.** $-\ln(e^{-x} + 3) + C$ *(6-2)*   **44.** $-(e^x + 2)^{-1} + C$ *(6-2)*

45. $y = f(x) = 3\ln|x| + x^{-1} + 4$ *(6-2, 6-3)*   **46.** $y = 3x^2 + x - 4$ *(6-3)*

47. $L_5 = 480$ ft; $R_5 = 320$ ft; $A_5 = 400$ ft; error bound for L_5 and $R_5 = 160$; error bound for $A_5 = 80$ *(6-4)*

48. The height of each rectangle represents an instantaneous rate, and the base of each rectangle represents a time interval; rate times time is distance. *(6-4)*

49. $n > 400$ *(6-4)*   **50.** Height $= \int_0^5 (160 - 32t)\,dt = 400$ ft *(6-4)*

51. (A) Average $f(x) = 6.5$ (B) *(6-5)*

52. $\frac{1}{3}(\ln x)^3 + C$ *(6-2)*

53. $\frac{1}{8}x^8 - \frac{2}{5}x^5 + \frac{1}{2}x^2 + C$ *(6-2)*

54. $\frac{2}{3}(6 - x)^{3/2} - 12(6 - x)^{1/2} + C$ *(6-2)*

55. $\frac{1{,}234}{15} \approx 82.267$ *(6-5)*

56. $\frac{64}{15} \approx 4.267$ *(6-5)*

57. $y = 3e^{x^3} - 1$ *(6-3)*

58. $N = 800e^{0.06t}$ *(6-3)*

59. *(6-5)*

60. $M_5 = 0.748\,05$ *(6-5)*   **61.** $f''(x) = (4x^2 - 2)e^{-x^2}$ $\boxed{|f''(x)| \le 2}$ *(6-5)*

62. $I = \int_0^1 e^{-x^2} \, dx = M_5 \pm 0.003\ 34 = 0.748\ 05 \pm 0.003\ 34\ (6\text{-}5)$ **63.** $n \geq 13\ (6\text{-}5)$ **64.** Limited growth

$(6\text{-}3)$

65. Exponential decay   **66.** Unlimited growth   **67.** Logistic growth

$(6\text{-}3)$

$(6\text{-}3)$

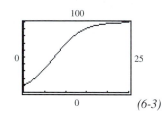

$(6\text{-}3)$

68. 1.167 *(6-5)*   **69.** 99.074 *(6-5)*   **70.** −0.153 *(6-5)*

71. $L_2 = \$180{,}000; R_2 = \$140{,}000; \$140{,}000 \leq \int_{200}^{600} C'(x) \, dx \leq \$180{,}000\ (6\text{-}4)$

72. The height of the rectangle, $C'(x)$, represents the marginal cost at a production level of x units—that is, the approximate cost per unit at that production level. The width of the rectangle represents the number of units involved in the increase in production over $x = 200$. Thus, the cost per unit times the number of units equals the increase in production costs. The approximation improves as n increases. *(6-4)*

73. $\displaystyle\int_{200}^{600} \left(600 - \frac{x}{2} \right) dx = \$160{,}000\ (6\text{-}5)$ **74.** $\displaystyle\int_{10}^{40} \left(150 - \frac{x}{10} \right) dx = \$4{,}425\ (6\text{-}5)$

75. $P(x) = 100x - 0.01x^2; P(10) = \$999\ (6\text{-}3)$ **76.** $\int_0^{15} (60 - 4t) \, dt = 450$ thousand barrels *(6-5)*

77. 109 items *(6-5)*   **78.** $16e^{2.5} - 16e^2 - 8 \approx \$68.70\ (6\text{-}5)$

79. (A) $A_5 = 725$; error bound $= 75$ (B)

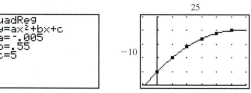

(C) 729 *(6-4, 6-5)*

80. Useful life $= 10 \ln \frac{20}{3} \approx 19$ yr; total profit $= 143 - 200e^{-1.9} \approx 113.086$ or \$113,086 *(6-5)*

81. $S(t) = 50 - 50e^{-0.08t}; 50 - 50e^{-0.96} \approx \$31$ million; $-(\ln 0.2)/0.08 \approx 20$ mo *(6-3)*   **82.** 6.5 ppm *(6-5)*

83. 1 cm² *(6-3)*   **84.** 800 gal *(6-5)*   **85.** (A) 145 million   (B) About 46 years *(6-3)*

86. $\dfrac{-\ln 0.04}{0.000\ 123\ 8} \approx 26{,}000$ yr *(6-3)*   **87.** $N(t) = 95 - 70e^{-0.1t}; N(15) \approx 79$ words/min *(6-3)*

CHAPTER 7

Exercise 7-1

1. $\int_a^b g(x) \, dx$ **3.** $\int_a^b [-h(x)] \, dx$

5. Since the shaded region in Figure C is below the x axis, $h(x) \leq 0$; thus, $\int_a^b h(x) \, dx$ represents the negative of the area of the region.

7. 20   **9.** $\frac{7}{3} \approx 2.333$ **11.** 9   **13.** 7.021   **15.** 0.693   **17.** $\int_a^b [-f(x)] \, dx$ **19.** $\int_b^c f(x) \, dx + \int_c^d [-f(x)] \, dx$

21. $\int_c^d [f(x) - g(x)] \, dx$ **23.** $\int_a^b [f(x) - g(x)] \, dx + \int_b^c [g(x) - f(x)] \, dx$

25. Find the intersection points by solving $f(x) = g(x)$ on the interval $[a, d]$ to determine b and c. Then observe that $f(x) \geqslant g(x)$ over $[a, b]$, $g(x) \geqslant f(x)$ over $[b, c]$, and $f(x) \geqslant g(x)$ over $[c, d]$. Thus,

area $= \int_a^b [f(x) - g(x)]\, dx + \int_b^c [g(x) - f(x)]\, dx + \int_c^d [f(x) - g(x)]\, dx.$

27. 2.5   **29.** 7.667   **31.** 23.667   **33.** 15   **35.** 32   **37.** 36   **39.** 9   **41.** 2.832   **43.** 18   **45.** 1.858

47. 17   **49.** 8   **51.** 101.75   **53.** 8   **55.** 17.979   **57.** 5.113   **59.** 8.290   **61.** 3.166   **63.** 1.385

65. Total production from the end of the fifth year to the end of the 10th year is $50 + 100 \ln 20 - 100 \ln 15 \approx 79$ thousand barrels.

67. Total profit over the 5-yr useful life of the game is $20 - 30e^{-1.5} \approx 13.306$ or \$13,306.

69. 1935: 0.412; 1947: 0.231; income was more equally distributed in 1947.

71. 1963: 0.818; 1983: 0.846; total assets were less equally distributed in 1983.

73. (A) $f(x) = 0.3125x^2 + 0.7175x - 0.015$ (B) 0.104   **75.** Total weight gain during the first 10 hr is $3e - 3 \approx 5.15$ g.

77. Average number of words learned during the second 2 hr is $15 \ln 4 - 15 \ln 2 \approx 10$.

Exercise 7-2

1. 4.12   **3.** 509.14   **5.** (A) 10.72   (B) 3.28   (C) 10.72   **7.** (A) .75   (B) .11   (C)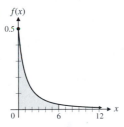

9. 8 yr   **11.** (A) .11   (B) .10   **13.** $P(t \geqslant 12) = 1 - P(0 \leqslant t \leqslant 12) = .89$

15. \$12,500   **17.**   If $f(t)$ is the rate of flow of a continuous income stream, then the total income produced from 0 to 5 yr is the area under the graph of $y = f(t)$ from $t = 0$ to $t = 5$.

19. $8,000(e^{0.15} - 1) \approx \$1,295$ **21.**   If $f(t)$ is the rate of flow of a continuous income stream, the total income produced from 0 to 3 yr is the area under the graph of $y = f(t)$ from $t = 0$ to $t = 3$.

23. \$255,562; \$175,562   **25.** $20,000(e^{0.25} - e^{-0.08}) \approx \$7,218$ **27.** \$875

29. Clothing store: $FV = 120,000(e^{0.5} - 1) \approx \$77,847$; computer store: $FV = 200,000(e^{0.5} - e^{0.25}) \approx \$72,939$; the clothing store is the better investment.

31. Bond: $FV = 10,000e^{0.4} \approx \$14,918$; business: $FV = 25,000(e^{0.4} - 1) \approx \$12,296$; the bond is the better investment.

33. \$55,230

35. $\dfrac{k}{r}(e^{rT} - 1)$

37. \$625,000   **39.**   The shaded area is the consumers' surplus and represents the total savings to consumers who are willing to pay more than \$150 for a product but are still able to buy the product for \$150.

41. $9,900   **43.**

The area of the region *PS* is the producers' surplus and represents the total gain to producers who are willing to supply units at a lower price than $67 but are still able to supply the product at $67.

45. $CS = \$3,380; PS = \$1,690$ **47.** $CS = \$6,980; PS = \$5,041$ **49.** $CS = \$7,810; PS = \$8,336$

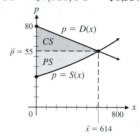

51. $CS = \$8,544; PS = \$11,507$ **53.** (A) $\bar{x} = 21.457; \bar{p} = \$6.51$ (B) $CS = 1.774$ or $\$1,774; PS = 1.087$ or $\$1,087$

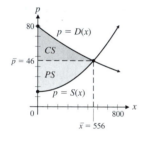

Exercise 7-3

1. $\frac{1}{3}xe^{3x} - \frac{1}{9}e^{3x} + C$ **3.** $\frac{x^3}{3}\ln x - \frac{x^3}{9} + C$ **5.** $u = x + 2;$ $\dfrac{(x + 2)(x + 1)^6}{6} - \dfrac{(x + 1)^7}{42} + C$ **7.** $-xe^{-x} - e^{-x} + C$

9. $\frac{1}{2}e^{x^2} + C$ **11.** $(xe^x - 4e^x)\big|_0^1 = -3e + 4 \approx -4.1548$ **13.** $(x \ln 2x - x)\big|_1^3 = (3 \ln 6 - 3) - (\ln 2 - 1) \approx 2.6821$

15. $\ln(x^2 + 1) + C$ **17.** $(\ln x)^2/2 + C$ **19.** $\frac{2}{3}x^{3/2} \ln x - \frac{4}{9}x^{3/2} + C$

21.

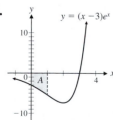

The integral represents the negative of the area between the graph of $y = (x - 3)e^x$ and the *x* axis from $x = 0$ to $x = 1$.

23.

The integral represents the area between the graph of $y = \ln 2x$ and the *x* axis from $x = 1$ to $x = 3$.

25. $(x^2 - 2x + 2)e^x + C$ **27.** $\dfrac{xe^{ax}}{a} - \dfrac{e^{ax}}{a^2} + C$ **29.** $\left(-\dfrac{\ln x}{x} - \dfrac{1}{x}\right)\bigg|_1^e = -\dfrac{2}{e} + 1 \approx 0.2642$

31. $6 \ln 6 - 4 \ln 4 - 2 \approx 3.205$ **33.** $xe^{x-2} - e^{x-2} + C$ **35.** $\frac{1}{2}(1 + x^2) \ln(1 + x^2) - \frac{1}{2}(1 + x^2) + C$

37. $(1 + e^x) \ln(1 + e^x) - (1 + e^x) + C$ **39.** $x(\ln x)^2 - 2x \ln x + 2x + C$

41. $x(\ln x)^3 - 3x(\ln x)^2 + 6x \ln x - 6x + C$ **43.** 1.56 **45.** 34.98 **47.** $\int_0^5 (2t - te^{-t}) \, dt = \$24$ million

49.

The total profit for the first 5 yr (in millions of dollars) is the same as the area under the marginal profit function, $P'(t) = 2t - te^{-t}$, from $t = 0$ to $t = 5$.

51. \$3,278   **53.** 0.264

55.

The area bounded by $y = x$ and the Lorenz curve $y = xe^{x-1}$, divided by the area under the graph of $y = x$ from $x = 0$ to $x = 1$ is the index of income concentration. The closer this index is to 0, the more equally distributed income is; the closer this index is to 1, the more concentrated income is in a few hands.

57. $S(t) = 1,600 + 400e^{0.1t} - 40te^{0.1t}$; 15 mo   **59.** \$977

61.

The area bounded by the price–demand equation, $p = 9 - \ln(x + 4)$, and the price equation, $y = \overline{p} = 2.089$, from $x = 0$ to $x = \overline{x} = 1,000$, represents the consumers' surplus. This is the amount consumers who are willing to pay more than \$2,089 save.

63. 2.1388 ppm   **65.** $N(t) = -4te^{-0.25t} - 40e^{-0.25t} + 80$; 8 wk; 78 words/min   **67.** 20,980

Exercise 7-4

1. $\ln \left| \dfrac{x}{1 + x} \right| + C$ **3.** $\dfrac{1}{3 + x} + 2 \ln \left| \dfrac{5 + 2x}{3 + x} \right| + C$ **5.** $\dfrac{2(x - 32)}{3} \sqrt{16 + x} + C$ **7.** $-\ln \left| \dfrac{1 + \sqrt{1 - x^2}}{x} \right| + C$

9. $\dfrac{1}{2} \ln \left| \dfrac{x}{2 + \sqrt{x^2 + 4}} \right| + C$ **11.** $\frac{1}{3}x^3 \ln x - \frac{1}{9}x^3 + C$ **13.** $x - \ln|1 + e^x| + C$ **15.** $9 \ln \frac{3}{2} - 2 \approx 1.6492$

17. $\frac{1}{2} \ln \frac{12}{5} \approx 0.4377$ **19.** $\ln 3 \approx 1.0986$ **21.** $-\dfrac{\sqrt{4x^2 + 1}}{x} + 2 \ln|2x + \sqrt{4x^2 + 1}| + C$

23. $\frac{1}{2} \ln|x^2 + \sqrt{x^4 - 16}| + C$ **25.** $\frac{1}{6}(x^3 \sqrt{x^6 + 4} + 4 \ln|x^3 + \sqrt{x^6 + 4}|) + C$ **27.** $-\dfrac{\sqrt{4 - x^4}}{8x^2} + C$

29. $\dfrac{1}{5} \ln \left| \dfrac{3 + 4e^x}{2 + e^x} \right| + C$ **31.** $\frac{2}{3}(\ln x - 8) \sqrt{4 + \ln x} + C$ **33.** $\frac{1}{5}x^2e^{5x} - \frac{2}{25}xe^{5x} + \frac{2}{125}e^{5x} + C$

35. $-x^3e^{-x} - 3x^2e^{-x} - 6xe^{-x} - 6e^{-x} + C$ **37.** $x(\ln x)^3 - 3x(\ln x)^2 + 6x \ln x - 6x + C$

39. $\frac{64}{3}$ **41.** $\frac{1}{2} \ln \frac{9}{5} \approx 0.2939$ **43.** $\dfrac{-1 - \ln x}{x} + C$

45. $\sqrt{x^2 - 1} + C$

47. 31.38   **49.** 5.48   **51.** $3,000 + 1,500 \ln \frac{1}{3} \approx \$1,352$ **53.**

55. $C(x) = 200x + 1{,}000 \ln(1 + 0.05x) + 25{,}000$; 608; $198,773   **57.** $100{,}000e - 250{,}000 \approx \$21{,}828$ **59.** 0.1407

61.

As the area bounded by the two curves gets smaller, the Lorenz curve approaches $y = x$ and the distribution of income approaches perfect equality—all persons share equally in the income available.

63. $S(t) = 1 + t - \dfrac{1}{1 + t} - 2 \ln|1 + t|$; $24.96 - 2 \ln 25 \approx \$18.5$ million

65.

The total sales (in millions of dollars) over the first 2 yr (24 mo) is the area under the graph of $y = S'(t)$ from $t = 0$ to $t = 24$.

67. $P(x) = \dfrac{2(9x - 4)}{135}(2 + 3x)^{3/2} - 2{,}000.83$; 54; $37,932   **69.** $100 \ln 3 \approx 110$ ft   **71.** $60 \ln 5 \approx 97$ items

73.

The area under the graph of $y = N'(t)$ from $t = 0$ to $t = 12$ represents the total number of items learned in that time interval.

Review Exercise

1. $\int_a^b f(x)\, dx$ *(7-1)*   **2.** $\int_b^c [-f(x)]\, dx$ *(7-1)*   **3.** $\int_a^b f(x)\, dx + \int_b^c [-f(x)]\, dx$ *(7-1)*   **4.** Area $= 1.153$

(7-1)

5. $\frac{1}{4}xe^{4x} - \frac{1}{16}e^{4x} + C$ *(7-3, 7-4)*   **6.** $\frac{1}{2}x^2 \ln x - \frac{1}{4}x^2 + C$ *(7-3, 7-4)*   **7.** $\dfrac{(\ln x)^2}{2} + C$ *(6-2)*   **8.** $\dfrac{\ln(1 + x^2)}{2} + C$ *(7-2)*

9. $\dfrac{1}{1 + x} + \ln\left|\dfrac{x}{1 + x}\right| + C$ *(7-4)*   **10.** $-\dfrac{\sqrt{1 + x}}{x} - \dfrac{1}{2}\ln\left|\dfrac{\sqrt{1 + x} - 1}{\sqrt{1 + x} + 1}\right| + C$ *(7-4)*   **11.** $\int_a^b [f(x) - g(x)]\, dx$ *(7-1)*

12. $\int_b^c [g(x) - f(x)]\, dx$ *(7-1)*   **13.** $\int_b^c [g(x) - f(x)]\, dx + \int_c^d [f(x) - g(x)]\, dx$ *(7-1)*

14. $\int_a^b [f(x) - g(x)]\, dx + \int_b^c [g(x) - f(x)]\, dx + \int_c^d [f(x) - g(x)]\, dx$ *(7-1)*

15. Area $= 20.833$

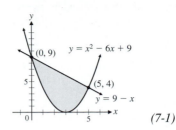
$(0, 9)$
$y = x^2 - 6x + 9$
$(5, 4)$
$y = 9 - x$
$(7-1)$

16. 1 *(7-3, 7-4)*
17. $\frac{15}{2} - 8 \ln 8 + 8 \ln 4 \approx 1.955$ *(7-4)*
18. $\frac{1}{6}(3x \sqrt{9x^2 - 49} - 49 \ln|3x + \sqrt{9x^2 - 49}|) + C$ *(7-4)*
19. $-2te^{-0.5t} - 4e^{-0.5t} + C$ *(7-3, 7-4)*
20. $\frac{1}{3}x^3 \ln x - \frac{1}{9}x^3 + C$ *(7-3, 7-4)*
21. $x - \ln|1 + 2e^x| + C$ *(7-4)*

22. (A) Area $= 8$

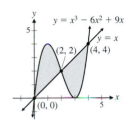
$y = x^3 - 6x^2 + 9x$
$y = x$
$(2, 2)$
$(4, 4)$
$(0, 0)$

(B) Area $= 8.38$

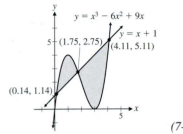
$y = x^3 - 6x^2 + 9x$
$y = x + 1$
$(1.75, 2.75)$
$(4.11, 5.11)$
$(0.14, 1.14)$
$(7-1)$

23. $\frac{1}{3}(\ln x)^3 + C$ *(6-2)*   **24.** $\frac{1}{2}x^2(\ln x)^2 - \frac{1}{2}x^2 \ln x + \frac{1}{4}x^2 + C$ *(7-3, 7-4)*   **25.** $\sqrt{x^2 - 36} + C$ *(6-2)*
26. $\frac{1}{2} \ln|x^2 + \sqrt{x^4 - 36}| + C$ *(7-4)*   **27.** $50 \ln 10 - 42 \ln 6 - 24 \approx 15.875$ *(7-3, 7-4)*
28. $x(\ln x)^2 - 2x \ln x + 2x + C$ *(7-3, 7-4)*   **29.** $-\frac{1}{4}e^{-2x^2} + C$ *(6-2)*   **30.** $-\frac{1}{2}x^2e^{-2x} - \frac{1}{2}xe^{-2x} - \frac{1}{4}e^{-2x} + C$ *(7-3, 7-4)*
31. 1.703 *(7-1)*   **32.** (A) $.189$ (B) $.154$ *(7-2)*

33.

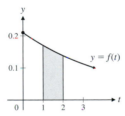
0.2
0.1
$y = f(t)$

The probability that the product will fail during the second year of warranty is the area under the probability density function $y = f(t)$ from $t = 1$ to $t = 2$. *(7-2)*

34. $R(x) = 65x - 6[(x + 1) \ln(x + 1) - x]$; 618/wk; \$29,506 *(7-3)*

35. (A)

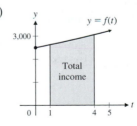
$y = f(t)$
$3,000$
Total income

(B) \$8,507 *(7-2)*   **36.** (A) \$20,824   (B) \$6,623 *(7-2)*

37. (A)

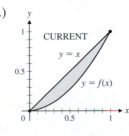
CURRENT
$y = x$
$y = f(x)$

PROJECTED
$y = x$
$y = g(x)$

(B) More equitably distributed, since the area bounded by the two curves will have decreased.
(C) Current $= 0.3$; projected $= 0.2$; income will be more equitably distributed 10 years from now. *(7-1)*

38. (A) CS = \$2,250; PS = \$2,700   (B) CS = \$2,890; PS = \$2,278

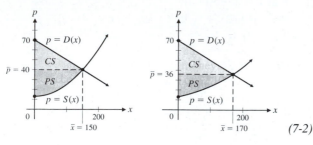

$(7-2)$

39. (A) 25.403 or 25,403 lb
  (B) PS = 121.6 or \$1,216 *(7-2)*
40. 4.522 mL; 1.899 mL *(6-5, 7-4)*
41.

$y = R(t)$

$(6-5, 7-1)$

42. .667; .333 *(7-2)*   **43.**

$y = f(t)$

The probability that the doctor will spend more than an hour with a randomly selected patient is the area under the probability density function $y = f(t)$ from $t = 1$ to $t = 3$. *(7-2)*

44. 45 thousand *(6-5, 7-1)*   **45.** .368 *(7-2)*

CHAPTER 8

Exercise 8-1

1. −8   **3.** −29   **5.** 12   **7.** 19   **9.** −448   **11.** 6   **13.** 150   **15.** 16π **17.** 791   **19.** 0.192   **21.** 118
23. 36,095.07   **25.** −1.926, 0.599   **27.** $2x + h$ **29.** $2y^2$ **31.** $E(0, 0, 3); F(2, 0, 3)$
33. (A) In the plane $y = c, c$ any constant, $z = x^2$.
  (B) The y axis; the horizontal line parallel to the y axis and passing through the point $(1, 0, 1)$; the horizontal line parallel to the y axis and passing through the point $(2, 0, 4)$
  (C) A parabolic "trough" lying on top of the y axis
35. (A) Upper semicircles whose centers lie on the y axis
  (B) Upper semicircles whose centers lie on the x axis
  (C) The upper hemisphere of radius 6 with center at the origin
37. (A) $a^2 + b^2$ and $c^2 + d^2$ both equal the square of the radius of the circle.
  (B) Bell-shaped curves with maximum values of 1 at the origin
  (C) A bell, with maximum value 1 at the origin, extending infinitely far in all directions.
39. \$4,400; \$6,000; \$7,100   **41.** $R(p, q) = -5p^2 + 6pq - 4q^2 + 200p + 300q; R(2, 3) = \$1,280; R(3, 2) = \$1,175$
43. 30,065 units   **45.** (A) \$272,615.08   (B) 12.2%   **47.** $T(70, 47) \approx 29$ min; $T(60, 27) = 33$ min
49. $C(6, 8) = 75; C(8.1, 9) = 90$ **51.** $Q(12, 10) = 120; Q(10, 12) \approx 83$

Exercise 8-2

1. −2   **3.** 5   **5.** $6x^2 + 9y^2$ **7.** $18y$ **9.** 72   **11.** $10x - 6y - 4$ **13.** 4   **15.** −6   **17.** 10
19. $12x^2y + \dfrac{y}{x}$ **21.** $32 + \ln 2$ **23.** $12x^2 + \dfrac{1}{x}$ **25.** $8xe^{4x^2 + 5y}$ **27.** $40xe^{4x^2 + 5y}$ **29.** $40e^4$ **31.** $8e^5$
33. $f_x(x, y) = 6x(x^2 - y^3)^2; f_y(x, y) = -9y^2(x^2 - y^3)^2$ **35.** $f_x(x, y) = 24xy(3x^2y - 1)^3; f_y(x, y) = 12x^2(3x^2y - 1)^3$

37. $f_x(x, y) = 2x/(x^2 + y^2); f_y(x, y) = 2y/(x^2 + y^2)$ **39.** $f_x(x, y) = y^4 e^{xy^2}; f_y(x, y) = 2xy^3 e^{xy^2} + 2y e^{xy^2}$
41. $f_x(x, y) = 4xy^2/(x^2 + y^2)^2; f_y(x, y) = -4x^2 y/(x^2 + y^2)^2$
45. $f_{xx}(x, y) = 2y^2 + 6x; f_{xy}(x, y) = 4xy = f_{yx}(x, y); f_{yy}(x, y) = 2x^2$
47. $f_{xx}(x, y) = -2y/x^3; f_{xy}(x, y) = (-1/y^2) + (1/x^2) = f_{yx}(x, y); f_{yy}(x, y) = 2x/y^3$
49. $f_{xx}(x, y) = (2y + xy^2)e^{xy}; f_{xy}(x, y) = (2x + x^2 y)e^{xy} = f_{yx}(x, y); f_{yy}(x, y) = x^3 e^{xy}$ **51.** $x = 2$ and $y = 4$
53. $x = 1.200$ and $y = -0.695$ **55.** (A) $-\frac{13}{3}$ (B) The function $f(0, y)$, for example, has values less than $-\frac{13}{3}$.
57. (A) $c = 1.145$ (B) $f_x(c, 2) = 0; f_y(c, 2) = 92.021$
59. $f_{xx}(x, y) + f_{yy}(x, y) = (2y^2 - 2x^2)/(x^2 + y^2)^2 + (2x^2 - 2y^2)/(x^2 + y^2)^2 = 0$ **61.** (A) $2x$ (B) $4y$
63. $P_x(1{,}200, 1{,}800) = 24$; profit will increase approx. \$24 per unit increase in production of type A calculators at the
$(1{,}200, 1{,}800)$ output level; $P_y(1{,}200, 1{,}800) = -48$; profit will decrease approx. \$48 per unit increase in production of
type B calculators at the $(1{,}200, 1{,}800)$ output level
65. $\partial x/\partial p = -5$: a \$1 increase in the price of brand A will decrease the demand for brand A by 5 lb at any price level (p, q);
$\partial y/\partial p = 2$: a \$1 increase in the price of brand A will increase the demand for brand B by 2 lb at any price level (p, q)
67. (A) $f_x(x, y) = 7.5x^{-0.25}y^{0.25}; f_y(x, y) = 2.5x^{0.75}y^{-0.75}$
(B) Marginal productivity of labor $= f_x(600, 100) \approx 4.79$; Marginal productivity of capital $= f_y(600, 100) \approx 9.58$
(C) Capital
69. Competitive **71.** Complementary
73. (A) $f_w(w, h) = 6.65w^{-0.575}h^{0.725}; f_h(w, h) = 11.34\ w^{0.425}h^{-0.275}$
(B) $f_w(65, 57) = 11.31$: for a 65 lb child 57 in. tall, the rate of change in surface area is 11.31 in.² for each pound gained in
weight (height is held fixed); $f_h(65, 57) = 21.99$: for a child 57 in. tall, the rate of change in surface area is 21.99 in.²
for each inch gained in height (weight is held fixed)
75. $C_W(6, 8) = 12.5$: index increases approx. 12.5 units for 1 in. increase in width of head (length held fixed) when $W = 6$ and
$L = 8$; $C_L(6, 8) = -9.38$: index decreases approx. 9.38 units for 1 in. increase in length (width held fixed) when $W = 6$
and $L = 8$

Exercise 8-3

1. $f_x(x, y) = 4; f_y(x, y) = 5$; the functions $f_x(x, y)$ and $f_y(x, y)$ never have the value 0.
3. $f_x(x, y) = -1.2 + 4x^3; f_y(x, y) = 6.8 + 0.6y^2$; the function $f_y(x, y)$ never has the value 0.
5. $f(-2, 0) = 10$ is a local maximum **7.** $f(-1, 3) = 4$ is a local minimum **9.** f has a saddle point at $(3, -2)$
11. $f(3, 2) = 33$ is a local maximum **13.** $f(2, 2) = 8$ is a local minimum **15.** f has a saddle point at $(0, 0)$
17. f has a saddle point at $(0, 0); f(1, 1) = -1$ is a local minimum
19. f has a saddle point at $(0, 0); f(3, 18) = -162$ and $f(-3, -18) = -162$ are local minima
21. The test fails at $(0, 0); f$ has saddle points at $(2, 2)$ and $(2, -2)$ **23.** f has a saddle point at $(0.614, -1.105)$
25. $f(x, y)$ is nonnegative and equals 0 when $x = 0$, so f has a local minimum at each point of the y axis.
27. (B) Local minimum **29.** 2,000 type A and 4,000 type B; Max $P = P(2, 4) = \$15$ million
31. (A) When $p = \$10$ and $q = \$12, x = 56$ and $y = 16$; when $p = \$11$ and $q = \$11, x = 6$ and $y = 56$.
(B) A maximum weekly profit of \$288 is realized for $p = \$10$ and $q = \$12$.
33. $P(x, y) = P(4, 2)$ **35.** 8 in. by 4 in. by 2 in. **37.** 20 in. by 20 in. by 40 in.

Exercise 8-4

1. Max $f(x, y) = f(3, 3) = 18$ **3.** Min $f(x, y) = f(3, 4) = 25$
5. $F_x = -3 + 2\lambda = 0$ and $F_y = 4 + 5\lambda = 0$ have no simultaneous solution.
7. Max $f(x, y) = f(3, 3) = f(-3, -3) = 18$; Min $f(x, y) = f(3, -3) = f(-3, 3) = -18$
9. Maximum product is 25 when each number is 5 **11.** Min $f(x, y, z) = f(-4, 2, -6) = 56$
13. Max $f(x, y, z) = f(2, 2, 2) = 6$; Min $f(x, y, z) = f(-2, -2, -2) = -6$ **15.** Max $f(x, y) = f(0.217, 0.885) = 1.055$
17. $F_x = e^x + \lambda = 0$ and $F_y = 3e^y - 2\lambda = 0$ have no simultaneous solution.
19. Maximize $f(x, 5)$, a function of just one independent variable.
21. (A) Max $f(x, y) = f(0.707, 0.5) = f(-0.707, 0.5) = 0.47$
23. 60 of model A and 30 of model B will yield a minimum cost of \$32,400 per week
25. (A) 8,000 units of labor and 1,000 units of capital; Max $N(x, y) = N(8{,}000, 1{,}000) \approx 263{,}902$ units
(B) Marginal productivity of money ≈ 0.6598; increase in production $\approx 32{,}990$ units
27. 8 in. by 8 in. by $\frac{8}{3}$ in. **29.** $x = 50$ ft and $y = 200$ ft; maximum area is 10,000 ft²

Exercise 8-5

1.

$y = 0.7x + 1$

3.

$y = -2.5x + 10.5$

5.

$y = x + 2$

7. $y = -1.5x + 4.5$; $y = 0.75$ when $x = 2.5$ **9.** $y = 2.12x + 10.8$; $y = 63.8$ when $x = 25$
11. $y = -1.2x + 12.6$; $y = 10.2$ when $x = 2$ **13.** $y = -1.53x + 26.67$; $y = 14.4$ when $x = 8$

15.

$y = 0.75x^2 - 3.45x + 4.75$

21. (A) $y = 1.52x - 0.16$; $y = 0.73x^2 - 1.39x + 1.30$ (B) The quadratic function.
23. The normal equations form a system of 4 linear equations in the 4 variables $a, b, c,$ and d, which can be solved using Gauss–Jordan elimination.
25. (A) $y = 110.5x + 1,834$ (B) 4,044,000
27. (A) $y = -0.48x + 4.38$ (B) $6.56 per bottle
29. (A) $P = -0.66T + 48.8$ (B) 11.18 beats/min
31. (A) $y = 0.0086x + 56.48$ (B) 58.19°F
33. (A) $D = -3.1 A + 54.6$ (B) 45%
35. (A) $y = 0.0871x + 10.80$ (B) 21.25 ft

Exercise 8-6

1. (A) $3x^2y^4 + C(x)$ (B) $3x^2$ **3.** (A) $2x^2 + 6xy + 5x + E(y)$ (B) $35 + 30y$

5. (A) $\sqrt{y + x^2} + E(y)$ (B) $\sqrt{y + 4} - \sqrt{y}$ **7.** (A) $\dfrac{\ln x \ln y}{x} + C(x)$ (B) $\dfrac{2 \ln x}{x}$ **9.** 9   **11.** 330

13. $(56 - 20\sqrt{5})/3$ **15.** 1   **17.** 16   **19.** 49   **21.** $\frac{1}{8} \int_1^5 \int_{-1}^1 (x + y)^2 \, dy \, dx = \frac{32}{3}$

23. $\frac{1}{15} \int_1^4 \int_2^7 (x/y) \, dy \, dx = \frac{1}{2} \ln \frac{7}{2} \approx 0.6264$ **25.** $\frac{4}{3}$ cubic units   **27.** $\frac{32}{3}$ cubic units

29. $\int_0^1 \int_1^2 xe^{xy} \, dy \, dx = \frac{1}{2} + \frac{1}{2}e^2 - e$ **31.** $\int_0^1 \int_{-1}^1 \dfrac{2y + 3xy^2}{1 + x^2} \, dy \, dx = \ln 2$

35. (A) $\frac{1}{3} + \frac{1}{4}e^{-2} - \frac{1}{4}e^2$ (B)

(C) Points to the right of the graph in part (B) are greater than 0; points to the left of the graph are less than 0.

37. $\dfrac{1}{0.4}\displaystyle\int_{0.6}^{0.8}\int_{5}^{7}\dfrac{y}{1-x}\,dy\,dx = 30\ln 2 \approx \$20.8$ billion

39. $\frac{1}{10}\displaystyle\int_{10}^{20}\int_{1}^{2}x^{0.75}y^{0.25}\,dy\,dx = \frac{8}{175}(2^{1.25}-1)(20^{1.75}-10^{1.75}) \approx 8.375$ or $8{,}375$ units

41. $\frac{1}{192}\displaystyle\int_{-8}^{8}\int_{-6}^{6}\left[10-\frac{1}{10}(x^2+y^2)\right]dy\,dx = \frac{20}{3}$ insects/ft^2 **43.** $\frac{1}{8}\displaystyle\int_{-2}^{2}\int_{-1}^{1}[100-15(x^2+y^2)]\,dy\,dx = 75$ ppm

45. $\frac{1}{10{,}000}\displaystyle\int_{2{,}000}^{3{,}000}\int_{50}^{60}0.000\,013\,3xy^2\,dy\,dx \approx 100.86$ ft   **47.** $\frac{1}{16}\displaystyle\int_{8}^{16}\int_{10}^{12}100\dfrac{x}{y}\,dy\,dx = 600\ln 1.2 \approx 109.4$

Review Exercise

1. $f(5,10)=2{,}900$; $f_x(x,y)=40$; $f_y(x,y)=70$ *(8-1, 8-2)*   **2.** $\partial^2 z/\partial x^2 = 6xy^2$; $\partial^2 z/\partial x\,\partial y = 6x^2 y$ *(8-2)*
3. $2xy^3 + 2y^2 + C(x)$ *(8-6)*   **4.** $3x^2 y^2 + 4xy + E(y)$ *(8-6)*   **5.** 1 *(8-6)*
6. $f_x(x,y)=5+6x+3x^2$; $f_y(x,y)=-2$; the function $f_y(x,y)$ never has the value 0. *(8-3)*
7. $f(2,3)=7$; $f_y(x,y)=-2x+2y+3$; $f_y(2,3)=5$ *(8-1, 8-2)*
8. $(-8)(-6)-(4)^2=32$ *(8-2)*   **9.** $(1,3,-\frac{1}{2})$, $(-1,-3,\frac{1}{2})$ *(8-4)*
10. $y=-1.5x+15.5$; $y=0.5$ when $x=10$ *(8-5)*   **11.** 18 *(8-6)*
12. $f_x(x,y)=2xe^{x^2+2y}$; $f_y(x,y)=2e^{x^2+2y}$; $f_{xy}(x,y)=4xe^{x^2+2y}$ *(8-2)*
13. $f_x(x,y)=10x(x^2+y^2)^4$; $f_y(x,y)=80xy(x^2+y^2)^3$ *(8-2)*
14. $f(2,3)=-25$ is a local minimum; f has a saddle point at $(-2,3)$ *(8-3)*   **15.** Max $f(x,y)=f(6,4)=24$ *(8-4)*
16. Min $f(x,y,z)=f(2,1,2)=9$ *(8-4)*   **17.** $y=\frac{116}{165}x+\frac{100}{3}$ *(8-5)*   **18.** $\frac{27}{5}$ *(8-6)*
19. 4 cubic units *(8-6)*   **20.** 0 *(8-6)*   **21.** (A) 12.56 (B) No *(8-6)*
22. $F_x=12x^2+3\lambda=0$, $F_y=-15y^2+2\lambda=0$, and $F_\lambda=3x+2y-7=0$ have no simultaneous solution. *(8-4)*
23. (A) $P_x(1,3)=8$; profit will increase $8,000 for a 100 unit increase in product A if production of product B is held fixed at an output level of $(1,3)$.
    (B) For 200 units of A and 300 units of B, $P(2,3)=\$100$ thousand is a local maximum. *(8-2, 8-3)*
24. 8 in. by 6 in. by 2 in. *(8-3)*   **25.** $y=0.63x+1.33$; profit in sixth year is $5.11 million *(8-4)*
26. (A) Marginal productivity of labor ≈ 8.37; marginal productivity of capital ≈ 1.67; management should encourage increased use of labor.
    (B) 80 units of labor and 40 units of capital; Max $N(x,y)=N(80,40)\approx 696$ units; marginal productivity of money ≈ 0.0696; increase in production ≈ 139 units
    (C) $\dfrac{1}{1{,}000}\displaystyle\int_{50}^{100}\int_{20}^{40}10x^{0.8}y^{0.2}\,dy\,dx = \dfrac{(40^{1.2}-20^{1.2})(100^{1.8}-50^{1.8})}{216} = 621$ items *(8-4)*
27. $T_x(70,17)=-0.924$ min/ft increase in depth when $V=70$ ft^3 and $x=17$ ft *(8-2)*
28. $\frac{1}{16}\displaystyle\int_{-2}^{2}\int_{-2}^{2}[100-24(x^2+y^2)]\,dy\,dx = 36$ ppm *(8-6)*   **29.** 50,000 *(8-1)*
30. $y=\frac{1}{2}x+48$; $y=68$ when $x=40$ *(8-5)*
31. (A) $y=0.4933x+25.20$ (B) 84.40 people/mi^2
    (C) 89.30 people/mi^2; 97.70 people/mi^2 *(8-5)*
32. (A) $y=1.069x+0.522$ (B) 64.68 yr (C) 64.78 yr; 64.80 yr *(8-5)*

CHAPTER 9

Exercise 9-1

7. (C)   **9.** (A)   **11.**   **13.**

15. $y = 3 + 3x + e^x$ **17.** $y = xe^{-x}$ **19.** $y = 2x - \dfrac{1}{x}$ **25.** $y = \sqrt{9 - x^2}$ **27.** $y = 2 - e^x$

29. (A) $y = 2 - e^{-x}$
(B) $y = 2$
(C) $y = 2 + e^{-x}$

31. (A) $y = 2 - e^x$
(B) $y = 2$
(C) $y = 2 + e^x$

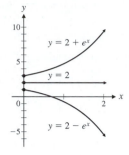

33. $dN/dt = 650; t = $ Time (in years), $N = N(t) = $ Number of employees at time t, $dN/dt = N' = $ Rate at which the number of employees is changing

35. $dT/dt = -k(T - 72); t = $ Time, $T = T(t) = $ Temperature of pizza at time t, $k = $ Positive constant of proportionality, $dT/dt = T' = $ Rate at which the temperature of the pizza is changing (dT/dt is negative because the temperature of the pizza is decreasing)

37. (A) $y = \dfrac{10}{1 + 9e^{-x}}$
(B) $y = 10$
(C) $y = \dfrac{10}{1 - 0.5e^{-x}}$

39. (A) $y = Cx^3 + 2$ for any C (B) No particular solution exists.
(C) $y = 2 - x^3$

41. (A)

(B) Each graph decreases to a local minimum and then increases, approaching the line $y = x$ as x approaches ∞.

(C)

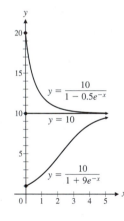

(D) Each graph is increasing for all x and approaches the line $y = x$ as x approaches ∞.

43. (A) $\bar{p} = 5$
(B) $p(t) = 5 - 4e^{-0.1t}$
$p(t) = 5 + 5e^{-0.1t}$

45. (B) $A(t) = 2{,}500e^{0.08t} - 2{,}500$
$A(t) = 3{,}500e^{0.08t} - 2{,}500$

47. (A) $\overline{N} = 200$

(B) $N(t) = 200 - 150e^{-0.5t}$
$N(t) = 200 + 100e^{-0.5t}$

49. (A) $\overline{N} = C$

(B) $N(t) = 100e^{2-2e^{-0.5t}}$
$N(t) = 200e^{2-2e^{-0.5t}}$

Exercise 9-2

1. $dy/dt = 100{,}000$ **3.** $dy/dt = k(10{,}000 - y)$

5. The annual sales now are \$2 million, and sales are increasing at a rate proportional to the annual sales.

7. In a community of 5,000 people, a single person began the spread of a rumor that is spreading at a rate proportional to the product of the number of people who have heard the rumor and the number who have not.

9. $\dfrac{1}{y^2} y' = \dfrac{1}{x}$ **11.** $\dfrac{1}{y} y' = \dfrac{3 - x}{x}$

13. General solution: $y = x^3 + C$; particular solution: $y = x^3 - 1$

15. General solution: $y = 2 \ln x + C$; particular solution: $y = 2 \ln x + 2$

17. General solution: $y = Ce^x$; particular solution: $y = 10e^x$

19. General solution: $y = 25 - Ce^{-x}$; particular solution: $y = 25 - 20e^{-x}$

21. General solution: $y = Cx$; particular solution: $y = 5x$

23. General solution: $y = (3x + C)^{1/3}$; particular solution: $y = (3x + 24)^{1/3}$

25. General solution: $y = Ce^{e^x}$; particular solution: $y = 3e^{e^x}$

27. General solution: $y = \ln(e^x + C)$; particular solution: $y = \ln(e^x + 1)$

29. General solution: $y = Ce^{x/2} - 1$; particular solution: $y = 3e^{x/2} - 1$

31. General solution: $y = 2 - 1/(e^x + C)$; particular solution: $y = 2 - e^{-x}$ **33.** $y + \frac{1}{3}y^3 = x + \frac{1}{3}x^3 + C$

35. $\frac{1}{2}\ln|1 + y^2| = \ln|x| + (x^2/2) + C$ or $\ln(1 + y^2) = \ln(x^2) + x^2 + C$ **37.** $\ln(1 + e^y) = \frac{1}{2}x^2 + C$

39. $y = \sqrt{1 + (\ln x)^2}$ **41.** $y = \frac{1}{4}[x + \ln(x^2) + 3]^2$ **43.** $y = \sqrt{\ln(x^2 + e)}$

45. The differential equation cannot be written in the form $f(y)y' = g(x)$.

47. The integrand e^{x^2} does not have an elementary antiderivative.

49. (A) $M = 3/(1 - e^{-k})$; $M = 4/(1 - e^{-1.5k})$ (B) 7.3   **51.** $5{,}000\, e^{0.275} \approx \$6{,}582.65$

53. (A) $\dfrac{7 \ln 0.5}{\ln 0.8} \approx 22$ days   (B) $100{,}000(1 - e^{0.5 \ln 0.5}) \approx 29{,}300$ people

55. $\dfrac{30 \ln 0.1}{\ln 0.5} \approx 100$ days   **57.** $\dfrac{4 \ln 0.2}{\ln 0.8} \approx 29$ yr   **59.** $k = 0.2$; $M = \$11$ million

61. $\dfrac{2 \ln(5/12)}{\ln(5/6)} \approx 9.6$ min   **63.** $25 + 300e^{4 \ln(2/3)} \approx 84.26°F$

65. (A) $100e^{5 \ln 1.4} \approx 538$ bacteria   (B) $\dfrac{\ln 10}{\ln 1.4} \approx 6.8$ hr

67. (A) $\dfrac{50{,}000}{1 + 499e^{2 \ln(99/499)}} \approx 2{,}422$ people   (B) $\dfrac{10 \ln(1/499)}{\ln(99/499)} \approx 38.4$ days

69. $I = As^k$

71. (A) $\dfrac{1{,}000}{1 + 199e^{7 \ln(99/199)}} \approx 400$ people   (B) $\dfrac{\ln(3/3{,}383)}{\ln(99/199)} \approx 10$ days

Exercise 9-3

1. Yes; $f(x) = -k$, $g(x) = 0$ **3.** No; the equation has a y^2 term, so it is not first-order linear.

5. $\int (x^3 y' + 3x^2 y)\, dx = x^3 y$ **7.** $\int (e^{-3x} y' - 3e^{-3x} y)\, dx = e^{-3x} y$ **9.** $\int (x^4 y' + 4x^3 y)\, dx = x^4 y$

11. $\int (e^{-0.5x}y' - 0.5e^{-0.5x}y)\,dx = e^{-0.5x}y$ **13.** $I(x) = e^{2x};\ y = 2 + Ce^{-2x};\ y = 2 - e^{-2x}$

15. $I(x) = e^x;\ y = -e^{-2x} + Ce^{-x};\ y = -e^{-2x} + 4e^{-x}$ **17.** $I(x) = e^{-x};\ y = 2xe^x + Ce^x;\ y = 2xe^x - 4e^x$

19. $I(x) = e^x;\ y = 3x^3e^{-x} + Ce^{-x};\ y = 3x^3e^{-x} + 2e^{-x}$ **21.** $I(x) = x;\ y = x + (C/x);\ y = x$

23. $I(x) = x^2;\ y = 2x^3 + (C/x^2);\ y = 2x^3 - (32/x^2)$ **25.** $I(x) = e^{x^2/2};\ y = 5 + Ce^{-x^2/2}$

27. $I(x) = e^{-2x};\ y = -2x - 1 + Ce^{2x}$ **29.** $I(x) = x;\ y = e^x - (e^x/x) + (C/x)$

31. $I(x) = x;\ y = \frac{1}{2}x\ln x - \frac{1}{4}x + (C/x)$ **33.** $I(x) = x^{3/2};\ y = 4x + Cx^{-3/2}$

35. (B) The solution is wrong. The term $1/x$ in front of the integral and the term x inside the integral cannot be canceled.
   (C) $y = \frac{1}{4}x^3 + \frac{2}{3}x^2 + \frac{1}{2}x + (C/x)$

37. (B) The solution is wrong. A constant of integration is omitted.
   (C) $y = \frac{1}{2}e^{-x} + Ce^{-3x}$

39. $y = 1 + (C/x)$ **41.** $y = C(1 + x^2) - 1$ **43.** $y = Ce^{x^2} - 1$ **45.** $y = Ce^{kt}$ **47.** $y = (b/a) + Ce^{-ax}$

49. $A(t) = 62{,}500 - 42{,}500e^{0.048t}$; 8.035 years, \$24,105   **51.** \$15,124.60   **53.** $A(t) = 35{,}000e^{0.07t} - 30{,}000$; \$19,667.36

55. 5.23%   **57.** $p(t) = 20 + 10e^{-3t};\ \bar{p} = 20$ **59.** 622 lb; $\dfrac{622}{250} \approx 2.5$ lb/gal

61. $600 - 200e^{-0.5} \approx 479$ lb; $3 - e^{-0.5} \approx 2.4$ lb/gal   **63.** 6.2 hr

65. $120 + 40e^{-0.15} \approx 154$ lb; $-(\ln \frac{3}{4})/0.005 \approx 58$ days; weight will approach 120 lb if the diet is maintained for a long period of time

67. $17.5\left(\dfrac{125 - 130e^{-0.15}}{1 - e^{-0.15}}\right) \approx 1{,}647$ cal

69. Student A: $0.9 - 0.8e^{-4.8} \approx 0.8934$ or 89.34%; student B: $0.7 - 0.3e^{-4.8} \approx 0.6975$ or 69.75%

Review Exercise

3. (B) *(9-1)*   **4.** (A) *(9-1)*

5.

(9-1)

6.

(9-1)

7. $dy/dt = -k(y - 5),\ k > 0$ *(9-2)*   **8.** $dy/dt = ky,\ k > 0$ *(9-2)*

9. A single person began the spread of a disease that is spreading at a rate proportional to the product of the number of people who have contracted the disease and the number who have not. *(9-2)*

10. The amount of radioactive material now is 100 g, and the amount is decreasing at a rate proportional to the amount present. *(9-2)*

11. First-order linear form: $y' + (3x^2 - e^x)y = (1 - x)e^x - 9$ *(9-2, 9-3)*   **12.** Neither form *(9-2, 9-3)*

13. Separation of variables form: $(y + 2)^{-1}y' = x - 3$; first-order linear form: $y' + (3 - x)y = 2x - 6$ *(9-2, 9-3)*

14. Separation of variables form: $y^2e^y y' = 4xe^x$ *(9-2, 9-3)*   **15.** $y = C/x^4$ *(9-2)*

16. $y = \frac{1}{6}x^2 + (C/x^4)$ *(9-3)*   **17.** $y = -1/(x^3 + C)$ *(9-2)*

18. $y = e^x + Ce^{2x}$ *(9-3)*   **19.** $y = \frac{1}{2}x^7 + Cx^5$ *(9-3)*   **20.** $y = C(2 + x) - 3$ *(9-2)*

21. $y = 10 - 10e^{-x}$ *(9-2, 9-3)*   **22.** $y = x - 1 + e^{-x}$ *(9-3)*   **23.** $y = e^2e^{-2e^{-x}}$ *(9-2)*

24. $y = (x^2 + 4)/(x + 4)$ *(9-3)*   **25.** $y = \sqrt{x^2 + 16} - 4$ *(9-2)*

26. $y = \frac{1}{3}x\ln x - \frac{1}{9}x + \frac{19}{9}(1/x^2)$ *(9-3)*   **27.** $y = \sqrt{1 + 2x^2}$ *(9-2)*

28. $y = (2x + 1)e^{-x^2}$ *(9-3)*   **29.** $y = -2 + Cx^4$ *(9-2, 9-3)*

32. $y = 2 + Cx^5$ (A) $y = 2 + Cx^5$ for any constant C
   (B) No particular solution exists.   (C) $y = 2 - x^5$ *(9-3)*

33. $y = \sqrt{x^2 + 16}$ *(9-2)*

34. (A)

(B) The graphs are increasing and cross the x axis only at $x = 0$.

(C)

(D) Each graph has a local maximum, a local minimum, and crosses the x axis three times. *(9-1)*

35. (A) $M = 7/(1 - e^{-5k})$, $M = 4/(1 - e^{-2k})$ **(B)** 9.2 *(9-2)*   **36.** $500e^{-0.25 \ln 20} \approx \$236.44$ *(9-2)*
37. (A) 5 yr **(B)** \$74,000 *(9-2)*
38. (A) $p = 50 + Ae^{-t}$ **(B)** $\bar{p} = 50$ **(C)**

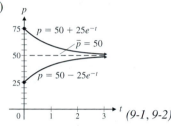

(9-1, 9-2)

39. $A(t) = 100,000 - 40,000e^{0.05t}$; $(\ln 2.5)/0.05 \approx 18.326$ yr; \$91,630 *(9-3)*
40. 4.82% *(9-3)*   **41.** $y = 100 + te^{-t} - 100e^{-t}$ *(9-3)*
42. (A) 211.1 lb **(B)** 10.3 hr *(9-3)*   **43. (A)** 337 birds **(B)** 357 birds *(9-2)*
44. (A) $200 - 199e^{(5/2)\ln(190/199)} \approx 23$ people **(B)** $\dfrac{2 \ln(100/199)}{\ln(190/199)} \approx 30$ days *(9-2)*

CHAPTER 10

Exercise 10-1

1. $24/(2 - x)^5$ **3.** $16e^{-2x}$ **5.** $1 - x + \frac{1}{2}x^2 - \frac{1}{6}x^3 + \frac{1}{24}x^4$ **7.** $1 + 3x + 3x^2 + x^3$ **9.** $2x - 2x^2 + \frac{8}{3}x^3$
11. $1 + \frac{1}{3}x - \frac{1}{9}x^2 + \frac{5}{81}x^3$ **13. (A)** $p_3(x) = -1$; $-0.562 < x < 0.562$ **(B)** $p_4(x) = x^4 - 1$; for all x
15. $1 + (x - 1) + \frac{1}{2}(x - 1)^2 + \frac{1}{6}(x - 1)^3 + \frac{1}{24}(x - 1)^4$ **17.** $1 + 3(x - 1) + 3(x - 1)^2 + (x - 1)^3$
19. $3(x - \frac{1}{3}) - \frac{9}{2}(x - \frac{1}{3})^2 + 9(x - \frac{1}{3})^3$ **21.** $1 - 2x + 2x^2 - \frac{4}{3}x^3$; 0.604 166 67   **23.** $4 + \frac{1}{8}x - \frac{1}{512}x^2$; 4.123 046 9
25. $1 + \frac{1}{2}(x - 1) - \frac{1}{8}(x - 1)^2 + \frac{1}{16}(x - 1)^3 - \frac{5}{128}(x - 1)^4$; 1.095 437 5   **27.** $n! (4 - x)^{-(n+1)}$ **29.** $3^n e^{3x}$
31. $-(n - 1)!(6 - x)^{-n}$ **33.** $\dfrac{1}{4} + \dfrac{1}{4^2}x + \dfrac{1}{4^3}x^2 + \dfrac{1}{4^4}x^3 + \cdots + \dfrac{1}{4^{n+1}}x^n$ **35.** $1 + 3x + \dfrac{3^2}{2!}x^2 + \dfrac{3^3}{3!}x^3 + \cdots + \dfrac{3^n}{n!}x^n$
37. $-\dfrac{1}{6}x - \dfrac{1}{6^2 2}x^2 - \dfrac{1}{6^3 3}x^3 - \cdots - \dfrac{1}{6^n n}x^n$ **39.** $-1 - (x + 1) - (x + 1)^2 - (x + 1)^3 - \cdots - (x + 1)^n$
41. $(x - 1) - \dfrac{1}{2}(x - 1)^2 + \dfrac{1}{3}(x - 1)^3 - \dfrac{1}{4}(x - 1)^4 + \cdots + \dfrac{(-1)^{n+1}}{n}(x - 1)^n$
43. $\dfrac{1}{e^2} - \dfrac{1}{e^2}(x - 2) + \dfrac{1}{2!e^2}(x - 2)^2 - \dfrac{1}{3!e^2}(x - 2)^3 + \cdots + \dfrac{(-1)^n}{n!e^2}(x - 2)^n$
45. $3 \ln 3 + (1 + \ln 3)(x - 3) + \dfrac{1}{3 \cdot 2}(x - 3)^2 - \dfrac{1}{3^2 2 \cdot 3}(x - 3)^3$

$\qquad + \dfrac{1}{3^3 3 \cdot 4}(x - 3)^4 - \dfrac{1}{3^4 4 \cdot 5}(x - 3)^5 + \cdots + \dfrac{(-1)^n}{3^{n-1}(n-1) \cdot n}(x - 3)^n$

47. $p_1(x) = x, p_2(x) = x - \frac{1}{2}x^2, p_3(x) = x - \frac{1}{2}x^2 + \frac{1}{3}x^3$

x	$p_1(x)$	$p_2(x)$	$p_3(x)$	$f(x)$
-0.2	-0.2	-0.22	$-0.222\ 667$	$-0.223\ 144$
-0.1	-0.1	-0.105	$-0.105\ 333$	$-0.105\ 361$
0	0	0	0	0
0.1	0.1	0.095	$0.095\ 333$	$0.095\ 31$
0.2	0.2	0.18	$0.182\ 667$	$0.182\ 322$

x	$|p_1(x) - f(x)|$	$|p_2(x) - f(x)|$	$|p_3(x) - f(x)|$
-0.2	$0.023\ 144$	$0.003\ 144$	$0.000\ 477$
-0.1	$0.005\ 361$	$0.000\ 361$	$0.000\ 028$
0	0	0	0
0.1	$0.004\ 69$	$0.000\ 31$	$0.000\ 023$
0.2	$0.017\ 678$	$0.002\ 322$	$0.000\ 345$

49.

51. $-1.323 < x < 1.168$

53. $e^a \left[1 + (x - a) + \frac{1}{2!}(x - a)^2 + \frac{1}{3!}(x - a)^3 + \cdots + \frac{1}{n!}(x - a)^n \right] = e^a \sum_{k=0}^{n} \frac{1}{k!}(x - a)^k$

55. $\frac{1}{a} - \frac{1}{a^2}(x - a) + \frac{1}{a^3}(x - a)^2 - \frac{1}{a^4}(x - a)^3 + \cdots + \frac{(-1)^n}{a^{n+1}}(x - a)^n = \sum_{k=0}^{n} \frac{(-1)^k}{a^{k+1}}(x - a)^k$

57. $n \geq$ degree of f **59.** $p_2(x) = 10 - 0.0005x^2$; $9.85   **61.** $p_2(x) = 8 - \frac{3}{40}(x - 60) - \frac{1}{1024}(x - 60)^2$; $7.12

63. $p_2(t) = 10 - 0.8t^2$; approx. $18 million   **65.** $p_2(t) = -3 + \frac{3}{25}t^2$; 6.32 cm^2

67. $p_2(x) = 120 + 30(x - 5) + \frac{3}{4}(x - 5)^2$; 126.25 ppm   **69.** $p_2(t) = 6 - 0.06t^2$; 27.5 wpm

71.

73.

75.

77.
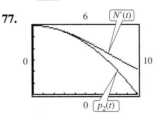

Exercise 10-2

1. $|x| < 1$ or $-1 < x < 1$ **3.** $|x| < \frac{1}{2}$ or $-\frac{1}{2} < x < \frac{1}{2}$ **5.** $|x| < 1$ or $-1 < x < 1$ **7.** $|x| < 5$ or $-5 < x < 5$

9. $|x| < 1$ or $-1 < x < 1$ **11.** $-\infty < x < \infty$ **13.** (A) 6   (B) Yes   **15.** $|x + 3| < 1$ or $-4 < x < -2$

17. $-\infty < x < \infty$ **19.** $|x - 2| < 4$ or $-2 < x < 6$ **21.** $|x - 8| < 2$ or $6 < x < 10$

23. (A)   (B) $-\infty < x < \infty$

25. $p_n(x) = 1 + 4x + \dfrac{4^2}{2!}x^2 + \cdots + \dfrac{4^n}{n!}x^n; \ 1 + 4x + \dfrac{4^2}{2!}x^2 + \cdots + \dfrac{4^n}{n!}x^n + \cdots; -\infty < x < \infty$

27. $p_n(x) = 2x - \dfrac{2^2}{2}x^2 + \dfrac{2^3}{3}x^3 - \cdots + \dfrac{(-1)^{n-1}2^n}{n}x^n; \ 2x - \dfrac{2^2}{2}x^2 + \dfrac{2^3}{3}x^3 - \cdots + \dfrac{(-1)^{n-1}2^n}{n}x^n + \cdots; |x| < \tfrac{1}{2} \text{ or } -\tfrac{1}{2} < x < \tfrac{1}{2}$

29. $p_n(x) = 1 + \dfrac{1}{2}x + \dfrac{1}{2^2}x^2 + \cdots + \dfrac{1}{2^n}x^n; \ 1 + \dfrac{1}{2}x + \dfrac{1}{2^2}x^2 + \cdots + \dfrac{1}{2^n}x^n + \cdots; |x| < 2 \text{ or } -2 < x < 2$

31. $p_n(x) = -(x - 1) - \dfrac{1}{2}(x - 1)^2 - \dfrac{1}{3}(x - 1)^3 - \cdots - \dfrac{1}{n}(x - 1)^n;$

$\quad - (x - 1) - \dfrac{1}{2}(x - 1)^2 - \dfrac{1}{3}(x - 1)^3 - \cdots - \dfrac{1}{n}(x - 1)^n - \cdots; |x - 1| < 1 \text{ or } 0 < x < 2$

33. $p_n(x) = -1 + (x - 2) - (x - 2)^2 + \cdots + (-1)^{n-1}(x - 2)^n;$
$\quad -1 + (x - 2) - (x - 2)^2 + \cdots + (-1)^{n-1}(x - 2)^n + \cdots; |x - 2| < 1 \text{ or } 1 < x < 3$

35. (A) $|x| < 4$ or $-4 < x < 4$
$\quad$ (B) The coefficients of the odd powers of x are 0.
$\quad$ (C) $|x| < 2$ or $-2 < x < 2$

37. (A) $\lim\limits_{n \to \infty} \left| \dfrac{a_{n+1}}{a_n} \right|$ does not exist $\quad$ (B) $|x| < 1$ or $-1 < x < 1$

Exercise 10-3

1. $1 + 2x + \dfrac{1}{2}x^2 + \cdots + \left[1 + \dfrac{(-1)^{n-1}}{n} \right] x^n + \cdots, -1 < x < 1$ $\quad$ **3.** $x^3 + x^4 + x^5 + \cdots + x^{n+3} + \cdots, -1 < x < 1$

5. $x + x^3 + \dfrac{1}{2!}x^5 + \cdots + \dfrac{1}{n!}x^{2n+1} + \cdots, -\infty < x < \infty$ $\quad$ **7.** $1 + x^3 + x^6 + \cdots + x^{3n} + \cdots, -1 < x < 1$

9. $1 - x^2 + x^4 - \cdots + (-1)^n x^{2n} + \cdots, -1 < x < 1$ $\quad$ **11.** $\dfrac{1}{2} + \dfrac{1}{2^2}x + \dfrac{1}{2^3}x^2 + \cdots + \dfrac{1}{2^{n+1}}x^n + \cdots, -2 < x < 2$
13. $1 + 8x^3 + 8^2x^6 + \cdots + 8^n x^{3n} + \cdots, -\tfrac{1}{2} < x < \tfrac{1}{2}$

15. $10^x = 1 + (\ln 10)x + \dfrac{(\ln 10)^2}{2!}x^2 + \dfrac{(\ln 10)^3}{3!}x^3 + \cdots + \dfrac{(\ln 10)^n}{n!}x^n + \cdots; -\infty < x < \infty$

17. $-\dfrac{1}{\ln 2}x - \dfrac{1}{2 \ln 2}x^2 - \dfrac{1}{3 \ln 2}x^3 - \cdots - \dfrac{1}{n \ln 2}x^n - \cdots; -1 < x < 1$

19. $\dfrac{1}{4} - \dfrac{1}{4^2}x^2 + \dfrac{1}{4^3}x^4 - \cdots + \dfrac{(-1)^n}{4^{n+1}}x^{2n} + \cdots, -2 < x < 2$

21. (A) $x \geq 0$ $\quad$ (B) No; some powers of x are not nonnegative integers.

23. (A) $1 + x^2 + x^4 + x^6 + \cdots + x^{2n} + \cdots, -1 < x < 1$
$\quad$ (B) $2x + 4x^3 + 6x^5 + \cdots + 2nx^{2n-1} + \cdots, -1 < x < 1$

25. $x - \dfrac{1}{3}x^3 + \dfrac{1}{5}x^5 - \cdots + \dfrac{(-1)^n}{2n + 1}x^{2n+1} + \cdots, -1 < x < 1$ $\quad$ **27.** $5 - \dfrac{1}{4}x^4 - \dfrac{1}{10}x^5 - \cdots - \dfrac{1}{n(n - 3)}x^n - \cdots; -1 < x < 1$

29. (A) $\qquad$ (B) Differentiating the Taylor series for $\sinh x$ twice yields the series for $\sinh x$.

31. $1 + (x - 3) + (x - 3)^2 + \cdots + (x - 3)^n + \cdots, 2 < x < 4$

33. $(x - 1) - \dfrac{1}{2}(x - 1)^2 + \dfrac{1}{3}(x - 1)^3 - \cdots + \dfrac{(-1)^{n-1}}{n}(x - 1)^n + \cdots, 0 < x < 2$

35. $1 + 3(x - 1) + 3^2(x - 1)^2 + \cdots + 3^n(x - 1)^n + \cdots, \tfrac{2}{3} < x < \tfrac{4}{3}$

37. $1 + 3x + 6x^2 + \cdots + \dfrac{n(n - 1)}{2}x^{n-2} + \cdots, -1 < x < 1$

39. (A) $1 + x^2 + x^4 + \cdots + x^{2n} + \cdots, -1 < x < 1$ **(B)** f' and $\dfrac{1}{1 - x^2}$ have identical Taylor series at 0.

41. $1 + 2x + 2x^2 + \cdots + 2x^n + \cdots, -1 < x < 1$

39. $x + \dfrac{1}{3}x^3 + \dfrac{1}{5}x^5 + \cdots + \dfrac{1}{2n + 1}x^{2n+1} + \cdots, -1 < x < 1$ **45.** $1 + \dfrac{1}{2!}x^2 + \dfrac{1}{4!}x^4 + \cdots + \dfrac{1}{(2n)!}x^{2n} + \cdots, -\infty < x < \infty$

47. $-2 + 3x + \dfrac{1}{6}x^3 - \dfrac{1}{24}x^4 + \cdots + \dfrac{(-1)^{n+1}}{n(n - 1)(n - 2)}x^n + \cdots; -1 < x < 1$

49. $\ln a + \dfrac{1}{a}(x - a) - \dfrac{1}{2a^2}(x - a)^2 + \cdots + \dfrac{(-1)^{n-1}}{na^n}(x - a)^n + \cdots, 0 < x < 2a$

51. $\dfrac{1}{1 - ab} + \dfrac{b}{(1 - ab)^2}(x - a) + \dfrac{b^2}{(1 - ab)^3}(x - a)^2 + \cdots + \dfrac{b^n}{(1 - ab)^{n+1}}(x - a)^n + \cdots; |x - a| < \left|\dfrac{1 - ab}{b}\right|$

Exercise 10-4

1. $0.4; |R_1(0.6)| \leqslant 0.18$ **3.** $0.544; |R_3(0.6)| \leqslant 0.0054$ **5.** $0.255; |R_2(0.3)| \leqslant 0.009$
7. $0.261\ 975; |R_4(0.3)| \leqslant 0.000\ 486$ **9.** $0.818\ 667; |R_3(0.2)| \leqslant 0.000\ 067$ **11.** $0.970\ 446; |R_3(0.03)| \leqslant 0.000\ 000\ 033$
13. $0.492; |R_3(0.6)| \leqslant 0.0324$ **15.** $0.058\ 272; |R_3(0.06)| \leqslant 0.000\ 003\ 24$ **17.** $0.6704; n = 4$ **19.** $0.9608; n = 2$
21. $0.095; n = 2$ **23.** $0.01; n = 1$ **25.** 0.1973 **27.** 0.0656 **29.** 0.0193 **31.** $n = 9$ **33.** $|x| < 1.817$
35. $-0.427 < x < 0.405$ **37.** $0.745; |R_2(0.3)| \leqslant 0.0135$ **39.** $1.051\ 25; |R_2(0.05)| \leqslant 0.000\ 063$
41. $4.123\ 047; |R_2(1)| \leqslant 0.000\ 061$
43. The second computation is correct. The use of the Taylor series for f at 0 is invalid because the lower limit of integration lies outside the interval of convergence of the series.

45. (A) $p_2(x) = 1 + x + \dfrac{1}{2!}x^2$

```
QuadReg
 y=ax²+bx+c
 a=.5004168056
 b=1.0016675
 c=1
 R²=1
```

(B) Max $|f(x) - p_2(x)| = 0.000\ 171$
    Max $|f(x) - q_2(x)| = 0.000\ 065$
(C) $q_2(x)$

47. $\displaystyle\int_0^1 \dfrac{10x^6}{4 + x^2}dx \approx 0.302$; index of income concentration ≈ 0.698
49. $S(4) \approx 2.031$ within ± 0.004, or \$2,031,000 within $\pm$\$4,000 **51.** 2 yr; \$1,270
53. $30.539°C$ **55.** $A(2) \approx 6.32\ cm^2$ within ± 0.031 **57.** $N(5) \approx 27.5$ wpm within ± 0.1875

Review Exercise

1. $-6/(x + 5)^4$ *(10-1)* **2.** $1 + \frac{1}{3}x - \frac{1}{9}x^2 + \frac{5}{81}x^3; 1.003\ 322$ *(10-1)*
3. $2 + \frac{1}{4}(x - 3) - \frac{1}{64}(x - 3)^2 + \frac{1}{512}(x - 3)^3; 1.974\ 842$ *(10-1)* **4.** $3 + \frac{1}{6}x^2; 3.001\ 667$ *(10-1)* **5.** $-\frac{1}{4} < x < \frac{1}{4}$ *(10-2)*
6. $1 < x < 11$ *(10-2)* **7.** $-1 < x < 1$ *(10-2)* **8.** $-\infty < x < \infty$ *(10-2)* **9.** $(-1)^n 9^n e^{-9x}$ *(10-1)*
10. $\dfrac{1}{7} + \dfrac{1}{7^2}x + \dfrac{1}{7^3}x^2 + \cdots + \dfrac{1}{7^{n+1}}x^n + \cdots, -7 < x < 7$ *(10-2)*

11. $\ln 2 + \dfrac{1}{2}(x - 2) - \dfrac{1}{8}(x - 2)^2 + \cdots + \dfrac{(-1)^{n-1}}{n2^n}(x - 2)^n + \cdots, 0 < x < 4$ *(10-2)*

12. $\dfrac{1}{10} - \dfrac{1}{10^2}x + \dfrac{1}{10^3}x^2 - \cdots + \dfrac{(-1)^n}{10^{n+1}}x^n + \cdots, -10 < x < 10$ *(10-3)*

13. $\dfrac{1}{4}x^2 + \dfrac{1}{4^2}x^4 + \dfrac{1}{4^3}x^6 + \cdots + \dfrac{1}{4^{n+1}}x^{2n+2} + \cdots, -2 < x < 2$ *(10-3)*

14. $x^2 + 3x^3 + \dfrac{3^2}{2!}x^4 + \cdots + \dfrac{3^n}{n!}x^{n+2} + \cdots, -\infty < x < \infty$ *(10-3)*

15. $x + \dfrac{1}{e}x^2 - \dfrac{1}{2e^2}x^3 + \cdots + \dfrac{(-1)^{n-1}}{ne^n}x^{n+1} + \cdots, -e < x < e$ *(10-3)*

16. $\dfrac{1}{2} + \dfrac{1}{2^2}(x - 2) + \dfrac{1}{2^3}(x - 2)^2 + \cdots + \dfrac{1}{2^{n+1}}(x - 2)^n + \cdots, 0 < x < 4$ *(10-3)*

17. (A) The coefficients of the odd powers of x are 0. (B) $-\sqrt{5}/5 < x < \sqrt{5}/5$ *(10-2, 10-3)*

18. (A) $0 \leq x < 1$ (B) No; some powers of x are not nonnegative integers. *(10-2, 10-3)*

19. $f(x) = \dfrac{1}{2} + \dfrac{1}{2^2}x + \dfrac{1}{2^3}x^2 + \dfrac{1}{2^4}x^3 + \cdots + \dfrac{1}{2^{n+1}}x^n + \cdots, -2 < x < 2$

$g(x) = \dfrac{1}{2^2} + \dfrac{1}{2^2}x + \dfrac{3}{2^4}x^2 + \cdots + \dfrac{n}{2^{n+1}}x^{n-1} + \cdots, -2 < x < 2$ *(10-3)*

20. $f(x) = x^2 - x^4 + x^6 - \cdots + (-1)^n x^{2n+2} + \cdots, -1 < x < 1;$
$g(x) = 2x - 4x^3 + 6x^5 - \cdots + (2n+2)(-1)^n x^{2n+1} + \cdots, -1 < x < 1$ *(10-3)*

21. $\dfrac{1}{3 \cdot 9}x^3 - \dfrac{1}{5 \cdot 9^2}x^5 + \dfrac{1}{7 \cdot 9^3}x^7 - \cdots + \dfrac{(-1)^n}{(2n+3)9^{n+1}}x^{2n+3} + \cdots; -3 < x < 3$ *(10-3)*

22. $\dfrac{1}{5 \cdot 16}x^5 + \dfrac{1}{7 \cdot 16^2}x^7 + \dfrac{1}{9 \cdot 16^3}x^9 + \cdots + \dfrac{1}{(2n+5)16^{n+1}}x^{2n+5} + \cdots; -4 < x < 4$ *(10-3)*

23. (A) $g(x) = 2 - 3(x-1) + (x-1)^3; h(x) = 9(x+1) - 6(x+1)^2 + (x+1)^3$
(B) The two series represent the same polynomial function for $-\infty < x < \infty$. *(10-2)*

24. (A) f is not differentiable at 0. (B) If $a > 0, p_n(x) = x$ for $n \geq 1$. If $a < 0, p_n(x) = -x$ for $n \geq 1$. *(10-2)*

25. $1.78; |R_2(0.6)| \leq 0.108$ *(10-4)* **26.** $1.0618; |R_2(0.06)| \leq 0.000\ 108$ *(10-4)*

27. $0.261\ 975; n = 4$ *(10-4)* **28.** $0.03; n = 1$ *(10-4)*

29. $5 - \dfrac{1}{3}x^3 - \dfrac{1}{8}x^4 - \cdots - \dfrac{1}{n(n-2)}x^n - \cdots; -1 < x < 1$ *(10-3)*

30. $3 - 4x + \dfrac{1}{6}x^3 - \dfrac{1}{12}x^4 + \cdots + \dfrac{(-1)^{n+1}}{(n-3)!(n-1)n}x^n + \cdots; -\infty < x < \infty$ *(10-3)*

31. 0.0612 *(10-4)* **32.** 0.314 *(10-4)* **33.** $n = 7$ *(10-4)* **34.** $|x| < 0.6918$ *(10-4)*

35. $-0.616 < x < 0.616$ *(10-4)* **36.** $-0.488 < x < 0.488$ *(10-4)*

37. $p_2(x) = 5 - \frac{1}{1,000}x^2; \$4.93$ *(10-4)* **38.** $p_2(t) = 9 - 0.03t^2; \$80,000$ *(10-4)*

39. $\displaystyle\int_0^1 \dfrac{18x^3}{8 + x^2}dx \approx 0.516$; index of income concentration ≈ 0.484 *(10-4)*

40. $S(8) \approx 3.348$ within ± 0.001, or $\$3,348,000$ within $\pm\$1,000$ *(10-4)* **41.** $A(2) \approx 10.17$ cm^2 within ± 0.01 *(10-4)*

42. $\dfrac{1}{5}\displaystyle\int_0^5 \dfrac{5,000t^2}{10,000 + t^4}dt \approx 4.06$ *(10-4)* **43.** $\dfrac{1}{5}\displaystyle\int_0^5 (10 + 2t - 5e^{-0.01t^2})dt \approx 10.4$ *(10-4)*

CHAPTER 11

Exercise 11-1

1. $\frac{1}{3}$ **3.** Diverges **5.** 2 **7.** Diverges **9.** 1 **11.** Diverges **13.** Diverges **15.** 1

17. $\frac{14}{3}$ **19.** 10 **21.** Diverges **23.** True **25.** False

27. $F(b) = \dfrac{1}{3} - \dfrac{1}{3b^3}; \lim\limits_{b\to\infty} F(b) = \dfrac{1}{3}$ **29.** $F(b) = \dfrac{10^{-6}b^2}{2}; \lim\limits_{b\to\infty} F(b) = \infty$ **31.** $F(b) = 2 - 2e^{-b/2}; \lim\limits_{b\to\infty} F(b) = 2$

33. $F(b) = 2\sqrt{b} - 2$; $\lim_{b \to \infty} F(b) = \infty$ **35.** $F(b) = 1 - \dfrac{1}{b + 1}$; $\lim_{b \to \infty} F(b) = 1$

37. Yes   **39.** 1   **41.** Diverges   **43.** $\frac{1}{2}$ **45.** Diverges   **47.** Diverges   **49.** \$120,000   **51.** \$50,000

53. Increasing the interest rate to 6% decreases the capital value to \$100,000, while decreasing the interest rate to 4% increases the capital value to \$150,000. In general, increasing the interest rate decreases the amount of capital required to establish the income stream.

55. (A) 7.5 billion ft^3 (B) 6.14 yr   **57.** 500 gal   **59.** Approx. 8,000,000 immigrants

Exercise 11-2

1. $f(x) \geq 0$ from graph

$$\int_0^4 f(x)\, dx = 1$$

3. No   **5.** No

7. (A) $\displaystyle\int_1^3 \frac{1}{8}x\, dx = \frac{1}{2}$ (B) $\displaystyle\int_0^2 \frac{1}{8}x\, dx = \frac{1}{4}$ (C) $\displaystyle\int_3^4 \frac{1}{8}x\, dx = \frac{7}{16}$

9. (A) $\displaystyle\int_1^1 f(x)\, dx = 0$ (B) $\displaystyle\int_5^\infty f(x)\, dx = 0$ (C) $\displaystyle\int_{-\infty}^5 f(x)\, dx = 1$

11. $F(x) = \begin{cases} 0 & \text{if } x < 0 \\ \frac{1}{16}x^2 & \text{if } 0 \leq x \leq 4 \\ 1 & \text{if } x > 4 \end{cases}$

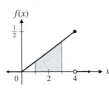

13. (A) $F(4) - F(2) = \frac{3}{4}$ (B) $F(2) - F(0) = \frac{1}{4}$ **15.** (A) 2 (B) $\frac{4}{3}$

17. $f(x) \geq 0$ from graph

$$\int_0^\infty \frac{2}{(1 + x)^3}\, dx = 1$$

19. (A) $\displaystyle\int_1^4 \frac{2}{(1 + x)^3}\, dx = .21$

(B) $\displaystyle\int_3^\infty \frac{2}{(1 + x)^3}\, dx = \frac{1}{16}$

(C) $\displaystyle\int_0^2 \frac{2}{(1 + x)^3}\, dx = \frac{8}{9}$

21. $F(x) = \begin{cases} 0 & \text{if } x < 0 \\ 1 - \dfrac{1}{(1 + x)^2} & \text{if } x \geq 0 \end{cases}$

23. (A) 1 (B) 3

25. $F(x) = \begin{cases} 0 & \text{if } x < 0 \\ \frac{3}{4}x^2 - \frac{1}{4}x^3 & \text{if } 0 \leqslant x \leqslant 2 \\ 1 & \text{if } x > 2 \end{cases}$

27. $F(x) = \begin{cases} 0 & \text{if } x < -1 \\ \frac{3}{8} + \frac{1}{2}x + \frac{1}{8}x^4 & \text{if } -1 \leqslant x \leqslant 1 \\ 1 & \text{if } x > 1 \end{cases}$

29. $x \approx 0.57$ **31.** $x \approx 0.44$ **33.** No such constant exists since $\int_1^\infty f(x)\, dx$ diverges   **35.** 2   **37.** No

39. 0.683; 0.954; 0.997   **41.** $f(x) = \begin{cases} 2x & \text{if } 0 \leqslant x \leqslant 1 \\ 0 & \text{otherwise} \end{cases}$ **43.** $f(x) = \begin{cases} 12x - 24x^2 + 12x^3 & \text{if } 0 \leqslant x \leqslant 1 \\ 0 & \text{otherwise} \end{cases}$

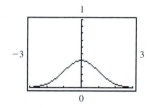

45. $F(x) = \begin{cases} 0 & \text{if } x < 0 \\ \frac{1}{2}x^2 & \text{if } 0 \leqslant x < 1 \\ 2x - \frac{1}{2}x^2 - 1 & \text{if } 1 \leqslant x \leqslant 2 \\ 1 & \text{if } x > 2 \end{cases}$

47. (A) The probability that the daily demand is between 2 million and 6 million kWh is .48.

(B) $\int_0^8 (.2 - .02x)\, dx = .96$ (C) $\int_5^{10} (.2 - .02x)\, dx = .25$

49. (A) The probability that it takes the computer more than 5 and less than 10 sec to respond is .239.

(B) $\int_0^1 \frac{1}{10}e^{-x/10}\, dx = 1 - e^{-1/10} \approx .0952$ (C) $\int_4^\infty \frac{1}{10}e^{-x/10}\, dx = e^{-2/5} \approx .6703$

51. (A) $\int_4^{10} .003x\sqrt{100 - x^2}\, dx = \frac{(84)^{3/2}}{1,000} \approx .7699$ (B) $\int_0^8 .003x\sqrt{100 - x^2}\, dx = .784$ (C) $\sqrt{100 - (100)^{2/3}} \approx 8{,}858$ lb

53. (A) $\int_7^{10} \frac{1}{5{,}000}(10x^3 - x^4)\, dx = .47178$ (B) $\int_0^5 \frac{1}{5{,}000}(10x^3 - x^4)\, dx = \frac{3}{16} = .1875$

55. (A) $\int_0^{20} \frac{800x}{(400 + x^2)^2}\, dx = .5$ (B) $\int_{15}^\infty \frac{800x}{(400 + x^2)^2}\, dx = .64$ (C) 10 days

57. (A) $\int_{30}^\infty \frac{1}{20}e^{-x/20}\, dx = e^{-1.5} \approx .223$ (B) $\int_{80}^\infty \frac{1}{20}e^{-x/20}\, dx = e^{-4} \approx .018$

Exercise 11-3

1. $\frac{4}{3}; \frac{2}{9}; .471$ **3.** 3.5; .75; .866   **5.** $\frac{4}{3}; \frac{1}{18}; .236$ **7.** $1/\sqrt{2} \approx .707$ **9.** $\sqrt{10} \approx 3.162$ **11.** $4 - 2\sqrt{2} \approx 1.172$

13. X_1 has mean 1; X_2 has mean 0.   **15.** X_1 and X_2 both have median 1.   **17.** $\frac{4}{3}; \frac{2}{9}; .471$ **19.** $\frac{8}{3}; \frac{8}{9}; .943$

21. 1.00; 0.25; 0.50   **23.** $e^{.5} \approx 1.649$ **25.** $\frac{2}{3}$ **27.** 1   **29.** $(\ln 2)/2 \approx .347$

31. $F(x) = \frac{1}{2}$ for any x satisfying $0.5 \leqslant x \leqslant 2.5$. Median is not unique.

33. $F(x) = \begin{cases} 0 & \text{if } x < 0 \\ x - \frac{1}{2}x^2 & \text{if } 0 \leq x < 1 \\ \frac{1}{2} & \text{if } 1 \leq x \leq 2 \\ \frac{1}{2}x^2 - 2x + \frac{5}{2} & \text{if } 2 < x \leq 3 \\ 1 & \text{if } x > 3 \end{cases}$; $F(x) = \frac{1}{2}$ for any x satisfying $1 \leq x \leq 2$. Median is not unique.

35. $a\mu + b$ **37.** $x_1 = 1; x_2 = \sqrt{2} \approx 1.414; x_3 = \sqrt{3} \approx 1.732$ **39.** $x_1 = 1; x_2 = 3; x_3 = 9$ **41.** .54 **43.** 1.47

45. (A) $\frac{50}{3} \approx \$16.667$, or $\$16,667$ (B) $20 - 5(\sqrt{2}/2) \approx \$16.464$, or $\$16,464$ **47.** $3 \ln 2 \approx 2.079$ min

49. 1 million gal **51.** $\frac{20}{3} \approx 6.7$ min **53.** 20 days **55.** 1.8 hr

Exercise 11-4

1. $f(x) = \begin{cases} \frac{1}{2} & \text{if } 0 \leq x \leq 2 \\ 0 & \text{otherwise} \end{cases}$ **3.** $f(x) = \begin{cases} 2e^{-2x} & \text{if } x \geq 0 \\ 0 & \text{otherwise} \end{cases}$

$F(x) = \begin{cases} 0 & \text{if } x < 0 \\ x/2 & \text{if } 0 \leq x \leq 2 \\ 1 & \text{if } x > 2 \end{cases}$ $\qquad F(x) = \begin{cases} 1 - e^{-2x} & \text{if } x \geq 0 \\ 0 & \text{otherwise} \end{cases}$

5. $\mu = 3; m = 3; \sigma = 2/\sqrt{3} \approx 1.155$ **7.** $\mu = 5; m = 5 \ln 2 \approx 3.466; \sigma = 5$

9. .3413 **11.** .4987 **13.** .3159 **15.** .4932 **17.** .4332 **19.** .4995

21. .1915 **23.** .2881 **25.** X_1 has standard deviation 1; X_2 has standard deviation .722.

27. X_1 has median 1; X_2 has median .693. **29.** $\mu = 2; \frac{1}{2}$

31. $\mu = 1; 1 - e^{-1} \approx .632$ **33.** $\mu = 0; \sigma = 5/\sqrt{3} \approx 2.887; 1/\sqrt{3} \approx .577$

35. $\mu = 6; \sigma = 6; 1 - e^{-2} \approx .865$ **37.** .7888 **39.** .5328 **41.** .0122 **43.** .1056

45. (A) .6827 (B) .6827 (C) .6827 The probability of being within 1 standard deviation of the mean is always .6827.

47. (A) .9973 (B) .9973 (C) .9973 The probability of being within 3 standard deviations of the mean is always .9973.

53. .2; .139; .2 **55.** $\mu = p/(p - 1)$, provided $p > 1$. The mean does not exist if $0 \leq p \leq 1$.

57. $m = \sqrt[p]{2}$ **59.** $\frac{3}{8} = .375$ **61.** $1 - e^{-2/3} \approx .487$

63. $1 - (1/\sqrt{2}) \approx .293$ **65.** 2.28% **67.** 1.247% **69.** (A) $-1/(\ln .7) \approx 2.8$ yr (B) $e^{-1} \approx .368$

71. 0.82% **73.** $e^{-2.5} \approx .082$ **75.** A's, 80.2 or greater; B's, 74.2–80.2; C's, 65.8–74.2; D's, 59.8–65.8; F's, 59.8 or lower

Review Exercise

1. $\frac{1}{2}$ *(11-1)* **2.** Diverges *(11-1)* **3.** 8 *(11-1)* **4.** .75

5. $\mu = \frac{2}{3} \approx 0.6667; V(X) = \frac{2}{9} \approx .2222;$ $\sigma = \sqrt{2}/3 \approx .4714$ *(11-3)*

(11-2)

6. $F(x) = \begin{cases} 0 & \text{if } x < 0 \\ x - \frac{1}{4}x^2 & \text{if } 0 \leq x \leq 2 \\ 1 & \text{if } x > 2 \end{cases}$

(11-2)

7. $2 - \sqrt{2} \approx .5858$ *(11-3)* **8.** .4641 *(11-4)*

9. $f(x) = \begin{cases} \frac{1}{10} & \text{if } 5 \leq x \leq 15 \\ 0 & \text{otherwise} \end{cases}$

$F(x) = \begin{cases} 0 & \text{if } x < 5 \\ \frac{1}{10}(x - 5) & \text{if } 5 \leq x \leq 15 \\ 1 & \text{if } x > 15 \end{cases}$ *(11-4)*

10. $f(x) = \begin{cases} 5e^{-5x} & \text{if } x \geq 0 \\ 0 & \text{otherwise} \end{cases}$

$F(x) = \begin{cases} 1 - e^{-5x} & \text{if } x \geq 0 \\ 0 & \text{otherwise} \end{cases}$ *(11-4)*

11. No; the area under the graph of f is not 1. *(11-2)*   **12.** No; f is negative for some values of x. *(11-2)*

13. $\frac{31}{32} \approx .9688$ **14.** $\mu = \frac{5}{3} \approx 1.6667$; $V(X) = \frac{20}{9} \approx 2.2222$;   **15.** $F(x) = \begin{cases} 1 - x^{-5/2} & \text{if } x \geq 1 \\ 0 & \text{otherwise} \end{cases}$

$\sigma = \frac{2}{3}\sqrt{5} \approx 1.4907$ *(11-3)*

(11-2)

(11-2)

16. $2^{2/5} \approx 1.3195$ *(11-3)*   **17.** $f(x) = \begin{cases} \frac{1}{2}e^{-x/2} & \text{if } x \geq 0 \\ 0 & \text{otherwise} \end{cases}$ *(11-4)*   **18.** $1 - e^{-1} \approx .6321$ *(11-2)*

19. $F(x) = \begin{cases} 1 - e^{-x/2} & \text{if } x \geq 0 \\ 0 & \text{otherwise} \end{cases}$ *(11-4)*   **20.** $\mu = 2$; $\sigma = 2$; $m = 2 \ln 2 \approx 1.3863$ *(11-4)*

21. (A) .9104   (B) .0668 *(11-4)*   **22.** (A) .3345   (B) .9970 *(11-4)*   **23.** 1 *(11-1)*   **24.** $\frac{1}{3}$ *(11-1)*   **25.** Yes *(11-1)*

26. $k = 10$ *(11-2)*   **27.** No such constant exists because $\int_0^\infty f(x)\,dx$ diverges. *(11-2)*

28. $P(X \leq 3) = 1 - e^{-1} \approx .6321$ *(11-4)*   **29.** $\mu = 5$; $m = 5\sqrt{2} - 5 \approx 2.0711$ *(11-3)*   **30.** 1.68 *(11-3)*

31. 1 *(11-1)*   **32.** $a\sigma^2 + a\mu^2 + b\mu + c$ *(11-3)*

33. In comparison with f_1, the graph of f_2 is shifted to the right, so X_2 has the greater mean. *(11-3)*

34. In comparison with f_1, the area under the graph of f_2 is more spread out, so X_2 has the greater variance. *(11-3)*

35. X_2 has mean 6; X_1 has mean 4. *(11-3)*   **36.** X_2 has variance 12; X_1 has variance 8. *(11-3)*

37. (A) 20,000,000 barrels   (B) 4.09 yr *(11-1)*

38. (A) The probability that the weekly demand is between 40 and 100 pounds is .36.   (B) $\frac{3}{4} = .75$ (C) 80 lb *(11-2)*

39. \$40,000 *(11-1)*   **40.** (A) .896   (B) 50%   (C) 50% *(11-2, 11-3)*

41. (A) $e^{-1} \approx .3679$ (B) $1 - e^{-.25} \approx .2212$ *(11-4)*   **42.** .0228 *(11-4)*

43. (A) 60.46%   (B) 7.35% *(11-4)*

44. (A) The probability that the shelf-life of the drug is between 2 and 8 months is $\frac{5}{18} \approx .2778$.

(B) $\frac{2}{3} \approx .6667$ (C) 10 months *(11-2, 11-3)*

45. (A) $e^{-4} \approx .0183$ (B) $\frac{1}{2}$ month *(11-2, 11-3)*   **46.** (A) 25 ml   (B) 6.93 hr *(11-1)*

47. 1.22% *(11-4)*   **48.** 45,000; 50,000 *(11-1)*

CHAPTER 12

Exercise 12-1

1. $\pi/3$ rad   **3.** $\pi/4$ rad   **5.** $\pi/6$ rad   **7.** II   **9.** IV   **11.** I   **13.** 1   **15.** 0   **17.** -1 **19.** 60°   **21.** 45°
23. 30°   **25.** $\sqrt{3}/2$ **27.** $\frac{1}{2}$ **29.** $-\sqrt{3}/2$ **31.** 0.1411   **33.** -0.6840 **35.** 0.7970   **37.** $3\pi/20$ rad   **39.** 15°

41. 1   **43.** 2   **45.** $1/\sqrt{3}$ or $\sqrt{3}/3$ **49.**

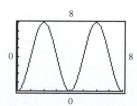

51.

53. (A) $P(13) = 5, P(26) = 10, P(39) = 5, P(52) = 0$

(B) $P(30) \approx 9.43, P(100) \approx 0.57$; thus, 30 weeks after January 1 the profit on a week's sales of bathing suits is $943, and 100 weeks after January 1 the profit on a week's sales of bathing suits is $57.

(C)

55. (A) $V(0) = 0.10, V(1) = 0.45, V(2) = 0.80, V(3) = 0.45, V(7) = 0.45$

(B) $V(3.5) \approx 0.20, V(5.7) \approx 0.76$; thus, the volume of air in the lungs of a normal seated adult 3.5 sec after exhaling is approximately 0.20 L, and 5.7 sec after exhaling is approx. 0.76 L.

(C)

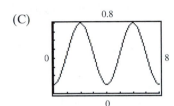

57. (A) $-5.6°$ (B) $-4.7°$

Exercise 12-2

1. $-\sin t$ **3.** $3x^2 \cos x^3$ **5.** $t \cos t + \sin t$ **7.** $(\cos x)^2 - (\sin x)^2$ **9.** $5(\sin x)^4 \cos x$ **11.** $\dfrac{\cos x}{2\sqrt{\sin x}}$

13. $-\dfrac{x^{-1/2}}{2} \sin \sqrt{x} = \dfrac{-\sin \sqrt{x}}{2\sqrt{x}}$ **15.** $f'\left(\dfrac{\pi}{6}\right) = \cos\dfrac{\pi}{6} = \dfrac{\sqrt{3}}{2}$

17.

$f(x)$

Increasing on $[-\pi, 0]$; decreasing on $[0, \pi]$; concave upward on $[-\pi, -\pi/2]$ and $[\pi/2, \pi]$; concave downward on $[-\pi/2, \pi/2]$; local maximum at $x = 0$; $f'(x) = -\sin x$; $f(x) = \cos x$.

19. $\dfrac{(\cos x)^2 + (\sin x)^2}{(\cos x)^2} = \dfrac{1}{(\cos x)^2} = (\sec x)^2$ **21.** $\dfrac{x \cos \sqrt{x^2 - 1}}{\sqrt{x^2 - 1}}$ **23.** $2e^x \cos x$

25.

27.

29.

31. (A) $P'(t) = \dfrac{5\pi}{26} \sin \dfrac{\pi t}{26}, 0 < t < 104$

(B) $P'(8) = \$0.50$ hundred or $\$50$ per week; $P'(26) = \$0$ per week; $P'(50) = -\$0.14$ hundred or $-\$14$ per week

(C)

t	$P(t)$	
26	\$1,000	Local maximum
52	\$0	Local minimum
78	\$1,000	Local maximum

(D)

t	$P(t)$	
0	\$0	Absolute minimum
26	\$1,000	Absolute maximum
52	\$0	Absolute minimum
78	\$1,000	Absolute maximum
104	\$0	Absolute minimum

(E) Same answer as for part (C)

33. (A) $V'(t) = \dfrac{0.35\pi}{2} \sin \dfrac{\pi t}{2}, 0 \le t \le 8$ **(B)** $V'(3) = -0.55$ L/sec; $V'(4) = 0.00$ L/sec; $V'(5) = 0.55$ L/sec

(C)

t	$V(t)$	
2	0.80	Local maximum
4	0.10	Local minimum
6	0.80	Local maximum

(D)

t	$V(t)$	
0	0.10	Absolute minimum
2	0.80	Absolute maximum
4	0.10	Absolute minimum
6	0.80	Absolute maximum
8	0.10	Absolute minimum

(E) Same answer as for part (C)

Exercise 12-3

1. $-\cos t + C$ **3.** $\frac{1}{3} \sin 3x + C$ **5.** $\frac{1}{13}(\sin x)^{13} + C$ **7.** $-\frac{3}{4}(\cos x)^{4/3} + C$ **9.** $\frac{1}{3} \sin x^3 + C$ **11.** 1   **13.** 1
15. $\sqrt{3}/2 - \frac{1}{2} \approx 0.366$ **17.** 1.4161   **19.** 0.0678   **21.** $e^{\sin x} + C$ **23.** $\ln|\sin x| + C$ **25.** $-\ln|\cos x| + C$

27. (A) $f(x)$ **(B)** $M_6 = 0.532$ **(C)**   **(D)** $|I - M_6| \le 0.0625$

29. (A) \$520 hundred or \$52,000   **31. (A)** 104 tons
(B) \$106.38 hundred or \$10,638   **(B)** 31 tons

(C) $P(t)$ 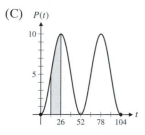   **(C)** $P(n)$

Review Exercise

1. (A) $\pi/6$ **(B)** $\pi/4$ **(C)** $\pi/3$ **(D)** $\pi/2$ *(12-1)*   **2. (A)** -1 **(B)** 0 **(C)** 1 *(12-1)*
3. $-\sin m$ *(12-2)*   **4.** $\cos u$ *(12-2)*
5. $(2x - 2)\cos(x^2 - 2x + 1)$ *(12-2)*   **6.** $-\frac{1}{3}\cos 3t + C$ *(12-3)*   **7. (A)** $30°$ **(B)** $45°$ **(C)** $60°$ **(D)** $90°$ *(12-1)*
8. (A) $\frac{1}{2}$ **(B)** $\sqrt{2}/2$ **(C)** $\sqrt{3}/2$ *(12-1)*   **9. (A)** -0.6543 **(B)** 0.8308 *(12-1)*   **10.** $(x^2 - 1)\cos x + 2x \sin x$ *(12-2)*
11. $6(\sin x)^5 \cos x$ *(12-2)*   **12.** $(\cos x)/[3(\sin x)^{2/3}]$ *(12-2)*   **13.** $\frac{1}{2}\sin(t^2 - 1) + C$ *(12-3)*   **14.** 2 *(12-3)*
15. $\sqrt{3}/2$ *(12-3)*   **16.** -0.243 *(12-3)*   **17.** $-\sqrt{2}/2$ *(12-2)*   **18.** $\sqrt{2}$ *(12-3)*

19. (A)   (B) $M_4 = 0.587$ (C)   (D) $|I - M_4| \leqslant 0.05$ *(6-5, 12-2, 12-3)*

20. $\pi/12$ *(12-1)*   **21.** (A) -1 (B) $-\sqrt{3}/2$ (C) $-\frac{1}{2}$ *(12-1)*   **22.** $1/(\cos u)^2 = (\sec u)^2$ *(12-2)*
23. $-2x(\sin x^2)e^{\cos x^2}$ *(12-2)*   **24.** $e^{\sin x} + C$ *(12-3)*   **25.** $-\ln|\cos x| + C$ *(12-3)*   **26.** 15.2128 *(12-3)*
27. *(12-2, 12-3)*   **28.** *(12-2, 12-3)*   **29.** *(12-2, 12-3)*

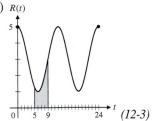

30. (A) $R(0) = \$5$ thousand; $R(2) = \$4$ thousand; $R(3) = \$3$ thousand; $R(6) = \$1$ thousand
   (B) $R(1) = \$4.732$ thousand, $R(22) = \$4$ thousand; thus, the revenue is $\$4,732$ for a month of sweater sales 1 month after January 1, and $\$4,000$ for a month of sweater sales 22 months after January 1. *(12-1)*

31. (A) $R'(t) = -\dfrac{\pi}{3}\sin\dfrac{\pi t}{6}, 0 \leqslant t \leqslant 24$

   (B) $R'(3) = -\$1.047$ thousand or $-\$1,047$/mo; $R'(10) = \$0.907$ thousand or $\$907$/mo; $R'(18) = \$0.000$ thousand

(C)

t	$P(t)$	
6	\$1,000	Local minimum
12	\$5,000	Local maximum
18	\$1,000	Local minimum

(D)

t	$P(t)$	
0	\$5,000	Absolute maximum
6	\$1,000	Absolute minimum
12	\$5,000	Absolute maximum
18	\$1,000	Absolute minimum
24	\$5,000	Absolute maximum

(E) Same answer as for part (C) *(12-2)*

32. (A) $\$72$ thousand or $\$72,000$ (B) $\$6.270$ thousand or $\$6,270$ (C) 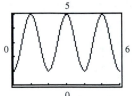 *(12-3)*

APPENDIX A

Self-Test on Basic Algebra

1. (A) T (B) T (C) F (D) T *(A-1)*   **2.** (A) $(y + z)x$ (B) $(2 + x) + y$ (C) $2x + 3x$ *(A-2)*
3. $x^3 + 3x^2 + 5x - 2$ *(A-3)*   **4.** $x^3 - 3x^2 - 3x + 22$ *(A-3)*   **5.** $3x^5 + x^4 - 8x^3 + 24x^2 + 8x - 64$ *(A-3)*
6. 3 *(A-3)*   **7.** 1 *(A-3)*   **8.** $14x^2 - 30x$ *(A-3)*.   **9.** $6x^2 - 5xy - 4y^2$ *(A-3)*   **10.** $4a^2 - 12ab + 9b^2$ *(A-3)*
11. $4xy - 2y^2$ *(A-3)*   **12.** $m^4 - 6m^2n^2 + n^4$ *(A-3)*   **13.** $x^3 - 6x^2y + 12xy^2 - 8y^3$ *(A-3)*
14. (A) 4.065×10^{12} (B) 7.3×10^{-3} *(A-6)*   **15.** (A) $255,000,000$ (B) $0.000\ 406$ *(A-6)*
16. (A) $\{2, 4, 5, 6\}$ (B) $\{5\}$ (C) $\{8\}$ (D) $\{2, 4\}$ *(A-1)*   **17.** (A) 28 (B) 5 (C) 4 (D) 10 *(A-1)*
18. (A) T (B) F *(A-2)*   **19.** 0 and -3 are two examples of infinitely many. *(A-2)*   **20.** $6x^5y^{15}$ *(A-6)*
21. $3u^4/v^2$ *(A-6)*   **22.** 6×10^2 *(A-6)*   **23.** x^6/y^4 *(A-6)*   **24.** $u^{7/3}$ *(A-7)*   **25.** $3a^2/b$ *(A-7)*   **26.** $\frac{5}{9}$ *(A-6)*
27. $x + 2x^{1/2}y^{1/2} + y$ *(A-7)*   **28.** $6x + 7x^{1/2}y^{1/2} - 3y$ *(A-7)*   **29.** $(3x - 1)(4x + 3)$ *(A-4)*

30. $(4x - 3y)(2x - 3y)$ *(A-4)*   **31.** Not factorable relative to the integers *(A-4)*
32. $3n(2n - 5)(n + 1)$ *(A-4)*   **33.** $(x - y)(7x - y)$ *(A-4)*

34. Not factorable relative to the integers *(A-4)*   **35.** $\dfrac{12a^3b - 40b^2 - 5a}{30a^3b^2}$ *(A-5)*   **36.** $\dfrac{7x - 4}{6x(x - 4)}$ *(A-5)*

37. $\dfrac{-8(x + 2)}{x(x - 4)(x + 4)}$ *(A-5)*   **38.** $\dfrac{y + 2}{y(y - 2)}$ *(A-5)*   **39.** $\dfrac{-1}{7(7 + h)}$ *(A-5)*   **40.** $\dfrac{xy}{y - x}$ *(A-7)*

41. (A) Subtraction (B) Commutative (+) (C) Distributive (D) Associative $(\cdot)$ (E) Negatives (F) Identity (+) *(A-2)*

42. (A) 90 (B) 45 *(A-1)*   **43.** $6x^{2/5} - 7(x - 1)^{3/4}$ *(A-7)*   **44.** $2\sqrt{x} - 3\sqrt[3]{x^2}$ *(A-7)*   **45.** $2 - \frac{3}{2}x^{-1/2}$ *(A-7)*

46. $\sqrt{3x}$ *(A-7)*   **47.** $\sqrt{x} + \sqrt{5}$ *(A-7)*   **48.** $\dfrac{1}{\sqrt{x - 5}}$ *(A-7)*   **49.** $\dfrac{1}{\sqrt{u + h} + \sqrt{u}}$ *(A-7)*   **50.** $x = 2$ *(A-8)*

51. $x = 0, 5$ *(A-9)*   **52.** $x = \pm\sqrt{7}$ *(A-9)*   **53.** $x = -4, 5$ *(A-9)*   **54.** $x = (3 \pm \sqrt{17})/4$ *(A-9)*

55. $x < 4$ or $(-\infty, 4)$ *(A-8)*   **56.** $x \geq \frac{9}{2}$ or $[\frac{9}{2}, \infty)$ *(A-8)*

57. $2 \leq x < 12$ or $[2, 12)$ *(A-8)*   **58.** $y = \frac{2}{3}x - 2$ *(A-8)*   **59.** $y = 3/(x - 1)$ *(A-8)*   **60.** Yes *(A-1)*

61. $3.1487 \times 10^4 = \$31,487$ per person *(A-6)*   **62.** \$20,000 at 8%; \$40,000 at 14% *(A-8)*   **63.** 4,000 tapes *(A-8)*

Exercise A-1

1. T   **3.** T   **5.** T   **7.** T   **9.** $\{1, 2, 3, 4, 5\}$ **11.** $\{3, 4\}$ **13.** $\varnothing$ **15.** $\{2\}$ **17.** $\{-7, 7\}$
19. $\{1, 3, 5, 7, 9\}$ **21.** $A' = \{1, 5\}$ **23.** 40   **25.** 60   **27.** 60   **29.** 20   **31.** 95   **33.** 40
35. (A) $\{1, 2, 3, 4, 6\}$ (B) $\{1, 2, 3, 4, 6\}$ **37.** $\{1, 2, 3, 4, 6\}$
39. True
41. False
43. False
45. False
47. False
49. (A) 2
   (B) 4
   (C) 8
   (D) 16
51. No   **53.** 800   **55.** 200   **57.** 200   **59.** 800   **61.** 200   **63.** 200   **65.** 6   **67.** A+, AB+
69. A−, A+, B+, AB−, AB+, O+   **71.** O+, O−   **73.** B−, B+   **75.** Everybody in the clique relates to each other.

Exercise A-2

1. vu **3.** $(3 + 7) + y$ **5.** $u + v$ **7.** T   **9.** T   **11.** T   **13.** T   **15.** T   **17.** T   **19.** T   **21.** F
23. T   **25.** T   **27.** No   **29.** (A) F (B) T (C) T   **31.** $\sqrt{2}$ and π are two examples of infinitely many.
33. (A) N, Z, Q, R (B) R (C) Q, R (D) Q, R
35. (A) T (B) F, since, for example, $(8 - 4) - 2 \neq 8 - (4 - 2)$ (C) T
   (D) F, since, for example, $(8 \div 4) \div 2 \neq 8 \div (4 \div 2)$.
37. $\frac{1}{11}$ **39.** (A) 2.166 666 666... (B) 4.582 575 69... (C) 0.437 500 000... (D) 0.261 261 261...

Exercise A-3

1. 3   **3.** $x^3 + 4x^2 - 2x + 5$ **5.** $x^3 + 1$ **7.** $2x^5 + 3x^4 - 2x^3 + 11x^2 - 5x + 6$ **9.** $-5u + 2$
11. $6a^2 + 6a$ **13.** $a^2 - b^2$ **15.** $6x^2 - 7x - 5$ **17.** $2x^2 + xy - 6y^2$ **19.** $9y^2 - 4$ **21.** $6m^2 - mn - 35n^2$
23. $16m^2 - 9n^2$ **25.** $9u^2 + 24uv + 16v^2$ **27.** $a^3 - b^3$ **29.** $16x^2 + 24xy + 9y^2$ **31.** 1   **33.** $x^4 - 2x^2y^2 + y^4$
35. $5a^2 + 12ab - 10b^2$ **37.** $-4m + 8$ **39.** $-6xy$ **41.** $u^3 + 3u^2v + 3uv^2 + v^3$
43. $x^3 - 6x^2y + 12xy^2 - 8y^3$ **45.** $2x^2 - 2xy + 3y^2$ **47.** $8x^3 - 20x^2 + 20x + 1$ **49.** $4x^3 - 14x^2 + 8x - 6$

51. $m + n$ **53.** No change   **55.** $(1 + 1)^2 \neq 1^2 + 1^2$; either a or b must be 0
57. $0.09x + 0.12(10{,}000 - x) = 1{,}200 - 0.03x$ **59.** $10x + 30(3x) + 50(4{,}000 - x - 3x) = 200{,}000 - 100x$
61. $0.02x + 0.06(10 - x) = 0.6 - 0.04x$

Exercise A-4

1. $3m^2(2m^2 - 3m - 1)$ **3.** $2uv(4u^2 - 3uv + 2v^2)$ **5.** $(7m + 5)(2m - 3)$ **7.** $(a - 4b)(3c + d)$
9. $(2x - 1)(x + 2)$ **11.** $(y - 1)(3y + 2)$ **13.** $(x + 4)(2x - 1)$ **15.** $(w + x)(y - z)$ **17.** $(a - b)(m + n)$
19. $(3y + 2)(y - 1)$ **21.** $(u - 5v)(u + 3v)$ **23.** Not factorable   **25.** $(wx - y)(wx + y)$ **27.** $(3m - n)^2$
29. Not factorable   **31.** $4(z - 3)(z - 4)$ **33.** $2x^2(x - 2)(x - 10)$ **35.** $x(2y - 3)^2$ **37.** $(2m - 3n)(3m + 4n)$
39. $uv(2u - v)(2u + v)$ **41.** $2x(x^2 - x + 4)$ **43.** $(r - t)(r^2 + rt + t^2)$ **45.** $(a + 1)(a^2 - a + 1)$
47. $[(x + 2) - 3y][(x + 2) + 3y]$ **49.** Not factorable   **51.** $(6x - 6y - 1)(x - y + 4)$
53. $(y - 2)(y + 2)(y^2 + 1)$ **55.** $a^2(3 + ab)(9 - 3ab + a^2b^2)$ **57.** True   **59.** False

Exercise A-5

1. $8d^6$ **3.** $\dfrac{15x^2 + 10x - 6}{180}$ **5.** $\dfrac{15m^2 + 14m - 6}{36m^3}$ **7.** $\dfrac{1}{x(x - 4)}$ **9.** $\dfrac{x - 6}{x(x - 3)}$ **11.** $\dfrac{x - 5}{(x - 1)^2(x + 1)}$
13. $\dfrac{2}{x - 1}$ **15.** $\dfrac{5}{a - 1}$ **17.** $\dfrac{x^2 + 8x - 16}{x(x - 4)(x + 4)}$ **19.** $\dfrac{7x^2 - 2x - 3}{6(x + 1)^2}$ **21.** $-\dfrac{1}{x}$ **23.** $\dfrac{-17c + 16}{15(c - 1)}$ **25.** $\dfrac{1}{x - 3}$
27. $\dfrac{-1}{2x(x + h)}$ **29.** $\dfrac{x - y}{x + y}$ **31.** (A) Incorrect (B) $x + 1$ **33.** (A) Incorrect (B) $2x + h$
35. (A) Incorrect (B) $\dfrac{x^2 - x - 3}{x + 1}$ **37.** (A) Correct   **39.** $\dfrac{-2x - h}{3(x + h)^2 x^2}$ **41.** x

Exercise A-6

1. $2/x^9$ **3.** $3w^7/2$ **5.** $2/x^3$ **7.** $1/w^5$ **9.** 5   **11.** $1/a^6$ **13.** y^6/x^{12} **15.** 8.23×10^{10} **17.** 7.83×10^{-1}
19. 3.4×10^{-5} **21.** $40{,}000$ **23.** 0.007 **25.** $61{,}710{,}000$ **27.** $0.000\ 808$ **29.** 1   **31.** 10^{14} **33.** $y^6/25x^4$
35. 4×10^2 **37.** $4y^3/3x^5$ **39.** $\frac{7}{4} - \frac{1}{4}x^{-3}$ **41.** $\frac{3}{4}x - 1x^{-1} - \frac{1}{4}x^{-3}$ **43.** $\dfrac{x^2(x - 3)}{(x - 1)^3}$ **45.** $\dfrac{2(x - 1)}{x^3}$
47. 2.4×10^{10}; $24{,}000{,}000{,}000$ **49.** 3.125×10^4; $31{,}250$ **51.** 64   **55.** uv **57.** $\dfrac{bc(c + b)}{c^2 + bc + b^2}$
59. (A) 6.674×10^{11}
(B) 1.8241
(C) 0.5482
61. (A) $\$20{,}445$
(B) $\$1{,}346$
(C) 6.58%
63. (A) 9×10^{-6} (B) $0.000\ 009$ (C) 0.0009%
65. $1{,}531{,}000$

Exercise A-7

1. $6\sqrt[5]{x^3}$ **3.** $\sqrt[5]{(4xy^3)^2}$ **5.** $\sqrt{x^2 + y^2}$ (not $x + y$)   **7.** $5x^{3/4}$ **9.** $(2x^2y)^{3/5}$ **11.** $x^{1/3} + y^{1/3}$ **13.** 5   **15.** 64
17. -6 **19.** Not a rational number (not even a real number)   **21.** $\frac{8}{125}$ **23.** $\frac{1}{27}$ **25.** $x^{2/5}$ **27.** m **29.** $2x/y^2$
31. $xy^2/2$ **33.** $2/3x^{7/12}$ **35.** $2x + 3$ **37.** $6x^3$ **39.** 2   **41.** $12x - 6x^{35/4}$ **43.** $3u - 13u^{1/2}v^{1/2} + 4v$
45. $25m - n$ **47.** $9x - 6x^{1/2}y^{1/2} + y$ **49.** $\frac{1}{2}x^{1/3} + x^{-1/3}$ **51.** $\frac{2}{3}x^{-1/4} + \frac{1}{3}x^{-2/3}$ **53.** $\frac{1}{2}x^{-1/6} - \frac{1}{4}$ **55.** $4n\sqrt{3mn}$
57. $\dfrac{2\sqrt{x} - 2}{x - 2}$ **59.** $7(x - y)(\sqrt{x} + \sqrt{y})$ **61.** $\dfrac{1}{xy\sqrt{5xy}}$ **63.** $\dfrac{1}{\sqrt{x + h} + \sqrt{x}}$ **65.** $\dfrac{1}{\sqrt{t} + \sqrt{x}}$
67. $x = y = 1$ is one of many choices.   **69.** $x = y = 1$ is one of many choices.   **71.** False   **73.** True   **75.** True
77. False   **79.** $\dfrac{x + 8}{2(x + 3)^{3/2}}$ **81.** $\dfrac{x - 2}{2(x - 1)^{3/2}}$ **83.** $\dfrac{x + 6}{3(x + 2)^{5/3}}$ **85.** 103.2 **87.** 0.0805 **89.** $4{,}588$
91. (A) and (E); (B) and (F); (C) and (D)

Exercise A-8

1. $m = 5$ **3.** $x < -9$ **5.** $x \leq 4$ **7.** $x < -3$ or $(-\infty, -3)$

9. $-1 \leq x \leq 2$ or $[-1, 2]$

11. $y = 8$ **13.** $x > -6$ **15.** $y = 8$ **17.** $x = 10$

19. $y \geq 3$ **21.** $x = 36$ **23.** $m < 3$ **25.** $x = 10$ **27.** $3 \leq x < 7$ or $[3, 7)$

29. $-20 \leq C \leq 20$ or $[-20, 20]$

31. $y = \frac{3}{4}x - 3$

33. $y = -(A/B)x + (C/B) = (-Ax + C)/B$ **35.** $C = \frac{5}{9}(F - 32)$ **37.** $B = A/(m - n)$

39. $-2 < x \leq 1$ or $(-2, 1]$

41. (A) and (C): $a > 0$ and $b > 0$, or $a < 0$ and $b < 0$ (B) and (D): $a > 0$ and $b < 0$, or $a < 0$ and $b > 0$
43. (A) $>$ (B) $<$ **45.** Negative   **47.** True   **49.** False   **51.** True   **53.** 3,500 \$15 tickets; 4,500 \$25 tickets
55. \$7,200 at 10%; \$4,800 at 15%   **57.** \$21,005   **59.** 5,851 books   **61.** (B) 6,180 books   (C) At least \$11.50
63. 5,000   **65.** 12.6 yr

Exercise A-9

1. $\pm\sqrt{11}$ **3.** $-1, 3$ **5.** $-2, 6$ **7.** $0, 2$ **9.** $3 \pm 2\sqrt{3}$ **11.** $-2 \pm \sqrt{2}$ **13.** $0, 2$ **15.** $\pm\frac{3}{2}$ **17.** $\frac{1}{2}, -3$
19. $(-1 \pm \sqrt{5})/2$ **21.** $(3 \pm \sqrt{3})/2$ **23.** No real solution   **25.** $-4 \pm \sqrt{11}$ **27.** $\pm\sqrt{3}$
29. $-\frac{1}{2}, 2$ **31.** $(x - 2)(x + 42)$ **33.** Not factorable in the integers   **35.** $(2x - 9)(x + 12)$
37. $(4x - 7)(x + 62)$ **39.** $r = \sqrt{A/P} - 1$
41. If $c < 4$, there are two distinct real roots; if $c = 4$, there is one real double root; and if $c > 4$, there are no real roots.
43. 1,575 bottles at \$4 each   **45.** 0.2, or 20%   **47.** 8 ft/sec; $4\sqrt{2}$ or 5.66 ft/sec

APPENDIX B

Exercise B-1

1. $5, 7, 9, 11$ **3.** $\frac{3}{2}, \frac{4}{3}, \frac{5}{4}, \frac{6}{5}$ **5.** $9, -27, 81, -243$ **7.** 23 **9.** $\frac{101}{100}$ **11.** $1 + 2 + 3 + 4 + 5 + 6 = 21$
13. $5 + 7 + 9 + 11 = 32$ **15.** $1 + \frac{1}{10} + \frac{1}{100} + \frac{1}{1,000} = \frac{1,111}{1,000}$ **17.** 3.6 **19.** 82.5 **21.** $\frac{1}{2}, -\frac{1}{4}, \frac{1}{8}, -\frac{1}{16}, \frac{1}{32}$
23. $0, 4, 0, 8, 0$ **25.** $1, -\frac{3}{2}, \frac{9}{4}, -\frac{27}{8}, \frac{81}{16}$ **27.** $a_n = n - 3$ **29.** $a_n = 4n$ **31.** $a_n = (2n - 1)/2n$
33. $a_n = (-1)^{n+1}n$ **35.** $a_n = (-1)^{n+1}(2n - 1)$ **37.** $a_n = (\frac{2}{5})^{n-1}$ **39.** $a_n = x^n$ **41.** $a_n = (-1)^{n+1}x^{2n-1}$
43. $1 - 9 + 25 - 49 + 81$ **45.** $\frac{4}{7} + \frac{8}{9} + \frac{16}{11} + \frac{32}{13}$ **47.** $1 + x + x^2 + x^3 + x^4$ **49.** $x - \dfrac{x^3}{3} + \dfrac{x^5}{5} - \dfrac{x^7}{7} + \dfrac{x^9}{9}$

51. (A) $\displaystyle\sum_{k=1}^{5}(k + 1)$ (B) $\displaystyle\sum_{j=0}^{4}(j + 2)$ **53.** (A) $\displaystyle\sum_{k=1}^{4}\dfrac{(-1)^{k+1}}{k}$ (B) $\displaystyle\sum_{j=0}^{3}\dfrac{(-1)^j}{j + 1}$ **55.** $\displaystyle\sum_{k=1}^{n}\dfrac{k + 1}{k}$

57. $\displaystyle\sum_{k=1}^{n}\dfrac{(-1)^{k+1}}{2^k}$ **59.** False
61. True   **63.** $2, 8, 26, 80, 242$ **65.** $1, 2, 4, 8, 16$
67. $1, \frac{3}{2}, \frac{17}{12}, \frac{577}{408}; a_4 = \frac{577}{408} \approx 1.414\ 216, \sqrt{2} \approx 1.414\ 214$

69. $1, 1, 2, 3, 5, 8, 13, 21, 34, 55$

Exercise B-2

1. (A) Arithmetic, with $d = -5; -26, -31$ (B) Geometric, with $r = 2; -16, 32$ (C) Neither
   (D) Geometric, with $r = \frac{1}{3}; \frac{1}{54}, \frac{1}{162}$ **3.** Geometric; 1   **5.** Neither   **7.** Arithmetic; 127.5
   9. $a_2 = 11, a_3 = 15$ **11.** $a_{21} = 82, S_{31} = 1{,}922$ **13.** $S_{20} = 930$ **15.** $a_2 = -6, a_3 = 12, a_4 = -24$ **17.** $S_7 = 547$
   19. $a_{10} = 199.90$ **21.** $r = 1.09$ **23.** $S_{10} = 1{,}242, S_\infty = 1{,}250$ **25.** 2,706   **27.** -85 **29.** 1,120
   31. (A) Does not exist   (B) $S_\infty = \frac{8}{5} = 1.6$ **33.** 2,400   **35.** 0.999
   37. Use $a_1 = 1$ and $d = 2$ in $S_n = (n/2)[2a_1 + (n-1)d]$.   **39.** $S_n = na_1$ **41.** No   **43.** Yes
   45. $\$48 + \$46 + \cdots + \$4 + \$2 = \$600$
   47. About $\$11{,}670{,}000$ **49.** $\$1{,}628.89; \$2{,}653.30$

Exercise B-3

1. 720   **3.** 10   **5.** 1,320   **7.** 10   **9.** 6   **11.** 1,140   **13.** 10   **15.** 6   **17.** 1   **19.** 816
21. $C_{4,0}a^4 + C_{4,1}a^3b + C_{4,2}a^2b^2 + C_{4,3}ab^3 + C_{4,4}b^4 = a^4 + 4a^3b + 6a^2b^2 + 4ab^3 + b^4$
23. $x^6 - 6x^5 + 15x^4 - 20x^3 + 15x^2 - 6x + 1$ **25.** $32a^5 - 80a^4b + 80a^3b^2 - 40a^2b^3 + 10ab^4 - b^5$ **27.** $3{,}060x^{14}$
29. $5{,}005p^9q^6$ **31.** $264x^2y^{10}$ **33.** $C_{n,0} = \dfrac{n!}{0!n!} = 1; C_{n,n} = \dfrac{n!}{n!0!} = 1$ **35.** 1 5 10 10 5 1; 1 6 15 20 15 6 1

Exercise B-4

1. $\Delta x = 3; \Delta y = 45; \Delta y/\Delta x = 15$ **3.** 12   **5.** 12   **7.** $dy = (24x - 3x^2)\, dx$

9. $dy = \left(2x - \dfrac{x^2}{3}\right) dx$ **11.** $dy = -\dfrac{295}{x^{3/2}}\, dx$

13. (A) $12 + 3\,\Delta x$ (B) 12   **15.** $dy = 6(2x + 1)^2\, dx$ **17.** $dy = \dfrac{9 - x^2}{(x^2 + 9)^2}\, dx$

19. $dy = 1.4; \Delta y = 1.44$ **21.** $dy = -3; \Delta y = -3\frac{1}{3}$ **23.** 120 in.^{3}

25. (A)   (B)

27. (A)   (B)

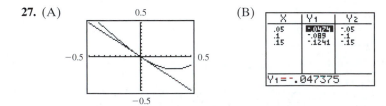

29. True   **31.** False   **33.** $dy = \dfrac{6x - 2}{3(3x^2 - 2x + 1)^{2/3}}\, dx$ **35.** $dy = 3.9; \Delta y = 3.83$

37. 40 unit increase; 20 unit increase   **39.** $-\$2.50; \$1.25$ **41.** $-1.37/\text{min}; -0.58/\text{min}$ **43.** 1.26 mm^2 **45.** 3 wpm

47. (A) 2,100 increase   (B) 4,800 increase   (C) 2,100 increase

Exercise B-5

1. $\frac{8}{3}$ **3.** $\frac{1}{2}$ **5.** 1   **7.** 4   **9.** 0   **11.** ∞ **13.** ∞ **15.** 5   **17.** ∞ **19.** 8   **21.** 0   **23.** ∞ **25.** ∞

27. $\frac{1}{3}$ **29.** -2 **31.** $-\infty$ **33.** 0   **35.** 0   **37.** 0   **39.** $\frac{1}{4}$ **41.** $\frac{1}{3}$ **43.** 0   **45.** 0   **47.** ∞

49. $\displaystyle\lim_{x\to\infty} \frac{\sqrt{1+x^2}}{x} = 1$ **51.** $\displaystyle\lim_{x\to-\infty} \frac{\sqrt[3]{x^3+1}}{x} = 1$

Exercise B-6

1. $R = \{(x,y)|0 \leq y \leq 4 - x^2, \quad 0 \leq x \leq 2\}$ **3.** R is a regular x region:

$R = \{(x,y)|0 \leq x \leq \sqrt{4-y}, \quad 0 \leq y \leq 4\}$ $R = \{(x,y)|x^3 \leq y \leq 12 - 2x, \quad 0 \leq x \leq 2\}$

   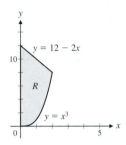

5. R is a regular y region: $R = \{(x,y)|\frac{1}{2}y^2 \leq x \leq y + 4, \quad -2 \leq y \leq 4\}$

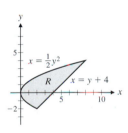

7. $\frac{1}{2}$ **9.** $\frac{39}{70}$ **11.** R consists of the points on or inside the rectangle with corners $(\pm 2, \pm 3)$; both.

13. R is the arch-shaped region consisting of the points on or inside the rectangle with corners $(\pm 2, 0)$ and $(\pm 2, 2)$ that are not inside the circle of radius 1 centered at the origin; regular x region.

15. $\frac{56}{3}$ **17.** $-\frac{3}{4}$ **19.** $\frac{1}{2}e^4 - \frac{5}{2}$

21. $R = \{(x,y)|0 \leq y \leq x + 1, \quad 0 \leq x \leq 1\}$

$\displaystyle\int_0^1 \int_0^{x+1} \sqrt{1 + x + y}\, dy\, dx = (68 - 24\sqrt{2})/15$

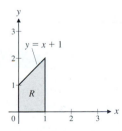

23. $R = \{(x,y)|0 \leq y \leq 4x - x^2, \quad 0 \leq x \leq 4\}$

$\displaystyle\int_0^4 \int_0^{4x-x^2} \sqrt{y + x^2}\, dy\, dx = \frac{128}{5}$

25. $R = \{(x, y)| 1 - \sqrt{x} \leq y \leq 1 + \sqrt{x}, \quad 0 \leq x \leq 4\}$

$\int_0^4 \int_{1-\sqrt{x}}^{1+\sqrt{x}} x(y - 1)^2 \, dy \, dx = \frac{512}{21}$

27. $\int_0^3 \int_0^{3-y} (x + 2y) \, dx \, dy = \frac{27}{2}$ **29.** $\int_0^1 \int_0^{\sqrt{1-y}} x\sqrt{y} \, dx \, dy = \frac{2}{15}$

31. $\int_0^1 \int_{4y^2}^{4y} x \, dx \, dy = \frac{16}{15}$ **33.** $\int_0^4 \int_0^{4-x} (4 - x - y) \, dy \, dx = \frac{32}{3}$

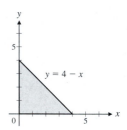

35. $\int_0^1 \int_0^{1-x^2} 4 \, dy \, dx = \frac{8}{3}$ **37.** $\int_0^4 \int_0^{\sqrt{y}} \dfrac{4x}{1 + y^2} \, dx \, dy = \ln 17$ **39.** $\int_0^1 \int_0^{\sqrt{x}} 4ye^{x^2} \, dy \, dx = e - 1$

41. $R = \{(x, y)| x^2 \leq y \leq 1 + \sqrt{x}, \quad 0 \leq x \leq 1.49\}$ **43.** $R = \{(x, y)| y^3 \leq x \leq 1 - y, \quad 0 \leq y \leq 0.68\}$

$\int_0^{1.49} \int_{x^2}^{1+\sqrt{x}} x \, dy \, dx \approx 0.96$ $\int_0^{0.68} \int_{y^3}^{1-y} 24xy \, dx \, dy \approx 0.83$

45. $R = \{(x, y)| e^{-x} \leq y \leq 3 - x, \quad -1.51 \leq x \leq 2.95\}$ Regular x region
$R = \{(x, y)| -\ln y \leq x \leq 3 - y, \quad 0.05 \leq y \leq 4.51\}$ Regular y region
$\int_{-1.51}^{2.95} \int_{e^{-x}}^{3-x} 4y \, dy \, dx = \int_{0.05}^{4.51} \int_{-\ln y}^{3-y} 4y \, dx \, dy \approx 40.67$

47. 323.402   **49.** −3.354

Exercise B-7

1. (A) $p(x) = a_0 + a_1(x - 1) + a_2(x - 1)(x - 3)$
 (B) $a_0 = 2, a_1 = 2, a_2 = 1$
 (C) $p(x) = 2 + 2(x - 1) + (x - 1)(x - 3)$

3. (A) $p(x) = a_0 + a_1(x + 1) + a_2(x + 1)x + a_3(x + 1)x(x - 2)$
 (B) $a_0 = 6, a_1 = -1, a_2 = 2, a_3 = -2$
 (C) $p(x) = 6 - (x + 1) + 2(x + 1)x - 2(x + 1)x(x - 2)$

5.

x_k	y_k	1ST D.D.	2ND D.D.
1	4		
		4	
2	8		1
		6	
3	14		

$p(x) = 4 + 4(x - 1) + (x - 1)(x - 2)$

7.

x_k	y_k	1ST D.D.	2ND D.D.	3RD D.D.
-1	-3			
		4		
0	1		-1	
		2		1
1	3		2	
		6		
2	9			

$p(x) = -3 + 4(x + 1) - (x + 1)x + (x + 1)x(x - 1)$

9.

x_k	y_k	1ST D.D.	2ND D.D.	3RD D.D.
-2	25			
		-5		
1	10		3	
		7		-1
2	17		-3	
		-2		
4	13			

$p(x) = 25 - 5(x + 2) + 3(x + 2)(x - 1) - (x + 2)(x - 1)(x - 2)$

13. $p(x) = -64 + 24(x + 4) - 4(x + 4)x + (x + 4)x(x - 4)$; (A) 8 (B) 56

15. $p(x) = \frac{1}{4}(x + 1)x(x - 1)$; (A) 1.5 (B) 6

17. $p(x) = 24 - 11(x + 4) + \frac{5}{2}(x + 4)(x + 2) - \frac{1}{2}(x + 4)(x + 2)x + \frac{1}{8}(x + 4)(x + 2)x(x - 2)$;
 (A) $\frac{57}{8} = 7.125$ (B) $-\frac{23}{8} = -2.875$

19. $p(x) = -24 + 18(x + 3) - 6(x + 3)(x + 2) + (x + 3)(x + 2)(x + 1)$; (A) $\frac{3}{8} = 0.375$ (B) $\frac{105}{8} = 13.125$

21. $p(x) = 2 - (x + 2) + \frac{1}{2}(x + 2)x = \frac{1}{2}x^2$ **23.** $p(x) = -4 + 2x$

25. $p(x) = 2(x + 1) - (x + 1)x = 2 + x - x^2$

27. $p(x) = 1 + 4(x + 2) - 3(x + 2)(x + 1) + (x + 2)(x + 1)x = x^3 - 3x + 3$

29. $p(x) = -3 + 3(x + 2) + (x + 2)(x + 1) - 2(x + 2)(x + 1)x + (x + 2)(x + 1)x(x - 1) = 5 - 6x^2 + x^4$

31. $p(x) = x^3 - 3x + 3$; the two polynomials are identical.

33. $p(x) = x^4 - 6x^2 + 5$; the two polynomials are identical.

35. $p(x) = 1 + \frac{1}{3}(x - 1) - \frac{1}{60}(x - 1)(x - 4)$

x	1	2	3	4	5	6	7	8	9
$p(x)$	1	1.4	1.7	2	2.3	2.5	2.7	2.9	3
$\sqrt{x}$	1	1.4	1.7	2	2.2	2.4	2.6	2.8	3

37. $p(x) = -4 - (x + 2) + 3(x + 2)(x + 1) - (x + 2)(x + 1)x = 6x - x^3$ **39.** $y = y_2 + (y_1 - y_2)x^2/x_1^2$

41. (A) $p(t) = 2 + 7.5t - 0.75t(t - 4) = 2 + 10.5t - 0.75t^2$ (B) $C(6) \approx 38$ or \$38,000   (C) approx. \$15,500

43. (A) $p(x) = 0.2x + 0.75x(x - 0.2) + 1.25x(x - 0.2)(x - 0.8) = 0.25x - 0.5x^2 + 1.25x^3$ (B) 0.1146

45. (A) $p(x) = 24.4 + 5.8(x - 2) - 1.6(x - 2)(x - 4) = 15.4x - 1.6x^2$ (B) $R(5) \approx 37$ or \$37,000
   (C) approx. 4,813 lamps

47. (A) $p(t) = 14 - t + 2t(t - 1) - t(t - 1)(t - 2) = 14 - 5t + 5t^2 - t^3$ (B) approx. 14.7°C

49. (A) $p(t) = 450 - 130t + 20t(t - 2) = 450 - 170t + 20t^2$ (B) approx. 4.25 days

51. (A) $p(t) = 10,000 + 350t + 15t(t - 10) - t(t - 10)(t - 20)$
$$= 10,000 + 45t^2 - t^3$$
   (B) approx. 14,000 registered voters

Index

Applications Index

Business & Economics